A-Level PHYSICS

FOURTH EDITION

Roger Muncaster

BSc PhD

Formerly Head of Physics
Bury Metropolitan College of Further Education

First published in 1981
Second edition 1985
Third edition 1989
Fourth edition 1993

Published by:
Nelson Thornes Ltd
Delta Place
27 Bath Road
Cheltenham
GL53 7TH
United Kingdom

04 05 / 10

A catalogue record for this book is available from the British Library.

ISBN 07487 1584 3 (Flexi-cover students' edition)
ISBN 07487 (Hard-cover library edition)

Related titles by the same author:

A-Level Physics:
Nuclear Physics and Fundamental Particles
ISBN 07487 1805 2

A-Level Physics:
Relativity and Quantum Physics
ISBN 07487 1799 4

A-Level Physics:
Medical Physics
ISBN 07487 2324 2

A-Level Physics:
Astrophysics and Cosmology
ISBN 07487 2865 1

The cover photograph shows polarized light through the injection area of a polycarbonate moulding. The photograph is by courtesy of the Paul Brierley Photo Library, Harlow, Essex.

Typeset by Tech-Set, Gateshead, Tyne & Wear.
Printed & Bound in Croatia by Zrinski

Contents

Preface

This book is intended to cover the NEAB, London, and AEB A-level syllabuses in Physics. It will also be found to cover the bulk of all the other syllabuses for A-level Physics, including those used overseas. Students following BTEC National courses involving Physics should also find the book useful, as should those university students who are studying Physics as a subsidiary subject. SI units are used throughout.

The aim has been to produce a book which is not so long that students are unlikely to read it. On the other hand, the book is not a set of 'revision notes' and it has been my intention to explain every topic thoroughly. It is hoped that the explanations are such that all students will understand them; at the same time, the content is intended to be such that the book will provide a proper basis for those students who are going on to study Physics at degree level.

The book has been arranged in seven main sections (A to G). Though there is no need to read the sections in the order in which they are presented, on the whole it is advisable to keep to the chapter sequence within any one section.

Practical details are given of those experiments which students are required to describe at length in examinations. The book contains many worked examples.

Chapters 9, 11 and 55 were extended for the second edition of the book; a chapter on thermodynamics was added at the same time. Sections on pressure, density, Archimedes' principle, reflection at plane surfaces, defects of vision, magnetic domains, U-values and impulse were added for the third edition. The treatments of various other topics were also revised and the number of experimental investigations was increased.

Since the advent of the GCSE examination and double science, students starting A-level courses tend to have less knowledge of Physics than they did previously. In the light of this, I have needed to make further additions to the book.

The number of worked examples has been greatly increased. Many of these are easier than was previously considered necessary. Questions have been added at relevant points in the text so that students can obtain an immediate test of their understanding of a topic. 'Consolidation' sections have been added at the ends of selected Chapters. These are designed to stress key points and, in some cases, to present an overview of a topic in a manner which would not be possible in the main text. Definitions and fundamental points are now highlighted – either by the use of screening or bold type.

At the end of each of the seven sections of the book there are questions, most of which are taken from past A-level papers. Over two hundred of these have been added for the fourth edition of the book.

A new edition gives me the opportunity to thank all those people who have suggested ways in which the book might be improved. I am particularly grateful to Jeni Davies for undertaking the laborious task of assisting with proof-reading, and for the invaluable suggestions she has made throughout the preparation of this edition.

Finally, I express my gratitude to the following examinations boards for permission to use questions from their past examination papers:

Associated Examining Board [AEB]
University of Cambridge Local Examinations Syndicate [C], reproduced by permission of University of Cambridge Local Examinations Syndicate
Cambridge Local Examinations Syndicate, Overseas Examinations [C(O)]
Northern Ireland Examinations Board [I]
Northern Examinations and Assessment Board (formerly the Joint Matriculation Board) [J]
Oxford and Cambridge Schools Examinations Board [O & C]
University of Oxford Delegacy of Local Examinations [O]
Southern Universities' Joint Board [S]
University of London Examinations and Assessment Council (formerly the University of London School Examinations Board) [L]
Welsh Joint Education Committee [W].

Where only part of the original question has been used, this is indicated by an asterisk in the acknowledgement to the board concerned thus [L*].

R. MUNCASTER
Helmshore

SECTION A

MECHANICS

1

VECTORS

1.1 VECTORS AND SCALARS

Vector quantities have both magnitude and direction; **scalar quantities** have magnitude only.

Examples of each type of quantity are given in Table 1.1.

Table 1.1
Examples of vectors and scalars

Scalars	*Vectors*
Distance	Displacement
Speed	Velocity
Mass	Force (weight)
Energy (work)	Acceleration
Volume	Momentum
Charge	Torque

Vectors can be represented by a line drawn in a particular direction. The length of the line represents the magnitude of the vector; the direction of the line represents the direction of the vector. In print, vector quantities are indicated by using bold type (e.g. $\boldsymbol{F}$) or by using an arrow (e.g. $\vec{F}$). The same symbol without the use of either bold type or an arrow (e.g. F) represents the magnitude of the vector.

Two vector quantities are equal only if they have the same magnitude <u>and</u> direction.

1.2 DISPLACEMENT

The displacement of a body may be defined as being the length and direction of the imaginary line joining it to some reference point.

Displacement is therefore a vector; the magnitude of the displacement is equal to the distance from the reference point.

Suppose a body moves from O to Y along the path OXY (Fig. 1.1). When the body is at Y its displacement from O is the vector, **OY**. The magnitude of the displacement is simply the length of OY. This is quite clearly less than the path length OXY, illustrating that the magnitude of the displacement of a body is not necessarily equal to the distance the body has actually moved.

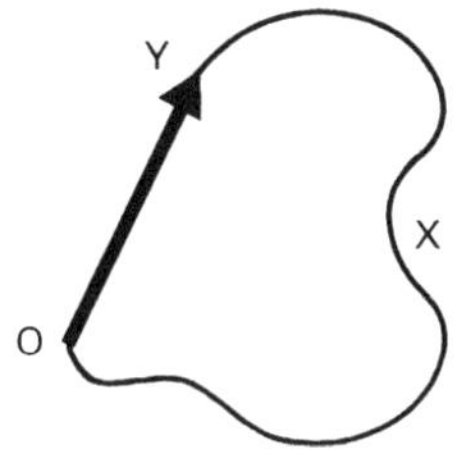

Fig. 1.1
To illustrate the difference between displacement and distance

1.3 SOME DEFINITIONS

Velocity is the rate of change of displacement, i.e. the rate of change of distance in a given direction.

Speed is the rate of change of distance.

Momentum is the product of mass and velocity.

Acceleration is the rate of change of velocity.

Velocity, momentum and acceleration are vectors

Note A body moving along a circular path may have constant speed but, because its direction is changing, it cannot have a constant velocity. It follows that if a body is moving around a circle, even if it has constant speed, it is being accelerated because its velocity is changing.

1.4 RELATIONSHIP BETWEEN SPEED AND VELOCITY

If a body moves along a straight line (without ever reversing its direction of motion), the distance it moves is equal to the magnitude of its displacement from the starting point. It follows, therefore, that since

$$\text{Speed} = \frac{\text{Distance moved}}{\text{Time taken}}$$

and

$$\text{Magnitude of velocity} = \frac{\text{Magnitude of displacement}}{\text{Time taken}}$$

then

$$\text{Speed} = \text{Magnitude of velocity}$$

It should be noted that this relationship is not necessarily true if the motion is not along a straight line, for then the magnitude of the displacement is less than the distance moved. The relationship does hold, though, if the time interval under consideration is infinitesimally short, for then the path length will also be infinitesimally short and therefore can be considered linear. Thus, for all types of motion

$$\text{Instantaneous speed} = \text{Magnitude of instantaneous velocity}$$

1.5 ADDITION AND SUBTRACTION OF VECTORS

Addition

The resultant of two or more vectors is the single vector which produces the same effect (in both magnitude and direction).

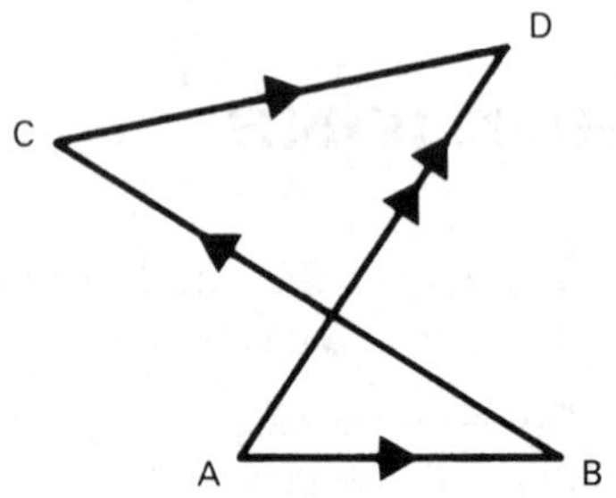

Fig. 1.2 To show that three separate displacements can be equivalent to one

Consider a body undergoing three separate displacements, from A to B, B to C and C to D (Fig. 1.2). Clearly the same result could have been achieved by the single displacement from A to D, and the vector **AD** is therefore the resultant of the other three,

i.e. **AD** = **AB** + **BC** + **CD**

It should be noted that the resultant, **AD**, is that displacement which makes a closed figure out of that obtained by setting out the individual displacements. The result will obviously hold for any number of displacements; it is also true no matter what the vectors are! Consider two forces, $\boldsymbol{P}$ and $\boldsymbol{Q}$, acting at a point, O (Fig. 1.3). From what has ben said, the resultant will be the vector $\boldsymbol{R}$ of Fig. 1.4.

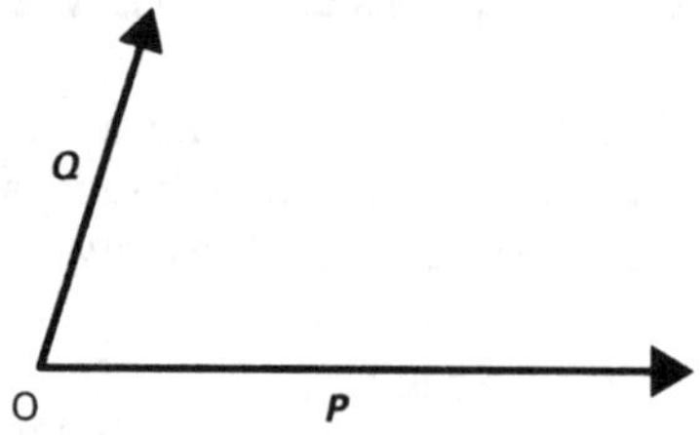

Fig. 1.3 Two forces acting at a point

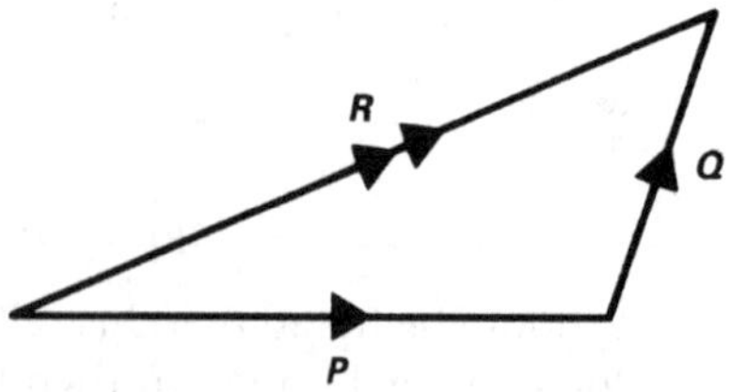

Fig. 1.4 The resultant of two forces

Clearly, $\boldsymbol{R}$ is a diagonal of the parallelogram of which $\boldsymbol{P}$ and $\boldsymbol{Q}$ are two sides. This provides a useful rule – **the parallelogram rule of vector addition**:

If two vectors, **OA** and **OB**, are represented in magnitude and direction by the sides OA and OB of a parallelogram OACB, then **OC** represents their resultant.

Fig. 1.5
The parallelogram rule

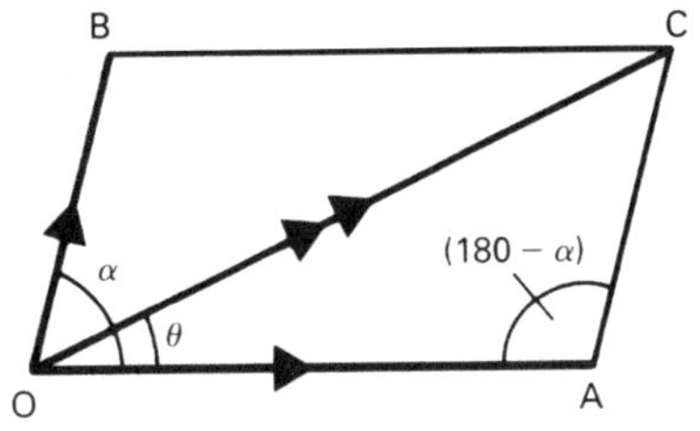

Thus, in Fig. 1.5,

$$\mathbf{OA} + \mathbf{OB} = \mathbf{OC}$$

Subtraction

This can be achieved by adding a vector of the same magnitude as that being subtracted but which acts in the opposite direction. For example

$$\mathbf{OA} - \mathbf{OB} = \mathbf{OA} + \mathbf{BO}$$

$$\therefore \quad \mathbf{OA} - \mathbf{OB} = \mathbf{BC} + \mathbf{BO}$$

$$\text{i.e.} \quad \mathbf{OA} - \mathbf{OB} = \mathbf{BA}$$

EXAMPLE 1.1

A force of 3 N acts at 90° to a force of 4 N. Find the magnitude and direction of their resultant, ***R***.

Solution

Fig. 1.6
Diagram for Example 1.1

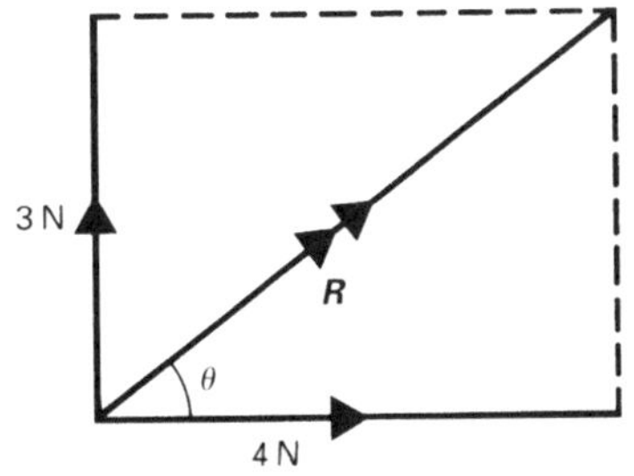

Refer to Fig. 1.6. By Pythagoras

$$R^2 = 3^2 + 4^2 = 25 \qquad \therefore\ R = 5\,\text{N}$$

Also

$$\tan\theta = \tfrac{3}{4} \qquad \therefore\ \theta = \tan^{-1}\left(\tfrac{3}{4}\right) = 37^\circ$$

The resultant is therefore a force of 5 N acting at 37° to the 4 N force and at 53° to the 3 N force.

EXAMPLE 1.2

A force of 3 N acts at 60° to a force of 5 N. Find the magnitude and direction of their resultant, $\boldsymbol{R}$.

Solution

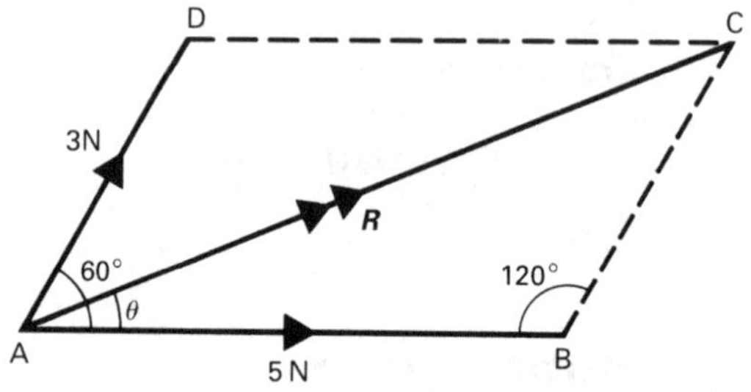

Fig. 1.7
Diagram for Example 1.2

Refer to Fig. 1.7. Applying the cosine rule (Appendix 3.7) to △ABC gives

$$R^2 = 5^2 + 3^2 - 2{\cdot}5{\cdot}3 \cos 120^\circ$$
$$= 25 + 9 + 15 = 49 \qquad \therefore \quad R = 7\,\text{N}$$

Applying the sine rule (Appendix 3.7) to △ABC gives

$$\frac{R}{\sin 120^\circ} = \frac{3}{\sin\theta}$$

$$\therefore \quad \sin\theta = \frac{3\sin 120^\circ}{R} = \frac{3\sin 120^\circ}{7} = 0.3712$$

$$\therefore \quad \theta = \sin^{-1} 0.3712 = 21.8^\circ \text{ or } 180^\circ - 21.8^\circ = 158.2^\circ$$

Your calculator will give you $\sin^{-1}$ 0.3712 as 21.8° but in general $\sin\alpha = \sin(180^\circ - \alpha)$ and therefore 158.2° is also a possibility.

It is obvious from the diagram that θ must be acute and therefore the required value is 21.8°. The resultant is therefore a force of 7 N acting at 21.8° to the 5 N force and at 38.2° to the 3 N force.

EXAMPLE 1.3

A particle which is moving due east at 4 m s^{-1} changes direction and starts to move due south at 3 m s^{-1}. Find the change in velocity.

Solution

The change in velocity is the 'new' velocity minus the 'old' velocity, just as a change in temperature, for example, would be the 'new' temperature minus the 'old' temperature. Therefore

$$\text{Change in velocity} = 3\text{ m s}^{-1}\text{ (S)} - 4\text{ m s}^{-1}\text{ (E)}$$
$$= 3\text{ m s}^{-1}\text{ (S)} + 4\text{ m s}^{-1}\text{ (W)}$$

The change in velocity is therefore the vector $\boldsymbol{R}$ of Fig. 1.8.

Fig. 1.8
Diagram for Example 1.3

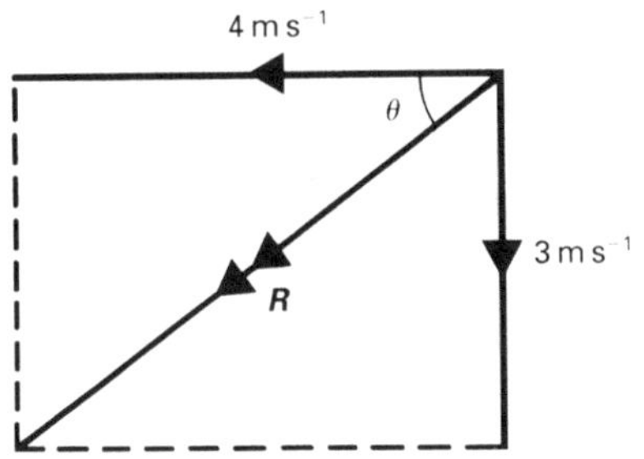

$$R^2 = 3^2 + 4^2 = 25 \qquad \therefore \quad R = 5\text{ m s}^{-1}$$

$$\tan\theta = \tfrac{3}{4} \qquad \therefore \quad \theta = \tan^{-1}\left(\tfrac{3}{4}\right) = 37^\circ$$

i.e. Change in velocity = 5 m s^{-1} at 37° S of W.

Alternatively, we can say that the velocity has increased by 5 m s^{-1} in the direction 37° S of W.

Note The parallelogram rule can also be used to obtain the resultant of more than two vectors. For example, suppose that the resultant of three vectors is required. The procedure is to use the rule to find the resultant of any two of them, and then to use it again to add this to the remaining vector.

QUESTIONS 1A

1. Find the magnitude and direction of the resultant of each of the following pairs of forces.
   (a) 7 N at 90° to 24 N,
   (b) 20 N at 60° to 30 N,
   (c) 40 N at 110° to 50 N,
   (d) 60 N at 150° to 20 N.

2. Find the resultant of a displacement of 30 m due east followed by a displacement of 70 m due south.

3. Find: **(a)** the increase in speed, **(b)** the increase in velocity when a body moving south at 20 m s^{-1} changes direction and moves north at 30 m s^{-1}.

4. Find the magnitude and direction of the increase in velocity when a body which has been moving due S at 6.0 m s^{-1} changes direction and moves NW at 8.0 m s^{-1}.

1.6 COMPONENTS OF VECTORS

It follows from the parallelogram rule that any vector can be treated as if it is the sum of a pair of vectors. There is an infinite number of these pairs and three are shown in Fig. 1.9. A perpendicular pair such as ***P*** and ***Q*** is the most useful.

Fig. 1.9
Components of a vector

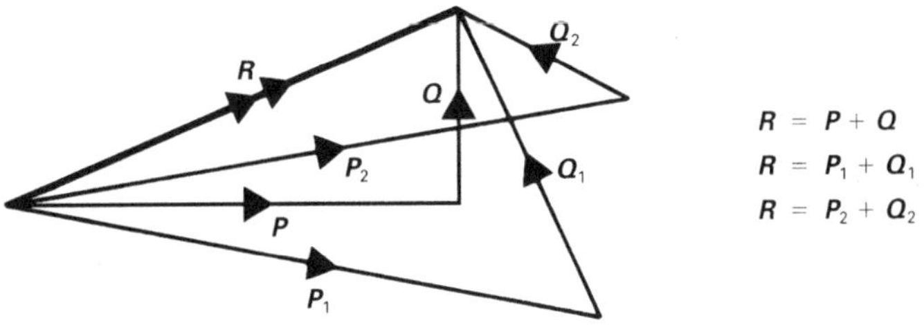

Consider a vector, $\boldsymbol{F}$, resolved into two perpendicular vectors of magnitudes AB and AD (Fig. 1.10). From simple trigonometry, AB $= F\cos\theta$ and AD $= F\sin\theta$, and therefore $\boldsymbol{F}$ can be resolved into two perpendicular vectors (called **the perpendicular components of $\boldsymbol{F}$**) of magnitudes $F\sin\theta$ and $F\cos\theta$ (Fig. 1.11).

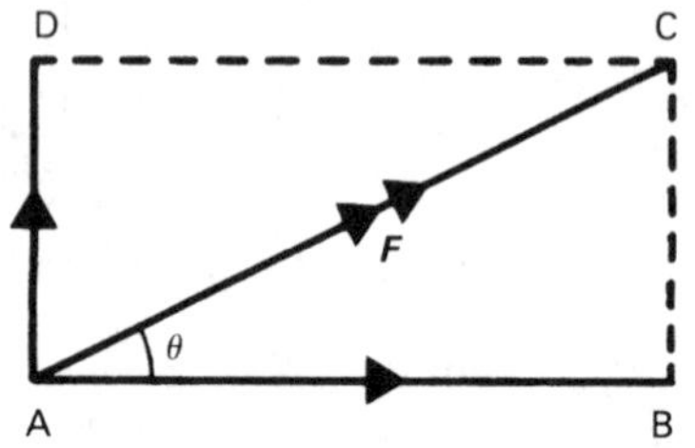

Fig. 1.10
Resolving a vector into two perpendicular components

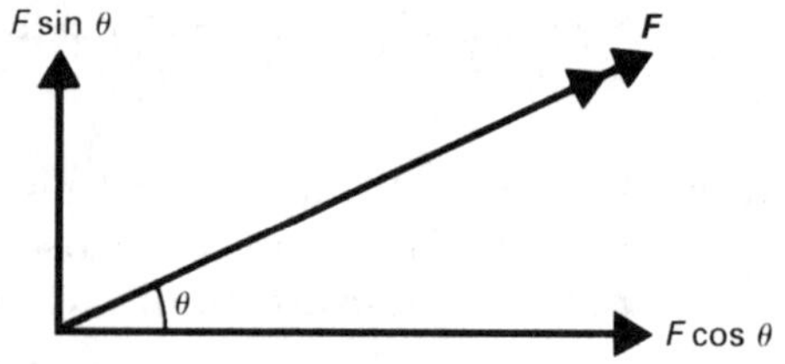

Fig. 1.11
The perpendicular components of a vector

EXAMPLE 1.4

Calculate the horizontal and vertical components of a force of 50 N which is acting at 40° to the horizontal.

Solution

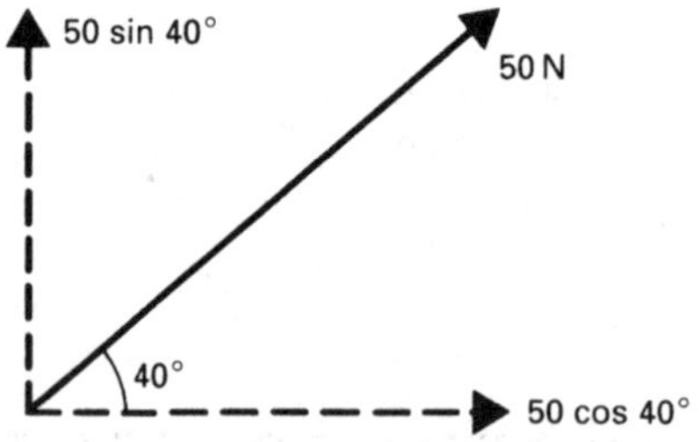

Fig. 1.12
Diagram for Example 1.4

Refer to Fig. 1.12.

Horizontal component $= 50\cos 40^\circ = 38\,\text{N}$

Vertical component $= 50\sin 40^\circ = 32\,\text{N}$

EXAMPLE 1.5

A body of weight 100 N rests on a plane which is inclined at 30° to the horizontal. Calculate the components of the weight parallel and perpendicular to the plane.

Solution

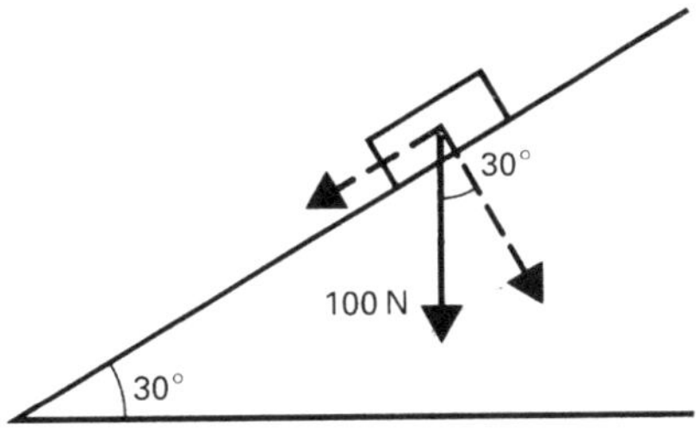

Fig. 1.13
Diagram for Example 1.5

Refer to Fig. 1.13.

Component parallel to plane = 100 sin 30° = 50.0 N

Component perpendicular to plane = 100 cos 30° = 86.6 N

EXAMPLE 1.6

Find the resultant of the system of forces shown in Fig. 1.14.

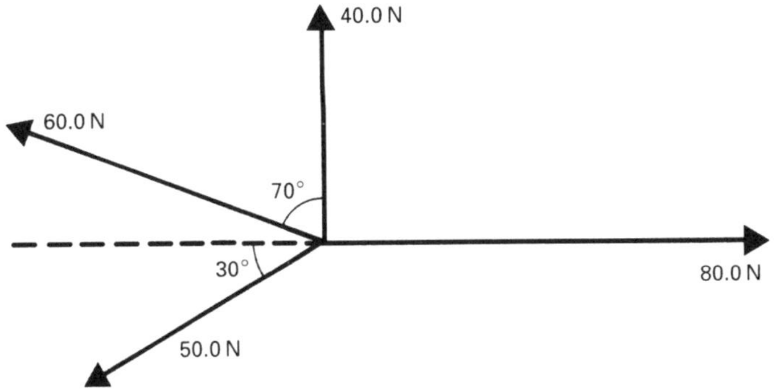

Fig. 1.14
Diagram for Example 1.6

Solution

Total upward force = 40.0 + 60.0 cos 70° − 50.0 sin 30° = 35.52 N

Total force to right = 80.0 − 60.0 sin 70° − 50.0 cos 30° = −19.68 N

The minus sign implies that the horizontal force is to the left. The resultant, ***R***, is as shown in Fig. 1.15.

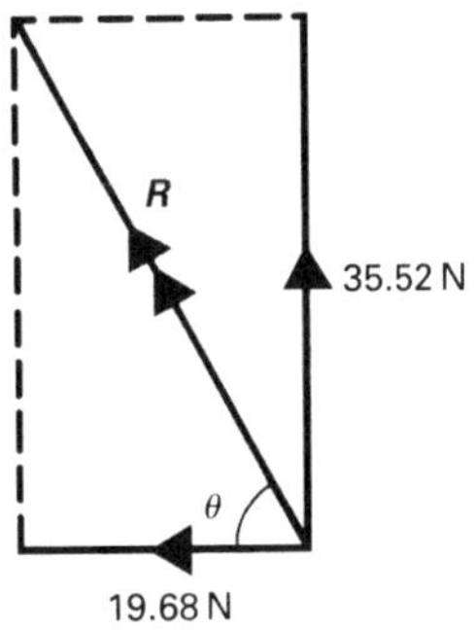

Fig. 1.15
Diagram for Example 1.6

$$R^2 = 35.52^2 + 19.68^2 \qquad \therefore \quad R = 40.6\ \text{N}$$

$$\tan\theta = \frac{35.52}{19.68} \qquad \therefore \quad \theta = 61.0^\circ$$

The resultant is therefore a force of 40.6 N acting at 61.0° to the horizontal.

QUESTIONS 1B

1. Find the horizontal and vertical components of: **(a)** a force of 30.0 N acting at 30° to the horizontal, **(b)** a velocity of 50.0 m s^{-1} at 60° to the horizontal.

2. A particle at weight 200 N rests on a plane inclined at 50° to the horizontal. What are the components of the weight: **(a)** parallel, **(b)** perpendicular to the plane?

3. Find the resultant of the system of forces shown below.

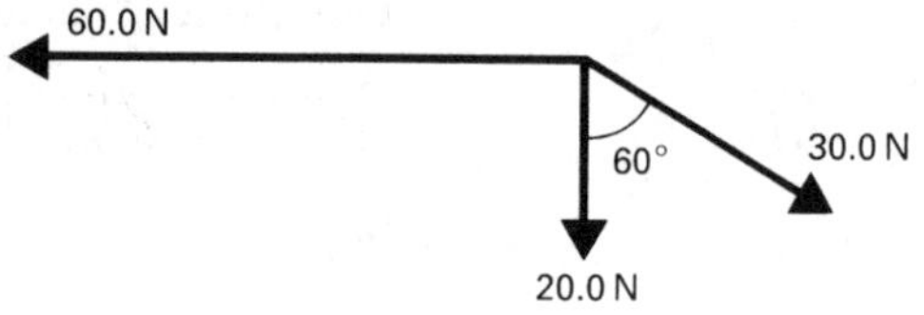

CONSOLIDATION

Vectors have both magnitude and direction; **scalars** have magnitude only.

Displacement and **velocity** are vectors; **distance** and **speed** are scalars.

Distance from reference point = magnitude of displacement.

Instantaneous speed = magnitude of instantaneous velocity

Vectors can be added and subtracted by using the **parallelogram rule.**

Components of Vectors

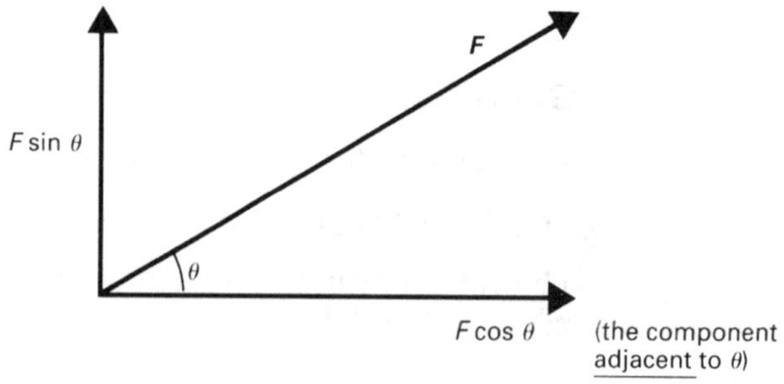

2

MOTION

2.1 NEWTON'S LAWS OF MOTION

In 1687 Sir Isaac Newton published his *Philosophiae Naturalis Principia Mathematica* (*The Mathematical Principles of Natural Science*), in which he stated the three laws on which the science of mechanics is based.

Newton's First Law

Every body continues in its state of rest or of uniform (unaccelerated) motion in a straight line unless acted on by some external force.

This law expresses the concept of inertia. The **inertia** of a body can be described as being its reluctance to start moving, or to stop moving once it has started.

Events often seem to contradict the first law, for it is our natural experience that there are many familiar examples of motion in which moving objects come to rest when (apparently) left to their own devices. Closer examination of the circumstances, however, reveals that in every case there is some sort of retarding force acting. Such forces are often due to friction between solid surfaces or to air resistance.

A body of large mass requires a large force to change its speed or its direction by a noticeable amount, i.e. the body has a large inertia. Thus, **the mass of a body is a measure of its inertia**.

Newton's Second Law

The rate of change of momentum of a body is directly proportional to the external force acting on the body and takes place in the direction of the force.

In mathematical terms the second law may be written as

$$F \propto \frac{\mathrm{d}}{\mathrm{d}t}(mv)$$

where F = the applied force, and

$$\frac{\mathrm{d}}{\mathrm{d}t}(mv) = \text{the rate of change of momentum.}$$

Introducing a constant of proportionality, k, this becomes

$$F = k\frac{\mathrm{d}}{\mathrm{d}t}(mv)$$

The SI unit of force (the newton) is defined in such a way that $k = 1$ provided that the rate of change of momentum is also expressed in the relevant SI unit (kg m s^{-2}), in which case

$$F = \frac{\mathrm{d}}{\mathrm{d}t}(mv) \qquad [2.1]$$

If the mass is constant, equation [2.1] becomes

$$F = m\frac{\mathrm{d}v}{\mathrm{d}t}$$

i.e.

$$F = ma \qquad [2.2]$$

where $a =$ the acceleration that results from the application of the force.

Equations [2.1] and [2.2] are the forms in which Newton's second law is normally used, but it should be remembered that they are valid only if a consistent set of units is used, and that equation [2.2] applies only in the case of a constant mass.

Equation [2.2] is used to define the newton. Thus:

> **The newton** (N) is that force which produces an acceleration of 1 m s^{-2} when it acts on a mass of 1 kg.

The experimental investigation of $F = ma$ is dealt with in section 2.15.

Newton's Third Law

> If a body A exerts a force on a body B, then B exerts an equal and oppositely directed force on A.

This law is often misinterpreted as meaning that the two forces cancel each other out because they are of equal strength and act in opposite directions. There is, in fact, no possibility of this, because the two forces each act on different bodies.

Thus, if a man pushes on a large stationary crate, the crate pushes back on the man with a force of exactly the same size. Whether or not the crate starts to move, has nothing to do with the force that it exerts on the man. In accordance with Newton's second law, the crate will start to move if the force exerted by the man is greater than any forces which are acting on the crate in such a way as to resist its motion (e.g. friction between the crate and the ground).

The third law implies that forces always occur in pairs – some examples are given below.

(i) The Earth exerts a gravitational force of attraction on the Moon; the Moon exerts a force of the same size on the Earth.

(ii) A rocket moves forward as a result of the push exerted on it by the exhaust gases which the rocket has pushed out.

(iii) When a man jumps off the ground it is because he has pushed down on the Earth and the Earth, in accordance with Newton's third law, has pushed up on him. It should not be overlooked that the other result of this is that the Earth moves down.

(iv) If a car is accelerating forward, it is because its tyres are pushing backward on the road and the road is pushing forward on the tyres. Note that if the car is moving forward and slowing down, the tyres push forward and the road pushes backward.

2.2 MASS AND WEIGHT

The weight of a body is the force acting on its mass due to the gravitational attraction of the Earth.*

In accordance with Newton's second law, a body acquires an acceleration whenever there is a net force acting on it. The acceleration that results from the effect of gravity (i.e. that results from its weight) is known as the acceleration due to gravity, g. By equation [2.2], the weight of a body of mass m is given by

$$\text{Weight} = mg \quad [2.3]$$

The force exerted by gravity is such that, at any given point in a gravitational field (and therefore at any given point on the Earth's surface), the acceleration due to gravity is the same for all bodies, no matter what their masses (see Chapter 8).

It follows that two bodies dropped from the same point above the surface of the Earth reach the ground at the same time even if their masses are different. (Note that this statement ignores the effect of air resistance; when the viscous drag of the air is significant, for example if one of the bodies is a feather or is falling by parachute, it is not even approximately true.) The acceleration due to gravity varies slightly from place to place on the Earth's surface, but it is normally sufficiently accurate to use a value of $9.8\,\text{m s}^{-2}$ everywhere. Thus, from equation [2.3] the weight (in newtons) of a body which has a mass m (in kg) is given by

$$\text{Weight} = m \times 9.8$$

Another unit, the **kilogram force** (kgf), is often used as a measure of weight. It is defined such that a mass of 1 kg has a weight of 1 kgf. This is not an SI unit and must not be used in any equation where it is not possible to use it on both sides of the equation. If in doubt, it is best to convert kilograms force to newtons by making use of

$$1\,\text{kgf} = 9.8\,\text{N}$$

Summary of Differences between Mass and Weight

(i) The mass of a body is a measure of its resistance to acceleration (i.e. it is a measure of the inertia of the body). The weight of a body is the force exerted on its mass by gravity.

(ii) In SI units mass is measured in kilograms, weight is measured in newtons.

(iii) The mass of a body is the same everywhere. The weight of a body on the surface of the Earth has a slight dependence on where it is, and would have considerably different values at other places in the Universe.

*The weight of a body on the Moon is the force exerted on its mass by the gravitational attraction of the Moon.

EXAMPLE 2.1

A body of mass 7.0 kg rests on the floor of a lift. Calculate the force, R, exerted on the body by the floor when the lift: (a) has an upward acceleration of $2.0\ \text{m s}^{-2}$, (b) has a downward acceleration of $3.0\ \text{m s}^{-2}$, (c) is moving down with a constant velocity. (Assume $g = 10\ \text{m s}^{-2}$.)

Solution

Fig. 2.1
Diagram for Example 2.1

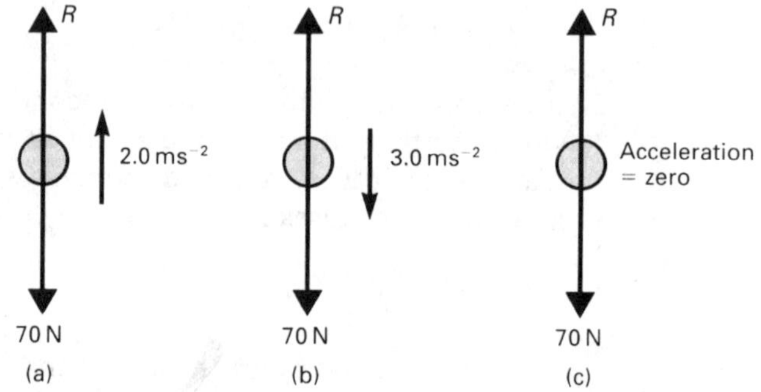

(a) Refer to Fig. 2.1(a) (When using $F = ma$ the direction of F must be the same as that of a. The body has an upward acceleration and therefore we require the resultant upward force.)

By Newton's second law (equation [2.2])

$$\underbrace{R - 70}_{\text{upward force}} = 7.0 \times \underset{\text{upward acceleration}}{2.0}$$

$$\therefore \quad R - 70 = 14 \qquad \text{i.e.} \quad R = 84\ \text{N}$$

(b) Refer to Fig. 2.1(b) (The acceleration is downward and therefore we require the resultant downward force.)

By equation [2.2].

$$\underbrace{70 - R}_{\text{downward force}} = 7.0 \times \underset{\text{downward acceleration}}{3.0}$$

$$\therefore \quad 70 - R = 21 \qquad \text{i.e.} \quad R = 49\ \text{N}$$

(c) Refer to Fig. 2.1(c). There is no acceleration and therefore, by equation [2.2], no resultant force, in which case

$$R = 70\ \text{N}$$

Notes

(i) By Newton's third law, R is equal and opposite to the force exerted by the body on the floor of the lift. It follows that if the body were resting on a bathroom scale rather than directly on the floor of the lift, the scale would register its weight as 84 N, 49 N and 70 N in situations (a), (b) and (c) respectively. Thus the body appears heavier than it actually is when it has an upward acceleration and lighter when it has a downward acceleration. In case (c), where it has no acceleration, it appears neither heavy nor light.

(ii) These results do not depend on the direction in which the lift is moving. For example in (a) the lift has an upward directed acceleration and therefore may be moving up with increasing speed or moving down with decreasing speed.

EXAMPLE 2.2

A body of mass 5.0 kg is pulled up a smooth plane inclined at 30° to the horizontal by a force of 40 N acting parallel to the plane. Calculate the acceleration of the body and the force exerted on it by the plane. (Assume $g = 10\,\mathrm{m\,s^{-2}}$.)

Solution

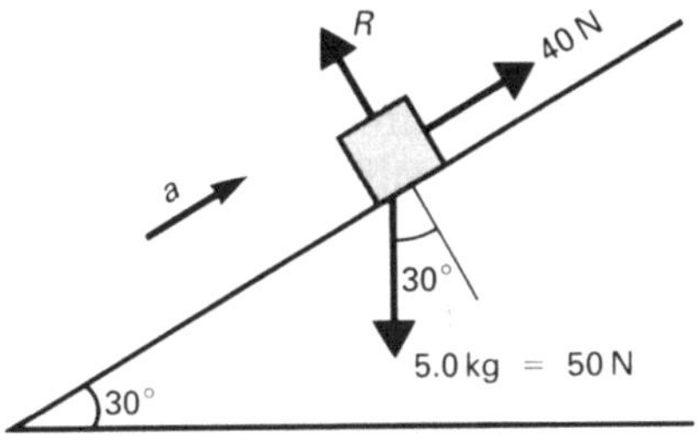

Fig. 2.2
Diagram for Example 2.2

The diagram shows the acceleration of the body and all the forces acting on it

Refer to Fig. 2.2. The plane is smooth and therefore the only force it exerts on the body is the normal reaction R. Let the acceleration of the body be a.

Consider the motion parallel to the plane. The weight has a component of $50 \sin 30°$ acting parallel to the plane (downwards) and therefore by Newton's second law (equation [2.2])

$$\underbrace{40 - 50 \sin 30°}_{\text{resultant force up the plane}} = 5.0 \times \underbrace{a}_{\text{acceleration up the plane}}$$

$$\therefore \quad 40 - 25 = 5.0a \qquad \text{i.e.} \quad a = 3.0\,\mathrm{m\,s^{-2}}$$

Note that the resultant force and the acceleration are in the same direction – up the plane.

Consider the motion perpendicular to the plane. The weight has a component of $50 \cos 30°$ perpendicular to the plane and in the opposite direction to R. There is no acceleration perpendicular to the plane and therefore no resultant force, in which case

$$R = 50 \cos 30° \qquad \text{i.e. } R = 43\,\mathrm{N}$$

EXAMPLE 2.3

A train is moving along a straight horizontal track. A pendulum suspended from the roof of one of the carriages of the train is inclined at 4° to the vertical. Calculate the acceleration of the train. (Assume $g = 10\,\mathrm{m\,s^{-2}}$.)

Solution

Suppose that the mass of the pendulum bob is m. The forces acting on the bob are its weight, mg, which acts vertically downwards and the tension, T, in the string. (Fig. 2.3.)

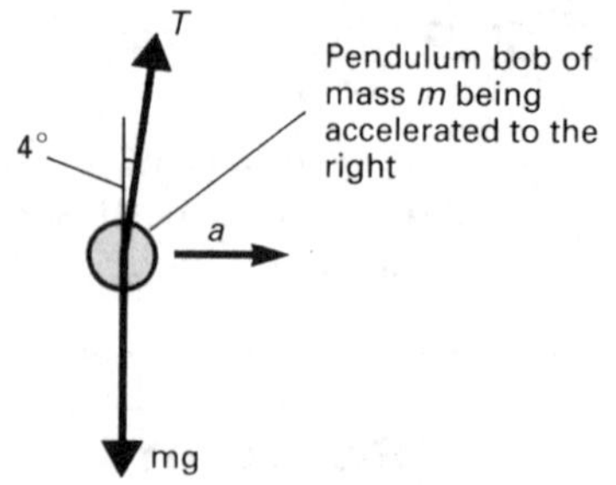

Fig. 2.3
Diagram for Example 2.3

Consider the horizontal motion. The pendulum bob is at rest with respect to the train and therefore it too has a horizontal acceleration, a (to the right). The horizontal component of the tension is $T \sin 4°$ and therefore by equation [2.2]

$$T \sin 4° = ma \qquad [2.4]$$

Consider the vertical motion. There is no vertical component of acceleration and therefore

$$T \cos 4° = mg \qquad [2.5]$$

Dividing equation [2.4] by equation [2.5] gives

$$\tan 4° = a/g$$

$$\therefore \quad a = g \tan 4° \qquad \text{i.e.} \quad a = 0.70 \text{ m s}^{-2}$$

EXAMPLE 2.4

Two blocks, A of mass m and B of mass $3m$, are side by side and in contact with each other. They are pushed along a smooth floor under the action of a constant force F applied to A. Find: (a) the acceleration of the blocks, (b) the force exerted on B by A.

Solution

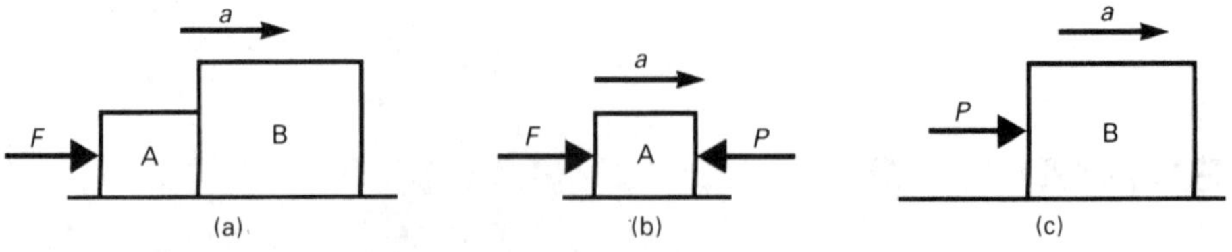

Fig. 2.4
The horizontal force(s):
(a) on the whole system
(b) on A, (c) on B

(a) Let the acceleration of the blocks be a. Consider the motion of the whole system (Fig. 2.4(a)). By Newton's second law (equation [2.2])

$$F = (m + 3m)a \qquad \text{i.e.} \quad a = \frac{F}{4m}$$

(b) Let the force on B due to A be P. By Newton's third law there will be an equal and opposite force on A (Fig. 2.4(b) and (c)). Applying Newton's second law to the motion of B gives

$$P = 3m \times a$$

$$\therefore \quad P = 3m \times \frac{F}{4m} \quad \text{i.e.} \quad P = \frac{3F}{4}$$

The reader should confirm that considering the motion of A gives the same result, though a little less easily.

QUESTIONS 2A

1. The resultant force on a body of mass 4.0 kg is 20 N. What is the acceleration of the body?

2. A body of mass 6.0 kg moves under the influence of two oppositely directed forces whose magnitudes are 60 N and 18 N. Find the magnitude and direction of the acceleration of the body.

3. Two forces, of magnitudes 30 N and 40 N and which are perpendicular to each other, act on a body of mass 25 kg. Find the magnitude and direction of the acceleration of the body.

4. Is the motion of the train in Example 2.3: (A) to the right, (B) to the left, or (C) is it impossible to tell?

5. A body of mass 3.0 kg slides down a plane which is inclined at 30° to the horizontal. Find the acceleration of the body: **(a)** if the plane is smooth, **(b)** if there is a frictional resistance of 9.0 N.
($g = 10\,\text{m}\,\text{s}^{-2}$.)

6. A railway truck of mass 6.0 tonnes moves with an acceleration of $0.050\,\text{m}\,\text{s}^{-2}$ down a track which is inclined to the horizontal at an angle α where $\sin\alpha = 1/120$. Find the resistance to motion, assuming that it is constant.
($g = 10\,\text{m}\,\text{s}^{-2}$, 1 tonne $= 1.0 \times 10^3$ kg.)

7. A body hangs from a spring-balance which is suspended from the ceiling of a lift. What is the mass of the body if the balance registers a reading of 70 N when the lift has an upward acceleration of $4.0\,\text{m}\,\text{s}^{-2}$?
($g = 10\,\text{m}\,\text{s}^{-2}$.)

8. What is the apparent weight during take-off of an astronaut whose actual weight is 750 N if the resultant upward acceleration is $5g$?

9. A body of mass 5.0 kg is pulled along smooth horizontal ground by means of a force of 40 N acting at 60° above the horizontal. Find: **(a)** the acceleration of the body, **(b)** the force the body exerts on the ground.
($g = 10\,\text{m}\,\text{s}^{-2}$.)

10. A railway engine of mass 100 tonnes is attached to a line of trucks of total mass 80 tonnes. Assuming there is no resistance to motion, find the tension in the coupling between the engine and the leading truck when the train: **(a)** has an acceleration of $0.020\,\text{m}\,\text{s}^{-2}$, **(b)** is moving at constant velocity.
(1 tonne $= 1.0 \times 10^3$ kg.)

11. A car of mass 1000 kg tows a caravan of mass 600 kg up a road which rises 1 m vertically for every 20 m of its length. There are constant frictional resistances of 200 N and 100 N to the motion of the car and to the motion of the caravan respectively. The combination has an acceleration of $1.2\,\text{m}\,\text{s}^{-2}$ with the engine exerting a constant driving force. Find: **(a)** the driving force, **(b)** the tension in the tow-bar.
($g = 10\,\text{m}\,\text{s}^{-2}$.)

Examples 2.1 to 2.4 are concerned with Newton's second law in the form '$F = ma$' (equation [2.2]). The examples that follow use the law in the form, 'Force = Rate of change of momentum' (equation [2.1]).

EXAMPLE 2.5

Water emerges at $2\,\mathrm{m\,s^{-1}}$ from a hose pipe and hits a wall at right angles. The pipe has a cross-sectional area of $0.03\,\mathrm{m^2}$. Calculate the force on the wall assuming that the water does not rebound. (Density of water $= 1000\,\mathrm{kg\,m^{-3}}$.)

Note In solving problems of this type we determine the mass of substance that has its momentum changed in one second. We then find the change in momentum of this mass and so obtain the change in momentum per second, i.e. the rate of change of momentum.

Solution

In one second the volume of water that hits the wall is that which has left the pipe in one second, i.e. that which was contained in a cylinder of length 2 m and cross-sectional area $0.03\,\mathrm{m^2}$, namely $2 \times 0.03 = 0.06\,\mathrm{m^3}$. The mass of water hitting the wall in one second is therefore $0.06 \times 1000 = 60\,\mathrm{kg}$. When the water hits the wall its speed changes from $2\,\mathrm{m\,s^{-1}}$ to zero, and therefore the rate of change of momentum is $60 \times 2 = 120\,\mathrm{N}$.

By Newton's second law, force = rate of change of momentum, and therefore the force exerted by the wall = 120 N. Therefore, by Newton's third law, the force exerted by the water = 120 N.

EXAMPLE 2.6

A helicopter of mass $1.0 \times 10^3\,\mathrm{kg}$ hovers by imparting a downward velocity v to the air displaced by its rotating blades. The area swept out by the blades is $80\,\mathrm{m^2}$. Calculate the value of v. (Density of air $= 1.3\,\mathrm{kg\,m^{-3}}$, $g = 10\,\mathrm{m\,s^{-2}}$.)

Solution

The volume of air displaced in one second $= 80\,v$, and therefore the mass of air displaced in one second $= 1.3 \times 80\,v = 104\,v$. It follows that in one second the momentum of the air increases by $104\,v^2$. By Newton's second law, rate of change of momentum = force, and therefore the force exerted on the air by the blades $= 104\,v^2$. By Newton's third law, the upward force on the helicopter is also $104\,v^2$. Since the helicopter is hovering, the upward force is equal to the weight of the helicopter, and therefore

$$104\,v^2 = 1.0 \times 10^3 g$$

i.e. $$v = 9.8\,\mathrm{m\,s^{-1}}$$

EXAMPLE 2.7

Sand falls onto a conveyor belt at a constant rate of $2\,\mathrm{kg\,s^{-1}}$. The belt is moving horizontally at $3\,\mathrm{m\,s^{-1}}$. Calculate: (a) the extra force required to maintain the speed of the belt, (b) the rate at which this force is doing work, (c) the rate at which the kinetic energy of the sand increases.

Account for the fact that the answers to (b) and (c) are different.

Solution

(a) Every second 2 kg of sand acquire a horizontal velocity of 3 m s^{-1}, and therefore the rate of increase of horizontal momentum = $2 \times 3 = 6$ N. By Newton's second law, force = rate of change of momentum, and therefore the extra force required to maintain the speed of the belt = 6 N.

(b) In one second the force moves 3 m, and therefore (by equation [5.1] or [5.7]) the rate at which the force is working = $6 \times 3 = 18$ W.

(c) Kinetic energy = $\frac{1}{2}mv^2$ (see section 5.3), and therefore the rate at which the kinetic energy of the sand is increasing = $\frac{1}{2} \times 2 \times 3^2 = 9$ W.

A finite time elapses before the sand acquires the speed of the belt. During this period the belt is slipping past the sand and therefore work has to be done to overcome friction between the sand and the belt. The rate at which work is done by the force is equal to the rate at which it is doing work against friction plus the rate at which it is doing work to increase the kinetic energy of the sand – hence the difference between (b) and (c). (**Note**. The rate at which work is done against friction is equal to the rate at which work is done to increase the kinetic energy of the sand no matter what the speed of the belt and no matter what the rate at which sand is falling onto the belt.)

QUESTIONS 2B

1. Water is squirting horizontally at 4.0 m s^{-1} from a burst pipe at a rate of 3.0 kg s^{-1}. The water strikes a vertical wall at right angles and runs down it without rebounding. Calculate the force the water exerts on the wall.

2. A machine gun fires 300 bullets per minute horizontally with a velocity of 500 m s^{-1}. Find the force needed to prevent the gun moving backwards if the mass of each bullet is 8.0×10^{-3} kg.

3. Coal is falling onto a conveyor belt at a rate of 540 tonnes every hour. The belt is moving horizontally at 2.0 m s^{-1}. Find the extra force required to maintain the speed of the belt. (1 tonne = 1000 kg.)

4. The rotating blades of a hovering helicopter sweep out an area of radius 4.0 m imparting a downward velocity of 12 m s^{-1} to the air displaced. Find the mass of the helicopter. (g = 10 m s^{-2}, density of air = 1.3 kg m^{-3}.)

5. Find the force exerted on each square metre of a wall which is at right angles to a wind blowing at 20 m s^{-1}. Assume that the air does not rebound. (Density of air = 1.3 kg m^{-3}.)

6. Hailstones with an average mass of 4.0 g fall vertically and strike a flat roof at 12 m s^{-1}. In a period of 5.0 minutes six thousand hailstones fall on each square metre of roof and rebound vertically at 3.0 m s^{-1}. Calculate the force on the roof if it has an area of 30 m^2.

7. The speed of rotation of the blades of the helicopter in question 4 is increased so that the air now has a downward velocity of 13 m s^{-1}. Find the (upward) acceleration of the helicopter.

2.3 THE EQUATIONS OF MOTION FOR UNIFORM ACCELERATION

Equations [2.6]–[2.9] describe the motion of bodies which are moving with constant (uniform) acceleration.

$$v = u + at \quad [2.6]$$

$$v^2 = u^2 + 2as \quad [2.7]$$

$$s = ut + \tfrac{1}{2}at^2 \quad [2.8]$$

$$s = \tfrac{1}{2}(u + v)t \quad [2.9]$$

where u = the velocity when $t = 0$,

v = the velocity at time t,

a = the constant acceleration,

s = the distance from the starting point at time t, (this is not necessarily the distance moved).

When using these equations it is necessary to bear in mind that u, v, a and s are vectors. If, say, the positive direction is taken to be up, then:

(i) the velocity of a body which is moving down is negative,

(ii) points below the starting point have negative values of s,

(iii) downward directed accelerations are negative.

Notes (i) An acceleration produces retardation whenever it acts in the opposite direction to the velocity, irrespective of whether the acceleration itself is being taken to be positive or negative.

(ii) The equations of motion can be deduced from the definitions of velocity and acceleration and therefore do not introduce any new ideas; equation [2.8], however, highlights the important result that when a body moves from rest $s \propto t^2$.

(iii) For a body moving at constant velocity $a = 0$ and equations [2.6] and [2.7] reduce to $v = u$. Substituting for u in equation [2.8] (with $a = 0$) or in equation [2.9] gives

$$s = vt \qquad \text{at constant velocity}$$

Derivation of the Equations of Motion for Uniform Acceleration

Suppose that a body is moving with constant acceleration a and that in a time interval t its velocity increases from u to v and its displacement increases from 0 to s. Then, since

Acceleration = Rate of change of velocity

$$a = \frac{v - u}{t}$$

i.e. $v = u + at$ [2.6]

The average velocity is $\frac{1}{2}(u + v)$ and therefore, since

Displacement = Average velocity × time

$$s = \tfrac{1}{2}(u + v)t \quad [2.9]$$

Eliminating t between equations [2.6] and [2.9] leads to equation [2.7], and eliminating v between any two of the three equations that have now been derived leads to equation [2.8].

EXAMPLE 2.8

A ball is thrown vertically upwards with a velocity of $20\,\mathrm{m\,s^{-1}}$. Calculate:

(a) the maximum height reached,

(b) the total time for which the ball is in the air.
(Assume $g = 10\,\mathrm{m\,s^{-2}}$.)

Solution

(a) We shall take the upward direction to be positive. In the notation of this section:

$u = 20\,\mathrm{m\,s^{-1}}$ (the velocity with which the ball leaves the thrower's hand)

$v = 0$ (at the maximum height)

$a = -10\,\mathrm{m\,s^{-2}}$ (the minus sign is necessary because 'up' has been taken to be positive)

$s = h$ (where h is the maximum height)

From equation [2.7]

$$0^2 = 20^2 + 2(-10)h$$

i.e. $h = 20\,\mathrm{m}$

(b) $u = 20\,\mathrm{m\,s^{-1}}$

$a = -10\,\mathrm{m\,s^{-2}}$

$t = t$ (where t is the time the ball is in the air)

$s = 0$ (since the ball is back on the ground)

From equation [2.8]

$$0 = 20\,t + \tfrac{1}{2}(-10)\,t^2$$

i.e. $t = 0$ or $t = 4\,\mathrm{s}$

The required solution is $t = 4\,\mathrm{s}$. The other solution, $t = 0$, refers to the fact that the height of the ball was also zero when it was first projected.

QUESTIONS 2C

Take $g = 10\,\mathrm{m\,s^{-2}}$ where necessary.

1. A particle is moving in a straight line with a constant acceleration of $6.0\,\mathrm{m\,s^{-2}}$. As it passes a point, A, its speed is $20\,\mathrm{m\,s^{-1}}$. What is its speed $10\,\mathrm{s}$ after passing A?

2. A particle which is moving in a straight line with a velocity of $15\,\mathrm{m\,s^{-1}}$ accelerates uniformly for $3.0\,\mathrm{s}$, increasing its velocity to $45\,\mathrm{m\,s^{-1}}$. What distance does it travel whilst accelerating?

3. A car starts to accelerate at a constant rate of $0.80\,\text{m s}^{-2}$. It covers 400 m whilst accelerating in the next 20 s. What was the speed of the car when it started to accelerate?

4. A body of mass 3.0 kg, initially at rest, moves along a smooth horizontal surface under the effect of a horizontal force of 12 N. **(a)** Find the acceleration of the body. **(b)** Find the speed of the body after 5.0 s.

5. A car moving at $30\,\text{m s}^{-1}$ is brought to rest with a constant retardation of $3.6\,\text{m s}^{-2}$. How far does it travel whilst coming to rest?

6. A stone is dropped from the top of a cliff which is 80 m high. How long does it take to reach the bottom of the cliff?

7. A particle is projected vertically upwards at $30\,\text{m s}^{-1}$. Calculate: **(a)** how long it takes to reach its maximum height, **(b)** the two times at which it is 40 m above the point of projection, **(c)** the two times at which it is moving at $15\,\text{m s}^{-1}$.

8. A stone is fired vertically upwards from a catapult and lands 5.0 s later. What was the initial velocity of the stone? For how long was the stone at a height of 20 m or more?

9. A hot-air balloon is 21 m above the ground and is rising at $8.0\,\text{m s}^{-1}$ when a sandbag is dropped from it. How long does it take the sandbag to reach the ground?

10. A stone is thrown vertically upwards at $10\,\text{m s}^{-1}$ from a bridge which is 15 m above a river. **(a)** What is the speed of the stone as it hits the river? **(b)** With what speed would it hit the river if it were thrown downwards at $10\,\text{m s}^{-1}$?

11. A bullet of mass 8.00×10^{-3} kg moving at $320\,\text{m s}^{-1}$ penetrates a target to a depth of 16.0 mm before coming to rest. Find the resistance offered by the target, assuming it to be uniform.

2.4 MOTION UNDER GRAVITY

A body that is projected at an angle to the vertical moves along a curved (parabolic) path. In order to solve problems involving motion of this type, we consider the horizontal and vertical components of the motion separately. This is justified because the horizontal motion has no effect on the vertical motion and vice versa. To appreciate this, consider two bodies, A and B, projected horizontally off the edge of a table, and suppose that the velocity with which A is projected is greater than that of B (Fig. 2.5). Both A and B reach the ground at the same time even

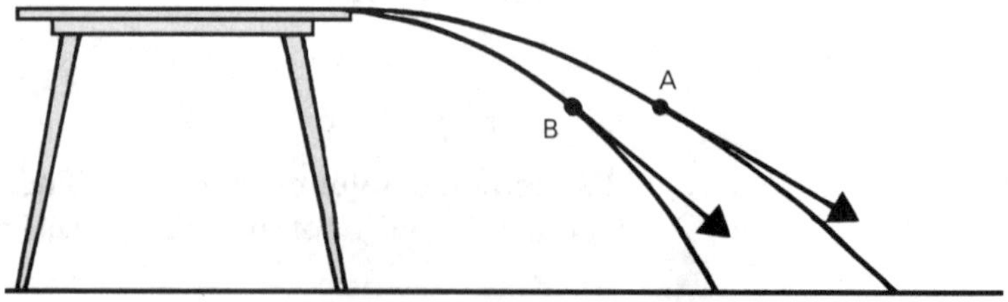

Fig. 2.5
To show the motion of two bodies projected horizontally under gravity

though their velocities of projection were different. This is because, initially, neither body had any vertical component of velocity (they were projected horizontally). The downward motions of both A and B are due to the effect of gravity, and this accelerates each at the same rate ($9.8\,\text{m s}^{-2}$). Since they both start from rest (in terms of the vertical motion) and travel the same vertical distance, they reach the ground at the same time. In the absence of air resistance, each body retains its original horizontal component of velocity for the whole of its motion. The horizontal distance travelled by A is therefore greater than that travelled by B.

EXAMPLE 2.9

A body is projected with a velocity of $200\,\mathrm{m\,s^{-1}}$ at an angle of $30°$ above the horizontal. Calculate:

(a) the time taken to reach the maximum height,

(b) its velocity after 16 s.

(Assume $g = 10\,\mathrm{m\,s^{-2}}$ and ignore air resistance.)

Fig. 2.6
Diagrams for Example 2.9

(a)

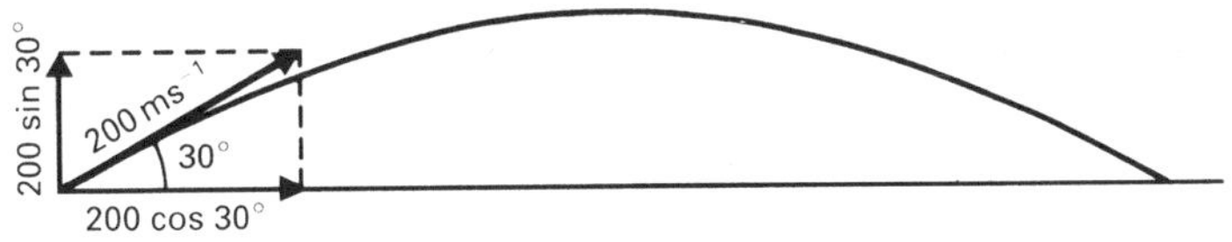

(b)

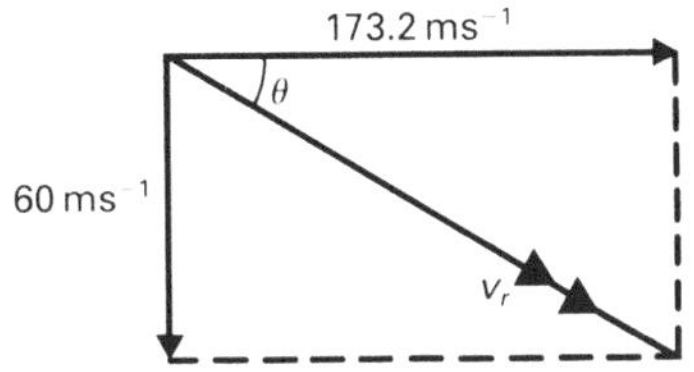

Solution

(a) Consider the vertical motion. In the notation of the last section:

$u = 200 \sin 30° = 100\,\mathrm{m\,s^{-1}}$

$v = 0$ (at the maximum height)

$a = -10\,\mathrm{m\,s^{-2}}$ (minus sign because 'up' has been taken to be positive)

$t = t$ (where t is the time taken to reach the maximum height)

From equation [2.6]

$$0 = 100 + (-10)t$$

i.e. $t = 10\,\mathrm{s}$

(b) Considering the vertical component of the motion:

$u = 100\,\mathrm{m\,s^{-1}}$

$v = v_y$

$a = -10\,\mathrm{m\,s^{-2}}$

$t = 16\,\mathrm{s}$

From equation [2.6]

$$v_y = 100 + (-10)16$$

i.e. $v_y = -60\,\mathrm{m\,s^{-1}}$

(The minus sign indicates that the body is moving downwards.)

The horizontal component of the velocity will still be $200 \cos 30°$ ($= 173.2\,\mathrm{m\,s^{-1}}$) since, in the absence of air resistance, there is no horizontal component of acceleration. The actual velocity, v_r, is therefore as shown in Fig. 2.6(b), from which

$$v_r^2 = 60^2 + 173.2^2 \qquad \text{i.e.} \quad v_r = 183\,\text{m s}^{-1}$$

Also,

$$\tan\theta = 60/173.2 \qquad \text{i.e.} \quad \theta = 19.1^\circ$$

2.5 PARABOLIC MOTION

A body projected with a velocity v at an angle α above the horizontal has a vertical component of velocity of $v \sin\alpha$. Its vertical displacement, y, after time t is given by equation [2.8] as

$$y = (v \sin\alpha)t - \tfrac{1}{2}gt^2 \qquad [2.10]$$

At the same time, its horizontal displacement, x, due to its constant horizontal component of velocity of $v \cos\alpha$, is given by $s = vt$ as

$$x = (v \cos\alpha)t \qquad [2.11]$$

Eliminating t between equations [2.10] and [2.11] leads to

$$y = x \tan\alpha - g\frac{\sec^2\alpha}{2v^2}x^2$$

This is the equation of a parabola and it follows, therefore, that a body moving under the influence of gravity travels along a parabolic path. The path of a charged particle in a uniform electric field is also a parabola (see section 50.1).

Points to Bear in Mind when Attempting Questions 2D

To Find Time of Flight

Use $s = ut + \frac{1}{2}at^2$ for the vertical motion with $s = 0$.

To Find Time to Maximum Height

Use $v = u + at$ for the vertical motion with $v = 0$.

To Find Maximum Height

Use $v^2 = u^2 + 2as$ for the vertical motion with $v = 0$.

To Find Range

Find the time of flight, t, then substitute for t in $s = vt$ for the horizontal motion.

To Find Direction of Motion

Use $\tan\theta = v_y/v_x$ where θ is the angle the direction of motion makes with the horizontal, and v_y and v_x are the vertical and horizontal components of velocity respectively.

QUESTIONS 2D

1. A particle is projected with a speed of 25 m s^{-1} at 30° above the horizontal. Find: **(a)** the time taken to reach the highest point of the trajectory, **(b)** the magnitude and direction of the velocity after 2.0 s.

2. A particle is projected with a velocity of 30 m s^{-1} at an angle of 40° above a horizontal plane. Find: **(a)** the time for which the particle is in the air, **(b)** the horizontal distance it travels.

3. A pebble is thrown from the top of a cliff at a speed of 10 m s^{-1} and at 30° above the horizontal. It hits the sea below the cliff 6.0 s later. Find: **(a)** the height of the cliff, **(b)** the distance from the base of the cliff at which the pebble falls into the sea.

4. A pencil is accidentally knocked off the edge of a (horizontal) desk top. The height of the desk is 64.8 cm and the pencil hits the floor a horizontal distance of 32.4 cm from the edge of the desk. What was the speed of the pencil as it left the desk?

5. A particle is projected from level ground in such a way that its horizontal and vertical components of velocity are 20 m s^{-1} and 10 m s^{-1} respectively. Find: **(a)** the maximum height of the particle, **(b)** its horizontal distance from the point of projection when it returns to the ground, **(c)** the magnitude and direction of its velocity on landing.

6. An aeroplane moving horizontally at 150 m s^{-1} releases a bomb at a height of 500 m. The bomb hits the intended target. What was the horizontal distance of the aeroplane from the target when the bomb was released?

2.6 GRAPHICAL REPRESENTATION OF MOTION IN A STRAIGHT LINE

Graphs can be used to represent the motion of a body which is moving in a straight line. (The motion must be in a straight line because there is no means of representing more than two directions, e.g. forwards and backwards, on a graph.) The method is particularly useful when the body under consideration has a non-uniform acceleration, for the equations of motion (section 2.3) do not apply in such cases and even calculus methods are of no use if the acceleration varies with time in such a way that it cannot be expressed mathematically.

Displacement–Time Graphs

By definition, velocity is rate of change of displacement and therefore **the slope of a graph of displacement against time represents velocity**. Suppose that the displacement–time graph shown in Fig. 2.7 refers to the motion of a shunting engine. Bearing in mind that the slope of the graph represents velocity, we can make the following analysis of the motion of the engine:

Fig. 2.7
A displacement–time graph

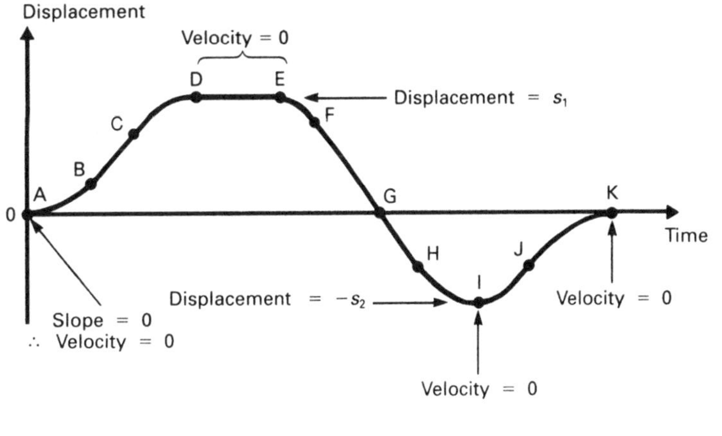

At A	Stationary
A–B	Accelerating (slope increasing)
B–C	Moving with constant velocity (slope constant)
C–D	Decelerating (slope decreasing)
D–E	Stationary
E–F	Accelerating and moving back towards the starting point
F–G	Moving with constant velocity
At G	Momentarily at the starting point
G–H	Moving away from the starting point with constant velocity in the opposite direction to the original direction
H–I	Decelerating
At I	Momentarily stationary
I–J	Accelerating and moving back towards the starting point
J–K	Decelerating
At K	Stationary at the starting point.

Note At the end of the period under consideration the engine is back at its starting point and therefore has zero displacement; the distance it has travelled, however, is $2s_1 + 2s_2$.

Velocity–Time Graphs

By definition, acceleration is rate of change of velocity and therefore **the slope of a graph of velocity against time represents acceleration. The area under such a graph represents distance.** We shall illustrate this by referring to the velocity–time graph in Fig. 2.8.

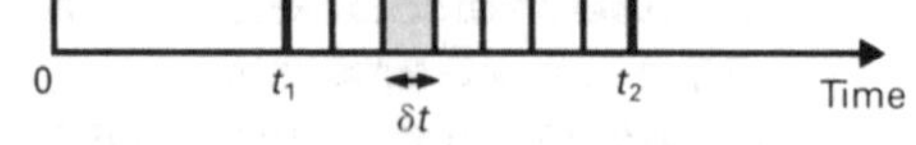

Fig. 2.8
To show that the area under a velocity–time graph represents distance

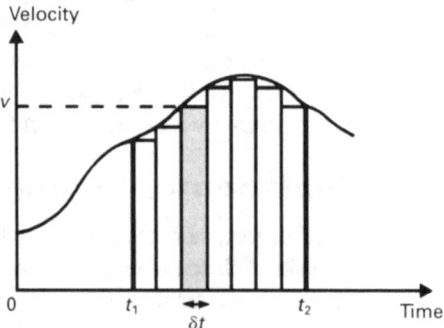

For a body which is moving with constant velocity, distance moved = velocity × time. It follows that if the velocity had the constant value of v during the time interval δt, the distance moved would be $v\,\delta t$. This is the area of the shaded strip, and therefore if the velocity varied with time according to the stepped line, the total distance moved in the interval from t_1 to t_2 would be the sum of the areas of the strips. By considering narrower and narrower strips we can make the stepped line follow the actual curve more and more closely. In the limit of infinitesimally narrow strips the sum of the areas of the strips is exactly equal to the area under the curve between t_1 and t_2, i.e. **the area under the curve between t_1 and t_2 represents the distance moved in the interval from t_1 to t_2.**

Suppose that a body moves in the manner represented by the velocity–time graph in 2.9. Bearing in mind that the slope of the graph represents acceleration, we can make the following analysis of the motion of the body:

Fig. 2.9
A velocity–time graph

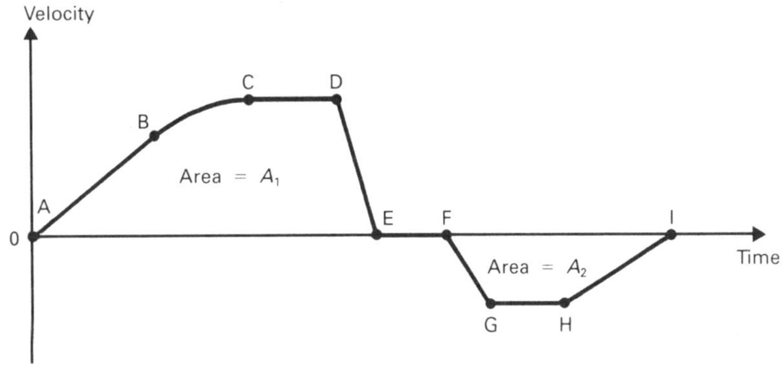

A–B	Moves from rest with a constant acceleration
B–C	Velocity still increasing, acceleration decreasing
C–D	Moving with constant velocity
D–E	Decelerating at a constant rate. Comes to rest
E–F	Stationary
F–G	Moving in the opposite direction to the original direction. Acceleration constant
G–H	Constant velocity
H–I	Decelerating at a constant rate. Comes to rest.

Note Total distance moved $= A_1 + A_2$. Net distance moved (i.e. magnitude of displacement $= A_1 - A_2$.

2.7 THE CONSERVATION OF LINEAR MOMENTUM

Suppose that two bodies, A and B, are involved in a collision (Fig. 2.10) and that there are no external forces acting. The force on A due to B, F_A, is, by Newton's third law, equal (in magnitude) to the force on B due to A, F_B. Therefore, by Newton's second law, each body experiences the same rate of change of momentum. Each force obviously acts for the same length of time as the other (i.e. for the duration of the collision), and therefore since the only forces that are acting are the internal forces F_A and F_B, the magnitudes of the changes of

Fig. 2.10
Collision of two bodies

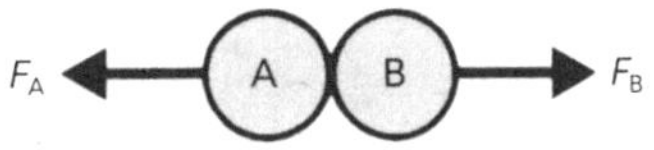

momentum of the two bodies will be the same. The changes in momentum, however, are oppositely directed and therefore the total change in momentum is zero. The result can be extended to any number of bodies in any situation where the bodies interact only with themselves, i.e. where there are no external forces. It is known as the **principle of conservation of linear momentum** and can be stated as:

> **The total linear momentum of a system of interacting (e.g. colliding) bodies, on which no external forces are acting, remains constant.**

The experimental investigation of the conservation of linear momentum is dealt with in section 2.14.

EXAMPLE 2.10

A body, A, of mass 4 kg moves with a velocity of 2 m s^{-1} and collides head-on with another body, B, of mass 3 kg moving in the opposite direction at 5 m s^{-1}. After the collision the bodies move off together with velocity v. Calculate v.

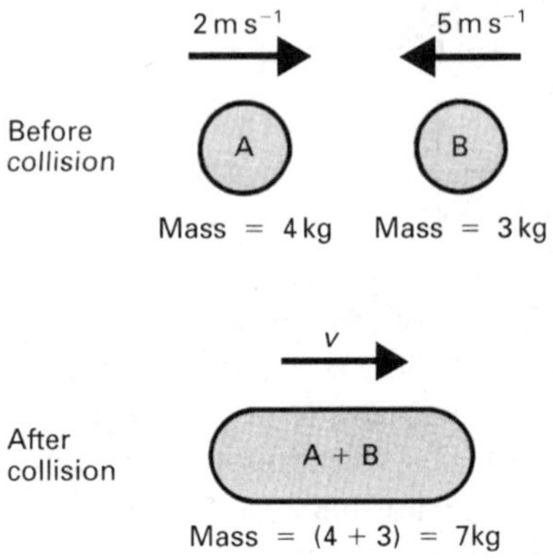

Fig. 2.11 Diagram for Example 2.10

Solution

Referring to Fig. 2.11 and taking momentum directed to the right to be positive, we find that

$$\text{Momentum of A before the collision} = 4 \times 2 = 8 \text{ kg m s}^{-1}$$

$$\text{Momentum of B before the collision} = 3 \times (-5) = -15 \text{ kg m s}^{-1}$$

$$\therefore \quad \text{The total momentum before the collision} = -7 \text{ kg m s}^{-1}$$

$$\text{Momentum of (A + B) after the collision} = 7v$$

By the principle of conservation of momentum

$$-7 = 7v$$

i.e. $\quad v = -1 \text{ m s}^{-1}$

The minus sign indicates that the bodies move to the left (i.e. in the original direction of B) after the collision.

EXAMPLE 2.11

A bullet of mass 6.0×10^{-3} kg is fired from a gun of mass 0.50 kg. If the muzzle velocity of the bullet is 300 m s^{-1}, calculate the recoil velocity of the gun.

Solution

Initially, both the bullet and the gun are at rest and their total momentum is zero. After firing, the momentum of the bullet $= 6.0 \times 10^{-3} \times 300 = 1.8$ kg m s^{-1}. By the principle of conservation of linear momentum, the total momentum after firing is equal to that before firing, and therefore the gun must have a momentum of

$1.8\,kg\,m\,s^{-1}$ in the opposite direction to that of the bullet. If the recoil velocity of the gun is v, then

$$0.50\,v = 1.8$$

i.e. $$v = 3.6\,m\,s^{-1}$$

QUESTIONS 2E

1. A body of mass 6 kg moving at $8\,m\,s^{-1}$ collides with a stationary body of mass 10 kg and sticks to it. Find the speed of the composite body immediately after the impact.

2. A bullet of mass m is fired horizontally from a gun of mass M. Find the recoil velocity of the gun if the velocity of the bullet is v.

3. A flat truck of mass 400 kg is moving freely along a horizontal track at $3.0\,m\,s^{-1}$. A man moving at right angles to the track jumps on to the truck causing its speed to decrease by $0.50\,m\,s^{-1}$. What is the mass of the man?

4. A kitten of mass 0.60 kg leaps at 30° to the horizontal out of a toy truck of mass 1.2 kg causing it to move over horizontal ground at $4.0\,m\,s^{-1}$. At what speed did the kitten leap?

5. A particle of mass $5\,m$ moving with speed v explodes and splits into two pieces with masses of $2m$ and $3m$. The lighter piece continues to move in the original direction with speed $5v$ relative to the heavier piece. What is the actual speed of the lighter piece?

2.8 ELASTIC COLLISIONS

Whenever two bodies collide, their total momentum is conserved unless there are external forces acting on them. The total kinetic energy (see section 5.3), however, usually decreases, since the impact converts some of it to heat and/or sound and/or permanently distorts the bodies leaving them with an increased amount of potential energy.

A collision in which some kinetic energy is lost is known as an **inelastic collision**. A **completely inelastic** collision is one in which the bodies stick together on impact. A collision is **elastic** if there is no loss of kinetic energy.

2.9 NEWTON'S EXPERIMENTAL LAW OF IMPACT

The relative velocity with which two bodies separate from each other, after a collision, is related to their relative velocity of approach and a constant known as the **coefficient of restitution**, e, of the two bodies. The relationship is known as Newton's experimental law of impact and can be expressed as

$$\text{Speed of separation} = e \times \text{Speed of approach} \qquad [2.12]$$

The coefficient of restitution of the two bodies is defined by equation [2.12] and depends on their elastic properties and the natures of their surfaces. These same properties determine whether a collision is elastic, inelastic or completely inelastic and therefore it is possible to classify a collision according to the value of e that is associated with it (Table 2.1).

Table 2.1
Classification of collisions

Type of collision	e
Elastic	1
Inelastic	<1
Completely inelastic	0

EXAMPLE 2.12

A body, A, of mass 6 kg and moving at 9 m s^{-1} collides head-on with another body, B, of mass 3 kg and moving in the same direction as A at 4 m s^{-1}. If the velocities of A and B after the collision are respectively v_A and v_B and the coefficient of restitution of the bodies is 0.8, calculate v_A and v_B. Assume that no external forces act on the system.

Fig. 2.12
Diagram for Example 2.12

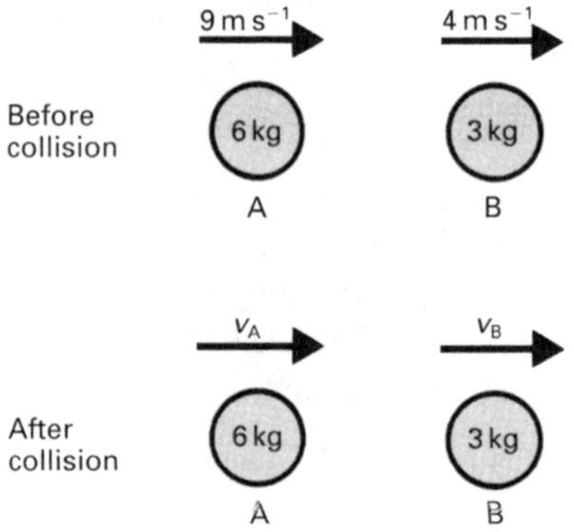

Solution

Refer to Fig. 2.12. There are no external forces acting on the system, in which case momentum is conserved and we may put

$$(6 \times 9) + (3 \times 4) = 6v_A + 3v_B$$

i.e. $$22 = 2v_A + v_B \qquad [2.13]$$

Using Newton's experimental law of impact (equation [2.12]) we have

$$v_B - v_A = 0.8\,(9 - 4)$$

i.e. $$v_B - v_A = 4 \qquad [2.14]$$

Solving equations [2.13] and [2.14] simultaneously gives

$$v_A = 6\,\text{m s}^{-1} \text{ and } v_B = 10\,\text{m s}^{-1}$$

QUESTIONS 2F

1. A sphere, A, of mass 3.0 kg moving at 8.0 m s^{-1} collides directly with another sphere, B, of mass 5.0 kg moving in the opposite direction to A at 4.0 m s^{-1}. Find the velocities of the spheres immediately after the impact if $e = 0.30$.

2. A sphere of mass m moving with velocity u is involved in an elastic collision with a sphere of mass $2m$ moving along the same line with velocity $-u$. Find the velocities of the spheres immediately after the impact.

3. A ball is dropped onto horizontal ground from a height of 9.0 m. Find the height to which the ball rises: **(a)** on the first bounce, **(b)** on the second bounce. ($e = 0.70$.)

2.10 IMPULSE

The impulse of a constant force, F, acting for a time, Δt, is defined by

$$\text{Impulse} = F\,\Delta t \qquad [2.15]$$

Impulse is a vector quantity; its direction is the same as that of the force. It follows from equation [2.15] that the unit of impulse is the newton second (N s). Note that $1\ \text{N s} = 1\ \text{kg m s}^{-1}$.

Suppose that a force, F, causes the momentum of a body to change by $\Delta(mv)$ in a time Δt. By Newton's second law, force = rate of change of momentum, and therefore

$$F = \frac{\Delta(mv)}{\Delta t}$$

i.e. $$F\,\Delta t = \Delta(mv)$$

Therefore by equation [2.15]

$$\text{Impulse} = \text{Change in momentum} \qquad [2.16]$$

It can be shown that equation [2.16] applies to variable forces too.

The definition of impulse imposes no limit on the length of time for which the force may act. Nevertheless, the concept of impulse is normally used only in situations where a large variable force is acting for only a short time, for example a golf-club striking a ball or the blow of a hammer on a nail. Forces such as these are known as impulsive forces.

When a batsman strikes a cricket ball he 'follows through' in order to keep the bat in contact with the ball for as long a time as possible. It follows from equation [2.15] that this increases the impulse and therefore, by equation [2.16], produces a larger change in momentum and so increases the speed at which the ball leaves the bat.

Suppose now that the ball is caught by a fielder. In catching it the fielder has to reduce the momentum of the ball to zero. It follows from equation [2.16] that the impulse on his hand will be the same no matter how he catches the ball. However, by equation [2.15], he can reduce the force he feels by drawing his hands backwards to increase the time taken to effect the catch. Not only is this less painful, but it also reduces the likelihood of the ball bouncing out of his hands.

The impulse of a variable force, F, acting for a time, t, is defined by

$$\text{Impulse} = \int_0^t F\,\mathrm{d}t \qquad [2.17]$$

Measuring the change in momentum that a variable force produces is usually much easier than measuring the way in which it varies with time. In practice, therefore, an impulse is more likely to be evaluated on the basis of equation [2.16] than equation [2.17].

2.11 FORCE–TIME GRAPHS

It follows from equation [2.17] that **the area under a graph of force against time represents impulse** (Fig. 2.13). It follows from equation [2.16] that **it also represents change in momentum**.

Fig. 2.13
Force–time graphs

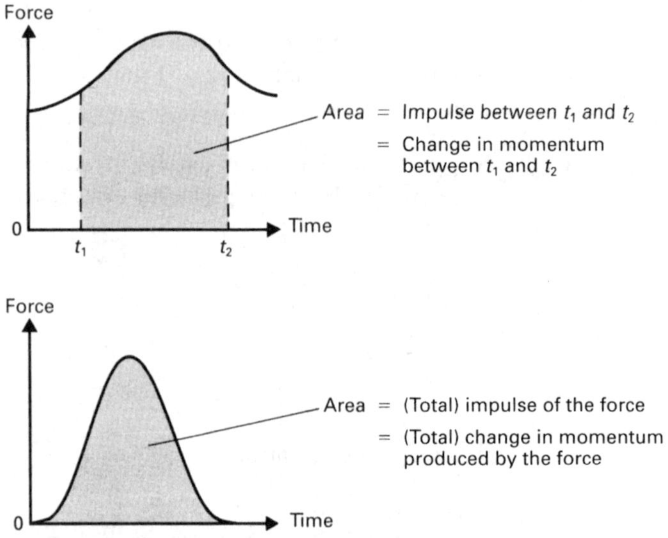

EXAMPLE 2.13

A body of mass 4 kg is moving at 5 m s^{-1} when it is given an impulse of 8 N s in the direction of its motion. (a) What is the velocity of the body immediately after the impulse? (b) If the impulse acts for 0.02 s, what is the average value of the force exerted on the body?

Solution

Fig. 2.14
Diagram for Example 2.13

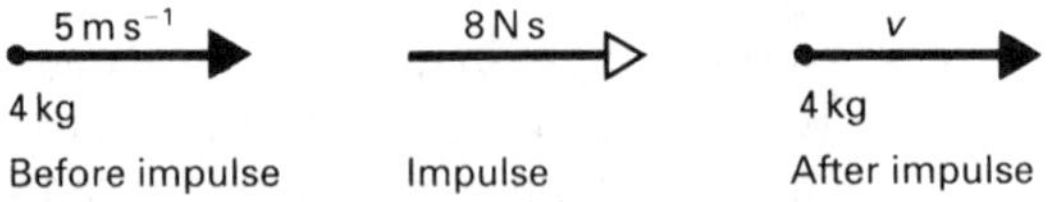

(a) Refer to Fig. 2.14. Let v = velocity of body immediately after the impulse.

$$\text{Impulse} = \text{Change in momentum}$$

$$\therefore \quad 8 = 4v - 4 \times 5$$

$$\therefore \quad 8 = 4v - 20 \quad \text{i.e.} \quad v = 7\,\text{m s}^{-1}$$

(b) Let F = average force

$$\text{Impulse} = F\,\Delta t$$

$$\therefore \quad 8 = F \times 0.02 \quad \text{i.e.} \quad F = 4 \times 10^2\,\text{N}$$

QUESTIONS 2G

1. A particle of mass 6.0 kg moving at $8.0\,\text{m s}^{-1}$ due N is subjected to an impulse of 30 N s. Find the magnitude and direction of the velocity of the particle immediately afterwards if the direction of the impulse is: **(a)** due N, **(b)** due S.

2. A ball of mass 6.0×10^{-2} kg moving at $15\,\text{m s}^{-1}$ hits a wall at right angles and bounces off along the same line at $10\,\text{m s}^{-1}$. **(a)** What is the magnitude of the impulse of the wall on the ball? **(b)** The ball is estimated to be in contact with the wall for 3.0×10^{-2} s, what is the average force on the ball?

3. A body of mass 2.0 kg and which is at rest is subjected to a force of 200 N for 0.20 s followed by a force of 400 N for 0.30 s acting in the same direction. Find: **(a)** the total impulse on the body, **(b)** the final speed of the body.

4. Find the final speed of the body in question 3 by using $F = ma$ and $v = u + at$.

2.12 FRICTION

Static Friction

When the surface of a body moves or tends to move over that of another, each body experiences a frictional force. The frictional forces act along the common surface, and each is in such a direction as to oppose the relative motion of the surfaces.

Fig. 2.15 illustrates an arrangement which can be used to investigate frictional forces. Small masses are added, one at a time, to the scale-pan in order to increase P. At first P is small and the block does not move, but as more masses are added, eventually a point is reached at which the block starts to slide. This is interpreted by supposing that for small values of P the frictional force F is equal to P but that there is a maximum frictional force which can be brought into play. This is called the **limiting frictional force** and its value is equal to the value of P at which the block starts to move. The way in which the frictional force depends on the normal reaction R can be investigated by placing weights on the block. The effect of the area of contact can be studied by repeating the experiment with different faces of the block in contact with the table.

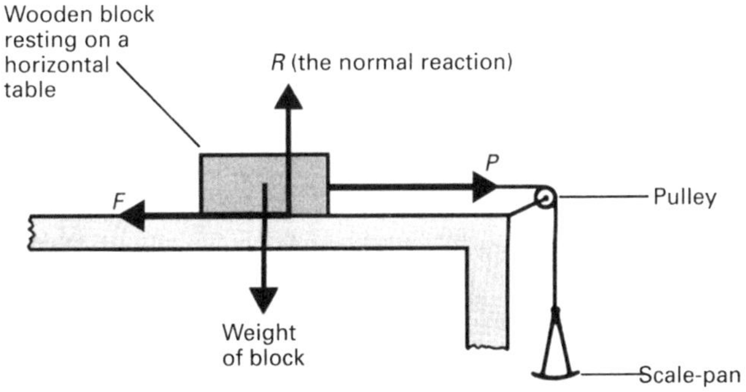

Fig. 2.15
Investigation of frictional forces

Sliding Friction

The frictional force which exists between two adjacent surfaces which are in relative motion is usually slightly less than the limiting frictional force between the surfaces and is called the **sliding** (or **dynamic** or **kinetic**) **frictional force**. This can be demonstrated by using the apparatus of Fig. 2.15 and giving the block a slight push each time a mass is added to the scale-pan. The value of P at which the block continues to move with constant velocity after being pushed is the value of the sliding frictional force and is less than the force required to produce motion when the block is not pushed.

The Laws of Friction

The results of experiments of the type described in Static Friction and Sliding Friction above are summarized in the laws of friction.

(i) The frictional force between two surfaces opposes their relative motion or attempted motion.

(ii) Frictional forces are independent of the area of contact of the surfaces.

(iii) For two surfaces which have no relative motion the limiting frictional force is directly proportional to the normal reaction.

For two surfaces which have relative motion the sliding frictional force is directly proportional to the normal reaction and is approximately independent of the relative velocity of the surfaces.

The Coefficients of Friction

The coefficient of limiting friction μ and the coefficient of sliding friction μ' are defined by

$$\mu = \frac{F}{R} \quad \text{and} \quad \mu' = \frac{F'}{R}$$

where F and F' are the limiting and sliding frictional forces respectively and R is the normal reaction. Both μ and μ' depend on the nature and the condition of the surfaces which are in contact but are independent of the area of contact. For steel on steel $\mu \approx 0.8$; for Teflon on Teflon $\mu \approx 0.04$. (The values given are approximate because even a mono-molecular layer of some surface impurity affects the experimental results.) If two surfaces are assumed to be perfectly smooth, there is no frictional force and $\mu = \mu' = 0$.

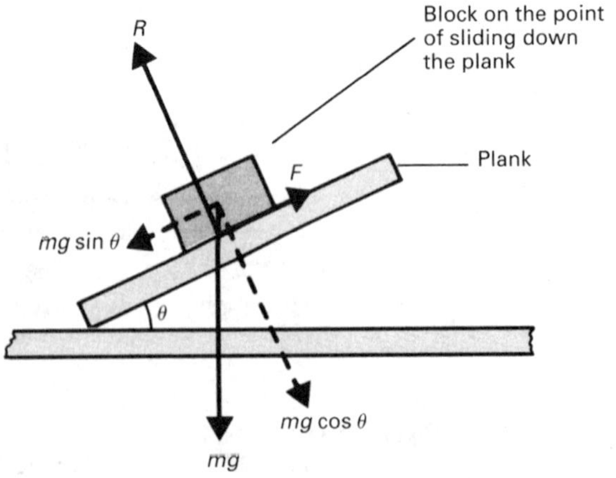

Fig. 2.16 Determination of the coefficient of limiting friction

The coefficient of limiting friction can be determined by carrying out an experiment of the type described in Static Friction above and measuring R and the minimum value of P that produces motion. The arrangement shown in Fig. 2.16 provides an alternative method. One end of the plank is raised gradually and the value of θ (the **angle of friction**) at which the block is on the point of slipping is measured. When the block is about to slip $\mu = F/R$, and therefore since

$$mg \sin \theta = F \quad \text{and} \quad mg \cos \theta = R$$

$$\mu = \frac{mg \sin \theta}{mg \cos \theta}$$

i.e. $\mu = \tan \theta$

An Explanation of the Laws of Friction

On a microscopic level, even a highly polished surface has bumps and hollows. It follows that when two surfaces are put together the actual area of contact is less than the apparent area of contact (Fig. 2.17).

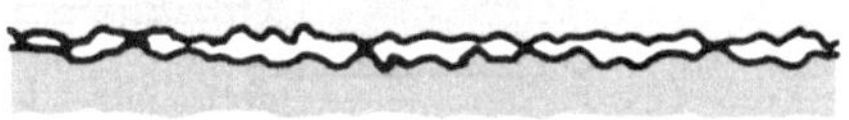

Fig. 2.17
Magnified cross-section through two surfaces in contact

For example, it has been estimated that for steel on steel, the actual contact area can be as little as one ten-thousandth of the apparent area. The pressures at the contact points are very high, and it is thought that the molecules are pushed into such close proximity that the attractive forces between them weld the surfaces together at these points. These welds have to be broken before one surface can move over the other. Clearly, therefore, no matter in which direction the motion occurs there is a force which opposes it. This explains law (i).

If the apparent area of contact of a body is decreased by turning the body so that it rests on one of its smaller faces, the number of contact points is reduced. Since the weight of the body has not altered, there is increased pressure at the contact points and this flattens the bumps so that the total contact area and the pressure return to their original values. Thus, although the apparent area of contact has been changed, the actual area of contact has not. This explains law (ii).

The extent to which the bumps are flattened depends on the weight of the body. Therefore the greater the weight, the greater the actual area of contact. This explains law (iii), because the weight is equal to the normal reaction.

2.13 DETERMINATION OF THE ACCELERATION DUE TO GRAVITY (*g*) BY FREE FALL

The apparatus is shown in Fig. 2.18. The principle of the method is to measure the time, t, for a ball-bearing to fall from rest through a measured distance, h.

The circuitry is such that switching on the electronic timer automatically cuts off the current to the electromagnet and releases the ball-bearing. The bearing falls freely until it strikes the hinged metal plate. The impact causes the plate to swing downwards, breaking the electrical connection at X and stopping the timer. The timer therefore automatically registers the time of fall.

Once h has been measured (with an extending rule, say) the acceleration due to gravity, g, can be calculated. It follows from $s = ut + \frac{1}{2}at^2$ (equation [2.8]) with $s = h$, $u = 0$, $a = g$ and $t = t$, that $g = 2h/t^2$, hence g.

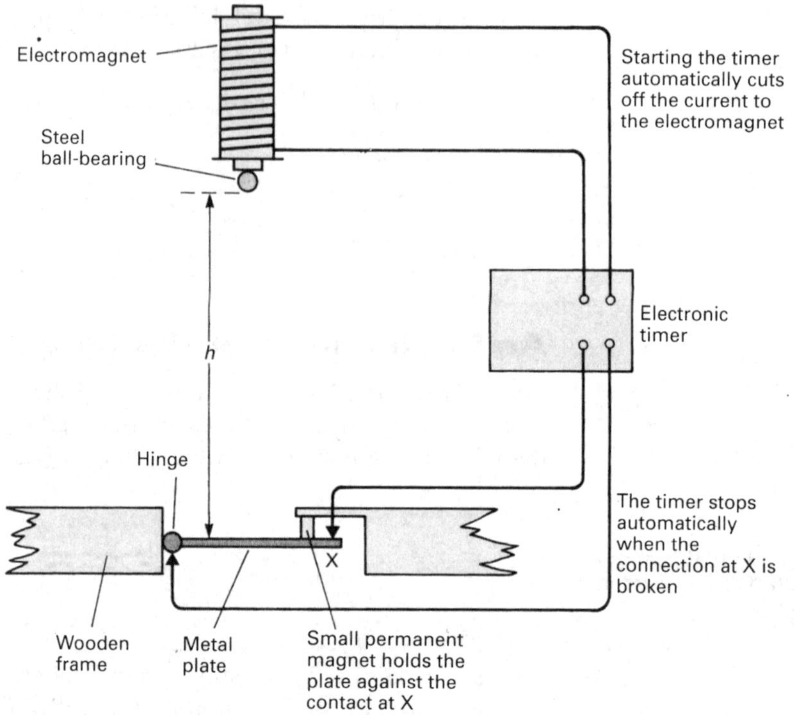

Fig. 2.18
Apparatus to determine ***g*** by free fall

Notes

(i) The timer should be capable of registering t with an uncertainty of ± 0.01 s or less.

(ii) h is measured from the bottom of the ball-bearing.

(iii) There may be a delay in releasing the ball-bearing due to residual magnetism in the electromagnet. The likelihood of this can be reduced by arranging that the bearing is held only weakly by the electromagnet to start with. This can be done by reducing the magnetizing current to the minimum that will hold the bearing, or by placing a piece of paper or thin card between the bearing and the electromagnet.

(iv) The experiment should be repeated a number of times and the average value of g found. Alternatively, the times of fall may be measured for a number of different values of h. Since $g = 2h/t^2$, $\sqrt{h} = (\sqrt{g/2})t$ and therefore the gradient of a graph of $\sqrt{h}$ against t is $\sqrt{g/2}$, allowing g to be found graphically. This has the advantage that the effect of any constant error in t (e.g. that due to the bearing not being released immediately the timer is started) is eliminated. (If there is an error of this type, the graph will not pass through the origin but the gradient will be unaffected.)

2.14 EXPERIMENTAL INVESTIGATION OF THE PRINCIPLE OF CONSERVATION OF LINEAR MOMENTUM

The principle of conservation of linear momentum can be investigated by means of two plastic vehicles riding on the cushion of air above a linear air-track (Fig. 2.19). The track is a hollow tube of triangular cross-section through which air is blown; the air emerges through holes in each side of the track. It has adjustable feet

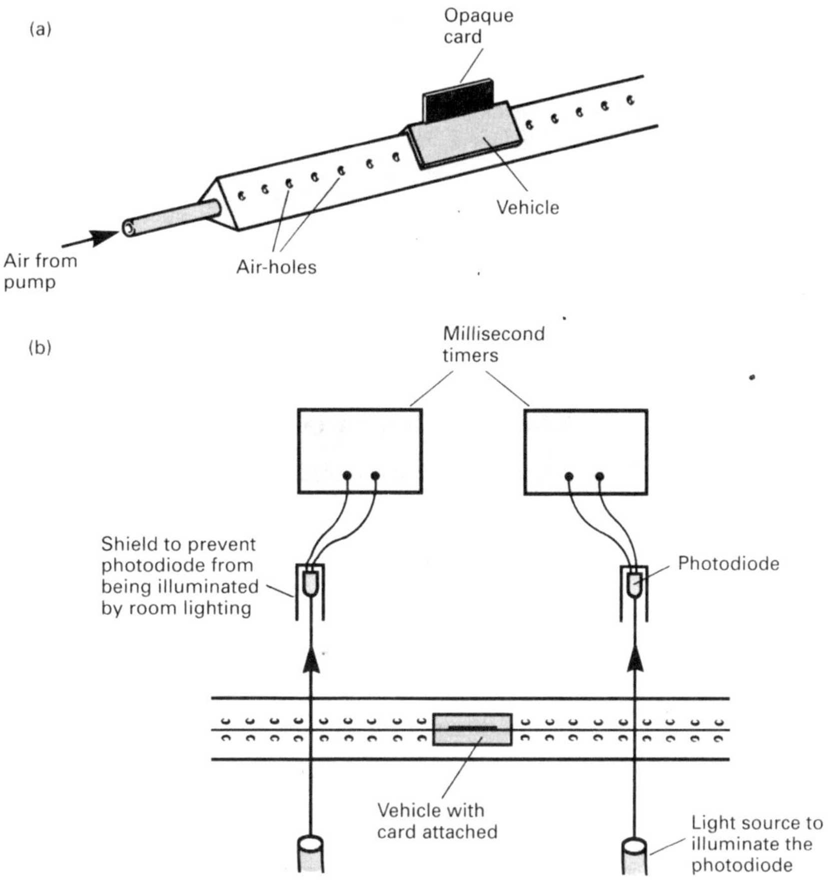

Fig. 2.19
(a) Vehicle on an air-track, (b) timing arrangement

allowing it to be made accurately horizontal so that the vehicles have no tendency to drift along it in either direction. A number of small (e.g. 50 g) masses may be attached to the vehicles. Each vehicle can carry an opaque card of known length (e.g. 10 cm) which is arranged to interrupt a beam of light falling on a photodiode. The circuitry is such that each of the millisecond timers is inoperative whilst light is falling on the photodiode to which it is connected. When a light beam is broken by the leading edge of a card the associated timer switches on and remains operative for as long as the card is in the beam. The timer therefore records the time for the vehicle to travel a distance equal to the length of the card and so allows the speed to be found.

Completely Inelastic Collision

Refer to Fig. 2.20. A is pushed towards B, which is stationary and has no card attached. A interrupts beam X and therefore its speed (u_A) before impact can be found. A pin on the front of A sticks in a small piece of plasticine on the back of B, and the vehicles then move together. The card on A interrupts beam Y allowing the (common) speed (v_{AB}) of A and B to be found.

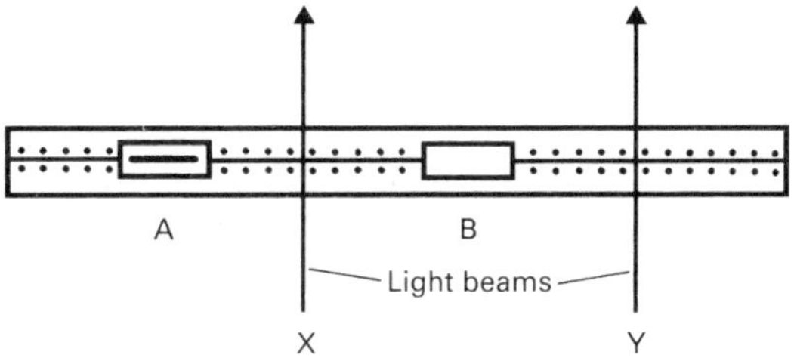

Fig. 2.20
Initial arrangement for inelastic collision

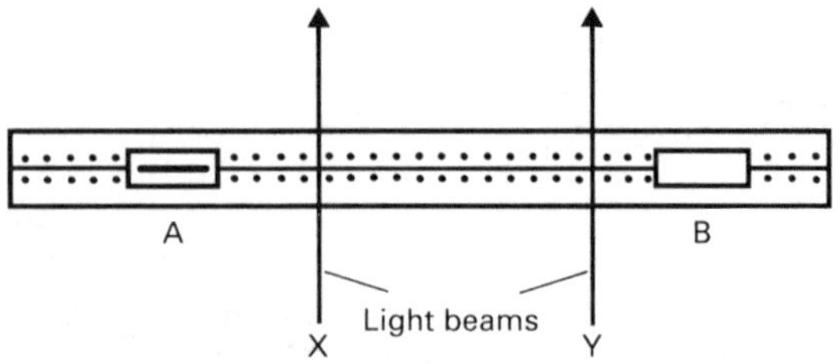

Fig. 2.21
Initial arrangement for elastic collision

Suppose the masses of A and B are m_A and m_B respectively. Momentum is conserved if $m_A u_A = (m_A + m_B) v_{AB}$. The experiment should be repeated for a number of different values of m_A, m_B and u_A.

Elastic Collision

Refer to Fig. 2.21. A and B (each carrying a card) are pushed towards each other so that they collide in the region between the beams. Since A will have passed through beam X and B will have passed through beam Y, their speeds before the collision can be found. Each vehicle has a stretched rubber band attached to its front end, and these act as buffers so that the collision is almost (perfectly) elastic (see section 2.8). It can be arranged that each vehicle reverses its direction of motion on impact. Since A then passes back through beam X and B passes through beam Y, their speeds after the collision can be found. The experiment requires two people – one to observe each timer.

Suppose the masses of A and B are m_A and m_B respectively, and their speeds are u_A and u_B before collision, and v_A and v_B after collision. Taking left to right as positive, the initial momentum is $m_A u_A - m_B u_B$ and the momentum after impact is $m_B v_B - m_A v_A$. Momentum is conserved if, within experimental error, $m_A u_A - m_B u_B = m_B v_B - m_A v_A$. The experiment should be repeated for a number of different values of m_A, m_B, u_A and u_B.

2.15 EXPERIMENTAL INVESTIGATION OF *F* = *ma*

Newton's second law in the form $F = ma$ can be investigated using the apparatus shown in Fig. 2.22. To compensate for friction, the slope of the runway is adjusted so that the trolley, when given a slight push, runs down it at constant speed (dots equally spaced on ticker-tape). The accelerating force is provided by means of an elastic thread attached to the rear of the trolley. The experimenter pulls on the

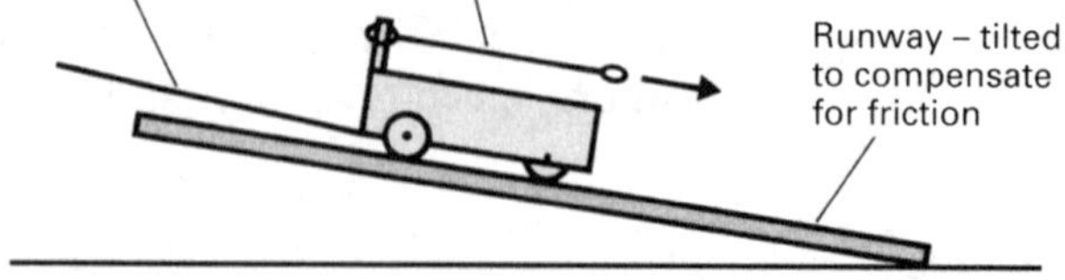

Fig. 2.22
Apparatus for investigating ***F*** = ***ma***

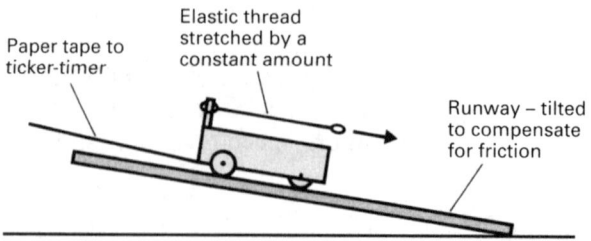

thread and walks along keeping the length of the thread constant (equal to the length of the trolley, say). The effects of friction have been compensated for by tilting the track and therefore the net force on the trolley is that provided by the stretched thread. Since the thread is stretched by a constant amount, the trolley is being accelerated by a constant force. The acceleration of the trolley is found by analysing the spacings of the dots on the ticker-tape. (The dots are produced at intervals of $\frac{1}{50}$ s, from which the velocity, and hence the acceleration, can be calculated.)

The effect of doubling (or tripling) the accelerating force is investigated by using two (or three) identical threads in parallel with each other and stretched by the same amount as in the first experiment. The effect of doubling (or tripling) the mass is investigated by stacking two (or three) identical trolleys on top of each other.

The accelerating force is proportional to the number of threads and the mass is proportional to the number of trolleys. A graph of acceleration against (number of threads/number of trolleys) can therefore be expected to be a straight line through the origin (i.e. $a \propto F/m$).

Note The wheels of the trolleys are made from a low-density material so that very little of the accelerating force is 'wasted' in providing the angular acceleration of the wheels.

CONSOLIDATION

Newton's first law Every body continues in a state of rest or of uniform (unaccelerated) motion in a straight line unless acted on by some external force.

Newton's second law The rate of change of momentum of a body is directly proportional to the external force acting on the body and takes place in the direction of the force.

$$F = \frac{\mathrm{d}}{\mathrm{d}t}(mv)$$

becomes

$$F = ma \qquad \text{for constant mass}$$

The newton (N) is defined as that force which produces an acceleration of 1 m s^{-2} when it acts on a mass of 1 kg.

Newton's third law If A exerts a force on B, then B exerts an equal and oppositely directed force on A

$$s = vt \qquad \text{for constant velocity}$$

$$\left.\begin{aligned} v &= u + at \\ v^2 &= u^2 + 2as \\ s &= ut + \tfrac{1}{2}at^2 \\ s &= \tfrac{1}{2}(u + v)t \end{aligned}\right\} \quad \text{for constant acceleration}$$

Displacement–Time Graphs

Gradient = velocity

Velocity–Time Graphs

Gradient = acceleration

Area under graph = distance

The principle of conservation of linear momentum The total linear momentum of a system of interacting (e.g. colliding) bodies, on which no external forces are acting, remains constant.

An elastic collision is one in which there is no loss of kinetic energy.

Law of Impact

Speed of separation = $e \times$ Speed of approach

Impulse

Impulse of constant force = $F\,\Delta t$

Impulse of variable force = $\int_0^t F\,\mathrm{d}t$

Impulse = Change in momentum (for both constant and variable forces)

3

TORQUE

3.1 DEFINITION OF TORQUE

Consider a force acting on a rigid body (Fig. 3.1) so as to cause it to turn about an axis which is perpendicular to the paper and passes through O. The effect of the force is determined by its turning moment, a quantity which depends not only on the size and direction of the force but also on where it acts. **The turning moment (or torque)** is defined by

$$T = Fd \tag{3.1}$$

where

T = torque (or turning moment) (N m)

F = the magnitude of the force (N)

d = the perpendicular distance of the line of action of the force from the axis (m).

Fig. 3.1
Definition of torque

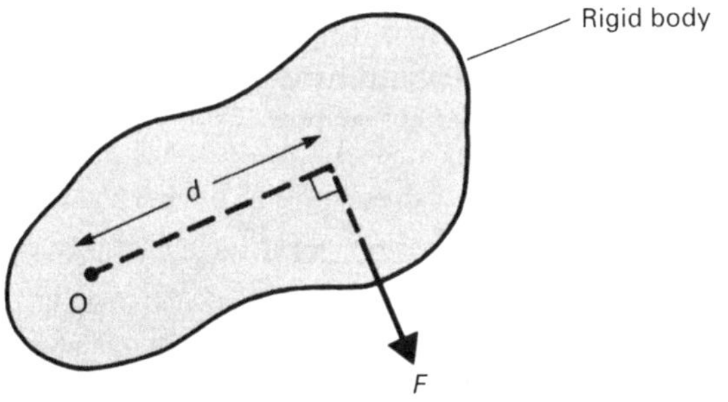

EXAMPLE 3.1

Find the moment of the 10 N force about the axis through O and perpendicular to the paper in each of the three situations shown in Fig. 3.2.

Fig. 3.2
Diagram for Example 3.1

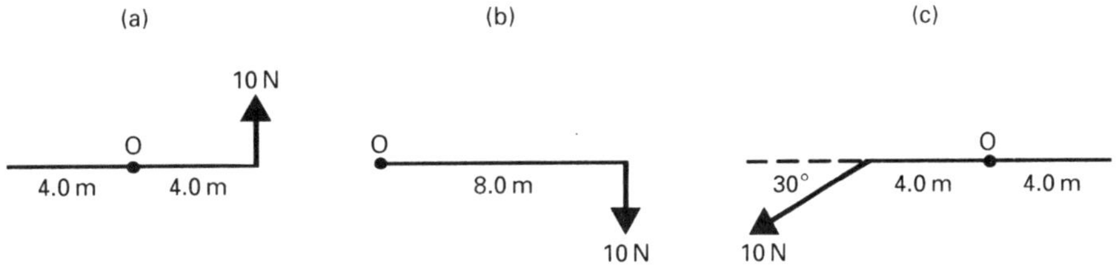

Solution

(a) Moment about O = $10 \times 4.0 = 40$ N m (anti-clockwise)

(b) Moment about O = $10 \times 8.0 = 80$ N m (clockwise)

(c) Refer to Fig. 3.3(a). Perpendicular distance of line of action of 10 N force from O = OB = OA sin 30° = 2.0 m

$\therefore$ Moment about O = $10 \times 2.0 = 20$ N m (anti-clockwise)

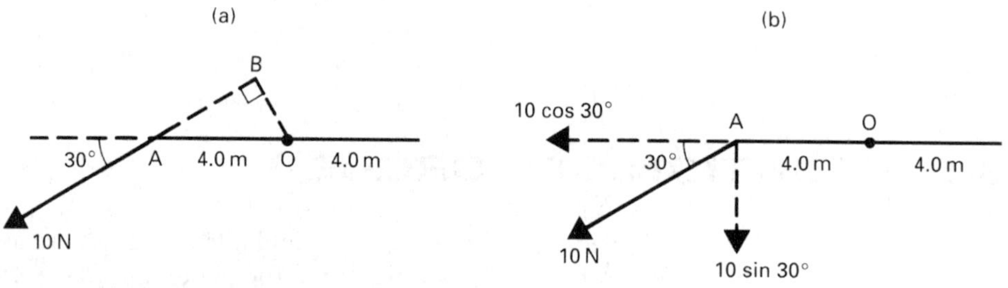

Fig. 3.3 Diagram for solution of Example 3.1(c)

Alternative Method

The 10 N force has components of 10 sin 30° and 10 cos 30° perpendicular and parallel to AO respectively (Fig. 3.3(b)).

Moment about O of perpendicular component = $10 \sin 30° \times 4.0$
= 20 N m
(anti-clockwise)

Moment about O of parallel component = 0

$\therefore$ Total moment about O = 20 N m (anti-clockwise)

QUESTIONS 3A

1. Find the moment of the 20 N force about axes perpendicular to the paper and through: **(a)** A, **(b)** B, **(c)** C, **(d)** D, **(e)** O where O is the centre of the rectangle

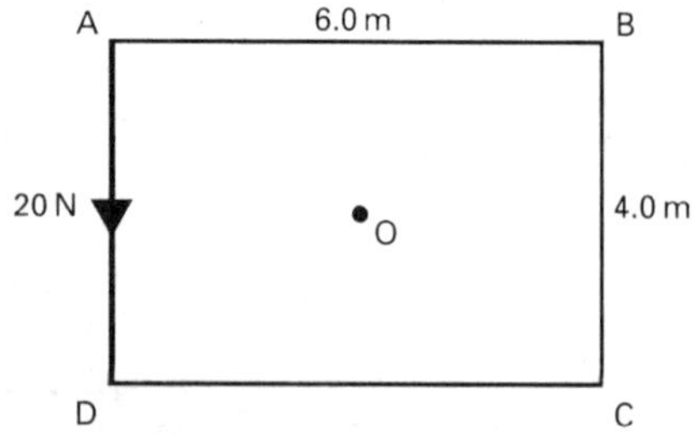

2. By resolving the 40 N force into two suitable components, or otherwise, find its moment about an axis perpendicular to the paper and through: **(a)** A, **(b)** B, **(c)** C.

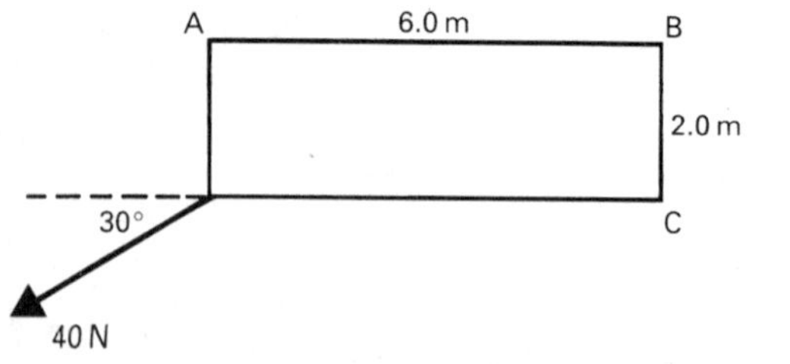

3.2 COUPLES

Two forces which are equal in magnitude and which are anti-parallel constitute a couple (Fig. 3.4).

Notes (i) There is no direction in which a couple can give rise to a resultant force, and therefore **a couple can produce a turning effect only** – it cannot produce translational motion.

(ii) Since a single force is bound to produce translation, it follows that **a couple cannot be represented by a single force**.

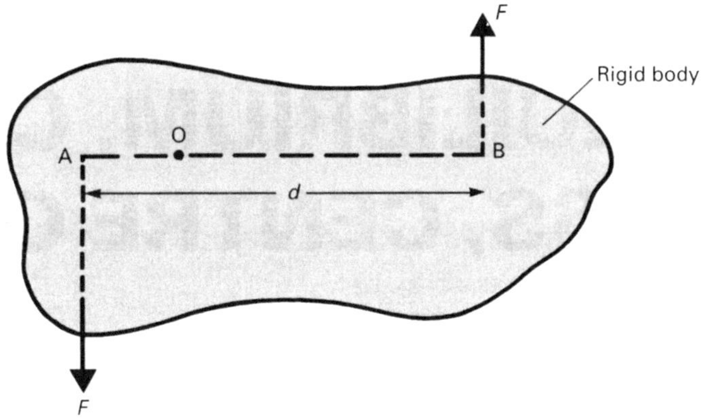

Fig. 3.4
Definition of a couple

3.3 TORQUE DUE TO A COUPLE

In Fig. 3.4,

$$\begin{aligned}\text{Total torque about O} &= F \times \text{OA} + F \times \text{OB}\\ &= F(\text{OA} + \text{OB})\\ &= Fd\end{aligned}$$

Thus, the torque about O does not depend on the position of O and therefore it follows that:

The torque due to a couple is the same about any axis and is given by

$$\text{Torque due to a couple} = \text{One force} \times \text{Separation of forces} \quad [3.2]$$

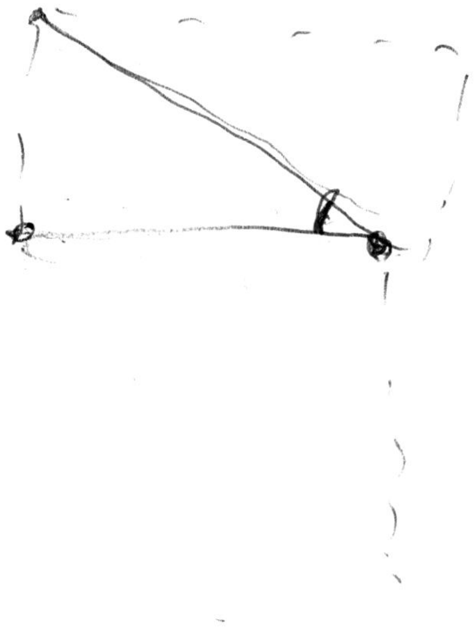

4

EQUILIBRIUM, CENTRE OF MASS, CENTRE OF GRAVITY

4.1 THE CONDITIONS FOR EQUILIBRIUM

A body is in equilibrium if:

(a) the acceleration of its centre of mass is zero in all directions, and

(b) its angular acceleration is zero.

Neither of these conditions requires that the body is at rest – a body may move with constant velocity and rotate with constant angular velocity and still be in equilibrium!

It follows from (a) and (b) that

A body is in equilibrium if:

(i) the resultant force on its centre of mass is zero, **and**

(ii) the total torque about all axes is zero.

Statements (i) and (ii) are often referred to as **the conditions for equilibrium** and are more useful in problem solving than (a) or (b). It can be shown that for a body subject to coplanar forces only, condition (i) will have been fulfilled if the resultant force in any two directions in the plane of the forces is zero. Condition (ii) will have been fulfilled if the total torque about any one axis which is perpendicular to the plane of the forces is zero. Therefore

To prove that a system of coplanar forces is in equilibrium it is sufficient to show that:

1 the resultant force in any two directions in the plane of the forces is zero, **and**

2 the total torque about any one axis which is perpendicular to the plane of the forces is zero.

Neither one of these conditions is sufficient on its own to show that a body is in equilibrium. On the other hand, if a body is known to be in equilibrium, then we may make use of 1 or 2 or both 1 and 2. It also follows that

If a body is in equilibrium:

I the resultant force is zero in all directions, and

II the total torque is zero about any axis.

Notes (i) Conditions 1 and 2 are known as sufficient conditions because together they form the minimum set of conditions which is sufficient to ensure equilibrium under the action of coplanar forces*. Conditions I and II are known as necessary conditions, in the sense that each is necessarily true, rather than that it is necessary to show them to be true.

(ii) Statement II is sometimes called **the principle of moments**, and can also be expressed as

If a body is in equilibrium, the total clockwise moment about any axis is equal to the total anti-clockwise moment about the same axis.

(iii) Statement I and statement II (in both its forms) also apply when the equilibrium is due to non-coplanar forces.

(iv) To prove that a body acted on by non-coplanar forces is in equilibrium it is sufficient to show that:

the resultant force in any three mutually perpendicular directions is zero, and

the total torque about each of any three mutually perpendicular axes is zero.

(v) When solving problems in which a system of coplanar forces is known to be in equilibrium we may choose two directions and apply condition 1 in each direction, and we may choose one axis and apply condition 2. Thus we resolve twice and take moments once. This gives three independent equations and allows us to find the values of three unknowns. There are two alternatives – we may resolve once and take moments twice, or we may take moments about three axes which are not in line with each other. It is not possible to obtain more than three independent equations and therefore there is no point in, for example, resolving twice and taking moments twice.

Concurrent Forces

Concurrent forces are forces whose lines of action intersect at a single point. A little thought should convince the reader that it is impossible for such a system of forces to produce a torque about any axis if their resultant is zero. It follows that

Concurrent forces are in equilibrium if their resultant is zero.

To prove that concurrent coplanar forces are in equilibrium it is sufficient to show that 1 is true. If we know that a system of concurrent coplanar forces is in equilibrium, we use 1 alone when solving problems – there is no point using 2.

Notes (i) If a body is in equilibrium under the action of three non-parallel coplanar forces, the forces must be concurrent. (See section 4.2.)

*This is not the only set of minimum conditions, but it is the one most commonly used.

(ii) **A particle** is an object which has mass but which is small enough to be regarded as a point. It follows that a set of forces acting on a particle must be concurrent forces.

EXAMPLE 4.1

The system of forces in Fig. 4.1 is in equilibrium. Find P and Q.

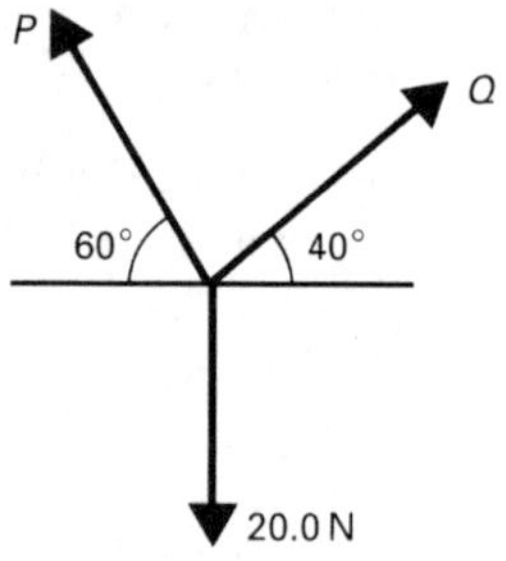

Fig. 4.1
Diagram for Example 4.1

Solution

We make use of condition 1 in the horizontal and vertical directions. Refer to Fig. 4.1

Resolving horizontally:

$$P \cos 60^\circ = Q \cos 40^\circ \qquad [4.1]$$

Resolving vertically:

$$P \sin 60^\circ + Q \sin 40^\circ = 20.0 \qquad [4.2]$$

By equation [4.1]

$$P = \frac{Q \cos 40^\circ}{\cos 60^\circ} \quad \text{i.e.} \quad P = 1.532\ Q \qquad [4.3]$$

Substituting for P in equation [4.2] gives

$$1.532\ Q \sin 60^\circ + Q \sin 40^\circ = 20.0$$

$$\therefore \quad 1.970\ Q = 20.0 \quad \text{i.e.} = 10.2\ \text{N}$$

Substituting for Q in equation [4.3] gives

$$P = 15.6\ \text{N}$$

Note We have resolved horizontally and vertically. It would have been quite reasonable to resolve perpendicular to P and perpendicular to Q. The advantage of this is that it gives an equation for Q which does not involve P and an equation for P which does not involve Q. The main disadvantage is that it is necessary to work out the angles that the forces make with these directions and although this is trivial, it leads to a rather messy diagram. It is by far the best method, though, when the unknown forces are at 90° to each other.

EXAMPLE 4.2

A uniform plank AB which is 6 m long and has a weight of 300 N is supported horizontally by two vertical ropes at A and B. A weight of 150 N rests on the plank at C where AC = 2 m. Find the tension in each rope.

Solution

The plank is uniform and therefore its weight acts at its mid-point, G, say. Let the tensions in the ropes at A and B be T_A and T_B respectively. Refer to Fig. 4.2.

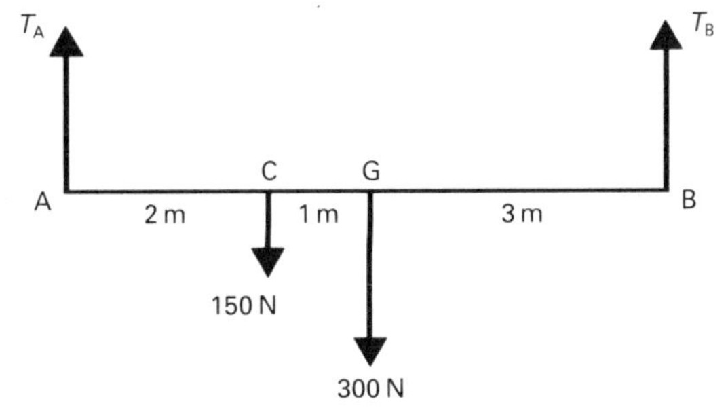

Fig. 4.2
Diagram for Example 4.2

The plank is in equilibrium and therefore the clockwise moment about any point is equal to the anti-clockwise moment about the same point. (Condition 2.)

Taking moments about A gives

$$T_B \times 6 = 150 \times 2 + 300 \times 3$$

$$\therefore \quad 6T_B = 1200 \quad \text{i.e.} \quad T_B = 200 \text{ N}$$

Resolving vertically gives

$$T_A + T_B = 150 + 300$$

$$\therefore \quad T_A + 200 = 450 \quad \text{i.e.} \quad T_A = 250 \text{ N}$$

Notes (i) As an alternative to resolving vertically, we could have taken moments about B to find T_A.

(ii) It is usually good policy to take moments about points where unknown forces are acting because this reduces the number of unknowns in each of the resulting equations.

EXAMPLE 4.3

A uniform ladder which is 5 m long and has a mass of 20 kg leans with its upper end against a smooth vertical wall and its lower end on rough ground. The bottom of the ladder is 3 m from the wall. Calculate the frictional force between the ladder and the ground. ($g = 10 \text{ m s}^{-2}$.)

Solution

Refer to Fig. 4.3. The ladder is uniform and therefore its weight, $20 \times 10 = 200$ N, acts at its mid-point G, a distance of 1.5 m from the wall. The wall is smooth and therefore the only force acting at the top of the ladder is the normal reaction R. By Pythagoras the point A at which the ladder makes contact with the wall is 4 m above the ground. The forces acting at the bottom of the ladder are the normal reaction S and the frictional force F. If the ladder were to slip, its bottom end would move to the right; it follows that F acts to the left as shown.

Fig. 4.3
Diagram for Example 4.3

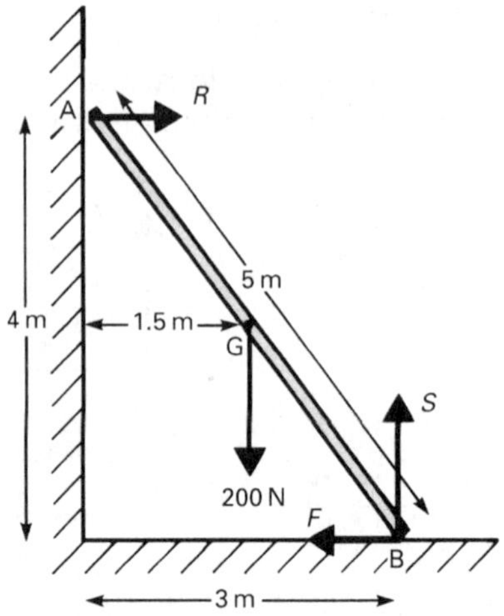

The ladder is in equilibrium and therefore there can be no resultant force in any direction. In particular there is no resultant vertical force, in which case

$$S = 200 \text{ N} \tag{4.4}$$

Because the ladder is in equilibrium the total torque about any point is zero. In particular, the total (net) torque about A is zero and therefore

$$(F \times 4) + (200 \times 1.5) = S \times 3$$

i.e. $4F + 300 = 3S$

Therefore by equation [4.4]

$$4F + 300 = 600$$

i.e. $F = 75$ N

Note The reason that we have chosen to consider the torque about A, rather than some other point, is that this automatically excludes R – a force in which we have no interest. The reader is advised to convince himself that considering the torque about G and/or B and making use of the fact that $F = R$ also gives $F = 75$ N.

Points to Bear in Mind when Attempting Questions 4A

(a) Draw a clear diagram showing all the forces acting on the particle (or body) whose equilibrium is being considered.

(b) Draw diagrams in which the angles look something like the angles they represent. There is no need to use a protractor, but an angle of 30°, say, should look more like 30° than 45° or 60°.

(c) A smooth surface can exert a force only at right angles to itself – the **normal reaction**.

(d) The tension is the same in each section of a light string which passes over a smooth pulley or a smooth peg, or which passes through a smooth hole or a smooth ring.

(e) There is no point in resolving in more than two directions.

(f) It is often an advantage to resolve perpendicular to an unknown force.

(g) It is often an advantage to take moments about points where unknown forces are acting.

QUESTIONS 4A

1. Solve the problem in Example 4.1 by resolving perpendicular to P and/or Q.

2. Two forces, P and Q, act NW and NE respectively. They are in equilibrium with a force of 50.0 N acting due S and a force of 20.0 N acting due E. Find P and Q.

3. A particle whose weight is 50.0 N is suspended by a light string which is at 35° to the vertical under the action of a horizontal force F. Find: **(a)** the tension in the string, **(b)** F.

4. A particle of weight W rests on a smooth plane which is inclined at 40° to the horizontal. The particle is prevented from slipping by a force of 50.0 N acting parallel to the plane and up a line of greatest slope. Calculate: **(a)** W, **(b)** the reaction due to the plane.

5. Two light strings are perpendicular to each other and support a particle of weight 100 N. The tension in one of the strings is 40.0 N. Calculate the angle this string makes with the vertical and the tension in the other string.

6. A uniform pole AB of weight $5W$ and length $8a$ is suspended horizontally by two vertical strings attached to it at C and D where AC = DB = a. A body of weight $9W$ hangs from the pole at E where ED = $2a$. Calculate the tension in each string.

7. AB is a uniform rod of length 1.4 m. It is pivoted at C, where AC = 0.5 m, and rests in horizontal equilibrium when weights of 16 N and 8 N are applied at A and B respectively. Calculate: **(a)** the weight of the rod, **(b)** the magnitude of the reaction at the pivot.

8. A uniform rod AB of length $4a$ and weight W is smoothly hinged at its upper end, A. The rod is held at 30° to the horizontal by a string which is at 90° to the rod and attached to it at C where AC = $3a$. Find: **(a)** the tension in the string, **(b)** the vertical component of the reaction at A, **(c)** the horizontal component of the reaction at A.

9. A sphere of weight 40 N and radius 30 cm rests against a smooth vertical wall. The sphere is supported in this position by a string of length 20 cm attached to a point on the sphere and to a point on the wall. Find: **(a)** the tension in the string, **(b)** the reaction due to the wall. (If you require a hint, turn to the answer.)

4.2 THE TRIANGLE OF FORCES

Suppose that a body is in equilibrium under the action of three non-parallel coplanar forces, ***P***, ***Q***, and ***R*** (Fig. 4.4). In order to satisfy condition (i) (p. 44), each force must be equal and opposite to the resultant of the other two. The system therefore reduces to one in which there are only two equal and opposite forces, (***R*** and ***R***′, say, where ***R***′ is the resultant of ***P*** and ***Q***). Furthermore, these two forces (***R*** and ***R***′) must be in line with each other, otherwise there would be a couple acting on the system and condition (ii) would not be satisfied. It follows that ***P*, *Q* and *R* must be concurrent**.

Fig. 4.4
Body acted on by three forces

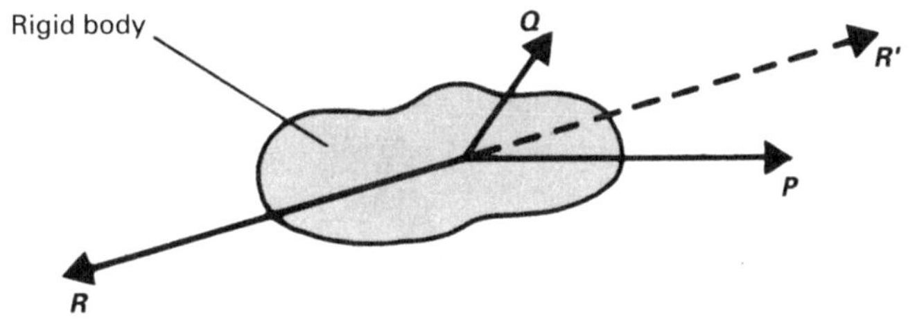

Bearing in mind that ***R***′ is the resultant of ***P*** and ***Q*** and that ***R*** = −***R***′, leads to Figs. 4.5(a), (b) and (c). It follows from Fig. 4.5 that:

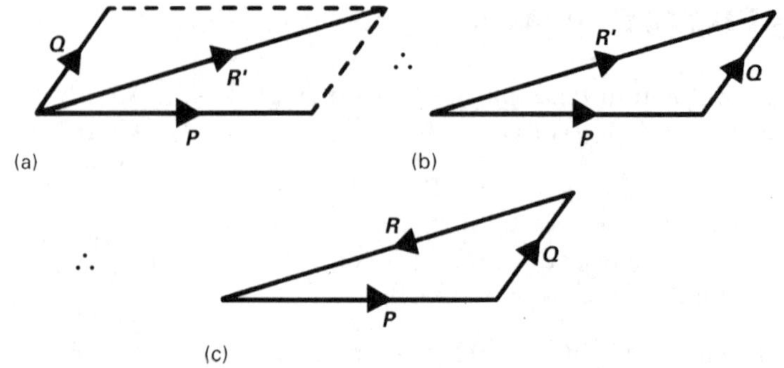

Fig. 4.5
The triangle of forces

> If a body is in equilibrium under the action of three coplanar forces, then the forces can be represented in magnitude and direction by the sides of a triangle taken in order. This is known as **the triangle of forces**.

QUESTIONS 4B

1. Solve the problem in example 4.1 by using a triangle of forces.

4.3 THE POLYGON OF FORCES

The triangle of forces can easily be extended to any number of forces, in which case:

> If a body is in equilibrium under the action of any number of forces, then the forces can be represented in magnitude and direction by the sides of a polygon taken in order

4.4 TYPES OF EQUILIBRIUM

There are three types of equilibrium and these are illustrated by the cone shown in Fig. 4.6.

(i) **Stable equilibrium** A body is in stable equilibrium if it returns to its equilibrium position after it has been displaced slightly (Fig. 4.6(a)).

(ii) **Unstable equilibrium** A body is in unstable equilibrium if it does not return to its equilibrium position and does not remain in the displaced position after it has been displaced slightly (Fig. 4.6(b)).

(iii) **Neutral equilibrium** A body is in neutral equilibrium if it stays in the displaced position after it has been displaced slightly (Fig. 4.6(c)).

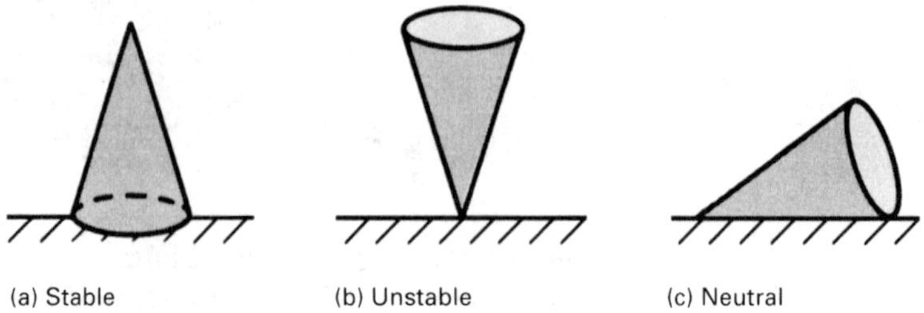

Fig. 4.6
Types of equilibrium

4.5 CENTRE OF MASS

The entire mass of a body can be considered to act at a single point, known as the **centre of mass** of the body. If a body is symmetrical and of uniform composition, the centre of mass is at the geometric centre of the body.

If a single force acts on a body and the line of action of the force passes through the centre of mass, the body will have a linear acceleration but no angular acceleration. Thus, a body which is accelerated from rest by such a force will move in a straight line without any rotation. As an example of this, imagine a stationary hammer resting on a frictionless surface. If forces such as ***P*** and ***Q*** are applied to the hammer (Figs 4.7(a) and (b)), it will move without rotation as shown.

Fig. 4.7
Effects of forces at centre of mass of a hammer

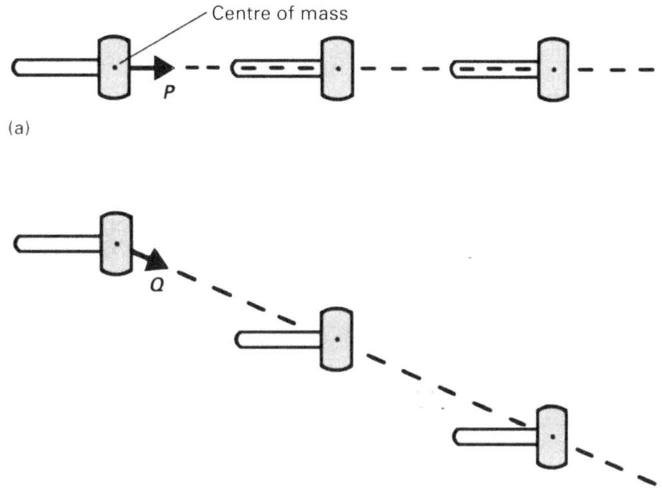

However, if a force such as ***R*** is applied to the hammer, its subsequent motion involves rotation because ***R*** does not act through the centre of mass (Fig. 4.8). Note that even when the body is rotating, the centre of mass moves along a straight line, i.e. the rotation takes place about the centre of mass. Thus, **in the absence of an actual pivot (e.g. an axle) a body behaves as if it is pivoted at its centre of mass and only at its centre of mass**.

Fig. 4.8
Effect of forces at centre of mass of a hammer

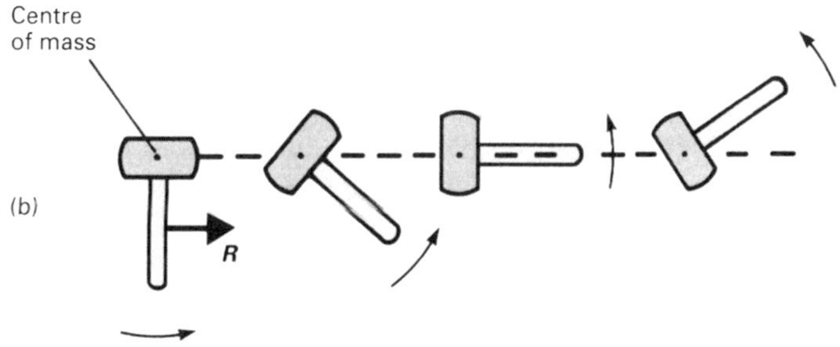

The motion of the centre of mass of a body cannot be affected by internal forces. Suppose that a space-ship, which is initially moving with uniform speed along a straight line, breaks into a number of pieces as a result of an explosion on board. No external force has acted on the mass of the space-ship and therefore the mass as a whole cannot (by Newton's second law) acquire an acceleration. Since the mass can be taken to be at the centre of mass, there can be no acceleration of the centre of mass. The pieces therefore move apart in such a way that the centre of mass continues to move with the original speed in the original direction.

4.6 CENTRE OF GRAVITY

The centre of gravity of a body is the single point at which the entire weight of the body can be considered to act. In uniform gravitational fields (such as that of the Earth on a small body) the centre of gravity coincides with the centre of mass.

Since the weight of a body acts at its centre of gravity, **a freely suspended body hangs in such a way that its centre of gravity is vertically below the pivot.** This is the basis of the usual experimental determination of the position of the centre of gravity of a body (Fig. 4.9).

Centres of gravity (and therefore centres of mass) can also be located by calculation (Examples 4.4 to 4.7).

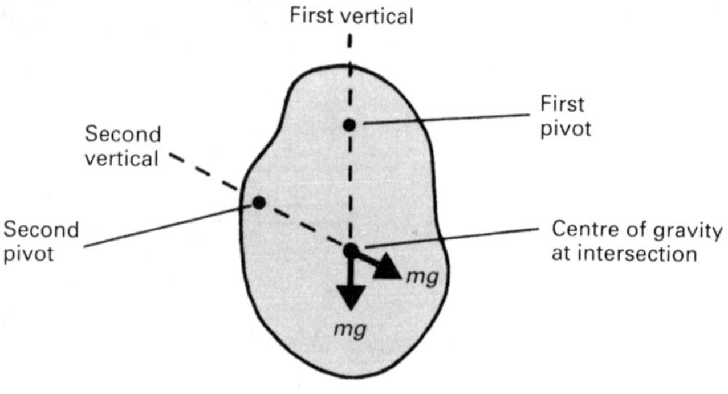

Fig. 4.9
Determination of centre of gravity

EXAMPLE 4.4

Calculate the position of the centre of gravity of a body which comprises two small spheres whose centres are connected by a straight rod of length L. The masses of the spheres are m_1 and m_2. The mass of the rod is very small and may be ignored.

Solution

By symmetry, the centre of gravity of the system is at a point on the line joining the centres of gravity of two spheres. Since the centre of gravity of each sphere is at its centre, the centre of gravity of the whole system is at a point such as G (Fig. 4.10).

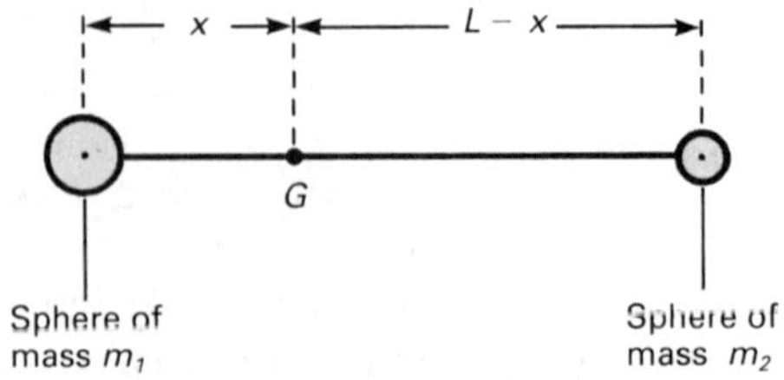

Fig. 4.10
Diagram for Example 4.4

The centre of gravity of a body is the point at which its weight acts, and therefore if the body were to be pivoted at its centre of gravity, there would be no gravitational torque about that point. Therefore,

$$\text{Torque about G due to } m_1 = \text{Torque about G due to } m_2$$

i.e. $m_1 g x = m_2 g (L - x)$

$\therefore$ $x(m_1 + m_2) = m_2 L$

$$\therefore \quad x = \frac{m_2 L}{m_1 + m_2} \qquad [4.5]$$

The method used in Example 4.4 can be extended in order to determine the positions of the centres of gravity of more complex structures (Example 4.5).

EXAMPLE 4.5

Calculate the position of the centre of gravity of a system of three particles each of mass m and located at the vertices of an equilateral triangle of side L.

Solution

The arrangement is as shown in Fig. 4.11. On the basis of Example 4.4 the centre of gravity of the particles at A and B is at G′ (the mid-point of AB). The system is therefore equivalent to one with a particle of mass $2m$ at G′ and a particle of mass m at C. The centre of gravity of the complete system is therefore on CG′ at G where G is a distance x from G′ given by equation 4.5 as

$$x = \frac{m \times CG'}{m + 2m}$$

i.e. $x = CG'/3$

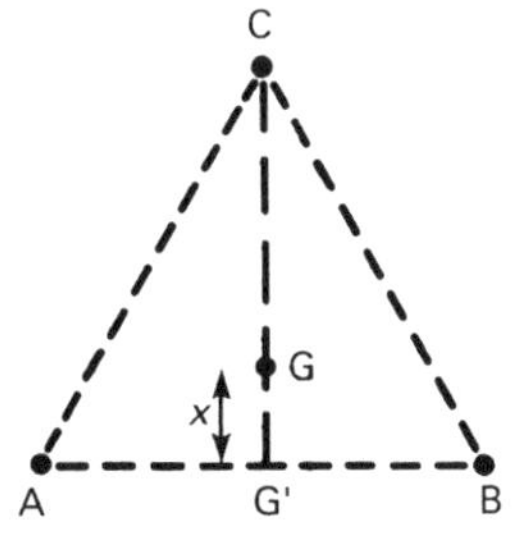

Fig. 4.11
Diagram for Example 4.5

EXAMPLE 4.6

MNOPQRST is a uniform lamina whose dimensions are as shown in Fig. 4.12(a). Find the distance of its centre of gravity from MN and from MT.

Solution

By symmetry the centre of gravity must lie on XY (Fig. 4.12(b)), and is therefore 3 cm from MN. It remains to find its distance from MT. We do this by regarding the lamina as a rectangle MNST and a square OPQR, the centres of gravity of

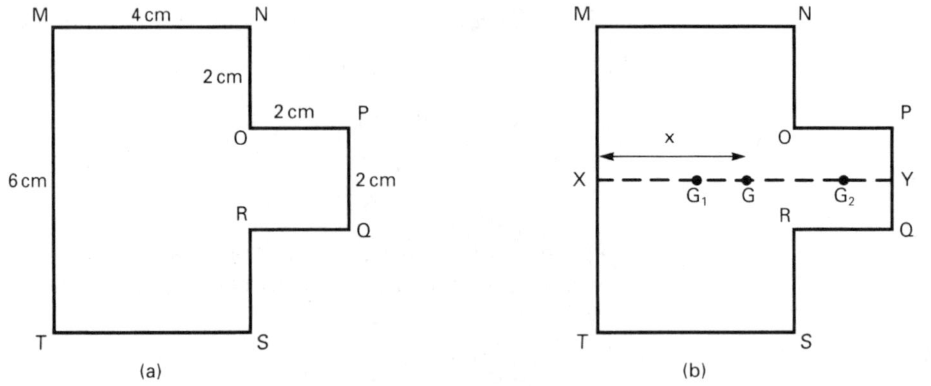

Fig. 4.12
Diagram for Example 4.6

which are at their centres G_1 and G_2 respectively. The centre of gravity of the whole lamina must lie between G_1 and G_2. Let it be at G, a distance x from MT. Let $w =$ the weight per unit area of the lamina.

Section	*Weight*	*Distance of centre of gravity from MT*
MNST	$24w$	2
OPQR	$4w$	5
MNPQRST	$28w$	x

$$\text{Moment of whole about MT} = \text{Sum of moments of parts about MT}$$

$$\therefore \quad 28wx = 24w \times 2 + 4w \times 5$$

$$\therefore \quad 28wx = 68w$$

$$\therefore \quad x = \frac{68w}{28w} \qquad \text{i.e.} \quad x = 2.4 \text{ cm}$$

The centre of gravity is therefore 3 cm from MN and 2.4 cm from MT.

EXAMPLE 4.7

MNOPQRST is a uniform lamina whose dimensions are as shown in Fig. 4.13(a). Find the distance of its centre of gravity from MN and from MT.

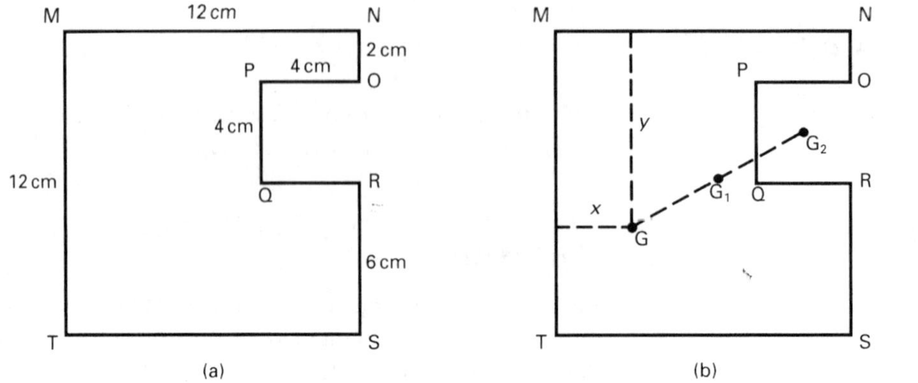

Fig. 4.13
Diagram for Example 4.7

Solution

We regard the lamina as a square MNST from which a smaller square OPQR has been removed. The centres of gravity at these squares are at their centres G_1 and G_2 (Fig. 4.13(b)). Let the centre of gravity of the lamina be at G, a distance x from MT and a distance y from MN. Let w = the weight per unit area of the lamina.

Section	Weight	Distance of centre of gravity from MT	Distance of centre of gravity from MN
MNST	$144w$	6	6
OPQR	$16w$	10	4
MNOPQRST	$128w$	x	y

Moment of whole about MT = Sum of moments of parts about MT

$$\therefore \quad 144w \times 6 = 16w \times 10 + 128wx$$

$$\therefore \quad 864w = 160w + 128wx$$

$$\therefore \quad 704w = 128wx$$

$$\therefore \quad x = \frac{704w}{128w} \quad \text{i.e.} \quad x = 5.5 \text{ cm}$$

Moment of whole about MN = Sum of moments of parts about MN

$$\therefore \quad 144w \times 6 = 16w \times 4 + 128wy$$

$$\therefore \quad 864w = 64w + 128wy$$

$$\therefore \quad 800w = 128wy$$

$$\therefore \quad y = \frac{800w}{128w} \quad \text{i.e.} \quad y = 6.25 \text{ cm}$$

The centre of gravity is therefore 5.5 cm from MT and 6.25 cm from MN.

QUESTIONS 4C

1. A light square frame ABCD of side $10a$ has particles of mass m, $2m$, $3m$ and $4m$ at A, B, C and D respectively. Find the distance of the centre of gravity: **(a)** from AB, **(b)** from AD.

2. A non-uniform rod AB of weight 40 N and length 20 cm is supported by a pivot at C where AC = 14 cm. The rod rests in horizontal equilibrium when a weight of 30 N is attached to it at B. Find the distance of the centre of gravity of the rod from A.

3. MNOPQR is a uniform lamina. Find the distance of its centre of gravity: **(a)** from MR, **(b)** from MN.

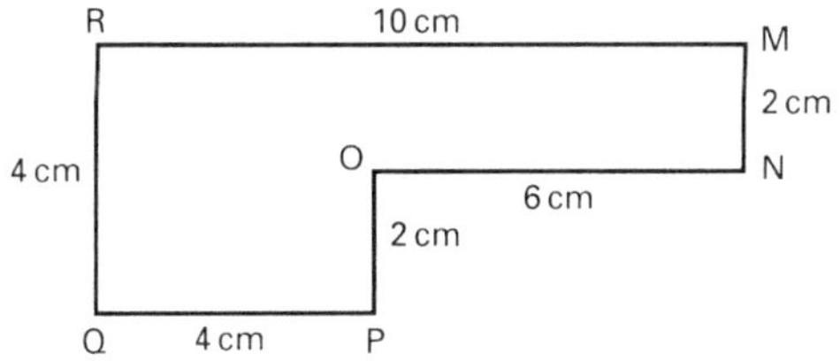

4. A circular plate of uniform thickness and radius 12 cm has a circular hole of radius 4 cm cut out of it. The centre of the hole is 2 cm from the centre, O, of the plate. Find the distance of the centre of gravity from O.

CONSOLIDATION

If three coplanar forces are in equilibrium, the forces are bound to be concurrent.

If a particle is in equilibrium, the resultant force on it is zero in which case it must be at rest or moving in a straight line at constant speed.

If a body is in equilibrium, the resultant force on it is zero and the resultant torque is zero in which case it must be at rest or moving in a straight line at constant speed and if it is rotating, it must be doing so with a constant angular velocity.

To solve problems in which concurrent coplanar forces are known to be in equilibrium resolve in (up to) two directions and make use of the fact that the resultant force in each direction is zero.

To solve problems in which non-concurrent coplanar forces are known to be in equilibrium resolve twice and take moments once, or resolve once and take moments twice, or take moments three times.

The centre of gravity of a body is the point at which its weight can be taken to act.

5

WORK, ENERGY, POWER

5.1 WORK

If a body moves as a result of a force being applied to it, the force is said to be doing work on the body. The work done is given by

$$W = Fs \qquad [5.1]$$

where

W = the work done (joules, J)

F = the constant applied force (N)

s = the distance moved in the direction of the force (m).

It follows from equation [5.1] that a force is doing no work if it is merely preventing a body moving, because in such a circumstance $s = 0$. Thus, if a man lifts some object, he is doing work whilst actually lifting it; but he does no work in holding it above his head, say, once he has lifted it into that position. The man would, of course, become tired if he were to hold a heavy object for a long time but this is because he is having to keep his muscles under tension; it is not because he is doing work on the object.

Suppose that a constant force, F, acts on a body so as to move it in a direction other than its own (Fig. 5.1). The component of $\boldsymbol{F}$ in the direction of motion is $F\cos\theta$, in which case the work done, W, is given by

$$W = Fs\cos\theta$$

Fig. 5.1
Force at angle to motion

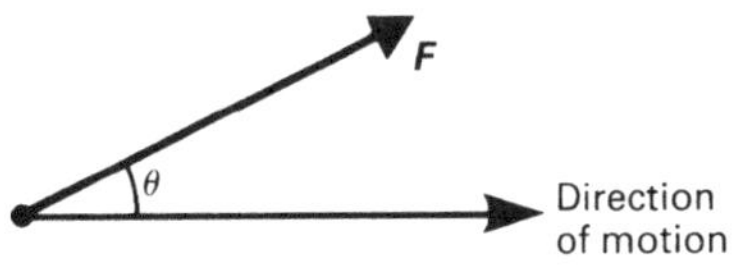

This situation can occur only if there is some other force preventing motion taking place in the direction of $\boldsymbol{F}$. For example, consider a man pulling a garden roller in the manner shown in Fig. 5.2. For convenience, the man is holding the handle at an angle θ to the horizontal and exerts a force F in the direction shown. The other force that acts on the roller is its weight, mg, and this of course, acts vertically downwards. The upward directed component of $\boldsymbol{F}$ will be less than the weight. Therefore there is no vertical motion and no work is done by the upward directed component of $\boldsymbol{F}$.

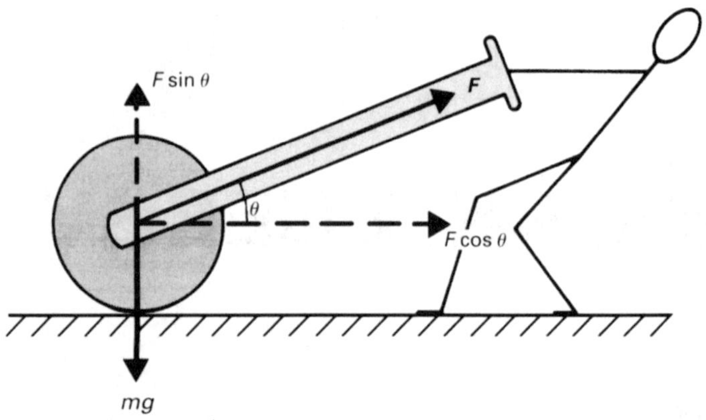

Fig. 5.2
Force diagram for a man pulling a roller

5.2 ENERGY

A body which is capable of doing work is said to possess energy. The amount of energy that a body has is equal to the amount of work that it can do (or what amounts to the same thing, the amount of work that must have been done on it to give it that energy).

Although it is often convenient to classify energy as being chemical energy or nuclear energy or heat energy, etc., there are basically only two types of energy – kinetic energy (KE) and potential energy (PE).

5.3 KINETIC ENERGY

The energy which a body possesses solely because it is moving is called kinetic energy.

The kinetic energy of a body can be defined as the amount of work it can do in coming to rest, or what amounts to the same thing, the amount of work that must have been done on it to increase its velocity from zero to the velocity it has. On this basis, if a body of mass m is moving with velocity v, then

$$\text{Kinetic energy} = \tfrac{1}{2} mv^2$$

Kinetic energy is a positive, scalar quantity

To Show that Kinetic Energy $= \frac{1}{2} mv^2$ (Variable Force)

Suppose that a body of mass m moves a small distance δs under the action of a force F. Suppose also that, though the force may be varying, δs is so small that the force can be considered constant over the distance δs. The work done δW is given by equation [5.1] as

$$\delta W = F\,\delta s$$

If the force increases the velocity of the body from zero to v, the total work done W is given by

$$W = \int_{v=0}^{v=v} F\,\mathrm{d}s$$

Using Newton's second law (equation [2.2]) we can write F as

$$F = m\frac{\mathrm{d}v}{\mathrm{d}t}$$

where $\mathrm{d}v/\mathrm{d}t$ is the acceleration of the body. Therefore

$$W = \int_{v=0}^{v=v} m\frac{\mathrm{d}v}{\mathrm{d}t}\,\mathrm{d}s$$

Bearing in mind that $v = \mathrm{d}s/\mathrm{d}t$, we can write

$$W = \int_0^v mv\,\mathrm{d}v$$

and therefore

$$W = [\tfrac{1}{2}mv^2]_0^v$$

i.e. $$W = \tfrac{1}{2}mv^2$$

By definition, the work done is the kinetic energy of the body, and therefore

$$\text{Kinetic energy} = \tfrac{1}{2}mv^2$$

Note The kinetic energy of a body depends only on its mass and its velocity and as such, the kinetic energy is independent of the way in which the body acquired this velocity. In view of this, the result that has just been derived could have been obtained more simply by specifying that the body was accelerated by a constant force. This will now be done.

To Show that Kinetic Energy $= \frac{1}{2}mv^2$ (Constant Force)

If a body of mass m moves a distance s under the action of a constant force F, the work done W by the force is given by equation [5.1] as

$$W = Fs$$

If the (constant) acceleration is a, then from Newton's second law $F = ma$ and therefore

$$W = mas \qquad [5.2]$$

If the body has been accelerated from rest to some velocity v, then from equation [2.7]

$$v^2 = 0^2 + 2as$$

i.e. $$as = \frac{v^2}{2}$$

Therefore from equation [5.2]

$$W = \tfrac{1}{2}mv^2$$

and therefore by definition

$$\text{Kinetic energy} = \tfrac{1}{2}mv^2$$

5.4 POTENTIAL ENERGY

The energy which a body possesses due to its position or to the arrangement of its component parts is called potential energy.

A brick which is suspended above the ground has energy because it could do work by falling to the ground. Its energy is due to its position and therefore it is potential energy. Furthermore, because it is due to its position in the Earth's gravitational field, it is more completely described as being gravitational potential energy. A charged particle situated in an electric field has electrical potential energy (section 39.6). The potential energy of a stretched bow string results from the elastic properties of the string and is therefore referred to as elastic potential energy. There are many forms of potential energy but this chapter is concerned primarily with gravitational potential energy.

5.5 GRAVITATIONAL POTENTIAL ENERGY

The potential energy of a body can be defined as the amount of work that was done on it to give it that energy. On this basis, if a body of mass m is at a height h, then

$$\text{Gravitational potential energy} = mgh$$

where

h = the height of the body above some arbitrary reference level (e.g. the ground or a bench top) where the potential energy is taken to be zero.

Alternatively, if a body of mass m is moved upwards, through a height Δh

$$\text{Increase in gravitational potential energy} = mg\Delta h$$

Note Because h is measured from an arbitrary level, the gravitational potential energy of a body is not an absolute property of the body and its position but depends on which reference level has been chosen. This is of no consequence though, because in the final analysis we are always concerned with the change in potential energy that occurs when a body moves.

To Show that Gravitational Potential Energy = *mgh*

Consider a body of mass m at some arbitrary height above the ground and moving upwards with some arbitrary velocity. If the body is just to maintain this velocity, there must be an upward directed force acting on it which is equal in magnitude to its weight mg. If the body moves upwards a further distance h under the influence of this (constant) force, the work done W by the force is given by equation [5.1] as

$$W = mgh$$

There can have been no change in the kinetic energy of the body because its velocity has been the same throughout. The work done on the body has therefore been used only to increase its potential energy. Thus in raising a body of mass m

through a distance h the increase in potential energy is mgh. It follows that if the potential energy of the body is taken to be zero when it is on the ground, then its potential energy at a height h is mgh, i.e.

$$\text{Gravitational potential energy} = mgh$$

5.6 CONSERVATION OF MECHANICAL ENERGY

The principle of conservation of mechanical energy can be stated as:

> **In a system in which the only forces acting are associated with potential energy (e.g. gravitational and elastic forces) the sum of the kinetic and potential energies is constant**

i.e. $$\text{KE} + \text{PE} = \text{a constant} \qquad [5.3]$$

Note that, in particular, equation [5.3] does not apply when there are frictional forces present.

As an example of the application of equation [5.3], we shall use it to obtain an expression for the velocity acquired by a body of mass m in falling freely from rest at a height h in a vacuum (Fig. 5.3). As the body falls it loses gravitational potential energy and gains kinetic energy. It follows from equation [5.3] that

$$\text{KE gained} = \text{PE lost}$$

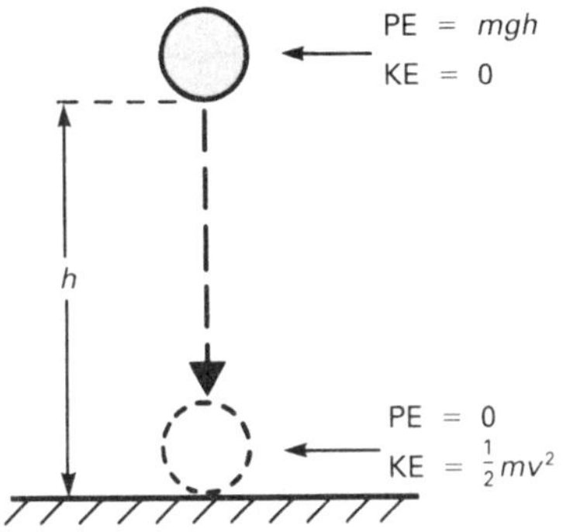

Fig. 5.3 Conversion of potential energy to kinetic energy

and therefore if the velocity of the body after it has fallen a distance h is v, then

$$\tfrac{1}{2}mv^2 = mgh$$

i.e. $$v = \sqrt{2gh}$$

The body comes to rest (at least momentarily) very soon after making contact with the ground. It does so because the Earth has exerted a force on it. The force is due to the solidity of the Earth, rather than to its gravitational properties. At the same time, the body exerts a force on the Earth, and both the body and the Earth become deformed. It is the kinetic energy which the body had immediately before the impact that has been used to produce these deformations. If they are permanent, the energy which created them is dissipated as heat and sound, and the body remains at rest on the ground. On the other hand, if the body and the Earth regain their original shapes, then they lose the elastic potential energy which they acquired at the impact and the body bounces. Some energy is bound to be dissipated as heat, and therefore the body has less than its original amount of kinetic energy and therefore does not reach its original height.

When friction is involved, and when work is done by external forces (i.e. forces other than those associated with potential energy) we make use of **the work–energy principle**:

$$\begin{pmatrix}\text{Work done by}\\ \text{external force}\end{pmatrix} = \begin{pmatrix}\text{Increase in}\\ \text{KE} + \text{PE}\end{pmatrix} + \begin{pmatrix}\text{Work done}\\ \text{against friction}\end{pmatrix} \qquad [5.4(a)]$$

If work is done against external forces, equation [5.4(a)] becomes

$$\begin{pmatrix}\text{Decrease in}\\ \text{KE} + \text{PE}\end{pmatrix} = \begin{pmatrix}\text{Work done}\\ \text{against}\\ \text{external forces}\end{pmatrix} + \begin{pmatrix}\text{Work done}\\ \text{against friction}\end{pmatrix} \qquad [5.4(b)]$$

In problems where there are sudden changes in velocity, e.g. where two bodies collide or where there is a sudden increase in the tension in a string (i.e. a jerk) some mechanical energy is converted to heat and/or sound*. In such circumstances, using the principle of conservation of mechanical energy or the work–energy principle (equations 5.4(a) and (b)) allows us to do no more than find out just how much energy has been converted in this way. For example if we know the height to which a bouncing ball rebounds, we can calculate the amount of mechanical energy converted to heat, etc., as a result of the impact, but energy considerations alone do not allow us to calculate the height to which the ball rebounds in the first place.

EXAMPLE 5.1

A car of mass 800 kg and moving at 30 m s^{-1} along a horizontal road is brought to rest by a constant retarding force of 5000 N. Calculate the distance the car moves whilst coming to rest.

Solution

If the car travels a distance s in coming to rest, then by equation [5.1] the work done by the car against the retarding force

$$= 5000s$$

The kinetic energy ($\frac{1}{2}mv^2$) lost by the car in coming to rest

$$= \tfrac{1}{2} \times 800 \times 30^2$$

$$= 360\,000 \text{ J}$$

The work done against the retarding force is equal to the kinetic energy lost by the car, and therefore

$$5000s = 360\,000$$

i.e. $s = 72$ m

Alternatively, the solution could have been obtained by using Newton's second law (equation [2.2]) to calculate the value of the retardation which could then be used in equation [2.7] to find s.

*This is not true in the special case of an elastic collision – see section 2.8.

EXAMPLE 5.2

A small block (Fig. 5.4) is released from rest at A and slides down a smooth curved track. Calculate the velocity of the block when it reaches B, a vertical distance h below A.

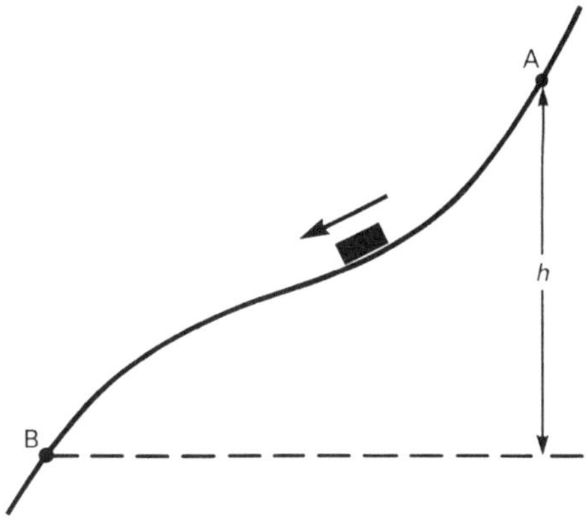

Fig. 5.4
Diagram for Example 5.2

Solution

Suppose that the mass of the block is m. The gravitational potential energy lost by the block in moving from A to B is mgh. If the velocity of the block on reaching B is v, then the kinetic energy gained by the block is $\frac{1}{2}mv^2$.

The track is smooth and therefore no work is done against friction, in which case

$$\text{KE gained} = \text{PE lost}$$

$$\therefore \quad \tfrac{1}{2}mv^2 = mgh$$

$$\text{i.e.} \quad v = \sqrt{2gh}$$

The problem has been solved by making use of the principle of conservation of mechanical energy. Unlike Example 5.1, it could not have been solved by using $F = ma$ and $v^2 = u^2 + 2as$ because the acceleration is not constant and therefore $v^2 = u^2 + 2as$ does not apply. A solution based on $F = ma$ is possible provided the equation of the curve is known, but it involves using calculus and is much more difficult than the solution given here.

Note that the speed at B does not depend on the particular shape of the curve. However the time to reach B does, and cannot be found by using energy considerations.

EXAMPLE 5.3

A car of mass 1.0×10^3 kg increases its speed from $10\,\text{m}\,\text{s}^{-1}$ to $20\,\text{m}\,\text{s}^{-1}$ whilst moving 500 m up a road inclined at an angle α to the horizontal where $\sin\alpha = \frac{1}{20}$. There is a constant resistance to motion of 300 N. Find the driving force exerted by the engine, assuming that it is constant. (Assume $g = 10\,\text{m}\,\text{s}^{-2}$.)

Solution

The work done by the engine is used to increase both the KE and PE of the car and to overcome the resistive force.

In moving 500 m along the road the car gains a vertical height of 500 sin $\alpha = 25$ m. Therefore

$$\text{PE gained} = 1.0 \times 10^3 \times 10 \times 25 = 2.5 \times 10^5 \text{ J}$$

$$\text{KE gained} = \tfrac{1}{2} \times 1.0 \times 10^3 \times 20^2 - \tfrac{1}{2} \times 1.0 \times 10^3 \times 10^2$$
$$= 1.5 \times 10^5 \text{ J}$$

$$\text{Work done against resistance} = 300 \times 500 = 1.5 \times 10^5 \text{ J}$$

By the work–energy principle,

$$\text{Work done by engine} = \text{Increase in PE} + \text{Increase in KE} + \text{Work against resistance}$$

$$\therefore \quad \text{Work done by engine} = 2.5 \times 10^5 + 1.5 \times 10^5 + 1.5 \times 10^5$$
$$= 5.5 \times 10^5 \text{ J}$$

If the driving force of the engine is F, then (by $W = Fs$)

$$5.5 \times 10^5 = F \times 500 \quad \text{i.e.} \quad F = 1.1 \times 10^3 \text{ N}$$

Alternatively, as with Example 5.1, the solution could have been obtained by using $F = ma$ and $v^2 = u^2 + 2as$.

QUESTIONS 5A

Questions 1 to 9 should be solved by using energy considerations (Assume $g = 10\,\text{m s}^{-2}$ where necessary.)

1. A car of mass 1.2×10^3 kg moves 300 m up a road which is inclined to the horizontal at an angle α where $\sin \alpha = \frac{1}{15}$. By how much does the gravitational PE of the car increase?

2. A particle is projected with speed v at an angle α to the horizontal. Find the speed of the particle when it is at a height h.

3. A car of mass 800 kg moving at $20\,\text{m s}^{-1}$ is brought to rest by the application of the brakes in a distance of 100 m. Calculate the work done by the brakes and the force they exert assuming that it is constant and that there is no other resistance to motion.

4. The speed of a dog-sleigh of mass 80 kg and moving along horizontal ground is increased from $3.0\,\text{m s}^{-1}$ to $9.0\,\text{m s}^{-1}$ over a distance of 90 m. Find: **(a)** the increase in the KE of the sleigh, **(b)** the force exerted on the sleigh by the dogs, assuming that it is constant and that there is no resistance to motion.

5. A simple pendulum consisting of a small heavy bob attached to a light string of length 40 cm is released from rest with the string at 60° to the downward vertical. Find the speed of the pendulum bob as it passes through its lowest point.

6. A car of mass 900 kg accelerates from rest to a speed of $20\,\text{m s}^{-1}$ whilst moving 80 m along a horizontal road. Find the tractive force (i.e. the driving force) exerted by the engine, assuming that it is constant and that there is a constant resistance to motion of 250 N.

7. A child of mass 20 kg starts from rest at the top of a playground slide and reaches the bottom with a speed of $5.0\,\text{m s}^{-1}$. The slide is 5.0 m long and there is a difference in height of 1.6 m between the top and the bottom. Find: **(a)** the work done against friction, **(b)** the average frictional force.

8. Two particles of masses 6.0 kg amd 2.0 kg are connected by a light inextensible string passing over a smooth pulley. The system is released from rest with the string taut. Find the speed of the particles when the heavier one has descended 2.0 m.

9. A ball of mass 50 grams falls from a height of 2.0 m and rebounds to a height of 1.2 m. How much kinetic energy is lost on impact?

5.7 POWER

The power of a machine is the rate at which it does work (alternatively, it is the rate at which it supplies* energy). The unit of power is the **watt** (W).

Thus

$$P = \frac{dW}{dt} \qquad [5.5]$$

where

P = the instantaneous power (W)

$\frac{dW}{dt}$ = the rate of working ($J\,s^{-1}$). Thus $1\,W = 1\,J\,s^{-1}$

If a machine is working at a steady rate,

$$\text{Power} = \frac{\text{Work done}}{\text{Time taken}} \qquad [5.6]$$

When the rate of working is not steady, equation [5.6] gives the average power.

Another useful expression for power can be obtained by combining equations [5.1] and [5.5]. Thus from equation [5.5]

$$P = \frac{dW}{dt}$$

Therefore, from equation [5.1]

$$P = \frac{d}{dt}(Fs)$$

If the force is constant

$$P = F\frac{ds}{dt}$$

i.e. $$P = Fv$$

where P is the power output of a machine which is doing work by exerting a force F and moving the point of application of the force with velocity v. Equation [5.7] is useful in, say, calculating the force exerted by a car engine when the car is moving at a known velocity and the power being produced by the engine is also known.

*The machine has not, of course, actually produced the energy, it has merely converted it from another form.

EXAMPLE 5.4

A pump raises water through a height of 3.0 m at a rate of 300 kilograms per minute and delivers it with a velocity of 8.0 m s^{-1}. Calculate the power output of the pump. (Assume $g = 10\,\mathrm{m\,s^{-2}}$.)

Solution

The work done by the pump is used to increase both the PE and the KE of the water. In one second the pump delivers $300/60 = 5.0$ kg of water. Therefore

$$\text{Increase in PE each second} = 5.0 \times 10 \times 3.0 = 150\ \mathrm{J}$$

$$\text{Increase in KE each second} = \tfrac{1}{2} \times 5.0 \times 8.0^2 = 160\ \mathrm{J}$$

Therefore

$$\text{Work done each second} = 150 + 160 = 310\ \mathrm{J}$$

Since work done per second is power, the power output of the pump is 310 W.

QUESTIONS 5B

Assume $g = 10\,\mathrm{m\,s^{-2}}$ where necessary.

1. A man of mass 75 kg climbs 300 m in 30 minutes. At what rate is he working?

2. A pump with a power output of 600 W raises water from a lake through a height of 3.0 m and delivers it with a velocity of $6.0\,\mathrm{m\,s^{-1}}$. What mass of water is removed from the lake in one minute?

3. What is the power output of a cyclist moving at a steady speed of $5.0\,\mathrm{m\,s^{-1}}$ along a level road against a resistance of 20 N.

4. What is the maximum speed at which a car can travel along a level road when its engine is developing 24 kW and there is a resistance to motion of 800 N?

5. A crane lifts an iron girder of mass 400 kg at a steady speed of $2.0\,\mathrm{m\,s^{-1}}$. At what rate is the crane working?

6. A man of mass 70 kg rides a bicycle of mass 15 kg at a steady speed of $4.0\,\mathrm{m\,s^{-1}}$ up a road which rises 1.0 m for every 20 m of its length. What power is the cyclist developing if there is a constant resistance to motion of 20 N?

6

CIRCULAR MOTION AND ROTATION

6.1 ANGULAR VELOCITY

Suppose that a particle (Fig. 6.1) moves from A to P along the arc AXP at a constant speed* in a time interval t. **The angular velocity**, ω, of the particle is given by

$$\omega = \frac{\theta}{t} \qquad [6.1]$$

where θ = the angle turned through in radians. (**The radian** (rad) is the SI unit of angle and is the angle subtended at the centre of a circle by an arc of the circumference equal in length to the radius of the circle.)

ω = the angular velocity of the particle about O (rad s^{-1})

t = the time taken (s).

The period, T, of the rotational motion is the time taken for the particle to complete one revolution (i.e. to turn through 2π radians) and is given by equation [6.1] as

$$\omega = \frac{2\pi}{T}$$

i.e.

$$T = \frac{2\pi}{\omega} \qquad [6.2]$$

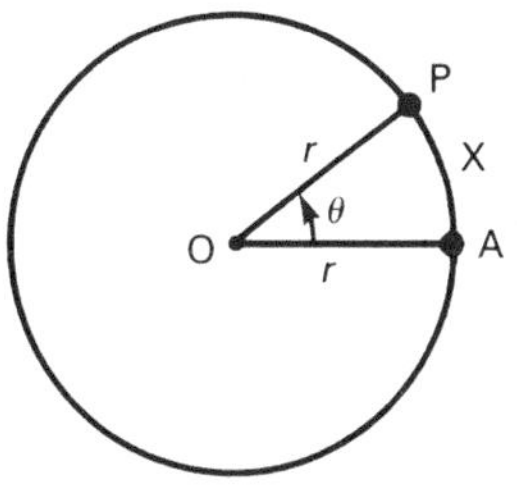

Fig. 6.1
Definition of angular velocity

*Note the use of the word 'speed' and not 'velocity'. The particle cannot have a constant velocity because its direction of motion is changing.

Note If the angular velocity is not constant, equation [6.1] is replaced by

$$\omega = \frac{d\theta}{dt}$$

where ω is the instantaneous angular velocity.

Referring again to Fig. 6.1, we can see that in time t the distance moved by the particle is the arc length AP and therefore its linear speed v is given by

$$v = \frac{\text{Arc length AP}}{t}$$

i.e. $$v = \frac{r\theta}{t}$$

Therefore, from equation [6.1]

$$v = \omega r \qquad [6.3]$$

6.2 CENTRIPETAL FORCE

If a body is moving along a circular path, there must be a force acting on it, for if there were not, it would move in a straight line in accordance with Newton's first law. Furthermore, if the body is moving at a constant speed, this force cannot (at any stage) have a component which is in the direction of motion of the body, for if it did it would be bound to either increase or decrease the speed of the body. The force that acts on the body must, therefore, be perpendicular to the direction of motion of the body and must therefore be directed towards the centre of the circular path. The force is known as a **centripetal force**.

If a brick is being whirled in a circle on one end of a piece of string, the centripetal force is provided by the tension in the string. If the string were to break, there would be no centripetal force and the brick would fly off at a tangent.

The centripetal force on an orbiting planet is gravitational; that on an electron moving round a nucleus is electrostatic.

6.3 CENTRIPETAL ACCELERATION

Because there is a resultant force on a body which is describing a circular path, the body must (by Newton's second law) have an acceleration. This acceleration must be in the same direction as the force, i.e. toward the centre of the circle. It is known as a **centripetal acceleration**. For a body which is moving with constant angular velocity, ω, along a circular path of radius, r, the magnitude of the centripetal acceleration can be shown to be given by

$$a = \omega^2 r \qquad [6.4]$$

where

a = the centripetal acceleration (m s^{-2}).

If the linear speed of the particle is v, then by equation [6.3]

$$a = \frac{v^2}{r} \qquad [6.5]$$

To Show that the Centripetal Acceleration $= \dfrac{v^2}{r}$

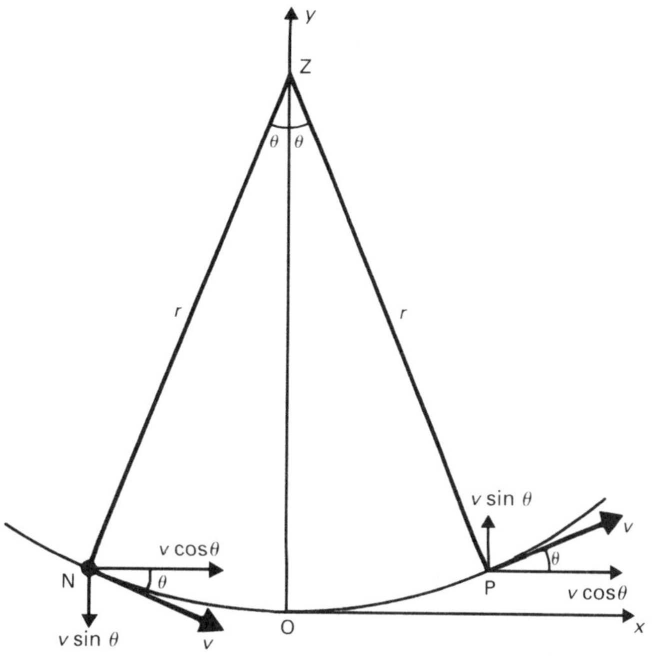

Fig. 6.2
To calculate the centripetal acceleration of a particle moving in a circle

Consider a particle moving with constant speed v along an arc NOP (Fig. 6.2). The x-component of velocity of the particle has the same value at P as at N and therefore its x-component of acceleration, a_x, is zero, i.e.

$$a_x = 0$$

As the particle moves from N to P its y-component of velocity changes by $2v \sin\theta$. If this takes place in a time interval, t, its y-component of acceleration, a_y, is given by

$$a_y = \frac{2v \sin\theta}{t} \qquad [6.6]$$

The speed of the particle along the arc is v, and therefore

$$t = \frac{\text{Arc length NOP}}{v}$$

i.e. $$t = \frac{2\theta r}{v}$$

Therefore from equation [6.6]

$$a_y = \frac{2v \sin\theta}{2\theta r/v}$$

i.e. $$a_y = \frac{v^2}{r}\frac{\sin\theta}{\theta}$$

If N and P are now taken to be coincident at O, then $\theta = 0$ and $\sin\theta/\theta$ has its limiting value of 1*, in which case

$$a_y = \frac{v^2}{r}$$

Thus at O, $a_x = 0$ and $a_y = v^2/r$ and therefore the acceleration is directed (entirely) along OZ, i.e. towards the centre of the circle. This result does not depend on the position of O and therefore in general:

The acceleration of a particle moving with constant speed v along a circular path of radius r is v^2/r and is directed towards the centre of the circle.

Note This result is not approximate, it does not depend on the approximate relationship $\sin\alpha \approx \alpha$, but on the limiting value of $\sin\alpha/\alpha$ as α tends to zero, and this is exactly equal to unity.

EXAMPLE 6.1

A particle of mass 3.0 kg is attached to a point O on a smooth horizontal table by means of a light inextensible string of length 0.50 m. The string is fully extended and the particle moves on the table in a circular path about O with a constant angular velocity of 8.0 radians per second. Calculate the tension in the string.

Solution

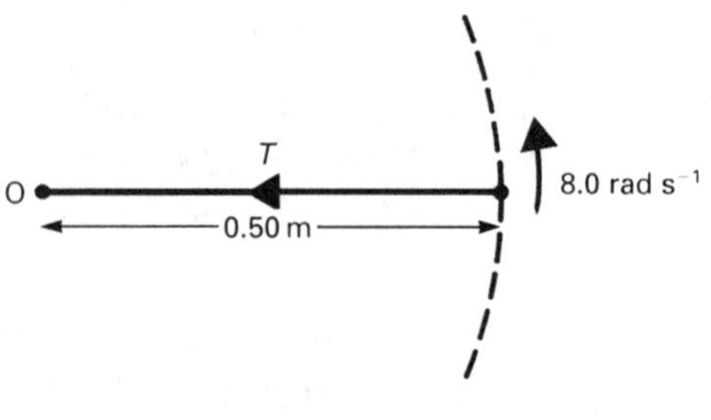

Fig. 6.3
Diagram for Example 6.1

Refer to Fig. 6.3. By Newton's second law

$$\text{Force} = \text{mass} \times \text{acceleration}$$

The 'acceleration' is the centripetal acceleration, $\omega^2 r$. The 'force' is the centripetal force and is provided by the tension, T, in the string. Therefore

$$T = 3.0 \times 8.0^2 \times 0.50$$

i.e. $$T = 96\text{ N}$$

*It is a general result that for a small angle α

$$\sin\alpha \approx \alpha \quad \text{measured in radians}$$

i.e. $$\frac{\sin\alpha}{\alpha} \approx 1$$

In the limit as α tends to zero $(\sin\alpha)/\alpha = 1$

i.e. $$\lim_{\alpha \to 0}\left(\frac{\sin\alpha}{\alpha}\right) = 1$$

EXAMPLE 6.2

A small bead of mass m is threaded on a smooth circular wire of radius r and centre O, and which is fixed in a vertical plane. The bead is projected with speed u from the highest point, A, of the wire. Find the reaction on the bead due to the wire when the bead is at P, in terms of m, g, r, u and θ where $\theta = \mathrm{A\hat{O}P}$.

Solution

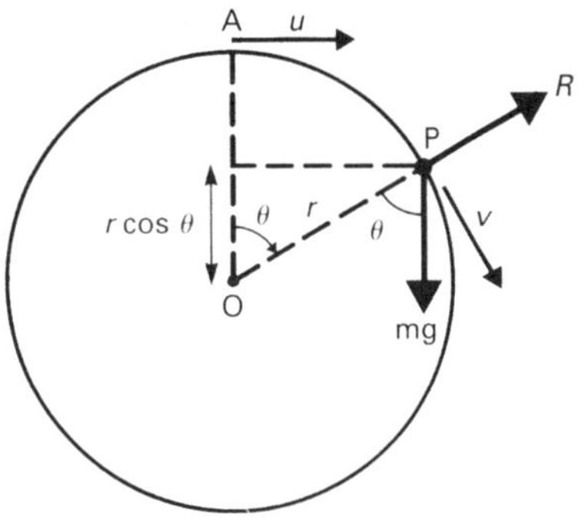

Fig. 6.4
Diagram for example 6.2

Refer to Fig. 6.4. Let the speed of the bead at $P = v$; let the reaction $= R$. Consider the motion from A to P. Since the wire is smooth, no work is done against friction and therefore

$$\text{Increase in KE} = \text{Decrease in PE}$$

$$\therefore \quad \tfrac{1}{2}mv^2 - \tfrac{1}{2}mu^2 = mg\,(r - r\cos\theta)$$

$$\therefore \quad v^2 = u^2 + 2gr\,(1 - \cos\theta)$$

Applying Newton's second law to the motion along PO gives

$$mg\cos\theta - R = m\frac{v^2}{r}$$

$$\therefore \quad mg\cos\theta - R = \frac{m}{r}[u^2 + 2gr\,(1 - \cos\theta)]$$

$$\therefore \quad R = mg\,(3\cos\theta - 2) - \frac{mu^2}{r}$$

QUESTIONS 6A

1. A particle moves along a circular path of radius 3.0 m with an angular velocity of 20 rad s^{-1}. Calculate: **(a)** the linear speed of the particle, **(b)** the angular velocity in revolutions per second, **(c)** the time for one revolution, **(d)** the centripetal acceleration.

2. A particle of mass 0.2 kg moves in a circular path with an angular velocity of 5 rad s^{-1} under the action of a centripetal force of 4 N. What is the radius of the path?

3. What force is required to cause a body of mass 3 g to move in a circle of radius 2 m at a constant rate of 4 revolutions per second?

4. An astronaut, as part of her training, is spun in a horizontal circle of radius 5 m. If she can withstand a maximum acceleration of 8g, what is the maximum angular velocity at which she can remain conscious?

5. A particle of mass 80 g rests at 16 cm from the centre of a turntable. If the maximum frictional force between the particle and the turntable is 0.72 N, what is the maximum angular velocity at which the turntable could rotate without the particle slipping?

6. The gravitational force on a satellite of mass m at a distance r from the centre of the Earth is $4.0 \times 10^{14}\ m/r^2$. Assuming that the Earth is a sphere of radius 6.4×10^3 km, find the period of revolution (in hours) of a satellite moving in a circular orbit at a height of 3.6×10^4 km above the Earth's surface.

7. A small bead is threaded on a smooth circular wire of radius r which is fixed in a vertical plane. the bead is projected from the lowest point of the wire with speed $\sqrt{6gr}$. Find the speed of the bead when it has turned through: **(a)** 60°, **(b)** 90°, **(c)** 180°, **(d)** 300°.

8. An aeroplane loops the loop in a vertical circle of radius 200 m, with a speed of 40 m s^{-1} at the top of the loop. The pilot has a mass of 80 kg. What is the tension in the strap holding him into his seat when he is at the top of the loop?

9. A bucket of water is swung in a vertical circle of radius r in such a way that the bucket is upside down when it is at the top of the circle. What is the minimum speed that the bucket may have at this point if the water is to remain in it?

6.4 VEHICLES GOING ROUND BENDS

The centripetal force required to cause a car to go round a bend on a level surface has to be provided by the frictional force exerted on the tyres by the road. The need to rely on friction is removed if the road is suitably 'banked' (Fig. 6.5). The normal reaction, $\boldsymbol{R}$, of the road on the car acquires a horizontal component ($R \sin\theta$) as a result of the banking. If the mass of the car is m and it is moving with constant speed v around a bend of radius r, the centripetal force needs to provide an acceleration of v^2/r, and therefore by Newton's second law (equation [2.2])

$$R \sin\theta = \frac{mv^2}{r} \qquad [6.7]$$

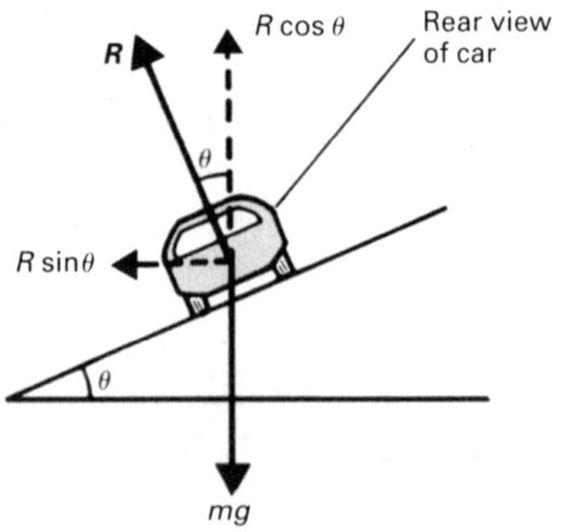

Fig. 6.5 Car going round a bend on a banked corner

Also, since there is no vertical acceleration,

$$R \cos\theta = mg \qquad [6.8]$$

Dividing equation [6.7] by equation [6.8] leads to

$$\tan\theta = \frac{v^2}{gr} \qquad [6.9]$$

If a railway train rounds a bend on a level track, the centripetal force is provided by the push of the outer rail on the flanges of the wheels. This causes a certain amount of wear which could be avoided by banking the track. Equation [6.9] obviously applies to this situation too.

Equation [6.9] also gives the angle at which an aircraft should be banked in order to turn (Fig. 6.6).

Fig. 6.6
Aircraft banking

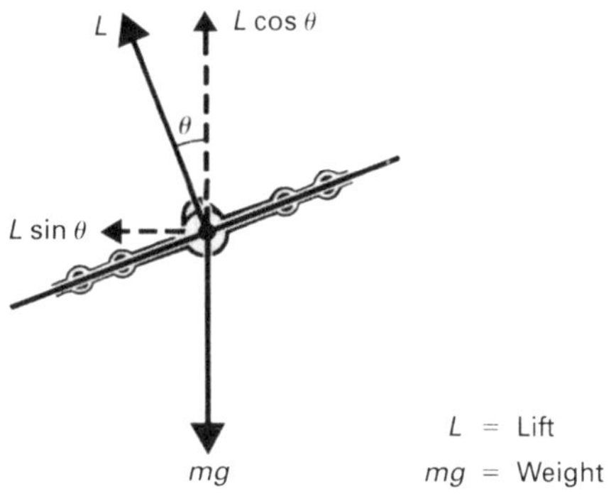

6.5 THE CONICAL PENDULUM

If a pendulum bob is displaced sideways and then given the appropriate velocity in a direction at right angles to its displacement, it will move in a horizontal circle and the string will sweep out a cone. Such an arrangement is shown in Fig. 6.7. There are two forces acting on the pendulum bob. These are:

(i) its weight (mg), and

(ii) the tension in the string (F).

Fig. 6.7
Action of a conical pendulum

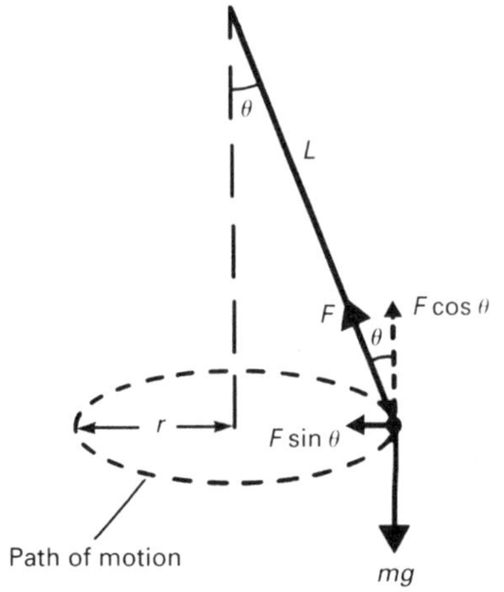

The horizontal component of the tension ($F \sin \theta$) provides the necessary centripetal force. If the radius of the circular path is r and the speed of the bob is v, then from Newton's second law

$$F \sin \theta = \frac{mv^2}{r} \qquad [6.10]$$

There is no vertical acceleration, and therefore

$$F \cos \theta = mg \qquad [6.11]$$

Dividing equation [6.10] by equation [6.11] leads to

$$\tan\theta = \frac{v^2}{gr}$$

Note that this is the same expression (equation [6.9]) that governs the angle at which a turning vehicle must lean. This is not surprising – the forces acting in Fig. 6.7 have the same relationship with each other as those in Fig. 6.5.

Bearing in mind that $v = \omega r$ (equation [6.3]), we obtain

$$\tan\theta = \frac{\omega^2 r^2}{gr}$$

i.e. $$\tan\theta = \frac{\omega^2 r}{g}$$

Referring to Fig. 6.7, $r = L\sin\theta$, and therefore

$$\tan\theta = \frac{\omega^2 L\sin\theta}{g}$$

i.e. $$\frac{\sin\theta}{\cos\theta} = \frac{\omega^2 L\sin\theta}{g}$$

i.e. $$\omega = \sqrt{\frac{g}{L\cos\theta}}$$

Therefore, from equation [6.2], the period, T, is given by

$$T = 2\pi\sqrt{\frac{L\cos\theta}{g}} \qquad [6.12]$$

EXAMPLE 6.3

A pendulum bob of mass 2.0 kg is attached to one end of a string of length 1.2 m. The bob moves in a horizontal circle in such a way that the string is inclined at 30° to the vertical. Calculate:

(a) the tension in the string,

(b) the period of the motion.

(Assume $g = 10\ \text{m s}^{-2}$.)

Solution

The forces acting on the bob are shown in Fig. 6.8; m is the mass of the bob and F is the tension in the string. Since there is no vertical acceleration,

$$F\cos 30^\circ = mg$$

i.e. $$F = \frac{mg}{\cos 30^\circ}$$

i.e. $$F = \frac{2.0 \times 10}{0.8660} = 23.1$$

i.e. Tension = 23 N

Fig. 6.8
Diagram for Example 6.3

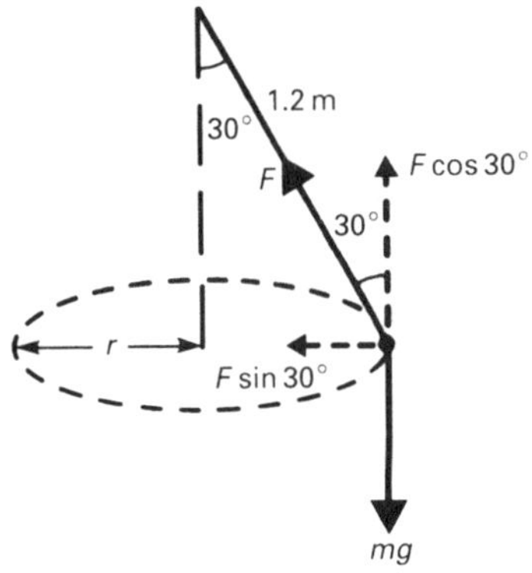

Applying Newton's second law to the horizontal motion gives

$$F \sin 30^\circ = m\omega^2 r$$

where ω is the angular velocity of the bob and r is the radius of the circular path. Noting that $r = 1.2 \sin 30^\circ$ leads to

$$F \sin 30^\circ = m\omega^2 \times 1.2 \sin 30^\circ$$

i.e. $$F = 1.2\, m\omega^2$$

Therefore, since $m = 2.0\,\text{kg}$ and $F = 23.1\,\text{N}$,

$$\omega = \sqrt{\frac{23.1}{1.2 \times 2}}$$

i.e. $$\omega = 3.103\ \text{rad s}^{-1}$$

From equation [6.2], period $= 2\pi/\omega$, and therefore

$$\text{Period} = 2\pi/3.103 = 2.03$$

i.e. $$\text{Period} = 2.0\ \text{s}$$

The same result could have been obtained (though less satisfyingly) by substituting the relevant values into equation [6.12].

QUESTIONS 6B

1. A particle of mass 0.20 kg is attached to one end of a light inextensible string of length 50 cm. The particle moves in a horizontal circle with an angular velocity of $5.0\,\text{rad s}^{-1}$ with the string inclined at θ to the vertical. Find the value of θ.

2. A particle is attached by means of a light, inextensible string to a point 0.40 m above a smooth, horizontal table. The particle moves on the table in a circle of radius 0.30 m with angular velocity ω. Find the reaction on the particle in terms of ω. Hence find the maximum angular velocity for which the particle can remain on the table.

3. A particle of mass 0.25 kg is attached to one end of a light inextensible string of length 3.0 m. The particle moves in a horizontal circle and the string sweeps out the surface of a cone. The maximum tension that the string can sustain is 12 N. Find the maximum angular velocity of the particle.

4. A particle of mass 0.30 kg moves with an angular velocity of $10\,\text{rad s}^{-1}$ in a horizontal circle of radius 20 cm inside a smooth hemispherical bowl. Find the reaction of the bowl on the particle and the radius of the bowl.

6.6 ROTATIONAL KINETIC ENERGY AND MOMENT OF INERTIA

When a body rotates it possesses energy which is due to the rotation. Since it has this energy because of its motion rather than its position, it is kinetic energy. It is distinct from the additional kinetic energy that it would have if it were also undergoing translational motion, and will be referred to as rotational kinetic energy. The wheels on a moving car rotate as they are moving along and therefore have both types of kinetic energy. It can be shown (see below) that

$$\text{Rotational KE} = \tfrac{1}{2} I\omega^2$$

where

ω = the angular velocity of the rotation (rad s^{-1})

I = a constant **for a given axis of rotation** known as **the moment of inertia** of the rotating body (kg m^2).

The moment of inertia of a body is a measure of the way in which its mass is distributed in relation to the axis about which it is rotating. As such it depends on the mass of the body, its size, its shape, and which axis is being considered (see section 6.7).

To Show that Rotational KE $= \frac{1}{2} I\omega^2$

Consider a rigid body rotating with constant angular velocity ω about a fixed axis which is perpendicular to the paper and which passes through O (Fig. 6.9). Suppose that the body is made up of particles such as that at P which has mass m_1 and is at a perpendicular distance r_1 from the axis.

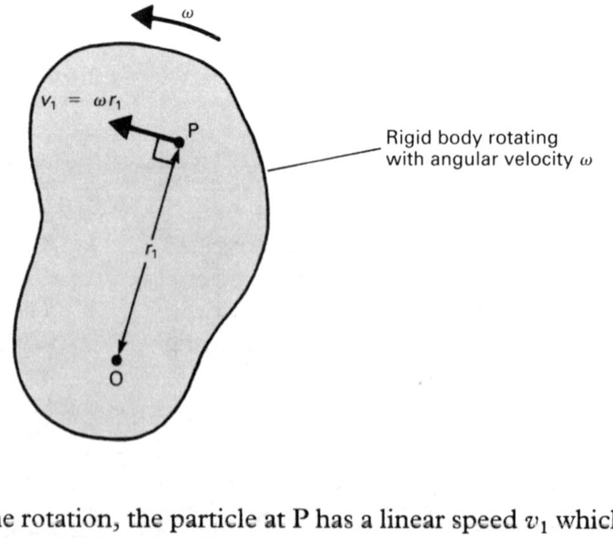

Fig. 6.9
Rotation of a rigid body with constant angular velocity

As a result of the rotation, the particle at P has a linear speed v_1 which is given by equation [6.3] as ωr_1. (Note that all the particles of the body have the same angular velocity, ω.) The (translational) kinetic energy of the particle at P is

$$\tfrac{1}{2} m_1 (\omega r_1)^2$$

If the rest of the particles of the body have masses $m_2, m_3, \ldots$, and their distances from the axis are $r_2, r_3, \ldots$, respectively, the total kinetic energy due to the rotation is given by

$$\text{Rotational KE} = \tfrac{1}{2} m_1(\omega r_1)^2 + \tfrac{1}{2} m_2(\omega r_2)^2 + \ldots$$

i.e. $$\text{Rotational KE} = \tfrac{1}{2} \omega^2(m_1 r_1^{\,2} + m_2 r_2^{\,2} + \ldots)$$

For a given body rotating about a given axis, the term within the brackets is constant and is called the **moment of inertia,** I, of the body about that axis. Thus,

$$\text{Rotational KE} = \tfrac{1}{2} I\omega^2$$

6.7 DEFINITION OF MOMENT OF INERTIA

The moment of inertia, I, of a body about an axis is defined by

$$\left.\begin{aligned} I &= m_1 r_1^{\,2} + m_2 r_2^{\,2} + \ldots \\ \text{i.e.} \quad I &= \sum m_i r_i^{\,2} \qquad i = 1, 2, 3, \ldots \end{aligned}\right\} \qquad [6.13]$$

where r_i is the perpendicular distance from the axis of a particle of mass m_i and the summation is taken over the whole of the body.

The mass of a real body is distributed continuously throughout the entire body, in which case equations [6.13] are replaced by

$$I = \int r^2 \, \mathrm{d}m \qquad [6.14]$$

where the integration is taken over the whole of the body.

Moments of inertia can be determined experimentally (see section 6.13) or, in simple cases, by evaluating equation [6.13] or [6.14]. The results of some such calculations are shown in Figs. 6.10 and 6.11.

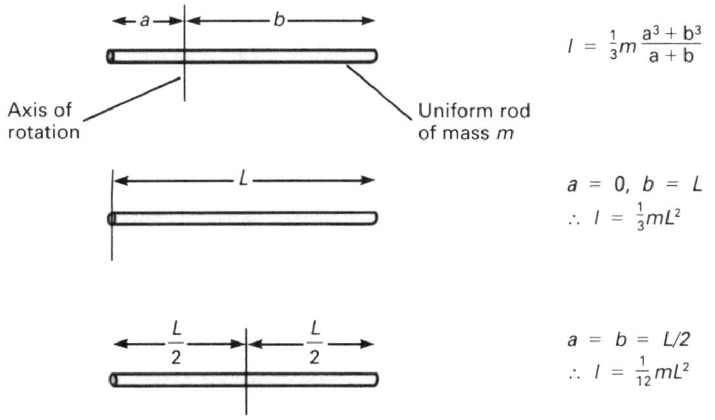

Fig. 6.10
Moments of inertia of a rod about various axes

Note It follows from equation [6.13] that the moment of inertia of a single particle of mass m about an axis is mr^2, where r is the perpendicular distance of the particle from the axis.

Fig. 6.11
Moments of inertia of cylindrical bodies

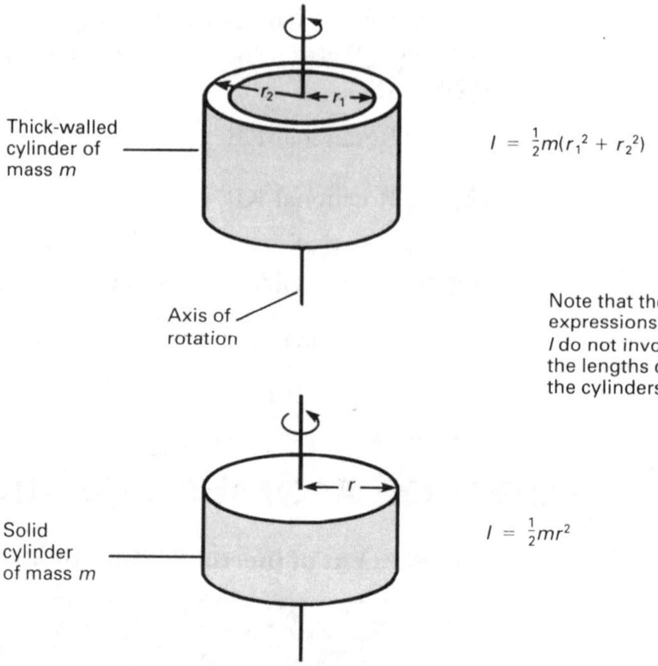

Note that the expressions for I do not involve the lengths of the cylinders

6.8 ANGULAR MOMENTUM

Consider a rigid body rotating with angular velocity ω about an axis which is perpendicular to the paper and which passes through O (Fig. 6.9). Consider a particle of the body which is at P and which has mass m_1 and is at a perpendicular distance r_1 from the axis. The body is rigid and therefore all of its particles have the same angular velocity, ω. Therefore, by equation [6.3] the linear velocity of P is ωr_1 and its linear momentum is $m_1 \omega r_1$.

> **The angular momentum** of a particle about an axis is the product of its linear momentum and the perpendicular distance of the particle from the axis.

Therefore the angular momentum about O of the particle at P is

$$m_1 \omega r_1^2$$

If the rest of the body is made up of particles of masses $m_2, m_3, \ldots$, whose distances from the axis are $r_2, r_3, \ldots$, respectively, the total angular momentum of the body about O is given by

$$\begin{aligned}\text{Angular momentum} &= m_1 \omega r_1^2 + m_2 \omega r_2^2 + \ldots \\ &= \omega(m_1 r_1^2 + m_2 r_2^2 + \ldots)\end{aligned}$$

Therefore (from equation [6.13])

$$\text{Angular momentum} = I\omega \qquad [6.15]$$

The Principle of Conservation of Angular Momentum

The linear momentum of a body moving along a straight line stays constant as long as no resultant external force acts on it (section 2.7). On the other hand, if a body is rotating, it is its angular momentum that is conserved. This is known as **the principle of conservation of angular momentum** and it can be stated as:

The total angular momentum of a system is constant unless an external torque acts on it.

The principle is readily demonstrated by a spinning skater. If she brings her arms in close to her body, her moment of inertia decreases (because some of her mass is now closer to her axis of rotation than it was previously) and her angular velocity increases to such an extent that her angular momentum ($I\omega$) is unchanged. It is left as an exercise for the reader to show that this results in the skater's rotational kinetic energy increasing in the same ratio as her angular velocity. The reader might also like to give some thought to what has provided this increase in energy.

6.9 THE ROTATIONAL FORM OF NEWTON'S SECOND LAW

If a torque is applied to a rigid body which is at rest, the body will start to rotate and will rotate with an ever increasing angular velocity, i.e. the application of the torque causes the body to have an angular acceleration. **Angular acceleration,** α, is defined as rate of change of angular velocity, i.e.

$$\alpha = \frac{d\omega}{dt}$$

where

α = angular acceleration (rad s^{-2}).

Thus, whereas in linear motion a force produces an acceleration which is related to the force through Newton's second law, in rotational motion a torque gives rise to an angular acceleration. The reader will not be surprised to learn therefore, that there is a rotational form of Newton's second law which relates torque and angular acceleration. It may be written as

$$T = I\alpha \qquad [6.16]$$

where

T = the applied torque (N m)

I = the moment of inertia of the body that is rotating (kg m^2)

α = the angular acceleration of the body (rad s^{-2}).

In a more general situation in which the moment of inertia is not constant this becomes

$$T = \frac{d}{dt}(I\omega) \qquad [6.17]$$

where $\frac{d}{dt}(I\omega)$ is the rate of change of angular momentum.

Equations [6.16] and [6.17] are the rotational forms of equations [2.2] and [2.1] respectively. Note that when there is no exernal torque (i.e. when $T = 0$) equation [6.17] reduces to $I\omega$ = a constant. Thus **the principle of conservation of angular momentum is merely a special case of the rotational form of Newton's second law**.

6.10 THE EQUATIONS OF ROTATIONAL MOTION

Equations [6.18]–[6.21] describe the motion of bodies which are moving with constant (uniform) angular acceleration.

$$\omega = \omega_0 + \alpha t \qquad [6.18]$$

$$\omega^2 = \omega_0{}^2 + 2\alpha\theta \qquad [6.19]$$

$$\theta = \omega_0 t + \tfrac{1}{2}\alpha t^2 \qquad [6.20]$$

$$\theta = \tfrac{1}{2}(\omega_0 + \omega)t \qquad [6.21]$$

where

ω_0 = the angular velocity when $t = 0$

ω = the angular velocity at time t

α = the constant angular acceleration

θ = the angle turned through in time t. (Note that if the direction of motion reverses, θ is the net angle turned through.)

These equations are analogous to those which govern uniformly accelerated linear motion (equations [2.6]–[2.9]), with u, v, a and s replaced by ω_0, ω, α and θ respectively.

EXAMPLE 6.4

A flywheel is mounted on a horizontal axle which has a radius of 0.06 m. A constant force of 50 N is applied tangentially to the axle. If the moment of inertia of the system (flywheel + axle) is 4 kg m^2, calculate:

(a) the angular acceleration of the flywheel,

(b) the number of revolutions that the flywheel makes in 16 s assuming that it starts from rest.

Solution

The arrangement is shown in Fig. 6.12.

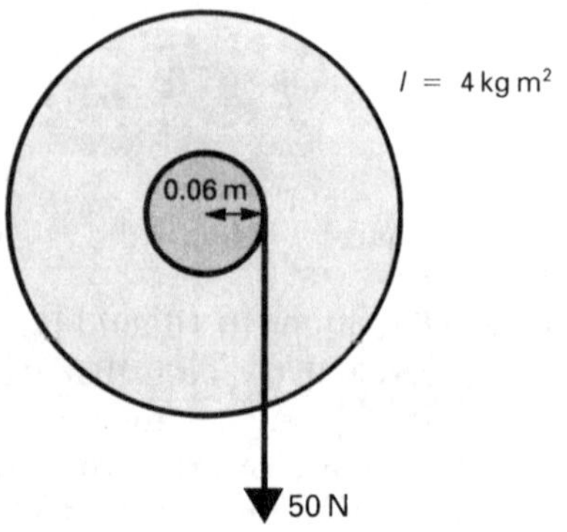

Fig. 6.12
Diagram for Example 6.4

(a) The torque, T, is given by

$$T = 50 \times 0.06 \qquad \text{(equation [3.1])}$$

i.e. $T = 3\,\text{N m}$

If $I(= 4\,\text{kg m}^2)$ is the moment of inertia of the system and α is the angular acceleration, then

$$T = I\alpha \qquad \text{(equation [6.16])}$$

$\therefore \quad 3 = 4\alpha$

i.e. $\alpha = 0.75\,\text{rad s}^{-2}$

(b) If θ is the angle turned through in time $t(= 16\,\text{s})$ and $\omega_0(= 0)$ is the initial angular velocity, then

$$\theta = \omega_0 t + \tfrac{1}{2}\alpha t^2 \qquad \text{(equation [6.20])}$$

$\therefore \quad \theta = 0 + \frac{1}{2} \times 0.75 \times 16^2$

i.e. $\theta = 96\,\text{rad}$

1 revolution $= 2\pi\,\text{rad}$

$\therefore$ Number of revolutions $= 96/2\pi \approx 15$

QUESTIONS 6C

1. A wheel of moment of inertia $0.30\,\text{kg m}^2$ mounted on a fixed axle accelerates uniformly from rest to an angular velocity of $60\,\text{rad s}^{-1}$ in 12 s. Find: **(a)** the angular acceleration, **(b)** the torque causing the wheel to accelerate, **(c)** the number of revolutions in this 12 s period.

2. A flywheel with a moment of inertia of $5.0\,\text{kg m}^2$ moves from rest under the action of a torque of 3.0 N m. Find: **(a)** the angular acceleration, **(b)** the angular velocity after 10 revolutions.

6.11 WORK DONE BY A TORQUE

Consider a rigid body turning through an angle θ as a result of a force F being applied to it (Fig. 6.13). Suppose that the axis of rotation passes through O and is

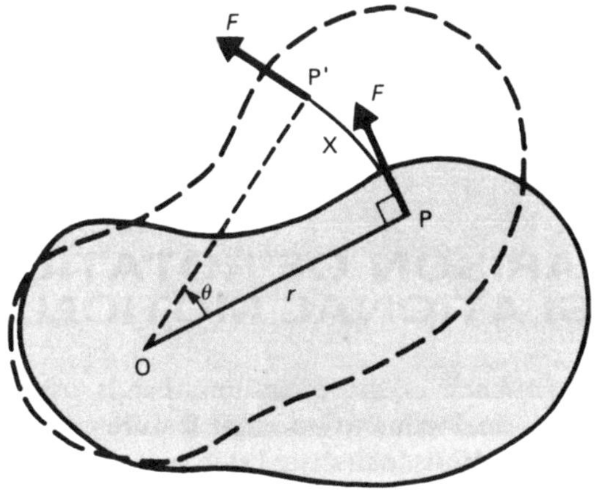

Fig. 6.13 Calculation of work done by a torque

perpendicular to the paper. Suppose also, that the perpendicular distance of the line of action of the force from the axis is constant and is equal to r. The force therefore gives rise to a constant torque, T, given by equation [3.1] as

$$T = Fr$$

As the body turns, the point of application of the force moves from P to P′ along the arc PXP′, and so moves a distance $r\theta$. The work done, W, is given by equation [5.1] as

$$W = Fr\theta$$

i.e.

$$W = T\theta \qquad [6.22]$$

where

W = the work done (J)

T = the constant torque (N m)

θ = the angle turned through (rad).

Notes (i) The work done can have increased the rotational kinetic energy of the body, and/or have been used to overcome any resistive forces (e.g. friction) that are present.

(ii) If the torque is not constant, equation [6.22] is replaced by

$$W = \int T \, d\theta$$

QUESTIONS 6D

1. A constant force of 30 N is applied tangentially to the rim of a wheel mounted on a fixed axle and which is initially at rest. The wheel has a moment of inertia of 0.20 kg m^2 and a radius of 15 cm.
(a) What is the torque acting on the wheel?
(b) Find the work done on the wheel in 10 revolutions.
(c) Assuming that no work is done against friction, use energy considerations to find the angular velocity of the wheel after 10 revolutions.

2. A disc and a hoop roll down a slope. They have the same mass and the same radius.
(a) Which has the greater moment of inertia?
(b) Does one loose more PE than the other?
(c) Which acquires the greater speed?

6.12 COMPARISON OF ROTATIONAL AND TRANSLATIONAL MOTION

Each of the quantities that is used in the treatment of rotational motion is analogous to one that features in the description of translational motion. The various 'pairs' are listed in Table 6.1

Table 6.1
Rotational analogues of translational quantities

Translational quantity	*Rotational analogue*
m	I
v	ω
F	T
a	α
s	θ

It follows that there are rotational analogues of the equations and expressions of translational motion, and that these can be obtained from the latter simply by replacing each quantity with its analogue from Table 6.1. This is done in Table 6.2.

Table 6.2
Rotational analogues of translational expressions and equations

Translation	*Rotation*
$\frac{1}{2}mv^2$	$\frac{1}{2}I\omega^2$
mv	$I\omega$
$F = ma$	$T = I\alpha$
$F = \frac{\mathrm{d}}{\mathrm{d}t}(mv)$	$T = \frac{\mathrm{d}}{\mathrm{d}t}(I\omega)$
$W = Fs$	$W = T\theta$

6.13 EXPERIMENTAL DETERMINATION OF THE MOMENT OF INERTIA OF A FLYWHEEL

The apparatus is shown in Fig. 6.14. It consists of a mass m attached to one end of a string, the other end of which passes through a hole in the axle of the flywheel whose moment of inertia I is to be determined. The mass is released so that it falls to the floor and drives the flywheel round. The string is of such a length that it is fully unwound just as the mass reaches the floor, and because of this and the way it is attached, it falls off the axle as soon as the mass completes its descent. The flywheel, though no longer driven, continues to rotate; it comes to rest only when its energy has been dissipated in doing work against frictional forces.

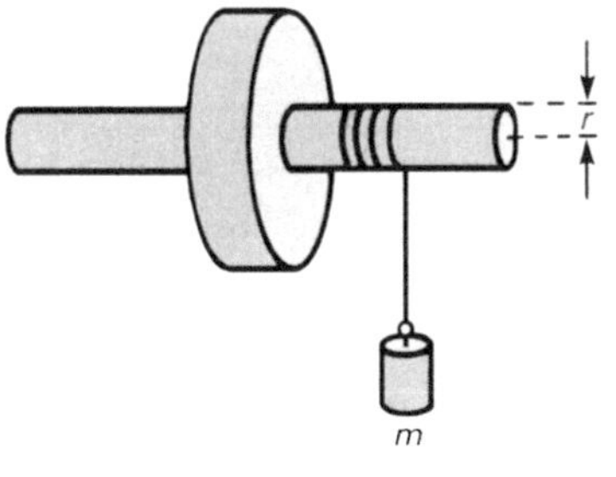

Fig. 6.14
Rotating flywheel

The potential energy lost by the falling mass is equal to the total gain in kinetic energy plus the work done against friction during the descent. Therefore if:

h = the distance the mass falls,

v = the speed of the mass on reaching the floor,

ω = the angular velocity of the flywheel when the mass reaches the floor,

n_1 = the number of revolutions of the flywheel whilst the mass is falling, and

f = the work done against friction during each revolution, then

$$mgh = \tfrac{1}{2}mv^2 + \tfrac{1}{2}I\omega^2 + n_1 f \qquad [6.23]$$

If the flywheel makes a further n_2 revolutions before coming to rest, the work done against friction is $n_2 f$. This is done at the expense of the kinetic energy of the flywheel, and therefore

$$\tfrac{1}{2}I\omega^2 = n_2 f$$

Substituting for f in equation [6.23] gives

$$mgh = \tfrac{1}{2}mv^2 + \tfrac{1}{2}I\omega^2 + \frac{n_1}{n_2}\left(\tfrac{1}{2}I\omega^2\right) \qquad [6.24]$$

The velocity with which the mass hits the floor is twice its average velocity, and therefore

$$v = \frac{2h}{t} \quad \text{and} \quad \omega = \frac{2h}{rt}$$

where t is the time the mass takes to reach the ground and r is the radius of the axle. Substituting for v and ω in equation [6.24] and rearranging gives

$$I = mr^2\left(\frac{gt^2}{2h} - 1\right)\left(\frac{n_2}{n_1 + n_2}\right)$$

The mass m is found by using a balance; r and h are measured with vernier calipers and a metre rule respectively; t is measured with a stopwatch; g is known; and n_1 and n_2 are counted – hence I.

7

SIMPLE HARMONIC MOTION

7.1 DEFINITION, EQUATIONS, EXPLANATIONS

There are many types of vibration but perhaps the most common is that which is known as simple harmonic motion (SHM). It is important not only because there are many examples of it but also because all other vibrations can be treated as if they are composed of simple harmonic vibrations. It is the way in which the acceleration of a body depends on its position that determines the particular type of vibration a body is undergoing.

If a body moves in such a way that its acceleration is directed towards a fixed point in its path and is directly proportional to its distance from that point, the body is moving with simple harmonic motion.

It follows that the 'fixed point' is the equilibrium position, i.e. the position at which the body would come to rest if it were to lose all of its energy.

If a body is vibrating with simple harmonic motion, its motion can be described by an equation of the form

$$\frac{d^2x}{dt^2} = -\omega^2 x \qquad [7.1]$$

where

$\frac{d^2x}{dt^2}$ = the acceleration of the body (m s^{-2})

x = the displacement of the body from its equilibrium position, i.e. from the 'fixed point' in its path (m)

ω^2 = a positive constant (s^{-2})

Notes (i) The minus sign in equation [7.1] ensures that the acceleration is always directed towards the equilibrium position, as required by the definition.

(ii) The constant of proportionality is written as ω^2 (rather than ω) because of the connection with circular motion – see section 7.2.

Integrating equation [7.1] leads to

$$v = \pm\,\omega\sqrt{a^2 - x^2} \qquad [7.2]$$

and

$$x = a \cos \omega t \qquad [7.3]$$

where

v = the velocity of the body at time t (m s^{-1})

a = **the amplitude** of the motion, i.e. the maximum displacement from the equilibrium position (m).

Note Equation [7.3] requires that $x = a$ when $t = 0$. An alternative expression for x is $x = a \sin \omega t$; this requires that $x = 0$ when $t = 0$, i.e. that the motion is taken to start from the equilibrium position rather than the point of maximum positive displacement.

Timing may commence when the body is at any point of its oscillation. To take account of this the expression

$$x = a \sin(\omega t + \varepsilon)$$

is used, in which ε, the **initial phase angle** or **epoch**, is a constant expressed in radians and given by $\sin \varepsilon = x_0/a$, where x_0 is the value of x at $t = 0$. The reader should confirm that this reduces to $x = a \sin(\omega t + \pi/2) = a \cos \omega t$ when $x = a$ at $t = 0$, and reduces to $x = a \sin(\omega t + 0) = a \sin \omega t$ when $x = 0$ at $t = 0$.

The period T of the motion (i.e. the time for one complete oscillation) is given by

$$T = 2\pi/\omega \qquad [7.4]$$

Note that **for any particular system the period is independent of the amplitude**. For example, if the amplutide of oscillation of a simple pendulum is increased, its average speed increases and there is no change in the time it takes to complete an oscillation.*

Fig. 7.1 illustrates the positional dependence of some of the parameters which are associated with the motion of a body whose oscillations are simple harmonic. The positive direction of x is, as usual, toward the right and therefore whenever the body is moving to the left its velocity is negative. Also, the acceleration is negative when it is directed towards the left. (Note that a negative acceleration must not be confused with retardation; the body slows down only when its acceleration is in the opposite direction to its velocity.)

As the body moves from A towards its equilibrium position, O, its speed increases and reaches a maximum at O. During this time its acceleration decreases to zero. Although there is no force acting on the body when it is at O, its inertia carries it through to B. From O to B there is a retarding force; the speed of the body decreases and is momentarily zero at B. As the body moves back to O its speed increases, and then decreases again from O to A.

Bearing in mind that the acceleration of a body of constant mass is proportional to the force acting on it, we see that a body which is moving with simple harmonic motion does so because there is a force acting on it which is proportional to the displacement of the body from its equilibrium position and is directed towards that

*The motion of a simple pendulum is not exactly simple harmonic and therefore this statement is only approximately true.

Fig. 7.1
Characteristics of SHM

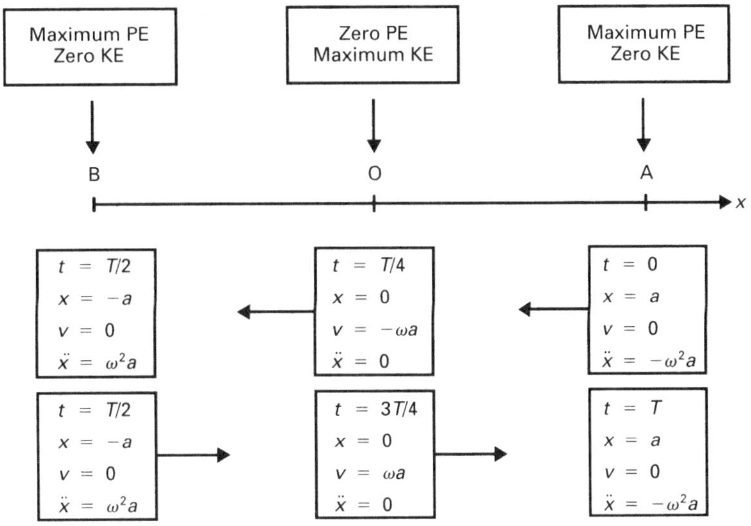

(*Note:* $\ddot{x}$ = acceleration)

position. Such is the case, for example, of a body oscillating on the end of a spring which obeys Hooke's law (section 11.1). If the body is pulled down below its equilibrium position and then released, a net upward force acts on it because the tension in the spring is greater than the weight of the body. The greater the downward displacement, the greater the tension and therefore the greater the upward directed force. When the body is above its equilibrium position its weight is greater than the tension and so the resultant force is downwards.

EXAMPLE 7.1

A particle is moving with SHM of period 8.0 s and amplitude 5.0 m. Find: (a) the speed of the particle when it is 3.0 m from the centre of its motion, (b) the maximum speed, (c) the maximum acceleration.

Solution

The equations for speed and acceleration involve ω; our first step therefore is to find ω.

$$T = 2\pi/\omega$$

$$\therefore \quad \omega = 2\pi/T = 2\pi/8.0 = 0.785\ s^{-1}$$

(a)
$$v = \pm\omega\sqrt{a^2 - x^2}$$

$$\therefore \quad v = \pm 0.785\sqrt{5.0^2 - 3.0^2} = \pm 3.14\,\text{m s}^{-1}$$

i.e. Speed 3.0 m from centre $= 3.1\,\text{m s}^{-1}$

(b) It follows from equation [7.2] with $x = 0$ that the maximum speed v_{max} is given by

$$v_{max} = \omega a$$

$$\therefore \quad v_{max} = 0.785 \times 5.0 = 3.93\,\text{m s}^{-1}$$

i.e. Maximum speed $= 3.9\,\text{m s}^{-1}$

(c) It follows from equation [7.1] that the magnitude of the acceleration is greatest when $x = \pm a$ and is given by

$$\text{maximum acceleration} = \omega^2 a$$

$$= (0.785)^2 \times 5.0 = 3.08\,\text{m s}^{-2}$$

i.e. $\text{Maximum acceleration} = 3.1\,\text{m s}^{-2}$

EXAMPLE 7.2

A particle is moving with SHM of period 24 s between two points, A and B (Fig. 7.2). Find the time taken for the particle to travel from: (a) A to B, (b) O to B, (c) O to C, (d) D to B, (e) C to E.

Fig. 7.2
Diagram for Example 7.2

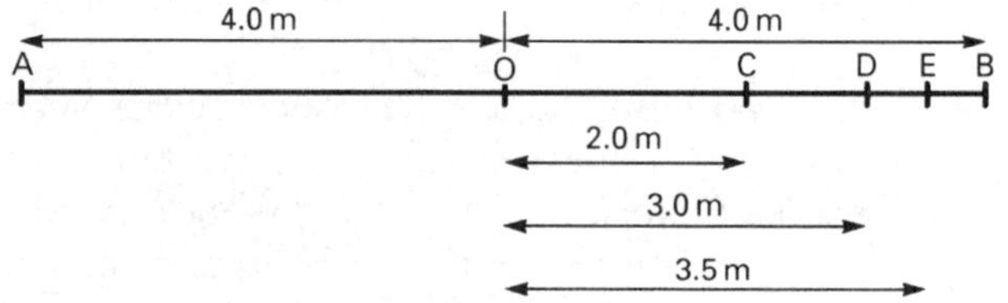

Solution

Let t_{AB} = time from A to B, t_{BC} = time from B to C, etc.

(a) t_{AB} = time for half an oscillation = 12 s

(b) t_{OB} = time for quarter of an oscillation = 6 s

(c) Amplitude (a) = OA = OB = 4.0 m. $\omega = 2\pi/T = 2\pi/24 = 0.262\,\text{s}^{-1}$. Therefore by

$$x = a \sin \omega t$$

$$OC = 4.0 \sin (0.262\, t_{OC})$$

$$\therefore \quad 2.0 = 4.0 \sin (0.262\, t_{OC})$$

$$\therefore \quad \sin (0.262\, t_{OC}) = 0.5$$

(Note the use of radians, *not* degrees.)

$$\therefore \quad 0.262\, t_{OC} = \sin^{-1} (0.5) = \frac{\pi}{6} = 0.524 \text{ rad}$$

$$\therefore \quad t_{OC} = 2.0\,\text{s}$$

(d) $t_{DB} = t_{BD}$. Therefore by

(x is the displacement from O, it is not the distance from D to B.)

$$x = a \cos \omega t$$

$$OD = 4.0 \cos (0.262\, t_{DB})$$

$$\therefore \quad 3.0 = 4.0 \cos (0.262\, t_{DB})$$

$$\therefore \quad \cos (0.262\, t_{DB}) = 0.75$$

$$\therefore \quad 0.262\, t_{DB} = \cos^{-1} (0.75) = 0.723 \text{ rad}$$

$$\therefore \quad t_{DB} = 2.8\,\text{s}$$

(e) $t_{CE} = t_{OE} - t_{OC}$. Since t_{OC} has already been found, it remains to find t_{OE}. By

$$x = a \sin \omega t$$

$$OE = 4.0 \sin (0.262\, t_{OE})$$

$$\therefore \quad 3.5 = 4.0 \sin (0.262\, t_{OE})$$

$$\therefore \quad \sin (0.262\, t_{OE}) = 0.875$$

$$\therefore \quad 0.262\, t_{OE} = \sin^{-1} (0.875) = 1.065 \text{ rad}$$

$$\therefore \quad t_{OE} = 4.1 \text{ s}$$

$$\therefore \quad t_{CE} = t_{OE} - t_{OC} = 4.1 - 2.0 = 2.1 \text{ s}$$

QUESTIONS 7A

1. A particle is moving with SHM of period 16 s and amplitude 10 m. Find the speed of the particle when it is 6.0 m from its equilibrium position.

2. How far is the particle in question 1 from its equilibrium position 1.5 s after passing through it? What is its speed at this time?

3. A tuning fork has a frequency of 256 Hz. What is the maximum speed of the tips of the prongs if they each oscillate with SHM of amplitude 0.40 mm. (Assume that the tips of the prongs move in straight lines.)

4. A particle moves with SHM of period 4.0 s and amplitude 4.0 m. Its displacement from the equilibrium position is x. Find the time taken for it to travel: **(a)** from $x = 4.0$ m to $x = 3.0$ m, **(b)** from $x = -4.0$ m to $x = 3.0$ m, **(c)** from $x = 0$ to $x = 3.0$ m, **(d)** from $x = 1.0$ m to $x = 3.0$ m.

5. A particle moving with SHM has a speed of 8.0 m s^{-1} and an acceleration of 12 m s^{-2} when it is 3.0 m from its equilibrium position. Find: **(a)** the amplitude of the motion, **(b)** the maximum velocity, **(c)** the maximum acceleration.

7.2 RELATIONSHIP BETWEEN SHM AND CIRCULAR MOTION

Consider a particle P moving with constant angular velocity ω around the circumference of a circle of radius a (Fig. 7.3). Consider, in particular, the motion of N, the point at which the perpendicular from P meets the diameter AOB.

The acceleration of P is $\omega^2 a$ (see section 6.3) and is directed towards O. It follows that the x-component of the acceleration of P is

$$-\omega^2 a \cos \theta$$

(The acceleration is directed towards the left, i.e. it is in the negative direction of x, hence the inclusion of the minus sign.) Since N is always vertically below P, its acceleration $\mathrm{d}^2x/\mathrm{d}t^2$ is equal to the x-component of the acceleration of P and therefore is given by

$$\frac{\mathrm{d}^2x}{\mathrm{d}t^2} = -\omega^2 a \cos \theta \qquad [7.5]$$

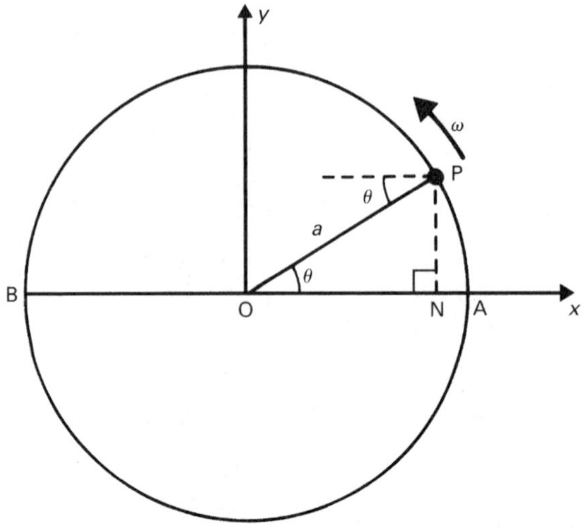

Fig. 7.3
To illustrate the relationship between SHM and circular motion

But $\cos\theta = \mathrm{ON}/a$, and ON is equal to x, the displacement of N, and therefore $\cos\theta = x/a$. Substituting for $\cos\theta$ in equation [7.5] gives

$$\frac{\mathrm{d}^2x}{\mathrm{d}t^2} = -\omega^2 x$$

Since this is the equation of motion of a particle which is moving with simple harmonic motion, the motion of N is simple harmonic. N completes one cycle in the time it takes P to complete one revolution, and therefore

$$\begin{matrix}\text{Period of} \\ \text{rotation of P}\end{matrix} = \begin{matrix}\text{Period of} \\ \text{oscillation of N}\end{matrix} = \frac{2\pi}{\omega}$$

7.3 THE SIMPLE PENDULUM

Consider a pendulum bob, P, of mass m attached to a light, inextensible string of length L. Suppose that the pendulum is suspended from a fixed point, O, and that when the string is at an angle θ to the vertical the velocity of the bob is v (Fig. 7.4).

The forces acting on the bob are its weight, mg, and the tension, F, in the string.

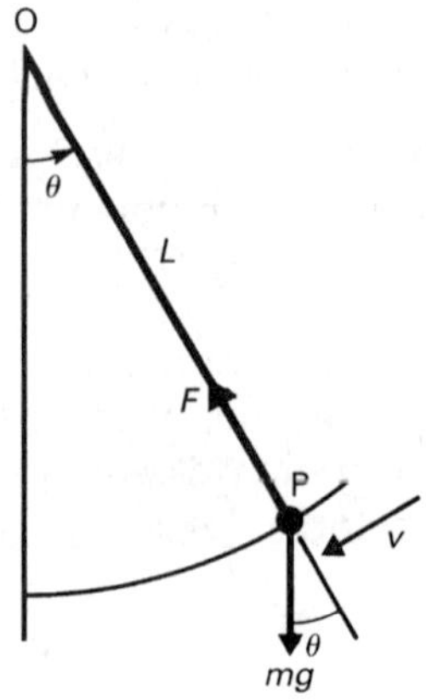

Fig. 7.4
To determine the period of oscillation of a simple pendulum

Consider the motion perpendicular to PO. By Newton's second law

$$mg \sin \theta = m \frac{dv}{dt}$$

The velocity, v, can be written in terms of the angular velocity $d\theta/dt$ as

$$v = -L \frac{d\theta}{dt}$$

(The minus sign is necessary because θ is measured from the vertical and therefore v is in the direction of θ decreasing.) Substituting for v gives

$$mg \sin \theta = -mL \frac{d^2\theta}{dt^2}$$

i.e.

$$\frac{d^2\theta}{dt^2} = -\frac{g}{L} \sin \theta$$

If the amplitude of oscillation is small, θ is small and therefore $\sin \theta \approx \theta$, in which case, to a reasonable approximation

$$\frac{d^2\theta}{dt^2} = -\frac{g}{L} \theta \qquad [7.6]$$

Since both g and L are positive constants, so also is g/L and therefore equation [7.6] is of the same form as equation [7.1]. It follows that, to a reasonable approximation, the motion of a simple pendulum is simple harmonic. Further, the positive constant ω^2 of equation [7.1] is equal to g/L and therefore, by equation [7.4], the period T of a simple pendulum is given by

$$T = 2\pi \sqrt{\frac{L}{g}} \qquad [7.7]$$

Notes

(i) Equation [7.7] does not involve m and therefore **the period of oscillation of a simple pendulum is independent of its mass**. This can be shown to be true no matter what the amplitude of oscillation.

(ii) Even for amplitudes of oscillation of as much as 15° the period calculated on the basis of equation [7.7] is accurate to within $\frac{1}{2}$%

7.4 DETERMINATION OF *g* USING A SIMPLE PENDULUM

A reasonably accurate determination of the acceleration due to gravity g can be made by measuring the period of oscillation and the length of a simple pendulum.

The pendulum, in the form of a small lead sphere attached to a suitable length (about 1 m) of sewing thread, should be suspended in the manner shown in Fig. 7.5. The wooden blocks should have well-defined right-angled edges at X so that there is no possibility of the pendulum swinging about more than one point.

Once the apparatus has been assembled the procedure outlined below should be followed.

(i) Measure the length L. (Note that the measurement is made to the centre of the bob.) Take care not to stretch the thread.

Fig. 7.5
Apparatus for the determination of ***g***

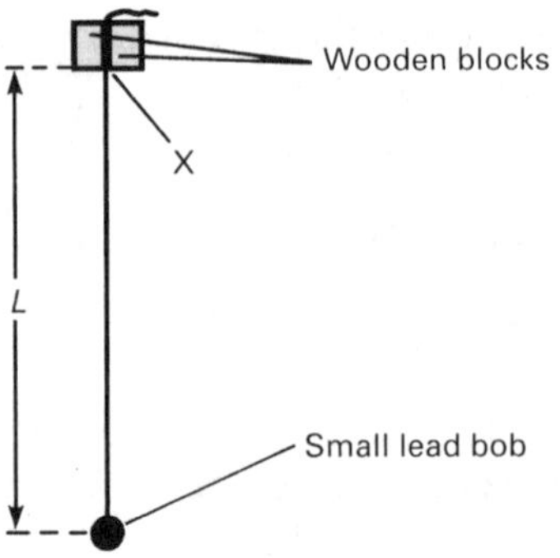

(ii) Displace the pendulum through about 5° and then release it so that it executes oscillations of small amplitude.

(iii) Determine the period T by using a stop-watch to time 50 oscillations. By using 50 oscillations, rather than one, the error that arises through being unable to start and stop the watch when the pendulum is exactly in the intended position is greatly reduced. The error which remains is minimized by making the timings to the mid-point of the motion because that is where the speed of the pendulum is greatest.

(iv) Repeat (i), (ii) and (iii) for about five more values of L.

From the theory of the simple pendulum

$$T = 2\pi\sqrt{\frac{L}{g}}$$

$$\therefore \quad T^2 = \frac{4\pi^2}{g}L$$

It follows that the gradient of a graph of T^2 against L is $4\pi^2/g$, in which case g can be determined by plotting such a graph and measuring its gradient.

Notes (i) A graph of T against $\sqrt{L}$ has a gradient of $2\pi/\sqrt{g}$ and therefore such a graph could have been used to determine g. The reason for choosing to plot T^2 against L is that the graph is linear and its gradient is $4\pi^2/g$ even if there is an error in the measurement of L, providing it is a constant error.

(ii) The approximation made in deriving equation [7.7] leads to an error of less than 0.05% if the amplitude of oscillation does not exceed 5°. This is insignificant compared with the errors that are likely to be involved in the measurements of L and T.

7.5 A BODY ON A SPRING

The extension of a spring which obeys Hooke's law (section 11.1) is proportional to the tension which has produced it. Therefore

$$\text{Tension} = k \times \text{extension} \qquad [7.8]$$

where

k = a constant of proportionality which is known as the **spring constant**. It is equal to the tension required to produce unit extension ($N\,m^{-1}$).

Suppose that a suspended spring which obeys Hooke's law has a body of mass m attached to its lower end. In Fig. 7.6(a) the body is at rest in its equilibrium position. There can be no resultant force acting on the body and therefore the tension F_0 is given by

$$F_0 = mg$$

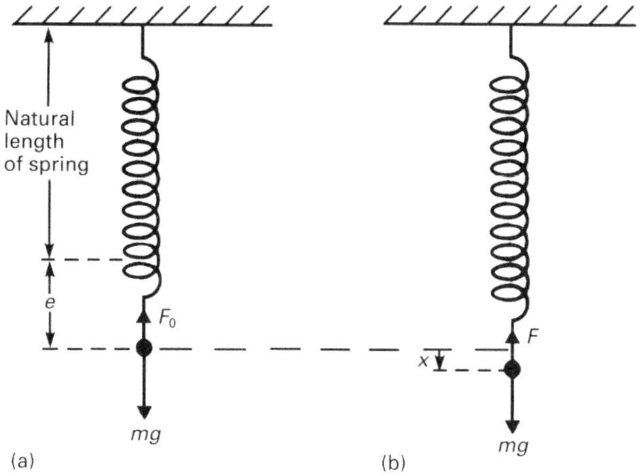

Fig. 7.6
Oscillation of a body on a spring

It follows from equation [7.8] that since the extension is e

$$mg = ke \tag{7.9}$$

Suppose now that the body is displaced downwards through a distance x (Fig. 7.6(b)). The body is no longer in equilibrium and feels a net upward force of $(F - mg)$, where F is the instantaneous value of the tension in the spring. Therefore by Newton's second law

$$F - mg = -m\frac{d^2x}{dt^2} \tag{7.10}$$

(The minus sign is present because the resultant force on the body is directed upwards and therefore acts so as to decrease x.) By equation [7.8], since the total extension is now $(e + x)$

$$F = k(e + x)$$

Therefore from equation [7.10]

$$k(e + x) - mg = -m\frac{d^2x}{dt^2}$$

But, from equation [7.9], $mg = ke$, and therefore

$$kx = -m\frac{d^2x}{dt^2}$$

i.e.

$$\frac{d^2x}{dt^2} = -\frac{k}{m}x \tag{7.11}$$

Since both k and m are positive constants, so also is k/m and therefore equation [7.11] may be written as

$$\frac{d^2x}{dt^2} = -\omega^2 x$$

where ω^2 is a positive constant equal to k/m. This equation represents simple harmonic motion and therefore the motion of the body is simple harmonic. Since $\omega^2 = k/m$, equation[7.4] gives the period of the motion as

$$\text{Period} = 2\pi\sqrt{\frac{m}{k}} \qquad [7.12]$$

Except in the idealized case of a spring of zero mass, it is necessary to take account of the fact that the spring itself oscillates. It can be shown that m needs to be replaced by $(m + m_s)$, where m_s is a constant known as the **effective mass** of the spring. (Note that m_s is less than the actual mass of the spring because it is the lowest coil which oscillates with the full amplitude of the suspended body.) With this modification then, equation [7.12] becomes

$$\text{Period} = 2\pi\sqrt{\frac{m + m_s}{k}} \qquad [7.13]$$

7.6 DETERMINATION OF *g* BY USING A MASS ON AN OSCILLATING SPRING

From equation [7.9], $m = ke/g$. Substituting for m in equation [7.13] and squaring leads to

$$T^2 = 4\pi^2\left(\frac{ke/g + m_s}{k}\right)$$

where T is the period of oscillation. Removing the brackets gives

$$T^2 = \frac{4\pi^2}{g}e + \frac{4\pi^2 m_s}{k} \qquad [7.14]$$

Thus a graph of T^2 against e is linear (Fig. 7.7) and has a gradient of $4\pi^2/g$, and therefore enables g to be determined. Such a graph can be obtained by adding a number of different masses to the spring and measuring the static extension e which each produces, together with the corresponding period of oscillation T.

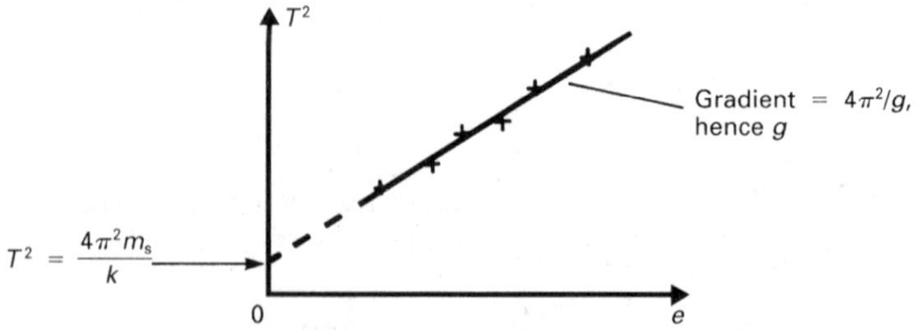

Fig. 7.7
T^2 against e for a body oscillating on a spring

Note When $e = 0$, $T^2 = 4\pi^2 m_s/k$, and therefore m_s can be found providing k is known. The value of k is found by plotting m against e since, by equation [7.9], the gradient of such a graph is k/g.

7.7 THE WORK DONE IN STRETCHING A SPRING

The tension, F, in a spring whose extension is x and which obeys Hooke's law is given by

$$F = kx$$

where k is the spring constant. If the extension is increased by δx where δx is so small that F can be considered constant, then (by equation [5.1]) the work done, δW, is given by

$$\delta W = F\,\delta x$$

i.e. $$\delta W = kx\,\delta x$$

The total work done in increasing the extension from 0 to x, i.e. the **elastic potential energy** stored in the spring when its extension is x, is given by W, where

$$W = \int_0^x kx\,\mathrm{d}x$$

i.e. $$W = \tfrac{1}{2}kx^2$$

Substituting for k from $F = kx$ gives $W = \frac{1}{2}Fx$; substituting for x gives $W = F^2/(2k)$, i.e.

$$W = \tfrac{1}{2}kx^2 = \tfrac{1}{2}Fx = \frac{F^2}{2k}$$

CONSOLIDATION

A body is moving with SHM if its acceleration is directed towards a fixed point in its path and is directly proportional to its distance from that point.

The amplitude of the motion (a) is the maximum displacement from the equilibrium position.

If a body is moving with SHM, its motion can be described by an equation of the form

$$\frac{\mathrm{d}^2x}{\mathrm{d}t^2} = -\omega^2 x \qquad [7.1]$$

The converse is also true, and therefore if we are required to show that a body is moving with SHM, it is sufficient to show that its motion is described by an equation of the same form as equation [7.1].

$$v = \pm\omega\sqrt{a^2 - x^2} \qquad [7.2]$$

$$x = a\cos\omega t \qquad (\text{if } x = a \text{ when } t = 0)$$

$$x = a\sin\omega t \qquad (\text{if } x = 0 \text{ when } t = 0)$$

$$T = 2\pi/\omega \qquad f = 1/T$$

Magnitude of maximum acceleration $= \omega^2 a$ (at $x = \pm a$)

Magnitude of maximum velocity $= \omega a$ (at $x = 0$)

To obtain v in terms of t substitute for x in equation [7.2].

The period (T) is independent of the amplitude (a).

The first step in solving many SHM problems is to find ω.

$$T = 2\pi\sqrt{\frac{L}{g}} \qquad \text{(Simple pendulum)}$$

$$T = 2\pi\sqrt{\frac{m + m_s}{k}} \qquad \text{(Mass on a spring)}$$

$$W = \tfrac{1}{2}kx^2 = \tfrac{1}{2}Fx = \frac{F^2}{2k}$$

8

GRAVITATION AND GRAVITY

8.1 KEPLER'S LAWS

Throughout the last few decades of the sixteenth century, Tycho Brahe made precise measurements of the positions of the planets and various other bodies in the Solar System. Johannes Kepler made a detailed analysis of the measurements, and by 1619 had announced three laws which describe planetary motion.

1. The orbit of each planet is an ellipse which has the Sun at one of its foci.
2. Each planet moves in such a way that the (imaginary) line joining it to the Sun sweeps out equal areas in equal times.
3. The squares of the periods of revolution of the planets about the Sun are proportional to the cubes of their mean distances from it.

Fig. 8.1 illustrates law 2 but gives an exaggerated idea of the eccentricity of most planetary orbits. With the exceptions of Mercury and Pluto, the planets follow very nearly circular paths.

Fig. 8.1
Illustration of Kepler's second law

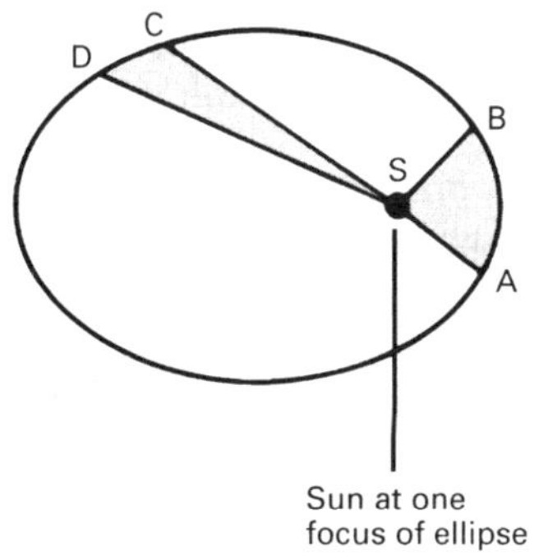

The average speed of the planet between A and B is greater than between C and D and in such a way that

Area ABS = Area CDS

8.2 NEWTON'S LAW OF UNIVERSAL GRAVITATION

About fifty years after Kepler's laws had been announced, Isaac Newton showed that any body which moves about the Sun in accordance with Kepler's second law must be acted on by a force which is directed towards the Sun. He was able to show that if this force is inversely proportional to the square of the distance of the body from the Sun, then the body must move along a path which is a conic section (i.e. elliptical, circular, parabolic or hyperbolic). Newton then showed that when the path is elliptical or circular the period of revolution is given by Kepler's third law. Thus, a centrally directed inverse square law of attraction is consistent with all three of Kepler's laws. Newton proposed that the planets are held in their orbits by just such a force. He further proposed that it is the same type of force which maintains the Moon in its orbit about the Earth, and which the Earth exerts on a body when it causes it to fall to the ground. Extending these ideas, Newton proposed that every body in the Universe attracts every other with a force which is inversely proportional to the square of their separation. His next step was to turn his attention to the masses of the bodies involved.

According to Newton's third law, if the Earth exerts a force on a body, then that same body must exert a force of equal magnitude on the Earth. Newton knew that the force exerted on a body by the Earth is proportional to the mass of the body. He saw no reason why the body should behave any differently from the Earth, in which case the force exerted on the Earth by the body must be proportional to the mass of the Earth. Since the two forces are equal, a change in one must be accompanied by an equal change in the other. It follows that each force must be proportional to the product of the Earth's mass and the mass of the body.

The ideas of the last two paragraphs are summarised in **Newton's law of universal gravitation**.

> Every particle in the Universe attracts every other with a force which is proportional to the product of their masses and inversely proportional to the square of their separation.

Thus

$$F = G\frac{m_1 m_2}{r^2} \qquad [8.1]$$

where

F = the gravitational force of attraction between two particles whose masses are m_1 and m_2, and which are a distance r apart

G = a constant of proportionality known as the **universal gravitational constant** ($= 6.67 \times 10^{-11}\ \text{N m}^2\ \text{kg}^{-2}$).

Note Equation [8.1] is concerned with particles (i.e. point masses) but, in the circumstances listed below, it can also be used for bodies of masses m_1 and m_2 whose centres are a distance r apart.

(a) It is valid for two bodies of any size provided that they each have spherical symmetry. (The Sun and the Earth is a good approximation.)

(b) It is a good approximation when one body has spherical symmetry and the other is small compared with the separation of their centres (e.g. the Earth and a brick).

(c) It is a good approximation when neither body has spherical symmetry, but where both are small compared with the separation of their centres (e.g. two bricks a few metres apart).

8.3 TO SHOW THAT KEPLER'S THIRD LAW IS CONSISTENT WITH $F = G\dfrac{m_1 m_2}{r^2}$

Consider a planet of mass m moving about the Sun in a circular* orbit of radius r. Suppose that the mass of the Sun is m_s and that the angular velocity of the planet is ω (Fig. 8.2).

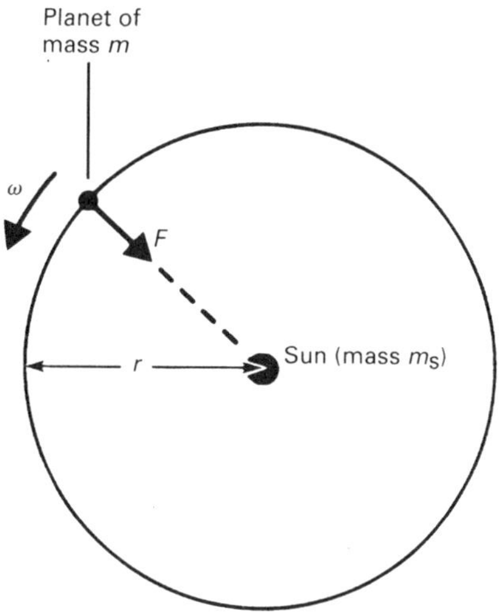

Fig. 8.2
Force on planet in circular motion around the Sun

The force F which provides the centripetal acceleration $\omega^2 r$ (section 6.3) is given by Newton's second law as

$$F = m\omega^2 r \qquad [8.2]$$

By Newton's law of universal gravitation (equation [8.1])

$$F = G\frac{mm_s}{r^2}$$

Therefore, from equation [8.2]

$$G\frac{mm_s}{r^2} = m\omega^2 r$$

But $\omega = 2\pi/T$, where T is the period of revolution of the planet, and therefore

$$G\frac{mm_s}{r^2} = m\frac{4\pi^2}{T^2}r$$

i.e. $$T^2 = \frac{4\pi^2}{Gm_s}r^3$$

Since G, m_s and π have the same values no matter which planet is being considered,

$$T^2 \propto r^3$$

This is Kepler's third law; it has been derived on the basis of Newton's law of universal gravitation and therefore the two laws are consistent.

*The mathematics required to treat the general case of an elliptical orbit is beyond the scope of this book but leads to the same result.

8.4 NEWTON'S TEST OF THE INVERSE SQUARE LAW

The last two sections make it clear that Newton's law of universal gravitation is consistent with Kepler's laws of planetary motion. However, the forces which hold the planets in their orbits are due, in every case, to the Sun. In order to show that gravitational attraction is universal, Newton needed to test it in circumstances which did not involve the Sun. The obvious test was to apply his ideas to the Earth–Moon system.

If a body of mass m is at the surface of the Earth, the force acting on the body is its weight mg. This same force is given by the law of universal gravitation as

$$G\frac{mm_E}{r_E^2}$$

where m_E and r_E are respectively the mass and radius of the Earth. Therefore

$$G\frac{mm_E}{r_E^2} = mg$$

i.e. $$G = \frac{gr_E^2}{m_E} \quad [8.3]$$

The law of universal gravitation gives the force exerted by the Earth on the Moon in its orbit as

$$G\frac{m_M m_E}{r_M^2}$$

where m_M is the mass of the Moon and r_M is the radius of its orbit. It is this force which provides the Moon's centripetal acceleration $\omega^2 r_M$, and therefore

$$G\frac{m_M m_E}{r_M^2} = m_M\,\omega^2\,r_M$$

i.e. $$G\frac{m_E}{r_M^2} = \omega^2\,r_M$$

But ω, the angular velocity of the Moon, is equal to $2\pi/T$ where T is its period of revolution about the Earth, and therefore

$$G\frac{m_E}{r_M^2} = \frac{4\pi^2\,r_M}{T^2}$$

Substituting for G from equation [8.3] leads to

$$g = \frac{4\pi^2\,r_M^3}{T^2\,r_E^2} \quad [8.4]$$

The value of r_E which was available to Newton was poor by present-day standards. Even so, equation [8.4] gave a value for g that was sufficiently close to the accepted value for Newton to conclude that the Earth exerted the same type of force on the Moon as the Sun did on the planets.

QUESTIONS 8A

1. Find the gravitational force of attraction between two 10 kg particles which are 5.0 cm apart.
($G = 6.7 \times 10^{-11}\,\text{N}\,\text{m}^2\,\text{kg}^{-2}$.)

2. The average orbital radii about the Sun of the Earth and Mars are 1.5×10^{11} m and 2.3×10^{11} m respectively. How many (Earth) years does it take Mars to complete its orbit?

3. Calculate the mass of the Earth by considering the force it exerts on a particle of mass m at its surface. (Radius of Earth = 6.4×10^3 km, $g = 9.8\,\text{m}\,\text{s}^{-2}$, $G = 6.7 \times 10^{-11}\,\text{N}\,\text{m}^2\,\text{kg}^{-2}$.)

8.5 THE MASS OF THE EARTH

In 1798, a hundred and twenty one years after Newton had proposed the law of universal gravitation, Henry Cavendish made the first laboratory determination of the value of G.* Once this had been done, it was possible to obtain a value for the mass of the Earth on the basis of equation [8.3]. (Newton had worked the calculation in the opposite direction. He estimated G by using a value for m_E which was based on a guess at the density of the Earth.)

8.6 DEPENDENCE OF THE ACCELERATION DUE TO GRAVITY ON DISTANCE FROM THE CENTRE OF THE EARTH

The density of the Earth varies with depth but is largely independent of direction. It follows that the Earth can be treated as being made up of a large number of spherical shells of uniform density. This is useful, for it can be shown that:

(i) the acceleration due to gravity **outside** a spherical shell of uniform density is the same as it would be if the entire mass of the shell were concentrated at its centre, and

(ii) the acceleration due to gravity at all points **inside** a spherical shell of uniform density is zero.

These results will now be used to obtain expressions for the acceleration due to gravity both above and below the surface of the Earth.

Outside the Earth (i.e. $r > r_E$)

The acceleration due to gravity, g, at the surface of the Earth is given by rearranging equation [8.3] as

$$g = G\frac{m_E}{r_E^2} \qquad [8.5]$$

where m_E and r_E are respectively the mass and the radius of the Earth.

It follows from (i) that the acceleration due to gravity at a point outside the Earth has the value it would have if the entire mass of the Earth were at its centre. Therefore, by analogy with equation [8.5], the acceleration due to gravity g' at a distance r from the centre of the Earth when $r > r_E$ is given by

$$g' = G\frac{m_E}{r^2} \qquad [8.6]$$

*Cavendish had two lead spheres attached (one) to each end of a horizontal beam suspended by a silvered copper wire. When two larger spheres were brought up to the smaller ones, they deflected under the action of the gravitational force and twisted the suspension. The wire had been calibrated previously so that the strength of the force could be determined by measuring the angle through which the wire was twisted. An improved version of the experiment was performed by Boys in 1895.

Dividing equation [8.6] by equation [8.5] leads to

$$g' = \frac{r_E{}^2}{r^2} g \qquad [8.7]$$

Inside the Earth (i.e. $r < r_E$)

Consider a point P (Fig. 8.3) which is inside the Earth and at a distance r from its centre. From (ii) the acceleration due to gravity g' at P is due only to the sphere of radius r. If the mass of this sphere is m, then by analogy with equation [8.5]

$$g' = G\frac{m}{r^2} \qquad [8.8]$$

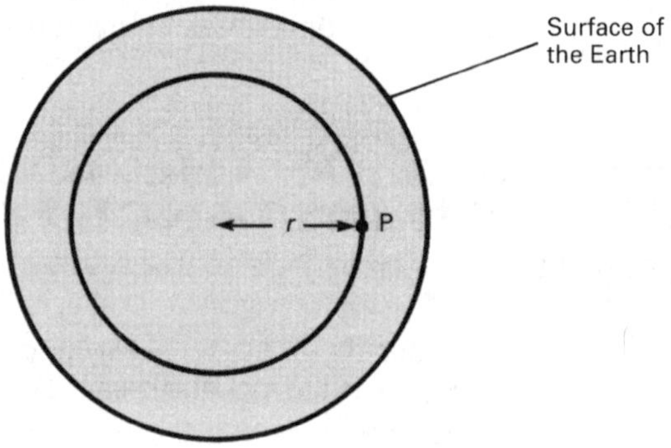

Fig. 8.3
To calculate the acceleration due to gravity inside the Earth

If the Earth is assumed to have uniform density*, ρ, then

$$m = \tfrac{4}{3}\pi r^3 \rho \quad \text{and} \quad m_E = \tfrac{4}{3}\pi r_E{}^3 \rho$$

and therefore

$$\frac{m}{m_E} = \frac{r^3}{r_E{}^3}$$

Substituting for m in equation [8.8] gives

$$g' = G\frac{m_E r}{r_E{}^3}$$

Replacing $Gm_E/r_E{}^2$ by g (equation [8.5]) gives

$$g' = \frac{r}{r_E} g \qquad [8.9]$$

The variation of g' as a function of r on the basis of equation [8.7] and equation [8.9] is shown in Fig. 8.4. Note that each of these equations reduces to $g' = g$ when $r = r_E$.

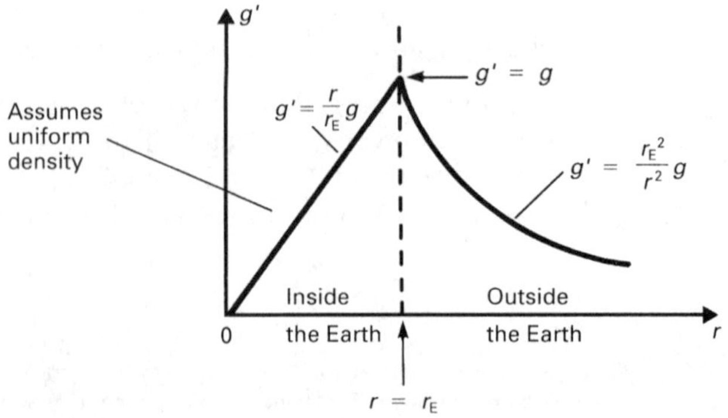

Fig. 8.4
Variation of ***g*** with distance from Earth's core

*As has been stated at the beginning of this section, the density of the Earth is not uniform. It is, however, normal practice, at this level, to assume that it is in carrying out this calculation.

8.7 ESCAPE VELOCITY

If a ball is thrown upwards from the surface of the Earth, its speed decreases from the moment it is projected due to the retarding effect of the Earth's gravitational field. The height which the ball ultimately attains depends on the speed with which it is projected – the greater the speed, the greater the height. If the ball were to be required to escape from the Earth, it would have to be projected with a velocity which is at least great enough for the ball to reach infinity before coming to rest. The **minimum** velocity that achieves this is known as the **escape velocity**.

Consider a body of mass m being projected upwards with velocity v from the Earth's surface. When the body is at a distance r from the centre of the Earth it will feel a gravitational force of attraction, F, due to the Earth and given by equation [8.1] as

$$F = G\frac{mm_E}{r^2}$$

where m_E is the mass of the Earth. In moving a small distance δr against this force, the work done δW at the expense of the kinetic energy of the body is given by equation [5.1] as

$$\delta W = G\frac{mm_E}{r^2}\,\delta r$$

Therefore the total work done, W, in moving from the Earth's surface (where $r = r_E$) to infinity (where $r = \infty$) is given by

$$W = \int_{r_E}^{\infty} G\frac{mm_E}{r^2}\,\mathrm{d}r$$

i.e. $$W = Gmm_E\left[\frac{-1}{r}\right]_{r_E}^{\infty}$$

i.e. $$W = G\frac{mm_E}{r_E}$$

If the body is to be able to do this amount of work (and so escape), it needs to have at least this amount of kinetic energy at the moment it is projected. The escape velocity v is therefore given by

$$\tfrac{1}{2}mv^2 = G\frac{mm_E}{r_E}$$

i.e. $$v = \sqrt{\frac{2Gm_E}{r_E}} \qquad [8.10]$$

Substituting known values into equation [8.10] leads to $v \approx 11\ \mathrm{km\,s^{-1}}$.

Notes

(i) This calaculation applies only to bodies which are not being driven, i.e. to projectiles. A body which is in powered flight does not have to rely on its initial kinetic energy to overcome the Earth's gravitational attraction, and therefore need never reach the escape velocity.

(ii) The escape velocity does not depend on the direction of projection. This is because the kinetic energy a body loses in reaching any particular height depends only on the height concerned and not on the path taken to reach it.

8.8 SATELLITE ORBITS

Consider a satellite of mass m orbiting the Earth with speed v along a circular path of radius r. The centripetal force, mv^2/r, is provided by the gravitational attraction of the Earth, and is given by Newton's law of universal gravitation as Gmm_E/r^2, where m_E is the mass of the Earth. Therefore

$$\frac{mv^2}{r} = G\frac{mm_E}{r^2}$$

i.e. $$v = \sqrt{\frac{Gm_E}{r}} \qquad [8.11]$$

The orbital period, T, is given by $T = 2\pi r/v$, and therefore by equation [8.11]

$$T = 2\pi\sqrt{\frac{r^3}{Gm_E}} \qquad [8.12]$$

Since both G and m_E are constants, it follows from equations [8.11] and [8.12] that **both the speed and the orbital period of an Earth satellite depend only on the radius of its orbit**. Two situations are of particular interest.

An Orbit Close to the Earth's Surface

To a good approximation, for a satellite which is less than about 200 km above the Earth we may put $r = r_E$, where r_E is the radius of the Earth. Substituting for r in equations [8.11] and [8.12] gives

$$v = \sqrt{\frac{Gm_E}{r_E}} \quad \text{and} \quad T = 2\pi\sqrt{\frac{r^3}{Gm_E}}$$

Taking $G = 6.7 \times 10^{-11}\,\text{N m}^2\,\text{kg}^{-2}$, $m_E = 6.0 \times 10^{24}\,\text{kg}$ and $r_E = 6.4 \times 10^6\,\text{m}$, we find

$$v = 7.9\,\text{km s}^{-1} \quad \text{and} \quad T = 85\,\text{minutes}$$

Geostationary (synchronous) Orbit

A satellite with an orbital period of 24 hours will always be at the same point above the Earth's surface (providing, of course, it is above the equator and is moving in the same direction as the Earth is rotating). Satellites of this type can be used to relay television signals and telephone messages (by radio link) from one point on the Earth's surface to another. Examples of these communications satellites are Syncom 2, Syncom 3 and Early Bird.

Substituting $G = 6.7 \times 10^{-11}\,\text{N m}^2\,\text{kg}^{-2}$, $m_E = 6.0 \times 10^{24}\,\text{kg}$ and $T = 24\,\text{hours}$ ($= 8.64 \times 10^4\,\text{s}$) in equation [8.12] gives

$$r = 42\,400\,\text{km}$$

Having calculated r we can use equation [8.11] (or $2\pi r = vT$) to calculate v. We find

$$v = 3.1\,\text{km s}^{-1}$$

Since $r_E = 6.4 \times 10^6\,\text{m}$, the height above the Earth's surface of the geostationary orbit is

$$42\,400 - 6400 = 36\,000\,\text{km}$$

Notes (i) Many artificial satellites actually move in elliptical orbits. For example, the orbit of Sputnik 1 took it from less than 250 km to over 900 km above the surface of the Earth. This is not so far from circular as it might at first seem. The radius of the Earth is about 6400 km, and therefore the semi-major axis of the ellipse was only about 10% bigger than the semi-minor axis.

(ii) The Moon, of course, is an Earth satellite and therefore equations [8.11] and [8.12] also apply to the Moon.

(iii) When a satellite is to be placed in orbit it is first carried to the desired height by rocket. It is then given the necessary tengential velocity (v) by firing rocket engines which are aligned parallel to the Earth's surface. If the satellite is still moving upwards when it reaches orbital height, it needs also to be given a downward directed thrust at this stage.

8.9 THE VARIATION OF *g* WITH LATITUDE

The acceleration due to gravity at the equator is $9.78\ \mathrm{m\,s^{-2}}$, whereas at the poles it is $9.83\ \mathrm{m\,s^{-2}}$. There are two main causes of this variation.

(i) The equatorial radius of the Earth is greater than the polar radius. Therefore a body at the equator is slightly further away from the centre of the Earth and consequently feels a smaller gravitational attraction. This accounts for $0.02\ \mathrm{m\,s^{-2}}$ of the observed difference of $0.05\ \mathrm{m\,s^{-2}}$.

(ii) Because the Earth rotates, its gravitational pull on a body at the equator has to provide the body with a centripetal acceleration of $0.03\ \mathrm{m\,s^{-2}}$. This does not apply at the poles.

8.10 WEIGHTLESSNESS

Consider an object of mass m hanging from a spring balance which is itself hanging from the roof of a lift (Fig. 8.5). The body is subjected to a downward directed force mg due to the Earth, and an upward directed force T, say, due to the tension in the spring. The net downward force is $(mg - T)$, and therefore by Newton's second law

$$mg - T = ma \qquad [8.13]$$

where a is the downward directed acceleration of the body.

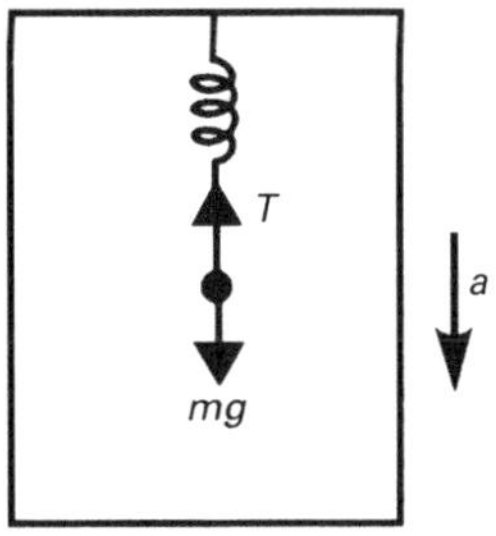

Fig. 8.5 Object suspended by a spring in a lift

If the lift is stationary or is moving with constant speed, $a = 0$ and therefore, by equation [8.13], $T = mg$, i.e. the balance registers the weight of the body as mg. However, if the lift is falling freely under gravity, both it and the body have a downward directed acceleration of g, i.e. $a = g$. It follows from equation [8.13] that $T = 0$, i.e. the balance registers the weight of the body as zero. It is usual to refer to a body in such a situation as being **weightless**. The term should be used with care; a gravitational pull of magnitude mg acts on the body whether it is in free fall or not, and therefore, in the strictest sense it has weight even when in free fall. The reason it is said to be weightless is that, whilst falling freely, it exerts no force on its support. Similarly, a man standing on the floor of a lift would exert no force on the floor if the lift were in free fall. In accordance with Newton's third law, the floor of the lift would exert no upward push on the man and therefore he would not have the sensation of weight.

An astronaut in an orbiting spacecraft has a centripetal acceleration equal to g', where g' is the acceleration due to gravity at the height of the orbit. The spacecraft has the same centripetal acceleration. The astronaut therefore has no acceleration relative to his spacecraft, i.e. he is weightless.

Note A body is weightless in the strictest sense only at a point where there is no gravitational field. An example of such a point is to be found between the Earth and the Moon where the two gravitational fields cancel.

8.11 GRAVITATIONAL POTENTIAL AND POTENTIAL ENERGY

The gravitational potential at a point in a gravitational field is defined as being numerically equal to the work done in bringing a unit mass from infinity (where the potential is zero) to that point.

Thus

$$U = \frac{W}{m} \qquad [8.14]$$

where

U = the gravitational potential at some point (J kg^{-1})

W = the work done in bringing a mass m from infinity to that point.

It has been shown in section 8.7, that the work which has to be done to take a mass m from the surface of the Earth to infinity is Gmm_E/r_E, where m_E and r_E are respectively the mass and the radius of the Earth. The work required to accomplish the reverse process, i.e. to bring the same body from infinity to the surface of the Earth, is therefore $-Gmm_E/r_E$. It follows from equation [8.14] that the gravitational potential U_E at the surface of the Earth is given by

$$U_E = -G\frac{m_E}{r_E} \qquad [8.15]$$

Notes (i) The minus sign in equation [8.15] indicates that the gravitational potential at the surface of the Earth is less than that at infinity. It follows that a body at infinity would 'fall' towards the Earth; a body on the Earth does not 'fall' to infinity.

(ii) It follows from equation [8.15] that in general

$$U = -G\frac{m}{r}$$

where U is **the gravitational potential** due to a body of mass m at a point outside the body and at a distance r from its centre. (This assumes that the body has spherical symmetry and/or is small compared with r.)

(iii) **The gravitational potential energy** of a body of mass m at a point where the gravitational potential is U is given by

$$\text{PE} = mU$$

This follows from the definition of gravitational potential and because the potential at infinity is zero.

8.12 GRAVITATIONAL FIELD STRENGTH

The gravitational field strength at a point in a gravitational field is defined as the force per unit mass acting on a mass placed at that point.

Thus

$$g = \frac{F}{m}$$

where

g = gravitational field strength ($\text{N kg}^{-1} = \text{m s}^{-2}$)

F = the force acting on a mass m

The same symbol (g) is used for gravitational field strength as for acceleration due to gravity. This is because they are one and the same thing, i.e. the field strength at a point in a gravitational field is equal to the gravitational acceleration of any mass placed at that point. We shall illustrate this for the particular case of the Earth. The gravitational force acting on a mass m at the Earth's surface is its weight, mg, where g is the acceleration due to gravity, and therefore the force per unit mass, i.e. the gravitational field strength, is mg/m – which is also g.

Note Gravitational field strength is a vector quantity. Its direction is that in which a mass would move under the influence of the field, i.e. towards a point of lower gravitational potential.

The gravitational field of the Earth is shown in Fig. 8.6. Note that it is a radial field directed towards the Earth and that it is stronger (field lines closer together) close to the Earth's surface than it is farther away. Over a limited area of the Earth's surface (an area that can be considered flat) the field can be considered to be uniform (field lines equally spaced) – see Fig. 8.7.

Field Strength Due to a Point Mass

The force on a point mass m' at a distance r from a point mass m is given by Newton's law of universal gravitation as F where

$$F = G\frac{mm'}{r^2}$$

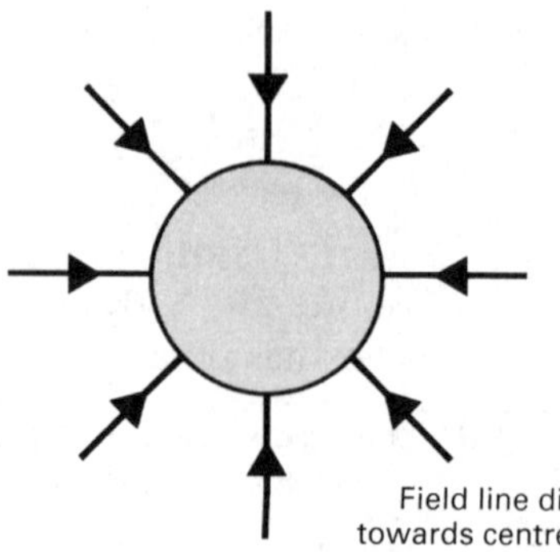

Fig. 8.6
The gravitational field of the Earth

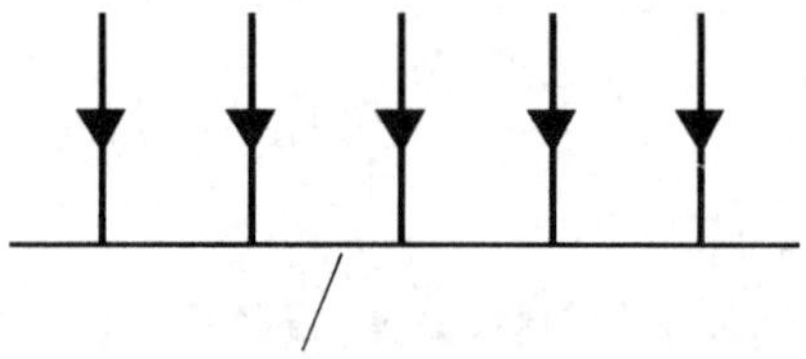

Fig. 8.7
The gravitational field of the Earth over a small area

The force per unit mass, i.e. the gravitational field strength, g, at a distance r from the point mass m is therefore given by

$$g = \frac{F}{m'}$$

i.e.

$$g = G\frac{m}{r^2} \quad [8.16]$$

Notes (i) Equation [8.16] also applies outside an extended body of mass m at a distance r from its centre providing the body has spherical symmetry and/or is small compared with r.

(ii) When $m =$ the mass of the Earth, m_E, and $r =$ the radius of the Earth, r_E, equation [8.16] becomes

$$g = G\frac{m_E}{r_E^2} \quad [8.17]$$

where g is the field strength (acceleration due to gravity) at the surface of the Earth. We arrived at this same result in section 8.6 (equation [8.5]) by thinking of g as the acceleration due to gravity. Equation [8.17] is particularly useful for it is often necessary when solving examination questions to express g in terms of G and vice versa.

8.13 THE ANALOGY BETWEEN GRAVITY AND ELECTRICITY

Electric field strength is force per unit charge – gravitational field strength is force per unit mass. The definition of electrical potential involves the work done in bringing a unit (positive) charge from infinity – that of gravitational potential involves the work done in bringing a unit mass from infinity. The reader may find it useful to compare results (i) and (ii) of section 8.6 with the way in which the electric field of a hollow sphere varies with distance from the centre of the sphere (Fig. 39.8).

Table 8.1. Gravitational and electrical quantities compared

Gravitational quantity	Electrical quantity
$U = \frac{W}{m}$	$V = \frac{W}{Q}$
$g = \frac{F}{m}$	$E = \frac{F}{Q}$
$g = -\frac{dU}{dx}$	$E = -\frac{dV}{dx}$

Table 8.2. Comparison of effects of point masses and point charges

Gravitational quantity	Electrical quantity
$U = -G\frac{m}{r}$	$V = \frac{1}{4\pi\varepsilon_0}\frac{Q}{r}$
$g = G\frac{m}{r^2}$	$E = \frac{1}{4\pi\varepsilon_0}\frac{Q}{r^2}$
$F = G\frac{m_1 m_2}{r^2}$	$F = \frac{1}{4\pi\varepsilon_0}\frac{Q_1 Q_2}{r^2}$

In Table 8.1 expressions for gravitational potential and field strength are given together with the analogous electrical expressions. Table 8.2 compares the gravitational field strength and the gravitational potential at a distance r from a point mass m with the analogous electrical quantities at a distance r from a point charge Q. Note also the similarity between the expression for the gravitational force between two point masses with that for the electrical force between two point charges in vacuum.

Notes

(i) There is no gravitational analogue of electrical permittivity, i.e. the gravitational force between two masses does not depend on the medium in which they are situated.

(ii) The gravitational force, unlike the electrical force, is always attractive.

8.14 TO SHOW THAT $g = -dU/dx$

Suppose that a particle of mass m is moved by a force F from A to B in a gravitational field of strength g (Fig. 8.8). Suppose also that AB $= \delta x$, where δx is so small that F can be considered constant between A and B.

The work done δW in going from A to B is given by

$$\delta W = F\delta x$$

By definition $g = -F/m$ (the minus sign is necessary because g and F are oppositely directed), therefore

$$\delta W = -mg\delta x \qquad [8.18]$$

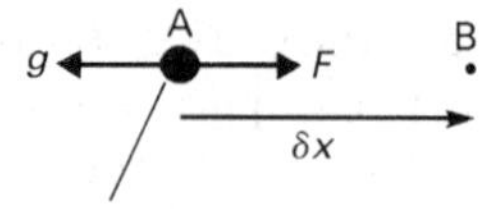

Fig. 8.8
To establish the relationship between ***g*** and ***U***

By definition, the increase in gravitational potential, δU, in moving the mass m from A to B is given by

$$\delta U = \delta W/m$$

Therefore, by equation [8.18]

$$\delta U = -g\delta x$$

Therefore, in the limit

$$g = -\mathrm{d}U/\mathrm{d}x$$

EXAMPLE 8.1

Use the following notation: m_{E} = mass of Earth, r_{E} = radius of Earth, g = acceleration due to gravity at surface of Earth, G = universal gravitational constant.

(a) Write down an expression for the gravitational potential at the Earth's surface.

(b) By how much would the gravitational PE of a body of mass m increase if it were moved from the Earth's surface to infinity.

(c) Hence find an expression for the minimum velocity with which a body could be projected from the Earth's surface and never return: (i) which involves G, (ii) which involves g

Solution

(a)
$$\text{Gravitational potential} = -G\frac{m_{\mathrm{E}}}{r_{\mathrm{E}}}$$

(b) The gravitational potential at infinity is zero and therefore 'in moving to infinity' the potential increases by $Gm_{\mathrm{E}}/r_{\mathrm{E}}$. Therefore

$$\text{Increase in PE of mass } m = G\frac{mm_{\mathrm{E}}}{r_{\mathrm{E}}}$$

(c) (i) If a projectile of mass m is to have just sufficient KE to reach infinity from the Earth's surface, then (since decrease in KE = increase in PE) its velocity v on leaving the surface must be given by

$$\tfrac{1}{2}mv^2 = G\frac{mm_{\mathrm{E}}}{r_{\mathrm{E}}} \qquad \text{i.e.} \quad v = \sqrt{\frac{2Gm_{\mathrm{E}}}{r_{\mathrm{E}}}}$$

Since the projectile has just sufficient KE to reach infinity, this is the minimum velocity that will prohibit its return, i.e.

$$\text{Minimum velocity} = \sqrt{\frac{2Gm_{\mathrm{E}}}{r_{\mathrm{E}}}} \qquad [8.19]$$

(ii) The acceleration due to gravity, g, at the surface of the Earth is given by

$$g = G\frac{m_E}{r_E^2} \qquad [8.17]$$

$$\therefore \quad G\frac{m_E}{r_E} = gr_E$$

Substituting in equation [8.19] gives

$$\text{Minimum velocity} = \sqrt{2gr_E}$$

EXAMPLE 8.2

Using the notation of Example 8.1, find expressions for: (a) the gravitational PE of a satellite of mass m orbiting the Earth at a distance r from its centre, (b) the KE of the satellite, (c) the total energy of the satellite. (d) Explain how each of these quantities would change if the orbit of the satellite were so low that it encountered a considerable amount of air resistance.

Solution

(a) At a distance r from the centre of the Earth

$$\text{Gravitational potential} = -G\frac{m_E}{r}$$

$$\therefore \quad \text{Gravitational PE of satellite} = -G\frac{m_E m}{r} \qquad [8.20]$$

(b) By the law of universal gravitation and Newton's second law

$$G\frac{m_E m}{r^2} = m\frac{v^2}{r}$$

$$\therefore \quad \text{KE of satellite } (= \tfrac{1}{2}mv^2) = G\frac{m_E m}{2r} \qquad [8.21]$$

(c)

$$\text{Total energy} = \text{PE} + \text{KE}$$

$$= -G\frac{m_E m}{r} + G\frac{m_E m}{2r}$$

$$\text{i.e.} \quad \text{Total energy} = -G\frac{m_E m}{2r} \qquad [8.22]$$

(d) If the satellite encountered air resistance, it would do work against friction and therefore its total energy would decrease, i.e. become more negative. It follows from equation [8.22] that r would decrease and therefore by equation [8.20] the PE would decrease (i.e. become more negative); it follows from equation [8.2] that the KE would increase.

CONSOLIDATION

Newton's law of universal gravitation Every particle in the universe attracts every other with a force which is proportional to the product of their masses and inversely proportional to the square of their separation.

$$F = G\frac{m_1 m_2}{r^2}$$

When dealing with bodies rather than particles r represents the separation of centres.

Gravitational potential (U) at a point is defined as being numerically equal to the work done in bringing a unit mass from infinity (where the potential is zero) to that point, i.e.

$$U = \frac{W}{m}$$

Also

$$U = -G\frac{m}{r} \qquad \left(\begin{array}{l}\text{Outside a body of mass } m \text{ and} \\ \text{a distance } r \text{ from its centre,} \\ \text{or at } r \text{ from a point mass } m\end{array}\right)$$

$$\text{Gravitational PE} = mU$$

Gravitational field strength at a point is defined as force per unit mass and is equal to the acceleration due to gravity at the point.

$$\text{Field strength } (g) = -\frac{dU}{dx} \qquad \text{(In general)}$$

$$\text{Field strength } (g) = G\frac{m}{r^2} \qquad \left(\begin{array}{l}\text{Outside a body of mass } m \text{ and} \\ \text{at a distance } r \text{ from its centre,} \\ \text{or at } r \text{ from a point mass } m\end{array}\right)$$

Gravitational fields are directed towards points of lower gravitational potential.

Satellites

$$T^2 \propto r^3$$

$$\left.\begin{array}{l}\text{PE} = -G\dfrac{m_E m}{r} \\ \text{KE} = G\dfrac{m_E m}{2r}\end{array}\right\} \quad \begin{array}{l}\text{For a body of mass } m \\ \text{at a distance } r \text{ from} \\ \text{the centre of the Earth}\end{array}$$

Escape velocity (from Earth) $= \sqrt{\dfrac{2Gm_E}{r_E}} = \sqrt{2gr_E}$

QUESTIONS ON SECTION A

Assume $g = 10\,\mathrm{m\,s^{-2}} = 10\,\mathrm{N\,kg^{-1}}$ unless otherwise stated.

MECHANICS (Chapters 1–5)

A1 Find, both graphically and by calculation, the horizontal and vertical components of a force of 50 N which is acting at 40° to the horizontal.

A2 The horizontal and vertical components of a force are respectively 20 N and 30 N. Calculate the magnitude and direction of the force.

A3 Calculate the magnitude and direction of the resultant of the forces shown in the figure below.

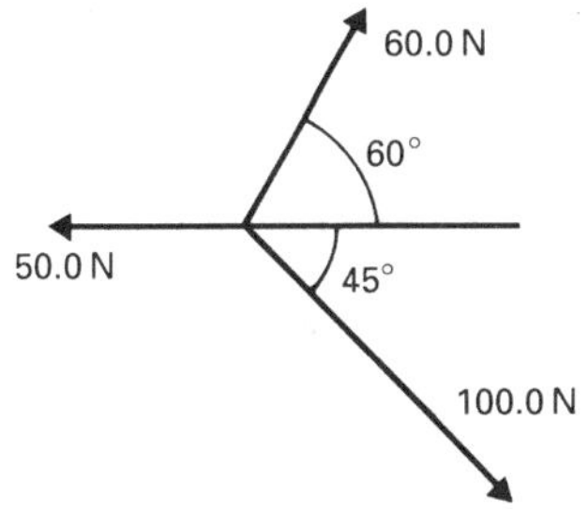

A4 The diagrams below show a sphere, S, resting on a table and a free-body diagram on which the forces acting on the sphere have been marked.

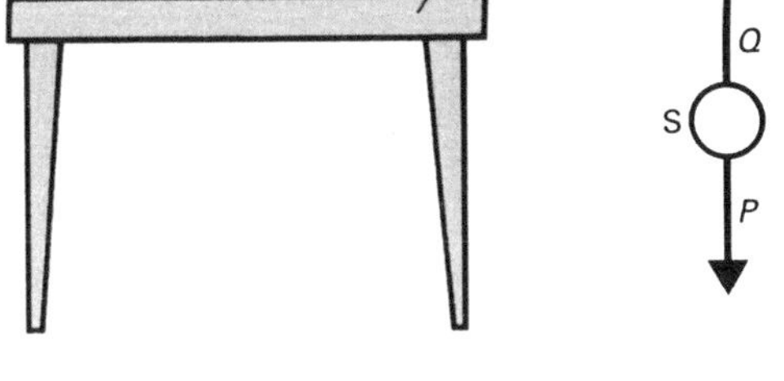

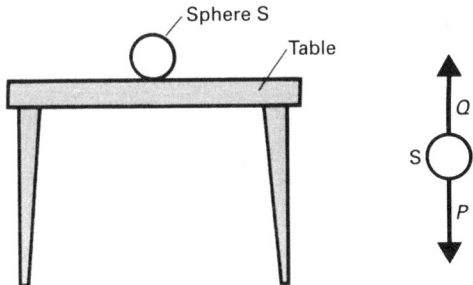

We know from Newton's third law of motion that forces occur in equal and opposite pairs. On which body does the force which pairs with force P act? Give its direction.

On which body does the force which pairs with force Q act? Give its direction.

A force F_1 acts on an object O and this same force F_1 forms a Newton's third law pair with a second force, F_2. State *two* ways in which F_1 and F_2 are similar and *two* ways in which they differ. [L, '91]

A5 Raindrops of mass 5×10^{-7} kg fall vertically in still air with a uniform speed of $3\,\mathrm{m\,s^{-1}}$. If such drops are falling when a wind is blowing with a speed of $2\,\mathrm{m\,s^{-1}}$, what is the angle which the paths of the drops make with the vertical? What is the kinetic energy of a drop? [S]

A6 A ship initially at rest accelerates steadily on a perfectly smooth sea. How would you attempt to estimate the value of the acceleration from within the ship? You cannot see out, but you have available all the apparatus normally found in a school laboratory. (You may assume that the acceleration is not less than $1\,\mathrm{m\,s^{-2}}$.) [W]

A7 A light string carrying a small bob of mass 5.0×10^{-2} kg hangs from the roof of a *moving* vehicle.

(a) What can be said about the motion of the vehicle if the string hangs vertically?

(b) The vehicle moves in a horizontal straight line from left to right, with a constant acceleration of $2.0\,\mathrm{m\,s^{-2}}$.

(i) Show in a sketch the forces acting on the bob.

(ii) By resolving horizontally and vertically or by scale drawing, determine the angle which the string makes with the vertical.

(c) The vehicle moves down an incline making an angle of 30° with the horizontal with a constant acceleration of $3.0\,\mathrm{m\,s^{-2}}$. Determine the angle which the string makes with the vertical. [J, '92]

A8 A hose with a nozzle 80 mm in diameter ejects a horizontal stream of water at a rate of $0.044\,m^3\,s^{-1}$. With what velocity will the water leave the nozzle? What will be the force exerted on a vertical wall situated close to the nozzle and at right-angles to the stream of water, if, after hitting the wall,
(a) the water falls vertically to the ground,
(b) the water rebounds horizontally?
(Density of water = $1000\,kg\,m^{-3}$.) [AEB, '79]

A9 What is the connection between force and momentum? A helicopter of total mass 1000 kg is able to remain in a stationary position by imparting a uniform downward velocity to a cylinder of air below it of effective diameter 6 m. Assuming the density of air to be $1.2\,kg\,m^{-3}$, calculate the downward velocity given to the air. [J]

A10 An astronaut is outside her space capsule in a region where the effect of gravity can be neglected. She uses a gas gun to move herself relative to the capsule. The gas gun fires gas from a muzzle of area $160\,mm^2$ at a speed of $150\,m\,s^{-1}$. The density of the gas is $0.800\,kg\,m^{-3}$ and the mass of the astronaut, including her space suit, is 130 kg. Calculate:
(a) the mass of gas leaving the gun per second,
(b) the acceleration of the astronaut due to the gun, assuming that the change in mass is negligible. [J, '92]

A11 Sand is poured at a steady rate of $5.0\,g\,s^{-1}$ on to the pan of a direct reading balance calibrated in grams. If the sand falls from a height of 0.20 m on to the pan and it does not bounce off the pan then, neglecting any motion of the pan, calculate the reading on the balance 10 s after the sand first hits the pan. [W, '92]

A12 A pebble is dropped from rest at the top of a cliff 125 m high. How long does it take to reach the foot of the cliff, and with what speed does it strike the ground? With what speed must a second pebble be thrown vertically downwards from the cliff top if it is to reach the bottom in 4 s? (Ignore air resistance.) [S]

A13 A stone thrown horizontally from the top of a vertical cliff with velocity $15\,m\,s^{-1}$ is observed to strike the (horizontal) ground at a distance of 45 m from the base of the cliff. What is **(a)** the height of the cliff, **(b)** the angle the path of the stone makes with the ground at the moment of impact? [S]

A14 A ball is thrown vertically upwards and caught by the thrower on its return. Sketch a graph of *velocity* (taking the upward direction as positive) against *time* for the whole of its motion, neglecting air resistance. How, from such a graph, would you obtain an estimate of the height reached by the ball? [L]

A15 A bus travelling steadily at $30\,m\,s^{-1}$ along a straight road passes a stationary car which, 5 s later, begins to move with a uniform acceleration of $2\,m\,s^{-2}$ in the same direction as the bus.
(a) How long does it take the car to acquire the same speed as the bus?
(b) How far has the car travelled when it is level with the bus? [W, '92]

A16 A cricketer throws a ball of mass 0.20 kg directly upwards with a velocity of $20\,m\,s^{-1}$, and catches it again 4.0 s later.
Draw labelled sketch graphs to show
(a) the velocity,
(b) the kinetic energy,
(c) the height
of the ball against time over the stated 4.0 s period. Your graphs must show numerical values of the given quantities. [S]

A17 A 'hammer' thrown in athletics consists of a metal sphere, mass 7.26 kg, with a wire handle attached, the mass of which can be neglected. In a certain attempt it is thrown with an initial velocity which makes an angle of 45° with the horizontal and its flight takes 4.00 s.

Find the horizontal distance travelled and the kinetic energy of the sphere just before it strikes the ground, stating any assumptions and approximations you make in order to do so. [S]

A18 The graph shown represents the variation in vertical height with time for a ball thrown upwards and returning to the thrower.

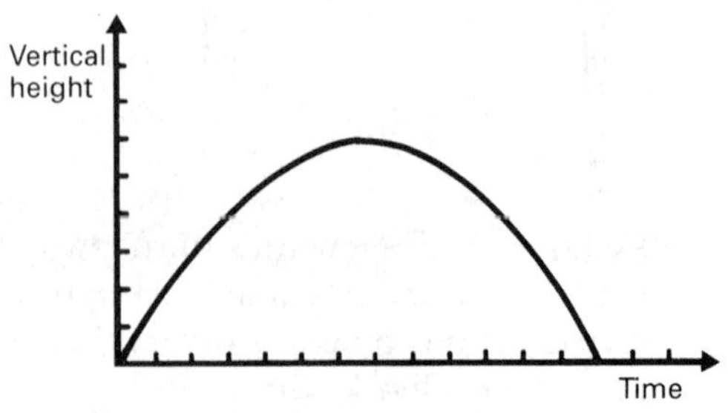

From this graph how could a *velocity* against *time* graph be constructed? Sketch the likely form of such a graph. [L]

A19 The diagram shows the speed–time graph for a swimmer performing one complete cycle of the breast stroke.

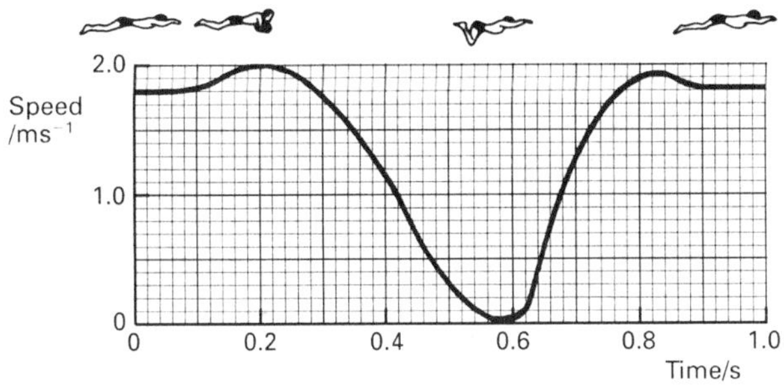

(a) Determine the maximum acceleration of the swimmer. Explain how you arrive at your answer.
(b) Without making any further calculations *sketch* a labelled acceleration–time graph for the same time interval as that shown in the diagram.
(c) Use the graph to estimate the distance travelled in one complete cycle of the stroke. Show your working clearly. [AEB, '89]

A20 The graph shows how the speed of a car varies with time as it accelerates from rest.

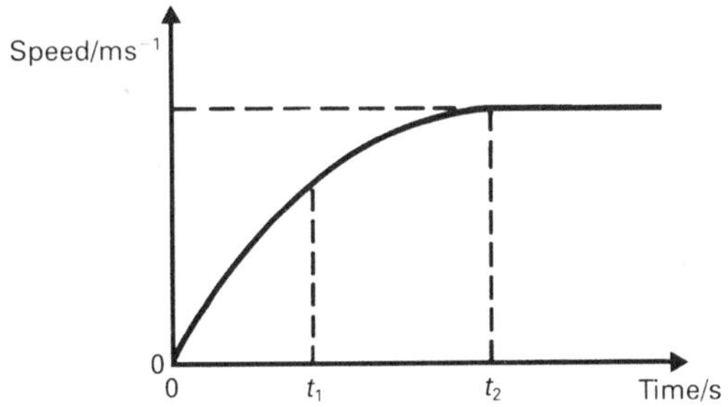

(a) State:
(i) the time at which the acceleration is a maximum,
(ii) how you could use the graph to find the distance travelled between times t_1 and t_2.
(b) The driving force produced by the car engine can be assumed to be constant. Explain, in terms of the forces on the car, why
(i) the acceleration is not constant,
(ii) the car eventually reaches a constant speed.

(c) The diagram shows a simple version of an instrument used to measure acceleration, in which a mass is supported between two springs in a box so that when one spring is extended, the other is compressed. At rest, the mass is in the position shown.

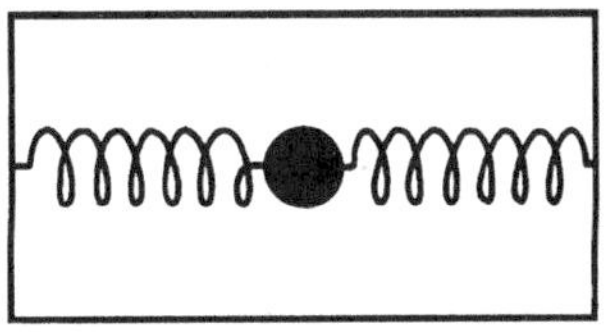

Redraw the diagram showing the position of the mass when the box accelerates to the right. Explain why the mass takes up this position. [J, '91]

A21 **(a)** A ball is thrown vertically upwards from the surface of the Earth with an initial velocity u. Neglecting frictional forces, sketch a graph to show the variation of the velocity v of the ball with time t as the ball rises and then falls back to Earth. What information contained in the graph enables you to determine: **(i)** the gravitational acceleration, **(ii)** the maximum height to which the ball rises?
(b) If the frictional forces in the air were not negligible, how, in the above situation, would: **(i)** *the initial deceleration* of the ball, **(ii)** *the maximum height* reached by the ball, be affected? [AEB, '79]

A22 **(a)** Define *acceleration*. Explain how it is possible for a body to be undergoing an acceleration although its speed remains constant.
(b) A ball is placed at the top of a slope as shown in Fig. 1.

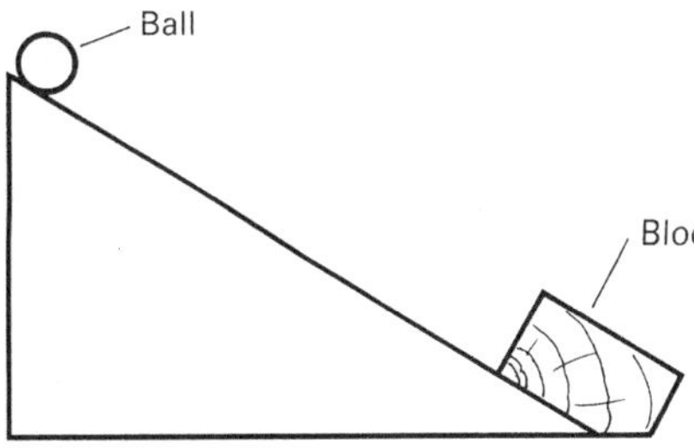

Fig. 1

A block is fixed rigidly to the lower end of the slope. The ball of mass 0.70 kg is released at time $t = 0$ from the top of the incline and v, the velocity of the ball down the slope, is found to vary with t as shown in Fig. 2.

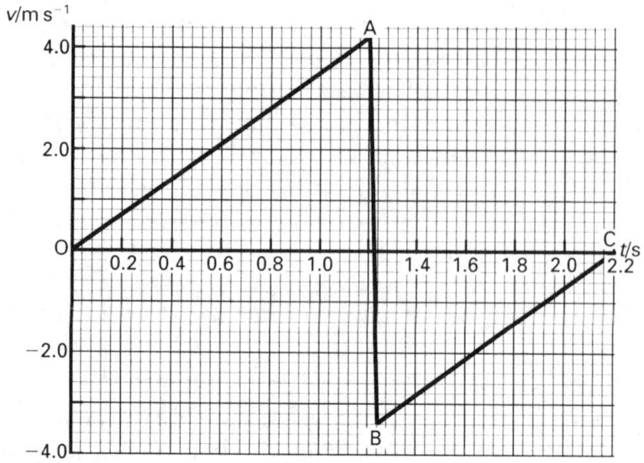

Fig. 2

(i) Describe qualitatively the motion of the ball during the periods OA, AB and BC.

(ii) Calculate: **(1)** the acceleration of the ball down the incline, **(2)** the length of the incline, **(3)** the mean force experienced by the ball during impact with the block.

(iii) Discuss whether the collision between the block and the ball is elastic. [C, '92]

A23 A rocket is caused to ascend vertically from the ground with a constant acceleration, a. At a time, t_s, after leaving the ground the rocket motor is shut off.

(a) Neglecting air resistance and assuming the acceleration due to gravity, g, is constant, sketch a graph showing how the velocity of the rocket varies with time from the moment it leaves the ground to the moment it returns to ground. In your sketch represent the ascending velocity as positive and the descending velocity as negative. Indicate on your graph **(i)** t_s, **(ii)** the time to reach maximum height, t_h, **(iii)** the time of flight, t_f.

(b) Account for the form of each portion of the graph and explain the significance of the area between the graph and the time axis from zero time to **(i)** t_s, **(ii)** t_h, **(iii)** t_f.

(c) Either by using the graph or otherwise, derive expressions in terms of a, g and t_s for **(i)** t_h, **(ii)** the maximum height reached, **(iii)** t_f. [J]

A24 A sphere of mass 3 kg moving with velocity $4\,\text{m}\,\text{s}^{-1}$ collides head-on with a stationary sphere of mass 2 kg, and imparts to it a velocity of $4.5\,\text{m}\,\text{s}^{-1}$. Calculate the velocity of the 3 kg sphere after the collision, and the amount of energy lost by the moving bodies in the collision. [S]

A25 The diagram shows a body of mass 2 kg resting in a frictionless horizontal gully in which it is constrained to move. It is acted upon by the force shown for 5 s after which time it strikes and sticks to the body B of mass 3 kg, the force being removed at this instant. What will the speed of the combined masses be? [L]

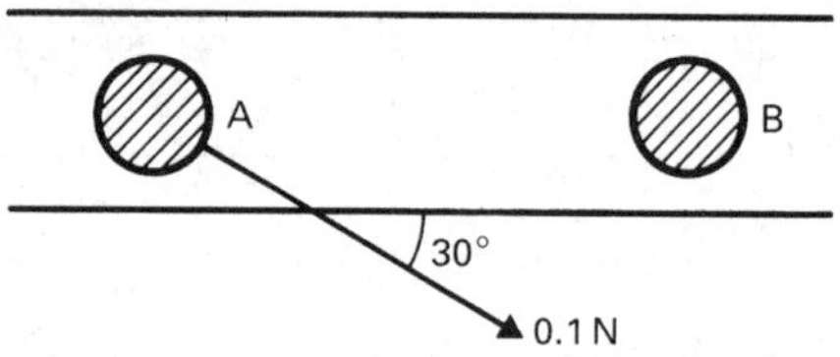

A26 A railway truck of mass 4×10^4 kg moving at a velocity of $3\,\text{m}\,\text{s}^{-1}$ collides with another truck of mass 2×10^4 kg which is at rest. The couplings join and the trucks move off together. What fraction of the first truck's initial kinetic energy remains as kinetic energy of the two trucks after the collision? Is energy conserved in a collision such as this? Explain your answer briefly. [L]

A27 **(a)** **(i)** Define *linear momentum*.

(ii) State *the principle of conservation of linear momentum* making clear the condition under which it can be applied.

(b) A spacecraft of mass 20 000 kg is travelling at $1500\,\text{m}\,\text{s}^{-1}$. Its rockets eject hot gases at a speed of $1200\,\text{m}\,\text{s}^{-1}$ relative to the spacecraft. During one burn, the rockets are fired for a 5.0 s period. In this time the speed of the spacecraft increases by $3.0\,\text{m}\,\text{s}^{-1}$.

(i) What is the acceleration of the spacecraft?

(ii) Assuming that the mass of fuel ejected is negligible compared with the mass of the spacecraft determine the distance travelled during the burn. Give your answer to four significant figures.

(iii) What is the thrust produced by the rocket?

(iv) Determine the mass of gas ejected by the rocket during the burn.

[AEB, '92]

A28 A bullet of mass 0.020 kg is fired horizontally at 150 m s^{-1} at a wooden block of mass 2.0 kg resting on a smooth horizontal plane. The bullet passes through the block and emerges undeviated with a velocity of 90 m s^{-1}. Calculate:
(a) the velocity acquired by the block.
(b) the total kinetic energy before and after penetration and account for their difference. [W, '91]

A29 Distinguish between an *elastic collision* and an *inelastic collision*.
A particle A of mass m moving with an initial velocity u makes a 'head-on' collision with another particle B of mass $2m$, B being initially at rest. In terms of u, calculate the final velocity of A if the collision is **(i)** elastic, **(ii)** inelastic, (assume that the two particles adhere on collision). [AEB, '79]

A30 A puck collides perfectly inelastically with a second puck originally at rest and of three times the mass of the first puck. What proportion of the original kinetic energy is lost, and where does it go? [W]

A31 State the law of conservation of linear momentum.

A proton of mass 1.6×10^{-27} kg travelling with a velocity of 3×10^{7} m s^{-1} collides with a nucleus of an oxygen atom of mass 2.56×10^{-26} kg (which may be assumed to be at rest initially) and rebounds in a direction at 90° to its incident path. Calculate the velocity and direction of motion of the recoil oxygen nucleus, assuming the collision is elastic and neglecting the relativistic increase of mass. [O & C*]

A32 **(a)** In an experiment to investigate the nature of different types of collision, a trolley of mass 1.6 kg was given a push towards a second trolley of mass 0.8 kg travelling more slowly but in the same direction. The speeds of both trolleys before and after collision were measured. The results for two different types of collisions were as follows:

Type A collision	*Speed before*	*Speed after*
1.6 kg trolley	0.70 m s^{-1}	0.30 m s^{-1}
0.8 kg trolley	0.10 m s^{-1}	0.89 m s^{-1}
Type B collision		
1.6 kg trolley	0.60 m s^{-1}	0.37 m s^{-1}
0.8 kg trolley	0.10 m s^{-1}	0.57 m s^{-1}

Describe a technique for measuring the speed of a trolley before and after a collision, showing how the speed is calculated.

Show that the results given above are consistent with the principle of conservation of linear momentum. Why should the speeds be measured *immediately* before and after the collisions?

(b) Distinguish clearly between *elastic* and *inelastic* collisions. Determine the nature of each of the collisions in (a) above, supporting your choice with appropriate calculations.

Describe how: **(i)** an elastic collision, **(ii)** an inelastic collision, could be simulated experimentally using the two trolleys and any necessary additional apparatus. [L]

A33 **(a)** A linear air-track is a length of metal track along which objects (gliders) can move with negligible friction, supported on a cushion of air. A glider of mass 0.40 kg is stationary near one end of a level air-track and an air-rifle is mounted close to the glider with its barrel aligned along the track. A pellet of mass 5.0×10^{-4} kg is fired from the rifle and sticks to the glider which acquires a speed of 0.20 m s^{-1}. Calculate the speed with which the pellet struck the glider.
(b) Describe an experimental arrangement you would use to verify the above result.
(c) A student using an air-track fails to level it correctly. The speed of a glider along the track near its centre is 0.20 m s^{-1} and when it has moved a further distance of 0.90 m it is 0.22 m s^{-1}. Determine the angle made by the track to the horizontal.
(d) A glider reaches the end of a level air-track and rebounds from a rubber band stretched across the track. Assuming that, whilst in contact with the band, the force exerted on the glider is proportional to the displacement of the band from the point of impact, sketch a graph showing how the velocity of the glider varies with time. Explain the shape of the graph. [J]

A34 **(a)** The law of conservation of momentum suggests momentum is conserved in any collision. A tennis ball dropped onto a hard floor rebounds to about 60% of its initial height. State how momentum is conserved in this event.

(b) **(i)** A top class tennis player can serve the ball, of mass 57 g, at an initial horizontal speed of $50\,\mathrm{m\,s^{-1}}$. The ball remains in contact with the racket for 0.050 s. Calculate the average force exerted on the ball during the serve.

(ii) *Sketch* a graph showing how the horizontal acceleration of the ball might possibly vary with time during the serve, giving the axes suitable scales.

(iii) Explain how this graph would be used to show that the speed of the ball on leaving the racket is $50\,\mathrm{m\,s^{-1}}$.

[AEB, '90]

A35 **(a)** **(i)** What is meant by the term *linear momentum*?

(ii) State the law of conservation of linear momentum.

(iii) Explain how force is related to linear momentum.

(b) Consider the following:

(i) a vehicle in space changes its direction by firing a rocket motor,

(ii) a dart is thrown at a board and sticks to the board,

(iii) a ball is dropped to the floor and rebounds.

In each case discuss how the law of conservation of linear momentum may be applied.

(c) A student devises the following experiment to determine the velocity of a pellet from an air rifle.

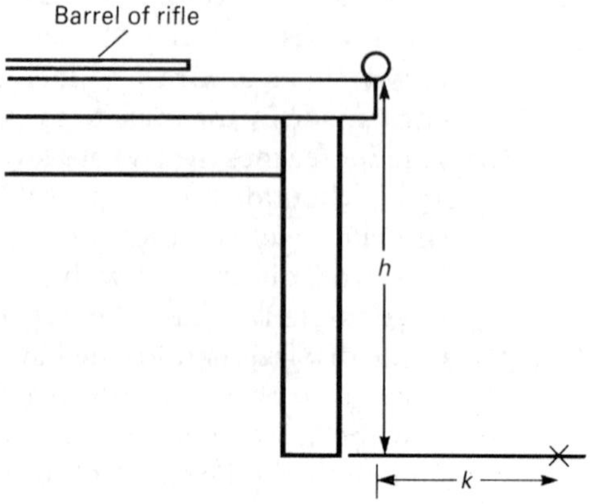

A piece of plasticine of mass M is balanced on the edge of a table such that it just fails to fall off. A pellet of mass m is fired horizontally into the plasticine and remains embedded in it. As a result the plasticine reaches the floor a horizontal distance k away. The height of the table is h.

(i) Show that the horizontal velocity of the plasticine with pellet embedded is $k\left(\frac{g}{2h}\right)^{1/2}$

(ii) Obtain an expression for the velocity of the pellet before impact with the plasticine. [S]

A36 A particle A is suspended as shown in the figure by two strings, which pass over smooth pulleys, and are attached to particles B and C. At A, the strings are at right angles to each other, and make equal angles with the horizontal. If the mass of B is 1 kg, what are the masses of A and C? [S]

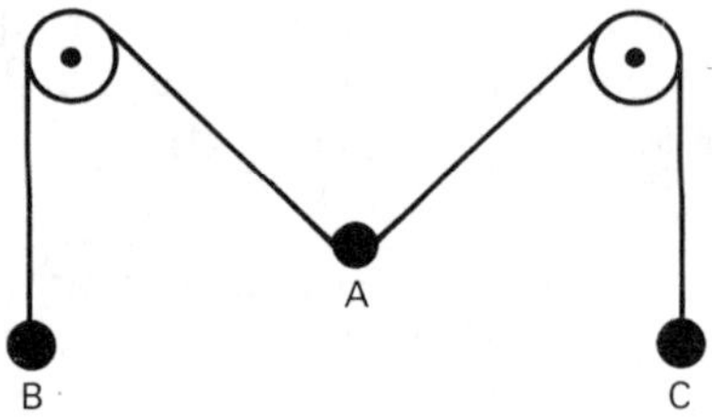

A37 State the conditions that a rigid body may be in equilibrium under the action of three forces.

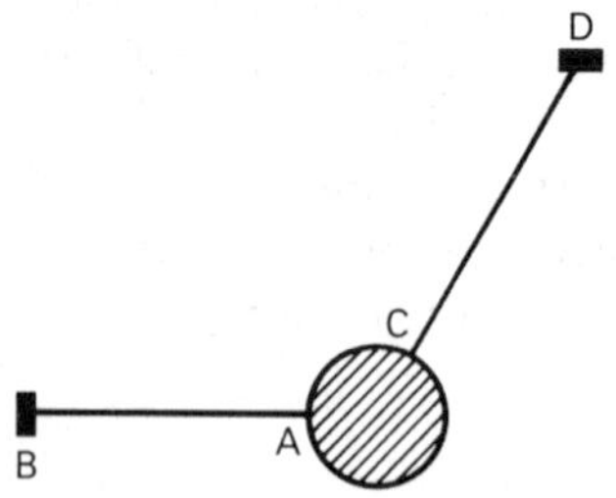

A uniform sphere of mass 2 kg is kept in position as shown by two strings AB and CD. AB is horizontal, and CD is inclined at 30° to the vertical. Calculate the tension in each string. [S]

A38 A uniform beam AB of length 5 m and which weighs 200 N is supported horizontally by two vertical ropes X and Y at A and B respectively. Calculate the tensions in the ropes if a man weighing 700 N stands on the beam at a distance of 2 m from A.

A39 A uniform rod AB of weight W and which is 20 cm long is suspended by two vertical springs X and Y attached to the rod at A and B respectively. The upper ends of the springs are attached to a horizontal beam. When the springs are unextended they have the same length. The tension T_X in X is given by $T_X = kx$ and the tension T_Y in Y is given by $T_Y = 3ky$, where k is a constant and x and y are the extensions of X and Y respectively. At what distance from A must a body of weight $5W$ be attached to the rod if the rod is to be horizontal?

A40 Weights of 1 N, 2 N, 3 N and 4 N are attached to a light wire which has been bent to form a circle of radius $5\sqrt{2}$ cm. The weights are equally spaced around the circle and the 1 N weight is diametrically opposite the 3 N weight. Calculate the distance of the centre of mass from the centre of the circle.

A41 A uniform ladder which is 10 m long and weighs 300 N leans with its upper end against a smooth vertical wall and its lower end on rough horizontal ground. The bottom of the ladder is 6 m from the base of the wall. A man weighing 700 N stands on the ladder at a point 6 m above the ground. Calculate the magnitudes and directions of the forces exerted on the ladder by **(a)** the wall, **(b)** the ground.

A42 **(a)** State the conditions for the equilibrium of a body which is acted upon by a number of forces.

(b) A student holds a uniform metre rule at one end in two different ways, as shown in Figs. 1 and 2.

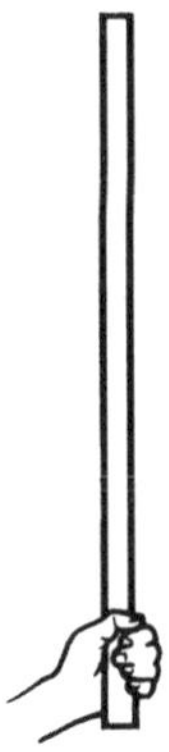

Fig. 1

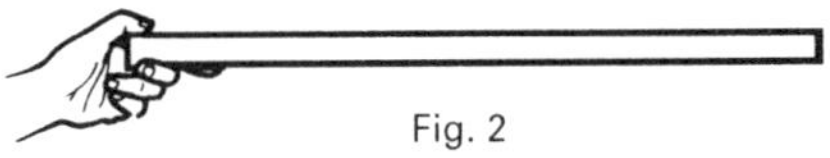

Fig. 2

On a copy of Fig. 1 draw and label an arrow to represent the weight W of the metre rule and an arrow to represent the force F provided by the student's hand. What is the relationship between magnitudes of F and W?

In Fig. 2, the rule is held horizontally between the thumb and first finger. On a copy of Fig. 2, draw and label all the forces acting on the metre rule. List these forces in order of increasing magnitude.

[C, '91]

A43 Fig. 1 shows the bone and muscle structure of a person's arm supporting a 5.0 kg mass in

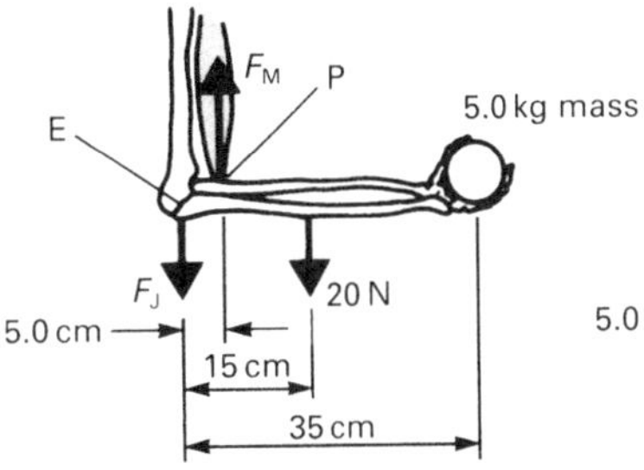

Fig. 1

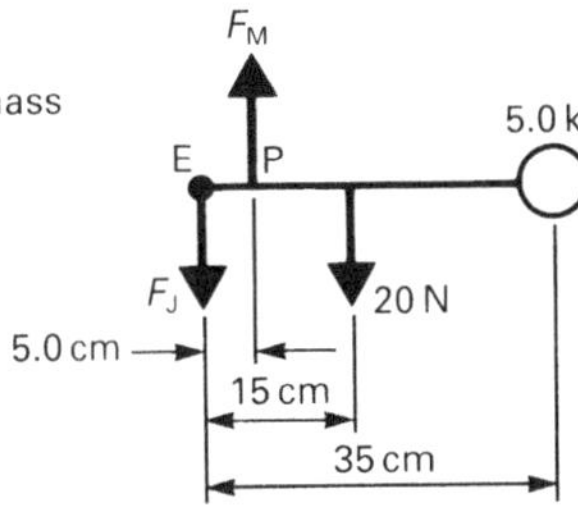

Fig. 2

equilibrium. The forearm is horizontal and is at right angles to the upper arm. Fig. 2 shows the equivalent mechanical system.

F_M is the force exerted by the biceps muscle and F_J is the force at the elbow joint. The perpendicular distance between the lines of action of F_M and F_J is 5.0 cm.

(a) **(i)** Explain why the 20 N force has been included.

(ii) State the conditions which must be met by the forces when the arm is in equilibrium.

(b) **(i)** Calculate the magnitude of the force F_M.

(ii) Show that F_M has the same magnitude when the forearm is at 45° to the horizontal with the upper arm still vertical.

(c) In many athletes the distance between the elbow joint, E, and the muscle attachment, P, is greater than 5.0 cm. Explain how this is an advantage in lifting and throwing events. [AEB, '89]

A44 The diagram shows the driving wheel of a car with a torque of 125 N m being applied to the axle of the wheel. The car is travelling at a constant velocity of 30 m s^{-1}.

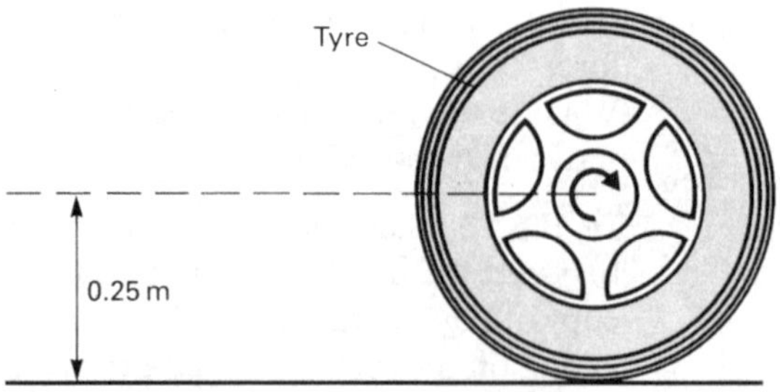

Copy the diagram and show on it

(a) the direction in which the car is travelling.

(b) Draw an arrow on the diagram representing F, the horizontal component of the force which the road exerts on the tyre.

(c) Since the car is not accelerating, what can be deduced about the resultant torque on the wheel?

(d) Find F.

(e) What power is being supplied to the wheel? [C, '92]

A45 **(a)** State the conditions for equilibrium for a body which is acted upon by a system of coplanar forces.

(b) The diagram shows a safety valve which releases steam when the pressure rises above a pre-determined value. The mass of the rod and the valve cover are negligible.

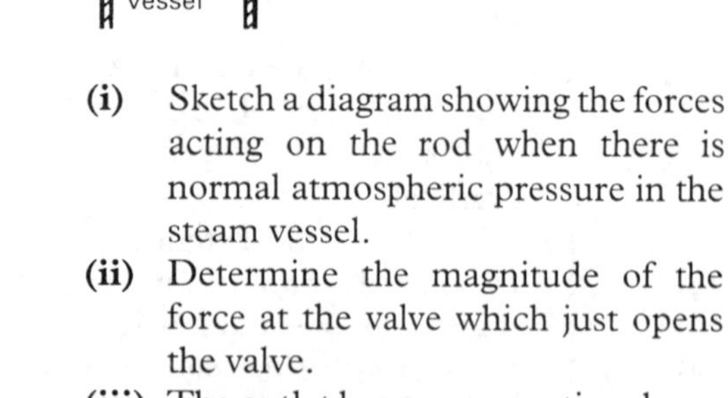

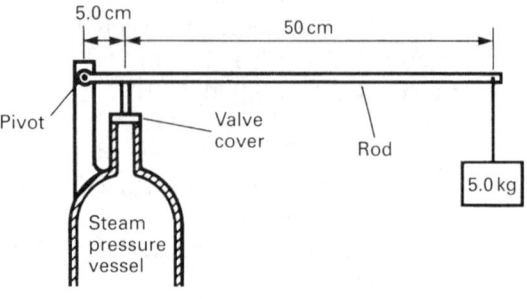

(i) Sketch a diagram showing the forces acting on the rod when there is normal atmospheric pressure in the steam vessel.

(ii) Determine the magnitude of the force at the valve which just opens the valve.

(iii) The outlet has a cross-sectional area of $2.0\,\text{cm}^2$. Using the force calculated in (b)(ii) determine the excess pressure above atmospheric pressure in the vessel when the valve just opens. Give your answer in pascals (Pa). [AEB, '90]

A46 Derive an expression for the kinetic energy of a particle of mass m which has momentum p.
A stationary radioactive nucleus disintegrates into an α particle of relative atomic mass 4, and a residual nucleus of relative atomic mass 144. If the kinetic energy of the α particle is $3.24 \times 10^{-13}\,\text{J}$, what is the kinetic energy of the residual nucleus? [S]

A47 As shown in the diagram, two trolleys P and Q of masses 0.50 kg and 0.30 kg respectively are held together on a horizontal track against a spring which is in a state of compression. When the spring is released the trolleys separate freely and P moves to the left with an initial velocity of $6\,\text{m}\,\text{s}^{-1}$. Calculate:

(a) the initial velocity of Q,

(b) the initial total kinetic energy of the system.

Calculate also the initial velocity of Q if trolley P is held still when the spring under the same compression as before is released. [J]

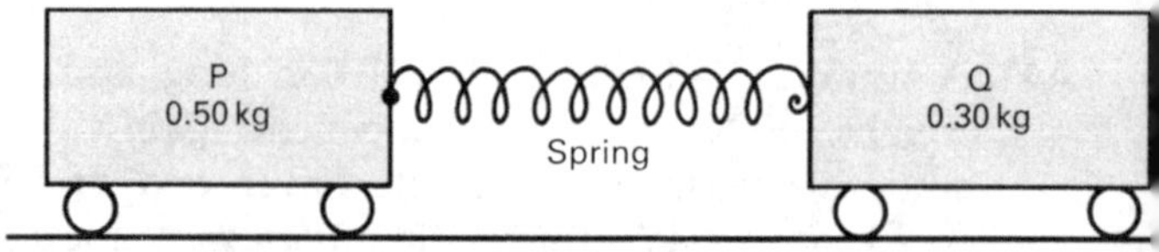

A48 A bullet of mass $2.0 \times 10^{-3}\,\text{kg}$ is fired horizontally into a free-standing block of wood of mass $4.98 \times 10^{-1}\,\text{kg}$, which it knocks forward with an initial speed of $1.2\,\text{m}\,\text{s}^{-1}$.

(a) Estimate the speed of the bullet.

(b) How much kinetic energy is lost in the impact?

(c) What becomes of the lost kinetic energy? [S]

A49 A block of wood of mass 1.00 kg is suspended freely by a thread. A bullet, of mass 10 g, is fired horizontally at the block and becomes embedded in it. The block swings to one side, rising a vertical distance of 50 cm. With what speed did the bullet hit the block? [L]

A50 In an experiment to investigate collisions on a level airtrack., eight optical sensors are positioned 0.10 m apart close to the track, as shown in the diagram. When a marker on a

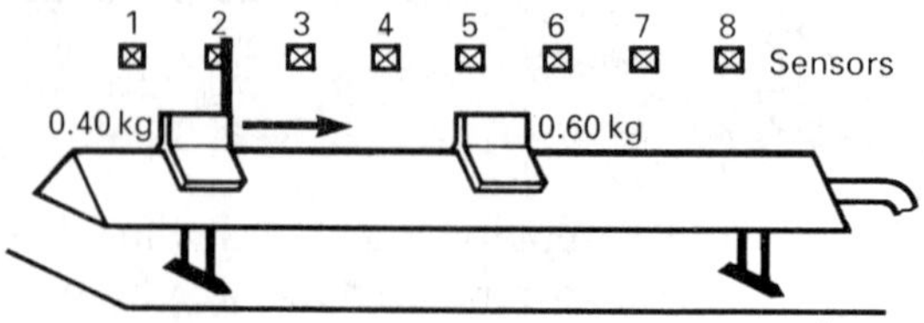

moving glider passes sensor number 1, an electronic timer is started. As the glider passes each sensor, the time taken for the glider to travel from sensor number 1 is recorded. A glider of mass 0.40 kg is given a push along the track. As it passes sensor number 5, it collides with, and sticks to, a stationary glider of mass 0.60 kg. The recorded times at each sensor are shown below.

Sensor number	1	2	3	4	5	6	7	8
Time s	0	0.66	1.32	1.98	2.64	4.31	5.98	7.65

(a) Calculate the speed just before and the speed just after the collision.
(b) Show that momentum is conserved in the collision.
(c) Calculate the kinetic energy before the collision and the kinetic energy after the collision.
Account for the difference. [J, '91]

A51 A motor car collides with a crash barrier when travelling at 100 km h^{-1} and is brought to rest in 0.1 s. If the mass of the car and its occupants is 900 kg calculate the average force on the car by a consideration of momentum.

Because of the seat belt, the movement of the driver, whose mass is 80 kg, is restricted to 0.20 m, relative to the car. By a consideration of energy calculate the average force exerted by the belt on the driver. [J]

A52 This question is about the design of experiments to measure the speed of an air-gun pellet.

The speed of the pellet is known to be about 40 m s^{-1} and the mass of the pellet is about 0.5 g.

(a) One student suggests that momentum ideas might be used. It is proposed that the pellet be fired into a trolley of mass M and that the speed v of the trolley after impact be determined by finding the time it takes a card to cross the path of a light beam. The light beam illuminates a photodiode which controls a timer. The timer can record the time the light beam is cut off to the nearest 0.01 s. The system is shown in Fig. 1.

(i) Explain how the speed of the pellet could be obtained from the proposed measurements.

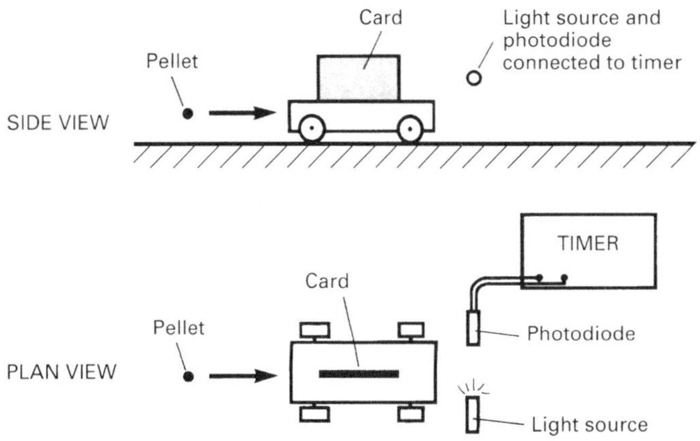

Fig. 1

(ii) It is decided that the final speed should be about 0.1 m s^{-1}. Deduce a suitable mass for the trolley.
(iii) Assuming that the final speed is to be determined to a precision of about 2% give a suitable length for the card. Explain how you arrived at your answer.
(iv) It is suggested that the answer is not accurate because the trolley will not be 'friction-free'. How will this affect the final result?
(v) State and explain whether you would expect the result to be more accurate or less accurate if the card were made longer. [AEB, '91]

A53 Explain what is meant by *kinetic energy*, and show that for a particle of mass m moving with velocity v, the kinetic energy is $\frac{1}{2}mv^2$.
A steel ball is:
(a) projected horizontally with velocity v, at a height h above the ground,
(b) dropped from a height h and bounces on a fixed horizontal steel plate.
Neglecting air resistance, and using suitable sketch graphs, explain how the kinetic energy of the ball varies in (a) with its height above the ground, and in (b) with its height above the plate. [J]

A54 A model railway truck P, of mass 0.20 kg and a second truck, Q, of mass 0.10 kg are at rest on two horizontal straight rails, along which they can move with negligible friction. P is acted on by a horizontal force of 0.10 N which makes an angle of 30° with the track. After P has travelled 0.50 m, the force is removed and P then collides with and sticks to Q. Calculate:

(a) the work done by the force,
(b) the speed of P before the collision,
(c) the speed of the combined trucks after the collision. [J, '90]

A55 A lorry of mass 3.5×10^4 kg attains a steady speed v while climbing an incline of 1 in 10

with its engine operating at 175 kW. Find v. (Neglect friction.) [W, '90]

A56 A particle A of mass 2 kg and a particle B of mass 1 kg are connected by a light elastic string C, and initially held at rest 0.9 m apart on a smooth horizontal table with the string in tension. They are then simultaneously released. The string releases 12 J of energy as it contracts to its natural length. Calculate the

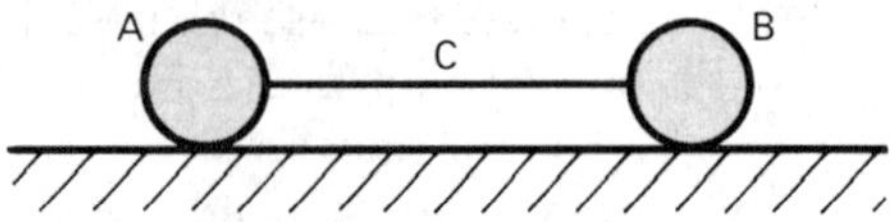

velocity acquired by each of the particles. Where do the particles collide? [S]

A57 State Newton's laws of motion, and show how the principle of conservation of linear momentum may be derived from them.
A particle of mass 3 kg and a particle Q of mass 1 kg are connected by a light elastic string and initially held at rest on a smooth horizontal table with the string in tension. They are then simultanously released. The string gives up 24 J of energy as it contracts to its natural length. Calculate the velocity acquired by each of the particles, assuming no energy is lost.
A helicopter of mass 810 kg supports itself in a stationary position by imparting a downward velocity v to all the air in a circle of area 30 m^2. Given that the density of air is 1.20 kg m^{-3}, calculate the value of v. What is the power needed to support the helicopter in this way, assuming no energy is lost? [S]

A58 The acceleration–time graph above is drawn for a body which starts from rest and moves in a straight line. The body is of mass 10 kg. Use the graph to find:

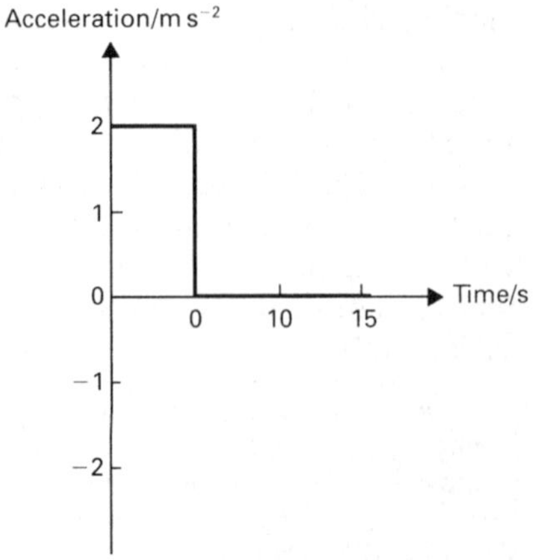

(a) the distance travelled in 15 s,
(b) the average force acting over the whole 15 s period. [L]

A59 A stone of mass 80 g is released at the top of a vertical cliff. After falling for 3 s, it reaches the foot of the cliff, and penetrates 9 cm into the ground. What is:
(a) the height of the cliff,
(b) the average force resisting penetration of the ground by the stone? [S]

A60 **(a)** A car of mass 1000 kg is initially at rest. It moves along a straight road for 20 s and then comes to rest again. The speed–time graph for the movement is:

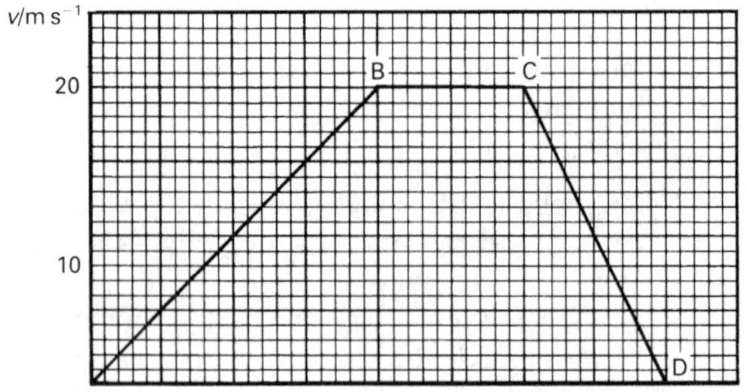

(i) What is the total distance travelled?
(ii) What resultant force acts on the car during the part of the motion represented by CD?
(iii) What is the momentum of the car when it has reached its maximum speed? Use this momentum value to find the constant resultant accelerating force.

(iv) During the part of the motion represented by OB on the graph, the constant resultant force found in (iii) is acting on the moving car although it is moving through air. Sketch a graph to show how the driving force would have to vary with time to produce this constant acceleration. Explain the shape of your graph.

(b) If, when travelling at this maximum speed, the 1000 kg car had struck and remained attached to a stationary vehicle of mass 1500 kg, with what speed would the interlocked vehicles have travelled immediately after collision?

Calculate the kinetic energy of the car just prior to this collision and the kinetic energy of the interlocked vehicles just afterwards. Comment upon the values obtained.

Explain how certain design features in a modern car help to protect the driver of a car in such a collision. [L]

A61 A large cardboard box of mass 0.75 kg is pushed across a horizontal floor by a force of 4.5 N. The motion of the box is opposed by **(i)** a frictional force of 1.5 N between the box and the floor, and **(ii)** an air resistance force kv^2, where $k = 6.0 \times 10^{-2}\ \mathrm{kg\,m^{-1}}$ and v is the speed of the box in $\mathrm{m\,s^{-1}}$.

Sketch a diagram showing the directions of the forces which act on the moving box. Calculate maximum values for

(a) the acceleration of the box,
(b) its speed. [L]

A62 A stone is projected vertically upwards and eventually returns to the point of projection. Ignoring any effects due to air resistance draw sketch graphs to show the variation with time of the following properties of the stone: **(i)** velocity, **(ii)** kinetic energy, **(iii)** potential energy, **(iv)** momentum, **(v)** distance from point of projection, **(vi)** speed. [AEB, '82]

A63 **(a)** State Newton's laws of motion. Explain how the *newton* is defined from these laws.

(b) A rocket is propelled by the emission of hot gases. It may be stated that both the rocket and the emitted hot gases each gain kinetic energy and momentum during the firing of the rocket.

Discuss the significance of this statement in relation to the laws of conservation of energy and momentum, explaining the essential difference between these two quantities.

(c) A bird of mass 0.50 kg hovers by beating its wings of effective area $0.30\ \mathrm{m^2}$.

(i) What is the upward force of the air on the bird?
(ii) What is the downward force of the bird on the air as it beats its wings?
(iii) Estimate the velocity imparted to the air, which has a density of $1.3\ \mathrm{kg\,m^{-3}}$, by the beating of the wings.

Which of Newton's laws is applied in each of (i), (ii) and (iii) above? [L]

A64 A horizontal force of 2000 N is applied to a vehicle of mass 400 kg which is initially at rest on a horizontal surface. If the total force opposing motion is constant at 800 N, calculate:

(a) the acceleration of the vehicle,
(b) the kinetic energy of the vehicle 5 s after the force is first applied,
(c) the total power developed 5 s after the force is first applied. [AEB, '85]

A65 On a linear air-track the gliders float on a cushion of air and move with negligible friction. One such glider of mass 0.50 kg is at rest on a level track. A student fires an air rifle pellet of mass 1.5×10^{-3} kg at the glider along the line of the track. The pellet embeds itself in the glider which recoils with a velocity of $0.33\ \mathrm{m\,s^{-1}}$.

(a) State the principle you will use to calculate the velocity at which the pellet struck the glider.

Calculate the velocity at which the pellet struck.

(b) Another student repeats the experiment with the air-track inclined at an angle of 2° to the horizontal. Initially, the glider is at rest at the bottom of the track. After the impact the glider recoils with the same initial velocity but slows down and stops momentarily further along the track.

Explain *in words* how to calculate how far along the air track the glider moves before stopping instantaneously.

Calculate how far the glider moves along the air track before stopping momentarily. [O & C, '92]

A66 Two ski slopes are of identical length and vertical height. Slope A, Fig. 1 is concave whilst slope B, Fig. 2 is partly convex. Two skiers, of equal weight, start from rest at the top of each slope. Assume that the effects of friction on the skis and of air resistance on the moving skiers' motion are negligible.

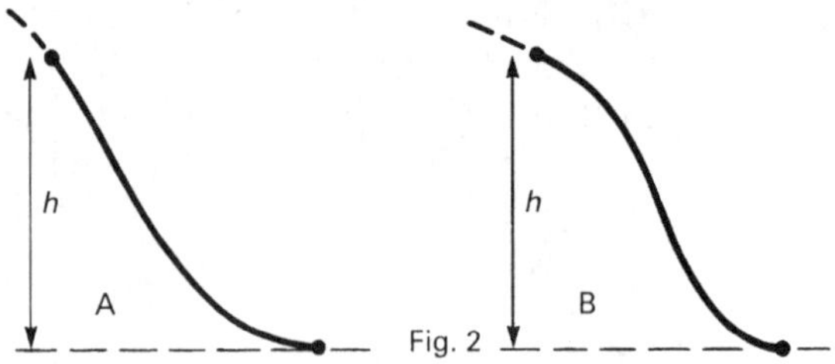

Fig. 1 Fig. 2

(a) At the bottom of the ski-run, will the skier on slope A be moving faster, at the same speed, or slower than the skier on slope B? Justify your answer.

(b) Will the skier on slope A take longer, the same time, or less time than the skier on slope B to complete the ski run? Justify your answer.

(c) A heavier skier joins the first skier on slope A. He also starts from rest at the top of the slope. Will he take longer, the same time, or less time than the first skier to complete the ski run? Justify your answer.

(d) Which of the three skiers will have most energy at the bottom of the ski run? Explain. [O & C, '90]

A67 **(a)** What do you understand by the *principle of conservation of energy*?

(b) Explain how the principle applies to: **(i)** an object falling from rest in vacuo, **(ii)** a man sliding from rest down a vertical pole, if there is a constant resistive force opposing the motion. Sketch graphs, using one set of axes for **(i)** and another set for **(ii)**, showing how each form of energy you consider varies with time, and point out the important features of the graphs.

(c) A motor car of mass 600 kg moves with constant speed up an inclined straight road which rises 1.0 m for every 40 m travelled along the road. When the brakes are applied with the power cut off, there is a constant resistive force and the car comes to rest from a speed of $72\,\mathrm{km\,h^{-1}}$ in a distance of 60 m. By using the principle of conservation of energy, calculate the resistive force and the deceleration of the car. [J]

A68 A vehicle has a mass of 600 kg. Its engine exerts a tractive force of 1500 N, but motion is resisted by a constant frictional force of 300 N. Calculate:

(a) the acceleration of the vehicle,

(b) its momentum 10 s after starting to move,

(c) its kinetic energy 15 s after starting to move. [S]

A69 A typical escalator in the London Underground rises at an angle of 30° to the horizontal. It lifts people through a vertical height of 15 m in 1.0 minute. Assuming all the passengers stand still whilst on the escalator, 75 people can step on at the bottom and off at the top in each minute. Take the average mass of a passenger to be 75 kg.

(a) Find the power needed to lift the passengers when the escalator is fully laden. For this calculation assume that any kinetic energy given to the passengers by the escalator is negligible.

(b) The frictional force in the escalator system is 1.4×10^4 N when the escalator is fully laden. Calculate the power needed to overcome the friction. Hence find the power input for the motor driving the fully laden escalator, given that the motor is only 70% efficient.

(c) When the passengers walk up the moving escalator, is more or less power required by the motor to maintain the escalator at the same speed? Explain your answer. [O & C, '90]

A70 A muscle exerciser consists of two steel ropes attached to the ends of a strong spring contained in a telescopic tube. When the ropes are pulled sideways in opposite directions, as shown in the simplified diagram (on p. 125), the spring is compressed.

The spring has an uncompressed length of 0.80 m. The force F (in N) required to compress the spring to a length x (in m) is calculated from the equation

$$F = 500\,(0.80 - x)$$

The ropes are pulled with equal and opposite forces, P, so that the spring is compressed to a length of 0.60 m and the ropes make an angle of 30° with the length of the spring.

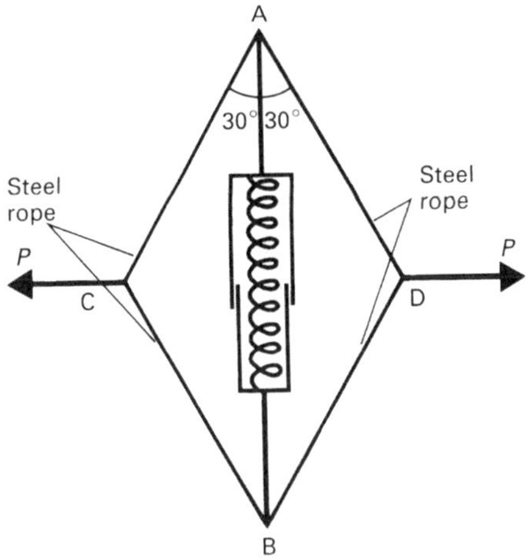

(a) Calculate
  (i) the force, F,
  (ii) the work done in compressing the spring.
(b) By considering the forces at A or B, calculate the tension in each rope.
(c) By considering the forces at C or D, calculate the force, P [J, '91]

A71 The blades of a large wind turbine, designed to generate electricity, sweep out an area of $1400\,m^2$ and rotate about a horizontal axis which points directly into a wind of speed $15\,m\,s^{-1}$, as illustrated in the diagram.

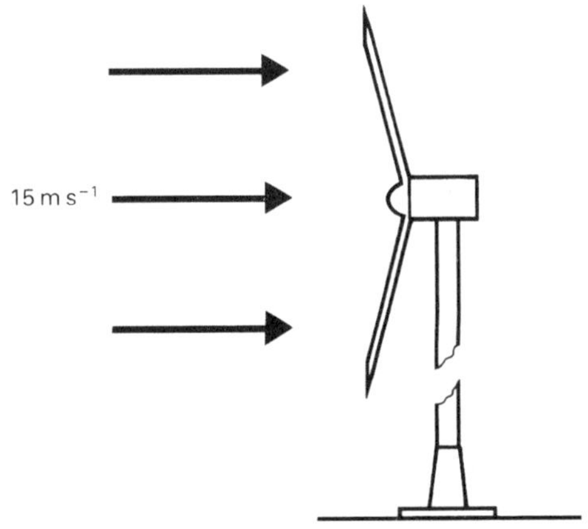

(a) Calculate the mass of air passing per second through the area swept out by the blades.
(Take the density of air to be $1.2\,kg\,m^{-3}$.)
(b) The mean speed of the air on the far side of the blades is reduced to $13\,m\,s^{-1}$. How much kinetic energy is lost by the air per second?
(c) How many turbines, operating with 70% efficiency, would be needed to equal the power output of a single conventional 1000 MW power station?
(d) Suggest *two* advantages, and *two* disadvantages, of wind turbines as a source of energy. [O, '92]

A72 The thrust F exerted on a rocket by the jet of expelled gases depends on the cross-sectional area, A, of the jet, the density, ρ, of the mixture of gases and the velocity, v, at which they are ejected. The following relationships have been suggested between these quantities, in each of which k is a dimensionless constant:
(a) $F = kA\rho v$
(b) $F = kA\rho v^2$
(c) $F = kA^2\rho v^2$.
Use the method of dimensions to show for each whether it is possible. [S]

CIRCULAR MOTION AND ROTATION (Chapter 6)

A73 What force is necessary to keep a mass of 0.8 kg revolving in a horizontal circle of radius 0.7 m with a period of 0.5 s? What is the direction of this force?
(Assume that $\pi^2 = 10$.) [L]

A74 Use Newton's laws of motion to explain why a body moving with uniform speed in a circle must experience a force towards the centre of the circle.

An aircraft of mass 1.0×10^4 kg is travelling at a constant speed of $0.2\,km\,s^{-1}$ in a horizontal circle of radius 1.5 km.
(a) What is the angular velocity of the aircraft?
(b) Show on a sketch the forces acting on the aircraft in the vertical plane containing the aircraft and the centre of the circle. Find the magnitude and direction of their resultant.
(c) Explain why a force is exerted on a passenger by the aircraft. In what direction does this force act? [C]

A75 A spaceman in training is rotated in a seat at the end of a horizontal rotating arm of length 5 m. If he can withstand accelerations up to $9g$, what is the maximum number of revolutions per second permissible? [L]

A76 A simple pendulum, suspended from a fixed point, consists of a light cord of length 500 mm and a bob of weight 2.0 N. The bob is made to move in a horizontal circular path. If the maximum tension which the cord can withstand is 5.0 N show whether or not it is possible for the radius of the path of the bob to be 300 mm. [L]

A77 Explain why there must be a force acting on a particle which is moving with uniform speed in a circular path. Write down an expression for its magnitude.

A conical pendulum consists of a small massive bob hung from a light string of length 1 m and rotating in a horizontal circle of radius 30 cm. With the help of a diagram indicate what forces are acting on the bob. How do they account for the motion of the bob? Deduce the speed of rotation in revolutions per minute. [J]

A78 A particle of mass m travels at constant speed, v, in a vacuum along a path consisting of two straight lines connected by a semicircle, AB, of diameter d as shown in the diagram.

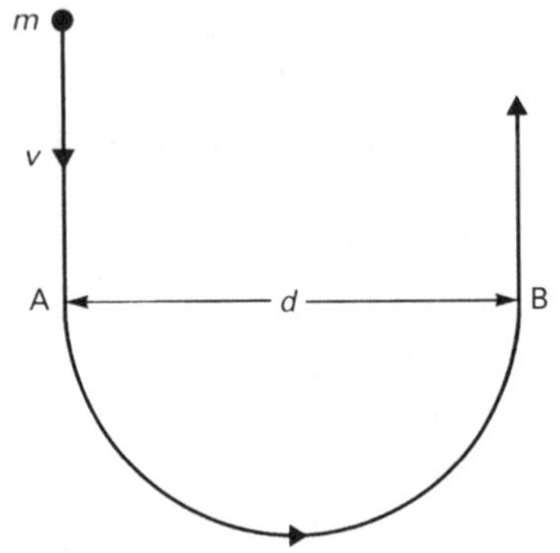

For the section of the path from A to B, find:

(a) the time taken,
(b) the change in the momentum of the particle,
(c) the force acting on the particle at any point along the semicircular path AB,
(d) the work done on the particle by this force. [J]

A79 The diagram shows a section of a curtain track in a vertical plane. The curved section, CDE, forms a circular arc of radius of curvature

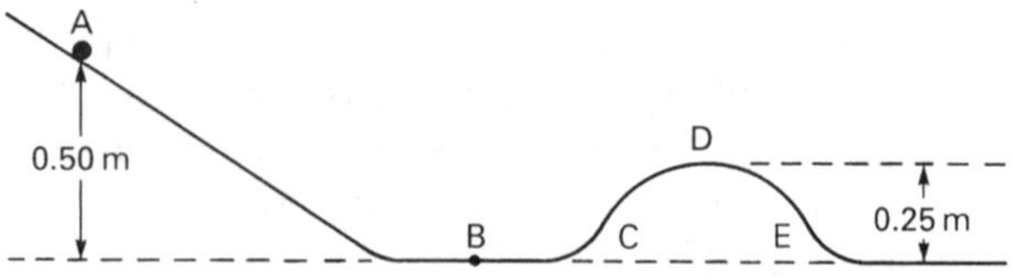

0.75 m and the point D is 0.25 m higher than B. A ball-bearing of mass 0.060 kg is released from A, which is 0.50 m higher than B. Assume that rotational and frictional effects can be ignored and that the ball-bearing remains in contact with the track throughout the motion.

(a) Calculate the speed of the ball-bearing **(i)** at B, **(ii)** at D.
(b) Draw a diagram showing the forces acting on the ball-bearing when it is at D and calculate the reaction between the track and the ball-bearing at this point. [J]

A80 A compressed spring is used to propel a ball-bearing along a track which contains a circular loop of radius 0.10 m in a vertical plane. The spring obeys Hooke's law and requires a force of 0.20 N to compress it 1.0 mm.

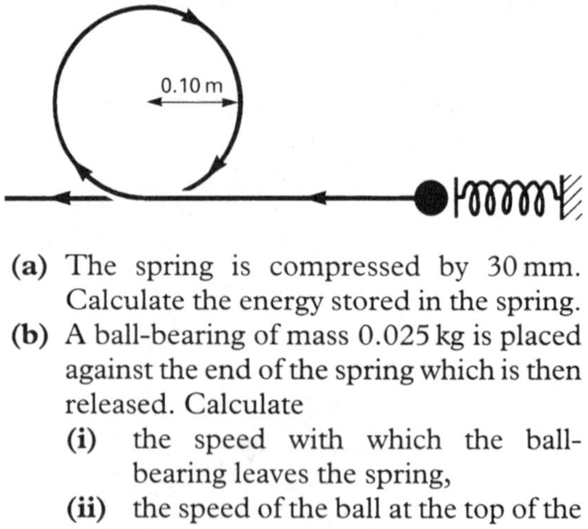

(a) The spring is compressed by 30 mm. Calculate the energy stored in the spring.
(b) A ball-bearing of mass 0.025 kg is placed against the end of the spring which is then released. Calculate
  (i) the speed with which the ball-bearing leaves the spring,
  (ii) the speed of the ball at the top of the loop,
  (iii) the force exerted on the ball by the track at the top of the loop.

Assume that the effects of friction can be ignored. [J, '89]

A81 A special prototype model aeroplane of mass 400 g has a control wire 8 m long attached to its body. The other end of the control line is attached to a fixed point. When the aeroplane flies with its wings horizontal in a horizontal circle, making one revolution every 4 s, the

control wire is elevated 30° above the horizontal. Draw a diagram showing the forces exerted on the plane and determine:
(a) the tension in the control wire,
(b) the lift on the plane.
(Assume that $\pi^2 = 10$.) [AEB, '79]

A82 A small mass of 5 g is attached to one end of a light inextensible string of length 20 cm and the other end of the string is fixed. The string is held taut and horizontal and the mass is released. When the string reaches the vertical position, what are the magnitudes of:
(a) the kinetic energy of the mass,
(b) the velocity of the mass,
(c) the acceleration of the mass,
(d) the tension in the string?
(Neglect air friction.) [J]

A83 A boy ties a string around a stone and then whirls the stone so that it moves in a horizontal circle at constant speed.
(a) Draw a diagram showing the forces acting on the stone, assuming that air resistance is negligible. Use your diagram to explain
(i) why the string cannot be horizontal,
(ii) the direction of the resultant force on the stone and
(iii) the effect that the resultant force has on the path of the stone.
(b) The mass of the stone is 0.15 kg and the length of the string between the stone and the boy's hand is 0.50 m. The period of rotation of the stone is 0.40 s. Calculate the tension in the string.
(c) The boy now whirls the stone in a vertical circle, but the string breaks when it is horizontal. At this instant, the stone is 1.0 m above the ground and rising at a speed of 15 m s^{-1}. Describe the subsequent motion of the stone until it hits the ground and calculate its maximum height. [O & C, '91]

A84 Derive an expression for the magnitude of the acceleration of a particle moving with speed v in a circle of radius r.

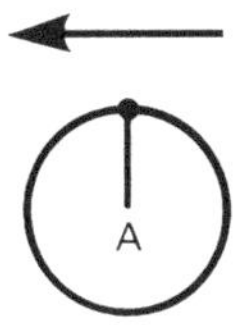

A particle of mass m moves in a circle in a vertical plane, being attached to a fixed point A by a string of length r. The motion of the mass is such that the string is just fully extended at the highest point. Determine:
(a) the minimum speed v at the highest point for this to happen,
(b) the speed V of the particle, and the tension in the string when the particle is at its lowest point.
What is the component of the acceleration of the particle along the tangent to the circle at the instant when the string makes an angle θ with the vertical?

If the particle was initially suspended at rest vertically below A, and was set in motion as described above by an impact with a particle of mass $2m$, determine the velocity u of this particle on the assumption that no energy is lost in the collision. [S]

A85 **(a)** In problems involving linear motion the following equations are often used:
(i) Force = mass × acceleration,
(ii) Kinetic energy = $\frac{1}{2}$ × mass × (velocity)^2,
(iii) Work = force × distance
Using words, write down the corresponding equations for rotational motion.
(b) A couple of torque 5 N m is applied to a flywheel initially at rest. Calculate its kinetic energy after it has completed 5 revolutions. Ignore friction. [W, '92]

A86 A gramophone record A is dropped on to a turntable B which is rotating freely at 10 revolutions per second. The mass of A is 0.25 kg and the mass of B is 0.50 kg. The radius of A is 0.05 m and the radius of B is 0.10 m. What is the final speed of rotation (in rev s^{-1}) of the record and turntable together? (The moment of inertia of a disc is given by $I = \frac{1}{2} MR^2$.) [W, '90]

A87 A swivel chair consists of a seat mounted on a screw-threaded column in such a way that when the seat is given a clockwise rotation it rises vertically. The seat, of moment of inertia about its rotation axis I and mass M, is given an initial clockwise rotation of angular velocity ω. How far does the seat rise, assuming no friction?

What change of angular momentum has occurred during the rise?

Explain the apparent violation of the law of conservation of angular momentum. [W]

A88 **(a)** For a rigid body rotating about a fixed axis, explain with the aid of a suitable diagram what is meant by *angular velocity*, *kinetic energy* and *moment of inertia*.

(b) In the design of a passenger bus, it is proposed to derive the motive power from the energy stored in a flywheel. The flywheel, which has a moment of inertia of $4.0 \times 10^2\,\text{kg m}^2$, is accelerated to its maximum rate of rotation of 3.0×10^3 revolutions per minute by electric motors at stations along the bus route.

(i) Calculate the maximum kinetic energy which can be stored in the flywheel.

(ii) If, at an average speed of 36 kilometres per hour, the power required by the bus is 20 kW, what will be the maximum possible distance between stations on the level? [J]

A89 A cylindrical rocket of diameter 2.0 m develops a spinning motion in space of period 2.0 s about the axis of the cylinder. To eliminate this spin two jet motors which are attached to the rocket at opposite ends of a diameter are fired until the spinning motion ceases. Each motor turns the rocket in the same direction and provides a constant thrust of $4.0 \times 10^3\,\text{N}$ in a direction tangential to the surface of the rocket and in a plane perpendicular to its axis. If the moment of inertia of the rocket about its cylindrical axis is $6.0 \times 10^5\,\text{kg m}^2$, calculate the number of revolutions made by the rocket during the firing and the time for which the motors are fired. [J]

A90 **(a)** **(i)** Explain what is meant by the *moment of inertia* of a body.

(ii) Why is there no unique value for the moment of inertia of a given body?

(iii) A rigid body rotates about an axis with an angular velocity ω. If the relevant moment of inertia of the body is I, show that its rotational kinetic energy is $\frac{1}{2}I\omega^2$.

(b) **(i)** A motor car is designed to run off the rotational kinetic energy stored in a flywheel in the car. The flywheel is to be accelerated up to some maximum rotational speed by electric motors placed at various stations along the route. If the flywheel has a moment of inertia of $300\,\text{kg m}^2$ and is accelerated to 4200 revolutions per minute at a station, calculate the kinetic energy stored in the flywheel. Assuming that at an average speed of $54\,\text{km h}^{-1}$ the power required by the car is 15 kW, what is the maximum possible distance between stations on the car's route?

(ii) What assumption did you make in the last calculation? Comment on the feasibility of the design.

[W, '90]

A91 **(a)** A rigid body is rotating with angular velocity ω about a fixed axis O. Considering a small particle of the body of mass m, at distance r from the axis, state the linear velocity of the particle at any instant, the linear momentum of the particle at that instant, the angular momentum of the particle, and the kinetic energy of the particle.

Write down expressions for **(i)** the angular momentum about the axis O, **(ii)** the kinetic energy, of the whole body, regarded as an assemblage of individual particles, and hence explain the meaning and the importance of the idea of moment of inertia.

(b) Describe how you would determine experimentally the moment of inertia of a flywheel about its usual axis of rotation. An outline of the method only is required, and no formulae need be proved.

(c)

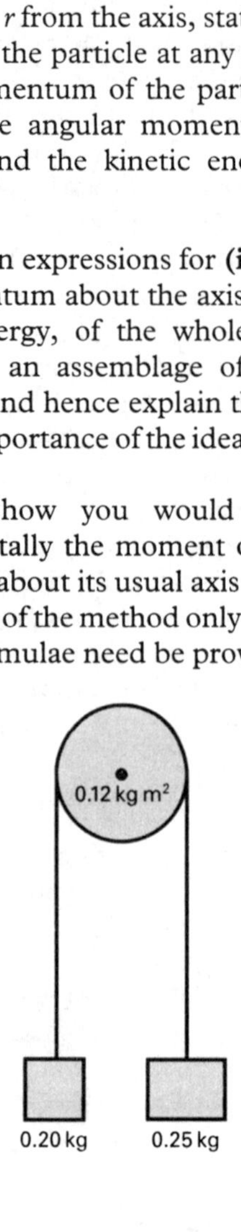

Masses 0.20 kg and 0.25 kg are suspended as in the figure on p. 128 from a light cord which passes over a wheel of radius 0.15 m and moment of inertia 0.12 kg m^2. Initially, the two masses are held at the same horizontal level. Explain what happens when they are released from rest if the cord does not slip on the wheel. Assuming that the wheel rotates freely about its axis, calculate the angular velocity of the wheel and the speed of each mass when the vertical distance between the masses is 0.3 m. [O]

A92 The turntable of a record player rotates at a steady angular speed of 3.5 rad s^{-1}. A record is dropped from rest on to the turntable. Initially, the record slips but eventually it moves with the same angular speed as the turntable.

(a) The angle the turntable turns thrrough while the record is slipping on its surface is 0.25 rad. Find the average angular acceleration of the record while it is attaining the steady speed of the turntable.

(b) The moment of inertia of the record about its axis of rotation is 1.1×10^{-3} kg m^2. What additional torque must be applied by the turntable motor to maintain the constant angular speed of the turntable while the record is accelerating? [C]

A93 **(a)** Explain what is meant by *moment of inertia* and *angular momentum*. State the relationship between them.

(b) Explain the following observations:

(i) Flywheels are often large diameter wheels with heavy rims rather than disc type wheels of constant thickness.

(ii) A pirouetting skater rotates faster as she draws her arms closer to her body.

(c) In a single cylinder petrol engine, energy from the combustion process is supplied to the crankshaft (the axle of the engine) once every two rotations of the crankshaft, namely during each power stroke , and a flywheel is attached to the crankshaft to smooth the motion.

In such an engine each power stroke produces mechanical energy $E = 1$ kJ, and its flywheel has moment of inertia $I = 0.5$ kg m^2.

(i) If the engine starts from rest, determine the angular velocity after 20 revolutions. Neglect the effect of friction.

(ii) Show, that in the case of zero friction, the crankshaft angular acceleration α is given by

$$\alpha = \frac{E}{4\pi I}$$

(iii) In the practical case where there is friction, calculate the frictional torque when the engine is operating at its highest speed. [W, '92]

SIMPLE HARMONIC MOTION (Chapter 7)

A94 **(a)** Define simple harmonic motion (SHM) for a particle moving in a straight line.

(b) Use your definition to explain how SHM can be represented by the equation

$$\frac{\mathrm{d}^2 x}{\mathrm{d}t^2} = -\omega^2 x$$

(c) A mechanical system is known to perform SHM. What quantity must be measured in order to determine ω for the system? [J]

A95 A body of mass 200 g is executing simple harmonic motion with an amplitude of 20 mm. The maximum force which acts upon it is 0.064 N. Calculate:

(a) its maximum velocity,

(b) its period of oscillation. [L]

A96 **(a)** State the conditions for an oscillatory motion to be considered *simple harmonic*.

(b) A body of mass 0.30 kg executes simple harmonic motion with a period of 2.5 s and an amplitude of 4.0×10^{-2} m. Determine:

(i) the maximum velocity of the body,

(ii) the maximum acceleration of the body,

(iii) the energy associated with the motion. [S]

A97 A particle moves with simple harmonic motion in a straight line with amplitude 0.05 m and period 12 s. Find:

(a) the maximum speed,

(b) the maximum acceleration, of the particle.

Write down the values of the constants P and Q in the equation

$$x/m = P \sin [Q(t/s)]$$

which describes its motion. [C]

A98 A sinusoidal voltage is applied to the Y plates of a cathode ray oscilloscope which has a calibrated time base. A stationary trace, with an amplitude of 4.0 cm and a wavelength of 1.5 cm, is obtained when the time base is set at 1.0 cm ms^{-1}. The time base is then switched off and the trace becomes a vertical line. Calculate the maximum speed of the spot of light on the end of the tube when producing the vertical line. [L]

A99 A small piece of cork in a ripple tank oscillates up and down as ripples pass it. If the ripples travel at 0.20 m s^{-1}, have a wavelength of 15 mm and an amplitude of 5.0 mm, what is the maximum velocity of the cork? [L]

A100 A body moving with simple harmonic motion has velocity v and acceleration a when the displacement from its mean position is x. Sketch graphs of a against x, and v against x. [L]

A101 The displacement y of a particle vibrating with simple harmonic motion of angular speed ω is given by

$$y = a \sin \omega t \quad \text{where } t \text{ is the time}$$

What does a represent?

Sketch a graph of the *velocity* of the particle as a function of time starting from $t = 0$ s.

A particle of mass 0.25 kg vibrates with a period of 2.0 s. If its greatest displacement is 0.4 m what is its maximum kinetic energy? [L]

A102 The displacement–time equation for a particle moving with simple harmonic motion is

$$x = a \sin (\omega t + \varepsilon)$$

(a) Explain what each of the symbols represents, illustrating your answer with a rough graph showing how x varies with t.

(b) Write down the velocity–time equation, and draw a corresponding graph showing how the velocity v varies with t.

(c) If m is the mass of the particle, the kinetic energy at displacement x is $\frac{1}{2}m\omega^2(a^2 - x^2)$. Write down the expressions for the potential energy at displacement x, and the total energy.

(d) The total energy of an atom oscillating in a crystal lattice at temperature T is, on average, $3kT$, where k is the Boltzmann constant 1.38×10^{-23} J K^{-1}. Assuming that copper atoms, each of mass 1.06×10^{-25} kg, execute simple harmonic motion of amplitude 8×10^{-11} m at 300 K, calculate the corresponding frequency. [O]

A103 The bob of a simple pendulum moves simple harmonically with amplitude 8.0 cm and period 2.00 s. Its mass is 0.50 kg. The motion of the bob is undamped.

Calculate maximum values for

(a) the speed of the bob, and

(b) the kinetic energy of the bob. [L]

A104 The following statements refer to a body in simple harmonic motion along a straight line. Write each reference letter (A, B, etc.) on a new line and state whether the corresponding statement is correct or incorrect. If you consider a statement to be incorrect, make a short comment pointing out the error.

A. The displacement of the body must be small.

B. The kinetic energy of the body is constant.

C. The period is constant.

D. The amplitude varies sinusoidally with time.

E. At certain instants, the acceleration is zero.

F. The acceleration of the body can be greater than the acceleration due to gravity.

Show that the motion of a simple pendulum is simple harmonic, and obtain an expression for the period, stating any assumptions made.

How would you obtain experimentally the relationship between period and length? Explain how you would use your results to obtain the value of the acceleration due to gravity.

(If the acceleration of a body is related to its position x by a relationship of the type

$a = -kx$, you may assume that the subsequent motion is simple harmonic of period $2\pi/k^{1/2}$.)

Explain why the tension in the string of a simple pendulum is not constant as it swings. At what points does the tension have its maximum and minimum values? Consider whether these values are greater or less than that when the pendulum hangs stationary. [W]

A105 A light spring is suspended from a rigid support and its free end carries a mass of 0.40 kg which produces an extension of 0.060 m in the spring. The mass is then pulled down a further 0.060 m and released causing the mass to oscillate with simple harmonic motion.

(a) Potential energy is stored in two ways in this arrangement: explain briefly what they are.

(b) Calculate the kinetic energy of the mass as it passes through the mid-point of its motion. [L]

A106 (a) The displacement x, in m, from the equilibrium position of a particle moving with simple harmonic motion is given by

$$x = 0.05 \sin 6t$$

where t is the time, in s, measured from an instant when $x = 0$.

(i) State the amplitude of the oscillations.

(ii) Calculate the time period of the oscillations and the maximum acceleration of the particle.

(b) A mass hanging from a spring suspended vertically is displaced a small amount and released. By considering the forces on the mass at the instant when the mass is released, show that the motion is simple harmonic and derive an expression for the time period. Assume that the spring obeys Hooke's law. [J, '89]

A107 A small mass suspended from a light helical spring is drawn down 15 mm from its equilibrium position and released from rest. After 3 seconds the mass reaches this position once more. Find values for the constants a, ω and ε, in the equation $x = a \sin(\omega t + \varepsilon)$ which describes the motion of the mass. Here x measures the distance from the equilibrium position and t the elapsed time since release. [W]

A108 (a) Define *simple harmonic motion.*

(b) A light helical spring, for which the force necessary to produce unit extension is k, hangs vertically from a fixed support and carries a mass M at its lower end. Assuming that Hooke's law is obeyed and that there is no damping, show that if the mass is displaced in a vertical direction from its equilibrium position and released, the subsequent motion is simple harmonic. Derive an expression for the time period in terms of M and k.

(c) If $M = 0.30$ kg, $k = 30$ N m^{-1} and the initial displacement of the mass is 0.015 m, calculate:

(i) the maximum kinetic energy of the mass,

(ii) the maximum and minimum values of the tension in the spring during the motion.

(d) Sketch graphs showing how **(i)** the kinetic energy of the mass, **(ii)** the tension in the spring vary with displacement from the equilibrium position.

(e) If the same spring with the same mass attached were taken to the Moon, what would be the effect, if any, on the time period of the oscillations? Explain your answer. [J]

A109 (a) Define *simple harmonic motion.*

(b) The displacement of a body undergoing SHM is given by $y = A \sin \omega t$.

(i) Explain what A and ω represent.

(ii) Draw a graph showing how y varies with t.

(iii) Underneath this, and using the same scales for t, sketch graphs showing how the velocity v and the acceleration a vary with t.

(c) A mass m hangs on a string of length l from a rigid support. The mass is pulled aside, so that the string makes an angle θ with the vertical, and then released.

(i) Show that the mass executes SHM, stating any assumptions made.

(ii) Prove that the period T of this SHM is given by $T = 2\pi\sqrt{\dfrac{l}{g}}$.

(iii) A student times a simple pendulum to determine T. Does it matter how many oscillations are counted? Does it matter from where the counts are taken – the end or the middle of the swing? Give reasons.

(d) A piston in a car engine performs SHM. The piston has a mass of 0.50 kg and its amplitude of vibration is 45 mm. The revolution counter in the car reads 750 revolutions per minute. Calculate the maximum force on the piston. [W, '90]

A110 **(a)** A body of mass m is suspended from a vertical, light, helical spring of force constant k, as in Fig. 1. Write down an expression for the period T of vertical oscillations of m.

(b) Two such identical springs are now joined as in Fig. 2 and support the same mass m. In terms of T, what is the period of vertical oscillations in this case?

(c) The identical springs are now placed side by side as in Fig. 3, and m is supported symmetrically from them by means of a weightless bar. In terms of T, what is the period of vertical oscillations in this case?

Fig. 1

Fig. 2

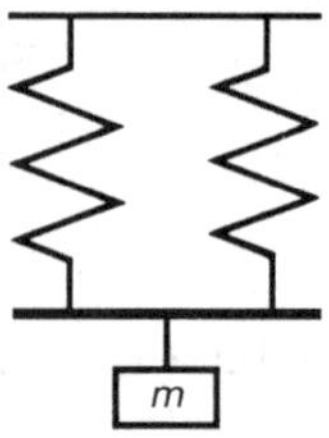

Fig. 3

[W, '91]

A111 **(a)** Define simple harmonic motion. Use your definition to explain what relationship must exist at any instant between the force acting on a body performing such a motion in a straight line and its distance from a fixed point. At what point(s) in the motion is **(i)** the velocity, **(ii)** the acceleration, a maximum?

(b) One end of a spring is attached to a fixed point and the other end carries a body of small mass m which produces a static extension a. Show that, if the body is displaced vertically through a further small distance, it will oscillate with simple harmonic motion. (The mass of the spring may be neglected.)

Given that the time period, T, of the body performing simple harmonic motion is given by the expression

$$T = 2\pi\sqrt{\frac{\text{Mass of the body}}{\text{Force on the body per unit displacement}}}$$

derive an expression for the period of oscillation of the body on the spring.

(c) Suggest how you would investigate experimentally whether the bob of a simple pendulum, when oscillating through a small angle, was executing simple harmonic motion. [L]

A112 A 100 g mass is suspended vertically from a light helical spring and the extension in equilibrium is found to be 10 cm. The mass is now pulled down a further 0.5 cm and is then released from rest. Stating any assumptions you make, show that the subsequent motion of the mass is simple harmonic motion. Calculate:

(a) the period of oscillation,

(b) the maximum kinetic energy of the mass. [J]

A113 A light platform is supported by two identical springs, each having spring constant $20\,\text{N}\,\text{m}^{-1}$, as shown in the diagram.

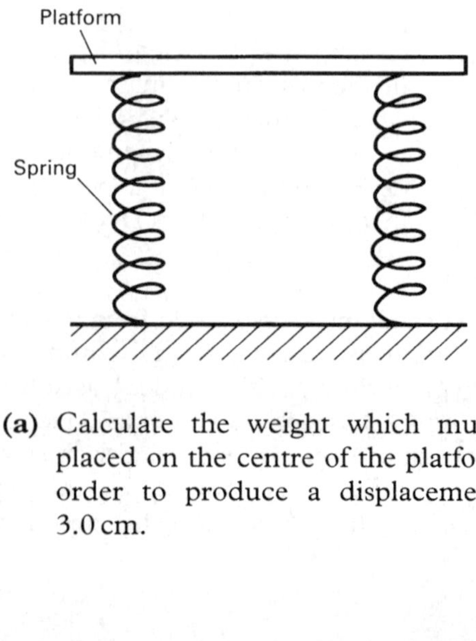

(a) Calculate the weight which must be placed on the centre of the platform in order to produce a displacement of 3.0 cm.

(b) The weight remains on the platform and the platform is depressed a further 1.0 cm and then released. **(i)** What is the frequency of oscillation of the platform? **(ii)** What is the maximum acceleration of the platform? [C, '91]

A114 Define *simple harmonic motion.*

An extension of 2.5 cm is produced when a mass is hung from the lower end of a light helical spring which is fixed at the top end and to which Hooke's law may be assumed to apply. If the mass is depressed slightly and then released, show that the vertical vibrations executed are simple harmonic and calculate their time period.

If the mass of the spring is taken into account, the oscillating mass M may be considered increased to $(M + m)$. Give reasons why m is less than the actual mass of the spring and describe an experiment in which a series of known masses is used to determine the value of m. [J]

A115 Define *simple harmonic motion* and state where the magnitude of the acceleration is **(a)** greatest, **(b)** least.

Some sand is sprinkled on a horizontal membrane which can be made to vibrate vertically with simple harmonic motion. When the amplitude is 0.10 cm, the sand just fails to make continuous contact with the membrane. Explain why this phenomenon occurs and calculate the frequency of vibration. [J]

A116 Define *simple harmonic motion*, and explain what is meant by the *amplitude* and *period* of such a motion.

Show that the vertical oscillations of a mass suspended by a light helical spring are simple harmonic, and obtain an expression for the period.

A small mass rests on a scale-pan supported by a spring; the period of vertical oscillations of the scale-pan and mass is 0.5 s. It is observed that when the amplitude of the oscillation exceeds a certain value, the mass leaves the scale-pan. At what point in the motion does the mass leave the scale-pan, and what is the minimum amplitude of the motion for this to happen? [S]

A117 **(a)** The displacemenmt y of a body moving with SHM is given by

$$y = A \sin \omega t.$$

(i) Sketch the variation of y with t.
(ii) With reference to your sketch explain what is meant by A and ω.
(iii) Sketch on the same axes the variation of velocity v with time.
(iv) Copy and complete the expression for the velocity:

$$v = A\omega \sin (\omega t \,..........)$$

(b) A light spring of force constant k is attached to a solid support and a mass m is fixed to its lower end as shown below.

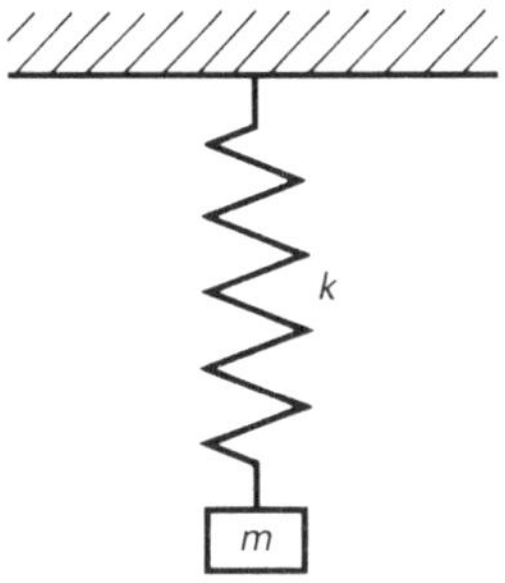

Prove that when displaced vertically and released, the mass moves with SHM of period

$$T = 2\pi \sqrt{\frac{m}{k}}.$$

(c) The following system may be used commercially to measure mass (or weight).

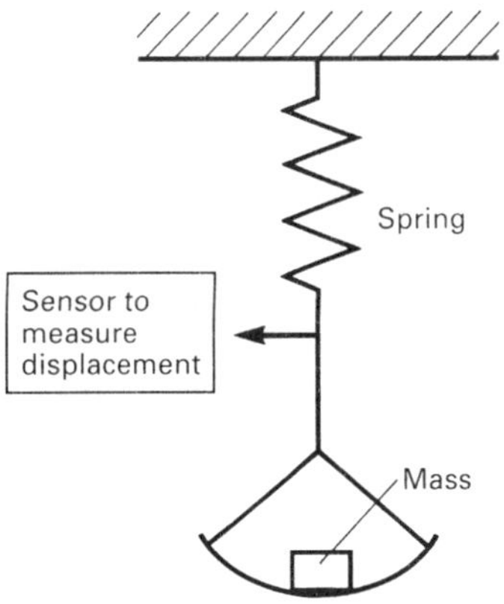

(i) If, at the sensor, the maximum measurable deflection is 10 mm and $k = 10^5\,\mathrm{N\,m^{-1}}$, calculate the maximum measurable mass, M_{max}.

(ii) Calculate the frequency of oscillation when M_{max} is applied.
(iii) Suggest an improvement to the system to enable many measurements of mass to be made in rapid succession. [W, '92]

A118 A mass hangs from a light spring. The mass is pulled down 30 mm from its equilibrium position and then released from rest. The frequency of oscillation is 0.50 Hz.
(a) Calculate:
(i) the angular frequency, ω, of the oscillation
(ii) the magnitude of the acceleration at the instant it is released from rest.
(b) Sketch a graph of the acceleration of the mass against time during the first 4.0 s of its motion. Put a scale on each axis.
(c) After a few oscillations half of the mass becomes detached when it is at the lowest point of its motion. The act of detachment still leaves the remaining half instantaneously at rest.

Is the period of the subsequent oscillation the same, shorter or longer than the original period? Account for your answer. [O & C, '91]

A119 **(a)** **(i)** Define *simple harmonic motion.*
(ii) Show that the equation

$$y = a \sin(\omega t + \varepsilon)$$

represents such a motion and explain the meaning of the symbols y, a, ω and ε.
(iii) Draw with respect to a common time axis graphs showing the variation with time t of the displacement, velocity and kinetic energy of a heavy particle that is describing such a motion.
(b) When a metal cylinder of mass 0.2 kg is attached to the lower end of a light helical spring the upper end of which is fixed, the spring extends by 0.16 m. The metal cylinder is then pulled down a further 0.08 m.
(i) Find the force that must be exerted to keep it there, if Hooke's law is obeyed.
(ii) The cylinder is then released. Find the period of vertical oscillations, and the kinetic energy the cylinder possesses when it passes through its mean position. [O]

GRAVITATION (Chapter 8)

A120 Explorer 38, a radio-astronomy research satellite of mass 200 kg, circles the Earth in an orbit of average radius $3R/2$ where R is the radius of the Earth. Assuming the gravitational pull on a mass of 1 kg at the Earth's surface to be 10 N, calculate the pull on the satellite. [L]

A121 A satellite of mass 66 kg is in orbit round the Earth at a distance of $5.7R$ above its surface, where R is the value of the mean radius of the Earth. If the gravitational field strength at the Earth's surface is $9.8\ \text{N kg}^{-1}$, calculate the centripetal force acting on the satellite.

Assuming the Earth's mean radius to be 6400 km, calculate the period of the satellite in orbit in hours. [L]

A122 **(a)** Define *acceleration*. An object is thrown vertically upwards from the surface of the Earth. Air resistance can be neglected. Sketch labelled graphs on the same axes to show how **(i)** the velocity, **(ii)** the acceleration of the object vary with time. Mark on the graphs the time at which the object reaches maximum height and the time at which it returns to its original position.
(b) Modern gravity meters can measure g, the acceleration of free fall, to a high degree of accuracy. The principle on which they work is of measuring t, the time of fall of an object through a known distance h in a vacuum. Assuming that the object starts from rest, deduce the relation between g, t and h.
(c) State Newton's law of gravitation relating the force F between two point objects of masses m and M, their separation r and the gravitational constant G.

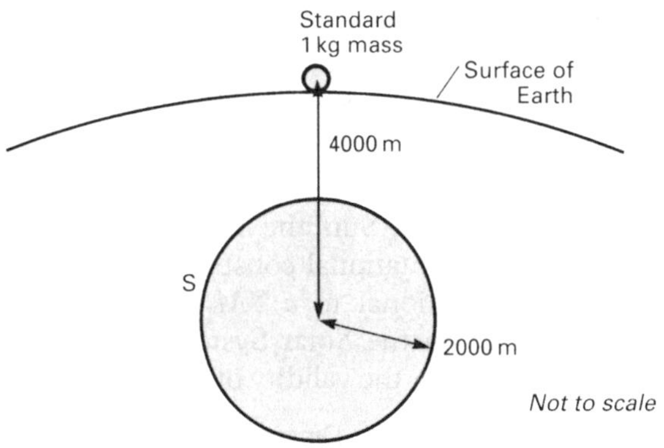

(d) The diagram shows a standard kilogram mass at the surface of the Earth and a spherical region S of radius 2000 m with its centre 4000 m from the surface of the Earth. The density of the rock in this region is 2800 kg m^{-3}. What force does the matter in region S exert on the standard mass?

(e) If region S consisted of oil of density 900 kg m^{-3} instead of rock, what difference would there be in the force on the standard mass?

(f) Suggest how gravity meters may be used in oil prospecting. Find the uncertainty within which the acceleration of free fall needs to be measured if the meters are to detect the (rather large) quantity of oil stated in (e).

($G = 6.67 \times 10^{-11}$ N m^2 kg^{-2}.) [C, '91]

A123 A communication satellite is placed in an orbit such that it remains directly above a fixed point on the Earth's surface at all times.

(a) What is the period of this satellite?

(b) Explain why the satellite must be in orbit above the equator.

(c) Show that the correct height for the orbit does not depend upon the mass of the satellite. [S]

A124 The gravitational force acting on an astronaut travelling in a space vehicle in low Earth orbit is only slightly less than if he were standing on Earth.

(a) Explain why the force is only slightly less.

(b) Explain why, when travelling in the space vehicle, the astronaut appears to be 'weightless'. [L]

A125 **(a)** State the Kepler law of planetary motion which relates period to orbit radius. Show that it is consistent with an inverse square law of force between massive bodies.

(b) When a space shuttle is in an orbit at a mean height of 0.33×10^6 m above the surface of the Earth, it requires 91 minutes to complete one orbit. Use this information to obtain a value for the mass of the Earth.

(c) Describe a laboratory experiment to measure the acceleration of free fall. Explain carefully how the value is obtained from the measurements made and comment upon the accuracy you would expect.

(d) Explain why an astronaut inside the shuttle of part (b) feels weightless even though the intensity of the Earth's gravitational field at that height is approximately 9 N kg^{-1}.

(Mean Earth radius = 6.37×10^6 m; Universal Gravitational constant = 6.67×10^{-11} N m^2 kg^{-2}.) [S]

A126 An artificial satellite travels in a circular orbit round the Earth. Explain why its speed would have to be greater for an orbit of small radius than for one of large radius. [L]

A127 A man is able to jump vertically 1.5 m on Earth. What height might he be expected to jump on a planet of which the density is one third that of the Earth but of which the radius is one half that of the Earth? [L]

A128 Assuming the Earth to be a sphere of radius 6×10^6 m, estimate the mass of the Earth, given that the acceleration of free fall is 10 m s^{-2} and that the gravitational constant G is 7×10^{-11} N m^2 kg^{-2}. [C(O)]

A129 The Moon-rover used by astronauts on the Moon breaks down. Explain whether or not the force required **(a)** to lift it, **(b)** to start it moving horizontally with a given acceleration would be more or less than on Earth. Frictional forces may be considered to be negligible.

While engaged in lifting the vehicle an astronaut lets drop simultaneously a spanner and a piece of paper. Describe and explain the fall of these two objects compared with what would be observed on Earth. [AEB, '79]

A130 The diagram shows a binary star system consisting of two stars each of mass 4.0×10^{30} kg separated by 2.0×10^{11} m. The stars rotate about the centre of mass of the system.

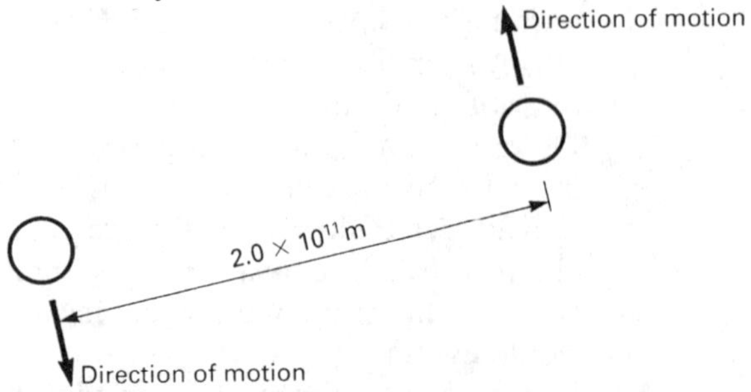

(a) (i) Copy the diagram and, on your diagram, label with a letter L a point where the gravitational field strength is zero. Explain why you have chosen this point.

(ii) Determine the gravitational potential at L. ($G = 6.7 \times 10^{-11}$ N m^2 kg^{-2}.)

(b) (i) Calculate the force on each star due to the other.

(ii) Calculate the linear speed of each star in the system.

(iii) Determine the period of rotation. [AEB, '92]

A131 Kepler's third law of planetary motion, as simplified by taking the orbits to be circles round the Sun, states that if r denotes the radius of the orbit of a particular planet and T denotes the period in which that planet describes its orbit, then r^3/T^2 has the same value for all the planets.

The orbits of the Earth and of Jupiter are very nearly circular with radii of 150×10^9 m and 778×10^9 m respectively, while Jupiter's period round the Sun is 11.8 years.

(a) Show that these figures are consistent with Kepler's third law.

(b) Taking the value of the gravitational constant, G, to be 6.67×10^{-11} N m^2 kg^{-2}, estimate the mass of the Sun. [O*]

A132 (a) State Newton's law of gravitation and derive the dimensions of the gravitational constant G.

(b) If a planet is assumed to move around the sun in a circular orbit of radius r with periodic time T, derive an expression for T in terms of r and other relevant quantities. [J]

A133 Explain what is meant by the gravitational constant G, and derive its dimensions in terms of mass M, length L and time T.

Assuming that the period of rotation t of a planet in its orbit depends only on its distance d from the Sun, the mass M_s of the Sun and the gravitational constant G, show that t^2 is proportional to d^3/M_s. Use the following data on the Solar System to test, as far as possible, the validity of this result.

Planet	*Distance from Sun*/km	*Period*/days
Mercury	0.53×10^8	88
Earth	1.49×10^8	365
Mars	2.28×10^8	687
Jupiter	7.78×10^8	4333
Uranus	28.7×10^8	30690

The distance of the Moon from the Earth is 3.8×10^5 km and its period of rotation is 27.3 days. Deduce the ratio of the mass of the Sun to that of the Earth. [O & C]

A134 Assuming that the Earth (mass m) describes a circular orbit of radius R at angular velocity ω round the Sun (radius r, mass M) due to gravitational attraction:

(a) write down the Earth's equation of motion,

(b) obtain the mean density of the Sun, given $\omega = 2.0 \times 10^{-7}$ rad s^{-1}; $R/r = 200$; $G = 6.7 \times 10^{-11}$ kg^{-1} m^3 s^{-2}; volume of a sphere $= \frac{4}{3}\pi r^3$. [S]

A135 Explain how the mass, M, of the Sun can be calculated from a knowledge of the following:
R, distance from Earth to Sun,
r, distance from Earth to Moon,
T, orbital period of Earth,
t, orbital period of Moon,
m, the mass of the Earth. [L]

A136 Explain what is meant by the universal gravitational constant G. Derive the relationship between G and the acceleration of free fall, g, at the surface of the Earth (neglecting rotation of the Earth and assuming that it is spherical).

Explain why the rotation of the Earth about its axis affects the value of g at the equator.

Calculate the percentage change in g between the poles and the equator (again assuming that the Earth is spherical).

The orbit of the Moon is approximately a circle of radius 60 times the equatorial radius of the Earth. Calculate the time taken for the Moon to complete one orbit, neglecting the rotation of the Earth.

(Acceleration of free fall at the poles of the Earth = $9.8\,\mathrm{m\,s^{-2}}$. Equatorial radius of the Earth = 6.4×10^6 m. 1 day = 8.6×10^4 seconds.) [L]

A137 Describe the circumstances under which a body can be said to be *weightless*. [C]

A138 Show that $\mathrm{N\,kg^{-1}}$ is a valid unit for g, the acceleration due to gravity.

Draw a graph showing how g varies with distance from the Earth's centre. Start your graph from the Earth's surface and assume that $g = 10\,\mathrm{m\,s^{-2}}$ at the surface. Take as your unit along your distance axis the Earth's radius (6.4×10^6 m) and extend the axis to six radii.

Estimate from your graph the loss in potential energy as a body of mass 1 kg falls from 2.56×10^7 m to 1.92×10^7 m from the Earth's centre.

Determine (*not* from your graph) the distance from the Earth to the Moon, and the value of the Earth's g at the Moon. (You may assume that 1 lunar month = 28 days.) [W]

A139 Distinguish between the gravitational constant G and the acceleration due to gravity g.

Assuming that the Earth is a uniform homogeneous sphere of radius R and density Δ obtain expressions for the acceleration due to gravity:

(a) at a pole of the Earth,
(b) at a height h above the Earth at the pole,
(c) at a point on the equator. [J]

A140 **(a)** **(i)** Define gravitational field strength.

(ii) Show that gravitational field strength is equal to g (the acceleration due to gravity).

(b) Explain carefully the distinction between weight and mass.

(c) How are weight and mass each measured? (*One sentence on each is expected.*) [W, '90]

A141 **(a)** **(i)** Explain what is meant by *gravitational potential* and *gravitational potential energy*.

(ii) Use your explanations to show that the difference in potential energy between a point on the Earth's surface and one at a height h above it is, to a close approximation, equal to mgh where m is the mass of the body under consideration and g is the gravitational field strength at the Earth's surface.

(b) The base of a mountain is at sea level where the gravitational field strength is $9.810\,\mathrm{N\,kg^{-1}}$. The value of the gravitational field strength at the top of the mountain is $9.790\,\mathrm{N\,kg^{-1}}$. Calculate the height of the mountain above sea level.

(c) Outline a method of measuring the gravitational field strength to the accuracy required in **(b)** above.

(Radius of the Earth = 6000 km.) [O & C, '92]

A142 **(a)** Define *gravitational field strength* and *gravitational potential*, stating the relationship between them. Explain what is meant by the term *uniform field* and discuss to what extent the gravitational field of the Earth can be considered to be uniform by considering two points on the surface **(i)** separated by a distance of about 10 km, **(ii)** at opposite ends of a diameter. Assume that the Earth is a homogeneous sphere.

(b) Write down an expression for the gravitational potential at the surface of the Earth in terms of its mass M, radius R and the gravitational constant G. Sketch a graph showing the variation of potential with position along a line passing through the centre of the Earth and point out the important features of the graph. (Only consider points external to the surface and in one direction only.)

(c) Derive an expression for the escape velocity, v, at the surface of a planet in terms of the radius, r, of the planet and the acceleration of free fall, g_p, at the surface of the planet. [J]

A143 What are the gravitational potentials at a point on the Earth's surface due to **(a)** the Earth, **(b)** the Sun?

(Mass of Earth = 6.0×10^{24} kg; radius of Earth = 6.4×10^{6} m; mass of Sun = 2.0×10^{30} kg; radius of Earth's orbit = 1.5×10^{11} m; $G = 6.7 \times 10^{-11}$ N m^2 kg^{-2}.) [C]

A144 **(a)** **(i)** State Newton's law of gravitation. Give the meaning of any symbol you use.

(ii) Define *gravitational field strength.*

(iii) Use your answers to **(i)** and **(ii)** to show that the magnitude of the gravitational field strength at the Earth's surface is

$$\frac{GM}{R^2}$$

where M is the mass of the Earth, R is the radius of the Earth and G is the gravitational constant.

(b) Define *gravitational potential.* Use the data below to show that its value at the Earth's surface is approximately -63 MJ kg^{-1}.

(c) A communications satellite occupies an orbit such that its period of revolution about the earth is 24 hr. Explain the significance of this period and show that the radius, R_0, of the orbit is given by

$$R_0 = \sqrt[3]{\frac{GMT^2}{4\pi^2}}$$

where T is the period of revolution and G and M have the same meanings as in (a) (iii).

(d) Calculate the least kinetic energy which must be given to a mass of 2000 kg at the Earth's surface for the mass to reach a point a distance R_0 from the centre of the Earth. Ignore the effect of the Earth's rotation.

($G = 6.7 \times 10^{-11}$ N m^2 kg^{-2}, $M = 6.0 \times 10^{24}$ kg, $R = 6.4 \times 10^{6}$ m.) [J, '89]

A145 **(a)** Define *gravitational potential* at a point.

(b) As a spacecraft falls towards the Earth, it loses gravitational potential energy. What becomes of the lost potential energy

(i) when the spacecraft is falling freely towards the Earth well away from the Earth's atmosphere,

(ii) when the spacecraft is falling through the Earth's atmosphere at constant speed?

(c) **(i)** Calculate the gravitational potential difference between a point on the Earth's surface and a point 1600 km above the Earth's surface.

(ii) Calculate the minimum energy required to project a spacecraft of mass 2.0×10^{6} kg from the surface of the Earth so that it escapes completely from the influence of the Earth's gravitational field.

(Radius of Earth=6400 km; Universal Gravitational constant = 6.7×10^{-11} N m^2 kg^{-2}; mass of the Earth = 6.0×10^{24} kg.) [AEB, '87]

A146 What do you understand by the term *gravitational field*; define *gravitational field strength.*

Show that the radius R of a satellite's circular orbit about a planet of mass M is related to its period as follows:

$$R^3 = \frac{GM}{4\pi^2} T^2$$

where G is the universal gravitational constant.

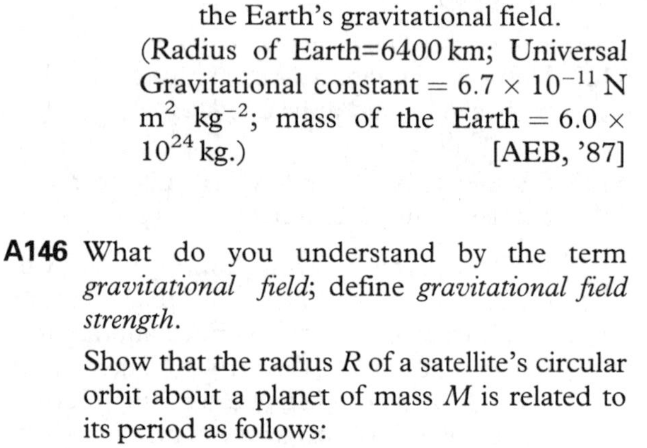

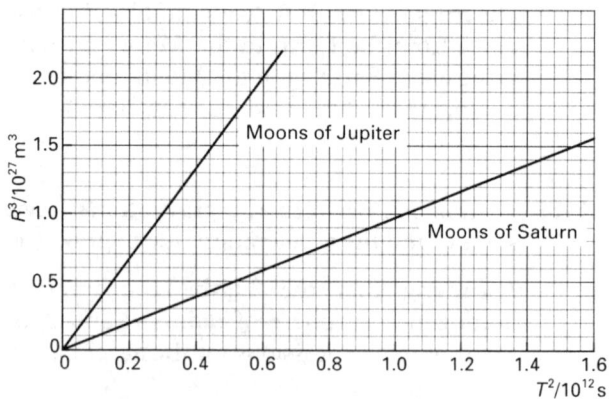

The diagram shows two graphs of R^3 against T^2; one is for the moons of Jupiter and the other is for the moons of Saturn. R is the mean distance of a moon from a planet's centre and T is its period.

The orbits are assumed to be circular.

The mass of Jupiter is 1.90×10^{27} kg.

(a) Why are the lines straight?

(b) Find a value for the mass of Saturn.

(c) Find a value for the universal gravitational constant G. [L*]

A147 Explain what is meant by the statement: *The gravitational potential at one Earth's radius above the Earth's surface is* $-31.3 \times 10^6\,\mathrm{J\,kg^{-1}}$.

(a) The table below gives the gravitational potential V_g at distances from the centre of the Earth in units of Earth radii. The radius of the Earth $R = 6.38 \times 10^6$ m.

V_g/MJ kg^{-1}	−62.7	−31.3	−20.9	−15.7
Distance/R	1.0	2.0	3.0	4.0

(i) Use these data to determine the gravitational potential at distances from the Earth's centre of 10×10^6 m and 15×10^6 m. Indicate how you determined the values.

(ii) A spacecraft of mass 4.0×10^4 kg has its motors switched off. It slows down as it moves away from 10×10^6 m above the Earth's centre to 15×10^6 m. Find the loss of kinetic energy of the craft and the average force acting on the craft.

(iii) A slow-moving meteorite is captured by the Earth's gravitational field. Determine the speed with which it will crash into the surface on the assumption that it is not slowed by air resistance.

(b) The Space Shuttle, with its engines shut down, is moving in the same circular orbit above the Earth and at the same speed as a satellite that it is trying to capture. The two craft are separated by a distance of a few kilometres. The Shuttle can catch up the satellite by using its engines in reverse for a few seconds to slow it down. The Shuttle falls into a lower orbit and passes the satellite. Using its engines to accelerate for a few seconds it returns to the original orbit just in front of the satellite. Use your knowledge of gravitational forces and uniform motion in a circular orbit to explain the physics of this procedure.

[O & C, '90]

A148 **(a)** Write down an expression for the gravitational potential difference between a point P on the Earth's surface and a distant point Q as shown below.

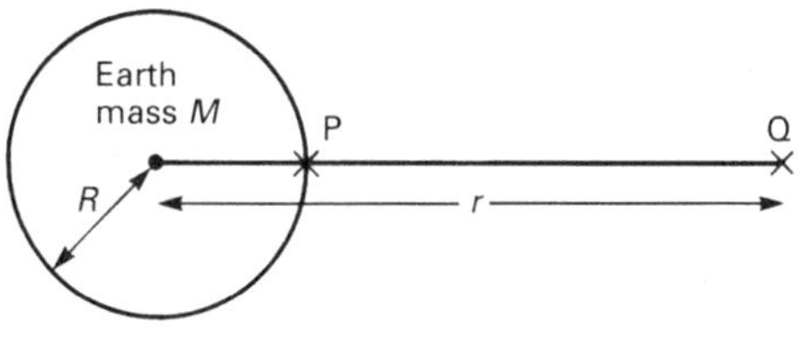

Show that if r is only slightly greater than R, the gravitational potential difference becomes $g(r - R)$ where g is the gravitational field strength on the Earth's surface.

(b) The graph shows how the gravitational potential difference between a point on the Earth's surface and a distant point, distance x from the Earth's surface, changes near to the Moon's surface. The Moon's surface is 384 000 km from the Earth's surface.

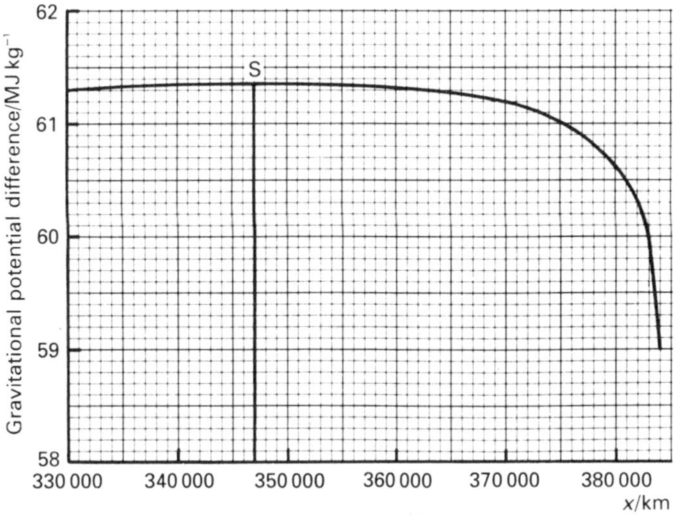

The graph shows the gravitational potential difference first increasing, then achieving a maximum value and finally decreasing to a smaller value on the Moon's surface.

(i) Use the graph to determine the amount of potential energy released as a mass of 200 kg falls to the surface of the Moon from a height of 14 000 km. At what speed will it hit the surface?

(ii) What feature of the graph justifies the assumption that the potential energy of a body measured with respect to the Moon's surface is proportional to its height above that surface? Obtain from the graph the height to which this assumption is true.

(iii) The net force acting on a body moving from the Earth's surface to the surface of the Moon is the resultant of two components, one due to the attraction of the body towards the Earth and the other due to its attraction towards the Moon.

Explain how the net force changes in going from the Earth's surface to the point S, shown on the graph, where the gravitational potential difference is a maximum.

What is the value of the net force at the point where the gravitational potential difference is a maximum? Give a reason for your answer.

Explain why the gravitational potential difference is a maximum at this point. [L]

A149 (a) (i) Explain why a force is required for a mass to travel at a constant speed in a circular path. State the direction of this force and give an equation for its magnitude, defining any terms used.

(ii) State how this force is provided in the case of a satellite orbitting the Earth.

(iii) Show that the speed of a satellite in orbit close to the Earth is given by $(gR)^{\frac{1}{2}}$ where g is the acceleration of free fall and R is the radius of the Earth.

(iv) Calculate the speed of the satellite and the period of the orbit given that $g = 9.8\,\text{m}\,\text{s}^{-2}$ and $R = 6.4 \times 10^3\,\text{km}$.

(b) The most useful communication satellites are those in geostationary orbits. A satellite in geostationary orbit remains above the same point on the Earth's

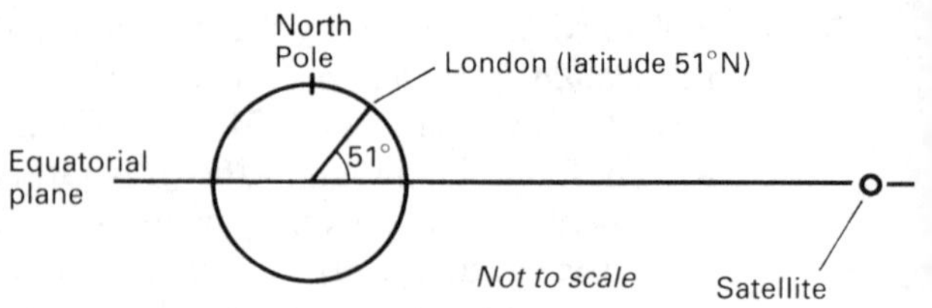

surface at all times. This is only possible when the satellite is in an orbit in the equatorial plane. Only 'line of sight' communication is possible in satellite communications. This means that communication can only occur provided there is no obstruction between the transmitter, the satellite and the receiver.

(i) The relationship between the period T and radius R of an orbit is

$$T^2 = kR^3$$

where k is a constant.

Using your answer to (a) (iv) determine the radius of the orbit for a geostationary satellite.

(ii) Estimate the delay between the transmission and reception of a signal using the satellite. Show how you arrive at your answer.

(iii) Calculate the most northerly latitude for which satellite communication is possible.

(iv) State with reasons how many satellites are needed to provide communication between all places on the equator and indicate on a diagram how this can be achieved.

(v) State and explain the advantages of communicating using geostationary satellites compared with those whose position relative to the Earth's surface is continually changing. [AEB, '89]

SECTION B

STRUCTURAL PROPERTIES OF MATTER

9

SOLIDS AND LIQUIDS

9.1 INTRODUCTION

The kinetic theory accounts for all three states of matter (solid, liquid and gas) by assuming that matter is made up of molecules which are in continual motion. This motion exists at all temperatures above absolute zero, and the kinetic energy associated with it is often referred to as **thermal energy**. The molecules* exert forces of attraction on each other, and so they also possess potential energy. The forces are due to the electrostatic interactions of the electrons and nuclei of the molecules. The force between a pair of molecules depends upon the spatial distribution of the electrons and the separation of the molecules. At very small separations the net force must become repulsive, for if the attractive force were to exist right down to zero separation, all matter would collapse in on itself. The force must be negligible at large separations in order to account for the properties of gases. The kinetic energy, on the other hand, depends only on temperature; in fact **temperature is the outward manifestation of kinetic energy**. It is the relative magnitude of the kinetic and potential energies which determines whether a substance is in the solid, the liquid or the gaseous state.

9.2 INTERMOLECULAR FORCE AND POTENTIAL ENERGY

Consider two isolated molecules whose separation is such that they are exerting attractive forces on each other. If one of the molecules were to be removed to infinity, work would have to be done on it in order to overcome the attractive force, and therefore its potential energy would increase. However, it is convenient to regard the potential energy of each molecule as being zero when their separation is infinite (because at such a separation they have no influence on each other), and therefore when two molecules are attracting each other their potential energy is negative.

9.3 TO SHOW THAT $F = -dE/dr$

Consider two molecules exerting forces of attraction on each other (Fig. 9.1). If the force F on A moves it a small distance δr (so that F can be considered constant) to the right, then the work done δW on A is given by

$$\delta W = F\,\delta r \qquad [9.1]$$

*We shall not distinguish between atoms and molecules.

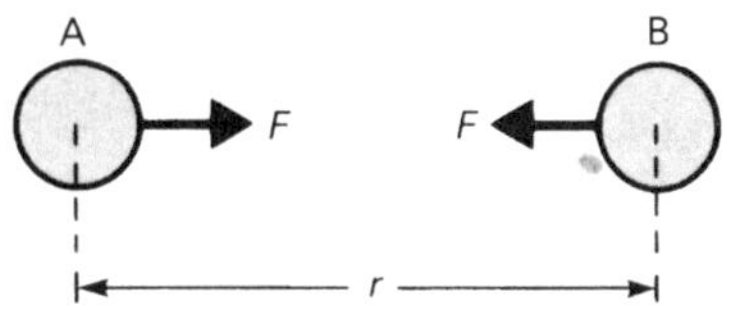

Fig. 9.1
Mutually attracting molecules

If δE is the resulting change in the potential energy of A, then

$$\delta E = -\delta W \qquad [9.2]$$

The minus sign is present because as A moves towards B, under the influence of the attractive force, its potential energy decreases. By equations [9.1] and [9.2],

$$\delta E = -F\,\delta r$$

and therefore in the limit

$$F = -\mathrm{d}E/\mathrm{d}r$$

9.4 THE INTERMOLECULAR POTENTIAL ENERGY AND FORCE CURVES

The potential energy E of a pair of molecules (or atoms), due to the electrostatic force F between them, varies as a function of their separation r as shown in Fig. 9.2. Since $F = -\mathrm{d}E/\mathrm{d}r$, a plot of F against r is, in fact, a plot of the negative of the gradient of the energy curve against r. Such a plot is shown in Fig. 9.3.

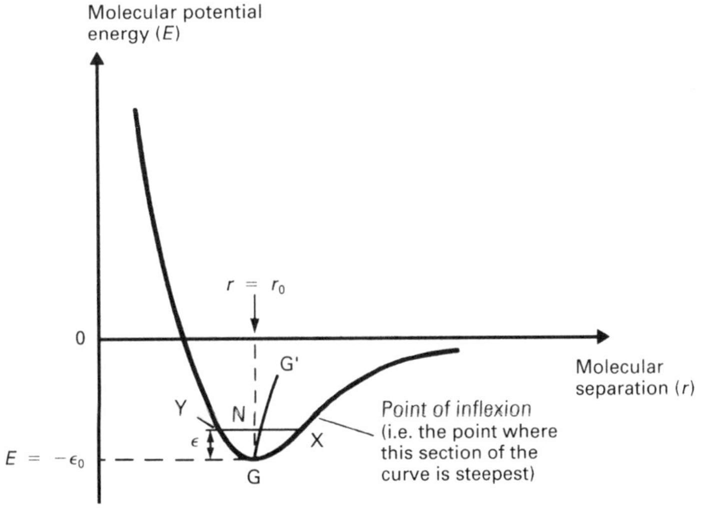

Fig. 9.2
The potential energy of a pair of molecules as a function of their separation

When $r = r_0$ there is no net force between the molecules and their potential energy (Fig. 9.2) has its minimum value. Thus if two molecules have a separation of r_0 they are at their equilibrium separation. Any increase or decrease in their separation would require energy, since work would have to be done against the net attractive or the net repulsive force respectively. The equilibrium is stable because an increase in r leads to an attractive force which restores r to r_0; similarly a decrease in r produces a repulsive force which again restores r to r_0. (The value of r_0 depends on the particular solid but it is often $\sim 3 \times 10^{-10}$ m.)

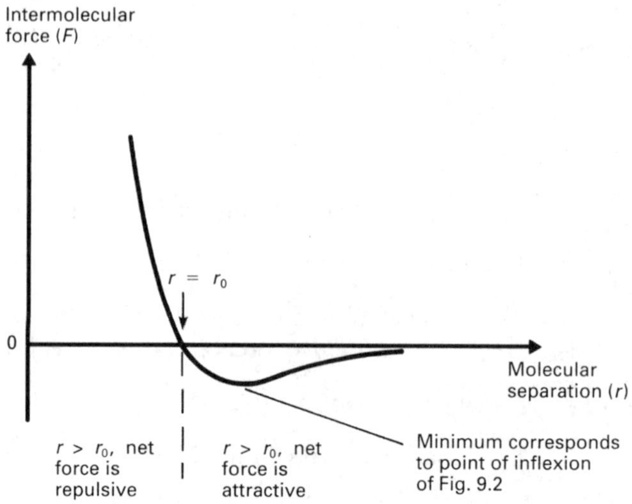

Fig. 9.3
The force between a pair of molecules as a function of their separation

9.5 SOLIDS

Solids have fixed shapes and fixed volumes. Consider a solid at absolute zero; the molecules would have no kinetic energy and therefore would be stationary at their equilibrium separation r_0. At higher temperatures the molecules would possess kinetic energy and could use this to oppose the intermolecular forces. Suppose a pair of molecules has an amount of kinetic energy ε. By exchanging this kinetic energy for potential energy, they would be able to increase their separation such that their situation became that represented by the point X on the energy curve (Fig. 9.2); or decrease it to Y. At X the molecules would feel an attractive force which would restore them to their equilibrium separation r_0. On reaching r_0 they would once again have kinetic energy ε but would now be moving toward each other; they would therefore decrease their separation to the state represented by Y. At Y their directions of motion would again reverse. Thus, **the molecules of a solid at temperatures above absolute zero oscillate about their equilibrium positions**. Because their kinetic energy is low compared with their potential energy ($\varepsilon < 0.1\varepsilon_0$), the molecules of solids can merely vibrate about fixed positions. They are therefore locked into a geometrically ordered array, and as a consequence a solid has both a fixed volume and a fixed shape.

At temperatures above absolute zero the mean separation is not necessarily r_0. In Fig. 9.2 XN is greater than YN, and therefore the mid-point of XY, the point on which the oscillation is centred, corresponds to a separation which is greater than r_0. As the temperature increases from absolute zero, the mid-point moves from G towards G′. Thus, the mean separation of the two molecules (and consequently of all the pairs of molecules within the solid) increases with temperature, i.e. the curve shown represents the normal situation, that of a solid which expands on heating.

The linear expansivity α of a solid is defined by

$$\alpha = \frac{\delta L}{L\,\delta\theta} \qquad [9.3]$$

where

δL = the increase in length brought about by a small increase in temperature $\delta\theta$

L = the original length of the specimen.

The value of α depends on the temperature at which it is measured. However, for temperature increases of less than about 100°C the variation is slight and equation [9.3] can be replaced by the more useful expression

$$L_1 = L_0 + \alpha L_0(\theta_1 - \theta_0) \quad [9.4]$$

where

α = the mean linear expansivity of the solid in the temperature range θ_o to θ_1 (unit = $°C^{-1}$ or K^{-1})

L_0 = the length of the specimen at θ_0

L_1 = the length of the specimen at θ_1.

9.6 LIQUIDS

A liquid has a fixed volume but no fixed shape. The molecules of liquids, like those of solids, vibrate. In liquids though, each molecule has a particular set of nearest neighbours for only a short time. This occurs because the molecules of liquids have greater average kinetic energies than those of solids. (Note that the molecules of liquids, like those of solids and gases, have a range of kinetic energies, and the energy of any particular molecule is constantly changing due to intermolecular collisions.) The increased kinetic energy results in larger amplitudes of vibration, and therefore there is more likelihood of a molecule being able to pass through the gaps between the molecules surrounding it. There is, therefore, a continual molecular migration superimposed upon the vibrational motion, and this accounts for the ability of a liquid to adopt the shape of its container. The molecules, however, are close together and a change in volume would require that the intermolecular forces were overcome – liquids, therefore, have fixed volumes.

9.7 BROWNIAN MOTION

This was first observed in 1827 by Robert Brown, a Scottish botanist, while using a microscope to look at a suspension of pollen grains in water. He noticed that the pollen grains were in a state of continual motion. The motion was both random and jerky. Brownian motion can be observed when small particles of any kind are suspended in a fluid (e.g. smoke particles suspended in air). The motion can be made more pronounced by:

(i) increasing the temperature of the fluid, and/or

(ii) decreasing the size of the suspended particles.

Brownian motion is now regarded as strong evidence that fluids are composed of molecules in a state of unceasing random motion. For example, we consider that a smoke particle suspended in air is constantly being bombarded by air molecules. At any one time, though, if the air molecules move randomly, the smoke particle is likely to receive a bigger impact on one side than on the opposite side. Because the smoke particle is small, this statistical imbalance will be significant and therefore the particle will speed up or slow down and/or change direction. The smoke particle is in thermal equilibrium with the air molecules, in which case its average kinetic energy will be equal to that of an air molecule, i.e.

$$\left(\tfrac{1}{2}M\overline{C^2}\right)_{\text{smoke particle}} = \left(\tfrac{1}{2}m\overline{c^2}\right)_{\text{air molecule}}$$

Since $M \gg m$, then $\overline{C^2} \ll \overline{c^2}$, i.e. the mean square of speed of the smoke particle is much less than that of an air molecule.

The spot on the scale of a sensitive light-beam galvanometer quivers as a result of Brownian motion of the mirror and sets a limit on the sensitivity of the instrument.

9.8 LATENT HEAT

It can be seen from Fig. 9.2 that if two atoms are to become completely free of each other (i.e. to acquire zero potential energy), then they need to gain an amount ε_0 of energy (typically 0.01 to 0.1 eV).* Thus, ε_0 can be considered to be the **binding energy** of such a pair of atoms.

Consider a solid in which each atom has n nearest neighbours (i.e. a coordination number of n). Since interatomic forces are short-range forces, it can be assumed that each atom interacts only with these n atoms, i.e. that each atom is involved in n bonds. In order for such a bond to be broken, the amount of energy that each of the two atoms involved has to acquire is $\varepsilon_0/2$. Since the number of atoms in one mole (see section 14.3) is numerically equal to N_A (where N_A is the Avogadro constant) and each atom is involved in n bonds, the total energy required to separate the atoms of one mole of solid completely at absolute zero is $nN_A\,\varepsilon_0/2$. Thus,

$$L_s = \tfrac{1}{2}nN_A\,\varepsilon_0 \qquad [9.5]$$

where

L_s = the latent heat of sublimation† of one mole of solid at zero kelvin.

The latent heat of sublimation at temperatures above zero kelvin would be slightly less because the average potential energy of the atoms of such a solid would be greater than $-\varepsilon_0$. The latent heat of vaporization would be even smaller.

9.9 FREE SURFACE ENERGY

Consider a liquid whose atoms have a coordination number of n. To a first approximation, those atoms which form part of the surface will have a coordination number of $n/2$. (This is because those atoms which are above the surface are in the vapour phase, and therefore only an insignificant number of them is within range of the surface atoms.) Suppose that the surface area is increased by unit area, and that there are A atoms per unit surface area. In order that this can happen, A atoms have to leave the interior and enter the surface. Thus, A atoms that have previously been involved in n bonds each, are now involved in only $n/2$ bonds each, i.e. A atoms have been removed from $n/2$ bonds. The removal of one atom from one bond requires that it gains an amount $\varepsilon_0/2$ of energy. Therefore the

*1 eV = 1.6×10^{-19} J (see section 47.1).

† Sublimation is the direct conversion of a solid into its vapour.

energy involved in removing A atoms from $n/2$ bonds is $A \times \frac{n}{2} \times \frac{\varepsilon_0}{2}$. The energy needed to create unit surface area is called **free surface energy** σ (see section 10.17), and therefore

$$\sigma = \tfrac{1}{4}nA\varepsilon_0 \qquad [9.6]$$

The relationship is approximate because in addition to the approximation made above:

(i) the liquid will not be at zero kelvin and therefore the binding energy of a single pair of atoms will be less than ε_0,

(ii) the atoms of the vapour will have some effect on the coordination number of the surface atoms.

9.10 TYPES OF SOLID

Crystalline Solids*

In a crystalline solid the atoms (or ions or molecules) are arranged in a regular three-dimensional array. There is a basic unit, called the **unit cell**, which is repeated throughout the structure in all three dimensions. Some unit cells are shown in Figs. 9.12 to 9.14 (see pp. 155–6).

The way in which the atoms are arranged (i.e. the particular crystal form) has a marked bearing on the physical properties of the solid concerned. For example, though both graphite and diamond are crystalline forms of carbon, graphite is soft and is used in pencils and as a solid lubricant, whereas diamond is one of the hardest materials known. Diamond is more dense that graphite. It is transparent and is an electrical insulator; graphite is opaque and conducts electricity well (Fig. 9.4).

When atoms are arranged in a regular fashion their total potential energy is less than it would be if they were packed irregularly. It is not surprising, therefore, that most solids are crystalline (or polycrystalline – see following paragraphs). The regularity of the arrangement ensures that all parts of it require the same energy to bring about melting. Crystals therefore have definite melting points.

In a so-called 'single crystal' equivalent planes of atoms are all parallel to each other. It is often necessary to take special precautions to produce such a crystal.

Polycrystalline samples are much more common. These consist of large numbers of tiny crystals packed together and orientated randomly with respect to each other. These tiny crystals are called **grains** or **crystallites**, and the boundaries between them are known as **grain boundaries**.

The physical properties of single crystals depend on the directions in which they are measured relative to the crystal axes – a property known as **anisotropy**. For example, the electrical resistivity of a single crystal of tin is $1.3 \times 10^{-7}\,\Omega\,\text{m}$ parallel to the main axis but only $1.0 \times 10^{-7}\,\Omega\,\text{m}$ perpendicular to it. The coefficients of linear expansivity are respectively $3.1 \times 10^{-5}\,\text{K}^{-1}$ and $1.6 \times 10^{-5}\,\text{K}^{-1}$.

Any particular substance normally forms crystals of a characteristic general shape. Thus, crystals of ammonium alum are octahedral; those of quartz are hexagonal prisms capped by hexagonal pyramids (Fig. 9.5). This must not be taken to mean that all crystals of the same substance are identical – normally some faces will have

*Crystal structures are discussed in section 9.11.

Fig. 9.4
The structure of
(a) diamond, (b) graphite

(a)

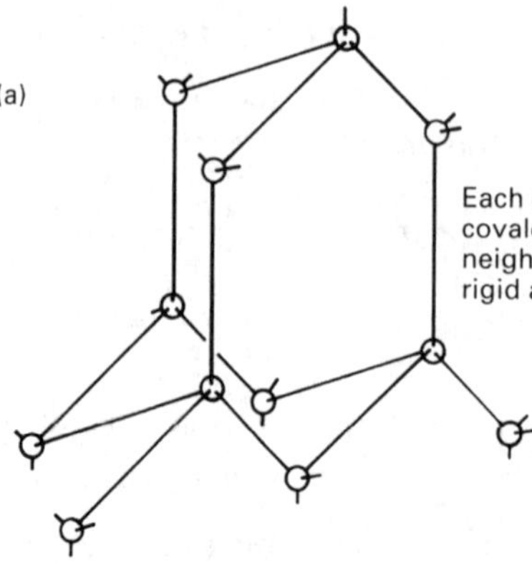

Each carbon atom is locked into position, covalently bonded to its four nearest neighbours. Diamond is therefore both rigid and strong.

(b)

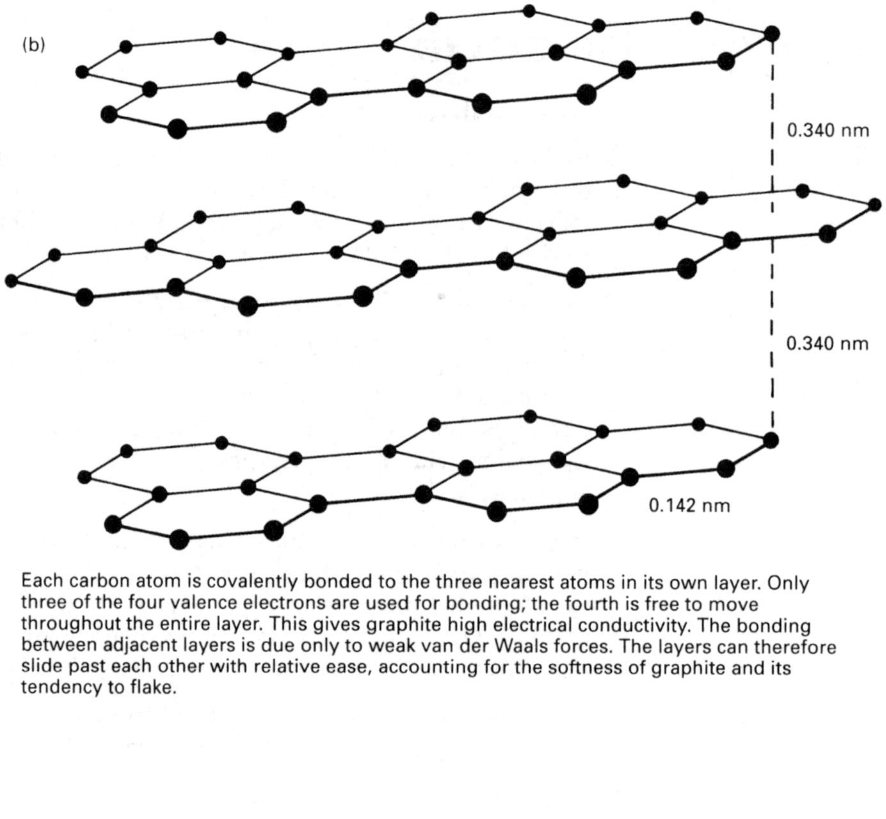

Each carbon atom is covalently bonded to the three nearest atoms in its own layer. Only three of the four valence electrons are used for bonding; the fourth is free to move throughout the entire layer. This gives graphite high electrical conductivity. The bonding between adjacent layers is due only to weak van der Waals forces. The layers can therefore slide past each other with relative ease, accounting for the softness of graphite and its tendency to flake.

Fig. 9.5
Ideal crystals of
(a) ammonium alum,
(b) quartz

(a)

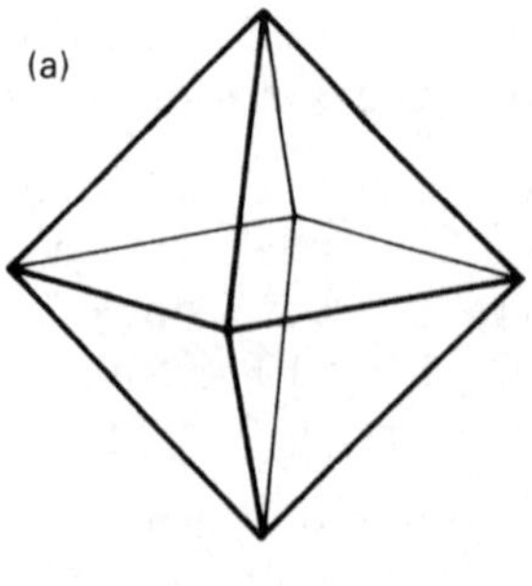

(b)

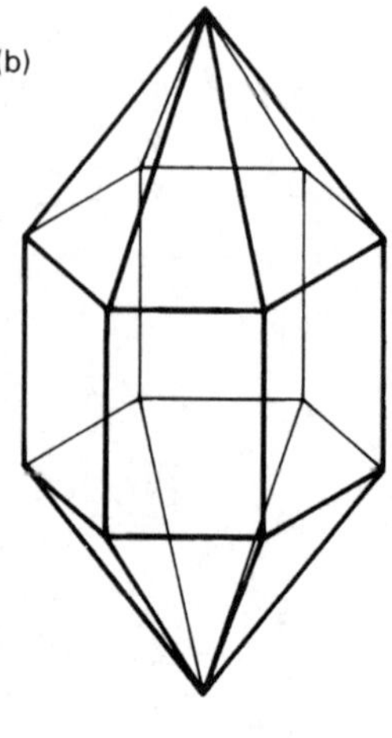

grown at greater rates than others, producing a distorted version of the ideal shape. This would happen, for example, if during its period of growth, the crystal were always resting on the same face, or if it came into contact with other crystals. When we say that crystals of the same substance have the same general shape, we mean that the angles betwen corresponding faces are always the same. Thus, although an ideally grown crystal of sodium chloride is a perfect cube (Fig. 9.6(a)), one grown under non-ideal conditions may have the form shown in Fig. 9.6(b). Fig. 9.7 shows cross-sections of ideal and distorted crystals of quartz.

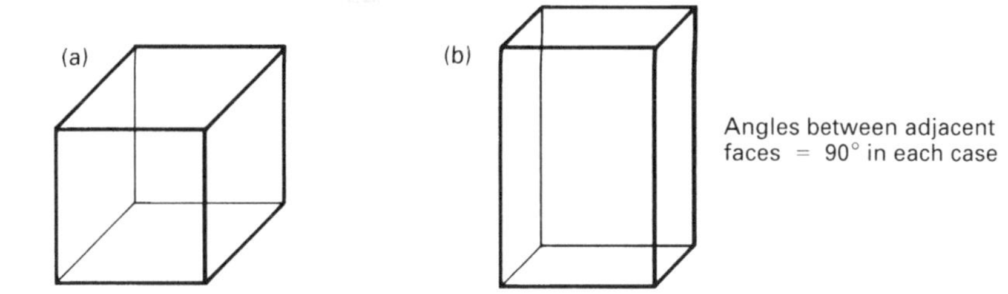

Fig. 9.6
(a) Ideal, and
(b) distorted crystals of sodium chloride

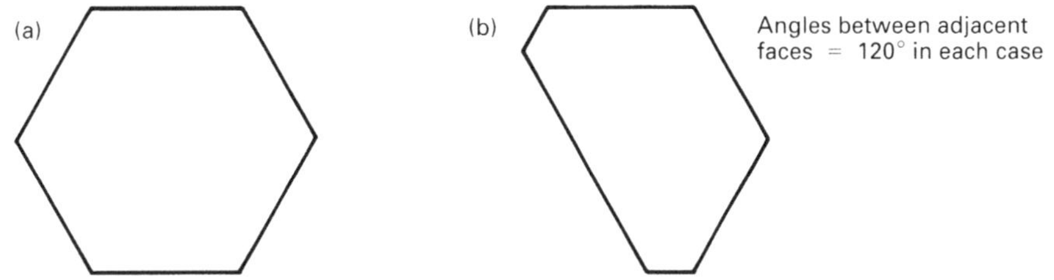

Fig. 9.7
Cross-sections of
(a) ideal, and
(b) distorted crystals of quartz

If a crystal is struck so that it fractures, it will often break along one of a small number of well-defined planes – **cleavage planes**. These are normally parallel to crystal faces. The reason for the ease of fracture is usually that the bonding between the atoms (or ions) on each side of a cleavage plane is weaker than that across other planes. Mica has only one cleavage plane and so splits easily into thin sheets. Sodium chloride has three mutually perpendicular cleavage planes.

Amorphous Solids

An amorphous solid (e.g. glass, wax) has no definite melting point. Rather, as it is heated it becomes progressively more like a liquid, i.e. it becomes softer and more fluid. Amorphous solids have no regular outlines, no directional properties, no cleavage planes and no long-range regular arrangements of atoms. Though they exhibit the properties of solids, they can be regarded as liquids with very high viscosities. For example, a glass window which has been in place for a few hundred years may be noticeably thicker towards the bottom as a result of flow under the moderate influence of its own weight. Indeed, in general there is no clear distinction between solids and liquids. For example, indium and gold, two crystalline solids, flow when subjected to very high pressures and are used in sealing-gaskets between components in high-vacuum systems.

If an amorphous solid is cooled slowly from a temperature at which it is soft, its atoms may have time to arrange themselves into a form which produces crystallization on further cooling.

Glasses

There are many types of glass. The most common is that used in windows and in the manufacture of bottles. It is called **soda-lime glass** and is made from a mixture of silica, calcium oxide and sodium oxide (SiO_2, CaO and Na_2O). The relative proportions of these oxides depend on the exact purpose for which the glass is to be used. There may also be small amounts of other substances present, for example as colouring agents. Another common group of glasses is the borosilicates (basically a combination of SiO_2 and B_2O_3). These are highly resistant to thermal shock because they have low coefficients of expansion, and are used in the manufacture of ovenware (e.g. Pyrex). They were originally developed to overcome the problems caused by (cold) rain falling on (hot) railway signal lights. Borosilicates are also resistant to attack by a wide variety of chemicals, making them ideal for laboratory glassware.

Glasses are amorphous materials and so do not have definite melting points. They soften on heating, and the maximum temperature at which a glass can be used depends on the type of glass. The maximum is normally between 400°C and 700°C, but fused quartz* can be used at over 1000°C. At room temperature glasses are hard and brittle. Though the common glasses are transparent, some are translucent (the so-called **opal glasses**) and some are opaque. Most glasses absorb ultraviolet radiation, but fused quartz and some specially developed glasses are transparent to it. Glass transmits infrared radiation provided its wavelength is not too long. The common glasses are electrical insulators. All glasses are inorganic. They are produced by melting mixtures of their raw materials, and then cooling the melt in such a way that crystallization does not occur.

Polymers

Polymers have large molecules in the form of long chains. Each chain consists of a large number of small molecules, known as **monomers**, joined together by strong (covalent) bonds. They are usually organic, and may be either natural (e.g. rubber, wool, cellulose and proteins) or synthetic (e.g. polythene and the various forms of nylon). The monomers in a chain may all be identical, or may be of two or more different types. Polythene, for example, has only one type, and is produced by polymerizing ethene ($CH_2{=}CH_2$) to produce chains containing thousands of CH_2 groups (Fig. 9.8). The naturally occurring polymers tend to be more complex.

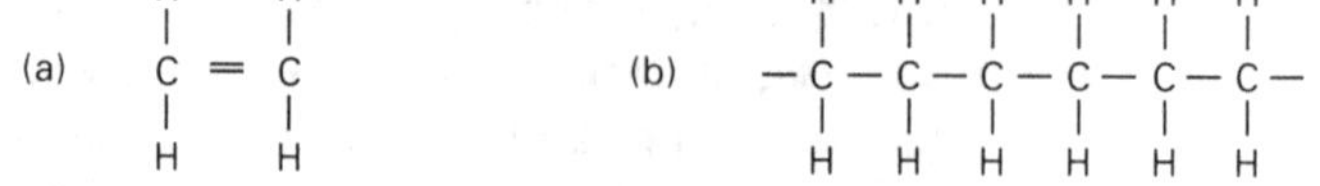

Fig. 9.8
To illustrate the effect of polymerization: (a) the monomer (ethene), (b) the polymer (polyethene, commonly called polythene)

The chains may be **linear**, **branched** or **cross-linked**; the three types are represented schematically in Fig. 9.9. The chain type has a strong bearing on the properties of the polymer. A material in which the chains are linear is flexible because the chains can easily slide past each other. Branching reduces this ability and so increases the rigidity of the material. Chains of this type cannot pack closely together, thus causing reductions in density, tensile strength, melting point, and crystallinity (see p. 153). Cross-linked polymers are very rigid because the chains are incapable of any sliding. There are no crystalline regions in cross-linked polymers.

*A vitreous (glassy) form of silica made by melting (fusing) pure quartz and then cooling it in such a way that it does not recrystallize. It is extremely resistant to thermal shock.

Fig. 9.9
Different types of long-chain molecule: (a) linear, (b) branched chain, (c) cross-linked chains

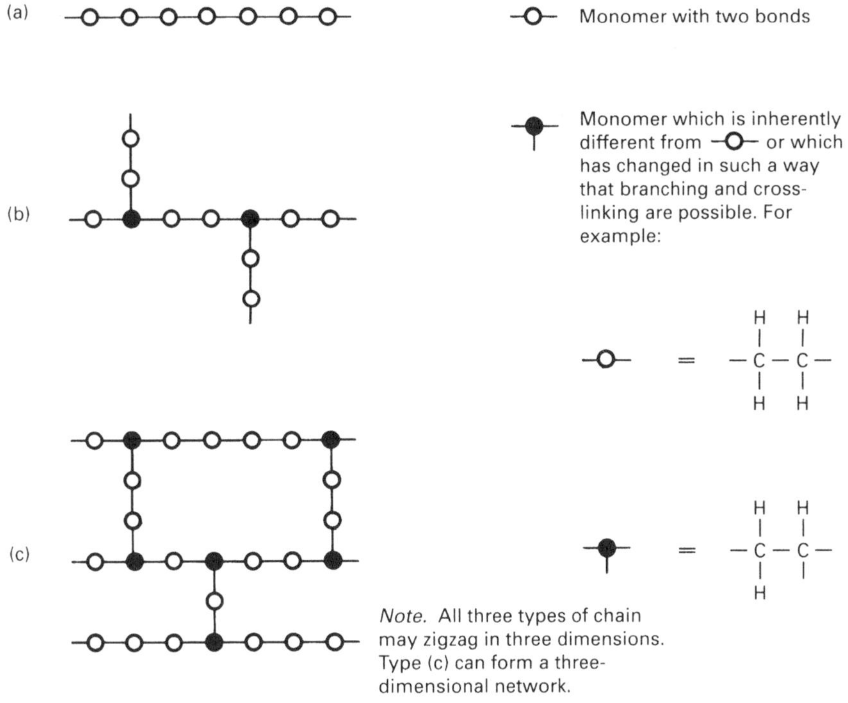

Many polymers may be classified as being either thermoplastics or thermosetting plastics. Some polymers are described as elastomers, some as fibres. These four classifications are discussed in the sections that follow and examples are given in Table 9.1.

Thermoplastics

These soften and become more flexible on heating; they regain their previous rigidity on cooling. They can be moulded while warm and retain their moulded form when they cool. There are usually only weak forces (e.g. van der Waals forces) between the chains. Heating overcomes these, and the chains can then slide past each other so that the material takes up the shape of the mould. Since the bonds are weak, the amount of heat required is not so great that the polymer decomposes. The bonds reform and restore the rigidity on cooling.

Thermosetting Plastics

These are cross-linked polymers and are more brittle and more rigid than the thermoplastics. They do not soften on heating and can withstand higher temperatures than thermoplastics because more energy is needed to break the relatively strong bonds between the chains. If the temperature is increased to the extent that the bonds break, the material decomposes.

Thermosetting plastics are moulded before polymerization is complete. They are then heated to produce further cross-linking, so setting the material, irreversibly, in its moulded form*.

*With some materials, epoxy resins for example, polymerization can be completed at room temperature.

Table 9.1
Some common polymers

	Chain type	*Percentage crystallinity*	*Uses*
Thermoplastics			
Low-density polythene	Branched	60	Polythene bags
High-density polythene	Linear	95	Buckets
PTFE	Linear	90	Non-stick cookware
Polystyrene	Linear*	0	Electrical insulators, packaging
Acrylics	Linear*	0	Headlamp lenses, baths
PVC	Linear*	0	Drainpipes, curtain rails
Fibres			
Nylons (polyamides)	Linear	Various	Can also be used in non-fibre form, e.g. nylon gears.
Terylene (polyester)	Linear	60	Non-fibre forms of polyester are used, e.g. electrical plugs, recording tape, but these forms are not called Terylene.
Elastomers			
Raw rubber (latex)	Linear	0	
Vulcanized rubber	Linear + cross-links	0	Tyres, vibration dampers
Neoprene	Linear + cross-links	0	Vacuum seals, petrol hoses
Thermosetting plastics			
Bakelite	Cross-linked	0	Ashtrays
Ebonite	Cross-linked	0	
Epoxy resins	Cross-linked	0	Adhesives. Also with glass fibre for boat hulls, etc.
Melamine formaldehyde	Cross-linked	0	Light fittings, tableware, kitchen surfaces

*Bulky side groups (e.g. chlorine in the case of PVC) prevent crystallinity.

Elastomers

These are materials which can be stretched considerably and still return to their original lengths when the stresses are removed. They are linear chain polymers with a degree of cross-linking between the chains. The cross-links are few enough to allow the chains to slide past each other during stretching, but numerous enough to pull the chains back into position once the stress is removed. Thus it is the presence of the cross-links which accounts for the elasticity. The ability of elastomers to undergo very large extensions is for quite a different reason. It is because the chains are tangled and have relatively large amounts of open space between them – stretching them simply straightens out the chains.

Raw rubber (latex) is an elastomer, but it has so few cross-links that it may easily be stretched beyond its elastic limit (see section 11.1) and so is of little use. In the process known as **vulcanization**, the rubber is heated together with sulphur; the sulphur atoms bond covalently with carbon atoms to form extra cross-links between the linear chains, thus producing a more useful material. If large amounts of sulphur are added, the number of cross-links becomes so great that it becomes the rigid, brittle material known as ebonite – a thermosetting plastic.

Fibres

These are linear chain polymers in which the chains have been aligned along the length of the fibre, and in which there are reasonably strong bonds between the chains (hydrogen bonding in the case of nylons, dipole–dipole bonding in the case of Terylene). Synthetic fibres are thermoplastic materials; many of them can be used in their non-fibre forms (e.g. nylon). Cellulose is a natural polymeric fibre.

Crystallinity in Polymers

In some polymers there are regions in which the chains are close together and parallel to each other. There is therefore a degree of long-range order in these regions, and they are said to be **crystalline**. At the other extreme are the so-called **amorphous polymers** in which the chains criss-cross in a random way like tangled strands of spaghetti. Linear chain polymers may be either crystalline or amorphous. Crystallinity tends not to occur in polymers with highly branched chains because the chains cannot pack sufficiently closely. Highly cross-linked polymers are completely amorphous.

We have seen that increased rigidity can be produced by increasing the amount of cross-linking between the chains. It can also be produced by creating crystalline regions in the polymer. The rigidity is due to the forces between individual atoms in adjacent chains in the crystalline regions. Although these forces are usually weak (e.g. van der Waals forces), the side-by-side arrangement of the chains means that there are large numbers of these 'bonds', making it difficult for the chains to slide past each other.

The greater the crystallinity, the higher the melting point and the higher the density. The effects of crystallinity are illustrated by the two forms of polythene* (see Table 9.2).

Table 9.2
A typical low-density and a typical high-density polythene compared

	Low-density polythene	*High-density polythene*
Crystallinity	50%	76%
Density	920 kg m^{-3}	960 kg m^{-3}
Melting point	110 °C	135 °C
Tensile strength at yield	12×10^6 Pa	31×10^6 Pa
Chain type	Branched	Linear

(Data kindly supplied by BP Chemicals Limited)

General Properties of Polymers

The main bonds in polymers are covalent; this accounts for their low thermal and electrical conductivities. Polymers are less dense than both metals and ceramics. They are usually resistant to water and acids but may be attacked by organic solvents. Production costs for plastics are much less than those for metals – a polythene bucket is much cheaper than a metal one.

Plastics often have other materials incorporated with them. The purpose of these additives may be to increase flexibility, increase strength, improve weathering properties, provide better insulation characteristics, add colour or simply to

*The two forms are produced by employing different conditions during polymerization.

reduce cost. For example, glass fibres can be added to epoxy and polyester resins to increase strength. The addition of mica to some thermosetting plastics makes them even better electrical insulators. Lead compounds can be added to PVC to prevent it decomposing in strong sunlight.

9.11 CRYSTAL STRUCTURES

Early crystallographers suspected that the external regularity of crystals was due to their atoms being arranged in regular three-dimensional arrays. This was confirmed in 1912 when von Laue and his students, Friedrich and Knipping, showed that crystals could be used to diffract X-rays, and therefore must be acting as three-dimensional diffraction gratings. The diffraction pattern they obtained was too complicated for them to use it to determine the structure of the crystal that had produced it, but one year later, using a simplified version of the technique, W.H. and W.L. Bragg succeeded. X-ray diffraction has since proved to be the single most important method of determining crystal structures.

The atoms (or ions or molecules) in a crystal are arranged in such a way that the total potential energy of the structure is as small as possible, in which case the structure is as stable as possible. The way in which this is achieved depends on the type of crystal concerned. Some examples are discussed below.

Metal Crystals

The valence electrons of metals are free to move throughout the whole of the metal, and therefore metals can be regarded as an array of positive ions in a 'sea' of electrons. For the purpose of this discussion we may consider the ions to be incompressible, equal-sized spheres. There is no directional bonding (as there is, for example, in diamond). In these circumstances the most stable arrangement is that in which the spheres occupy the minimum possible volume. This arrangement is known as **close packing**. The spheres are arranged in layers, where each sphere is surrounded by a hexagonal ring of six others in contact with it (Fig. 9.10). Fig. 9.11 shows the way in which two of these layers must fit together in order to fulfil the requirement of close packing. There are two types of hollows in layer B – those marked by crosses and those marked by dots. There are therefore two ways in which a third layer may be added to the first two.

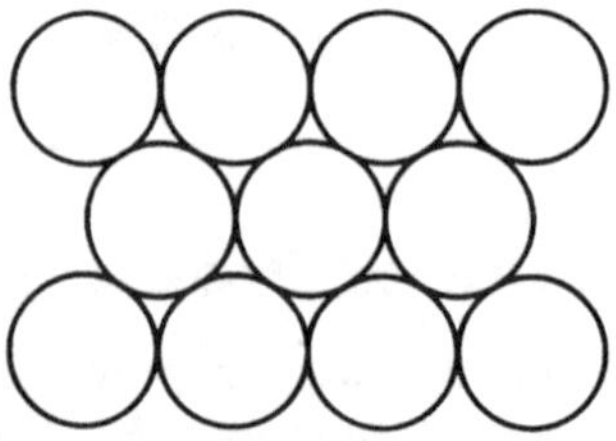

Fig. 9.10
Spheres packing together to occupy the minimum space

If the spheres of the third layer occupy the hollows marked by crosses, they are directly above the spheres in layer A. When this is so, the fourth layer is always a repeat of layer B, and the overall sequence is ABAB, etc. – a structure known as **hexagonal close packing** (Fig. 9.12).

If the spheres of the third layer occupy the hollows marked by dots, the layer is different from both A and B; we shall call it layer C. In this situation the overall sequence is always ABCABC, etc. – a structure known as **cubic close packing**.

Fig. 9.11
Close-packed layers

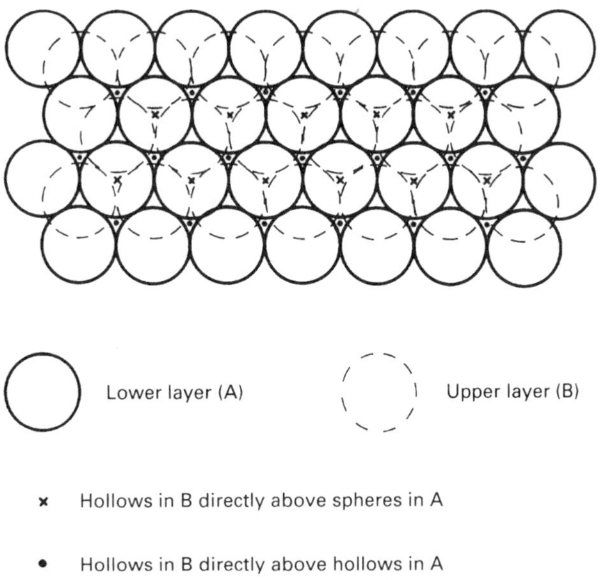

Fig. 9.12
Unit cell of hexagonal close-packed structure

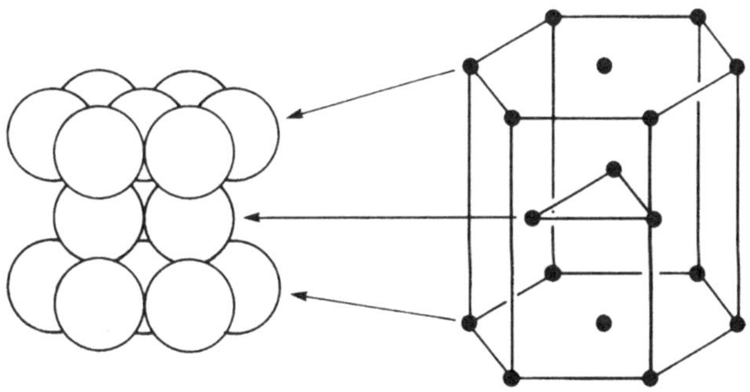

Coordination number = 12
Utilizes 74% of available space
Examples: Mg, Zn

The unit cell contains ions from four layers and is known as a **face-centred** cube (Fig. 9.13) because there is an ion at the centre of each face. Fig. 9.13(a) shows the relationship between the layers and the unit cell.

A more open structure than the two discussed so far is that known as **body-centred cubic** (Fig. 9.14). The least stable metals (e.g. lithium, sodium, potassium) tend to crystallize in this form. Close packing cannot occur because the thermal vibrations of the ions are able to overcome the relatively weak cohesive forces in these metals.

Fig. 9.13
Unit cell of cubic close-packed (face-centred cubic) structure

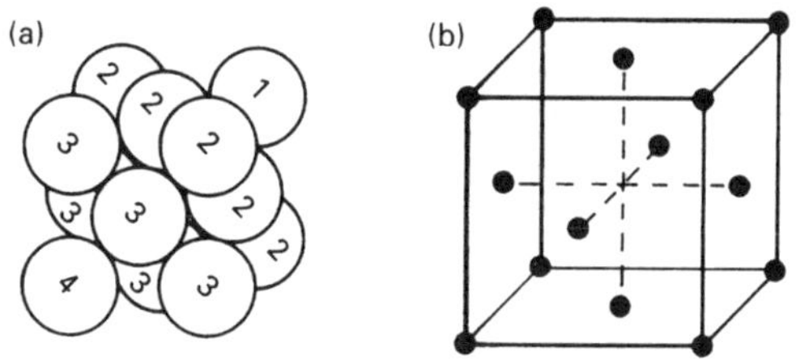

Coordination number = 12
Utilizes 74% of available space
Examples: Cu, Ag, Au, Al;
Fe between 906°C and 1401°C

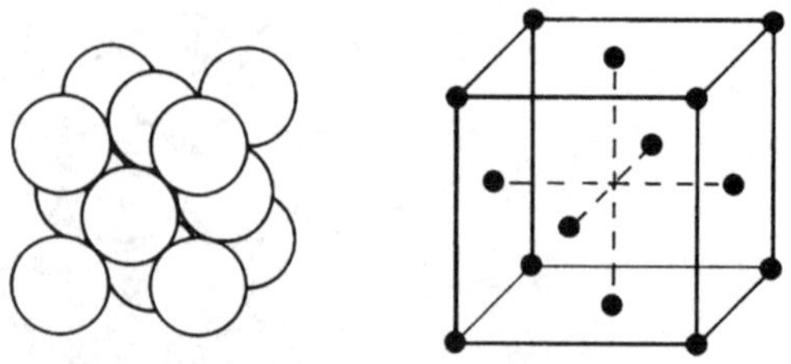

Fig. 9.14
Unit cell of body-centred cubic structure

Coordination number = 8
Utilizes 68% of available space
Examples: Li, Na, K; Fe below 906°C

Ionic Crystals

As with metals, the ions tend to pack closely together because the bonding is non-directional. We may still regard the ions as incompressible spheres, but we need to take account of there being both positively charged and negatively charged ions present. The arrangement shown in Fig. 9.15(a) is more stable than that in Fig. 9.15(b) because in Fig. 9.15(b) there are much stronger repulsive forces between the negative ions. Consideration must also be given to the fact that in practice the two types of ion are normally of different sizes. Fig. 9.16 illustrates the effect of the relative size of the central ion and those around it. Each arrangement makes maximum use of the available space, and each does so without allowing the negative ions to be in contact with each other.

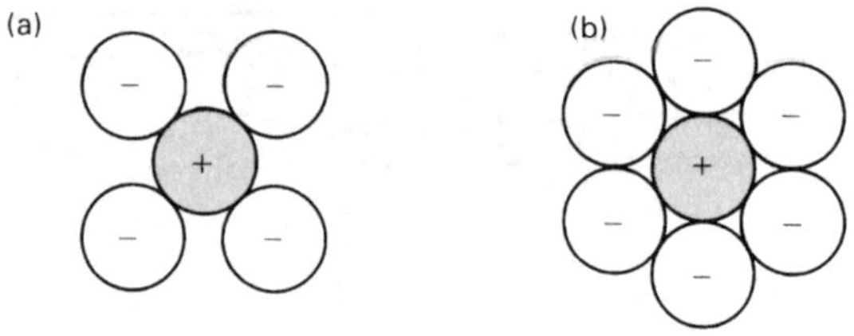

Fig. 9.15
(a) Stable, and
(b) unstable arrangements of negative ions around a positive ion

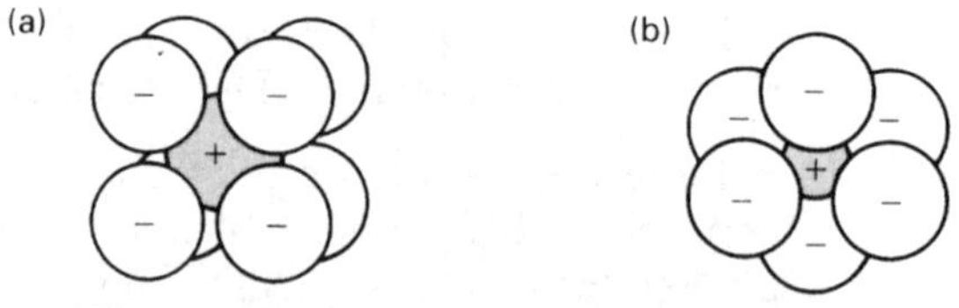

$0.73 < r_+/r_- < 1.00$
Example: CsCl

$0.41 < r_+/r_- < 0.73$
Example: NaCl

Fig. 9.16
The effect of the relative size (r_+/r_-) of the central ion on those around it

Caesium chloride crystallizes in the form of Fig. 9.16(a); the extended structure is shown in Fig. 9.17. There are eight Cl^- ions around each Cs^+ ion, and eight Cs^+ ions around each Cl^- – the so-called 8 : 8 coordination. Though the lattice resembles that shown in Fig. 9.14, it is not known as body-centred cubic because there are two types of ion present.

Sodium chloride crystallizes in the form of Fig. 9.16(b); the extended structure is shown in Fig. 9.18. It can be regarded as two interpenetrating face-centred cubic structures. Each Na^+ ion has six Cl^- ions as its nearest neighbours, and each Cl^- ion has six Na^+ ions as nearest neighbours – 6 : 6 coordination.

There are two crystalline forms of zinc sulphide – zinc blende and wurtzite. The radius ratio of Zn^{2+} to S^{2-} is 0.48, and therefore zinc sulphide would be expected to have the sodium chloride structure. However, both crystalline forms of zinc

Fig. 9.17
Caesium chloride lattice showing 8:8 coordination

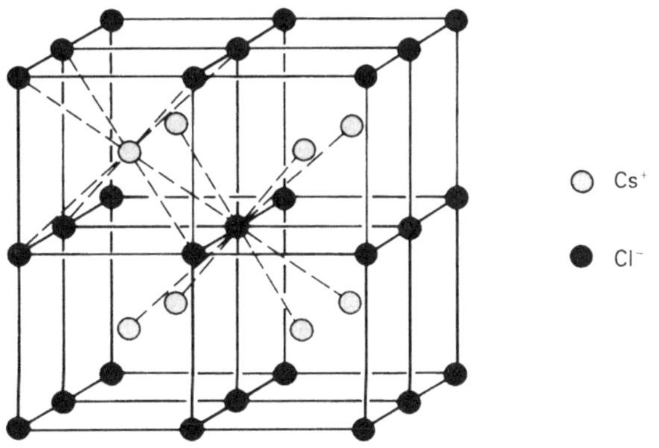

Fig. 9.18
Sodium chloride lattice

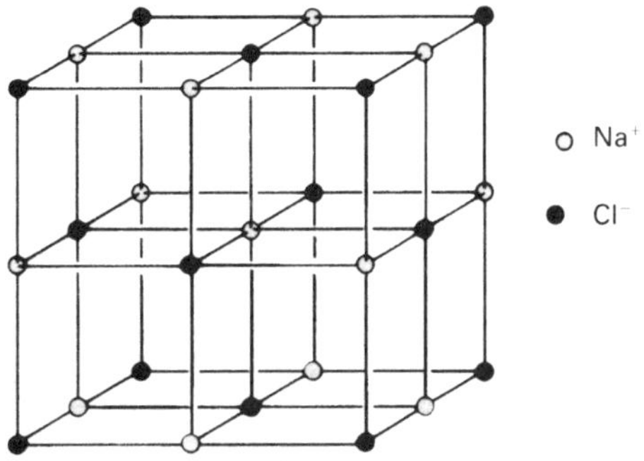

sulphide are such that each Zn^{2+} ion is surrounded tetrahedrally by four S^{2-} ions and vice versa. The reason for this is thought to be that the Zn—S bond has a degree of covalent character which imposes directional constraints on the crystal form adopted.

Diamond Structure

Diamond is one of the two crystalline forms of carbon (see section 9.10). Each carbon atom is covalently bonded to four others. Covalent bonds are highly directional, and we can no longer think in terms of spheres being packed together as closely as possible. The four bonds on each carbon atom point towards the vertices of a regular tetrahedron (Fig. 9.19(a)). Fig. 9.19(b) shows how the tetrahedron can be orientated so that its vertices are at four corners of a cube. The extended structure of diamond is shown in Fig. 9.4(a).

Fig. 9.19
Tetrahedral structure in diamond

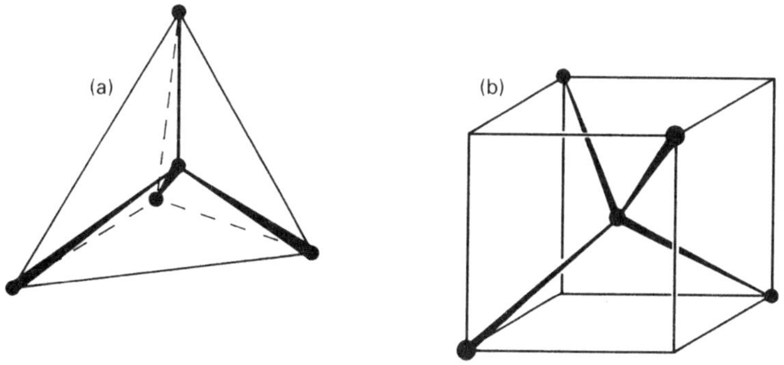

9.12 MEASUREMENT OF LINEAR EXPANSIVITY

Fig. 9.20 shows an apparatus for determining the linear expansivity of a material in the form of a rod about 50 cm long. The length, L, of the rod is measured at room temperature. The rod is then placed in the apparatus with one end against the frame, and the micrometer is screwed up until it is in contact with the other end. The micrometer reading, and the temperature, θ_1, are noted. The micrometer is now screwed back to allow the rod room to expand, and steam is passed through the jacket. When the thermometer reading has stopped increasing, the temperature, θ_2, is recorded. The micrometer is brought back into contact with the rod and the reading is noted. The difference, ΔL, between the two micrometer readings is the amount by which the length of the rod has increased. The linear expansivity, α, is calculated from

$$\alpha = \frac{\Delta L}{L(\theta_2 - \theta_1)}$$

Fig. 9.20
Apparatus for determining linear expansivity

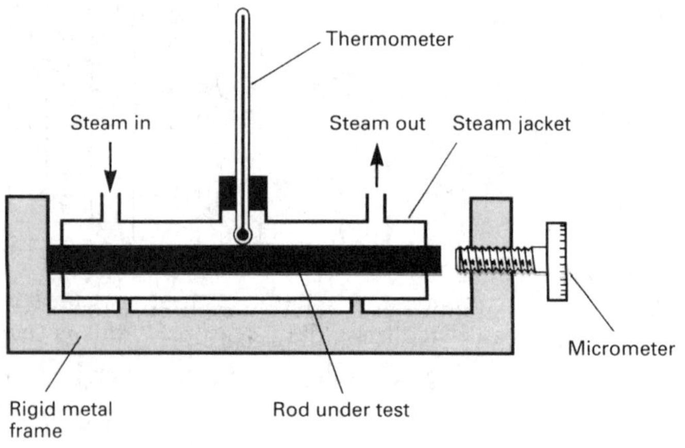

9.13 THE OIL FILM EXPERIMENT

If a small drop of olive oil is placed on the surface of some clean water, the oil spreads to form a large circular film. If it is assumed that the oil spreads until the film is only one molecule thick (i.e. a monomolecular layer), an estimate of the size of an oil molecule can be made by determining the thickness of the film.

It is important that the surface of the water is clean. To this end, water is poured into a large shallow tray until it is overflowing. The surface is then cleaned by drawing two waxed rods across it from the centre outwards. Lycopodium powder is now sprinkled onto the surface so that when the film is formed its edges may be seen easily. A small, spherical drop of oil is obtained on a V-shaped fine wire by dipping it into the oil. The diameter of the drop is measured by holding it in front of a millimetre scale and viewing it through a magnifying glass, or by using a travelling microscope. The drop is then touched onto the water surface. The oil spreads and pushes the lycopodium powder outwards to leave a clear film of oil whose diameter can be measured.

$$\text{Volume of oil drop} = \text{Volume of film}$$

$$\therefore \quad \tfrac{4}{3}\pi\left(\frac{d}{2}\right)^3 = \pi\left(\frac{D}{2}\right)^2 t$$

where

$$d = \text{diameter of oil drop}$$
$$D = \text{diameter of film}$$
$$t = \text{thickness of film}$$

Rearranging gives

$$t = \frac{2}{3}\frac{d^3}{D^2}$$

hence t.

A drop with a diameter of 0.5 mm produces a film with a diameter of about 200 mm, and gives a value for t of approximately 2 nm. Oil molecules are long and thin, and 'stand on end' on water. It follows that this figure of 2 nm represents the length of the molecule.

CONSOLIDATION

Kinetic energy of molecules depends on temperature.

Potential energy of each of two molecules is taken to be zero at infinite separation because they can have no influence on each other at infinite separation. The PE is negative at the equilibrium separation because work has to be done (and therefore the PE has to be increased) to separate the molecules to infinity.

Intermolecular PE and Force Curves

Minimum on PE curve corresponds to zero on force curve.

Point of inflexion on PE curve corresponds to minimum on force curve.

Force = – (gradient of PE curve)

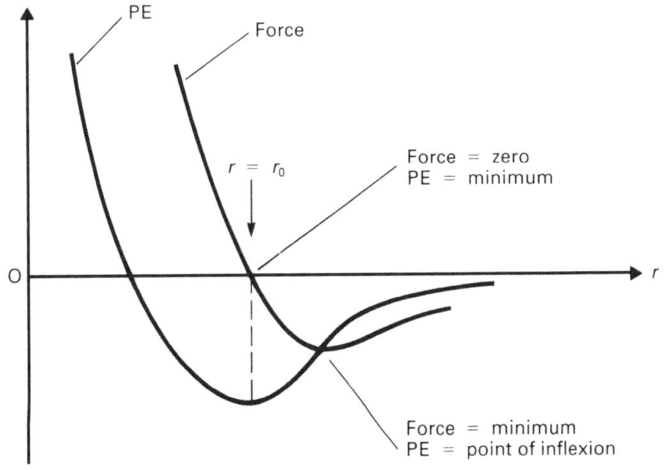

Solids

Fixed volume, fixed shape.

Molecules vibrate about fixed positions.

$$L_1 = L_0 + \alpha L_0 (\theta_1 - \theta_0)$$

Liquids

Fixed volume, no fixed shape.

Molecules vibrate about non-fixed positions.

10

FLUIDS AT REST

10.1 INTRODUCTION

This chapter is concerned with fluids. A fluid is a substance that can flow; it follows that **both liquids and gases are fluids**.

An important concept in connection with fluids is that of pressure. The pressure in a fluid depends on its density, and we shall begin the chapter by discussing density.

10.2 DENSITY

The density of a substance is defined by

$$\rho = \frac{m}{V} \qquad [10.1]$$

where

ρ = density of substance ($kg\,m^{-3}$)

m = mass of substance (kg)

V = volume of substance (m^3)

The relative density of a substance is defined by

$$\text{Relative density} = \frac{\text{Density of substance}}{\text{Density of water (at 4°C)}} \qquad [10.2]$$

Relative density has no units.

The specific volume of a substance is the reciprocal of its density, i.e. it is the volume of unit mass of the substance. Unit = $m^3\,kg^{-1}$. (Note the use of the word 'specific' to denote unit mass, as it does in specific heat capacity, etc.)

Methods of determining densities by experiment are summarized in section 10.14.

10.3 PRESSURE

The pressure acting on a surface is defined as the force per unit area acting at right angles to the surface, i.e.

$$p = \frac{F}{A} \qquad [10.3]$$

where

p = pressure on surface (SI unit = **the pascal** (Pa). 1 Pa = 1 N m^{-2}.)

F = the force acting at right angles to the surface (N)

A = the area over which the force is acting (m^2).

Note The SI unit of pressure is the pascal. Other units in common use are the atmosphere (atm), the millimetre of mercury (mmHg) and the bar. None of these is an SI unit. Standard atmospheric pressure is 1.01×10^5 Pa (3 sig. fig.) and in these various other units it is 1 atm (exactly), 760 mmHg and 1.01 bar.

10.4 PRESSURE IN FLUIDS

(a) The pressure in a fluid increases with depth. All points at the same depth in the fluid are at the same pressure.

(b) Any surface in a fluid experiences a force due to the pressure of the fluid.

(i) The force is perpendicular to the surface no matter what the orientation of the surface.

(ii) The magnitude of the force is independent of the orientation of the surface.

This final statement is illustrated in Fig. 10.1 and is often stated as **'pressure acts equally in all directions'**.

Fig. 10.1
'Pressure acts equally in all directions'

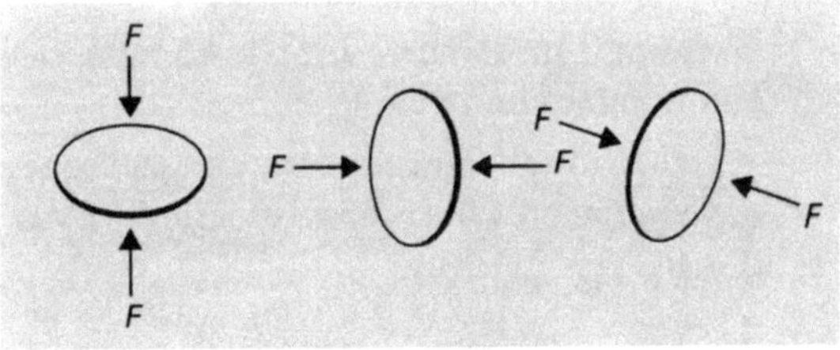

$F = p\delta A$ in each case where p = pressure in fluid and δA = area of surface. (The surfaces are assumed to be small so that variation of pressure with depth may be ignored.)

Note Though the force associated with the pressure at a point is a vector quantity, the pressure itself is a scalar, i.e. pressure has no direction. (The statement that pressure acts equally in all directions can be misleading in this respect!) Consider the pressure at a point in a fluid.

We cannot assign a direction to the pressure – all we can do is assign a direction to the force that the pressure creates on some surface placed in the fluid, and this depends on the orientation of the surface.

Fig. 10.2
To calculate pressure as a function of depth

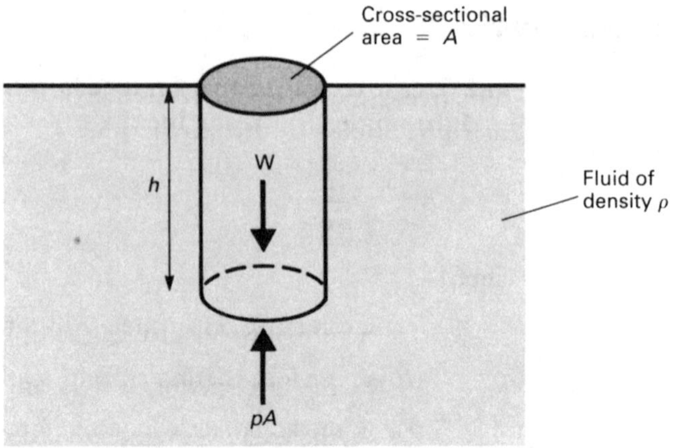

Pressure Variation with Depth

Consider a cylindrical region of cross-sectional area A and height h in a fluid of density ρ (Fig. 10.2). The top of the cylinder is at the surface of the fluid, and the (vertical) forces acting on it are its weight, W, and an upward directed force of pA due to the pressure, p, at the bottom of the cylinder. The cylinder is in equilibrium and therefore

$$\begin{aligned} pA &= W \\ &= \text{mass of cylinder} \times g \\ &= \text{volume of cylinder} \times \rho g \\ &= hA\rho g \end{aligned}$$

$$\therefore \quad p = h\rho g \qquad [10.4]$$

where p is the pressure due to the fluid at a depth h below the surface.

Notes

(i) Equation [10.4] is not valid in the case of gases when h is large. The density of a gas decreases with height, and the equation has been derived on the assumption that the density is constant. The equation is a reasonable approximation when h is small. However, the densities of gases are low, and when h is small the pressure variation with depth is also small and is usually ignored.

(ii) We have derived equation [10.4] by considering a cylindrical region within the fluid. The same result would have been obtained whatever shape the region had been taken to be.

(iii) A little thought should convince the reader that the difference in pressure, Δp, between two points separated by a vertical distance h in a fluid of density ρ is given by

$$\Delta p = h\rho g$$

10.5 WHY THE SURFACE OF A LIQUID IS HORIZONTAL

Consider two points, M and N, on the same horizontal level in a stationary liquid. Consider also a cylindrical region of cross-sectional area A whose end faces are centred on M and N (Fig. 10.3).

Fig. 10.3
To show that all points on the same level are at the same pressure

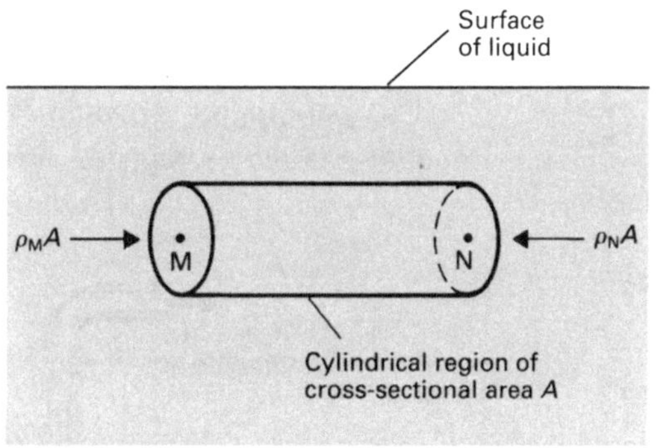

The horizontal forces acting on the cylinder are $p_M A$ and $p_N A$ as shown, where p_M and p_N are the fluid pressures at M and N respectively. The cylinder is in equilibrium and therefore $p_M A = p_N A$. It follows that $p_M = p_N$ and therefore that **all points on the same horizontal level are at the same pressure**.

All points in the liquid surface must be at atmospheric pressure, p_A, and it follows from equation [10.4], therefore, that they must all be at the same height above MN. Since MN is horizontal, the surface must also be horizontal.

10.6 THE U-TUBE MANOMETER

This consists of a U-shaped tube containing a liquid. It is used to measure pressure. The pressure to be measured (that of a gas, say) is applied to one arm of the manometer; the other arm is open to the atmosphere (Fig. 10.4).

Fig. 10.4
The (open) U-tube manometer

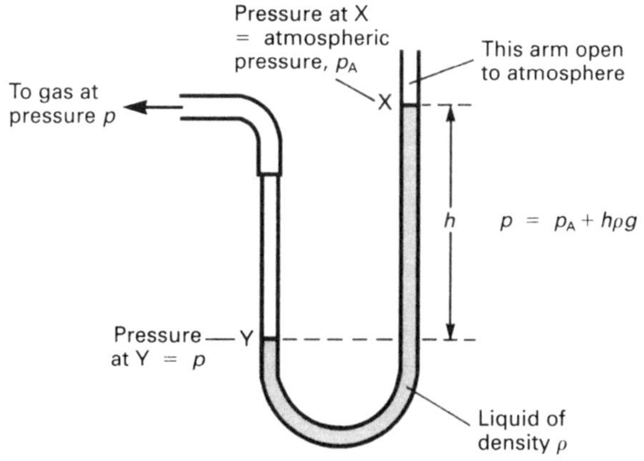

The liquid surface at Y is a vertical distance h below that at X. Therefore, by equation [10.4],

$$p = p_A + h\rho g$$

hence p.

Notes

(i) Manometers can be used to measure pressures both above and below atmospheric pressure.

(ii) Mercury is used as the manometer liquid unless the pressure being measured is close to atmospheric pressure, in which case a liquid of lower density (e.g. oil or water) is more suitable.

(iii) The pressure registered by the manometer, $h\rho g$, is known as the **gauge** pressure. The actual pressure, $p_A + h\rho g$, is called the **absolute** pressure. The manometer shown in Fig. 10.5, in which the arm on the right is closed and evacuated, registers absolute pressure directly.

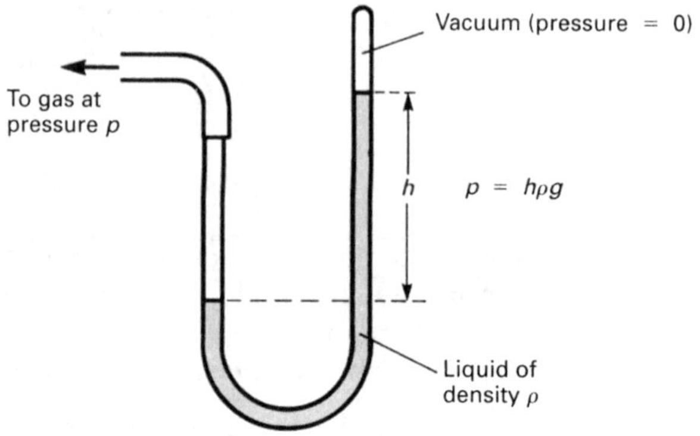

Fig. 10.5 U-tube manometer for absolute pressure

EXAMPLE 10.1

Refer to Fig. 10.6. Calculate the pressure of the gas in the bulb. (Atmospheric pressure = 1.01×10^5 Pa, density of mercury = 1.36×10^4 kg m^{-3}, $g = 9.81$ m s^{-2}.)

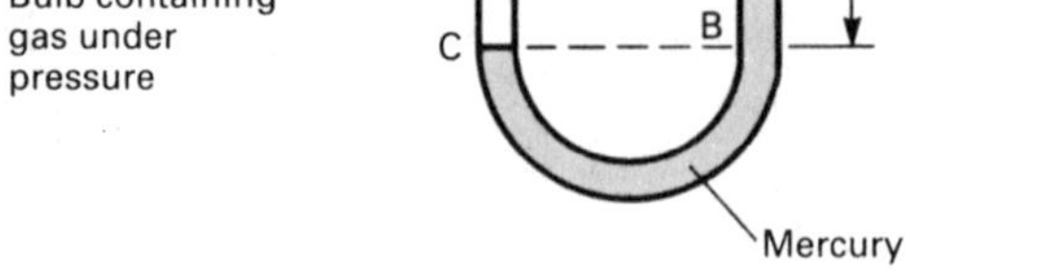

Fig. 10.6 Diagram for Example 10.1

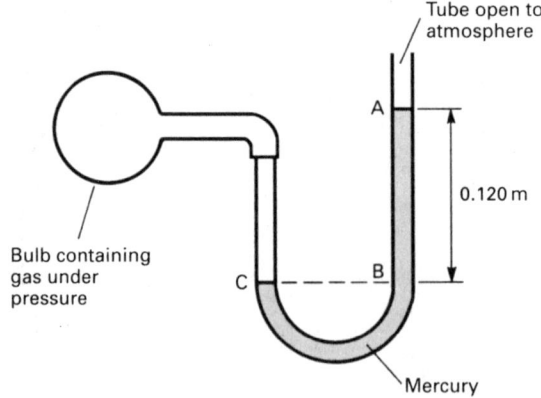

Solution

$$\text{Pressure at A} = \text{atmospheric pressure} = 1.01 \times 10^5 \text{ Pa}$$

Since pressure increases with depth,

$$\begin{aligned}\text{Pressure at B} &= 1.01 \times 10^5 + 0.120 \times 1.36 \times 10^4 \times 9.81 \\ &= 1.01 \times 10^5 + 0.16 \times 10^5 \\ &= 1.17 \times 10^5\end{aligned}$$

Since C is at the same level as B,

$$\text{Pressure at C} = 1.17 \times 10^5$$

i.e. $\text{Pressure of gas} = 1.17 \times 10^5 \text{ Pa}$

EXAMPLE 10.2

Refer to Fig. 10.7. Calculate the pressure of the gas in the bulb. (Atmospheric pressure = 760 mmHg.)

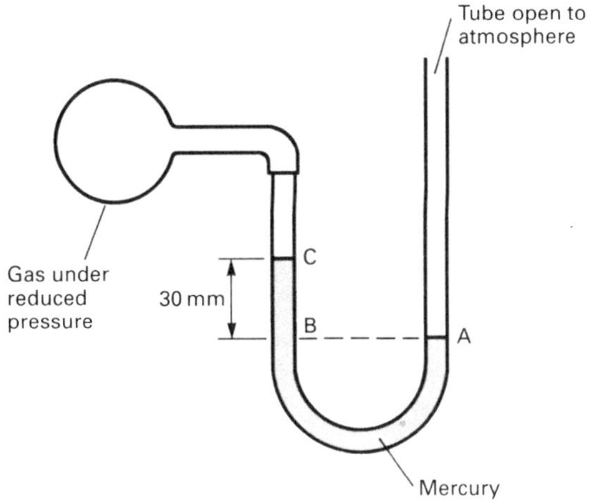

Fig. 10.7 Diagram for Example 10.2

Solution

	Pressure at A = 760 mmHg
∴	Pressure at B = 760 mmHg
∴	Pressure at C = 760 − 30 = 730 mmHg
i.e.	Pressure of gas = 730 mmHg

QUESTIONS 10A

Where necessary and unless otherwise stated use the following data: atmospheric pressure = 760 mmHg, density of mercury = 1.36×10^4 kg m^{-3}, $g = 9.81$ m s^{-2}.

1. An open U-tube manometer containing an oil of density 897 kg m^{-3} is used to measure the pressure of a gas. The oil level in the open tube is 25.0 cm higher than that in the limb connected to the gas. Calculate **(a)** the gauge pressure, **(b)** the absolute pressure of the gas. (Atmospheric pressure = 9.98×10^4 Pa.)

2. What is the atmospheric pressure (in pascals) on a day when a mercury barometer is reading 772 mmHg?

3. A beaker of cross-sectional area 60 cm^2 contains 600 cm^3 of mercury. Find the pressure on the inner surface of the base of the beaker.

4. The pressure on the upper surface of a submerged submarine is 1.20×10^6 Pa; the pressure on the base of the hull is 1.40×10^6 Pa. Calculate the height of the submarine. (Density of seawater = 1.04×10^3 kg m^{-3}.)

5. Find the pressure of the enclosed gas in each of the following situations.

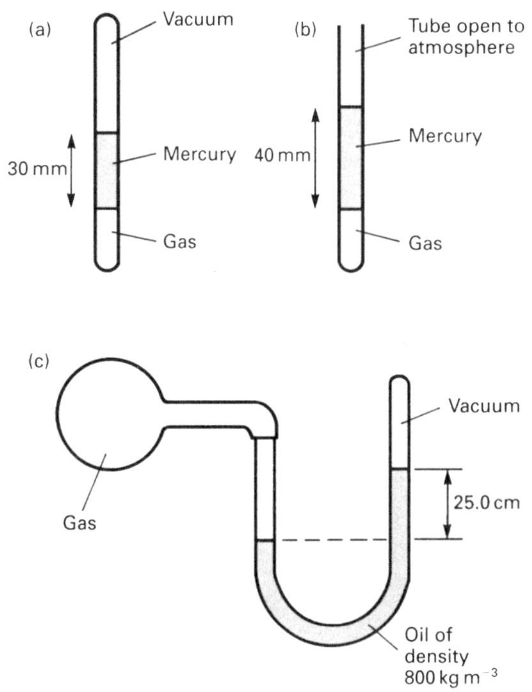

10.7 THE BOURDON GAUGE

The Bourdon gauge (Fig. 10.8) has a curved metal tube which is closed at one end and of elliptical cross-section. The closed end is linked to a pointer. If the pressure in the tube increases, the tube straightens slightly and moves the pointer over the scale.

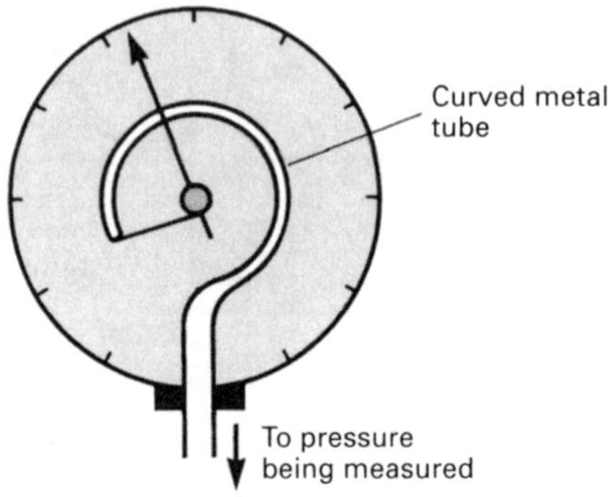

Fig. 10.8
The Bourdon gauge

The gauge detects the difference in pressure between the inside and outside of the curved tube, i.e. it detects the difference between the pressure being measured and the atmospheric pressure prevailing at the time. In this sense, then, it behaves like the open tube manometer shown in Fig. 10.4.

The scale is calibrated in some suitable unit of pressure (e.g. Pa, mmHg, atm). Some gauges are calibrated in such a way that when they are open to the atmosphere the scale reading is zero, i.e. the pointer 'starts' from zero. The absolute (actual) pressure is obtained by adding the value of the prevailing atmospheric pressure to the scale reading. Others have the pointer offset so that it 'starts' at a reading of one atmosphere (or its equivalent, e.g. 760 mmHg or 1.0×10^5 Pa) and gives a reading which is (approximately) equal to the actual value of the pressure being measured. The reading is approximate because the gauge cannot take account of variations in atmospheric pressure.

Bourdon gauges with an extensive variety of pressure ranges are available, and they can be used to measure pressures below atmospheric pressure as well as above it. Some gauges are used to measure (actual) pressures of as little as one millimetre of mercury, whilst others have ranges extending up to a few thousand atmospheres.

10.8 BALANCING COLUMNS

Fig. 10.9 shows a U-tube containing two immiscible liquids (i.e. liquids that do not mix with each other, such as paraffin and water).

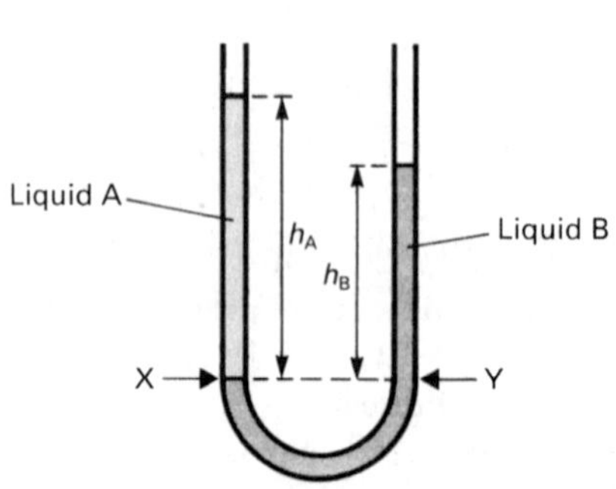

Fig. 10.9
Balancing columns

The pressure, p_X, at X is equal to atmospheric pressure, p, plus the pressure exerted by the head, h_A, of liquid A, i.e.

$$p_X = p + h_A \rho_A g$$

where ρ_A is the density of liquid A. Similarly, the pressure, p_Y, at Y is given by

$$p_Y = p + h_B \rho_B g$$

where ρ_B is the density of liquid B. Since X and Y are at the same level in liquid B, $p_X = p_Y$, and therefore

$$p + h_A \rho_A g = p + h_B \rho_B g$$

$$\therefore \quad h_A \rho_A g = h_B \rho_B g$$

i.e. $$\frac{\rho_A}{\rho_B} = \frac{h_B}{h_A}$$

The ratio of the densities of the liquids can therefore be found by measuring h_A and h_B. If liquid B is water, h_B/h_A is the relative density of liquid A.

10.9 THE HYDRAULIC JACK. PASCAL'S PRINCIPLE

Fig. 10.10 illustrates the operating principle of a hydraulic jack in which a downward directed force, F_X, is being used to balance a much larger force, F_Y. The pressure in the liquid at both X and Y is p. The liquid therefore exerts upward directed forces of pA_X and pA_Y on the pistons at X and Y respectively.

Fig. 10.10
The hydraulic jack

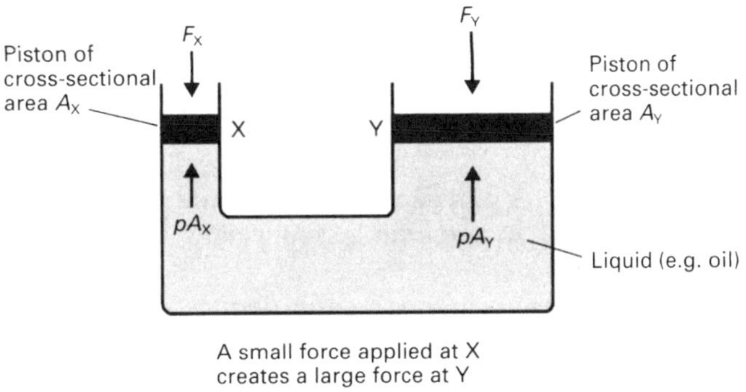

It follows that

$$F_X = pA_X \quad \text{and} \quad F_Y = pA_Y$$

Eliminating p between these equations gives

$$\frac{F_X}{F_Y} = \frac{A_X}{A_Y}$$

Thus, the ratio of the forces is equal to the ratio of the areas of the respective pistons. Bearing in mind that the areas of the pistons are proportional to the squares of their diameters, it follows that if the diameter of the piston at Y is ten times that of X, then a moderate effort of 100 N, say, at X could move a load of 10 000 N at Y. However, (assuming that the liquid is total incompressible) the effort has to move one hundred times further than the load is moved. Hydraulic braking systems and hydraulic presses work in a similar fashion.

Suppose that the pressure at X and Y in the absence of the forces F_X and F_Y is p_0. When the forces are applied the pressure at both X and Y increases by $(p - p_0)$. In fact, the pressure at every point of the liquid increases by $(p - p_0)$. This is an illustration of **Pascal's principle**, which can be stated as:

Any pressure applied to an enclosed fluid is transmitted undiminished to every part of the fluid and to the walls of its container regardless of its shape.

Notes

(i) Although Pascal's principle applies to both liquids and gases, a gas cannot be used as the working fluid in a hydraulic jack because gases are compressible. Most of the effort would go into compressing the gas rather than into moving the load.

(ii) Pascal's principle illustrates an important difference between fluids and solids, namely that **a fluid transmits pressure (unchanged), whereas a solid transmits force (unchanged)**. Consider Fig. 10.11. If a force, F, is applied to the (smaller) left-hand face, X, of the solid, the solid (assuming that it does not move) exerts the same force on anything in contact with its right-hand face, Y, even though the faces are not the same size. Thus, when the force at X increases, the force at Y increases by the same amount. However, the increase in pressure at X is greater than that at Y.

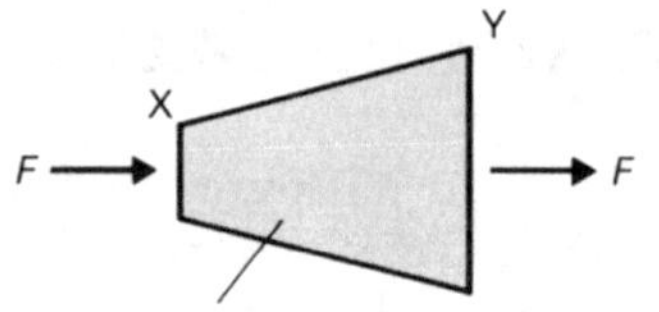

Fig. 10.11
Solids transmit force

10.10 ARCHIMEDES' PRINCIPLE

A body immersed in a fluid (totally or partially) experiences an upthrust (i.e. an apparent loss of weight) which is equal to the weight of fluid displaced.

The principle is easily deduced. Consider a cylinder of height h and cross-sectional area A a distance h_0 below the surface of a fluid of density ρ (Fig. 10.12).

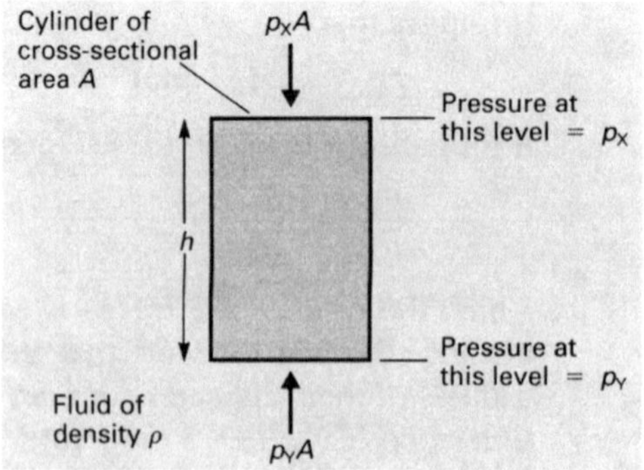

Fig. 10.12
To deduce Archimedes' principle

$$\begin{aligned} \text{Volume of fluid displaced} &= \text{Volume of cylinder} \\ &= Ah \\ \therefore \quad \text{Mass of fluid displaced} &= Ah\rho \\ \therefore \quad \text{Weight of fluid displaced} &= Ah\rho g \end{aligned} \qquad [10.5]$$

The fluid exerts forces of $p_X A$ and $p_Y A$ on the top and bottom faces of the cylinder. The upthrust (i.e. the resultant upward force due to the fluid) is therefore given by

$$\begin{aligned}\text{Upthrust} &= p_Y A - p_X A \\ &= (h + h_0)\rho g A - h_0 \rho g A \qquad \text{(by equation [10.4])} \\ &= h\rho g A\end{aligned}$$

Therefore, by equation [10.5],

$$\text{Upthrust} = \text{Weight of fluid displaced}$$

which is Archimedes' principle. (Note, the result clearly does not depend on the fact that we have considered a cylinder.) Archimedes' principle can be verified by experiment (see section 10.12).

If a body is more dense than the fluid in which it is immersed, then its weight is greater than the weight of the fluid it displaces. By Archimedes' principle, therefore, its weight is greater than the upthrust, and it falls down through the fluid unless it is supported in some way. A body which is less dense than the fluid around it, on the other hand, experiences a net upward force and rises up through the fluid.

When a body floats the upthrust on it must be equal to its weight for it moves neither up nor down. It follows from Archimedes' principle, therefore, that the weight of the body is equal to the weight of the fluid displaced. This is known as the **principle of flotation**. As we have just seen, it is a special case of Archimedes' principle. It can be stated as:

A floating body displaces its own weight of fluid.

The principle of flotation, like Archimedes' principle itself, applies to both partially immersed bodies (e.g. ships) and totally immersed bodies (e.g. submarines and airships).

EXAMPLE 10.3

An object is weighed with a spring balance, first in air and then whilst totally immersed in water. The readings on the balance are 0.48 N and 0.36 N respectively. Calculate the density of the object. (Density of water = $1.0 \times 10^3\,\text{kg}\,\text{m}^{-3}$.)

Solution

The object has the same volume as the water it displaces and therefore

$$\frac{\text{Density of object}}{\text{Density of water}} = \frac{\text{Mass of object}}{\text{Mass of water displaced}}$$

$$= \frac{\text{Weight of object}}{\text{Weight of water displaced}}$$

By Archimedes' principle, weight of water displaced = upthrust, and therefore

$$\frac{\text{Density of object}}{\text{Density of water}} = \frac{\text{Weight of object}}{\text{Upthrust in water}} \qquad [10.6]$$

$$\therefore \quad \text{Density of object} = \frac{0.48}{0.12} \times 1.0 \times 10^3$$
$$= 4.0 \times 10^3\,\text{kg}\,\text{m}^{-3}$$

10.11 MEASUREMENT OF DENSITY USING ARCHIMEDES' PRINCIPLE

Solids

Weigh the solid in air and in water, and then use equation [10.6].

Liquids

The density of a liquid can be found by determining the upthrust on some suitable object when it is immersed in the liquid and then when it is immersed in water.

By analogy with equation [10.6]

$$\frac{\text{Density of object}}{\text{Density of liquid}} = \frac{\text{Weight of object}}{\text{Upthrust in liquid}} \qquad [10.7]$$

Dividing equation [10.6] by equation [10.7] gives

$$\frac{\text{Density of liquid}}{\text{Density of water}} = \frac{\text{Upthrust in liquid}}{\text{Upthrust in water}}$$

from which the density of the liquid can be found.

10.12 VERIFICATION OF ARCHIMEDES' PRINCIPLE BY EXPERIMENT

Suspend a glass stopper from a spring balance to obtain the weight of the stopper in air. Gently lower the stopper into a displacement can filled to the spout with water. The difference between the two spring balance readings is the upthrust on the stopper. Collect the water that runs out of the can in a previously weighed beaker. Weigh the beaker with the water in it to find the weight of the water displaced by the stopper. If the weight of the water is equal to the upthrust, Archimedes' principle has been verified.

10.13 THE HYDROMETER

The hydrometer provides a quick method of measuring the relative densities of liquids.

In accordance with the principle of flotation, whenever the hydrometer floats in a liquid the weight of the liquid it displaces is equal to its own weight. It follows that it sinks further into water, say, than it does into a liquid of higher density (Fig. 10.13).

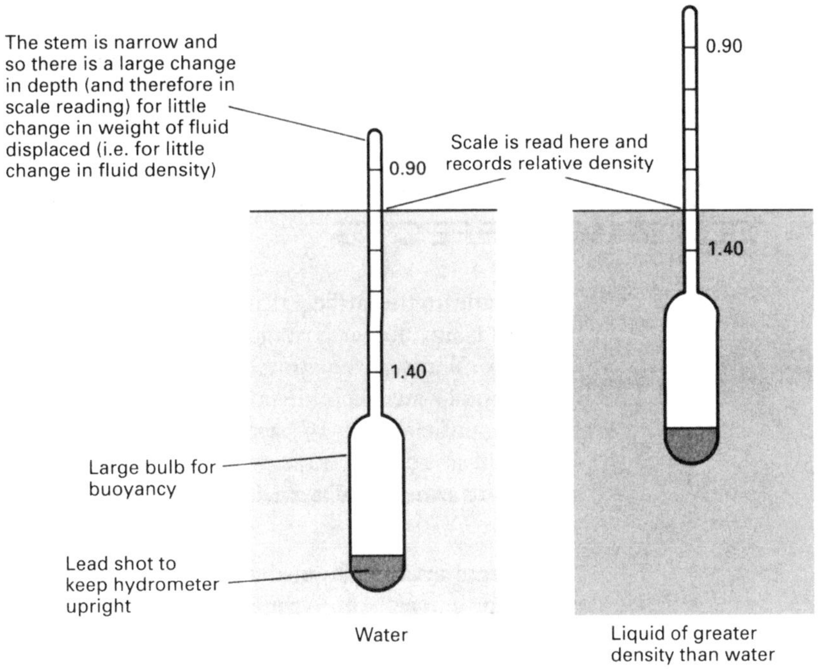

Fig. 10.13
The hydrometer

10.14 SUMMARY OF METHODS OF DETERMINING DENSITIES AND RELATIVE DENSITIES

(i) Measure the mass (m) and volume (V) and then use $\rho = m/V$ to obtain the density.

The mass should be found by using a beam balance or top-pan balance rather than a spring balance – a spring balance measures weight. The volume of a solid may be found by measuring its dimensions or by a displacement method. The volume of a liquid may be found by using a measuring cylinder, pipette or burette. The volume of a gas may be found by enclosing it in a container of known volume.

(ii) The method of 'balancing columns' can be used for liquids. (See section 10.8.)

(iii) The hydrometer can be used for liquids. (See section 10.13.)

(iv) The method based on Archimedes' principle can be used for liquids and solids. (See section 10.11.)

(v) Relative density bottle can be used for liquids and fine powders. (See GCSE texts.)

10.15 SURFACE TENSION

A steel needle can be caused to float on water even though steel is more dense than water. A liquid spilled on to a surface that it does not wet tends to form into small drops, rather than spread into a continuous film. These are two examples of phenomena which suggest that the surface of a liquid behaves like an elastic skin in a state of tension. This is indeed the case and can be understood by a consideration of the effects of intermolecular forces.

10.16 MOLECULAR EXPLANATION OF SURFACE TENSION EFFECTS

A molecule in the surface of a liquid is subject to intermolecular forces from below, but not from above (providing the effects of the molecules of the vapour are ignored). Thus, if the coordination number of the molecules of the interior is n, then that of a surface molecule will be $n/2$. Therefore, if a molecule of the interior has a potential energy of (say) -0.4 eV, then a surface molecule, being involved in only half as many bonds, will have a potential energy of -0.2 eV. Thus, **the potential energy of a molecule in the surface exceeds that of one in the interior**.

All systems arrange themselves in such a way that they have the minimum possible potential energy. In order that the potential energy associated with the intermolecular forces (the surface tension forces) can be a minimum, the number of molecules which reside in the surface has to be a minimum. Therefore:

(i) liquids have the smallest possible surface area, and

(ii) the average separation of the molecules in the surface of a liquid is greater than that of molecules in the interior.

The requirement that the surface area is a minimum means that a liquid subject to surface tension forces only, will assume the shape of a sphere. (This is because a sphere is the shape which allows a given volume of material to have the smallest possible surface area.) Liquids are normally subject to gravitational forces in addition to surface tension forces, in which case the adopted shape is that which minimizes the total potential energy. Small drops of liquid are nearly spherical, and become more so as the drop decreases in size. This is because the ratio of the surface area (which is proportional to r^2) to the weight (which is proportional to r^3) and therefore of surface tension force to gravitational force, increases as r decreases. Soap bubbles are almost perfect spheres because they have large surface areas and negligible masses. The effect of gravity can be eliminated by using two immiscible liquids of the same density (Fig. 10.14). The phenylamine (aniline) and water are at such a temperature that their densities are equal, in which case the upthrust on each phenylamine globule is exactly equal to its weight, and therefore the globule is not subject to any net gravitational force. Drops of liquid which are falling freely under gravity are also spherical. This is because every part of the drop is being accelerated to the same extent and the acceleration cannot, therefore, affect the shape of the drop.

Fig. 10.14
Eliminating the effects of gravity

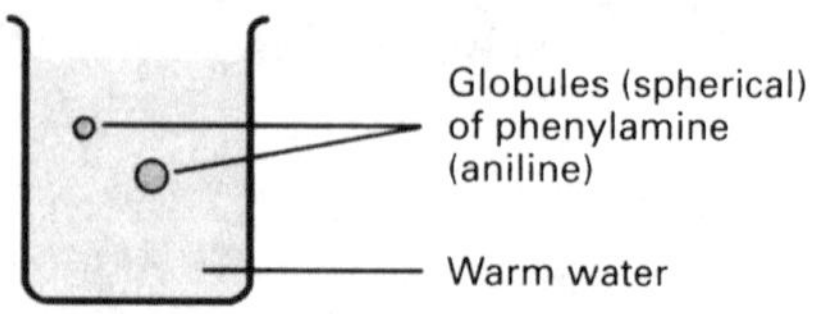

The molecules in the interior of a liquid are, of course, at their equilibrium separation, and therefore the attractive forces of their neighbours are balanced by the repulsive forces. This is not true of the surface molecules, the separation of these is greater than the equilibrium separation (requirement (ii)), and therefore they exert a net attractive force on each other. Thus, at any point in the surface of a liquid there is a net force away from that point due to the attractions of the molecules around it. **The surface therefore behaves like an elastic skin in a state of tension.**

10.17 SURFACE TENSION AND FREE SURFACE ENERGY

The surface tension γ of a liquid is defined as the force per unit length acting in the surface and perpendicular to one side of an imaginary line drawn in the surface. (Unit = $N\,m^{-1}$.)

Free surface energy σ is defined as the work done in isothermally creating unit area of new surface. (Unit = $J\,m^{-2} = N\,m^{-1}$.)

Whenever the surface area of a given volume of liquid is increased, work has to be done against the surface tension forces. Alternatively, one may think of the work being necessary to provide the extra energy needed to have an increased number of molecules in the surface.

Consider stretching a thin film of liquid on a horizontal frame (Fig. 10.15). Since the film has both an upper and lower surface, the force F on AB due to surface tension is given by

$$F = 2L\gamma$$

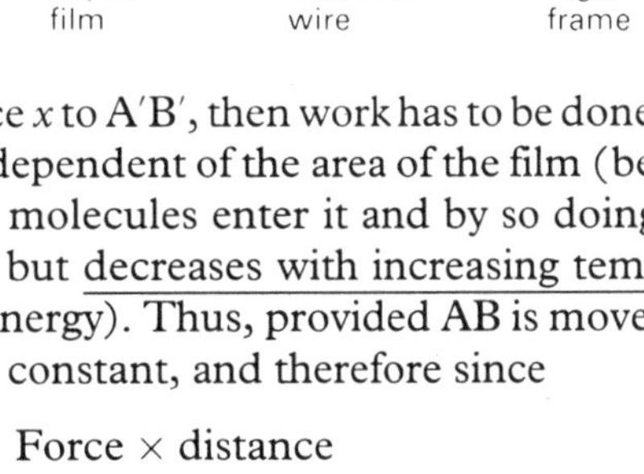

Fig. 10.15
A thin film of liquid being stretched

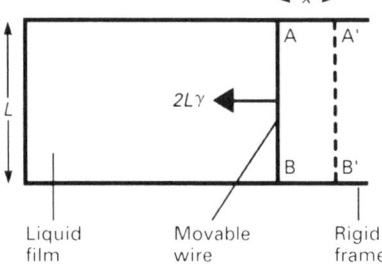

If AB is moved a distance x to A′B′, then work has to be done against this force. The surface tension, γ, is independent of the area of the film (because as the size of the surface increases more molecules enter it and by so doing maintain the average molecular separation), but decreases with increasing temperature (because this decreases the binding energy). Thus, provided AB is moved isothermally to A′B′, the force on AB will be constant, and therefore since

$$\text{Work done} = \text{Force} \times \text{distance}$$

$$\text{Work done} = 2L\gamma x$$

The increase in surface area is $2Lx$ (upper and lower surfaces), and therefore the work done per unit area increase (the free surface energy σ) is given by

$$\sigma = \frac{2L\gamma x}{2Lx}$$

i.e. $$\sigma = \gamma$$

Thus, the free surface energy σ is equal to the surface tension γ. This provides a second definition of γ:

The surface tension γ is the work done in isothermally increasing the surface area of the liquid by unit area. (Unit $= \mathrm{J\,m^{-2}} = \mathrm{N\,m^{-1}}$.)

10.18 SOME SURFACE TENSION PHENOMENA

Floating Needle

The needle (Fig. 10.16) creates a depression in the liquid surface so that the surface tension forces F (which act in the surface) now have an upward directed component which is capable of supporting the weight of the neєdle.

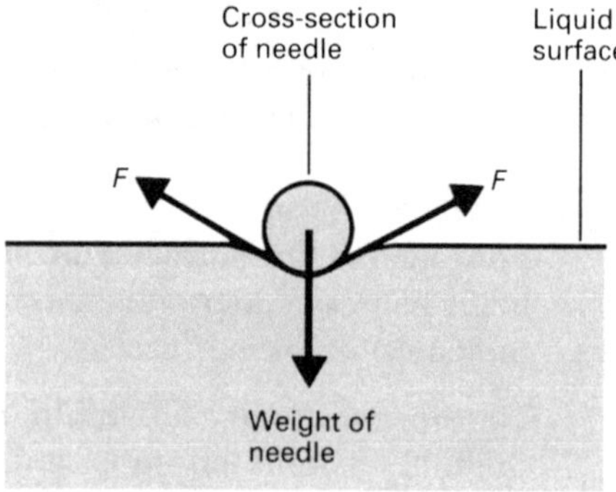

Fig. 10.16
A needle supported by surface tension forces

Thread on a Soap Film

In Fig. 10.17(a) there are equal and opposite forces on each side of the thread and therefore it stays where it has been placed. If the film is broken in the region bounded by the thread (Fig. 10.17(b)), there are forces on the outside of the thread only. The thread is therefore pulled into a circle (the shape with the maximum area for a given perimeter) and therefore the liquid film has the minimum possible area.

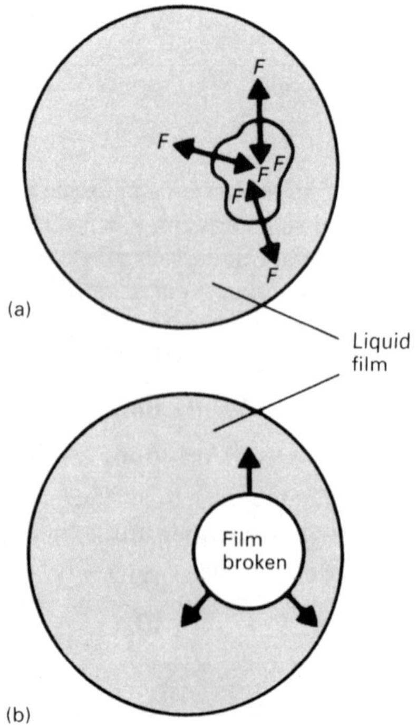

Fig. 10.17
To show that a liquid attains the minimum possible surface area

Camphor Boat

The camphor (Fig. 10.18) sublimes and interacts with the water at the back of the boat, reducing the surface tension there, so that F' is less than F. There is therefore a net forward force which drags the boat through the water.

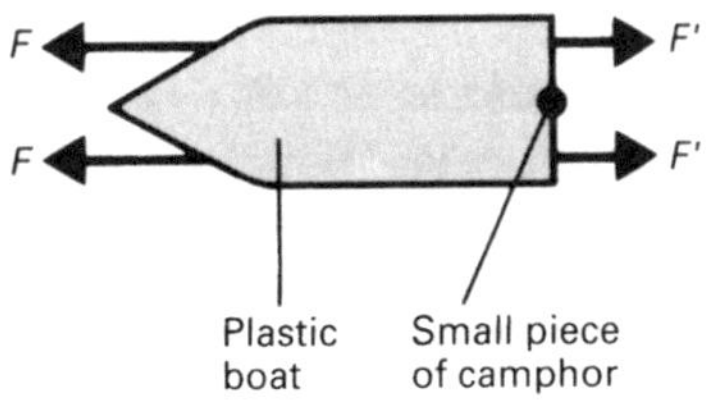

Fig. 10.18
Camphor boat

10.19 ANGLE OF CONTACT

The surface of a liquid is usually curved where it is in contact with a solid. The particular form that this curvature takes is determined by the relative strengths of what are called the cohesive and adhesive forces.

The cohesive force is the attractive force exerted on a liquid molecule by the neighbouring liquid molecules.

The adhesive force is the attractive force exerted on a liquid molecule by the molecules in the surface of the solid.

Consider a liquid in a container with vertical sides. If the adhesive force is large compared with the cohesive force, the liquid tends to stick to the wall and so has a concave meniscus (Fig. 10.19(a)). On the other hand, if the adhesive force is small compared with the cohesive force, the liquid surface is pulled away from the wall and the meniscus is convex (Fig. 10.19(b)). Whether the meniscus is concave or convex depends on the liquid concerned and on the solid with which it is in contact. For example, water has a concave meniscus when in contact with glass and a convex meniscus when in contact with wax; mercury has a convex meniscus with (clean) glass.

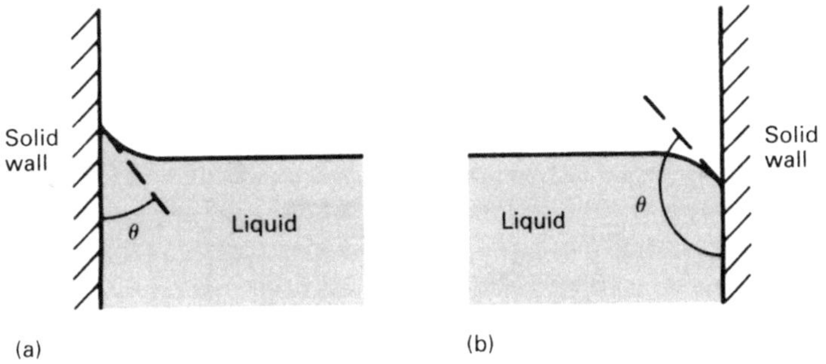

Fig. 10.19
(a) Concave, and
(b) convex menisci

The angle of contact θ is defined as the angle between the solid surface and the tangent plane to the liquid surface at the point where it touches the solid; the angle is measured through the liquid. It can be seen from Fig. 10.19 that the meniscus is concave when θ is less than 90° and is convex when θ is greater than 90°. A liquid is said to 'wet' a surface with which its angle of contact is less than 90°. The angle of contact between water and clean glass is zero, that between mercury and clean glass is 137°. Thus water 'wets' clean glass, mercury does not.

The zero angle of contact between water and clean glass is due to the adhesive force between water and glass being very much larger than the cohesive force between the water molecules themselves. This explains why water tends to spread into a thin continuous film when splashed on a horizontal clean glass surface. Mercury, on the other hand, forms into little drops; water on the roof of a freshly waxed car behaves in a similar fashion.

The addition of a detergent to a liquid lowers its surface tension and reduces the contact angle. Water-proofing agents have the opposite effect.

10.20 CAPILLARY RISE. MEASUREMENT OF γ

Water in a capillary tube rises above the level of the water outside. The effect is known as capillary rise and is most marked with narrow tubes. The ability of blotting paper to soak up ink is due to the same effect; the spaces between the fibres act as fine capillary tubes. A liquid whose angle of contact is greater than 90° suffers capillary depression. Both capillary rise and capillary depression are caused by surface tension and provide a means by which the surface tension γ of a liquid may be measured.

Suppose that a capillary tube is held vertically in a liquid which has a concave meniscus (Fig. 10.20). Surface tension forces cause the liquid to exert a downward directed force on the walls of the tube. In accordance with Newton's third law, the tube exerts an equal and opposite force on the liquid and it rises in the tube. At equilibrium the weight of the liquid which has been lifted up is equal to the vertical component of the force exerted by the tube. The mass of the raised liquid is the product of its density ρ and its volume $\pi r^2 h$, and therefore its weight is

$$\rho \pi r^2 hg$$

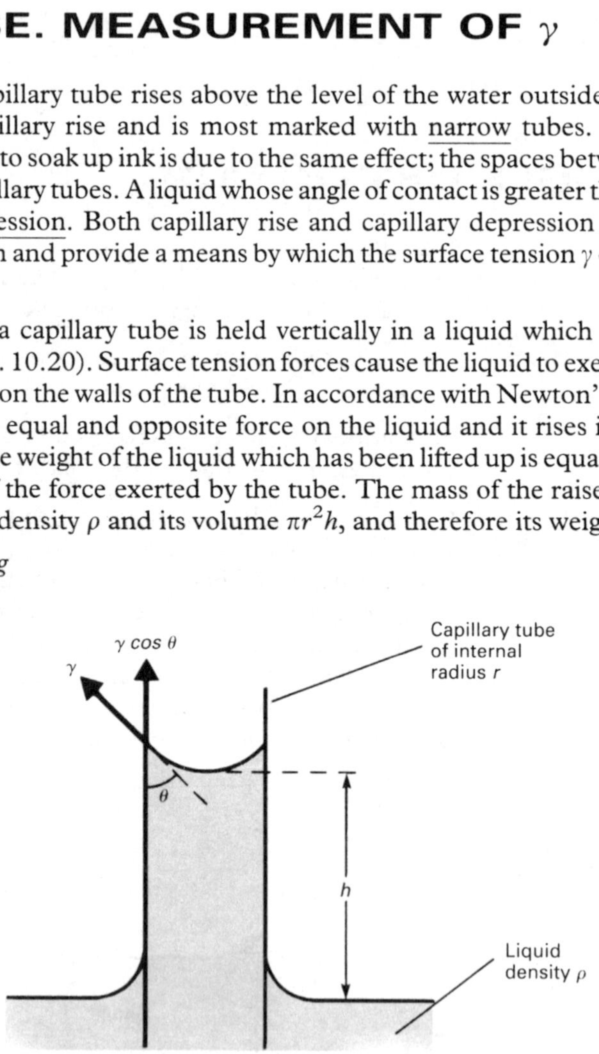

Fig. 10.20
Liquid in a capillary tube (not to scale)

The length of the liquid surface in contact with the tube is equal to the circumference $2\pi r$ of the tube, and therefore the vertical component of the force exerted by the tube is

$$2\pi r\gamma \cos\theta$$

Therefore at equilibrium

$$2\pi r\gamma \cos\theta = \rho \pi r^2 hg$$

i.e.

$$\gamma = \frac{\rho rhg}{2\cos\theta} \qquad [10.8]$$

from which γ can be determined.

Notes (i) The weight of the small quantity of liquid in the meniscus has been ignored in deriving equation [10.8].

(ii) h and r are normally measured with a travelling microscope. The tube should be broken at the level of the meniscus in order to measure r.

(iii) θ and ρ are found from tables or measured in separate experiments.

(iv) Equation [10.8] also holds for capillary depression.

10.21 PRESSURE DIFFERENCE ACROSS A SPHERICAL INTERFACE

The pressure inside a soap bubble is greater than the pressure of the air outside the bubble. If this were not so, the combined effect of the external pressure and the surface tension forces in the soap film would cause the bubble to collapse. Similarly, the pressure inside an air bubble in a liquid exceeds the pressure in the liquid, and the pressure inside a mercury drop is greater than that outside it.

In order to derive an expression for the excess pressure inside an air bubble in a liquid we shall consider the forces acting on one half of such a bubble (Fig. 10.21). Suppose that the radius of the bubble is r and that the surface tension of the liquid is γ. The half not shown exerts a surface tension force around the rim of the half we are considering. This force is directed to the left and, since the length of the rim is $2\pi r$, is of magnitude $2\pi r\gamma$. The resultant force due to the pressure p_o outside is also to the left and is acting perpendicular to an area πr^2 (the area of the flat face of the hemisphere) and is therefore of magnitude $p_o \pi r^2$ since p_o is the force per unit area. The resultant force due to the internal pressure p_i is to the right and its magnitude is $p_i \pi r^2$. The hemisphere is in equilibrium under the action of these forces, and therefore

$$p_i \pi r^2 = p_o \pi r^2 + 2\pi r\gamma$$

i.e.

$$p_i - p_o = \frac{2\pi r\gamma}{\pi r^2}$$

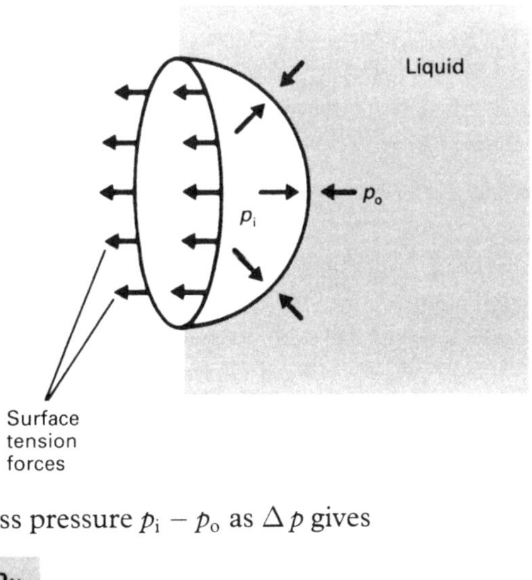

Fig. 10.21
Forces on air bubble in a liquid

Writing the excess pressure $p_i - p_o$ as Δp gives

$$\Delta p = \frac{2\gamma}{r} \quad \text{(for air bubbles and spherical drops)} \quad [10.9]$$

Note The smaller the bubble, the greater the excess pressure.

The excess pressure inside a spherical drop of mercury is given by the same expression if γ and r are taken to represent the surface tension of mercury and the radius of the drop respectively.

A soap film has two surfaces and therefore the excess pressure Δp inside a soap bubble is given by

$$\Delta p = \frac{4\gamma}{r} \quad \text{(for a soap bubble)}$$

where γ is the surface tension of the soap solution and r is the radius of the bubble. (**Note**. This assumes that the inner and outer surfaces have the same radius of curvature – a reasonable approximation.)

10.22 EXPERIMENTAL DETERMINATION OF γ BY JAEGER'S METHOD

The apparatus is shown in Fig. 10.22(a). When the tap is opened water drips into the large container and increases the pressure in the system. An air bubble starts to form at the lower end of the narrow tube. As more water drips into the container the bubble grows, and as it does so its radius of curvature decreases (see Fig. 10.22(b)).

Suppose that when the radius of the bubble is r the head of the liquid in the manometer is h_1, in which case the pressure inside the bubble is

$$h_1\rho_1 g + A$$

Fig. 10.22 Apparatus for the determination of γ

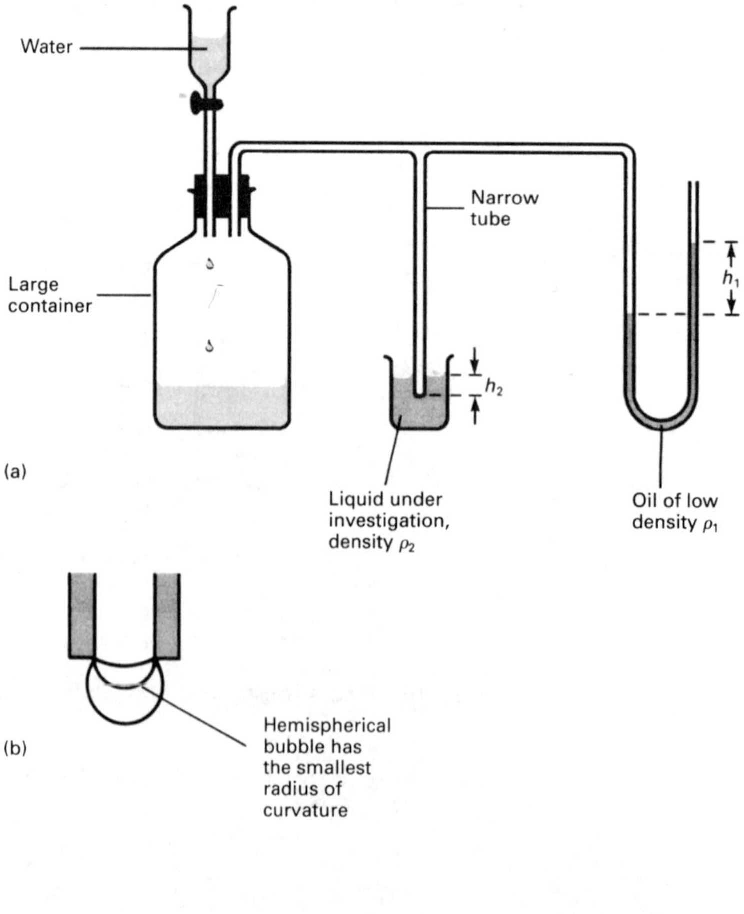

where ρ_1 is the density of the manometer liquid, g is the acceleration due to gravity and A is the atmospheric pressure. The pressure outside the bubble is

$$h_2\rho_2 g + A$$

where h_2 is the depth at which the bubble is formed and ρ_2 is the density of the liquid whose surface tension is being measured. The excess pressure in the bubble is therefore

$$h_1\rho_1 g - h_2\rho_2 g$$

The excess pressure in a bubble of this type is given by equation [10.9] as $2\gamma/r$, and therefore

$$\frac{2\gamma}{r} = h_1\rho_1 g - h_2\rho_2 g \qquad [10.10]$$

where γ is the surface tension of the liquid under test. The only variables in this equation are r and h_1, and therefore it follows that h_1 will have its maximum value when r has its minimum value. The bubble has its smallest possible radius of curvature when it is hemispherical, for if it were to grow any larger, its radius would increase. It follows that when h_1 has its maximum value the bubble is hemispherical and its radius of curvature is equal to the internal radius of the tube.

In practice the bubble becomes unstable and breaks away from the end of the tube as soon as its size increases beyond the stage where the bubble is hemispherical. When this happens the pressure in the system falls to atmospheric and another bubble begins to form as more water drips from the funnel. The tap is set so that bubbles form slowly (about one per second). Once a suitable rate has been achieved the maximum value of h_1 is recorded. A travelling microscope can be used to measure r (the internal radius of the lower end of the narrow tube) and h_2. The values of, ρ_1, ρ_2 and g can be obtained from tables and used in equation [10.10], together with the measured values of h_1, h_2 and r, to calculate γ.

Note When a bubble breaks away its radius is not exactly equal to that of the tube. This limits the accuracy to which absolute determinations of γ may be made by Jaeger's method. It does, however, provide a reliable means of investigating the temperature dependence of surface tension. Providing bubbles are formed at the same rate for each measurement, the relative values of γ are very accurate.

CONSOLIDATION

$$\text{Density} = \frac{\text{Mass}}{\text{Volume}}$$

$$\text{Relative density} = \frac{\text{Density of substance}}{\text{Density of water (at } 4^\circ\text{C)}}$$

$$\text{Specific volume} = \frac{\text{Volume}}{\text{Mass}}$$

The pressure on a surface is defined as the force per unit area acting at right angles to the surface. Pressure is a scalar.

Pressure in Fluids

(a) Increases with depth ($p = h\rho g$)

(b) 'Acts equally in all directions' (Strictly, it is the force due to the pressure that acts equally in all directions.)

Archimedes' Principle A body immersed in a fluid (totally or partially) experiences an upthrust (i.e. an apparent loss of weight) which is equal to the weight of the fluid displaced.

The Principle of Flotation A floating body displaces its own weight of fluid.

The surface tension of a liquid is defined as the force per unit length acting in the surface and perpendicular to one side of an imaginary line drawn in the surface. (Unit = $N\,m^{-1}$.)

Free surface energy is defined as the work done in isothermally creating unit area of new surface. (Unit = $J\,m^{-2} = N\,m^{-1}$.)

Surface tension = Free surface energy

A liquid is said to 'wet' (i.e. stick to) a surface with which its angle of contact is less than $90°$.

Solids transmit force; fluids transmit pressure.

11

ELASTICITY

11.1 DEFINITIONS

If forces are applied to a material in such a way as to deform it, then the material is said to be being stressed. As a result of the stress the material becomes strained. Initially we shall be concerned only with solids, and with stress which results in an increase in length (tensile stress) or a decrease in length (compressive stress).

Stress — Force per unit area of cross-section. Unit $= \mathrm{N\,m^{-2}} =$ pascal (Pa)

Strain — $\dfrac{\text{Change in length}}{\text{Original length}}$ (pure number)

Elasticity — A material is said to be elastic if it returns to its original size and shape when the load which has been deforming it is removed.

Hooke's law — Up to some maximum load (known as the **limit of proportionality**) the extension of a wire (or spring) is proportional to the applied load.

Elastic limit — This is the maximum load which a body can experience and still regain its original size and shape once the load has been removed. (The elastic limit sometimes coincides with the limit of proportionality.)

Yield point — If the stress is increased beyond the elastic limit, a point is reached at which there is a marked increase in extension. This is the yield point. The internal structure of the material has changed – the crystal planes have (effectively)* slid across each other. The material is said to be showing **plastic** behaviour. Few materials exhibit a yield point – mild steel is one that does.

Strength — This relates to the maximum force which can be applied to a material without it breaking.

Breaking stress — This is also called **ultimate tensile strength** and is the maximum stress which can be applied to a material.

Stiffness — This relates to the resistance which a material offers to having its size and/or shape changed.

Ductility — A ductile material is one which can be permanently stretched.

*See section 11.12.

Brittleness A brittle material cannot be permanently stretched; it breaks soon after the elastic limit has been reached. Brittle materials are often very strong in compression.

The stress-strain curve of a (hypothetical) elastic material is shown in Fig. 11.1. The corresponding load–extension curve is slightly different because as the length of the specimen increases its cross-sectional area decreases. The curve shown represents a ductile material; for a brittle material section EB is very short.

If the stress is removed at a point such as C, which is beyond the elastic limit, the body has a permanent strain equal to OO′.

Fig. 11.1
Stress–strain curve for an elastic material

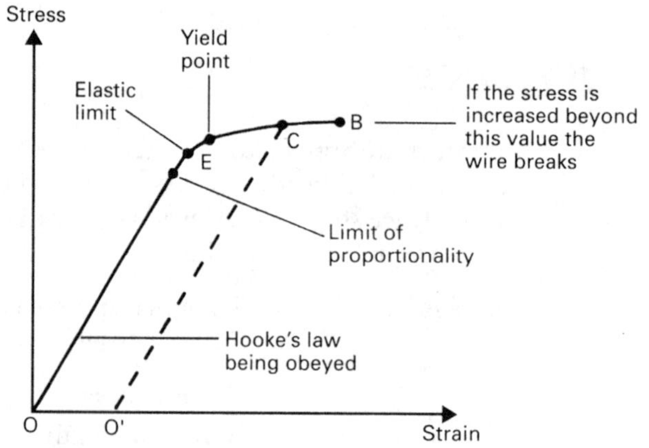

11.2 YOUNG'S MODULUS

Provided the stress is not so high that the limit of proportionality has been exceeded, the ratio stress/strain is a constant for a given material and is known as **Young's modulus**. Thus

$$E = \frac{\text{Tensile (or compressive) stress}}{\text{Tensile (or compressive) strain}} \qquad [11.1]$$

where

E = Young's modulus (N m^{-2} = Pa).

Young's modulus is clearly a measure of a material's resistance to changes in length. For example:

E (natural rubber)* = $1 \times 10^6\ \text{N m}^{-2}$
E (mild steel) = $2 \times 10^{11}\ \text{N m}^{-2}$

Note Bending a beam involves both tensile and compressive stress – the outer surface is stretched, the inner surface is compressed.

11.3 MOLECULAR EXPLANATION OF HOOKE'S LAW

Consider a plot of intermolecular force, F, against intermolecular separation, r, for a solid (Fig. 11.2). When the stress is zero the mean separation of the molecules is r_0. A tensile stress acts in opposition to the attractive forces between the molecules,

*This is an average value. Rubber does not obey Hooke's law and therefore the ratio stress/strain depends on the stress applied.

Fig. 11.2
Molecular explanation of Hooke's law

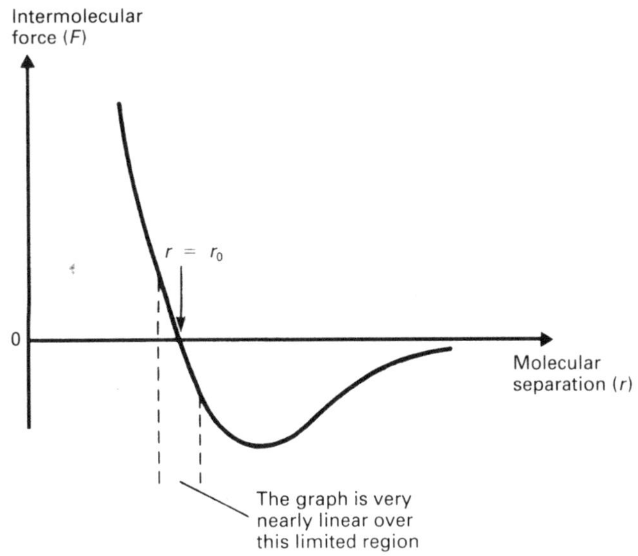

and is therefore capable of increasing their separation. For values of r close to r_0 the graph can be considered to be linear, and therefore, providing the stress is not so large as to take r out of this region, equal increases in tensile stress will produce equal increases in extension (Hooke's law). Note that Hooke's law applies also to compressive stress.

The work done in stretching a wire is stored as elastic potential energy (see section 11.5). On a molecular level this corresponds to the increased potential energy of the molecules which results from their increased separation.

11.4 EXPERIMENTAL DETERMINATION OF YOUNG'S MODULUS

Consider the experimental arrangement shown in Fig. 11.3. When Q is loaded there is a tendency for its support to sag. The errors that would result if this were to happen are avoided by carrying the reference scale on a second wire, P, suspended

Fig. 11.3
Apparatus for investigating the extension of a wire

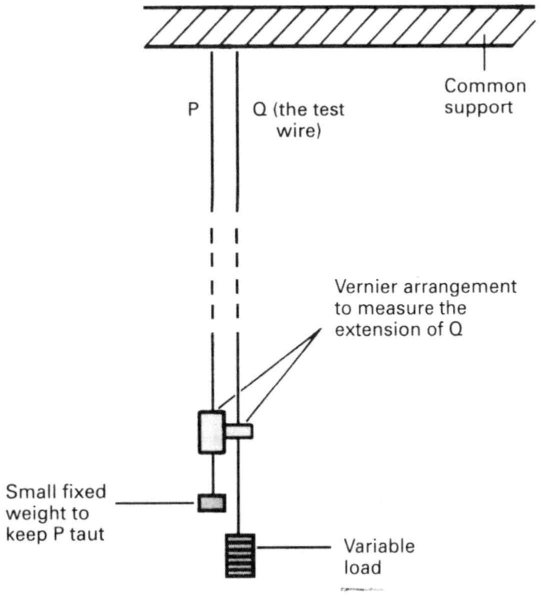

from the same beam as Q. Both P and Q are made from the same material and are of the same length; errors due to expansion, as a result of temperature changes during the experiment, are therefore avoided. The test wire is loaded (typically up to 100 N in 5 N steps), and the resulting extension is measured as a function of the load. The wires are as long as is convenient (typically 2 m) and thin in order to obtain as large an extension as possible; even so a vernier arrangement is needed to measure the extension (typically 1 mm). If the test wire is free of kinks at the start and the limit of proportionality is not exceeded, the measurement can be used to produce a plot similar to that in Fig. 11.4.

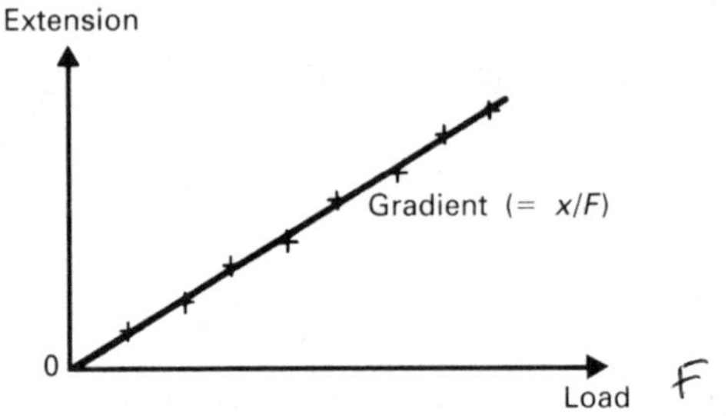

Fig. 11.4 Typical results of extending a wire

From equation [11.1]

$$E = \frac{\text{Stress}}{\text{Strain}}$$

i.e. $$E = \frac{F/A}{x/L} = \frac{L}{A\,(x/F)}$$

where

F = applied load (N)

A = area of cross-section of wire (m^2)

x = extension (m)

L = original length (m).

Bearing in mind that x/F is the gradient of the graph, we have

$$E = \frac{L}{A \times \text{gradient}}$$

The gradient is measured from the graph, often as the mean of the results obtained with an increasing load and a decreasing load. L can be measured with an extending ruler or metre rule. A is obtained by determining the diameter of the wire at several places with a micrometer.

11.5 THE WORK DONE IN STRETCHING A WIRE (STRAIN ENERGY)

Consider a wire whose extension is x when the force on it is F. If the extension is increased by δx, where δx is so small that F can be considered constant, then (by equation [5.1]) the work done, δW, is given by

$$\delta W = F\delta x$$

The total work done in increasing the extension from 0 to x, i.e. the **elastic potential energy** stored in the wire (the **strain energy**) when its extension is x, is given by W, where

$$W = \int_0^x F\mathrm{d}x \qquad [11.2]$$

If the wire obeys Hooke's law, we may put

$$F = kx$$

where k is a constant, and therefore by equation [11.2]

$$W = \int_0^x kx\,\mathrm{d}x$$

i.e.

$$W = \tfrac{1}{2}kx^2 \qquad [11.3]$$

Alternatively, since $F = kx$, substituting for k gives

$$W = \tfrac{1}{2}Fx \qquad [11.4]$$

A wire of length L and cross-sectional area A has a volume of AL and therefore by equation [11.4]

$$\text{Strain energy per unit volume} = \frac{\tfrac{1}{2}Fx}{AL}$$

$$= \frac{1}{2}\frac{F}{A} \times \frac{x}{L}$$

i.e.

Strain energy per unit volume = ½ stress × strain [11.5]

Notes (i) For a wire of a material with Young's modulus E it follows from equation [11.1] that $F = (EA/L)\,x$, i.e. $k = EA/L$ and therefore by equation [11.3]

$$W = \frac{EAx^2}{2L} \qquad [11.6]$$

(ii) Equations [11.3] to [11.6] apply only as long as Hooke's law is obeyed. If the extension is so great that the limit of proportionality is exceeded or the wire does not obey Hooke's law anyway, the work done can be found from a graph of force against extension (Fig. 11.5). **The strain energy per unit volume is the area under a graph of stress against strain.**

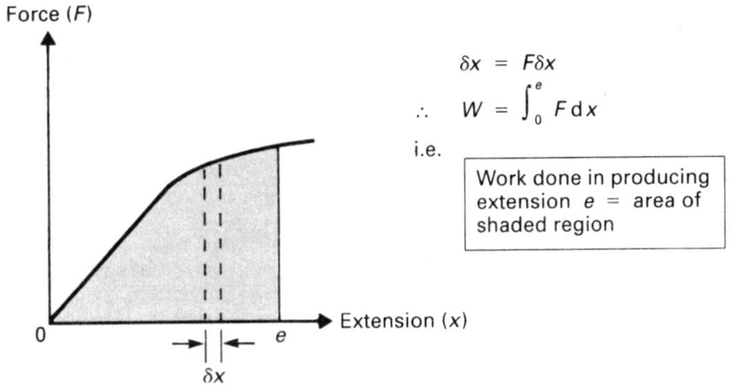

Fig. 11.5
Work done in stretching a wire

EXAMPLE 11.1

A steel wire, AB, of length 0.60 m and cross-sectional area $1.5 \times 10^{-6}\,\text{m}^2$ is attached at B to a copper wire, BC, of length 0.39 m and cross-sectional area $3.0 \times 10^{-6}\,\text{m}^2$. The combination is suspended vertically from a fixed point at A, and supports a weight of 250 N at C. Find the extension of each section of the wire. (Young's modulus of steel = 2.0×10^{11} Pa, Young's modulus of copper = 1.3×10^{11} Pa.)

Solution

Each section of the wire is subject to the full force of 250 N. Let x_1 = extension of AB; let x_2 = extension of BC.

$$E = \frac{\text{Stress}}{\text{Strain}}$$

Therefore for steel

$$2.0 \times 10^{11} = \frac{250/1.5 \times 10^{-6}}{x_1/0.60} = \frac{1.00 \times 10^8}{x_1}$$

$$\therefore \quad x_1 = \frac{1.00 \times 10^8}{2.0 \times 10^{11}} = 5.0 \times 10^{-4}\,\text{m} = 0.50\,\text{mm}$$

For copper

$$1.3 \times 10^{11} = \frac{250/3.0 \times 10^{-6}}{x_2/0.39} = \frac{3.25 \times 10^7}{x_2}$$

$$\therefore \quad x_2 = \frac{3.25 \times 10^7}{1.3 \times 10^{11}} = 2.5 \times 10^{-4}\,\text{m} = 0.25\,\text{mm}$$

EXAMPLE 11.2

A steel rod of length 0.60 m and cross-sectional area $2.5 \times 10^{-5}\,\text{m}^2$ at 100 °C is clamped so that when it cools it is unable to contract. Find the tension in the rod when it has cooled to 20 °C. (Young's modulus of steel = 2.0×10^{11} Pa, linear expansivity of steel = $1.6 \times 10^{-7}\,°\text{C}^{-1}$.)

Solution

It follows from equation [9.4] that if the rod were allowed to contract, its length would decrease by

$$0.60 \times 1.6 \times 10^{-7}\,(100 - 20) = 7.68 \times 10^{-6}\,\text{m}$$

The extension of the clamped rod at 20 °C is therefore 7.68×10^{-6} m.

$$E = \frac{\text{Stress}}{\text{Strain}}$$

$$\therefore \quad 2.0 \times 10^{11} = \frac{\text{Stress}}{7.68 \times 10^{-6}/0.60}$$

i.e. $\quad \text{Stress} = 2.56 \times 10^6\,\text{Pa}$

$$\text{Stress} = \frac{\text{Tension}}{\text{Cross-sectional area}}$$

$$\therefore \quad 2.56 \times 10^6 = \frac{\text{Tension}}{2.5 \times 10^{-5}}$$

i.e. $\text{Tension} = 64\,\text{N}$

QUESTIONS 11A

1. An aluminium wire of length 0.35 m and radius 0.20 mm is stretched by 1.4 mm. Young's modulus of aluminium is 7.0×10^{10} Pa.
   (a) Find the strain in the wire.
   (b) Find the stress in the wire.
   (c) Find the cross-sectional area of the wire.
   (d) Find the tension in the wire.

11.6 BULK MODULUS AND SHEAR MODULUS

So far we have been concerned only with stress which results in a change in length. Two other types of stress will now be considered. The associated moduli of elasticity are called the **bulk modulus** and the **shear modulus**. The latter is sometimes referred to as the **rigidity modulus**.

Shear (Rigidity) Modulus

A shear stress is one which changes the shape of a body; the strain which results is called a **shear strain**. Fig. 11.6 illustrates a solid block WXYZ whose lower face is fixed. A force F acts on the block tangential to its upper face. The force provides a shear stress which distorts the block so that its new shape is WX′Y′Z. **The shear modulus** G is defined by

$$G = \frac{\text{Shear stress}}{\text{Shear strain}} \qquad (\text{Unit} = \text{N m}^{-2} = \text{Pa})$$

where

$$\text{Shear stress} = \text{Tangential force per unit area} = F/A$$

and

$$\text{Shear strain} = \text{Tangent of angle of shear} = \tan\alpha = \Delta x/y$$

i.e.

$$G = \frac{F/A}{\Delta x/y} \qquad [11.7]$$

(**Note**: twisting a wire involves shear stress.)

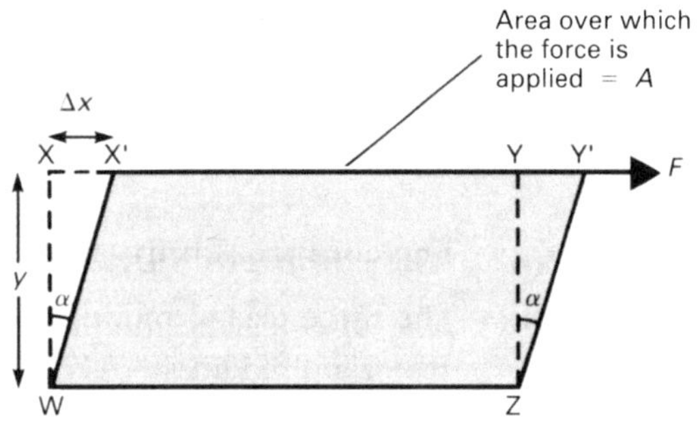

Fig. 11.6
Block subjected to a shear stress

Bulk Modulus

This refers to situations in which the volume of a substance is changed by the application of an external stress. Unlike the shear modulus and Young's modulus, which refer to solids only, bulk moduli are possessed by solids, liquids and gases.

In Fig. 11.7 the application of a force ΔF, which is everywhere normal to the surface of a spherical body, has changed its volume by ΔV. The **bulk modulus** K is defined by

$$K = \frac{\text{Bulk stress}}{\text{Bulk strain}} \qquad (\text{Unit} = \text{N m}^{-2} = \text{Pa})$$

where

$$\text{Bulk stress} = \text{Increased force per unit area} = \Delta F/A$$

and

$$\text{Bulk strain} = \frac{\text{Change in volume}}{\text{Original volume}} = \frac{\Delta V}{V}$$

i.e. $$K = -\frac{F/A}{\Delta V/V}$$

Fig. 11.7
Sphere subjected to a radial stress

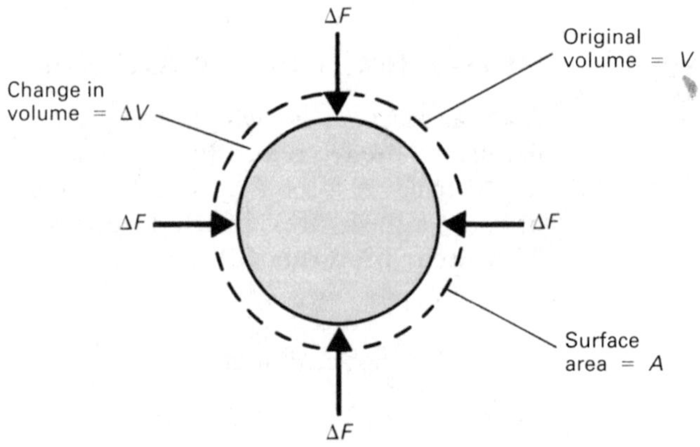

Note When ΔF is positive ΔV is negative, and therefore it has been necessary to include the minus sign in order to make K a positive constant.

$\Delta F/A$ is the change in pressure Δp, and therefore

$$K = -\frac{\Delta p}{\Delta V/V}$$

which in the limit as $\Delta p \to 0$ becomes

$$K = -V\frac{\mathrm{d}p}{\mathrm{d}V}$$

Notes (i) **The compressibility** κ of a substance is given by $\kappa = 1/K$.

(ii) The three elastic moduli have the same order of magnitude for any one material, and apply only in the region where the ratio of stress to strain is constant.

11.7 PLASTICITY

A **perfectly plastic** material is one which shows no tendency to return to its original size and shape when the load which has been deforming it is removed (plasticine is a good example). In this sense a perfectly plastic material is the opposite of an elastic material. The application of a load to a plastic material causes **dislocations** (i.e. gaps in the crystal lattice – see section 11.12) to move. This produces the same effect as planes of atoms sliding past each other.

11.8 ELASTIC HYSTERESIS

Fig. 11.8 shows the force–extension curve of a sample of rubber for both loading and unloading. The extension due to any given force is greater during unloading than during loading, i.e. the unloading extension lags behind the loading extension. The effect is called **elastic hysteresis**, and the region enclosed by the two curves is called a **hysteresis loop**. Metals also exhibit hysteresis, but to a much smaller extent.

Fig. 11.8
The effect of loading and unloading a sample of rubber

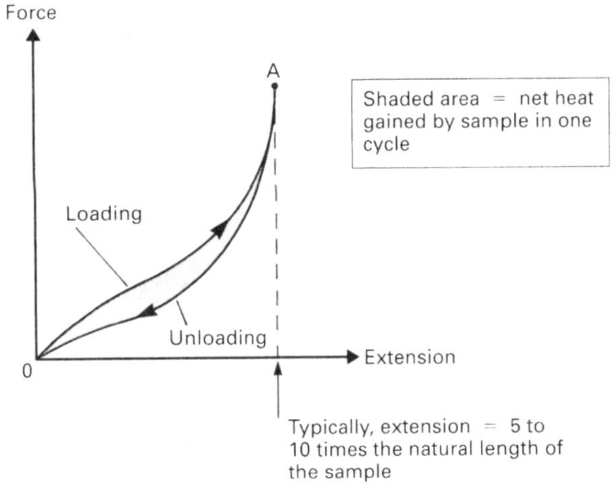

When rubber is stretched it becomes warmer. When the stress is released its temperature falls but it remains a little warmer than it was initially. The net increase in the heat content of the sample during the cycle is equal to the area of the hysteresis loop.

11.9 SOME PROPERTIES OF RUBBER

(i) Samples of some types of rubber can be stretched as much as 10 times their natural lengths and still regain their original sizes when the stresses are removed. A typical metal, on the other hand, can be subjected to only about 1/10 000 of this extension before its elastic limit is exceeded.

(ii) Rubber does not obey Hooke's law, i.e. the value of the ratio stress/strain depends on the particular stress at which it is measured – see Fig. 11.8. The sample stretches easily at first, but has become very stiff (steep slope) by the time the extension corresponding to point A has been reached. At A the extension is such that the long-chain molecules of the rubber (see 'Elastomers' in section 9.10) have become fully straightened out. Any further extension can be achieved only by stretching the bonds between the carbon atoms in the chains.

(iii) Some types of rubber have particularly large hysteresis loops, and so are useful as vibration absorbers. If a block of such a rubber is placed between a piece of vibrating machinery and the floor, much of the energy of the mechanical vibration is converted to heat energy in the rubber, and so is not transmitted to the floor. The rubber used in the manufacture of tyres has a small hysteresis loop, for it is clearly desirable that as little heat as possible is generated in a tyre.

(iv) Heating a stretched* rubber band causes it to contract. The higher temperature produces increased lateral bombardment of the long-chain molecules causing them to kink and so shorten.

Fig. 11.9
Stress–strain curves of typical samples of (a) copper, (b) glass, (c) rubber

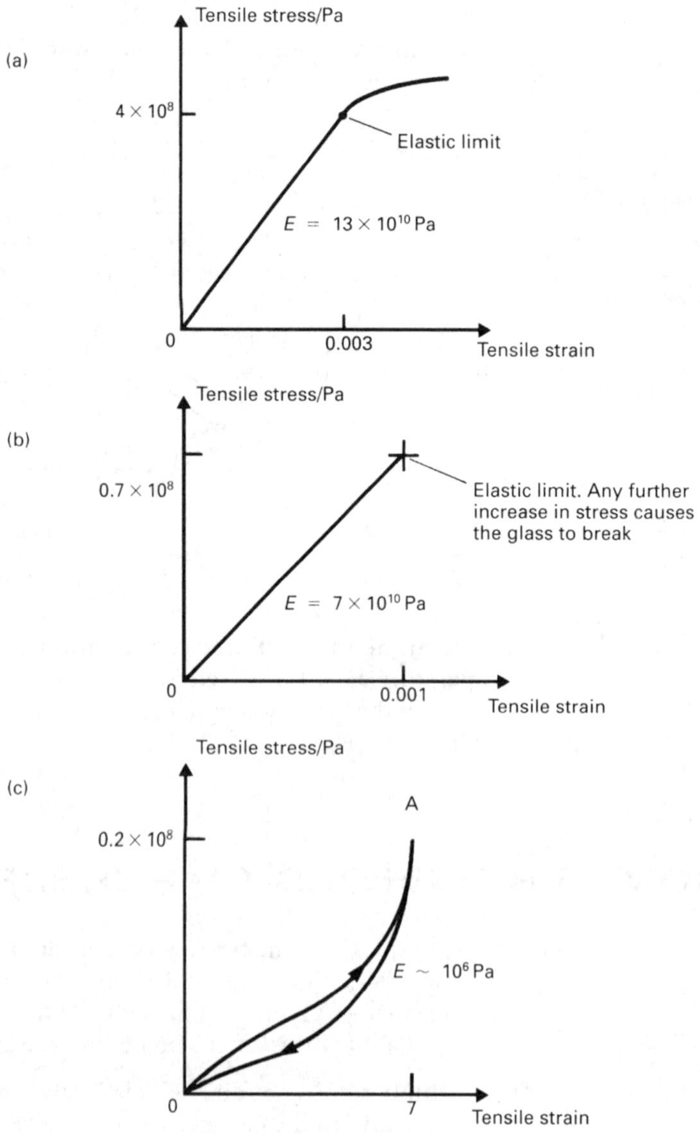

*An unstressed sample of rubber, on the other hand, behaves quite normally and expands on heating.

(v) For most materials the value of Young's modulus decreases with increasing temperature. In the case of rubber, though, the ratio stress/strain increases with increasing temperature. This is because the increased lateral bombardment of the long-chain molecules which occurs at higher temperatures makes it more difficult to straighten them.

(vi) If a rubber band is stretched rapidly, its temperature increases. This behaviour is opposite to that of metals and most other materials. Stretching the rubber produces greater alignment of its long-chain molecules. This increased amount of order is akin to crystallization in the sense that as a result of it the rubber is in a lower energy state than previously. The energy released in attaining this state heats the sample.

The stress/strain curve for rubber is compared with those of copper and glass in Fig. 11.9.

11.10 FATIGUE

If a material is repeatedly stressed and unstressed (or stressed first in one direction and then in another), it becomes weaker, i.e. the strain produced by a given amount of stress increases. If the repeated stressing is continued, the material may fracture even though the maximum stress applied in any of the stress cycles could have been sustained indefinitely if it had been applied steadily. The failure of a material under these circumstances is called **fatigue failure** or **fatigue fracture**. It has been estimated that about 90% of the failures which occur in aircraft components are due to fatigue.

Mild steel and many other ferrous metals can safely undergo an infinite number of stress cycles, provided that the maximum stress is kept below a particular value known as the **fatigue limit**. There is no such limit for non-ferrous materials. In such cases the maximum loading is kept below that which would cause failure within the time for which the component is required to last.

Fatigue fractures usually start in the surface at points of high stress, e.g. at sharp corners and around rivet holes. It is believed that each time the material is stressed a small amount of plastic strain is produced. Since it is plastic strain, the effects of repeated stressings are cumulative and eventually produce fracture.

11.11 CREEP

The term **creep** is used to describe the gradual increase in strain which occurs when a material is subjected to stress for a long period of time. Unlike fatigue it occurs even when the stress is constant. It is most marked at elevated temperatures and may be so severe that the material eventually fractures. The greater the stress, the more quickly this happens. The turbine blades in jet engines are particularly susceptible to creep because they are under high stress and are at high temperatures. Soft metals (e.g. lead) and most plastics show considerable creep even at room temperature. Fig. 11.10 shows a typical creep curve. Note the accelerated rate of creep just before fracture.

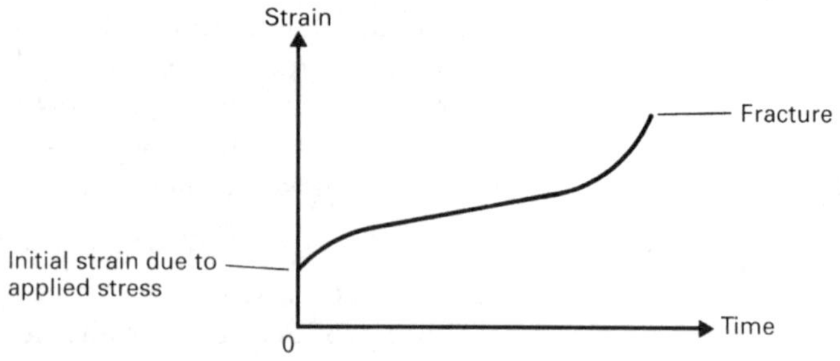

Fig. 11.10
Typical creep curve

11.12 DISLOCATIONS

The ductility of metals (i.e. their ability to undergo plastic strain) might be thought to be due to the various crystal planes which make up the structure slipping over each other to take up new positions (Fig. 11.11). According to this idea every atom in plane X has had to break a bond with an atom in plane Y, and then form a new one with a different atom in plane Y. However, calculations reveal that this process would require stresses which are about a hundred times greater than those which are needed to produce plastic strain in practice. Thus real metals are not as strong as this simple model would suggest. An explanation was offered in 1934 by G.I. Taylor who put forward the idea of **dislocations**. One type, an **edge dislocation**, is shown in Fig. 11.12. It takes the form of an incomplete plane of atoms (AB in Fig. 11.12(a)). Forces applied in the manner shown move atoms B and N closer together and eventually a bond forms between them at the expense of that between M and N (Fig. 11.12(b)).

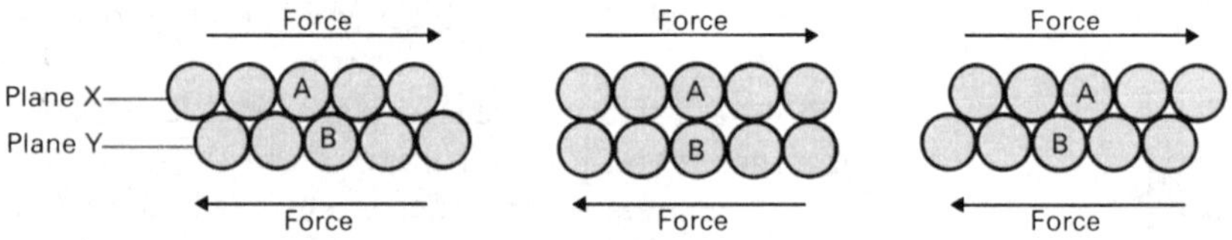

Fig. 11.11
Sliding of crystal planes

If the stress is maintained, M and Z bond together leaving plane XY incomplete (Fig. 11.12(c)). In this way, then, the dislocation moves from left to right through the crystal. The end result is the same as it would have been if rows 1 and 2 had slipped over row 3. However, it has been achieved much more easily for only one bond has been broken at a time, whereas the wholesale movement of the planes would require a large number of bonds to be broken at the same time. The process is commonly likened to the movement of a ruck in a carpet. A large force is required to drag a heavy carpet over a floor. However, if there is a ruck in the carpet, it can be moved by the almost effortless process of pushing the ruck from one side to the other (Fig. 11.13).

11.13 THE STRENGTHENING OF METALS

It follows from what has been said in section 11.12 that metals can be made stronger by impeding the movement of dislocations. This can be done in a number of ways.

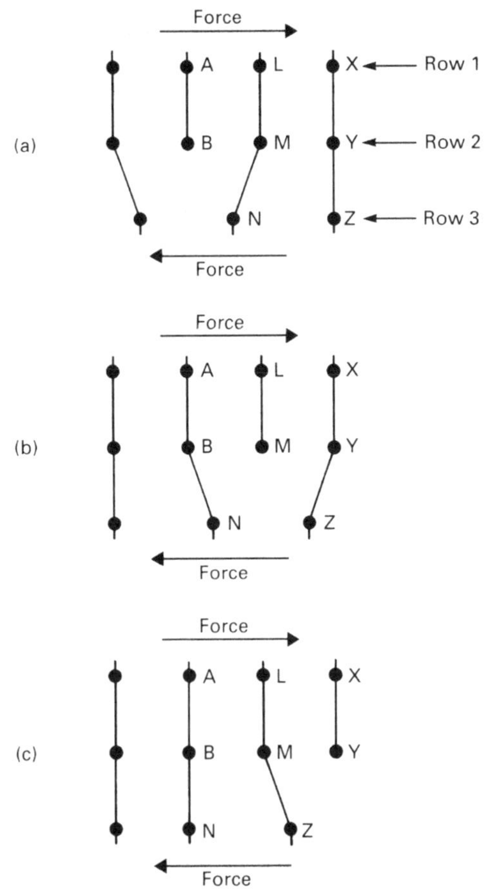

Fig. 11.12 Movement of a dislocation

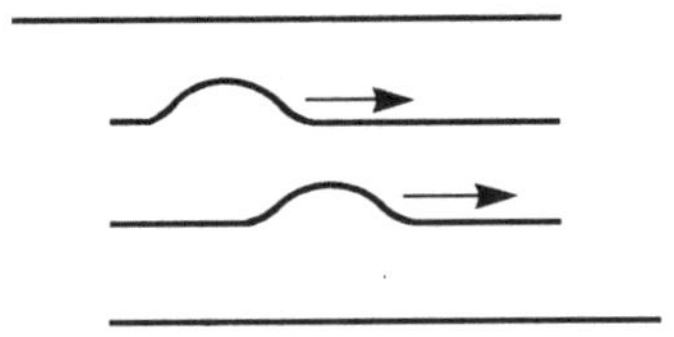

Fig. 11.13 Sliding a carpet by moving a ruck

(i) Increasing the number of dislocations, because the increased number makes it more likely that the various dislocations will obstruct each other in a 'log jam' effect. The number of dislocations can be increased by plastically deforming the metal repeatedly – this is the process known as **work hardening**.

(ii) Introducing 'foreign' atoms (e.g. carbon atoms in steel) into the structure. These disturb the regularity of the lattice and by so doing hinder the movement of dislocations.

(iii) Dislocations have difficulty in moving across grain boundaries and therefore samples in which the grain size is small (and which therefore have many grain boundaries) tend to be strong.

A metal in which there were no dislocations would, of course, be extremely strong. To date, though, such perfect crystals have been made only on a very small scale. They are known as 'whiskers' and are typically only a few micrometers thick, though a few millimetres long.

12

FLUID FLOW

12.1 TERMINOLOGY

Fluids

Both liquids and gases are fluids.

Viscosity

If a fluid is viscous then it offers a resistance to the motion through it of any solid body – or what amounts to the same thing, to its own motion past a solid body. In both these circumstances (except where the fluid is a gas of very low density) the layer of fluid in immediate contact with the solid surface is stationary with respect to that surface, and therefore the motion causes adjacent layers of fluid to move past each other. There exists a kind of internal friction which offers a resistance to the motion of one layer of fluid past another, and it is this that is the origin of the viscous force. In liquids the internal friction is due to intermolecular forces of attraction. In gases the viscous force arises as a result of the interchange of molecules that takes place between the different layers of the flowing gas. Thus, whenever molecules move from a fast-flowing layer into a more slowly moving layer, they increase the average speed of the molecules of that layer. It is as if the faster layer is dragging the slower layer along with it. At the same time, the random molecular motion means that molecules from the slower-moving layer move into the faster-moving layer, and therefore the average molecular speed of the faster-moving layer is reduced. Thus, the presence of an adjacent slow-moving layer slows down the fast-moving layer.

Steady Flow

If the flow of a fluid is steady (also known as **streamline flow**, **orderly flow** and **uniform flow**), then all the fluid particles that pass any given point follow the same path at the same speed (i.e. they have the same velocity). Thus, in steady flow no aspect of the flow pattern changes with time.

Turbulent Flow

This is also known as **disorderly flow**. In this type of flow the speed and direction of the fluid particles passing any point vary with time.

Line of Flow

The path followed by a particle of the fluid is called the line of flow of the particle.

Streamline

A streamline is a curve whose tangent at any point is along the direction of the velocity of the fluid particle at that point. Streamlines never cross.

For a fluid undergoing steady flow all the fluid particles that pass any given point follow the same path, i.e. all the particles passing any given point have the same line of flow. It follows that **in steady flow the streamlines coincide with the lines of flow**.

Laminar Flow*

This is a special case of steady flow in which the velocities of all the particles on any given streamline are the same, though the particles of different streamlines may move at different speeds (Fig. 12.1). As an example of laminar flow, consider a liquid flowing in an open channel of uniform cross-section. If the fluid is viscous, it flows as a series of parallel layers (laminae). The layer in contact with the base of the channel is at rest, and the speed of each layer is greater than the speeds of those below it. If the channel is wide, the drag effects of the side walls can be ignored, and therefore the velocities of all the particles within each layer are the same (Fig. 12.2).

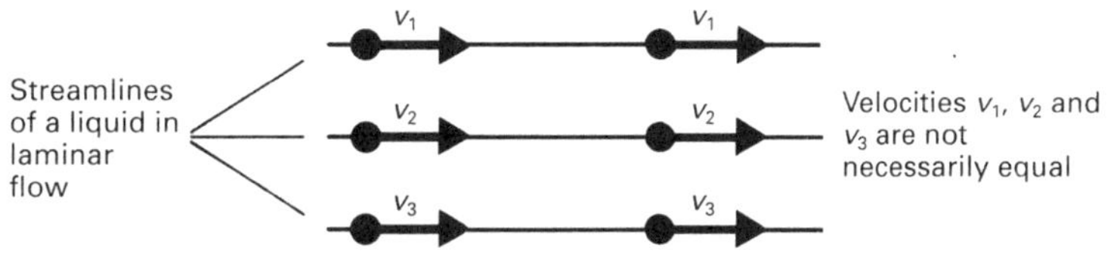

Fig. 12.1 Streamlines of a liquid in laminar flow

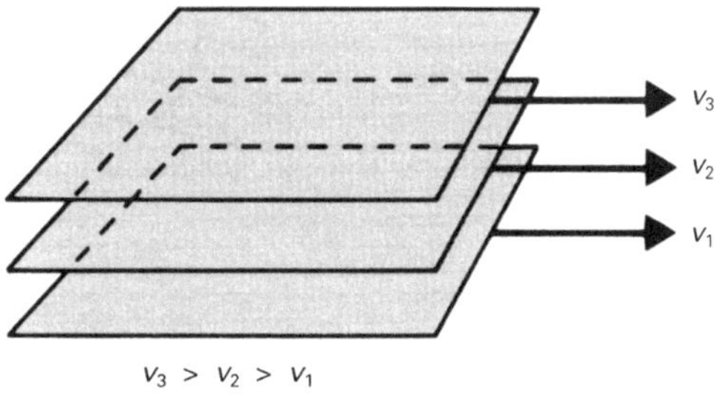

Fig. 12.2 To illustrate laminar flow

As a second example, consider a viscous fluid flowing in a pipe of uniform circular cross-section. In this case the fluid flows as a series of concentric cylinders. All the particles of fluid within such a cylinder flow at the same speed. The speed of the cylinder adjacent to the wall of the pipe is zero, and the speeds increase towards the centre.

Tube of Flow

This is a tubular region of a flowing fluid whose boundaries are defined by a set of streamlines.

Incompressible Fluid

This is a fluid in which changes in pressure produce no change in the density of the fluid. Liquids can be considered to be incompressible; gases subject only to small pressure differences can also be taken to be incompressible.

*The term 'laminar flow' is often used loosely as being synonymous with the less restricting term 'steady flow'.

12.2 THE EQUATION OF CONTINUITY

If a fluid is undergoing steady flow, then the mass of fluid which enters one end of a tube of flow must be equal to the mass that leaves at the other end during the same time. This must be so because in steady flow no fluid can leave the tube of flow through the side walls of the tube (streamlines do not cross each other), and therefore there would be a change in the mass within the tube if it were not so. A change in mass would mean a change in the number of fluid particles within the tube, in which case there would exist fluid particles where none had previously existed (or no particles where there had been some). This cannot happen under conditions of steady flow because the velocity at any point has to be unvarying.

Consider a fluid undergoing steady flow, and consider a section XY of a tube of flow within the fluid (Fig. 12.3). Let

A_X and A_Y be the cross-sectional areas of the tube of flow at X and Y respectively,

ρ_X and ρ_Y be the densities of the fluid at X and Y respectively,

v_X and v_Y be the velocities of the fluid particles at X and Y respectively.

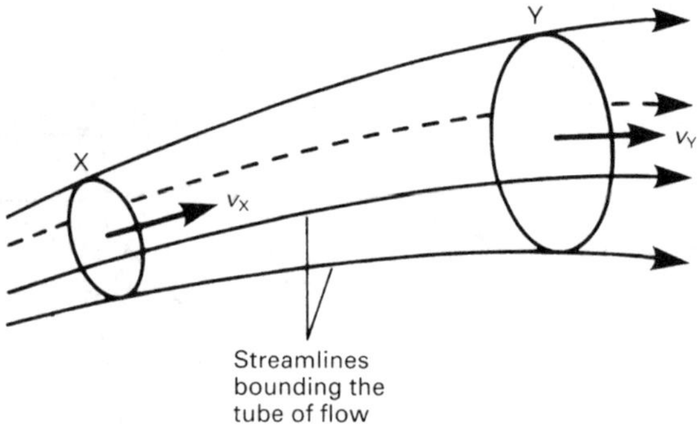

Fig. 12.3
Section of tube of flow

In a time interval Δt the fluid at X will move forward a distance $v_X \Delta t$. Therefore a volume $A_X v_X \Delta t$ will enter the tube at X. The mass of fluid entering at X in time Δt will therefore be

$$\rho_X A_X v_X \Delta t$$

Similarly, the mass leaving at Y in the same time will be

$$\rho_Y A_Y v_Y \Delta t$$

Since the mass entering at X is equal to the mass leaving at Y,

$$\rho_X A_X v_X \Delta t = \rho_Y A_Y v_Y \Delta t$$

i.e.

$$\rho_X A_X v_X = \rho_Y A_Y v_Y \qquad [12.1]$$

Equation [12.1] is known as the **equation of continuity**. For an incompressible fluid $\rho_X = \rho_Y$, and therefore the equation takes the form

$$A_X v_X = A_Y v_Y \qquad [12.2]$$

$A_X v_X$ is known as the **flow rate** (or **volume flux**) of the fluid at X.

12.3 BERNOULLI'S EQUATION

This states that **for an incompressible, non-viscous fluid undergoing steady flow, the pressure plus the kinetic energy per unit volume plus the potential energy per unit volume is constant at all points on a streamline.**

i.e. $$p + \frac{1}{2}\rho v^2 + \rho g h = \text{A constant}$$

where

p = the pressure within the fluid

ρ = the density of the fluid

v = the velocity of the fluid

g = the acceleration due to gravity, and

h = the height of the fluid (above some arbitrary reference line).

Proof Consider a tube of flow within a non-viscous, incompressible fluid undergoing steady flow (Fig. 12.4). Let

p_X and p_Y = pressures at X and Y

v_X and v_Y = velocities at X and Y

A_X and A_Y = areas of cross-section at X and Y

h_X and h_Y = average heights at X and Y.

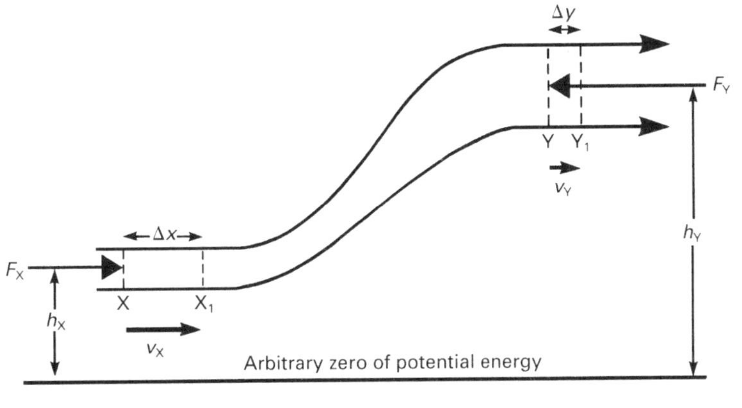

Fig. 12.4 Derivation of Bernoulli's equation

Let X_1 be close to X so that each of the parameters listed above has the same value at X_1 as at X. Let Y_1 be close to Y with a similar consequence. Since the fluid is incompressible, the density will be the same at all points; let this be ρ.

Consider the section of fluid which is between X and Y, moving to occupy the region between X_1 and Y_1. The fluid moves in this direction because the force F_X is greater than the force F_Y. The force F_X moves a distance Δx, and the fluid moves a distance Δy against the force F_Y.

The net work done on the fluid is therefore given by

$$\text{Work done on fluid} = F_X \Delta x - F_Y \Delta y$$

Since the fluid is undergoing steady flow, the mass of fluid that was originally between X and X_1 is equal to the mass which is now between Y and Y_1. Let this mass be m. Thus a mass m which originally had velocity v_X and average height h_X has been replaced by an equal mass with velocity v_Y and average height h_Y. Therefore,

$$\text{Gain in kinetic energy} = \tfrac{1}{2}mv_Y^2 - \tfrac{1}{2}mv_X^2$$

$$\text{Gain in potential energy} = mgh_Y - mgh_X$$

None of the work done on the fluid has been used to overcome internal friction because the fluid is non-viscous, and therefore by the principle of conservation of energy,

$$\text{Work done} = \text{Gain in KE} + \text{Gain in PE}$$

$$\therefore \quad F_X\Delta x - F_Y\Delta y = \tfrac{1}{2}mv_Y^2 - \tfrac{1}{2}mv_X^2 + mgh_Y - mgh_X$$

$$\text{i.e.} \quad p_XA_X\Delta x - p_YA_Y\Delta y = \tfrac{1}{2}mv_Y^2 - \tfrac{1}{2}mv_X^2 + mgh_Y - mgh_X$$

But, $A_X\Delta x$ = Volume between X and X_1 $= m/\rho$, and similarly $A_Y\Delta y = m/\rho$. Therefore,

$$p_X\frac{m}{\rho} - p_Y\frac{m}{\rho} = \tfrac{1}{2}mv_Y^2 - \tfrac{1}{2}mv_X^2 + mgh_Y - mgh_X$$

$$\text{Thus} \quad p_X - p_Y = \tfrac{1}{2}\rho v_Y^2 - \tfrac{1}{2}\rho v_X^2 + \rho gh_Y - \rho gh_X$$

$$\text{i.e.} \quad p_X + \tfrac{1}{2}\rho v_X^2 + \rho gh_X = p_Y + \tfrac{1}{2}\rho v_Y^2 + \rho gh_Y$$

Since X and Y were arbitrarily chosen points we may write

$$p + \tfrac{1}{2}\rho v^2 + \rho gh = \text{A constant}$$

In practice, Bernoulli's equation cannot apply exactly – real fluids are viscous and gases are easily compressed. Nevertheless, as long as the equation is used with care, it gives meaningful results and its qualitative implications are valid.

12.4 CONSEQUENCES OF BERNOULLI'S EQUATION

It follows from Bernoulli's equation that whenever a flowing fluid speeds up, there is a corresponding decrease in the pressure and/or the potential energy of the fluid. If the flow is horizontal, the whole of the velocity increase is accounted for by a decrease in pressure.

An aerofoil (e.g. an aircraft wing) is shaped so that air flows faster along the top of it than the bottom. There is, therefore, a greater pressure below the aerofoil than above it. It is this difference in pressure that provides the lift. A spinning ball experiences a similar effect. The spin drags air around with the ball (Fig. 12.5). The ball therefore has a resultant force acting on it towards the top of the page.

In accordance with the equation of continuity, fluids speed up at constrictions, and therefore there is a decrease in pressure at constrictions. This effect is made use of in such devices as filter pumps, Bunsen burners and carburettors.

The Venturi meter (Fig. 12.6) is a device which introduces a constriction into a pipe carrying a fluid, in order that the velocity of the fluid can be measured by measuring the resulting drop in pressure.

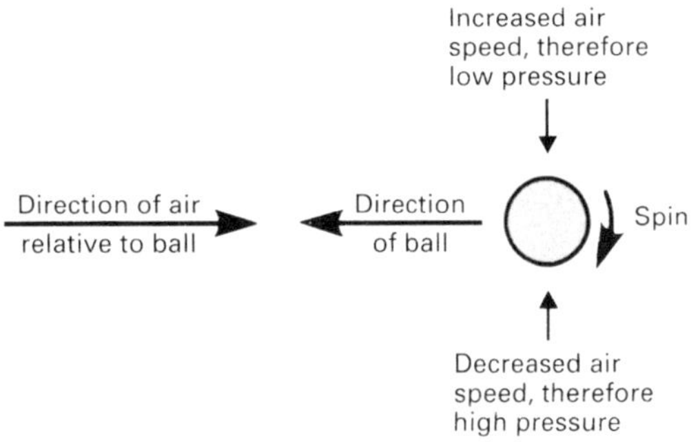

Fig. 12.5
To illustrate the effect of spin

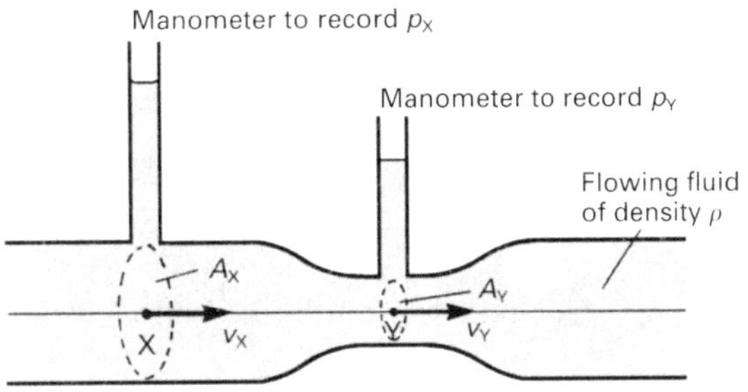

Fig. 12.6
The Venturi meter

Consider the fluid to be non-viscous, incompressible (of density ρ) and in horizontal steady flow. Let the pressure and velocity respectively be p_X and v_X at X, and be p_Y and v_Y at Y on the same streamline as X. Applying Bernoulli's equation at X and Y gives

$$p_X + \tfrac{1}{2}\rho v_X^2 = p_Y + \tfrac{1}{2}\rho v_Y^2$$

If the cross-sectional areas at X and Y are A_X and A_Y, then from the equation of continuity

$$A_X v_X = A_Y v_Y \quad \text{i.e.} \quad v_Y = \frac{A_X v_X}{A_Y}$$

$$\therefore \quad p_X + \tfrac{1}{2}\rho v_X^2 = p_Y + \tfrac{1}{2}\rho\left(\frac{A_X v_X}{A_Y}\right)^2$$

$$\therefore \quad p_X - p_Y = \tfrac{1}{2}\rho\left(\frac{A_X^2}{A_Y^2} - 1\right)v_X^2$$

Thus by measuring the pressures p_X and p_Y and knowing ρ, A_X and A_Y, it is possible to find the velocity v_X of the fluid in the unconstricted (main) section of the pipe.

EXAMPLE 12.1

Calculate the velocity with which a liquid emerges from a small hole in the side of a tank of large cross-sectional area if the hole is 0.2 m below the surface (Assume $g = 10\,\text{m}\,\text{s}^{-2}$.)

Solution

Refer to Fig. 12.7. We shall assume that we are dealing with a non-viscous, incompressible liquid in steady flow, in which case we may apply Bernoulli's equation to points X and Y on the streamline XY.

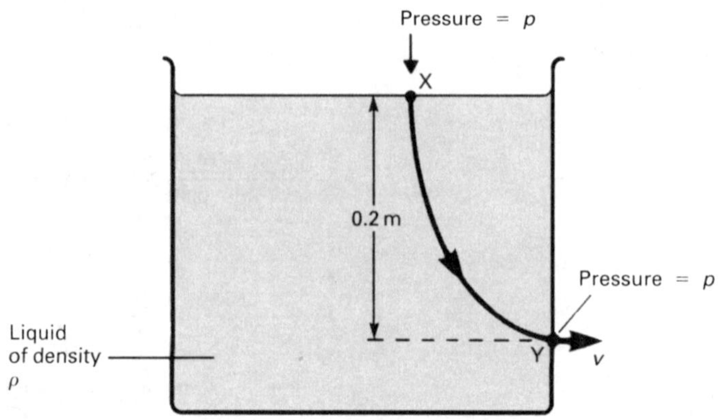

Fig. 12.7 Diagram for Example 12.1

Thus, taking the pressure, height and velocity at X to be p_X, h_X and v_X, and the pressure, height and velocity at Y to be p_Y, h_Y and v_Y, we may put

$$p_X + \rho g h_X + \tfrac{1}{2}\rho v_X{}^2 = p_Y + \rho g h_Y + \tfrac{1}{2}\rho v_Y{}^2 \qquad [12.3]$$

The pressure at both X and Y is atmospheric pressure p, and therefore

$$p_X = p_Y = p$$

Taking heights to be measured from the level of Y we have

$$h_X = 0.2\,\text{m} \qquad h_Y = 0$$

If we assume that the tank is wide enough for the rate at which the surface level falls to be negligible, then

$$v_X = 0 \qquad v_Y = v \text{ (the velocity of emergence)}$$

Substituting in equation [12.13] gives

$$p + \rho \times 10 \times 0.2 + 0 = p + 0 + \tfrac{1}{2}\rho v^2$$

i.e. $v = \sqrt{2 \times 10 \times 0.2} = 2\,\text{m}\,\text{s}^{-1}$

In general $v = \sqrt{2gh}$ and is equal to the velocity acquired by a body falling from rest through a height h – a result which is known as **Torricelli's theorem**. In practice v would be less than $2\,\text{m}\,\text{s}^{-1}$ because of viscous effects.

12.5 THE PITOT–STATIC TUBE

The Pitot–static tube is a device used to measure the velocity of a moving fluid. It consists of two manometer tubes – the Pitot tube and the static tube. The Pitot tube has its opening facing the fluid flow; the static tube has its opening at right angles to this.

When the Pitot–static tube is used to measure the velocity of a flowing liquid, the liquid itself can be used as the manometer liquid (Fig. 12.8). Providing the liquid has reached its equilibrium level in the Pitot tube, the liquid at Y will be stationary (i.e. Y is a **stagnation point**). Suppose that X is a point on the same streamline as Y, but sufficiently distant from it for the liquid there to have its full velocity, v. If the liquid is in steady flow and can be considered non-viscous and incompressible, we may apply Bernoulli's equation to X and Y. Bearing in mind that the flow is horizontal, this gives

$$p_X + \tfrac{1}{2}\rho v^2 = p_Y \qquad [12.4]$$

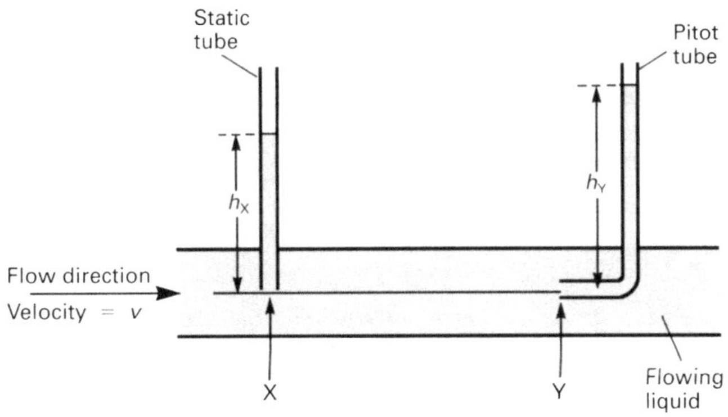

Fig. 12.8 Pitot–static tube to measure velocity of a liquid

where p_X and p_Y are the pressures in the liquid at X and Y respectively. Rearranging equation [12.4] gives

$$v = \sqrt{\frac{2}{\rho}(p_Y - p_X)} \qquad [12.5]$$

or

$$v = \sqrt{\frac{2}{\rho}\left(\begin{matrix}\text{Pressure at}\\ \text{stagnation point}\end{matrix} - \begin{matrix}\text{Pressure where}\\ \text{fluid velocity} = v\end{matrix}\right)} \qquad [12.6]$$

The pressure, p_Y, at Y is equal to the pressure exerted by the liquid in the Pitot tube plus atmospheric pressure, p_A. Therefore

$$p_Y = h_Y \rho g + p_A$$

The pressure, p_X, at X is equal to the pressure exerted by the liquid in the static tube plus atmospheric pressure, and therefore

$$p_X = h_X \rho g + p_A$$

Therefore

$$p_Y - p_X = \rho g(h_Y - h_X)$$

Therefore by equation [12.5]

$$v = \sqrt{2g(h_Y - h_X)} \qquad [12.7]$$

Gases cannot be used as manometer fluids, and therefore the type of Pitot–static tube used to measure gas velocities has the form shown in Fig. 12.9. The head of liquid in the manometer measures the difference ($h\rho_m g$) between the pressure at the stagnation point, Y, and the pressure at X, where the gas has velocity v. Therefore by equation [12.6]

$$v = \sqrt{\frac{2}{\rho}(h\rho_m g)}$$

The terms: 'static pressure', 'dynamic pressure' and 'total pressure' are often used in connection with flowing fluids.

Static Pressure

The static pressure at a point in a flowing fluid is the actual pressure at that point. As such, it is the pressure measured in such a way that the measurement does not affect, and is not affected by, the velocity of the fluid. One way of achieving this is with a manometer whose opening is parallel to the flow direction. (The static tube in Fig. 12.8 measures the static pressure at X.)

Fig. 12.9
Pitot–static tube to measure the velocity of a gas

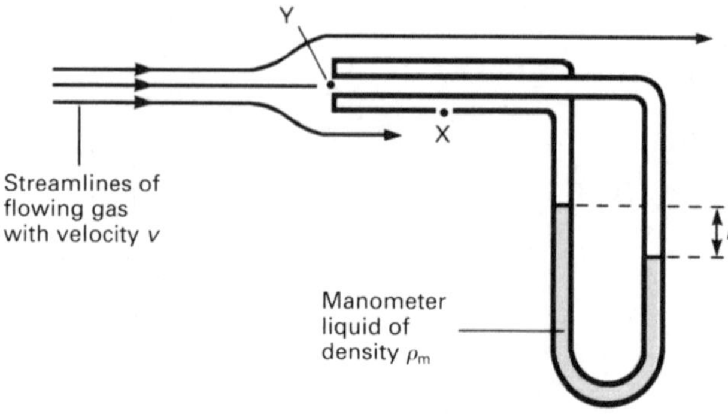

Dynamic Pressure

For a fluid of density ρ, moving with velocity v, the dynamic pressure is $\frac{1}{2}\rho v^2$.

Dynamic pressure is not a true pressure – simply a quantity which has the same dimensions (see Appendix 2) as pressure.

Total Pressure

The total pressure is the sum of the static and dynamic pressures. The pressure and the total pressure are equal to each other at a stagnation point.

12.6 THE COEFFICIENT OF VISCOSITY (η)

The coefficient of viscosity of a fluid is a measure of the degree to which the fluid exhibits viscous effects. The higher the coefficient of viscosity, the more viscous the fluid – the coefficient of viscosity of golden syrup at room temperature is about 10^5 times that of water at the same temperature. The coefficients of viscosity of most fluids have a marked temperature dependence; those of liquids decrease with increasing temperature, whereas those of gases increase with increasing temperature.

Viscous effects are due to the frictional force which exists between two adjacent layers of fluid which are in relative motion. Consider a viscous fluid undergoing laminar flow, and consider in particular two parallel layers of area A separated by a small distance δy and whose velocities are v and $v+\delta v$ (Fig. 12.10). It was suggested by Newton that the frictional force F between the layers is proportional to A and to the **velocity gradient** $\delta v/\delta y$, i.e.

$$F \propto A\frac{\delta v}{\delta y}$$

(This is the opposite of the situation with solids – the frictional force between the surfaces of two solids is independent of the area of contact and of the relative velocity. See section 2.12.) Introducing a constant of proportionality, η, we have

$$F = \eta A\frac{\delta v}{\delta y} \qquad [12.8]$$

Equation [12.8] is sometimes called **Newton's law of viscosity**. It holds for all gases and for many liquids. Such liquids are called **Newtonian liquids**; water

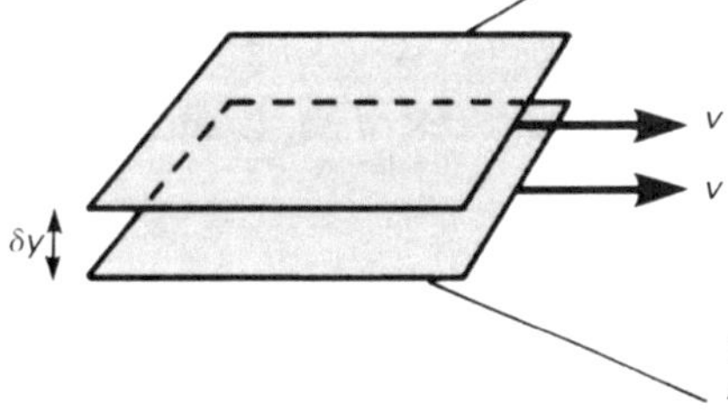

Fig. 12.10
Friction between successive layers of a liquid

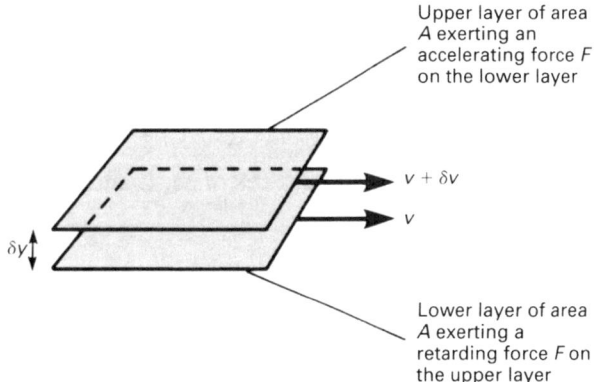

is an example. For any given values of A and $\delta v/\delta y$, F is large for those fluids which have high values of η, and therefore η is a measure of the viscosity of the fluid, and can meaningfully be called its **coefficient of viscosity**.

There are some liquids, called **non-Newtonian liquids**, for which F is not proportional to $\delta v/\delta y$ and which, therefore, do not have constant values of η – i.e. they do not have coefficients of viscosity in the normal sense. Oil-paint is an example of a non-Newtonian liquid.

Notes

(i) The units of η are $\mathrm{N\,s\,m^{-2}} = \mathrm{kg\,m^{-1}\,s^{-1}}$.

(ii) By rearranging equation [12.8] as

$$\eta = \frac{F/A}{\delta v/\delta y}$$

and then comparing it with equation [11.7] of section 11.6 we can draw an analogy between η and the shear modulus G of a solid. In each case F/A is the shear stress. In the case of a solid, though, the stress produces a fixed strain ($\Delta x/y$) proportional to the stress, whereas with a Newtonian fluid the strain increases without limit as long as the stress is applied and it is the rate of change of the strain ($\delta v/\delta y$) which is proportional to the stress.

12.7 POISEUILLE'S FORMULA

Consider a viscous liquid undergoing steady flow through a pipe of circular cross-section. Because of viscous drag the velocity varies from a maximum at the centre of the pipe to zero at the walls. We shall use dimensional analysis (see Appendix 2) to derive an expression for the volume V of liquid passing any section of the pipe in time t.

It can reasonably be supposed that the **rate of volume flow** V/t depends on (i) the coefficient of viscosity η of the liquid, (ii) the radius r of the pipe, and (iii) the pressure gradient p/l, where p is the pressure difference between the ends of the pipe and l is its length. If we express the relationship as

$$\frac{V}{t} = k\eta^x r^y \left(\frac{p}{l}\right)^z$$

where k is a dimensionless constant and x, y and z are unknown indices, then since each side of the equation must have the same dimensions,

$$[V/t] = [\eta^x][r^y][(p/l)^z]$$

$$\therefore \quad L^3T^{-1} = (ML^{-1}T^{-1})^x(L)^y(ML^{-2}T^{-2})^z$$

$$\text{i.e.} \quad L^3T^{-1} = M^{x+z}L^{y-x-2z}T^{-x-2z}$$

Equating the indices of M, L and T on both sides gives

$$0 = x + z \qquad \text{(for M)}$$
$$3 = y - x - 2z \qquad \text{(for L)}$$
$$-1 = -x - 2z \qquad \text{(for T)}$$

Solving gives $z = 1, x = -1, y = 4$. The relationship is therefore

$$\frac{V}{t} = k\frac{r^4}{\eta}\left(\frac{p}{l}\right)$$

The value of k cannot be found by using dimensional analysis, however, mathematical analysis shows that its value is $\pi/8$, and therefore

$$\frac{V}{t} = \frac{\pi r^4 p}{8\eta l} \qquad [12.9]$$

This is called **Poiseuille's formula** in recognition of Poiseuille who in 1844 made the first thorough experimental investigation of the steady flow of a liquid (water) through a pipe. The formula applies only to Newtonian fluids (see section 12.6) which are undergoing steady flow.

The speed of bulk flow is defined as the rate of volume flow divided by the cross-sectional area of the pipe. Steady flow occurs only when the speed of bulk flow is less than a certain critical value v_c. Since Poiseuille's formula applies only to steady flow, it does not hold when the speed of bulk flow exceeds v_c. Experiment shows that for cylindrical pipes

$$v_c \approx \frac{1100\,\eta}{r\rho}$$

where ρ and η are the density and coefficient of viscosity of the fluid and r is the radius of the pipe.

12.8 MEASUREMENT OF η BY USING POISEUILLE'S FORMULA

The method makes use of the apparatus shown in Fig. 12.11 and is suitable for liquids which flow easily (e.g. water). (For high-viscosity liquids see section 12.10.)

The liquid under test flows steadily through the capillary tube from a constant head device and the volume V of liquid which emerges in a known time t is measured. The pressure difference between the ends of the capillary tube is $h\rho g$ (where ρ is the density of the liquid and g is the acceleration due to gravity) and therefore from equation [12.9]

$$\frac{V}{t} = \frac{\pi r^4 h\rho g}{8\,\eta l}$$

Fig. 12.11
Apparatus for measuring η using Poiseuille's formula

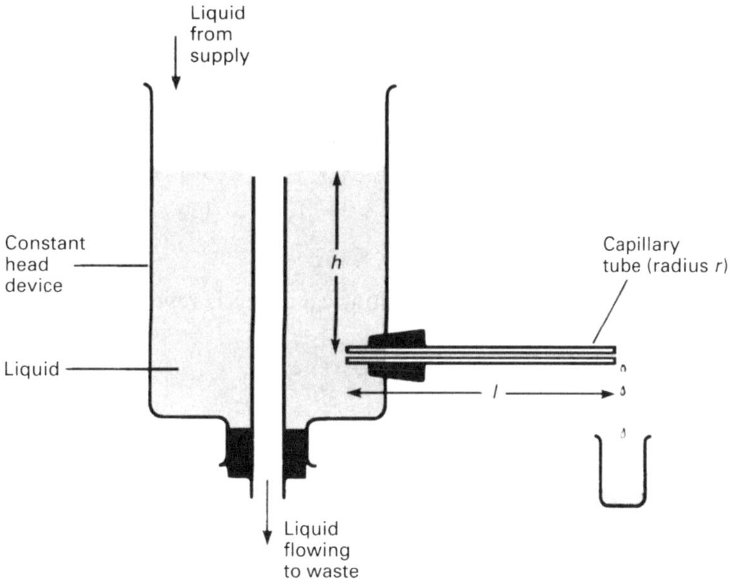

Poiseuille's formula applies only if the flow is steady. In order to check that this is the case the measurements are repeated for different values of h and a graph of (V/t) against h is plotted. The graph is linear providing h has been kept below the value at which the rate of flow is so high that turbulence sets in. The gradient of the graph is $\pi r^4 \rho g/(8\eta l)$, enabling η to be calculated once r and l have been measured (ρ and g are found from tables). The mean radius r of the tube can be found by measuring the length and mass of a mercury thread introduced into the tube.

Notes

(i) Great care is needed when measuring r because it appears in the calculation of η as r^4. This makes the percentage error in η due to an error in r four times the percentage error in r.

(ii) A capillary tube is used because r needs to be small so that h is large enough to be measured accurately.

12.9 STOKES' LAW AND TERMINAL VELOCITY

Derivation of Stokes' Law

Consider a sphere of radius r moving with velocity v through a fluid whose coefficient of viscosity is η. The sphere experiences a viscous force F which acts in the opposite direction to that in which the sphere is moving. We shall use dimensional analysis (see Appendix 2) to obtain an expression for F.

It can reasonably be supposed that F depends only on r, η and v. (Though the mass of the sphere and the density of the fluid have a bearing on how the velocity varies under the effect of an applied force, they have no direct influence on the drag force.) If we express the relationship as

$$F = kr^x\eta^y v^z$$

where k is a dimensionless constant and x, y and z are unknown indices, then since each side of the equation must have the same dimensions

$$[F] = [r^x][\eta^y][v^z]$$

$$\therefore \quad \mathrm{MLT^{-2}} = (\mathrm{L})^x(\mathrm{ML^{-1}T^{-1}})^y(\mathrm{LT^{-1}})^z$$

$$\text{i.e.} \quad \mathrm{MLT^{-2}} = \mathrm{M}^y\mathrm{L}^{x+z-y}\mathrm{T}^{-y-z}$$

Equating the indices of M, L and T on both sides gives

$$1 = y \quad \text{(for M)}$$

$$1 = x + z - y \quad \text{(for L)}$$

$$-2 = -y - z \quad \text{(for T)}$$

Solving gives $y = 1$, $z = 1$, $x = 1$. The relationship is therefore

$$F = kr\eta v$$

A full mathematical analysis reveals that $k = 6\pi$, and therefore

$$F = 6\pi r\eta v \qquad [12.10]$$

Equation [12.10] was first derived by Stokes and is known as **Stokes' law**.

Notes

(i) Strictly the law applies only to a fluid of infinite extent.

(ii) Stokes' law does not hold if the sphere is moving so fast that conditions are not streamline.

Terminal Velocity

Consider a sphere falling from rest through a viscous fluid. The forces acting on the sphere are its weight W, the upthrust U due to the displaced fluid, and the viscous drag F (see Fig. 12.12). Initially the downward force W is greater than the upward force, $U + F$, and the sphere accelerates downwards. As the velocity of the sphere increases so too does the viscous drag, and eventually $U + F$ is equal to W. The sphere continues to move downwards but, because there is now no net force acting on it, its velocity has a constant maximum value known as its **terminal velocity** v_t.

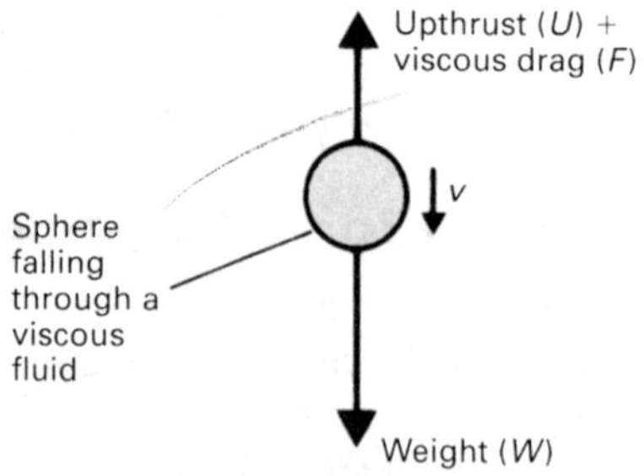

Fig. 12.12
Sphere falling through a viscous fluid

If ρ_f and ρ_s are the densities of the fluid and the sphere respectively, then

$$W = \tfrac{4}{3}\pi r^3 \rho_s g$$

and

$$U = \tfrac{4}{3}\pi r^3 \rho_f g$$

At the terminal velocity

$$U + F = W \qquad [12.11]$$

and

$$F = 6\pi r\eta v_t$$

where η is the coefficient of viscosity of the fluid. Substituting for W, U and F in equation [12.10] gives

$$\tfrac{4}{3}\pi r^3 \rho_f g + 6\pi r \eta v_t = \tfrac{4}{3}\pi r^3 \rho_s g$$

i.e. $$6\pi r \eta v_t = \tfrac{4}{3}\pi r^3 (\rho_s - \rho_f) g$$

i.e. $$v_t = \frac{2r^2(\rho_s - \rho_f)g}{9\eta} \qquad [12.12]$$

12.10 MEASUREMENT OF η BY USING STOKES' LAW

The method is suitable for liquids of high viscosity such as glycerine and treacle, and makes use of equation [12.12]. (For low-viscosity liquids see section 12.8.) The liquid whose coefficient of viscosity η is being determined is contained in a large measuring cylinder (Fig. 12.13). A small ball-bearing of radius r is dropped gently into the liquid. The time taken for the ball to fall from mark A to mark B is determined. Providing A is sufficiently far below the surface, the bearing will have reached its terminal velocity v_t before reaching A, in which case $v_t = AB/t$. If ρ_f and ρ_s are the densities of the liquid and the sphere respectively, then from equation [12.12]

$$\frac{AB}{t} = \frac{2r^2(\rho_s - \rho_f)g}{9\eta}$$

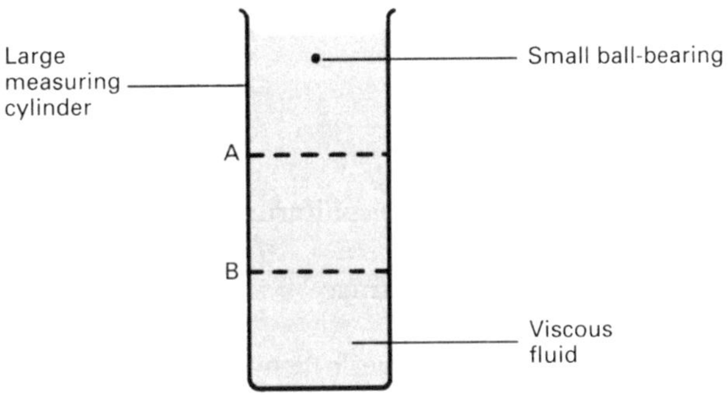

Fig. 12.13 Apparatus for measuring η using Stoke's law

A micrometer can be used to measure r; ρ_s and ρ_f are found from tables or are determined in additional experiments; hence η can be deduced.

Notes

(i) Stokes' law applies strictly only when the fluid is of infinite extent. The error due to the impossibility of fulfilling this condition is reduced by using a measuring cylinder which is wide compared with the diameter of the ball-bearing, and by having B well away from the bottom.

(ii) If the velocity of the bearing is so large that it produces turbulence, Stokes' law does not hold and equation [12.12] is not applicable. Using a highly viscous liquid and a small ball-bearing avoids this problem and also makes t large enough to be measured accurately.

QUESTIONS ON SECTION B

Assume $g = 10\,\mathrm{m\,s^{-2}} = 10\,\mathrm{N\,kg^{-1}}$ unless otherwise stated.

SOLIDS AND LIQUIDS (Chapter 9)

B1 The graph shows the potential energy of a pair of atoms at different distances apart.

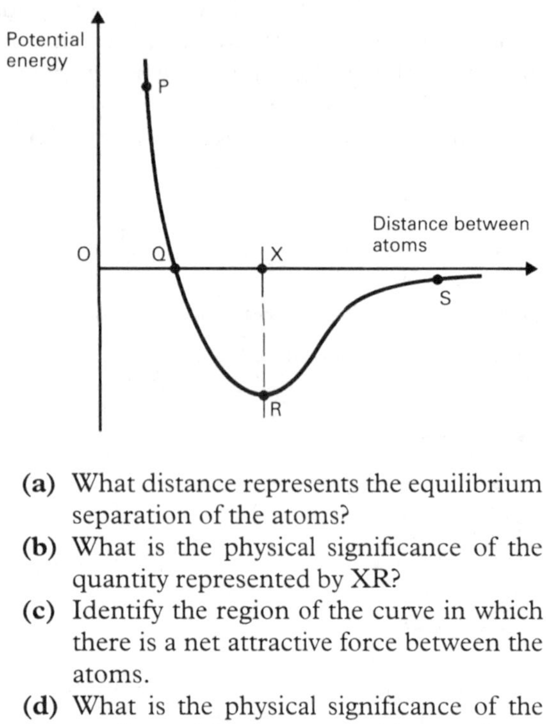

(a) What distance represents the equilibrium separation of the atoms?
(b) What is the physical significance of the quantity represented by XR?
(c) Identify the region of the curve in which there is a net attractive force between the atoms.
(d) What is the physical significance of the gradient of the tangent to this curve?

[J, '92]

B2 A certain molecule consists of two identical atoms, each of mass 1.7×10^{-27} kg. The equilibrium separation of the atoms in the molecule is x_0. The figure above shows the way in which the force F of repulsion between the atoms varies with their separation x.
(a) Account for the general shape of the graph and use it to find x_0.
(b) Sketch a graph of the potential energy V of the molecule as a function of x, marking the position of x_0 on the x-axis. How is V related to F?

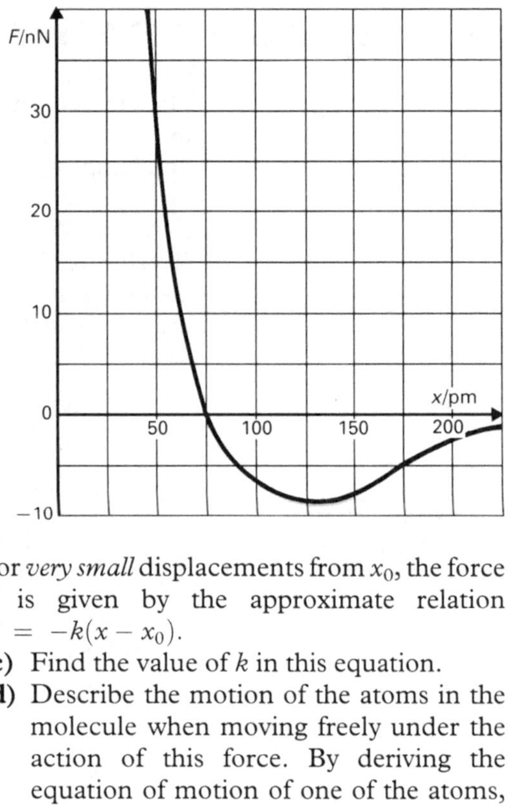

For *very small* displacements from x_0, the force F is given by the approximate relation $F = -k(x - x_0)$.
(c) Find the value of k in this equation.
(d) Describe the motion of the atoms in the molecule when moving freely under the action of this force. By deriving the equation of motion of one of the atoms, or otherwise, find the frequency of the motion. [C]

B3 The very simplified curves (p. 209) represent, for two adjacent atoms or molecules, the variation with the separation r between their centres of the potential energy V_r due to the interaction between them, and the force F_r between them.
(a) Explain the general relation between the F_r curve and the V_r curve, and the significance of the broken lines A and B.
(b) With reference to the V_r curve (a), explain how:
  (i) the lower part is consistent with molecules in a solid oscillating about a mean position,
  (ii) the effect of a rise in temperature could be represented,

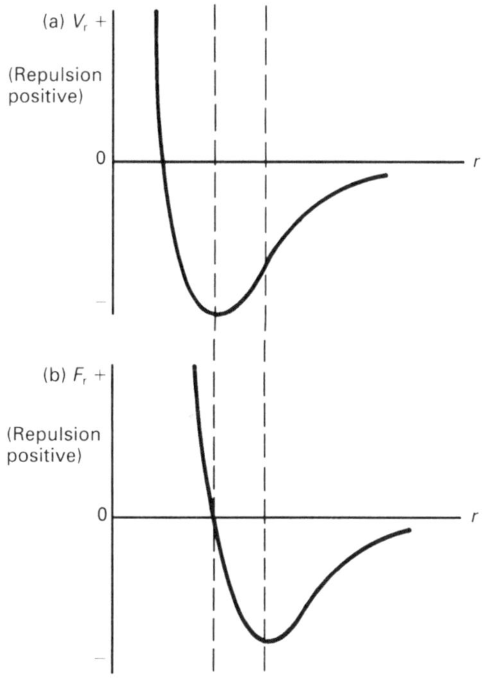

(iii) the thermal expansion of a solid on heating is accounted for,
(iv) the latent heat of vaporization (or sublimation) per atom can be estimated.

(c) With reference to the F_r curve (b), explain how:
(i) the property of elasticity is represented,
(ii) Hooke's law is accounted for.

(d) How can it be forecast that a solid will rupture under a large enough stress, and will melt at a high enough temperature? [O]

B4 The specific latent heat of vaporization of a particular liquid is $2.0 \times 10^5\,\text{J kg}^{-1}$, its relative molecular mass is 30 and its coordination number is 10. On the basis of this data, and given that the Avogadro constant $= 6 \times 10^{23}\,\text{mol}^{-1}$, obtain a value for the binding energy of a pair of (adjacent) molecules of the liquid.

On freezing, the coordination number increases to 12. Estimate the specific latent heat of fusion.

B5 **(a)** Calculate the potential energy, in eV, per pair of atoms of a solid for which the latent heat of sublimation is $1.3 \times 10^4\,\text{J mol}^{-1}$ and the number of neighbours per atom is 6. The Avogadro constant, $N_A = 6.0 \times 10^{23}\,\text{mol}^{-1}$ and $1\,\text{eV} = 1.6 \times 10^{-19}\,\text{J}$.

(b) For a pair of atoms, sketch a graph showing how the potential energy per atom pair varies with the distance between the atoms. Show on your graph **(i)** the equilibrium separation, r_0, **(ii)** the value of the energy calculated in (a).

(c) Mark on your sketch graph a point P corresponding to a separation *other than the equilibrium value* and explain how you would determine, from the graph, the force between the atoms at P. Indicate whether you consider the force at P to be attractive or repulsive. [J]

B6 The latent heat of vaporization of water is $4 \times 10^4\,\text{J mol}^{-1}$ at the boiling point, and each water molecule has, on average, 10 near neighbours. Estimate the binding energy for a pair of water molecules.

Ignore the work done in expansion in your calculation, but explain whether this assumption leads to an overestimate or an underestimate of the binding energy.

(The Avogadro constant $N_A = 6 \times 10^{23}\,\text{mol}^{-1}$.) [W]

B7 Calculate the average volume occupied by a single molecule of a solid whose density is $1.2 \times 10^3\,\text{kg m}^{-3}$ and whose relative molecular mass is 90. (The Avogadro constant $= 6 \times 10^{23}\,\text{mol}^{-1}$.)

Hence, stating any assumption that you make, estimate the distance between the centres of two adjacent molecules of the solid.

B8 **(a)** Estimate the diameter of a water molecule given that the relative molecular mass of water is 18 and its density is $1000\,\text{kg m}^{-3}$.

(b) Using the value obtained in part (a), estimate the binding energy of water molecules, given that the surface tension of water is $0.072\,\text{N m}^{-1}$ and that the number of near neighbours in the water is 10.

(Assume that $\gamma = N_s z\varepsilon/4$. Take $N_A = 6.0 \times 10^{23}\,\text{mol}^{-1}$) [W, '91]

B9 An alloy contains two metals, X and Y, of densities $3.0 \times 10^3\,kg\,m^{-3}$ and $5.0 \times 10^3\,kg\,m^{-3}$ respectively. Calculate the density of the alloy
(a) if the volume of X is twice that of Y, and
(b) if the mass of X is twice that of Y.

B10 An alloy of two metals, X and Y, has a volume of $5.0 \times 10^{-4}\,m^3$ and a density of $5.6 \times 10^3\,kg\,m^{-3}$. The densities of X and Y are $8.0 \times 10^3\,kg\,m^{-3}$ and $4.0 \times 10^3\,kg\,m^{-3}$ respectively. Find the mass of X and the mass of Y.

FLUIDS AT REST (Chapter 10)

B11 An open U-tube manometer containing mercury is used to measure the pressure of a gas. The mercury level in the open tube is 600 mm higher than that in the limb which is in contact with the gas. What is the pressure (in pascals) of the gas?

(Density of mercury $= 1.36 \times 10^4\,kg\,m^{-3}$, atmospheric pressure $= 1.01 \times 10^5\,Pa$, $g = 9.81\,m\,s^{-2}$.)

B12 The diagram shows a mercury manometer recording a pressure of 150 kPa. The atmospheric pressure is 100 kPa.
(Take the density of mercury as $13\,600\,kg\,m^{-3}$.)

What is the height difference h of the mercury surfaces? [O, '91*]

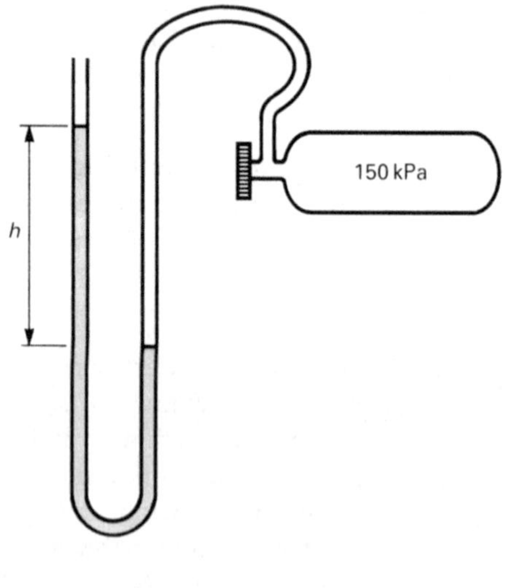

B13 A body has a weight of 160 N when weighed in air and a weight of 120 N when totally immersed in a liquid of relative density 0.8. What is the relative density of the body?

B14 An object is suspended from a force meter ('spring balance') capable of reading forces to within $\pm 0.01\,N$. The object is found to have a weight of 4.92 N in air and 3.87 N when immersed in water.
(a) Calculate the density of the material from which the object is made.
(b) Discuss the reading which could be obtained if the object were suspended from the force meter within an evacuated enclosure.

(Density of air $= 1.3\,kg\,m^{-3}$, density of water $= 1.0 \times 10^3\,kg\,m^{-3}$.) [S]

B15 A tank contains a liquid of density $1.2 \times 10^3\,kg\,m^{-3}$. A body of volume $5.0 \times 10^{-3}\,m^3$ and density $9.0 \times 10^2\,kg\,m^{-3}$ is totally immersed in the liquid and is attached by a thread to the bottom of the tank. What is the tension in the thread?

B16 A ball with a volume of $32\,cm^3$ floats on water with exactly half of the ball below the surface. What is the mass of the ball? (Density of water $= 1.0 \times 10^3\,kg\,m^{-3}$.)

B17 An object floats in a liquid of density $1.2 \times 10^3\,kg\,m^{-3}$ with one quarter of its volume above the liquid surface. What is the density of the object?

B18 A hot-air balloon has a volume of $500\,m^3$. The balloon moves upwards at a *constant* speed in air of density $1.2\,kg\,m^{-3}$ when the density of the hot air inside it is $0.80\,kg\,m^{-3}$.
(a) What is the combined mass of the balloon and the air inside it?
(b) What is the upward acceleration of the balloon when the temperature of the air inside it has been increased so that its density is $0.7\,kg\,m^{-3}$?

B19 An object with a volume of $1.0 \times 10^{-5}\,m^3$ and density $4.0 \times 10^2\,kg\,m^{-3}$ floats on water in a tank of cross-sectional area $1.0 \times 10^{-3}\,m^2$.
(a) By how much does the water level drop when the object is removed?
(b) Show that this decrease in water level reduces the force on the base of the tank by an amount equal to the weight of the object.

(Density of water $= 1.0 \times 10^3\,kg\,m^{-3}$.)

B20 Derive, explaining the meaning of the terms on the right-hand side, the approximate relationship

$$\sigma \approx \tfrac{1}{4} n A \varepsilon_0$$

where σ is the work done in isothermally creating unit surface area of a liquid. Explain why the relationship is approximate.

B21 Show that σ, the work done in isothermally creating unit surface area of a liquid, is equal to γ, the force per unit length acting in the surface of the liquid at right-angles to one side of an imaginary line drawn in the surface.

B22 The specific latent heat of vaporization and the surface tension of a particular liquid are $7.5 \times 10^4\,\mathrm{J\,kg^{-1}}$ and $4.0 \times 10^{-2}\,\mathrm{N\,m^{-1}}$ respectively. The relative molecular mass of the liquid is 40. Estimate the number of molecules in $1\,\mathrm{cm^2}$ of the liquid surface. (The Avogadro constant $= 6 \times 10^{23}\,\mathrm{mol^{-1}}$.)

B23 Explain briefly, with the aid of a diagram, what you would expect to happen to a nearly spherical water droplet resting on a clean horizontal surface if a tiny amount of detergent were added to it.

How do you account for the change that might occur? [L]

B24 The velocity v of surface waves on a liquid may be related to their wavelength λ, the surface tension of the liquid σ and its density ρ by the following equation

$$v = k\lambda^{\alpha}\sigma^{\beta}\rho^{\gamma}$$

where k is a dimensionless constant.

Find values for α, β and γ by a dimensional argument. [W]

B25 A spherical drop of mercury of radius 2 mm falls to the ground and breaks into 10 smaller drops of equal size. Calculate the amount of work that has to be done. (Surface tension of mercury $= 4.72 \times 10^{-1}\,\mathrm{N\,m^{-1}}$.)

What is the minimum speed with which the original drop could have hit the ground? (Density of mercury $= 1.36 \times 10^4\,\mathrm{kg\,m^{-3}}$.)

B26 Two soap bubbles have radii of 3 cm and 4 cm. The bubbles are in a vacuum and they combine to form a single larger bubble. Calculate the radius of this bubble. (You may assume that the surface tension of soap solution is constant throughout.)

B27 A glass barometer tube has an internal radius of 3 mm. Calculate the actual atmospheric pressure on a day when the height of the mercury column is 760.2 mm. (Surface tension of mercury $= 4.72 \times 10^{-1}\,\mathrm{N\,m^{-1}}$, angle of contact of mercury with glass $= 137^\circ$, density of mercury $= 1.36 \times 10^4\,\mathrm{kg\,m^{-3}}$, acceleration due to gravity $= 9.81\,\mathrm{m\,s^{-2}}$.)

B28 A soap bubble whose radius is 12 mm becomes attached to one of radius 20 mm. Calculate the radius of curvature of the common interface.

B29 Define the terms *surface tension*, *angle of contact*.

The end of a clean glass capillary tube, having internal diameter 0.6 mm, is dipped into a beaker containing water, which rises up the tube to a vertical height of 5.0 cm above the water surface in the beaker. Calculate the surface tension of water. (Density of water $= 1000\,\mathrm{kg\,m^{-3}}$.)

What would be the difference if the tube were not perfectly clean, so that the water did not wet it, but had an angle of contact of 30° with the tube surface? [S]

B30 Define *surface tension*. Give a concise explanation of the origin of surface tension in terms of intermolecular forces.

The pressure difference on the two sides of a spherical liquid–gas interface is $2\gamma/R$; what do γ and R represent? (*You may use this expression in* (a) *if you so wish, but you are advised to use it in parts* (b) *and* (c).)

(a) Derive an expression for the height of the liquid column in a vertical, uniform capillary tube. (Neglect any correction for the mass of the meniscus and assume that the angle of contact is zero.) Describe the experimental determination of the surface tension of water by the capillary rise method giving with reasons, a suitable value for the radius of the tube.

(b) The two vertical arms of a manometer, containing water, have different internal radii of $10^{-3}\,\mathrm{m}$ and $2 \times 10^{-3}\,\mathrm{m}$ respectively. Determine the difference in height of the two liquid levels when the arms are open to the atmosphere.

(c) Explain why the pressure difference is not constant across the meniscus of the liquid column in a capillary tube, and discuss the general shape of the meniscus.

The surface tension and density of water are $7 \times 10^{-2}\,\mathrm{N\,m^{-1}}$ and $10^3\,\mathrm{kg\,m^{-3}}$ respectively. [W]

B31 By considering the work done per unit area in increasing the surface area of a bubble blown in a liquid, or otherwise, derive an expression for the excess pressure p inside a bubble of radius r.

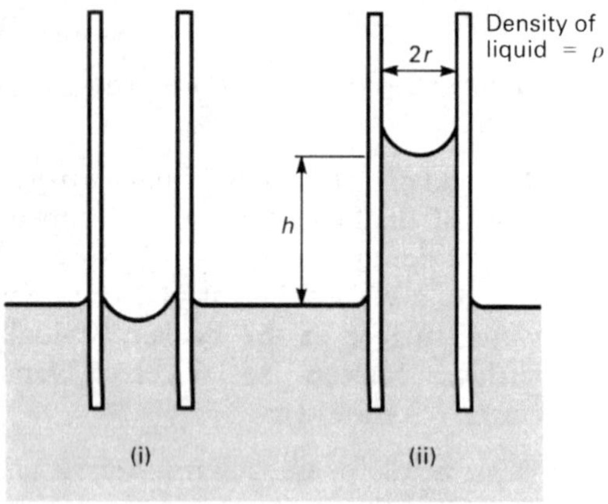

The diagrams above represent glass capillary tubes dipping into a liquid. Explain why the situation represented by (i) is unstable while that in (ii) is stable.

Use the data given in diagram (ii) to derive an expression for the height h to which the liquid rises, given that the angle of contact between the liquid and glass is zero.

By considering intermolecular forces explain why the surface of a liquid is different from the bulk of the liquid.

Suggest why there might be a connection between the surface energies (surface tensions) of liquids and their normal boiling points. [L]

B32 The pressure difference p across a spherical surface of radius r between air and a liquid, where γ is the surface tension of the liquid, is given by

$$p = \frac{2\gamma}{r}$$

(a) Show that this expression is consistent with γ being measured in $\mathrm{N\,m^{-1}}$. It can be shown that γ is also equal to the energy stored per unit area in the surface. Show that this is also consistent with γ being measured in $\mathrm{N\,m^{-1}}$.

(b) Describe a method for measuring γ which is based on measuring the excess pressure in a bubble.

(c) Using the energy definition of γ given above calculate the energy stored in the surface of a soap bubble 2.0 cm in radius if its surface tension is $4.5 \times 10^{-2}\,\mathrm{N\,m^{-1}}$. If the thickness of the surface is 6.0×10^{-7} m and the density of the soap solution is $1000\,\mathrm{kg\,m^{-3}}$, calculate the speed with which the liquid fragments will fly apart when the bubble is burst. What assumptions have you made in your calculation? [L]

B33 **(a)** Draw and label a diagram of apparatus suitable for measuring the surface tension of water by Jaeger's method.

Assume that the pressure p within the apparatus *when it is assembled* equals the pressure p_0 of the atmosphere outside. Sketch a graph which shows how the pressure difference $p - p_0$ changes with time from the instant p begins to increase until the moment a bubble is about to break away from the bottom of the capillary for the third time.

How are the pressure differences shown in the graph related to **(i)** the position of the liquid meniscus in the capillary and **(ii)** the radius of the bubble formed at the bottom of the capillary?

State which quantities you would measure if you were using this apparatus to determine the surface tension of water and describe how you would measure them.

(b) The diagram, (p. 213) which is not to scale, shows two capillary tubes of uniform bore fitting tightly into a short length of rubber tubing. AB and CD are two threads of water. The capillary tube containing CD is kept horizontal while that containing AB is raised through an angle θ until the water surface at D is both flat and vertical.

(i) Calculate the surface tension of water given that θ is 10.5°, AB is 11.4 cm,

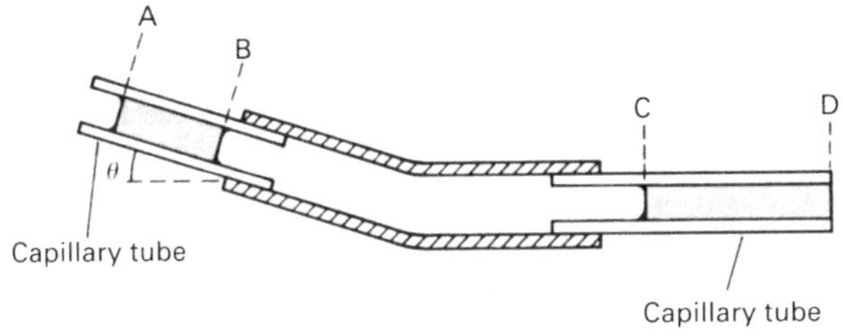

the radius of the capillary tube at C is 0.72 mm and the density of water is $1.00 \times 10^3\ \text{kg m}^{-3}$. The angle of contact between water and glass is zero. (You may assume the relation $\Delta p = 2\gamma/r$, and that $g = 9.8\ \text{m s}^{-2}$.)

(ii) Suggest an experimental procedure to determine when the water surface at D is flat. [L]

ELASTICITY (Chapter 11)

B34 Define *tensile stress*, *tensile strain*, *Young's modulus*.

A mass of 11 kg is suspended from the ceiling by an aluminium wire of length 2 m and diameter 2 mm. What is:

(a) the extension produced,

(b) the elastic energy stored in the wire?

The Young's modulus of aluminium is 7×10^{10} Pa (N m^{-2}). [S]

B35 The maximum upward acceleration of a lift of total mass 2500 kg is $0.5\ \text{m s}^{-2}$. The lift is supported by a steel cable, which has a maximum safe working stress of 1.0×10^8 Pa. What minimum area of cross-section of cable should be used? [C(O)]

B36 An elastic string of cross-sectional area $4\ \text{mm}^2$ requires a force of 2.8 N to increase its length by one tenth. Find Young's modulus for the string. If the original length of the string was 1 m, find the energy stored in the string when it is so extended. [W]

B37 **(a)** For an elastic wire under tension there is, under certain conditions, a simple relation between the *applied stress* and the *strain produced*. Explain the meaning of the terms in italics, state the relation and indicate the conditions that must be fulfilled.

(b) A long thin vertical steel wire is fixed at the upper end. Describe, giving reasons for the design of the apparatus used, how you would measure the extensions caused by the addition of various loads at the lower end.

(c) A massive stone pillar 20 m high and of uniform cross-section rests on a rigid base and supports a vertical load of 5.0×10^5 N at its upper end. State, with reasons, where in the pillar the maximum compressive stress occurs. If the compressive stress in the pillar is not to exceed $1.6 \times 10^6\ \text{N m}^{-2}$, what is the minimum cross-sectional area of the pillar?

Density of the stone $= 2.5 \times 10^3\ \text{kg m}^{-3}$. [J]

B38 A cylindrical copper wire and a cylindrical steel wire, each of length 1.000 m and having equal diameters are joined at one end to form a composite wire 2.000 m long. This composite wire is subjected to a tensile stress until its length becomes 2.002 m. Calculate the tensile stress applied to the wire.

(The Young modulus for copper $= 1.2 \times 10^{11}$ Pa and for steel $= 2.0 \times 10^{11}$ Pa.) [W, '91]

B39 **(a)** A heavy rigid bar is supported horizontally from a fixed support by two vertical wires, A and B, of the same initial length and which experience the same extension. If the ratio of the diameter of A to that of B is 2 and the ratio of Young's modulus of A to that of B is 2, calculate the ratio of the tension in A to that in B.

(b) If the distance between the wires is *D*, calculate the distance of wire A from the centre of gravity of the bar. [J]

B40 **(a)** Define *stress*, *strain* and the *Young modulus*.

(b) **(i)** Describe an experiment to determine the Young modulus for a material in the form of a wire.

(ii) Which measurement requires particular care, from the point of view of accuracy, and why?

(c) **(i)** Derive an expression for the potential energy stored in a stretched wire.

(ii) A steel wire of diameter 1 mm and length 1.5 m is stretched by a force of

50 N. Calculate the potential energy stored in the wire.
(Young modulus of steel = 2×10^{11} Pa.

(iii) The wire is further stretched to breaking. Where does the stored energy go? [W, 90]

B41 Define Young's modulus and describe a method to measure its value for a uniform elastic wire. State the precautions necessary to ensure an accurate result.
The ends of a uniform wire of cross-sectional area 10^{-6} m^2 and negligible mass are attached to fixed points A and B which are 1 m apart in the same horizontal plane. The wire is initially straight and unstretched. A mass of 0.5 kg is attached to the mid-point of the wire and hangs in equilibrium with the mid-point at a distance 10 mm below AB. Calculate the value of Young's modulus for the wire. [O & C]

B42 **(a)** Describe an experiment using two long, parallel, identical wires to determine the Young modulus for steel. Explain why it is necessary to use two such wires. Indicate what quantities you would measure and what measuring instrument you would use in each case. State what graph you would plot, and show how it is used to calculate the Young modulus.

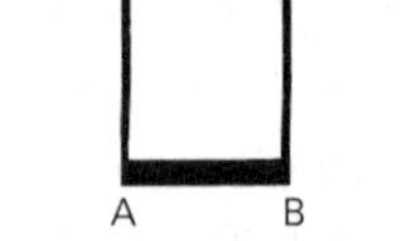

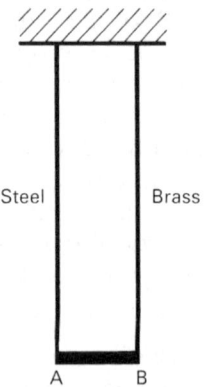

(b) A light rigid bar is suspended horizontally from two vertical wires, one of steel and one of brass, as shown in the diagram. Each wire is 2.00 m long. The diameter of the steel wire is 0.60 mm and the length of the bar AB is 0.20 m. When a mass of 10.0 kg is suspended from the centre of AB the bar remains horizontal.

(i) What is the tension in each wire?

(ii) Calculate the extension of the steel wire and the energy stored in it.

(iii) Calculate the diameter of the brass wire.

(iv) If the brass wire were replaced by another brass wire of diameter 1.00 mm, where should the mass be suspended so that AB would remain horizontal?
(The Young modulus for steel = 2.0×10^{11} Pa, the Young modulus for brass = 1.0×10^{11} Pa.) [J, '91]

B43 **(a)** For moderate loads, most metals are elastic. What is meant by the term *elastic*?

(b) The graph shows a stress–strain diagram for a steel wire, of cross-section area 0.80×10^{-6} m^2, that is stretched to its elastic limit L.

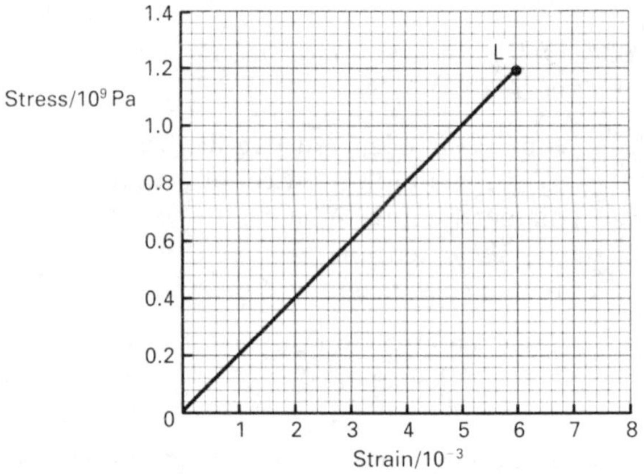

Use this graph to estimate:

(i) the Young modulus of steel;

(ii) the tension in the wire at its elastic limit L;

(iii) the maximum elastic strain energy that can be stored in unit volume (1.0 m^3) of steel. [O, '91]

B44 Define *stress*, *strain*, and *Young's modulus* of an elastic material.

Describe an experiment for measuring the Young's modulus of a material in the form of a wire.

A rubber cord has a diameter of 5.0 mm, and an unstretched length of 1.0 m. One end of the cord is attached to a fixed support A. When a mass of 1.0 kg is attached to the other end of the cord, so as to hang vertically below A, the cord is observed to elongate by 100 mm. Calculate the Young's modulus of rubber.

If the 1 kg mass is now pulled down a further short distance and then released, what is the period of the resulting oscillations? [S]

B45 Explain the term *Young's modulus.*

A nylon guitar string 62.8 cm long and 1 mm diameter is tuned by stretching it 2.0 cm. Calculate **(a)** the tension, **(b)** the elastic energy stored in the string.

Young's modulus of nylon $= 2 \times 10^9$ Pa. [S]

B46 Two copper wires A and B, of the same known areas of cross-section, are subjected to measured stretching forces and the corresponding extensions are measured. The results, on a force–extension graph, are shown in the diagram below.

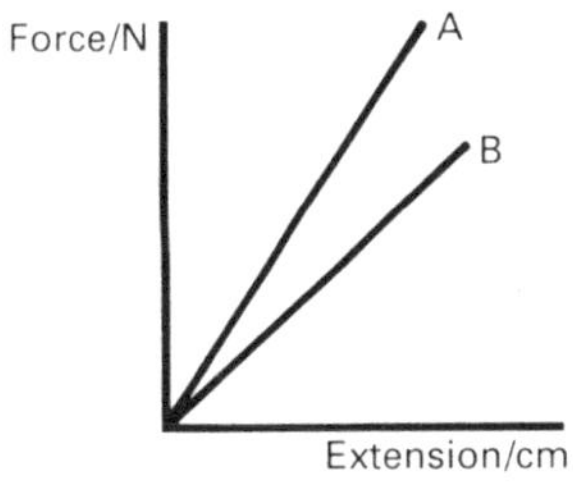

Explain what deduction you could make about the difference between the two wires.
Define the quantities which you would plot to get the same graph for both wires. How would you use this second graph to evaluate an important physical constant of copper? [L]

B47 A submerged wreck is lifted from a dock basin by means of a crane to which is attached a steel cable 10 m long of cross-sectional area 5 cm^2 and Young's modulus 5×10^{10} N m^{-2}. The material being lifted has a mass 10^4 kg and mean density 8000 kg m^{-3}. Find the change in extension of the cable as the load is lifted clear of the water.

Assume that at all times the tension in the cable is the same throughout its length.

Density of water $=$ 1000 kg m^{-3}. [J]

B48 A copper wire LM is fused at one end, M, to an iron wire MN. The copper wire has length 0.900 m and cross-section 0.90×10^{-6} m^2. The iron wire has length 1.400 m and cross-section 1.30×10^{-6} m^2. The compound wire is stretched; its total length increases by 0.0100 m.

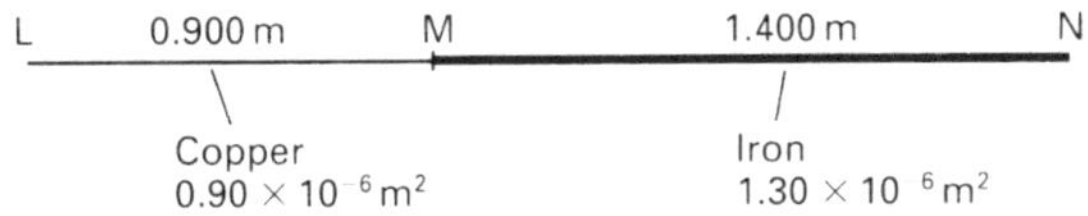

Calculate:
(a) the ratio of the extensions of the two wires,
(b) the extension of each wire,
(c) the tension applied to the compound wire.
(Young's modulus for copper $= 1.30 \times 10^{11}$ N m^{-2}; Young's modulus for iron $= 2.10 \times 10^{11}$ N m^{-2}.) [L]

B49 Explain the terms tensile *stress* and tensile *strain* as applied to a specimen of material and explain the meaning of the word *tensile.*

A car breaks down and the driver asks a friend to tow it using a piece of nylon rope which is 10.00 m long and has a diameter of 10.0 mm. The rope obeys Hooke's law and has a Young modulus of 3.0×10^9 N m^{-2}. The mass of the car and driver is 750 kg.

(a) When towing on a level road at a constant speed, it is found that the rope extends by 0.025 m. Calculate the tension in the rope and hence the net resistive force acting on the towed car.
(b) The two cars now ascend a slope which rises 1.0 m vertically for every 15.0 m travelled on the road. They maintain the same speed as in part (a). What is the new length of the towrope?
(c) How much elastic energy is stored in the rope while the cars are climbing?
(d) A stretched towrope must be regarded as dangerous because of the energy released should it break or become detached. This danger can be reduced by careful choice of towrope.

By comparison with the original rope in each case, state and explain how the energy stored in the rope could be reduced by using a rope with a different
(i) Young modulus,
(ii) area of cross-section. [O & C, '91]

B50 The graph (p. 216) represents the tension–extension graph for a copper wire of length 1.2 m and cross-sectional area 1.5×10^{-6} m^2.

The tension is gradually increased from zero to a maximum value, and then reduced back to zero.

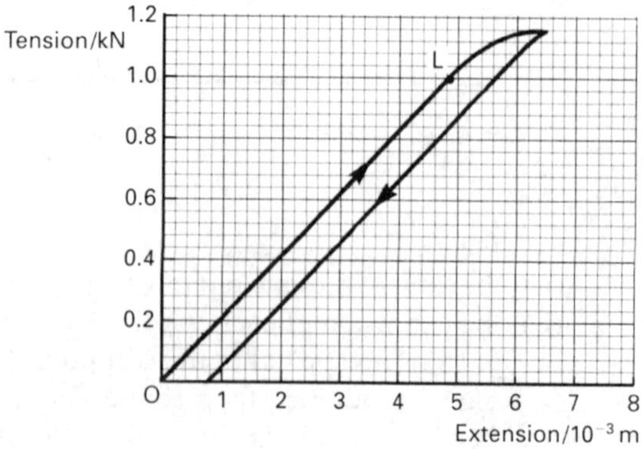

(a) Use the region OL of the graph to find the Young modulus for the material of the wire.
(b) Why is the unloading curve displaced from the loading curve?
(c) Shade the area of the graph which represents the energy lost as heat during the loading–unloading cycle. [O, '92*]

B51 A 20 m length of continuous steel railway line of cross-sectional area $8.0 \times 10^{-3}\,m^2$ is welded into place after heating to a uniform temperature of 40 °C.

(Take Young's modulus for steel to be 2.0×10^{11} Pa, its linear expansivity to be $12 \times 10^{-6}\,K^{-1}$, its density to be $7800\,kg\,m^{-3}$, and its specific heat capacity to be $500\,J\,kg^{-1}\,K^{-1}$.)

Calculate, for normal operating conditions at 15 °C:
(a) the tensile strain,
(b) the tensile stress,
(c) the elastic strain energy in the rail.
How much heat would be required to return the rail to 40 °C? Explain briefly why your answer is not the same as that of (c). [O*]

B52 The diagrams show an apparatus designed to demonstrate the resistance to shear of a new material.

One end X of the steel bar is fixed. The other end has a hole of diameter 6 mm drilled in it. When the room temperature is 20 °C, the distance between the fixed end of the bar and the nearer edge of the hole is 2.0 m as shown in Fig. 1. At this temperature half of the hole protrudes beyond the restraining pillars.

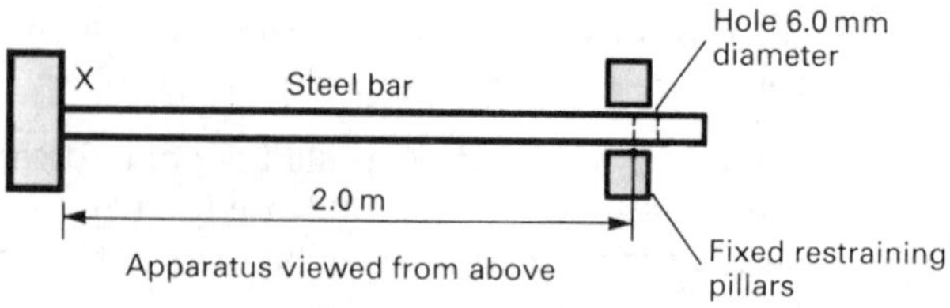

Fig. 1

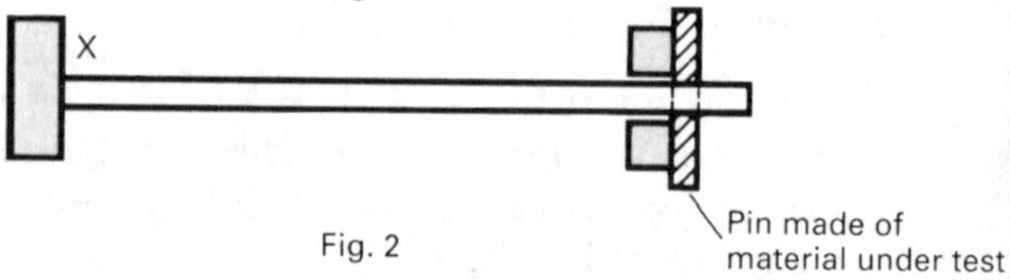

Fig. 2

The bar is heated in a constant temperature enclosure until the hole just clears the restraining pillars. A pin, which just fits the hole and made of the material under test, is then inserted through the hole as shown in Fig. 2.
(a) Calculate the temperature of the enclosure.
(b) Given that the bar does not extend beyond its limit of proportionality, calculate the tensile stress in the steel bar when the temperature returns to 20 °C.
(Young modulus for steel = 1.2×10^{11} Pa, linear expansivity of steel = $1.5 \times 10^{-6}\,K^{-1}$.)
[AEB, '90]

B53 **(a)** Give the meanings of the terms *tensile stress* and *Young's modulus*. Define the quantity which relates these terms.
(b) When measuring Young's modulus of a material it is common to use a specimen which is **(i)** very long, and **(ii)** very thin. Give the reasons for this.

Describe how you would measure accurately the extension of such a wire under an applied load.
(c) The graph (p. 217) shows how the extension of a wire varies with the load applied to it. The wire used has a length 3.00 m and a diameter 5.0×10^{-4} m.
(i) Calculate the tensile stress produced by a load of 50 N.
(ii) Find the energy stored in the wire when this load is acting.

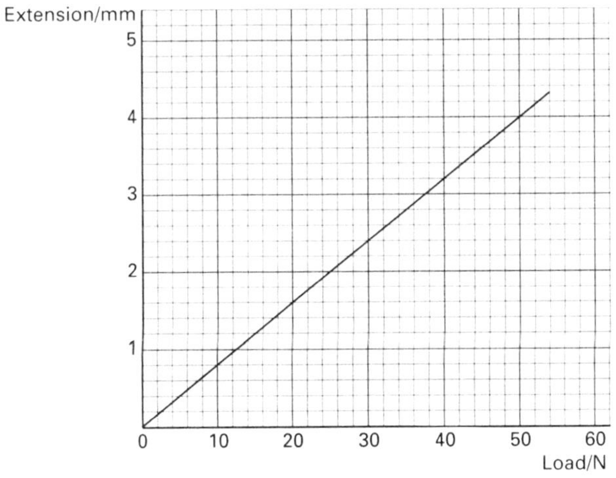

(iii) Calculate the reduction in gravitational potential energy of a 5.0 kg mass used to provide the load.

(iv) Suggest why the answers to (ii) and (iii) above are different.

(v) Calculate Young's modulus for the metal of the wire. [S]

B54 The sketch shows, approximately, how the resultant force between adjacent atoms in a solid depends on *r*, their distance apart.

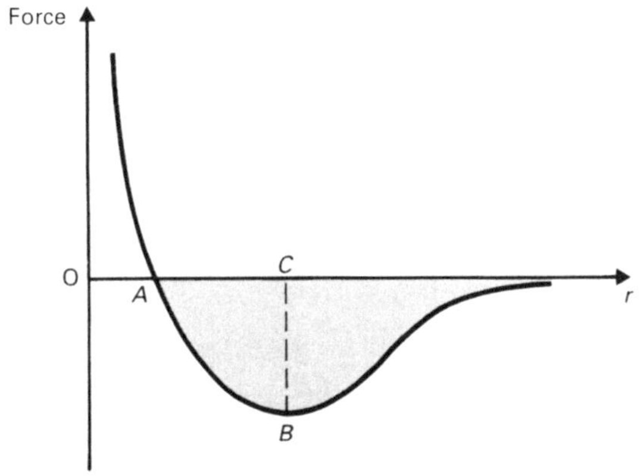

(a) Which distance on the graph represents the equilibrium separation of the atoms? Briefly justify your answer.

(b) What is the significance of the shaded area?

(c) Use the graph to explain why you would expect the solid to obey Hooke's Law for *small* extensions and compressions.

[W, '92]

B55 **(a)** The graphs represent stress–strain curves for two different materials, A and B. F_A and F_B are the respective points at which each material fractures.

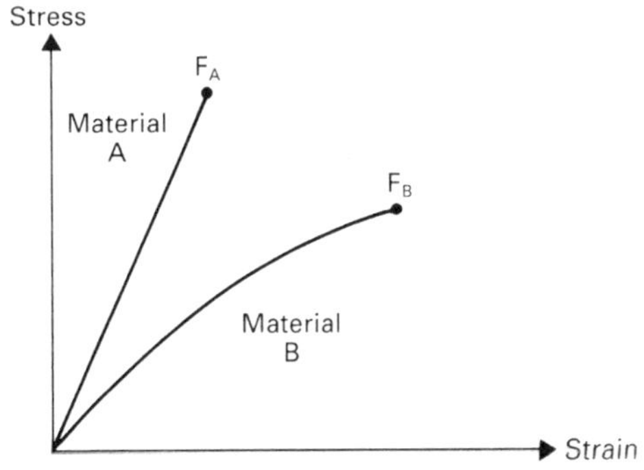

State, giving your reasons, which material, A or B,

(i) obeys Hooke's law up to the point of fracture,

(ii) is the weaker,

(iii) has the greater value of Young's modulus.

(b) A thin steel wire initially 1.5 m long and of diameter 0.50 mm is suspended from a rigid support. Calculate **(i)** the final extension and **(ii)** the energy stored in the wire when a mass of 3.0 kg is attached to the lower end. Assume that the material obeys Hooke's law.

(Young's modulus for steel = $2.0 \times 10^{11}\,\mathrm{N\,m^{-2}}$.) [J]

B56 In the model of a crystalline solid the particles are assumed to exert both attractive and repulsive forces on each other. Sketch a graph of the potential energy between two particles as a function of the separation of the particles. Explain how the shape of the graph is related to the assumed properties of the particles.

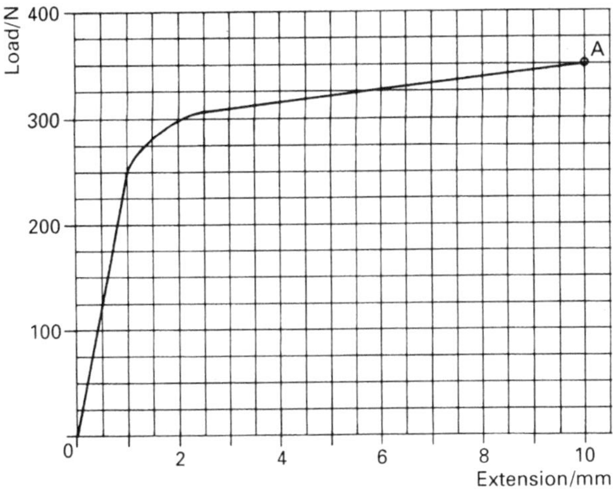

The force F, in N, of attraction between two particles in a given solid varies with their separation d, in m, according to the relation

$$F = \frac{7.8 \times 10^{-20}}{d^2} - \frac{3.0 \times 10^{-96}}{d^{10}}$$

State, giving a reason, the resultant force between the two particles at their equilibrium separation. Calculate a value for this equilibrium separation.

The graph (p. 217) displays a load against extension plot for a metal wire of diameter 1.5 mm and original length 1.0 m. When the load reached the value at A the wire broke. From the graph deduce values of

(a) the stress in the wire when it broke,
(b) the work done in breaking the wire,
(c) the Young modulus for the metal of the wire.

Define *elastic* deformation. A wire of the same metal as the above is required to support a load of 1.0 kN without exceeding its elastic limit. Calculate the minimum diameter of such a wire. [O & C]

B57 **(a) (i)** Define *stress* and *strain* as related to the extension of a wire.
(ii) A rubber cord and a steel wire are each subjected to linear stress. Draw sketch graphs showing how the resultant strain of each sample depends on the applied stress, and point out any important differences between the graphs.
(iii) For some materials, the strain–stress curve obtained when the tension applied to the specimen is being increased may differ significantly from that when the tension is being decreased, even though no permanent extension has been caused. How may this phenomenon be interpreted?
(b) (i) Describe the important features of the structure of a polymeric solid, such as rubber.
(ii) Making reference to the curve you have drawn for rubber in (a)(ii) above, account for the behaviour of rubber under linear stress in terms of changes which may occur in the internal structure of the polymer.
[I*]

B58 The end X of a uniform cylindrical rod XY is clamped in a fixed horizontal position. The free end Y is depressed under the action of the weight of the rod by a small amount d. The rod projects a distance l from the point of clamping X. The depression d is found to be directly proportional to the ratio g/A where g is the acceleration due to gravity and A the cross-sectional area of the rod. Also, d depends on l and the density ρ and the Young modulus E of the material of the rod. Use the method of dimensions to determine how d might depend on l, ρ and E.

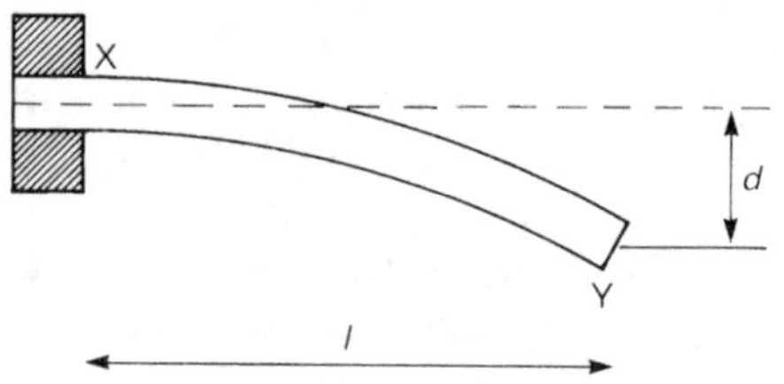

How would you show experimentally the way in which d varies with the length and radius of the rod? [O & C*]

B59 **(a)** In order to determine *Young's modulus* for the material of a wire in a school laboratory, it is usual to apply a *tensile stress* to the wire and to measure the *tensile strain* produced. Explain the meanings of the terms in italics and state the relationship between them.

(b)

Additional load/kg	Scale reading/mm
0	2.8
2.0	3.8
4.0	4.5
6.0	5.1
8.0	5.7
10.0	6.3
12.0	6.9
14.0	7.5
16.0	8.1

The table shows readings obtained when stretching a wire supported at its upper end by suspending masses from its lower end. The unstretched length of the wire was 2.23 m and its diameter 0.71 mm. Using a graphical method, determine a value for Young's modulus for the material of the wire.

(c) Describe suitable apparatus for obtaining the readings shown and explain the important features of the design.

(d) A student noticed that when a mass of 10.0 kg suspended from a wire identical to that described above was pushed downwards and released, it executed vertical oscillations of small amplitude. Use the graph to explain briefly why you would expect the oscillations to be simple harmonic. [J]

B60 **(a)** When materials are stretched their behaviour may be either *elastic* or *plastic*. Distinguish carefully between these terms.

(b) Whilst stretching a length of thin copper wire it is noticed that

(i) at first a fairly strong pull is needed to stretch it by a small amount and that it stretches uniformly,

(ii) beyond a certain point the wire extends by a very much larger amount for no further increase in the pull,

(iii) finally the wire breaks.

Sketch a force–extension graph to illustrate the behaviour of this wire. Mark on it the region where the behaviour is elastic and the region where it is plastic. [L]

B61 The *force constant k* of a spring is the constant of proportionality in the Hooke's law relation $T = ke$ between tension T and extension e.

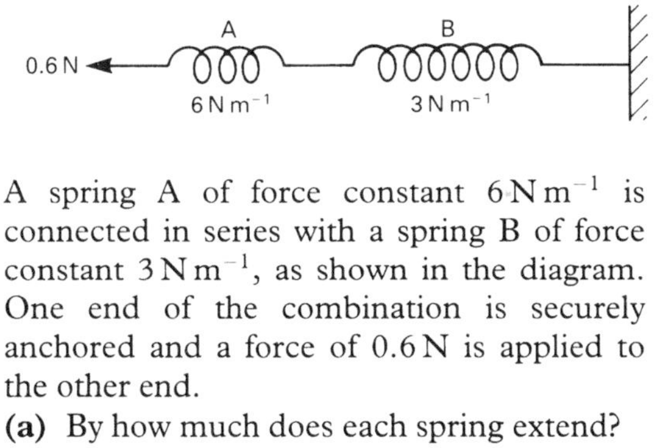

A spring A of force constant $6\,N\,m^{-1}$ is connected in series with a spring B of force constant $3\,N\,m^{-1}$, as shown in the diagram. One end of the combination is securely anchored and a force of 0.6 N is applied to the other end.

(a) By how much does each spring extend?

(b) What is the force constant of the combination? [C]

B62 **(a)** **(i)** Distinguish between *elastic* and *plastic* deformation of a material.

(ii) Sketch a graph to show how the extension x of a copper wire varies with F, the applied load. Mark on your sketch the region where the wire obeys Hooke's law.

(b) **(i)** A force is required to cause an extension of a spring. Explain why this causes energy to be stored in the spring.

(ii) A spring of spring constant k undergoes an elastic change resulting in an extension x. Deduce that W, its strain energy, is given by

$$W = \tfrac{1}{2}kx^2$$

(c) A toy train, mass m, travels along a track at speed v and is brought to rest by two spring buffers which are shown below.

Each buffer has spring constant k.

(i) By considering the energy transfer, derive an expression to show how the maximum compression of the buffers varies with the initial speed of the train.

(ii) Calculate the maximum compression of the buffers for a train of mass $m = 1.2$ kg travelling with an initial speed $v = 0.45\,m\,s^{-1}$ when the spring constant k of each buffer is $4.8 \times 10^3\,N\,m^{-1}$.

State and explain a reason why, in practice, spring buffers of this design are not used. [C, '92]

FLUID FLOW (Chapter 12)

B63 **(a)** Explain the terms *lines of flow* and *streamlines* when applied to fluid flow and deduce the relationship between them in laminar flow.

(b) State Bernoulli's equation, define the physical quantities which appear in it and the conditions required for its validity.

(c) The depth of water in a tank of large cross-sectional area is maintained at 20 cm and water emerges in a continuous stream out of a hole 5 mm in diameter in the base. Calculate:

(i) the speed of efflux of water from the hole,

(ii) the rate of mass flow of water from the hole.

Density of water = $1.00 \times 10^3\,\text{kg m}^{-3}$.

[J]

B64 **(a)** Distinguish between *static pressure*, *dynamic pressure* and *total pressure* when applied to streamline (laminar) fluid flow and write down expressions for these three pressures at a point in the fluid in terms of the flow velocity v, the fluid density ρ, pressure p, and the height h, of the point with respect to a datum.

(b) Describe, with the aid of a labelled diagram, the Pitot–static tube and explain how it may be used to determine the flow velocity of an incompressible, non-viscous fluid.

(c) The static pressure in a horizontal pipeline is 4.3×10^4 Pa, the total pressure is 4.7×10^4 Pa, and the area of cross-section is 20 cm^2. The fluid may be considered to be incompressible and non-viscous and has a density of $10^3\,\text{kg m}^{-3}$. Calculate:

(i) the flow velocity in the pipeline,

(ii) the volume flow rate in the pipeline.

[J]

B65 Air flows over the upper surfaces of the wings of an aeroplane at a speed of $120.0\,\text{m s}^{-1}$, and past the lower surfaces of the wings at $110.0\,\text{m s}^{-1}$. Calculate the 'lift' force on the aeroplane if it has a total wing area of $20.0\,\text{m}^2$.

(Density of air = $1.29\,\text{kg m}^{-3}$.)

B66 A large tank contains water to a depth of 1.0 m. Water emerges from a small hole in the side of the tank 20 cm below the level of the surface. Calculate:

(a) the speed at which the water emerges from the hole,

(b) the distance from the base of the tank at which the water strikes the floor on which the tank is standing.

If a second hole were to be drilled in the wall of the tank vertically below the first hole, at what height above the base of the tank would this second hole have to be if the water issuing from it were to hit the floor at the same point as that from the first hole?

B67 **(a)** By considering the flow of an incompressible fluid along a horizontal pipe as shown in Fig. 1, derive Bernoulli's equation using the conservation of energy principle.

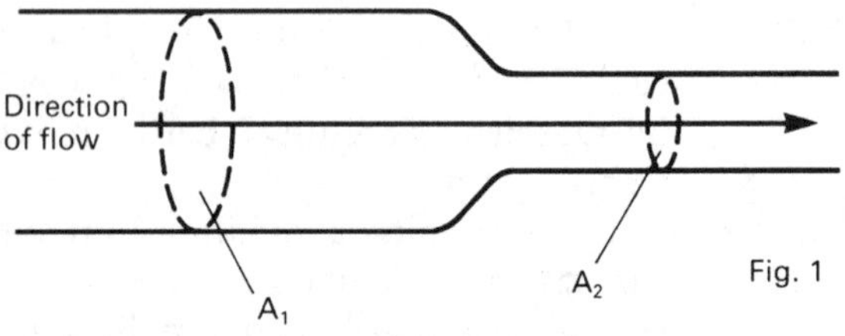

Fig. 1

(b) The water aspirator is a laboratory device used for the partial evacuation of air from a vessel. A jet of running water from a pipe constricted at X, as shown in Fig. 2, is directed into the expanded opening of a funnel at Y and passes out into the drain.

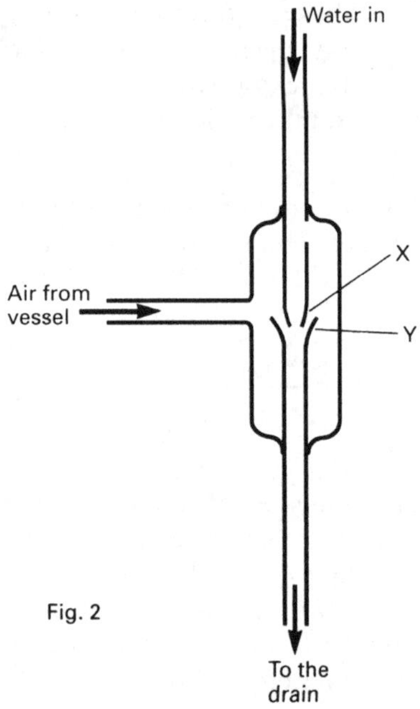

Fig. 2

(i) By considering the effect of the water jet on the air in the region of the jet, explain why this is a practical example of the Bernoulli effect.

(ii) Calculate the maximum reduction in pressure that could be achieved using this pump if the jet diameter is 2.0 mm and volume rate of flow of water is $1.3 \times 10^{-4}\,\text{m}^3\,\text{s}^{-1}$.

(Density of air = $1.3\,\text{kg m}^{-3}$.) [J, '91]

B68 **(a)** State the equation of continuity for a *compressible* fluid flowing through a pipe.

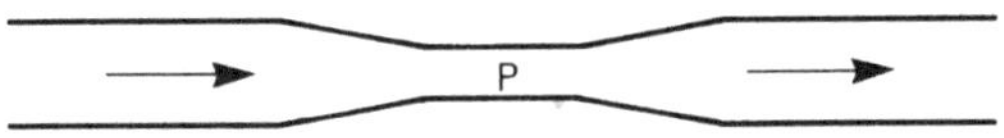

A horizontal pipe of diameter 36.0 cm tapers to a diameter of 18.0 cm at P. An ideal gas at a pressure of 2.00×10^5 Pa is moving along the wider part of the pipe at a speed of $30.0\ \mathrm{m\,s^{-1}}$. The pressure of the gas at P is 1.80×10^5 Pa. Assuming that the temperature of the gas remains constant calculate the speed of the gas at P.

(b) State Bernoulli's equation for an incompressible fluid, giving the meanings of the symbols in the equation.

(c) For the gas in (a) recalculate the speed at P on the assumption that it can be treated as an *incompressible* fluid, and use Bernoulli's equation to calculate the corresponding value for the pressure at P. Assume that in the wider part of the pipe the gas speed is still $30.0\ \mathrm{m\,s^{-1}}$, the pressure is still 2.00×10^5 Pa and at this pressure the density of the gas is $2.60\ \mathrm{kg\,m^{-3}}$.

(d) Draw a labelled diagram to show how you would use the change in pressure discussed in (c), treating the gas as an incompressible fluid, to obtain a value for the speed of the gas in the pipe. Show how the result is calculated. [J]

B69 **(a)** A cylinder of large cross-sectional area, containing water, stands on a horizontal bench. The water surface is at a height h above the bench. Water emerges horizontally from a hole in the side of the cylinder, at a height y above the bench.

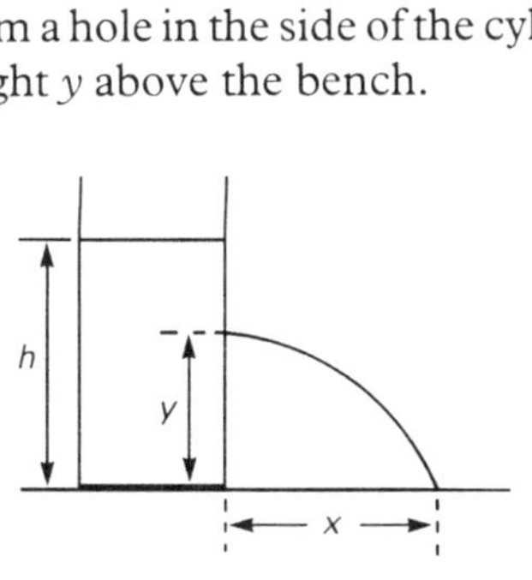

(i) Use Bernoulli's equation to derive expressions for the speed at which the water emerges from the hole, and the speed at which it hits the bench.

(ii) Derive expressions for the time for the water to travel from the hole to the bench, and for x, the horizontal distance the water travels from the cylinder.

(b) Draw a diagram of a Pitot–static tube, and with reference to Bernoulli's equation explain how the tube may be used to measure the speed of a boat in sea water. [J]

B70 **(a)** Derive Bernoulli's equation for an incompressible fluid.

(b) State what you understand by *static pressure* and *dynamic pressure* and state how they are related to terms which appear in Bernoulli's equation.

(c) **(i)** Non viscous oil flows from the bottom of a tank in a horizontal pipe. State how you would measure the static pressure and the dynamic pressure of the oil in the pipe.

(ii) If the density of the oil is ρ and it is moving at a speed v in the pipe at a depth d below the surface of the oil in the tank derive from Bernoulli's equation an expression showing how the static pressure as measured in (i) is related to the atmospheric pressure, p_0, acting on the surface of the oil in the tank. (You may neglect any motion of the oil in the tank.) [J]

B71 **(a)** What do you understand by the *equation of continuity* as applied to a fluid in motion?

(b) Derive Bernoulli's equation for an incompressible fluid.

(c) A simple garden syringe used to produce a jet of water consists of a piston of area $4.00\ \mathrm{cm^2}$ which moves in a horizontal cylinder which has a small hole of area $4.00\ \mathrm{mm^2}$ at its end. If the force on the piston is 50.0 N calculate a value for the speed at which the water is forced out of the small hole, assuming the speed of the piston is negligible.
The density of water is $1.00 \times 10^3\ \mathrm{kg\,m^{-3}}$.

(d) Explain why the speed of the piston may be ignored. [J]

B72 **(a)** Explain the meaning of the term *laminar flow*. Describe how, for a liquid flowing in a horizontal pipe, it can be shown whether or not laminar flow occurs.

(b) (i) State both the equation of continuity and Bernoulli's equation for incompressible fluids.

(ii) Draw and label a diagram of a Venturi meter suitable for measuring the velocity of flow of a liquid in a horizontal pipe. Use the equations of (i) to obtain an expression from which the velocity of flow of the liquid in the Venturi meter can be calculated. What measurements must be made when the Venturi meter is used?

(iii) Explain how assumptions made in your derivation in (ii) could limit the usefulness of a Venturi meter.

[J]

B73 (a) (i) The Pitot tube shown in the figure below is used to measure the speed of flow of a gas in a pipe.

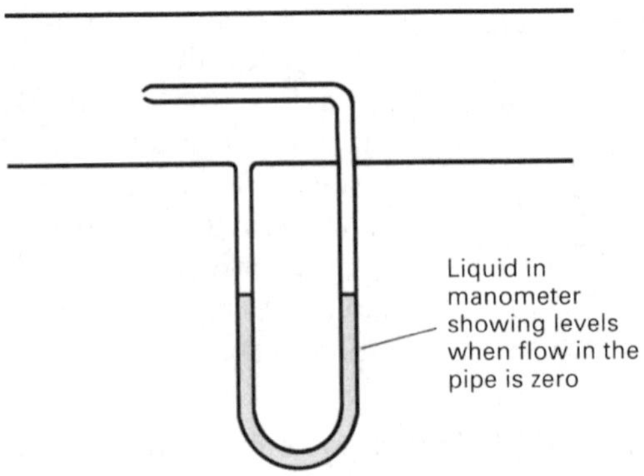

Redraw the diagram showing the direction of flow of the gas and the corresponding levels of liquid in the manometer.

(ii) By considering Bernoulli's equation show that the difference h in the levels of the liquid in the manometer is given by

$$h = \frac{\rho v^2}{2\rho_0 g}$$

where ρ is the density of the gas, ρ_0 is the density of the liquid in the manometer, v is the speed of the gas along the pipe, g is the acceleration due to gravity.

(b) The figure below shows a variation of the Pitot tube used to indicate the speed of aircraft.

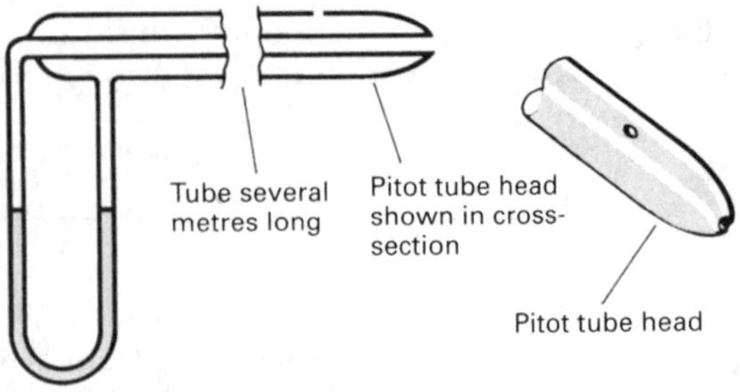

(i) State *one* factor that would have to be taken into consideration when choosing where to position the Pitot tube head on the exterior of an aircraft if the true airspeed is to be indicated.

(ii) Would water be a suitable liquid for the manometer if airspeeds of up to 600 km h^{-1} are to be indicated in the cockpit of an aircraft? Justify your answer.

(c) The figure below shows a Venturi meter used to indicate the speed of flow of a liquid through a pipe.

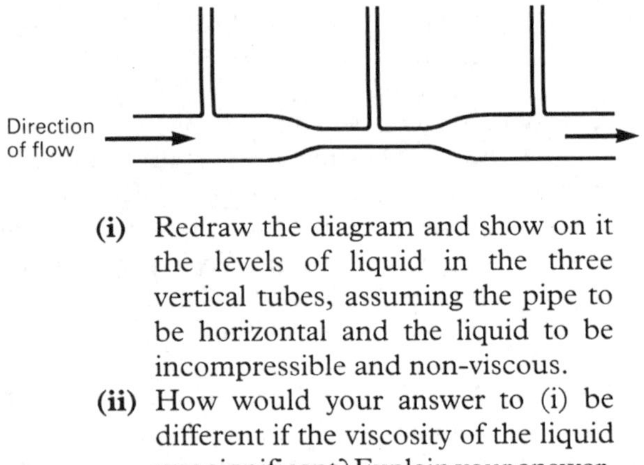

(i) Redraw the diagram and show on it the levels of liquid in the three vertical tubes, assuming the pipe to be horizontal and the liquid to be incompressible and non-viscous.

(ii) How would your answer to (i) be different if the viscosity of the liquid was significant? Explain your answer.

[J, '92]

B74 (a) Give an account of an experiment which makes use of Poiseuille's formula to measure the viscosity of water.

(b) An empty vessel which is open at the top has a horizontal capillary tube of length 20 cm and internal radius 1.0 mm protruding from one of its side walls immediately above the base. Water flows into the vessel at a constant rate of 1.5 cm^3 s^{-1}. At what depth does the water level stop rising?
(You may assume that the flow is steady. Coefficient of viscosity of water = 1.0 ×

10^{-3} N s m^{-2}, density of water = 1.0×10^3 kg m^{-3}, acceleration due to gravity = 10 m s^{-2}.)

B75 A liquid flows steadily through two pipes, A and B, which are joined end to end and whose internal radii are r and $2r$ respectively. If B is 8 times longer than A and the pressure difference between the ends of the composite pipe is 9000 N m^{-2}, what is the pressure difference across A?

B76 **(a)** Define coefficient of viscosity, η, and show that its dimensions in M, L and T are $ML^{-1}T^{-1}$. What is meant by laminar flow?

(b) Poiseuille's formula for the volume of liquid V flowing in time t through a uniform capillary of radius r under laminar conditions is

$$\frac{V}{t} = \frac{\pi r^4 p}{8\eta l},$$

where p/l is the pressure gradient along the tube.

(i) Show that this equation is dimensionally consistent.

(ii) Describe how you would apply the equation to measure η for water at room temperature.

(iii) Laminar conditions should obtain provided that the value of

$$\frac{r^3 p\rho}{4\eta^2 l} < 1150,$$

where ρ is the density of the liquid. Taking η to be 1.2×10^{-3} Pa s (N s m^{-2}) and ρ to be 1000 kg m^{-3} for water, estimate the greatest head of water under which laminar flow should hold for a capillary of length 0.2 m and radius 0.7 mm. [O]

B77 **(a)** When a sphere of radius a moves slowly with a speed v through a fluid of viscosity η, Stokes' law tells us that the force F on the sphere due to viscous drag is given by the expression $F = 6\pi a\eta v$. Show that this expression is dimensionally correct.

(b) In an experiment to compare the viscosities of two oils, small spheres are allowed to fall through long columns of the liquids. What conditions are necessary in order that Stokes' law may be applied?

State what measurements are necessary to find the *terminal* speed of the spheres. [L*]

B78 Frictional forces and viscous drag both oppose relative motion. Suggest some similarities and differences between them.

Explain why a small sphere, falling through liquid in a deep tank, eventually moves with a constant speed (the terminal velocity). Sketch a graph showing how the acceleration of the sphere varies with time after its release at the surface of the liquid.

Describe how you would measure such a terminal velocity, explaining how you would use your measurements to ensure that your result was the true *terminal* velocity. [C*]

B79 The viscous force on a sphere, of radius r, moving through a fluid with velocity v can be expressed as $6\pi\eta r v$, where η is the coefficient of viscosity of the fluid. What is the limitation on the use of this expression? [L]

B80 **(a)** Stokes' law may be represented by the equation shown below.

$$F = 6\pi\eta r v$$

(i) State the physical quantities represented by the symbols F, η, r and v and the conditions under which the relationship is valid.

(ii) Tiny spherical particles of alumina, having a wide range of radii, are stirred up in a beaker of water 8.0 cm deep. Draw a diagram showing the forces acting on *one* such particle, including the force of upthrust (equal to the weight of water displaced by the particle), shortly after stirring has ceased and the water has achieved a still condition. Hence determine the radius of the largest particle to remain in suspension after 24 hours. You may assume that the particles fall through the water with terminal velocity.

(Density of water = 1.0×10^3 kg m^{-3}, density of alumina = 2.7×10^3 kg m^{-3}, (coefficient of viscosity of water = 1.0×10^{-3} N s m^{-2}.)

(b) **(i)** State Bernoulli's equation for an incompressible fluid.

(ii) Sketch a section through an aircraft wing and explain how the movement of such a wing through the air results in an upward force (lift) on the wing.

(iii) A particular aircraft design calls for a lift of about 1.2×10^4 N on each square metre of the wing when the speed of the aircraft through the air is 100 m s^{-1}. Assuming that the air flows past the wing with streamline flow and the flow past the lower surface is equal to the speed of the aircraft, what is the required speed of the air over the upper surface of the wing? (Density of air = 1.3 kg m^{-3}.) [J, '90]

B81 The stress σ between two planes of molecules in a moving liquid is given by

$$\sigma = \frac{\eta v}{x}$$

where v is the difference in the velocities of the planes, x their distance apart and η a constant for the liquid.

(a) Show that the dimensions of η are $ML^{-1}T^{-1}$.

(b) The force F acting on a sphere moving through a liquid is known to depend upon

(i) the radius r of the sphere,

(ii) the speed u of the sphere,

(iii) the constant η for the liquid.

Find how F depends on r, u and η. [W, '91]

B82 A body moving through air at a high speed v experiences a retarding force F given by

$$F = kA\rho v^x$$

where A is the surface area of the body, ρ is the density of the air and k is a numerical constant. Deduce the value of x.

A sphere of radius 50 mm and mass 1.0 kg falling vertically through air of density 1.2 kg m^{-3} attains a steady velocity of 11.0 m s^{-1}. If the above equation then applies to its fall what is the value of k in this instance? [L]

B83 The drag force F exerted on a vehicle due to its motion through still air is given by

$$F = \frac{\rho D v^2}{2}$$

where ρ is the density of air, v is the speed of the car and D is the drag factor.

(a) Write down the units of F, ρ and v and hence determine the unit of D.

(b) The magnitude of the drag factor of a particular car is 0.33. Calculate the speed when the rate at which energy is dissipated in overcoming air resistance is 3.0 kW. (The density of air = 1.3 kg m^{-3}.)

(c) State what happens to the energy dissipated in overcoming air resistance.

(d) Some cars are streamlined like the one in the diagram.

State and explain the effect of the shape of the car on the vertical forces acting on the vehicle when it starts from rest and accelerates. [AEB, '91]

B84 When a sphere of radius a and density d, falls through oil contained in a tank, it descends with uniform velocity v. The relation between v, a, and d is

$$v/a^2 = Ad - B$$

where A and B are constants.

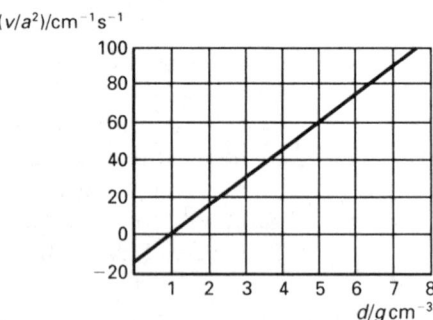

The above graph shows the results of some experiments. Determine from the graph the numerical values of A and B. What is the radius of a steel sphere of density 7.5 g cm^{-3} which falls through the oil with velocity 3.9 cm s^{-1}? [S]

B85 **(a)** Draw diagrams to show the forces acting on an object falling through a viscous liquid

(i) at the instant of release,
(ii) when it has reached its terminal velocity.

Write down an equation for the forces acting on the object in (ii). Describe and explain the motion of an object projected downwards through a viscous medium, assuming that the projection velocity of the object is greater than its terminal velocity.

(b) (i) Describe how the terminal velocity of a small sphere falling through motor oil could be measured.

(ii) In an experiment to determine the coefficient of viscosity of motor oil the following measurements were made.

Mass of glass sphere	1.2×10^{-4} kg
Diameter of sphere	4.0×10^{-3} m
Terminal velocity of sphere	5.4×10^{-2} m s^{-1}
Density of oil	860 kg m^{-3}

Calculate the coefficient of viscosity of the oil. [J]

B86 (a) (i) State Newton's law of viscosity and hence deduce the dimensions of the coefficient of viscosity.

(ii) The rate of volume flow, $\frac{dV}{dt}$, of liquid of viscosity η, through a pipe of internal radius r and length l, is given by the equation

$$\frac{dV}{dt} = \frac{\pi p r^4}{8\eta l}$$

where p is the pressure difference between the ends of the pipe. Show that this equation is dimensionally correct.

(b) The figure shows a tank containing a light lubricating oil. The oil flows out of the tank through a horizontal pipe of length 0.10 m and internal diameter 4.0 mm.

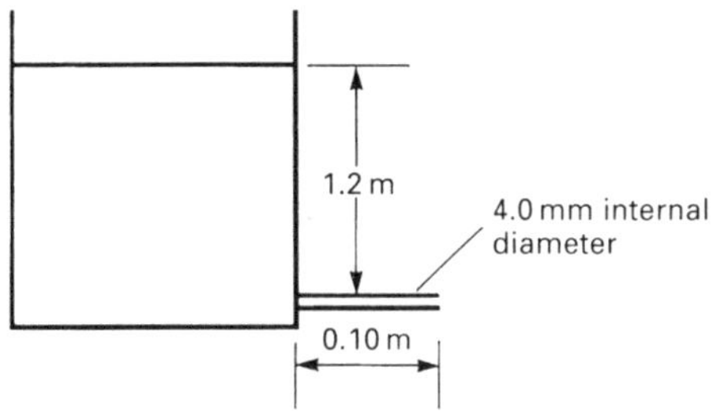

(i) Calculate the volume of oil which flows through the pipe in one minute when the level of oil in the tank is 1.2 m above the pipe and does not significantly alter during this time.

Density of oil $= 9.2 \times 10^2$ kg m^{-3}
Coefficient of viscosity of oil $= 8.4 \times 10^{-2}$ N s m^{-2}

(ii) It is found that the volume flow is greater at higher temperatures. Assuming that density changes can be ignored, suggest an explanation for this effect in terms of the nature of the viscous force.

(c) Discuss how the lubricating properties of an oil are affected by:
(i) the coefficient of viscosity of the oil,
(ii) its variation with temperature. [J]

SECTION C

THERMAL PROPERTIES OF MATTER

13

THERMOMETRY AND CALORIMETRY

13.1 TEMPERATURE

The temperature of a body is its degree of hotness (or coldness). Thus, temperature is a measure of how hot (or cold) a body is, and should not be confused with the amount of heat it contains.

13.2 TEMPERATURE SCALES

There are many types of thermometer, but each makes use of a particular thermometric property (i.e. a property whose value changes with temperature) of a particular thermometric substance. For example: a mercury-in-glass thermometer makes use of the change in length of a column of mercury confined in a capillary tube of uniform bore; a platinum resistance thermometer makes use of the increase in the electrical resistance of platinum with increasing temperature.

In order to establish a temperature scale it is necessary to make use of **fixed points**: A fixed point is the single temperature at which it can confidently be expected that a particular physical event (e.g. the melting of ice under specific conditions) always takes place. Three such points are defined below.

The ice point is the temperature at which pure ice can exist in equilibrium with water at **standard atmospheric pressure** (i.e. at a pressure of 760 mm of mercury).

The steam point is the temperature at which pure water can exist in equilibrium with its vapour at standard atmospheric pressure.

The triple point of water is that unique temperature at which pure ice, pure water and pure water vapour can exist together in equilibium.

The triple point is particularly useful, since there is only one pressure at which all three phases (solid, liquid and gas) can be in equilibrium with each other.

The SI unit of temperature is the kelvin (K). **An interval of one kelvin is defined as being 1/273.16 of the temperature of the triple point of water as measured on the thermodynamic scale of temperature** (see later in this section and in section 16.6). The triple point of water is the fixed point of the scale and is assigned the value of 273.16 K. On this basis absolute zero is 0 K, the ice point is 273.15 K, and the steam point is 373.15 K.

Another unit, the **degree Celsius** (°C), is often used and is defined by

$$\theta = T - 273.15 \qquad [13.1]$$

where

θ = temperature in °C, and

T = temperature in K.

The Celsius scale was originally defined by using the ice and steam points as fixed points of the scale, and designating them as 0 °C and 100 °C respectively. Bearing in mind that these temperatures are respectively 273.15 K and 373.15 K, we can easily see that the more recent definition (equation [13.1]) is consistent with this. It also follows from equation [13.1] that **a temperature change of 1 K is exactly equal to a temperature change of 1 °C.**

A mercury-in-glass thermometer could be calibrated by marking the positions of the mercury when the thermometer is at the ice point and the steam point, and then dividing the interval between these two marks (designated 0 °C and 100 °C respectively) into a hundred equal divisions. If this procedure were to be adopted, the Celsius temperature θ corresponding to a length l_θ of the mercury column would be given by

$$\theta = \frac{l_\theta - l_0}{l_{100} - l_0} \times 100 \qquad [13.2]$$

where l_0 and l_{100} are the lengths of the mercury column at 0 °C and 100 °C respectively. Such a calibration regards equal increases in the length of the mercury column as being due to equal increases in temperature. There is of course no valid reason for making this assumption, and so if such a thermometer is used, it is important to stress that the measured temperatures are according to the mercury-in-glass scale of temperature. If a platinum resistance thermometer were to be calibrated by making an equivalent assumption, i.e. that equal increases in temperature produce equal increases in the resistance of platinum, then temperatures measured by this thermometer would be according to the platinum resistance scale. These two scales coincide only at the fixed points (0 °C and 100 °C), because, as might be expected, the volume of mercury and the resistance of platinum do not vary in the same way.

The thermodynamic scale of temperature is totally independent of the properties of any particular substance and is therefore an absolute scale of temperature. Although this scale is theoretical, it can be shown (see section 16.6) that it is identical with the scale based on the pressure variation of an ideal gas (see Chapter 14) at constant volume. The fixed point of both scales is the triple point of water (273.16 K) and **the kelvin temperature T on both the ideal gas scale and the thermodynamic scale can be found from**

$$T = \frac{p_T}{p_{\mathrm{Tr}}} \times 273.16 \qquad [13.3]$$

where p_T is the pressure of an ideal gas at temperature T, and p_{Tr} is the pressure of the same volume of the gas at the triple point of water. Ideal gases do not exist, but real gases at low pressures are a good approximation to them. This means that results obtained using constant-volume gas thermometers incorporating real gases can be adjusted to coincide exactly with the theoretically correct temperatures of the thermodynamic scale. (The unknown temperature is estimated on the basis of equation [13.3] at a number of different (low) pressures. The results are then extrapolated to what would be obtained at zero pressure if such a measurement were possible, because at zero pressure a real gas would behave like an ideal gas.) In practice, therefore, the various types of thermometer are calibrated in terms of the constant-volume gas thermometer. As a result, the measured value of any particular temperature is the same (within the limits of accuracy of the instrument being used) no matter what type of thermometer is used to measure it.

13.3 LIQUID-IN-GLASS THERMOMETERS

These are simple to use and cheap to buy, but cannot be used for accurate work because:

(i) parallax errors prevent the scale being read to better than about 0.1 °C;

(ii) non-uniform bore limits the accuracy to about 0.1 °C;

(iii) the glass expands and contracts and can take many hours to reach its correct size, and therefore spoils the calibration;

(iv) the accuracy of the calibration depends on whether or not the thermometer is upright, and on how much of the stem is exposed.

This type of thermometer is easily adjusted to the constant-volume gas thermometer scale by suitably spacing the degree markings on the glass. Liquid-in-glass thermometers have relatively large heat (thermal) capacities, and this limits their use in two distinct ways:

(i) they cannot be used to follow rapidly changing temperatures; and

(ii) they can considerably affect the temperature of the body whose temperature they are being used to measure.

The majority of liquid-in-glass thermometers use mercury as the thermometer liquid. This is because:

(i) mercury is opaque and therefore easily seen;

(ii) mercury is a good conductor of heat and therefore can rapidly take up the temperature of its surroundings;

(iii) mercury does not wet (i.e. stick to) the glass.

The range of such a thermometer is from −39 °C (the freezing point of mercury) to something below its normal boiling point of 357 °C. This upper limit can, however, be extended by filling the thermometer with an inert gas such as nitrogen; this increases the pressure on the mercury so that its boiling point can be increased to about 800 °C. Ordinary soda-lime glass or Pyrex would soften at such a temperature, and therefore the thermometer would probably be made from fused quartz. If the mercury is replaced by ethyl alcohol, temperatures as low as −114.9 °C (the freezing point of alcohol) can be measured. Alcohol is also more sensitive to temperature change than mercury but its expansion is very non-linear. The use of liquid pentane can reduce the lower limit even more, to about −200 °C.

EXAMPLE 13.1

A particular resistance thermometer has a resistance of 30.00 Ω at the ice point, 41.58 Ω at the steam point and 34.59 Ω when immersed in a boiling liquid. A constant-volume gas thermometer gives readings of 1.333×10^5 Pa, 1.821×10^5 Pa and 1.528×10^5 Pa at the same three temperatures. Calculate the temperature at which the liquid is boiling: (a) on the scale of the gas thermometer, (b) on the scale of the resistance thermometer.

Solution

The Celsius temperature θ_g, according to the gas thermometer scale, is given by

$$\theta_g = \frac{p_\theta - p_0}{p_{100} - p_0} \times 100$$

where p_θ is the gas pressure at the temperature of the boiling liquid and p_0 and p_{100} are the gas pressures at 0 °C and 100 °C respectively. Thus

$$\begin{aligned}\theta_g &= \frac{1.528 \times 10^5 - 1.333 \times 10^5}{1.821 \times 10^5 - 1.333 \times 10^5} \times 100 \\ &= \frac{0.195}{0.488} \times 100 \\ &= 39.96\ ^\circ\text{C}\end{aligned}$$

The Celsius temperature θ_r according to the resistance scale is given by

$$\theta_r = \frac{R_\theta - R_0}{R_{100} - R_0} \times 100$$

where R_θ is the resistance at the temperature of the boiling liquid and R_0 and R_{100} are the resistance values at 0 °C and 100 °C respectively. Thus

$$\begin{aligned}\theta_r &= \frac{34.59 - 30.00}{41.58 - 30.00} \times 100 \\ &= \frac{4.59}{11.58} \times 100 \\ &= 39.64\ ^\circ\text{C}\end{aligned}$$

EXAMPLE 13.2

The resistance R_θ of a particular resistance thermometer at a Celsius temperature θ as measured by a constant-volume gas thermometer is given by

$$R_\theta = 50.00 + 0.1700\theta + 3.00 \times 10^{-4}\theta^2$$

Calculate the temperature as measured on the scale of the resistance thermometer which corresponds to a temperature of 60 °C on the gas thermometer.

Solution

A resistance R_θ corresponds to a temperature θ_r on the scale of the resistance thermometer which is given by

$$\theta_r = \frac{R_\theta - R_0}{R_{100} - R_0} \times 100$$

where R_0 and R_{100} are the resistances at 0 °C and 100 °C respectively. It follows that the resistance temperature which corresponds to a temperature of 60 °C on the gas thermometer scale is given by

$$\theta_r = \frac{R_{60} - R_0}{R_{100} - R_0} \times 100$$

where R_{60} is the resistance at 60 °C on the gas thermometer scale.

$$R_\theta = 50.00 + 0.1700\theta + 3.00 \times 10^{-4}\theta^2$$

$$\therefore \quad R_0 = 50.00\,\Omega$$

and $\quad R_{60} = 50.00 + 10.20 + 1.08 = 61.28\,\Omega$

and $\quad R_{100} = 50.00 + 17.00 + 3.00 = 70.00\,\Omega$

Therefore

$$\theta_r = \frac{61.28 - 50.00}{70.00 - 50.00} \times 100 = \frac{11.28}{20.00} \times 100 = 56.40\,°\text{C}$$

EXAMPLE 13.3

Derive equation [13.2].

Solution

If equal increases in the length of a mercury column are regarded as being due to equal increases in temperature, then a graph of length of column against temperature is a straight line (Fig. 13.1).

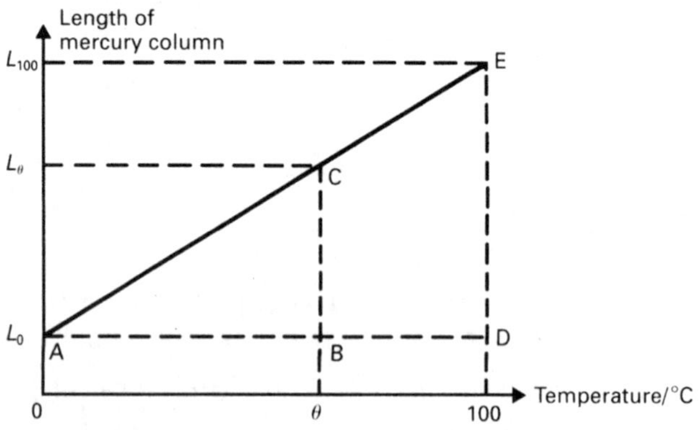

Fig. 13.1 Length of mercury column against temperature measured on the mercury-in-glass scale

Since △ABC and △ADE are similar,

$$\frac{AB}{AD} = \frac{BC}{DE}$$

$$\therefore \quad \frac{\theta - 0}{100 - 0} = \frac{l_\theta - l_0}{l_{100} - l_0}$$

i.e. $$\theta = \frac{l_\theta - l_0}{l_{100} - l_0} \times 100$$

QUESTIONS 13A

1. A resistance thermometer has a resistance of $21.42\,\Omega$ at the ice point, $29.10\,\Omega$ at the steam point and $28.11\,\Omega$ at some unknown temperature θ. Calculate θ on the scale of this thermometer.

2. A particular constant-volume gas thermometer registers a pressure of 1.937×10^4 Pa at the triple point of water and 2.618×10^4 Pa at the boiling point of a liquid. What is the boiling point of the liquid according to this thermometer?

3. The temperature measurement described in question 2 was repeated using the same thermometer but with a different quantity of (the same) gas. The readings on this occasion were 4.068×10^4 Pa at the triple point of water and 5.503×10^4 Pa at the boiling point of the liquid. **(a)** What is the boiling point of the liquid according to this measurement? **(b)** Which of the two values is the better approximation to the ideal gas temperature, and why? **(c)** Estimate the ideal gas temperature.

13.4 THERMOCOUPLES

Whenever two dissimilar metals are in contact an EMF is set up at the point of contact. The magnitude of this EMF depends on the temperature at the junction of the two metals, and therefore the effect (known as the **thermoelectric or Seebeck effect**) can be used in thermometry. The devices which are used in this way are called thermocouples, and at their simplest consist of two wires of different metals joined to each other and to a high-resistance millivoltmeter as shown in Fig. 13.2. The reading on the millivoltmeter increases as the temperature of the junction increases, due to the increased EMF at the junction.

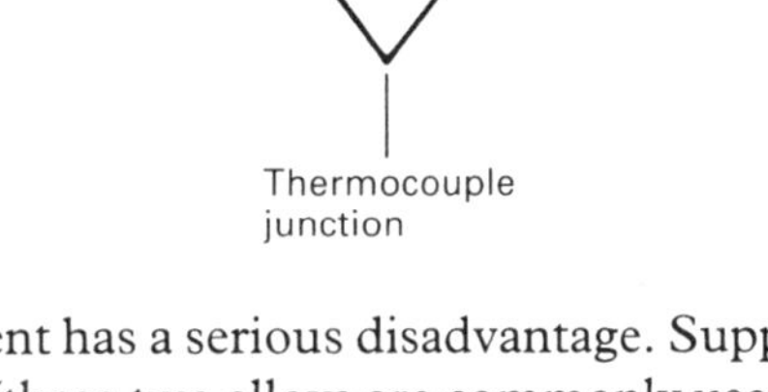

Fig. 13.2
Simple thermocouple

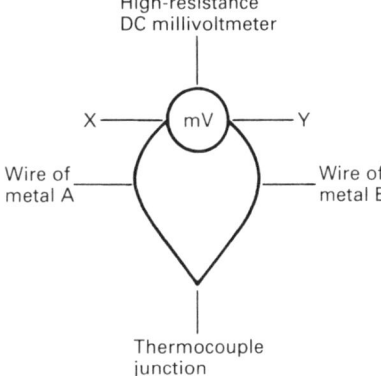

This simple arrangement has a serious disadvantage. Suppose metal A is chromel and metal B is alumel (these two alloys are commonly used in the manufacture of thermocouples), and that the terminal posts of the meter are brass. At X then, there is an EMF due to a chromel/brass thermocouple, and at Y there is a different EMF due to a brass/alumel thermocouple. The meter reading will be the algebraic sum of the three EMFs, and not the EMF of the actual chromel/alumel thermocouple which is required.

This difficulty can be overcome by using a second junction as shown in Fig. 13.3. With this arrangement, the EMFs produced at the meter terminals are equal and opposite, and therefore cancel each other. The extra junction that has been introduced, the so-called 'cold' junction, acts as a reference junction. The hot junction acts as the temperature measuring junction. The cold junction is normally placed in crushed ice and water so that it is always at 0 °C. The EMF at

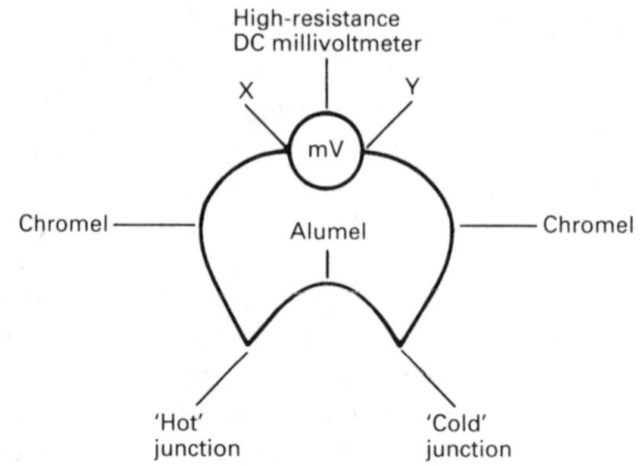

Fig. 13.3 Thermocouple with reference junction

the cold junction is therefore always the same, and so it is a simple matter to adjust the meter reading to allow for this EMF. (**Note**. The use of the terms 'hot junction' and 'cold junction' arises because thermocouples are normally used to measure temperatures above 0 °C, in which case the reference junction is the colder of the two.)

Thermocouples have very small heat capacities, and so have very little effect on the temperature of the body whose temperature they are measuring, and can measure rapidly fluctuating temperatures. In both these respects thermocouples are superior to other types of thermometer. In addition, they are cheap and easy to use, and are ideal for use with a pen-recorder.

The thermoelectric EMFs of many pairs of metals have been measured as a function of the hot junction temperature θ as measured by a constant-volume gas thermometer and expressed in degrees Celsius. In every case if the cold junction is maintained at 0 °C, it is found that to a good approximation the EMF E is given by

$$E = \alpha\theta + \beta\theta^2 \qquad [13.4]$$

where the values of α and β depend on the particular pair of metals concerned. This relationship is, of course, parabolic and therefore there exists a value of θ, known as the **neutral temperature**, θ_n, for which $dE/d\theta = 0$ (Fig. 13.4). It is clearly not desirable to use a thermocouple to measure temperatures close to its neutral temperature, because the variation of EMF with temperature is small and the thermometer is therefore insensitive in this region.

The particular pair of metals used depends on the temperature range for which the thermocouple is intended. Chromel/alumel thermocouples are normally used up to about 1100 °C, and produce a thermoelectric EMF of about 4 mV for every 100 °C difference in temperature between the hot and cold junctions. Above 1100 °C and up to about 1700 °C platinum/platinum–rhodium is used on account of the high melting points of platinum and platinum–rhodium. All these metals, particularly platinum and platinum–rhodium alloy, are readily available in states of high purity and so can be used to make thermocouples which give highly

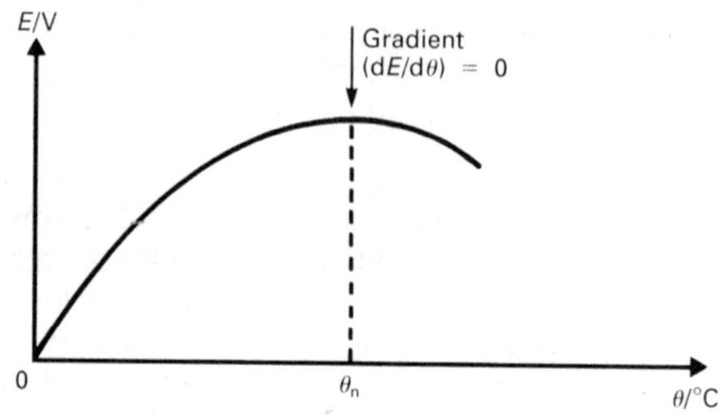

Fig. 13.4 Thermocouple EMF as a function of temperature

reproducible results. The disadvantage of platinum/platinum–rhodium is its relatively low thermoelectric EMF, about 1 mV per 100 °C.

For the most accurate work the millivoltmeter is replaced by a potentiometer (see Chapter 38). The use of a potentiometer, however, prevents the thermocouple being used to measure rapidly changing temperatures.

The values of α and β of equation [13.4] which are relevant to the commonly used thermocouple materials can be obtained from tables, and can be used in equation [13.4] to determine θ once E has been measured. Alternatively, calibration charts (plots of E against θ) are available.

13.5 RESISTANCE THERMOMETERS

Resistance thermometers rely on the fact that the resistances of metals are temperature-dependent, and therefore a measurement of resistance can be used as a measurement of temperature. They are usually made of platinum because of its high temperature coefficient of resistance and high melting point (1773 °C); features which make platinum resistance thermometers both sensitive and useful over large ranges of temperature. Also, platinum is readily available in a state of high purity, so that the measurements made with one particular platinum resistance thermometer are likely to match those made with another. The platinum is in the form of wire coiled on a suitable insulator such as mica or alumina. In use, the thermometer forms one arm of a Wheatstone bridge (see Chapter 37). This arrangement allows very slight changes in resistance, and therefore in temperature, to be measured. Platinum resistance thermometers are extremely accurate from −200 °C up to 1200 °C. The main disadvantage of thermometers of this type is that they have relatively large heat capacities. This means that they take a considerable time to come into thermal equilibrium with their surroundings, and therefore prevents them following rapidly changing temperatures. This is precluded anyway because a Wheatstone bridge has to be used.

When calibrated against constant-volume gas thermometers the resistance R of platinum is found to vary with Celsius temperature θ according to

$$R = R_0(1 + \alpha\theta + \beta\theta^2) \quad [13.5]$$

where R_0 is the resistance of the platinum at 0 °C and α and β are constants. The values of R_0, α and β pertaining to any particular thermometer are found by measuring its resistance at the ice point, the steam point and at the melting point of sulphur (444.6 °C), and inserting the three pairs of values of R and θ in equation [13.5]. Once R_0, α and β have been found equation [13.5] can be used to determine θ for any measured value of R.

13.6 THERMISTORS

These devices, like resistance thermometers, rely on their change of electrical resistance with temperature as a means of measuring temperature. Unlike resistance thermometers, however, they have negative temperature coefficients of resistance; their resistance decreasing approximately exponentially with increasing temperature. Thermistors are semiconducting devices cheaply manufactured out of several different mixtures of semiconducting oxide powders

($Fe_3O_4 + MgCr_2O_4$ is a common mixture). They are very robust. When a Wheatstone bridge circuit is used to measure their resistance they are about twenty times as sensitive as resistance thermometers. The resistance of the connecting wires is of no significance, since the devices themselves typically have a resistance of 1 kΩ. Thermistors have very small thermal capacities, and therefore respond quickly and have little effect on the temperature they are measuring. The range is typically −70 °C to 300 °C. They are less stable than resistance thermometers, and therefore less accurate.

13.7 THE CONSTANT-VOLUME GAS THERMOMETER

A simple constant-volume gas thermometer is shown in Fig. 13.5. When the thermometer is in use the bulb is placed inside the enclosure whose temperature is required. The gas in the bulb (air in the simplest versions) expands and forces mercury up the movable tube. The height of this tube is then adjusted to bring the mercury in the left-hand tube back to its original position at a fixed mark A. The gas now has its original volume. At this stage the head of mercury h is measured and the pressure p_θ of the gas is calculated from $p_\theta = p_A + h$ where p_A is the prevailing atmospheric pressure expressed in mm of mercury.

If p_0 and p_{100} are the pressures at 0 °C and 100 °C respectively, the temperature of the enclosure can be found from

$$\theta = \frac{p_\theta - p_0}{p_{100} - p_0} \times 100$$

where θ is the desired temperature in °C according to the constant-volume gas scale.

Fig. 13.5
Constant-volume gas thermometer

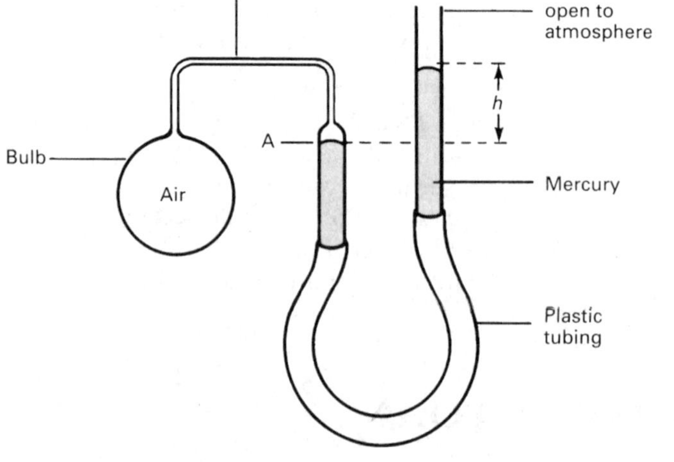

There are a number of sources of error:

(i) the bulb expands;

(ii) air is not an ideal gas;

(iii) the air in the capillary tube is not at the temperature being measured.

13.8 HEAT CAPACITY

The temperature rise produced by the addition of any given amount of heat to a body is determined by the mass of the body and the substance(s) of which it is composed.

The heat capacity* **(C) of a body is defined as being the heat required to produce unit temperature rise.**

It follows that if the temperature of a body whose heat capacity is C rises by $\Delta\theta$ when an amount of heat ΔQ is added to it, then

$$\Delta Q = C\Delta\theta \tag{13.6}$$

Unit of heat capacity $= \text{J K}^{-1} = \text{J}\,^\circ\text{C}^{-1}$ (see note (ii)).

The term **specific heat capacity** refers to the heat capacity of unit mass of a substance.

The specific heat capacity (c) of a substance is the heat required to produce unit temperature rise in unit mass of the substance.

It follows that if the temperature of a body of mass m and specific heat capacity c rises by $\Delta\theta$ when an amount of heat ΔQ is added to it, then

$$\Delta Q = mc\Delta\theta \tag{13.7}$$

Unit of specific heat capacity $= \text{J kg}^{-1}\,\text{K}^{-1} = \text{J kg}^{-1}\,^\circ\text{C}^{-1}$ (see note (ii)).

Notes

(i) The value of c depends on the temperature at which it is measured. However, over moderate changes in temperature, the variation is slight (except at low temperatures) and is normally ignored at this level.

(ii) Equations [13.6] and [13.7] involve only changes in temperature and so the numerical values of C and c when expressed in $\text{J}\,^\circ\text{C}^{-1}$ and $\text{J kg}^{-1}\,^\circ\text{C}^{-1}$ are the same as those expressed in J K^{-1} and $\text{J kg}^{-1}\,\text{K}^{-1}$ respectively.

QUESTIONS 13B

1. Calculate the quantity of heat required to raise the temperature of a metal block with a heat capacity of $23.1\,\text{J}\,^\circ\text{C}^{-1}$ by $30.0\,^\circ\text{C}$.

2. An electrical heater supplies 500 J of heat energy to a copper cylinder of mass 32.4 g. Find the increase in temperature of the cylinder.
(Specific heat capacity of copper $=$ $385\,\text{J kg}^{-1}\,^\circ\text{C}^{-1}$.)

3. How much heat must be removed from an object with a heat capacity of $150\,\text{J}\,^\circ\text{C}^{-1}$ in order to reduce its temperature from $80.0\,^\circ\text{C}$ to $20.0\,^\circ\text{C}$?

4. A metal block of heat capacity $36.0\,\text{J}\,^\circ\text{C}^{-1}$ at $70\,^\circ\text{C}$ is plunged into an insulated beaker containing 200 g of water at $18\,^\circ\text{C}$. The block and the water eventually reach a common temperature of $\theta\,^\circ\text{C}$. **(a)** Write down expressions in terms of θ for **(i)** the decrease in temperature of the block, **(ii)** the increase in temperature of the water. **(b)** Find in terms of θ, **(i)** the heat lost by the block, **(ii)** the heat gained by the water. **(c)** Assuming that no heat is used to heat the beaker and that no heat is lost to the surroundings, find the value of θ.
(Specific heat capacity of water $=$ $4.2 \times 10^3\,\text{J kg}^{-1}\,^\circ\text{C}^{-1}$.)

*Sometimes called **thermal capacity**.

13.9 THE COOLING CORRECTION

Experimental determinations of specific heat capacities usually involve some loss of heat to the surroundings. Losses due to conduction and convection can be reduced by lagging, or by surrounding the apparatus with a layer of still air, or by evacuating the region around the substance under test. Losses due to radiation are significant at high temperatures and can be reduced by using polished surfaces. One way of reducing the effect of the heat losses which remain is to apply a cooling correction.

Suppose that in some experiment the temperature is recorded both during heating and after heating has been discontinued, and that the temperature is found to vary with time as shown by the solid curve in Fig. 13.6. If there had been no heat loss, the maximum temperature would have been $\theta_m + \Delta\theta$. It can be shown that

$$\Delta\theta = \frac{A}{A'}\Delta\theta' \qquad [13.8]$$

Measuring $\Delta\theta'$ and the areas A and A' enables the correction ($\Delta\theta$) to be found.

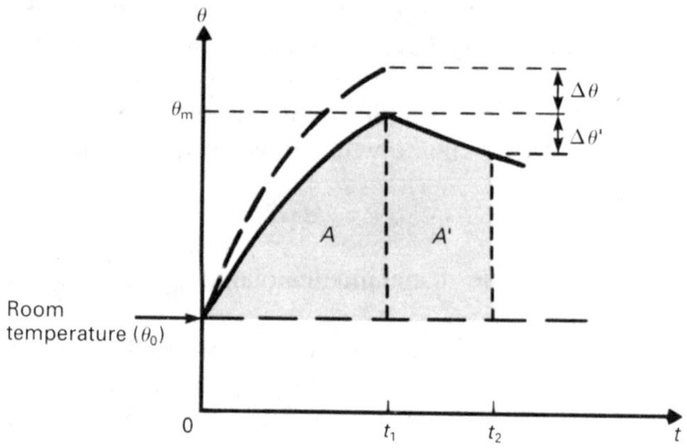

Fig. 13.6 Temperature against time with and without cooling

Theory of the Cooling Correction

The cooling correction is based on Newton's law of cooling (section 13.15). The reader should be familiar with this before proceeding.

If Newton's law of cooling applies, the rate of loss of heat to the surroundings, dQ/dt, both during heating and during cooling is given by

$$\frac{dQ}{dt} = k(\theta - \theta_0) \qquad [13.9]$$

where k is a constant of proportionality. The total loss of heat, Q, in the interval between $t = 0$ and $t = t_1$ is given by

$$Q = \int_0^{t_1} \frac{dQ}{dt}\,dt$$

Therefore by equation [13.9]

$$Q = k\int_0^{t_1} (\theta - \theta_0)\,dt$$

i.e. $\quad Q = k \times \text{Area } A$

Similarly, the loss of heat, Q', between $t = t_1$ and $t = t_2$ is given by

$$Q' = k \times \text{Area } A'$$

Therefore

$$\frac{Q}{Q'} = \frac{\text{Area } A}{\text{Area } A'}$$

But $Q = C\Delta\theta$ and $Q' = C\Delta\theta'$ where C is the heat capacity of the body, and therefore

$$\frac{\Delta\theta}{\Delta\theta'} = \frac{A}{A'}$$

i.e. $$\Delta\theta = \frac{A}{A'}\Delta\theta'$$

Note Calorimetry experiments are usually carried out in still air and it may seem surprising therefore that Newton's law of cooling (which applies to conditions of forced convection) is the basis of the cooling correction. It is used because it simplifies the theory, and is justified because the error it introduces is an error only in a correction term and is therefore of little significance overall.

13.10 ELECTRICAL METHODS OF MEASURING SPECIFIC HEAT CAPACITIES

Specific Heat Capacity of a Liquid

The apparatus is shown in Fig. 13.7. The rheostat should be adjusted to give a suitable current through the heating coil. The inner calorimeter contains a known mass of the liquid under test. The temperature θ_0 of the liquid is recorded. The switch is closed and the heater current and PD are recorded. The liquid is stirred continuously and its temperature is measued at one-minute intervals. Heating is continued until the temperature has risen by about 50 °C. The current and PD change slightly due to the increased resistance of the heating coil at higher temperatures, and their values should be recorded immediately before switching off the heater. The heater is switched off and the temperature is recorded until it has fallen to about 10 °C below its maximum value θ_m.

Fig. 13.7 Apparatus for determining the specific heat capacity of a liquid

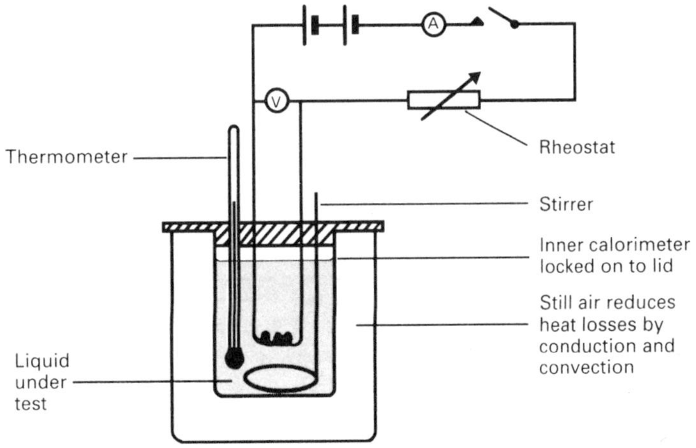

If the specific heat capacity of the liquid and the heat capacity of the inner calorimeter are c and C respectively, and $\Delta\theta$ is the cooling correction found from equation [13.8], then

$$VIt = (mc + C)(\theta_m + \Delta\theta - \theta_0)$$

where V and I are the average heater PD and current and t is the time for which heating is carried out; hence c.

Specific Heat Capacity of a Solid

The apparatus is shown in Fig. 13.8. The material under test is in the form of a solid cylinder of mass m, into which two holes have been drilled to accommodate a heater and a thermometer. The procedure is basically the same as that for a liquid. The specific heat capacity c is calculated from

$$VIt = mc(\theta_m + \Delta\theta - \theta_0)$$

Fig. 13.8
Apparatus for determining the specific heat capacity of a solid

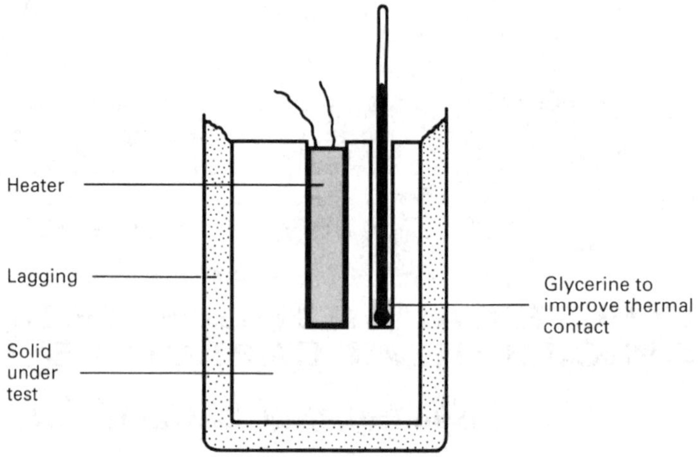

Nernst's Method for a Solid

The apparatus is shown in Fig. 13.9. A platinum heating coil is wound on paraffin-waxed paper around a cylindrical plug X of the metal under test. The paper insulates the coil from X so that its turns are not shorted out. The plug and coil are inserted into a cylindrical block Y of the same metal as X. A layer of paraffin wax around the coil insulates it from Y. The leads to the heating coil are used to

Fig. 13.9
Nernst's apparatus

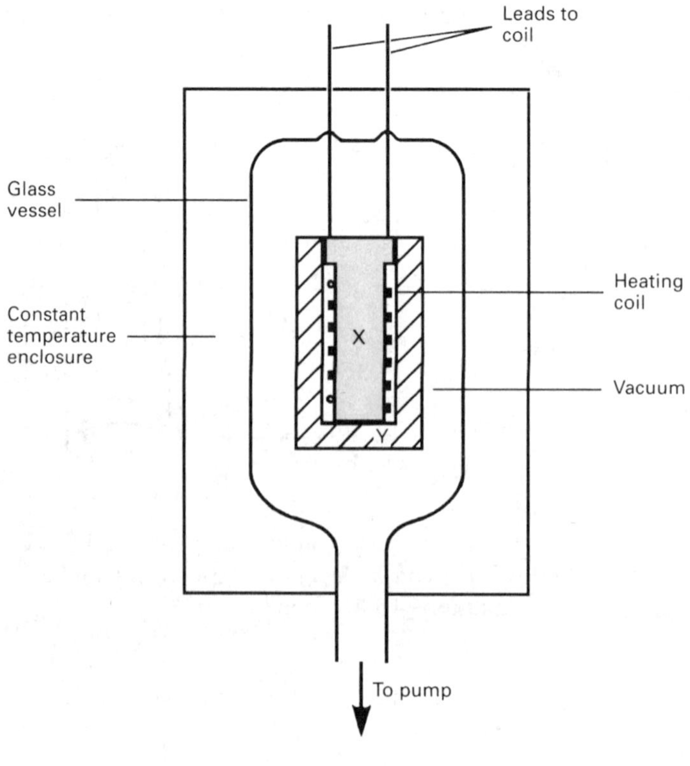

suspend the metal inside a glass vessel which can be evacuated. The apparatus is surrounded by a constant-temperature enclosure, the temperature of which is the temperature at which the specific heat measurement is required. (The specific heat capacity of a substance depends on the temperature at which it is measured.) The apparatus is left until the metal acquires this temperature and then the glass vessel is evacuated. Since the metal is in a vacuum and is at the temperature of its surroundings, heat losses are almost entirely eliminated.

The electrical energy used in a measured time t to raise the temperature of the specimen by a small amount $\Delta\theta$ is determined by measuring the current I through the coil and the PD V across it. The temperature rise is found by using the coil as a platinum resistance thermometer. In order to do this the resistance of the coil is measured immediately before the heater current is switched on, and immediately after it has been switched off. Temperature rises of as little as 10^{-3} K can be used.

13.11 THE CONTINUOUS FLOW METHOD FOR THE SPECIFIC HEAT CAPACITY OF A LIQUID

The method is due to Callendar and Barnes (1899).

Liquid is passed through the continuous flow calorimeter (Fig. 13.10) at a constant rate until all conditions are steady. At this stage the temperatures θ_X and θ_Y at X and Y, and the mass m_1 of liquid flowing through the calorimeter in time t are measured, together with the current I_1 through the heating coil and the PD V_1 across it. Under steady conditions none of the electrical energy which is being supplied is being used to heat the calorimeter, and therefore

$$V_1 I_1 t = m_1 c(\theta_Y - \theta_X) + Q \quad [13.10]$$

where Q is the heat lost to the surroundings in time t.

Fig. 13.10
Callender and Barnes' continuous flow calorimeter

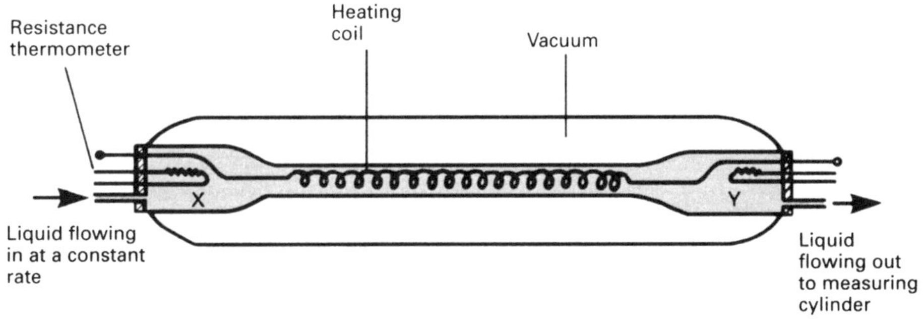

The rate of flow is altered so that the mass of liquid flowing in time t is m_2. The current and PD are adjusted (to I_2 and V_2) to bring the temperature at Y back to its original value θ_Y. The temperature at X is that of the tank supplying the liquid and is constant at θ_X. Since all temperatures are the same as they were with the initial flow rate, the heat lost in time t is again Q. Therefore

$$V_2 I_2 t = m_2 c(\theta_Y - \theta_X) + Q \quad [13.11]$$

Subtracting equation [13.10] from equation [13.11] gives

$$(V_2 I_2 - V_1 I_1)t = (m_2 - m_1)c(\theta_Y - \theta_X)$$

Hence c.

Advantages

(i) The presence of the vacuum prevents heat losses by convection, and the effect of losses due to conduction and radiation is eliminated.

(ii) The temperatures which are measured are steady and therefore can be determined accurately by using platinum resistance thermometers. This allows small temperature rises to be used (typically 2 °C) and the method is therefore suitable for determining the manner in which the specific heat capacity changes with temperature.

(iii) The calculation does not involve the heat capacities of the various parts of the apparatus and so there is no need to know their values.

Disadvantage

A large quantity of liquid is required.

Further Points

(i) The percentage error is least when the difference between the two flow-rates is large.

(ii) Continuous flow methods can also be used for gases.

13.12 LATENT HEAT

It is necessary to supply energy (heat) to a solid in order to melt it, even if the solid is already at its melting point. This energy is called **latent heat**. It is distinct from any heat that might have been used to bring the solid up to its melting point in the first place, and from that which might be used to raise the temperature of the liquid once the solid has melted.

The energy is used to provide the increased molecular potential energy of the liquid phase and, when the phase change results in expansion, to do external work in pushing back the atmosphere. The energy used to do external work is usually much less than that used to increase the potential energy of the molecules, and in the case of ice, which contracts on melting, is negative.

The conversion of a liquid to a vapour (vaporization) and the direct conversion of a solid to a vapour (sublimation) also require latent heat to be supplied. These two processes usually involve large changes in volume, and the proportion of the latent heat which is used to do external work is greater than in melting.

In terms of the first law of thermodynamics (section 14.15) melting (i.e. fusion), vaporization and sublimation are represented by

$$L = \Delta U + \Delta W$$

where

L = the latent heat supplied in order to cause the phase change

ΔU = the increase in internal potential energy which accompanies the phase change. (There is no change in temperature and therefore no change in kinetic energy.)

ΔW = the external work done as a result of the phase change. This term is positive for expansion and negative for contraction.

The specific latent heat (l) of fusion (or vaporization or sublimation) of a substance is defined as the energy required to cause unit mass of the substance to change from solid to liquid (or liquid to vapour, or solid to vapour) without temperature change. (Unit $= \mathrm{J\,kg^{-1}}$.)

Note The value of l depends on the temperature (and therefore the pressure) at which it is measured.

It follows that the heat, ΔQ, which must be added to change the phase of a mass, m, of substance is given by

$$\Delta Q = ml$$

where l is the specific latent heat of fusion, vaporization or sublimation according to the particular phase change which is taking place. For the reverse processes (liquid to solid, vapour to liquid, and vapour to solid) ΔQ represents the amount of heat that must be removed from the substance.

EXAMPLE 13.4

A calorimeter with a heat capacity of $80\,\mathrm{J\,{}^\circ C^{-1}}$ contains 50 g of water at 40 °C. What mass of ice at 0 °C needs to be added in order to reduce the temperature to 10 °C? Assume no heat is lost to the surroundings. (Specific heat capacity of water $= 4.2 \times 10^3\,\mathrm{J\,kg^{-1}\,{}^\circ C^{-1}}$, specific latent heat of ice $= 3.4 \times 10^5\,\mathrm{J\,kg^{-1}}$.)

Solution

Heat lost by calorimeter cooling to 10 °C

$$= 80(40 - 10) = 2400\,\mathrm{J}$$

Heat lost by water cooling to 10 °C

$$= 50 \times 10^{-3} \times 4.2 \times 10^3(40 - 10) = 6300\,\mathrm{J}$$

$\therefore$ Total heat lost $= 2400 + 6300 = 8700\,\mathrm{J}$

Let mass of ice $= m$

Heat used to melt ice at 0 °C

$$= m \times 3.4 \times 10^5 = 3.4 \times 10^5\,m$$

Heat used to increase temperature of melted ice to 10 °C

$$= m \times 4.2 \times 10^3(10 - 0) = 4.2 \times 10^4\,m$$

$\therefore$ Total heat used $= 3.4 \times 10^5\,m + 4.2 \times 10^4\,m = 3.82 \times 10^5\,m$

Since no heat is lost to the surroundings,

$$3.82 \times 10^5\,m = 8700$$

$\therefore$ $m = 0.0228\,\mathrm{kg}$

i.e. Mass of ice required $= 23\,\mathrm{g}$

QUESTIONS 13C

1. Calculate the heat required to melt 200 g of ice at 0 °C.
(Specific latent heat of ice $= 3.4 \times 10^5\,\mathrm{J\,kg^{-1}}$.)

2. Calculate the heat required to turn 500 g of ice at 0 °C into water at 100 °C.
(Specific latent heat of ice $= 3.4 \times 10^5\,\mathrm{J\,kg^{-1}}$, specific heat capacity of water $= 4.2 \times 10^3\,\mathrm{J\,kg^{-1}\,°C^{-1}}$.)

3. Calculate the heat given out when 600 g of steam at 100 °C condenses to water at 20 °C.
(Specific latent heat of steam $= 2.26 \times 10^6\,\mathrm{J\,kg^{-1}}$, specific heat capacity of water $= 4.2 \times 10^3\,\mathrm{J\,kg^{-1}\,°C^{-1}}$.)

13.13 EXPERIMENTAL DETERMINATION OF THE SPECIFIC LATENT HEAT OF VAPORIZATION OF A LIQUID

The method about to be described is a continuous flow method and makes use of a self-jacketing vaporizer.

The apparatus is shown in Fig. 13.11. The liquid under investigation is heated to boiling point and the vapour which is produced passes to the condenser by way of holes (H) in the inner wall of the vessel. Boiling is continued, and eventually the temperatures of all parts of the apparatus become steady. At this stage the condensed vapour is collected, over a measured time t, and its mass m_1

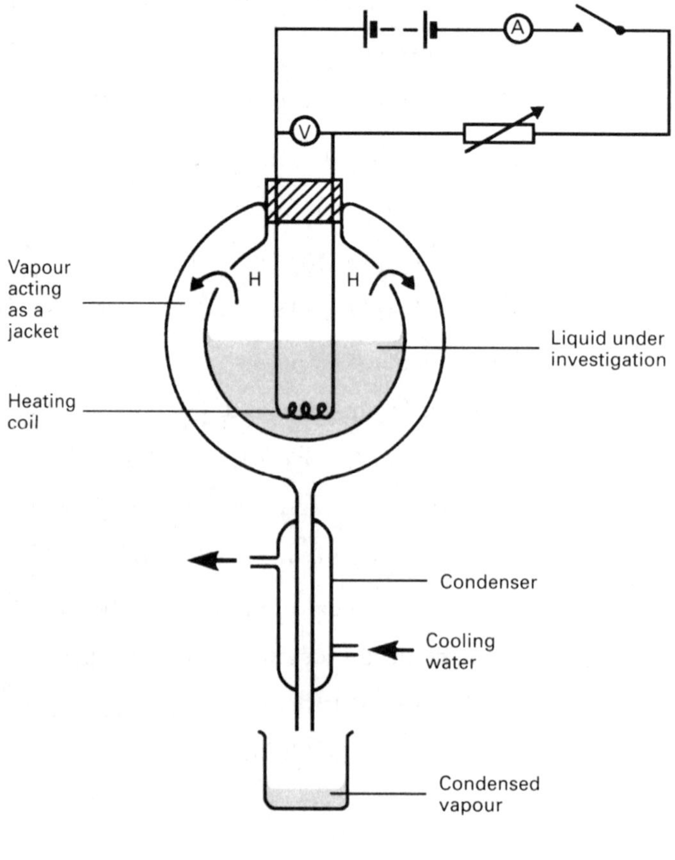

Fig. 13.11
Apparatus for determining the specific latent heat of vaporization of a liquid

determined. If V_1 and I_1 are the heater PD and current, then the electrical energy supplied in time t is V_1I_1t. Since the temperatures are steady, this energy is used only to vaporize the liquid and to offset heat losses, and therefore

$$V_1I_1t = m_1l + Q \qquad [13.12]$$

where l is the specific latent heat of vaporization of the liquid and Q is the heat lost to the surroundings in time t.

The heater PD and current are now changed to V_2 and I_2 and the new mass m_2 of vapour which condenses in the same time t is measured.

Each part of the apparatus is at the same temperature as it was with the initial rate of heating and the energy lost in time t is again Q. Therefore.

$$V_2I_2t = m_2l + Q \qquad [13.13]$$

Subtracting equation [13.12] from equation [13.13] gives

$$(V_2I_2 - V_1I_1)t = (m_2 - m_1)\,l$$

from which l can be determined.

Note The liquid which is being vaporized is surrounded by its vapour (hence self-jacketing vaporizer). Any heat lost by the vapour causes it to condense, not to cool, and therefore the liquid is surrounded by a constant temperature enclosure which is at its own temperature; this considerably reduces heat losses from the liquid.

13.14 EXPERIMENTAL DETERMINATION OF THE SPECIFIC LATENT HEAT OF FUSION OF ICE BY THE METHOD OF MIXTURES

A calorimeter of mass m_c is about two-thirds filled with water of mass m_w which is about 5 °C above room temperature. The water and the calorimeter are left for a short time until they reach the same temperature as each other. This temperature (θ_1) is measured using a sensitive ($\frac{1}{10}$ °C) thermometer.

A lump of melting ice (i.e. ice at 0 °C) is then dried with blotting paper and immediately added to the water. The mixture is then stirred gently until the lump has melted. This procedure is repeated with further lumps until the temperature of the mixture is approximately as far below room temperature as θ_1 was above. The lowest temperature attained (θ_2) is recorded.

The calorimeter and its contents are weighed to determine the mass (m_i) of the ice.

$$\begin{bmatrix}\text{Heat lost by}\\ \text{water cooling}\\ \text{from } \theta_1 \text{ to } \theta_2\end{bmatrix} + \begin{bmatrix}\text{Heat lost by}\\ \text{calorimeter}\\ \text{cooling from}\\ \theta_1 \text{ to } \theta_2\end{bmatrix} = \begin{bmatrix}\text{Heat used}\\ \text{to melt ice}\\ \text{at } 0^\circ\text{C}\end{bmatrix} + \begin{bmatrix}\text{Heat used to}\\ \text{increase}\\ \text{temperature}\\ \text{of melted ice}\\ \text{from } 0^\circ\text{C to } \theta_2\end{bmatrix}$$

Therefore

$$m_wc_w(\theta_1 - \theta_2) + m_cc_c(\theta_1 - \theta_2) = m_i\,l + m_ic_w(\theta_2 - 0)$$

from which the specific latent heat of fusion of ice (l) can be found, providing the specific heat capacities, c_w and c_c, of water and of the calorimeter material respectively, are known.

Notes (i) It is very important that the ice is dry when it is added to the water. If it is not, the mass of ice that is melted is less than m_i.

(ii) No cooling correction is applied because θ_1 is as far above room temperature as θ_2 is below it, and therefore, to a reasonable approximation, the mixture gains as much heat from the surroundings whilst it is below room temperature as it loses to them whilst it is above room temperature.

13.15 COOLING LAWS

Newton's Law of Cooling

This applies when a body is cooling under conditions of forced convection (i.e. when it is in a steady draught). It states that the rate of loss of heat of a body is proportional to the difference in temperature between the body and its surroundings, i.e.

$$\begin{pmatrix}\text{Rate of loss of}\\ \text{heat to surroundings}\end{pmatrix} \propto \begin{pmatrix}\text{Excess}\\ \text{temperature}\end{pmatrix}$$

or

$$\begin{pmatrix}\text{Rate of loss of}\\ \text{heat to surroundings}\end{pmatrix} = k(\theta - \theta_0)$$

where

θ = temperature of body

θ_0 = temperature of surroundings

k = a constant of proportionality whose value depends on both the nature and the area of the body's surface.

The law can be taken to be a good approximation for cooling under conditions of natural convection (a body cooling in still air for example) provided the excess temperature is not greater than about 30 °C. For higher excess temperatures than this the five-fourths power law should be used.

The Five-Fourths Power Law

This applies when a body is cooling under conditions of natural convection. It can be stated as

$$\begin{pmatrix}\text{Rate of loss of}\\ \text{heat to surroundings}\end{pmatrix} = k(\theta - \theta_0)^{5/4}$$

Experimental Investigation of Newton's Law of Cooling

The rate at which a body loses heat is proportional to the rate at which its temperature falls provided that its heat capacity does not vary with temperature (since $\Delta Q = C\Delta\theta$, where C = heat capacity). Therefore for a body cooling under conditions where Newton's law of cooling applies

$$\frac{d\theta}{dt} = -K(\theta - \theta_0)$$

where $d\theta/dt$ is (minus) the rate of fall of temperature of the body and K is a constant of proportionality whose value depends on the nature and the area of the body's surface, and on the heat capacity of the body. (Since the temperature is falling, the presence of the minus sign makes K a positive constant.) It follows that Newton's law of cooling can be verified by verifying this equation.

Method

A calorimeter containing hot water and standing on an insulating surface (e.g. a wooden block) is placed in the stream of air from an electric fan or an open window. The temperature of the water is measured at one-minute intervals using a $\frac{1}{10}$ °C thermometer. The water should be stirred gently prior to each temperature measurement. A graph of temperature (θ) versus time (t) is plotted (Fig. 13.12(a)). The gradient of this graph at any temperature θ is the rate of fall of temperature at that value of θ. The gradients are measured (by constructing tangents to the curve) at various values of θ and are then plotted against the corresponding excess temperature $(\theta - \theta_0)$ as in Fig. 13.12(b). If this plot is a straight line through the origin, Newton's law of cooling has been verified.

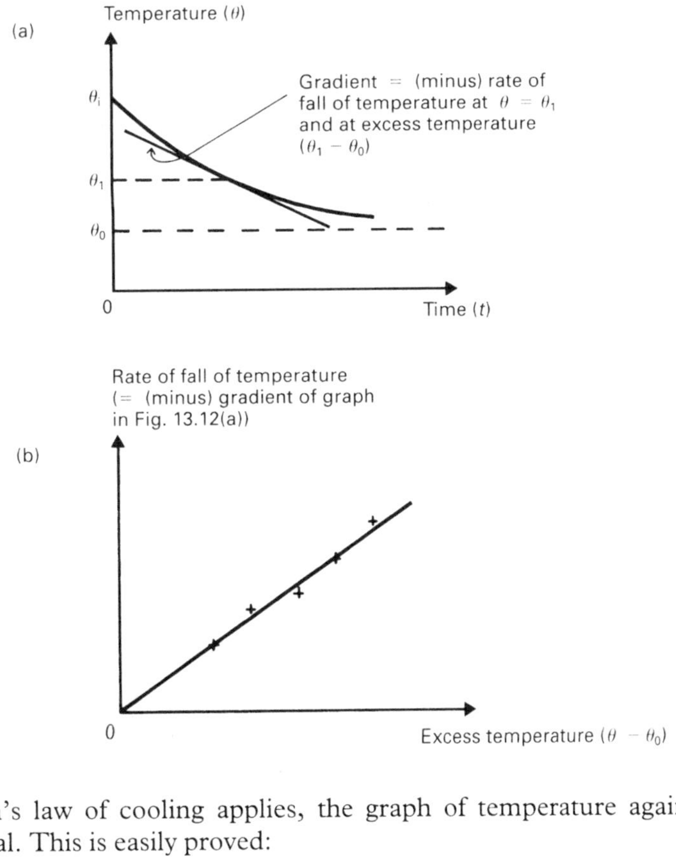

Fig. 13.12
Plot of (a) temperature against time during cooling, (b) rate of fall of temperature against excess temperature

If Newton's law of cooling applies, the graph of temperature against time is exponential. This is easily proved:

$$\frac{d\theta}{dt} = -K(\theta - \theta_0)$$

$$\therefore \quad \int_{\theta_i}^{\theta} \frac{d\theta}{(\theta - \theta_0)} = -K \int_0^t dt$$

where θ_i is the initial temperature.

$$\therefore \quad \left[\log_e(\theta - \theta_0)\right]_{\theta_i}^{\theta} = -K[t]_0^t$$

$$\therefore \quad \log_e\left(\frac{\theta - \theta_0}{\theta_i - \theta_0}\right) = -Kt$$

$$\therefore \quad (\theta - \theta_0) = (\theta_i - \theta_0)e^{-Kt}$$

$$\text{i.e.} \quad \left(\begin{array}{c}\text{Excess}\\ \text{temperature}\end{array}\right) = \left(\begin{array}{c}\text{Initial excess}\\ \text{temperature}\end{array}\right)e^{-Kt}$$

CONSOLIDATION

The temperature of a body is a measure of how hot it is, not how much heat it contains.

Celsius temperature, θ, based on some thermometric property, X, (e.g. the length of a column of mercury) is given by

$$\theta = \frac{X_\theta - X_0}{X_{100} - X_0} \times 100$$

Thermometers calibrated on the basis of this equation necessarily agree with each other only at the fixed points. They may agree at other temperatures too.

Kelvin temperature, T, on the thermodynamic scale and on the ideal gas scale is given by

$$T = \frac{p_T}{p_{\text{Tr}}} \times 273.16$$

where p_T and p_{Tr} are the pressures of a fixed volume of an ideal gas at temperature T and the triple point of water respectively.

The thermodynamic scale and the ideal gas scale are absolute scales, i.e. they do not depend on the properties of any particular substance.

An interval of one kelvin is defined as 1/273.16 of the temperature of the triple point of water.

Celsius temperature, θ, is defined by

$$\theta = T - 273.15$$

where T is the corresponding temperature in kelvins.

Heat capacity (C) is a property of a body. It is the heat required to produce unit temperature rise in the body. (Unit $=$ $\text{J}\,^\circ\text{C}^{-1}$ or $\text{J}\,\text{K}^{-1}$.)

Specific heat capacity (c) is a property of a substance. It is the heat required to produce unit temperature rise in unit mass of the substance. (Unit $=$ $\text{J}\,\text{kg}^{-1}\,^\circ\text{C}^{-1}$ or $\text{J}\,\text{kg}^{-1}\,\text{K}^{-1}$.)

For a change of temperature $\Delta Q = C\Delta\theta \qquad \Delta Q = mc\Delta\theta$

For a change of phase $\Delta Q = ml$

14

GASES

14.1 THE GAS LAWS

The experimental relationships between the pressures, volumes and temperatures of gases were investigated by various workers in the seventeenth and eighteenth centuries. These early experiments resulted in three laws – the so-called **gas laws**.

Boyle's Law

For a fixed mass of gas at constant temperature, the product of pressure and volume is constant.

On the basis of the other two laws the Kelvin scale of temperature was introduced, and these laws are stated below in terms of that scale.

Charles' Law

For a fixed mass of gas at constant pressure, the volume is directly proportional to the temperature measured in kelvins.

The Pressure Law

For a fixed mass of gas at constant volume, the pressure is directly proportional to the temperature measured in kelvins.

Representing pressure, volume, and temperature in kelvins by p, V and T respectively, we can formulate the three laws as:

At constant T	pV = a constant	or	$p \propto 1/V$
At constant p	V/T = a constant	or	$V \propto T$
At constant v	p/T = a constant	or	$p \propto T$

It should be noted that the three laws are not independent; any one of them can be derived from the other two. The experimental investigation of the gas laws is dealt with in sections 14.9 to 14.11.

14.2 CONCEPT OF AN IDEAL GAS AND THE IDEAL GAS EQUATION

No gas obeys the gas laws exactly. Nevertheless they provide a fairly accurate description of the way gases behave when they are at low pressures and are at temperatures which are well above those at which they liquefy. A useful concept is that of an **ideal (or perfect) gas** – a gas which obeys the gas laws exactly. The behaviour of such a gas can be accounted for by

$$pV = nRT \qquad [14.1]$$

where

p = the pressure of the gas ($N\,m^{-2}$ = pascals, Pa)

V = the volume of the gas (m^3)

n = the number of moles (see section 14.3) of gas (mol)

R = the universal molar gas constant (= $8.31\,J\,K^{-1}\,mol^{-1}$)

T = the temperature of the gas **in kelvins**.

Equation [14.1] is known as the equation of state of an ideal gas (or simply as the **ideal gas equation**); it embodies the three gas laws and Avogadro's law (section 14.6). It can be shown that a gas which obeys this equation exactly must be subject to the assumptions inherent in the kinetic theory of gases (section 14.4). In particular, there would be no forces between the molecules of such a gas and therefore the **internal energy** (i.e. the energy of the molecules) of such a gas would be entirely kinetic and would depend only on its temperature.

Summary

(i) An ideal gas obeys the gas laws and $pV = nRT$ exactly. No such gas exists.

(ii) The internal energy of an ideal gas is entirely kinetic and depends only on its temperature.

(iii) The behaviour of real gases and unsaturated vapours (see Chapter 15) can be described by $pV = nRT$ if they are at low pressures and are at temperatures which are well above those at which they liquefy.

Note For a gas at pressure p_1, volume V_1 and temperature T_1 equation [14.1] gives

$$p_1V_1 = nRT_1 \qquad \text{i.e.} \qquad p_1V_1/T_1 = nR$$

If the same sample of gas is at pressure p_2, volume V_2 and temperature T_2

$$p_2V_2 = nRT_2 \qquad \text{i.e.} \qquad p_2V_2/T_2 = nR$$

(The number of moles is the same in each case (n) because we are dealing with the same sample of gas, i.e. with a fixed mass of gas and therefore with a fixed number of moles.) Combining these equations gives

$$\frac{p_1V_1}{T_1} = \frac{p_2V_2}{T_2} \qquad \text{for a fixed mass of gas}$$

Any unit of pressure can be used for p_1 providing the same unit is used for p_2. Similarly, any unit of volume can be used for V_1 as long as the same unit is used for V_2, but **both T_1 and T_2 must be expressed in kelvins**.

EXAMPLE 14.1

A gas (which can be considered ideal) has a volume of $100\,\text{cm}^3$ at 2.00×10^5 Pa and 27 °C. What is its volume at 5.00×10^5 Pa and 60 °C?

Solution

$$p_1 = 2.00 \times 10^5\,\text{Pa} \qquad p_2 = 5.00 \times 10^5\,\text{Pa}$$

$$V_1 = 100\,\text{cm}^3 \qquad V_2 = V_2$$

$$T_1 = 27\,^\circ\text{C} = 300\,\text{K} \qquad T_2 = 60\,^\circ\text{C} = 333\,\text{K}$$

$$\frac{p_1 V_1}{T_1} = \frac{p_2 V_2}{T_2}$$

$$\therefore \quad \frac{2.00 \times 10^5 \times 100}{300} = \frac{5.00 \times 10^5 \times V_2}{333}$$

$$\therefore \quad V_2 = \frac{2.00 \times 10^5 \times 100 \times 333}{300 \times 5.00 \times 10^5} = 44.4\,\text{cm}^3$$

Note that V_1 was expressed in cm^3 and therefore V_2 is in cm^3.

EXAMPLE 14.2

Refer to Fig 14.1. Initially A contains $3.00\,\text{m}^3$ of an ideal gas at a temperature of 250 K and a pressure of 5.00×10^4 Pa, whilst B contains $7.20\,\text{m}^3$ of the same gas at 400 K and 2.00×10^4 Pa. Find the pressure after the connecting tap has been opened and the system has reached equilibrium, assuming that A is kept at 250 K and B is kept at 400 K.

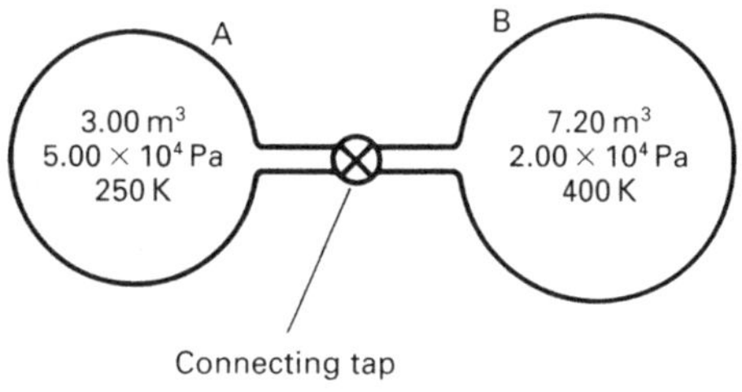

Fig. 14.1 Diagram for Example 14.2

Solution

On opening the tap some gas moves from A to B, reducing the pressure in A and increasing it in B. This continues until, at equilibrium, the pressure in A is equal to that in B. Let this final pressure be p. The 'trick' is to recognize that the total mass of gas, and therefore the total number of moles, is the same after the tap is opened as it was before.

Since $pV = nRT$, $n = \dfrac{pV}{RT}$ and therefore

$$\text{Number of moles initially in A} = \frac{5.00 \times 10^4 \times 3.00}{250R} = \frac{600}{R}$$

$$\text{Number of moles initially in B} = \frac{2.00 \times 10^4 \times 7.20}{400R} = \frac{360}{R}$$

$$\therefore \quad \text{Total number of moles initially} = 960/R$$

$$\text{Number of moles finally in A} = \frac{p \times 3.00}{250R} = \frac{1.20 \times 10^{-2}}{R}p$$

$$\text{Number of moles finally in B} = \frac{p \times 7.20}{400R} = \frac{1.80 \times 10^{-2}}{R}p$$

$$\therefore \quad \text{Total number of moles finally} = 3.00 \times 10^{-2}p/R$$

The total number of moles does not change and therefore

$$\frac{3.00 \times 10^{-2}}{R}p = \frac{960}{R} \qquad \text{i.e.} \quad p = 3.20 \times 10^4 \text{ Pa}$$

14.3 USEFUL RELATIONSHIPS

The mole is the unit of a quantity called 'amount of substance'.

1 mole (mol) of a substance is defined as the amount of it that contains the same number of elementary units (atoms or molecules, according to which is considered to be its elementary unit) as there are atoms in 12 g of carbon 12. This has been found by experiment to be 6.022×10^{23}. For example, a mole of molecular oxygen contains 6.022×10^{23} molecules; a mole of sodium vapour contains 6.022×10^{23} atoms, etc.

The Avogadro constant, N_A, is numerically equal to the number of atoms in one mole. $N_A = 6.022 \times 10^{23}\,\text{mol}^{-1}$. (Note, it is not a number, but a quantity measured in mol^{-1}.)

The relative molecular mass, M_r, of a substance is defined by

$$M_r = \frac{\text{Average mass of a molecule of the substance}}{\text{One twelfth the mass of a carbon 12 atom}}$$

Relative molecular mass is a number and therefore has no units.

The molar mass of a substance is defined as the mass per mole of the substance. (The SI unit is kg mol^{-1}, but molar mass is often expressed in g mol^{-1}.) Molar mass in g mol^{-1} and mass of one mole in g are numerically equal to relative molecular mass. For example, for oxygen in its molecular form, O_2, relative molecular mass $= 32$, molar mass $= 32\,\text{g mol}^{-1}$ and mass of one mole $= 32$ g.

The volume of 1 mole of a gas at STP is $22.4 \times 10^{-3}\,\text{m}^3$. **Note.** STP = **standard temperature and pressure,** i.e. 273 K and 760 mm of mercury ($1.013 \times 10^5\,\text{N m}^{-2}$).

This relationship can be obtained by substituting the relevant values of the parameters into equation [14.1]. Thus, rearranging the equation gives

$$V = \frac{nRT}{p}$$

Therefore, for 1 mole of gas at STP

$$V = \frac{(1)(8.31)(273)}{1.013 \times 10^5} = 22.4 \times 10^{-3}\,\text{m}^3$$

QUESTIONS 14A

1. What is the temperature of $19.0\,\text{m}^3$ of an ideal gas at a pressure of 600 mmHg if the same gas occupies $12.0\,\text{m}^3$ at 760 mmHg and 27 °C?

2. A gas has a volume of $60.0\,\text{cm}^3$ at 20 °C and 900 mmHg. What would its volume be at STP, i.e. at 273 K and 760 mmHg?

3. A cylinder contains $2.40 \times 10^{-3}\,\text{m}^3$ of hydrogen at 17 °C and 2.32×10^6 Pa. The relative molecular mass of hydrogen = 2, $R = 8.31\,\text{J}\,\text{K}^{-1}\,\text{mol}^{-1}$ and the Avogadro constant, $N_A = 6.02 \times 10^{23}\,\text{mol}^{-1}$. Calculate:
   (a) the number of moles of hydrogen in the cylinder,
   (b) the number of molecules of hydrogen in the cylinder,
   (c) the mass of the hydrogen,
   (d) the density of hydrogen under these conditions.

14.4 THE KINETIC THEORY OF GASES (DERIVATION OF $p = \frac{1}{3}\rho\overline{c^2}$)

This is an attempt to explain the experimentally observed properties of gases by considering the motion of the molecules (or atoms) of which they are composed. A number of assumptions are made.

(i) The molecules of a particular gas are identical.

(ii) Collisions between the molecules and with the container are (perfectly) elastic (see section 2.8).

(iii) The molecules exert no forces on each other except during impacts (which are assumed to have negligible duration anyway) and the effect of gravity is ignored so that:

 (a) between collisions the molecules move in straight lines at constant speed, and

 (b) the motion is random.

(iv) There is a sufficiently large number of molecules for statistics to be meaningfully applied.

(v) The size of the molecules is negligible compared to their separation.

(vi) The laws of Newtonian mechanics apply.

Some of these assumptions run through the entire analysis; others are used more specifically.

Consider a gas enclosed in a cubical container of side L (Fig. 14.2). Let each molecule of the gas have mass m (assumption (i)). Consider, initially, a single molecule which is moving towards wall X, and suppose that its x-component of velocity is u_1. This molecule will have an x-component of momentum mu_1 towards the wall. The molecule will eventually reverse the direction of its momentum by

Fig. 14.2
Derivation of $p = \frac{1}{3}\rho\overline{c^2}$

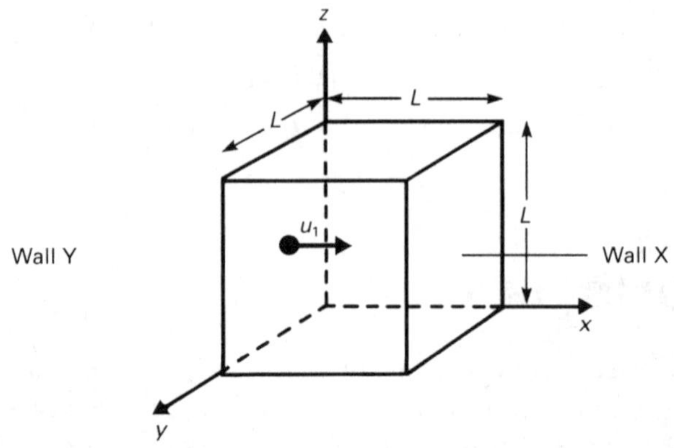

colliding with the wall. Since the collision will be elastic (assumption (ii)), it will rebound with the same speed so that its momentum will now be $-mu_1$. The change in the x-component of momentum is therefore $2mu_1$.

The molecule has to travel a distance $2L$ (from X to Y and back to X) before it next collides with wall X. The time for such a trip is $2L/u_1$, and therefore this molecule's rate of change of momentum due to collision with X will be

$$\frac{2mu_1}{2L/u_1} = \frac{mu_1{}^2}{L}$$

By Newton's second law, rate of change of momentum is equal to force, and therefore $mu_1{}^2/L$ is the force exerted on the molecule by the wall. By Newton's third law, the molecule exerts an equal but oppositely directed force on the wall, and therefore

$$\text{Force on X} = mu_1{}^2/L$$

Therefore

$$\text{Force per unit area on X} = \frac{mu_1{}^2/L}{L^2} \qquad (\text{since area of X} = L^2)$$

Therefore

$$\text{Pressure on X} = \frac{mu_1{}^2}{L^3}$$

If there are N molecules in the container and their x-components of velocity are u_1, $u_2, \ldots, u_N$, the total pressure, p, on wall X will be given by

$$p = \frac{m}{L^3}(u_1{}^2 + u_2{}^2 + \cdots + u_N{}^2)$$

Therefore

$$p = \frac{m}{L^3} N\,\overline{u^2} \qquad [14.2]$$

where

$\overline{u^2}$ is the mean square velocity in the x-direction.

Since mN is the total mass of gas in the container, mN/L^3 is the density, ρ, of the gas and therefore, by equation [14.2],

$$p = \rho\overline{u^2} \qquad [14.3]$$

If

c = the resultant velocity of a molecule whose x-, y- and z-components of velocity are u, v and w respectively, then

$$c^2 = u^2 + v^2 + w^2$$

Therefore

$$\overline{c^2} = \overline{u^2} + \overline{v^2} + \overline{w^2} \qquad [14.4]$$

where

$\overline{c^2}$ is the mean square velocity of the molecules

$\overline{v^2}$ is the mean square velocity in the y-direction

$\overline{w^2}$ is the mean square velocity in the z-direction.

Since there is a large number of molecules and they are moving randomly (assumptions (iv) and (iii)b)

$$\overline{u^2} = \overline{v^2} = \overline{w^2}$$

Therefore, from equation [14.4],

$$\overline{u^2} = \tfrac{1}{3}\overline{c^2}$$

Therefore, from equation [14.3]

$$p = \tfrac{1}{3}\rho\,\overline{c^2} \qquad [14.5]$$

14.5 RELATIONSHIP BETWEEN MOLECULAR KINETIC ENERGY AND TEMPERATURE

On the basis of the kinetic theory of gases,

$$p = \tfrac{1}{3}\rho\,\overline{c^2}$$

Therefore, for any volume V of gas,

$$pV = \tfrac{1}{3}\rho V\,\overline{c^2}$$

Therefore,

$$pV = \tfrac{1}{3}M\,\overline{c^2} \qquad [14.6]$$

where

M = the mass of volume V of the gas.

Equation [14.6] may be rewritten as

$$pV = \tfrac{2}{3}N(\tfrac{1}{2}m\,\overline{c^2}) \qquad [14.7]$$

where

N = the total number of molecules in volume V, and

m = the mass of one molecule.

The ideal gas equation for n moles of a gas of volume V and pressure p is

$$pV = nRT \qquad [14.8]$$

where

R = the universal molar gas constant, and

T = the temperature in kelvins.

Thus the predictions of the kinetic theory of gases (represented by equation [14.7]) are in agreement with idealized experimental observation (represented by equation [14.8]) if

$$\tfrac{2}{3}N(\tfrac{1}{2}m\overline{c^2}) = nRT$$

i.e. $$\tfrac{1}{2}m\overline{c^2} = \frac{3}{2}\frac{nR}{N}T$$

i.e. $$\tfrac{1}{2}m\overline{c^2} = \frac{3}{2}\frac{R}{N_A}T$$

(since N/n is the number of molecules per mole, i.e. N_A, the Avogadro constant). Both R and N_A are universal constants, and therefore so also is R/N_A; it is called **Boltzmann's constant,** k $(= 1.38 \times 10^{-23}\,\mathrm{J\,K^{-1}})$ and is the gas constant per molecule. The left-hand term is the average translational* kinetic energy of a single molecule, and therefore

$$\text{Average translational KE of a molecule} = \tfrac{3}{2}kT = \frac{3}{2}\frac{R}{N_A}T \quad [14.9]$$

Thus, in order to make the kinetic theory consistent with the ideal gas equation we need to accept the validity of equation [14.9], i.e. in addition to assumptions (i) to (vi) of section 14.4, we need to make the further assumption that **the average translational kinetic energy of a molecule is equal to (3/2) kT**. Such an assumption is reasonable, since putting heat energy into a gas increases its temperature and must also increase the kinetic energy of its molecules becuase there is no other way that the energy can be absorbed. (An ideal gas can have no potential energy because it has no intermolecular forces, and there is nothing other than molecules present.)

Note The three gas laws (section 14.1) can be combined as $pV \propto T$. In order to make the kinetic theory consistent with this expression, rather than with the more demanding $pV = nRT$, we need only make the assumption that $\frac{1}{2}m\overline{c^2}$ is proportional to T, for it then follows from equation [14.7] that $pV \propto T$.

QUESTIONS 14B

1. By what factor does **(a)** the mean square speed, **(b)** the root mean square speed of the molecules of a gas increase when its temperature is doubled?

2. The temperature of a gas is increased in such a way that its volume doubles and its pressure quadruples. If the root mean square speed of the molecules was originally 250 m s^{-1}, what is it at the higher temperature?

3. Find the value of the ratio

$$\frac{\text{Root mean square speed of hydrogen molecules}}{\text{Root mean square speed of oxygen molecules}}$$

(a) when the two gases are at the same temperature, **(b)** when the oxygen is at 100 °C and the hydrogen is at 30 °C.
(Relative molecular mass of hydrogen = 2, relative molecular mass of oxygen = 32.)

*In section 14.16 diatomic and polyatomic molecules are considered. These have both translational and rotational kinetic energies.

14.6 AVOGADRO'S LAW

Equal volumes of all gases at the same temperature and pressure contain the same number of molecules.

The law was announced by Avogadro in 1811 and was well-established before the kinetic theory was developed. It is embodied in the ideal gas equation, $pV = nRT$. In order to illustrate this we shall consider two gases distinguished by the subscripts $_1$ and $_2$. Applying the ideal gas equation gives

$$p_1V_1 = n_1RT_1 \quad \text{and} \quad p_2V_2 = n_2RT_2$$

For equal volumes at the same pressure $p_1V_1 = p_2V_2$, and therefore

$$n_1RT_1 = n_2RT_2 \qquad [14.10]$$

If the gases are at the same temperature, $T_1 = T_2$, and therefore from equation [14.10]

$$n_1 = n_2$$

Thus equal volumes of two gases which are at the same temperature and pressure contain the same number of moles. It follows from the definition of the mole (section 14.3) that the gases also contain the same number of molecules. Thus, under the conditions to which Avogadro's law relates, the number of molecules in each gas is the same, i.e. Avogadro's law is embodied in $pV = nRT$.

Note Avogadro's law can be applied to real gases which are at low pressures and are at temperatures well above those at which they liquefy.

14.7 DALTON'S LAW OF PARTIAL PRESSURES

The total pressure of a mixture of gases, which do not interact chemically, is equal to the sum of the partial pressures, i.e. to the sum of the pressures that each gas would exert if it alone occupied the volume containing the mixture.

Suppose that a volume V contains n_1 moles of a gas whose partial pressure is p_1 and n_2 moles of a gas whose partial pressure is p_2. If the temperature of the gases is T, then by equation [14.1]

$$p_1V = n_1RT \quad \text{and} \quad p_2V = n_2RT$$

Dividing gives

$$\frac{p_1}{p_2} = \frac{n_1}{n_2} \qquad [14.11]$$

From Dalton's law, the total pressure p is given by

$$p = p_1 + p_2$$

Substituting for p_2 from equation [14.11] gives

$$p_1 = \left(\frac{n_1}{n_1 + n_2}\right)p$$

and substituting for p_1 gives

$$p_2 = \left(\frac{n_2}{n_1 + n_2}\right)p$$

Note Dalton's law still applies if one or more of the components of the mixture is a vapour (saturated or unsaturated) – see Example 15.1.

14.8 THE MAXWELLIAN DISTRIBUTION OF MOLECULAR SPEEDS

It can be shown, on the basis of statistical mechanics, that the speeds of gas molecules are distributed as illustrated in Fig. 14.3. The curve, which is known as the Maxwellian distribution of molecular speeds, agrees well with that obtained by experiment. The quantity $N(c)$ is such that $N(c)\delta c$ is the number of molecules whose speeds are in the narrow range c to $c + \delta c$. Theoretically

$$c_0 : \bar{c} : \sqrt{\overline{c^2}} = 1 : 1.13 : 1.23$$

where

c_0 = most probable speed

$\bar{c}$ = mean speed

$\sqrt{\overline{c^2}}$ = root mean square speed.

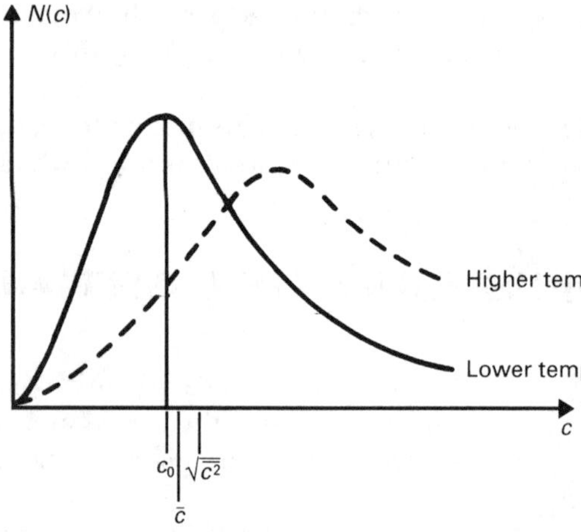

Fig. 14.3 Maxwellian distribution of molecular speeds

14.9 EXPERIMENTAL INVESTIGATION OF BOYLE'S LAW

The apparatus is shown in Fig. 14.4. The gas under investigation is the air trapped above the oil in the glass tube. The volume, V, of the air is read directly from the scale. It is compressed by using a foot pump to increase the pressure above the oil in the reservoir. The pressure, p, of the trapped air is the same as that of the air in the oil reservoir. (Whether or not this can be read directly from the Bourdon gauge depends on the particular type of gauge being used – see section 10.7.) The pressure is increased in stages, allowing a number of pairs of values of p and V to be taken. Compressing the air warms it slightly – it should be allowed to cool to room temperature (indicated by a steady volume reading) before each measurement is made.

A graph of V against $1/p$ is plotted. If the graph is a straight line through the origin, Boyle's law has been verified for the particular temperature and range of pressures investigated.

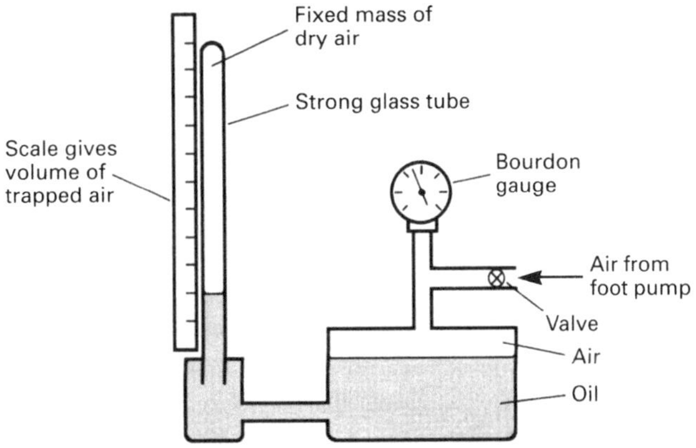

Fig. 14.4
Boyle's law apparatus

14.10 EXPERIMENTAL INVESTIGATION OF CHARLES' LAW

The apparatus is shown in Fig. 14.5. A column of air is trapped inside a capillary tube by a short thread of concentrated sulphuric acid. The reason for using the acid (rather than mercury, say) is that it absorbs any water that might be in the air and so allows meaningful results to be obtained. The tube has a uniform bore and therefore the volume of the trapped air is proportional to the length of the air column. The water is heated and the length, l, of the air column is measured for a number of different temperatures, θ. The water should be heated slowly, and stirred before each reading, to allow the air to reach the temperature of the water. The pressure of the air throughout the experiment is constant (equal to atmospheric pressure plus the pressure exerted by the acid thread).

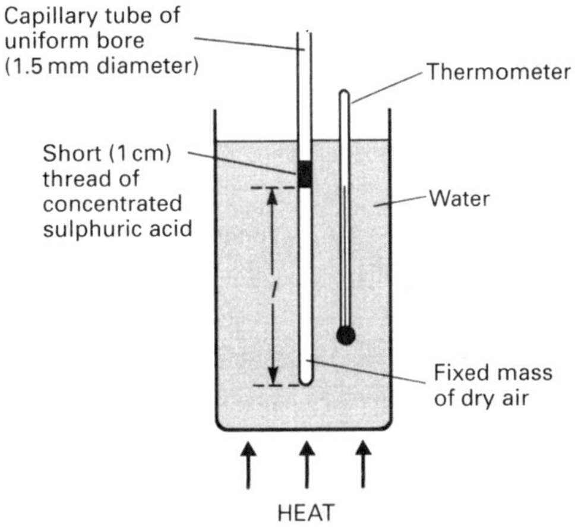

Fig. 14.5
Charles' law apparatus

A graph of l against Celsius temperature, θ, is plotted. The graph will be a straight line (Fig. 14.6(a)) showing that (for the particular pressure and range of temperatures investigated) the volume of a fixed mass of dry air at a constant pressure increases uniformly with temperature. This is one form of Charles' law.

Alternatively, a graph of l against Kelvin temperature, T, where $T = \theta + 273$, could be plotted (Fig. 14.6(b)). This graph passes through the origin and therefore verifies that the volume of a fixed mass of gas (dry air) at constant pressure is directly proportional to the temperature measured in Kelvins. This is the form of Charles' law given in section 14.1.

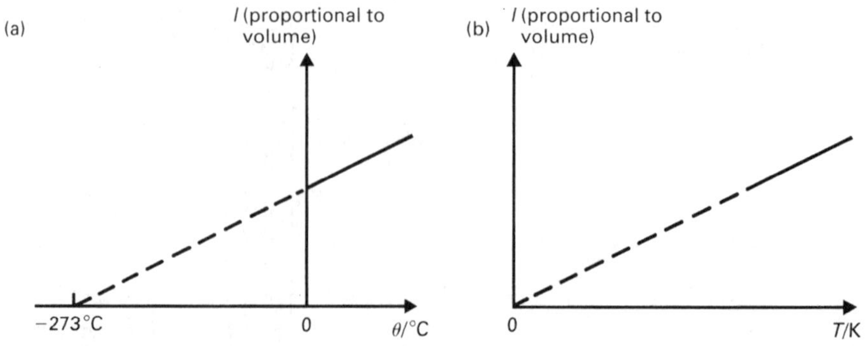

Fig. 14.6
Graphs for Charles' law investigation

Note Small values of l should be avoided; otherwise the rounded end of the capillary tube introduces a significant error into the assumption that the volume of the trapped air is proportional to l.

14.11 EXPERIMENTAL INVESTIGATION OF THE PRESSURE LAW

The apparatus is shown in Fig. 14.7.* It enables the pressure variation with temperature of a fixed mass of dry air at constant volume to be investigated.

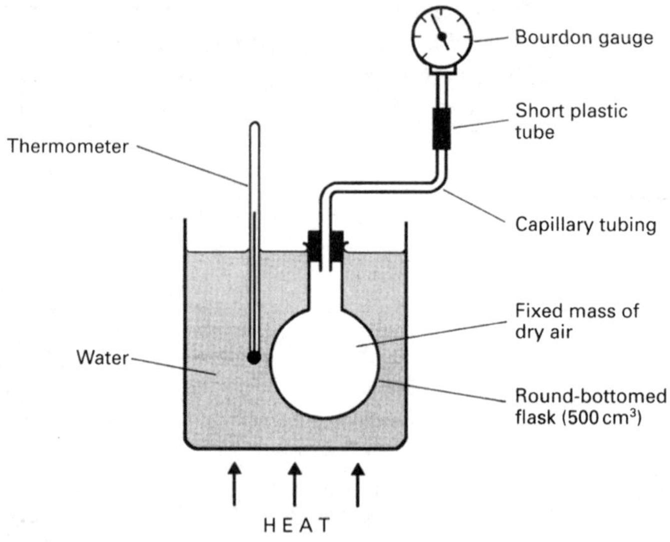

Fig. 14.7
Pressure law apparatus

The water is heated, and the pressure, p, of the air in the flask is recorded for a number of different temperatures, θ. (Whether or not the pressure can be read directly from the Bourdon gauge depends on the particular type of gauge being used—see section 10.7.) The water should be heated slowly, and stirred before each reading, to allow the air in the flask to reach the temperature of the water.

The air in the Bourdon gauge and connecting tube is not at the same temperature as that in the flask. Using a large flask and capillary tubing reduces the significance of the error that this causes.

*An alternative form of apparatus is shown in Fig. 13.5.

A graph of p against Celsius temperature, θ, is plotted. The graph will be a straight line (Fig. 14.8(a)) showing that (for the particular volume and range of temperatures investigated) the pressure of a fixed mass of dry air at constant volume increases uniformly with temperature. This is one form of the pressure law.

Fig. 14.8
Graphs for pressure law investigation

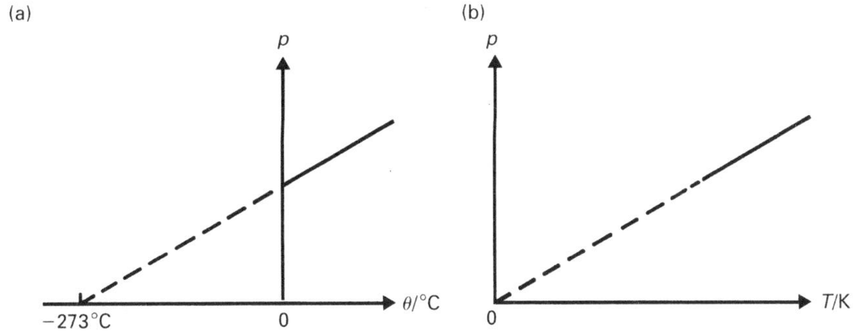

Alternatively, a graph of p against Kelvin temperature, T, where $T = \theta + 273$, could be plotted (Fig. 14.8(b)). This graph passes through the origin and therefore verifies that the pressure of a fixed mass of gas (dry air) at constant volume is directly proportional to the temperature measured in Kelvins. This is the form of the pressure law given in section 14.1.

14.12 VOLUME AND PRESSURE COEFFICIENTS OF GASES

The expansivity of a gas at constant pressure (or **volume coefficient**) α_p is defined by

$$\alpha_p = \frac{V - V_0}{V_0\theta} \quad \text{i.e.} \quad V = V_0(1 + \alpha_p\theta)$$

where

V_0 = volume of gas at 0 °C

V = volume of gas at Celsius temperature θ.

It follows from equation [14.1] that for an ideal gas at a constant pressure p

$$V = nR(273 + \theta)/p \quad \text{and} \quad V_0 = nR(273)/p$$

$$\therefore \quad \alpha_p = \frac{[nR(273 + \theta)/p] - [nR(273)/p]}{[nR(273)/p]\theta}$$

i.e. $\alpha_p = 1/273\,\mathrm{K}^{-1}$ (or $^\circ\mathrm{C}^{-1}$)

The coefficient of pressure increase at constant volume (or pressure coefficient) α_v is defined by

$$\alpha_v = \frac{p - p_0}{p_0\theta} \quad \text{i.e.} \quad p = p_0(1 + \alpha_v\theta)$$

where

p_0 = pressure of gas at 0 °C

p = pressure of gas at Celsius temperature θ.

It can be shown (by the method used for α_p) that

$$\alpha_v = 1/273\,\text{K}^{-1} \quad (\text{or}\,^\circ\text{C}^{-1})$$

14.13 REVERSIBLE PROCESSES

If, at every stage, a process can be made to go in the reverse direction by an infinitesimal change in the conditions which are causing it to take place, it is said to be a reversible process.

It follows that when the state of a system is changed reversibly:

(i) the system is in **thermodynamic equilibrium** (i.e. all parts of the system are at the same temperature and pressure) at every instant, and

(ii) at the completion of the process the system could be returned to its initial state by passing through the intermediate states in reverse order, and without there being any net change in the rest of the Universe.

In practice, it is impossible to produce a perfect reversible change. However, processes which take place very slowly and which do not involve friction are often good approximations to reversible changes. The slow compression of a gas by the movement of a light, frictionless piston in a non-conducting cylinder is an example of an approximately reversible process, because a slight decrease in the force on the piston would allow the gas to expand and no energy will have been dissipated as heat. Other examples include the changes of pressure, volume and temperature which are associated with the passage of a sound wave through air, and the movement of a pendulum about a frictionless support in a vacuum.

14.14 EXTERNAL WORK DONE BY AN EXPANDING GAS

Consider a gas enclosed in a cylinder by a frictionless piston of cross-sectional area A (Fig. 14.9). Suppose that the piston is in equilibrium under the action of the force pA exerted by the gas and an external force F. Suppose now, that the gas

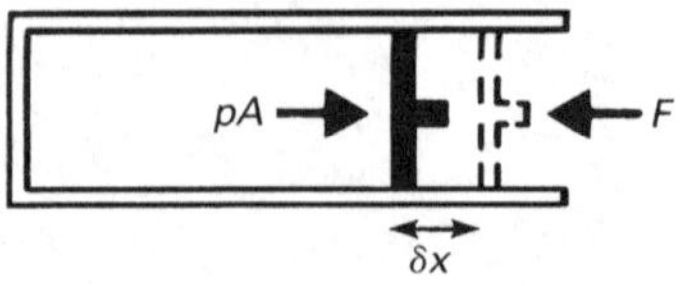

Fig. 14.9 Gas expanding in a cylinder

expands and moves the piston outwards through a distance δx, where δx is so small that p can be considered to be constant. The external work done δW by the expansion is given, by equation [5.1], as

$$\delta W = pA\,\delta x$$

i.e. $$\delta W = p\delta V \qquad [14.12]$$

where δV is the small increase in volume of the gas. The total work done W by the gas if its volume changes by a finite amount from V_1 to V_2 is therefore given by

$$W = \int_{V_1}^{V_2} p\,\mathrm{d}V \qquad [14.13]$$

Equation [14.13] holds no matter what the relationship between p and V. For the particular case of an **isobaric process** (i.e. one in which p is constant)

$$W = \int_{V_1}^{V_2} p\,\mathrm{d}V = p\int_{V_1}^{V_2} \mathrm{d}V$$

i.e. $W = p(V_2 - V_1)$ [14.14]

In the general case, if a plot of p against V is available (known as an **indicator diagram**), the work done can be obtained graphically. Suppose that the pressure of a gas varies with volume as shown in Fig. 14.10. The work done W by the gas as its volume changes from V_1 to V_2 is given by

$$W = \int_{V_1}^{V_2} p\,\mathrm{d}V = \text{Area of shaded region}$$

Fig. 14.10 Indicator diagram for a gas

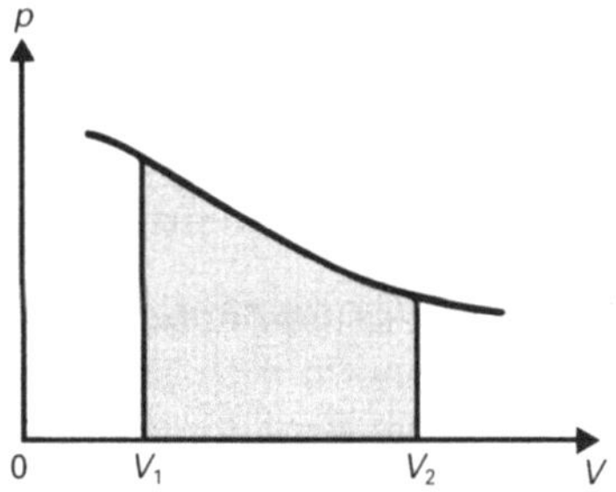

Notes (i) Equation [14.13] also applies when a gas is compressed, in which case work is being done on the gas.

(ii) Strictly, equation [14.13] can be applied only if the change takes place reversibly – if it does not, the values of pressure and temperature at any instant will be different in different regions of the gas.

(iii) Equation [14.13] also applies to solids and liquids. In these cases, though, the increases in volume are small and therefore the amounts of external work done are small compared with increases in internal energy.

14.15 THE FIRST LAW OF THERMODYNAMICS

Thermodynamics is the study of the relationship between heat and other forms of energy. When the principle of conservation of energy is stated with reference to heat and work it is known as the **first law of thermodynamics**.

The heat energy (ΔQ) supplied to a system is equal to the increase in the internal energy (ΔU) of the system plus the work done (ΔW) by the system on its surroundings.

i.e. $\Delta Q = \Delta U + \Delta W$ [14.15]

The internal energy of a system is the sum of the kinetic and potential energies of the molecules of the system. It follows from equation [14.15] that it may be increased by:

(i) putting heat energy into the system, and/or

(ii) doing work on the system.

When the internal energy of a system changes the change depends only on the initial and final states of the system, and not on how the change was brought about. This is equivalent to saying that the internal energy of a system depends only on the state that it is in, and not on how it reached that state. (Note, a system is said to have changed 'state' if some observable property of the system, e.g. its temperature, pressure, or phase, has changed.)

It is not possible to determine the absolute value of the internal energy of a real system*. This is no real problem, though, because we are always concerned with changes in internal energy, and these can be determined. To do so we make use of equation [14.15] – either directly or indirectly. Either we find the values of ΔQ and ΔW and use them (directly) in equation [14.15] to calculate ΔU, or we use the measured values of the quantities in an equation which is based on equation [14.15] – equation [14.16] for example.

An **isolated system** is one which is cut off from any form of external influence. In particular, no work can be done on it or by it (i.e. $\Delta W = 0$), and no heat can enter it or leave it (i.e. $\Delta Q = 0$). It follows from equation [14.15] that $\Delta U = 0$, and therefore that **the internal energy of an isolated system is constant**.

When a system undergoes an **adiabatic process** (see section 14.21) $\Delta Q = 0$, and equation [14.15] reduces to $\Delta U = -\Delta W$. Bearing in mind that ΔW represents work done by the system, $(-\Delta W)$ represents work done on the system. Thus, **when a system undergoes an adiabatic process the increase in internal energy of the system is equal to the work done on it.**

Notes

(i) We stated immediately after equation [14.15] that the internal energy of a system is the sum of the kinetic and potential energies of the molecules of the system. This should not be taken to imply that we are defining internal energy in this way. Absolute values of internal energy are not defined at all for real systems; changes in internal energy are defined by equation [14.15].

(ii) The internal energy of an ideal gas is due entirely to the kinetic energy of the molecules. It therefore follows from equation [14.9] that the internal energy, U, of one mole of an ideal monatomic gas at kelvin temperature T is given by

$$U = \tfrac{3}{2}RT$$

The increase in internal energy, ΔU, due to an increase in temperature, ΔT, is given by

$$\Delta U = \tfrac{3}{2}R\Delta T$$

*It is possible for an ideal gas – see Note (ii).

14.16 THE PRINCIPAL MOLAR HEAT CAPACITIES OF A GAS

The molar heat capacity of a substance is the heat required to produce unit temperature rise in one mole of the substance.

A change in temperature involves a change in pressure and/or volume. With solids and liquids such changes are small and are normally ignored. Large changes occur with gases, and in order to define the heat capacity of a gas it is necessary to specify the particular conditions of pressure and volume. Two cases are of special interest: (i) when the pressure is constant, (ii) when the volume is constant. The heat capacities measured under these conditions are called the principal heat capacities.

The molar heat capacity of a gas at constant pressure (C_p) is the heat required to produce unit temperature rise in one mole of the gas when the pressure remains constant.

The molar heat capacity of a gas at constant volume (C_v) is the heat required to produce unit temperature rise in one mole of the gas when the volume remains constant.

When a gas is heated at constant pressure it expands, and therefore some of the heat which is supplied to the gas is used:

(i) to do external work, and (in the case of a real gas)

(ii) to increase the potential energy of its molecules.

When a gas is heated at constant volume, on the other hand, all of the heat which is supplied to it is used to increase the temperature. It follows that the amount of heat required to raise the temperature of a gas at constant pressure is greater than that required to raise its temperature by the same amount at constant volume. In particular, C_p is greater than C_v.

Note The principal heat capacities for unit mass of gas are called the principal specific heat capacities at constant pressure and constant volume and are denoted by c_p and c_v respectively.

14.17 TO SHOW THAT $C_p - C_v = R$ FOR AN IDEAL GAS

Suppose that one mole of an ideal gas is heated so that its temperature increases by ΔT at constant volume. It follows from the definition of C_v that the heat supplied ΔQ is given by

$$\Delta Q = C_v \Delta T$$

Since there is no change in volume, the external work done ΔW is zero. From the first law of thermodynamics (equation [14.15])

$$\Delta Q = \Delta U + \Delta W$$

$$\therefore \quad C_v \Delta T = \Delta U \qquad [14.16]$$

where ΔU is the increase in internal energy of the gas. It is important to note that the internal energy of an ideal gas depends only on its temperature, and therefore equation [14.16] holds whenever the temperature of one mole of an ideal gas increases by ΔT, it is not restricted to situations in which the temperature increase occurs at constant volume.

Suppose now that one mole of the same gas is heated so that its temperature increases by the same amount ΔT at constant pressure. It follows from the definition of C_p that the heat supplied ΔQ is given by

$$\Delta Q = C_p \Delta T$$

The external work done ΔW by the gas is given (by equation [14.14]) as

$$\Delta W = p\Delta V$$

where ΔV is the (non-zero) change in volume and p is the constant pressure. From the first law of thermodynamics

$$\Delta Q = \Delta U + \Delta W$$

$$\therefore \quad C_p \Delta T = \Delta U + p\Delta V$$

Substituting for ΔU from equation [14.16] gives

$$C_p \Delta T = C_v \Delta T + p\Delta V \qquad [14.17]$$

If the initial volume and temperature of the gas are V and T respectively, then since we are concerned with one mole of an ideal gas

$$pV = RT$$

and

$$p(V + \Delta V) = R(T + \Delta T)$$

Subtracting gives

$$p\Delta V = R\Delta T$$

Substituting for $p\Delta V$ in equation [14.17] gives

$$C_p \Delta T = C_v \Delta T + R\Delta T$$

i.e.

$$C_p - C_v = R \qquad [14.18]$$

14.18 CALCULATION OF C_p/C_v FOR AN IDEAL MONATOMIC GAS

The internal energy of an ideal gas is entirely kinetic. The moment of inertia of a monatomic molecule can be considered to be zero and therefore the kinetic energy of such a molecule is associated with its translational motion only*. It follows that the average total kinetic energy of a monatomic molecule is given by equation [14.9], and that the internal energy U of one mole of such a gas is given by

$$U = \tfrac{3}{2} N_A kT$$

i.e.

$$U = \tfrac{3}{2} RT$$

If the temperature changes by ΔT, the corresponding change in internal energy ΔU is given by

$$\Delta U = \tfrac{3}{2} R\Delta T$$

*If the moment of inertia were not zero, the molecule would have additional kinetic energy due to its rotational motion.

By equation [14.16]

$$\Delta U = C_v \Delta T$$

$$\therefore \quad C_v \Delta T = \tfrac{3}{2} R \Delta T$$

i.e. $C_v = \tfrac{3}{2} R$

By equation [14.18]

$$C_p - C_v = R$$

$$\therefore \quad \frac{C_p - C_v}{C_v} = \frac{R}{\tfrac{3}{2} R}$$

i.e. $\dfrac{C_p}{C_v} - 1 = \tfrac{2}{3}$

i.e. $\dfrac{C_p}{C_v} = \tfrac{5}{3} = 1.67$

14.19 C_p/C_v FOR DIATOMIC AND POLYATOMIC GASES

Molecules which contain more than one atom have non-negligible moments of inertia and therefore possess rotational kinetic energy in addition to translational kinetic energy. When this is taken into account it can be shown that

$$\frac{C_p}{C_v} = \tfrac{7}{5} = 1.40 \quad \text{for a diatomic gas, and}$$

$$\frac{C_p}{C_v} = \tfrac{4}{3} = 1.33 \quad \text{for a polyatomic gas.}$$

The ratio C_p/C_v is denoted by γ, and therefore:

$\gamma = 1.67$ for a monatomic gas
$\gamma = 1.40$ for a diatomic gas
$\gamma = 1.33$ for a polyatomic gas

Note $C_p - C_v = R$ holds no matter what the atomicity of the gas.

14.20 ISOTHERMAL PROCESSES

An isothermal process is a process which takes place at constant temperature.

It follows from the ideal gas equation that when a gas expands or contracts isothermally

$$pV = \text{a constant} \qquad [14.19]$$

The internal energy of an ideal gas depends only on its temperature and therefore, for an ideal gas which is involved in an isothermal process, $\Delta U = 0$ and the first law of thermodynamics reduces to $\Delta Q = \Delta W$. Thus if the gas expands and does external work ΔW, an amount of heat ΔQ has to be supplied to the gas in order to maitain its temperature. Conversely, if the gas contracts, work is being done on it and an amount of heat ΔQ has to be allowed to leave the gas.

Any attempt to produce an isothermal change requires that the gas is contained in a vessel which has thin, good-conducting walls and which is surrounded by a constant temperature reservoir. In addition, the expansion or contraction must take place slowly. If these conditions are not fulfilled when, say, a gas expands, then the energy used by the gas in doing external work has to be provided at the expense of the kinetic energy of its molecules, and the temperature of the gas falls.

Note Equation [14.19] can be expressed as

$$p_1V_1 = p_2V_2$$

where p_1 and V_1 are the initial pressure and volume of the gas, and p_2 and V_2 are the pressure and volume after the isothermal change has taken place.

14.21 ADIABATIC PROCESSES

An adiabatic process is one which takes place in such a way that no heat enters or leaves the system during the process.

It can be shown that when an ideal gas undergoes a reversible adiabatic expansion or contraction

$$pV^{\gamma} = \text{a constant} \qquad [14.20]$$

where γ is the ratio of the principal heat capacities of the gas.

Since $\Delta Q = 0$, the first law of thermodynamics reduces to $\Delta U = -\Delta W$. Thus if the gas expands and does external work, its temperature falls. Conversely, an adiabatic compression causes the temperature of the gas to rise.

A truly adiabatic process is an ideal which cannot be realized. However, when a gas expands rapidly, the expansion is nearly adiabatic, particularly if the gas is contained in a vessel which has thick, badly conducting walls. Two examples of approximately adiabatic processes are:

(i) the rapid escape of air from a burst tyre,

(ii) the rapid expansions and contractions of air through which a sound wave is passing.

Notes (i) Equation [14.20] can also be expressed in the form

$$p_1V_1^{\gamma} = p_2V_2^{\gamma} \qquad [14.21]$$

where p_1 and V_1 are the initial pressure and volume of the gas, and p_2 and V_2 are the pressure and volume after the adiabatic change has taken place.

(ii) $pV = nRT$ applies to any change of the state of an ideal gas and can be expressed as

$$\frac{p_1 V_1}{T_1} = \frac{p_2 V_2}{T_2} \qquad [14.22]$$

Dividing equation [14.21] by equation [14.22] gives

$$T_1 V_1^{(\gamma-1)} = T_2 V_2^{(\gamma-1)}$$

from which the final temperature T_2 can be calculated.

14.22 ISOTHERMAL AND ADIABATIC PROCESSES COMPARED

Fig. 14.11 illustrates isothermal and adiabatic expansions of an ideal gas which is initially at a pressure p_1 and volume V_1. The temperature fall which accompanies the adiabatic expansion results in a lower final pressure than that produced by the isothermal expansion. Note that the area under the isothermal is greater than that under the adiabatic, i.e. more work is done by the isothermal expansion than by the adiabatic expansion. Note also that the adiabatic through any point is steeper than the isothermal through that point.

Fig. 14.11
p–V curves for isothermal and adiabatic expansions

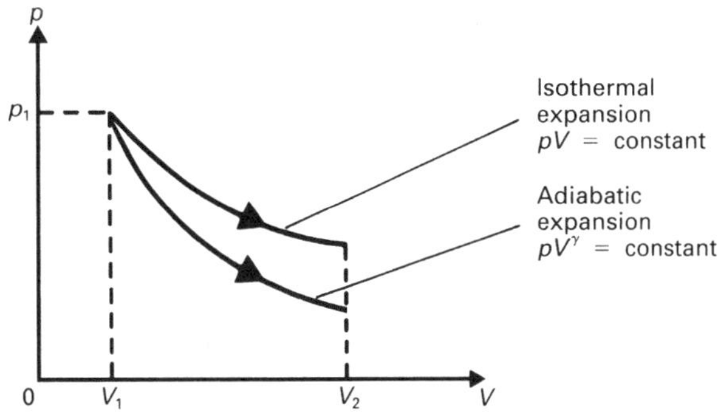

Fig. 14.12 illustrates the p–V curves of a gas which is expanded adiabatically from a volume V_1 to a volume V_2, and is then compressed isothermally to its original volume.

Fig. 14.12
p–V curves for adiabatic expansion followed by isothermal compression

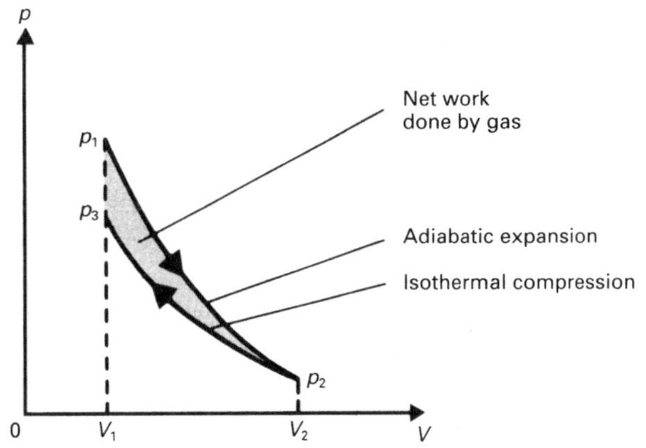

Work done by gas in expanding $=$ Area $V_1p_1p_2V_2$

Work done on gas in contracting $=$ Area $V_1p_3p_2V_2$

Net work done by gas $=$ Area $p_1p_2p_3$

14.23 MEASUREMENT OF γ FOR AIR BY CLÉMENT AND DÉSORMES' METHOD

The apparatus is shown in Fig. 14.13. The principal component is a vessel of large volume ($\sim$10 litres) containing air and a little phosphorus pentoxide to dry it.

Fig. 14.13
Clément and Désormes' apparatus

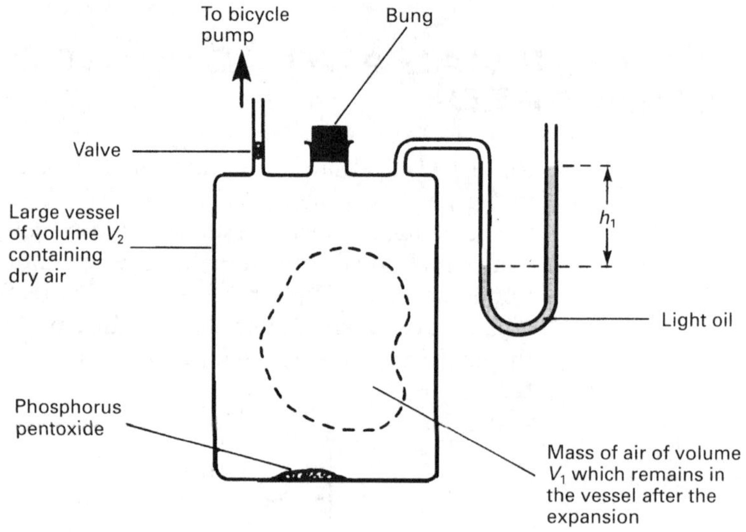

Procedure

(i) Air is pumped into the vessel until the pressure inside it is a little above atmospheric. The air is then allowed to cool to room temperature and the (now steady) manometer reading h_1 is recorded.

(ii) The bung is removed for about one second allowing the air to undergo an (approximately) reversible adiabatic expansion. The pressure falls to atmospheric and the air cools.

(iii) The air is now left to regain room temperature and the new (steady) manometer reading h_2 is recorded.

Theory

Suppose that the air which remains in the vessel initially occupied a volume V_1 at a pressure p_1. Immediately after the expansion this same mass of air is at atmospheric pressure p and now occupies the whole of the vessel so that its volume is V_2. Since the expansion is adiabatic

$$p_1V_1^{\gamma} = pV_2^{\gamma} \qquad [14.23]$$

Suppose that p_2 is the pressure of the air when it has regained room temperature. Thus, a mass of air which initially had volume V_1 and pressure p_1 at room temperature, now has volume V_2 and pressure p_2, also at room temperature, and therefore

$$p_1V_1 = p_2V_2 \qquad [14.24]$$

From equation [14.23]

$$\left(\frac{V_2}{V_1}\right)^\gamma = \frac{p_1}{p}$$

From equation [14.24]

$$\frac{V_2}{V_1} = \frac{p_1}{p_2}$$

$$\therefore \quad \left(\frac{p_1}{p_2}\right)^\gamma = \frac{p_1}{p}$$

Taking logs gives

$$\gamma \log(p_1/p_2) = \log(p_1/p)$$

Note. Using natural logs would give the same result.

i.e. $$\gamma = \frac{\log(p_1/p)}{\log(p_1/p_2)} \qquad [14.25]$$

If the density of the manometer liquid is ρ,

$$\left.\begin{aligned} p_1 &= p + h_1 \rho g \\ p_2 &= p + h_2 \rho g \end{aligned}\right\} \qquad [14.26]$$

Measured values of p and ρ can be used in equations [14.26] to find p_1 and p_2, the values of which can then be used in equation [14.25] together with the value of p to calculate γ.

When $h_1 \rho g \ll p$ and $h_2 \rho g \ll p$ equation [14.25] reduces to

$$\gamma = \frac{h_1}{h_1 - h_2}$$

from which γ can be calculated without there being any need to measure ρ and p.

Notes

(i) The walls of the vessel act as a source of heat and therefore the expansion is not truly adiabatic. The error is reduced by allowing the expansion to take place rapidly and by using a vessel of large volume.

(ii) Rapid expansion requires the bung to be large. Unfortunately, when the orifice is large the air oscillates as a result of the expansion, and this makes it impossible to know exactly when to replace the bung. A bung of optimum size should be used.

(iii) The apparatus should be shielded from draughts and direct sunlight.

14.24 REAL GASES

When real gases are subjected to pressures which are greater than a few atmospheres and/or when they are at temperatures near to those at which they liquefy, it is found that they no longer conform to the ideal gas equation (nor, therefore, to the gas laws). This is not surprising because (as shown in section 14.5) that equation is consistent with the kinetic theory of gases which assumes:

(i) that there are no intermolecular forces, and

(ii) that the volume of the molecules is negligible compared with their separation.

An increase in pressure or a decrease in temperature clearly reduces the validity of assumption (ii). Assumption (i) also becomes less valid because there is ample evidence (see Chapter 9) that the closely packed molecules of solids and liquids do exert forces on each other.

The extent of the departure from ideal gas behaviour varies from gas to gas, but of the common gases carbon dioxide shows considerable non-ideal characteristics.

14.25 ANDREWS' EXPERIMENTS ON CARBON DIOXIDE

The apparatus which Andrews used to investigate the behaviour of carbon dioxide is shown, schematically, in Fig. 14.14. By tightening the screws, Andrews was able to force water into the glass tubes and so increase the pressures and decrease the volumes of the gases trapped in the upper portions of the tubes. These tubes had been calibrated beforehand, so that it was a simple matter for Andrews to read off the volumes of the trapped gases by noting the positions of the tops of the mercury columns. By assuming that the nitrogen obeyed Boyle's law (a reasonable assumption at the pressures and temperatures involved as Andrews knew), he was able to calculate the pressure of the nitrogen once he had measured its volume. Since both gases were at the same pressure, this gave him the pressure of the carbon dioxide as desired. The capillary tubes were surrounded by a water bath, the purpose of which was to maintain the gases at a constant temperature. In this way then, Andrews measured the volume of the carbon dioxide as a function of its pressure at a fixed temperature. Altering the water bath temperature allowed him to obtain this information for a number of different temperatures. He presented his results as a series of **isothermals** (i.e. a series of plots of pressure against volume, each at a fixed temperature). Some of these curves are shown in Fig. 14.15.

Fig. 14.14
Andrews' apparatus

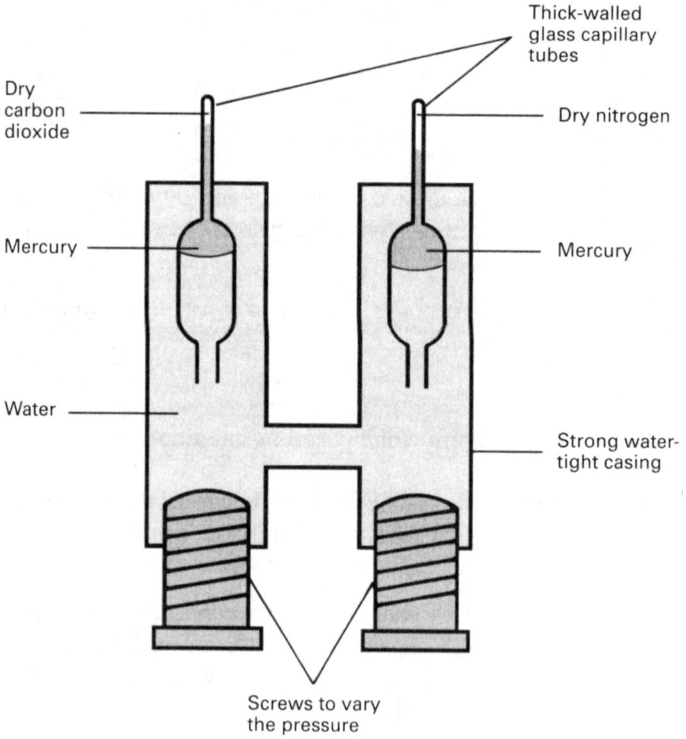

Fig. 14.15
Andrews' isothermals (p–V curves) for a fixed mass of carbon dioxide

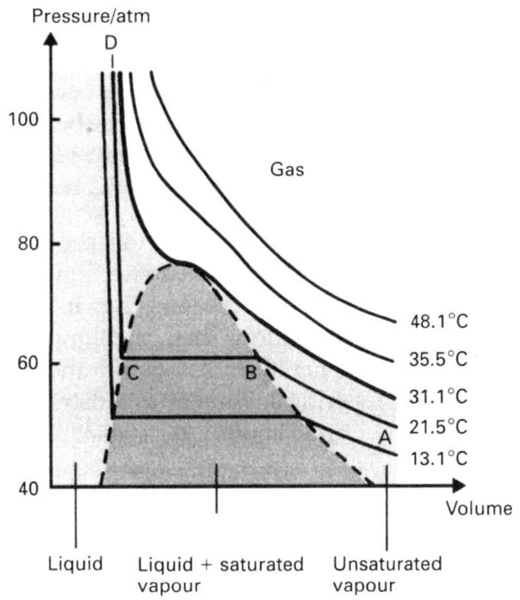

The diagram shows the critical nature of the 31.1 °C isothermal. Above 31.1 °C the carbon dioxide exists as a gas no matter how high the pressure, and the curves are approximately hyperbolic (the shape they would be if the carbon dioxide were an ideal gas). Below 31.1 °C the carbon dioxide can exist in both the gaseous state (as a vapour) and the liquid state. Consider the carbon dioxide to be in the state of pressure, volume and temperature that is represented by the point A on the 21.5 °C isothermal. In this state the carbon dioxide is an unsaturated vapour (see section 15.2), and if it is compressed, the p–V curve is very nearly hyperbolic until the pressure reaches that represented by B. At B the carbon dioxide begins to liquefy. Between B and C the volume decreases as the screws are turned in, but there is no increase in pressure. The decrease in volume is due to the fact that in moving from B to C more and more liquid forms, so that at C the carbon dioxide is entirely liquid. From C to D and beyond, large increases in pressure produce very little decrease in volume – as might be expected, since liquds are virtually incompressible.

14.26 TERMINOLOGY

It is now possible to define some useful terms.

Critical temperature (T_c) is the temperature above which a gas cannot be liquefied, no matter how great the pressure. (T_c = 31.1 °C for carbon dioxide.)

Critical pressure (p_c) is the minimum pressure that will cause liquefaction of a gas at its critical temperature. (p_c = 73 atm for carbon dioxide.)

Specific critical volume (V_c) is the volume occupied by 1 kg of a gas at its critical temperature and critical pressure.

Gas is the term applied to a substance which is in the gaseous phase and is above its critical temperature.

Vapour is the term applied to a substance which is in the gaseous phase and is below its critical temperature.

Thus, a vapour can be liquefied simply by increasing the pressure on it; a gas cannot.

Notes (i) Oxygen, nitrogen and hydrogen are traditionally called **permanent gases**, since it was originally thought that they could not be liquefied. This misconception arose because the early workers had no knowledge of the necessity for a gas to be below its critical temperature, and each of these gases has a critical temperature which is well below room temperature (−118 °C, −146 °C and −240 °C respectively).

(ii) It can be seen from the p–V curves of carbon dioxide (Fig. 14.15), for example, that when a liquid at its critical temperature (and critical pressure) becomes gaseous, then it does so without any change of volume. Under these conditions then, the liquid and its saturated vapour have the same density. Therefore, if a liquid and its saturated vapour are in equilibrium at their critical temperature, there is no meniscus, i.e. no distinction between liquid and vapour.

14.27 CURVES OF *pV* AGAINST *p*

A convenient way to show the departure from ideal gas behaviour at some temperature, is to plot pV against p at that temperature. For an ideal gas such a plot is, of course, a straight line parallel to the p axis, but for a fixed mass of real gas the curves typically have the form shown in Fig. 14.16.

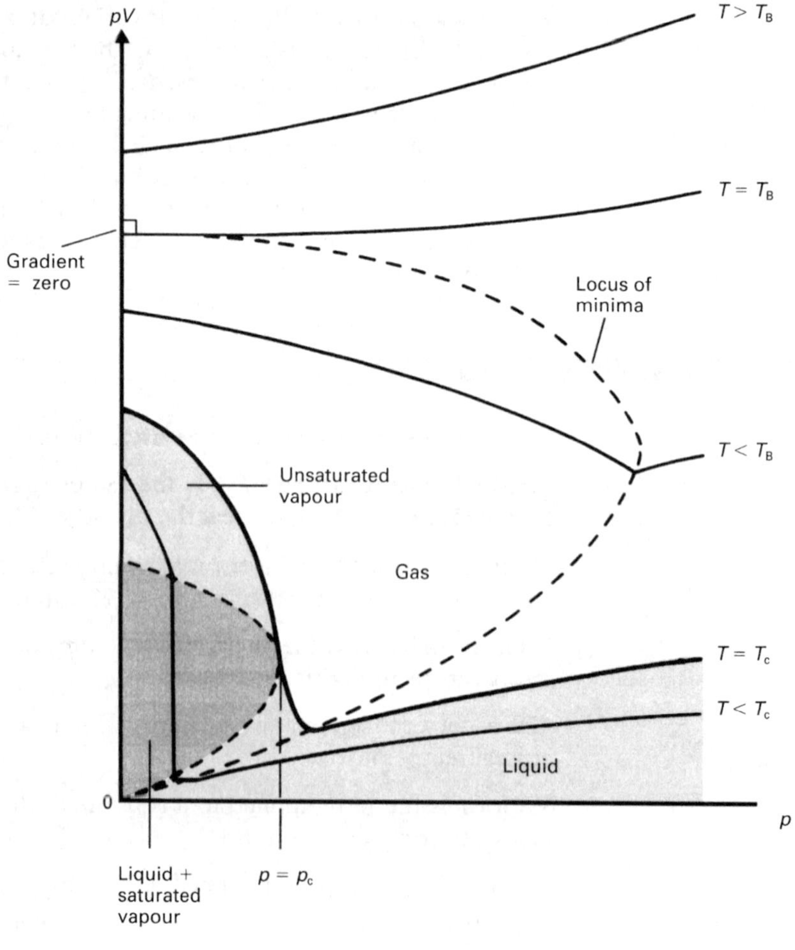

Fig. 14.16 Plots of ***pV*** against ***p*** for a typical real gas

At the **Boyle temperature** T_B the isothermal is horizontal at $p = 0$, and therefore very nearly so for other low pressures, i.e. **the Boyle temperature of a gas is the temperature at which the gas approximates best to an ideal gas**.

Kamerlingh Onnes showed that the behaviour of all real gases can be accounted for by

$$pV = A + Bp + Cp^2 + \ldots^{\star} \quad [14.27]$$

A, B, C, etc. are called **virial coefficients**, and for a fixed mass of any particular gas have values which depend only on temperature. On the whole, $|A| \gg |B| \gg |C|$, etc.

Two particular situations are of interest.

At $p = 0$

When the behaviour of any real gas is extrapolated to zero pressure it is found to behave like an ideal gas. Therefore, when $p = 0$, equation [14.27] must become identical with $pV = nRT$. Thus,

$$A + Bp + Cp^2 + \cdots = nRT \quad \text{when } p = 0$$

i.e. $$A = nRT \quad [14.28]$$

Since the virial coefficients are not functions of p, equation [14.28] is true for all pressures

At the Boyle Temperature

At the Boyle temperature the graph of pV against p is horizontal at $p = 0$, i.e.

$$\frac{\mathrm{d}(pV)}{\mathrm{d}p} = 0 \qquad \text{when } p = 0$$

Therefore by equation [14.27]

$$\frac{\mathrm{d}}{\mathrm{d}p}(A + Bp + Cp^2 + \cdots) = 0 \quad \text{when } p = 0$$

i.e. $$B + 2Cp + \cdots = 0 \qquad \text{when } p = 0$$

i.e. $$B = 0$$

$\star$Any function can be represented by an infinite polynomial such as equation [14.27], and therefore at first sight it might seem to be of little interest. The importance of the equation lies in:
(i) each term is less significant than the preceding one, and
(ii) B varies in the same way with temperature for all gases.

Thus at the Boyle temperature equation [14.27] becomes

$$pV = nRT + Cp^2 + \cdots$$

Since C and the succeeding coefficients are normally less significant than B (and B is actually zero here), to a good approximation, $pV = nRT$ at the Boyle temperature, as expected.

14.28 VAN DER WAALS' EQUATION OF STATE

There have been many attempts to find an equation which accurately represents the behaviour of real gases over wide ranges of pressure and temperature. One of the most useful is that suggested by van der Waals in 1873. Van der Waals' equation is a modification of the ideal gas equation which takes account of the finite size of the molecules and the attractive forces between them.

Because the molecules of a real gas have a finite size, the volume in which they are free to move is less than that of their container. It follows that the volume available to the molecules is $V - b$, where V is the volume of the container and b is a factor which is proportional to the volume of the molecules themselves. Values of b differ from gas to gas.

The molecules of a real gas exert attractive forces on each other. The effect of this is to reduce the force on the walls of the container so that the observed pressure p is less than that in the interior of the gas. It can be shown that the pressure in the interior (i.e. that which would be observed at the walls if there were no intermolecular forces) is given by $p + a/V^2$, where a is a constant whose value depends on the gas under consideration.

A little thought should convince the reader that what is meant by 'pressure' in the context of the ideal gas equation is the pressure that would act at the walls of the container if there were no intermolecular forces, and that what is meant by 'volume' is the volume in which the molecules are free to move. The ideal gas equation is therefore replaced by

$$\left(p + \frac{a}{V^2}\right)(V - b) = RT$$

which is **van der Waals' equation for one mole of gas**. At high pressures the volume is small, and therefore a/V^2 is significant in comparison with p, and b is significant in comparison with V.

Notes

(i) The values of a and b for any particular gas can be found by experiment. The units of a are $\text{N m}^4\,\text{mol}^{-2}$, the units of b are $\text{m}^3\,\text{mol}^{-1}$.

(ii) When the number of moles of gas is n the equation becomes $(p + n^2a/V^2)(V - nb) = nRT$.

(iii) The walls of the container exert attractive forces on the gas molecules, but this has no effect on the pressure. The momentum of a molecule which is approaching a wall is increased by the force due to the wall, and the wall itself acquires a momentum which is equal to that gained by the gas molecule but is oppositely directed to it. It follows that the momentum of the wall immediately after a gas molecule has collided with it is the same as it would have been had there been no attractive force.

CONSOLIDATION

Boyle's law At constant T, $pV = $ a constant or $p \propto 1/V$

Charles' law At constant p, $V/T = $ a constant or $V \propto T$

Pressure law At constant V, $p/T = $ a constant or $p \propto T$

An ideal gas obeys $pV = nRT$ exactly. The internal energy is entirely kinetic and depends only on temperature.

$$p = \tfrac{1}{3}\rho\overline{c^2}$$

$$\text{Average KE of a monatomic molecule} = \tfrac{3}{2}kT = \frac{3}{2}\frac{R}{N_A}T$$

Avogadro's law Equal volumes of all gases at the same temperature and pressure contain the same number of molecules.

Dalton's law of partial pressures The total pressure of a mixture of gases, which do not interact chemically, is equal to the sum of the partial pressures, i.e. to the sum of the pressures that each gas would exert if it alone occupied the volume containing the mixture.

A system is in **thermodynamic equilibrium** if all parts of it are at the same temperature and pressure.

$$W = \int_{V_1}^{V_2} p\,\mathrm{d}V = \text{area under } p\text{–}V \text{ curve}$$

$$W = p(V_2 - V_1) \quad \text{at constant pressure}$$

$$\Delta Q = \Delta U + \Delta W$$

where

$\Delta Q = $ heat put into system, $\Delta U = $ increase in internal energy of system, $\Delta W = $ work done by system.

$$\Delta U = C_v \Delta T \quad \text{for one mole of ideal gas}$$

and

$$\left.\begin{aligned} U &= \tfrac{3}{2}RT \\ \Delta U &= \tfrac{3}{2}R\Delta T \end{aligned}\right\} \quad \text{for one mole of an ideal monatomic gas}$$

For an Ideal Gas

$$C_p - C_v = R \quad \text{and} \quad \frac{C_p}{C_v} = \gamma$$

where

$\gamma = 1.67$ (monatomic)

$\gamma = 1.40$ (diatomic)

$\gamma = 1.33$ (polyatomic)

$$\frac{p_1 V_1}{T_1} = \frac{p_2 V_2}{T_2} \quad \text{for any change of state of an ideal gas}$$

$$p_1 V_1 = p_2 V_2 \quad \text{for an isothermal (i.e. constant temperature) change}$$

$$\left.\begin{aligned} p_1 V_1^{\gamma} &= p_2 V_2^{\gamma} \\ T_1 V_1^{(\gamma-1)} &= T_2 V_2^{(\gamma-1)} \end{aligned}\right\} \quad \text{for an adiabatic (i.e. constant heat) change}$$

15

VAPOURS

The distinction between a gas and a vapour is given in section 14.26.

15.1 EVAPORATION

Evaporation is the process by which a liquid* becomes a vapour. It can take place at all temperatures, but occurs at the greatest rate when the liquid is at its boiling point.

The kinetic theory supposes that the molecules of liquids are in continual motion and make frequent collisions with each other. Although the average kinetic energy of a molecule is constant at any particular temperature, it may gain kinetic energy as a result of collisions with other molecules. If a molecule which is near the surface and is moving towards the surface gains enough energy to overcome the attractive forces of the molecules behind it, it escapes from the surface. It follows that the rate of evaporation can be increassed by:

(i) increasing the area of the liquid surface;

(ii) increasing the temperature of the liquid (since this increases the average kinetic energy of all the molecules without increasing the strength of the intermolecular forces of attraction);

(iii) causing a draught to remove the vapour molecules before they have a chance to return to the liquid;

(iv) reducing the air pressure above the liquid (since this decreases the possibility of a vapour molecule rebounding off an air molecule).

When a liquid evaporates it loses those of its molecules which have the greatest kinetic energies, and therefore **when a liquid evaporates it cools.**

15.2 SATURATED AND UNSATURATED VAPOURS

Suppose that a container is partly filled with a liquid and then sealed. Some molecules escape from the liquid by the process of evaporation and exist as a vapour in the region above. The vapour molecules move about at random, and some of them return to the liquid. The rate of condensation (i.e. the rate at which molecules return to the liquid) is determined by the number of molecules in the vapour phase. Initially this is low, and the rate of evaporation exceeds the rate of condensation. There is, therefore, a net gain of molecules by the vapour, and eventually a **dynamic equilibrium** is established in which the rate at which

*Solids evaporate but the rate of evaporation of a solid is negligible unless it is close to its melting point.

molecules enter the vapour is equal to the rate at which they return to the iquid. The region above the iquid is said to be **saturated** with vapour, for it now contains the maximum possible number of molecules which the conditions will allow. (If the number of vapour molecules were to increase by some means, the rate of condensation would become greater than the rate of evaporation and the equilibrium would be re-established.) The pressure exerted by a saturated vapour is called the **saturated vapour pressure** (SVP) and its value depends only on temperature.

If the volume of the space above the liquid is increased, there is a momentary decrease in the density of the vapour, in particular, immediately above the liquid surface. This decreases the rate of condensation and restores the pressure to its previous value, i.e. SVP is independent of volume. If the increase in volume is continued, more and more liquid evaporates and eventually there is none left. Any further increase in volume causes the vapour to become unsaturated. Once this happens the pressure varies with volume in a manner which is approximately consistent with Boyle's law. A plot of pressure against volume at a fixed temperature is shown in Fig. 15.1. (Note that Andrews' isothermals in Fig. 14.15 for carbon dioxide at temperatures below its critical temperature are also of this form.)

Fig. 15.1
p–V curve for vapour in a sealed container

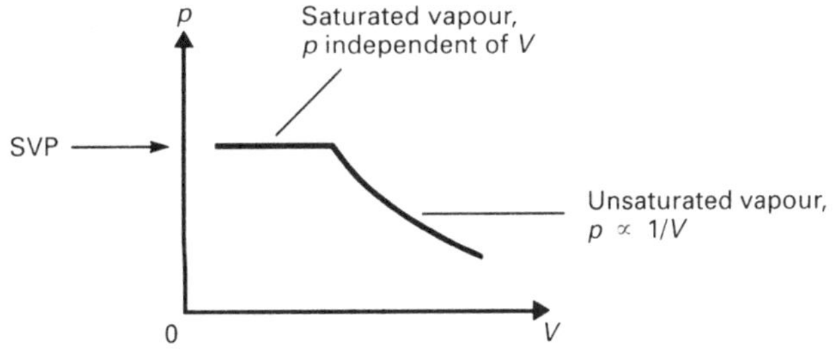

It follows from what has been said so far that **a saturated vapour can be defined as being a vapour which is in equilibrium with its own liquid.** If the temperature of such a system is increased there are two distinct consequences.

(i) The kinetic energy of the vapour molecules increases.

(ii) The rate of evaporation increases and therefore there is an increase in the number of molecules in the vapour phase.

If the volume of the system is constant, each of these effects produces an increase in pressure. The effect of (i) alone would be to give a pressure increase of the form predicted by the pressure law (approximately); the additional effect of (ii) means that the increase in pressure with increasing temperature is much more rapid than this (see Fig. 15.2).

If the temperature is increased at constant pressure, the volume increases, but because of (ii) it increases much more rapidly than required by Charles' law.

Fig. 15.2
p–θ curve for vapour in a sealed container

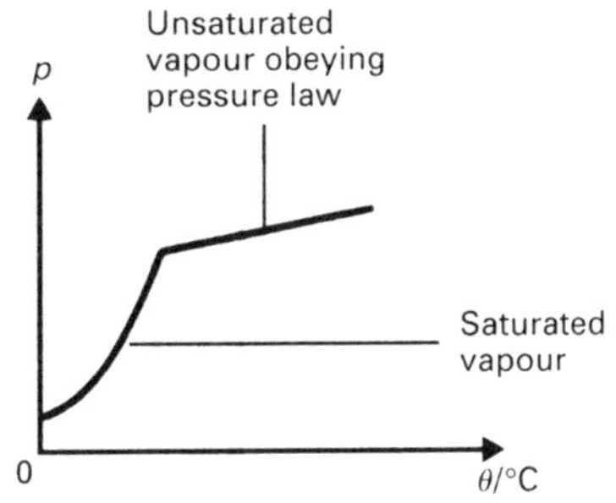

Summary

(i) A saturated vapour is a vapour which is in equilibrium with its own liquid.

(ii) The gas laws refer to fixed masses of gases. Changing the state of a saturated vapour involves condensation or evaporation and therefore changes its mass. It follows that saturated vapours do not obey the gas laws. In particular SVP depends only on temperature.

(iii) Unsaturated vapours, like real gases, obey the gas laws approximately. In carrying out calculations at this level unsaturated vapours can be taken to obey the gas laws exactly.

15.3 MIXTURES OF GASES AND SATURATED VAPOURS

Dalton's law of partial pressure applies. The total pressure is that of the gas plus that of the vapour. It must be borne in mind that the gas obeys the gas laws, the saturated vapour does not (see Example 15.1).

15.4 SUPERSATURATED VAPOURS

If the temperature of a saturated vapour is reduced suddenly, there is a brief period* during which the vapour contains more molecules than it should at the new temperature. Such a vapour is called a supersaturated vapour and it is not in equilibrium with its liquid.

15.5 BOILING

A liquid boils when its temperature is such that bubbles of vapour form throughout its volume. The pressure inside these bubbles is the SVP of the liquid at the temperature concerned, and must be at least as big as the pressure outside the bubbles otherwise they would collapse. Thus:

The boiling point of a liquid is that temperature at which its SVP is equal to the external pressure.

The external pressure is equal to

(i) the pressure of the atmosphere above the liquid, plus

(ii) the hydrostatic pressure due to the liquid itself, plus

(iii) the pressure due to surface tension effects.

The last two of these are normally ignored but, in particular, (ii) accounts for the lower part of a boiling liquid being hotter than the upper part.

*If there are no nucleating sites present (e.g. dust), the vapour may remain supersaturated for a long time.

If the pressure above a boiling liquid is increased, it stops boiling because the external pressure is now greater than the SVP. If the temperature of the liquid is increased, its SVP rises and eventually becomes equal to the new external pressure. Thus the boiling point of a liquid increases with pressure and a plot of external pressure against boiling point is identical to a plot of SVP against temperature. The SVP of water is shown as a function of temperature in Fig. 15.3.

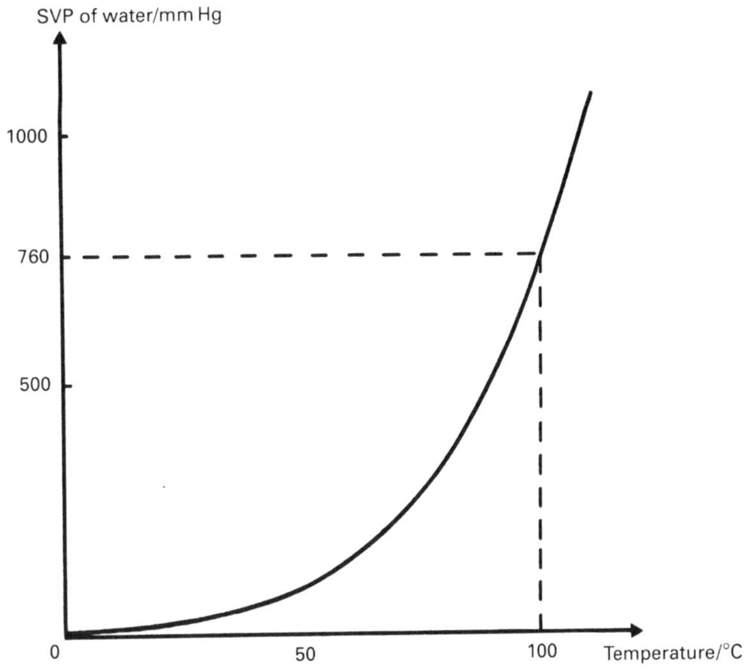

Fig. 15.3 Variation of saturated vapour pressure of water with temperature

Boiling differs from evaporation in that:

(i) boiling occurs throughout the volume of a liquid, whereas evaporation occurs only at the surface, and

(ii) for any given external pressure a liquid boils at a single temperature only, whereas evaporation takes place at all temperatures.

15.6 EXPERIMENTAL DETERMINATION OF SVP BY THE DYNAMIC METHOD

The apparatus is shown in Fig. 15.4. The pressure above the liquid is reduced to some desired value below atmospheric pressure by means of the vacuum pump. The liquid is then heated gently and it starts to boil at a temperature which is determined by the pressure inside the apparatus. The vapour is condensed and returned to the round-bottomed flask, thereby preventing a pressure build-up within the apparatus. The thermometer registers the temperature of the saturated vapour. The pressure p above the liquid is given by $p = (p_a - h\rho g)$ where p_a is atmospheric pressure and ρ is the density of the mercury. Since a liquid boils when its temperature is such that its SVP is equal to the external pressure, p is the SVP of the liquid at the temperature registered by the thermometer.

Replacing the vacuum pump by a bicyle pump allows SVPs above atmospheric pressure to be determined.

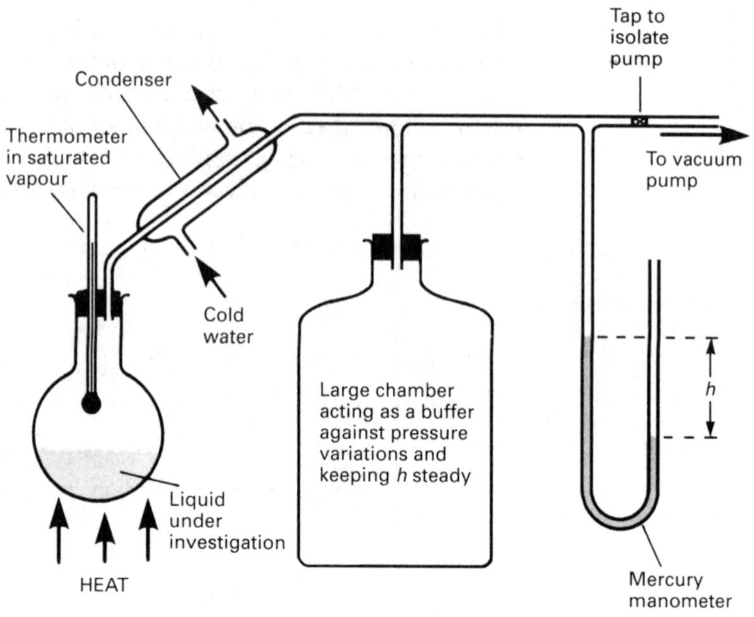

Fig. 15.4
Apparatus for the determination of SVP by the dynamic method

EXAMPLE 15.1

A closed vessel contains air saturated with water vapour at 77 °C. The total pressure in the vessel is 1000 mmHg. Calculate the new pressure in the vessel if the temperature is reduced to 27 °C. (The SVP of water at 77 °C = 314 mmHg; SVP of water at 27 °C = 27 mmHg.)

Solution

By Dalton's law of partial pressures, the pressure of the air at 77 °C (350 K) = 1000 − 314 = 686 mmHg. Treating the air as an ideal gas and assuming that its volume is V and is constant, we see that its pressure, p, at 27 °C (300 K) is (by equation [14.22]) given by

$$\frac{686 \times V}{350} = \frac{pV}{300}$$

i.e. $p = 588$ mmHg

The pressure of the water at 27 °C = 27 mmHg, and therefore the total pressure at 27 °C = 588 + 27 = 615 mmHg.

16

THERMODYNAMICS

The first law of thermodynamics is dealt with in Chapter 14.

16.1 THERMAL EQUILIBRIUM AND THE ZEROTH LAW OF THERMODYNAMICS

If two bodies are in thermal contact and there is no net flow of heat energy between them, the bodies are said to be in **thermal equilibrium** with each other. The bodies must possess some property which determines whether they are in thermal equilibrium – we call this property temperature. It follows that **heat can flow from one body to another only if they are at different temperatures.**

Experiment shows that two bodies which are separately in thermal equilibrium with a third body are also in thermal equilibrium with each other. This is known as the **zeroth law of thermodynamics.** It is called the zeroth law because although the other laws of thermodynamics inherently assume its validity (and therefore logically come after it) they had been established for many years before the first formal statement of it. The reader may feel that the zeroth law is merely a statement of the obvious – maybe it is, but the principle it embodies is fundamental to the whole of thermodynamics and therefore needs to be stated formally.

In order to see how we make use of the zeroth law, suppose we wish to discover whether two bodies, A and B, are at the same temperature, i.e. whether they are in thermal equilibrium with each other. We do this by first bringing A into thermal equilibrium with a third body – a thermometer – and then bringing B into thermal equilibrium with the same thermometer. If the thermometer gives the same reading in each case, by using the zeroth law we can say that A and B are at the same temperature. If, under these circumstances, A and B were not at the same temperature, i.e. if the zeroth law were not true, there would be no point in taking readings with thermometers.

16.2 ENTHALPY

The function $U + pV$ is involved in many applications of thermodynamics; it has therefore been found useful to give it a name – **enthalpy**. Thus the enthalpy, H, of a substance is defined by

$$H = U + pV \qquad [16.1]$$

where U is the internal energy of the substance (see section 14.15) when it is at a pressure p and has a volume V.

Many processes which produce changes in enthalpy take place at constant pressure (chemical reactions for example), and we shall be concerned only with enthalpy changes of this type. Suppose that in some constant pressure process U increases to $(U + \Delta U)$ and V increases to $(V + \Delta V)$. If, as a result of this, the enthalpy increases by ΔH to $(H + \Delta H)$, then from equation [16.1]

$$H + \Delta H = U + \Delta U + p(V + \Delta V) \qquad [16.2]$$

Subtracting equation [16.1] from equation [16.2] gives

$$\Delta H = \Delta U + p\Delta V \qquad \text{(at constant pressure)} \qquad [16.3]$$

Note Though p is always measured in $\mathrm{N\,m^{-2}}$, the units in which ΔH, ΔU and ΔV are measured depend on the amount of substance involved. For unit mass of substance, ΔH and ΔU are respectively called the specific enthalpy change and the specific internal energy change, and are measured in $\mathrm{J\,kg^{-1}}$. The corresponding value of ΔV is called the specific volume change and is measured in $\mathrm{m^3\,kg^{-1}}$. For 1 mole of substance ΔH and ΔU are measured in $\mathrm{J\,mol^{-1}}$ and ΔV is measured in $\mathrm{m^3\,mol^{-1}}$. When the amount of substance is neither 1 kg nor 1 mol, or is unspecified, ΔH and ΔU are measured in J and ΔV is measured in $\mathrm{m^3}$.

The term $p\,\Delta V$ in equation [16.3] is the work done by the substance as it expands against the constant pressure p. It therefore follows from the first law of thermodynamics (equation [14.15]) that

$$\Delta Q = \Delta U + p\Delta V \qquad [16.4]$$

where ΔQ is the heat supplied to the substance. Comparing equations [16.3] and [16.4] gives

$$\Delta H = \Delta Q \qquad \text{(at constant pressure)} \qquad [16.5]$$

i.e. **in a constant pressure process the enthalpy change is equal to the heat supplied.**

Chemical reactions in which heat is absorbed are called **endothermic reactions**; those in which heat is given out are called **exothermic reactions**. It follows from equation [16.5] that:

$\Delta H > 0$ for an **endothermic** reaction at constant pressure,

$\Delta H = 0$ for an **adiabatic** process at constant pressure,

$\Delta H < 0$ for an **exothermic** reaction at constant pressure.

Equation [16.5] provides two other useful relationships.

$$L_V = \Delta H_{LV} \qquad \text{and} \qquad L_F = \Delta H_{SL}$$

where

L_V = the specific latent heat of vaporization at some pressure and temperature

ΔH_{LV} = the specific enthalpy change when the substance goes from the liquid to the vapour phase at that pressure and temperature

L_F = the specific latent heat of fusion at some pressure and temperature

ΔH_{SL} = the specific enthalpy change when the substance goes from solid to the liquid phase at that pressure and temperature.

It follows from equation [16.5] and the definition of molar heat capacity at constant pressure, C_p, (section 14.16) that if the temperature of 1 mole of gas increases by ΔT, then its enthalpy increases by ΔH, where

$$\Delta H = C_p \Delta T$$

For unit mass of gas

$$\Delta H = c_p \Delta T$$

where c_p is the specific heat capacity of the gas at constant pressure.

EXAMPLE 16.1

Calculate (a) the increase in enthalpy, (b) the increase in internal energy when 4.000 kg of water at 100°C and a pressure of 1.013×10^5 Pa is turned into steam at the same temperature and pressure. (Specific enthalpy change for the conversion of water to steam at 100°C = 2.261×10^6 J kg^{-1}, specific volume of water at 100°C = 1.044×10^{-3} m^3 kg^{-1}, specific volume of steam at 100°C = 1.637 m^3 kg^{-1})

Solution

There are 4.000 kg of water, and therefore

$$\text{Increase in enthalpy } (\Delta H) = 4.000 \times 2.261 \times 10^6$$
$$= 9.044 \times 10^6 \text{ J}$$

The water is vaporized at constant pressure and therefore $\Delta H = \Delta U + p\Delta V$ (equation [16.3] applies. Here $\Delta H = 9.044 \times 10^6$ J, $p = 1.013 \times 10^5$ Pa and, bearing in mind that there are 4.000 kg of water

$$\Delta V = 4.000\,(1.637 - 1.044 \times 10^{-3}) = 6.544 \text{ m}^3$$

$$\therefore \quad 9.044 \times 10^6 = \Delta U + (1.013 \times 10^5)(6.544)$$

i.e. $\quad \Delta U = 8.381 \times 10^6$ J

i.e. $\quad$ increase in internal energy $= 8.381 \times 10^6$ J.

16.3 HEAT ENGINES

Consider a heating coil through which an electric current is being passed and which is immersed in a flowing liquid. Once a steady state (section 17.2) has been attained, the state of the coil does not change in any way, and all of the electrical energy goes into heating the liquid. Similarly, when mechanical work is done to overcome friction, it too can be done in such a way that all of the mechanical energy is converted to heat. In general, the transformation of work (of any kind) into heat can be accomplished with 100% efficiency and can be continued indefinitely.

However, the opposite process, the continual conversion of heat into work, is never 100% efficient. A device which converts heat into work is called a **heat engine** (internal combustion engines and steam engines are examples). If it is to be useful, the engine must be able to work continuously, and in order to do this it must work in a cycle. To illustrate this we shall consider a gas enclosed in a cylinder by a piston. If heat is supplied to the gas, it expands and pushes back the piston.

Thus heat has been converted to work. The process stops, however, as soon as the pressure of the gas becomes equal to the pressure outside the cylinder. Before there can be any further conversion of heat into work, the gas has to be returned to its initial (compressed) state. That is, if there is to be a continual conversion of heat into work, the gas has to undergo a cycle. Furthermore, the gas can be returned to its initial state only if some of the heat it initially absorbed is given up to a sink which is at a lower temperature than the source which provided the heat in the first place.

Thus, in practice we find that all heat engines operate by taking some working substance around a cycle, and:

(i) take in heat at a high termperature,

(ii) do work,

(iii) reject some of the heat at a lower temperature.

This is illustrated in Fig. 16.1. Since the engine rejects some of the heat it initially takes in, it has converted only part of it into work. This should not be taken to mean that the engine has violated the first law of thermodynamics. There has been no loss of energy; it is just that some of it is still in the form of heat.

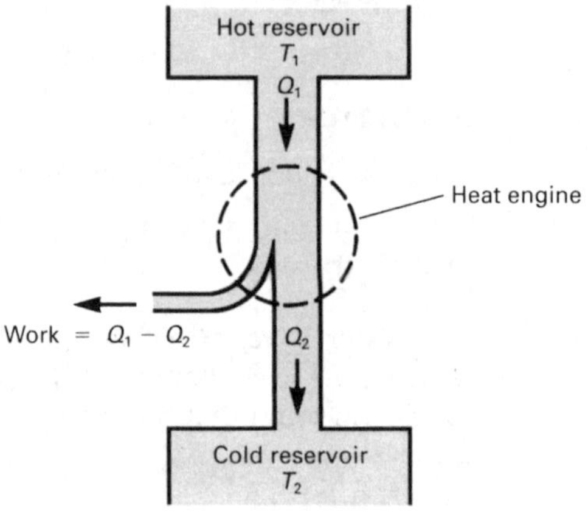

Fig. 16.1
The principle of a heat engine

16.4 THERMAL EFFICIENCY OF HEAT ENGINES

The thermal efficiency η of a heat engine is defined by

$$\eta = \frac{\text{Work done in one cycle}}{\text{Heat taken in at the higher temperature}} \qquad [16.6]$$

At the completion of the cycle the engine's working substance is in the same state as it was initially, and therefore there can have been no change in its internal energy. It follows from the first law of thermodynamics, therefore, that the work done is equal to the net quantity of heat absorbed, i.e.

$$\text{Work done in one cycle} = Q_1 - Q_2$$

where

Q_1 = the heat taken in at the higher temperature

Q_2 = the heat rejected at the lower temperature.

Therefore by equation [16.6]

$$\eta = \frac{Q_1 - Q_2}{Q_1} \quad \text{(for both reversible and irreversible engines)} \qquad [16.7]$$

It can be shown that if the cycle is carried out reversibly (see section 14.13), then

$$\eta = \frac{T_1 - T_2}{T_1} \quad \text{(for all reversible heat engines)} \qquad [16.8]$$

where T_1 and T_2 are the Kelvin temperatures (as measured on the ideal gas scale*) at which the heat is respectively absorbed and rejected.

Notes

(i) Equation [16.8] is valid for all reversible engines, regardless of the particular cycle and the particular working substance, as long as the heat is taken in entirely at the single temperature T_1 and is rejected entirely at the single temperature T_2.

(ii) It can be shown that no heat engine is more efficient than a reversible one working between the same two temperatures, and therefore no heat engine can possibly have an efficiency greater than that given by equation [16.8].

(iii) It follows from equation [16.8] that the efficiency of a heat engine can never be 100% (i.e. η cannot be equal to 1) because the reservoir to which the engine rejects heat would have to be at a temperature of zero kelvin (i.e. $T_2 = 0\,\text{K}$) and this, of course, is impossible.

(iv) Equation [16.8] can be rewritten as $\eta = 1 - T_2/T_1$, and therefore the efficiency is increased by decreasing T_2/T_1, i.e. **the efficiency is increased by taking in heat at as high a temperature as possible and rejecting heat at as low a temperature as possible.**

(v) The efficiency of a real heat engine is less than that given by equation [16.8] because of losses due to frictional effects, turbulence, etc., and because the heat is usually taken in over a range of temperatures and rejected over a range of temperatures.

16.5 THE SECOND LAW OF THERMODYNAMICS

Though there is nothing in the first law of thermodynamics to prevent it being otherwise, it is a matter of common experience that:

(i) no heat engine that works in a cycle completely converts heat into work, and

(ii) when a cold body and a hot body are brought into contact with each other, heat always flows from the hot body to the cold body – never from the cold body to the hot body.

The second law of thermodynamics is a formal statement of these observations. It can be stated in a number of different (but equivalent) ways. One such statement is:

> It is not possible to convert heat continuously into work without at the same time transferring some heat from a warmer body to a colder body.

Thus, whereas the first law tells us of the equivalence of heat and work, the second law is concerned with the circumstances in which heat can be converted into work.

If the second law were not true, it would be possible to run ships on heat extracted from the sea. It is not possible to do so, though, because the second law requires there to be a reservoir at a lower temperature than the sea into which some of the

*The significance of specifying the ideal gas scale will become apparent on reading section 16.6.

rejected heat can be discharged. There is no such reservoir, except perhaps the ship's cold-store, and this is cold only because refrigeration units are consuming energy to keep it so!

The experience that heat cannot be completely converted into work is associated with the fact that heat is fundamentally different from other forms of energy. The heat energy possessed by a body is the energy of the random motions of its molecules. This is quite distinct from, say, the kinetic energy the body has when it is moving. The kinetic energy of a moving body represents the ordered motion which its molecules have superimposed on their random motion. When we try to convert heat into work we are trying to change the random molecular motion into ordered motion. The reason we cannot accomplish this with 100% efficiency is that we cannot control the individual motions of the molecules.

16.6 THE THERMODYNAMIC SCALE OF TEMPERATURE

The efficiency of a reversible heat engine depends only on the temperatures of the source and the sink between which it is operating. Kelvin realized that if a temperature scale were defined in terms of the efficiency of such an engine, it would be independent of the properties of any particular substance – it would therefore be an absolute scale.

Kelvin suggested that the scale (now called the **thermodynamic scale**) should be such that the ratio of any two temperatures on it should be equal to the ratio of the quantities of heat taken in and rejected by a reversible heat engine operating between the same two temperatures. Thus if we represent temperatures on the thermodyanic scale by τ, then for a reversible engine taking in heat Q_1 at temperature τ_1 and rejecting heat Q_2 at a lower temperature τ_2,

$$\frac{\tau_2}{\tau_1} = \frac{Q_2}{Q_1} \qquad [16.9]$$

The efficiency of such an engine is given by equation [16.7] as

$$\eta = \frac{Q_1 - Q_2}{Q_1} \qquad \text{i.e.} \quad \eta = 1 - \frac{Q_2}{Q_1}$$

Therefore, by equation [16.9]

$$\eta = 1 - \frac{\tau_2}{\tau_1} \qquad [16.10]$$

If the temperatures between which the engine is operating had been measured on the **ideal gas scale** (see section 13.2) and had been found to be T_1 and T_2, then from what has been said in section 16.4, the efficiency η would be given by

$$\eta = \frac{T_1 - T_2}{T_1} \qquad \text{i.e.} \quad \eta = 1 - \frac{T_2}{T_1} \qquad [16.11]$$

Comparing equations [16.10] and [16.11] we see that

$$\frac{\tau_2}{\tau_1} = \frac{T_2}{T_1} \qquad [16.12]$$

That is to say, any two temperatures on the thermodynamic scale are in the same ratio as the same two temperatures measured on the ideal gas scale. Finally, by making the temperature of the triple point of water the fixed point of both the thermodynamic scale and the ideal gas scale, and assigning to it the same numerical value (273.16 K) in each case, the two scales become identical. For example, if the temperature of the sink to which the engine is discharging heat is taken to be the temperature of the triple point of water then $\tau_2 = T_2 = 273.16\,\text{K}$ and therefore by equation [16.12]

$$\frac{273.16}{\tau_1} = \frac{273.16}{T_1}$$

i.e. $\tau_1 = T_1$

Now we have established that the two scales are identical, there is no longer any need to distinguish between them. Therefore from now on the single symbol T should be taken to refer to either scale.

16.7 ENTROPY

The first law of thermodynamics is concerned with energy; the second law is concerned with a quantity called **entropy**. It can be defined by*

$$\delta S = \delta Q/T \quad \text{(for a reversible process only)} \qquad [16.13]$$

where

δS = the increase in entropy of some system when it undergoes a reversible change ($\text{J}\,\text{K}^{-1}$)

δQ = the heat absorbed by the system, and where δQ is so small that the process can be considered to take place at a constant temperature T measured in kelvins.

For the more general case of a reversible process in which the temperature is not necessarily constant and where a system changes from an initial state (1) to some other state (2)

$$\Delta S = \int_1^2 \frac{\text{d}Q}{T} \quad \text{(for a reversible process only)} \qquad [16.14]$$

where ΔS = the increase in entropy when the system changes reversibly from state 1 to state 2 ($\text{J}\,\text{K}^{-1}$)

$\int_1^2 \frac{\text{d}Q}{T}$ = the sum of the ratios of the quantities of heat absorbed at each point on the path from state 1 to state 2 to the temperatures at those points, i.e. the sum of the terms $\delta Q/T$ of equation [16.13].

Notes (i) Equations [16.13] and [16.14] are valid for reversible processes only.

(ii) The entropy of a system depends only on the state of the system. **When the entropy of a system changes the change depends only on the initial and final states of the system, not on the particular process by which it was accomplished, nor on whether it was reversible or irreversible.** At first this statement may seem to contradict note (i), but it does not. Though the changes in entropy are the same, $\int \text{d}Q/T$ for the reversible process is not equal to $\int \text{d}Q/T$ of the irreversible process.

*An alternative definition is given later.

(iii) Since T cannot be negative, it follows from equation [16.14] that **the entropy of a system increases when it absorbs heat and decreases when it rejects heat.**

(iv) For an adiabatic process there is no change in heat content and equation [16.14] reduces to

$$\Delta S = 0 \quad \text{(for any reversible adiabatic process)}$$

Processes which occur without change in entropy are called **isentropic** processes and therefore **reversible adiabatic processes are isentropic**.

(v) For a reversible isothermal process equation [16.4] reduces to

$$\Delta S = \frac{Q}{T} \quad \text{(for any reversible isothermal process)} \qquad [16.15]$$

where Q is the heat absorbed at the constant temperature T.

(vi) It follows from note (ii) that when a substance is taken through a complete cycle the net change in entropy is zero, i.e.

$\Delta S = 0$ for a working substance undergoing a complete cycle (for both reversible and irreversible processes)

When the cycle is carried out reversibly the entropy lost by the source is equal to that gained by the sink, and therefore the entropy change for the whole system (sink, source and working substance) is zero. For an irreversible cycle, though, the entropy lost by the source is less than that gained by the sink. Therefore even though there is no change in the entropy of the working substance, there is an increase in the entropy of the system as a whole.

EXAMPLE 16.2

Calculate the change in entropy of 3.00 kg of water at 100 °C when it is converted to steam at 100 °C. (Specific latent heat of vaporization of water = $2.26 \times 10^6\,\mathrm{J\,kg^{-1}}$ at 100 °C.)

Solution

The process is both isothermal and reversible, in which case the change in entropy, ΔS, is given by equation [16.5] as

$$\Delta S = Q/T$$

Here $Q = 3.00 \times 2.26 \times 10^6 = 6.78 \times 10^6\,\mathrm{J}$

$T = 373\,\mathrm{K}\ (= 100\,^\circ\mathrm{C})$

$$\therefore \quad \Delta S = \frac{6.78 \times 10^6}{373}$$

$$= 1.82 \times 10^4\,\mathrm{J\,K^{-1}}$$

EXAMPLE 16.3

Calculate the change in entropy of 5.00 kg of water when it is heated reversibly from 0 °C to 100 °C. (Specific heat capacity of water in the range 0 °C to 100 °C = $4.20 \times 10^3\,\text{J kg}^{-1}\,{}^\circ\text{C}^{-1}$.)

Solution

The change in entropy, ΔS, is given by equation [16.4] as

$$\Delta S = \int_{T=273\,\text{K}}^{T=373\,\text{K}} \frac{dQ}{T}$$

It follows from equation [13.7] that $dQ = mc\,dT$.

$$\Delta S = \int_{273}^{373} \frac{(5.00)\,(4.20 \times 10^3)}{T}\,dT$$

$$= 21.0 \times 10^3 [\log_e T]_{273}^{373}$$

$$= 21.0 \times 10^3 \log_e \left(\tfrac{373}{273}\right)$$

$$= 6.55 \times 10^3\,\text{J K}^{-1}$$

EXAMPLE 16.4

5.00 kg of water are heated from 0 °C to 100 °C by being placed in contact with a body which has a large heat capacity and which is itself at 100 °C. Calculate the changes in entropy of: (a) the water; (b) the body; (c) the Universe. (Specific heat capacity of water in the range 0 °C to 100 °C = $4.20 \times 10^3\,\text{J kg}^{-1}\,{}^\circ\text{C}^{-1}$.)

Solution

(a) This is identical to Example 16.3, and therefore

$$\Delta S_{\text{water}} = +6.55 \times 10^3\,\text{J K}^{-1}$$

(b) The body is of very large heat capacity and therefore, to a good approximation, we may assume that its temperature is constant at 100 °C (373 K). The entropy change of the body is therefore given (by equation [16.15]) as

$$\Delta S_{\text{body}} = -Q/373$$

where Q is the heat lost by the body. This is equal to the heat gained by the water, and therefore

$$Q = 5.00 \times 4.20 \times 10^3 \times 100 = 2.10 \times 10^6\,\text{J}$$

$$\therefore \quad \Delta S_{\text{body}} = -\frac{2.10 \times 10^6}{373}$$

i.e. $\Delta S_{\text{body}} = -5.63 \times 10^3\,\text{J K}^{-1}$

(c) The change in entropy of the Universe is equal to that of the whole system (i.e. the water and the body), and therefore

$$\Delta S_{\text{Universe}} = \Delta S_{\text{water}} + \Delta S_{\text{body}}$$

$$= 6.55 \times 10^3 - 5.63 \times 10^3$$

i.e. $$\Delta S_{\text{Universe}} = 920\,\text{J}\,\text{K}^{-1}$$

Note The transfer of heat from the body (at 100°C) to the water (initially at 0°C) is an irreversible process, for at no stage could it go in the opposite direction. In performing the calculations, though, we have used equations which govern reversible processes. We are justified in doing this because entropy changes depend only on the initial and final states of the system concerned, and not on the manner in which the changes occur. (See Note (ii) on p. 289.) The change in entropy of the water is therefore the same as it would be if its temperature were increased reversibly to 100°C. Likewise, the entropy lost by the body as a result of losing heat is the same as it would be if the heat had been lost reversibly. Note, though, that there is an overall increase in entropy (of 920 J K^{-1}) as, of course, there must be for an irreversible process.

16.8 TEMPERATURE – ENTROPY DIAGRAMS (*T–S* DIAGRAMS)

These are plots of temperature against entropy and are a useful alternative to *p–V* diagrams. Suppose that during some reversible process the temperature and entropy of a substance vary in the arbitrary manner shown in Fig. 16.2.

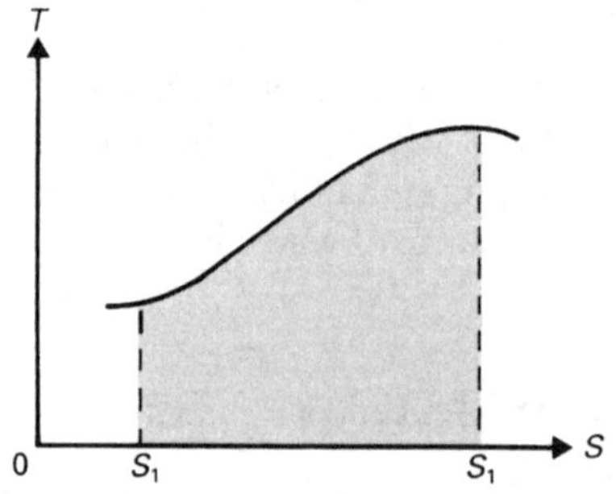

Fig. 16.2
To illustrate the significance of the area under a ***T–S*** curve

It follows from equation [16.13] that for a reversible process

$$\delta Q = T\delta S$$

The heat absorbed by the substance when its entropy changes from S_1 to S_2 is therefore given by Q, where

$$Q = \int_{S_1}^{S_2} T\text{d}S$$

i.e. **Heat absorbed = Area of shaded region** (for a reversible process only) [16.16]

When a substance is taken through a complete cycle it ends up in the same state as the one it started in; in particular it has the same temperature and the same entropy, and is therefore represented by a closed loop on a *T–S* diagram. Suppose a substance is taken reversibly through the (arbitrary) cycle shown in Fig. 16.3 from A to B and back to A, in the direction AXBYA.

In going from A to B the entropy is increasing, and therefore heat is being absorbed. Heat is rejected in going from B to A. From equation [16.16]

	Heat absorbed	=	Area under AXB
	Heat rejected	=	Area under AYB
∴	Net heat absorbed	=	Area of shaded region

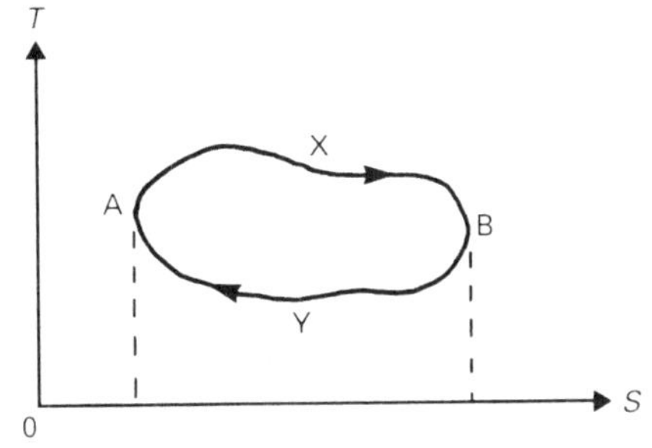

Fig. 16.3 To illustrate the significance of the area of a closed loop on a ***T–S*** diagram

At the completion of the cycle the substance is in its original state, and therefore there can have been no change in its internal energy, in which case it follows from the first law of thermodynamics that the net heat absorbed is equal to the work done by the substance, i.e.

Net heat absorbed in one cycle = Work done = Area of shaded region (for a reversible cycle only)

In order to illustrate some of these ideas we shall now consider the situation of a gas (not necessarily ideal) undergoing the particularly simple sequence of operations known as a **Carnot cycle**. The cycle is represented in Fig. 16.4 both by a p–V curve and by a T–S curve.

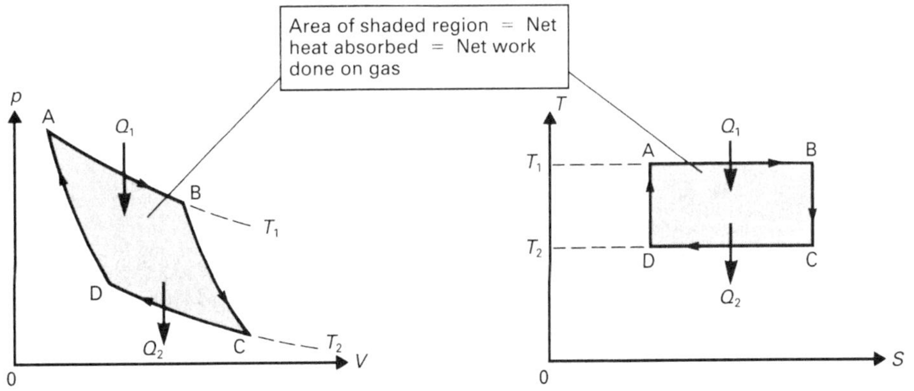

Fig. 16.4 Carnot cycle of a gas represented by (a) a ***p–V*** curve, (b) a ***T–S*** curve

Bearing in mind that $\Delta S = Q/T$ for a reversible isothermal process and that $\Delta S = 0$ for a reversible adiabatic process, we can make the following analysis, in which ΔS_{AB} represents the increase in entropy in going from A to B etc.

A→B Reversible isothermal expansion An amount of heat Q_1 is taken in by the gas at a constant temperature T_1:

$$\Delta S_{\mathrm{AB}} = Q_1/T_1$$

B→C Reversible adiabatic expansion:

$$\Delta S_{BC} = 0$$

C→D Reversible isothermal contraction. An amount of heat Q_2 is rejected by the gas at a constant temperature T_2:

$$\Delta S_{CD} = -Q_2/T_2$$

D→A Reversible adiabatic contraction:

$$\Delta S_{DA} = 0$$

$$\therefore \quad \text{Total increase in entropy} = \frac{Q_1}{T_1} - \frac{Q_2}{T_2}$$

The thermodynamic scale of temperature (section 16.6) is defined in such a way that $Q_1/T_1 = Q_2/T_2$, and therefore the total increase in entropy in going around the cycle is zero.

16.9 ENTROPY CHANGES IN IRREVERSIBLE PROCESSES

Whenever a system undergoes an irreversible process, though the entropy of some components of the system may decrease, there is an increase in the entropy of the system as a whole.

We shall illustrate this by considering some examples of irreversible processes.

Thermal Conduction

The conduction of heat from a hotter body to a colder body is an irreversible process because there is a finite temperature difference whilst the heat transfer is taking place, and the transfer can take place in one direction only. If the system is perfectly lagged, the heat lost by the hotter body will be equal to that gained by the cooler body; let it be Q. If the mean temperatures of the hotter and colder bodies are T_1 and T_2 respectively, then

$$\text{Decrease in entropy of hotter body} \approx Q/T_1$$

$$\text{Increase in entropy of cooler body} \approx Q/T_2{}^{\star}$$

But $T_2 < T_1$ and therefore $Q/T_2 > Q/T_1$, i.e. there has been a net increase in entropy.

Note It follows that if heat were to flow from the colder body to the hotter body there would be a decrease in entropy, and this would be in violation of the principle of increase of entropy (see section 16.10).

Processes Involving Friction

If two bodies are rubbed against each other, the mechanical work done in overcoming friction generates heat. The bodies absorb this heat and their entropy increases. There is no change in the entropy of the agency doing the work because it neither loses nor gains heat. There is therefore an overall increase in entropy.

⋆A non-approximate method for this type of calculation is given in Example 16.3.

Cooling of Food in a Refrigerator

As the food in a refrigerator cools, its entropy decreases because heat is being removed from it. But this heat goes into the surrounding air, and so the entropy of the air increases. In addition, electrical energy is being consumed, and this will probably have involved the burning of some fuel (coal or oil, for example). The entropy of the combustion products (hot gases, smoke, etc.) is greater than that of the original fuel. Calculations show that there is a net increase in entropy.

Irreversible Heat Engines

The efficiency of a heat engine, reversible or irreversible, which takes in heat Q_1 at a temperature T_1 and rejects heat Q_2 at a temperature T_2 is $(Q_1 - Q_2)/Q_1$ (see section 16.4). In the case of a reversible engine

$$\frac{Q_1 - Q_2}{Q_1} = \frac{T_1 - T_2}{T_1}$$

The efficiency of an irreversible engine is less than that of a reversible one, and therefore for an irreversible engine

$$\frac{Q_1 - Q_2}{Q_1} < \frac{T_1 - T_2}{T_1}$$

$$\therefore \quad 1 - \frac{Q_2}{Q_1} < 1 - \frac{T_2}{T_1}$$

$$\therefore \quad \frac{T_2}{T_1} < \frac{Q_2}{Q_1}$$

$$\therefore \quad \frac{Q_1}{T_1} < \frac{Q_2}{T_2}$$

Thus the entropy lost by the source (Q_1/T_1) is less than that gained by the sink (Q_2/T_2). Since the only other component of the system, the working substance, undergoes no entropy change in a complete cycle, there has been an overall increase in entropy.

16.10 PRINCIPLE OF INCREASE OF ENTROPY (ENTROPY VERSION OF THE SECOND LAW)

We have seen that when a system undergoes a reversible process there is no change in the entropy of the system, and that in an irreversible process there is always an increase in entropy. Reversible processes are an ideal that cannot be realized in practice, i.e. all real processes are irreversible. It follows that **all real processes occur in such a way that there is a net increase in entropy.** This is called **the principle of increase of entropy.** It is a consequence of the second law of thermodynamics, and in fact is one of the many ways in which the second law can be stated.

Every time entropy increases the opportunity to convert some heat into work is lost for ever. For example, there is an increase in entropy when hot and cold water are mixed. The warm water which results will never separate itself into a hot layer and a cold layer. There has been no loss of energy but some of the energy is no longer available for conversion into work. We can envisage a (distant) future in which the temperature of the Universe is the same throughout. The entropy of the Universe will then have reached its maximum value and all processes will cease – the so-called **'heat death' of the Universe**.

16.11 THE STATISTICAL SIGNIFICANCE OF ENTROPY

Imagine a glass container in which there are a thousand grains of salt, and then imagine that a thousand grains of black pepper are carefully placed on top of them. If the container is shaken, the mixture will become uniformly grey. Continued shaking will keep redistributing the grains at random, but we would not expect that the original distribution would ever return. Thus the system has gone from a highly organized state with salt at the bottom and pepper on top, into a highly disorganized state where there is complete uniformity. The reader should realize that if we were to label the grains in some way (by numbering them, say), then the chance that any particular distribution would occur (all the odd numbers being on top for example) would be just as unlikely as that with all the pepper at the top – no matter how we numbered the grains. The point is, of course, that the grains are not labelled. The system has gone from a statistically unlikely state (salt and pepper separate) to one of a very large number of indistinguishable (uniformly grey) states in which there are approximately five hundred grains of salt and five hundred grains of pepper in each half of the mixture.

This has been just one example of the common experience that in all natural processes (involving large, and therefore statistically meaningful, numbers) the amount of disorder tends to increase up to some maximum value. We saw in section 16.10 that whenever some natural process takes place there is an increase in entropy. Thus natural processes increase both disorder and entropy. This is no coincidence; entropy and disorder are related, and it can be shown that entropy is in fact a measure of disorder. This is not too surprising, for we stated in section 16.5 that when work is converted into heat, ordered motion is being changed into disordered motion, and later saw that increases in heat content are brought about by increases in entropy.

16.12 HEAT PUMPS AND REFRIGERATORS

Both heat pumps and refrigerators (we shall explain the difference in the next paragraph) act like heat engines working in reverse, i.e. they take in heat at a low temperature and reject heat at a higher temperature. In order that they can do this, some external agency (an electric motor for example) has to do work on the working substance of the device. Fig. 16.5 compares the action of a heat engine operating between temperatures T_1 and T_2 with that of a heat pump or refrigerator operating between the same two temperatures.

The purpose of a refrigerator is to cool whatever is inside it, i.e. to remove heat from the low temperature reservoir. The purpose of a heat pump, on the other hand, is to supply heat to the high temperature reservoir. For example, a heat pump might be used to heat a house in winter by taking heat from a (cold) river nearby. The effectiveness of refrigerators and heat pumps is measured by a quantity called the **coefficient of performance**. It is respectively the ratio of the heat extracted or supplied, to the work done by the external agency. Thus

$$\text{Coefficient of performance of a refrigerator} = \frac{Q_2}{W} = \frac{Q_2}{Q_1 - Q_2} = \frac{T_2}{T_1 - T_2} \quad \left(\text{for a reversible refrigerator}\right)$$

$$\text{Coefficient of performance of a heat pump} = \frac{Q_1}{W} = \frac{Q_1}{Q_1 - Q_2} = \frac{T_1}{T_1 - T_2} \quad \left(\text{for a reversible heat pump}\right)$$

Heat pumps provide a cheap form of heating, because the heat supplied (Q_1) is greater than the work done by the external agency ($Q_1 - Q_2$). Suppose that a heat pump working reversibly extracts heat from a river at 7 °C (280 K) and delivers it to a room at 21 °C (294 K). The pump is reversible, and therefore

$$\frac{Q_1}{Q_1 - Q_2} = \frac{T_1}{T_1 - T_2}$$

$$= \frac{294}{294 - 280} = 21$$

Thus the coefficient of performance is 21, i.e. 21 joules of heat would be provided with the consumption of only one joule of work! Compare this with a conventional electric fire where one joule of electrical energy can (at best) supply one joule of heat.

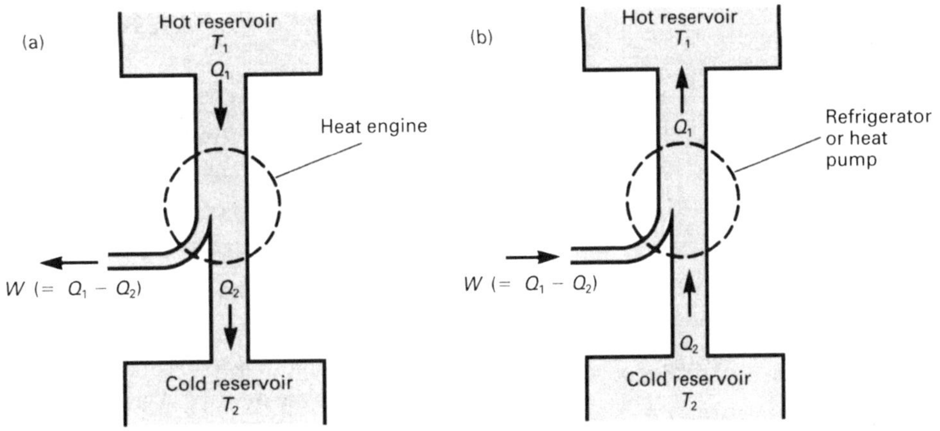

Fig. 16.5
(a) The action of a heat engine, compared with (b) that of a heat pump or refrigerator

16.13 THE PETROL ENGINE CYCLE (OTTO CYCLE)

Fig. 16.6(a) shows the p–V curve for the cycle of operations known as an Otto cycle. **The Otto cycle** is an idealized form of the cycle that occurs in a petrol engine. Refer also to Fig. 16.6(b).

A→A′ The inlet valve opens and the exhaust valve closes.

A′→A **Induction stroke**. A mixture of typically 7% petrol vapour and 93% air (by weight) at about 50 °C is drawn into the cylinder (through the inlet valve) as the piston moves down.

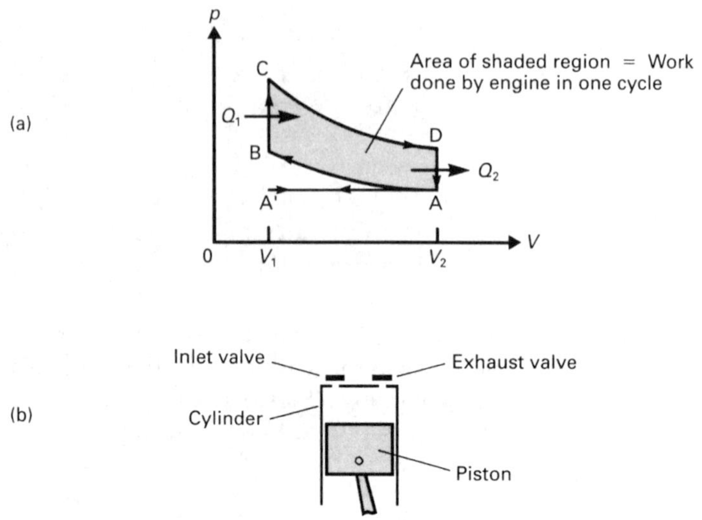

Fig. 16.6
(a) ***p–V*** diagram for idealized petrol engine cycle, (b) piston and cylinder

At A — The inlet valve closes.

A→B — **Compression stroke**. The piston moves up, compressing the gas adiabatically. The temperature rises to about 300 °C

B→C — A spark ignites the air-petrol mixture at B, supplying heat Q_1 and increasing the pressure at constant volume. The temperature rises to about 2000 °C.

C→D — **Power stroke**. The increased pressure pushes the piston down and the gas expands, adiabatically, decreasing both the pressure and temperature.

D→A — The exhaust valve opens at D, and most of the burnt gas rushes out of the cylinder, removing an amount of heat Q_2. The pressure and temperature of the gas that remains in the cylinder decrease.

A→A′ — **Exhaust stroke**. The rest of the burnt gas is expelled as the piston moves up.

At A′ — The exhaust valve closes and the inlet valve opens. The cycle starts again.

Notes

(i) Each cycle consists of four strokes of the piston: A′ to A (down), A to B (up), C to D (down) and A to A′ (up). It is therefore referred to as a four-stroke cycle.

(ii) The fuel is burnt inside the cylinder; it is therefore an **internal combustion engine**.

(iii) The expansion (C to D) and compression (A to B) of the gas are adiabatic because the piston moves at high speed.

(iv) If the working substance is assumed to be air behaving as an ideal gas, then the thermal efficiency η can be shown to be given by

$$\eta = 1 - \frac{1}{(v_2/v_1)^{\gamma-1}} \qquad [16.17]$$

where

(v_2/v_1) = **the compression ratio** of the engine, i.e. the ratio of the maximum volume of the gas to its minimum volume (see Fig. 16.6(a)). The compression ratio is typically 9:1.

γ = the ratio of the principal specific heat capacities of the gas – approximately 1.4 (see section 14.19).

It follows from equation [16.17] that increasing the compression ratio increases the efficiency of the engine. There is an upper limit to the compression ratio that can be achieved in practice because the temperature rise that accommmpanies the compression (A to B) can become so high that the mixture is ignited before it is sparked. This is called pre-ignition and it reduces the efficiency because it tends to push the piston back down the cylinder before it has completed its stroke. The use of a high-octane fuel allows a high compression ratio to be utilized.

(v) The efficiency of an actual engine is less than that given by equation [16.17] because the working substance is not air, and it does not behave as an ideal gas. The efficiency is further reduced by frictional effects, turbulence, loss of heat to the cylinder walls and the fact that the exhaust valve takes a finite time to open and close. The actual efficiency is typically 28%; the theoretical efficiency (η) is typically 58%.

(vi) The p–V curve of an actual petrol engine cycle has rounded 'corners'. This reduces the area enclosed by the curve and so accounts for the reduction in the work done per cycle. Also, the actual cycle has A→A′ slightly above A′→A.

16.14 THE DIESEL CYCLE

Fig. 16.7(a) shows the p–V curve of an idealized four-stroke Diesel cycle. Refer also to Fig. 16.7(b).

Fig. 16.7
(a) ***p–V*** diagram for idealized Diesel cycle, (b) piston and cylinder

(a)

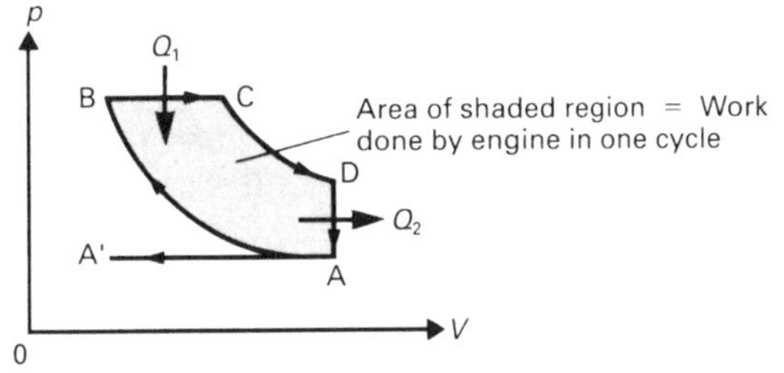

(b)

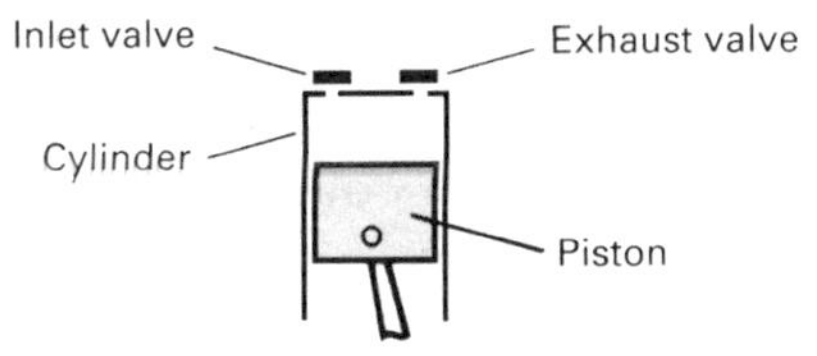

At A′ The inlet valve opens and the exhaust valve closes.

A′→A **Induction stroke**. Air is drawn into the cylinder at atmospheric pressure as the piston moves down.

At A The inlet valve closes.

A→B **Compression stroke**. The air is compressed adiabatically as the piston moves up. The temperature of the air rises to about 700 °C and is now hot enough to ignite the fuel.

B→C **First part of power stroke**. The fuel (diesel oil) is sprayed into the cylinder and is ignited by the hot air. The fuel enters at such a rate that as it burns (supplying heat Q_1) it forces the piston down at constant pressure.

C→D **Second part of power stroke**. The fuel supply is cut off at C and the burnt gas expands adiabatically and pushes the piston down. The temperature falls.

D→A The exhaust valve opens at D and most of the burnt gas rushes out of the cylinder, removing an amount of heat Q_2. The pressure and temperature of the gas remaining in the cylinder decrease.

A→A **Exhaust stroke**. The rest of the burnt gas is expelled from the cylinder as the piston moves up.

At A′ The exhaust valve closes and the inlet valve opens. The cycle starts again.

Notes

(i) Each cycle consists of four strokes of the piston: A′ to A (down), A to B (up), B to D (down) and A to A′ (up). It is therefore a four-stroke cycle.

(ii) The fuel is burnt inside the cylinder; it is therefore an **internal combustion engine**.

(iii) There is no fuel in the cylinder during compression (A to B) and therefore (unlike the case of the petrol engine) very high compression ratios (typically 16:1) can be utilized without any risk of pre-ignitiion. This makes Diesel engines more efficient than petrol engines.

Diesel engines have the added advantage of using a cheaper fuel. On the other hand, the higher working pressures of Diesel engines makes them more expensive to produce and they have lower power/weight ratios than petrol engines. The theoretical efficiency is typically 65%, but the efficiency of an actual engine is less than this (typically 36%) because of frictional effects, etc. (see note (v) of section 16.13).

(iv) Note (vi) of section 16.13 applies here too.

16.15 THE STEAM ENGINE CYCLE (RANKINE CYCLE)

Fig. 16.8 shows the p–V curve for the cycle of operations known as a Rankine cycle. **The Rankine cycle** is an idealized form of the cycle which occurs in a steam engine.

A→B Water is compressed adiabatically. There is very little change in volume and only a slight increase in pressure.

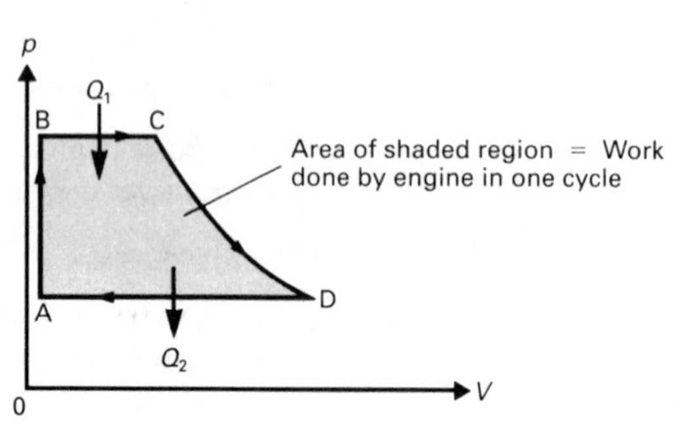

Fig. 16.8 Idealized steam engine cycle

B→C The water is heated (in the boiler) at constant pressure to its boiling point at the pressure of the boiler. As the heating continues the water vaporizes, at the same constant pressure, to form steam, which expands into the cylinder.

C→D The steam is now receiving no heat. It expands adiabatically and cools.

D→A The steam is condensed to water at constant pressure and temperature.

Notes

(i) Steam engines are known as **external combustion engines** because the fuel is burned outside the cylinder.

(ii) The theoretical efficiency is typically 30%; the actual efficiency is much less, typically 10%. A major cause of this large difference is the drop in pressure that occurs as the stream passes along the pipes leading from the boiler to the cylinder.

(iii) The theoretical efficiency is much less than that of both the Otto cycle and the Diesel cycle. This reflects the fact that the heat is supplied at a much lower temperature (about 250 °C) in the case of the steam engine.

17

HEAT TRANSFER

Heat may be transferred from one point to another by conduction, convection or radiation. This chapter is concerned mainly with conduction and radiation; convection is discussed briefly in sectiion 17.12.

17.1 THERMAL CONDUCTION

Conduction is the process by which heat flows from the hotter regions of a substance to the colder regions without there being any net movement of the substance itself. The mechanism by which conduction occurs depends on the nature of the material concerned; various mechanisms are described in section 17.5.

17.2 DEFINITION OF THERMAL CONDUCTIVITY (*k*)

We shall have cause to refer to steady state conditions. **A substance is in steady state when the temperatures at all points in it are steady.**

Consider a thin disc of some material of cross-sectional area A and thickness δx (Fig. 17.1). Suppose that the hotter face is maintained at a temperature θ and that the other face is maintained at $\theta - \delta\theta$.

Fig. 17.1
Definition of thermal conductivity

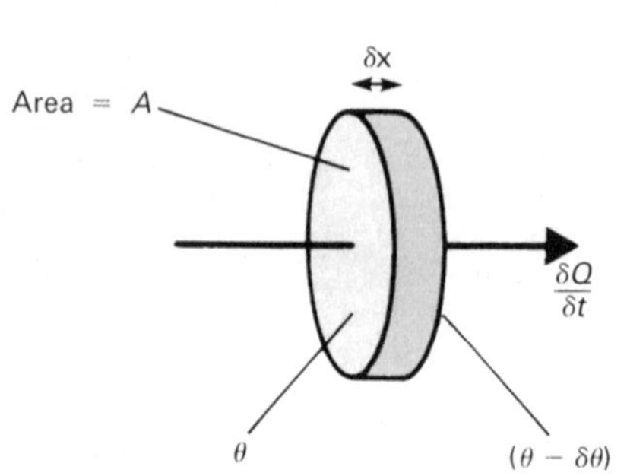

Suppose also that the rate of flow of heat from the hotter face to the colder face is $\delta Q/\delta t$. It can be shown by experiment that **if there are no heat losses from the sides and steady state conditions prevail,** then

$$\frac{\delta Q}{\delta t} \propto A\frac{\delta\theta}{\delta x}$$

With the introduction of a constant of proportionality k this can be written as

$$\frac{\delta Q}{\delta t} = -kA\frac{\delta\theta}{\delta x}$$

which in the limit as $\delta x \to 0$ becomes

$$\frac{\mathrm{d}Q}{\mathrm{d}t} = -kA\frac{\mathrm{d}\theta}{\mathrm{d}x} \qquad [17.1]$$

where

$\mathrm{d}Q/\mathrm{d}t$ is the rate of flow of heat from the hotter face to the colder face and is at right angles to the faces (unit = W)

$\mathrm{d}\theta/\mathrm{d}x$ is called the **temperature gradient** across the section concerned (unit = $\mathrm{K\,m^{-1}}$)

k is a constant whose value depends on the material of the disc. It is called the **coefficient of thermal conductivity** of the material (unit = $\mathrm{W\,m^{-1}\,K^{-1}}$). Values of k for some common materials are given in Table 17.1.

Notes (i) When heat is flowing in the positive direction of x (as in Fig. 17.1) the temperature gradient is negative, and therefore the presence of the minus sign in equation [17.1] makes k a positive constant.

(ii) It is the existence of the temperature gradient which causes the heat to flow. If it were not for the fact that the two faces are being maintained at their respective temperatures, the effect of the heat flow would be to destroy the temperature gradient by warming the cooler regions.

(iii) Equation [17.1] is used to define k. Thus:

The coeffcient of thermal conductivity of a material is the rate of flow of heat per unit area per unit temperature gradient when the heat flow is at right angles to the faces of a thin parallel-sided slab of the material under steady state conditions.

Table 17.1
Values of k for some common substances at room temperature

Substance	$k/\mathrm{W\,m^{-1}\,K^{-1}}$
Silver	418
Copper	385
Aluminium	238
Iron	80
Lead	38
Mercury	8
Glass (Pyrex)	1.1
Brick	~ 1
Rubber	0.2
Air	0.03

17.3 TEMPERATURE DISTRIBUTION ALONG A UNIFORM BAR

Perfectly Lagged Bar

Consider two infinitesimally thin sections X and Y of a perfectly lagged bar of uniform cross-sectional area A (Fig. 17.2), and suppose that steady state conditions have been attained. From equation [17.1]

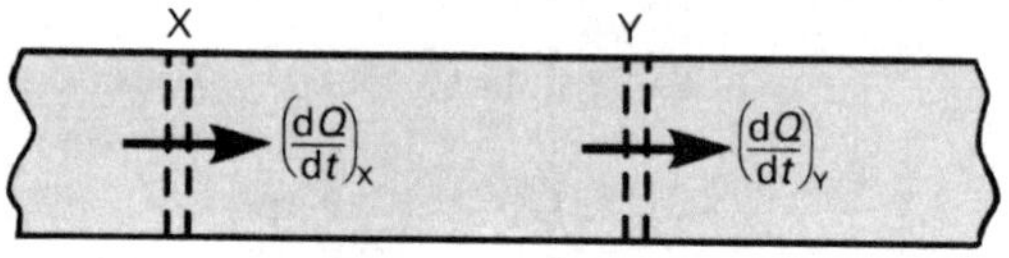

Fig. 17.2 Heat flow along a uniform bar

$$\text{At X:} \quad \left(\frac{dQ}{dt}\right)_X = -k_X A\left(\frac{d\theta}{dx}\right)_X \qquad [17.2]$$

$$\text{At Y:} \quad \left(\frac{dQ}{dt}\right)_Y = -k_Y A\left(\frac{d\theta}{dx}\right)_Y \qquad [17.3]$$

where k_X and k_Y are the thermal conductivities at X and Y and $(d\theta/dx)_X$ and $(d\theta/dx)_Y$ are the temperature gradients at X and Y. The bar is perfectly lagged, so no heat can escape from the sides, and therefore the rates of flow of heat across X and Y are equal, i.e.

$$\left(\frac{dQ}{dt}\right)_X = \left(\frac{dQ}{dt}\right)_Y$$

Therefore from equations [17.2] and [17.3]

$$k_X A\left(\frac{d\theta}{dx}\right)_X = k_Y A\left(\frac{d\theta}{dx}\right)_Y \qquad [17.4]$$

To a very good approximation thermal conductivity is independent of temperature, and therefore $k_X = k_Y$ and equation [17.4] reduces to

$$\left(\frac{d\theta}{dx}\right)_X = \left(\frac{d\theta}{dx}\right)_Y$$

Thus the temperature gradients at X and Y are equal, and since X and Y are any two sections, it follows that:

The temperature gradient is the same at all points along a perfectly lagged uniform bar.

Unlagged bar

Consider again sections X and Y, but suppose now that the bar is unlagged. Some of the heat which flows through X will flow out of the sides of the bar before reaching Y, in which case

$$\left(\frac{dQ}{dt}\right)_X > \left(\frac{dQ}{dt}\right)_Y$$

Therefore, from equations [17.2] and [17.3] and since $k_X = k_Y$

$$\left(\frac{d\theta}{dx}\right)_X > \left(\frac{d\theta}{dx}\right)_Y$$

and it follows that:

Temperature gradient decreases with distance from the hot end of an unlagged uniform bar.

Fig. 17.3 is based on these results and shows the steady state temperature distribution of a perfectly lagged uniform bar of length L, together with that of an identical unlagged bar. Each bar has its hot end maintained at a temperature θ_2 and its cold end at θ_1. The situations of the two bars are illustrated in terms of heat flow lines in Fig. 17.4.

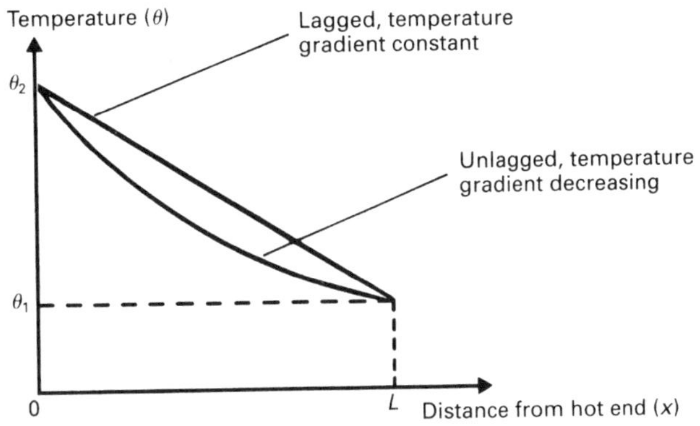

Fig. 17.3 Temperature distribution along a uniform bar

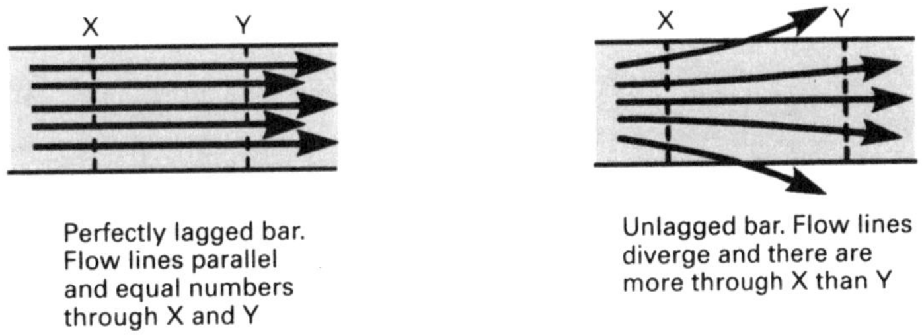
X
Y
Perfectly lagged bar. Flow lines parallel and equal numbers through X and Y
Unlagged bar. Flow lines diverge and there are more through X than Y

Fig. 17.4 Heat flow lines in a lagged and an unlagged bar

Since the temperature gradient of the lagged bar is constant

$$\frac{d\theta}{dx} = -\frac{(\theta_2 - \theta_1)}{L}$$

Therefore from equation [17.1]

$$\frac{dQ}{dt} = kA\frac{(\theta_2 - \theta_1)}{L} \quad \left(\begin{matrix}\text{for a perfectly} \\ \text{lagged bar}\end{matrix}\right) \qquad [17.5]$$

This is a particularly useful form of equation [17.1], but it is relevant only to the case of a perfectly lagged bar.

17.4 ANALOGY BETWEEN THERMAL AND ELECTRICAL CONDUCTION

The flow of heat is anlogous to the flow of electrical charge, and it is possible to think in terms of a **heat current** $\mathrm{d}Q/\mathrm{d}t$ in the same way as we think in terms of electrical current I. A heat current flows through a perfectly lagged uniform bar whenever there is a temperature difference $(\theta_2 - \theta_1)$ across it. This is equivalent to the way in which an electrical current is caused to flow by a potential difference V. Equation [17.5] can be rearranged as

$$\frac{\mathrm{d}Q}{\mathrm{d}t} = \frac{(\theta_2 - \theta_1)}{(L/kA)}$$

and for an electrical conductor of resistance R

$$I = \frac{V}{R}$$

Comparing these two equations we see that (L/kA) is equivalent to R, and therefore may be thought of as the **thermal resistance** of the bar. Electrical conductivity σ and resistivity ρ are related by $\sigma = 1/\rho$ (section 36.1), and therefore (from equation [36.2]) electrical resistance R is given by $R = L/(\sigma A)$. Since thermal resistance is equal to $L/(kA)$, it follows that k is equivalent to σ. Table 17.2 summarizes these results. Note also that the temperature gradient is equivalent to potential gradient.

Table 17.2
Electric quantities and their thermal analogues

Electrical quantity		*Analogous thermal quantity*	
Electric current	I	Heat current	$\frac{\mathrm{d}Q}{\mathrm{d}t}$
Potential difference	V	Temperature difference	$\theta_2 - \theta_1$
Electrical resistance	$R = \frac{L}{\sigma A}$	Thermal resistance	$\frac{L}{kA}$
Electrical conductivity	σ	Thermal conductivity	k
Potential gradient	$\frac{\mathrm{d}V}{\mathrm{d}x}$	Temperature gradient	$\frac{\theta_2 - \theta_1}{L}$

The analogy is made use of in Example 17.1

EXAMPLE 17.1

Two perfectly lagged metal bars, X and Y, are arranged (a) in series, (b) in parallel. When the bars are in series the 'hot' end of X is maintained at 90 °C and the 'cold' end of Y is mantained at 30 °C. When the bars are in parallel the 'hot' end of each is maintained at 90 °C and the 'cold' end of each is maintained at 30 °C. Calculate the ratio of the total rate of flow of heat in the parallel arrangement to that in the series arrangement. The length of each bar is L and the cross-sectional area of each is A. The thermal conductivity of X is $400\ \mathrm{W\,m^{-1}\,K^{-1}}$ and that of Y is $200\ \mathrm{W\,m^{-1}\,K^{-1}}$.

Solution

The bars are perfectly lagged, and therefore $\mathrm{d}Q/\mathrm{d}t = kA\,(\theta_2 - \theta_1)/L$ applies.

(a) When the bars are in series the rate of flow of heat is the same through each and therefore if the temperature at the juncton of the bars is θ,

$$400A\,\frac{90-\theta}{L} = 200A\,\frac{\theta-30}{L}$$

i.e.

$$\theta = 70\,^\circ\mathrm{C}$$

$$\therefore \quad \text{Rate of flow of heat} = 400A\,\frac{90-70}{L} \text{ or } 200A\,\frac{70-30}{L}$$

$$= \frac{8000A}{L}$$

(b) When the bars are in parallel the rates of flow of heat are given by

$$\left(\frac{\mathrm{d}Q}{\mathrm{d}t}\right)_\mathrm{X} = 400A\,\frac{90-30}{L} = \frac{24\,000A}{L}$$

and

$$\left(\frac{\mathrm{d}Q}{\mathrm{d}t}\right)_\mathrm{Y} = 200A\,\frac{90-30}{L} = \frac{12\,000A}{L}$$

$$\therefore \quad \text{Total rate of flow of heat} = \frac{24\,000A}{L} + \frac{12\,000A}{L} = \frac{36\,000A}{L}$$

The ratio of the rates of heat flow is therefore

$$\frac{36\,000A/L}{8000A/L} = 4.5$$

Alternative Method

The problem can be solved by making use of the analogy with electrical conduction. Suppose that R_X and R_Y are the thermal resistances of X and Y respectively. $R_\mathrm{X} = L/400A$ and $R_\mathrm{Y} = L/200A$

$$\therefore \quad R_\mathrm{Y} = 2R_\mathrm{X}$$

(a) When the bars are in series the total thermal resistance R is given by $R = R_\mathrm{X} + R_\mathrm{Y} = 3R_\mathrm{X}$. Therefore, by analogy with $V = IR$

$$(90-30) = \frac{\mathrm{d}Q}{\mathrm{d}t}(3R_\mathrm{X})$$

i.e.

$$\frac{\mathrm{d}Q}{\mathrm{d}t} = \frac{60}{3R_\mathrm{X}}$$

(b) When the bars are in parallel the total thermal resistance R is given by

$$\frac{1}{R} = \frac{1}{R_\mathrm{X}} + \frac{1}{R_\mathrm{Y}} = \frac{1}{R_\mathrm{X}} + \frac{1}{2R_\mathrm{X}}$$

i.e.

$$R = \frac{2R_\mathrm{X}}{3}$$

Therefore, by analogy with $V = IR$

$$90 - 30 = \frac{\mathrm{d}Q}{\mathrm{d}t}\left(\frac{2R_X}{3}\right)$$

i.e. $$\frac{\mathrm{d}Q}{\mathrm{d}t} = \frac{60 \times 3}{2R_X}$$

The ratio of the rates of flow of heat is therefore

$$\frac{60 \times 3}{2R_X} \bigg/ \frac{60}{3R_X} = 4.5$$

17.5 CONDUCTION MECHANISMS

The temperature of a substance is a manifestation of the kinetic energies of its molecules, and where there are free electrons (i.e. in metals), of the kinetic energies of the electrons. Bearing this in mind, we can account for the way in which heat is conducted.

Gases

Collisions between gas molecules, and the resulting energy transfer, tend to redistribute the energy in such a way that all regions eventually contain (on average) equally energetic molecules. There is, therefore, a flow of heat from regions at high temperature to those at low temperature. (The process may be compared with the way in which two initially separate gases eventually each occupy the whole of the space available to them.)

Non-metallic Solids and Liquids

In solids and liquids the molecules are, in comparison with those in gases, locked in postion in the lattice. Conduction occurs as a result of the energetic movements of the molecules in the high-temperature regions being transmitted, by way of collisions with intervening molecules, to the cooler regions.

Metals

In metals there are electrons which are free to move about the whole of the lattice, in which sense they behave like the molecules in a gas. They are therefore able to transfer energy in a similar manner to the molecules of a gas. The electrons, however, move about a thousand times faster than gas molecules, and this accounts for the high thermal conductivities of metals. Though the molecules also play a part in the conduction process, the role of the electrons greatly predominates.

The Wiedemann–Franz law expresses the observation that the ratio k/σ is nearly the same for all metals. This is a reflection of thermal conduction in metals being largely due to the movement of the same free electrons as those which are responsible for electrical conduction in metals.

17.6 EXPERIMENTAL DETERMINATION OF THERMAL CONDUCTIVITY

The methods which are about to be described make use of equation [17.5]. This equation is valid only if steady state conditions hold and the lines of heat flow through the specimen are parallel. In addition, it is necessary that both the rate of flow of heat and the temperature gradient are large enough to be measured with a reasonable degree of accuracy. For a good conductor the conditions are satisfied when the specimen is in the form of a long thin bar. (A typical bar is 20 cm long and has a diameter of 4 cm.) Specimens of poorly conducting materials should be in the form of a thin disc. (A typical disc is 2 mm thick and has a diameter of 10 cm.)

Good Conductor (Searle's Bar)

The apparatus is shown in Fig. 17.5. The heater is switched on and water is passed through the copper coil at a constant rate. If the bar is assumed to be perfectly lagged, then at steady state (i.e. when all four thermometers give steady readings) the rate of flow of heat between X and Y is given (by equation [17.5]) as

$$\frac{dQ}{dt} = kA\frac{\theta_2 - \theta_1}{x}$$

where k is the thermal conductivity of the material of the bar. Since the bar is assumed to be perfectly lagged and none of the heat is being used to increase its temperature (steady state), all of the heat which flows along the bar is being used to increase the temperature of the water. Therefore

$$\frac{dQ}{dt} = mc(\theta_4 - \theta_3)$$

where m is the mass of water flowing per unit time and c is the specific heat capacity of water. Equating the right-hand sides of these equations gives

$$kA\frac{\theta_2 - \theta_1}{x} = mc(\theta_4 - \theta_3)$$

hence k.

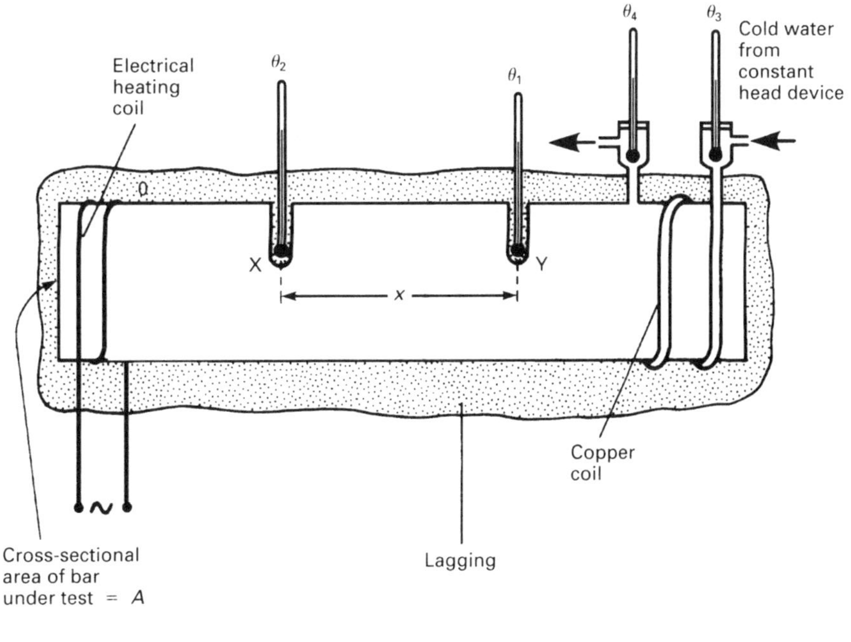

Fig. 17.5
Searle's bar

Note The holes at X and Y contain oil to ensure good thermal contact between the thermometers and the bar. If thermocouples are used there is no need to have such large holes and there is less disturbance of the ideal flow pattern.

Poor Conductor (Lees' Disc)

The apparatus is shown in Fig. 17.6(a). The sample (e.g. cardboard) is in the form of a thin disc and is sandwiched between the thick base X of a steam chest and a thick brass slab Y. The arrangement is suspended on three strings which are attached to Y.

Steam is passed through the chest and the apparatus is left to reach steady state. The sample is thin and therefore, to a good approximation, no heat is lost from its sides. It follows that at steady state: (i) equation [17.5] holds, and (ii) the rate at which heat is flowing through the sample is equal to the rate at which Y is losing heat to the surroundings. If the latter is dQ/dt, then

$$\frac{dQ}{dt} = kA\frac{\theta_2 - \theta_1}{x}$$

where k is the thermal conductivty of the sample and A is the area of one of its faces.

At this stage the sample is removed so that Y comes into direct contact with X and is heated by it. When the temperature of Y has risen by about 10 °C, X is removed and the sample is put back on top of Y (Fig. 17.6(b)). Since X is no longer present, Y cools and its temperature θ is recorded at one-minute intervals until it has dropped to about 10 °C below its steady state temperature θ_1. A graph of temperature against time is plotted (Fig. 17.6(c)). If m and c are respectively the

Fig. 17.6
Lee's disc: (a) full apparatus, (b) disc cooling, (c) cooling curve

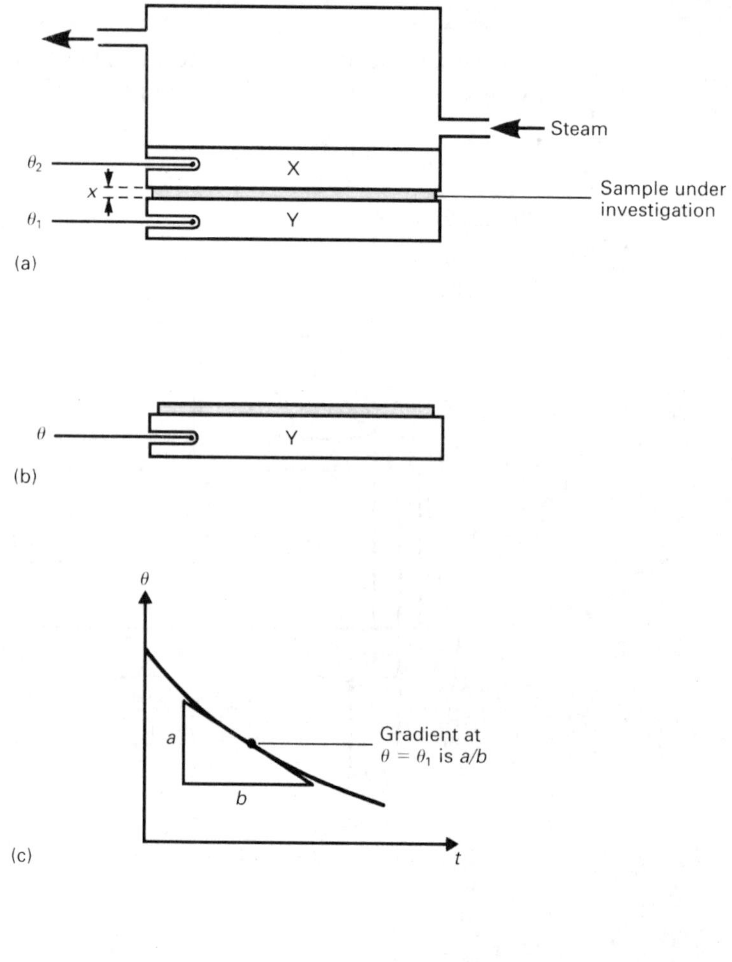

mass and specific heat capacity of Y, then the rate at which it is losing heat to the surroundings when its temperature is θ_1 is given by $mc(a/b)$, where a/b is the gradient of the graph (i.e. the rate of fall of temperature at θ_1). The conditions under which Y is losing heat are the same as those at steady state, and therefore

$$kA\frac{(\theta_2 - \theta_1)}{x} = mc\frac{a}{b}$$

from which k can be determined.

Notes (i) The upper and lower surfaces of the sample should be smeared with petroleum jelly (Vaseline) to give good thermal contact with X and Y.

(ii) The thermometers actually register the temperatures of X and Y, but since these are good conductors, the temperature gradients across them are small and therefore $(\theta_2 - \theta_1)$ is, to a good approximation, the temperature difference across the sample.

17.7 THERMAL RADIATION

We shall describe **thermal radiation** as being electromagnetic radiation emitted by a body solely on account of its temperature. The radiation spans a continuous range of wavelengths and the distribution of energy amongst these wavelengths depends on the temperature of the emitter. At temperatures below about 1000 °C the energy is associated almost entirely with infrared wavelengths; at higher temperatures visible and ultraviolet wavelengths are also involved. (These aspects are discussed more fully in section 17.10). Thermal radiation has all the general properties of electromagnetic waves. It can be reflected; its speed in a vacuum is $3 \times 10^8 \text{ m s}^{-1}$; it cannot be deflected by electric and magnetic fields; the intensity of the radiation produced by a point source falls off as the inverse square of the distance from the source; etc.

When thermal radiation is incident on a body some of the radiation may be reflected, some transmitted, and some may be absorbed and produce a heating effect. A substance which transmits the thermal radiation incident on it is said to be **diathermanous**, one which absorbs the radiation is said to be **adiathermanous**. (Equivalent respectively to substances which are transparent and substances whch are opaque to visible light.) The absorption of electromagnetic radiation of any wavelength may produce a heating effect. Thus, though X-radiation, for example, is not normally thought of as thermal radiation, heat is produced when X-rays are absorbed.

17.8 PRÉVOST'S THEORY OF EXCHANGES

According to this theory a body emits radiation at a rate which is determined only by the nature of its surface and its temperature, and absorbs radiation at a rate which is determined by the nature of its surface and the temperature of its surroundings.

Suppose that a body is suspended by a non-conductng thread inside an evacuated enclosure whose walls are maintained at a constant temperature T. Since the enclosure is evacuated, there is no possibility of conduction and convection and events are controlled only by radiative processes, i.e. Prévosts's theory applies. If the temperature of the body is greater than that of the surroundings, the body emits radiation at a greater rate than it absorbs it and its temperature falls, eventually becoming equal to T. Conversely, if the initial temperature of the body is less than that of the enclosure, the temperature of the body increases until it becomes equal to T. It is important to note that emission and absorption do not cease at this stage; instead there is a dynamic equilibrium in which the rate of emission is equal to the rate of absorption.

It follows that **if the surface of a body is such that the body is a good absorber of radiation, it must be an equally good emitter**, otherwise its temperature would rise above that of its surroundings. It also follows that a good emitter is a good absorber. These conclusions are confirmed by simple experiments (e.g. Leslie's cube). In particular, **matt black surfaces are the best absorbers and the best emitters of radiation; highly polished silver surfaces are both poor emitters and poor absorbers**.

17.9 THE BLACK BODY

A black body is a body which absorbs all the radiation which is incident on it.

The concept is an idealized one, but it can be very nearly realized in practice – Fig. 17.7 illustrates how. The inner wall of the enclosure is matt black so that most of any radiation which enters through the small hole is absorbed on reaching the wall. The small amount of radiation which is reflected has very little chance of escaping through the hole before it too is absorbed in a subsequent encounter with the wall.

Fig. 17.7
Approximate realization of a black body

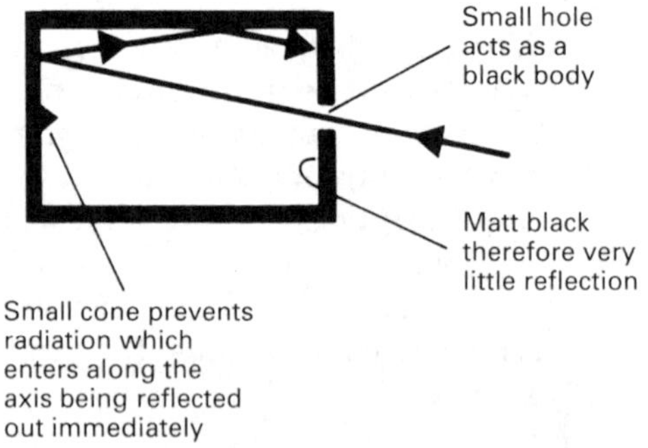
Small hole acts as a black body
Matt black therefore very little reflection
Small cone prevents radiation which enters along the axis being reflected out immediately

A black body radiator (or cavity radiator) is one which emits radiation which is characteristic of its temperature and, in particular, which does not depend on the nature of its surfaces.

A black body radiator can be made by surrounding the enclosure of Fig. 17.7 with a heating coil. The radiation which is emitted by any section of the wall is involved in many reflections before it eventually emerges from the hole. Any section which is a poor emitter absorbs very little of the radiation which is incident on it, and those sections which are good emitters absorb most of the radiation incident on them. This has the effect of mixing the radiations before they emerge, and of making the temperature the same at all points on the inner surface of the enclosure.

17.10 ENERGY DISTRIBUTION IN THE SPECTRUM OF A BLACK BODY

Fig. 17.8 illustrates the way in which the energy radiated by a black body is distributed amongst the various wavelengths. E_λ is such that $E_\lambda \delta\lambda$ represents the energy radiated per unit time per unit surface area of the black body in the wavelength interval λ to $\lambda + \delta\lambda$. It follows that **the area under any particular curve is the total energy radiated per unit time per unit surface area at the corresponding temperature**.

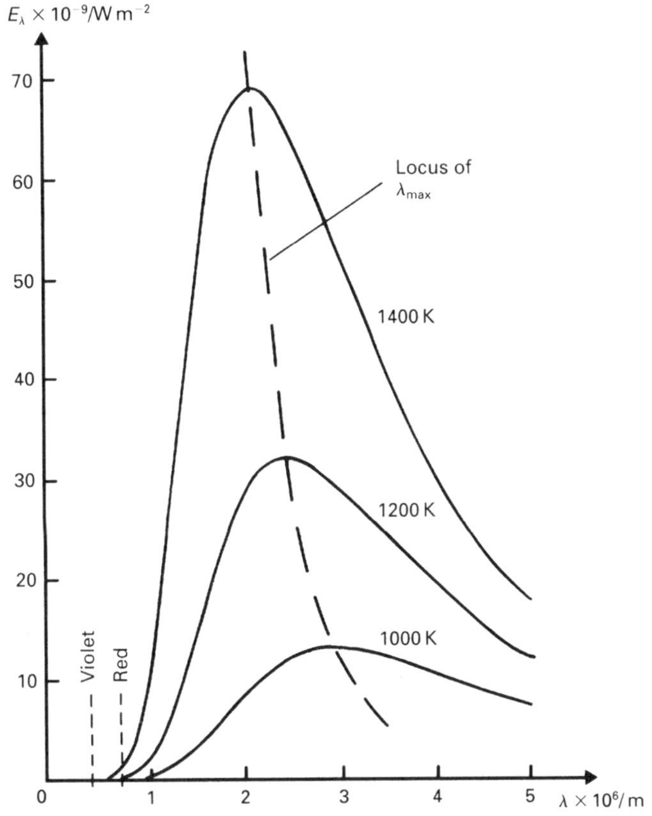

Fig. 17.8
Energy distribution of a black body

The curves embody two important laws.

Wien's Displacement Law

The wavelength λ_{max} at which the maximum amount of energy is radiated decreases with temperature and is such that

$$\lambda_{max} T = \text{a constant} \qquad [17.6]$$

where T is the temperature of the black body in kelvins. Equation [17.6] is known as **Wien's displacement law**. The value of the constant is found by experiment to be 2.9×10^{-3} m K.

The curves illustrate the well known observation that the colour of a body whch is hot enough to be emitting visible light depends on its temperature. At about 1200 K the visible wavelengths which are emitted lie predominately at the red end of the spectrum and a body at this temperature is said to be red-hot. At higher temperatures the proportions of the other spectral colours increase so that increasing temperatures cause the overall colour to change from red through yellow to white. The intensity distribution of the wavelengths emitted by the Sun is the same as that of a black body at about 6000 K, i.e. the temperature of the Sun's surface is about 6000 K. Some stars are much hotter than the Sun and appear blue.

Stefan's Law

The total energy radiated per unit time per unit surface area of a black body is proportional to the fourth power of the temperature of the body expressed in kelvins.

Thus

$$E = \sigma T^4 \qquad [17.7]$$

where

σ = a constant of proportionality known as **Stefan's constant**. Its value is $5.67 \times 10^{-8}\ \mathrm{W\,m^{-2}\,K^{-4}}$.

Note that the value of E at any temperature T is equal to the area under the corresponding curve, i.e. $E = \int_0^\infty E_\lambda \mathrm{d}\lambda$.

If a black body whose temperature is T is in an enclosure at a temperature T_0, the rate at which unit surface area of the black body is receiving radiation from the enclosure is $\sigma T_0{}^4$. The net rate of loss of energy by the black body is therefore given by E_{net} where

$$E_{\text{net}} = \sigma(T^4 - T_0{}^4) \qquad [17.8]$$

In the case of a non-black body equations [17.7] and [17.8] are replaced by

$$E = \varepsilon\sigma T^4$$

and

$$E_{\text{net}} = \varepsilon\sigma(T^4 - T_0{}^4)$$

where ε is called the **total emissivity** of the body. Its value depends on the nature of the surface of the body and lies between 0 and 1.

EXAMPLE 17.2

A 100 W electric light bulb has a filament which is 0.60 m long and has a diameter of 8.0×10^{-5} m. Estimate the working temperature of the filament if its total emissivity is 0.70. (Stefan's constant $= 5.7 \times 10^{-8}\ \mathrm{W\,m^{-2}\,K^{-4}}$.)

Solution

The surface area of the filament is that of a cylinder of diameter 8.0×10^{-5} m and length 0.60 m and is therefore $\pi \times 8.0 \times 10^{-5} \times 0.60 = 1.51 \times 10^{-4}\ \mathrm{m^2}$.

The bulb is rated at 100 W and therefore E, the energy radiated per unit time per unit surface area of the filament, is given by

$$E = \frac{100}{1.51 \times 10^{-4}} = 6.62 \times 10^5\ \mathrm{W\,m^{-2}}$$

But

$$E = \varepsilon\sigma T^4$$

$$\therefore \quad 6.62 \times 10^5 = 0.70 \times 5.7 \times 10^{-8} \times T^4$$

i.e. $$T^4 = 16.6 \times 10^{12}$$

$$\therefore \quad T = 2018\,\text{K} \approx 2.0 \times 10^3\,\text{K}$$

17.11 THE THERMOPILE

The essential features of a simple thermopile are shown in Fig. 17.9.

Fig. 17.9
Thermopile

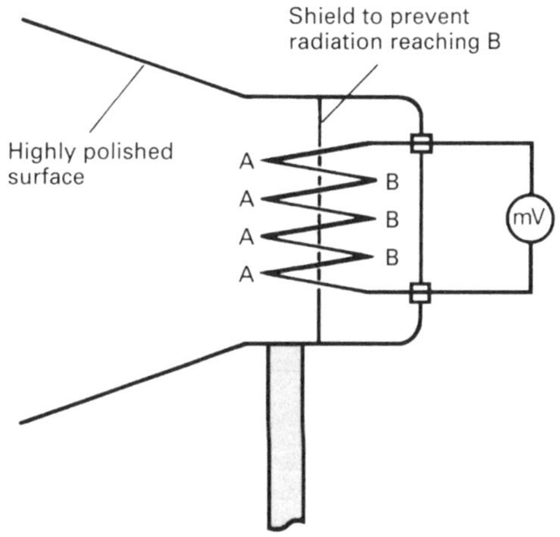

The device detects the presence of thermal radiation and consists of a number of thermocouples connected in series. (Thermocouples are described in section 13.4.) One set of thermocouple junctions (A) is exposed to the radiation and is heated by it; the other set (B) is shielded from the radiation. A highly polished metal cone concentrates the radiation on the exposed junctions; these junctions are coated with lamp-black to enhance the efficiency with which the radiation is absorbed.

More sensitive forms of the instrument have large numbers of junctions and often employ antimony and bismuth as the thermocouple materials.

17.12 CONVECTION

If a beaker containing a liqud is heated from underneath, the liquid at the bottom becomes warmer than that above it. If, like most liquids, it expands on heating, it is now less dense than the liquid above it. It therefore rises to the top, and cooler liquid from above moves downwards to take its place. On reaching the bottom this liquid also becomes heated and so it too moves upwards, and the process continues. The circulating current of liquid established in this way is called a convection current. The process itself is called convection, and though it can occur in both liquids and gases, it obviously cannot occur in solids. We can summarize by saying that convection is the process in which heat is transferred through a fluid by movement of the fluid itself.

Most fluids expand on heating, and are therefore heated from below but cooled from above – refrigerators have the freezing compartment at the top! The best known example of a fluid that contracts on heating is that of water between 0 °C and 4 °C – the so-called **anomalous expansion of water**. If the air temperature above a pond is below 4 °C, convection occurs until the temperature of all the water in the pond has fallen to 4 °C. Convection then ceases because any further cooling of the water in contact with the air causes it to expand and remain at the surface.

17.13 U-VALUES

U-values provide architects and building engineers with a simple means of estimating heat losses from buildings. They take account not only of heat lost by conduction, but of any lost by convection and/or radiation.

The U-value of a structure (e.g. a cavity wall or a window) is defined as the heat transferred per unit time through unit area of the structure when there is unit temperature difference across it.

Thus

$$\text{Rate of transfer of heat} = UA\,\Delta T \qquad [17.9]$$

where

U = U-value ($\mathrm{W\,m^{-2}\,K^{-1}}$)

A = area of structure ($\mathrm{m^2}$)

ΔT = temperature difference across structure (K).

The U-values of the various types of wall, window, roof, etc. that are commonly used in buildings are based on data obtained by experiment and are listed in tables. Some typical values are given in Table 17.3. (The table used by an architect would be much more detailed. The U-value of a cavity wall, for example, depends on the type and thickness of brick, the width of the cavity, the wind conditions on the outside of the wall, etc.)

Table 17.3
Typical U-values of some structures

Structure	*U-value*/$\mathrm{W\,m^{-2}\,K^{-1}}$
Single brick wall	3.3
Brick wall with air cavity	1.8
Brick wall with foam-filled cavity	0.6
Single window	5.5
Double-glazed window	1.9
Tiled roof	2.0
Tiled roof with insulation	0.5

The reader may be wondering why architects use U-values rather than coefficients of thermal conductivity when calculating heat losses from buildings. In order to answer we shall calculate: (a) on a U-value basis, and (b) on a thermal conductivity basis, the rate at which heat is lost from a room through a window (single-glazed) of area 2.0 $\mathrm{m^2}$ in which the glass is 10 mm thick, and where the air temperatures inside and outside the room are 20 °C and 0 °C respectively. We shall take the U-value of the window to be 5.5 $\mathrm{W\,m^{-2}\,K^{-1}}$ and the coeffcient of thermal conductivity of glass to be 1.1 $\mathrm{W\,m^{-1}\,K^{-1}}$.

(a) Rate of loss of heat $= UA\,\Delta T$ (equation [17.9])

$$= 5.5 \times 2.0 \times 20$$

$$= 2.2 \times 10^2\,\text{W}$$

(b) Rate of loss of heat $= kA\dfrac{(\theta_2 - \theta_1)}{L}$ (equation [17.5])

$$= \frac{1.1 \times 2.0 \times 20}{10 \times 10^{-3}}$$

$$= 4.4 \times 10^3\,\text{W}$$

The thermal conductivity calculation has given a rate of loss of heat which is twenty times greater than the U-value calculation! The U-value estimate is reasonable; the thermal conductivity estimate (4.4 kW) is ridiculously high. The error has arisen because the inner surface of the glass is at a lower temperature than the air inside the room, and the outer surface is at a higher temperature than the air outside. The U-value calculation takes account of this; the thermal conductivity calculation does not. It is as if there is an insulating layer of air a few millimetres in thickness in contact with each surface of the glass. The window should therefore be regarded not as just a single piece of glass, but as a composite structure – a sheet of glass sandwiched between two layers of air. In the U-value calculation ΔT represents the temperature difference across the structure, and this is 20 °C. In the thermal conductivity calculation $(\theta_2 - \theta_1)$ represents the temperature difference across the glass, and we should not have taken its value to be 20 °C.

Architects base their calculations on U-values rather than on coefficients of thermal conductvity because they are concerned with the air temperatures inside and outside a room, not with the temperatures on the surfaces of a piece of glass, or of a wall, etc.

The reader may now be wondering why the surface of a window pane is not at the same temperature as the air a few millimetres away. The transfer of heat from the interior of a room to a window is due primarily to convection, as is that from the outer surface to the air outside. Convection causes air to move past the window, but because air is a viscous fluid, the air within a few millimetres of the window pane moves much more slowly than that which is further away. Indeed, the air in immediate contact with the glass is stationary. Heat transfer through this region is therefore governed more by conduction than it is by convection (Fig 17.10). Since air is a poor conductor of heat, there is a relatively large temperature change across the layer, and the surface of the glass is at a different temperature from that of the bulk of the air around it. U-values are based on data from experiments that take account of these insulating layers of air.

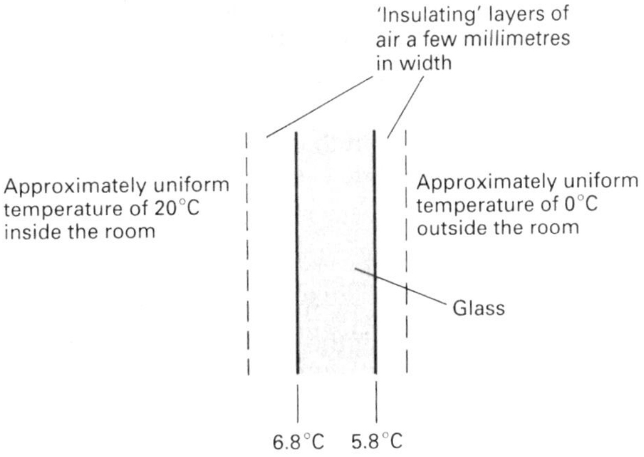

Fig. 17.10
Temperature change across a window

It is possible to estimate the temperatures on the surfaces of the window pane in our example (see later in this section). Such an estimate gives 6.8 °C for the inner surface and 5.8 °C for the outer surface. There is therefore a temperature difference of 1 °C across the glass. The reader should not be surprised by this value – it is one twentieth of the value we used in our original calculation, and that gave an estimate of the heat flow rate which was twenty times too high!

Thermal Resistance Coefficient

The thermal resistance coefficient of a material is the thermal resistance of unit area of the material and is defined by

$$X = \frac{L}{k} \qquad [17.10]$$

where

X = thermal resistance coefficient ($m^2\,K\,W^{-1}$)

L = thickness of material (m)

k = coefficient of thermal conductivity ($W\,m^{-1}\,K^{-1}$).

The thermal resistance coefficient of a structure which consists of a number of different components in series is the sum of the thermal resistance coefficients of the individual components. In Fig. 17.11, for example, the thermal resistance coefficient, X_S, of the structure (for heat transfer between the outer surfaces of A and C is given by

$$X_S = X_A + X_B + X_C$$

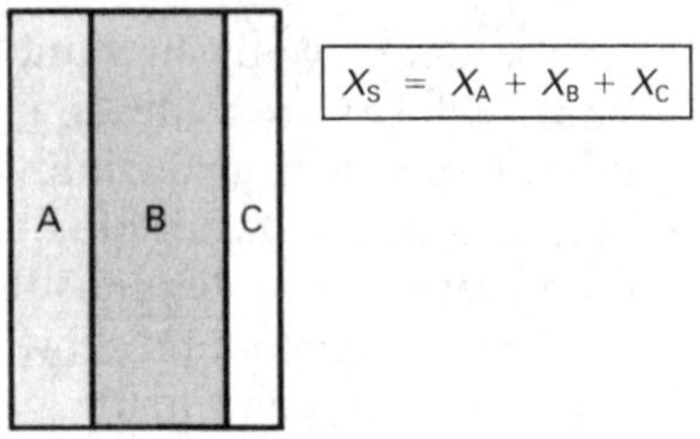

Fig. 17.11
To calculate thermal resistance coefficient of a composite structure

where X_A, X_B and X_C are respectively the thermal resistance coefficients of A, B and C. The thermal resistance coefficient, X_W, of the window in Fig. 17.10 is given by

$$X_W = X_i + X_g + X_o$$

where X_g is the thermal resistance coefficient of the glass and is calculated on the basis of equation [17.10], and X_i and X_o are respectively the effective thermal resistance coefficients of the 'layers' of air on the inner and outer surfaces of the glass. Equation [17.10] cannot be used to calculate these, but it is found by experiment that $X_i = 0.120\,m^2\,K\,W^{-1}$ and $X_o = 0.053\,m^2\,K\,W^{-1}$. The U-value of the window is the reciprocal of its thermal resistance coefficient, i.e.

$$U = \frac{1}{X_W}$$

We are now in a position to show how we estimated the temperatures on the surfaces of the glass in the window. Equation [17.5] can be rewritten as

$$\frac{dQ}{dt} = \frac{A}{X}(\theta_2 - \theta_1)$$

The rate of flow of heat ($\mathrm{d}Q/\mathrm{d}t$) through the window has already been found to be 2.2×10^2 W, and its area (A) is $2.0\ \mathrm{m}^2$. If the temperature on the inner surface of the glass is taken to be θ, then considering the heat flow across the inner layer of air gives

$$2.2 \times 10^2 = \frac{2.0}{0.120}(20 - \theta) \qquad \text{i.e.} \quad \theta = 6.8\ {}^\circ\mathrm{C}$$

We leave it as an exercise for the reader to show that the temperature on the outer surface of the glass is approximately 5.8 °C.

CONSOLIDATION

A substance is in **steady state** when the temperatures at all points in it are steady.

$$\frac{\mathrm{d}Q}{\mathrm{d}t} = kA\frac{\mathrm{d}\theta}{\mathrm{d}x} \quad \text{at steady state}$$

$$\frac{\mathrm{d}Q}{\mathrm{d}t} = \frac{kA(\theta_2 - \theta_1)}{L} \quad \text{at steady state when there are no heat losses from the sides}$$

$$\frac{\mathrm{d}Q}{\mathrm{d}t} = \frac{\theta_2 - \theta_1}{(L/kA)} \quad \text{is analogous to} \quad I = \frac{V}{R}$$

i.e.

$$\text{Rate of flow of heat} = \frac{\text{Temperature difference}}{\text{Thermal resistance}}$$

is analogous to

$$\text{Rate of flow of charge} = \frac{\text{Potential difference}}{\text{Electrical resistance}}$$

QUESTIONS ON SECTION C

THERMOMETRY (Chapter 13)

C1 A bath of oil is maintained at a steady temperature of about 180 °C, which is measured both with a platinum resistance thermometer and a mercury-in-glass thermometer. Explain why you would expect the temperatures indicated by the two thermometers to be different. At what temperatures would the two thermometers show the same value? [J]

C2 Explain why two thermometers, using different thermometric properties and calibrated at two fixed points, would not necessarily show the same termperature except at the fixed points.

Why is the constant volume gas thermometer chosen as a standard?

What type of thermometer is recommended to measure accurately a temperature of **(a)** about 15 K, and **(b)** 2000 K? (No details required.) [W]

C3 **(a)** Explain how a temperature scale is defined.
(b) Discuss the relative merits of **(i)** a mercury-in-glass thermometer, **(ii)** a platinum resistance thermometer, **(iii)** a thermocouple, for measuring the temperature of an oven which is maintained at about 300°C. [J]

C4 The resistance of the element in a platinum resistance thermometer is 6.750 Ω at the triple point of water and 7.166 Ω at room temperature. What is the temperature of the room on the scale of the resistance thermometer? The triple point of water is 273.16 K. State one assumption you have made. [L]

C5 Describe the structure of a simple constant volume gas thermometer. Discuss how it would be used to establish a scale of temperature.

Explain why the same temperature measured on two different scales need not have the same value.

Discuss the circumstances in which: **(a)** a gas thermometer, and **(b)** a thermocouple might be used.

Why is it generally not sensible to use a thermoelectric EMF as the physical property used to *define* a scale of temperature? [L]

C6 The graph below shows the variation in resistance of a piece of platinum wire with the temperature being measured on the ideal gas scale.

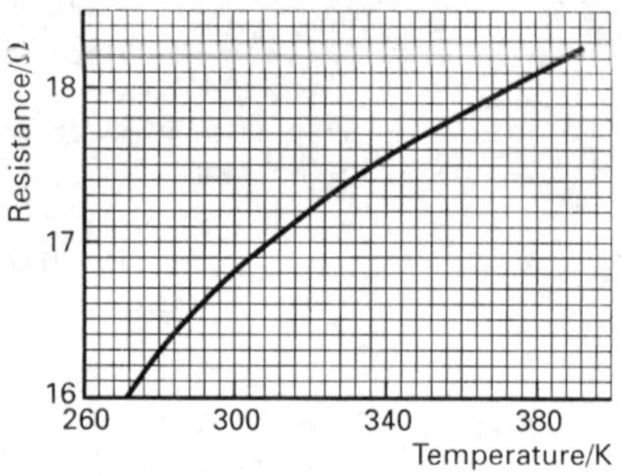

What is the *Celsius* temperature corresponding to a resistance of 17 Ω on
(a) the ideal gas scale, and
(b) the platinum resistance scale? [L]

C7

	Resistance of resistance thermometer	*Pressure recorded by constant volume gas thermometer*
Steam point 100 °C	75.000 Ω	1.10×10^7 N m^{-2}
Ice point 0 °C	63.000 Ω	8.00×10^6 N m^{-2}
Room temperature	64.992 Ω	8.51×10^6 N m^{-2}

Using the above data, which refer to the observations of a particular room temperature using two types of thermometer, calculate the room temperature on the scale of the resistance thermometer and on the scale of the constant volume gas thermometer.

Why do these values differ slightly? [L]

C8 The value of the property X of a certain substance is given by

$$X_t = X_0 + 0.50t + (2.0 \times 10^{-4})t^2,$$

where t is the temperature in degrees Celsius measured on a gas thermometer scale. What would be the Celsius temperature defined by the property X which corresponds to a temperature of 50 °C on this gas thermometer scale? [L]

C9 **(a)** What is meant by a thermometric property? What qualities make a particular property suitable for use in a practical thermometer?

A Celsius temperature scale may be defined in terms of a thermometric property X by the following equation:

$$\theta = \frac{X - X_0}{X_{100} - X_0} \times 100\,°\mathrm{C} \qquad (1)$$

where X_0 is the value of the property at the ice point, X_{100} at the steam point, and X at some intermediate temperature. If X is plotted against θ a straight line always results no matter what thermometric property is chosen. Explain this.

(b) On the graph, line A shows how X varies with θ (following equation (1) above), line B shows how a second thermometric property Q varies with θ, the temperature measured on the X scale.

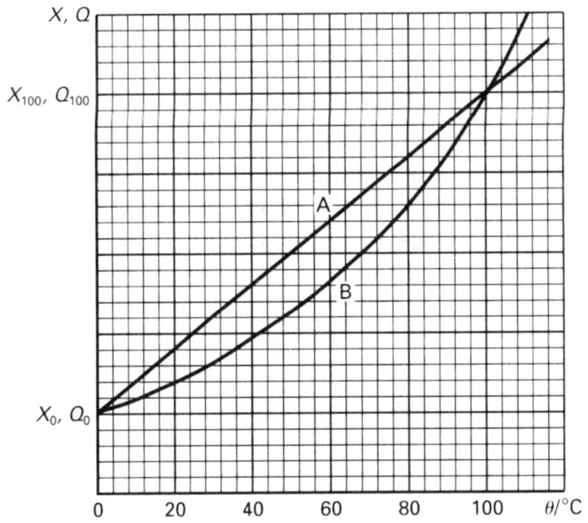

(i) Describe, in principle, how you would conduct an experiment to obtain line B.

(ii) If $\theta = 40\,°\mathrm{C}$ recorded by an X-scale thermometer, what temperature would be recorded by a Q-scale thermometer?

(iii) At what two temperatures will the X and Q scales coincide?

(c) The ideal gas scale of temperature is one based on the properties of an ideal gas. What is the particular virtue of this scale? Describe very briefly how readings on such a scale can be obtained using a thermometer containing a real gas. [L]

C10 A temperature T can be defined by $T = T_\mathrm{f}(X/X_\mathrm{f})$, where T_f is the assigned temperature of a fixed point and X and X_f are the values of a thermometric property of a substance at T and T_f respectively. On the ideal-gas scale, the fixed point is the triple point of water and $T_\mathrm{f} = 273.16\,\mathrm{K}$.

(a) List *four* thermometric properties which are used in thermometry. Explain why certain thermometric properties of a gas are taken as standard.

(b) Explain what is meant by a fixed point and by the triple point of water.

(c) Sketch and label the simple form of constant-volume gas thermometer found in school laboratories, and describe how it is used to determine the boiling point of a liquid on the ideal-gas scale.

(d) For a thermometer which is not based on the properties of gases, explain how you would calibrate it in terms of the ideal-gas scale.

(e) Compare the advantages and disadvantages of the constant-volume gas thermometer with those of any *two* other types of thermometers.

(f) The pressures recorded in a certain constant-volume gas thermometer at the triple point of water and at the boiling point of a liquid were 600 mm of Hg and 800 mm of Hg respectively. What is the apparent temperature of the boiling point? However, it was found that the volume of the thermometer increased by 1% between the two temperatures. Obtain a more accurate value of the boiling point. [W]

CALORIMETRY (Chapter 13)

C11 Describe how you might measure, by an electrical method, the specific heat capacity of copper provided in the form of a cylinder 4 cm long and 1 cm in diameter.

Describe the procedure you would use to make an allowance for heat loss and explain how you would derive the specific heat capacity from your measurements.

When a metal cylinder of mass 2.0×10^{-2} kg and specific heat capacity $500\,J\,kg^{-1}\,K^{-1}$ is heated by an electrical heater working at constant power, the initial rate of rise of temperature is $3.0\,K\,min^{-1}$. After a time the heater is switched off and the initial rate of fall of temperature is $0.3\,K\,min^{-1}$. What is the rate at which the cylinder gains heat energy immediately before the heater is switched off? [J]

C12 Define *specific heat capacity* of a substance. Describe in detail a method for the determination of this quantity for a liquid and explain how the result is calculated from the observations.

What are the particular advantages of a continuous flow method? A copper block has a conical hole bored in it into which a conical copper plug just fits. The mass of the block is 376 g and that of the plug is 18 g. The block and plug are initially at room temperature 10 °C and almost completely surrounded by a layer of insulating material. The plug is removed from the block, cooled to a temperature of −196°C and then quickly inserted into the block again. The temperature of the block falls to 3°C and then slowly rises. Calculate the value of the mean specific heat capacity of copper (in the range −196 °C to 3 °C) obtained by ignoring heat flow into the block from the surroundings.

Suggest how the experiment could be improved so as to minimize the error arising from this heat flow from the surroundings into the block.

(Assume that the mean specific heat capacity of copper in the temperature range 3 °C to 10 °C is $380\,J\,kg^{-1}\,K^{-1}$.) [L]

C13 In a constant flow calorimeter, being used for measuring the specific heat capacity of a liquid, a PD of 4.0 V was applied to the heating coil. The rate of flow of liquid was now doubled and, by adjusting the applied PD, the same inlet and outlet temperatures were obtained. Assuming heat losses to be negligible calculate the new value of the applied PD. [L]

C14 In an electrical constant flow experiment to determine the specific heat capacity of a liquid, heat is supplied to the liquid at a rate of 12 W. When the rate of flow is $0.060\,kg\,min^{-1}$ the temperature rise along the flow is 2.0 K. Use these figures to calculate a value for the specific heat capacity of the liquid.
If the true value of the specific heat capacity is $5400\,J\,kg^{-1}\,K^{-1}$, estimate the percentage of heat lost in the apparatus. Explain briefly how in practice you would reduce or make allowance for this heat loss. [L]

C15 When water was passed through a continuous-flow calorimeter the rise in temperature was from 16.0 to 20.0 °C, the mass of water flowing was 100 g in one minute, the potential difference across the heating coil was 20 V and the current was 1.5 A. Another liquid at 16.0 °C was then passed through the calorimeter and to get the same change in temperature the potential difference was changed to 13 V, the current to 1.2 A and the rate of flow to 120 g in one minute. Calculate the specific heat of the liquid if the specific heat of water is assumed to be $4.2 \times 10^3\,J\,kg^{-1}\,°C^{-1}$.

State *two* advantages of the continuous flow method of calorimetry. [J]

C16 **(a)** Describe how you would determine the specific heat capacity of a liquid by the continuous flow method.
(b) What are the chief advantages and disadvantages of this method?
(c) What special difficulties would you expect to meet in attempting to use this method for **(i)** saturated brine and **(ii)** glycerol?
(d) With a certain liquid, the inflow and outflow temperatures were maintained at 25.20 °C and 26.51 °C respectively. For a PD of 12.0 V and current 1.50 A, the rate of flow was 90 g per minute; with 16.0 V and 2.00 A, the rate of flow was 310 g per minute. Find the specific heat capacity of the liquid, and also the power lost to the surroundings. [O]

C17 In a constant flow calorimeter experiment to find the specific heat capacity of a liquid, the potential difference across the heating coil is doubled. By what factor must the rate of flow of liquid be changed if the inlet and outlet temperatures are to remain the same? Neglect heat losses in the calculation.

When this experiment is performed what purpose is served by making the mean temperature of the liquid the same on the two occasions? [AEB, '79]

C18 **(a)** When bodies are in thermal equilibrium, their temperatures are the same. Explain in energy terms the condition for two bodies to be in thermal equilibrium with one another.

(b) The temperature of a beaker of water is to be measured using a mercury-in-glass thermometer.

(i) Why is it necessary to wait before taking the reading?

(ii) Explain briefly how you might estimate the heat capacity (energy required per unit temperature rise) of a mercury-in-glass thermometer.

(iii) If the beaker contains 120 g of water at 60°C, what temperature would be recorded by the mercury-in-glass thermometer if it was initially at 18°C and had a heat capacity of $30\,\mathrm{J\,K^{-1}}$?

(Assume the specific heat capacity of water to be $4200\,\mathrm{J\,kg^{-1}\,K^{-1}}$ and ignore the heat losses to the beaker and surroundings while the temperature is being taken.)

(iv) Why, if a more accurate value of the temperature were required in this case, might you use a thermocouple?

(v) Describe briefly how you would calibrate a thermocouple and use it to measure the temperature of the water. Show how you would calculate the temperature of the water from your readings. [L]

C19 Define *specific heat capacity*, and *specific latent heat*.

1 kg of vegetables, having a specific heat capacity $2200\,\mathrm{J\,kg^{-1}\,K^{-1}}$, at a temperature 373 K, are plunged into a mixture of ice and water at 273 K. How much ice is melted?

(Specific latent heat of fusion of ice = $3.3 \times 10^5\,\mathrm{J\,kg^{-1}}$.) [S]

C20 3 kg of molten lead (melting point 600 K) is allowed to cool down until it has solidified. It is found that the temperature of the lead falls from 605 K to 600 K in 10 s, remains constant at 600 K for 300 s, and then falls to 595 K in a further 8.4 s. Assuming that the rate of loss of energy remains constant, and that the specific heat capacity of solid lead is $140\,\mathrm{J\,kg^{-1}\,K^{-1}}$, calculate:

(a) the rate of loss of energy from the lead,
(b) the specific latent heat of fusion of lead,
(c) the specific heat capacity of liquid lead. [S]

C21 0.020 kg of ice and 0.10 kg of water at 0 °C are in a container. Steam at 100 °C is passed in until all the ice is just melted. How much water is now in the container?

(Specific latent heat of steam = $2.3 \times 10^6\,\mathrm{J\,kg^{-1}}$.
Specific latent heat of ice = $3.4 \times 10^5\,\mathrm{J\,kg^{-1}}$.
Specific heat capacity of water = $4.2 \times 10^3\,\mathrm{J\,kg^{-1}\,K^{-1}}$.) [L]

C22 **(a)** In terms of the kinetic theory of matter explain why energy must be supplied to a liquid in order to vaporize it.

(b) Describe, with the aid of a labelled diagram, an electrical method for the determination of the specific latent heat of vaporization of a liquid. Explain how the result is derived from the readings taken.

(c) When a piece of ice of mass 6.00×10^{-4} kg at a temperature of 272 K is dropped into liquid nitrogen boiling at 77 K in a vacuum flask, $8.00 \times 10^{-4}\,\mathrm{m^3}$ of nitrogen, measured at 294 K and 0.75 m of mercury pressure, are produced. Calculate the mean specific heat capacity of ice between 272 K and 77 K. Assume that the specific latent heat of vaporization of the nitrogen is $2.13 \times 10^5\,\mathrm{J\,kg^{-1}}$ and that the density of nitrogen at STP is $1.25\,\mathrm{kg\,m^{-3}}$. [J]

C23 Wet clothing at a temperature of 0 °C is hung out to dry when the air temperature is 0 °C and there is a dry wind blowing. After some time, it is found that some of the water has evaporated and the remainder has frozen. What is the source of the energy required to evaporate the water? Estimate the proportion of the water originally in the clothing which remains as ice. State any assumptions you make.

(Specific latent heat of fusion of ice at 273 K = 333 kJ kg^{-1}; specific latent heat of vaporization of water at 273 K = 2500 kJ kg^{-1}.) [S]

C24 Describe with the aid of a labelled diagram a method of measuring the latent heat of vaporization of a liquid.

In a factory heating system water enters the radiators at 60 °C and leaves at 38 °C. The system is replaced by one in which steam at 100 °C is condensed in the radiators, the condensed steam leaving at 82 °C. What mass of steam will supply the same heat as 1.00 kg of hot water in the first instance?

(The latent heat of vaporization of water is 2.260×10^6 J kg^{-1} at 100 °C. The specific heat of water is 4.2×10^3 J kg^{-1} °C^{-1}.) [J]

C25 Describe how you would determine the specific latent heat of vaporization of a liquid by the continuous flow method.

What becomes of the energy used to change a liquid into a vapour at the same temperature?

A beaker containing ether at a temperature of 13 °C is placed in a large vessel in which the pressure can be reduced so that the ether boils; this results in a cooling of the remaining ether. What proportion of the ether has evaporated when the temperature of the remainder has been reduced to 0 °C? (Assume no interchange of heat between the ether and its surroundings.)

(Mean specific heat capacity of ether over the temperature range 0–13 °C = 2.4×10^3 J kg^{-1} K^{-1}.

Mean specific latent heat of vaporization of ether in temperature range 0–13 °C = 3.9×10^5 J kg^{-1}.) [S]

C26 A domestic kettle is marked 250 V, 2.3 kW and the manufacturer claims that it will heat a pint of water to boiling point in 94 s.

(a) Test this claim by calculation and state any simplifying assumptions you make.

(b) If the kettle is left switched on after it boils, how long will it take to boil away half a pint of water measured from when it first boils?

(c) Estimate the work done against an atmospheric pressure of 100 kPa when 1 cm^3 of water evaporates at 100°C, producing 1600 cm^3 of steam. Express this as a percentage of the total energy required to evaporate 1 cm^3 of water at 100°C.

(Specific heat capacity of water = 4.2×10^3 J kg^{-1} K^{-1}, specific latent heat of vaporisation of water = 2.3×10^6 J kg^{-1}, density of water = 1.0 g cm^{-3}, 1 pint = 570 cm^3.) [J, '92]

C27 The graph refers to an experiment in which an initially solid specimen of nitrogen absorbs heat at a constant rate. Nitrogen melts at 63 K, and the specific heat capacity of solid nitrogen is 1.6×10^3 J kg^{-1} K^{-1}.

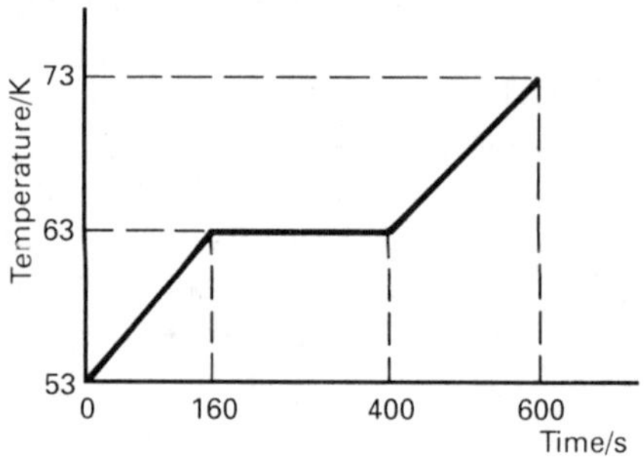

Calculate the specific latent heat of fusion of nitrogen.

Calculate the specific heat capacity of liquid nitrogen. [S]

C28 **(a)** In an espresso coffee machine, steam at 100 °C is passed into milk to heat it. Calculate

(i) the energy required to heat 150 g of milk from room temperature (20 °C) to 80 °C,

(ii) the mass of steam condensed.

(b) A student measures the temperature of the hot coffee as it cools. The results are given below:

Time/min	0	2	4	6	8
Temp/°C	78	66	56	48	41

A friend suggests that the rate of cooling is exponential.

(i) Show quantitatively whether this suggestion is valid.

(ii) Estimate the temperature of the coffee after a total of 12 min.

(Specific heat capacity of milk = 4.0 kJ kg^{-1} K^{-1}, specific heat capacity of water = 4.2 kJ kg^{-1} K^{-1}, specific latent heat of steam = 2.2 MJ kg^{-1}.) [O & C, '91]

GASES AND VAPOURS (Chapters 14 and 15)

C29 A uniform capillary tube, closed at one end, contained air trapped by a thread of mercury 85 mm long. When the tube was held horizontally the length of the air column was 50 mm; when it was held vertically with the closed end downwards, the length was 45 mm. Find the atmospheric pressure.

(Take $g = 10\,\text{m s}^{-2}$; density of mercury = $14 \times 10^3\,\text{kg m}^{-3}$.) [C(O)]

C30 A uniform, vertical glass tube, open at the lower end and sealed at the upper end, is lowered into sea-water, thus trapping the air in the tube. It is observed that when the tube is submerged to a depth of 10 m, the sea-water has entered the lower half of the tube. To what depth must the tube be lowered so that the sea-water fills three-quarters of the tube?

(The length of the glass tube is negligible compared with the depth to which it is submerged. The vapour pressure of sea-water may be neglected.) [S]

C31 A mercury barometer tube, with a scale attached, has a little air above the mercury. The top of the tube is 1.00 m above the level of the mercury in the reservoir. When the tube is vertical the height of the mercury column is 700 mm. When the tube is inclined at 60° to the vertical the reading of the mercury level on the scale is 950 mm. To what height would the mercury have risen in the vertical tube had there not been any air in it? [L]

C32 **(a)** Explain briefly what is meant by the term *ideal gas*.

(b) A volume of 0.23 m^3 contains nitrogen at a pressure of 0.50×10^5 Pa and temperature 300 K. Assuming that the gas behaves ideally, calculate the amount in mol of nitrogen present.

(c) Calculate the root mean square speed of nitrogen molecules at a temperature of 300 K

(Molar mass of nitrogen = 0.028 kg mol^{-1}, molar gas constant = 8.3 J K^{-1} mol^{-1}.) [O & C, '92]

C33 Two vessels A, B of equal volume are connected by a narrow tube of negligible internal volume. Initially, the whole system is

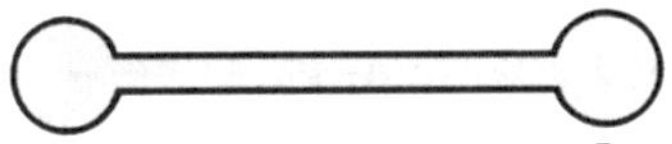

filled with 3 g of dry air at a pressure of 10^5 Pa and temperature 300 K. The temperature of the vessel B is now raised to 600 K, the temperature of A remaining 300 K. What is:
(a) the new pressure in the system,
(b) the mass of air in A and in B? [S]

C34 An industrial firm supplies compressed air cylinders of volume 0.25 m^3 filled to a pressure of 20 MPa at 17°C.

Calculate the contents of the cylinder expressed in: **(a)** moles (mol), **(b)** kilograms (kg).

(Take the gas constant R as 8.3 J K^{-1} mol^{-1} and the molar mass of air as 0.029 kg mol^{-1}.) [O, '91]

C35 A bicycle tyre has a volume of $1.2 \times 10^{-3}\,\text{m}^3$ when fully inflated. The barrel of a bicycle pump has a working volume of $9 \times 10^{-5}\,\text{m}^3$. How many strokes of this pump are needed to inflate the completely flat tyre (i.e. zero air in it) to a total pressure of 3.0×10^5 Pa, the atmospheric pressure being 1.0×10^5 Pa? (Assume the air is pumped in slowly, so that its temperature does not change.)

Explain why the barrel of the bicycle pump becomes hot when the tyre is being inflated quickly. [S]

C36 **(a)** The equation relating pressure p, volume V and thermodynamic temperature T of an ideal gas is

$$pV = nRT$$

Identify the terms n and R.

(b) Nitrogen gas under an initial pressure of 5.0×10^6 Pa at 15 °C is contained in a cylinder of volume 0.040 m^3. After a period of three years the pressure has fallen to 2.0×10^6 Pa at the same temperature because of leakage.

(Assume molar mass of nitrogen = $0.028\,\text{kg mol}^{-1}$, $R = 8.3\,\text{J mol}^{-1}\,\text{K}^{-1}$, Avogadro constant = $6.0 \times 10^{23}\,\text{mol}^{-1}$.)

Calculate:

(i) the mass of gas originally present in the cylinder,

(ii) the mass of gas which escaped from the cylinder in three years,

(iii) the average number of nitrogen molecules which escaped from the cylinder per second.

(Take one year to be equal to 3.2×10^7 s.) [O, '92]

C37 A cylinder containing 19 kg of compressed air at a pressure 9.5 times that of the atmosphere is kept in a store at 7 °C. When it is moved to a workshop where the temperature is 27 °C a safety valve on the cylinder operates, releasing some of the air. If the valve allows air to escape when its pressure exceeds 10 times that of the atmosphere, calculate the mass of air that escapes. [L]

C38 A mole of an ideal gas at 300 K is subjected to a pressure of 10^5 Pa and its volume is $0.025\,\text{m}^3$. Calculate:

(a) the molar gas constant R,

(b) the Boltzmann constant k,

(c) the average translational kinetic energy of a molecule of the gas.

($N_A = 6.0 \times 10^{23}\,\text{mole}^{-1}$.) [W, '90]

C39 A vessel of volume $1.0 \times 10^{-3}\,\text{m}^3$ contains helium gas at a pressure of 2.0×10^5 Pa when the temperature is 300 K.

(a) What is the mass of helium in the vessel?

(b) How many helium atoms are there in the vessel?

(c) Calculate the r.m.s. speed of the helium atoms.

(Relative atomic mass of helium = 4, the Avogadro constant = $6.0 \times 10^{23}\,\text{mol}^{-1}$, the molar gas constant $R = 8.3\,\text{J mol}^{-1}\,\text{K}^{-1}$.)

[W, '92]

C40 Use a simple treatment of the kinetic theory of gases, stating any assumptions made, to derive the expression $\overline{c^2} = 3p/\rho$ for the mean square speed of the molecules in terms of the density and pressure of the gas.

What would be the total kinetic energy of the atoms of 1 kg of neon gas at a pressure of 10^5 Pa and temperature 293 K, given that the density of neon under these conditions is $828\,\text{g m}^{-3}$. What would be the total kinetic energy of the atoms of 1 kg of neon gas at 300 K? Hence determine the specific heat capacity of neon at constant volume. [S]

C41 **(a)** State Avogadro's law.

(b) The pressure p of an ideal gas is given by:

$$p = \tfrac{1}{3}\,nm\langle c^2\rangle$$

where n is the number of molecules per unit volume, m is the mass of one molecule and $\langle c^2\rangle$ is the mean square speed of the gas molecules.

Use the above equation to deduce Avogadro's law. [W, '91]

C42 **(a)** State the assumptions made in the kinetic theory of gases and prove $p = \frac{1}{3}\rho\overline{c^2}$, in the usual notation. Hence derive **(i)** Boyle's law, and **(ii)** the perfect gas law, assuming that the average kinetic energy of a molecule is proportional to the absolute temperature.

(b) Consider whether the assumptions of the kinetic theory are likely to be true for real gases.

(c) At room temperature, $\sqrt{\overline{c^2}}$ of a gas molecule is typically about $10^2\,\text{m s}^{-1}$. Explain why, if a gas is released at one side of a room, it may be several minutes before it can be detected on the other side of the room.

(d) At a certain instant of time, ten molecules have the following speeds: 100, 300, 400, 400, 500, 600, 600, 600, 700, 900 m s^{-1} respectively. Calculate $\sqrt{\overline{c^2}}$. [W]

C43 **(a)** One mole of an ideal gas at pressure p and Celsius temperature θ occupies a volume V. Sketch a graph showing how the product pV varies with θ. What information can you obtain from the gradient of the graph and the intercept on the temperature axis?

(b) Some helium (molar mass of which = $0.004\,\text{kg mol}^{-1}$) is contained in a vessel of volume $8.0 \times 10^{-4}\,\text{m}^3$ at a temperature of 300 K. The pressure of the gas is 200 kPa. Calculate

(i) the mass of helium present
(ii) the internal energy (the translational kinetic energy of the gas molecules).
(Molar gas constant = $8.3\,J\,K^{-1}\,mol^{-1}$.)
[O & C, '90]

C44 A cubical container of volume $0.10\,m^3$ contains uranium hexafluoride gas at a pressure of 1.0×10^6 Pa and a temperature of 300 K.
(a) Assuming that the gas is ideal determine:
(i) the number of moles of gas present, given that the universal gas constant, $R = 8.3\,J\,K^{-1}\,mol^{-1}$,
(ii) the mass of gas present, given that its relative molecular mass is 352,
(iii) the density of the gas,
(iv) the r.m.s. speed of the molecules.
(b) A student suggests that since the molecules are so massive, the density of the gas at the bottom of the container would be significantly greater than the density at the top. Explain whether you agree or disagree. [AEB, '86]

C45 Explain what is meant by the *root mean square speed* of the molecules of a gas.

State four assumptions which are made in the simple kinetic theory of an ideal gas. Derive an expression for the pressure, p, of an ideal gas in terms of its density, ρ, and the mean square speed of its molecules.

How is temperature interpreted in terms of the theory?

The speed of sound in air is $\sqrt{\dfrac{1.4p}{\rho}}$. Show that the speed of sound in air is of the same order of magntidue as the root mean square speed of the air molecules. Explain in physical terms why you consider this to be a reasonable result. [L]

C46 The graph illustrates the distribution of molecular speeds in oxygen at room temperature.
(a) Copy this graph and show on it a second curve to show the effect of increasing the temperature of the gas on the distribution of molecular speeds. Label the second curve A.
(b) A sealed vessel has a volume of $1.5 \times 10^{-3}\,m^3$ and contains oxygen at a pressure of 1.0×10^4 Pa and a temperature of 300 K.

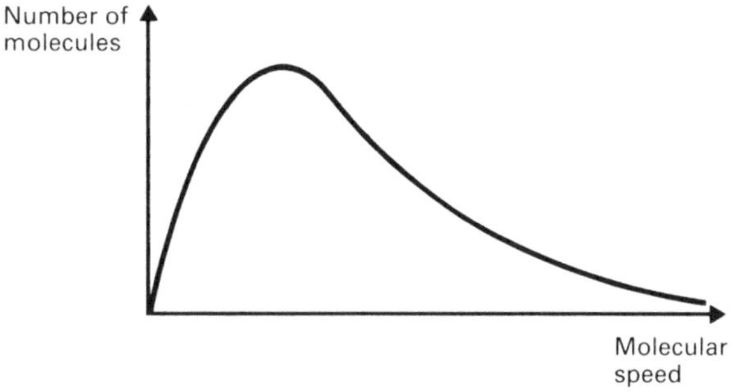

Given that the molar gas constant, $R = 8.3\,J\,mol^{-1}\,K^{-1}$, the Avogadro constant, $N_A = 6.0 \times 10^{23}\,mol^{-1}$, and the molar mass of oxygen = $32 \times 10^{-3}\,kg\,mol^{-1}$, determine:
(i) the number of moles of oxygen in the vessel,
(ii) the number of molecules in the vessel,
(iii) the root mean square speed of the molecules in the vessel. [AEB, '89]

C47 Helium gas is contained in a cylinder by a gas-tight piston which can be assumed to move without friction. The gas occupies a volume of $1.0 \times 10^{-3}\,m^3$ at a temperature of 300 K and a pressure of 1.0×10^5 Pa.
(a) Calculate
(i) the number of helium atoms in the container,
(ii) the total kinetic energy of the helium atoms.
(b) Energy is now supplied to the gas in such a way that the gas expands and the temperature remains constant at 300 K.

State and explain what changes, if any, will have occurred in the following quantities:
(i) the internal energy of the gas,
(ii) the r.m.s. speed of the helium atoms,
(iii) the density of the gas.
(The Boltzmann constant = $1.4 \times 10^{-23}\,J\,K^{-1}$.)
[J, '90]

C48 The kinetic theory of gases leads to the equation

$$p = \tfrac{1}{3}\rho\overline{c^2}$$

where p is the *pressure*, ρ is the *density* and $\overline{c^2}$ is the *mean square molecular speed*. Explain the meaning of the terms in italics and list the simplifying assumptions necessary to derive this result. Discuss how this equation is related to Boyle's Law.

Air may be taken to consist of 80% nitrogen molecules and 20% oxygen molecules of relative molecular masses 28 and 32 respectively. Calculate:

(a) the ratio of the root mean square speed of nitrogen molecules to that of oxygen molecules in air,

(b) the ratio of the partial pressures of nitrogen and oxygen molecules in air, and

(c) the ratio of the root mean square speed of nitrogen molecules in air at 10 °C to that at 100 °C. [O & C]

C49 Show that, for an ideal gas, the coefficient of pressure increase at constant volume and the coefficient of cubic expansivity at constant pressure are equal in value. [AEB, '79]

C50 **(a)** A flask is filled with water vapour at 30°C and sealed. The velocity of any particular water vapour molecule in the flask may vary randomly in two different ways. What are these two ways?

Describe, with the aid of a diagram, how the motion of one of the water vapour molecules could change during a time interval in which it has six collisions with other molecules.

(b) Explain why a small increase in pressure will do more work on a gas than on a liquid. [L, '91]

C51 **(a)** **(i)** Write down the equation which defines a temperature on the Kelvin scale in terms of the properties of an ideal gas. Explain the symbols you use.

(ii) A simple form of gas thermometer consists of a capillary tube sealed at one end and containing a thread of mercury which traps a mass of dry air. Describe how you would calibrate it on the gas scale and use it to determine the boiling point of a liquid known to be about 350 K. Explain how the temperature is calculated from the readings and state any assumptions you make.

(b) A cylinder fitted with a piston which can move without friction contains 0.050 mol of a monatomic ideal gas at a temperature of 27 °C and a pressure of 1.0×10^5 Pa. Calculate:

(i) the volume,

(ii) the internal energy of the gas.

(c) The temperature of the gas in **(b)** is raised to 77°C, the pressure remaining constant. Calculate:

(i) the change in internal energy,

(ii) the external work done,

(iii) the total heat energy supplied.

(Molar gas constant = $8.3\,\mathrm{J\,mol^{-1}\,K^{-1}}$.) [J]

C52 At a temperature of 100 °C and a pressure of 1.01×10^5 Pa, 1.00 kg of steam occupies $1.67\,\mathrm{m^3}$ but the same mass of water occupies only $1.04 \times 10^{-3}\,\mathrm{m^3}$. The specific latent heat of vaporization of water at 100 °C is $2.26 \times 10^6\,\mathrm{J\,kg^{-1}}$. For a system consisting of 1.00 kg of water changing to steam at 100 °C and 1.01×10^5 Pa, find:

(a) the heat supplied to the system,

(b) the work done by the system,

(c) the increase in internal energy of the system. [C]

C53 **(a)** State the *first law of thermodynamics*.

(b) Give *one* practical example of each of the following:

(i) a process in which heat is supplied to a system without causing an increase in temperature,

(ii) a process in which no heat enters or leaves a system but the temperature changes. [C]

C54 Some gas, assumed to behave ideally, is contained within a cylinder which is surrounded by insulation to prevent loss of heat, as shown below.

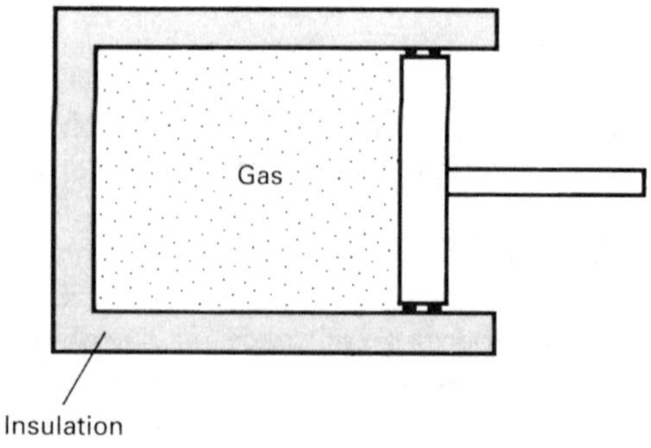

Initially the volume of gas is $2.9 \times 10^{-4}\,\mathrm{m^3}$, its pressure is 1.04×10^5 Pa and its temperature is 314 K.

(a) Use the equation of state for an ideal gas to find the amount, in moles, of gas in the cylinder.
(b) The gas is then compressed to a volume of $2.9 \times 10^{-5}\,\text{m}^3$ and its temperature rises to 790 K. Calculate the pressure of the gas after this compression.
(c) The work done on the gas during the compression is 91 J. Use the first law of thermodynamics to find the increase in the internal energy of the gas during the compression.
(d) Explain the meaning of *internal energy*, as applied to this system, and use your result in **(c)** to explain why a rise in the temperature of the gas takes place during the compression.
(Molar gas constant $= 8.3\,\text{J K}^{-1}\,\text{mol}^{-1}$.)

C55 The diagram shows curves (not to scale) relating pressure, p, and volume, V, for a fixed mass of an ideal monatomic gas at 300 K and 500 K. The gas is in a container fitted with a piston which can move with negligible friction.

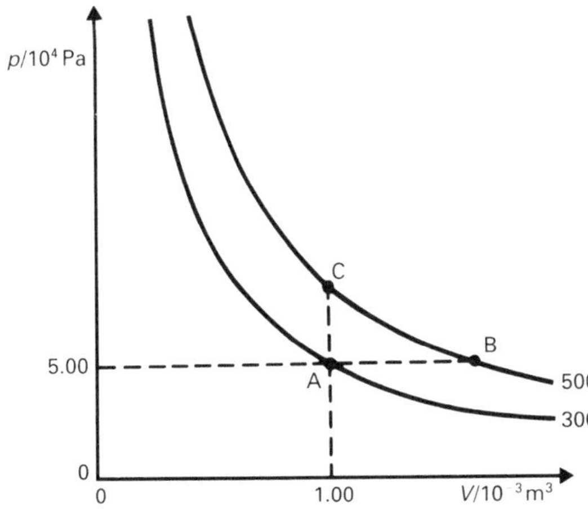

(a) Give the equation of state for n moles of an ideal gas, defining the symbols used. Show by calculation that:
(i) the number of moles of gas in the container is 2.01×10^{-2},
(ii) the volume of the gas at B on the graph is $1.67 \times 10^{-3}\,\text{m}^3$.
Molar gas constant, $R = 8.31\,\text{J mol}^{-1}\,\text{K}^{-1}$
(b) The kinetic theory gives the equation $p = \frac{1}{3}\rho\overline{c^2}$ where ρ is the density of the gas.
(i) Explain what is meant by $\overline{c^2}$.
(ii) Use the equation to derive an expression for the total internal energy of one mole of an ideal monatomic gas at kelvin temperature T.
Calculate the total internal energy of the gas in the container at point A on the graph.
(c) State the first law of thermodynamics as applied to a fixed mass of an ideal gas when heat energy is supplied to it so that its temperature rises and it is allowed to expand. Define any symbols used.
(d) Explain how the first law of thermodynamics applies to the changes represented on the graph by **(i)** A to C and **(ii)** A to B. Calculate the heat energy absorbed in each case. [J, '89]

C56 **(a)** The first law of thermodynamics is represented by the equation

$$Q = \Delta U + W$$

Explain each term in this equation.
(b) An engine (shown below) burns a mixture of petrol vapour and air. When the engine is running it makes 25 power strokes per second and develops a mean power of 18 kW.

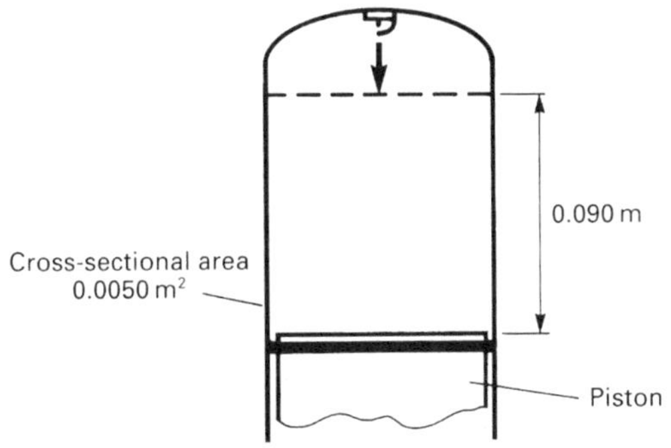

Neglecting losses in the engine due to friction and other causes, calculate the work done in each power stroke.
(c) The burning starts when the piston is at the top of its stroke and the resulting high pressure drives the piston downwards through a distance of 0.090 m. The cylinder has a cross-sectional area of $0.0050\,\text{m}^2$.
Calculate:
(i) the mean force on the piston head during the power stroke;
(ii) the mean pressure of the hot gas.
[O, '92]

C57 Why is the energy needed to raise the temperature of a given mass of gas by a certain amount greater if the pressure is kept constant than if the volume is kept constant?
[L]

C58 **(a)** Define the molar heat capacity of a gas at
(i) constant volume ($C_{v,\text{ molar}}$),
(ii) constant pressure ($C_{p,\text{ molar}}$).
(b) C_p is greater than C_v. Explain carefully why this is so.
(c) Show that for an ideal gas

$$C_{p,\text{ molar}} - C_{v,\text{ molar}} = R$$

where R is the molar gas constant.
[W, '92]

C59 **(a)** Explain what is meant by the terms *internal energy* and *molar heat capacity*.
(b) Explain why the molar heat capacity of a gas at constant pressure is different from the molar heat capacity at constant volume. Which of these is the larger? Explain your answer.
(c) What additional fact may be stated about the internal energy of a gas if the gas is ideal?
(d) **(i)** A quantity of 0.200 mol of air enters a diesel engine at a pressure of 1.04×10^5 Pa and at a temperature of 297 K. Assuming that air behaves as an ideal gas find the volume of this quantity of air.
(ii) The air is then compressed to one twentieth of this volume, the pressure having risen to 6.89×10^6 Pa. Find the new temperature.
(iii) Heating of the air then takes place by burning a small quantity of fuel in it to supply 6150 J. This is done at a constant pressure of 6.89×10^6 Pa as the volume of air increases and the temperature rises to 2040 K. Find
(1) the molar heat capacity of air at constant pressure,
(2) the volume of air after burning the fuel,
(3) the work done by the air during this expansion,
(4) the change in the internal energy of the air during this expansion.
(Molar gas constant = $8.31\,\text{J}\,\text{K}^{-1}\,\text{mol}^{-1}$.)

C60 What happens to the energy added to an ideal gas when it is heated **(a)** at constant volume, and **(b)** at constant pressure?

Show from this that a gas can have a number of values of specific heat capacity. Deduce an expression for the difference between the specific heat capacities of a gas at constant pressure and at constant volume.

If the ratio of the principal specific heat capacities of a certain gas is 1.40 and its density at STP is $0.090\,\text{kg}\,\text{m}^{-3}$ calculate the values of the specific heat capacity at constant pressure and at constant volume.
(Standard atmospheric pressure = $1.01 \times 10^5\,\text{N}\,\text{m}^{-2}$.)
[L]

C61 What are the two principal specific heat capacities of a gas, and why are they different? Show that the difference between these specific heat capacities is given by $c_p - c_v = p/\rho T$, where p, ρ, T are the pressure, density and temperature respectively.

A steel pressure vessel of volume $2.2 \times 10^{-2}\,\text{m}^3$ contains 4.0×10^{-2} kg of a gas at a pressure of 1.0×10^5 Pa and temperature 300 K. An explosion suddenly releases 6.48×10^4 J of energy, which raises the pressure instantaneously to 1.0×10^6 Pa. Assuming no loss of heat to the vessel, and ideal gas behaviour, calculate:
(a) the maximum temperature attained,
(b) the two principal specific heat capacities of the gas.
What is the velocity of sound in this gas at a temperature of 300 K?
[S]

C62 Explain why an ideal gas can have an infinite number of molar heat capacities and define the principal values.

What is meant by **(a)** an *isothermal* change, **(b)** an *adiabatic* change?
[AEB, '79]

C63 Derive an expression for the difference between the molar heat capacities of a perfect gas at constant pressure and at constant volume.
A thermally-insulated tube through which a gas may be passed at constant pressure contains an electric heater and thermometers for measuring the temperature of the gas as it

enters and as it leaves the tube. $3.0 \times 10^{-3}\ m^3$ of gas of density $1.8\ kg\ m^{-3}$ flows into the tube in 90 s and, when electrical power is supplied to the heater at a rate of 0.16 W, the temperature difference between the outlet and inlet is 2.5 K.

(a) Calculate a value for the specific heat capacity of the gas at constant pressure.

(b) In what way would your result have been different if the viscous drag on the gas had been significant?

(c) What features of the constant-flow method make it particularly suitable for the accurate determination of the specific heat capacities of fluids? [C]

C64 Explain clearly and concisely why, for a fixed mass of perfect gas:

(a) the internal energy remains constant when the gas expands isothermally,

(b) the heat capacity at constant pressure is greater than the heat capacity at constant volume. [W]

C65 **(a)** State the *first law of thermodynamics.*

Explain in simple kinetic theory terms what happens to the energy that enters the substance as heat in each of the following cases:

(i) an ideal gas maintained at constant volume,

(ii) an ideal gas maintained at constant pressure,

(iii) a crystalline solid at a temperature remote from its melting point,

(iv) a crystalline solid at its melting point.

(b) **(i)** Show that, for an ideal monatomic gas, the ratio $\gamma = C_p/C_v$ of the two principal molar heat capacities should be equal to 5/3.

(ii) Describe one experimental method for determining the value of γ for air. [O]

C66 What is meant by **(a)** an *adiabatic* and **(b)** an *isothermal* change of state of a gas?

A gas is contained in a thin-walled metal cylinder and compressed by a piston moving with constant velocity. Explain why the change is approximately adiabatic or isothermal according as the piston moves with a high or low velocity. [J]

C67 Air is contained in a cylinder by a frictionless gas-tight piston.

(a) Find the work done by the gas as it expands from a volume of $0.015\ m^3$ to a volume of $0.027\ m^3$ at a constant pressure of 2.0×10^5 Pa.

(b) Find the final pressure if, starting from the same initial conditions as in (a), and expanding by the same amount, the change occurs **(i)** isothermally, **(ii)** adiabatically.

(γ for air = 1.40) [W, '91]

C68 **(a)** Prove that the work done by a gas in expanding through a small volume δV at pressure p is $p\delta V$. Show that for large changes in volume, the work done is given by the area under the p–V curve.

(b) **(i)** Write down an expression for the first law of thermodynamics, defining *each* term in it.

(ii) How does the expression written down in (i) become modified in the case of

(I) an adiabatic change,

(II) an isothermal change, in the case of an ideal gas?

(iii) Briefly describe how such conditions are achieved in practice.

(c) In a particular automated process, an action by a piston R is transferred by means of moving air to a second piston S which is maintained in equilibrium by a variable force F as shown below:

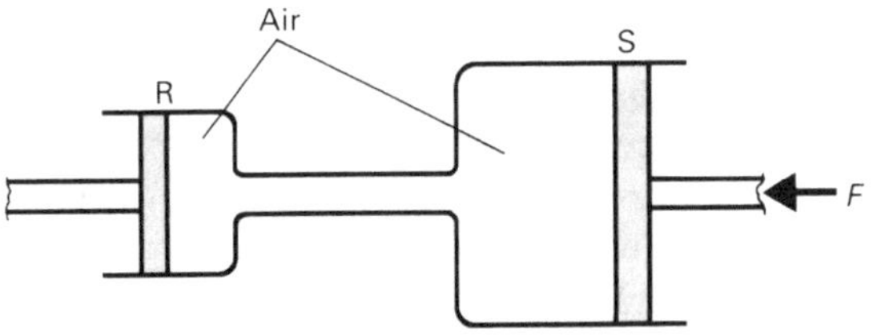

Area of piston R = $10^{-3}\ m^2$
Area of piston S = $2 \times 10^{-3}\ m^2$

After being at rest for a long time, the volume and pressure of the air between R and S are $10^{-4}\ m^3$ and 2×10^5 Pa respectively. Piston R is then moved a distance of 50 mm towards S, and this causes S to move a distance of 10 mm.

(i) Calculate the force exerted on S by the gas before R is moved.

(ii) Calculate the value of F after R is moved, assuming that the change has taken place isothermally.
(iii) Calculate the value of F after R is moved assuming that the change has taken place adiabatically
(The value of γ for air is 1.4.) [W, '92]

C69 (a) Explain what is meant by an *isothermal change* and by an *adiabatic change*, giving, in each case for an ideal gas, the equation relating the initial and final pressures and volumes. Explain qualitatively why the temperature of an ideal gas decreases during an adiabatic expansion.
(b) A fixed mass of gas is in an initial state A given by p_1, v_1, T_1. The gas expands adiabatically to a state B given by p_2, v_2, T_2. It is then heated up at constant volume until it reaches the initial temperature, i.e. state C given by p_3, v_2, T_1.
(i) On a graph of p against v, sketch the isothermals for temperatures T_1 and T_2.
(ii) On the same graph draw lines showing the transition from state A, via state B, to state C.
(iii) If $T_1 = 300$ K and $v_2 = 4v_1$, find the value of T_2; assume that γ, the ratio of the principal molar heat capacities of the gas, is 1.5.
(iv) Show that γ is given by $(p_1/p_3)^\gamma = (p_1/p_2)$.
(c) It is generally stated that an adiabatic change must be carried out 'rapidly'. What is meant by 'rapidly' in this context? [W]

C70 A vessel of volume 1.0×10^{-2} m^3 contains an ideal gas at a temperature of 300 K and pressure 1.5×10^5 Pa. Calculate the mass of gas, given that the density of the gas at temperature 285 K and pressure 1.0×10^5 Pa is 1.2 kg m^{-3}.

750 J of heat energy is suddenly released in the gas, causing an instantaneous rise of pressure to 1.8×10^5 Pa. Assuming ideal gas behaviour, and no loss of heat to the containing vessel, calculate the temperature rise, and hence the specific heat capacity at constant volume of the gas. [S]

C71 (a) Define (i) the *Avogadro constant*, (ii) the *atomic mass unit*. If the numerical value of the Avogadro constant is y and that of the atomic mass unit expressed in grams is x, show that $yx = 1$.
(b) Use the following data to calculate the root-mean-square speed of helium molecules at 2000 °C.
Mass of one mole of helium = 4.00 g.
Molar gas constant = 8.31 J mol^{-1} K^{-1}. [J]

C72 The cylinder in Figure 1 holds a volume $V_1 = 1000$ cm^3 of air at an initial pressure $p_1 = 1.10 \times 10^5$ Pa and temperature $T_1 = 300$ K. Assume that air behaves like an ideal gas.

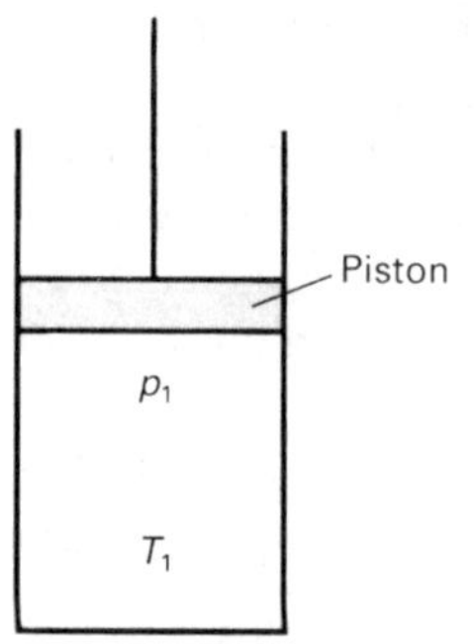

Fig. 1

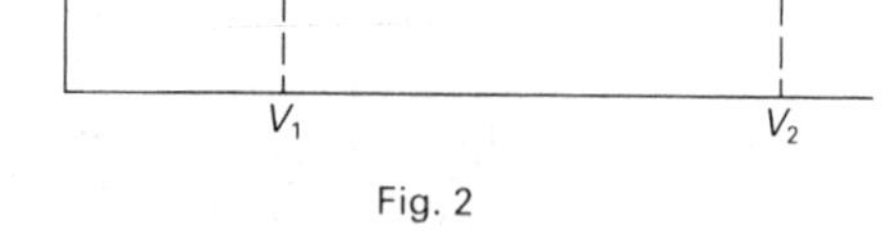

Fig. 2

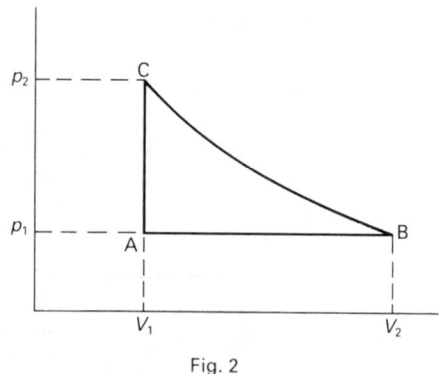

Figure 2 shows a sequence of changes imposed on the air in the cylinder.
(a) AB – the air is heated to 375 K at constant pressure. Calculate the new volume, V_2.
(b) BC – the air is compressed isothermally to volume V_1. Calculate the new pressure, p_2.
(c) CA – the air cools at constant volume to pressure p_1. State how a value for the work done on the air during the full sequence of changes may be found from the graph in Figure 2. [L]

C73 What is an *adiabatic* change?

A vessel of volume $8.00 \times 10^{-3}\,m^3$ contains an ideal gas at a pressure of 1.14×10^5 Pa. A stopcock in the vessel is opened and the gas expands adiabatically, expelling some of its original mass, until its pressure is equal to that outside the vessel (1.01×10^5 Pa). The stopcock is then closed and the vessel is allowed to stand until the temperature returns to its original value; in this equilibrium state, the pressure is 1.06×10^5 Pa.

(a) Explain why there was a temperature change as a result of the adiabatic expansion.

(b) Find the volume which the mass of gas finally left in the vessel occupied under the original conditions.

(c) Sketch a graph showing the way in which the pressure and volume of the mass of gas finally left in the vessel changed during the operations described above. (It is *not* necessary to plot exact numerical values of p and V.)

(d) What is the value of γ, the ratio of the principal heat capacities of the gas?

(e) What can you deduce about the molecules of the gas? Give your reasons. [C]

C74 **(a)** **(i)** Prove the work done by a gas in expanding through δV at pressure p is $p\delta V$ and show that for large changes in volume the work is the area under the p–V curve.

(ii) State the first law of thermodynamics.

(iii) Deduce the relationship $C_p - C_v = R$ for the difference between the principal molar heat capacities (C_p and C_v) for an ideal gas.

(b) The diagram represents an energy cycle whereby a mole of an ideal gas is firstly cooled at constant pressure (A→B) then heated at constant volume (B→C) and then returned to its original state (C→A).

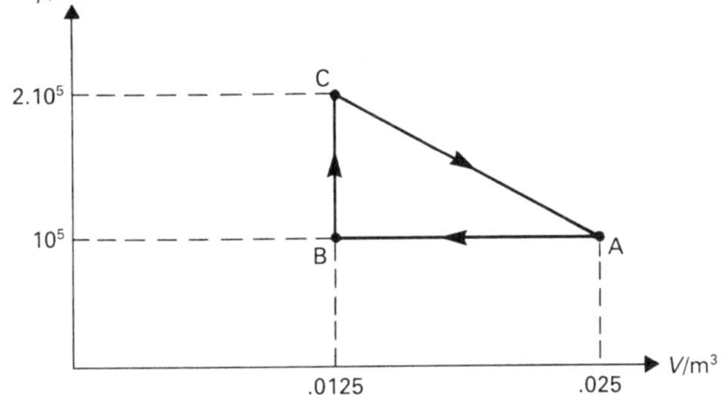

(i) Calculate the temperature of the gas at A, at B, and at C.

(ii) Calculate the heat given out by the gas in the process A→B.

(iii) Calculate the heat absorbed in the process B→C.

(iv) Calculate the net amount of work done in the cycle.

(v) Calculate the net amount of heat transferred in the cycle.

($R = 8.3\,J\,mol^{-1}\,K^{-1}$; $C_v = \frac{5}{2}R$.) [W]

C75 State the first law of thermodynamics and explain what is meant by the *internal energy* of a system. What constitutes the internal energy of an ideal gas? Starting from the expression $P = \frac{1}{3}\rho \overline{c^2}$, show that the internal energy of an ideal *monatomic* gas is $\frac{3}{2}PV$, and discuss the interpretation of temperature in the kinetic theory.

Explain the following observations:

(a) when pumping up a bicycle tyre the pump barrel gets warm, and

(b) when a gas at high pressure in a container is suddenly released, the container cools.

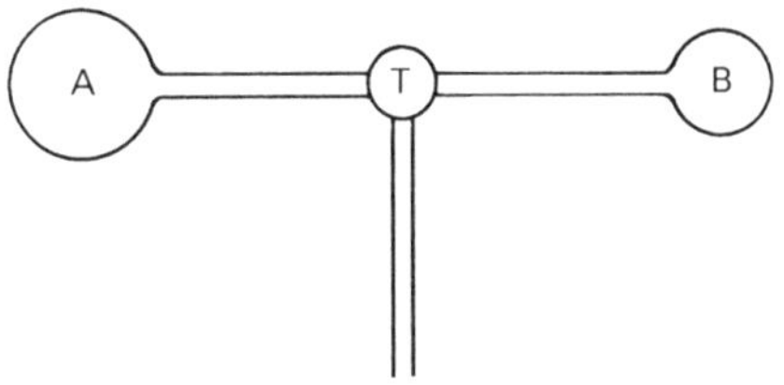

Two bulbs, A of volume $100\,cm^3$ and B of volume $50\,cm^3$, are connected to a three way tap T which enables them to be filled with gas or evacuated. The volume of the tubes may be neglected.

(c) Initially bulb A is filled with an ideal gas at 10 °C to a pressure of 3.0×10^5 Pa. Bulb B is filled with an ideal gas at 100 °C to a pressure of 1.0×10^5 Pa. The two bulbs are connected with A maintained at 10 °C and B at 100 °C. Calculate the pressure at equilibrium.

(d) Bulb A, filled at 10 °C to a pressure of 3.0×10^5 Pa, is connected to a vacuum pump with a cylinder of volume $20\,cm^3$. Calculate the pressure in A after one inlet stroke of the pump. The air in the pump is now expelled into the atmosphere. Calculate the pressure in A after the second inlet stroke. Calculate the number of strokes of the pump to reduce the pressure in A to 1.0×10^3 Pa. The whole system is maintained at 10 °C throughout the process [O & C]

C76 The specific latent heat of vaporization of a particular liquid at 130 °C and a pressure of 2.60×10^5 Pa is $1.84 \times 10^6 \, J\,kg^{-1}$. The specific volume of the liquid under these conditions is $2.00 \times 10^{-3} \, m^3\,kg^{-1}$, and that of the vapour is $5.66 \times 10^{-1} \, m^3\,kg^{-1}$. Calculate:
(a) the work done, and
(b) the increase in internal energy when 1.00 kg of the vapour is formed from the liquid under these conditions.

C77 **(a)** Explain what is meant by a *reversible* change.
(b) State the *first law of thermodynamics*, and discuss the experimental observations on which it is based.
(c) A mass of 0.35 kg of ethanol is vaporized at its boiling point of 78 °C and a pressure of 1.0×10^5 Pa. At this temperature, the specific latent heat of vaporization of ethanol is $0.95 \times 10^6 \, J\,kg^{-1}$, and the densities of the liquid and vapour are $790 \, kg\,m^{-3}$ and $1.6 \, kg\,m^{-3}$ respectively. Calculate:
(i) the work done by the system;
(ii) the change in internal energy of the system.
Explain in molecular terms what happens to the heat supplied to the system. [O]

C78 **(a)** State *four* of the basic assumptions made in developing the simple kinetic theory for an ideal gas.
(b) The theory derives the formula $p = \frac{1}{3}\rho \overline{c^2}$ where p is the pressure of the gas, ρ is the density of the gas and $\overline{c^2}$ is the mean square speed of the molecules. Explain more fully what is meant by $\overline{c^2}$ and explain its significance in relation to the temperature of a gas.
(c) Describe briefly the experiments which Andrews performed on carbon dioxide. (A detailed description of the apparatus is *not* required.)
(i) Draw graphs to show the pressure–volume relationship which Andrews obtained for various temperatures. Indicate on your diagram the various states of the carbon dioxide.
(ii) Use your graphs to explain the meaning of critical temperature. What is its significance in connection with the liquefaction of gases? [AEB, '79]

C79 The model of a gas as a large number of elastic bodies moving about in a random manner is the basic idea of the kinetic theory of gases. In terms of this model, explain:
(a) what is meant by an ideal gas,
(b) how a gas exerts a pressure when enclosed in a container,
(c) why the atmospheric pressure decreases with height,
(d) how the atmosphere, which is not in a container, exerts a pressure at all.
Which of the assumptions made to develop a quantitative expression for the pressure of an ideal gas require to be modified to explain the behaviour of a real gas? Illustrate your answer by considering a p–V isothermal for an ideal gas and a p–V isothermal for a real gas at a temperature below its critical value.

A series of experiments was performed by Andrews to obtain p–V isothermals for carbon dioxide. Sketch a set of p–V isothermals for water, noting particularly any dissimilarities between the curves for water and carbon dioxide. [L]

C80 Explain what is meant by the *critical temperature* of a real gas (such as carbon dioxide), and describe, with the aid of pressure–volume diagrams, the behaviour of a real gas at a temperature **(a)** above the critical temperature, **(b)** equal to the critical temperature, **(c)** below the critical temperature.

Either: Describe, with the aid of a diagram, an experiment by which the departure of a real gas from ideal gas behaviour may be studied.

Or: Explain how van der Waals attempted to produce an equation which would describe the behaviour of a real gas. [S]

C81 What are the conditions under which the equation $pV = RT$ gives a reasonable description of the relationship between the pressure p, the volume V and the temperature T of a real gas?

Sketch p–V isothermals for the gas–liquid states and indicate the region in which $pV = RT$ applies. Indicate the state of the substance in the various regions of the p–V diagram. Mark and explain the significance of the critical isothermal.

Discuss a way in which the equation $pV = RT$ may be modified so that it can be applied more generally. Explain and justify on a molecular basis the additional terms introduced. Discuss the success of this modification. [L]

C82 **(a)** State the conditions under which the behaviour of a *real* gas will deviate significantly from that expected of an *ideal* gas.

(b) **(i)** On a pV against p diagram sketch an isotherm for a real gas at the Boyle temperature. On the same set of axes sketch isotherms for temperatures just above and just below the Boyle temperature, labelling the isotherms clearly.

(ii) Explain how the properties of the atoms or molecules of a gas give rise to the shape of the isotherm you have drawn *below* the Boyle temperature.

(c) A quantity of oxygen gas occupies 0.20 m^3 at a temperature of 27 °C and pressure of 10 atmospheres. If it were to be liquefied, what volume of liquid oxygen, density 1.1×10^3 kg m^{-3}, would be produced? The oxygen gas in its initial state may be considered to behave as an ideal gas.

What condition must be met before the gas can be liquefied by the increase of pressure alone?

(1 atmosphere = 1.0×10^5 Pa, relative molecular mass of oxygen = 32, molar gas constant = 8.3 J mol^{-1} K^{-1}.) [J, '89]

C83 The equation of state for one mole of a real gas is

$$\left(p + \frac{a}{V^2}\right)(V - b) = RT$$

where p is the pressure of the gas, V is the volume and T is the absolute temperature of the gas. Determine the dimensions of **(i)** a, **(ii)** b, **(iii)** R. [W, '90]

C84 **(a)** Write down van der Waals' equation of state for a real gas and explain how the assumptions of the simple kinetic theory of gases are modified in the derivation of the equation.

(b) Carbon dioxide has a density of 344 kg m^{-3} at its critical pressure 7.5×10^6 Pa and critical temperature 304 K.

(i) By considering the mass of 1 mol of CO_2 show that the value of the critical volume is 1.28×10^{-4} m^3.

(ii) Hence calculate the van der Waals' constants a and b given that the critical volume $V_c = 3b$.

(Molar mass of $CO_2 = 4.4 \times 10^{-2}$ kg, molar mass constant = 8.3 J mol^{-1} K^{-1}.)

(c) Sketch, on the same axes, P–V isotherms for a fixed mass of CO_2 at

(i) the critical temperature,

(ii) a temperature below the critical temperature.

Mark on the same axes points corresponding to the critical volume and critical pressure.

(d) For the isotherm you have drawn in (c) (ii) give the state (or states) of the CO_2 when

(i) it has the critical volume,

(ii) it is at the critical pressure. [J, '91]

C85 **(a)** Which two assumptions of the kinetic theory of ideal gases are unlikely to be valid for real gases at high pressure?

(b) The equation of state for one mole of an ideal gas is $pV = RT$.

(i) Write down Van der Waals' equation for one mole of a real gas.

(ii) Explain the reasons for the modifications made.

(c) The following data refer to nitrogen gas.

Critical pressure = 3.4×10^6 Pa

Critical volume = 9.0×10^{-5} m^3 mol^{-1}

Van der Waals' constant,

$a = 1.4 \times 10^{-1}$ Pa m^6 mol^{-2}

Van der Waals' constant,

$b = 3.9 \times 10^{-5}$ m^3 mol^{-1}

(i) Use the given data to calculate the critical temperature of nitrogen.

(ii) Calculate the temperature of an ideal gas with the same pressure and volume per mole as given in the data.

(d) Sketch a graph of pV against p for 1 mole of nitrogen at its critical temperature. On the same axes, sketch the graph for 1 mole of an ideal gas at the temperature calculated in (c) (ii).

Calculate the values of the intercept of each graph on the pV axis.

(Molar gas constant, $R = 8.3\,\mathrm{J\,mol^{-1}\,K^{-1}}$.) [J]

C86 **(a)** The pressure exerted by an ideal gas can be written in the form

$$p = (m/M)RT/V$$

Summarize briefly the large-scale properties of a fixed mass of such a gas.

(b) The pressure exerted by an ideal gas can also be expressed in the form

$$p = \tfrac{1}{3}\rho\overline{c^2}$$

Summarize briefly the assumptions made about an ideal gas when deriving this result.

(c) A cylinder of volume $0.080\,\mathrm{m^3}$ contains oxygen at a temperature of 280 K and a pressure of 90 kPa. Calculate:

(i) the mass of oxygen in the cylinder,

(ii) the number of oxygen molecules in the cylinder,

(iii) the RMS speed of the oxygen molecules.

(The Avogadro constant, $N_A = 6.0 \times 10^{23}\,\mathrm{mol^{-1}}$; molar gas constant, $R = 8.3\,\mathrm{J\,K^{-1}\,mol^{-1}}$; molar mass of oxygen, $M = 0.032\,\mathrm{kg\,mol^{-1}}$.)

(d) In practice real gases deviate in behaviour from an ideal gas.

(i) Under what conditions does a real gas behave much like an ideal gas?

(ii) By reference to Andrews' experiments with carbon dioxide describe briefly the ways in which that gas differs in behaviour from an ideal gas. [L]

C87 **(a)** **(i)** Explain how the molecules of a gas cause pressure.

(ii) When a fixed mass of gas at constant volume is heated its pressure rises. Give the reasons for this, in terms of the behaviour of gas molecules.

(b) **(i)** State the two principal assumptions made in the simple kinetic theory of gases.

(ii) Explain how the equation of state for an ideal gas can be modified for a real gas in conditions where the above assumptions are not valid.

(c) A quantity of ideal gas whose ratio of principal molar heat capacities is $5/3$ has temperature 300 K, volume $64 \times 10^{-3}\,\mathrm{m^3}$, and pressure 243 kPa.

It is made to undergo the following three changes in order:

A: reversible adiabatic compression to a volume $27 \times 10^{-3}\,\mathrm{m^3}$,

B: reversible isothermal expansion back to $64 \times 10^{-3}\,\mathrm{m^3}$,

C: a return to its original state.

(i) Calculate the pressure on completion of process A.

(ii) Calculate the temperature at which process B occurs.

(iii) Describe process C.

(iv) Show on a sketch graph of pressure (y-axis) against volume the changes described. [S]

C88 Give an account of how the behaviour of real gases differs from that of an ideal gas, illustrating your answer with sketch graphs relating pressure, volume and temperature. Explain what is meant by *critical temperature* and *critical pressure*.

State van der Waals' equation and explain the physical significance of the various terms. Sketch and discuss the general nature of the p–V relation predicted by van der Waals' equation at a temperature **(a)** above and **(b)** below the critical temperature.

Outline a method of liquefying oxygen (critical temperature 155 K; critical pressure 50 atmospheres). [W]

C89 **(a)** In his experiments on carbon dioxide how did Andrews measure the pressure and volume of the carbon dioxide at high pressure?

(b) Measurements of the pressure p and the volume V of a fixed mass of gas over a wide range of pressures were obtained for different temperatures. On the same axes sketch graphs to illustrate the relation between pV and p for a real gas under isothermal conditions.

(i) at a temperature below its critical temperature,

(ii) at its critical temperature and

(iii) at a temperature above its critical temperature.

On the same axes sketch a graph to illustrate the relation for an ideal gas at one temperature.

(c) Explain why some of the assumptions of the kinetic theory of an ideal gas may have to be modified for real gases. Hence explain why a real gas may deviate from Boyle's law. [J]

C90 (a) The partly labelled diagram below shows the apparatus used by Andrews in his experiments on carbon dioxide.

State what A is and explain its purpose. Explain why each of the water baths, P, Q and R were used during the experiment.

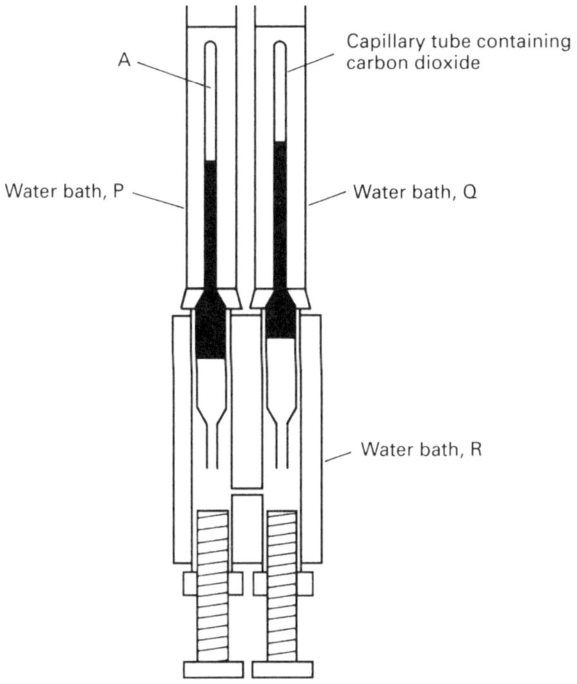

(b) State the meanings of *critical temperature* and *critical pressure.*

(c) Some carbon dioxide initially at a temperature above its critical temperature is subjected to the following changes.

(i) It is compressed isothermally to a pressure above its critical pressure.

(ii) Then at this pressure it is cooled at constant pressure until the temperature is well below its critical temperature.

(iii) Then at this temperature it is expanded isothermally until all the carbon dioxide is again a gas.

Sketch a graph of pressure against volume to illustrate these changes, and discuss the associated changes of state. [J]

C91 (a) In terms of simple kinetic theory, explain qualitatively how a gas exerts a pressure. If the pressure of an ideal gas is given by $p = \frac{1}{3}\rho \overline{c^2}$ where ρ is the density of the gas and $\overline{c^2}$ is the mean square speed of the molecules, explain any change in the pressure that may occur if the gas is:

(i) allowed to expand while the temperature is kept constant,

(ii) heated while the volume is kept constant.

(b) Sketch isothermal curves to show how the pressure of a fixed mass of substance (e.g. carbon dioxide) varies with volume over a wide range of temperature and pressure. Indicate on your sketch the regions where the substance is in *the liquid phase, the saturated vapour phase, the unsaturated vapour phase* and *the gas phase.*

(c) An unsaturated vapour of mass 5×10^{-4} kg and at a temperature of 20 °C is compressed isothermally until, at a volume $V_1 = 9 \times 10^{-5}\,\text{m}^3$ and a pressure 6×10^6 Pa, the vapour first becomes saturated. Further compression of the vapour causes the formation of liquid until, when the volume is V_2, the substance is changed completely to liquid. If V_2 is negligible compared with V_1 and the temperature remains constant throughout the process, calculate:

(i) the work that must be performed during the compression from V_1 to V_2,

(ii) the amount of thermal energy that must be supplied to, or removed from, the substance during the same compression.

(Assume that the specific latent heat of vaporization of the liquid at 20 °C is $1.2 \times 10^5\,\text{J kg}^{-1}$.) [AEB, '79]

C92 Sketch a graph to show how the saturated vapour pressure of a liquid varies with temperature. Give a qualitative explanation of the shape of the graph. [C]

C93 In terms of the kinetic theory of matter explain:

(a) what is meant by *saturated vapour* and *saturation vapour pressure,*

(b) how the saturation vapour pressure varies with temperature.

Describe an experiment to measure the saturation vapour pressure of water vapour at 300 K (27 °C). Discuss one practical difficulty in using the apparatus you describe to measure the saturation vapour pressure of water vapour at 275 K (2 °C).

State, with reasons, *two* advantages of using mercury in a barometer. [J]

C94 State the relation between pressure and volume at constant temperature for **(a)** an ideal gas, **(b)** a saturated vapour.

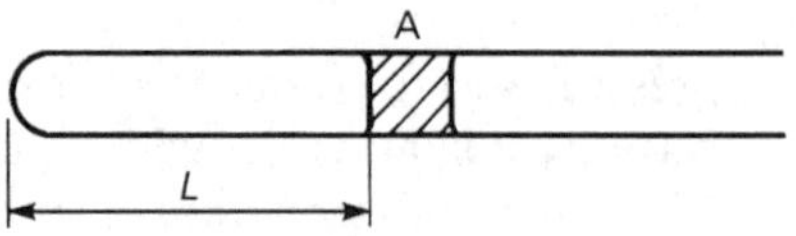

A long uniform horizontal capillary tube, sealed at one end, and open to the air at the other, contains air trapped behind a short column of water A. The length L of the trapped air column at temperatures 300 K and 360 K is 10 cm and 30 cm respectively. Given that the vapour pressures of water at the same temperatures are 4 kPa and 62 kPa respectively, calculate the atmospheric pressure. [S]

C95 A closed space contains a very small quantity of a liquid and its vapour. Describe what happens and illustrate by sketch graphs how the pressure in the space changes when:
(a) the volume of the space is slowly increased at constant temperature,
(b) the temperature is slowly raised at constant volume. [J]

C96 A sealed vessel contains a mixture of air and water vapour in contact with water. The total pressures in the vessel at 27 °C and 60 °C are respectively 1.0×10^5 Pa and 1.3×10^5 Pa. If the saturated vapour pressure of water at 60 °C is 2.0×10^4 Pa what is its value at 27 °C? ($1\ \text{Pa} = 1\ \text{N m}^{-2}$.) [L]

C97 State the relation between pressure and volume at constant temperature for **(a)** an ideal gas, **(b)** a saturated vapour.

The saturation vapour pressure of water is $6 \times 10^4\ \text{N m}^{-2}$ at temperature 360 K and $0.3 \times 10^4\ \text{N m}^{-2}$ at temperature 300 K. A vessel contains only water vapour at a temperature of 360 K and pressure $2 \times 10^4\ \text{N m}^{-2}$. It may be assumed that unsaturated water vapour behaves like an ideal gas. If the vapour were to remain unsaturated what would be the pressure in the vessel at 300 K? What is the actual pressure at this temperature, and what fraction, if any, of the vapour has condensed? [S]

C98 Show by sketch graphs the dependence on temperature of:
(a) the volume of an ideal gas, the pressure remaining constant,
(b) the pressure of a saturated vapour.

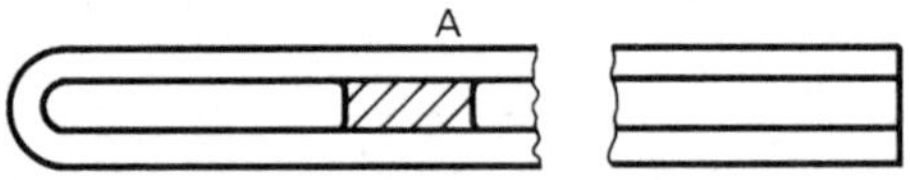

A long uniform capillary tube, sealed at one end and open to the air at the other, contains air trapped behind a short column of water A. Show by a sketch graph how the length of the air column depends on the temperature of the system, paying particular attention to the form of the curve where the temperature is
(i) low enough for the vapour pressure of water to be negligible,
(ii) approaching the boiling point of water. [S]

C99 **(a)** Describe an experiment to determine the saturation vapour pressure of water at various temperatures in the range 75 °C to 110 °C. Sketch a graph showing the results which would be obtained from the experiment.
[SVP of water at 75 °C = 38 kPa (kN m^{-2}); standard atmospheric pressure = 101 kPa (kN m^{-2}).]
(b) By considering the effect of temperature on the SVP of water vapour, explain why it is essential to have a safety valve on the water boiler of a central heating system.
(c) In a Boyle's law experiment using damp air, the following results were obtained:
Initial pressure (air unsaturated) = 8.5 kPa (kN m^{-2})
Pressure when volume reduced to $\frac{1}{2}$ of initial volume = 16.0 kPa (kN m^{-2})
Pressure when volume reduced to $\frac{1}{3}$ of initial volume = 23.0 kPa (kN m^{-2}).
(i) Show that the vapour exerts its saturation pressure when the volume is reduced to half its initial value.

(ii) Calculate the saturation vapour pressure at the temperature of the experiment.
(iii) Calculate the initial pressure of the water vapour. [AEB, '79]

C100 **(a)** **(i)** Explain what is meant by a *saturated vapour*.
(ii) State Dalton's law of partial pressures.
(b) Suggest an experiment to investigate the variation with temperature of the saturated vapour pressure of water vapour over the range of 0 °C to 100 °C. Sketch the apparatus you would use, list the measurements you would make, and describe how the results would be obtained.
(c) The saturated vapour pressure of water at 20 °C is 18 mmHg (= 2.4 kPa). Draw sketch-graphs showing how the pressure, p, of 1 m^3 of water vapour (with a small amount of water present throughout) will vary when:
(i) the vapour is compressed isothermally to a volume of 0.2 m^3,
(ii) the vapour (and the water) are heated at constant volume to the boiling point of water (100 °C).
In each case, show on your graph the final vapour pressure exerted by the water vapour.
(d) In pure atmospheric air it may be assumed that 80% of the molecules present are nitrogen (molar mass = 0.028 kg) and that 20% are oxygen (molar mass = 0.032 kg).
(Take atmospheric pressure as 100 kPa, temperature to be 17 °C, and the molar gas constant to be 8.3 $J\,K^{-1}\,mol^{-1}$.)
Showing all stages in your working, calculate:
(i) the partial pressure exerted by each gas,
(ii) the density of the oxygen present,
(iii) the density of the air. [O]

C101 In an experiment to determine the specific latent heat of vaporization of benzene, it was found that when the electrical power input to the heater was 82 W, 10.0 g of benzene was evaporated in 1 minute; when the power input was reduced to 30 W, the rate of evaporation was 2.0 g per minute. Calculate the specific latent heat of vaporization of benzene.

The saturation vapour pressure of benzene is 1.0×10^5 Pa at a temperature of 80 °C; at the same temperature, the saturation vapour pressure of acetone (propanone) is 1.8×10^5 Pa. Which of these two compounds has the higher boiling point, and why?
(Atmospheric pressure = 1.0×10^5 Pa.) [S]

THERMODYNAMICS (Chapter 16)

C102 **(a)** Explain what is meant by the statement that two bodies are in thermal equilibrium.
(b) State the zeroth law of thermodynamics. Explain why it is so called and its relevance in the use of a thermometer to measure temperature.

C103 The specific latent heat of vaporization of a particular liquid at 30 °C and 1.20×10^5 Pa is $3.20 \times 10^5\ J\,kg^{-1}$. Under the same conditions of temperature and pressure the specific volume of the liquid is $1.00 \times 10^{-3}\ m^3\,kg^{-1}$, and that of its vapour is $4.51 \times 10^{-1}\ m^3\,kg^{-1}$. If 5.00 kg of the liquid is vaporized at 30 °C and 1.20×10^5 Pa, what is:
(a) the increase in enthalpy?
(b) the increase in internal energy?

C104 What is the maximum theoretical efficiency of a heat engine which takes in heat at 25.0 °C and rejects it at 10.0 °C?

C105 **(a)** When a system is taken from A to C via B it absorbs 180 J of heat and does 130 J of work. How much heat does the system absorb in going from A to C via D, if it performs 40 J of work in doing so?
(b) The decrease in internal energy in going from D to A is 30 J. Calculate the heat absorbed by the system in going from:
(i) A to D,
(ii) D to C.

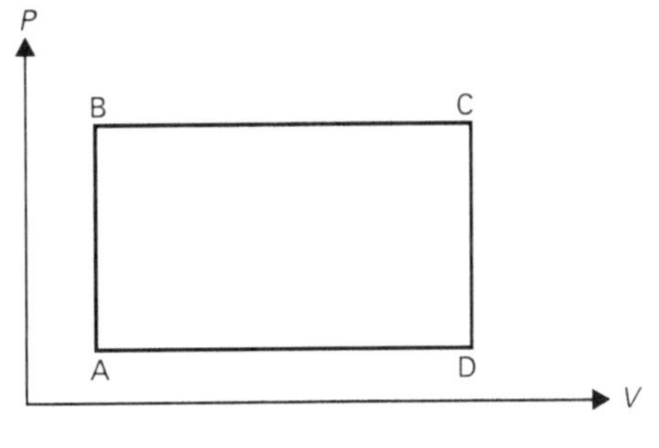

C106 Calculate the change in entropy of 1.00 kg of ice at 0 °C when it is melted and converted to water at 0 °C. (Specific latent heat of fusion of ice at 0 °C = 3.34×10^5 J kg^{-1}.)

C107 An electrical resistor which is consuming energy at 200 W is kept at a constant temperature of 27 °C by a stream of cooling water. In a time interval of 1 minute what is the increase in entropy of
(a) the resistor,
(b) the cooling water?

C108 **(a)** Distinguish between the operation of a *heat engine* and a *heat pump*.
(b) The maximum efficiency, E_{max}, of a heat engine is given by

$$E_{max} = \frac{T_H - T_C}{T_H} \quad \text{or} \quad 1 - \frac{T_C}{T_H}$$

where T_H is the temperature of the heat reservoir and T_C is the tempeature of the heat sink.
(i) Write down an expression for the efficiency of a heat engine in terms of heat transfer into and out of the engine.
(ii) Use one of the above expressions and your answer to (i) to show that, in the ideal case, the total entropy change is zero.
(c) A heat engine uses a heat sink which is at a temperature of 27 °C and steam at normal atmospheric pressure as the source.

Determine the maximum efficiency of the engine. [AEB, '90]

C109 The graph relates the pressure and volume of a fixed mass of an ideal gas which is first compressed isothermally from A to B and then allowed to expand adiabatically from B to C.

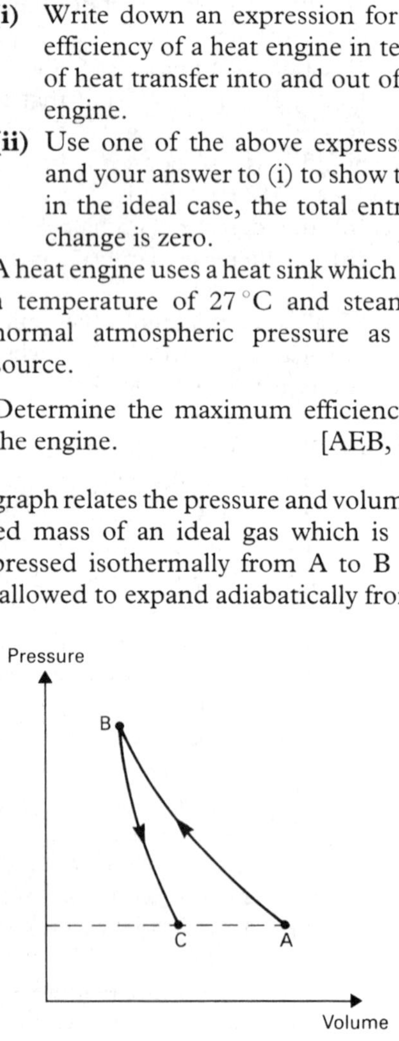

(a) For each of the changes shown on the graph, state and explain whether:
(i) the temperature of the gas changes,
(ii) there is heat transfer to or from the gas,
(iii) work is done on or by the gas.
(b) What single process will bring the gas back to its initial condition?
(c) Sketch the corresponding entropy–temperature graph for the changes A to B and B to C. [AEB, '87]

C110 The diagram shows the *T–S* curve of a gas undergoing a reversible cyclic process. Calculate the heat absorbed by the gas in going from:
(a) A to B
(b) B to C.

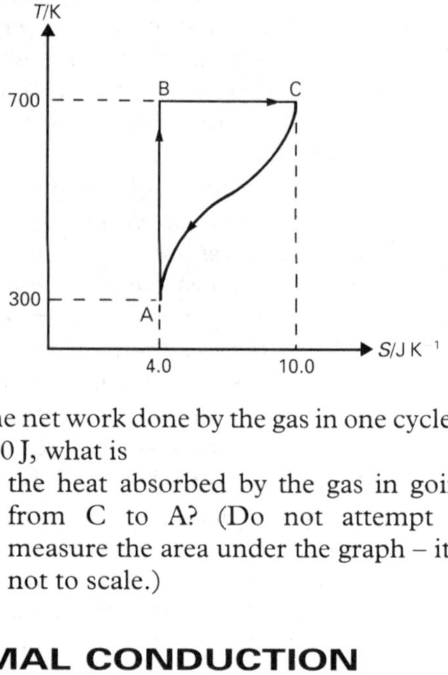

If the net work done by the gas in one cycle is 1200 J, what is
(c) the heat absorbed by the gas in going from C to A? (Do not attempt to measure the area under the graph – it is not to scale.)

THERMAL CONDUCTION (Chapter 17)

C111 Explain what is meant by *temperature gradient*.

An ideally lagged compound bar 25 cm long consists of a copper bar 15 cm long joined to an aluminium bar 10 cm long and of equal cross-sectional area. The free end of the copper is maintained at 100 °C and the free end of the aluminium at 0 °C. Calculate the temperature gradient in each bar when steady state conditions have been reached.
(Thermal conductivity of copper
= 390 W m^{-1} °C^{-1}.
Thermal conductivity of aluminium
= 210 W m^{-1} °C^{-1}.) [J]

C112 If a copper kettle has a base of thickness 2.0 mm and area $3.0 \times 10^{-2}\,m^2$, estimate the steady difference in temperature between inner and outer surfaces of the base which must be maintained to enable enough heat to pass through so that the temperature of 1.000 kg of water rises at the rate of $0.25\,K\,s^{-1}$. Assume that there are no heat losses, the thermal conductivity of copper = $3.8 \times 10^2\,W\,m^{-1}\,K^{-1}$ and the specific heat capacity of water = $4.2 \times 10^3\,J\,kg^{-1}\,K^{-1}$.

After reaching the temperature of 373 K the water is allowed to boil under the same conditions for 120 seconds and the mass of water remaining in the kettle is 0.948 kg. Deduce a value for the specific latent heat of vaporization of water (neglecting condensation of the steam in the kettle). [J]

C113 A cubical container full of hot water at a temperature of 90 °C is completely lagged with an insulating material of thermal conductivity $6.4 \times 10^{-2}\,W\,m^{-1}\,°C^{-1}$. The edges of the container are 1.0 m long and the thickness of the lagging is 1.0 cm. Estimate the rate of flow of heat through the lagging if the external temperature of the lagging is 40 °C. Mention any assumptions you make in deriving your result.

Discuss qualitatively how your result will be affected if the thickness of the lagging is increased considerably assuming that the temperature of the surrounding air is 18 °C. [J]

C114 A thin-walled hot-water tank, having a total surface area $5\,m^2$, contains $0.8\,m^3$ of water at a temperature of 350 K. It is lagged with a 50 mm thick layer of material of thermal conductivity $4 \times 10^{-2}\,W\,m^{-1}\,K^{-1}$. The temperature of the outside surface of the lagging is 290 K. What electrical power must be supplied to an immersion heater to maintain the temperature of the water at 350 K? (Assume the thickness of the copper walls of the tank to be negligible.)

What is the justification for the assumption that the thickness of the copper walls of the tank may be neglected? (Thermal conductivity of copper = $400\,W\,m^{-1}\,K^{-1}$.)

If the heater were switched off, how long would it take for the temperature of the hot water to fall 1 K?
(Density of water = $1000\,kg\,m^{-3}$; specific heat capacity of water = $4170\,J\,kg^{-1}\,K^{-1}$.) [S]

C115 The diagram shows a lagged copper bar acting as a thermal link between a bath of boiling water and an ice–water mixture.

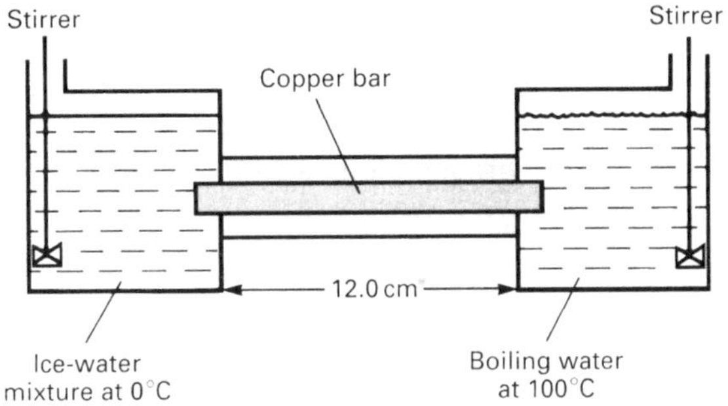

Calculate the energy flow per second through the bar. Hence calculate the mass of ice which should melt during a 15 s period. Write down one important assumption that you make.
In practice the amount of ice that melts per second is likely to be different from the calculated amount. Give a reason why the calculated amount might be
(a) higher, and
(b) lower than the amount melted in practice. (Thermal conductivity of copper = $385\,W\,m^{-1}\,K^{-1}$, area of cross-section of copper bar = $1.50\,cm^2$, specific latent heat (specific enthalpy change) of fusion of ice = $3.34 \times 10^5\,J\,kg^{-1}$.) [L, '91]

C116 **(a)** Describe and contrast convection and conduction as mechanisms of heat transfer.
(b) In many experiments in physics it is necessary to reduce the rate of transfer of thermal energy between the sample under investigation and any surrounding container as much as possible. Give a brief account of the ways in which this may be achieved.
(c) Define *thermal conductivity* and show that it has an SI unit $W\,m^{-1}\,K^{-1}$.
(d) A small greenhouse consists of $34\,m^2$ of glass of thickness 3.0 mm and $9.0\,m^2$ of concrete wall of thickness 0.080 m. On a sunny day, the interior of the

greenhouse receives a steady 25 kW of solar radiation. Estimate the difference in temperature between the inside and outside of the greenhouse. The temperatures inside and outside may be assumed uniform and heat transfer downwards into the ground inside the greenhouse may be neglected.
(Thermal conductivity of glass = $0.85\,\mathrm{W\,m^{-1}\,K^{-1}}$, thermal conductivity of concrete = $1.5\,\mathrm{W\,m^{-1}\,K^{-1}}$.)
[O & C, '92]

C117 Sketch graphs to illustrate the temperature distribution along a metal bar heated at one end when the bar is **(a)** lagged, and **(b)** unlagged. In each case assume that temperature equilibrium has been reached. Explain the difference between the two graphs.

By considering the relative increase in surface area explain why asbestos lagging of thickness 20 mm will be more effective in reducing the *total* heat losses from a copper pipe carrying stream at 100 °C if the pipe has a diameter of about 60 mm than if the pipe has a much smaller diameter.

A window pane consists of a sheet of glass of area $2.0\,\mathrm{m^2}$ and thickness 5.0 mm. If the surface temperatures are maintained at 0 °C and 20 °C calculate the rate of flow of heat through the pane assuming a steady state is maintained. The window is now double-glazed by adding a similar sheet of glass so that a layer of air 10 mm thick is trapped between the two panes. Assuming that the air is still, calculate the ratio of the rate of flow of heat through the window in the first case to that in the second.

Why in practice, would the ratio be much smaller than this?
(Conductivity of glass = $0.80\,\mathrm{W\,m^{-1}\,K^{-1}}$.
Conductivity of air = $0.025\,\mathrm{W\,m^{-1}\,K^{-1}}$.)[L]

C118 Certain special glasses have been developed for use in the manufacture of oven-to-table utensils. Discuss whether high or low values of thermal conductivity, expansivity and specific heat capacity are desirable for these glasses. [C]

C119 If a bar of copper of uniform area of cross-section is well lagged and its ends maintained at different temperatures, the temperature gradient along the bar is uniform. Explain why this is so.

Discuss the effect on the temperature gradient if the cross-sectional area of the bar had not been uniform but had increased uniformly, being greater at the hot end of the bar. Illustrate your answer with a graph. [L]

C120 **(a)** Explain what is meant by *temperature gradient*.
(b) The ends of a perfectly lagged bar of uniform cross-section are maintained at steady temperatures θ_1 and θ_2. Sketch a graph showing how the temperature varies along the bar and account for its shape.
(c) Describe an experimental arrangement which attempts to fulfil the conditions given in (b) and outline the measurements you would make in order to determine the thermal conductivity of the material of the bar. Show how to calculate the thermal conductivity from your measurements.
(d) An iron pan containing water boiling steadily at 100°C stands on a hot-plate and heat conducted through the base of the pan evaporates 0.090 kg of water per minute. If the base of the pan has an area of $0.04\,\mathrm{m^2}$ and a uniform thickness of 2.0×10^{-3} m, calculate the surface temperature of the underside of the pan.
(The thermal conductivity of iron = $66\,\mathrm{W\,m^{-1}\,K^{-1}}$, and the specific latent heat of vaporization of water at 100 °C = $2.2 \times 10^6\,\mathrm{J\,kg^{-1}}$.) [J]

C121 **(a)** A sheet of glass has an area of $2.0\,\mathrm{m^2}$ and a thickness 8.0×10^{-3} m. The glass has a thermal conductivity of $0.80\,\mathrm{W\,m^{-1}\,K^{-1}}$. Calculate the rate of heat transfer through the glass when there is a temperature difference of 20 K between its faces.
(b) A room in a house is heated to a temperature 20 K above that outside. The room has $2\,\mathrm{m^2}$ of windows of glass similar to the type used in (a). Suggest

why the rate of heat transfer through the glass is much less than the value calculated above.

(c) Explain why two sheets of similar glass insulate much more effectively when separated by a thin layer of air than when they are in contact. [AEB, '86]

C122 A double-glazed window consists of two panes of glass each 4 mm thick separated by a 10 mm layer of air. Assuming the thermal conductivity of glass to be 50 times greater than that of air calculate the ratios:

(a) temperature gradient in the glass to temperature gradient in the air gap,

(b) temperature difference across one pane of the glass to temperature difference across the air gap.

Sketch a graph showing how the temperature changes between the surface of the glass in the room and the surface of the glass outside, i.e. across the double-glazed window, if there is a large temperature difference between the room and the outside.

Explain why, in practice, the value of the ratio calculated in (a) is too high. [L]

C123 Outline an experiment to measure the thermal conductivity of a solid which is a poor conductor, showing how the result is calculated from the measurements.

Calculate the theoretical percentage change in heat loss by conduction achieved by replacing a single glass window by a double window consisting of two sheets of glass separated by 10 mm of air. In each case the glass is 2 mm thick. (The ratio of the thermal conductivities of glass and air is 3 : 1.)

Suggest why, in practice, the change would be much less than that calculated. [L]

C124 **(a)** Explain what is meant by a *temperature gradient*. Write down an expression for the rate of heat flow along a cylinder in terms of the temperature gradient and the constants of the cylinder, defining the constants.

In measuring the thermal conductivity of a good conductor, the specimen is normally long and the sides are heavily lagged. For a poor conductor, the specimen is usually very thin and there is often no lagging. What are the reasons for this?

(b) The two ends of a metal bar are maintained at different constant temperatures. On a single graph, sketch curves to show the variation of temperature along the bar when its surface is **(i)** perfectly lagged and **(ii)** unlagged. Explain the shapes of the two curves. [W]

C125 Describe how you would measure the thermal conductivity of a metal.

The two ends of an iron bar of uniform cross-section are maintained at temperatures of 100 °C and 20 °C respectively. Sketch, and explain, the variation of temperature along the bar:

(a) if it is lagged so that no heat can escape through the side,

(b) if it is not lagged.

Explain how aluminium, which is a good conductor of heat, can be used for heat insulation provided it is *very thin*, *crumpled*, and *polished*. [S]

C126 **(a)** The thermal conductivity, λ, of a material may be defined by reference to the expression

$$\frac{q}{t} = -\lambda A \frac{\Delta T}{\Delta x}$$

State the two conditions that must be satisfied before this expression may be applied.

(b) The thermal conductivities of glass and copper, when measured at room temperature, are found to be $0.6\,\mathrm{W\,m^{-1}\,K^{-1}}$ and $400\,\mathrm{W\,m^{-1}\,K^{-1}}$, respectively. Account for the large differences in these values in terms of the mechanisms of thermal conduction.

(c) You have been asked to design some apparatus to determine the thermal conductivity of copper. Describe the shape you would choose for the specimen of copper, giving reasons for your choice.

C127 At low temperatures, the thermal conductivity k of aluminium varies with temperature as illustrated. If heat flows steadily along a lagged aluminium bar of uniform cross section, how does the temperature θ vary with the distance s? Answer by means of a carefully drawn graph of θ against s.

Give your reasoning.

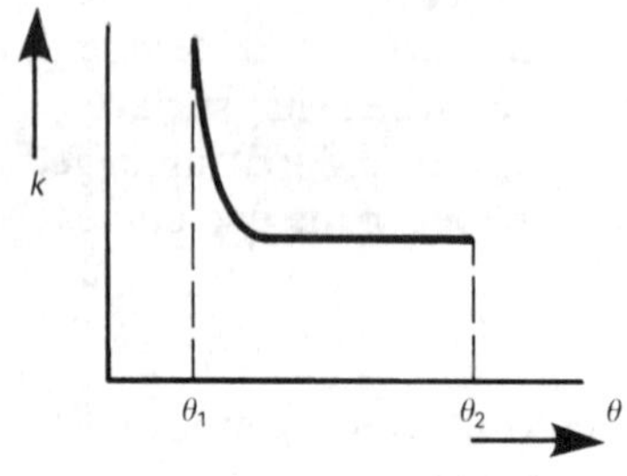

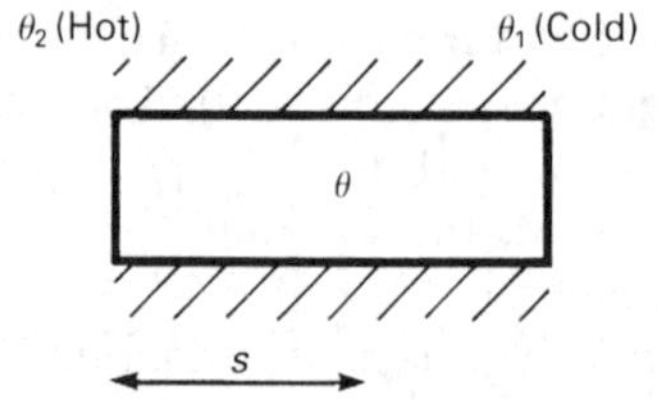

[S]

C128 **(a)** A well-lagged solid copper cylinder of cross section 5.0 cm^2 has one of its ends maintained at a constant temperature well above room temperature. Graph P in the diagram shows how the temperature, θ, of the cylinder varies along part of its length when steady state conditions are reached. Graph Q shows how the temperature changes in an identical unlagged cylinder in the same conditions.

(i) Explain why the temperature gradient is constant along the well-lagged cylinder. Draw a diagram, suitably labelled, showing the lines of heat flow in this cylinder.

(ii) Use graph P to calculate the temperature gradient along the well-lagged cylinder. Assuming the thermal conductivity of copper to be 390 W m^{-1} K^{-1}, calculate the rate of heat flow along the cylinder.

(iii) Describe how the temperature gradient changes along the unlagged cylinder by referring to graph Q. Account for the change.

Where, along the unlagged cylinder, is the temperature gradient equal to that along the lagged cylinder?

(b) Draw a fully labelled diagram of the apparatus you would use to measure the thermal conductivity of a *poor* conductor and list the measurements that you would make. [L]

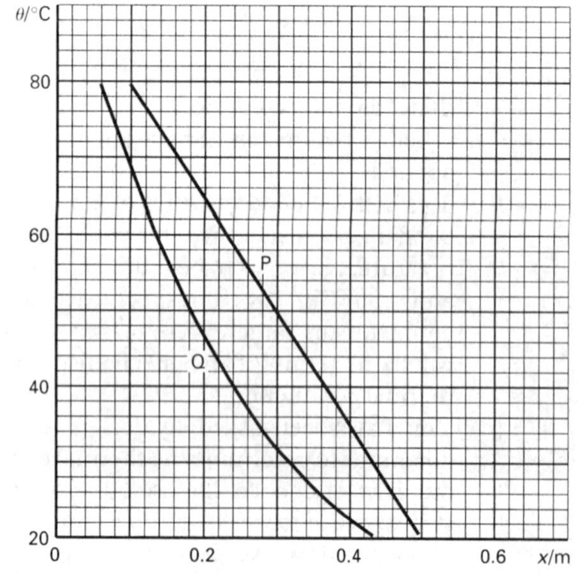

C129 **(a)** Explain what is meant by *temperature gradient*.

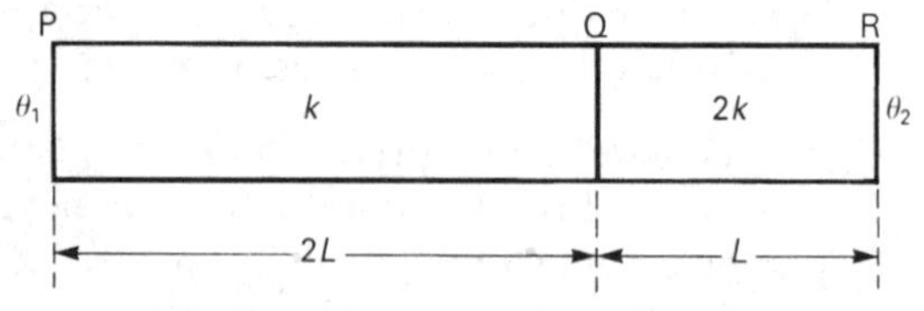

(b) The diagram shows a perfectly lagged composite bar PQR, both components having the same cross-sectional area. The material of QR has twice the thermal conductivity of PQ. Assuming that there is a steady flow of heat along the bar and that the ends P and R are at temperatures θ_1 and θ_2 respectively ($\theta_1 > \theta_2$), sketch a graph showing the variation of temperature along the length PR and account for its shape. Calculate the temperature midway between Q and R when $\theta_1 = 100\,°C$ and $\theta_2 = 0\,°C$.

(c) (i) For a thermal conductor along which heat is flowing at a steady rate, the *thermal resistance*, R_θ, can be defined in a similar way to the electrical resistance of an electrical conductor carrying a steady current. By comparing the equations for heat and current flow, show that the thermal resistance of a conductor of length l, cross-sectional area A and thermal conductivity k is given by

$$R_\theta = \frac{l}{kA}$$

(ii) Calculate the thermal resistance of a pane of window glass 6 mm thick and of area 2.0 m^2. Determine the heat flow rate through the glass when there is a temperature difference of 3.0 °C between its surfaces. A second identical pane of glass is now added to the first to make a double-glazed window with a 12 mm air gap between the glass panes. Calculate the new heat flow rate due to conduction for the same temperature difference between the outer surfaces.
(Thermal conductivity of glass $= 1.0\,\mathrm{W\,m^{-1}\,K^{-1}}$,
Thermal conductivity of air $= 0.02\,\mathrm{W\,m^{-1}\,K^{-1}}$.) [J]

C130 Ice is forming on the surface of a pond. When it is 4.6 cm thick, the temperature of the surface of the ice in contact with the air is 260 K, whilst the surface in contact with the water is at temperature 273 K. Calculate the rate of loss of heat per unit area from the water.

Hence determine the rate at which the thickness of the ice is increasing.

(Thermal conductivity of ice $= 2.3\,\mathrm{W\,m^{-1}\,K^{-1}}$,
Density of water $= 1000\,\mathrm{kg\,m^{-3}}$,
Specific latent heat of fusion of ice $= 3.25 \times 10^5\,\mathrm{J\,kg^{-1}}$.) [S]

C131 The rate of flow of heat energy through a perfectly lagged metal bar of area of cross-section X, length d, and thermal conductivity k, may be considered to be analogous to the rate of flow of charge through an electrical conductor of area of cross-section A, length l, and resistivity ρ. Show that the 'thermal resistance' of the metal bar which corresponds to the electrical resistance of the conductor is d/kX. Extending the analogy to thermal conductors in series show that the effective thermal conductivity K of a composite wall consisting of two parallel-sided layers of material of thickness d_1 and d_2 and thermal conductivities k_1 and k_2 respectively is given by the expression

$$K = \frac{d_1 + d_2}{\dfrac{d_1}{k_1} + \dfrac{d_2}{k_2}}$$

[AEB, '83]

C132 (a) Why is a good conductor of heat also a good conductor of electricity?

(b) Explain how heat is conducted through
(i) a good conductor,
(ii) a poor conductor.

(c) Write down equations for
(i) the rate of flow of charge (current) through a conductor,
(ii) the rate of flow of heat through a substance.
Name the symbols in these equations.

(d) By comparing the equations in part (c), find an expression for the quantity which is the thermal equivalent of electrical resistance.

(e) Using this idea of 'thermal' resistance, or otherwise, calculate the heat passing per second through 1 m^2 of glass of thickness 2 mm when its faces are maintained at 20 °C and 5 °C respectively.
(Thermal conductivity of glass $= 1.2\,\mathrm{W\,m^{-1}\,K^{-1}}$.)

(f) Two such sheets of glass are now placed 4 mm apart and sealed so as to trap air in the space between them forming a 'sandwich' of thickness 8 mm. Given that the thermal conductivity of air is $0.024\,\mathrm{W\,m^{-1}\,K^{-1}}$, calculate the rate of heat conduction per m^2 when the outside faces of the glass are again maintained at 20 °C and 5 °C respectively.

(g) Give *one* application of such a 'sandwich' and briefly explain why, in practice, the rate of heat transfer will be different from that calculated. [W, '91]

C133 The rate of thermal energy transfer dQ/dt perpendicular to the faces of a well-insulated rectangular slab is given by

$$\frac{dQ}{dt} = -kA\frac{d\theta}{dx}$$

where k is the thermal conductivity of the slab, A is its cross sectional area and $d\theta/dx$ is the temperature gradient.

(a) **(i)** What are the analogous quantities to Q, k and $d\theta/dx$ in the electrical conductivity equation?

(ii) Write down an equation for the thermal resistance in terms of k, A, and L, the thickness of the conductor.

(b) The diagram shows a furnace wall which is constructed of two types of brick. The temperatures of the inner and outer surfaces of the wall are 600 °C and 460 °C respectively, as shown in the diagram. The value of the thermal conductivity, k_1, for the inner layer of the furnace wall is $0.8\,W\,m^{-1}\,K^{-1}$ and that of the outer layer, k_2, is $1.6\,W\,m^{-1}\,K^{-1}$.

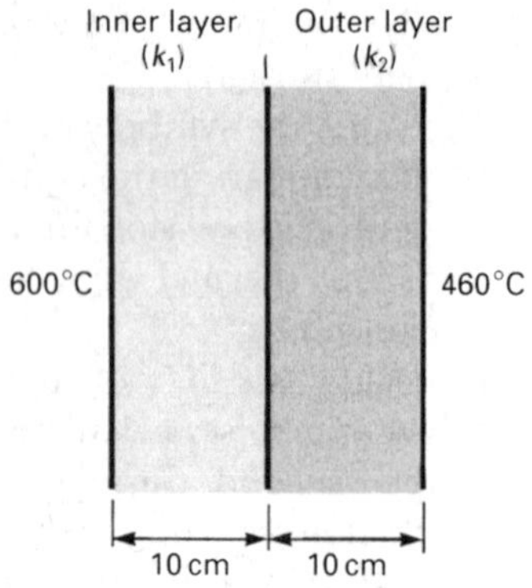

(i) Explain why, in the steady state, the rate of thermal energy transfer must be the same in both layers.

(ii) Determine the temperature at the interface, I, between the layers.

(iii) Sketch and label a graph which shows the variation of temperature with distance across the wall.

[AEB, '91]

THERMAL RADIATION (Chapter 17)

C134 State Prévost's theory of exchanges.

An enclosure contains a black body A which is in equilibrium with it. A second black body B, at a higher temperature than A, is then also placed in the enclosure. If the enclosure is maintained at constant temperature and all heat exchange is by radiation state and explain how the temperatures of A and B will change with time.

Describe an electrical instrument which can be used to detect heat radiation and explain its action.

A solid copper sphere, of diameter 10 mm, is cooled to a temperature of 150 K and is then placed in an enclosure maintained at 290 K. Assuming that all interchange of heat is by radiation, calculate the initial rate of rise of temperature of the sphere. The sphere may be treated as a black body.

(Density of copper = $8.93 \times 10^3\,kg\,m^{-3}$, specific heat capacity of copper = $3.70 \times 10^2\,J\,kg^{-1}\,K^{-1}$, the Stefan constant = $5.70 \times 10^{-8}\,W\,m^{-2}\,K^{-4}$.) [L]

C135 Sketch graphs showing the distribution of energy in the spectrum of black body radiation at three temperatures, indicating which curve corresponds to the highest temperature. If such a set of graphs were obtained experimentally, how would you use information from them to attempt to illustrate Stefan's Law? [L]

C136 Draw a graph showing the distribution of energy in the spectrum of a black body. Explain what quantity is plotted against the wavelength.

By considering how this energy distribution varies with temperature explain the colour changes which occur when a piece of iron is heated from cold to near its melting point.

[L]

C137 As the temperature of a black body rises what changes take place in **(a)** the total energy radiated from it and **(b)** the energy distribution amongst the wavelengths radiated? Illustrate (b) by suitable graphs and explain how the information required in (a) could be obtained from these graphs.

Use your graphs to explain how the appearance of the body changes as its temperature rises and discuss whether or not it is possible for a black body to radiate white light.

The element of an electric fire, with an output of 1.0 kW, is a cylinder 25 cm long and 1.5 cm in diameter. Calculate its temperature when in use, if it behaves as a black body. (Stefan's constant = $5.7 \times 10^{-8}\,\text{W m}^{-2}\,\text{K}^{-4}$.) [L]

C138 The silica cylinder of a radiant wall heater is 0.6 m long and has a radius of 5 mm. If it is rated at 1.5 kW estimate its temperature when operating. State *two* assumptions you have made in making your estimate. (Stefan's constant, $\sigma = 6 \times 10^{-8}\,\text{W m}^{-2}\,\text{K}^{-4}$.) [L]

C139 Explain what is meant by *black body radiation.*

State Stefan's law, and draw graphs to show how the energy depends on wavelength in the spectra of the radiation emitted by a black body at two different temperatures. Indicate which of your two graphs corresponds to the higher temperature.

A blackened metal sphere of diameter 10 mm is placed at the focus of a concave mirror of diameter 0.5 m directed towards the Sun. If the solar power incident on the mirror is $1600\,\text{W m}^{-2}$, calculate the maximum temperature which the sphere can attain. State the assumptions you make.
(Stefan's constant = $6 \times 10^{-8}\,\text{W m}^{-2}\,\text{K}^{-4}$.) [S]

C140 **(a)** The temperature of a piece of wire is gradually increased. Discuss the variation in character of the radiation emitted. Sketch graphs to illustrate this variation. (Assume that the wire behaves like a black body.)

Suggest how you might investigate experimentally the variation in the *total* radiation emitted by the wire at the various temperatures. In what way would this total radiation be related to the graphs already sketched?

(b) If the mean equilibrium temperature of the Earth's surface is T and the total rate of energy emission by the Sun is E show that

$$T^4 = \frac{E}{16\sigma\pi R^2}$$

where σ is the Stefan constant and R is the radius of the Earth's orbit around the Sun. (Assume that the Earth behaves like a black body.) [L]

C141 An unlagged, thin-walled copper pipe of diameter 2.0 cm carries water at a temperature of 40 K above that of the surrounding air. Estimate the power loss per unit length of the pipe if the temperature of the surroundings is 300 K and the Stefan constant, σ, is $5.67 \times 10^{-8}\,\text{W m}^{-2}\,\text{K}^{-4}$.

State *two* important assumptions you have made. [L]

C142 **(a)** **(i)** Give the meaning of the term *black body radiation.*

(ii) State Stefan's law of black body radiation.

(iii) A simple solar heating panel is made from a central heating radiator which has been given a dull black surface. Water is pumped slowly through the panel. Discuss the possibility of raising the water temperature significantly above the temperature of the surroundings. Physical reasons should be given for any conclusions that you draw.

(b) Two kinds of material are available to a householder for insulating the loft of the house. One material is in the form of loose chippings, the other in the form of a roll of matting. Both should be laid to the same thickness. Devise a simple experiment to test in a laboratory which should be the more effective. [S*]

C143 The diagram shows how E_λ, the energy radiated per unit area per second per unit

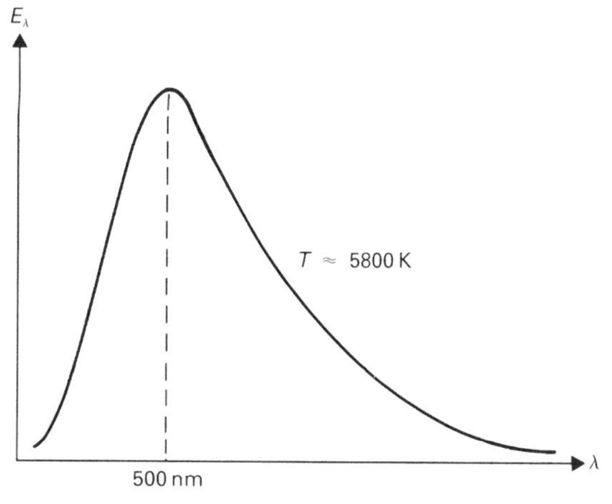

wavelength interval, varies with wavelength λ for radiation from the Sun's surface.

Calculate the wavelengths λ_{max} at which the corresponding curves peak for:

(a) radiation in the Sun's core where the temperature is approximately 15×10^6 K, and

(b) radiation in interstellar space which corresponds to a temperature of approximately 2.7 K

(You may use the relation $\lambda_{max} \times T =$ constant.)

Name the part of the electromagnetic spectrum to which the calculated wavelength belongs in each case. [L]

C144 Describe the structure of a thermopile.

A thermopile was connected to a centre-zero microammeter and it was found that, when a can of hot water was placed near, the microammeter deflected to the right. After removing the hot water the meter was allowed to return to zero and a can containing water and ice was brought near. The microammeter now deflected to the left. Explain this.

If a thermopile fitted with a horn is placed near a radiator of large surface area and then gradually moved away, the microammeter readings change very little until a certain distance is reached and, beyond this, the readings begin to fall. Explain this.

The solar radiation falling normally on the surface of the Earth has an intensity $1.40\,\text{kW m}^{-2}$. If this radiation fell normally on one side of a thin, freely suspended blackened metal plate and the temperature of the surroundings was 300 K, calculate the equilibrium temperature of the plate. Assume that all heat interchange is by radiation.

(Stefan's constant $= 5.67 \times 10^{-8}\,\text{W m}^{-2}\,\text{K}^{-4}$.) [L]

SECTION D

GEOMETRICAL OPTICS

18

REFRACTION

18.1 THE LAWS OF REFRACTION

(i) The incident ray, the refracted ray and the normal at the point of incidence are all in the same plane (Fig. 18.1).

(ii) At the boundary between any two given materials, the ratio of the sine of the angle of incidence to the sine of the angle of refraction is constant for rays of any particular wavelength (Fig. 18.1). This is known as **Snell's law**.

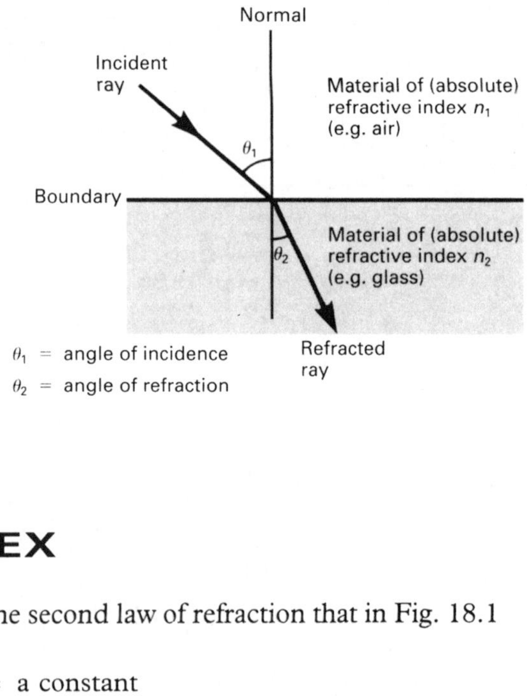

Fig. 18.1
Refraction at a plane surface

18.2 REFRACTIVE INDEX

It follows from the second law of refraction that in Fig. 18.1

$$\frac{\sin\theta_1}{\sin\theta_2} = \text{a constant}$$

The constant is known as **the refractive index of material (2) with respect to material (1)**, ${}_1n_2$*, and Snell's law can be written as

$$\frac{\sin\theta_1}{\sin\theta_2} = {}_1n_2 \qquad [18.1]$$

*Refractive index values depend on the wavelength (colour) of the light – see section 18.7.

It can be shown (see section 24.4) that

$$ {}_1n_2 = \frac{\text{Velocity of light in material (1)}}{\text{Velocity of light in material (2)}} \quad [18.2]$$

Note When light is travelling from material (2) to material (1) we use the refractive index of (1) with respect to (2), ${}_2n_1$. It follows from equation [18.2] that

$$ {}_2n_1 = \frac{1}{{}_1n_2} $$

EXAMPLE 18.1

Find the angle of refraction: (a) when a ray of light is travelling from air to glass at an angle of incidence of 40°, (b) when a ray of light is travelling from glass to air at an angle of incidence of 20°. (Refractive index of glass with respect to air = 1.50.)

Solution

(a) By Snell's law

$$\frac{\sin\theta_1}{\sin\theta_2} = {}_1n_2$$

The angle of incidence (θ_1) is 40° and therefore

$$\frac{\sin 40^\circ}{\sin\theta_2} = 1.50$$

$$\therefore \quad \sin\theta_2 = \frac{\sin 40^\circ}{1.50} = 0.4285$$

$$\therefore \quad \theta_2 = 25.4^\circ$$

(b) The angle of incidence (θ_1) is 20°. In this case light is travelling from glass to air and therefore we require the refractive index of air with respect to glass. This is the reciprocal of the refractive index of glass with respect to air and therefore

$$\frac{\sin 20^\circ}{\sin\theta_2} = \frac{1}{1.50}$$

$$\therefore \quad \sin\theta_2 = 1.50\sin 20^\circ = 0.5130$$

$$\therefore \quad \theta_2 = 30.9^\circ$$

It is convenient to define the **absolute refractive index**, n, of a material as

$$n = \frac{\text{Velocity of light in vacuum}}{\text{Velocity of light in material}}$$

By analogy, the absolute refractive index n_1 of material (1) is given by

$$n_1 = \frac{\text{Velocity of light in vacuum}}{\text{Velocity of light in material (1)}}$$

and the absolute refractive index n_2 of material (2) is given by

$$n_2 = \frac{\text{Velocity of light in vacuum}}{\text{Velocity of light in material (2)}}$$

Therefore

$$\frac{n_2}{n_1} = \frac{\text{Velocity of light in material (1)}}{\text{Velocity of light in material (2)}}$$

i.e. $$\frac{n_2}{n_1} = {}_1n_2 \qquad \text{(by equation [18.2])}$$

It is now possible to write equation [18.1] in terms of absolute refractive indices, i.e.

$$\frac{\sin\theta_1}{\sin\theta_2} = \frac{n_2}{n_1}$$

i.e. $$n_1 \sin\theta_1 = n_2 \sin\theta_2 \qquad [18.3]$$

The absolute refractive index of vacuum is, by definition, equal to 1, and that of air at normal atmospheric pressure and 20 °C is 1.0003. This difference is so slight that it is normally ignored.

It follows from equation [18.3] that if $n_2 > n_1$ (i.e. if material (2) has a greater **optical density** than material (1)), then $\theta_2 < \theta_1$. Thus, **when light travels from a material into one which has a greater absolute refractive index (i.e. a greater optical density), it bends towards the normal** (Fig. 18.2).

Fig. 18.2
Refraction by a glass block

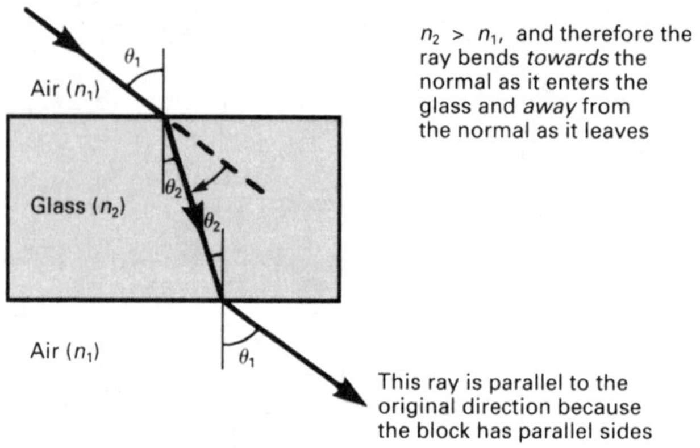

θ_1
Air (n_1)
Glass (n_2)
θ_2
θ_2
Air (n_1)
θ_1
$n_2 > n_1$, and therefore the ray bends *towards* the normal as it enters the glass and *away* from the normal as it leaves
This ray is parallel to the original direction because the block has parallel sides

EXAMPLE 18.2

A ray of light is incident in glass on a glass–water boundary. The angle of incidence is 50°. Calculate the angle of refraction. (Refractive index of glass = 1.50, refractive index of water = 1.33.)

Solution

$$n_1 \sin\theta_1 = n_2 \sin\theta_2$$

$$\therefore \quad 1.50 \sin 50^\circ = 1.33 \sin\theta_2$$

$$\therefore \quad \sin\theta_2 = \frac{1.50}{1.33}\sin 50^\circ = 0.8640$$

$$\therefore \quad \theta_2 = 59.8^\circ$$

Note We have worked with absolute refractive indices. If we had been given the refractive indices of glass and of water with respect to air, we could have proceeded as follows.

$$n_1 \sin\theta_1 = n_2 \sin\theta_2$$

$$\therefore \quad \frac{n_1}{n_a}\sin\theta_1 = \frac{n_2}{n_a}\sin\theta_2$$

where n_a is the (absolute) refractive index of air. It follows that

$${}_a n_1 \sin\theta_1 = {}_a n_2 \sin\theta_2 \qquad \text{etc.}$$

where ${}_a n_1$ is the refractive index of glass with respect to air and ${}_a n_2$ is the refractive index of water with respect to air.

18.3 TOTAL INTERNAL REFLECTION AND CRITICAL ANGLE

When light travels from an optically more dense material to an optically less dense material (e.g. from glass to air) it is possible for the angle of incidence to be such that the angle of refraction is 90° (Fig. 18.3(b)).

The angle of incidence at which this happens is called the **critical angle**, c. From equation [18.3]

$$n_1 \sin 90^\circ = n_2 \sin c$$

Therefore, since $\sin 90^\circ = 1$

$$\sin c = \frac{n_1}{n_2} = \frac{1}{{}_1 n_2} \qquad (n_2 > n_1) \qquad [18.4]$$

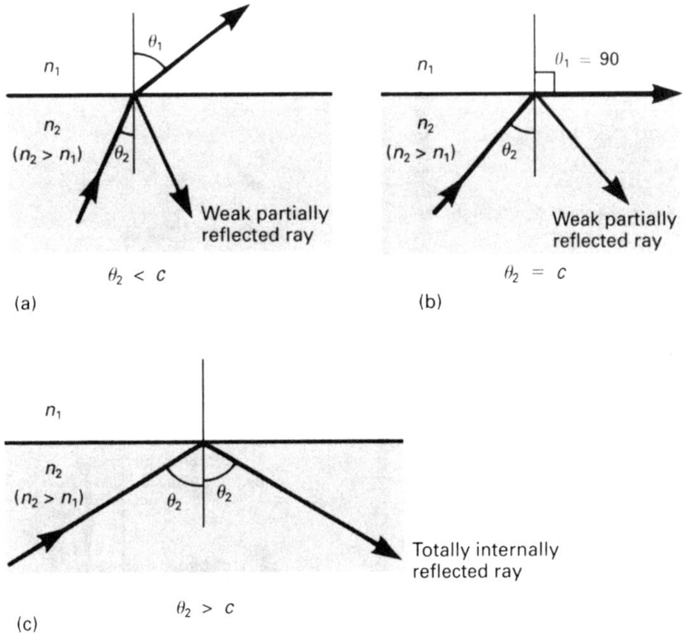

Fig. 18.3
Three possibilities when light travels towards a less dense medium

If $\theta_2 < c$, the light is refracted in the normal way (Fig. 18.3(a)). If $\theta_2 > c$, the light is totally reflected back into material (2) (Fig. 18.3(c)) – this is known as **total internal reflection**. Total internal reflection cannot occur when the light is travelling towards an optically more dense material.

QUESTIONS 18A

1. A ray of light is incident in air on the surface of a glass block. The angle of incidence is 60°. Calculate the angle of refraction.
(Refractive index of glass with respect to air = 1.50.)

2. A ray of light is incident in water: **(a)** on a plane water–air boundary at an angle of incidence of 45°, **(b)** on a plane water–glass boundary at an angle of incidence of 30°. Calculate the angle of refraction in each case.
(Refractive index of glass = 1.50, refractive index of water = 1.33.)

3. The refractive index of diamond with respect to air is 2.42. Calculate the critical angle for a diamond–air boundary.

18.4 TOTALLY REFLECTING PRISMS

The critical angle for a glass/air boundary is about 42°. Thus, whenever light which is travelling in glass is incident on such a boundary at an angle of more than 42°, it undergoes total internal reflection (Fig. 18.4). Totally reflecting prisms find applications in optical instruments and are superior to silvered glass mirrors in two respects.

(i) Mirrors absorb some of the incident light, whereas all the incident light is reflected when total internal reflection occurs.

(ii) In order that the silvered surface does not tarnish, glass mirrors are usually silvered on the back – the finite thickness of glass in front of the reflecting surface produces ghost images (Fig. 18.5).

Fig. 18.4
Totally reflecting prisms

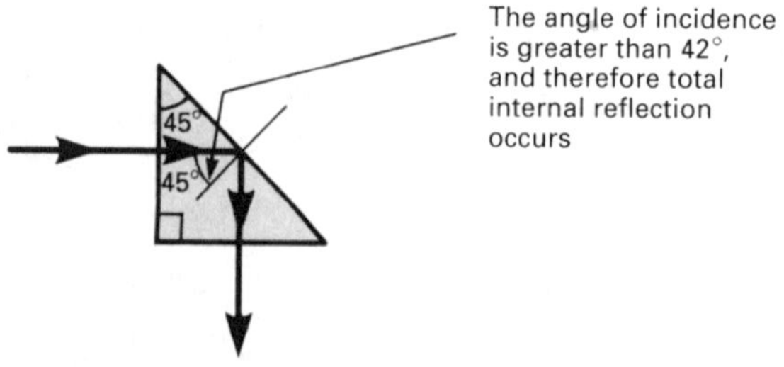
The angle of incidence is greater than 42°, and therefore total internal reflection occurs
45°
45°

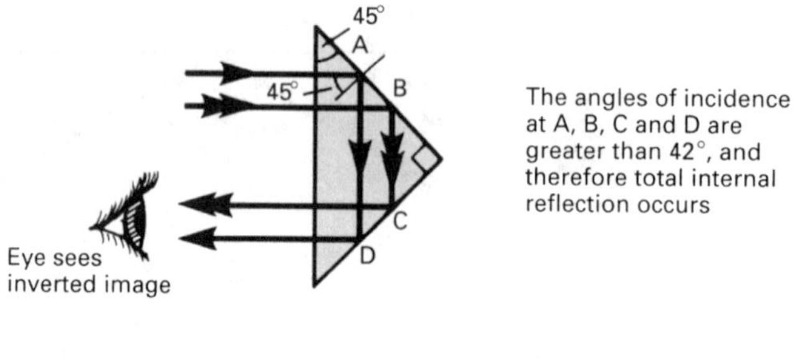
45°
A
45°
B
C
D
The angles of incidence at A, B, C and D are greater than 42°, and therefore total internal reflection occurs
Eye sees inverted image

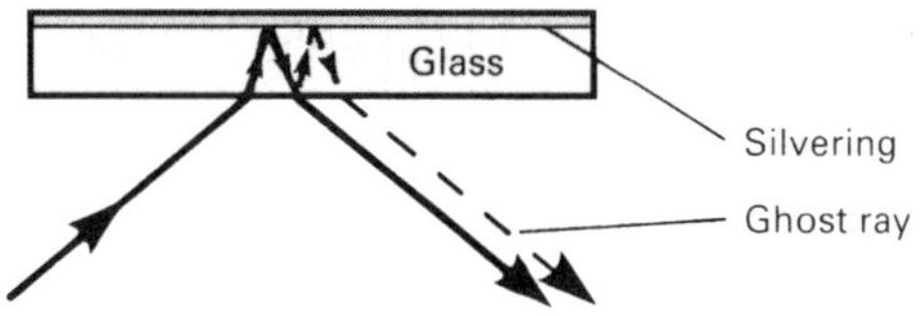

Fig. 18.5
Production of a ghost ray

Total internal reflection is made use of in optical fibres (light pipes) – see page 874, and is also responsible for the formation of mirages.

18.5 THE MEASUREMENT OF REFRACTIVE INDEX

Ray Tracing

The refractive index of a solid, in the form of a rectangular block, can be determined by passing a ray of light (from air) into the block and measuring the angle of incidence (θ_1) and the angle of refraction (θ_2). Then, from equation [18.3]

$$n_2 = n_1 \frac{\sin\theta_1}{\sin\theta_2}$$

where n_2 is the required absolute refractive index, and n_1 (the refractive index of air) can be taken to be unity.

Real and Apparent Depth Method

This method can be used for a solid in the form of a rectangular block or for a liquid.

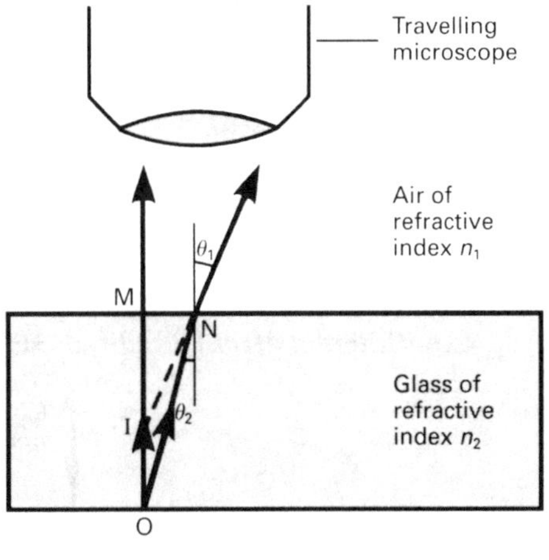

Fig. 18.6
Real and apparent depth method of measuring refractive index

In Fig. 18.6, light from O is refracted on emerging from the glass block at N so that it appears to have come from I. From equation [18.3]

$$n_1 \sin\theta_1 = n_2 \sin\theta_2$$

By simple geometry

$$\widehat{MIN} = \theta_1 \quad \text{and} \quad \widehat{MON} = \theta_2$$

$$\therefore \quad n_1 \sin\widehat{MIN} = n_2 \sin\widehat{MON}$$

$$\therefore \quad n_1 \frac{MN}{IN} = n_2 \frac{MN}{ON}$$

$$\text{i.e.} \quad \frac{n_2}{n_1} = \frac{ON}{IN}$$

The rays which enter the travelling microscope are confined to a narrow cone, in which case ON ≈ OM and IN ≈ IM and therefore

$$\frac{n_2}{n_1} = \frac{\text{OM}}{\text{IM}}$$

Since n_1 is the refractive index of air, it is equal to unity, and therefore

$$n_2 = \frac{\text{OM}}{\text{IM}}$$

i.e.
$$n_2 = \frac{\text{Real depth}}{\text{Apparent depth}} \qquad [18.5]$$

Procedure

(i) Focus the travelling microscope on O (say a mark on a piece of paper) before the block has been placed over it. Suppose that the scale position is (a).

(ii) Put the block in position and move the travelling microscope upwards so that it is focused on I, in which case the mark will once again be in focus. Suppose that the scale position is now (b).

(iii) Move the travelling microscope upwards again until the top of the block (M) is in focus. Suppose that the scale position is now (c).

Since OM $= (c - a)$ and IM $= (c - b)$, equation [18.5] gives

$$n_2 = \frac{c - a}{c - b}$$

The refractive index of a liquid can be found by focusing on, say, a grain of sand on the bottom, inside surface of some suitable, empty container (a). Some of the liquid is then put into the container, and the travelling microscope is refocused on the sand (b). Finally the travelling microscope is focused on the liquid surface (c).

The Air Cell Method

The refractive index of a liquid can be determined by using an air cell (i.e. two parallel-sided glass plates cemented so as to enclose a thin layer of air). The experimental arrangement, as seen from above, is shown in Fig. 18.7.

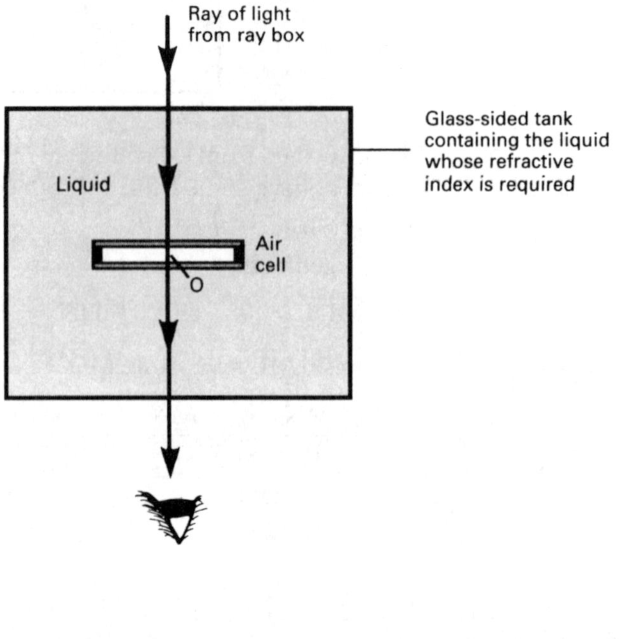

Fig. 18.7
Air cell method of measuring refractive index

When the air cell is in the orientation shown, the light from the ray box passes straight through to the observer. If the air cell is turned through a small angle, about an axis through O and perpendicular to the paper, the slit in the ray box appears to be displaced but is still visible to the observer. If the angle of rotation of the air cell is increased, there comes a point when the image of the slit suddenly disappears. This happens when the light is incident on the cell at such an angle that the light in the glass wall of the cell (Fig. 18.8) is incident on the glass/air boundary at the critical angle and therefore does not pass through the air film to the other side of the cell.

Fig. 18.8
Critical angle in an air cell

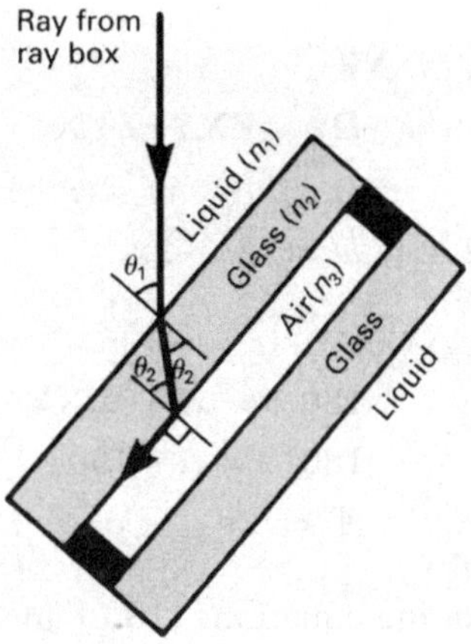

Applying equation [18.3] to the liquid/glass boundary gives

$$n_1 \sin \theta_1 = n_2 \sin \theta_2$$

Applying the same equation to the glass/air boundary gives

$$n_2 \sin \theta_2 = n_3 \sin 90^\circ$$

$$\therefore \quad n_1 \sin \theta_1 = n_3 \sin 90^\circ$$

But $n_3 = 1$ (the refractive index of air) and $\sin 90^\circ = 1$, therefore

$$n_1 = \frac{1}{\sin \theta_1}$$

Thus, n_1, the required refractive index, can be determined by measuring θ_1. There is, of course, a cut-off position on each side of the normal. In practice, the cell is turned from one of these positions to the other, and the angle turned through $(2\theta_1)$ is measured.

The Spectrometer Method

See section 21.12.

18.6 DEVIATION BY A PRISM

The deviation produced by a prism depends on the angle at which the light is incident on the prism. It can be shown, both theoretically and by experiment, that the deviation is a minimum when the light passes symmetrically through the prism. This situation is shown in Fig. 18.9, where D is the **minimum deviation**.

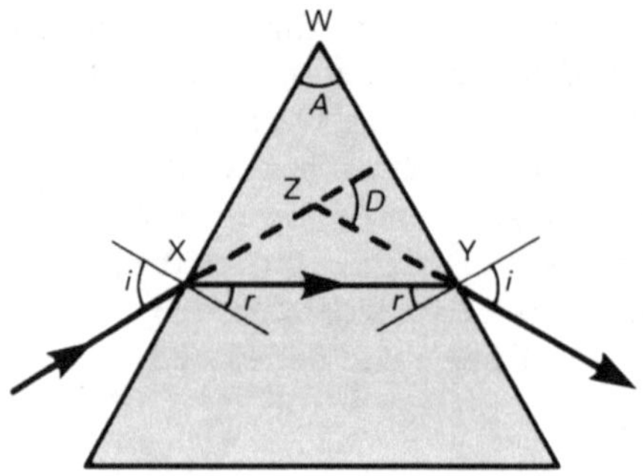

Fig. 18.9
Minimum deviation in a prism

Note that the internal ray is parallel to the base of the prism

In △XYZ

$$D = Z\hat{X}Y + Z\hat{Y}X$$

$$\therefore \quad D = (i - r) + (i - r)$$

$$\text{i.e.} \quad D = 2i - 2r \qquad [18.6]$$

In △WXY

$$180^\circ = A + W\hat{X}Y + W\hat{Y}X$$

$$\therefore \quad 180^\circ = A + (90^\circ - r) + (90^\circ - r)$$

$$\text{i.e.} \quad A = 2r \qquad [18.7]$$

Adding equations [18.6] and [18.7] leads to

$$i = \frac{A + D}{2}$$

From equation [18.7]

$$r = \frac{A}{2}$$

From equation [18.3]

$$n_1 \sin i = n_2 \sin r$$

$$\therefore \quad n_1 \sin \frac{A + D}{2} = n_2 \sin \frac{A}{2}$$

If the light is incident from air, $n_1 = 1$, and therefore

$$n_2 = \frac{\sin\frac{A + D}{2}}{\sin\frac{A}{2}} \qquad [18.8]$$

where n_2 is the refractive index of the material of the prism.

18.7 DISPERSION

When a narrow beam of white light is refracted by a prism the light spreads into a band of colours – the **spectrum** of the light. The effect was first explained by Newton. He identified the colours as ranging from red at one side of the band, through orange, yellow, green, blue and indigo, to violet at the other. When Newton isolated any one of the colours and passed it through a second prism there was no further colour change. He concluded that the colours had not been introduced by the prism, but that they were components of the white light – the prism had merely separated them from each other. The separation of the white light into its component colours in this way is called **dispersion**.

Fig. 18.10 illustrates the effect for a single ray of white light. For clarity only the extreme rays, red and violet, are shown and their angular separation is exaggerated. The red and violet rays, as components of the white light, are incident on the prism at the same angle. On entering the prism the violet ray is refracted through a larger angle than the red ray. It follows, therefore, that the refractive index of the prism material is greater for violet light than it is for red light. The difference is about 1% for glass, but depends on the type of glass.

Fig. 18.10
Dispersion due to a prism

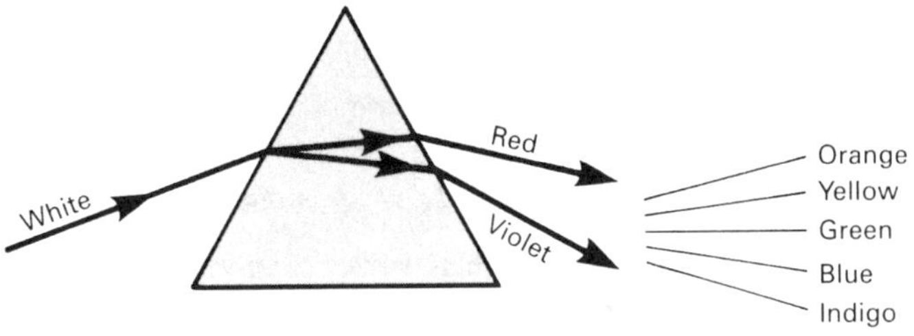

We now know that the spectrum of white light covers a continuous range of wavelengths. What Newton identified as red is in fact a range of different reds corresponding to the band of wavelengths which we see as red; what he identified as blue is a range of different blues corresponding to another band of wavelengths, and so on. The various wavelengths corresponding to any one colour are refracted by different amounts. It follows that for an accurate measurement of refractive index to be meaningful, it is necessary to know the wavelength at which it has been measured.

In a vacuum (and to a good approximation in air) all electromagnetic waves travel at the same speed. In particular, then, red light and violet light travel at the same speed in a vacuum. In glass, however, they travel at different speeds, and this is why the refractive index of glass for violet light is different from that for red light. In section 18.2 we defined the absolute refractive index n of a material by

$$n = \frac{\text{Velocity of light in vacuum}}{\text{Velocity of light in material}}$$

It follows that the absolute refractive index of glass for violet light n_V is given by

$$n_V = \frac{\text{Velocity of light in vacuum}}{\text{Velocity of violet light in glass}}$$

and the absolute refractive index of glass for red light n_R is given by

$$n_R = \frac{\text{Velocity of light in vacuum}}{\text{Velocity of red light in glass}}$$

Since n_R is less than n_V, it follows that the velocity of red light in glass is greater than that of violet light in glass. Neither red light nor violet light (nor any other colour) suffers a change in frequency on entering glass; their reduced velocities are due to their wavelengths being less in glass than they are in vacuum. It follows that

$$n_V = \frac{\text{Wavelength of violet light in vacuum}}{\text{Wavelength of violet light in glass}}$$

and

$$n_R = \frac{\text{Wavelength of red light in vacuum}}{\text{Wavelength of red light in glass}}$$

CONSOLIDATION

Light bends towards the normal when it travels from an optically less dense material to an optically more dense material, i.e. towards a material which has a greater refractive index.

$$\text{Refractive index of (2) w.r.t. (1)} = {}_1n_2 = \frac{\text{Velocity of light in (1)}}{\text{Velocity of light in (2)}}$$

$$\text{Absolute refractive index} = n = \frac{\text{Velocity of light in vacuum}}{\text{Velocity of light in material}}$$

$$\frac{\sin\theta_1}{\sin\theta_2} = {}_1n_2 \qquad \text{(For light going from (1) to (2))}$$

$$n_1 \sin\theta_1 = n_2 \sin\theta_2$$

Total internal reflection can occur only when light travels from an optically more dense material towards an optically less dense material (e.g. glass to air).

The critical angle (c) is the angle of incidence which causes the angle of refraction to be 90°.

$$\sin c = \frac{n_1}{n_2} = \frac{1}{{}_1n_2} \qquad (n_2 > n_1)$$

When light crosses the boundary between materials which have different refractive indices there is a change in velocity and in wavelength, but not in frequency.

19

LENSES

19.1 BASIC PROPERTIES OF LENSES

Lenses may be either converging or diverging. Examples of each type are shown in Fig. 19.1. A converging lens is so called because it causes rays of light incident on it to converge more (or diverge less) after refraction. A diverging lens does the opposite.

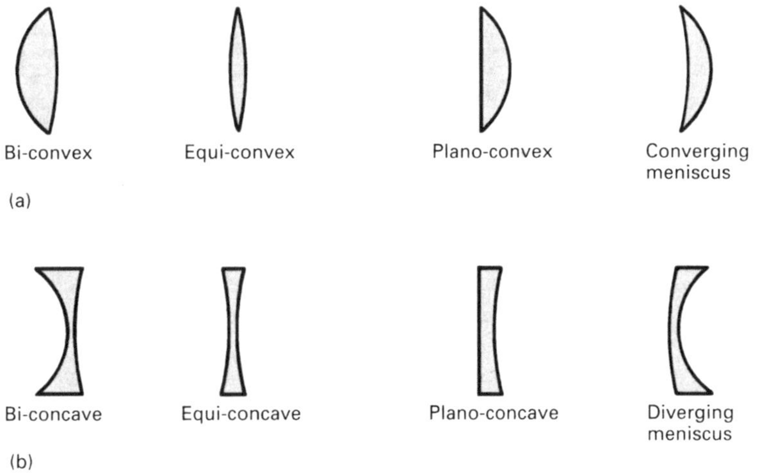

Fig. 19.1
(a) Converging lenses (thicker in middle).
(b) Diverging lenses (thinner in middle)

The focal point (principal focus) of a converging lens is that point, on the principal axis of the lens, to which rays of light which are parallel and close to the axis converge after refraction at the surfaces of the lens (Fig. 19.2).

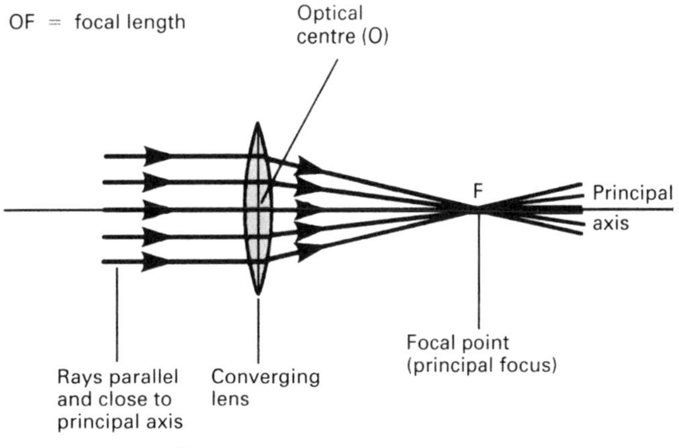

Fig. 19.2
Focal point of a converging lens

The principal axis of a lens (either converging or diverging) is the line which passes through the centres of curvature of the lens surfaces (Figs. 19.2 and 19.3).

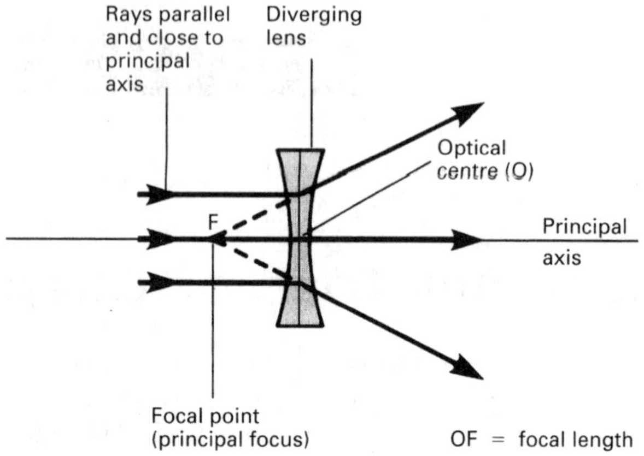

Fig. 19.3
Focal point of a diverging lens

The focal point (principal focus) of a diverging lens is that point, on the principal axis of the lens, from which rays of light which are parallel and close to the axis appear to diverge after refraction at the surfaces of the lens (Fig. 19.3).

The power of a lens is defined by

$$\text{Power} = \frac{1}{\text{Focal length in metres}}$$

Unit = metre^{-1}. (This is sometimes called the **dioptre**, but the name has not been adopted by SI.)

The power of a converging lens is positive; that of a diverging lens is negative. A converging lens of focal length 25 cm therefore has a power of 4 m^{-1}; a diverging lens of focal length 20 cm has a power of -5 m^{-1}.

Note Refraction actually takes place at the surfaces of a lens but it is normal practice when drawing ray diagrams (section 19.3) to treat it as occurring at the middle line – a reasonable approximation in the case of a thin lens.

19.2 IMAGES

Each type of lens can be used to produce both real and virtual images. (In order to obtain a real image with a diverging lens it is necessary to use a virtual object – see section 19.8.)

A real image is one through which rays of light actually pass, and it can therefore be formed on a screen. **A virtual image** is one from which rays of light only appear to have come, and it therefore cannot be formed on a screen.

Whenever we 'look at an object through a lens', the rays of light that enter our eyes do not come directly from the object but come (or appear to have come) from its image (Fig. 19.4).

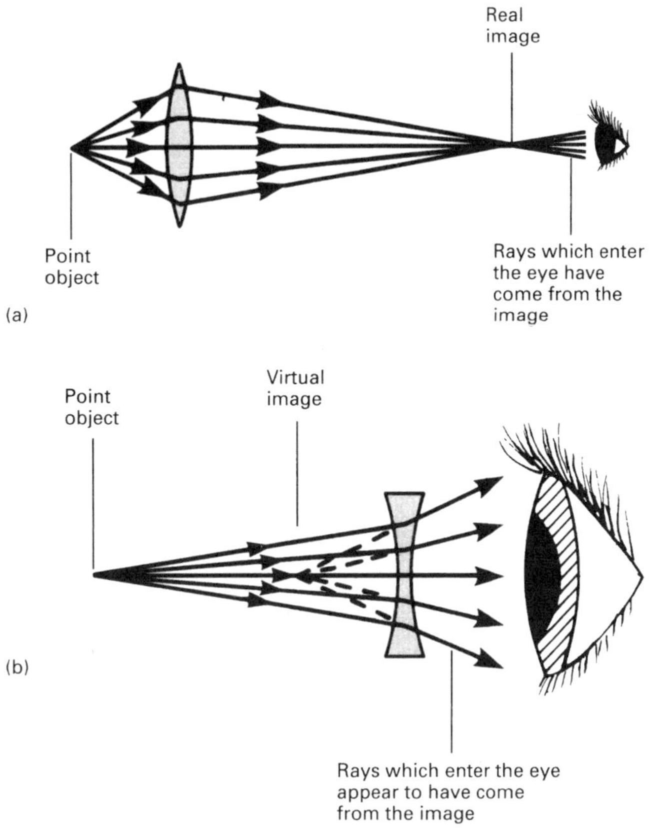

Fig. 19.4
Viewing an image produced by (a) a converging lens, (b) a diverging lens

Notes (i) When a lens (of either type) forms an image, the image is on the opposite side to the object if both object and image are real or both are virtual.

(ii) The object and image are on the same side of the lens if one is real and the other is virtual.

19.3 IMPORTANT RAY PATHS

All the rays of light that leave a single point on an object and go into making up an image pass through (or appear to have come from) a single point on the image. It follows that if the paths of any two rays from a single point on an object are known, the position of its image can be located. For each type of lens, there are three rays (Fig. 19.5) whose paths can be determined, and any two of them can be used.

In each case:

(i) ray I, through the optical centre (O) of the lens, is undeviated;

(ii) ray II, parallel to the principal axis, passes through (or appears to have come from) F;

(iii) ray III, through (or heading towards) F′, emerges parallel to the principal axis.

Rays of light which are incident on a lens and which are parallel to each other but not to the principal axis, converge to (or appear to diverge from) a point in the **focal plane** (Fig. 19.6) of the lens. The position of this point is located by recognizing that any ray which is incident on the optical centre of a lens passes through without deviation.

Fig. 19.5
Ray paths through
(a) a converging lens,
(b) a diverging lens

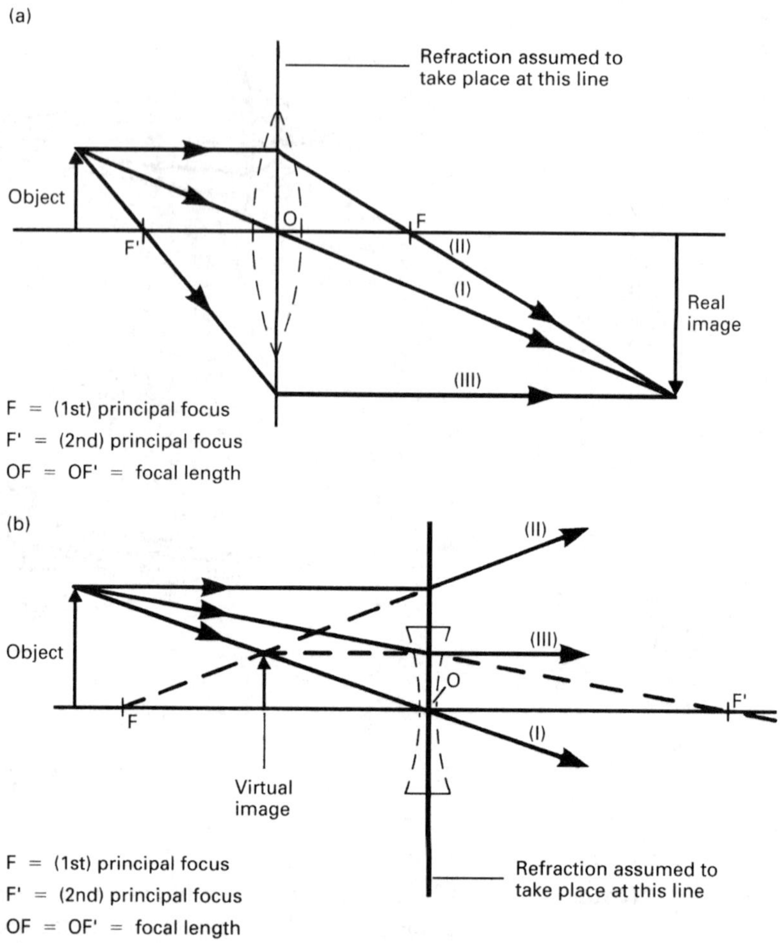

Fig. 19.6
Focal plane of
(a) a converging lens,
(b) a diverging lens

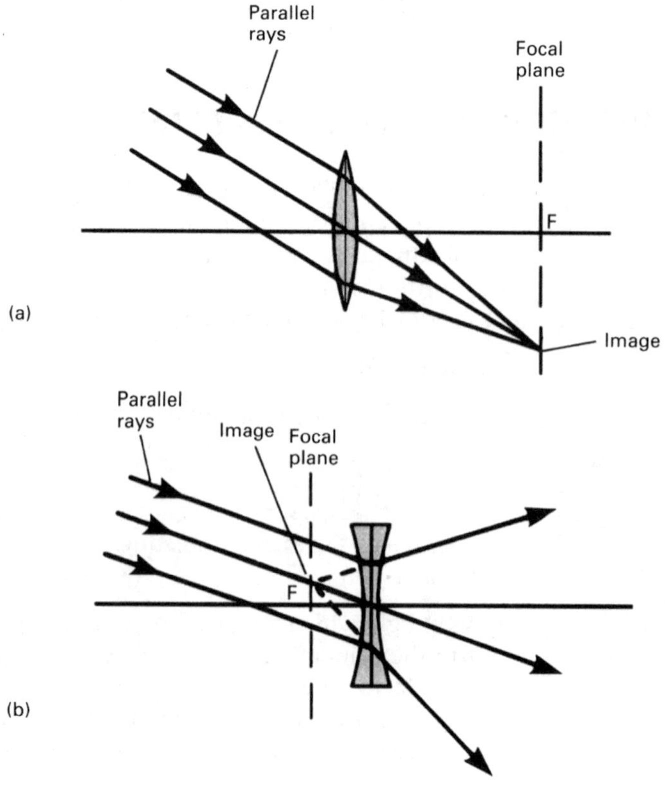

Note Rays which are parallel to each other have effectively come from a single point at infinity and therefore pass through (or appear to have come from) a single image point.

Fig. 19.7 shows the action of a lens on the cone of light incident on the lens from a single point on an object. Two of the rays which can be used to locate the image are also shown.

Fig. 19.7
Action of a converging lens on the cone of light from a point

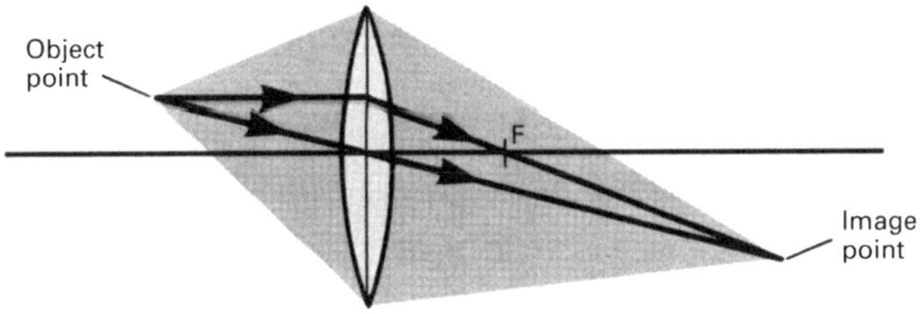

19.4 THE LENS FORMULA

$$\frac{1}{u}+\frac{1}{v}=\frac{1}{f} \qquad [19.1]$$

where

u = object distance
v = image distance
f = focal length of lens
(see Fig. 19.8)

Fig. 19.8
To illustrate the meanings of ***u***, ***v*** and ***f***

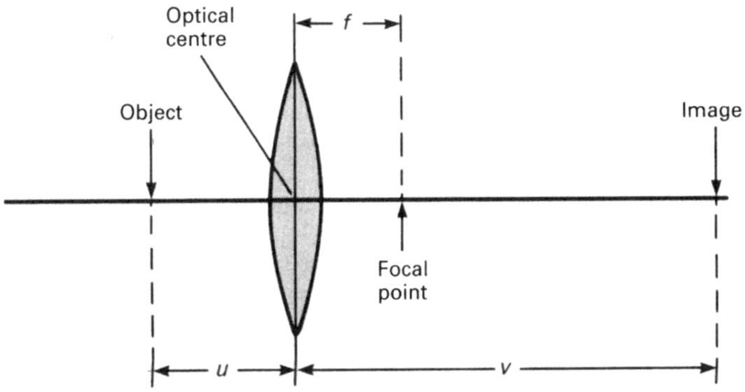

In order to distinguish between real and virtual images, and between converging and diverging lenses, it is necessary to employ a sign convention. This book uses the **Real is Positive** convention, in which:

(i) the focal lengths of converging lenses are positive, those of diverging lenses are negative;

(ii) distances from lenses to real objects and real images are positive, whereas distances to virtual objects and virtual images are negative.

EXAMPLE 19.1

An object is placed 20 cm from a converging lens of focal length 30 cm. Calculate the position of the image.

Solution

$$u = +20\text{ cm} \qquad \text{(real object)}$$

$$v = v$$

$$f = +30\text{ cm} \qquad \text{(converging lens)}$$

$$\frac{1}{u} + \frac{1}{v} = \frac{1}{f}$$

$$\therefore \quad \frac{1}{20} + \frac{1}{v} = \frac{1}{30}$$

$$\therefore \quad \frac{1}{v} = \frac{1}{30} - \frac{1}{20} = \frac{2-3}{60} = -\frac{1}{60}$$

i.e. $v = -60\text{ cm}$

Since the image distance has turned out to be negative, the image is virtual. The image is therefore 60 cm from the lens and on the same side as the object.

QUESTIONS 19A

1. Calculate the distance of the image from the lens in the following situations. In each case state whether the image is real or virtual.

	Object distance/cm	*Focal length*/cm	*Type of lens*
(a)	30.0	40.0	Converging
(b)	40.0	20.0	Diverging
(c)	15.0	20.0	Diverging
(d)	25.0	20.0	Converging
(e)	50.0	20.0	Converging

19.5 TWO THIN LENSES IN CONTACT

The images produced by lenses are often coloured at the edges as a result of dispersion (section 18.7). The defect is called **chromatic aberration,** and it can be reduced by using a lens which is a combination of two lenses made from different types of glass and cemented together (section 19.10). The focal length of the combination is related to the focal lengths of the component lenses.

Consider two thin lenses which are in contact with each other. Suppose that their focal lengths are f_1 and f_2 and that that of the combination is f (Fig. 19.9). It is assumed throughout that since the lenses are thin, CC_1 and CC_2 are not significant in comparison with u, v and v'.

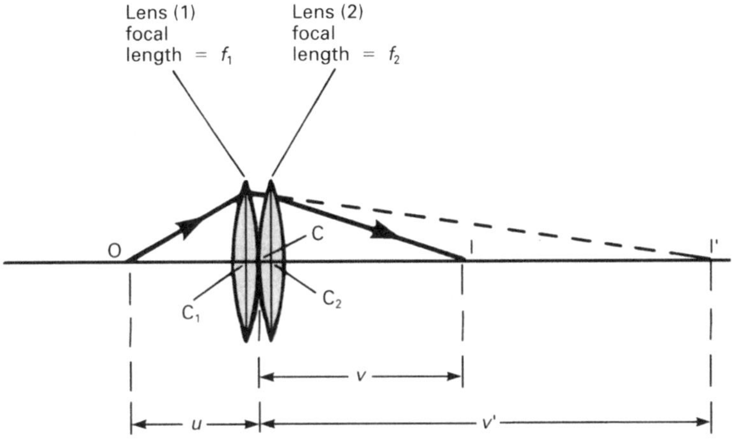

Fig. 19.9
Effect of two converging lenses in contact

If lens (2) were not present, lens (1) would produce an image of an object at O at I′. Therefore for lens (1)

$$\frac{1}{u}+\frac{1}{v'}=\frac{1}{f_1} \qquad [19.2]$$

Lens (2) produces an image, at I, of light which is originally converging to I′, i.e. a virtual object at I′ gives rise to a real image at I. Bearing in mind that virtual object distances are negative, we have for lens (2)

$$-\frac{1}{v'}+\frac{1}{v}=\frac{1}{f_2} \qquad [19.3]$$

Adding equations [19.2] and [19.3] leads to

$$\frac{1}{u}+\frac{1}{v}=\frac{1}{f_1}+\frac{1}{f_2}$$

But, for the combination,

$$\frac{1}{u}+\frac{1}{v}=\frac{1}{f}$$

Therefore

$$\frac{1}{f}=\frac{1}{f_1}+\frac{1}{f_2} \qquad [19.4]$$

19.6 THE PRINCIPLE OF REVERSIBILITY

According to the **principle of reversibility**, a ray of light which travels along any particular path from some point A to another point B, travels by the same path when going from B to A. Thus, if a ray of light were to leave O (Fig. 19.10) and travel to I, then a ray could travel from I to O along the same path. If I is the position of the image of an object which is at O, then every ray which leaves O and passes through the lens must also pass through I. Since each of these rays could make the journey in the opposite direction, it follows that an object at I would give rise to an image at O. This result applies to any object and its image, i.e. whenever an object at some point A gives rise to an image at some other point B, an object at B would give rise to an image at A.

Fig. 19.10
Principle of reversibility

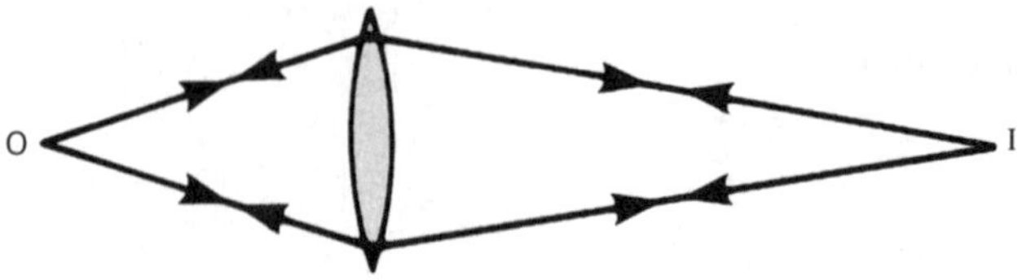

19.7 EXPERIMENTAL DETERMINATIONS OF FOCAL LENGTHS OF CONVERGING LENSES

Distant Object Method

A rough estimate of the focal length of a converging lens can be made by positioning the lens so that it produces a sharply focused image of a distant window on a screen. To a reasonable approximation, the rays incident on the lens from the window are parallel rays and therefore the distance between the lens and the screen can be taken to be the focal length of the lens. (Estimate the error in the case of a lens of focal length 20 cm when using a window which is 5 m from the lens.)

Lens Formula Method

By using an illuminated object and a screen, it is possible to obtain a number of values of u and the corresponding values of v. The best estimate of the focal length of the lens is the average of the values obtained by substituting each pair of values of u and v into the lens formula (equation [19.1]). Alternatively, a graph of $1/u$ against $1/v$ could be plotted (Fig. 19.11). Each intercept of such a graph is equal to $1/f$, hence f.

Fig. 19.11
Graph to determine f for a converging lens

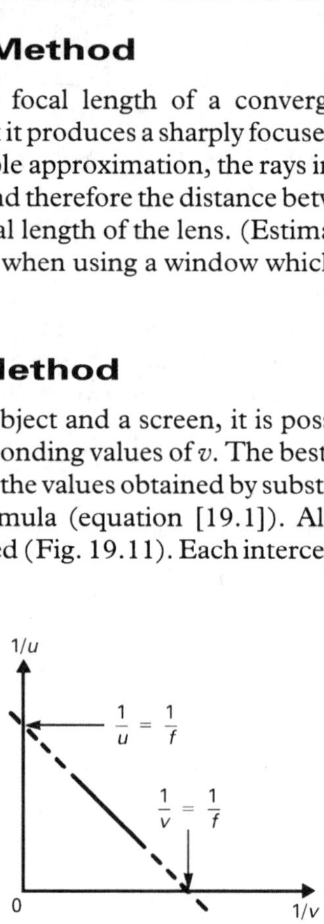

A no-parallax technique (see section 20.3) with two pins can be used instead of the object and screen. If this is done, virtual image positions can be located in addition to those of real images.

Plane Mirror Method (Self-conjugate Foci)

Consider a converging lens resting on a plane mirror (Fig. 19.12) and suppose that there is an object at the focal point (F) of the lens. Since the object is at the focal point, rays of light from it are parallel to the principal axis when they emerge from the lens. As a consequence, rays strike the mirror at 90° to its face and are reflected back along their original paths to F. Thus, the location of the focal point of the lens can be found by discovering the position at which some object is coincident with its own image. Once the position of the focal point has been found, its distance from the optical centre of the lens can be measured in order to obtain the focal length.

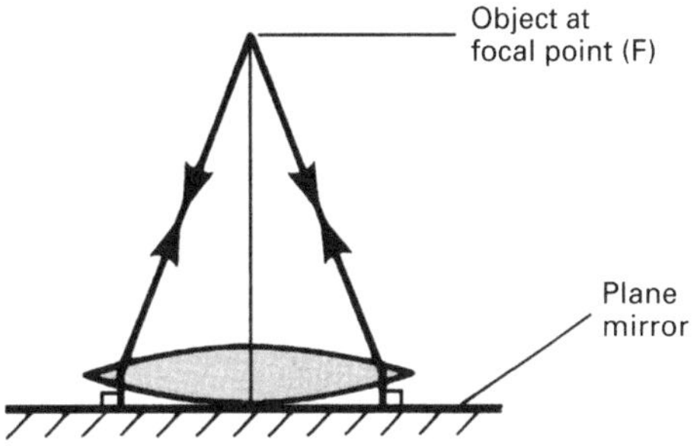

Fig. 19.12 Use of a mirror to determine ***f*** for a converging lens

In practice, the position of a pin suspended above the lens is adjusted so that there is no parallax between the tip of the pin and its image. Alternatively, an illuminated object (e.g. a cross-hair) set in a screen can be used.

The Displacement Method

In Fig. 19.13, a converging lens at L_1 produces an image at I of an object at O. It follows from the principle of reversibility (section 19.6) that an object at I would give rise to an image at O, i.e. an object at a distance v from the lens would give rise to an image whose distance from the lens is u. This situation could be achieved by keeping the object at O and moving the lens to L_2.

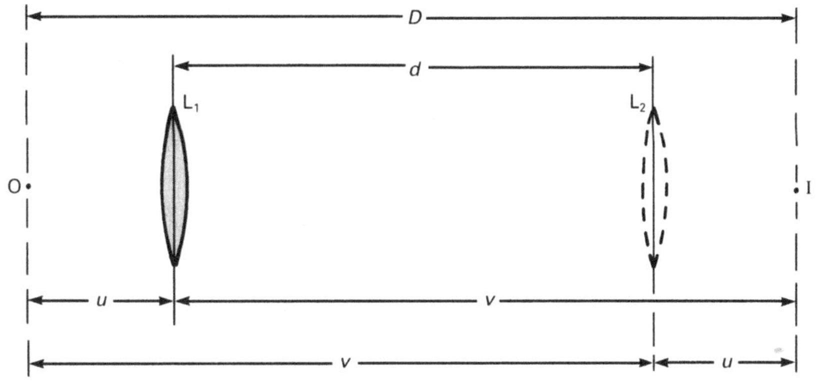

Fig. 19.13 Displacement method to determine ***f*** for a converging lens

From Fig. 19.13

$$2u = D - d \qquad \therefore \quad u = \frac{D-d}{2}$$

$$v = d + u = d + \frac{D-d}{2} \qquad \text{i.e.} \quad v = \frac{D+d}{2}$$

Substituting these values of u and v into the lens formula (equation [19.1]) leads to

$$\frac{2}{D-d} + \frac{2}{D+d} = \frac{1}{f}$$

$$\therefore \qquad \frac{2D + 2d + 2D - 2d}{(D-d)(D+d)} = \frac{1}{f}$$

$$\therefore \qquad \frac{4D}{D^2 - d^2} = \frac{1}{f}$$

i.e. $$f = \frac{D^2 - d^2}{4D} \qquad [19.5]$$

Equation [19.5] provides a convenient method of determining the focal length of a lens. It is particularly useful in the case of an inaccessible lens because, unlike many methods, there is no necessity to make measurements to the optical centre of the lens.

It can be shown that **the minimum distance between a real object and its real image is 4*f***, and that at this separation the object and image are equi-distant from the lens. Thus, if $D = 4f$, the two lens positions coincide and the displacement, *d*, is zero.

19.8 EXPERIMENTAL DETERMINATIONS OF FOCAL LENGTHS OF DIVERGING LENSES

Lens Formula Method

A diverging lens cannot produce a real image of a real object, and therefore it is not possible to use the simple arrangement: real object, lens, screen. There are, however, two other methods of obtaining pairs of values of u and v.

Method 1

A pin acts as an object for the diverging lens, and a second pin is used to locate the image position by the no-parallax technique.

Method 2

A converging lens (Fig. 19.14) is used to produce, on a screen, a real image (I_1) of an illuminated object (O). The position of this image is recorded. The diverging lens, whose focal length is required, is placed between the converging lens and the screen so that I_1 acts as a virtual object for the diverging lens. If the position of the diverging lens is taken to be A, $u = -AI_1$. The screen is moved away from the lens until it is at I and there is once again a properly focused image on it. This position is recorded, and v is found from $v = AI$.

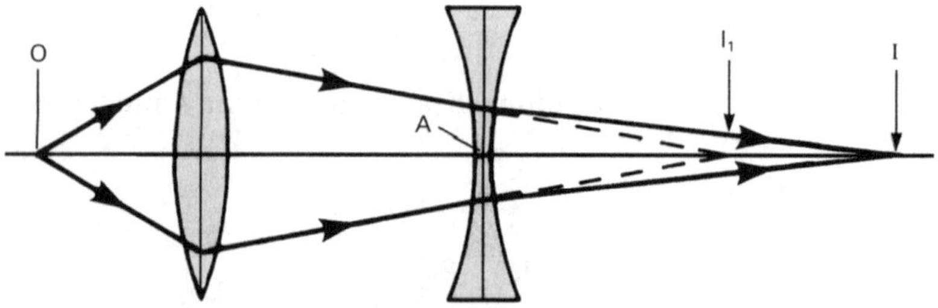

Fig. 19.14
Use of a converging lens to determine ***f*** for a diverging lens

The best estimate of the focal length of the diverging lens is the average of the values obtained by substituting each pair of values of u and v (from either method) into the lens formula (equation [19.1]). Alternatively, a graph of $1/u$ against $1/v$ could be plotted (Fig. 19.15). Each intercept is equal to $1/f$, hence f.

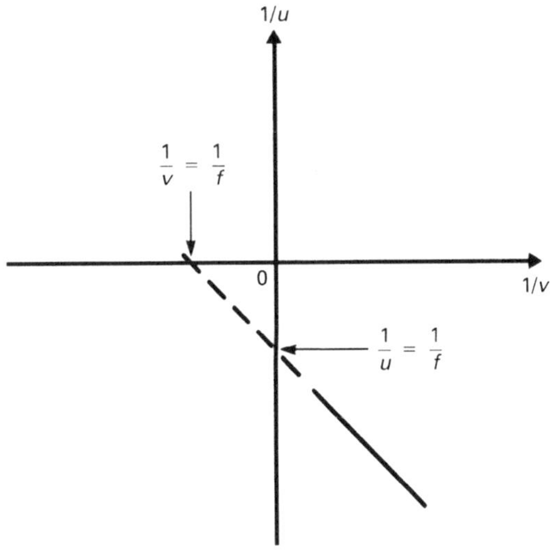

Fig. 19.15
Graph to determine ***f*** for a diverging lens

Converging Lens in Contact

The diverging lens, whose focal length is required, is placed in contact with a converging lens. If the converging lens is the more powerful of the two (i.e. if it has a shorter focal length), the combination acts as a converging lens. The focal length, f, of the combination is found by using any of the methods of section 19.7. The diverging lens is then removed, and the focal length, f_1, of the converging lens acting alone is determined. The focal length, f_2, of the diverging lens is given by equation [19.4] as

$$\frac{1}{f_2} = \frac{1}{f} - \frac{1}{f_1}$$

19.9 LATERAL (TRANSVERSE) MAGNIFICATION

Lateral magnification, m, is defined by

$$m = \frac{\text{Height of image}}{\text{Height of object}} \qquad [19.6]$$

In Fig. 19.16, △OAB and △ICB are similar, and therefore

$$\frac{h_i}{h_o} = \frac{v}{u}$$

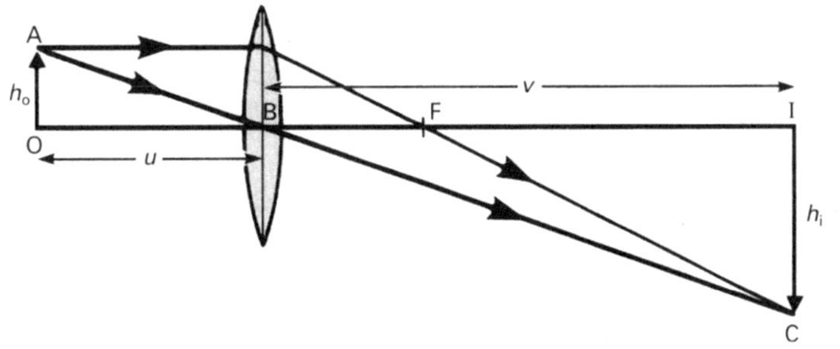

Fig. 19.16
To illustrate lateral (transverse) magnification

Therefore from equation [19.6]

$$m = \frac{h_i}{h_o} = \frac{v}{u} \qquad [19.7]$$

Notes (i) Both u and v are taken to be positive when using equation [19.7]; they represent the lengths OB and BI respectively.

(ii) Both equation [19.6] and equation [19.7] also apply in the case of diverging lenses, and mirrors.

EXAMPLE 19.2

An object of height 3.0 cm is placed 20 cm from a diverging lens of focal length 30 cm. Calculate the height of the image.

Solution

$u = +20\text{ cm}$ (real object)

$v = v$

$f = -30\text{ cm}$ (diverging lens)

$$\frac{1}{u} + \frac{1}{v} = \frac{1}{f}$$

$$\therefore \quad \frac{1}{20} + \frac{1}{v} = -\frac{1}{30}$$

$$\therefore \quad \frac{1}{v} = -\frac{1}{30} - \frac{1}{20} = \frac{-2-3}{60} = -\frac{5}{60}$$

$$\therefore \quad v = -12\text{ cm}$$

$$\frac{h_i}{h_o} = \frac{v}{u}$$

$$\therefore \quad \frac{h_i}{3.0} = \frac{12}{20}$$ (Note that the minus sign in $v = -12\text{ cm}$ has been ignored)

$$\therefore \quad h_i = 1.8\text{ cm}$$

i.e. Height of image $= 1.8\text{ cm}$

QUESTIONS 19B

1. Find the magnification produced when an object is placed: **(a)** 30 cm from a converging lens of focal length 20 cm, **(b)** 10 cm from a diverging lens of focal length 15 cm.

2. A particular lens produces a magnification of 3× when an object is placed 20 cm from it. **(a)** How far from the lens is the image? **(b)** Find both possible focal lengths of the lens.

3. How far from a converging lens of focal length 45 cm must an object be placed if the image is to be three times as big as the object
(a) if the image is real,
(b) if the image is virtual?

19.10 CHROMATIC ABERRATION

Because of dispersion (section 18.7) a simple (single-component) lens has slightly different focal lengths for the various colours which make up white light. It follows that when green light (say) is in focus the other colours are slightly out of focus and the edges of the image are coloured. The defect, which is called **chromatic aberration**, can be reduced by using an **achromatic doublet**. This is a combination of two lenses made from different types of glass and cemented together with Canada balsam. A common type is shown in Fig. 19.17. The crown glass component of the doublet is a converging lens and the deviation it produces is in the opposite direction to that produced by the flint glass, diverging lens. Each component deviates violet light more than red light and because the deviation produced by one component is in the opposite direction to that produced by the other, it is possible to arrange that the dispersion between red and violet produced by one component cancels that produced by the other. For any given deviation, crown glass produces less dispersion than flint glass. In order for the dispersions to cancel, the crown glass component has to produce more deviation than the flint glass component, and therefore since the crown glass lens is converging, the overall combination is also converging.

Fig. 9.17
An achromatic doublet

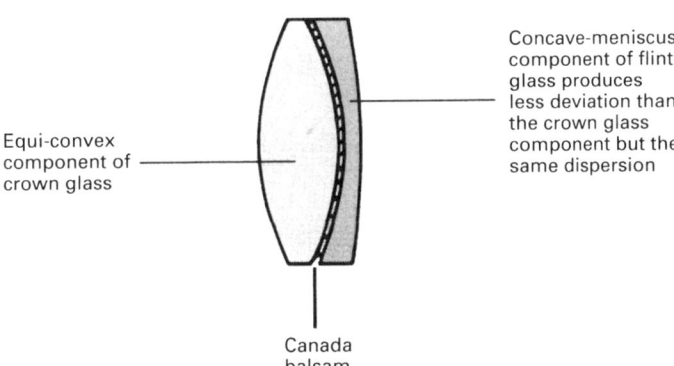

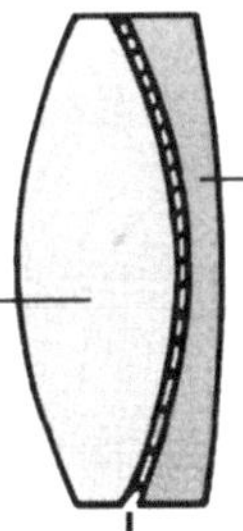
Equi-convex
component of
crown glass
Concave-meniscus
component of flint
glass produces
less deviation than
the crown glass
component but the
same dispersion
Canada
balsam

Although an achromatic doublet has the same focal length for red light as it does for violet light, its focal lengths for the other spectral colours are slightly different. These differences though, are considerably less than for a single-component lens and the doublet produces very little chromatic aberration.

CONSOLIDATION

Converging lenses are thick in the middle and have positive focal lengths.

Diverging lenses are thin in the middle and have negative focal lengths. They cannot produce real images of real objects.

$$\text{Power of a lens} = \frac{1}{\text{Focal length in metres}} \qquad (\text{Unit} = \text{m}^{-1} \text{ or dioptre.})$$

Power is positive for converging lenses, negative for diverging lenses.

A real image is one through which rays of light actually pass. It can be formed on a screen.

A virtual image is one from which rays of light only appear to have come. It cannot be formed on a screen.

A virtual object is one towards which rays of light are heading but which are intercepted by a lens (or mirror).

Lens Calculations

$$\frac{1}{u} + \frac{1}{v} = \frac{1}{f}$$

u is negative for virtual objects

v is negative for virtual images

f is negative for diverging lenses

$$m = \frac{h_i}{h_o} = \frac{v}{u}$$

Minus signs are ignored when using this equation – magnification cannot be negative.

20

MIRRORS

20.1 THE LAWS OF REFLECTION

(i) The reflected ray is in the same plane as the incident ray and the normal to the reflecting surface at the point of incidence.

(ii) The angle of reflection is equal to the angle of incidence.

The situation is illustrated in Fig. 20.1. It follows from the first law that since the incident ray and the normal are in the plane of the paper, then so too is the reflected ray. It follows from the second law that $i = r$.

Fig. 20.1
Reflection

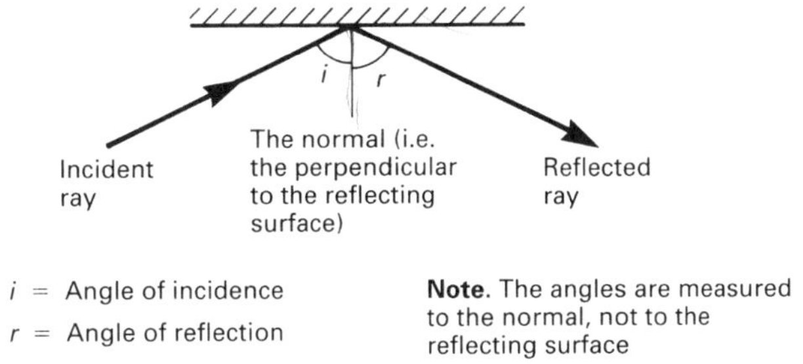

Note The laws apply whenever light is reflected, regardless of whether the reflecting surface is a plane mirror, a curved mirror or a diffuse reflector such as a piece of paper.

20.2 IMAGES PRODUCED BY PLANE MIRRORS

Fig. 20.2 illustrates the formation of an image by a plane mirror. Rays of light from a (real) point object at O are reflected by the mirror and enter the eye in such a way that they appear to have come from I. It follows that the image of the object is at I, and that it is a virtual (see section 19.2) image.

The straight line (not shown) joining the object and image is perpendicular to the mirror. Furthermore, the image is as far behind the mirror as the object is in front of it. These two statements are easily proved (see next page). Alternatively, their validity can be demonstrated by experiment (see, for example, section 20.3).

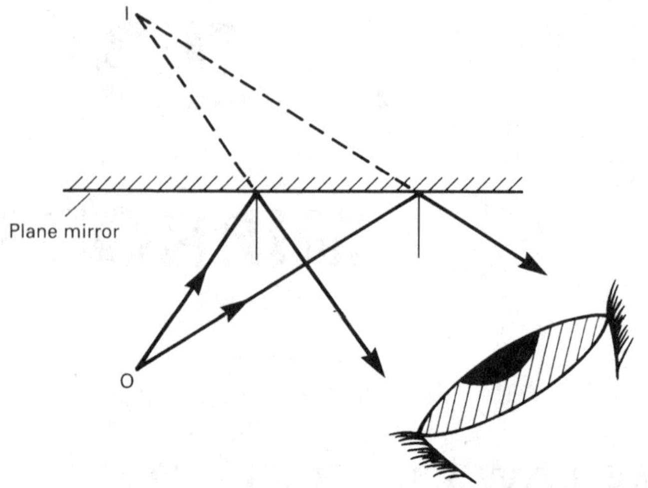

Fig. 20.2
Image of a point object in a plane mirror

To Show that the Line Joining a Point Object and its Image is Perpendicular to the Mirror

Refer to Fig. 20.3 in which I is the image of O, and ON is perpendicular to the mirror.

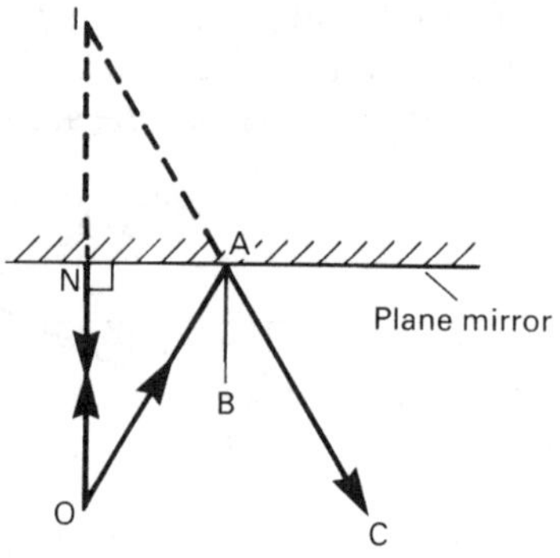

Fig. 20.3
To show that ONI is perpendicular to the mirror and that NO = NI

It follows from the second law that a ray incident along ON must be reflected back along NO. Since this reflected ray must appear to have come from the image (I), it follows that ONI is a straight line and therefore OI is perpendicular to the mirror, i.e. the line joining object and image is perpendicular to the mirror.

To Show that the Image is as Far Behind the Mirror as the Object is in Front

Refer to Fig. 20.3.

$$N\hat{O}A = O\hat{A}B \quad \text{(alternate angles)}$$
$$O\hat{A}B = C\hat{A}B \quad \text{(second law)}$$
$$C\hat{A}B = N\hat{I}A \quad \text{(corresponding angles)}$$
$$\therefore \quad N\hat{O}A = N\hat{I}A$$
$$\therefore \quad \tan N\hat{O}A = \tan N\hat{I}A$$
$$\therefore \quad \frac{NA}{NO} = \frac{NA}{NI}$$
$$\therefore \quad NO = NI$$

The image is therefore as far behind the mirror as the object is in front.

Fig. 20.4 shows a finite-sized object, AB, in front of a plane mirror.

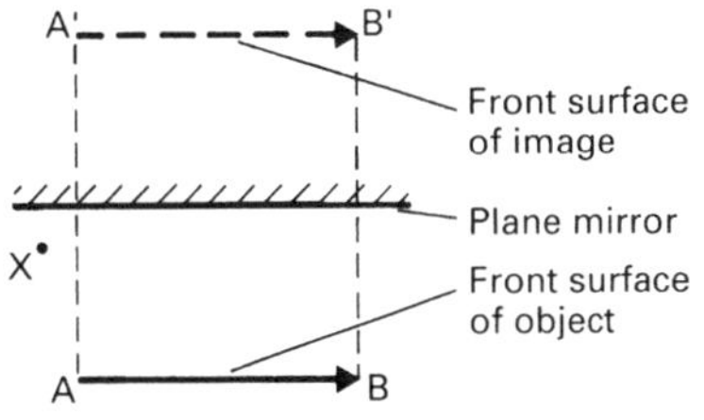

Fig. 20.4
Image of a finite-sized object in a plane mirror

From what has been said about the images of point objects, the image of A is at A′ and that of B is at B′, where AA′ and BB′ are at right angles to the mirror. It follows that A′B′ = AB, i.e. the image, A′B′, is the same size as the object, AB.

A person at X looking at the object would see the arrowhead on the left. However, if the same person were to look at the image, he or she would see the arrowhead on the right. This is known as **lateral inversion** – F appears as ꟻ. Note that although left becomes right (and vice versa), top does not become bottom – why?

Summary

(i) The image of a real object is virtual.

(ii) A line joining a point on the object to the corresponding point on the image is perpendicular to the mirror.

(iii) The image is as far behind the mirror as the object is in front.

(iv) The image is the same size as the object.

(v) The image is laterally inverted.

We have seen that a plane mirror produces a virtual image of a real object. It follows from the principle of reversibility (section 19.6) that a plane mirror must produce a real image of a virtual object – see Fig. 20.5.

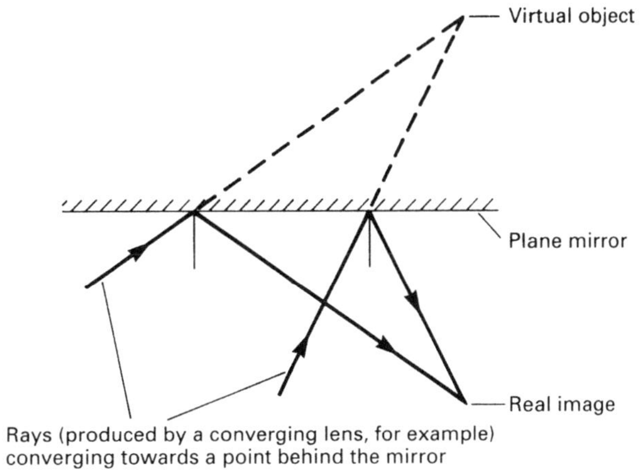

Fig. 20.5
Plane mirror producing a real image of a virtual object

20.3 IMAGE LOCATION BY NO PARALLAX

The method of no parallax can be used to locate the positions of both real and virtual images. The illuminated object and screen method (see, for example, section 19.7) can be used for real images only. In order to illustrate the no-parallax technique, we shall describe how it is used to locate the position of a real object produced by a plane mirror.

An object pin, O, is stood in front of a vertical plane mirror (Fig. 20.6).

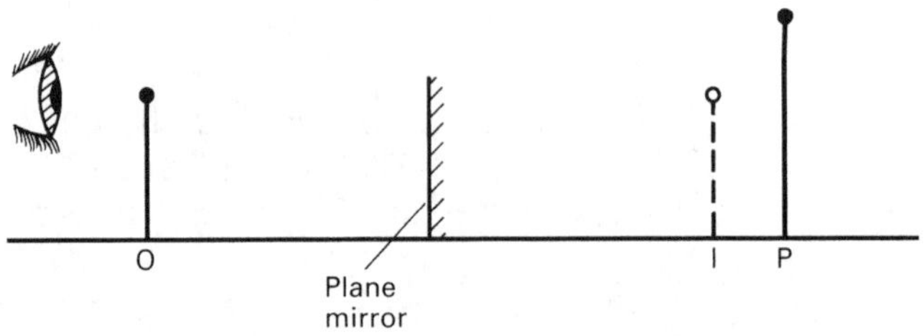

Fig. 20.6
Image location by no parallax

A second pin, P, is now placed behind the mirror. This pin should be large enough for the observer to see part of it over the top of the mirror. The observer looks at both P and the image, I, of O and then moves P until it appears to coincide with I. If P and I are merely in the observer's line of sight, rather than actually in the same place, they appear to move relative to each other when the observer moves his or her head from side to side (see Fig. 20.7(a)). There is said to be **parallax** between P and I.

(a)

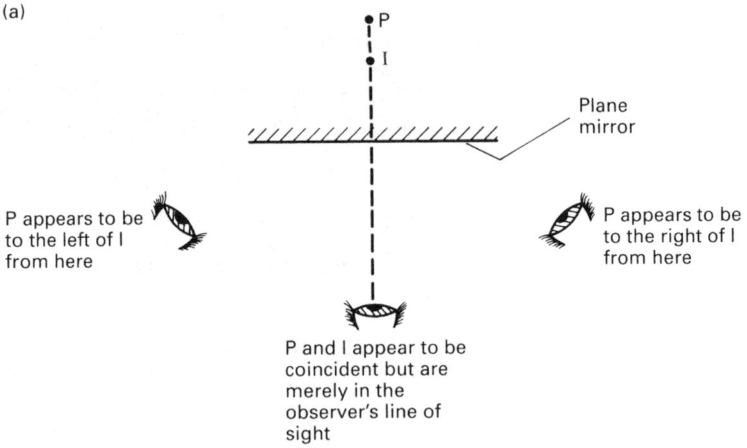

(b)

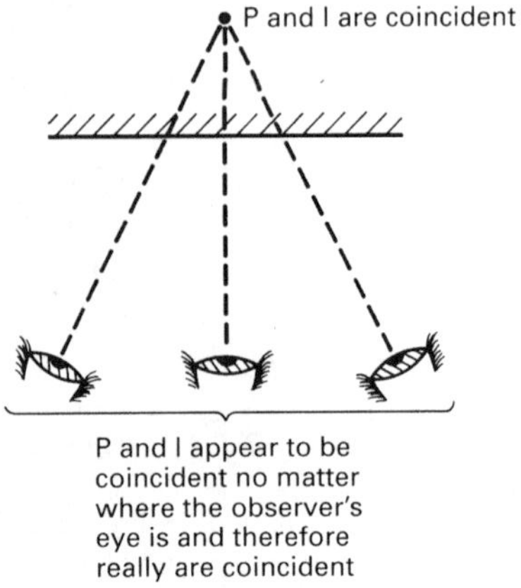

Fig. 20.7
(a) Parallax between P and I. (b) No parallax between P and I

The observer continues to move P until a position is found in which P and I move together on moving the head from side to side (Fig. 20.7(b)). There is now **no parallax** between P and I, i.e. they are coincident and P is at the position of the image. The image postion has therefore been found.

The no-parallax method can be used to locate the positions of images produced by curved mirrors and by lenses. The observer must always be in a position to see the image and therefore must be on the same side as the (real) object in the case of a mirror, but on the opposite side to the object in the case of a lens. The observer must also be able to see at least part of the locating pin. Some situations are illustrated in Fig. 20.8.

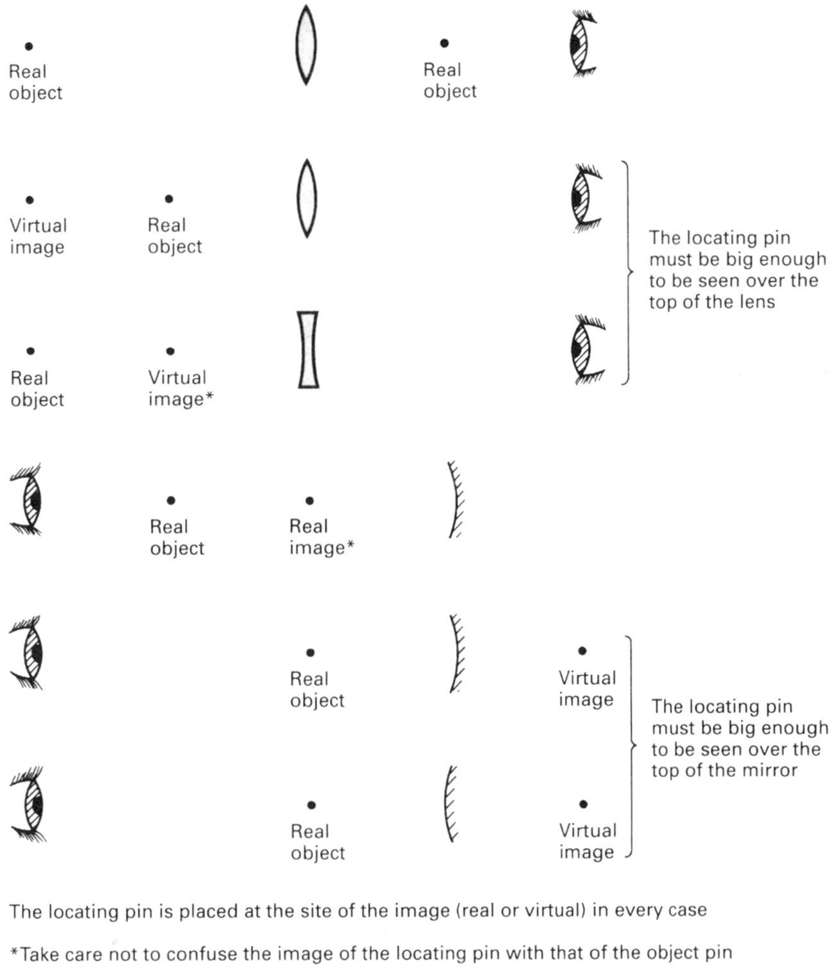

Fig. 20.8
Relative positions of object, observer and lens or mirror when obtaining image positions by no parallax

20.4 SPHERICAL MIRRORS. DEFINITIONS

Spherical mirrors may be concave (converging) or convex (diverging). The basic features of each type of mirror are illustrated by Fig. 20.9.

In each case:

(i) C is the **centre of curvature** of the mirror; it is the centre of the sphere of which the mirror's surface forms part;

(ii) P is the **pole** of the mirror;

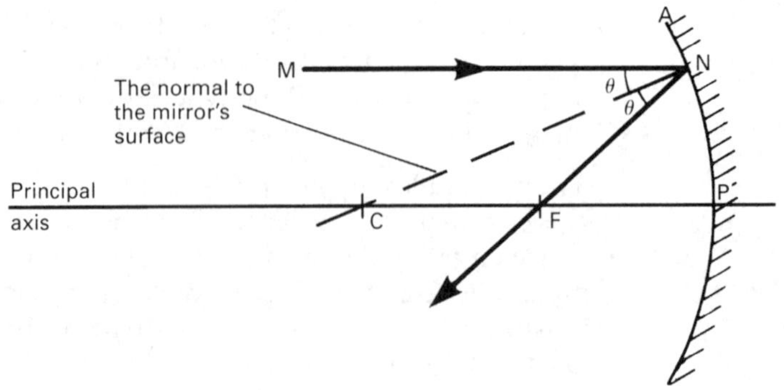

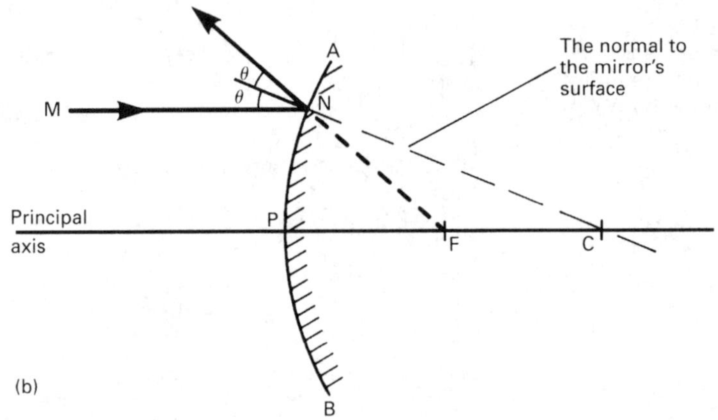

Fig. 20.9 Spherical mirrors: (a) converging, (b) diverging

(iii) CP is the **radius of curvature,** r, of the mirror;

(iv) the line through CP is the **principal axis** of the mirror;

(v) AB is the **aperture** of the mirror.

A ray of light which is parallel and close to the principal axis of a concave mirror (e.g. MN in Fig. 20.9(a)) is reflected through the **principal focus (focal point),** F, of the mirror.

A ray of light which is parallel and close to the principal axis of a convex mirror (e.g. MN in Fig. 20.9(b)) is reflected in such a way that it appears to have come from the **principal focus (focal point),** F, of the mirror.

In each case FP is the **focal length,** f, of the mirror, and

$$f = \frac{r}{2} \qquad [20.1]$$

20.5 IMPORTANT RAY PATHS

As with lenses (section 19.3), if the paths of any two rays from a single point on an object are known, its image position can be located. For each type of mirror, there are three rays (Fig. 20.10) whose paths can be determined, and any two of them can be used.

Notes

(i) Reflection is treated as if it takes place at the line which passes through P and is perpendicular to CP.

(ii) Since $f = r/2$ (equation [20.1]), F is the mid-point of CP.

Fig. 20.10
Ray paths for (a) a concave, (b) a convex mirror

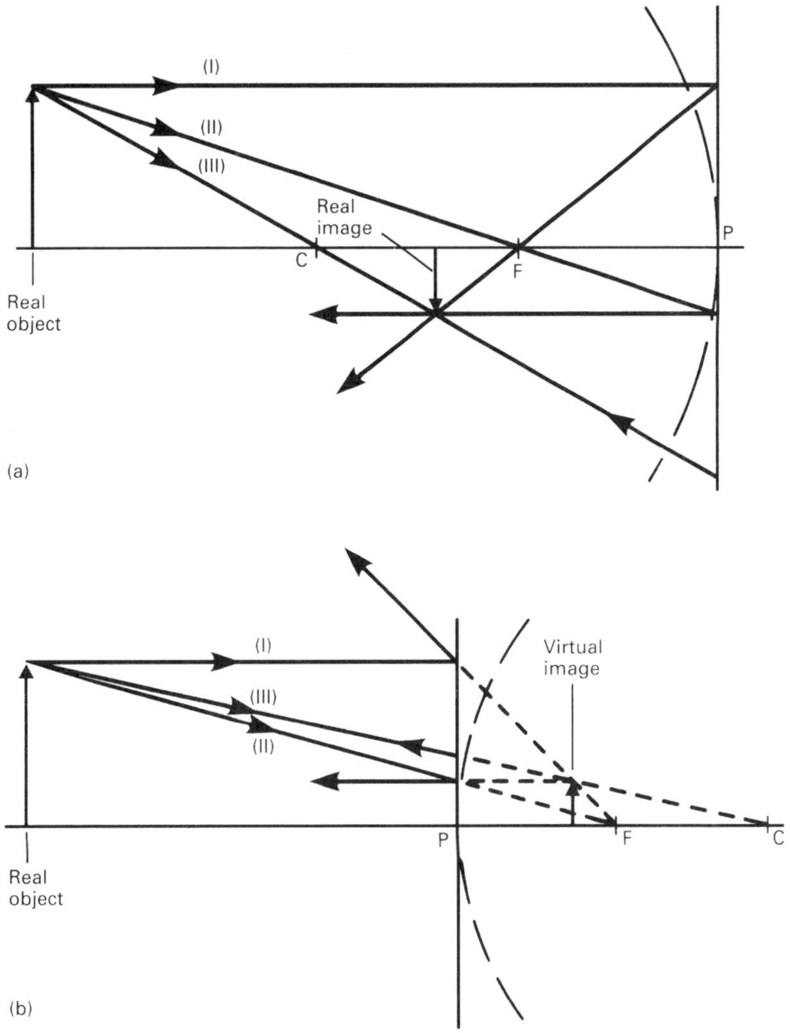

(iii) In each case:

(a) the incident ray (I) parallel to the principal axis is reflected through (or appears to have come from) F;

(b) the incident ray (II) passing through (or travelling towards) F is reflected back parallel to the principal axis;

(c) the incident ray (III) passing through (or travelling towards) C is reflected back along its original path.

20.6 THE MIRROR FORMULA

$$\frac{1}{u}+\frac{1}{v}=\frac{1}{f}=\frac{2}{r} \qquad [20.2]$$

where

u = object distance
v = image distance
f = focal length
r = radius of curvature

All distances are measured to the pole of the mirror.

In order to distinguish between real and virtual images, and the two types of spherical mirror, it is necessary to employ a sign convention. This book uses the **Real is Positive** convention, in which:

(i) the focal lengths and radii of curvature of concave mirrors are positive, and those of convex mirrors are negative;

(ii) distances from mirrors to real objects and real images are positive, whereas distances to virtual objects and virtual images are negative.

EXAMPLE 20.1

A convex mirror whose radius of curvature is 30 cm forms an image of a real object which has been placed 20 cm from the mirror. Calculate the position of the image and the magnification produced.

Solution

$$u = +20\text{ cm} \quad \text{(real object)}$$

$$v = v$$

$$r = -30\text{ cm} \quad \text{(convex mirror)}$$

From equation [20.2]

$$\frac{1}{u}+\frac{1}{v} = \frac{2}{r}$$

$$\therefore \quad \frac{1}{20}+\frac{1}{v} = \frac{2}{-30}$$

$$\text{i.e.} \quad \frac{1}{v} = \frac{-2}{30}-\frac{1}{20} = \frac{-4}{60}-\frac{3}{60} = \frac{-7}{60}$$

$$\therefore \quad v = \frac{-60}{7} = -8.6\text{ cm}$$

Since the image distance has turned out to be negative, the image is virtual. The image is therefore 8.6 cm behind the mirror.

Equation [19.7] applies, and therefore the magnification, m, is given by

$$m = \frac{v}{u} = \frac{60/7}{20} = \frac{3}{7}$$

Note The minus sign in $v = -60/7$ cm has been ignored – see note (i) in section 19.9. Thus, the magnification is $\frac{3}{7}$, i.e. the image is three-sevenths the size of the object.

QUESTIONS 20A

1. A concave mirror with a radius of curvature of 40 cm forms an image of a real object which has been placed 25 cm from the mirror. **(a)** What is the focal length of the mirror? **(b)** Calculate the distance of the image from the mirror and state whether it is real or virtual.

2. **(a)** What is the focal length of a mirror which forms a 4× magnified real image of an object which has been placed 20 cm from the mirror?

(b) State whether the mirror is concave or convex.

20.7 THE CAUSTIC CURVE AND PARABOLOIDAL MIRRORS

If rays of light are incident on a spherical concave mirror and are parallel to the principal axis of the mirror, those rays which are not close to the axis are reflected in such a way that they subsequently cross the axis at points which are closer to the mirror than its principal focus. (The discussion, so far, has assumed that the mirrors have been of small aperture, in which case it is a sufficiently good approximation to ignore this.) Because of this effect, the light is reflected into a region which is bounded by a surface known as the **caustic** (Fig. 20.11). The two-dimensional version of this surface, known as the **caustic-curve**, can often be seen on the surfaces of liquids contained in vessels of circular cross-section (e.g. on a cup of tea).

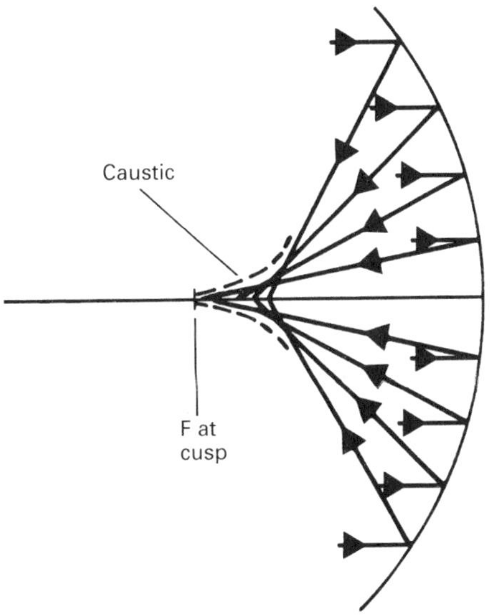

Fig. 20.11
Caustic curve produced by a spherical concave mirror

A **paraboloidal** concave mirror, on the other hand, focuses all rays which are parallel to its principal axis at its focus, no matter how wide the beam. It follows from the principle of reversibility that if a lamp is situated at the focus of such a mirror, an accurately parallel beam will be reflected from it. Paraboloidal mirrors therefore find application as search-light reflectors.

20.8 EXPERIMENTAL DETERMINATIONS OF FOCAL LENGTHS OF MIRRORS

Spherometer Method

A spherometer can be used to determine the radii of curvature, r, of both concave and convex mirrors. The focal lengths, f, are then found from $f = r/2$ (equation [20.1]).

Mirror Formula Method

Concave Mirrors

Concave mirrors behave like converging lenses in that they can produce both real and virtual images of real objects. Consequently, the techniques described in the 'Lens Formula Method' of section 19.7 can be used to determine the focal lengths of concave mirrors.

Convex Mirrors

When convex mirrors are used with real objects the images that result are virtual. The techniques of Method 1 in section 19.8 are suitable.

CONSOLIDATION

The Laws of Reflection – see page 375

Plane Mirrors – see summary on page 377

Curved Mirrors

Concave mirrors are converging and have positive focal lengths and positive radii of curvature.

Convex mirrors are diverging and have negative focal lengths and negative radii of curvature. They cannot produce real images of real objects.

$$\frac{1}{u} + \frac{1}{v} = \frac{1}{f} = \frac{2}{r}$$

u is negative for virtual objects

v is negative for virtual images.

If an object and image move relative to each other when an observer moves his/her head from side to side, there is said to be **parallax** between them – they are in different positions.

If an object and image move together when an observer moves his/her head from side to side, there is said to be **no parallax** between them – they are in the same position as each other.

21

OPTICAL INSTRUMENTS

21.1 THE VISUAL ANGLE AND ANGULAR MAGNIFICATION

The size that an object appears to be is determined by the size, b, of its image on the retina (Fig. 21.1). For small angles

$$b = a\theta$$

where a is the length of the eyeball and θ is the angle subtended at the eye by the object (i.e. the **visual angle**). Since a is constant, **the size of the image on the retina (and therefore the apparent size of the object) is proportional to the visual angle**. For example, a penny held at arm's length subtends a larger angle at the eye than the Moon does, and therefore can block it from view. Thus, although the penny is smaller than the Moon, it appears to be bigger because it is subtending a larger angle at the eye.

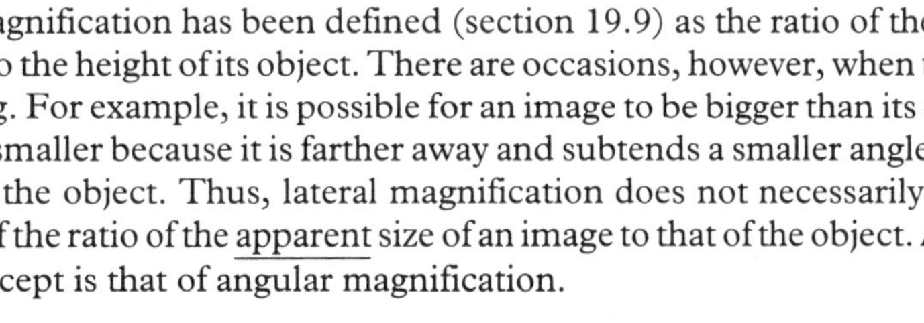

Fig. 21.1
Image formation by the eye

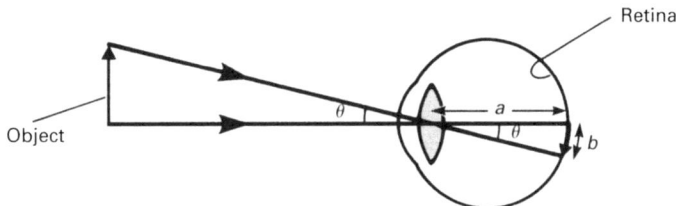

Lateral magnification has been defined (section 19.9) as the ratio of the height of an image to the height of its object. There are occasions, however, when this can be misleading. For example, it is possible for an image to be bigger than its object, yet to appear smaller because it is farther away and subtends a smaller angle at the eye than does the object. Thus, lateral magnification does not necessarily provide a measure of the ratio of the apparent size of an image to that of the object. A far more useful concept is that of angular magnification.

The angular magnification (also known as **magnifying power**), M, of an optical instrument is defined by

$$M = \frac{\beta}{\alpha} \qquad [21.1]$$

where

β = the angle subtended at the eye by the image

α = the angle subtended at the unaided eye by the object.

In order to determine α, it is necessary to specify the position of the object. Microscopes are used to view objects which are close by, and which, therefore, are seen to the best advantage by the naked eye when they are at the near point (see section 21.2). It is meaningful in the case of microscopes, therefore, to specify α as being the angle subtended by the object when it is at the near point. Those objects which are observed with telescopes are at large distances, in which case α is simply the angle subtended by the object wherever it happens to be.

21.2 THE NEAR POINT

The position at which an object (or an image which is acting as an object for the lens of the eye) is seen most clearly is called the **near point**. The near point varies from one individual to another but (by convention) is taken to be a distance of 25 cm from the eye. This distance is known as the **least distance of distinct vision** (D). An object whose distance from the eye is less than D appears blurred; one which is farther away appears smaller than when at the near point.

21.3 THE MAGNIFYING GLASS (SIMPLE MICROSCOPE)

When a converging lens is caused to produce a virtual, upright and enlarged image it is being used as a magnifying glass. The position of the image depends on the position of the object relative to that of the lens. Two (limiting) situations are examined.

Image at the Near Point

This arrangement (Fig. 21.2) produces the greatest angular magnification.

Fig. 21.2 Magnifying glass with image at near point

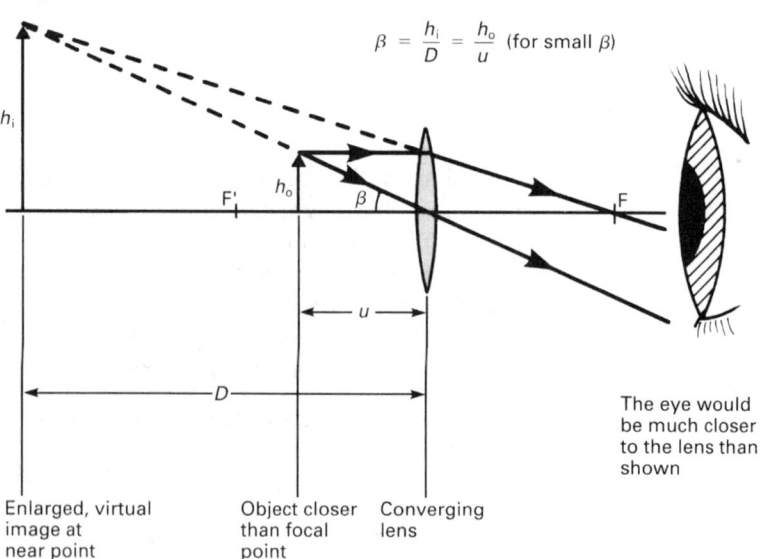

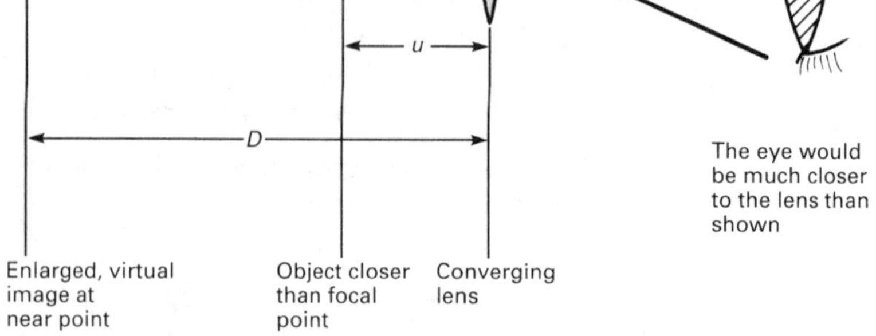

The angular magnification, M, of a microscope (see section 21.1) is given by

$$M = \frac{\beta}{\alpha}$$

where

β = the angle subtended at the eye by the image

α = the angle subtended at the unaided eye by the object when it is at the near point.

From Fig. 21.2

$$\beta = \frac{h_i}{D}$$

(This assumes that β is a small angle and that the eye is close to the lens.)*

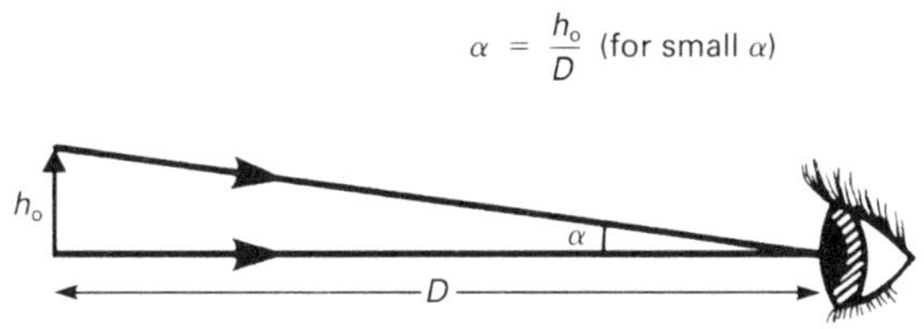

Fig. 21.3 Angle subtended by an object at the near point

If the object were at the near point, it would subtend the small angle h_o/D (Fig. 21.3), in which case

$$\alpha = \frac{h_o}{D}$$

Therefore, since $M = \beta/\alpha$

$$M = \frac{h_i/D}{h_o/D}$$

i.e. $$M = \frac{h_i}{h_o}$$

But h_i/h_o is, by definition, the **lateral magnification,** m, and therefore, in this case

$$M = m = \frac{h_i}{h_o}$$

It is possible to derive an alternative expression for M. Again, we shall assume that the angles are small and that the eye is close to the lens. From Fig. 21.2

$$\beta = \frac{h_o}{u}$$

From Fig. 21.3

$$\alpha = \frac{h_o}{D}$$

Therefore, since $M = \beta/\alpha$

$$M = \frac{h_o/u}{h_o/D}$$

i.e. $$M = \frac{D}{u} \qquad [21.2]$$

* (i) If the eye is close to the lens, the angle subtended at the eye (β) is a good approximation to the angle subtended at the lens and therefore β is as shown in Fig. 21.2.

(ii) Strictly, $\tan\beta = h_i/D$, but if β is small, $\beta = \tan\beta$ is a good approximation and therefore we may put $\beta = h_i/D$.

In the usual notation

$$\frac{1}{u}+\frac{1}{v}=\frac{1}{f} \quad \text{(equation [19.1])}$$

Here $u = u$ and $v = -D$ (the image is virtual and therefore $v = -25\,\text{cm}$), in which case

$$\frac{1}{u}-\frac{1}{D}=\frac{1}{f}$$

Multiplying by $(-D)$ leads to

$$\frac{D}{u}=\frac{D}{f}+1$$

Therefore, from equation [21.2]

$$M = \left(\frac{D}{f}+1\right) \qquad [21.3]$$

where

$D = +25\,\text{cm}$

f = the focal length of the lens.

Image at Infinity

This arrangement (Fig. 21.4) produces a smaller angular magnification than when the image is at the near point, but has the advantage that the eye is relaxed (unaccommodated).

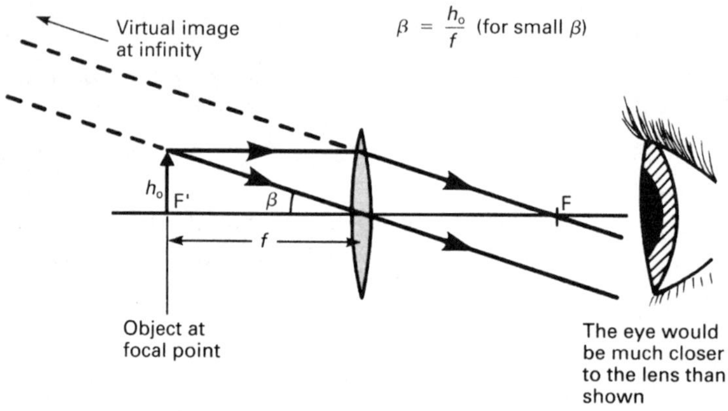

Fig. 21.4 Magnifying glass with image at infinity

As with the image at the near point we shall assume that the angles are small. From Fig. 21.4

$$\beta = \frac{h_o}{f}$$

(There is no need for the eye to be close to the lens – the image is at infinity and therefore subtends the same angle at the eye as it does at the eyepiece lens in any case.)

From Fig. 21.3

$$\alpha = \frac{h_o}{D}$$

Therefore, since $M = \beta/\alpha$

$$M = \frac{h_o/f}{h_o/D}$$

i.e. $$M = \frac{D}{f} \qquad [21.4]$$

where

$D = +25$ cm

f = the focal length of the lens.

It follows from both equation [21.3] and equation [21.4] that a lens of short focal length is required if the angular magnification is to be high. The surfaces of such a lens need to be highly curved. This sets an upper limit to the angular magnification that can be achieved using a single lens, because the extent to which images are distorted increases with curvature.

21.4 THE COMPOUND MICROSCOPE

Because it makes use of two lenses, the magnifying power of the compound microscope is much greater than that of the magnifying glass. The **objective lens** forms a real image of the object. This (intermediate) image acts as an object for the **eyepiece lens** which, behaving as a magnifying glass, produces an enlarged, virtual image of it. The focal length of each lens should be small in order to produce a high overall angular magnification. (Magnifying powers of 500× are not uncommon.)

Final Image at the Near Point

This arrangement (Fig. 21.5) produces the greatest angular magnification. The separation of the lenses is such that the intermediate image is formed inside F'_e so that the eyepiece lens acts as a magnifying glass.

Fig. 21.5 Compound microscope with image at near point

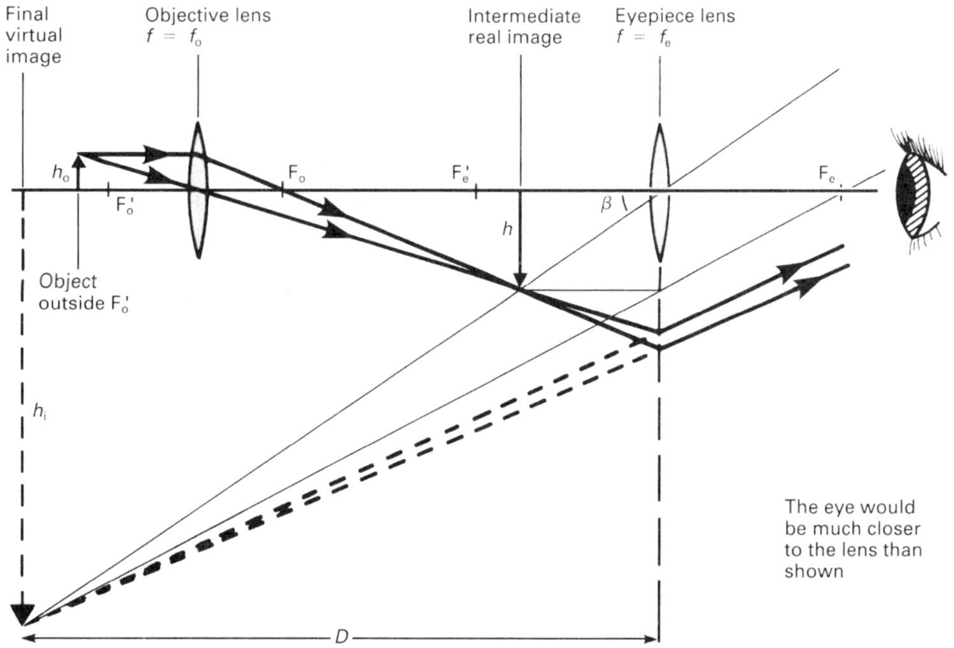

The angular magnification, M, of a microscope (see section 21.1) is given by

$$M = \frac{\beta}{\alpha}$$

where

β = the angle subtended at the eye by the image

α = the angle subtended at the unaided eye by the object when it is at the near point.

From Fig. 21.5

$$\beta = \frac{h_i}{D}$$

(This assumes that β is a small angle and that the eye is close to the lens.)* If the object were at the near point, it would subtend the small angle h_o/D (see Fig. 21.3), in which case

$$\alpha = \frac{h_o}{D}$$

Therefore, since $M = \beta/\alpha$

$$M = \frac{h_i/D}{h_o/D}$$

i.e. $$M = \frac{h_i}{ho}$$

But h_i/h_o is, by definition, the lateral magnification, m, and therefore, as for the magnifying glass with the image at the near point (section 21.3),

$$M = m = \frac{h_i}{h_o}$$

Note

$$M = \frac{h_i}{h_o} = \frac{h_i}{h} \times \frac{h}{h_o}$$

i.e. angular magnification = lateral magnification produced by eyepiece × lateral magnification produced by objective.

Image at Infinity

This arrangement (Fig. 21.6) produces a smaller angular magnification than when the image is at the near point, but has the advantage that the eye is relaxed (unaccommodated). The separation of the lenses is such that the intermediate image is formed at the focal point (F_e') of the eyepiece lens.

EXAMPLE 21.1

A compound microscope consists of two thin converging lenses. The focal length of the objective lens is 10 mm and that of the eyepiece lens is 20 mm. If an object is placed 11 mm from the objective lens, the instrument produces an image at infinity. Calculate the separation of the lenses and the magnifying power of the instrument.

* See footnote on page 387.

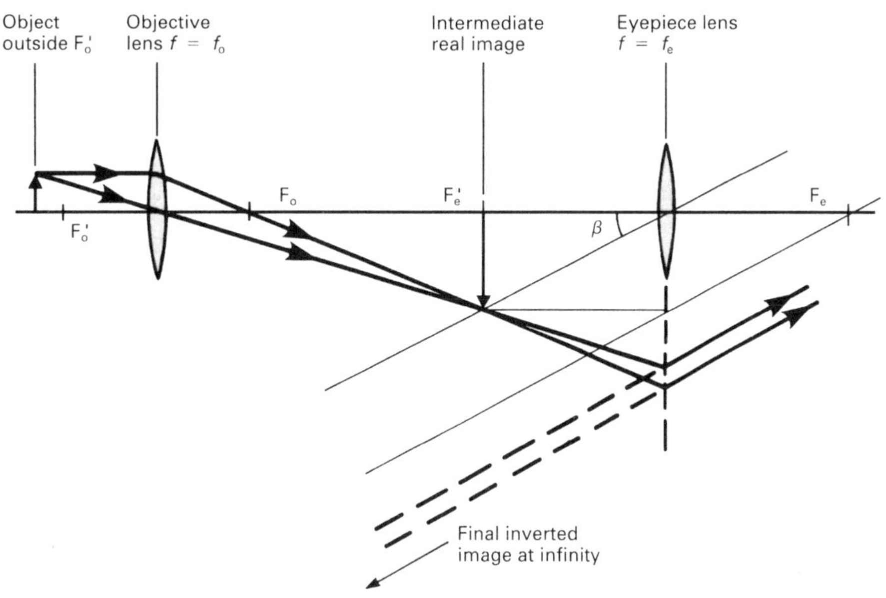

Fig. 21.6 Compound microscope with image at infinity

Solution

Since the final image is at infinity, the intermediate image must be at the focal point of the eyepiece lens. The situation is therefore as shown in Fig. 21.7. It is clear from Fig. 21.7 that the lens separation can be found once v, the distance of the intermediate image from the objective lens, is known. For this lens, employing the usual notation we have

$$u = 11\,\text{mm}$$

$$v = v$$

$$f = 10\,\text{mm}$$

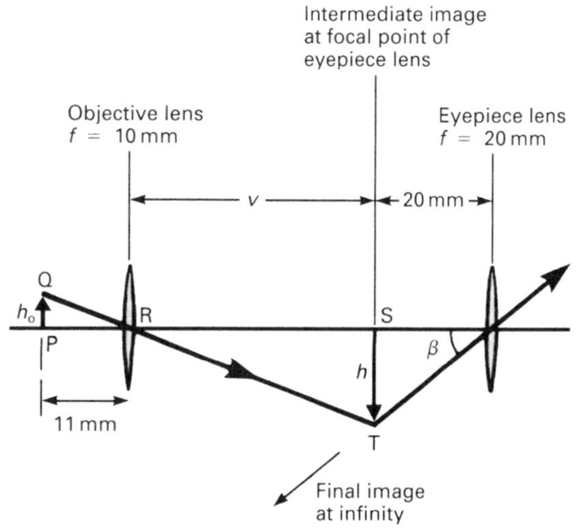

Fig. 21.7 Diagram for Example 21.1

Substituting these values into the lens formula (equation [19.1]) leads to

$$\frac{1}{11} + \frac{1}{v} = \frac{1}{10}$$

i.e. $v = 110\,\text{mm}$

Therefore, from Fig. 21.7, the lens separation is $110 + 20 = 130$ mm.

The magnifying power, M, is given by

$$M = \frac{\beta}{\alpha}$$

where

β = the angle subtended at the eye by the image

α = the angle subtended at the unaided eye by the object when it is at the near point.

Since the final image is at infinity, the angle subtended at the eye is the same as that subtended at the eyepiece lens and therefore β is as shown in Fig. 21.7, in which case, taking β to be a small angle we have

$$\beta = \frac{h}{20}$$

For a microscope $\alpha = h_o/D$ and, since this calculation is being worked in mm, $D = 250$ mm. Therefore

$$\alpha = \frac{h_o}{250}$$

Therefore, since $M = \beta/\alpha$

$$M = \frac{h/20}{h_o/250} = 12.5\frac{h}{h_o}$$

But, $\triangle$QPR is similar to $\triangle$TSR, and therefore

$$\frac{h}{h_o} = \frac{v}{11} = \frac{110}{11} = 10$$

Therefore

$$M = 12.5 \times 10 = 125$$

21.5 THE ASTRONOMICAL (REFRACTING) TELESCOPE

This consists of two converging lenses. The **objective lens** produces a real image of the object being viewed. This (intermediate) image acts as an object for the **eyepiece lens** which, behaving as a magnifying glass, produces a virtual image of it. If the magnifying power of the instrument is to be high, the focal length of the objective lens must be large and that of the eyepiece must be small.

Telescopes are used to view objects which are at great distances, and in each of the situations discussed here it will be assumed that the object is at infinity. Because the diameter of the objective lens is small compared with the distance of the object from the lens, all the rays reaching the lens from a single point on the object can be taken to be parallel to each other.

Final Image at Infinity (i.e. Normal Adjustment)

The arrangement is shown in Fig. 21.8. The object is at infinity, and therefore the intermediate image is in the focal plane of the objective lens. The separation of the lenses is such that their focal planes coincide, and therefore the eyepiece lens, acting as a magnifying glass, produces a final image which is at infinity. The eye is relaxed (unaccommodated).

Fig. 21.8
Astronomical telescope in normal adjustment

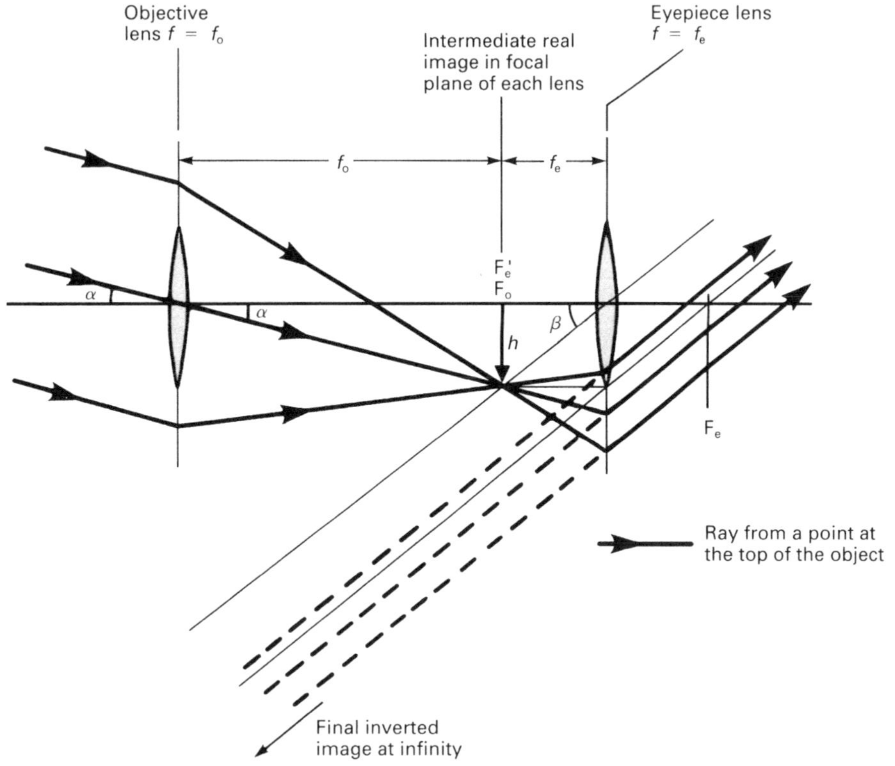

The angular magnification, M, of a telescope (see section 21.1) is given by

$$M = \frac{\beta}{\alpha}$$

where

β = the angle subtended at the eye by the image

α = the angle subtended at the unaided eye by the object.

Since both the object and the final image are at infinity, the angles they subtend at the eye are the same as those they subtend at the objective and at the eyepiece respectively. It follows that α and β are as shown in Fig. 21.8, from which

$$\beta = \frac{h}{f_e} \quad \text{and} \quad \alpha = \frac{h}{f_o}$$

(This assumes that α and β are small, in which case, to a good approximation, $\alpha = \tan \alpha$ and $\beta = \tan \beta$.)

Therefore, since $M = \beta/\alpha$

$$M = \frac{h/f_e}{h/f_o}$$

i.e.

$$M = \frac{f_o}{f_e} \qquad [21.5]$$

It is clear from this equation that, as stated previously, telescopes require long focal length objectives and short focal length eyepieces.

Final Image at the Near Point

The arrangement is shown in Fig. 21.9. The separation of the lenses is less than when the final image is formed at infinity. The intermediate image, though still in the focal plane of the objective lens, is now inside the focal point (F_e') of the eyepiece lens and in such a position that the final image is at the near point.

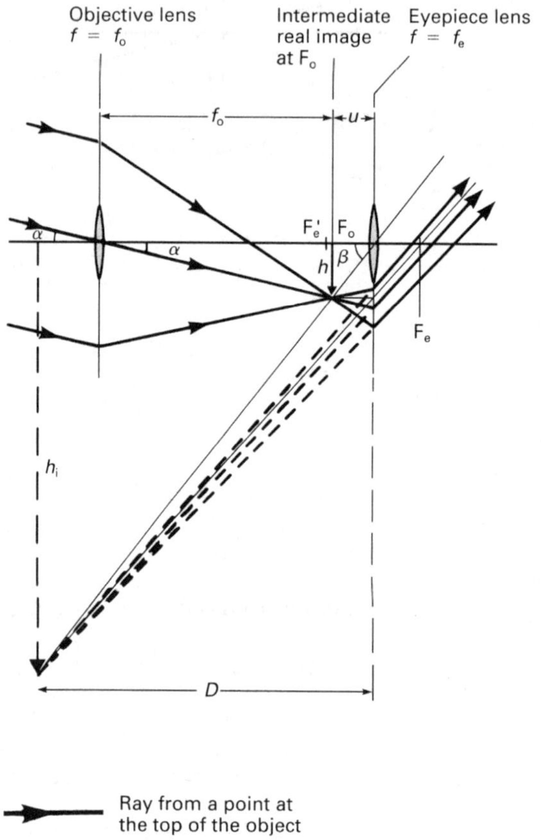

Fig. 21.9 Astronomical telescope with image at near point

QUESTIONS 21A

1. An astronomical telescope in normal adjustment has a total length of 78 cm and produces an angular magnification of 12. What is the focal length of the objective?

2. An astronomical telescope has an objective of focal length 80.0 cm and an eyepiece of focal length 5.0 cm. The telescope is in normal adjustment and a distant object subtends an angle of 5.0×10^{-3} radians at the objective. Find: **(a)** the length of the instrument, **(b)** the angular magnification it produces, **(c)** the size of the intermediate image.

3. A compound microscope has an objective of focal length 12.0 mm and an eyepiece of focal length 50.0 mm. The lenses have a separation of 90.0 mm and an object of height 0.30 mm is placed 15.0 mm from the objective. Calculate: **(a)** the distance of the intermediate image from the objective; **(b)** the distance of the final image from **(i)** the eyepiece, **(ii)** the objective; **(c)** the size of the intermediate image; **(d)** the angular magnification.

21.6 THE OBJECTIVE LENSES OF TELESCOPES

Ideally, the objective lens of a telescope should have a large diameter. There are two reasons for this.

(i) A telescope with a large diameter objective lens can resolve fine detail (see section 26.9).

(ii) The amount of light gathered by a telescope is proportional to the square of the diameter of its objective lens. This is a major consideration when the telescope is being used to view stars, which, because they are at great distances, appear dim.

The difficulties involved in obtaining large pieces of optical quality glass, and in supporting the lens, set an upper limit to this diameter – the world's largest refracting telescope has a 100 cm diameter objective.

The objective should be an **achromatic doublet** to reduce **chromatic aberration** (see section 19.10). In order that the image suffers the minimum possible **spherical aberration** (distortion) the deviations produced by each surface of the lens should be approximately the same. This is achieved when using a plano-convex lens by having the curved surface facing the object (Fig. 21.10).

Fig. 21.10
Telescope objective

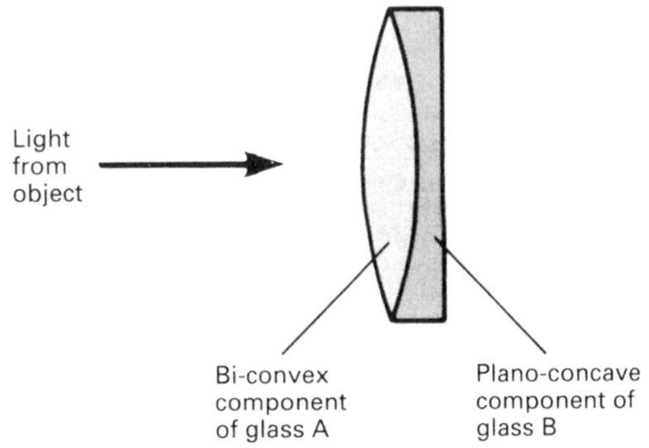

The lens surfaces are 'bloomed' (see section 25.8) to cut down reflections at the surfaces of the lens which would reduce the brightness of the final image.

21.7 THE TERRESTRIAL TELESCOPE

This uses one more lens that the astronomical refracting telescope (Fig. 21.11). The purpose of this (erecting) lens is to invert the intermediate image, and so make the final image erect.

Since the minimum separation of I_X and I_Y is $4f_X$ (see page 370), the introduction of the extra lens means that the length of the instrument exceeds that of the equivalent astronomical telescope by at least four times the focal length of the erecting lens. In order to keep this increase in length to the minimum, I_X and I_Y have to be equi-distant from the erecting lens and are therefore the same size as each other. It follows that when the final image is at infinity the angular magnification is f_o/f_e as in section 21.5. A further disadvantage of the erecting lens is that the extra reflections which occur at its surfaces decrease the intensity of the final image.

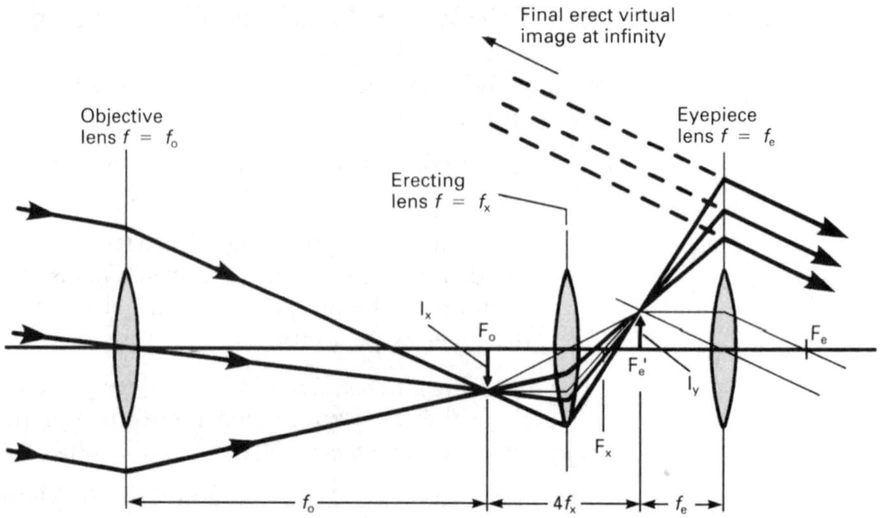

Fig. 21.11 Terrestrial telescope in normal adjustment

21.8 THE GALILEAN TELESCOPE

This uses a concave eyepiece lens in order to produce an erect final image without having to resort to the additional lens required by the terrestrial telescope. Fig. 21.12 shows a Galilean telescope in normal adjustment.

I_X acts as a virtual object for the concave eyepiece lens. Since I_X is at the focal point, F_e', of the eyepiece, the final image is at infinity.

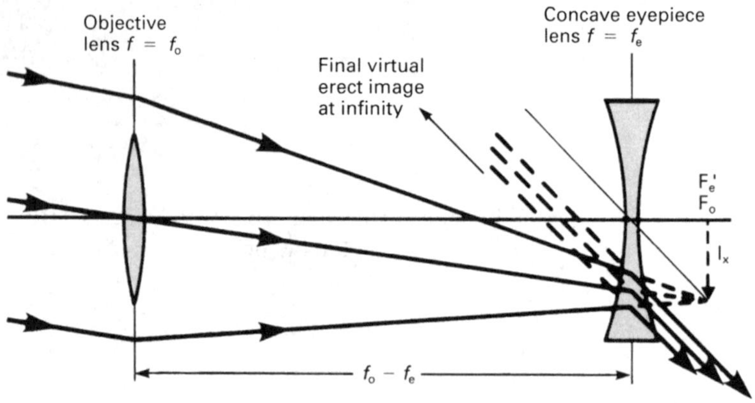

Fig. 21.12 Galilean telescope in normal adjustment

Advantages

(i) The final image is erect.

(ii) The telescope is shorter than both the astronomical telescope and the terrestrial telescope.

Disadvantages

(i) It has a small field of view.

(ii) It is impossible to place the eye in such a position that it collects all the light which has passed through the objective lens.

(iii) Telescopes often incorporate a cross-hair. This needs to be at the site of the intermediate image so that there is no parallax between the image of the cross-hair and the final image of the object being viewed. In the case of a

Galilean telescope it is impossible to meet this requirement, and therefore Galilean telescopes cannot be used to make precise determinations of the angular positions of the objects being viewed.

21.9 THE REFLECTING TELESCOPE

The purpose of the objective lens of a refracting telescope is to produce an image which can be examined by the eyepiece. In the reflecting telescope a concave mirror is used to the same end. Mirrors have a number of advantages over lenses.

(i) The diameter of a mirror can be much greater than that of a lens because a mirror can be supported over the whole of its non-reflecting surface. As a consequence, the world's largest telescopes (i.e. those that can be used to look at very faint objects and resolve very fine detail) are reflecting telescopes. The largest is the Russian telescope on Mt. Pastukhov – its mirror has a diameter of 6 m.

(ii) The glass which is used to make a mirror need only be free of air bubbles – there is no necessity for it to be homogeneous.

(iii) Mirrors cannot produce chromatic aberration.

(iv) A paraboloidal mirror is likely to produce less spherical aberration than a lens.

(v) There is only one surface to be ground.

In the **Cassegrain** reflecting telescope (Fig. 21.13) the intermediate image (I_1) formed by the concave objective acts as a virtual object for the (small) convex mirror. This produces a second intermediate image (I_2) in a convenient position to be magnified by the eyepiece.

Fig. 21.13 Cassegrain reflecting telescope

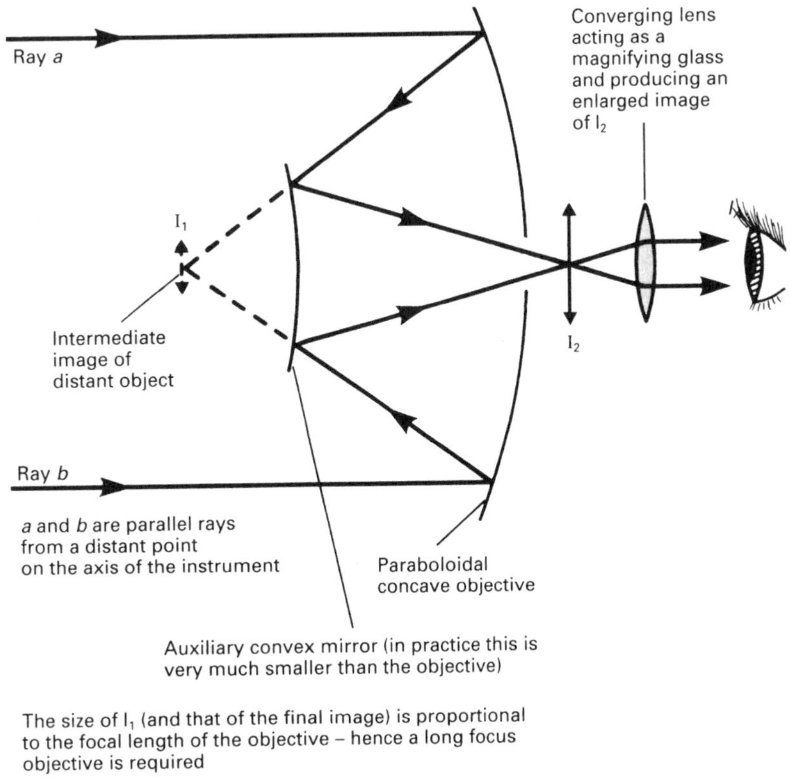
Ray a
Converging lens acting as a magnifying glass and producing an enlarged image of I_2
I_1
I_2
Intermediate image of distant object
Ray b
a and b are parallel rays from a distant point on the axis of the instrument
Paraboloidal concave objective
Auxiliary convex mirror (in practice this is very much smaller than the objective)
The size of I_1 (and that of the final image) is proportional to the focal length of the objective – hence a long focus objective is required

A **Newtonian** reflector is shown in Fig. 21.14.

Fig. 21.14
Newtonian reflecting telescope

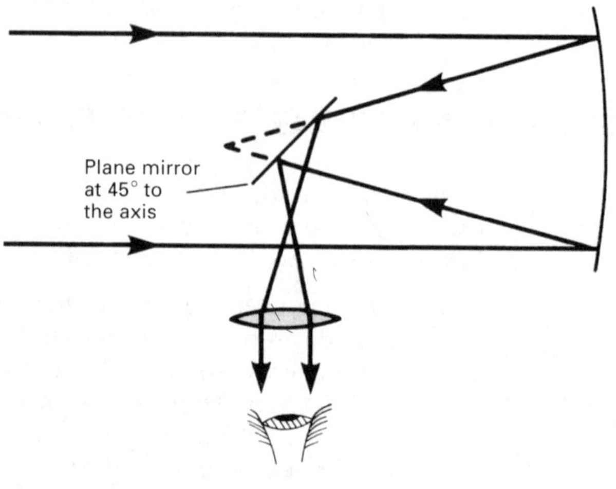

21.10 THE EYE-RING (EXIT-PUPIL)

The eye-ring of a telescope is a washer-like disc, positioned so that its circular aperture coincides with the image of the objective lens formed by the eyepiece lens (Fig. 21.15). Because of its position and because its aperture has the same diameter as this image, the eye-ring defines the **smallest** region through which passes all the light that has been refracted by both lenses. If an observer places his eye at the eye-ring, the maximum amount of light will enter his eye. **The position of the eye-ring is therefore the best position for the observer's eye.**

Fig. 21.15
The location of the eye-ring of a telescope

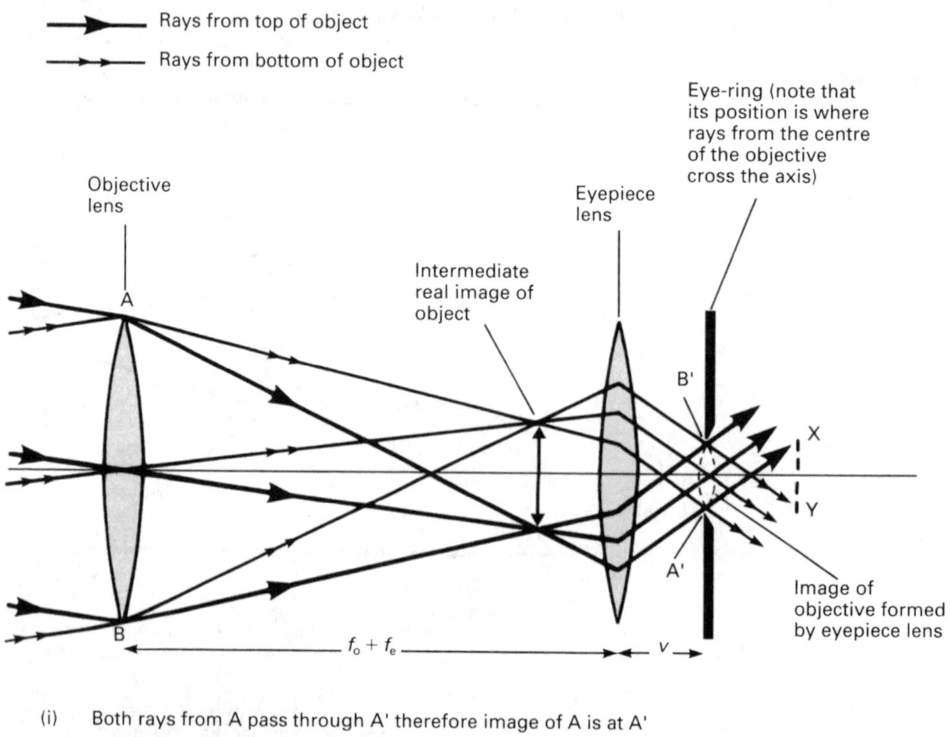

(i) Both rays from A pass through A' therefore image of A is at A'

(ii) A'B' is the smallest region through which *all* the light passes

(iii) XY represents the pupil of the observer's eye. Most of the light fails to enter the pupil when it is in the position shown. All the light enters the pupil when it is at A'B'

The diameter, D_e, of the eye-ring is equal to that of the image of the objective lens formed by the eye-piece lens. In normal adjustment the separation of these lenses is $f_o + f_e$ (where f_o and f_e are the respective focal lengths). Therefore, if v is the distance from the eyepiece lens to the image of the objective, and D_o is the diameter of the objective lens, then since by equation [19.7]

$$\frac{\text{Object size}}{\text{Image size}} = \frac{\text{Object distance}}{\text{Image distance}}$$

$$\frac{D_o}{D_e} = \frac{f_o + f_e}{v} \qquad [21.6]$$

Employing the usual notation, $u = f_o + f_e$, $v = v$ and $f = f_e$. Substituting these values into the lens formula (equation [19.1]) leads to

$$\frac{1}{f_o + f_e} + \frac{1}{v} = \frac{1}{f_e}$$

Multiplying by $(f_o + f_e)$ gives

$$1 + \frac{f_o + f_e}{v} = \frac{f_o + f_e}{f_e}$$

Therefore, from equation [21.6]

$$1 + \frac{D_o}{D_e} = \frac{f_o + f_e}{f_e}$$

i.e. $$\frac{D_o}{D_e} = \frac{f_o}{f_e}$$

In normal adjustment the angular magnification, M, is given by $M = f_o/f_e$, and therefore

$$M = \frac{D_o}{D_e}$$

21.11 THE SPECTROMETER

The spectrometer is an instrument which can be used to make accurate measurements of the deviation of a parallel beam of light which has passed through a prism or a diffraction grating. As such, the instrument provides a means of studying optical spectra. When used with a prism the spectrometer allows an accurate determination of refractive index to be made (see section 21.12). The use of a spectrometer with a diffraction grating is discussed in section 26.6.

The essential features of a spectrometer are shown in Fig. 21.16. The collimator is fixed, but the table and the telescope can each be rotated about a common (vertical) axis which passes through the centre of the table and is perpendicular to it. The instrument embodies a vernier scale arrangement so that the angles through which the table and the telescope have been turned can be measured to an accuracy of 1′ of arc. The collimator can be adjusted to provide a parallel beam of light when the slit is being illuminated. The collimator lens and the objective lens of the telescope are achromatic (see section 19.10).

Fig. 21.16 Spectrometer

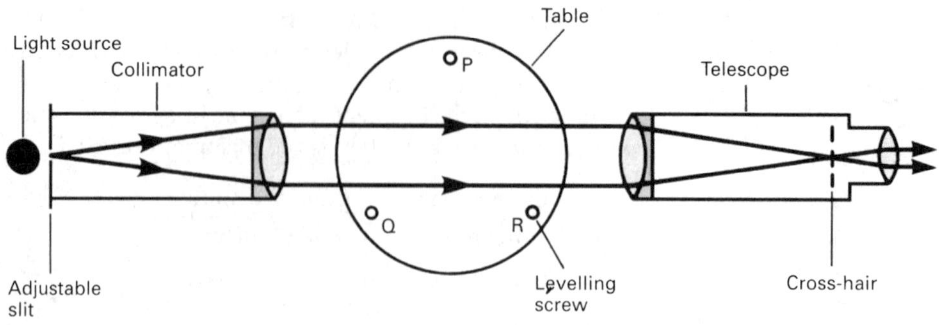

21.12 MEASUREMENT OF REFRACTIVE INDEX BY USING A SPECTROMETER

If the refracting angle A of a prism and the minimum deviation D produced by it are measured, the refractive index of the material of the prism can be found by making use of equation [18.8]. A spectrometer provides an accurate means of measuring A and D. Suppose that the refractive index of glass at the wavelength of sodium light is to be measured.

Adjusting the Spectrometer

(i) The telescope eyepiece is adjusted so that the cross-hairs are in sharp focus.

(ii) The telescope is focused on a distant object (by changing the position of the telescope objective relative to the eyepiece and the cross-hairs) in such a way that there is no parallax between the image and the cross-hairs. The significance of using a distant object is that, from now on, any parallel light which enters the telescope will be brought to a focus on the cross-hairs.

(iii) The collimator slit is illuminated with sodium light, and the telescope is turned so that it is in direct line with the slit. The slit is moved in or out of the collimator tube until the image of the slit, as seen through the telescope, is in sharp focus. Since the telescope has been set to focus parallel light, the collimator must now be providing parallel light. At this stage the width of the slit is optimized.

(iv) The prism is placed on the table in such a way that one face, XY say, is perpendicular to the line joining two of the levelling screws, Q and R say. The telescope is turned through 90° (by reference to the scale) and then the table is turned until XY reflects light into the telescope. The situation is now as shown in Fig. 21.17. If the plane of XY is not vertical, the centre of the image of the slit will be above or below the centre of the cross-hairs. If this is the case, Q is adjusted until the image is in the centre of the field of view. It is still possible that the other faces of the prism are not vertical and therefore, without moving the telescope, the table is turned so that XZ reflects light into the telescope. Levelling screw P is adjusted (if necessary) so that the image of the slit once more occupies the centre of the field of view. (The only effect on XY of adjusting P is to cause XY to move in its own plane and therefore leaves XY vertical.) The prism table is now level and the spectrometer is ready for use.

Fig. 21.17
Prism reflecting light from the collimator of the spectrometer into the telescope

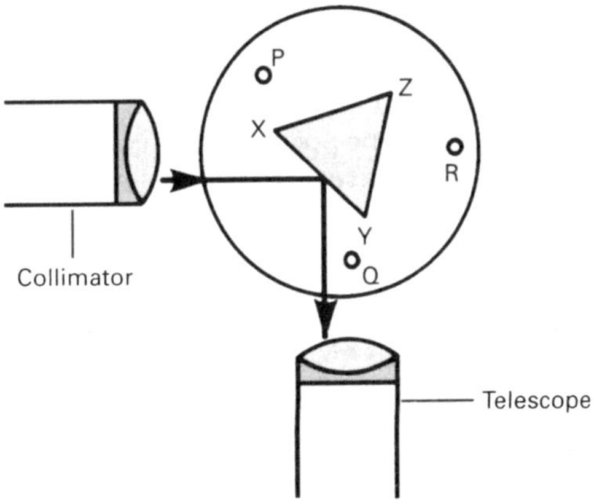

Measurement of *A*

The table is turned so that angle A faces the collimator and light is reflected by XY and XZ simultaneously (Fig. 21.18). The telescope is turned to T_1 so that an image of the slit (produced by reflection at XY) is centred on the cross-hairs. The position of T_1 is noted and the telescope is swung round to T_2 to receive light reflected from XZ. From simple geometry, the angle turned through in moving from T_1 to T_2 is equal to $2(\alpha + \beta) = 2A$; hence A.

Fig. 21.18
Measurement of angle ***A***

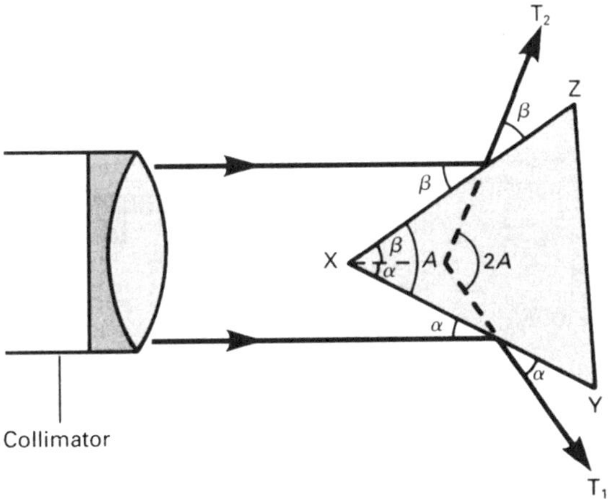

Measurement of *D*

The telescope and the table are rotated so that the telescope is receiving light which has passed approximately symmetrically through the prism (Fig. 21.19). The table is slowly rotated and, at the same time, the telescope is moved in such a way that the

Fig. 21.19
Measurement of angle ***D***

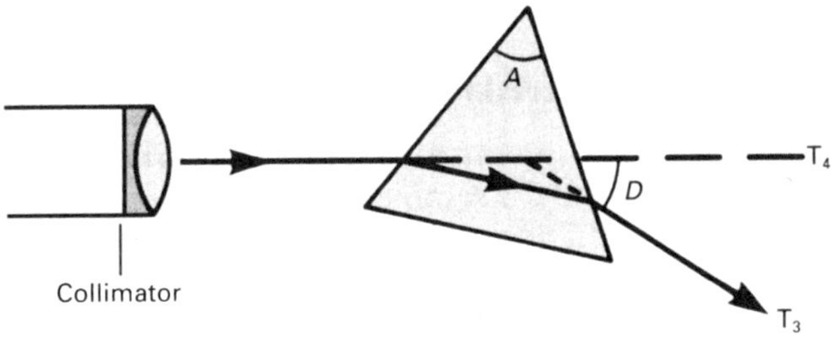

image remains centred on the cross-hairs. If the table is being rotated in the correct sense, a point is reached at which the image starts to move in the opposite direction to that in which it has been moving – this is the position of minimum deviation. The telescope position, T_3, is noted. The prism is removed and the telescope is turned to T_4 to receive the undeviated beam. The angle between T_3 and T_4 is D.

21.13 THE CAMERA

A simple camera is shown in Fig. 21.20. The lens can be moved in and out relative to the film so that light from objects at different distances may be focused on the film. The diameter, d, of the aperture can be altered by means of the diaphragm adjusting ring. When the camera is operated the shutter opens, for a predetermined time, and exposes the film to the light which has entered through the lens.

Fig. 21.20
Camera

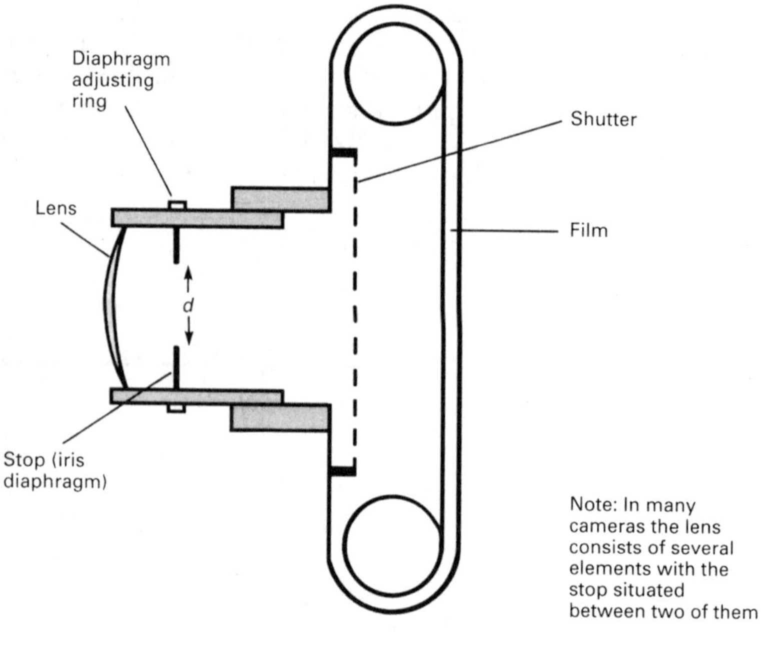
Diaphragm adjusting ring
Shutter
Lens
Film
d
Stop (iris diaphragm)
Note: In many cameras the lens consists of several elements with the stop situated between two of them

When the object being photographed is moving quickly, the exposure time has to be very short and therefore the diameter of the aperture has to be large in order that the amount of light which falls on the film is sufficient to expose it by the proper amount. When a slow-moving object is photographed, a long exposure can be used and the diameter of the aperture can be reduced. Using a small aperture has two advantages: (i) it reduces spherical aberration because the light which reaches the film has passed only through the centrel region of the lens, and (ii) it increases the **depth of field**.

f-number

The amount of light that reaches the film is proportional to the area of the aperture, i.e. to d^2. A camera is normally used to photograph objects whose distances from the lens are large compared with its focal length, f. It can be shown that the image of such an object covers an area which is approximately proportional to f^2. It follows that the amount of light per unit area of image is proportional to d^2/f^2 and that the required exposure time is proportional to f^2/d^2. f/d is called the **f-number** (or **relative aperture**). For a typical camera the available f-number settings are 2, 2.8, 4, 5.6, 8, 11, 16, 22. These numbers are such that the square of a number is (approximately) twice that of the number which precedes it. It follows that decreasing the f-number by one setting halves the exposure time.

Depth of Field

A camera is often required to photograph, simultaneously, objects which are at different distances from the lens. In Fig. 21.21(a) light from a point object at O is being brought to a focus at I on the film. Light from O_1 is out of focus and covers an area whose diameter is XY and which is known as a **circle of confusion**. Fig. 21.21(b) illustrates the effect of reducing the diameter of the aperture – the diameter X′Y′ of the circle of confusion is reduced and the photograph is clearer. The range of object distances for which the circles of confusion are so small that they are seen by the eye as points is called the **depth of field**. Thus, decreasing the aperture increases the depth of field.

Fig. 21.21
To illustrate depth of field

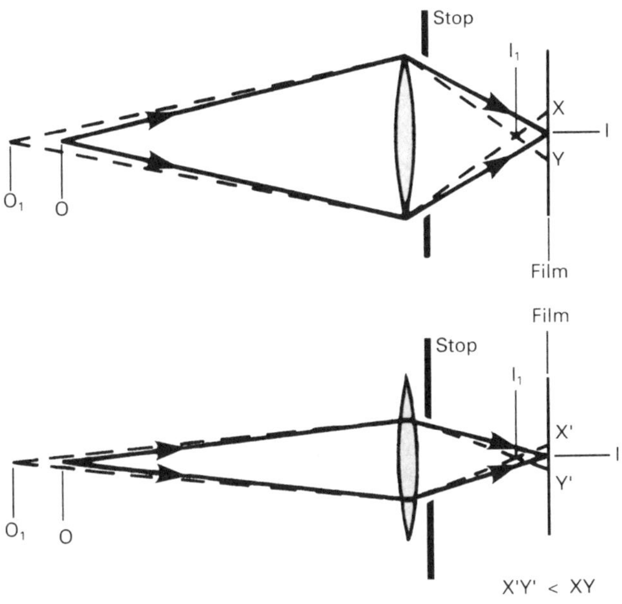

QUESTIONS 21B

1. A camera has an f-number of 5.6. When the camera is focused at infinity the centre of the lens is 4.9 cm from the film. What is the diameter of the lens?

2. When a camera is focused on an object 40.0 cm away the lens is 5.7 cm from the film. **(a)** What is the focal length of the lens? **(b)** How far would the lens need to be moved in order to focus the camera on an object at infinity?

21.14 THE EYE

Normal Eye

An eye (Fig. 21.22) produces a real, inverted image of the object being viewed. The image is produced on the **retina** – the light-sensitive region at the back of the eye. The shape, and therefore the focal length, of the eye lens can be altered by the action of the ciliary muscles attached to it. This makes it possible for light from

Fig. 21.22
Section through an eye

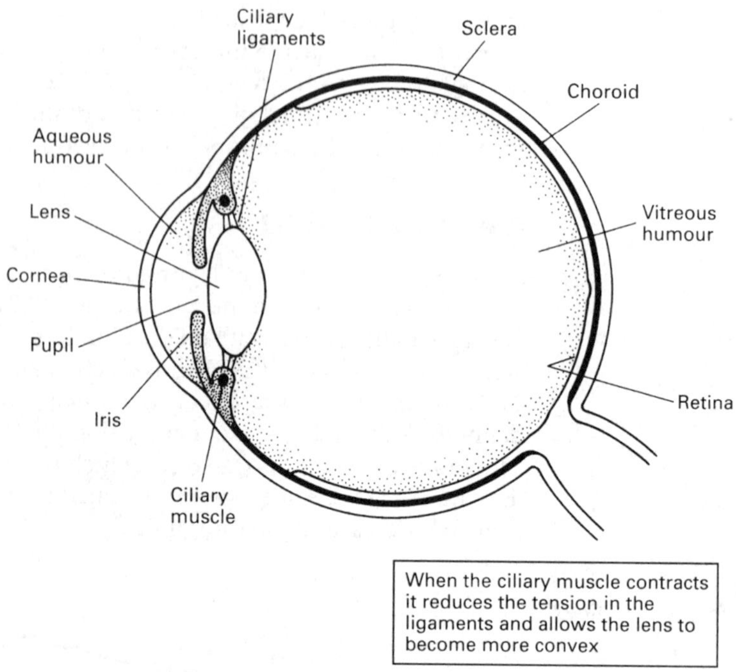

objects which are at different distances from the eye to be brought to a focus on the retina even though it is at a fixed distance from the lens. This ability of the eye is known as its **power of accommodation**. (Compare with this with the action of a camera, where objects at different distances are focused by altering the distance between the lens and the film.)

The closest point at which an eye can focus (comfortably) is called its **near point**; the most distant point is called the **far point**. For a normal eye these are at 25 cm and infinity, respectively. Light from an object close to the eye has to be deviated more than that from a more distant object. This requires a more powerful, and therefore more highly curved, lens. The lens is therefore at its most highly curved when focusing on the near point. An eye focused on infinity is relaxed and is said to be **unaccommodated**.

Note The refractive index of the lens is not very different from that of the aqueous humour or that of the vitreous humour. Consequently light undergoes very little deviation as a result of passing through the lens itself. The main function of the lens is to produce slight changes in deviation rather than large amounts of it. Most of the deviation occurs at the boundary between the air and the cornea.

Short Sight (Myopia)

A short-sighted person can see near objects clearly but not distant ones, i.e. his or her far point is closer than infinity (Fig. 21.23(a) and (b)). Light from a distant object is brought to a focus in front of the retina. Either the eyeball is too long or the eye, with the lens at its weakest, is too strong. The condition is corrected by using a suitable diverging lens to make parallel rays of light appear to have come from the (uncorrected) far point of the eye (Fig. 21.23(c)). It follows that the diverging lens must have a focal length which is equal to the uncorrected far point distance.

Fig. 21.23
(a) and (b) Short sight; (c) correction of short sight. (The eye lens is at its weakest in every case.)

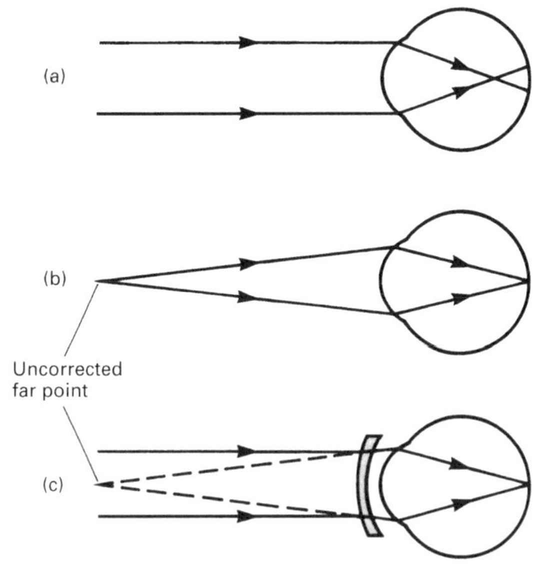

Long Sight (Hypermetropia)

A long-sighted person can see distant objects clearly but not those close by, i.e. his or her near point is greater than 25 cm from the eye (Fig. 21.24(a) and (b)). Light from a nearby object heads towards a focus behind the retina. Either the eyeball is too short or the eye, with the lens at its strongest, is too weak. The condition is corrected by using a suitable converging lens. The lens must produce a virtual image at the (uncorrected) near point of the eye of an object which is 25 cm from the eye (Fig. 21.24(c)). The focal length of the correcting lens is easily calculated. Suppose a long-sighted person has a near point distance of 150 cm. We see from Fig. 21.24(c) that an object 25 cm from the lens must give rise to a virtual image which is 150 cm from the lens, i.e. $u = 25$ cm, $v = -150$ cm and therefore by $1/u + 1/v = 1/f$, $f = 30$ cm. Thus a converging lens of focal length 30 cm is required.

Fig. 21.24
(a) and (b) Long sight; (c) correction of long sight. (The eye lens is at its strongest in every case.)

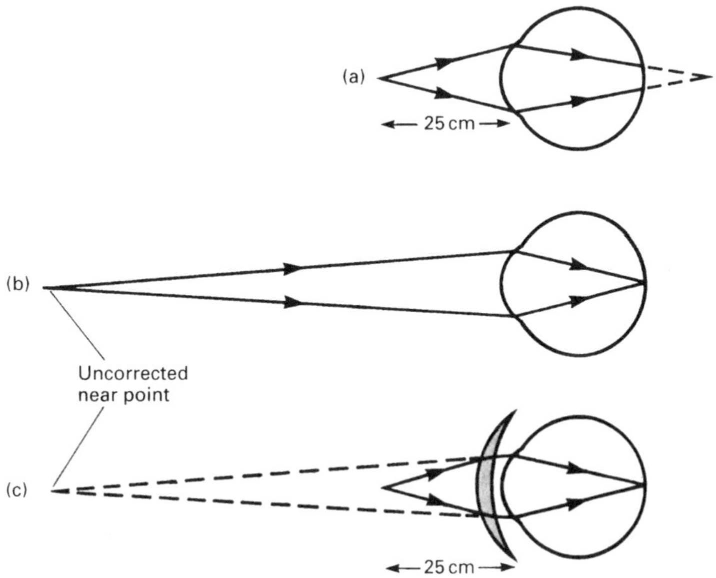

EXAMPLE 21.2

A man with a normal near point distance of 25.0 cm wears spectacles with diverging lenses of focal length 200 cm in order to correct his far point distance to infinity. What is his near point distance when wearing the spectacles?

Solution

The closest point on which the eye can focus whilst the spectacles are being worn is such that light from it appears to have come from a point 25.0 cm from the eye after refraction by the spectacle lens. Thus we need to find the object distance, u, for which the spectacle lens produces a virtual image distance of 25.0 cm. By

Note
v is minus 25.0 cm because it is a virtual image distance; f is minus 200 cm because the lens is diverging.

$$\frac{1}{u}+\frac{1}{v} = \frac{1}{f}$$

$$\frac{1}{u}+\frac{1}{-25.0} = \frac{1}{-200}$$

$$\therefore \quad \frac{1}{u} = -\frac{1}{200}+\frac{1}{25.0} = \frac{-1+8}{200} = \frac{7}{200}$$

$$\therefore \quad u = \frac{200}{7} = 28.6\text{ cm}$$

i.e. the near point distance with the spectacles is 28.6 cm.

QUESTIONS 21C

1. A girl can see clearly only those objects which are greater than 250 cm from her eyes. **(a)** What defect of vision does she have? **(b)** What type of lens (converging or diverging) would be used to correct the defect? **(c)** What would be the focal length of the spectacle lens that would allow her to see objects 25 cm from her eyes? **(d)** What would her far point distance be when wearing the spectacles?

CONSOLIDATION

The visual angle is the angle subtended at the eye by an object.

The near point is the position at which an object (or image) can be seen most clearly. An object (or image) closer than the near point appears blurred, one which is further away appears smaller than when at the near point.

Magnifying Glass (Simple Microscope)

$$M = \frac{\text{Angle subtended at eye by image}}{\text{Angle subtended at eye by object at the near point}}$$

The lens must have a short focal length for high angular magnification.

When the image is at the near point the angular magnification is at its largest and is equal to the lateral magnification. Also

$$M = \frac{h_i}{h_o} = \frac{D}{f}+1$$

When the image is at infinity the eye is relaxed (unaccommodated). Also

$$M = \frac{D}{f}$$

Compound Microscope

$$M = \frac{\text{Angle subtended at eye by image}}{\text{Angle subtended at eye by object at near point}}$$

Each lens must have a short focal length for high angular magnification.

The objective produces a real image which acts as an object for the eyepiece. The eyepiece acts as a magnifying glass and produces a virtual image.

When the final image is at the near point the angular magnification is at its largest and is equal to the lateral magnification.

When the final image is at infinity the eye is relaxed (unaccommodated).

Astronomical Telescope

A large objective can:

(i) collect a lot of light and so produce a bright image,

(ii) resolve fine detail.

$$M = \frac{\text{Angle subtended at eye by image}}{\text{Angle subtended at eye by object wherever it happens to be}}$$

Requires f_o large and f_e small for high angular magnification.

The objective produces a real image which acts as an object for the eyepiece. The eyepiece acts as a magnifying glass and produces a virtual image.

When the final image is at infinity the telescope is in **normal adjustment**. Also

$$M = \frac{f_o}{f_e}$$

Deriving Expressions for Angular Magnification

(i) It is necessary to specify that angles are small so that the approximations $\alpha = \tan\alpha$ and $\beta = \tan\beta$ may be used.

(ii) Except when the object or image is at infinity, it is necessary to specify that the eye is close to the lens concerned so that to a good approximation the angle subtended at the eye is equal to the angle subtended at the lens.

Short sight – nearby objects can be seen clearly, distant ones cannot. Corrected by using a diverging lens.

Long sight – distant objects can be seen clearly, nearby ones cannot. Corrected by using a converging lens.

22

EXPERIMENTAL DETERMINATION OF THE VELOCITY OF LIGHT

22.1 AN OUTLINE OF RÖMER'S ASTRONOMICAL METHOD

There have been many determinations of the velocity of light. The first of these was that of a Danish astronomer, Römer, in 1676. (Galileo had made an unsuccessful attempt in 1600.)

One effect of the Earth's motion about the Sun is to cause it to spend about six months of each year moving away from the planet Jupiter, and the remaining time moving towards it. Römer had noticed that the time intervals between Jupiter eclipsing one of its moons became progressively longer as the Earth moved away from the planet, and then progressively shorter as the earth moved back again. He reasoned that the eclipses actually take place at constant intervals, and that when the time intervals were increasing it was because the Earth was moving away from Jupiter and so the light had to travel a little farther each time. Knowing the diameter of the Earth's orbit about the Sun he was able to obtain a value for the velocity of light.

22.2 MICHELSON'S METHOD

In 1926 Michelson made use of an apparatus which involved a rotating octagonal steel prism (Fig. 22.1).

When the prism is stationary the light follows the path shown and an image of the source can be seen through the telescope. If the prism is rotated slowly, the image disappears because either face X is not in a suitable position to direct the outgoing beam to C, or face Y is unable to send the incoming beam to the telescope. However, if the speed of the prism is increased so that it turns through exactly one-eighth of a revolution in the same time that it takes light to travel from X to Y, then an image of the source is seen through the telescope. Michelson adjusted the speed of rotation until he was able to observe a stationary image of the source. This occurred when the prism was rotating at 530 rev s^{-1}. The experiment was carried out on Mt. Wilson (USA) and the concave reflector (C) was on another mountain 35 km away.

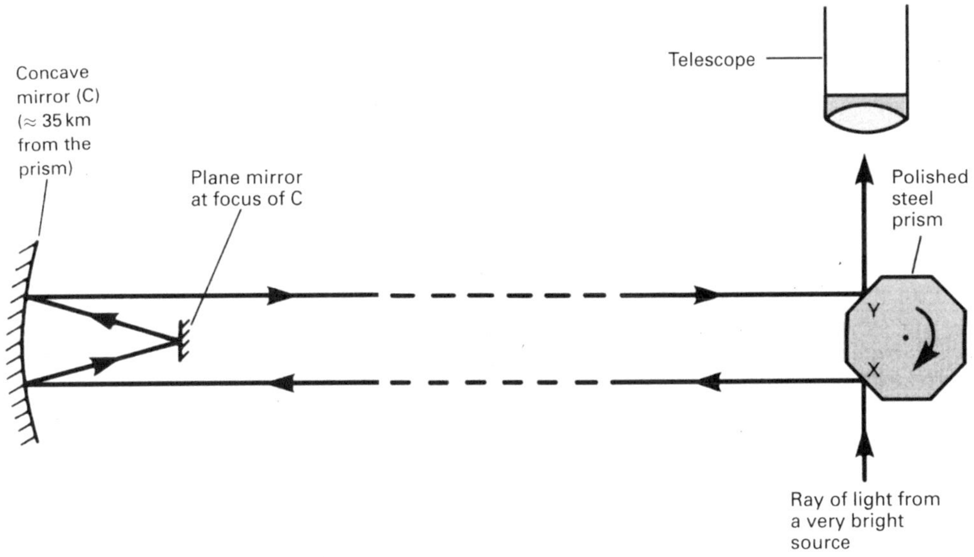

Fig. 22.1
Michelson's method for determining the velocity of light. (Note that the optical system used by Michelson was more complex than that shown here.)

The figures given here are approximate but can be used to indicate the procedure adopted by Michelson. Thus,

$$\begin{aligned}\text{Speed of light } (c) &= \frac{\text{Light path}}{\text{Time taken}} \\ &= \frac{2 \times 35 \times 10^3}{(1/8)(1/530)} \\ &\approx 3 \times 10^8 \text{ m s}^{-1}\end{aligned}$$

Michelson made a correction to take account of the fact that the light was travelling through air (rather than vacuum) and obtained a final value which was accurate to better than about one part in 10^5

QUESTIONS ON SECTION D

REFRACTION (Chapter 18)

D1 A ray of light is incident in air on the surface of a glass block. The angle of incidence is 30°. Calculate:

(a) the angle of refraction,

(b) the amount by which the ray is deviated on entering the glass.

(Refractive index of glass with respect to air = 1.5.)

D2 A ray of light is incident in glass on a plane glass–air boundary and makes an angle of 30° with the normal to the boundary. Calculate the angle through which the ray is deviated on entering the air.
(Refractive index of glass with respect to air = 1.5.)

D3 A ray of light is incident in air at an angle of 40° to the normal to one face of a 60° glass prism. Calculate the angle through which the ray has been deviated by the time it emerges from the prism.
(Refractive index of glass with respect to air = 1.50.)

D4 A ray of light is incident in water on a plane water–glass boundary. The angle of incidence is 30°. Calculate the angle of refraction.
(Refractive index of glass with respect to air = 3/2, refractive index of water with respect to air = 4/3.)

D5 A block of glass measures 5 cm × 5 cm × 8 cm. When the block is stood on one of its smaller faces and viewed from directly above it appears to be a cube. Calculate the refractive index of the glass.

D6 Calculate the critical angle for a glass–air surface if a ray of light which is incident in air on the surface is deviated through 15.5° when its angle of incidence is 40.0°.

D7 A narrow parallel pencil of monochromatic light is incident on a plane boundary between air and glass. Describe the phenomena which occur as the angle of incidence of the light is progressively increased from 0° to 90°, **(a)** in the air, **(b)** in the glass. [J]

D8 How would you use an air-cell to determine the refractive index of a liquid for sodium light? Point out the essential features of the structure of the cell, and give the theory of the method.

State and explain the difficulty encountered if white light is used in such an experiment.

A ray of monochromatic light is incident on one face of a prism of refracting angle 60°, made of glass of refractive index 1.50. Calculate the *least* angle of incidence for the ray to be transmitted through the second face. [J]

D9 **(a)** For light travelling in a medium of refractive index n_1 and incident on the boundary with a medium of refractive index n_2, explain what is meant by total internal reflection and state the circumstances in which it occurs.

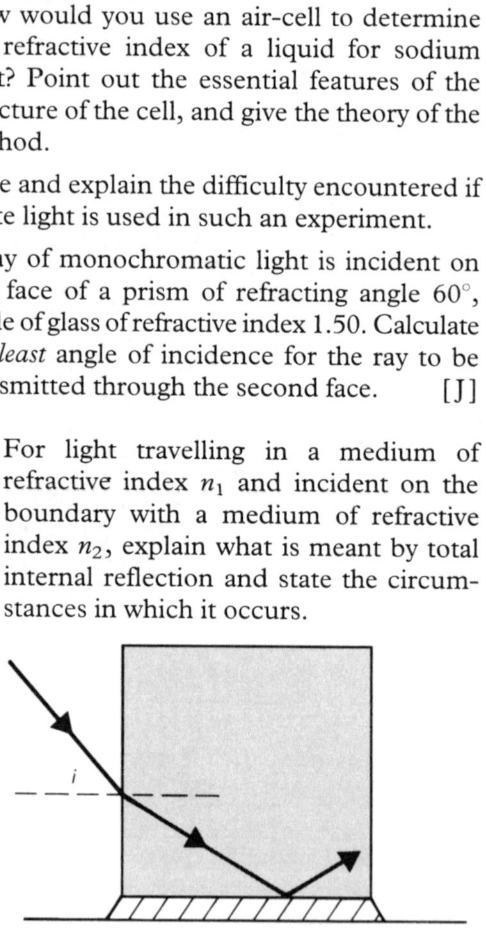

(b) A cube of glass of refractive index 1.500 is placed on a horizontal surface separated from the lower face of the cube by a film of liquid, as shown in the diagram. A ray of light from outside and in a vertical plane parallel to one face of the cube strikes another vertical face of the cube at an angle of incidence of $i = 48°\ 27'$ and, after refraction, is totally reflected at the critical angle at the glass-liquid interface. Calculate **(i)** the critical angle at the glass–liquid interface and **(ii)** the angle of emergence of the ray from the cube. [J]

D10 A plane mirror lies at the bottom of a long flat dish containing water, the mirror making an angle of 10° with the horizontal, as shown in the figure below.

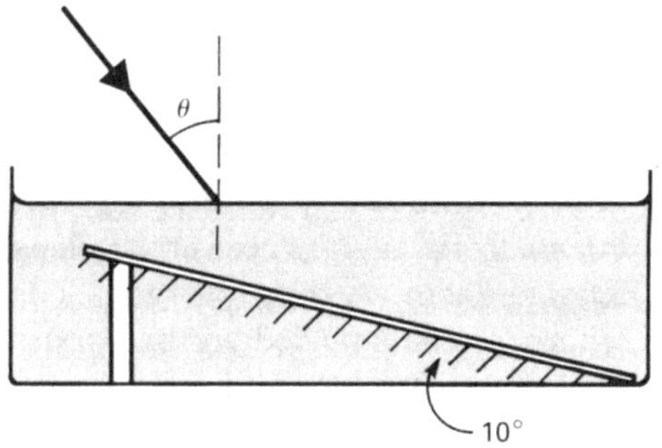

A narrow beam of monochromatic light falls on the surface of the water at an angle of incidence θ. If the refractive index of water is 4/3, determine the *maximum* value of θ for which light, after reflection from the mirror, would emerge from the upper surface of the water. [AEB, '79]

D11 Calculate the minimum deviation produced by a 60° glass prism if the refractive index of the glass is 1.50. What is the angle of incidence at which minimum deviation occurs?

D12 A ray of light strikes face AB of a triangular glass prism of section ABC, enters the prism and next strikes face AC. Angle BAC is 62.00°. What is the minimum angle of incidence on face AB which will allow light to emerge from face AC if the refractive index of the glass is 1.520?

What is the minimum deviation which this prism can cause for light entering face AB and leaving via face AC? [S]

D13 The diagram shows a narrow parallel horizontal beam of monochromatic light from a laser directed towards the point A on a vertical wall. A semicircular glass block G is placed symmetrically across the path of the light and with its straight edge vertical. The path of the light is unchanged.

The glass block is rotated about the centre, O, of its straight edge and the bright spot where the beam strikes the wall moves down from A to B and then disappears.

OA = 1.50 m   AB = 1.68 m

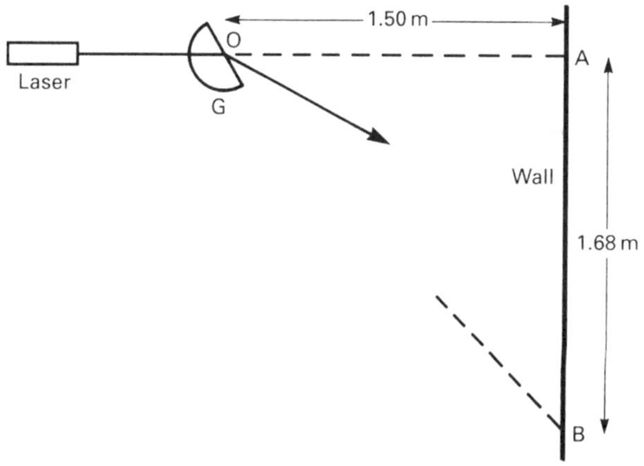

(a) Account for the disappearance of the spot of light when it reaches B.
(b) Find the refractive index of the material of the glass block G for light from the laser.
(c) Explain whether AB would be longer or shorter if a block of glass of higher refractive index was used. [L]

D14 Calculate the angular separation of the red and violet rays which emerge from a 60° glass prism when a ray of white light is incident on the prism at an angle of 45°. The glass has a refractive index of 1.64 for red light and 1.66 for violet light.

LENSES (Chapter 19)

D15 A real object which is 3.0 cm high is placed 60 cm from **(a)** a converging lens of focal length 20 cm, **(b)** a diverging lens of focal length 30 cm. Calculate the size and position of the image formed in each case.

Draw ray diagrams to illustrate the formation of the images.

D16 A convering lens forms a real image of a real object. If the image is twice the size of the object and 90 cm from it, calculate the focal length of the lens.

How far from the lens would the object have to be placed for the image to be the same size as the object?

D17 An object 3 cm high is placed 3 cm from a diverging lens of focal length 4 cm. The bottom of the object is on the principal axis of the lens. A converging lens of focal length 3 cm is placed coaxial with the diverging lens and 4 cm from it on the side remote from the object. Draw a scale diagram showing the paths through the system of two rays from the top of the object. Make it clear which lines in your diagram represent the paths of real rays, which represent virtual rays and which are construction lines. Measure the size of the final image.

D18 When an object is placed 50 cm from a thin converging lens a real image is formed at a distance of $33\frac{1}{3}$ cm from the lens. When a thin diverging lens is placed in contact with the first lens the image distance increases to 50 cm. What is the focal length of the diverging lens?

D19 An opaque disc P, 3 mm diameter, lies at the bottom of a glass beaker, and is illuminated from below by a source S. A converging lens L of focal length 10 cm, situated 15 cm above the disc, forms an image of this at Q. Where is Q situated, and what is the size of the image?

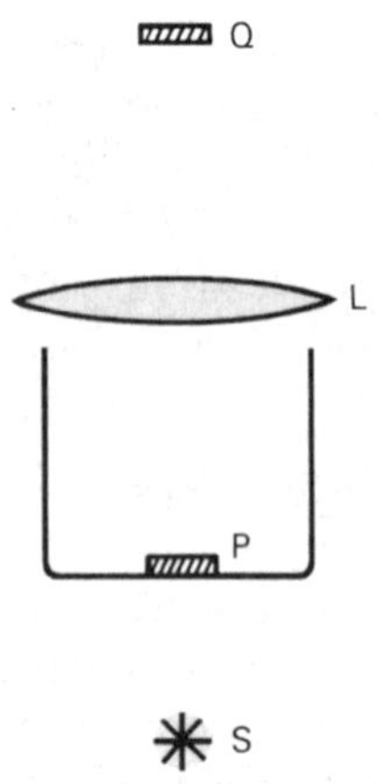

Explain qualitatively how the position and size of the image of the disc is changed when the beaker is filled with water. [S]

D20 A camera is fitted with a lens of focal length 50 mm. If a distant building subtends an angle of 0.1 rad at the camera, what is the size of its image on the film?

Over what distance must it be possible to move the lens so as to be able to focus sharply objects at distances between infinity and 550 mm?

If a disc of diameter 250 mm is photographed when placed 550 mm from the camera, what is the size of its image on the film? [S]

D21 Describe how you would determine the focal length (about 20 cm) of a diverging lens, being provided with a converging lens whose focal length is about 20 cm. Explain **(a)** the purpose of the converging lens, **(b)** why the two lenses are separated in the experiment.

Trace the paths of *two* rays from a point source off the axis of the arrangement you describe, marking on your diagram the positions of the principal foci, and distinguishing between rays and construction lines.

A point source is placed on the axis of, and 60 cm from, a thin converging lens of focal length 20 cm. A thin diverging lens of focal length 20 cm is placed on the opposite side of the converging lens and 5 cm from it. Calculate the position of the final image, and comment on the significance of the result with reference to the experiment you have described.

(At the beginning of your calculation state the sign convention you are using.) [J]

D22 Explain what is meant by **(a)** a virtual image, **(b)** a virtual object, in geometrical optics. Illustrate your answer by describing the formation of **(i)** a virtual image of a real object by a thin converging lens, **(ii)** a real image of a virtual object by a thin diverging lens. In each instance draw a ray diagram showing the passage of *two* rays through the lens for a non-axial object point. [J]

D23 A certain camera has a single converging lens of focal length 50 mm. What range of movement of the lens is necessary to produce a clear image of an object at any distance from the camera between 1.0 m and infinity? [C(O)]

D24 A slide projector is required to produce a real image 684 mm wide from an object 36 mm wide. If the distance of the object from the screen is to be 2000 mm, calculate:
(a) the distance of the lens from the object,
(b) the focal length of the lens required.
[AEB, '79]

D25 An object is placed 0.15 m in front of a converging lens of focal length 0.10 m. It is required to produce an image on a screen 0.40 m from the lens on the opposite side to the object. This is to be achieved by placing a second lens midway between the first lens and the screen. What type and of what focal length should this lens be? Sketch a diagram showing two rays from a non-axial point on the object to the final image. [L]

D26 **(a)** A student performs an experiment to measure the focal length of a converging lens. In the experiment a series of object and image distances (u and v) is obtained and then a graph is drawn of uv against $(u + v)$. This graph is shown.

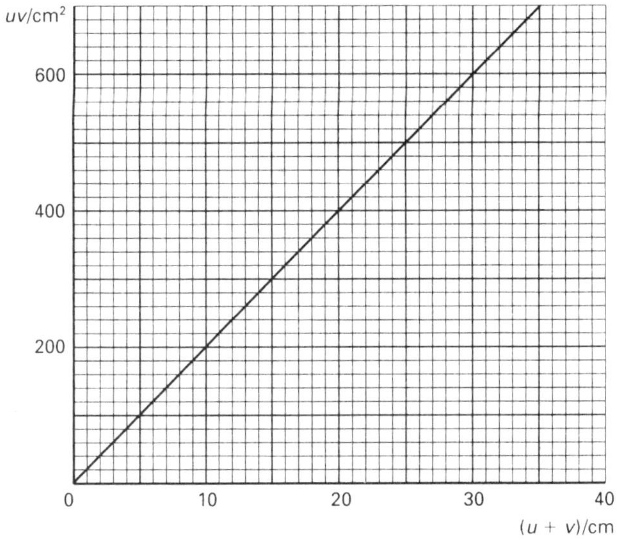

(i) Show that the slope of such a graph is equal to the focal length of the lens.
(ii) From the graph, obtain a value for this focal length.

(b) Explain why the images produced by this lens may be slightly coloured. [S]

MIRRORS (Chapter 20)

D27 A child is 1.4 m tall and her eyes are 10 cm below the top of her head. She wishes to see the whole length of her body in a vertical plane mirror whilst she herself is standing vertically.
(a) What is the minimum length of mirror that makes this possible?
(b) How far above the ground is the top of the mirror?

D28 When a plane mirror is turned through an angle α the reflected ray turns through an angle β. Derive the relationship between α and β.

D29 Two plane mirrors are parallel to, and facing, each other with a separation x. A point object lies between the mirrors at a distance y from one of them. What is the distance from the object of the image produced by two reflections in each mirror?

D30 A small luminous object is on the principal axis of a converging lens of focal length 40 cm. The object is 50 cm from the lens. Light from the object passes through the lens and is then reflected by a plane mirror. The mirror is 150 cm from the lens and perpendicular to its principal axis.
(a) What is the nature of the image produced by the mirror?
(b) How far from the lens is this image?

D31 An object is placed 4 cm from **(a)** a concave mirror which has a radius of curvature of 24 cm, **(b)** a convex mirror which has a radius of curvature of 40 cm. Calculate the position of each image and the magnification produced in each case.

Draw ray diagrams to illustrate the formation of the images.

D32 A rod which is 10 cm long is placed along the principal axis of a concave mirror such that the mid-point of the rod is 35 cm from the pole of the mirror. Calculate the radius of curvature of the mirror if it forms a real image of the rod which is 20 cm long.

OPTICAL INSTRUMENTS (Chapter 21)

D33 A thin converging lens of focal length 50 mm is to be used as a magnifying glass with the observer's eye close to the lens. If the observer can see images clearly anywhere between 250 mm from the lens and infinity, determine:
(a) the range of possible object distances,
(b) the corresponding range of magnifying powers. [J]

D34 Explain what is meant by the magnifying power of a magnifying glass. Derive expressions for the magnifying power of a magnifying glass when the image is **(a)** 25 cm from the eye and **(b)** at infinity. In each case draw the appropriate ray diagram. [J]

D35 Draw a ray diagram to illustrate the action of a thin lens used as a magnifying glass, the image being at the near point of the eye and the eye being close to the lens.

Distinguish clearly between the *magnification* of the image and the *magnifying power* of the lens. Explain why, in the case shown in your ray diagram, these quantities are numerically equal and derive an expression giving the magnifying power of the lens in terms of its focal length f and the nearest distance of distinct vision D. [J]

D36 A point object is placed on the axis of, and 3.6 cm from, a thin converging lens of focal length 3.0 cm. A second thin converging lens of focal length 16.0 cm is placed coaxial with the first and 26.0 cm from it on the side remote from the object. Find the position of the final image produced by the two lenses.

Why is this not a suitable arrangement for a compound microscope used by an observer with normal eyesight? For such an observer wishing to use the two lenses as a compound microscope with the eye close to the second lens decide, by means of a suitable calculation, where the second lens must be placed relative to the first. [J]

D37 The diagram shows two converging lenses arranged as a compound microscope giving a final virtual image 250 mm from the eye lens. The cross wires appear to the observer to be coincident with this image.

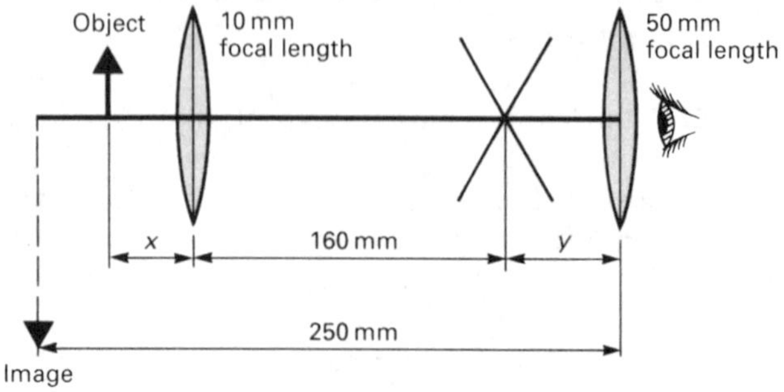

(a) Calculate y.
(b) Calculate the magnifying power of the instrument.
(c) Calculate x.
(**Note**. The diagram is *not* drawn to scale.) [S]

D38 The diagram shows the paths of two rays of light from the tip of an object B through the objective, O, and the eyelens, E, of a compound microscope. The final image is at the near point of an observer's eye when the eye is close to E. F_o and F_o' are the principal foci of O and F_E is one of the principal foci of E. The diagram is *not drawn to scale*.

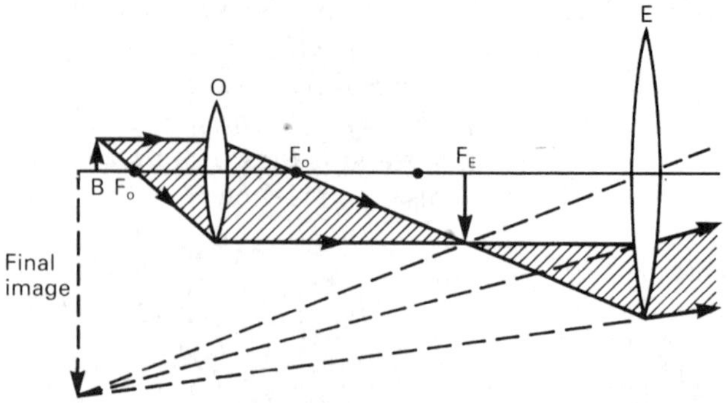

(a) Explain why:
(i) the object is placed just to the left of F_o,
(ii) the eyepiece is adjusted so that the intermediate image is to the right of F_E.
(b) In this arrangement, the focal lengths of O and E are 10 mm and 60 mm respectively. If B is 12 mm from O and the final image is 300 mm from E, calculate the distance apart of O and E. [J]

D39 Explain what is meant by the *least distance of distinct vision* and define the *magnifying power (angular magnification)* of a microscope.

Describe what is meant by **(a)** a simple microscope or magnifying lens and **(b)** a compound microscope. For each type of microscope, draw a ray diagram to show the passage through the microscope of two rays from a non-axial point object to produce an image at the least distance of distinct vision, and derive an expression for the magnifying power.

It is generally assumed that the eye can see clearly two points at the least distance of distinct vision when they are at least 10^{-3} m apart. Suggest suitable values for the focal lengths of the lenses and for their separation in a compound microsope which has just sufficient magnifying power for an observer to see particles which are 5×10^{-6} m apart. Assume that the least distance of distinct vision is 0.25 m. [W]

D40 **(a)** A thin converging lens can be used to form a real image or a virtual image of a small extended object at right angles to its principal axis. State how the nature of the image, and also the magnification (defined as *length of image/length of object*) depend on the position of the object relative to the principal focus on the object side.

(b) Defining the magnifying power of a simple magnifying glass to be

$$\frac{\text{Angle subtended at the eye by image in its actual situation}}{\text{Angle subtended at the unaided eye by object if it were placed at the eye's near point}}$$

show that the magnifying power of a magnifying glass of focal length f can range from d/f when the image is formed at infinity to $(1 + d/f)$ when the image is formed at the near-point distance d.

(c) A compound microscope consists of two thin lenses, an objective of focal length 20 mm and an eyepiece of focal length 50 mm, placed 220 mm apart. If the final image is viewed at infinity, calculate the distance of the object from the objective and also the magnifying power of the system when it is used by a man whose near-point distance is 250 mm.

(d) Draw a ray-trace diagram (not to scale) illustrating the passage of light through the microscope, showing rays starting from a non-axial point on the object. [O]

D41 A telescope is used to view a distant object, the final image being at infinity. Show on a diagram the paths through the telescope of two rays of light from each of two points on the object. Mark on your diagram the best position for the eye to be placed. [S]

D42 A telescope consists of two converging lenses: an objective lens of focal length 500 mm and an eyepiece lens of focal length 50 mm. When the telescope is in normal adjustment:

(a) what is the separation of the lenses,

(b) where is the final image located,

(c) is the image erect or inverted,

(d) what is the magnifying power,

(e) where should the pupil of the eye be placed to obtain the best view through the telescope? [S]

D43 Explain the term *angular magnification* as related to an optical instrument.

Describe, with the aid of a ray diagram, the structure and action of an astronomical telescope. Derive an expression for its angular magnification when used so that the final image is at infinity. With such an instrument what is the best position for the observer's eye? Why is this the best position?

Even if the lenses in such an instrument are perfect it may not be possible to produce clear separate images of two points which are close together. Explain why this is so. Keeping the focal lengths of the lenses the same, what could be changed in order to make the separation of the images more possible? [L]

D44 Define the *angular magnification* (or *magnifying power*) of a telescope. A telescope consists of two converging lenses. When in normal-adjustment, so that the image of a distant object is formed at infinity, the lenses are 450 mm apart, and the angular magnification is 8. What are the focal lengths of the two lenses? Is the image erect or inverted? [S]

D45 **(a)** What is meant by *normal adjustment* for an astronomical telescope? Why is it used in this way?

(b) An astronomical telescope in normal adjustment is required to have an angular magnification of 15. An objective lens of focal length 900 mm is available. Calculate the focal length of the eyepiece required and draw a ray diagram, not to scale, to show how the lenses should be arranged. The diagram should show three rays passing through the telescope from a non-axial point on a distant object. State the position of the final image, and whether or not it is inverted. [J, '89]

D46 Define *magnifying power* of an optical telescope.

Draw a ray diagram for an astronomical refracting telescope in normal adjustment, showing the paths through the instrument of three rays from a non-axial distant point object. Derive an expression for the magnifying power.

The magnifying power of an astronomical telescope in normal adjustment is 10. The real image of the *objective lens* produced by the eye lens has a diameter 0.40 cm. What is the diameter of the objective lens?

Discuss briefly the significance of the diameter of the objective lens on the optical performance of a telescope. [J]

D47 State what is meant by the *normal adjustment* in the case of an astronomical telescope.

Trace the paths of three rays from a distant non-axial point source through an astronomical telescope in normal adjustment.

Define the *magnifying power* of the instrument, and by reference to your diagram, derive an expression for its magnitude.

A telescope consists of two thin converging lenses of focal lengths 100 cm and 10 cm respectively. It is used to view an object 2000 cm from the objective. What is the separation of the lenses if the final image is 25 cm from the eye-lens? Determine the magnifying power for an observer whose eye is close to the eye-lens. [J]

D48 Draw a ray diagram showing the action of an astronomical telescope, consisting of two thin converging lenses, in normal adjustment, when forming separate images of two stars.

Explain why the images of the two stars formed on the retina of the eye will have a greater separation if the stars are viewed through the telescope than if they are viewed by the unaided eye.

The objective of an astronomical telescope in normal adjustment has a diameter of 150 mm and a focal length of 4.00 m. The eyepiece has a focal length of 25.0 mm. Calculate:
(a) the magnifying power of the telescope,
(b) the position of the eye ring (that is, the position of the image of the objective formed by the eyepiece),
(c) the diameter of the eye ring.
Give one advantage of placing the eye at the eye ring. [L]

D49 **(a)** **(i)** What is meant by the magnifying power M of an astronomical telescope?
(ii) Such a telescope can be made from two converging lenses. With the aid of a sketch, derive an expression for M in terms of the focal length of these lenses. Assume that the telescope is in normal adjustment.
(b) **(i)** Explain the significance of the eye-ring of a telescope.
(ii) The diameters of eye-rings seldom exceed 4 mm. Suggest a reason why.
(c) A telescope objective lens (of focal length 2.00 m) is used to photograph the Moon, which subtends an angle of 9.2 mrad (= 0.53°) at the Earth. The photographic film is placed in the principal focal plane of the lens, at right-angles to its principal axis.
(i) Calculate the diameter of the image of the Moon formed on the photographic film.
(ii) Suggest a reason why the complete telescope in normal adjustment is not used for taking the photograph.
[O, '92*]

D50 Draw a ray diagram to show how a converging lens produces an image of finite size of the Moon clearly focused on a screen. If the Moon subtends an angle of 9.1×10^{-3} radian at the centre of the lens, which has a focal length of 20 cm, calculate the diameter of this image.

With the screen removed, a second converging lens of focal length 5.0 cm is placed coaxial with the first and 24 cm from it on the side remote from the Moon. Find the position, nature and size of the final image. [J]

D51 A refracting telescope has an objective of focal length 1.0 m and an eyepiece of focal length 2.0 cm. A real image of the Sun, 10 cm in diameter, is formed on a screen 24 cm from the eyepiece. What angle does the Sun subtend at the objective? [L]

D52 Define *magnifying power* of a telescope. Show that, if a telescope is used to view a distant object and is adjusted so that the final image is at infinity, the magnifying power is given by the ratio of the focal lengths of the objective and eyepiece.

An astronomer has the choice of two telescopes of equal magnifying power but of different apertures. Explain what advantages he could obtain by choosing one rather than the other.

A telescope has two lenses of focal lengths 1.0 m and 0.10 m and it is adjusted to produce an image of a distant object on a screen. The object subtends an angle of 0.30° at the telescope objective. Calculate: **(a)** the linear size of the image formed on the screen 0.5 m from the eyepiece, and **(b)** the distance between the two lenses. Draw a diagram of the optical arrangement showing the paths of *two* rays through the lens system which come from a point on the object not on the axis. [L]

D53 Draw a ray diagram to show how the eye views the image formed by a convex lens when used as a magnifying lens. Define the *magnifying power* of a magnifying lens, and derive an expression for it when the image, seen by the eye, is at infinity.

Describe the optical system of an astronomical telescope and draw a ray diagram to show how the eye receives rays from a distant non-axial point, the final image being at infinity. Define the *angular magnification* of a telescope, and find an expression for it under the above conditions.

A surveyor looks through the telescope of a levelling instrument and would like to see the measuring stick, which is 0.6 km away, in the same detail as when he looks at it directly from a distance of 6 m. Determine the required angular magnification, and suggest suitable values for the focal lengths of the objective and eyepiece. [W]

D54 **(a)** Draw a ray diagram to show how a thin converging lens can form a magnified virtual image of an object.
What is meant by the term *angular magnification* and what is its value in this case?
Explain why the image appears magnified when both object and image subtend the same angle at the lens.

(b) A telescope is made from two lenses of focal lengths 100 cm and 5 cm. The instrument is adjusted so that a virtual image of the Moon is formed 25 cm from the more powerful lens.

(i) Draw a ray diagram of this arrangement.

(ii) Calculate the distance between the lenses and the angular magnification produced by the instrument.

(iii) Explain where the observer should position his eye in order to get the greatest field of view when using the telescope.

(c) The above telescope is now adjusted to form a real image of the Moon 20 cm from the more powerful lens.

(i) Calculate the distance between the lenses.

(ii) Suggest a possible use for a telescope adjusted in this manner. [S]

D55 A distant object subtending an angle of 0.10 minute of arc is viewed with a reflecting telecope whose objective is a concave mirror of focal length 1000 cm. The reflected light is intercepted by a convex mirror placed 950 cm from the pole of the concave mirror and a real image is formed in the vicinity of the pole of the concave mirror where there is a hole. This image is viewed with a convex lens of focal length 5 cm used as a magnifying glass and producing a final image at infinity. Draw a ray diagram (not to scale) for this arrangement using two rays from a non-axial point on the distant object which strike the objective at a small angle with the principal axis of the system.

Calculate:

(a) the size of the real image that would have been formed at the focus of the concave mirror,

(b) the size of the image formed by the convex mirror,

(c) the angle subtended by the final image at the optic centre of the convex lens.

Give *two* advantges of reflecting telescopes over refracting telescopes. (1 minute of arc $= 2.9 \times 10^{-4}$ radians.) [J]

D56 What is the function of a *collimator* in a spectrometer? Sketch the arrangement of the optical components of the collimator. [C]

D57 Define the *refractive index* of a medium, and state *Snell's law.*

Describe the construction and adjustment of a spectrometer. Explain, including the derivation of any equation used, how you would determine the refractive index of the material of a prism for light from a monochromatic source.

It has been suggested that the variation of refractive index n with wavelength λ is given by $n = A + B/\lambda$, where A and B are constants for a given material. Using a graphical method, or otherwise, investigate whether the following data fit this equation.

n	λ
1.633	4.5×10^{-7} m
1.630	5.0×10^{-7} m
1.627	5.5×10^{-7} m
1.625	6.0×10^{-7} m
1.623	6.5×10^{-7} m

[W]

D58 A camera has a lens with a focal length of 50 mm.

(a) When the camera is focused on a distant object, state

(i) the distance from the lens to the film,

(ii) *three* properties of the image formed by the lens.

(b) The camera is refocused on a near object which is 1000 mm from the lens. Calculate how far and in what direction the lens has been moved in order to produce an image at the film. [J, '90]

D59 Under certain light conditions a suitable setting for a camera is:

exposure time 1/125 second, aperture $f/5.6$.

If the aperture is changed to $f/16$ what would the new exposure time be in order to achieve the same film image density? What other effect would this change in f-number produce? [L]

D60 **(a)** Describe an experiment to determine the focal length of a thin converging lens.

(b) Describe the structure and operation of a simple camera. Discuss the advantages of a variable aperture lens in a camera.

(c) A camera is fitted with a lens having an f-number of 2.8. When the camera is focused at infinity the centre of the lens is 4.9 cm from the film. Calculate the effective diameter of the lens.

With the aperture fully open, the correct exposure for certain lighting conditions is $\left(\frac{1}{500}\right)$ s. What would be the f-number corresponding to $\frac{1}{4}$ of the full aperture and what would be the corresponding exposure time in order to obtain the same image brightness? [AEB, '79]

D61 A convex camera lens is used to form an image of an object 1.00 m away from it on a film 0.050 m from the lens. What is the focal length of the lens?

If the camera is used to photograph a distant object, how far from the film would the clear image be formed? What type of lens should be placed close to the first lens in order to enable the distant object to be focused on the film if the separation of the first lens and film cannot be changed in this camera? What is the focal length of this added lens? [L]

D62 A particular person cannot see objects clearly if they are more than 200 cm away. What spectacles are required to allow him to see distant objects clearly?

D63 An old lady has a near point of 150 cm. What spectacles does she need if she is to see objects which are only 25 cm from her eyes?

D64 A man wears spectacles whose lenses have a power of $+2.5\,\text{m}^{-1}$ in order to correct his near point to 25 cm. What is his near point distance when he is not wearing the spectacles?

D65 A boy has a near point of 40 cm and a far point of 400 cm.

(a) What spectacles are required to make his far point infinity?

(b) What is his range of vision when wearing the spectacles?

D66 A girl can see objects clearly only if they are between 60 cm and 600 cm away.

(a) What spectacles are required to allow her to see objects that are only 25 cm away?

(b) What is the most distant point on which she can focus whilst wearing the spectacles?

D67 **(a)** Using a labelled ray diagram in each case

(i) explain what is meant by *short sight*,

(ii) show how it is corrected.

(b) A man can see clearly only objects which lie between 0.50 m and 0.18 m from his eye.

(i) What is the power of the lens which when placed close to the eye would enable him to see distant objects clearly?

(ii) Calculate his least distance of distinct vision when using this lens. [J, '91]

D68 **(a)** Explain with the aid of a ray diagram the action of a spectacle lens in correcting the far point of an eye from 200 cm in front of the eye to infinity. Repeat the procedure for correcting the near point of an eye from 50 cm to 25 cm.

(b) In each case calculate the focal length of the spectacle lens, ignoring the separation of lens and eye. [J]

VELOCITY OF LIGHT (Chapter 22)

D69 **(a)** In an experiment to determine the speed of light in air, light from a point source is reflected from one face of a sixteen-sided mirror M, travels a distance d to a stationary mirror from which it returns and, after a second reflection at M, forms an image of the source on a screen. When M is rotated at certain speeds, the image is still seen in the same position. Explain how this can occur and show that, if the lowest speed of rotation for which the image remains in the same position is n (in revolutions per second), the speed of light, c, is given by

$$c = 32nd.$$

(b) Using the above arrangement, an image is seen on the screen when the speed of rotation is 900 revolutions per second. The speed of rotation is gradually increased until at 1200 revolutions per second the image is again seen. If $c = 3.00 \times 10^8 \text{m s}^{-1}$, calculate a value for d consistent with these figures. What is the lowest speed of rotation for which an image will be seen on the screen? [J]

D70 Describe a terrestrial method of measuring the speed of light. Explain precisely what observations are made, and how the speed is calculated from them.

How can the method you describe be adapted to show qualitatively that the speed of light is less in water than in air? Why would it be difficult to make a precise measurement of the speed of light in water?

What is the evidence that the speed of red and blue light is the same in vacuum, but that red light travels faster than blue light in water? [S]

D71 A plane mirror rotating at 35 rev s^{-1} reflects a narrow beam of light to a stationary mirror 200 metres away. The stationary mirror reflects the light normally so that it is again reflected from the rotating mirror. The light now makes an angle of 2.0 minutes of arc with the path it would travel if both mirrors were stationary. Calculate the velocity of light.

Give *two* reasons why it is important that an accurate value of the velocity of light should be known. [J]

SECTION E

WAVES AND THE WAVE PROPERTIES OF LIGHT

23

BASIC PROPERTIES OF WAVES

23.1 INTRODUCTION AND DEFINITIONS

A wave motion is a means of transferring energy from one point to another without there being any transfer of matter between the points. Waves may be classified as being either mechanical or electromagnetic. Mechanical waves (e.g. water waves, sound waves, waves in stretched strings) require a material medium for their propagation. Electromagnetic waves (e.g. light, radio, X-rays – see Chapter 28) can travel through a vacuum; their progress is impeded, to some extent, by the presence of matter.

When a mechanical wave travels from some point A to some other point B, it is because a disturbance of some kind at A has caused the particle there to become displaced. This particle drags its neighbour with it, so that it too becomes displaced and has a similar effect on the next particle, and so on until the disturbance reaches B. If the material is elastic, the particles oscillate about their rest positions. The motions of any pair of particles are the same, but that of the particle which is farther from the source occurs somewhat later and with reduced amplitude. If the disturbance at the source is of a repetitive nature, the wave is maintained. If it is not, the amplitude of vibration of each particle becomes progressively smaller and eventually the wave ceases to exist.

If the disturbance at the source is simple harmonic, a plot of the displacements of the particles at a single instant in time, as a function of distance from the source is sinusoidal (Fig. 23.1).

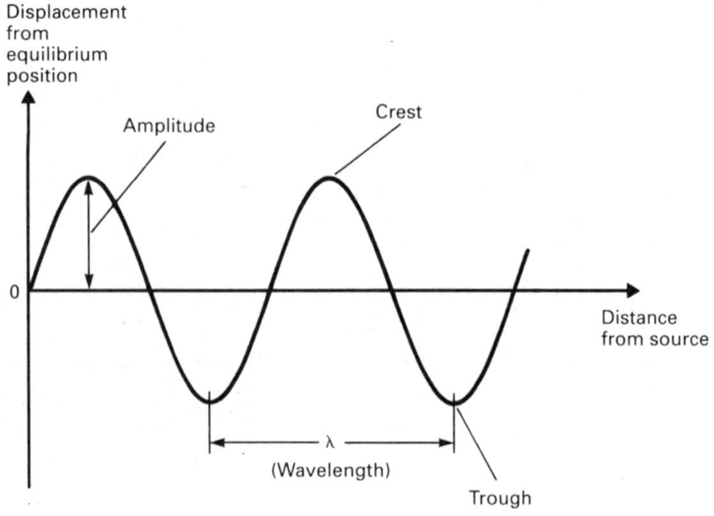

Fig. 23.1
Displacement against distance for particles in a simple harmonic wave

If, instead of considering the whole of the wave at one point in time, the displacement of a single particle is plotted as a function of time, the situation is as shown in Fig. 23.2.

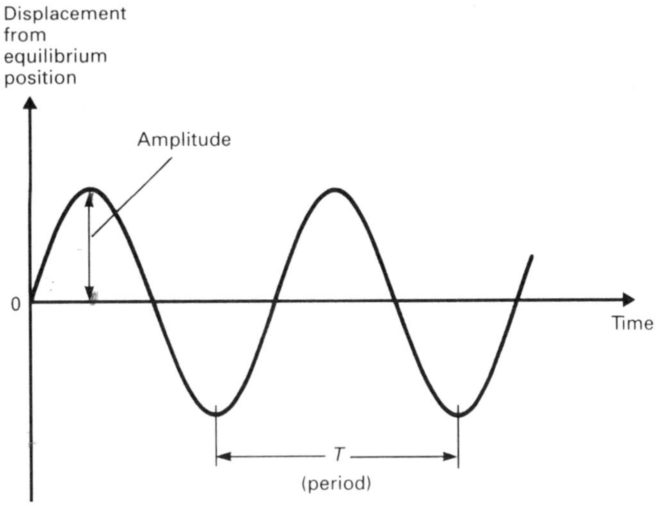

Fig. 23.2 Displacement against time for a particle in a simple harmonic wave

Notes

(i) Fig. 23.1 and Fig. 23.2 apply to both **transverse** waves (i.e. waves in which the displacements are perpendicular to the direction of travel) and to **longitudinal** waves (i.e. waves in which the displacements are parallel to the direction of travel). In the latter case it is necessary to adopt some (arbitrary) convention. For example, displacements to the right might be plotted above the axis.

(ii) The amplitude of vibration of a single particle is also the amplitude of the wave.

The wavelength (λ) is the distance between any particle and the nearest one which is at the same stage of its motion. In particular, it is the separation of two adjacent crests or troughs.

The amplitude (a) of the wave is the greatest displacement of any particle from its equilibrium position.

The period (T) is the time taken for any particle to undergo a complete oscillation. It is also the time taken for the wave to travel one wavelength.

The frequency (f) is the number of cycles that any particle undergoes in one second. It is also the number of wavelengths that pass a fixed point in one second. (Unit = hertz, symbol Hz.)

Derivation of $v = f\lambda$

Whenever the source of a wave motion undergoes one cycle the wave moves forward by one wavelength (λ). Since there are f such cycles each second, the wave progresses by $f\lambda$ in this time, and therefore the velocity (v) of the wave is given by

$$v = f\lambda \qquad [23.1]$$

It follows from the definitions of period and frequency that

$$f = \frac{1}{T} \qquad [23.2]$$

(This is easily seen by considering a numerical example. If $f = 10\,\text{Hz}$, 10 cycles occur in one second and therefore the period, T, is $\frac{1}{10}$ second.)

Equations [23.1] and [23.2] hold for all wave motions. Equation [23.2] also applies to any oscillatory motion.

An electromagnetic wave (see section 28.1) consists of a time-varying electric field accompanied by a time-varying magnetic field. The amplitude of such a wave is usually taken to be the maximum electric field strength associated with the wave. By plotting the electric field strength on the y-axes, we can use graphs of the kinds shown in Figs. 23.1 and 23.2 to represent electromagnetic waves.

23.2 THE PRINCIPLE OF SUPERPOSITION

The principle of superposition states that whenever two waves are travelling in the same region the total displacement at any point is equal to the vector sum of their individual displacements at that point.

The principle has been applied in producing Fig. 23.3, which represents two waves (A and B) of different amplitudes and frequencies being propagated along a single string.

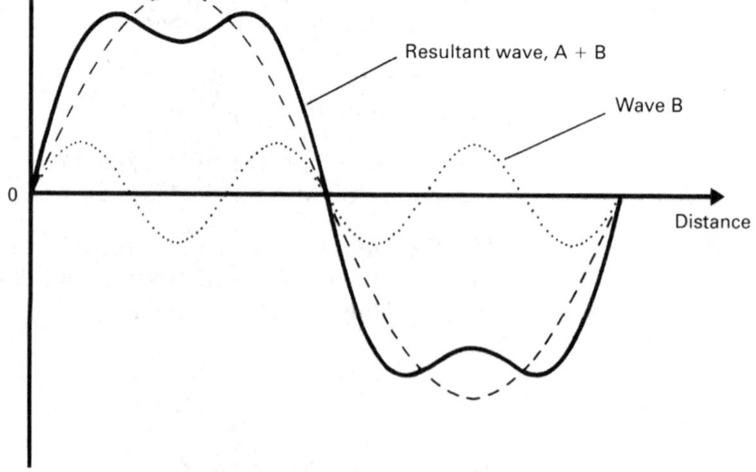

Fig. 23.3
To illustrate the principle of superposition

Whenever two waves are not travelling along the same line but merely cross at some point, they each emerge from the crossing point in the same form as they entered it. The principle of superposition applies at the point where they cross.

The phenomena of interference, diffraction, beats and stationary waves are consequences of the superposition of waves.

23.3 PHASE DIFFERENCE

Before proceeding the reader should compare Fig. 23.1 with a plot of sin θ against θ in order to appreciate that one wavelength corresponds to 2π radians, and that half a wavelength corresponds to π radians, etc.

When the crests of two waves of equal wavelength are together the waves are said to be **in phase** (i.e. they have a phase difference of zero). If a crest and a trough are together, the waves are **completely out of phase** (i.e. they have a phase difference of π radians). This occurs whenever one wave leads the other by half a wavelength (or, since a phase difference of $n \times 2\pi$, where n is an integer, is effectively a phase difference of zero, whenever one wave leads the other by an odd number of half-wavelengths).

QUESTIONS 23A

1. What is the phase difference between two waves, each of wavelength 12 cm, when one leads the other by: **(a)** 6 cm, **(b)** 3 cm, **(c)** 9 cm, **(d)** 12 cm, **(e)** 14 cm, **(f)** 18 cm, **(g)** 36 cm, **(h)** 39 cm.

23.4 THE EQUATION OF A PROGRESSIVE WAVE

The displacement, y, at some time, t, of a particle which is oscillating with simple harmonic motion of frequency, f, can be represented by

$$y = a \sin 2\pi f t \qquad [23.3]$$

Fig. 23.4
Sinusoidal wave

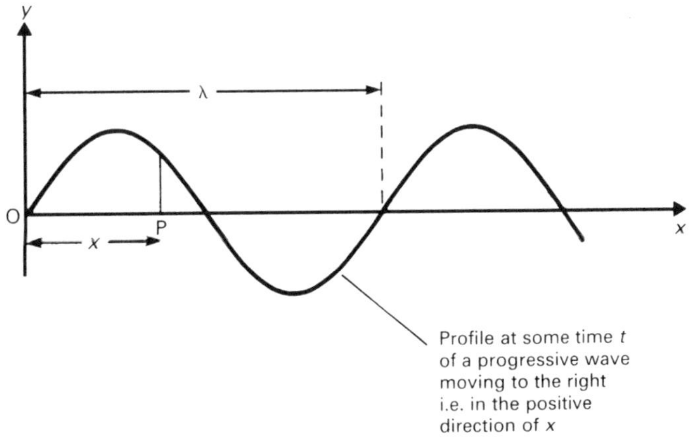

Fig. 23.4 shows a sinusoidal wave motion of wavelength λ and frequency f. Since it is a sinusoidal wave, all the particles contributing to it are oscillating sinusoidally and we may assume that the motion of the particle at the origin (O) is represented by equation [23.3]. Since the wave is moving to the right, the motion of the particle at

P will lag behind that of the particle at O by $(x/\lambda)T$, where T is the period of the motion. The displacement at P is therefore given by replacing t in equation [23.3] by $(t - (x/\lambda)T)$. Thus, the displacement, y, at any point, x, is given by

$$y = a \sin\left[2\pi f\left(t - \frac{xT}{\lambda}\right)\right]$$

But $T = 1/f$. Therefore

$$y = a \sin(2\pi f t - kx) \qquad [23.4]$$

where y is the displacement at time t and at a distance x from the origin, of a sinusoidal wave of frequency f travelling in the positive direction of the x-axis, and whose wavelength, λ, is given by $k = 2\pi/\lambda$.

24

HUYGENS' CONSTRUCTION

24.1 WAVEFRONTS AND RAYS

A wavefront is a line or surface, in the path of a wave motion, on which the disturbances at every point have the same phase.

If a point source generates two-dimensional waves in an isotropic medium (i.e. a medium in which the waves travel with the same speed in all directions), the wavefronts are circles centred on the source (Fig. 24.1). Wavefronts of this type can be seen when, for example, a small stone is thrown into a pond. The wavefronts associated with two-dimensional waves produced by a line source (e.g. a straight vibrator in a ripple tank) are straight lines. At large distances from a source of any kind which is producing two-dimensional waves, the wavefronts are straight lines (Fig. 24.1).

Fig. 24.1
Wavefronts and rays

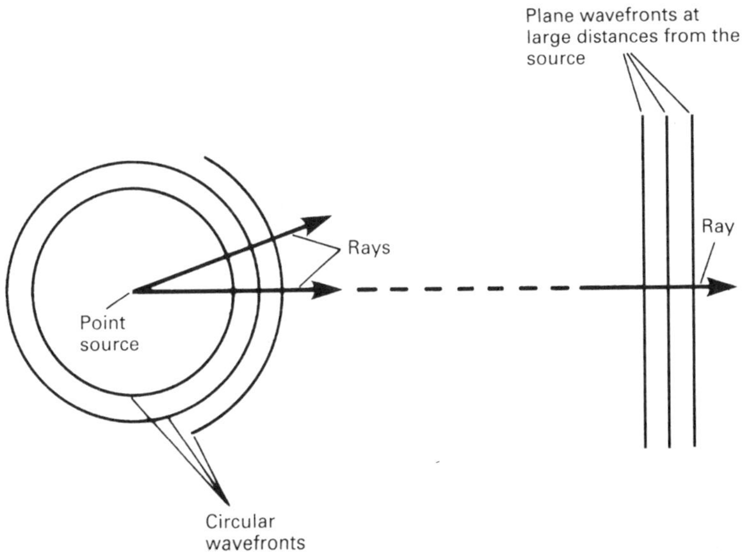

If a point source in an isotropic medium is sending out waves in three dimensions, the wavefronts are spheres centred on the source. A point source of light produces wavefronts of this type. At large distances from such a source the wavefronts are the surfaces of spheres of large radii, and therefore limited sections of them are essentially plane wavefronts. Indeed, at large distances from a source of any kind which is generating three-dimensional waves, the wavefronts are plane.

A ray is a line which represents the direction of travel of a wave; it is at right angles to the wavefronts (Fig. 24.1).

24.2 THE BASIS OF HUYGENS' CONSTRUCTION

If the present position of a wavefront is known, then Huygens' construction enables us to determine what its position will be at some later time. It can be used for all types of wave motion but finds particular application in solving problems in physical optics. In geometrical optics the location of image positions, etc. can usually be accomplished more easily by making use of ray diagrams.

Huygens postulated that each point on a wavefront could be regarded as being a secondary source of spherical wavelets and that the new wavefront is the surface which touches all of these secondary wavelets.

Fig. 24.2 illustrates how the construction is used to determine some subsequent position of a circular wavefront.

Fig. 24.2
Huygens' construction

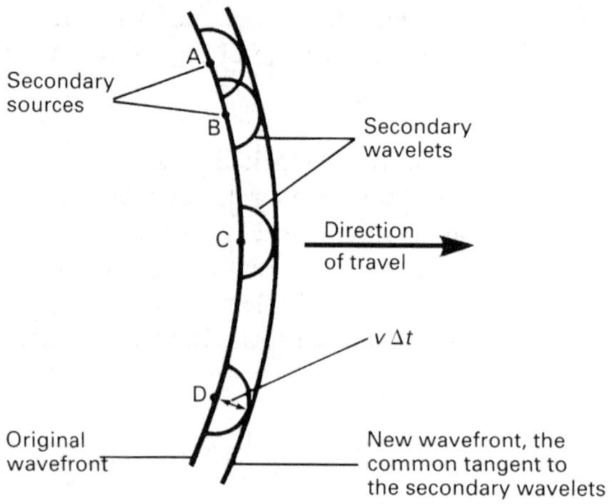

Notes (i) Whenever the construction is used, it is necessary to assume that the amplitude of the secondary wavelets is zero in the backward direction. This is done to prevent the construction predicting the existence of an additional wavefront behind the original one, since such a prediction would be contrary to observation.

(ii) Since it is a two-dimensional wave that is being considered, the secondary wavelets should be circular rather than spherical, and on account of (i) are shown as semicircles.

According to Huygens, every point on the original wavefront can be regarded as a secondary source; A, B, C and D have been arbitrarily chosen as four such points. The position of the wavefront at some time Δt later is found by constructing secondary wavelets centred on these points and of radius $v\Delta t$ (where v is the velocity of the wave motion). The new wavefront is the common tangent to these wavelets. Thus, in this case at least, Huygens' construction provides a result that is consistent with observation, i.e. that a circular wavefront continues as a circular wavefront.

24.3 HUYGENS' CONSTRUCTION APPLIED TO REFLECTION

Consider a parallel beam of light incident on a reflecting surface, XY, such that its direction of travel makes an angle i with the normal to the surface (Fig. 24.3). Consider also, that side A of an associated (plane) wavefront, AB, has just reached the surface. In the time that light from the other side, B, of the beam proceeds to C, a secondary wavelet of radius BC will be generated by A. Because XY is a reflecting

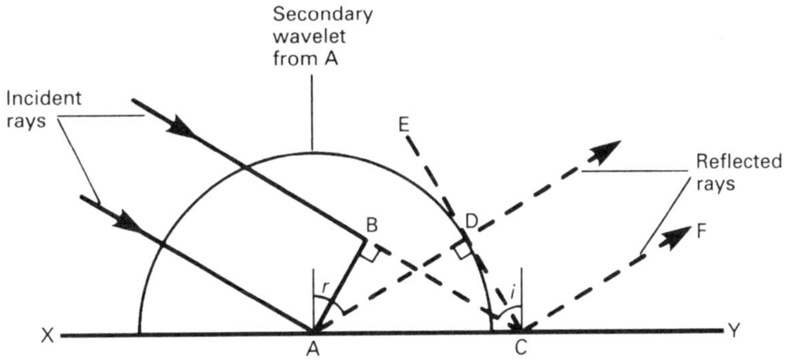

Fig. 24.3 Huygens' construction for reflection at a plane surface

surface, this secondary wavelet must be emitted into the region above XY, and is shown here as a semicircle. The new wavefront will be a tangent to this and must contain C, and (anticipating that it will be a plane wavefront) it will therefore be along CE as shown. The reflected beam must be at right angles to this wavefront and will be bounded by rays AD and CF as illustrated (where D is the point at which the tangent touches the semicircle and CF is parallel to AD).

In $\triangle$ABC and $\triangle$ADC, AC is common to each and

$$BC = AD \qquad \text{(construction)}$$

$$A\hat{B}C = A\hat{D}C \quad \text{(right angles)}$$

The two triangles are therefore congruent (RHS), and in particular

$$B\hat{C}A = D\hat{A}C$$

$$\therefore \quad (90^\circ - i) = (90^\circ - r)$$

$$\text{i.e.} \quad i = r$$

i.e. Angle of incidence = Angle of reflection

This is the second law of reflection.

Thus, Huygens' construction predicts that light is reflected in a manner which is consistent with observation. Although this result has been obtained by considering the behaviour of light, no properties which are specifically associated with light have been assumed and therefore it applies to any wave motion.

24.4 HUYGENS' CONSTRUCTION APPLIED TO REFRACTION

Consider a parallel beam of light incident on a refracting surface, XY, such that its direction of travel makes an angle θ_1 with the normal to the surface (Fig. 24.4). Consider also, that side A of an associated wavefront, AB, has just reached the surface. If light from the other side, B, of the beam subsequently travels to C in time t, then BC $= c_1 t$ (where c_1 is the velocity of light in the medium above XY). If c_2 is the velocity of light in the region below XY, then, in the same time, A will emit a secondary wavelet of radius $c_2 t$. It is shown here as a semicircle centred on A. The new wavefront will be a tangent to this and must contain C, and (anticipating that it will be a plane wavefront) it will therefore be along CE as shown. The refracted beam must be at right-angles to this wavefront and will be bounded by rays AD and CF as illustrated (where D is the point at which the tangent touches the semicircle and CF is parallel to AD).

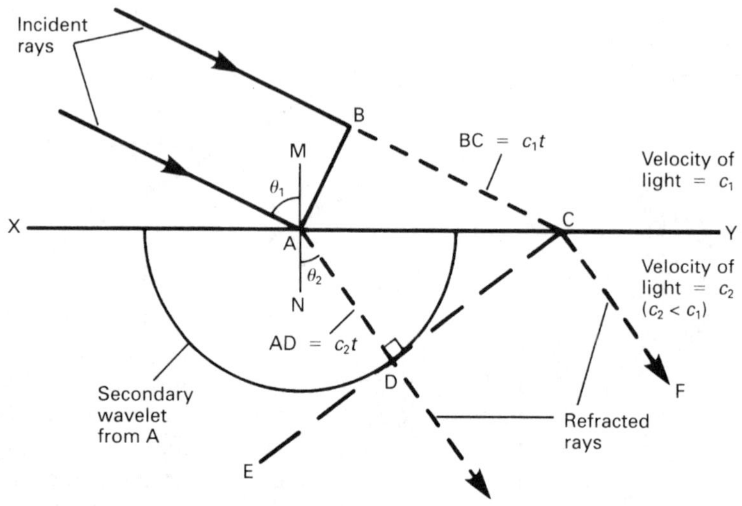

Fig. 24.4
Huygens' construction for refraction at a plane surface

Referring to $\triangle$ABC and $\triangle$ADC, we see that

$$\frac{\sin \mathrm{C\hat{A}B}}{\sin \mathrm{A\hat{C}D}} = \frac{\mathrm{BC/AC}}{\mathrm{AD/AC}} = \frac{\mathrm{BC}}{\mathrm{AD}} = \frac{c_1 t}{c_2 t} = \frac{c_1}{c_2} \qquad [24.1]$$

But

$$\mathrm{C\hat{A}B} + \mathrm{B\hat{A}M} = 90^\circ \quad \text{and} \quad \theta_1 + \mathrm{B\hat{A}M} = 90^\circ \quad \therefore \quad \mathrm{C\hat{A}B} = \theta_1$$

and

$$\mathrm{A\hat{C}D} + \mathrm{C\hat{A}D} = 90^\circ \quad \text{and} \quad \theta_2 + \mathrm{C\hat{A}D} = 90^\circ \quad \therefore \quad \mathrm{A\hat{C}D} = \theta_2$$

Therefore, from equation [24.1]

$$\frac{\sin \theta_1}{\sin \theta_2} = \frac{c_1}{c_2} \qquad [24.2]$$

Since (for any given wavelength) c_1 and c_2 are constants, equation [24.2] is consistent with Snell's law (section 18.1), i.e. Huygens' construction produces a result which is consistent with observation.

In the mid-seventeenth century there were two views as to the nature of light. According to Huygens light was a wave motion; according to Isaac Newton light rays were streams of particles – corpuscles. We have seen that Huygens' theory accounts for the laws of reflection and refraction – Newton's theory also accounts for these laws. The wave theory, on the basis of equation [24.2], predicts that light bends towards the normal when it enters a material in which its velocity is smaller (i.e. $\theta_2 < \theta_1$ when $c_2 < c_1$). **Newton's corpuscular theory,** on the other hand, predicts that the opposite happens, i.e. that light bends away from the normal on moving into a region in which its velocity is smaller.

In 1850, almost two centuries after the rival theories had first been proposed, Foucault measured the velocity of light in air and in water. He found that the velocity of light in water is less than it is in air. Since observation shows that light bends towards the normal in going from air to water, Foucault's result confirmed the prediction of the wave theory.

Equation [24.2] is valid for all wave motions. Water waves, for example, are refracted towards the normal when they enter a region in which they have reduced velocity. The effect can be observed by making use of a ripple tank, in one section of which the water is less deep than it is in the rest of the tank. (This can be achieved by putting a glass plate on part of the base of the tank.) Since water waves travel more slowly in the shallower water, they are refracted towards the normal as they enter the shallow region (Fig. 24.5).

Fig. 24.5
Refraction of water waves

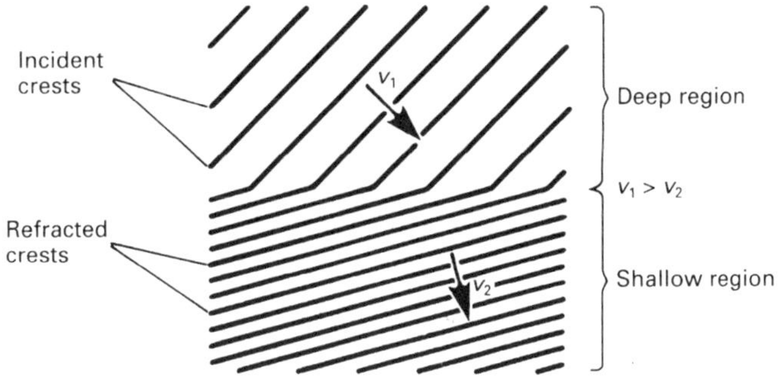

24.5 HUYGENS' CONSTRUCTION APPLIED TO TOTAL INTERNAL REFLECTION

Consider a parallel beam of light incident on the boundary, XY, between two media, such that its direction of travel makes an angle θ to the normal to XY (Fig. 24.6). Consider also, that side A of the beam has just reached XY. If light from the other side, B, of the beam subsequently travels to C in time t, then BC = c_2t (where c_2 is the velocity of light in the medium below XY). If c_1 is the velocity of light in the medium above XY, then, in the same time, A will emit a secondary wavelet of radius c_1t into the region above XY. If $c_1 > c_2$ (i.e. if the light is travelling towards an optically less dense medium), there is the possibility that c_1t is greater than AC. This would produce a secondary wavelet such as S. If this happens, refraction cannot occur because it is not possible to find a wavefront which is tangential to S and which passes through C. The only alternative is that the secondary wavelet from A remains in the region below XY, in which case its radius is c_2t. This leads to total internal reflection, and the usual Huygens' construction for reflection at a plane surface (section 24.3) then applies.

Fig. 24.6
Huygens' construction applied to total internal reflection

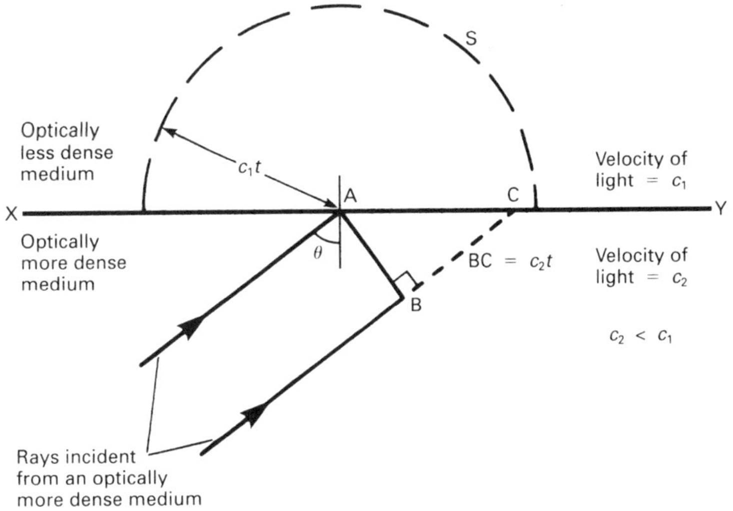

Thus, total internal reflection occurs if $c_1t > \mathrm{AC}$, i.e. if

$$c_1t > \frac{\mathrm{BC}}{\sin \mathrm{C\hat{A}B}}$$

But

$$\mathrm{BC} = c_2t \quad \text{and} \quad \mathrm{C\hat{A}B} = \theta$$

Therefore, total internal reflection occurs if

$$c_1t > \frac{c_2t}{\sin \theta}$$

i.e. if

$$\sin \theta > \frac{c_2}{c_1}$$

Therefore, the maximum value of θ for which total internal reflection does not occur (i.e. the critical angle c) is given by

$$\sin c = \frac{c_2}{c_1} = \frac{n_1}{n_2}$$

where n_1 and n_2 are the refractive indices of the regions above and below XY respectively ($n_1 < n_2$).

24.6 SECONDARY WAVELET TREATMENT OF YOUNG'S FRINGES

Young's fringes are discussed in section 25.3. Fig. 24.7 illustrates the secondary wavelet treatment of the formation of the fringes. S_1 and S_2 act as sources which are supposed to be emitting in phase with each other. Bright fringes occur in those

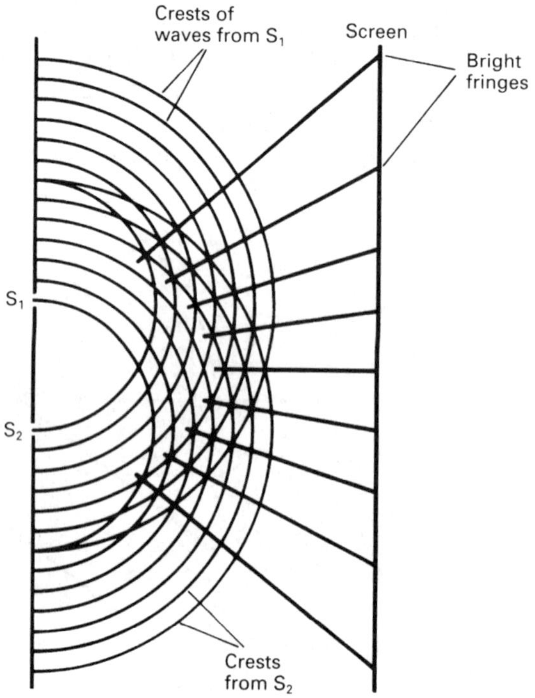

Fig. 24.7 Secondary wavelet treatment of Young's fringes

directions in which the light from S_1 interferes constructively with that from S_2, i.e. in those directions where crests are together. It is normal practice to consider that the fringes are equally spaced; they do not appear to be even approximately equally spaced here because the diagram is not to scale.

24.7 SECONDARY WAVELET TREATMENT OF LLOYD'S MIRROR

Lloyd's mirror is discussed in section 25.5. In Fig. 24.8 crests from a point source S are shown incident on a screen. Those sections of wavefronts which have been reflected undergo a phase change of π radians on reflection. The reflected rays appear to come from S′ and, because of the phase change, at points where crests from S meet the reflector there are troughs from S′. Destructive interference occurs where crests and troughs are together.

Fig. 24.8
Secondary wavelet treatment of Lloyd's mirror

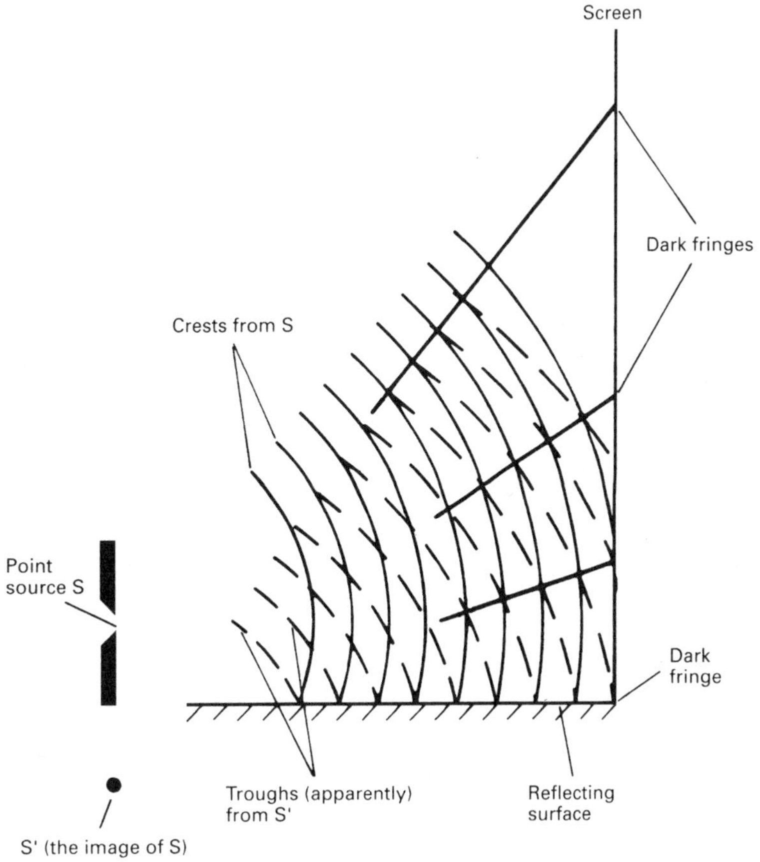

24.8 SECONDARY WAVELET TREATMENT OF THE DIFFRACTION GRATING

The diffraction grating is discussed in section 26.5. A secondary wavelet treatment is given in Fig. 24.9.

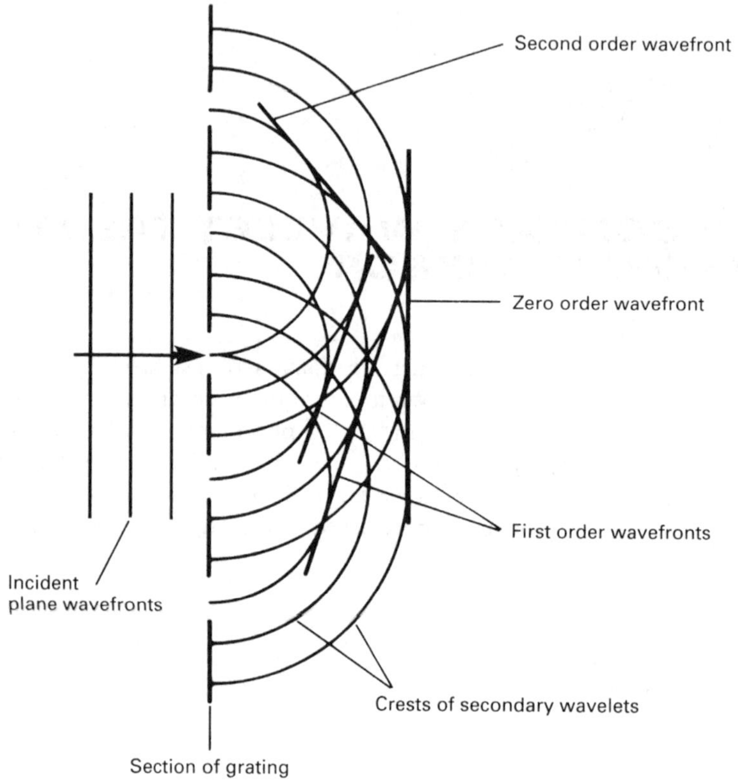

Fig. 24.9 Secondary wavelet treatment of a diffraction grating

24.9 DIFFRACTION OF WATER WAVES AT A GAP

Diffraction is discussed in detail in Chapter 26. The phenomenon can be demonstrated using water waves in a ripple tank. Fig. 24.10 illustrates what happens when plane waves (produced by a straight vibrator) are incident on a gap in a barrier. In Fig. 24.10(a) the width of the gap is small compared with the wavelength of the waves and there is considerable sideways spreading, i.e. considerable diffraction. In Fig. 24.10(b) the gap is much larger than the wavelength and there is very little diffraction – most of the energy associated with the waves is propagated in the same direction as the incident waves. Thus, **the greater the ratio of wavelength to gap width, the greater the spreading**.

Huygens' construction can be used to predict the shapes of the wavefronts. Fig. 24.10(c) illustrates this for the case of Fig. 24.10(b). Note that though the construction accounts for the shape of the wavefronts, it does not account for the reduced amplitude in directions other than the straight through direction.

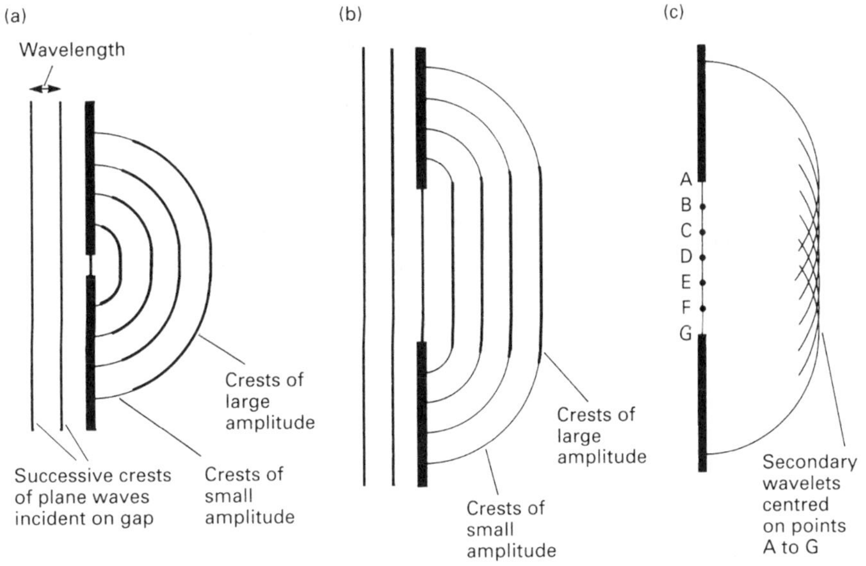

Fig. 24.10
Diffraction of water waves (a) at a narrow gap, (b) at a wide gap. (c) Huygens' construction for (b)

25

INTERFERENCE OF LIGHT WAVES

25.1 INTRODUCTION

If two waves are in the same place at the same time, they produce an effect which is equal to the combined effects of the two waves in accordance with the principle of superposition (section 23.2). The phenomenon is known as **interference**, and the two waves are said to interfere with each other. Interference occurs whenever two waves come together, but certain conditions need to be fulfilled if the effects of the interference are to be capable of being observed.

25.2 THE CONDITIONS FOR TWO SOURCES OF LIGHT TO PRODUCE OBSERVABLE INTERFERENCE

(i) The sources must be **coherent,** i.e. there must be a constant phase difference between them and therefore they must have the same frequency. (This phase difference may be zero but does not have to be.)

(ii) The waves that are interfering must have approximately the same amplitude. (Otherwise the resulting interference pattern lacks contrast.)

When light is emitted by a source it is as a result of electron transitions within the individual atoms of the source. These transitions occur randomly, and each gives rise to a short burst of radiation (a **photon**) that lasts, typically, for 10^{-9} s. (This is not true in the case of laser light.) Thus, since there must be a constant phase difference in order for interference to be observed, the two waves that are producing the interference must, in practice, have come from the same point on the same source, and must have done so within 10^{-9} s of each other. (**Note.** If this is not done, the interference pattern changes so rapidly that the impression is one of uniform illumination – nevertheless, the interference still occurs.)

25.3 YOUNG'S DOUBLE-SLIT EXPERIMENT*

The first demonstration of optical interference that was recognized as such, was provided by Thomas Young in 1801. The experiment gave strong support to the wave theory of light. The arrangement used by Young was similar to that shown in Fig. 25.1, a major difference being that Young used a white light source.

Fig. 25.1
Young's demonstration of interference

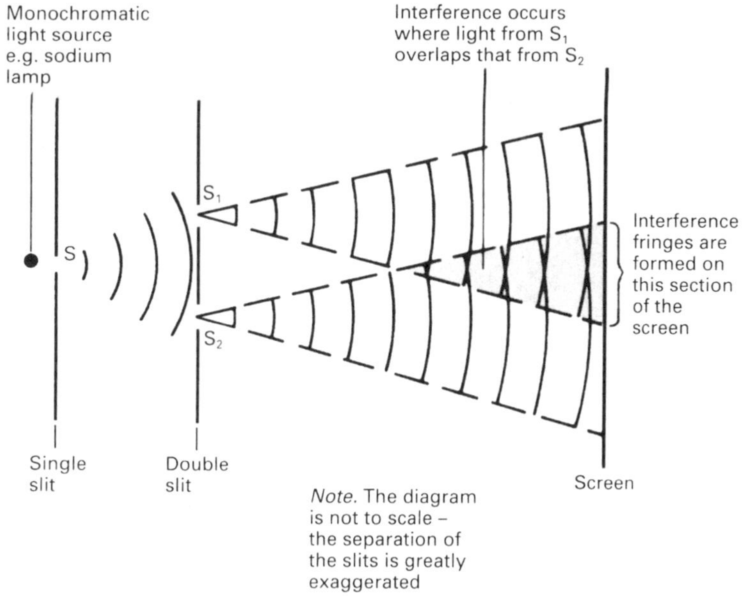

S, S_1 and S_2 are narrow slits which are parallel to each other. Because S is narrow, it diffracts (see Chapter 26) the light that falls on it and so illuminates both S_1 and S_2. Diffraction also takes place at S_1 and S_2, and interference occurs in the region where the light from S_1 overlaps that from S_2. Because S is narrow the light which emerges from S_1 comes from the same point as that which emerges from S_2. Furthermore, as long as the path difference is short, any two waves which interfere with each other will have left S within 10^{-9} s of each other. (The significance of this is explained in section 25.2.) Thus, S_1 and S_2 are coherent sources and the interference they produce is observable.

Fig. 25.2
Variation of intensity across Young's fringes

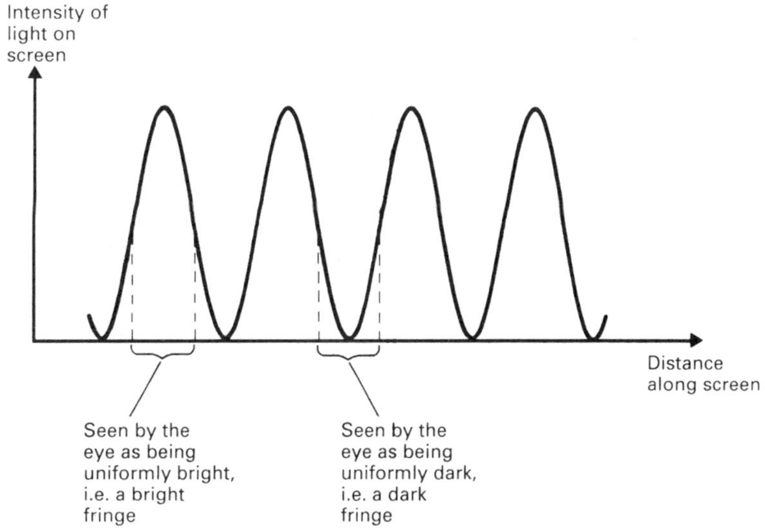

*A secondary wavelet treatment is given in section 24.6.

A series of alternately bright and dark bands (**interference fringes**), which are equally spaced and parallel to the slits, can be observed on a screen placed anywhere in the region of overlap. (**Note**. The actual variation in light intensity along the screen has the form shown in Fig. 25.2.)

Calculation of Fringe Separation

Fig. 25.3 represents the relative positions of the coherent sources, S_1 and S_2, and a point P on the screen. The perpendicular distance, D, from the plane of the slits to the screen is very much greater than the slit separation, a (typically, $D = 20$ cm, $a = 0.1$ cm). The path difference ($S_2P - S_1P$) between waves reaching P from S_2 and S_1 can be found by applying Pythagoras' theorem to $\triangle S_2NP$ and $\triangle S_1MP$.

Fig. 25.3
Optical geometry in production of Young's fringes

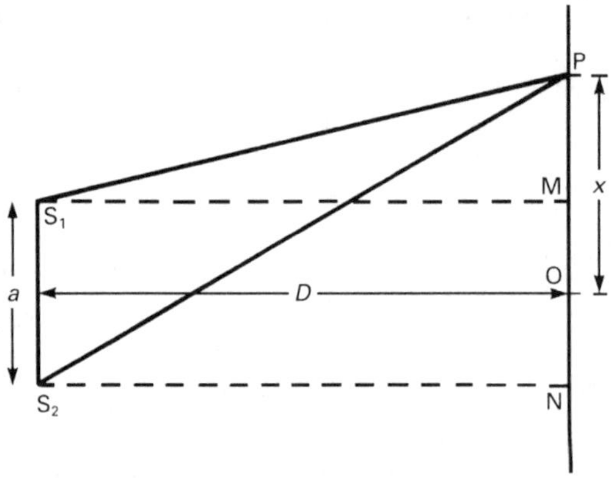

Thus,

$$\begin{aligned} S_2P^2 - S_1P^2 &= [S_2N^2 + NP^2] - [S_1M^2 + MP^2] \\ &= NP^2 - MP^2 \\ &= [NP + MP]\ [NP - MP] \\ &= \left[\left(x + \frac{a}{2}\right) + \left(x - \frac{a}{2}\right)\right] \left[\left(x + \frac{a}{2}\right) - \left(x - \frac{a}{2}\right)\right] \\ &= [2x]\ [a] \end{aligned}$$

$$\therefore \qquad S_2P - S_1P = \frac{2xa}{S_2P + S_1P}$$

Since $a \ll D$, ($S_2P + S_1P$) can be taken to be equal to $2D$ (as long as P is close to O). Therefore,

$$S_2P - S_1P = \frac{xa}{D} \qquad [25.1]$$

if a crest leaves S_1 at the same time as a crest leaves S_2,* there will be a bright fringe wherever the path difference ($S_2P - S_1P$) is equal to a whole number of wavelengths, i.e. for light of wavelength λ there is brightness if

$$S_2P - S_1P = n\lambda \quad (n = 0, 1, 2, 3, \ldots)$$

*Though this is unlikely to occur in practice, it does not invalidate the analysis which follows.

Therefore, from equation [25.1]

$$\frac{xa}{D} = n\lambda$$

i.e. $$x = n\frac{\lambda D}{a} \qquad [25.2]$$

Therefore, since $n = 0, 1, 2, 3, \ldots$ there are bright fringes at

$$x = 0, \frac{\lambda D}{a}, \frac{2\lambda D}{a}, \frac{3\lambda D}{a}, \ldots$$

Thus the difference in the values of x for any pair of adjacent bright fringes is $\lambda D/a$, i.e. the fringes are equally spaced with a separation of $\lambda D/a$.

At the sites of dark fringes the path differences are equal to odd numbers of half-wavelengths and a similar procedure reveals that the separation of these fringes is also equal to $\lambda D/a$. Thus

$$y = \frac{\lambda D}{a} \qquad \text{(For fringes close to O and } a \ll D) \qquad [25.3]$$

where

y = the separation of adjacent bright (or dark) fringes

Typically,

D = 20 cm = 0.2 m

a = 1 mm = 10^{-3} m

λ = 6×10^{-7} m

Therefore, from equation [25.3]

y = 1.2×10^{-4} m = 0.12 mm

Measurement of Wavelength

If a, y and D are measured, the wavelength of the light being used can be found from equation [25.3]. In practice the screen is replaced by a travelling microscope, and y is found by traversing about twenty bright fringes and making use of the fact that they are equally spaced. In order to locate the plane on which the microscope is focused (i.e. the plane of the fringes), a pin is moved around in front of the microscope until it is in sharp focus. The value of D can then be found by using a metre rule to measure the distance of the pin from the plane of the slits. The travelling microscope is also used to measure a.

The following points should be observed in the experiment.

(i) The separation of the fringes is increased by increasing D. (This follows from equation [25.3].) This decreases the error involved in measuring y but also reduces the intensity of the fringes.

(ii) The separation of the fringes is increased by decreasing a. (This follows from equation [25.3].)

(iii) Increasing the width of any of the three slits increases the intensity of the pattern but the fringes become more blurred.

(iv) Moving S closer to S_1 and S_2 increases the intensity of the fringes but does not affect their separation.

White Light Pattern

If white light is used, the fringes are coloured. The colour at any particular point is determined by the degree to which the various wavelengths are being destroyed at that point. Each wavelength produces its own fringe system. At O (Fig. 25.3), the path difference is zero for every wavelength and therefore each fringe system has a bright band at O. Since all of these bands are being produced in the same place, **the central band is white**. There is no other point where every wavelength produces constructive interference. On each side of this central band there is a region which gives the impression of being dark. Next to each of these dark bands there is a coloured band. In each case, the colour varies in spectral sequence from being predominantly blue nearest the central band to predominantly red. The pattern repeats, but the colours become less pronounced away from the centre and the impression is that of white light.

QUESTIONS 25A

1. The distance between the 1st bright fringe and the 21st bright fringe in a Young's double-slit arrangement was found to be 2.7 mm. The slit separation was 1.0 mm and the distance from the slits to the plane of the fringes was 25 cm. What was the wavelength of the light?

2. In a Young's double-slit experiment a total of 23 bright fringes occupying a distance of 3.9 mm were visible in the travelling microscope. The microscope was focused on a plane which was 31 cm from the double slit and the wavelength of the light being used was 5.5×10^{-7} m. What was the separation of the double slit?

3. For each of the situations **(a)** to **(e)** state which **two** of A to G would occur. The notation is that of Fig. 25.3.

(a) A filter is placed in front of S_1 so that it transmits only half the light intensity of S_2.
(b) The width of both S_1 and S_2 is halved.
(c) The wavelength of the light is reduced.
(d) D is increased.
(e) a is increased.

A The bright fringes become darker.
B The dark fringes become brighter.
C The fringe separation increases.
D The fringe separation decreases.
E The sharpness of the fringes increases.
F The sharpness of the fringes decreases.
G The bright fringes change colour.

25.4 DIVISION OF WAVEFRONT, DIVISION OF AMPLITUDE

In the double-slit experiment, the two waves that interfere with each other originate at different parts of the wavefront emitted by the source. This method of producing two coherent sources from a single source is known as **division of wavefront**. The arrangement known as Lloyd's mirror (section 25.5) also makes use of this class of interference.

Newton's rings (section 25.7) and the interference effects produced by the air wedge (section 25.6) and the parallel-sided thin film (section 25.8) are all examples of interference effects produced by **division of amplitude**. In this type of interference, the two waves that interfere originate at the same point on the wavefront produced by the source, each wave having part of the amplitude of the original. An extended (large) source, rather than a slit or pin-hole, can be used and this leads to easily observed interference effects.

25.5 LLOYD'S MIRROR*

The arrangement is shown in Fig. 25.4. S is an illuminated slit which is parallel to both the glass plate and the screen, and is very close to the plane of the plate. Light waves diffracted by S travel to a point such as P on the screen either directly (i.e. along SP) or after being reflected by the mirror (i.e. along SAP). Though S is the source of the 'direct' waves, the reflected waves appear to come from S′ (the virtual image of S), and therefore the interference can be thought of as being due to coherent sources at S and S′. Since SA = S′A, the path difference between the two sets of waves reaching P is (S′P − SP). There is no (geometrical) path difference between waves arriving at N, and therefore N might be expected to be the site of a bright fringe. However, **light undergoes a phase change of π radians whenever it is reflected at a more dense medium** (as is the case here), and therefore the fringe at N is dark. Thus, there are dark fringes where the geometrical path difference is a whole number of wavelengths and bright fringes where the geometrical path difference is an odd number of half-wavelengths.

Fig. 25.4
Lloyd's mirror

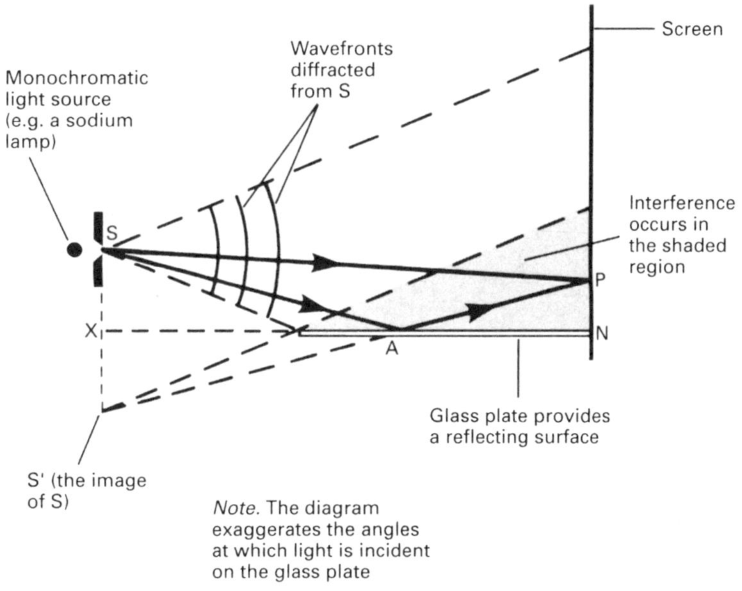

Calculation of Fringe Separation

The situation is equivalent to that of the double-slit experiment (section 25.3). The fringes are therefore equally spaced and the separation, y, of adjacent bright (or dark) fringes is given by equation [25.3] as

$$y = \frac{\lambda D}{a}$$

where

λ = the wavelength of the light being used

D = XN

a = SS′.

*A secondary wavelet treatment is given in section 24.7.

25.6 THE AIR WEDGE

An air wedge is a wedge-shaped film of air, such as could be produced by placing the edge of a razor blade or a piece of thin foil between two microscope slides at one end only (Fig. 25.5).

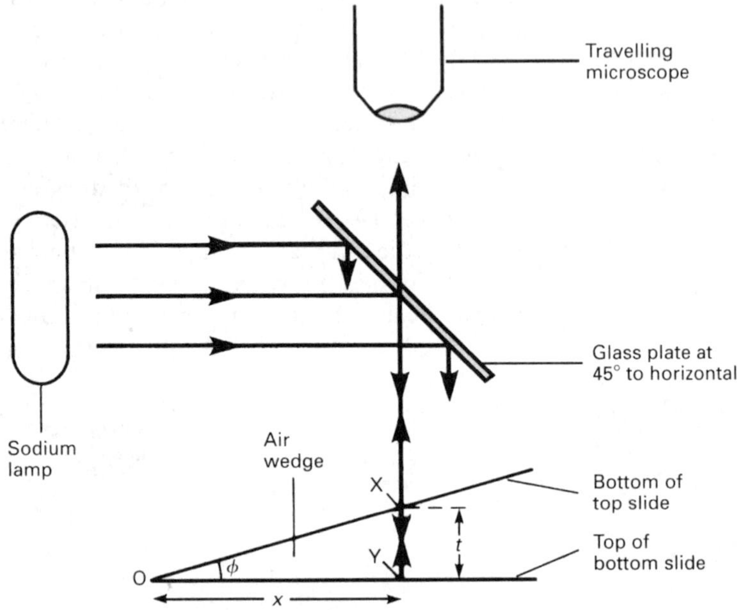

Fig. 25.5
Air wedge

Note. The diagram greatly exaggerates the angle of the wedge. ϕ is typically $< 1'$

Some of the light which is incident on X from above is reflected at X, whereas some of it crosses the air gap to Y, and is reflected at Y. Thus, waves travel upwards, from both X and Y, and since both sets of waves originate at X, X and Y act as two coherent sources which have been produced by division of amplitude. There is a geometrical path difference of $2t$ between the waves from X and those from Y. The waves which are reflected at Y are being reflected at a more dense medium and therefore undergo a phase change of π radians (see section 25.5). Waves reflected at X suffer no such phase change. Therefore there is darkness when $2t$ is equal to a whole number of wavelengths and brightness when $2t$ is equal to an odd number of half-wavelengths. Thus:

darkness if $\quad 2t = n\lambda \quad (n = 0, 1, 2, \ldots)$ [25.4]

brightness if $\quad 2t = (n + \frac{1}{2})\lambda \quad (n = 0, 1, 2, \ldots)$ [25.5]

The view through the microscope is of alternately bright and dark bands (interference fringes) which are equally spaced and parallel to the line along which the microscope slides make contact. In view of the phase change, **the fringe at O is dark**.

Calculation of Fringe Separation

Putting $n = 0, 1, 2, \ldots$ in equation [25.4] reveals that there are dark fringes if

$$2t = 0, \lambda, 2\lambda, \ldots$$

i.e. $$t = 0, \frac{\lambda}{2}, \frac{2\lambda}{2}, \ldots$$

But, from Fig. 25.5, $t = x \tan \phi$, and therefore there are dark fringes if

$$x \tan \phi = 0, \frac{\lambda}{2}, \frac{2\lambda}{2}, \ldots$$

i.e.
$$x = 0, \frac{\lambda}{2 \tan \phi}, \frac{2\lambda}{2 \tan \phi}, \ldots$$

Thus, the fringes are equally spaced and the separation, y, of adjacent dark (or bright) fringes is given by

$$y = \frac{\lambda}{2 \tan \phi} \qquad [25.6]$$

Determination of the Angle of the Wedge

The fringe separation, y, can be measured by traversing about twenty bright fringes with a travelling microscope and making use of the fact that the fringes are equally spaced. If λ is known, it is then possible to determine the angle, ϕ, of the air wedge by using equation [25.6].

Once the value of ϕ is known, it is possible to determine the thickness of the foil that is producing the wedge.

Thus, in Fig. 25.6

$$d = L \tan \phi$$

where

L = the length of the wedge. This can be measured with a metre rule.

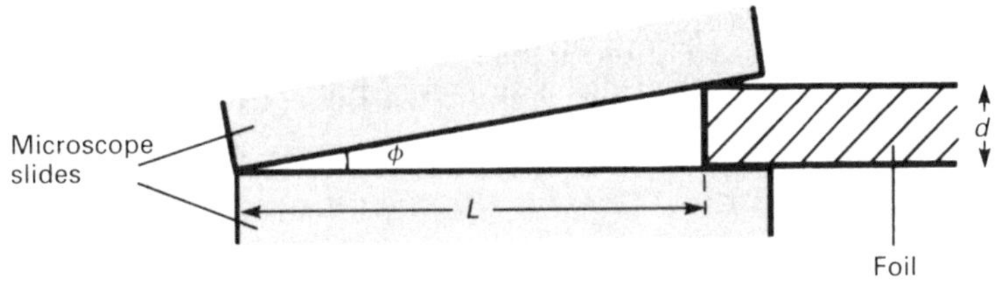

Fig. 25.6 Wedge angle and foil thickness (not to scale)

Notes

(i) The travelling microscope (or the eye) must be focused close to the upper surface of the air wedge because that is where the fringes are localized. The explanation of this is beyond the scope of this book, but it is associated with the fact that the interference is being produced by an extended source.

(ii) The angle, ϕ, of the wedge has to be small (typically 10^{-3} rad) in order that the fringes are sufficiently far apart to be observable. This follows from equation [25.6].

(iii) The fringes are clearer at the apex than at the thick end of the wedge. This is because the path differences near the thick end are many wavelengths in length, in which case the spread in wavelength of the light being used becomes significant. For example, suppose that the yellow light from a sodium lamp is being used to illuminate the air wedge. The light is not truly monochromatic – it consists of two wavelengths which, though close enough to allow the light to be considered to be monochromatic when small path differences are being employed, leads to blurring when the path differences are large. The two wavelengths are 5.890×10^{-7} m and 5.896×10^{-7} m. A path difference which is equal to 500 of the longer wavelengths is 500.5 of the shorter wavelengths. Therefore, where one component is producing a bright fringe, the other is producing a dark fringe, and vice versa.

Fringes Produced by Transmission

Fringes can also be viewed by transmission (Fig. 25.7). Wave B is reflected at both X and Y, and therefore it undergoes reflection at a more dense medium on two occasions. The second phase change that occurs cancels the effect of the first. Wave A does not undergo a phase change. Therefore the fringes are complementary to those produced by reflection, i.e. equation [25.4] gives the condition for brightness and equation [25.5] gives the condition for darkness. Wave A is considerably stronger than the twice reflected wave B, and therefore the fringes have much less contrast than those seen by reflected light.

Fig. 25.7
Transmission fringes

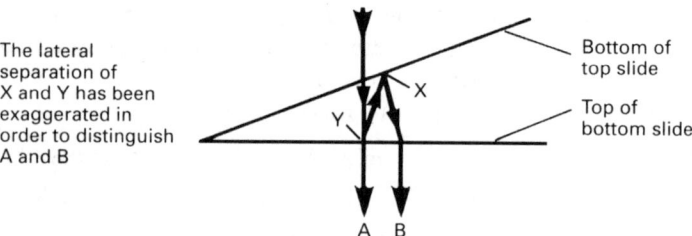

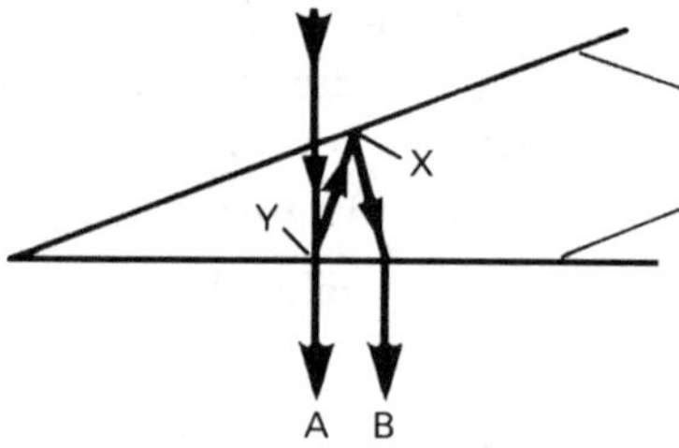

25.7 NEWTON'S RINGS

The interference effect known as Newton's rings was first studied, as the name suggests, by Isaac Newton. His explanation of their formation was based on the corpuscular theory of light. The effect was first correctly accounted for by Thomas Young on the basis of the wave theory. An arrangement for producing and viewing the rings is shown in Fig. 25.8.

Fig. 25.8
Apparatus to demonstrate Newton's rings

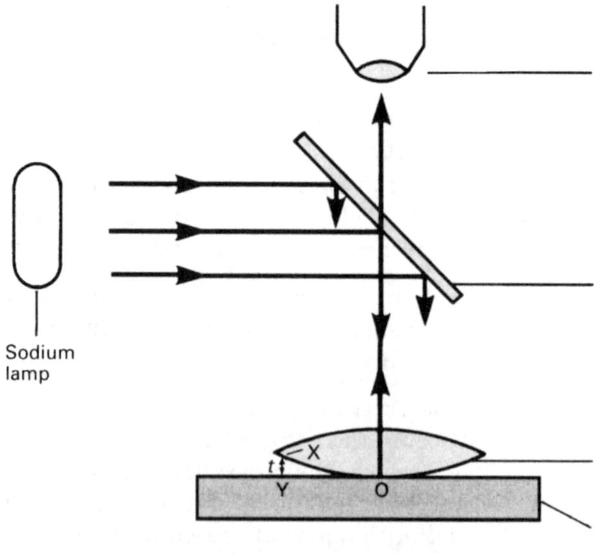

A convex lens of large radius of curvature ($\sim$ 100 cm) rests on an optically flat glass block illuminated from above. The interference is produced by division of amplitude and therefore an extended (large) source of light can be used.

The layer of air between the lens and the block acts as an air wedge (section 25.6). Light reflected from a point such as X on the lower surface of the lens, interferes with that reflected from a point (Y) vertically below it on the upper surface of the block. The 'fringes' take the form of concentric bright and dark rings centred on O. The reflection at Y produces a phase change of π radians (section 25.5), and the conditions for brightness and darkness are the same as those which produce fringes by reflection with an air wedge (section 25.6). Thus:

darkness if $2t = n\lambda \qquad (n = 0, 1, 2, \ldots)$ [25.7]

brightness if $2t = (n + \frac{1}{2})\lambda \qquad (n = 0, 1, 2, \ldots)$ [25.8]

Since a phase change of π radians occurs when light is reflected at the glass block, **the central spot** (for which $t = 0$) **is dark**. Young verified the phase change of π radians by placing an oil between the lens and the block. The oil was optically less dense than the block but more dense than the lens. Both reflections now took place at more dense media and therefore two phase changes occurred. The effect of one cancelled that of the other and the central spot became bright.

Newton's rings can be used to test the accuracy with which a lens has been ground – the rings are not circular if the surface of the lens is not spherical (or the block is not flat).

Measurement of Wavelength

The effect can be used to determine the wavelength, λ, of a light source if the radius of curvature, R, of the lower surface of the lens is known.

From Fig. 25.9 and the theorem of intersecting chords (see Appendix A3.11)

$$(2R - t)t = R_n^2$$

where R_n = the radius of the nth ring.

Fig. 25.9 Newton's rings – geometry

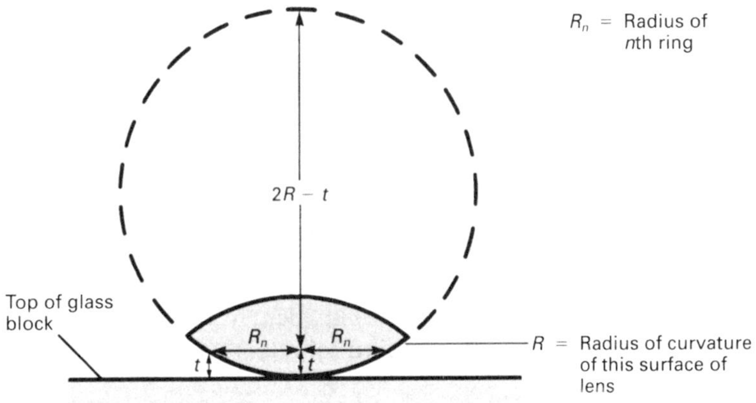

Since $t \ll R$, this reduces to

$$2Rt = R_n^2$$

i.e. $$2Rt = \frac{D_n^2}{4} \qquad [25.9]$$

where

D_n = the diameter of the nth ring.

But, by equation [25.7], at the sites of dark rings

$$2t = n\lambda \qquad (n = 0, 1, 2, \ldots)$$

Substituting in equation [25.9] leads to

$$n\lambda R = \frac{D_n^2}{4}$$

i.e. for dark rings

$$D_n^2 = 4R\lambda n \qquad (n = 0, 1, 2, \ldots) \qquad [25.10]$$

Similarly, from equations [25.8] and [25.9], for bright rings

$$D_n^2 = 4R\lambda\left(n + \tfrac{1}{2}\right) \qquad (n = 0, 1, 2, \ldots)$$

The diameters, D_n, of, say, the 5th, 10th, 15th, . . . , 30th dark rings are measured with a travelling microscope. It follows from equation [25.10] that if D_n^2 is plotted against n, the graph is a straight line of gradient $4R\lambda$ (Fig. 25.10). The value of R, the radius of curvature of the lens, can be measured with a spherometer and this value is used, together with that of the gradient, to calculate λ.

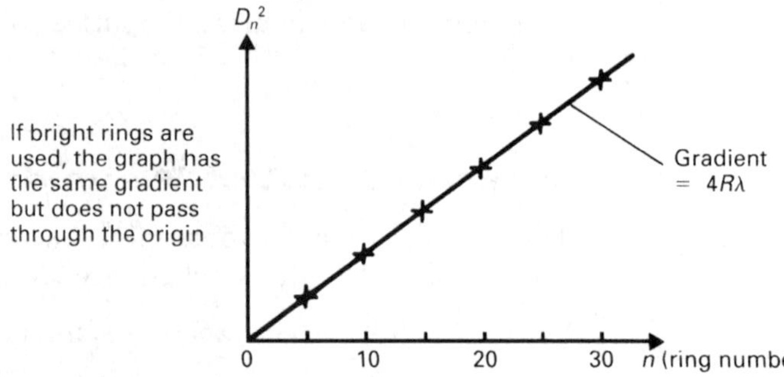

Fig. 25.10
Plot to determine λ from measurements on Newton's rings

Notes

(i) The diameters of the rings are used rather than their radii because it is difficult to locate the exact centre of the central spot.

(ii) A lens with a large radius of curvature is used so that the rings are sufficiently far apart to be observable. (This follows from equation [25.10].)

(iii) Note (i) of section 25.6 applies. Note (iii) also applies, but with obvious modifications.

25.8 PARALLEL-SIDED THIN FILMS

Consider waves incident in air at an angle i on a parallel-sided thin film of material (e.g. a soap film) of refractive index n_1 (Fig. 25.11).

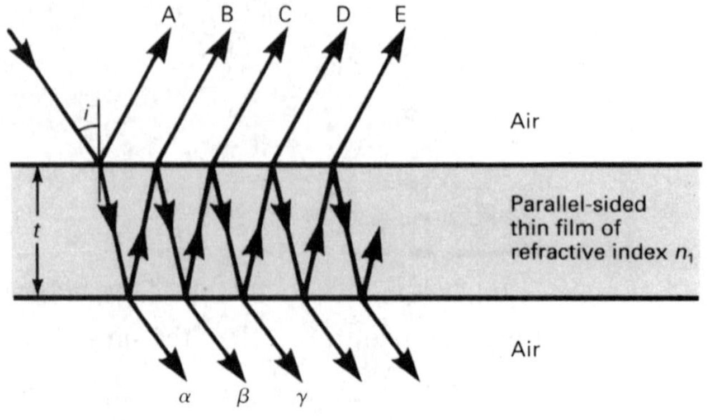

Fig. 25.11
Multiple reflection in a thin film

If i is small (i.e. nearly normal incidence), the geometrical path difference between ray A and ray B can be considered to be $2t$, where t is the thickness of the film. The geometrical path differences between B and C, C and D, and α and β are also equal to $2t$. The effects due to the transmitted light (rays $\alpha, \beta, \gamma, \ldots$) will be considered separately from those of the reflected light (rays A, B, C, ...).

Transmitted Light

If the geometrical path difference between adjacent rays is equal to a whole number of wavelengths, all the rays ($\alpha, \beta, \gamma, \ldots$) are in phase with each other and therefore interfere constructively if they are brought to a focus by a lens (e.g. that of the eye). If the geometrical path difference between adjacent rays is equal to an odd number of half-wavelengths, pairs of adjacent rays destroy each other. Thus:

brightness if $2t = n\lambda_1 \quad (n = 0, 1, 2, \ldots)$ [25.11]

darkness if $2t = (n + \frac{1}{2})\lambda_1 \quad (n = 0, 1, 2, \ldots)$ [25.12]

where

λ_1 = the wavelength of the light in the film.

It can be shown (see section 18.7) that $\lambda_1 = \lambda/n_1$, where λ is the wavelength of the light in air. Equations [25.11] and [25.12] can therefore be rewritten in terms of λ. Thus there is:

brightness if $\mathbf{2n_1t = n\lambda} \quad (n = 0, 1, 2, \ldots)$

darkness if $\mathbf{2n_1t = (n + \frac{1}{2})\lambda} \quad (n = 0, 1, 2, \ldots)$

Reflected Light

Ray A is reflected at a more dense medium and therefore undergoes a phase change of π radians. Therefore when $2n_1t = n\lambda$, B, C, D, ... are in phase with each other, but out of phase with A. It can be shown that the total intensity of B, C, D, ... is equal to that of A, and therefore there is

darkness if $\mathbf{2n_1t = n\lambda} \quad (n = 0, 1, 2, \ldots)$

If $2n_1t$ is equal to an odd number of half-wavelengths, the two strongest rays (A and B) are in phase with each other. Those which are out of phase with them (C, E, ...) are incapable of destroying them. Therefore there is

brightness if $\mathbf{2n_1t = (n + \frac{1}{2})\lambda} \quad (n = 0, 1, 2, \ldots)$

Blooming

When a lens is used to form an image, some of the incident light is reflected at the surfaces of the lens. This is undesirable because:

(i) it produces a background of unfocused light which reduces the contrast of the image, and

(ii) it reduces the brightness of the image.

If the lens is coated with a thin film of transparent material (magnesium fluoride is often used), it is possible to make the surfaces appreciably non-reflecting (Fig. 25.12). The process is known as 'blooming'. Magnesium fluoride is optically more

Fig. 25.12
Principle of blooming

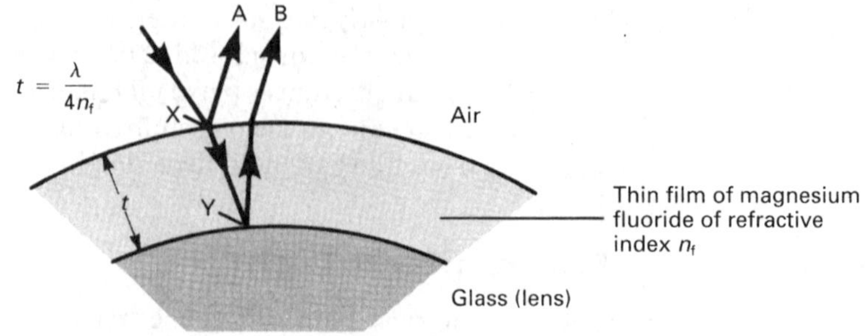

dense than air but optically less dense than glass. Therefore ray A undergoes a phase change of π radians on being reflected at X, and ray B suffers a similar phase change at Y. The effect of one phase change cancels that of the other, and therefore the two rays interfere destructively if the thickness, t, of the film is such that the optical path difference between the rays is equal to half a wavelength. For light which is incident at near normal incidence and has a wavelength in air of λ, this occurs when $2n_f t = \lambda/2$ where n_f is the refractive index of magnesium fluoride.

Lenses, of course, are usually illuminated by white light. This cannot be completely extinguished because destructive interference cannot occur for all the wavelengths of which the light is composed. However, if $2n_f t = \lambda/2$ holds for the average wavelength present (i.e. green), an appreciable amount of destructive interference occurs. This accounts for the purple appearance of bloomed lenses.

Colours in Thin Films

In Fig. 25.13, light from an extended source is incident on a parallel-sided thin film. If the angles of incidence of X and Y are large, the light that enters the eye from A has travelled a significantly greater distance in the film than that from B. If

Fig. 25.13
Production of colours

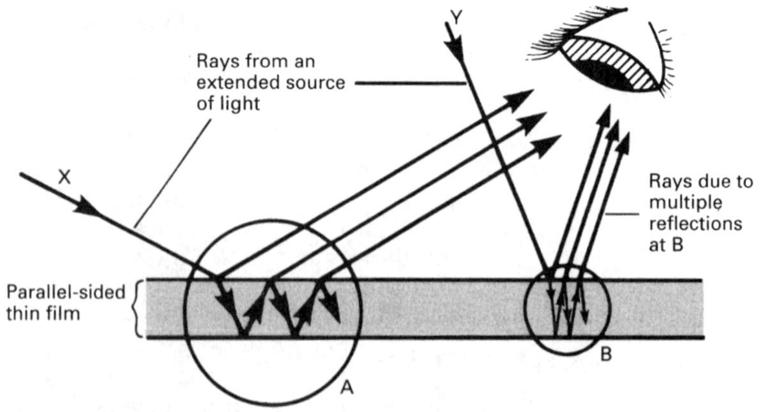

the incident light is white, the colours that reinforce each other in the light from A are unlikely to be the same as those which reinforce in the light from B. The light from A is brought to a focus at a different point on the retina from that coming from B, and therefore the eye sees the two regions as being different colours. (**Note**. The eye is focused on infinity because the rays from each region are parallel to each other.) The colours seen in thin films of oil on water and in soap bubbles are produced in this way.

25.9 VERTICAL SOAP FILMS

If a soap film is held with its plane vertical, it drains and becomes thicker at the bottom, forming an approximately wedge-shaped film. When such a film is illuminated by white light a series of horizontal coloured bands can be seen. (Reflected light gives greater contrast than transmitted light.) As the film continues to drain, the bands move downwards. The top of the wedge becomes thinner, and eventually the film breaks. If it is being viewed by reflection, the top of the film appears black just before it breaks. This is because the top of the film is so thin that there is effectively no geometrical path difference between the light coming from the two surfaces of the film, in which case the phase difference is due only to the phase change of π radians suffered by the light reflected at the front surface.

CONSOLIDATION

Two sources of waves are **coherent** if they have a constant phase difference. Since they must have a constant phase difference, they must have the same frequency. (**Note**. The phase difference may be zero, but it does not have to be.)

Light undergoes a phase change of π radians whenever it is reflected at a more dense medium.

Fringes produced by transmission are bright where fringes produced by reflection are dark, and vice versa.

26

DIFFRACTION OF LIGHT WAVES

26.1 INTRODUCTION

If a slit is placed in the path of a parallel beam of light, most of the light passes through the slit without changing direction, i.e. in a manner which is consistent with the idea that light travels in straight lines. However, careful observation of the situation reveals that some of the light spreads into regions which would be in shadow if light travelled only in straight lines. If the slit is wide, the spreading is so slight that it can normally be ignored. But if the width of the slit is comparable with the wavelength of the light, the effect is quite pronounced. The spreading of light in this way is called **diffraction**. Another example of the effect occurs when light falls on a small opaque disc. Light bends around the disc and produces brightness at the centre of its geometrical shadow. All types of wave motion exhibit diffraction effects. It is a common experience, for example, that sound waves spread round corners.

Diffraction can be accounted for on the basis of Huygens' construction. In Chapter 24 we accounted for the propagation of wavefronts by considering that every point on a wavefront emits secondary wavelets and by regarding the new position of the wavefront as being along the line which is the common tangent to all these secondary wavelets. The justification for choosing this line, rather than any other, is that each wavefront dealt with in Chapter 24 is an extensive wavefront, in which case it can be shown that this is the only line along which superposition of the secondary wavelets does not cause them to destroy each other. However, if a wavefront is limited in some way, there are additional lines where complete destruction does not occur. These lines also represent the positions of wavefronts. These additional wavefronts account for the spreading that occurs when an extended wavefront meets an obstruction and becomes a limited wavefront.

We now give a formal definition of diffraction.

> Diffraction occurs as a result of the superposition of secondary wavelets from a continuous section of wavefront that has been limited by an aperture or opaque object.

26.2 FRAUNHOFER DIFFRACTION

There are two general classes of diffraction – Fraunhofer diffraction and Fresnel diffraction. This book is concerned only with Fraunhofer diffraction, i.e. diffraction in which the wavefronts incident on a diffracting obstacle are plane, and give rise to diffracted wavefronts which are also plane. A distant source can be used to provide the incident wavefronts, but it is often more convenient to use a source which is at the focal point of a converging lens. Fig. 26.1 shows a practical arrangement for producing the Fraunhofer diffraction pattern of a single slit. An alternative arrangement omits L_2 and has the screen at a large distance from the diffracting slit.

Fig. 26.1
Fraunhofer diffraction

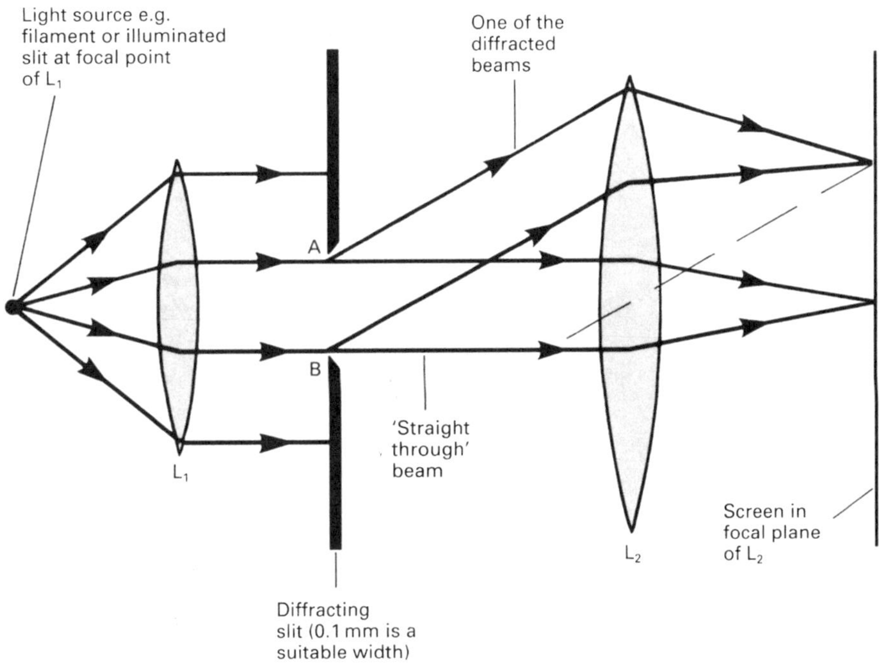

26.3 AN IMPORTANT RESULT

It can be shown that if light from a point is brought to a focus at some other point by means of a lens, then each of the ray paths involved contains the same number of wavelengths (and therefore the **optical path length** of each ray is the same). Thus in Fig. 26.2(a), there are the same number of wavelengths along OBB′ I as there are along the geometrically shorter path, OAA′ I. This is made possible by the fact that light has a shorter wavelength in glass than in air, and ray X travels farther in the glass than does ray Y. Fig. 26.2(b) shows spherical wavefronts and the way in which they are modified by the presence of a lens.

Fig. 26.2
(a) Optical paths of rays through a lens. (b) Action of lens on wavefronts

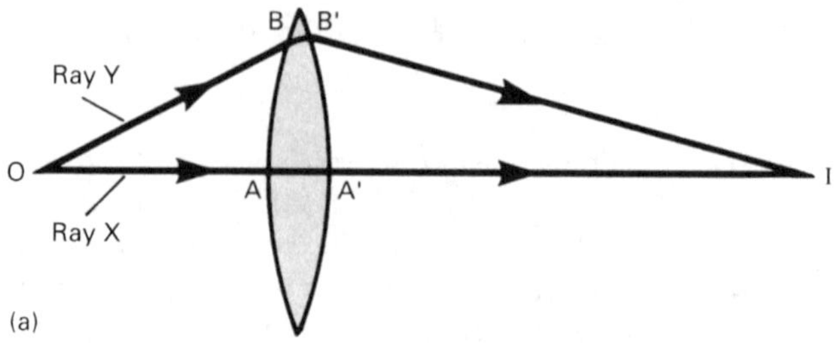

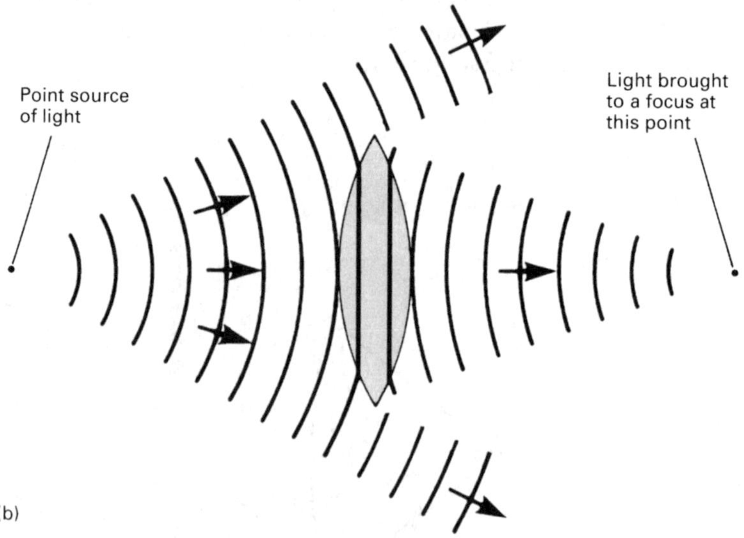

The result also holds, of course, for the particular case of a plane wavefront incident on a lens. Thus in Fig. 26.3, each of the ray paths shown contains the same number of wavelengths, and therefore the presence of the lens does not affect the phase relationship of the light travelling to P from Y with that of the light reaching P from X.

Fig. 26.3
Plane wavefront incident on a lens

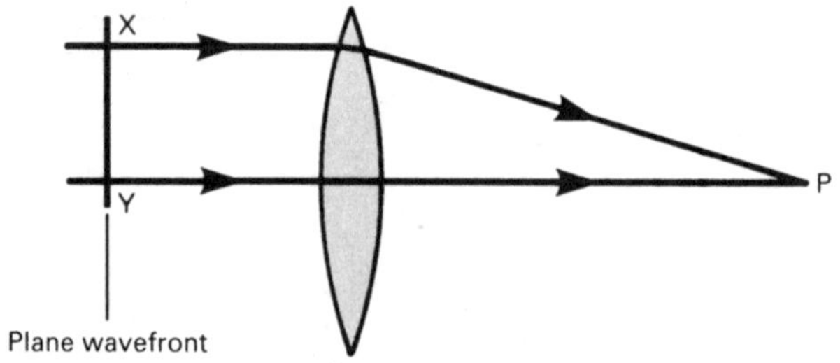

26.4 DIFFRACTION AT A SINGLE SLIT

A typical single slit diffraction pattern, such as that which might be produced by the arrangement of Fig. 26.1, is shown in Fig. 26.4(a). The intensity distribution is shown in Fig. 26.4(b). The pattern can be accounted for by considering the superposition of the secondary wavelets which are imagined to be emitted by each point on the wavefront between A and B (Fig. 26.1). A complete analysis of the situation is beyond the scope of this book, but we shall (i) derive an expression which gives the angular positions of the minima (e.g. X and Y in Fig. 26.4(b)), and (ii) account for the central maximum.

Fig. 26.4
(a) Diffraction pattern from a single slit.
(b) Corresponding variation of intensity

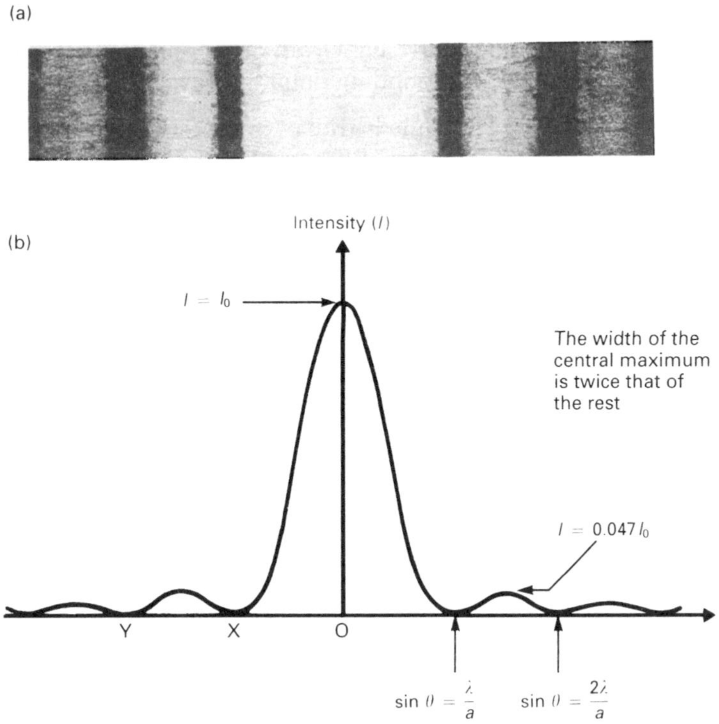

The Minima

In Fig. 26.5, AB is a slit of width a. Consider the slit to be split into $2n$ (where $n = 1, 2, 3, \ldots$) equal sections of width $a/2n$. Let these sections be AC, CD, DE, EF, etc. Consider some direction θ to the normal, and suppose that λ is the wavelength of the light being used. If θ is such that $\text{AN} = \lambda/2$, the wave from A will be completely out of phase with that from C. Thus the waves from A and C will destroy each other when they come together, at P, on the screen. (The presence of

Fig. 26.5
Diffraction at a single slit

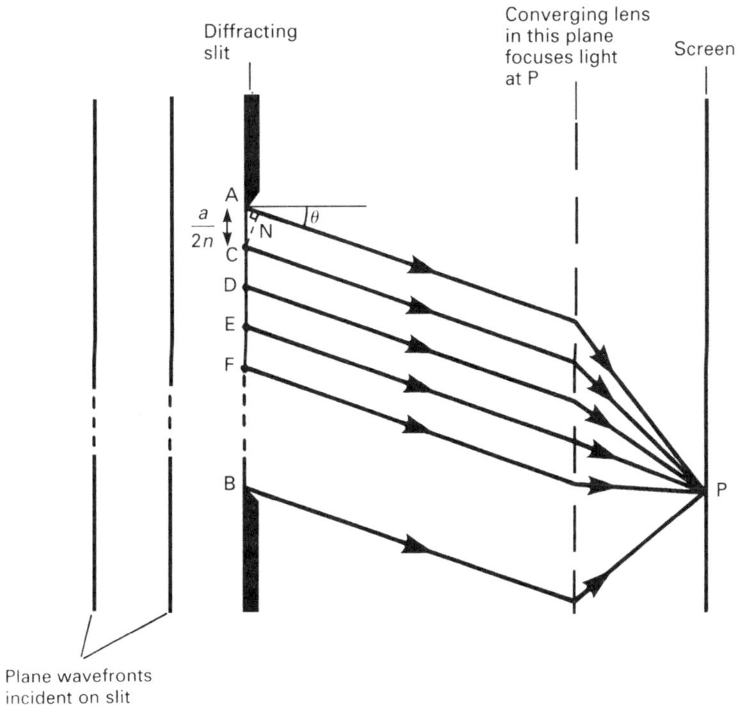

the lens does not affect the phase relationship of A and C – see section 26.3.) Similarly, light from each point between A and C will be destroyed by that from the corresponding point between C and D.

The same is true of every other pair of sections, such as DE and EF. Therefore, there is no diffracted light in those directions θ which are such that

$$\text{AN} = \frac{\lambda}{2}$$

i.e. $$\text{AC} \sin \text{A}\hat{\text{C}}\text{N} = \frac{\lambda}{2}$$

From simple geometry, $\text{A}\hat{\text{C}}\text{N} = \theta$, and therefore

$$\text{AC} \sin \theta = \frac{\lambda}{2}$$

i.e. $$\frac{a}{2n} \sin \theta = \frac{\lambda}{2}$$

i.e. the angular positions, θ, of the minima are given by

$$a \sin \theta = n\lambda \qquad (n = 1, 2, 3, \ldots) \qquad [26.1]$$

Notes

(i) It follows from equation [26.1] that the minima are not equally spaced. However, for values of θ which are less than about 10° (i.e. $\sin \theta \approx \theta$) it is often a sufficiently good approximation to consider that the minima are equally spaced.

(ii) The values of θ given by equation [26.1] apply to each side of the normal.

(iii) It follows from equation [26.1] that at the edges of the central band $\sin \theta = \lambda/a$ (see Fig. 26.4(b)). Therefore if $a \gg \lambda$, $\sin \theta$ is small and we may put $\sin \theta = \theta$, in which case $\theta = \lambda/a$.

The Central Maximum

In Fig. 26.6, all the waves arriving at O, at any one time, will have left the various points on the wavefront at AB at the same time as each other. Since all these waves are in phase with each other when they leave AB, they are still in phase when they reach O. (In making this statement we are again relying on the fact that the presence of the lens does not affect the phase relationships involved.) Thus, at O, the radiation from each point on AB enhances that from every other, and therefore there is brightness at O.

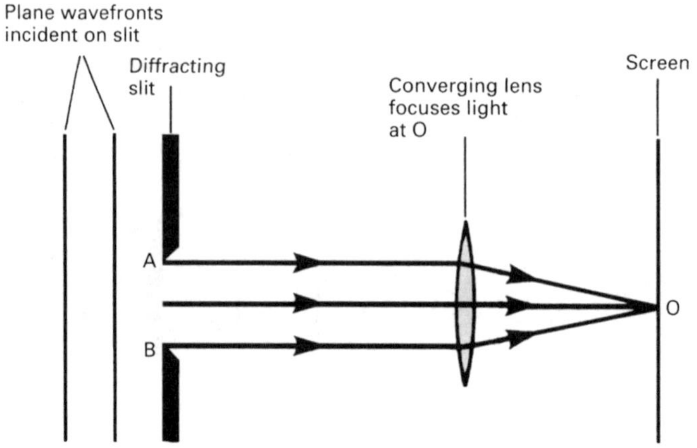

Fig. 26.6
Formation of the central maximum

The positions of the other maxima (i.e. the **subsidiary maxima**) are difficult to determine, but the points listed below are worthy of note.

(i) The subsidiary maxima are much less intense than the central maximum because O is the only point at which waves from each point on the wavefront interfere constructively with those from every other.

(ii) The subsidiary maxima lie approximately mid-way between the minima.

(iii) The width of the central maximum is twice that of each subsidiary maximum.

26.5 THE DIFFRACTION GRATING*

A diffraction grating is an arrangement of identical, equally-spaced diffracting elements. Normally, it consists of a large number of parallel lines of equal width ruled on glass (a transmission grating) or metal (a reflection grating). In each case the diffracting elements are the clear spaces between the rulings. There are, typically, 600 lines per millimetre. Diffraction gratings are used to produce optical spectra.

Fig. 26.7 represents a section of a (transmission) diffraction grating which is being illuminated normally by light of wavelength λ.

Fig. 26.7
Section of a diffraction grating

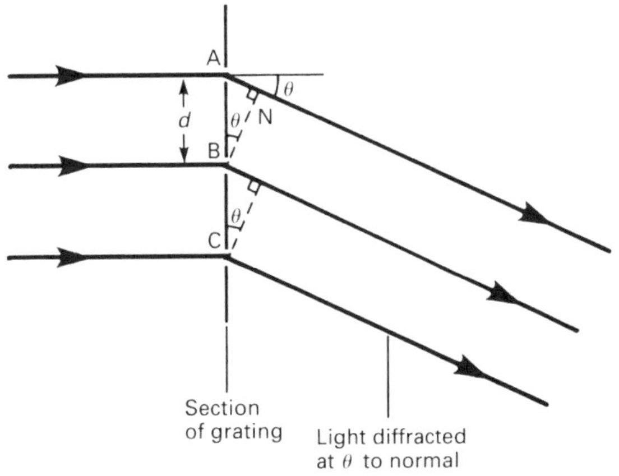

Each of the clear spaces (A, B, C, etc.) acts like a very narrow slit and diffracts the incident light to an appreciable extent in all the forward directions. Consider that light which is diffracted at some angle θ to the normal. The slits are equally spaced and therefore if θ is such that light from A is in phase with that from B, it is also in phase with that from every other slit, and it follows that light from each slit is in phase with that from every other. This happens whenever

$$AN = n\lambda \qquad (n = 0, 1, 2, \ldots)$$

i.e. $$d \sin\theta = n\lambda \qquad (n = 0, 1, 2, \ldots) \qquad [26.2]$$

*A secondary wavelet treatment is given in section 24.8.

Thus, if the light from all the slits is brought to a common focus (e.g. by the eye, or the telescope of a spectrometer), that from each slit interferes constructively with that from every other at those values of θ given by equation [26.2]. The effect of the grating, therefore, is to produce a series of bright images, known as **principal maxima**, the angular positions, θ, of which are given by equation [26.2]. Each value of θ applies to either side of the normal.

It can be shown that a grating which has a large number of slits produces what can be considered to be completely destructive interference in all directions other than those given by equation [26.2], and therefore gives rise to very sharp principal maxima.

The zero order principal maximum (i.e. that for which $n = 0$ in equation [26.2]) is given by equation [26.2] as having an angular position, θ, of zero for all values of d and λ (i.e. for any grating illuminated by light of any wavelength). The positions of the first order ($n = 1$), second order ($n = 2$), etc. principal maxima, however, depend on both d and λ. A typical grating has a **grating spacing,** d, of $1/600\,000$ m (corresponding to 600 lines per millimetre, i.e. 600 000 lines per metre). If such a grating is illuminated by light whose wavelength, λ, is 6×10^{-7} m, we find, by substituting these values in equation [26.2], that

$$\text{if } n = 1, \qquad \theta = 21.1^\circ$$

and

$$\text{if } n = 2, \qquad \theta = 46.1^\circ$$

But, when $n = 3$, equation [26.2] gives $\sin\theta = 1.08$ (which is impossible) and therefore this particular grating cannot produce a third order image when illuminated by light of this particular wavelength (see Fig. 26.8).

Fig. 26.8
Different orders of images produced by a diffraction grating

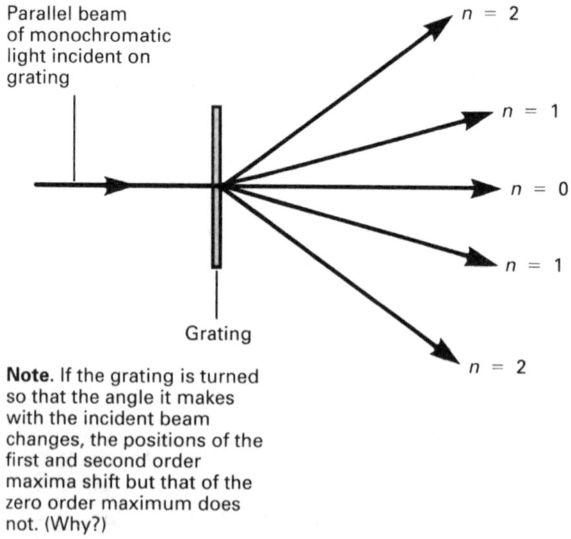
Parallel beam of monochromatic light incident on grating
n = 2
n = 1
n = 0
n = 1
n = 2
Grating
Note. If the grating is turned so that the angle it makes with the incident beam changes, the positions of the first and second order maxima shift but that of the zero order maximum does not. (Why?)

Note It follows from equation [26.2] that the number of orders of principal maxima that can be produced can be increased by increasing d (i.e. reducing the number of lines per metre). Reducing the total number of lines decreases the sharpness of the principal maxima and also gives rise to faint images, **subsidiary maxima,** in the regions between the principal maxima. (See section 26.8.)

EXAMPLE 26.1

When a grating with 300 lines per millimetre is illuminated normally with a parallel beam of monochromatic light a second order principal maximum is observed at 18.9° to the straight through direction. Find the wavelength of the light.

Solution

300 lines per millimetre = 3.00×10^5 lines per metre and therefore the grating spacing, d, is given by $d = 1/3.00 \times 10^5$ m. Since we are dealing with a second order maximum, $n = 2$ and therefore by

$$n\lambda = d \sin \theta$$

$$2\lambda = \frac{1}{3.00 \times 10^5} \sin 18.9^\circ = 1.080 \times 10^{-6}$$

$$\therefore \quad \lambda = 5.40 \times 10^{-7} \text{ m}$$

EXAMPLE 26.2

How many principal maxima are produced when a grating with a spacing of 2.00×10^{-6} m is illuminated normally with light of wavelength 6.44×10^{-7} m?

Solution

We first calculate the highest order possible.

$$n\lambda = d \sin \theta \qquad \therefore \qquad n = \frac{d}{\lambda} \sin \theta$$

Since $\sin \theta \leqslant 1$, $\quad n \leqslant \dfrac{d}{\lambda}$

$$\therefore \quad n \leqslant \frac{2.00 \times 10^{-6}}{6.44 \times 10^{-7}} \qquad \text{i.e. } n \leqslant 3.11$$

Since n must be an integer, the highest order possible is given by $n = 3$. The grating therefore produces 7 principal maxima. (The zero order maximum plus 3 on each side of it.)

QUESTIONS 26A

1. Light of wavelength 5.70×10^{-7} m is incident normally on a grating with a spacing of 2.00×10^{-6} m. What is the angle to the normal of: **(a)** a first order principal maximum, **(b)** a second order principal maximum?

2. A diffraction grating produces a second order principal maximum at 50.6° to the normal when being illuminated normally with light of wavelength 644 nm. Calculate the number of lines per millimetre of the grating.

Intensity Distribution of the Principal Maxima

The amount of light which is available to contribute to the constructive interference which occurs in any particular direction is determined by the amount of light which is diffracted in that direction by each slit. The intensity distribution is therefore that of a single slit diffraction pattern. This is illustrated by Fig. 26.9, which is assumed to represent the situation of a grating which has a large number of lines, and therefore no subsidiary maxima are shown.

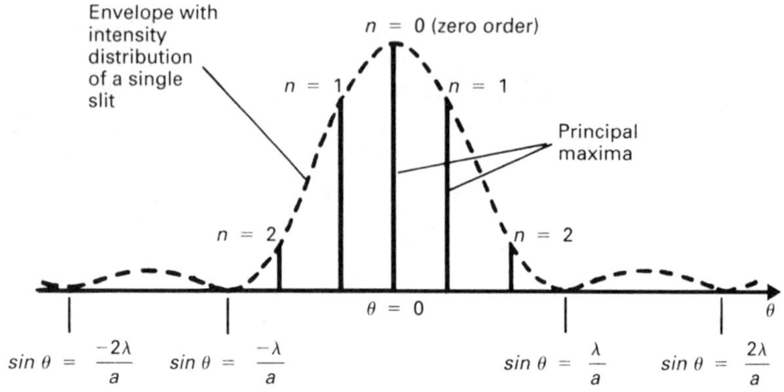

Fig.26.9
Intensity variation of the principal maxima

The angular positions of the principal maxima are determined by equation [26.2] and therefore depend on the grating spacing, d. The shape of the diffraction envelope, however, is governed by the width of the clear spaces. If the principal maxima are to be reasonably intense, they must all lie within the central band of the diffraction envelope. This sets an upper limit to the width of the clear spaces (see Example 26.3).

EXAMPLE 26.3

A particular transmission grating has clear spaces of width a, and opaque sections of width b. The grating is capable of producing third order principal maxima, but none of higher order, when illuminated normally by light of wavelength λ. Calculate the maximum value of the ratio a/b if all the principal maxima are to lie within the central band of the diffraction envelope.

Solution

If we take θ' to be the angular position of the edge of the central band of the single slit diffraction pattern, then we find, by putting $n = 1$ in equation [26.1], that

$$\sin\theta' = \frac{\lambda}{a}$$

If we take θ_3 to be the angular position of a third order principal maximum, then we find, by putting $n = 3$ and $d = (a + b)$ in equation [26.2], that

$$\sin\theta_3 = \frac{3\lambda}{a+b}$$

If the third order principal maximum is to lie within the central band of the diffraction envelope, then it is necessary that

$$\theta_3 < \theta'$$

i.e. $$\sin\theta_3 < \sin\theta'$$

i.e. $$\frac{3\lambda}{a+b} < \frac{\lambda}{a}$$

i.e. $$3a < a+b$$

i.e. $$2a < b$$

i.e. $$\frac{a}{b} < \frac{1}{2}$$

Thus, the maximum value of a/b is $\frac{1}{2}$, i.e. the opaque sections must be at least twice as wide as the clear sections. (Note that if a/b is exactly equal to $\frac{1}{2}$, the third order principal maxima are formed at the edges of the central band and are of zero intensity, i.e. missing.)

26.6 MEASUREMENT OF WAVELENGTH USING A DIFFRACTION GRATING AND SPECTROMETER

The wavelength, λ, of any line in a line spectrum (e.g. that of sodium) can be determined by using equation [26.2] if the angular position, θ, of one of the principal maxima produced by a grating is measured and the grating spacing, d, is known. (The value of d is normally supplied by the manufacturer.) The grating is used in conjunction with a **spectrometer**, so that θ may be measured to a high degree of accuracy (1′ of arc). The most significant error is often due to the value of d being different from the specified value as a result of expansion or contraction. The experimental arrangement is shown in Fig. 26.10.

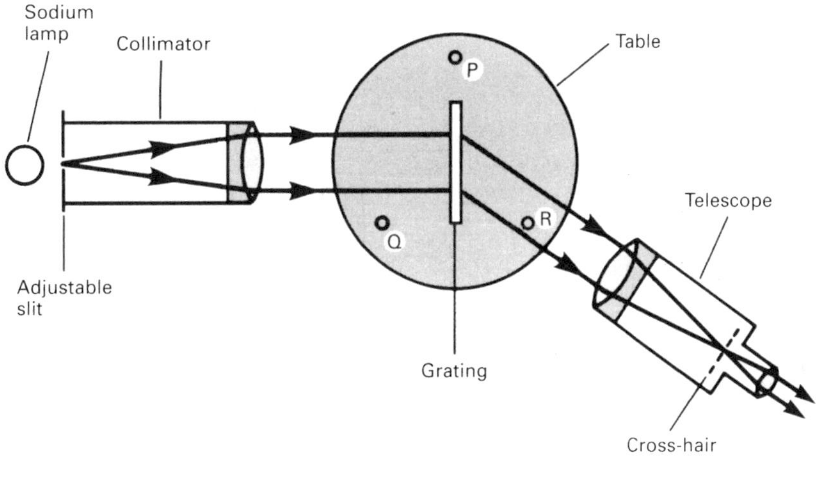

Fig. 26.10 Measurement of wavelength using a grating and spectrometer

Method

(i) The telescope eyepiece is adjusted so that the cross-hairs are in sharp focus.

(ii) The telescope is focused on a distant object (by changing the position of the telescope objective relative to the eyepiece and cross-hairs) in such a way that there is no parallax between the image and the cross-hairs. The significance of using an object which is distant is that, from now on, any parallel light which enters the telescope will be brought to a focus on the cross-hairs.

(iii) The slit of the collimator is illuminated (by a sodium lamp) and the telescope is turned so that it is in direct line with the slit. The slit is moved in or out of the collimator tube until the image of the slit, as seen through the telescope, is in sharp focus. Since the telescope has been set to focus parallel light, the collimator must now be providing parallel light. At this stage the width of the slit is optimized.

(iv) The grating is placed on the table so that its plane is perpendicular to two (say Q and R) of the levelling screws. The telescope is moved through exactly 90° (by reference to the scale) and then the table is turned until the grating reflects light onto the cross-hairs of the telescope. The situation is now as shown in Fig. 26.11, and the grating is at 45° to the light from the collimator. If the plane of the grating is not vertical, the image of the slit seen through the telescope, will be displaced either upwards or downwards. If this is the case, either Q or R is adjusted until the image occupies the centre of the field of view – the plane of the grating is now vertical.

(v) The table is turned through exactly 45° (by reference to the scale) so that the grating is now being illuminated normally, and the telescope is turned to a position such as T_1 (Fig. 26.12) where one of the first order principal

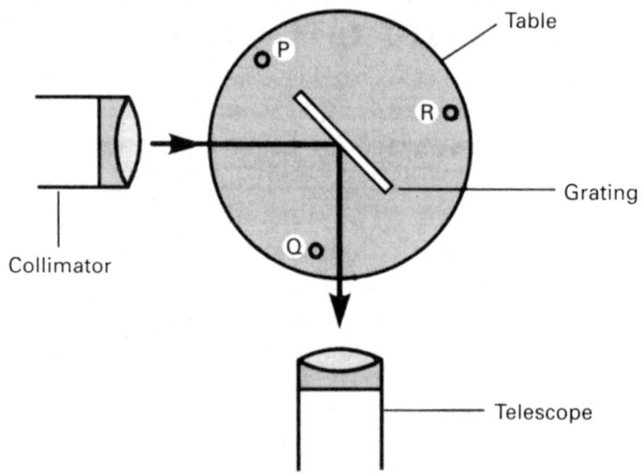

Fig. 26.11
Light from collimator reflected off grating at 45° into telescope

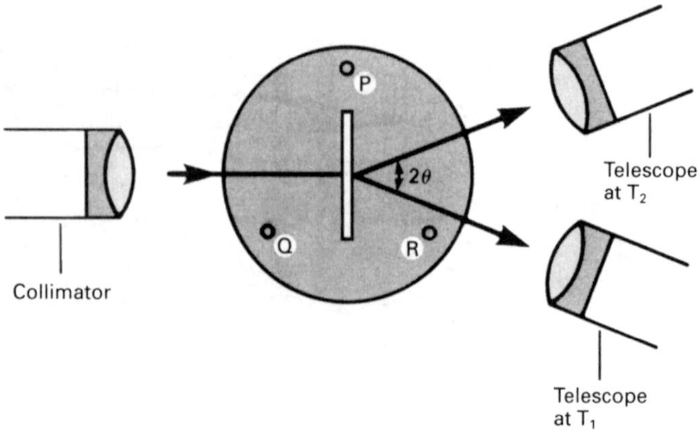

Fig. 26.12
Measurement of 2θ

maxima can be seen. (If the lines on the grating are not vertical, the image will not be in the centre of the field of view and the third screw, P, must be adjusted.) The angular position of the telescope (and therefore that of the first order image) is recorded.

(vi) The telescope is swung round to position T_2 (Fig. 26.12) where the other first order image is visible, and its position is recorded.

The angle between the two first order principal maxima (and therefore between the two recorded positions of the telescope) is 2θ, where, from equation [26.2] with $n = 1$,

$$d \sin \theta = \lambda$$

Thus, by halving the difference between the two angular positions of the telescope, λ can be found.

Note Measuring the double angle, 2θ, rather than θ, gives half the error.

26.7 GRATING SPECTRA

It follows from equation [26.2] that the angle at which any given principal maximum (except that for which $n = 0$) is formed by a grating depends on the wavelength of the light being used. Thus, a grating illuminated by white light produces a spectrum of the light. The most obvious difference between such a spectrum and that produced by a prism is that **a grating deviates violet light less than red**.

When $n = 0$, each wavelength produces a principal maximum at $\theta = 0$ and therefore **the central maximum is white**.

26.8 DIFFRACTION PRODUCED BY MULTIPLE SLITS

A typical diffraction grating (section 26.5) produces very sharp principal maxima. This is because it has a very large number of slits. (A grating with 600 lines per mm and which is 30 mm wide has 18 000 slits.) We shall now look more closely at how the diffraction pattern is affected by the number of slits.

Consider a parallel beam of monochromatic light of wavelength λ incident normally on a screen containing N slits, each of width a and separated by opaque sections of width b. The light diffracted by each slit interferes with that diffracted by the other $N - 1$ slits. The intensity distributions of the patterns so produced are shown in Fig. 26.13 for various values of N. The angular position, θ, of the principal maxima are given by equation [26.2], with $d = a + b$, because in deriving that equation we placed no limitation on the number of slits involved. Examination of the patterns reveals a number of points of interest.

(i) Increasing the value of N increases the intensities of the principal maxima compared with those of the subsidiary maxima.

(ii) Increasing the value of N increases the sharpness of the principal maxima. (It can be shown that the sharpness depends only on N and not on $a + b$ or a.)

(iii) The angular positions (θ) of the principal maxima do not depend on N.

(iv) Increasing the value of N increases the number of subsidiary maxima. (There are $N - 2$ subsidiary maxima between each adjacent pair of principal maxima.)

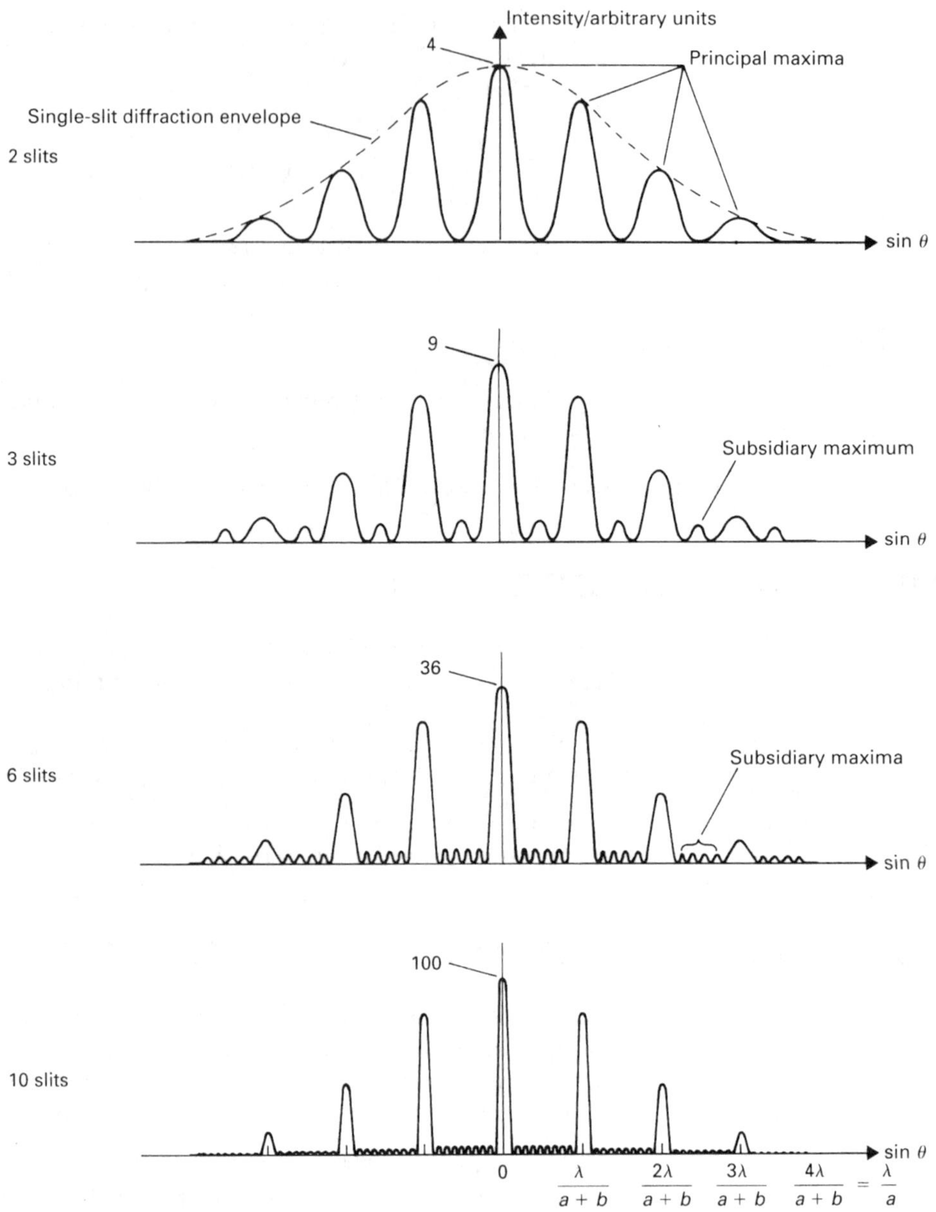

Fig. 26.13
To illustrate the effect on the diffraction pattern of varying ***N*** whilst keeping ***a*** and (***a + b***) fixed

(v) Increasing the number of slits allows more light through and therefore increases the absolute intensities of the diffraction patterns. (This is not immediately obvious from looking at Fig. 26.13, because it has been drawn in such a way that equivalent principal maxima are all the same size.) The mathematics involved is beyond the scope of this book, but it can be shown that the intensities of the principal maxima are proportional to N^2.

(vi) For any given value of N, altering $a + b$ alters the angular positions of the various principal maxima, and therefore also alters their relative intensities because they move to different parts of the diffraction envelope. Altering a has a similar effect because it changes the width of the diffraction envelope. When $(a + b)/a$ is an integer some principal maxima are missing because they occur exactly at the edges of the diffraction envelope and therefore have zero intensity. The patterns in Fig. 26.13 have been drawn for the case of $(a + b)/a = 4$, in which case the fourth-order principal maxima are missing. (See also Example 26.3.)

Why the Principal Maxima are Sharp when *N* is Large

At the site of a principal maximum θ is such that the light from each slit is in phase with that from every other. Suppose now that we change θ by a small amount so that the path difference, $(a+b)\sin\theta$, between the light from two adjacent slits changes by a small amount (Δx, say) where Δx is so small that it hardly affects the interference between light from adjacent slits. The maximum change in path difference that this creates is that between the two extreme slits and is equal to $(N-1)\Delta x$, where N is the total number of slits. If N is small, there is no possibility of the path difference between the light from any two slits changing by enough to produce a significant amount of destructive interference, and the overall intensity will still be reasonably high. However, if N is large, there is the likelihood of there being a slit which is far enough away from one of the end slits (say) for the path difference to change by $\lambda/2$, so that the light from one of these slits is destroyed by that from the other. The same will be true of the light from the two slits adjacent to these, and so on. Thus a slight change in θ will produce a marked decrease in intensity when N is large, i.e. a sharp principal maximum.

Origin of the Subsidiary Maxima

A complete mathematical analysis is beyond the scope of this book, but we can give some indication of how subsidiary maxima arise. Consider the case when $N=4$. Suppose that the four slits are A, B, C and D, in that order. If the path difference from adjacent slits is $\lambda/4$ or $3\lambda/4$, then the light from A destroys that from C, and the light from B destroys that from D, giving an overall intensity of zero. Also, if the path difference between the light from adjacent slits is $\lambda/2$, the light from A destroys that from B or D and that from C destroys that from D or B, and therefore the intensity is again zero. Thus there are three minima between the zero order and first order principal maxima. Between these three minima there are bound to be two maxima – these are the subsidiary maxima. They are not principal maxima, because to produce a principal maximum the light from each slit must be in phase with that from every other.

In the case of three slits there is only one subsidiary maximum between any adjacent pair of principal maxima. Consider the case of the subsidiary maximum between the zero order and first order principal maxima. It occurs when $(a+b)\sin\theta = \lambda/2$, for then light from one of the slits destroys that from a slit adjacent to it, so that the overall intensity is that of the light from the third slit. (The intensity of this subsidiary maximum is slightly less than $\frac{1}{9}$ (not $\frac{1}{3}$) that of the zero order principal maximum – why?) The minima occur at values of θ such that $(a+b)\sin\theta = \lambda/3$ or $2\lambda/3$, i.e. such that the path differences between the waves from adjacent slits are $\lambda/3$ and $2\lambda/3$ respectively. If the reader wishes to verify that this situation does in fact give an overall intensity of zero, he or she should try adding together (graphically or otherwise) three sine waves of equal amplitude such that each is displaced from one of the others by $\lambda/3$ or $2\lambda/3$, e.g. $\sin(\alpha-120^\circ)+\sin\alpha+\sin(\alpha+120^\circ)$ or $\sin(\alpha-240^\circ)+\sin\alpha+\sin(\alpha+240^\circ)$.

26.9 DIFFRACTION AT A CIRCULAR APERTURE

Whenever light is incident on an aperture, it is diffracted to some extent. The particular case of a circular aperture is important, because the amount of diffraction which takes place at such an aperture determines the abilities of the lenses of telescopes and other optical instruments to resolve fine detail.

The Fraunhofer diffraction pattern of a circular aperture is shown in Fig. 26.14, and takes the form of a bright central disc (known as **Airy's disç** and containing over 91% of the light), surrounded by a number of much less intense rings. The intensity distribution across a diameter of the pattern is almost the same as that shown in Fig. 26.4(b). The edge of the central disc makes an angle θ with the 'straight through' direction which is given by $\sin\theta = 1.22\,\lambda/a$, where a is the diameter of the aperture. When $a \gg \lambda$, this reduces to $\theta = 1.22\,\lambda/a$.

Fig. 26.14
Diffraction pattern due to a circular aperture

CONSOLIDATION

Diffraction at a Single Slit

The width of the central maximum is twice that of each of the subsidiary maxima.

Minima occur where

$$a \sin\theta = n\lambda \qquad (n = 1, 2, 3, \ldots)$$

Transmission Diffraction Grating

Consists of a **large number** of **identical** slits which are **equally spaced**.

Principal maxima occur at angles such that light from each slit is in phase with that from every other slit.

The sharpness of the principal maxima is increased by increasing the total number of slits.

Principal maxima occur where

$$d \sin\theta = n\lambda \qquad (n = 0, 1, 2, \ldots) \qquad [26.2]$$

It follows from equation [26.2] that increasing d increases the number of orders of principal maxima.

27

POLARIZATION OF LIGHT WAVES

27.1 THE PHENOMENON OF POLARIZATION

Wave motions are either longitudinal or transverse.

Longitudinal waves are waves in which the vibrations are along the direction of travel of the wave (e.g. sound waves, compression waves in springs).

Transverse waves are waves in which the vibrations are perpendicular to the direction of travel of the wave (e.g. electromagnetic waves, water waves). **Note**. A sound wave passing through water is, of course, a longitudinal wave.

These two types of waves may be distinguished in that **transverse waves can be polarized, longitudinal waves cannot be polarized**.

If all the vibrations of a transverse wave are in a single plane which contains the direction of propagation of the wave, the wave is said to be **plane-polarized** (**or linearly polarized**).

Fig. 27.1 depicts two ropes along which transverse waves are travelling. Wave A is plane-polarized in the *xy*-plane and wave B is plane-polarized in the *xz*-plane. Each wave can pass through its respective slit without any hindrance as long as the slits have the orientations shown. But, if slit A′, say, is rotated through 90° about O*x* so that it becomes parallel to O*z*, wave A is totally incapable of passing through. Such a procedure would have no effect whatsoever on a longitudinal wave.

Fig. 27.1
(a) Plane polarized waves

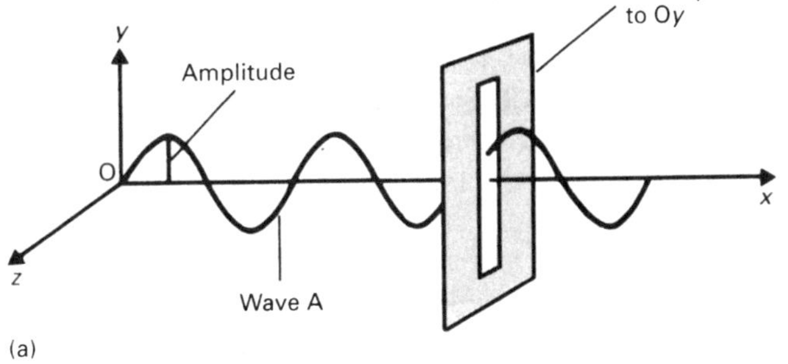

Fig. 27.1
(b) Plane polarized waves

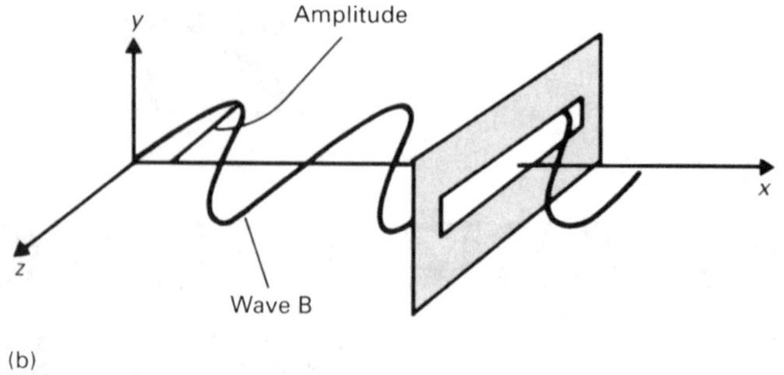

27.2 POLARIZATION AND LIGHT WAVES

The 'vibrations' of a light wave are a varying electric field, E, and a varying magnetic field, B, which are perpendicular to each other and which have the same frequency. Each of these fields is perpendicular to the direction of travel of the wave, and therefore light is a transverse wave motion.

Fig. 27.2 is a pictorial representation of a **plane-polarized** light wave in which the variation of E takes place exclusively in the xy-plane and that of B in the (perpendicular) xz-plane.

Fig. 27.2
To illustrate that ***E*** and ***B*** are mutually perpendicular

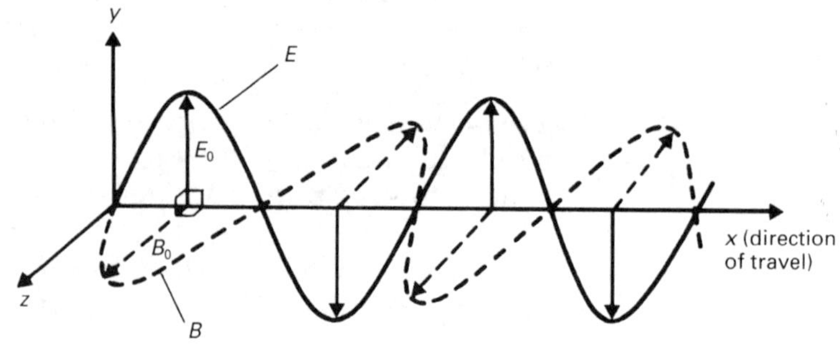

Notes

(i) E and B are in phase with each other, i.e. E has its maximum value E_0 at the same time as B has its maximum value B_0.

(ii) Though it is not apparent in Fig. 27.2, the magnitudes of E and B, at any one time, are related by $E/B = c$ (where $c =$ the velocity of light) and in particular $E_0/B_0 = c$.

When light interacts with matter, the effects of the electric field usually dominate those of the magnetic field. For example, it can be shown by experiment that it is the electric component of light which affects photographic film and which produces fluorescence. The plane of polarization of a light wave is therefore regarded as being that which contains E*. Thus, the wave shown in Fig. 27.2 is plane-polarized in the xy-plane.

Most sources of light emit waves whose planes of polarization vary randomly with time. These variations take place suddenly and at intervals of as little as 10^{-9} s. A wave of this type is said to be **unpolarized**, because over any reasonable interval of time its plane of polarization does not favour any one of the possible directions more than any other.

*From now on, whenever the term 'vibrations' is used in the context of a light wave it should be taken to refer to the variation of the associated electric field.

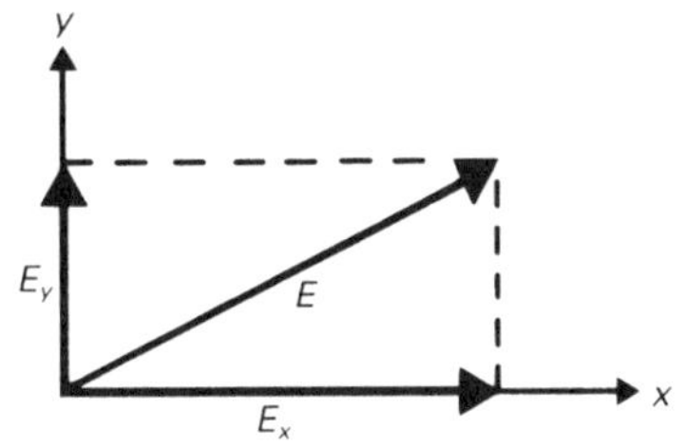

Fig. 27.3
Components of ***E***

Suppose that an unpolarized wave is travelling at right angles to the paper and that at some instant in time the electric field (E), at some point on the wave, is in the direction shown in Fig. 27.3, and therefore has components E_x and E_y as shown. After about 10^{-9} s the direction of E will change and therefore the values of E_x and E_y will also change. Over a reasonable period E will occupy a large number of orientations, each with equal probability, and therefore the average value of E_x will be equal to that of E_y. Thus, although the 'vibrations' of an unpolarized wave actually take place in every direction which is perpendicular to the direction of propagation, such a wave can be regarded as having vibrations only in two perpendicular directions, and such that the amplitude of vibration is the same in each of these directions. This gives rise to the commonly used pictorial representation of an unpolarized light wave which is shown in Fig. 27.4.

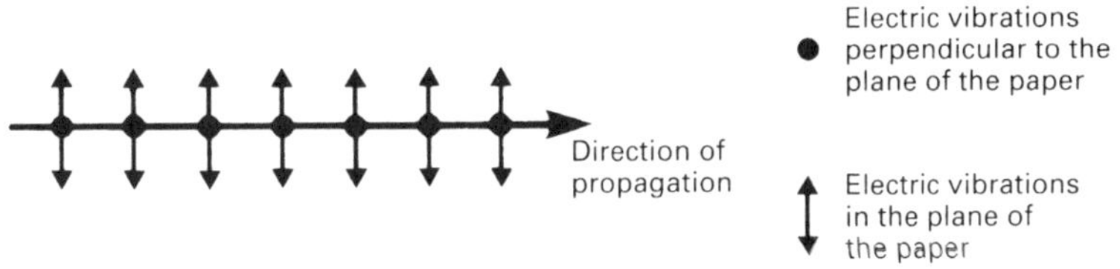

Fig. 27.4
Pictorial representation off an unpolarized light wave

27.3 PRODUCTION OF PLANE-POLARIZED LIGHT BY USING POLAROID

Polaroid is a material which transmits only those components, of any light which is incident on it, which are in a particular direction – we shall call this the 'reference' direction.

Consider a beam of unpolarized light incident on two sheets of Polaroid (P and Q) whose reference directions (indicated by ↑) are parallel to each other (Fig. 27.5). From what has been said in section 27.2, the 'vibrations' of the unpolarized light which is incident on P can be regarded as being in one or other of any two perpendicular directions which are at right angles to the direction of propagation. For convenience, we shall take one of these directions to be in the plane of the

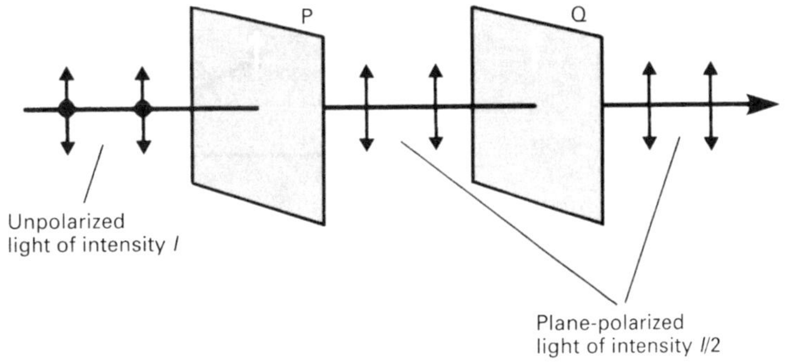

Fig. 27.5
Action of Polaroid on unpolarised light

paper, i.e. parallel to the reference directions of the Polaroids. (The 'vibrations' which are in this direction are indicated by ↕ in Fig. 27.5.) The direction of the other 'vibrations' must, therefore, be perpendicular to the plane of the paper, i.e. at right angles to the reference directions of the Polaroids. (These vibrations are represented by • in Fig. 27.5.)

The 'vibrations' which are perpendicular to the plane of the paper are completely absorbed by P. Therefore the beam that emerges from P has only half the intensity of the original (unpolarized) beam, and all of its 'vibrations' are in the plane of the paper. Thus, the beam that is incident on Q is plane-polarized in the plane of the paper, and when Q is in the orientation shown it transmits all of this beam. However, if Q is slowly rotated about the direction of propagation, the intensity of the light that emerges from it decreases and becomes zero when the reference direction of Q is perpendicular to that of P. (It can be shown that the intensity is proportional to $\cos^2\theta$, where θ is the angle between the reference directions of P and Q.)

The beam that emerges from P has half the intensity of the original beam no matter what the orientation of P. Rotating P about the direction of propagation does, however, alter the plane of polarization of the light that emerges from it. The effect of such a rotation on the intensity of the light that emerges from Q is therefore the same as that produced by rotating Q.

Polaroid is the trade name of sheets of nitrocellulose in which are embedded crystals of quinine iodosulphate. These crystals are **dichroic**, i.e. they transmit only those components of the 'vibrations' of light which are at a particular orientation to their axes. (This is called **selective absorption**.) The ability of Polaroid to exhibit the same property, lies in the fact that all the crystal axes are aligned within it. Dichroism is also exhibited by **tourmaline**—a naturally occurring crystalline material.

27.4 PRODUCTION OF PLANE-POLARIZED LIGHT BY REFLECTION

If an unpolarized beam of light is incident on a glass surface at an angle of about 57°, the light that is reflected from the surface is plane-polarized. (This can be checked by looking through a piece of Polaroid at light reflected in this way. If the Polaroid is slowly rotated about the line of vision, the intensity of the light reaching the eye varies from some maximum value to zero.) At angles of incidence other than 57°, the reflected light is partially plane-polarized.

When light is incident on glass at 57° the reflected ray is at right angles to the refracted ray. The significance of this will now be explained.

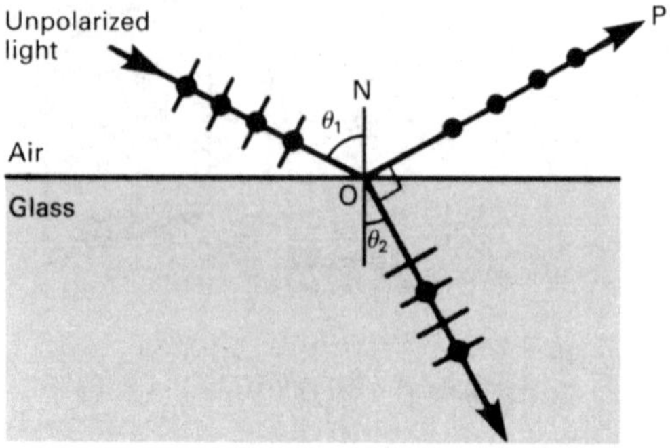

Fig. 27.6
Polarization by reflection

Suppose that unpolarized light is incident on a glass surface at such an angle that the reflected ray is at 90° to the refracted ray (Fig. 27.6). When the light enters the surface it causes the electrons in the surface to oscillate. The electrons radiate light as a result of these oscillations, and are the source of the reflected and refracted rays. It is as if all the light is refracted initially, and then some of it is immediately emitted as the reflected ray. Those 'vibrations' of the refracted ray which are in the plane of the paper cannot be emitted because they are parallel to the direction of propagation of the reflected ray. The reflected ray can therefore contain only those 'vibrations' which are perpendicular to the plane of the paper, i.e. it is plane-polarized. The refracted ray is partially plane-polarized, i.e. it contains more 'vibrations' which are in the plane of the paper than which are perpendicular to the plane of the paper.

When the refracted ray is not at 90° to the reflected ray, the 'vibrations' which are in the plane of the paper each have a component which is perpendicular to the direction of propagation of the reflected ray, and can, therefore, contribute to it and produce a reflected ray which is only partially plane-polarized.

If we take n to be the refractive index of glass, and apply Snell's law to Fig. 27.6, we obtain

$$\frac{\sin\theta_1}{\sin\theta_2} = n \qquad [27.1]$$

OP is the reflected ray and therefore $\mathrm{N\hat{O}P} = \theta_1$, in which case

$$\theta_2 = 90^\circ - \theta_1$$

Substituting for θ_2 in equation [27.1], we have

$$\frac{\sin\theta_1}{\sin(90^\circ - \theta_1)} = n$$

i.e. $$\frac{\sin\theta_1}{\cos\theta_1} = n$$

i.e. $$\boldsymbol{\tan\theta_1 = n} \qquad [27.2]$$

Equation [27.2] is known as **Brewster's law**, and θ_1, the angle of incidence at which the reflected ray is completely plane-polarized, is called the **angle of polarization** or **polarizing angle**. If the refractive index of glass is taken to be 1.54, equation [27.2] gives $\theta_1 = 57^\circ$.

27.5 DOUBLE REFRACTION

If an unpolarized ray is incident on a crystal of calcite (Iceland spar), it is split into two rays. These are known as the **ordinary ray** (the O-ray) and the **extraordinary ray** (the E-ray). **Each ray is plane-polarized in a direction which is perpendicular to that of the other.** The O-ray obeys the normal laws of refraction (hence the name); the E-ray does not. If an object is viewed through a crystal of calcite, two images are seen (Fig. 27.7). The phenomenon is known as **double refraction**.

The Nicol prism (Fig. 27.8) is a device which makes use of double refraction to produce an effect which is the same as that produced by Polaroid and tourmaline. Canada balsam (the transparent cement that holds the two pieces of calcite

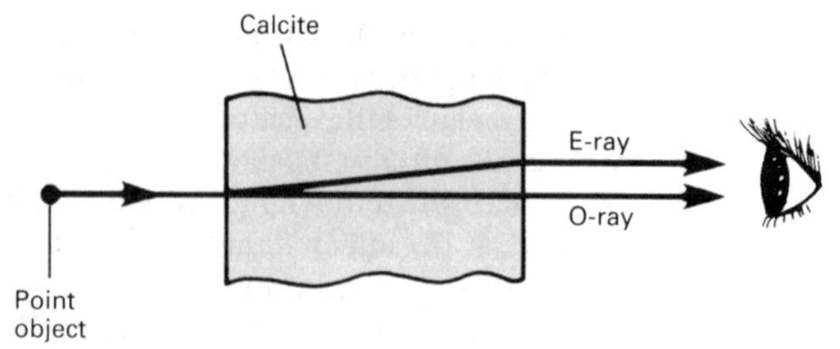

Fig. 27.7
Double refraction in calcite

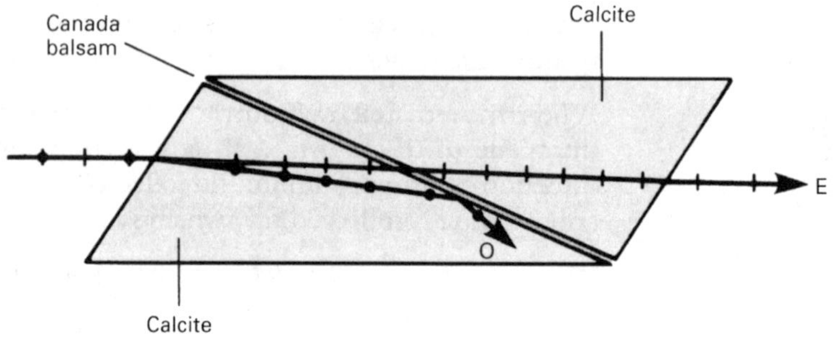

Fig. 27.8
Nicol prism

together) has a refractive index which makes it optically less dense than calcite as far the O-ray is concerned, but optically more dense than calcite for the E-ray. The crystal faces are angled in such a way that the O-ray undergoes total internal reflection at the boundary with the balsam. The E-ray passes straight through, providing a beam of plane-polarized light.

27.6 SCATTERING

If a beam of unpolarized light passes through a region which contains small particles (e.g. smoke particles or water droplets in the air, or a colloidal suspension of particles in a liquid), the intensity of the 'straight through' beam is reduced as a result of the scattering produced by the particles (**Tyndall effect**). The light that passes straight through is unpolarized, but the rest is plane-polarized to an extent which depends on the angle through which it has been scattered. Light scattered at 90° to the direction of incidence is completely plane-polarized.

Light is scattered out of the direct beam of the Sun by air molecules. Short wavelengths are scattered more effectively than long wavelengths and therefore the sky is blue.

27.7 SUNGLASSES

As explained in section 27.4, reflected light is plane-polarized to some extent. The glare (i.e. reflected light) from a wet road, for example, can be reduced by using Polaroid sunglasses, i.e. sunglasses which contain suitably oriented Polaroid.

27.8 STRESS ANALYSIS

If two sheets of Polaroid are placed one in front of the other, light cannot pass through the combination if their reference directions are at right angles to each other. If, however, a piece of glass which is under stress is placed between the Polaroids (Fig. 27.9), a pattern of interference fringes can be seen. Glass becomes doubly refracting when it is under stress and interference between the E-ray and the O-ray thus produced, produces a ray which is no longer plane-polarized at right angles to the analyser. The light intensity at any point in the field of view depends on the thickness of the glass and the amount of strain in the corresponding region of the specimen. The device is called a **strain viewer** and is used to detect regions of high strain which would be liable to failure.

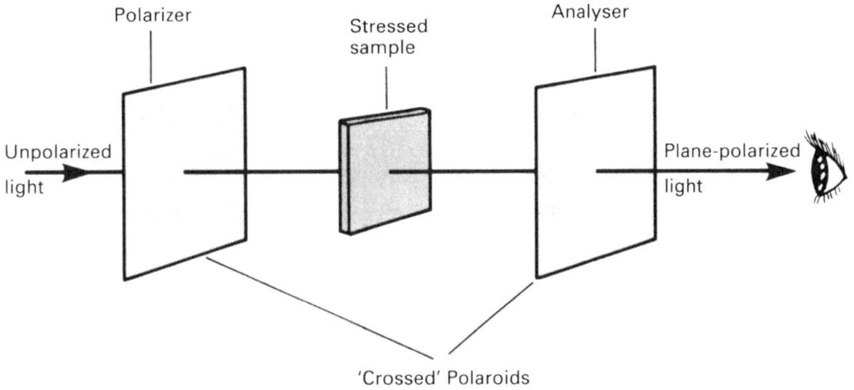

Fig. 27.9 Strain viewer

Many plastic materials also become doubly refracting when under stress. By making plastic models of bridges, etc. and examining them in a strain viewer, engineers are able to detect potential weak points.

CONSOLIDATION

Longitudinal waves are waves in which the vibrations are along the direction of travel of the wave.

Transverse waves are waves in which the vibrations are perpendicular to the direction of travel of the wave.

Transverse waves can be polarized; longitudinal waves cannot.

The vibrations of an unpolarized light wave actually take place in every direction which is perpendicular to the direction of propagation, but it can be regarded as having vibrations only in two perpendicular directions and such that the amplitude of vibration is the same in each.

Double refraction – the plane of polarization of the E-ray is at 90° to that of the O-ray.

28

ELECTROMAGNETIC WAVES. OPTICAL SPECTRA

28.1 THE ELECTROMAGNETIC SPECTRUM

The main divisions of the electromagnetic spectrum are shown in Fig. 28.1. Table 28.1 lists some of the properties of the various radiations and gives examples of how they are generated. Properties which are common to all types of electromagnetic waves are listed below and on p. 473 as (i) to (vi).

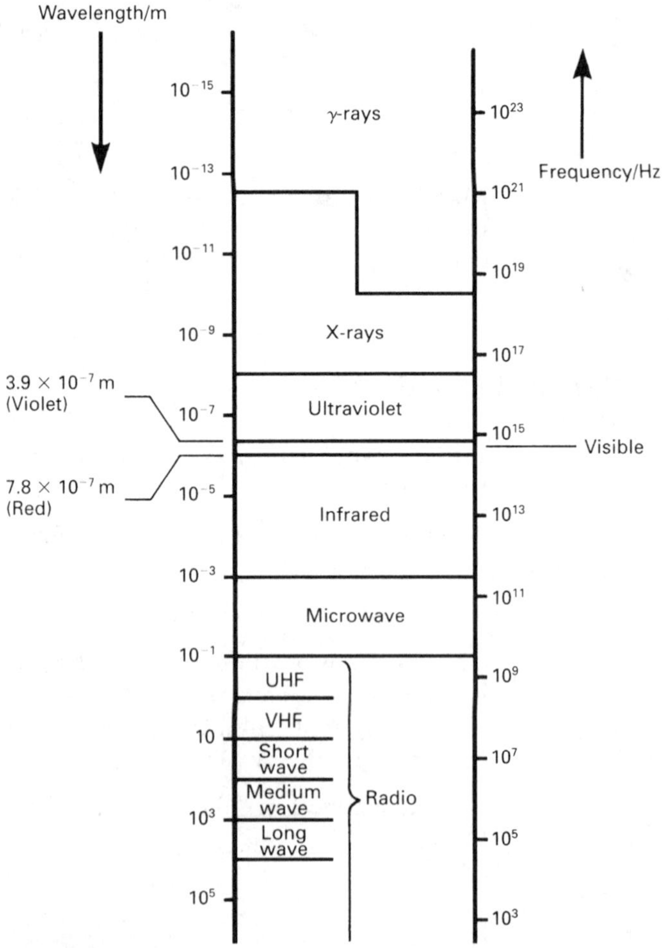

Fig. 28.1
The electromagnetic spectrum

(i) Electromagnetic waves consist of varying electric and magnetic fields. The two fields are perpendicular to each other and to the direction of travel of the wave. Each field vibrates at the same frequency, the frequency of the wave.

(ii) All electromagnetic waves travel at the same speed in vacuum, $2.998 \times 10^8\ \mathrm{m\,s^{-1}}$.

(iii) Electromagnetic waves are unaffected by electric and magnetic fields.

(iv) Electromagnetic waves travel in straight lines (within the limits set by diffraction).

(v) Since electromagnetic waves are transverse, they are capable of being polarized.

(vi) Electromagnetic waves can be caused to produce interference effects.

Table 28.1 Generation, properties and uses of electromagnetic radiation

Type	*Examples of generation*	*Main properties and uses*
γ-rays	Radioactive decay Nuclear fission and fusion reactions Interactions between elementary particles	See section 52.4
X-rays	Rapid deceleration of fast electrons (e.g. in X-ray tubes) Atomic transitions involving innermost electron orbits (e.g. in X-ray tubes)	See section 49.3
Ultraviolet	Atomic transitions (e.g. in carbon arc, mercury vapour lamp, the Sun)	Produces ionization and fluorescence Promotes chemical reactions Affects photographic film Produces photoelectric effect Absorbed by glass
Visible	Atomic transitions (e.g. in discharge tubes, incandescent lamps, lasers, flames)	Stimulates the retina Affects photographic film Initiates photosynthesis
Infrared	Atomic transitions Molecular vibrations	Produces heating Used in 'night sights'
Microwaves	Klystrons and Magnetrons Masers	Radar Telemetry Electron spin resonance
Radio waves	Electrical oscillations	Radio communication

Table 28.2 lists means of detecting the various radiations.

Table 28.2 Detection of electromagnetic radiation

Type	*Detection*
γ and X	Geiger-Müller tubes Ionization chambers Solid state detectors Scintillation counters Photographic film
Ultraviolet	Photoelectric cell Fluorescent materials Photographic film
Visible	Photoelectric cell The eye Photographic film
Infrared	Thermopile Special photographic film
Microwave	Crystal detectors (silicon and germanium)
Radio	Radio receivers

28.2 TYPES OF SPECTRA

Spectra are of two basic types: **emission** spectra and **absorption** spectra. An emission spectrum is the spectrum of radiations emitted by an object which is acting as a source of radiation. An absorption spectrum is the spectrum of the radiations transmitted by a substance which has absorbed some of the radiation incident on it.

Both emission spectra and absorption spectra can be further classified as being line spectra, band spectra or continuous spectra.

28.3 OPTICAL EMISSION SPECTRA

Optical emission spectra are produced when light from a luminous source is incident on a prism or a diffraction grating. The diffraction grating is normally preferred because:

(i) its ability to separate wavelengths which are very close together is better than that of the prism;

(ii) a prism distorts the spectrum by crowding together the wavelengths at the red end;

(iii) problems due to absorption by glass can be avoided by using reflection gratings.

A disadvantage of diffraction gratings is that they spread the energy of the incident light over the various orders of the spectra which they produce and thereby reduce the intensity of any one of them.

Line Spectra

A typical line spectrum (that of sodium vapour) is shown in Fig. 28.2(a). The bright lines are images of the slit of the spectrometer used to obtain the spectrum. Line spectra are produced by atoms which are so far apart that they are not interacting with each other. It follows that line spectra can be produced only by low-density monatomic gases and vapours. Sources which give rise to line spectra include low-pressure discharge tubes (e.g. sodium lamps), flames, gaseous nebulae (such as that in Orion) and the tails of comets.

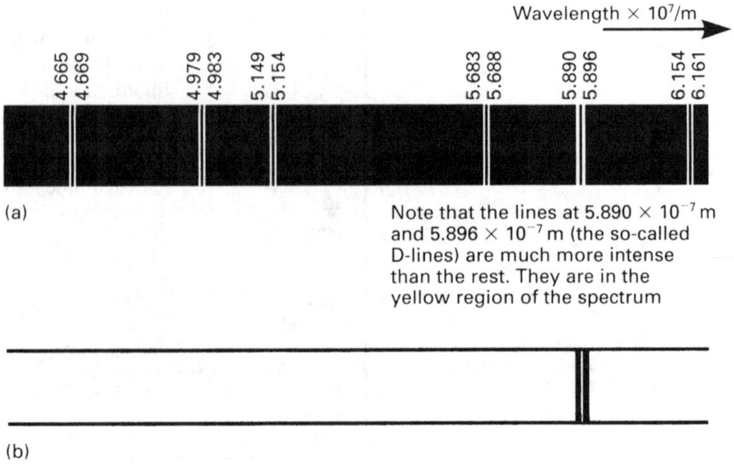

Fig. 28.2 Sodium vapour line spectra: (a) emission, (b) absorption

The electrons of an isolated atom can be considered to move in definite orbits about the nucleus. If an atom is involved in a violent collision, as it may be, for example, in a discharge tube or a hot gas, an electron may gain energy and move from its normal orbit to one which is farther from the nucleus. Such an atom is unstable and the electron soon returns to its original orbit. In doing so it gives up the energy it gained in the collision by emitting light. The wavelength of the emitted light depends on the two orbits concerned. When large numbers of atoms are involved, light of a number of different wavelengths may be produced, corresponding to electron transitions between different pairs of orbits. The various wavelengths constitute the line spectrum.

All atoms of any particular element have the same set of orbits and these are characteristic of the element concerned. It follows that **each element produces a unique line spectrum which may be used to identify the element**.

(The reader is advised to study sections 48.3 to 48.6 in order to gain a more complete understanding of the origins of line spectra.)

Band Spectra

Band spectra are composed of separate groups of lines known as bands; the lines within each band are closer at one side than the other (see Fig. 28.3). They are produced by gases and vapours whose molecules contain more than one atom (e.g. O_2, CO). The bands produced by heavy molecules are close together; those of light molecules are widely spaced. The bands in the spectrum of molecular hydrogen (H_2) are so widely spaced that the spectrum has the appearance of a line spectrum.

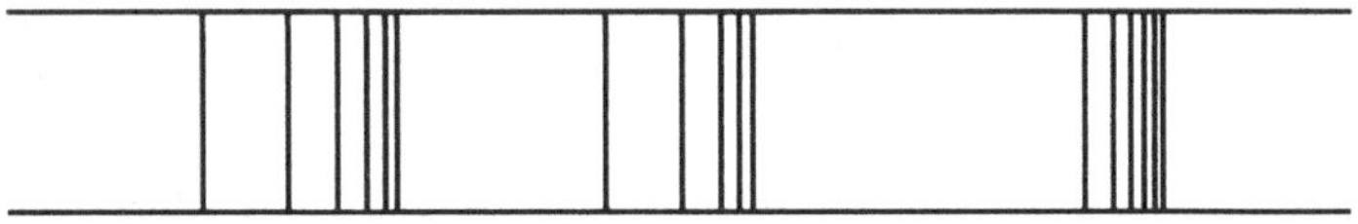

Fig. 28.3
Band spectra

Continuous Spectra

Continuous spectra are produced by hot solids and liquids, and by high-density gases such as the Sun. As the name suggests, spectra of this type consist of a continuous range of wavelengths. The atoms are so close together that they interact with each other. As a consequence some of the electrons have a continuous range of energies and the transitions which they undergo give rise to radiation of all wavelengths.

A continuous spectrum is not characteristic of the substance which produces it. Instead, the relative amounts of energy radiated at the various wavelengths are determined by the temperature of the emitter and the nature of its surface (see section 17.10).

28.4 OPTICAL ABSORPTION SPECTRA

Absorption spectra, like emission spectra, can be of all three types: line, band and continuous. The absorption spectrum which a substance produces is of the same type as its emission spectrum.

Suppose that white light which has passed through a cell containing sodium vapour is examined with a spectrometer. The spectrometer reveals a continuous spectrum from which certain wavelengths are missing (Fig. 28.2(b)). The missing wavelengths have been absorbed by the vapour, for the light of these wavelengths is of exactly the right energy to induce electron transitions within the sodium atoms. These transitions take place in the opposite direction to those which produce the emission spectrum. It is not surprising, therefore, to find that the dark lines in the absorption spectrum occur at the same wavelengths as the bright lines in the emission spectrum.

The radiation which the atoms have absorbed is emitted soon afterwards. The reader may now be at a loss to understand why there are any dark lines at all. There is no problem – the emission takes place in all directions and therefore very little can be emitted in the 'straight through' direction being examined with the spectrometer.

There are fewer lines in the absorption spectrum than the emission spectrum (Fig. 28.2). This is because we have supposed that the sodium vapour which has produced the absorption spectrum is colder than that which produced the emission spectrum. The electron transitions which give rise to the absorption must, therefore, each involve a low-lying energy level and so there are fewer possible transitions. (**Note**. A glowing gas can produce an absorption spectrum provided that the light it emits is less intense than that which it is to absorb.)

28.5 THE SOLAR SPECTRUM. FRAUNHOFER LINES

Careful examination of the Sun's spectrum reveals that it is crossed by a large number of dark lines. The lines were discovered by Fraunhofer in 1814 and, accordingly, are known as Fraunhofer lines. Most of the lines are formed by the relatively cool gases in the outer regions of the Sun absorbing certain of the wavelengths radiated by the interior.

The discovery of helium was due to the realization that some of the lines could not be attributed to any element known at the time.

The majority of the Fraunhofer lines exhibit Doppler broadening (see Chapter 35). The broadening can be accounted for as being due to the Sun's rotation, providing that the lines originate on the Sun. The remaining lines are the result of absorption by the Earth's atmosphere.

CONSOLIDATION

Line spectra are produced by low-density monatomic gases and vapours.

Band spectra are produced by gases and vapours whose molecules contain more than one atom.

Continuous spectra are produced by hot solids and liquids, and by high-density gases such as the Sun.

Line spectra and band spectra are characteristic of the elements which produce them; continuous spectra are not.

An absorption spectrum contains fewer lines than the corresponding emission spectrum.

29

FORCED VIBRATIONS AND RESONANCE

29.1 DAMPING

Unless it is maintained by some source of energy, the amplitude of vibration of any oscillatory motion becomes progressively smaller – the motion is said to be **damped** (see Fig. 29.1). The decrease in amplitude occurs because some of the energy of the oscillating system is used to overcome resistive forces. For example, the amplitude of vibration of a simple pendulum decreases because of air resistance and friction at the support. This is an example of natural damping. Many systems are artificially damped to cut down unwanted vibration – the shock absorbers on a car serve this purpose. (Electromagnetic damping and critical damping are discussed in section 42.12.)

Fig. 29.1
Damped oscillation

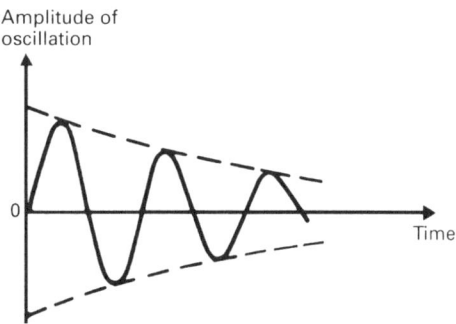

29.2 BARTON'S PENDULUMS

Refer to Fig. 29.2. Pendulums P, Q, R and S, though much lighter than A, are sufficiently massive for their motion to be considered undamped. (The significance of this will emerge on reading section 29.3.) Suppose that the heavy pendulum (A) is displaced so that it oscillates, at its natural frequency, in a plane which is perpendicular to the paper. Its vibrations are transferred through the common support string (XY) to the other pendulums and they start to oscillate. Since these pendulums are being forced to oscillate by A, they are said to be executing **forced oscillations**. Once the motion has settled down, observation shows that:

(i) All the pendulums vibrate with the same frequency (the natural frequency of A).

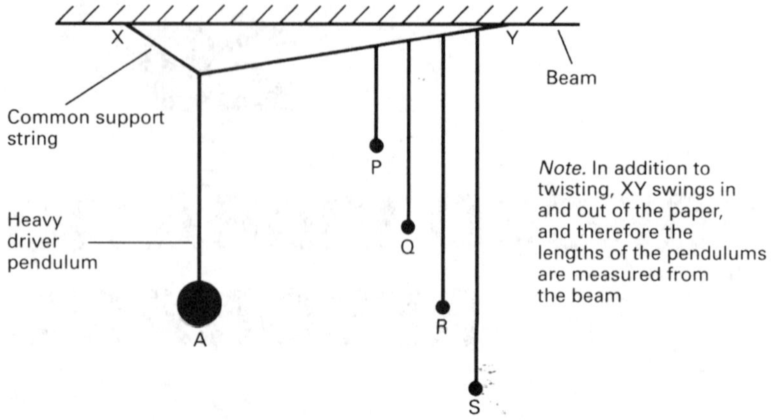

Fig. 29.2
Barton's pendulums

(ii) Pendulum R (whose natural frequency is equal to the forcing frequency because it has the same length as A) oscillates with greater amplitude than P, Q and S. R is said to be **resonating** with A.

(iii) The motion of R is a quarter of a period behind that of A. The shorter pendulums, P and Q, are approximately in phase with A. Pendulum S is almost half a period behind A.

29.3 A DETAILED INVESTIGATION OF FORCED VIBRATIONS

Barton's pendulums provide a useful demonstration experiment; the apparatus shown in Fig. 29.3 allows a more complete investigation of forced vibrations to be made. By using driver pendulums of different lengths, N is forced to vibrate at a number of different frequencies.

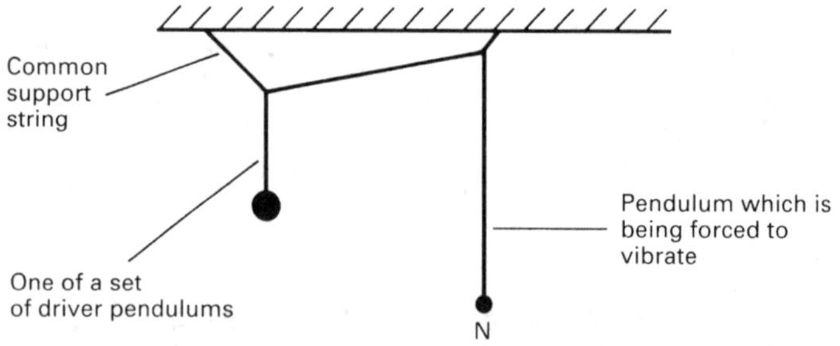

Fig. 29.3
Apparatus to investigate forced oscillations

As might be expected from what has been said in section 29.2, observation shows that:

(i) N vibrates at the same frequency as the driver pendulum, no matter which driver is being used;

(ii) the amplitude of vibration of N depends on the frequency of the driver pendulum;

(iii) the phase difference between N and the driver pendulum depends on which driver pendulum is being used.

Fig. 29.4 illustrates these results more completely, and shows the consequences of increasing the effects of damping on N. (This is achieved by replacing the bob of N

Fig. 29.4
Effect of damping on forced vibrations

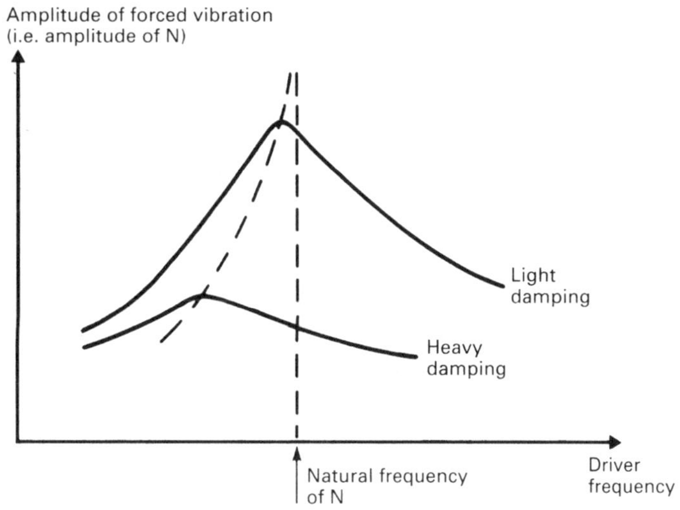

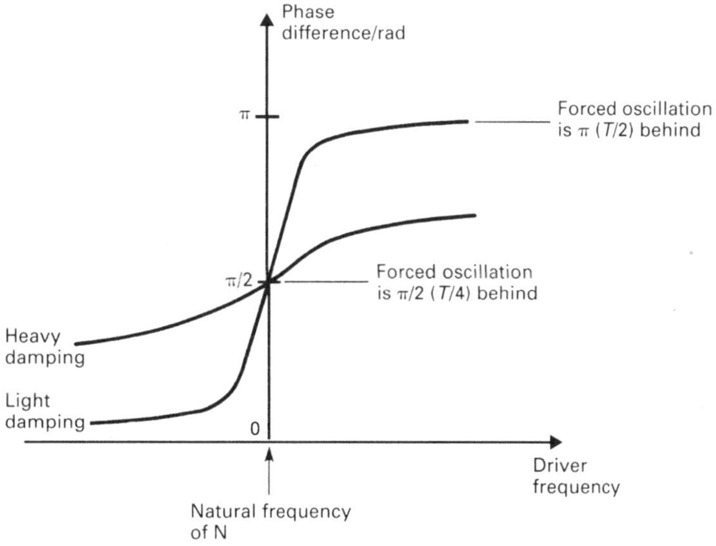

with a less massive one, or by attaching a piece of card so as to increase the air resistance.) Note, in particular, that when N is heavily damped the system resonates at a frequency which is somewhat less than the natural frequency of N.

If a periodic force is applied to any mechanical system, it will be set into forced vibration. If the driving force is simple harmonic, and the natural mode of oscillation of the system is also simple harmonic, the results of Fig. 29.4 apply.

29.4 SOME CONSEQUENCES OF RESONANCE

(i) Soldiers need to break step when crossing bridges. (Failure to do so caused the loss of over two hundred French infantrymen in 1850.)

(ii) Opera singers can shatter wine glasses by forcing them to vibrate at their natural frequencies.

(iii) A diver on a springboard builds up the amplitude of oscillation of the board by 'bouncing' on it at its natural frequency.

(iv) If a loose part in a car rattles when the car is travelling at a certain speed, it is likely that a resonant vibration is occurring.

(v) A column of air can be made to resonate to a particular note (see Chapter 33).

(vi) Electrical resonance is made use of to tune radio circuits.

(vii) Resonant vibrations of quartz crystals are used to control clocks and watches.

CONSOLIDATION

A system is undergoing **forced oscillations** if it is being forced to oscillate by some other system at the frequency of that system.

When a system is being forced to oscillate, but at its own natural frequency it is said to be **resonating** with the forcing system.

30

BEATS

30.1 THE PHENOMENON OF BEATS

If two notes which have nearly equal frequencies and similar amplitudes are sounded at the same time, then someone listening to them hears a single note which has the mean frequency of the original notes. The amplitude of this note repeatedly rises and falls (to zero if the amplitudes of the original notes are equal). This periodic increase and decrease in amplitude is a result of successive occurrences of constructive and destructive interference between the two notes, as they repeatedly become in phase and then out of phase with each other (see Fig. 30.1). The variations in amplitude (and intensity) are called **beats**, and can occur when two waves of any type are superposed. The number of times the sound reaches maximum intensity* in one second is called the **beat frequency**. Intensity maxima occur whenever the two waves are in phase with each other.

Fig. 30.1
Two waves producing beats

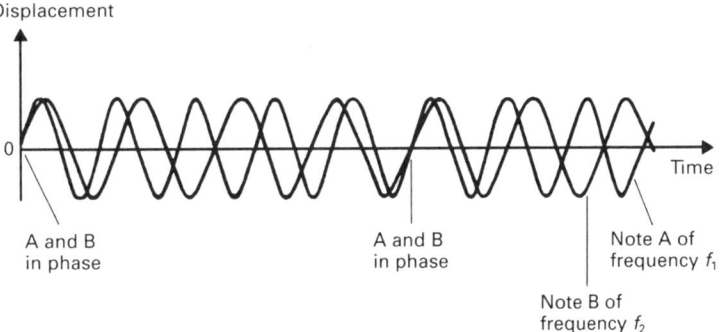

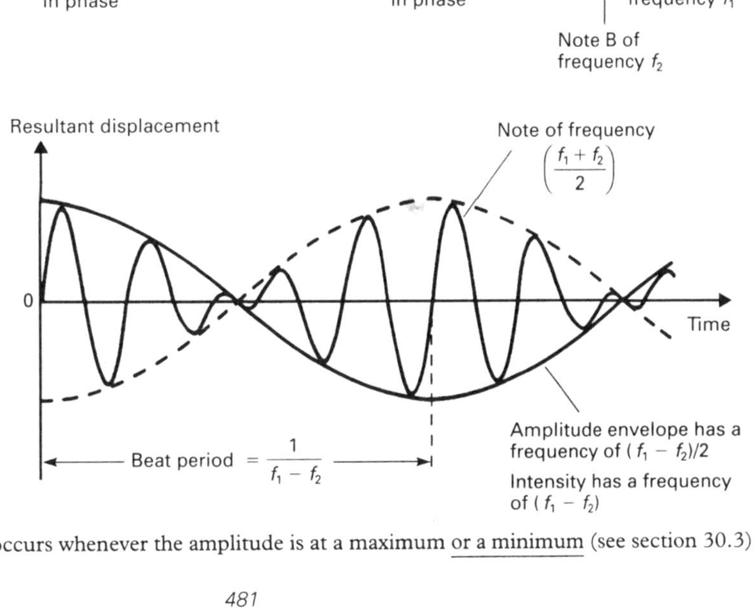

*This occurs whenever the amplitude is at a maximum or a minimum (see section 30.3).

Derivation of Beat Frequency $= f_1 - f_2$

Consider two notes, A and B, whose frequencies are f_1 and f_2 respectively. In some time t, A completes $f_1 t$ cycles and B completes $f_2 t$ cycles.

If t is such that A completes one more cycle than B, then

$$f_1 t - f_2 t = 1$$

i.e. $$t = \frac{1}{f_1 - f_2}$$

If the two notes are initially in phase, then (because t is the interval in which A completes one more cycle than B) t is the time that elapses before they are next in phase with each other. Thus, the **beat period** (i.e. the interval between successive in-phase situations) is given by

$$\text{Beat period} = \frac{1}{f_1 - f_2}$$

i.e. $$\text{Beat frequency} = f_1 - f_2 \quad (f_1 > f_2) \tag{30.1}$$

30.2 MEASURING AN UNKNOWN FREQUENCY

The phenomenon of beats can be used to determine the unknown frequency of some wave motion by causing the wave to beat with a wave of the same kind whose frequency is known.

Suppose that a note of unknown frequency, f_1, is made to produce beats with a note of known frequency, f_2. If f_2 is not very different from f_1, it is possible to count the number of beats that occur in some given time, and hence determine the beat frequency, f. Bearing in mind that we do not know which of f_1 and f_2 is the higher frequency, we obtain from equation [30.1]

$$f = f_1 - f_2 \quad \text{or} \quad f = f_2 - f_1$$

i.e. $$f_1 = f_2 \pm f \tag{30.2}$$

In order to discover which of f_1 and f_2 is the higher frequency (and therefore whether to make use of + or − in equation [30.2]), one of the frequencies, f_2 say, is changed slightly and the effect on the beat frequency is noted. For example, if the note of frequency f_2 is being produced by a tuning fork, its frequency could be reduced by loading one of the prongs of the fork with a small piece of plasticine. If this results in an increased beat frequency, f_1 must be higher than f_2, and therefore f_1 is found from $f_1 = f_2 + f$.

If a signal generator is available, it can be used as the source of the reference frequency, f_2, and has the advantage of being able to provide a continuously variable range of frequencies. It is possible to adjust the signal generator so that the beats are so far apart that they can no longer be heard. Suppose that an experimenter, who is capable of just discerning beats which are as much as 10 s apart, finds that he cannot hear beats when the signal generator frequency, f_2, is 1000 Hz. The beat frequency, f, must be less than 0.1 Hz (since he can detect beats up to 10 s apart), and therefore

$$f_1 = 1000 \pm 0.1\,\text{Hz}$$

Thus, the unknown frequency, f_1, can be determined to within 1 part in 10 000 – an extremely accurate measurement. (This assumes, of course, that the calibration of the signal generator is itself accurate.)

QUESTIONS 30A

1. When a tuning fork with a frequency of 512 Hz and a guitar string are sounded together beats can be heard with a frequency of 3 Hz. When the guitar string is tightened slightly (which increases its frequency) the beat frequency increases to 4 Hz. What was the original frequency of the string?

2. A black disc with a white spot on it rotates clockwise at 36 revolutions per second. When the disc is observed in light from a stroboscope whose frequency is close to the rotational frequency of the disc, the white spot appears to rotate clockwise at 2 revolutions per second. What is the frequency of the stroboscope? (A little thought should convince the reader that this question is concerned with beats!)

30.3 MATHEMATICAL TREATMENT OF BEATS

Suppose that y_1 and y_2 are the individual displacements of two sinusoidal wave motions whose frequencies are f_1 and f_2 respectively. If the amplitude of each wave is a, then it can be shown that, at a single point, the displacements vary with time according to

$$y_1 = a\sin(2\pi f_1 t) \qquad [30.3]$$

and

$$y_2 = a\sin(2\pi f_2 t)^{\star}$$

If the two waves are superposed, the resultant displacement, y, is given by

$$y = y_1 + y_2$$

i.e.
$$y = a\sin(2\pi f_1 t) + a\sin(2\pi f_2 t)$$

It is a general result that

$$\sin A + \sin B = 2\sin\left(\frac{A+B}{2}\right)\cos\left(\frac{A-B}{2}\right)$$

Therefore

$$y = 2a\sin\left(\frac{2\pi f_1 t + 2\pi f_2 t}{2}\right)\cos\left(\frac{2\pi f_1 t - 2\pi f_2 t}{2}\right)$$

i.e.
$$y = 2a\sin\left[2\pi\left(\frac{f_1+f_2}{2}\right)t\right]\cos\left[2\pi\left(\frac{f_1-f_2}{2}\right)t\right]$$

This can be rewritten as

$$y = A\sin\left[2\pi\left(\frac{f_1+f_2}{2}\right)t\right] \qquad [30.4]$$

where

$$A = 2a\cos\left[2\pi\left(\frac{f_1-f_2}{2}\right)t\right] \qquad [30.5]$$

⋆This assumes that the waves are in phase at $t = 0$, and simplifies the mathematics without invalidating the result.

Comparing equation [30.4] with equation [30.3] reveals that the resultant is a wave of frequency $(f_1 + f_2)/2$ whose amplitude is variable and is given by equation [30.5].* The intensity of a wave motion is proportional to the square of its amplitude, and therefore

$$\text{Intensity} \propto 4a^2 \cos^2 \left[2\pi\left(\frac{f_1 - f_2}{2}\right)t\right]$$

It is a general result that

$$2\cos^2\theta = 1 + \cos 2\theta$$

Therefore

$$\text{Intensity} \propto 2a^2\{1 + \cos[2\pi(f_1 - f_2)t]\}$$

A little consideration of this equation reveals that the intensity varies at a frequency of $f_1 - f_2$, i.e. the beat frequency is $f_1 - f_2$. Note that equation [30.5] gives the frequency of the amplitude variation as $(f_1 - f_2)/2$, i.e. the amplitude varies at half the rate of the intensity. This is because the intensity is a maximum when the amplitude has its maximum value ($2a$) and when it has its minimum value ($-2a$).

CONSOLIDATION

Beats are formed when two notes of nearly equal frequencies and similar amplitudes are sounded together.

The beat frequency f of two frequencies f_1 and f_2 is given by

$$f = f_1 - f_2 \qquad (f_1 > f_2)$$

*The cosine term is a slowly varying term, and it is this which makes it possible for it to be regarded as the amplitude.

31

STATIONARY (STANDING) WAVES

31.1 INTRODUCTION

In this chapter we shall concentrate on the formation and properties of stationary waves in general. The particularly important cases of stationary waves in strings and in columns of air are treated in Chapters 32 and 33.

31.2 THE FORMATION AND PROPERTIES OF STATIONARY WAVES

A stationary (standing) wave results when two waves which are travelling in opposite directions, and which have the same speed and frequency and approximately equal amplitudes, are superposed.

The superposition of two such waves results in points where the displacement is always zero – these points are called **nodes**. Mid-way between the nodes are points where the maximum displacements (i.e. the amplitudes of vibration) are greater than they are anywhere else – these points are called **antinodes**. The profile of a stationary wave does not travel and, though there is energy associated with the wave, energy does not pass along it.

Fig. 31.1 illustrates the formation of the stationary wave that results from the superposition of two (progressive) waves, A and B. Initially (i.e. at $t = 0$), A and B are in phase with each other and, in accordance with the principle of superposition (section 23.2), the resultant displacement is as shown in (a). One eighth of a period later (i.e. at $t = T/8$), A has moved to the left and B has moved to the right, so that their positions and their resultant are as shown in (b). At $t = T/4$, A and B are completely out of phase with each other and (momentarily) destroy each other at every point (c). The situations at $t = 3T/8$ and $t = T/2$ are shown in (d) and (e) respectively.

Fig. 31.2 shows how the resultant displacement changes between $t = 0$ and $t = T/2$. The displacement is always zero at N_1, N_2 and N_3, i.e. N_1, N_2 and N_3 are nodes. The point which is mid-way between N_1 and N_2 and that which is mid-way between N_2 and N_3 have larger amplitudes of vibration than the rest, i.e. these points are antinodes.

Fig. 31.1
Two waves producing a stationary wave

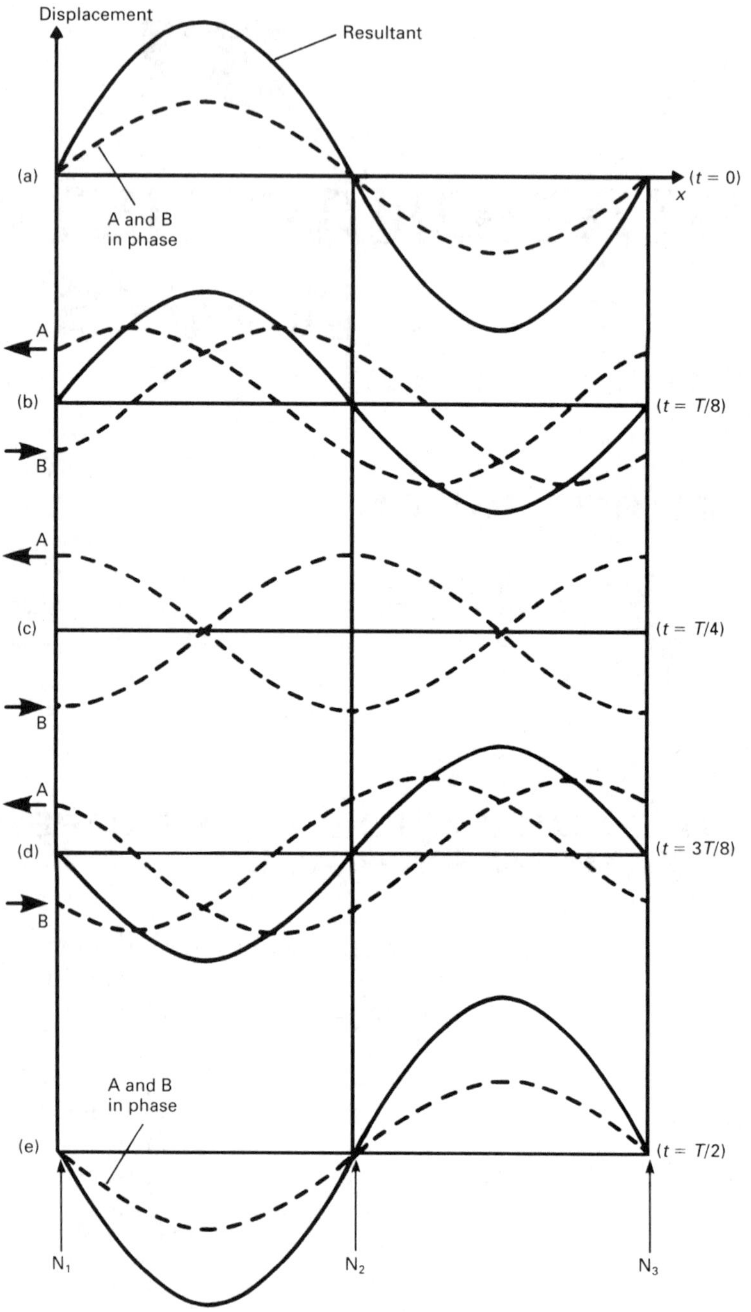

Notes

(i) $N_1N_2 = N_2N_3 = \lambda/2$, i.e. the separation of adjacent nodes $= \lambda/2$; and therefore the separation of adjacent antinodes $= \lambda/2$.

(ii) Unlike that of an (unattenuated) progressive wave, the amplitude of a stationary wave depends on position. It ranges from zero at the nodes to $2a$ at the antinodes, where a is the amplitude of either one of the progressive waves that have combined to produce the stationary wave.

(iii) At any one time, all the particles between two adjacent nodes are at the same phase of their motion. Each particle between the next pair of adjacent nodes is out of phase with these by π radians.

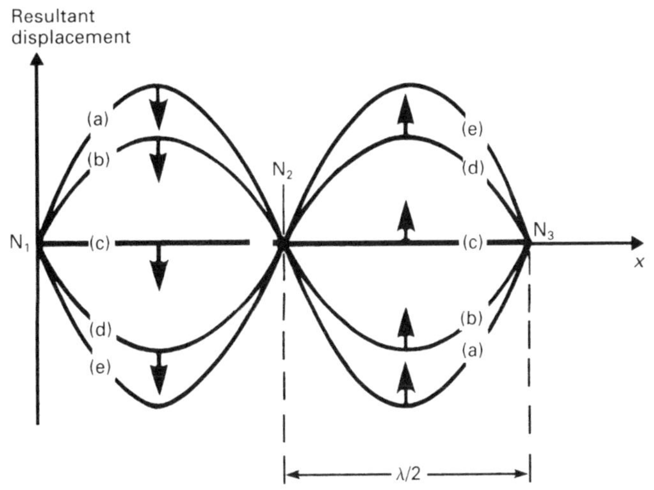

Fig. 31.2 Successive displacements of a stationary wave

31.3 MATHEMATICAL TREATMENT OF STATIONARY WAVES

The displacement y_1 of a (progressive) sinusoidal wave at time t and at a distance x from the origin is given by equation [23.4] as

$$y_1 = a \sin(2\pi f t - kx)$$

where

$$a = \text{the amplitude of the wave}$$

$$f = \text{the frequency of the wave}$$

$$k = 2\pi/\lambda \text{ where}$$

$$\lambda = \text{the wavelength of the wave.}$$

The displacement, y_2, of an identical wave travelling in the opposite direction is given by

$$y_2 = a \sin(2\pi f t + kx)$$

If these two waves are superposed, the resultant displacement, y, is given by

$$y = y_1 + y_2$$

i.e. $$y = a \sin(2\pi f t - kx) + a \sin(2\pi f t + kx)$$

It is a general result that

$$\sin(A - B) + \sin(A + B) = 2 \sin A \cos B$$

therefore

$$y = 2a \sin(2\pi f t) \cos(kx)$$

This can be rewritten as

$$y = A \sin(2\pi f t) \qquad [31.1]$$

where

$$A = 2a \cos(kx) \qquad [31.2]$$

Equation [31.1] represents a sinusoidal oscillation of frequency, f, and whose amplitude, A, depends on position and is given by equation [31.2]. Bearing in mind that $k = 2\pi/\lambda$, we see from equation [31.2] that:

(i) $A = 0$ when $x = \dfrac{\lambda}{4}, \dfrac{3\lambda}{4}, \dfrac{5\lambda}{4}, \cdots$

i.e. there are nodes at $x = \dfrac{\lambda}{4}, \dfrac{3\lambda}{4}, \dfrac{5\lambda}{4}, \cdots$

i.e. the separation of adjacent nodes is $\lambda/2$

(ii) $A = \pm 2a$ when $x = 0, \dfrac{\lambda}{2}, \lambda, \cdots$

i.e. there are antinodes at $x = 0, \dfrac{\lambda}{2}, \lambda, \cdots$

i.e. the separation of adjacent antinodes is $\lambda/2$.

It follows from equation [31.2] that the maximum displacement at an antinode is $2a$.

CONSOLIDATION

A stationary wave results when two waves which are travelling in opposite directions, and which have the same speed and frequency and approximately equal amplitudes are superposed.

A node is a point of zero displacement.

An antinode is a point of maximum displacement.

Separation of adjacent nodes $=$ Separation of adjacent antinodes $= \dfrac{\lambda}{2}$

All the particles between two adjacent nodes are in phase with each other and are completely out of phase with the particles between the next pair of adjacent nodes.

32

WAVES IN STRINGS

32.1 INTRODUCTION

If a transverse wave is caused to travel along a stretched string, the wave is reflected on reaching the ends of the string. The incident and reflected waves have the same speed, frequency and amplitude, and therefore their superposition results in a stationary wave.

If a stretched string is caused to vibrate by being plucked or struck, a number of different stationary waves are produced simultaneously. Only specific modes of vibration are possible, and these are considered in section 32.2.

32.2 THE MODES OF VIBRATION

The ends of a stretched string are fixed, and therefore the ends of the string must be displacement nodes. The three simplest modes of vibration which satisfy this condition in the case of a string of length L are shown in Fig. 32.1.

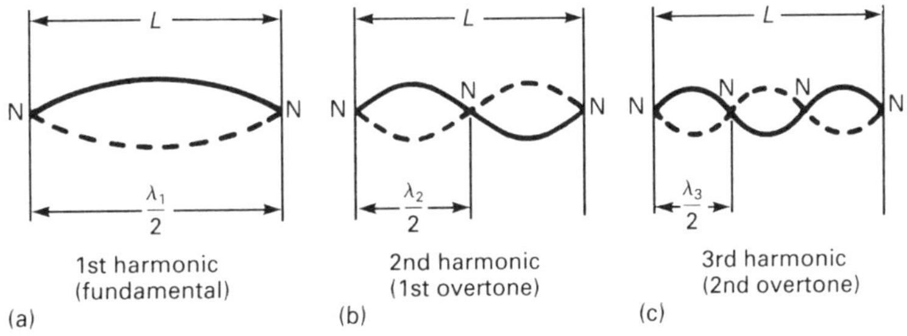

Fig. 32.1
Modes of vibration of a stretched string

The simplest mode of vibration (a) is called the **fundamental,** and the frequency at which it vibrates is called the **fundamental frequency**. The higher frequencies (e.g. (b) and (c)) are called **overtones**. (Note that the first overtone is the second harmonic, etc.) Representing the wavelengths of the first, second and third harmonics by λ_1, λ_2 and λ_3 respectively, and bearing in mind that the separation of adjacent nodes is equal to half a wavelength (section 31.2), we see from Fig. 32.1 that:

$$\frac{\lambda_1}{2} = L \qquad \frac{\lambda_2}{2} = \frac{L}{2} \qquad \frac{\lambda_3}{2} = \frac{L}{3}$$

Therefore, if λ_n is the wavelength of the nth harmonic,

$$\frac{\lambda_n}{2} = \frac{L}{n} \qquad [32.1]$$

The frequency, f_n, of the nth harmonic is given by equation [23.1] as

$$f_n = \frac{v}{\lambda_n} \qquad [32.2]$$

where

v = the velocity of either one of the progressive waves that have produced the stationary wave. (Note that the velocity is the same for all wavelengths.)

Therefore, from equations [32.1] and [32.2]

$$f_n = \frac{nv}{2L} \qquad [32.3]$$

The frequency, f_1, of the fundamental (i.e. the first harmonic) is given, by putting $n = 1$ in equation [32.3], as

$$f_1 = \frac{v}{2L}$$

Therefore, equation [32.3] can be rewritten as

$$f_n = nf_1$$

i.e. **the frequencies of the various overtones are whole-number multiples of the fundamental frequency.**

It can be shown that

$$v = \sqrt{\frac{T}{\mu}} \qquad [32.4]$$

where

T = the tension in the string (N)

μ = the **mass per unit length** of the string ($\mathrm{kg\,m^{-1}}$).

Therefore, by equations [32.2] and [32.4]

$$f_n = \frac{n}{2L}\sqrt{\frac{T}{\mu}} \qquad (n = 1, 2, 3, \ldots) \qquad [32.5]$$

QUESTIONS 32A

1. A wire of length 400 mm and mass 1.20×10^{-3} kg is under a tension of 120 N. What is: **(a)** the fundamental frequency of vibration, **(b)** the frequency of the third harmonic?

2. The fundamental frequency of vibration of a particular string is f. What would the fundamental frequency be if the length of the string were to be halved and the tension in it were to be increased by a factor of 4?

32.3 STRINGED INSTRUMENTS

When a guitar string is plucked or a piano string is struck, transverse waves travel along the string and are reflected on reaching its ends. The energy of any wave whose wavelength is such that it does not give rise to one of the allowed stationary waves is very quickly dissipated. The waves which remain have frequencies that are given by equation [32.5], and the string vibrates with all these frequencies simultaneously.

The largest amplitude of vibration, and therefore the predominant frequency, is that of the fundamental. The relative amplitudes of the various overtones depend on the particular instrument being played, and it is this that gives an instrument its characteristic sound (see section 34.2).

32.4 MELDE'S EXPERIMENT

If a string is caused to vibrate by being plucked or struck, it vibrates freely at all of its natural frequencies (i.e. the frequencies given by equation [32.5]). On the other hand, if a string is forced to vibrate at some particular frequency, it will vibrate with large amplitude only if the forcing frequency is one of the natural frequencies of the string. This can be very effectively demonstrated by the apparatus shown in Fig. 32.2, and is known as Melde's experiment.

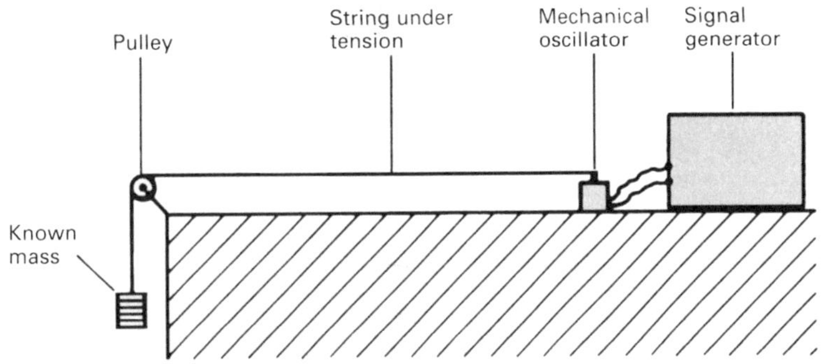

Fig. 32.2 Apparatus for Melde's experiment

The frequency of the signal generator is slowly increased and, at first, very little happens. Eventually though, a frequency f_1 (say) is reached at which the string vibrates with large amplitude in the form of a single loop (Fig. 32.3(a)). If the frequency is increased beyond this value, the amplitude of the vibrations dies away. When the forcing frequency reaches $2f_1$, the string again vibrates with large amplitude, but this time it vibrates as two loops (Fig. 32.3(b)). At $3f_1$ it vibrates as three loops, etc. Substituting the relevant values of L, T and μ in equation [32.5] confirms that the forcing frequencies, f_1, $2f_1$ and $3f_1$, are respectively equal to the frequencies of the first, second and third harmonics of the string. This, then, is an example of resonance – the string responds well only to those forcing frequencies which are equal to its natural frequencies of vibration.

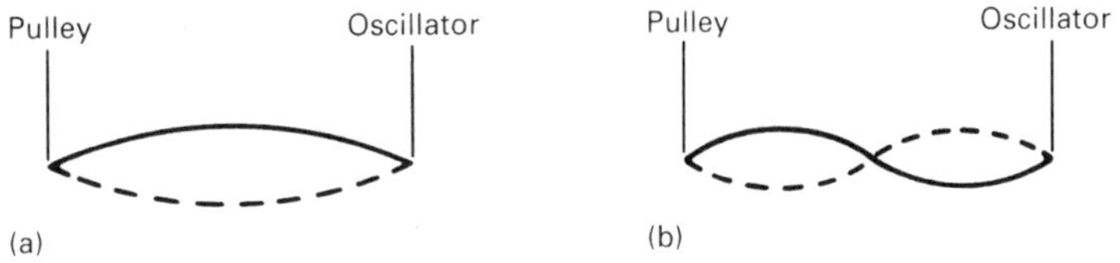

Fig. 32.3 Modes of vibration in Melde's experiment

Notes

(i) The amplitude of vibration of the oscillator is small in comparison with that of the string, and therefore the string behaves (almost) as if it is fixed at its point of attachment to the oscillator.

(ii) The reflected waves are not quite as 'strong' as the incident waves, and this prevents the displacements at the nodes being exactly zero.

(iii) The motion of the string can be 'frozen' if stroboscopic illumination is available. This demonstrates very convincingly that each section of the string is in anti-phase with that in an adjacent loop.

32.5 EXPERIMENTAL VERIFICATION OF $f_1 = \frac{1}{2L}\sqrt{\frac{T}{\mu}}$

The frequency, f_1, of the fundamental mode of vibration of a stretched string is given, by putting $n = 1$ in equation [32.5], as

$$f_1 = \frac{1}{2L}\sqrt{\frac{T}{\mu}}$$

It follows that:

(i) $f_1 \propto 1/L$ if T and μ are constant

(ii) $f_1 \propto \sqrt{T}$ if L and μ are constant

(iii) $f_1 \propto 1/\sqrt{\mu}$ if L and T are constant.

These relationships are sometimes referred to as **the laws of vibration of stretched strings**. They may be verified experimentally by using a sonometer (Fig. 32.4), as described below.

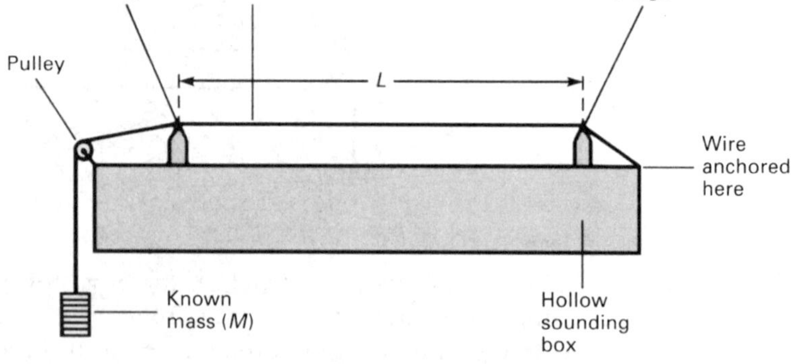

Fig. 32.4 Sonometer

To verify $f_1 \propto 1/L$

Having selected suitable values of T and μ, the position of the movable bridge is altered so that the vibrating length, L, of the wire is such that when the wire is plucked it produces the same note as a tuning fork of known frequency. If the experimenter is not sufficiently 'musical' to detect whether the two notes have the same pitch, he can make use of a resonance technique. A small piece of paper in the form of an inverted vee is placed on the centre of the wire, and the stem of a vibrating tuning fork is held against one of the bridges. This forces the wire to vibrate, and if its length is such that its fundamental frequency of vibration is equal to the frequency of the tuning fork, the wire vibrates with large amplitude and throws the paper off the wire. The procedure is repeated using tuning forks of other known frequencies, and without altering either T or μ. A graph of f_1 against $1/L$ is linear and passes through the origin, thus verifying the relationship.

To Verify $f_1 \propto \sqrt{T}$

With L kept constant at some suitable value, the mass, M, and therefore the tension $T(= Mg)$, is altered so that when the wire is plucked it produces the same note as a tuning fork of known frequency. The procedure is repeated using tuning forks of other known frequencies, and without changing either L or μ. A graph of f_1 against $\sqrt{T}$ is linear and passes through the origin, thus verifying the relationship.

To Verify $f_1 \propto 1/\sqrt{\mu}$

This relationship cannot be verified directly if tuning forks are used, because neither the frequencies of a set of tuning forks nor the masses per unit length of a set of wires are continuously variable. However, once it has been verified that $f_1 \propto 1/L$, it is sufficient to show that $L \propto 1/\sqrt{\mu}$ at constant T and constant f_1. First, the mass per unit length, μ, of a wire is determined by weighing. The length, L, of the wire is then adjusted so that when the wire is plucked it produces the same note as one of the tuning forks. The procedure is repeated using wires of different masses per unit length. Each wire must be under the same tension as the first wire, and in each case the length is adjusted until the wire vibrates at the same frequency as the tuning fork that was used with the first wire. A graph of L against $1/\sqrt{\mu}$ is linear and passes through the origin, thus verifying $f_1 \propto 1/\sqrt{\mu}$.

33

WAVES IN PIPES

33.1 INTRODUCTION

Suppose that the air at one end of a pipe is caused to vibrate. (In an organ pipe this is achieved by blasting air against a sharp edge. In the case of a clarinet air is blown across a reed.) The vibration produces a longitudinal wave which travels along the pipe and is reflected at its far end. Since the incident and reflected waves have the same speed, frequency and amplitude, a stationary wave results. The modes of vibration of an open pipe are different from those of a closed pipe; the two cases will be considered separately. (The reader will need to be familiar with the terms 'overtone' and 'harmonic' and may need to refer to Chapter 34 before continuing.)

33.2 CLOSED (STOPPED) PIPES

In Fig. 33.1, the air at Y cannot move and therefore Y must be the site of a displacement node. The pipe is open at X (the point at which the vibration is instigated), and therefore X must be the site of a displacement antinode. The three simplest modes of vibration which satisfy these conditions for a pipe of length L are shown in Fig. 33.2, where N and A represent nodes and antinodes respectively.

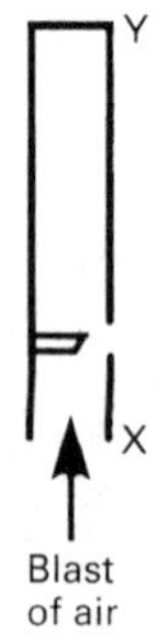

Fig. 33.1
Closed pipe

Representing the wavelengths in (a), (b) and (c) by λ_a, λ_b and λ_c respectively, and bearing in mind that the distance from a node to the next antinode is a quarter of a wavelength, we have

$$\frac{\lambda_a}{4} = L \qquad \frac{\lambda_b}{4} = \frac{L}{3} \qquad \frac{\lambda_c}{4} = \frac{L}{5}$$

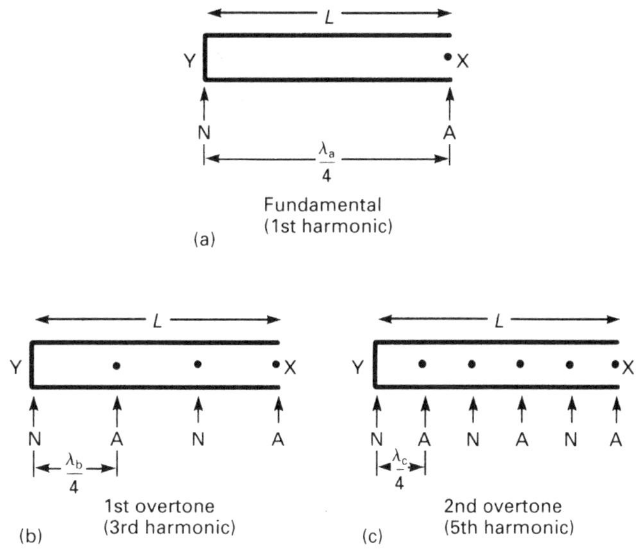

Fig. 33.2
Modes of vibration in a closed pipe

The velocity of a sound wave in air is independent of its wavelength and therefore has the same value for each of (a), (b) and (c). If, in each case, we represent the velocity by v, the respective frequencies, f_a, f_b and f_c are given by

$$f_a = \frac{v}{\lambda_a} \qquad f_b = \frac{v}{\lambda_b} \qquad f_c = \frac{v}{\lambda_c}$$

i.e. $$f_a = \frac{v}{4L} \qquad f_b = \frac{3v}{4L} \qquad f_c = \frac{5v}{4L}$$

Thus, $f_b = 3f_a$ and $f_c = 5f_a$. With the exception of f_a (the frequency of the fundamental) f_b and f_c are the lowest frequencies the pipe can produce (i.e. they are the first and second overtones respectively). It follows, therefore, that **a closed pipe can produce only odd harmonics**. Thus, for a closed pipe

$$f_n = \frac{nv}{4L} \qquad (n = 1, 3, 5, \ldots) \qquad [33.1]$$

where

f_n = the frequency of the nth harmonic.

33.3 OPEN PIPES

In Fig. 33.3, a longitudinal wave produced by a vibration at X is reflected by the free air at Y. The air is free to move at both X and Y and therefore there is a

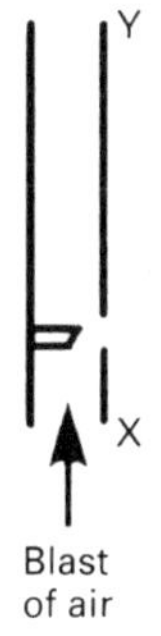

Fig. 33.3
Open pipe

displacement antinode at each of X and Y. The three simplest modes of vibration which satisfy this condition for a pipe of length L are shown in Fig. 33.4, where N and A represent nodes and antinodes respectively.

Representing the wavelengths in (a), (b) and (c) by λ_a, λ_b and λ_c respectively, and bearing in mind that the distance from a node to the next antinode is a quarter of a wavelength, we have

$$\frac{\lambda_a}{4} = \frac{L}{2} \qquad \frac{\lambda_b}{4} = \frac{L}{4} \qquad \frac{\lambda_c}{4} = \frac{L}{6}$$

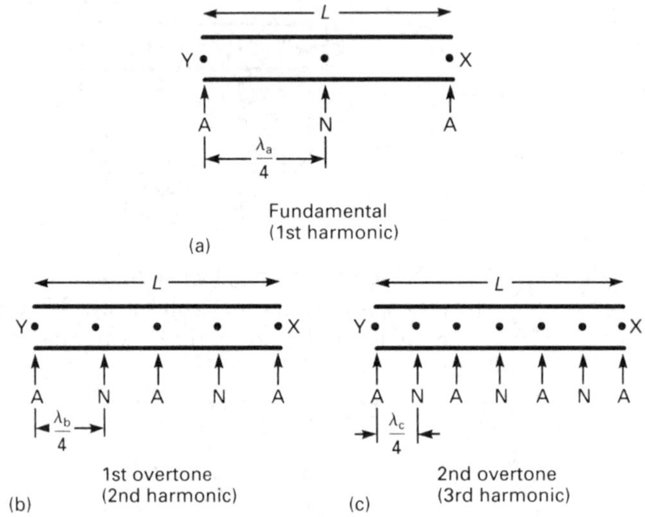

Fig. 33.4
Modes of vibration in an open pipe

The velocity of a sound wave in air is independent of its wavelength and therefore we may represent it by v for each of (a), (b) and (c), in which case the respective frequencies, f_a, f_b and f_c are given by

$$f_a = \frac{v}{\lambda_a} \qquad f_b = \frac{v}{\lambda_b} \qquad f_c = \frac{v}{\lambda_c}$$

i.e. $$f_a = \frac{v}{2L} \qquad f_b = \frac{2v}{2L} \qquad f_c = \frac{3v}{2L}$$

Thus, $f_b = 2f_a$ and $f_c = 3f_a$, i.e. the first and second overtones are the second and third harmonics respectively. It follows that **an open pipe can produce both odd and even harmonics**. Thus, for an open pipe

$$f_n = \frac{nv}{2L} \qquad (n = 1, 2, 3, \ldots) \qquad [33.2]$$

where

f_n = the frequency of the nth harmonic.

33.4 OPEN AND CLOSED PIPES COMPARED

It follows from equations [33.1] and [33.2] that:

(i) the fundamental frequency of an open pipe is twice that of a closed pipe of the same length;

(ii) a closed pipe of length $L/2$ produces the same fundamental as an open pipe of length L but, because both odd and even harmonics are present in the note from the open pipe, it gives a richer tone;

(iii) the velocity, v, of sound in air is proportional to $\sqrt{T}$ (where T is the temperature of the air **in kelvins**), and therefore, since v is involved in both equation [33.1] and equation [33.2], the frequencies produced by both closed and open pipes increase with temperature.

33.5 END CORRECTION

In practice, the vibrations at an open end of a sounding pipe extend into the free air just outside. The actual position of the associated displacement antinode is a short distance, c, beyond the end, and c is known as the **end-correction** (Fig. 33.5). The effective length of a closed pipe of length L is therefore $L + c$ and that of an open pipe of length L is $L + 2c$. It can be shown that

$$c = 0.6r \quad \text{approximately} \qquad [33.3]$$

where

r = the radius of the pipe

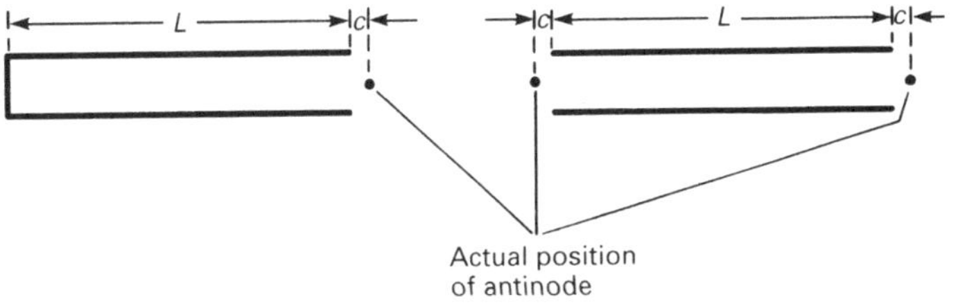

Fig. 33.5 End correction

It follows from equation [33.3] that the end-correction becomes more significant with increasing radius, and that when wide pipes are being considered it is necessary to replace L in equations [33.1] and [33.2] by $L + c$ and $L + 2c$ respectively.

QUESTIONS 33A

1. A closed (i.e. closed at one end) organ pipe has a fundamental of 400 Hz. What is: **(a)** the frequency of the first overtone; **(b)** the fundamental frequency of an open pipe of the same length?

2. Calculate the frequency of: **(a)** the fundamental, **(b)** the first overtone, produced by a pipe of length 40.0 cm which is closed at each end. (Velocity of sound in air = 340 m s^{-1}.)

33.6 PRESSURE VARIATIONS IN PIPES

At the sites of displacement antinodes the air in a sounding pipe moves to the maximum extent and the pressure variations are zero. Thus, there are pressure nodes at the sites of displacement antinodes. At a displacement node, the air is either coming toward the node from each side (in which case there is high pressure there) or it is moving away on each side (in which case there is low pressure there). There are therefore pressure antinodes at the sites of the displacement nodes.

There is, therefore, a pressure wave associated with the displacement wave, and the two waves have a phase difference of $\pi/2$ radians (i.e. a quarter of a wavelength).

33.7 DEMONSTRATION OF THE EXISTENCE OF NODES AND ANTINODES

Displacement Nodes and Antinodes

In Fig. 33.6 a thin paper disc with a few particles of fine sand on it is suspended by a fine silk thread inside a sounding organ pipe. If the disc is at a displacement antinode, the sand is buffeted by the motion of the air there. At displacement nodes the sand is motionless.

Fig. 33.6 Apparatus to demonstrate ***displacement*** nodes and antinodes

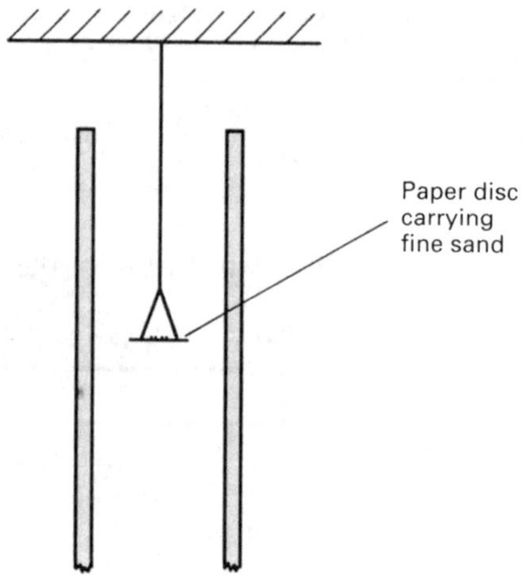

Pressure Nodes and Antinodes

Fig. 33.7 shows small microphones set into the wall of a sounding organ pipe. The planes of the microphone diaphragms are perpendicular to the direction of vibration of the air in the pipe and therefore are unaffected by the bulk motion of the air. Instead, the microphones respond to pressure variations. If the microphones are connected in turn to an oscilloscope, those which are at pressure antinodes are seen to produce waveforms of large amplitude. Those at pressure nodes give no signal.

Fig. 33.7 Apparatus to demonstrate ***pressure*** nodes and antinodes

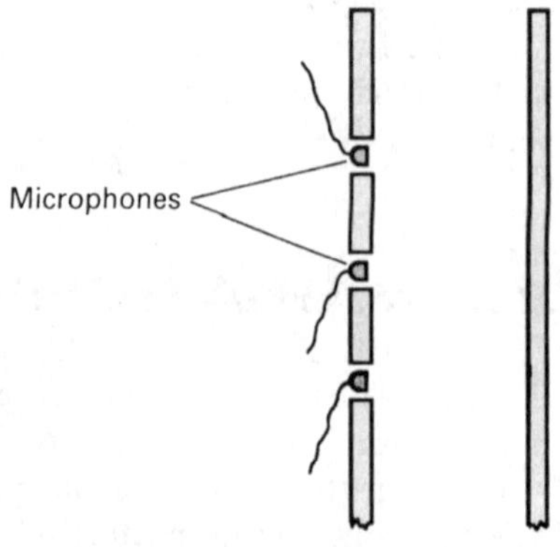

By using the same pipe to locate the pressure nodes and antinodes as that used to locate the displacement nodes and antinodes we can obtain direct confirmation that pressure antinodes are displacement nodes and vice versa.

CONSOLIDATION

A pipe open at one end produces only odd harmonics

$$f_n = \frac{nv}{4L} \qquad (n = 1, 3, 5, \ldots) \qquad [33.1]$$

A pipe open at both ends (or closed at both ends) produces both odd and even harmonics

$$f_n = \frac{nv}{2L} \qquad (n = 1, 2, 3, \ldots) \qquad [33.2]$$

End-corrections are significant with wide pipes and equation [33.1] and [33.2] become respectively,

$$f_n = \frac{nv}{4(L + c)} \qquad (n = 1, 3, 5, \ldots)$$

and

$$f_n = \frac{nv}{2(L + 2c)} \qquad (n = 1, 2, 3, \ldots)$$

34

MUSICAL NOTES AND SOUND

34.1 OVERTONES AND HARMONICS

The lowest frequency that a vibrating string or pipe can produce is called the **fundamental frequency**, and the corresponding note is called the **fundamental**.

A note whose frequency is n times that of the fundamental (where n is a whole number) is called the ***n*th harmonic**. (The first harmonic is therefore the fundamental.)

The overtones of a note are the notes of higher frequency which are actually produced with the fundamental. The first overtone is the harmonic whose frequency is the lowest of those which are present with the fundamental, the second overtone is the next higher harmonic which is present, etc.

34.2 THE CHARACTERISTICS OF MUSICAL NOTES

Any musical note is characterised by its **loudness**, its **pitch** and its **quality** (sometimes called its **timbre** or its **tone**).

Loudness

Loudness is a subjective quantity. A note which is regarded as being loud by one observer appears less loud to an observer whose hearing is poorer. Loudness increases with intensity, but the relationship between the two quantities is complex. Even when the situation is simplified by considering only a single observer, it is necessary to take account of the fact that the ear has different sensitivities to sounds of different pitch. (We cannot hear ultrasonic sounds, no matter how intense they are.)

The intensity of a sound wave is a measure of the amount of energy associated with it, and is proportional to the square of the amplitude of the wave. The energy, and therefore the intensity and loudness, of a note also depends on the mass of the medium or body which is vibrating in order to produce it. A loudspeaker cone, therefore, has a large surface area. A guitar string has only a small surface area, and if it were not for the fact that it is connected to the hollow sounding box of the instrument, its vibrations would be unlikely to be heard.

As a sound wave travels outwards from its source, its energy is spread over the surface of a sphere centred on the source. It follows, therefore, that the intensity, I, and the distance from the source, r, are related by

$$I \propto \frac{1}{r^2}$$

Pitch

The pitch of a note is determined solely by its frequency. A note of high pitch is due to a vibration of high frequency. If the frequency of a note is twice that of another, the former is said to be an **octave** higher than the latter.

Quality (Timbre or Tone)

The notes produced by musical instruments are not pure. When a guitar string sounds the note which is known as concert A, its fundamental frequency of vibration is 440 Hz and is accompanied by various harmonics, each of which is of a lower intensity than the fundamental. If a piano sounds concert A, the fundamental frequency is again 440 Hz, but the intensities of the various harmonics that accompany it are different from those produced by the guitar. The notes are said to have different qualities, i.e. **the quality of a note is determined by the relative strengths of its overtones**.

Note No overtones accompany the note produced by a tuning fork, unless it has been struck too hard.

34.3 EXPERIMENTAL DETERMINATION OF THE SPEED OF SOUND BY USING A RESONANCE TUBE

The apparatus is shown in Fig. 34.1. The vibrating tuning fork forces the air column to vibrate, and if the frequency of the tuning fork is equal to a natural frequency of vibration of the column, the tube **resonates** and a loud sound is heard.

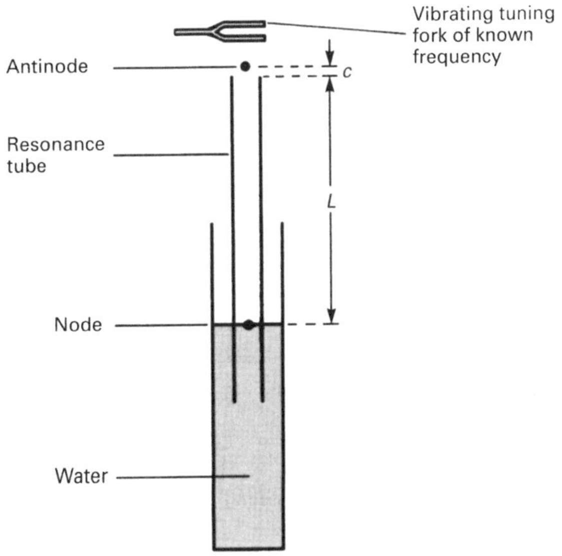

Fig. 34.1 Measurement of the speed of sound using a resonance tube

The water level (initially near the top of the tube) is lowered, by lowering the outer container, until the note is at its loudest. The water surface acts like the closed end of a pipe and is therefore the site of a displacement node. There is a displacement antinode a short distance c above the top of the tube, where c is the end-correction. The manner in which the resonance has been located ensures that the tube is vibrating in its fundamental mode, and therefore

$$\frac{\lambda}{4} = L + c$$

where

$$L = \text{the length of the air column at resonance}$$

$$\lambda = \text{the wavelength of the sound wave produced by the tuning fork.}$$

If the (known) frequency of the tuning fork is f, and v is the speed of sound, λ can be replaced by v/f, and therefore

$$\frac{v}{4f} = L + c \qquad [34.1]$$

i.e. $$L = \frac{v}{4} \cdot \frac{1}{f} - c \qquad [34.2]$$

The resonance lengths, L, corresponding to other known frequencies, f, are found. It follows from equation [34.2] that a graph of L against $1/f$ is linear and has a gradient of $v/4$ – hence v. Either intercept of such a graph provides a value for c, should it be required.

Notes (i) If only one tuning fork is available, the resonant length, L_1, corresponding to the first overtone (i.e. the third harmonic) should be found in addition to L. Then,

$$\frac{3\lambda}{4} = L_1 + c$$

therefore

$$\frac{3v}{4f} = L_1 + c \qquad [34.3]$$

Subtracting equation [34.1] from equation [34.3] gives

$$\frac{2v}{4f} = L_1 - L$$

i.e. $$v = 2f(L_1 - L)$$

Hence v.

(ii) The tuning forks could be replaced by a loudspeaker connected to a signal generator tuned to known frequencies. This has the advantage that the note does not fade, and therefore makes the resonance easier to detect. It may be necessary to calibrate the signal generator.

EXAMPLE 34.1

A small speaker emitting a note of 250 Hz is placed over the open upper end of a vertical tube which is full of water. When the water is gradually run out of the tube the air column resonates, initially when the water surface is 0.310 m below the top of the tube, and next when it is 0.998 m below the top. Find the speed of sound in air and the end-correction.

Solution

Since the tube was initially full of water, the first resonance that was heard must have been the fundamental, and therefore

$$\frac{\lambda}{4} = 0.310 + c$$

where c is the end-correction and λ is the wavelength of the note produced by the speaker. If the velocity of sound in air is v, it follows from $v = f\lambda$ that $\lambda = v/250$ and therefore

$$\frac{v}{4 \times 250} = 0.310 + c$$

i.e. $$v = 310 + 1000\,c \qquad [34.4]$$

Bearing in mind that there must be an antinode just above the open end of the tube and a node at the water surface, the next possible mode of vibration is the third harmonic (ANAN) and therefore

$$\frac{3\lambda}{4} = 0.998 + c$$

$$\therefore \quad \frac{3v}{4 \times 250} = 0.998 + c$$

i.e. $$3v = 998 + 1000\,c \qquad [34.5]$$

Subtracting equation [34.4] from equation [34.5] gives

$$2v = 688 \quad \text{i.e.} \quad v = 344\,\text{m}\,\text{s}^{-1}$$

Substituting for v in equation [34.4] gives

$$344 = 310 + 1000\,c$$

$$\therefore \quad 34 = 1000\,c \quad \text{i.e.} \quad c = 0.034\,\text{m}$$

34.4 EXPERIMENTAL DETERMINATION OF THE SPEED OF SOUND IN FREE AIR USING PROGRESSIVE WAVES

The reader should be familiar with the use of an oscilloscope to display Lissajous' figures (p. 795) before continuing.

The apparatus is arranged as shown in Fig. 34.2. The output from the signal generator is connected to the loudspeaker and to the X-plates of the oscilloscope, which has the time base off and the beam centralized. The signal generator is adjusted so that a note with a known frequency is produced by the loudspeaker. The note is received by the microphone and fed to the Y-plates of the oscilloscope.

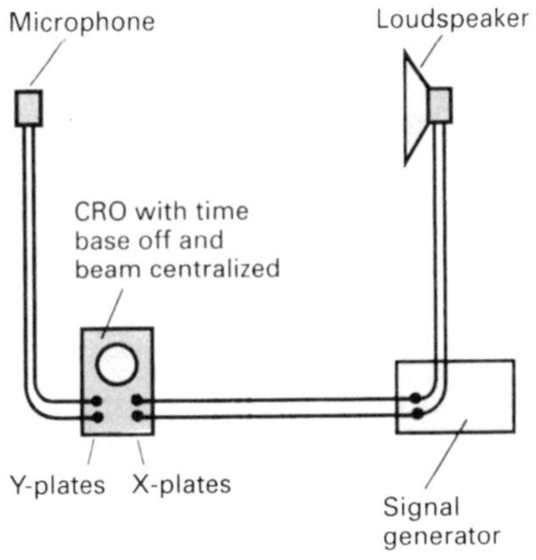

Fig. 34.2
Apparatus for measuring the speed of sound in free air

The oscilloscope therefore receives two signals of the same frequency (that of the signal generator). The phase difference between these two signals (and therefore the particular Lissajous' figure which is displayed on the screen) is determined by the time taken for the sound wave to travel from the loudspeaker to the microphone. The separation of the loudspeaker and the microphone is adjusted until a sloping line appears on the oscilloscope, indicating that the signal being received by the microphone is in phase with that at the speaker, and therefore that the microphone is a whole number of wavelengths away from the speaker. The microphone is now moved away from the speaker until a second line, sloping the same way as the first (e.g. top right to bottom left), appears on the oscilloscope. The distance moved by the microphone is equal to one wavelength (λ), in which case the speed of sound (v) can be determined from $v = f\lambda$, where f is the frequency of the signal generator. (If a suitable frequency, e.g. 3000 Hz, has been chosen, the microphone can be moved through a number of wavelengths in order to improve the accuracy of the experiment.)

34.5 THE SPEED OF SOUND IN GASES

It can be shown that the speed of sound in a gas is given by

$$v = \sqrt{\frac{\gamma p}{\rho}} \qquad [34.6]$$

where

v = the velocity of sound ($\mathrm{m\,s^{-1}}$)

γ = the ratio of the principal specific heats of the gas (i.e. $\gamma = C_p/C_v$ – see p. 267)

p = the pressure of the gas ($\mathrm{N\,m^{-2}}$)

ρ = the density of the gas ($\mathrm{kg\,m^{-3}}$).

Equation [34.6] might lead the reader to suppose that the speed of sound in a gas depends on the pressure of the gas. However, the density of a gas is proportional to its pressure, and therefore p/ρ in equation [34.6] is constant. Thus, **the speed of sound in a gas is independent of the pressure of the gas**.

If ρ in equation [34.6] is replaced by M/V, where M is the mass of some volume V of gas, we have

$$v = \sqrt{\frac{\gamma p V}{M}}$$

But, from equation [14.1]

$$pV = nRT$$

where

R = the universal molar gas constant ($\mathrm{J\,K^{-1}\,mol^{-1}}$)

n = the number of moles of gas in the mass M (mol)

T = the temperature of the gas in kelvins.

Therefore

$$v = \sqrt{\frac{\gamma nRT}{M}}$$

i.e. $$v = \sqrt{\frac{\gamma RT}{M_m}} \qquad [34.7]$$

where

M_m = the mass per mole of the gas ($kg\,mol^{-1}$).

Since R is a universal constant and γ and M_m are constants for any particular gas, it follows that for any given gas $v \propto \sqrt{T}$.

CONSOLIDATION

The fundamental (or first harmonic) is the lowest frequency note produced by a pipe or vibrating string.

The *n*th harmonic is a note whose frequency is n times that of the fundamental.

The first overtone is the note with the lowest frequency of those which are present with the fundamental – it is not necessarily the second harmonic.

The speed of sound in a gas is proportional to $\sqrt{T}$ (where T is the temperature in kelvins) but it is independent of the pressure.

35

THE DOPPLER EFFECT

35.1 INTRODUCTION

Whenever there is relative motion between a source of waves and an observer, the frequency of the wave motion as noted by the observer, is different from the actual frequency of the waves – the **Doppler effect**. The effect accounts for the sudden decrease in pitch (frequency) heard by a person standing in a railway station as a sounding train siren passes by.

35.2 DERIVATION OF EXPRESSION FOR FREQUENCY CHANGE

Consider a source of sound waves of frequency, f, and wavelength, λ, and suppose that the speed of the waves is c. Let the source be at S at $t = 0$ (Fig. 35.1) and suppose that it is moving along XY towards Y with velocity v_s. Suppose that the source emits a crest when it is at S. The next crest will be emitted $1/f$ seconds later, by which time (remembering that distance = velocity×time) the first crest will have travelled a distance $c(1/f)$ from S. That section of wavefront which is moving towards X will have reached point M, and that which is travelling towards Y will have reached N. During this same time the source itself will have travelled a distance $v_s(1/f)$, to T, say.

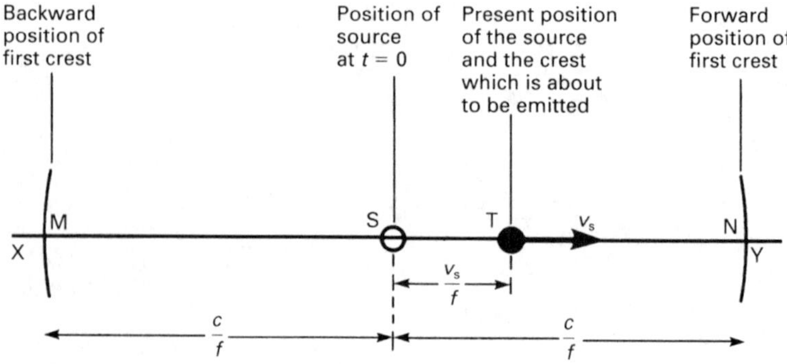

Fig. 35.1
To derive frequency change due to the Doppler effect

Forward of the Source

$$\text{Separation of successive crests} = \text{TN}$$
$$= \frac{c}{f} - \frac{v_s}{f}$$

But

$$\text{Separation of successive crests} = \lambda_f$$

where

λ_f = the wavelength forward of the source,

i.e. $$\lambda_f = \frac{c - v_s}{f} \qquad [35.1]$$

Behind the Source

$$\text{Separation of successive crests} = \text{TM}$$
$$= \frac{c}{f} + \frac{v_s}{f}$$

i.e. $$\lambda_b = \frac{c + v_s}{f} \qquad [35.2]$$

where

λ_b = the wavelength behind the source,

Bearing in mind that when the source is stationary the wavelength, λ, is given by $\lambda = c/f$, we see from equation [35.1] that $\lambda_f < \lambda$, i.e. the **waves 'bunch up' in front of a moving source**. An observer at Y receives crests more frequently than he would if the source were stationary, i.e. he hears a signal of higher frequency (given by c/λ_f). It is clear from equation [35.2] that $\lambda_b > \lambda$, and therefore an observer at X hears a note of lower frequency (given by c/λ_b) than he would if the source were stationary.

Suppose now that the observer (either at X or Y) is moving towards the source with velocity, v_o. The velocity of the waves relative to the observer is $(c + v_o)$. (**Note**. The velocity of the waves is not affected by the motion of their source.) The frequency forward of the source is therefore given by

$$f_f = \frac{c + v_o}{\lambda_f}$$

where

f_f = the frequency forward of the source

Therefore, by equation [35.1]

$$f_f = \left(\frac{c + v_o}{c - v_s}\right) f \qquad [35.3]$$

If

f_b = the frequency behind the source

$$f_b = \frac{c + v_o}{\lambda_b}$$

Therefore, by equation [35.2]

$$f_b = \left(\frac{c+v_o}{c+v_s}\right)f \qquad [35.4]$$

Notes

(i) Equations [35.3] and [35.4] can be used when only the source is moving, or when only the observer is moving, or when both source and observer are moving. Note that when $v_s = 0$, $f_f = f_b$, and therefore either equation can be used. This should not be unexpected, because when the source is stationary the terms 'forward' and 'backward' have no significance.

(ii) If the observer is moving away from the source, v_o is negative.

(iii) **The motion of the observer has no effect on the wavelength.** An observer moving towards a stationary source hears an increase in frequency because he intercepts the crests more frequently than he would if he were stationary.

(iv) Equations [35.3] and [35.4] apply to all types of mechanical waves. See section 35.3 for an explanation of the circumstances under which they may be applied to electromagnetic waves.

EXAMPLE 35.1

A cyclist and a railway train are approaching each other. The cyclist is moving at $10\,\text{m s}^{-1}$ and the train at $20\,\text{m s}^{-1}$. The engine driver sounds a warning siren at a frequency of 480 Hz. Calculate the frequency of the note heard by the cyclist (a) before, and (b) after the train has passed by.

(Speed of sound in air $= 340\,\text{m s}^{-1}$.)

Solution

(a) Before the train has passed by, the cyclist is forward of the source of sound and is moving towards it, in which case equation [35.3] applies with $c = 340\,\text{m s}^{-1}$, $f = 480\,\text{Hz}$, $v_s = 20\,\text{m s}^{-1}$ and $v_o = 10\,\text{m s}^{-1}$. The frequency f_f heard by the cyclist is therefore given by

$$f_f = \left(\frac{340+10}{340-20}\right)480 = 525\,\text{Hz}$$

(b) Once the train has passed by, the cyclist is behind the source of sound and is moving away from it, in which case equation [35.4] applies with $c = 340\,\text{m s}^{-1}$, $f = 480\,\text{Hz}$, $v_s = 20\,\text{m s}^{-1}$ and $v_o = -10\,\text{m s}^{-1}$ (the minus sign is necessary because the cyclist is moving away from the source). The frequency f_b heard by the cyclist is therefore given by

$$f_b = \left(\frac{340-10}{340+20}\right)480 = 440\,\text{Hz}$$

QUESTIONS 35A

1. A source of sound has a frequency of 500 Hz. Calculate the frequency heard by an observer in each of the situations shown in the table below. Take the speed of sound to be $340\,\mathrm{m\,s^{-1}}$ in each case.

2. What would be the wavelength of the sound waves at the site of the observer in each of situations **(a)** to **(h)** shown in the table below?

	Speed of source/$\mathrm{m\,s^{-1}}$	*Speed of observer*/$\mathrm{m\,s^{-1}}$	*Position of observer*
(a)	0	20 (towards source)	Not relevant
(b)	0	20 (away from source)	Not relevant
(c)	20	0	Forward of source
(d)	20	0	Behind source
(e)	30	50 (towards source)	Forward of source
(f)	30	50 (away from source)	Forward of source
(g)	30	50 (towards source)	Behind source
(h)	30	50 (away from source)	Behind source

35.3 THE DOPPLER EFFECT WITH LIGHT

The Doppler effect occurs not only with sound but also with light, and in this respect it has proved invaluable in astronomy.

When the light emitted by a star is examined spectrosocpically, it is found that each line in the spectrum of any particular element in the star occurs at a different wavelength from that of the corresponding line in the spectrum of the same element in the laboratory. With some stars all the spectral lines are at longer wavelengths than in the laboratory spectrum; with others they are at shorter wavelengths. The shifts in wavelengths are interpreted as being due to the Doppler effect. If the lines are at longer wavelengths (i.e. red shifted), the star is moving away from the Earth; if they are at shorter wavelengths (i.e. violet shifted), the star is approaching us. The extent of the shift in wavelength depends on the speed with which the star is moving relative to the Earth and can be used to calculate this speed.

The galaxies, apart from a few which are close to us, all exhibit a red shift and so must be moving away from the Earth. Measurements of the galactic red shifts reveal that the speed with which a galaxy is receding is proportional to its distance from us (**Hubble's law**). The galaxies are not only moving away from us, but also from each other, i.e. the Universe as a whole is expanding.

The Doppler effect has been used to measure the speed with which the Sun is rotating. The Fraunhofer lines originating at that side of the solar disc which is approaching us are shifted towards the violet, those from the other side of the disc are shifted towards the red. The rotational speeds of Saturn's rings have been determined in a similar manner.

The Doppler effect has also been used to detect the presence of double stars. A double star is two stars which are so close together that they appear as a single star even with a very large telescope. The stars rotate about their common centre of mass. When one of the pair is approaching us the other is moving away. The spectrum of what is apparently a single star shows both a red shift and a violet shift, revealing that there are in fact two stars.

Quasars (compact sources of radio waves that also emit light) show enormous red shifts, indicating recessional speeds in excess of $0.8c$ in some cases.

It is a consequence of the Special Theory of Relativity that it is impossible to distinguish the velocity of an observer towards a source of electromagnetic waves from that of the source towards the observer. It follows, therefore, that the quantities v_s and v_o have no significance when electromagnetic waves are being considered, and we should deal only with the relative velocity, v, of the source and observer. It turns out, however, that if $v \ll c$, equations [35.3] and [35.4] give valid results if one of v_s and v_o is put equal to v and the other is put equal to zero. (Each procedure gives the same result – this is a consequence of v being much less than c.)

Let us consider a situation in which the Earth and a star are separating with a relative velocity, v. Suppose that the star emits light of frequency, f. We shall assume that $v \ll c$, where c is the velocity of light, so that equations [35.3] and [35.4] apply.

Initially, we shall assume that the star is stationary and that the Earth is moving away from it. A person receiving light on the Earth is therefore considered to be an observer moving with velocity, v, away from a stationary source. The frequency, f', of the light received on the Earth is therefore given, by putting $v_s = 0$ and $v_o = -v$* in either equation [35.3] or equation [35.4], as

$$f' = \left(\frac{c - v}{c}\right) f$$

i.e. $$f' = \left(1 - \frac{v}{c}\right) f$$

Alternatively, we may consider that the Earth is stationary and that the star is moving away from it. In this case, a person on the Earth is to be considered to be a stationary observer situated behind a moving source. The frequency, f', of the light received on Earth is therefore given, by putting $v_s = v$ and $v_o = 0$ in equation [35.4], as

$$f' = \left(\frac{c}{c + v}\right) f$$

i.e. $$f' = f \Big/ \left(1 + \frac{v}{c}\right)$$

Expanding this by the binomial theorem as far as terms in v/c gives

$$f' = \left(1 - \frac{v}{c}\right) f$$

Thus, the two treatments give the same result as long as terms in $(v/c)^2$, $(v/c)^3$, etc. can be ignored (i.e. if $v \ll c$). Further, to this degree of accuracy, the result is the same as that obtained by a relativistic treatment.

Although it has been convenient to discuss the Doppler shifts mainly in terms of frequency, it is wavelengths that are usually measured. Accordingly, we shall now obtain an expression for the shift, $\Delta\lambda$, in the wavelength of the light emitted by a star which is receding from the Earth with a velocity, v, with respect to the Earth. In view of the previous discussion, we may assume that the Earth is stationary, in which case the observed wavelength, λ', of any particular spectral line in the light emitted by the star is given by equation [35.2] as

$$\lambda' = \frac{c + v}{f}$$

*The minus sign is present because the observer is moving away from the star – see note(ii) in section 35.2.

where f is the frequency of the emitted light. Replacing f by c/λ, where λ is the wavelength of the same spectral line in the laboratory, we have

$$\lambda' = \frac{c+v}{c/\lambda}$$

Rearranging gives

$$\lambda' = \lambda\left(1+\frac{v}{c}\right)$$

$$\therefore \quad \lambda' - \lambda = \frac{v}{c}\lambda$$

i.e. $$\Delta\lambda = \frac{v}{c}\lambda$$

If the wavelength shift, $\Delta\lambda$, is measured, this expression can be used to determine the speed, v, of the star as long as it is not so large that a relativistic expression is necessary. For a star which is approaching the Earth $\Delta\lambda = (v/c)\lambda$ still applies but the shift is to a shorter wavelength.

Note The Doppler effect cannot provide information about the transverse speeds of stars, it gives only the speeds of recession and approach.

QUESTIONS 35B

1. A star is moving away from the Earth at a speed of $4.12 \times 10^5\,\mathrm{m\,s^{-1}}$ relative to the Earth. Calculate the wavelength shift observed on the Earth for light of wavelength $5.150 \times 10^{-7}\,\mathrm{m}$ emitted by the star. ($c = 2.998 \times 10^8\,\mathrm{m\,s^{-1}}$.)

35.4 RADAR SPEED TRAPS

The speed of a moving car can be found by measuring the shift in frequency of microwaves reflected by it.

Consider a car moving with speed, v, towards a stationary source of microwaves of frequency, f. The car acts as an observer moving towards a stationary source, and the waves as received by the car have a frequency, f', given by putting $v_s = 0$ and $v_o = v$ in equation [35.3] (or equation [35.4]),* as

$$f' = \left(\frac{c+v}{c}\right)f \qquad [35.5]$$

Waves of this frequency (f') are reflected back to the source, so that the car is now acting as a source moving with velocity v, and the radar set is acting as a stationary observer forward of the source. The frequency, f'', of the waves on reaching the radar set is given, by putting $v_s = v$ and $v_o = 0$ in equation [35.3], as

$$f'' = \left(\frac{c}{c-v}\right)f'$$

*The use of these equations is justified because $v \ll c$ (see section 35.3).

Substituting for f' from equation [35.5] gives

$$f'' = \left(\frac{c}{c-v}\right)\left(\frac{c+v}{c}\right)f$$

i.e.
$$f'' = \left(\frac{c+v}{c-v}\right)f \qquad [35.6]$$

The fractional change in frequency, $\Delta f/f$, is given by

$$\frac{\Delta f}{f} = \frac{f'' - f}{f}$$

i.e.
$$\frac{\Delta f}{f} = \frac{f''}{f} - 1$$

Therefore, by equation [35.6]

$$\frac{\Delta f}{f} = \left(\frac{c+v}{c-v}\right) - 1$$

i.e.
$$\frac{\Delta f}{f} = \frac{2v}{c-v}$$

Since $v \ll c$, we can consider that the denominator is equal to c, and therefore

$$\frac{\Delta f}{f} = \frac{2v}{c}$$

i.e.
$$v = \frac{c\Delta f}{2f}$$

Δf is the beat frequency of the waves transmitted and received by the radar set. Thus by causing the incoming signal to beat with the transmitted signal, and knowing the frequency of the transmitted signal, we can find v.

35.5 SPECTRAL LINE BROADENING AND MEASUREMENT OF HIGH TEMPERATURES

Plasmas (gases) involved in nuclear fusion reactions have temperatures of millions of degrees Celsius. If the spectrum of such a gas is examined, it is found that the spectral lines are broadened. (It is of course necessary that some of the atoms which are present are not completely ionized.) The broadening is a consequence of the motion of the atoms which are emitting the light, and is significant because at temperatures as high as these the atoms are moving at very great speeds. Those atoms which are moving away from the observer emit longer wavelengths than they would if they were stationary; those moving towards the observer emit light of shorter wavelengths. The spectral lines are therefore broadened, and the extent of the broadening increases with the speed of the atoms. Since the speed is proportional to $\sqrt{T}$ (section 14.5), where T is the temperature in kelvins, T can be calculated by measuring the broadening.

CONSOLIDATION

The frequency f_f forward of a source moving with speed v_s and emitting waves of frequency f and speed c detected by an observer moving with speed v_o towards the source is given by

$$f_f = \left(\frac{c + v_o}{c - v_s}\right) f \qquad [35.3]$$

The frequency f_b behind a source moving with speed v_s and emitting waves of frequency f and speed c detected by an observer moving with speed v_o towards the source is given by

$$f_b = \left(\frac{c + v_o}{c + v_s}\right) f \qquad [35.4]$$

If the observer is moving away from the source, v_o is put into equations [35.3] and [35.4] as a negative number.

The motion of a source of waves does not affect the velocity of the waves.

The motion of an observer has no effect on wavelength.

Equations [35.3] and [35.4] can be used for electromagnetic waves providing that:

(i) one of v_o and v_s is put equal to zero and the other is taken to be the relative velocity of the source and the observer, and

(ii) the relative velocity of the source and observer is much less than the velocity of light.

QUESTIONS ON SECTION E

PROPERTIES OF WAVES (Chapter 23)

E1 When a sine-form voltage of frequency 1250 Hz is applied to the Y-plates of a cathode ray oscilloscope the trace on the tube is as shown in diagram (i).

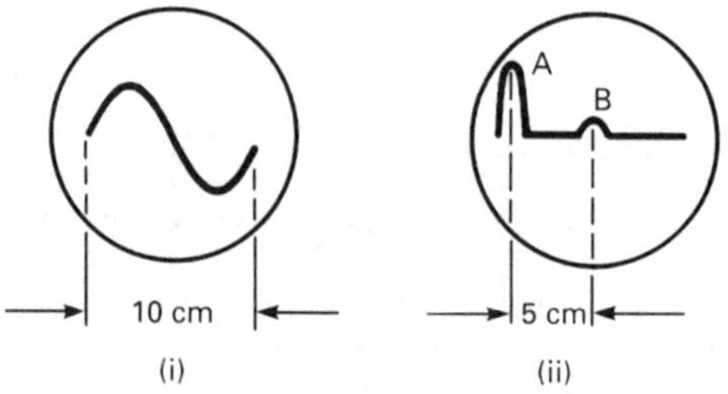

If a radar transmitter sends out short pulses, and at the same time gives a voltage to the Y-plates of the oscilloscope, with the time-base setting unchanged, the deflection A is produced as shown in diagram (ii). An object reflects the radar pulse which, when received at the transmitter and amplified, gives the deflection B. What is the distance of the object from the transmitter?

(Speed of radar waves $= 3 \times 10^8\,\mathrm{m\,s^{-1}}$.) [L]

E2 The equation for a transverse progressive simple harmonic wave moving in the positive direction of x may be expressed in the form

$$y = a \sin 2\pi(bt - cx)$$

where t represents time.

Justify this equation and explain the significance of the constants a, b, c, and the ratio b/c. [J]

E3 The equation $y = A \sin(\omega t + kx)$ represents a progressive wave.
(a) What do the quantities A, ω and k represent?
(b) In which direction is the wave moving? (Explain how you arrive at your conclusion.) [W, '90]

E4 The equation $y = a \sin(\omega t - kx)$ represents a plane wave travelling in a medium along the x-direction, y being the displacement at the point x at time t. Deduce whether the wave is travelling in the positive x-direction or in the negative x-direction.

If $a = 1.0 \times 10^{-7}\,\mathrm{m}$, $\omega = 6.6 \times 10^3\,\mathrm{s^{-1}}$ and $k = 20\,\mathrm{m^{-1}}$, calculate:
(a) the speed of the wave,
(b) the maximum speed of a particle of the medium due to the wave. [J]

HUYGENS' CONSTRUCTION (Chapter 24)

E5 Refraction occurs when waves pass from one medium into another of different refractive index. The table below gives data for sound waves and light waves.

Type of wave	*Speed in air/* $\mathrm{m\,s^{-1}}$	*Speed in water/* $\mathrm{m\,s^{-1}}$
Sound	340	1400
Light	3.00×10^8	2.25×10^8

(a) Plane waves travelling in air meet a plane water surface.
  (i) Use the data in the table above to calculate the angle of refraction in the water for light waves and for sound waves if the angle of incidence of the plane waves is 10°.
  (ii) Explain what happens in each case when the angle of incidence is 15°.
(b) Submarine detection helicopters are fitted with transmitters which send out sound waves. The reflected signals identify the location of the submarine. Explain why it is common for this transmitter to be suspended from the helicopter by a wire so that it is underwater. [J, '92]

E6 Explain what is meant by Huygens' principle.

Use the principle to show that a plane wave incident obliquely on a plane mirror is reflected:

(a) as a plane wave,

(b) so that the angle of incidence is equal to the angle of reflection. [J]

E7 Using Huygens' principle of secondary wavelets explain, making use of a diagram, how a refracted wavefront is formed when a beam of light, travelling in glass, crosses the glass–air boundary. Show how the sines of the angles of incidence and refraction are related to the speeds of light in air and glass. [L]

E8 A beam of light is refracted at the boundary of two media. Using Huygens' principle show how the ratio of the sine of the angle of incidence to the sine of the angle of refraction is related to the speeds of light in the two media. [AEB, '79]

INTERFERENCE (Chapter 25)

E9 An essential condition for interference to be observable between wavetrains originating from two sources is that the sources should be coherent. Explain what is meant by coherent in this context. [C]

E10 Draw a labelled diagram of Young's apparatus for producing and observing optical interference. Indicate clearly on your diagram the distances that need to be measured to enable you to determine the wavelength of the light. [W, '90]

E11 State the conditions which must be fulfilled if interference fringes are to be observed where two beams of light overlap.

Explain how it is that these conditions are fulfilled in interference experiments using **(a)** two slits, as in Young's experiment, **(b)** an extended source, as in Newton's rings experiment.

Describe, quoting the necessary formula, how you would determine the wavelength of monochromatic light by observation of Young's fringes. [S]

E12 Describe, quoting the necessary formula, how you would determine the wavelength of monochromatic light by observation of Young fringes.

Two whistles having the same frequency 2 kHz, and situated 3 m apart, are blown simultaneously. An observer moving along a line parallel to the line joining the whistles, and distant 20 m from it, observes minima of sound at a series of points spaced 1.14 m apart. Calculate the speed of sound in air.

Explain why it is possible to observe sound interference effects using two whistles, but not light interference using two lamps, even if the dimensions of the apparatus are suitably modified. [S]

E13 Microwave transmitters are situated at A and B as shown in the diagram and are 2.0 m apart. C, the mid-point of the line AB, is 100 m from the line PQR, which is parallel to ACB. The transmitters emit microwaves of wavelength 30 mm and of equal amplitude.

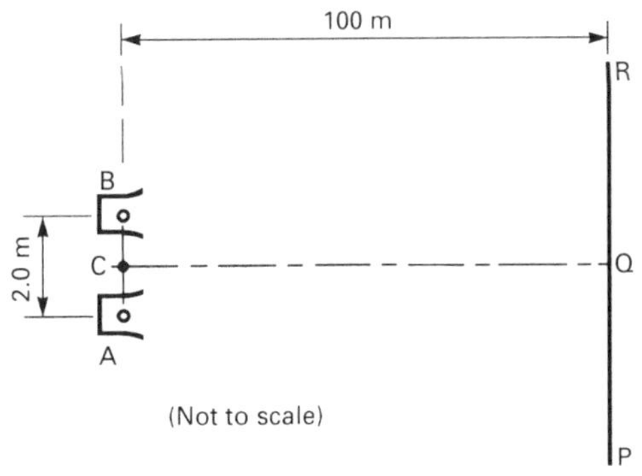

(a) An interference pattern is observed along PQR when the two transmitters emit in phase. Explain how the pattern arises and calculate the separation between adjacent maxima.

(b) The phase of transmitter A is changed so that there is a phase difference of π between the transmitters at A and B. The separation of the maxima along PQR does not change. Compare the new pattern with the original one.

(c) Half of a tube, with wall thickness 15 mm, is placed in front of the transmitter at A, as illustrated in the diagram on p. 516. The material of the wall is such that the wavelength of the radiation is reduced to 15 mm within it.

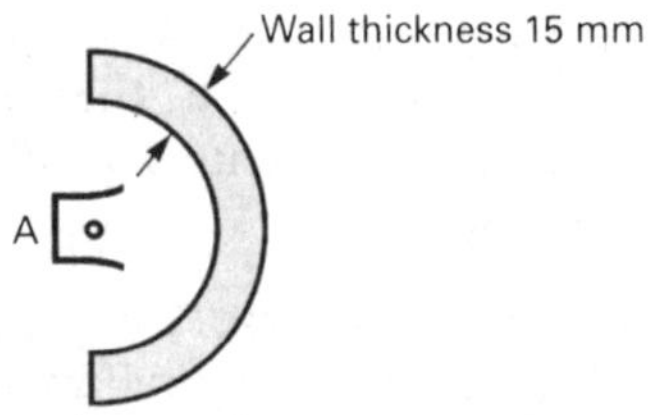

The transmitters at A and B emit in phase. Explain whether the pattern along PQR is like that for (a), (b) or neither. [O&C, '91]

E14 The only practical way to produce visible interference patterns with light is, in effect, to derive two sources from a single source. Explain why this is so.

Describe an experimental arrangement to observe interference in a wedge of air. How would you use this to determine a value for the wavelength of the light used?

A piece of wire of diameter 0.050 mm and two thin glass strips are available to produce the air wedge. If a total of 200 fringes are produced, what is the wavelength of the light used? [L]

E15 **(a)** A wedge-shaped film of air is formed between two, thin, parallel-sided, glass plates by means of a straight piece of wire. The two plates are in contact along one edge of the film and the wire is parallel to this edge.
- **(i)** Draw and label a diagram of the experimental arrangement you would use to observe and make measurements on interference fringes produced with light normally incident on the film.
- **(ii)** Explain the function of each part of the apparatus.

(b) In such an experiment using light of wavelength 589 nm, the distance between the seventh and one hundred and sixty-seventh dark fringes was 26.3 mm and the distance between the junction of the glass plates and the wire was 35.6 mm. Calculate the angle of the wedge and the diameter of the wire. [J]

E16 When monochromatic light is reflected from two flat glass plates, with a wedge-shaped air film of small angle between them, a pattern of bright and dark lines can be seen.

Explain in detail:
- **(a)** why this pattern is produced,
- **(b)** how the separation of the lines depends upon the angle of the wedge,
- **(c)** the effect of filling the space between the plates with a transparent liquid.

A glass slide is known to have an optically flat face. You wish to known whether or not another similar slide is also flat. Describe how you would find out, explaining how you would interpret your results. [L]

E17 Two optically flat glass plates are in contact along one edge and make a very small angle with each other. They are illuminated normally from above with yellow light of wavelength 600 nm and blue light of wavelength 450 nm. When viewed from above the wedge appears dark at the apex, and next appears dark at a distance of 6.00 mm from the apex. Explain why each of these dark areas occurs and calculate the angle between the two plates.

E18 A wedge-shaped film of air is formed between two thin parallel-sided optically flat glass plates as shown in the diagram below.

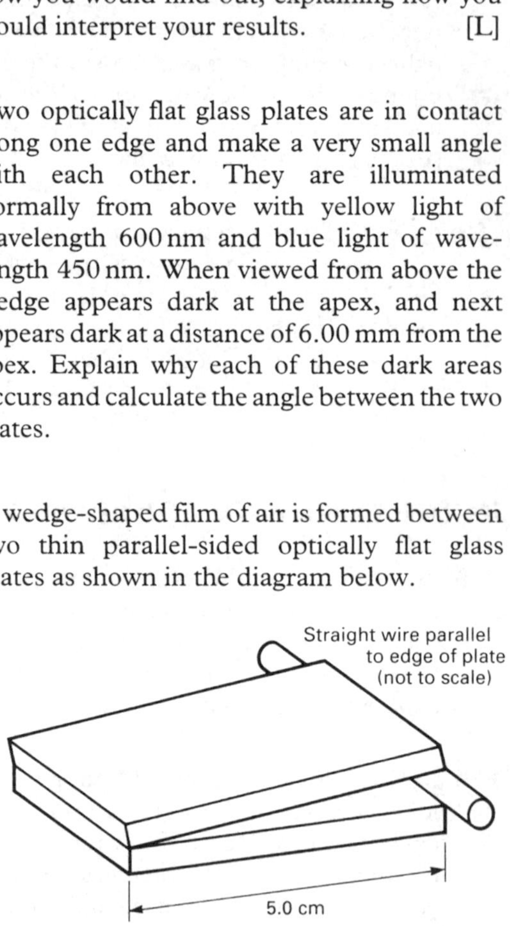

When the film is illuminated from above using monochromatic light of wavelength 550 nm a series of bright and dark fringes is observed.
- **(a)** Sketch the fringes and explain their formation, without calculation.
- **(b)** Calculate the distance between *adjacent* bright fringes when viewed from above for a wire of 0.05 mm diameter.
- **(c)** The wedge-shaped space between the plates is now filled with a transparent liquid of refractive index 1.2.
  - **(i)** What would be the new distance between alternate bright fringes?

(ii) If the refractive index of the glass plates is 1.6 discuss the possibility of using the above apparatus to determine the refractive indices of transparent liquids in the range 1.2 to 1.6. [J, '91]

E19 Explain why, for visible interference effects, it is normally necessary for the light to come from a single source and to follow different optical paths. Include a statement of the conditions required for complete destructive interference.

Explain *either* the colours seen in thin oil films *or* the colours seen in soap bubbles.

A wedge-shaped film of air between two glass plates gives equally spaced dark fringes, using reflected sodium light, which are 0.22 mm apart. When monochromatic light of another wavelength is used the fringes are 0.24 mm apart. Explain why the two fringe spacings are different. (The incident light falls normally on the air film in both cases.) Discuss what you would expect to see if the air film were illuminated by both sources simultaneously. Calculate the wavelength of the second source of light.

(Wavelength of sodium light = 589 nm (589×10^{-9} m).) [L]

E20 **(a)** A thin film of soap solution held in a circular wire frame mounted vertically is illuminated normally with sodium (yellow) light. When viewed from the same side as the source the film appears crossed by a series of horizontal black and yellow bands with a black band at the top.

(i) Explain the formation of the bands.
(ii) Explain why the band at the top of the frame is black.
(iii) If the sodium source is replaced by one emitting monochromatic red light describe and explain any changes which occur in the appearance of the bands.

(b) A film of soap solution of constant thickness 400 nm is illuminated normally with white light. By considering the condition necessary for constructive interference calculate the wavelengths of visible light at which there will be maximum intensity in the reflected beam.

You may assume that the refractive index for soap solution is the same for all wavelengths of white light and equal to 1.33. [J]

E21 **(a)** When monochromatic light is incident normally on a thin film, an interference pattern of bright and dark fringes is observed in light reflected from the film. Explain with the aid of a diagram why such an interference pattern is produced.

State what information about the film can be deduced from
(i) the shape of the fringes,
(ii) the separation of the fringes.

(b) An air wedge is made by separating two plane sheets of glass (microscope slides) by a fine wire at one end. When the wedge is illuminated normally by light of wavelength 5.9×10^{-7} m a fringe pattern is observed in the reflected light. The distance measured between the centre of the 1st bright fringe and the centre of the 11th bright fringe is 8.1 mm. Calculate the angle of the air wedge.

(c) As a soap film supported on a vertical frame slowly drains, patterns of horizontal alternate bright and dark fringes are observed both when viewed in reflected monochromatic light and when viewed in transmitted monochromatic light.

Explain why
(i) the contrast between fringes seen in transmitted light is less than that seen in reflected light,
(ii) a particular thickness of film which produces a dark fringe in reflected light produces a bright fringe in transmitted light. [J]

E22 A glass converging lens rests in contact with a horizontal plane sheet of glass. Describe how you would produce and view Newton's rings, using reflected sodium light.

Explain how the rings are formed and derive a formula for their diameters. Describe the measurements you would make in order to determine the radii of curvature of the faces of the lens, assuming that the wavelength of sodium light is known. Show how the result is derived from the observations.

How would the ring pattern change if:

(a) the lens were raised vertically one quarter of a wavelength,

(b) the space between the lens and the plate were filled with water? [J]

E23 In the interference of light what is meant by the requirement of coherency? How is this usually achieved in practice?

A thin spherical lens of long focal length is placed on a flat piece of glass. How, using this arrangement, would you demonstrate interference by reflection?

Explain how these fringes are formed and describe their appearance. What would be the effect if the spherical lens were replaced by a cylindrical one?

Such a system using a spherical lens is illuminated with light of wavelength 600 nm. When the lens is carefully raised from the plate 50 extra fringes appear and move towards the centre of the fringe system. By what distance was the lens raised? [L]

E24 **(a)** With the aid of a labelled diagram, describe an *experimental* arrangement for observing Newton's rings in reflected light. State *two* conditions that must be satisfied by the reflecting surfaces.
Sketch the appearance of the rings when viewed in monochromatic light.

(b) A glass disc of 10 mm radius with its lower surface specially shaped is placed on a plane glass surface and viewed from above in reflected light from a sodium lamp. A dark circular area, 1.5 mm radius, is seen surrounded by many alternate bright and dark rings of constant spacing, 0.30 mm, between successive bright rings.
Sketch the lower surface of the shaped plate, giving dimensions and angles. Explain why the central area of the fringe pattern is dark.
(Wavelength of sodium light = 5.9×10^{-7} m.) [J, '90]

E25 The amount of light reflected from the surface of a lens can be reduced by coating the surface of the lens with a transparent film of a substance such as magnesium fluoride. The refractive index of the film is less than that of the lens glass.

(a) Explain how the amount of reflected light can be reduced by the use of such a film.

(b) For monochromatic light state the conditions under which the fraction of incident light reflected by the lens would be reduced to zero. Determine the minimum thickness of coating that will produce a minimum amount of reflected light if the coating has a refractive index of 1.40 and the wavelength of the light is 550 nm. [J, '89*]

DIFFRACTION (Chapter 26)

E26 **(a)** Parallel monochromatic light is incident normally on a thin slit of width d and focused on to a screen. Derive the relationship between the wavelength λ and the angle of diffraction θ for the first minimum of intensity on the screen.

(b) If the light has a wavelength of 540 nm and is focused by a converging lens of focal length 0.50 m placed immediately in front of the slit which has a width of 0.10 mm calculate the distance from the centre of the intensity distribution to the first minimum.

(c) Explain how diffraction effects, similar to those referred to above, limit the sharpness of the image produced by a telescope. [J]

E27 The phenomenon of Fraunhofer diffraction may be demonstrated by illuminating a wide slit by a parallel beam of monochromatic light and focusing the light that passes through the slit on to a white screen. A diffraction pattern may then be observed on the screen.

(a) Sketch the intensity variation in the diffraction pattern as a function of distance across it.

(b) What would happen to the intensity variation if the width of the slit were halved? [C]

E28 **(a)** Draw a labelled diagram to show the essential features of a practical arrangement to demonstrate Fraunhofer diffraction of light due to a single narrow slit.

(b) **(i)** Sketch a graph of intensity against angular position showing the form of the Fraunhofer diffraction pattern obtained when monochromatic light is incident normally on a narrow single slit.

(ii) Explain, qualitatively, the formation of the central maximum and the first minimum in the diffraction pattern.

(iii) Light of wavelength 550 nm is incident normally on the slit. Determine the minimum width of the slit that would result in at least four minima appearing in the diffraction pattern within an angle of $\pm 5^\circ$ of the centre of the pattern.

(c) Explain what is meant by the *resolving power* of a telescope objective and discuss how a decrease in the diameter of an objective will affect its resolving power. [J, '92]

E29 An opaque card is pierced by two small holes 4.0 mm apart and strongly illuminated from one side. A lens on the other side of the card focuses images of the holes, 16.0 mm apart, on a screen 125 cm from the card. Find:
(a) the position of the lens,
(b) the focal length.

Why are measurements of the object and image sizes in this experimental arrangement unlikely to yield a reliable value for the focal length of the lens? [L]

E30 A narrow, parallel beam of light of wavelength 0.50 μm is incident normally on a grating with spacing 1.50 μm between the lines. The beams emerging from the grating fall on a screen placed parallel to the grating and distant 0.6 m from it. Draw a diagram to show where there will be light on the screen. Mark in the appropriate distances. [S]

E31 Light from a white source passes through a filter that transmits only the band of wavelengths from 400 nm to 600 nm. When this filtered light is incident normally on a certain diffraction grating, the 400 nm light in one order of the spectrum is diffracted at the same angle, 30°, as the 600 nm light in the adjacent order. Find the spacing between the lines in the grating. [C(O)]

E32 Describe in outline the steps you would take in setting up a diffraction grating in combination with a spectrometer to examine the spectrum of visible light. Explain the function of each part of the system and show on a clear diagram the passage of light rays from the source, through the system, to the obverver's eye.

When the spectrum of light containing violet and red components only is examined with a diffraction grating, it is found that the fourth line from the centre (not counting the zero-order line) is a mixture of red and violet. Explain this. If the grating has 500 lines per mm, and the diffraction angle for the composite line is 43.6°, find the wavelengths of the violet and red components.

What will be the fifth line in the spectrum and at what diffraction angle will it occur? [W]

E33 Light consisting of two wavelengths which differ by 160 nm passes through a diffraction grating with 2.5×10^5 lines per metre. In the diffracted light the third order of one wavelength coincides with the fourth order of the other. What are the two wavelengths and at what angle of diffraction does this coincidence occur? [L]

E34 Draw a diagram to show plane monochromatic waves falling normally on a grating and being diffracted. Using the diagram, explain in which direction diffracted maxima will be seen.

A diffraction grating is fitted on a spectrometer table with monochromatic light incident normally on it. The instrument is adjusted to observe the spectra produced. State *two* ways in which the spectra would be affected if a grating of greater spacing were used.

The slit of the collimator is now illuminated with light of wavelengths 5.890×10^{-5} cm and 6.150×10^{-5} cm. The grating has 6000 lines per cm and the telescope objective has a focal length of 20 cm. Calculate:
(a) the angle between the first-order diffracted waves leaving the grating,
(b) the separation between the centres of the two first-order lines formed in the focal plane of the objective. [J]

E35 A plane diffraction grating is illuminated by a source which emits two spectral lines of wavelengths 420 nm (420×10^{-9} m) and 600 nm (600×10^{-9} m). Show that the 3rd order line of one of these wavelengths is diffracted through a greater angle than the 4th order of the other wavelength. [L]

E36 Give a labelled sketch and a brief description of the essential features of a spectrometer incorporating a plane diffraction grating.

What part is played by **(a)** diffraction, and **(b)** interference, in the operation of the diffraction grating?

A source emitting light of two wavelengths is viewed through a grating spectrometer set at normal incidence. When the telescope is set at an angle of 20° to the incident direction, the second order maximum for one wavelength is seen superposed on the third order maximum for the other wavelength. The shorter wavelength is 400 nm. Calculate the longer wavelength and the number of lines per metre in the grating. At what other angles, if any, can superposition of two orders be seen using this source? [O&C]

E37 **(a)** Explain the action of a plane transmission grating on a plane wavefront incident normally on the grating.

(b) How may such a grating, in conjunction with a spectrometer, be used to determine the wavelength of light from a given monochromatic source? (You may assume that the initial adjustments to the spectrometer and grating have been made.)

(c) A parallel beam of light is incident normally on a diffraction grating having 5.00×10^5 lines per metre. After passing through the grating the light is focused on a screen using a lens of focal length 20.0 cm. If the wavelengths of the two lines in the spectrum are 500×10^{-9} m and 600×10^{-9} m respectively, calculate the separation on the screen of the two images in the first order spectrum.

[AEB, '79]

E38 With careful explanation, deduce the diffraction grating formula

$$n\lambda = d \sin \theta$$

An alternative name for the diffraction grating might be the 'interference grating'. Which name do you consider to be more appropriate in the light of your formula derivation? Give brief reasons.

A diffraction grating is of width L and consists of p equally spaced slits. What is the effect upon both the brightness of the diffraction lines and their angular positions of

(a) blocking alternate slits (keeping L constant),

(b) doubling L (thereby doubling p, but keeping the slit width constant)?

Explain your answers briefly.

The line spectrum of a certain substance consists of three prominent lines; blue (B), yellow (Y) and red (R). When the spectrum is examined with a diffraction grating having $d = 4.00 \times 10^{-6}$ m, it is found that the sequence of lines, moving from the centre, is B, Y, R, B, Y, B, R. Give a brief explanation for this.

Further, it is found that the diffraction angles θ of the fifth and seventh lines are 17.46° and 20.49° respectively, and that the sixth line is at an angular position exactly halfway between them. Find the wavelength of the blue line.

[W]

E39 **(a)** A spectrometer and diffraction grating are adjusted to view the spectrum of a source of light, the plane of the grating being normal to the incident parallel beam. The source emits four discrete wavelengths which are listed in the table below together with the settings of the telescope crosswires on the first-order diffraction maxima.

Wavelength/nm	*Telescope setting*	
	Left	*Right*
448	165.7°	194.4°
501	164.0°	196.0°
588	161.1°	198.9°
668	158.4°	201.6°

If the setting of the crosswires on the central diffraction maximum is 180.0°, use the data to draw a straight-line graph, and use the graph to determine the number of lines per metre ruled on the grating.

(b) **(i)** The source is replaced by a monochromatic one. The crosswire settings for the second-order diffraction maxima are 136.8° (left) and 223.2° (right). Calculate the wavelength of the light emitted by the source.

(ii) Excluding the zero order, what is the total number of diffraction maxima produced? Give the reason for your answer.

(c) The effective width of the grating is halved by masking the outer areas of the grating with masking-tape, and the observations in (b) are repeated. In what ways do the diffraction maxima now seen differ from those seen originally in (b) and in what ways are they similar? Explain your answer. [J]

E40 Draw a clearly labelled diagram, including wavelets originating from four adjacent gaps in a diffraction grating, to illustrate the formation of a second order spectrum for monochromatic light incident normally on the grating.

Calculate:
(a) the angular deviation of this spectrum for light of wavelength 589.0 nm, given that the grating has 2000 lines per centimetre,
(b) the greatest angular deviation this grating can produce for such light at normal incidence. [S]

E41 *List* the adjustments which have to be made to a spectrometer for use with a diffraction grating.

(Details of *how* the adjustments are carried out should *not* be given.) [W, '91]

POLARIZATION (Chapter 27)

E42 Explain why light can be polarized but sound cannot. Describe a method by which a plane-polarized beam of light can be distinguished from a partially plane-polarized beam.

E43 Describe two distinctly different methods of producing plane-polarized light.

E44 What is meant by **(a)** polarization by reflection, **(b)** the angle of polarization.

Calculate the angle of polarization for water of refractive index 1.33.

E45 How would you demonstrate that a beam of light is completely polarized?

A parallel beam of unpolarized light is incident at an angle of 58° on a plane glass surface. The reflected beam is completely polarized. What is:
(a) the refractive index of the glass,
(b) the angle of refraction of the transmitted beam? [S]

E46 Explain what is meant by double refraction. Describe how you could demonstrate experimentally that the two refracted beams produced from a single beam by a piece of calcite are plane-polarized at right angles to each other.

E47 Describe as fully as you can the nature of linearly polarized light.

When unpolarized light falls on the surface of a block of glass the reflected light is partially polarized. If the angle of incidence is $\tan^{-1} n$, where n is the refractive index of the glass, the reflected light is completely linearly polarized. Describe the apparatus you would use and the experiments you would perform in order to verify these statements for a sample of glass of known refractive index.

Why would it be necessary, if very accurate results were required, to use monochromatic light to verify the second statement?

Show that, when the condition for completely polarized light is satisfied, the reflected and refracted beams are at right angles to one another. [O&C]

E48 Explain why polaroid sunglasses are effective in reducing glare. [W, '90]

ELECTROMAGNETIC WAVES AND OPTICAL SPECTRA (Chapter 28)

E49 The graph refers to the output of a light source. It shows how the power output W of the light of a particular wavelength λ varies with the wavelength λ. Describe the spectrum of this source in terms of the phrases *continuous spectrum, line spectrum, emission spectrum, absorption spectrum.*

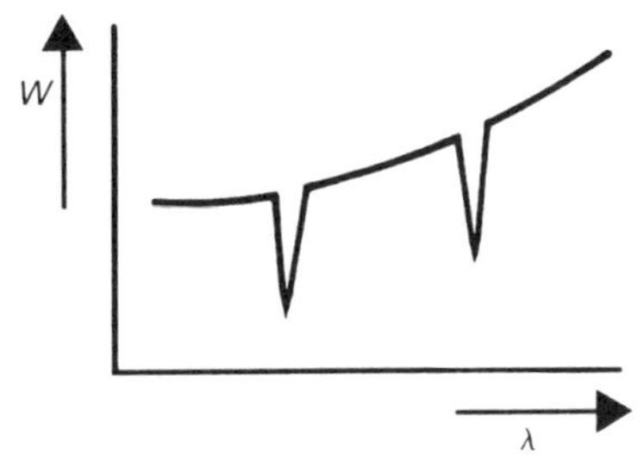

What would be the appearance of the spectrum of this source when viewed by an ordinary diffraction grating spectrometer? [S]

E50 Distinguish **(a)** between an emission spectrum and an absorption spectrum, **(b)** between a continuous spectrum and a line spectrum.

How do line spectra arise? [J]

E51 When a beam of white light is passed through a cloud of sodium vapour the emergent beam is found to contain less sodium yellow light than the incident beam. What changes occur in the sodium atoms to account for this? [J*]

E52 Sketch graphs showing the distribution of energy with wavelength in the visible spectrum (0.4–0.7 μm) of **(a)** the Sun, **(b)** a sodium street lamp.

These spectra may be studied by using a spectrometer with either a diffraction grating, or a prism. State *two* important differences between the spectra produced by these devices. [S]

E53 **(a)** Explain the presence of dark lines in the otherwise continuous spectrum of the Sun in terms of the atomic processes involved.
(b) In a complete solar eclipse the light from the main body of the Sun is cut off and only light from the corona of hot gases forming the outer layer reaches the Earth. State what kind of spectrum you would expect to see for this light. How would it compare with the spectrum referred to in (a)? [J, '91]

RESONANCE, BEATS AND WAVE PHENOMENA (Chapters 29–34)

E54 **(a)** Any system which is able to vibrate can be made to perform *forced vibrations* (oscillations) and may show *resonance*.
Explain the meanings of the italic terms.
(b) What is meant by *damping* and how does the amount of damping affect resonance? [W, '92]

E55 Explain how audible beats arise when two tuning forks of slightly different frequencies f_0 and f are sounded together. Derive an expression for the number of beats heard per second.

Describe an experiment using beats to determine f if f_0 is known.

A clock pendulum has a period of 2.00 s. A simple pendulum set up in front of it gains on the clock so that the two vibrate in phase at intervals of 22.0 s. Calculate:
(a) the period of the simple pendulum,
(b) the fractional change in length of the simple pendulum necessary for the two periods to be equal. [J]

E56 The equation $y = a\sin(\omega t - kx)$ represents a sound wave travelling along the x direction. What is the physical interpretation of the quantities a, y, ω, t, and k? Deduce whether the wave is travelling in the positive or negative x direction.

If $a = 2.0 \times 10^{-7}$ m, $\omega = 8 \times 10^3\,\text{s}^{-1}$ and $k = 25\,\text{m}^{-1}$ calculate **(a)** the speed of the wave, and **(b)** the maximum speed of a particle of the medium disturbed by the wave.

What is the equation of the reflected wave produced when such a wave strikes a rigid boundary normally? By considering the interference of the two waves, show that the resulting displacement is a stationary wave and deduce the distance between successive nodes. [L]

E57 Describe the difference between stationary waves and progressive waves. Outline an experimental arrangement to illustrate the formation of a stationary wave in a string.

Waves, of wavelength λ, from a source, S, reach a common point P, by two different routes.

At P the waves are found to have a phase difference of $3\pi/4$ rad.

Show graphically what this means. What is the minimum path difference between the two routes?

A string fixed at both ends is vibrating in the lowest mode of vibration for which a point a quarter of its length from one end is a point of maximum vibration. The note emitted has a frequency of 100 Hz. What will be the frequency emitted when it vibrates in the next mode such that this point is again a point of maximum vibration? [L]

E58 State the formula for the fundamental frequency of the note emitted by a stretched string which involves the tension in the string, giving the meaning of the symbols used.

A stretched string, of length l and under a tension T, emits a note of fundamental frequency f. The tension is now reduced to half of its original value and the vibrating length changed so that the frequency of the second harmonic is equal to f. What is the new length of the string in terms of the original length? [AEB, '79]

E59 **(a)** When a wire of length l under a tension T is set in transverse vibration in its fundamental mode, the frequency of vibration is n. Describe experiments (one in each instance) to determine the relation between
(i) n and l, when T is constant,
(ii) n and T, when l is constant,
and state the relations you would expect to obtain.

(b) A vertical wire of length L, cross-sectional area A and made of material of density ρ is fixed at its upper end and supports a mass M of volume V at the other. If the mass of the wire is negligible compared with M, use the following data to calculate the frequency of the fundamental mode of transverse vibration of the wire **(i)** before, **(ii)** after M is totally immersed in water.

Length $L = 0.500$ m
Area of cross-section $A = 7.50 \times 10^{-7}$ m^2.
Density $\rho = 8.00 \times 10^3$ kg m^{-3}.
Mass $M = 5.00$ kg.
Volume $V = 3.75 \times 10^{-4}$ m^3.
Density of water $= 1.00 \times 10^3$ kg m^{-3}.
[J]

E60 A sonometer wire of material having density of 8.0 g cm^{-3} is stretched so that its length is increased by 0.10%. The fundamental frequency of transverse vibrations of a part of the wire 50 cm long is then 150 Hz. Calculate:
(a) the velocity with which a transverse wave is transmitted along the stretched wire,
(b) the tension per unit area of cross-section of the wire,
(c) Young's modulus for the material of the wire. [J]

E61 The frequency of a wire vibrating in one loop is given by the formula

$$f = \frac{1}{2l}\sqrt{\frac{T}{m}}$$

where T is the tension and l the length. Explain the meaning of m and indicate how the formula would be altered if the wire vibrated in three loops. Describe an experiment to verify the relation between the length and the tension of a wire vibrating transversely in unison with a tuning fork of fixed frequency. Explain how you would use your observations to draw a linear graph, and how you would use this and any other necessary data to find the frequency of the fork. A tuning fork of frequency f and a wire of constant length vibrating transversely are sounded together. For different tensions beats are heard as follows:

Tension	T	$1.01T$	$0.99T$
Beats per second	2.0	0.7	3.3

Determine the frequency of the wire (in terms of f) when the tension is **(a)** T, **(b)** $1.01T$. When the tension is T, what percentage change in length of the wire is required so that the wire and fork sound in unison? [J]

E62 Describe the motion of the particles of a string under constant tension and fixed at both ends when the string executes transverse vibrations of **(a)** its fundamental frequency, **(b)** the first overtone (second harmonic). Illustrate your answer with suitable diagrams.

A horizontal sonometer wire of fixed length 0.50 m and mass 4.5×10^{-3} kg is under a fixed tension of 1.2×10^2 N. The poles of a horseshoe magnet are arranged to produce a horizontal transverse magnetic field at the midpoint of the wire, and an alternating sinusoidal current passes through the wire. State and explain what happens when the frequency of the current is progressively increased from 100 to 200 Hz. Support your explanation by performing a suitable calculation. Indicate how you would use such an apparatus to measure the fixed frequency of an alternating current. [J]

E63 **(a)** The velocity v of a transverse wave on a stretched string is given by $v = \sqrt{\frac{T}{\mu}}$ where T is the tension and μ is the mass per unit length. Show that the equation is dimensionally correct.

(b) **(i)** Derive an expression for the fundamental frequency of a vibrating wire.

(ii) Describe how you would use a sonometer to determine experimentally how the fundamental frequency of a vibrating wire depends upon *one* of the variables in the expression (b) (i) above.

(c) **(i)** Write down an expression for the frequency of the nth overtone of a vibrating wire of fundamental frequency f.

(ii) Why does the note sounded by a vibrating sonometer wire sound different when it is plucked in the middle from when it is plucked near one end?

(d) The mass of the vibrating length of a sonometer wire is 1.20 g and it is found that a note of frequency 512 Hz is produced when the wire is sounding its second overtone. If the tension in the wire is 100 N, calculate the vibrating length of the wire. [W, '90]

E64 A vibrator of variable frequency f is used to vibrate a horizontal wire which passes over a pulley and has a mass M attached to it as shown below. The length AB of the wire between the end of the vibrator and the pulley is 1.5 m.

When $f = 1000$ Hz a well-defined pattern of nodes and antinodes is seen. As f is gradually reduced this pattern disappears but the next well-defined pattern is seen when $f = 900$ Hz. If the mass of 1 m of the wire is 1.0×10^{-3} kg and assuming the points A and B are nodes then calculate

(a) the speed of the wave along the wire,

(b) the value of M

(Assume $g = 10.0\,\text{m s}^{-2}$.) [W, '92]

E65 When a tuning fork of frequency 256 Hz is sounded together with a sonometer wire emitting its fundamental frequency, 6 beats are heard every second. When the prongs of the tuning fork are lightly loaded, 4 beats are heard every second.

(a) What is the fundamental frequency of the sonometer wire?

(b) If the length of the sonometer wire is 25 cm and the mass per unit length of the material of the wire is $9.0 \times 10^{-3}\,\text{kg m}^{-1}$, calculate the tension in the sonometer wire. [W, '91]

E66 Explain what is meant by the *wavelength*, the *frequency*, and the *speed* of a sinusoidal travelling wave and derive a relation between them.

What is meant by a stationary wave? A stationary sinusoidal transverse wave of period T is set up on a stretched string so that there are nodes only at the two ends of the string and at its midpoint. The displacement of each point of the string has its maximum value at $t = 0$. Show on a single sketch the shape taken by the string at times $t = 0$, $T/8$, $T/4$, $3T/8$ and $T/2$.

A piano string 1.5 m long is made of steel of density $7.7 \times 10^3\,\text{kg m}^{-3}$ and Young's modulus $2 \times 10^{11}\,\text{N m}^{-2}$. It is maintained at a tension which produces an elastic strain of 1% in the string. What is the fundamental frequency of transverse vibration of the string? [O&C]

E67 **(a)** Beats are heard when a tuning fork and a sonometer wire vibrate simultaneously with frequencies f_1 and f_2 respectively, f_2 being slightly different from f_1.

(i) Explain what is meant by the term *beats* and explain how they are formed.

(ii) Show that the beat frequency is equal to the difference between f_1 and f_2.

(b) In an arrangement such as that described in (a), a sonometer wire vibrates under constant tension in its fundamental mode and two beats per second are heard when $f_1 = 512$ Hz. When the length of the wire is increased by 2 mm, beats are no longer heard. Calculate, to three significant figures, **(i)** the initial frequency of vibration, **(ii)** the initial length of the wire. [J]

E68 What is the frequency of the sound emitted by an open-ended organ pipe 1.7 m long when sounding its fundamental frequency?
(The speed of sound in air $= 340\,\text{m s}^{-1}$.)

What would be the effect (if any) on the frequency of the sound emitted of an increase of:
(a) the atmospheric pressure,
(b) the temperature of the air? [S]

E69 A source S emitting sound of frequency 210 Hz is mounted near the open end of a tube which is fitted with a movable piston. It is found that when the piston is moved slowly along the tube, the sound intensity reaches a

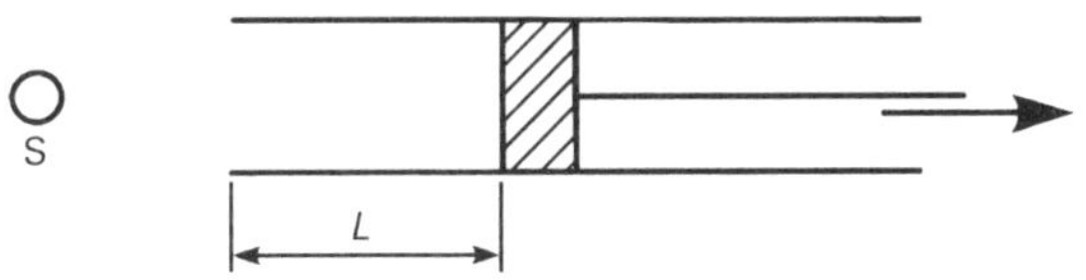

maximum when the length L of the tube between the piston and the open end is 41 cm, and a second maximum occurs when $L = 121$ cm. Explain this, and calculate the speed of sound in air. [S]

E70 Explain the conditions necessary for the creation of stationary waves in air.

Describe how **(i)** the displacement, **(ii)** the pressure vary at different points along a stationary wave in air and describe how these effects might be demonstrated experimentally.

A tube is closed at one end and closed at the other by a vibrating diaphragm which may be assumed to be a displacement node. It is found that when the frequency of the diaphragm is 2000 Hz a stationary wave pattern is set up in the tube and the distance between adjacent nodes is then 8.0 cm. When the frequency is gradually reduced the stationary wave pattern disappears but another stationary wave pattern reappears at a frequency of 1600 Hz. Calculate:
(a) the speed of sound in air,
(b) the distance between adjacent nodes at a frequency of 1600 Hz,
(c) the length of the tube between the diaphragm and the closed end,
(d) the next lower frequency at which a stationary wave pattern will be obtained. [J]

E71 Under what conditions are beats heard? Derive an expression for their frequency.

At 17°C, the air columns in two identical tubes resonate to a tuning fork of frequency 300 Hz. Find the frequency of the beats heard when the air columns are sounded simultaneously, the temperature in one of the air columns having been raised to 27°C. (You may neglect the expansion of the material of the tubes and assume that the columns vibrate in their fundamental modes.) [J]

E72 What do you understand by **(a)** forced vibrations, **(b)** free vibrations, and **(c)** resonance? Illustrate your answer by giving three distinct examples, one for each of (a), (b) and (c).

Explain how a stationary sound wave may be set up in a gas column and how you could demonstrate the presence of nodes and antinodes. State what measurements would be required in order to deduce the speed of sound in air from your demonstration, and show how you would calculate your result. [L]

E73 Two open-ended organ pipes are sounded together and 8 beats per second are heard. If the shorter pipe is of length 0.80 m what is the length of the other pipe? You may ignore any end corrections.
(Speed of sound in air = 320 m s^{-1}.) [L]

E74 **(a)** **(i)** The equation $y = A \sin(\omega t - kx)$ represents a sound wave travelling in the x-direction. Explain clearly what the letters y, A, ω and k represent for a sound wave.
(ii) State the connection between the above sound wave and one which has the equation $y = A \sin(\omega t + kx)$.
(b) Consider the following paragraph:
A sound wave in air is reflected by a solid wall placed at right angles to the direction of travel of the wave, so that the incident and reflected waves are *superimposed*. A system of *stationary waves* is set up and as a result there are regularly spaced displacement *nodes* and *antinodes*.
(i) Explain the meanings of the words in italics giving diagrams where appropriate.
(ii) If the sound wave is produced by a source of frequency 500 Hz, and the distance between adjacent nodes is 34.0 cm, calculate the speed of the sound wave.

(iii) The temperature of the air increases from 20 °C to 30 °C. Calculate the new distance between adjacent nodes.
(Take 0 °C = 273 K.) [W, '91]

E75 (a) Describe how you would use a resonance tube to determine the speed of sound in air inside the tube.
(b) For a vertical resonating column of air in a tube closed at one end, describe the motion of the air particles at various points along the axis of the tube when the tube is sounding **(i)** its fundamental note, **(ii)** its first overtone.
(c) Would you expect the speed of sound in a tube to be greater or less than the speed in the same gas in an open space? Give a reason for your answer. [J]

E76 For a resonance tube closed at one end, l is the shortest resonant length and f is the resonant frequency. When $f = 200$ Hz, $l = 402$ mm and when $f = 250$ Hz, $l = 318$ mm. Calculate the velocity of sound in air. [W, '90]

E77 (a) Distinguish between a progressive wave and stationary wave. Your answer should refer to energy, amplitude and phase.
(b) A small loudspeaker emitting a pure note is placed just above the open end of a vertical tube, 1.0 m long and a few centimetres in diameter, containing air. The lower end of the tube is closed. Describe in detail and explain what is heard as the frequency of the note emitted by the loudspeaker is gradually raised from 50 Hz to 500 Hz. (You may assume that the speed of sound in air is 340 m s^{-1} and need make only approximate calculations.)

The air temperature is now changed. It is found that as the frequency of the note emitted by the loudspeaker is raised from 50 Hz resonance first occurs for a frequency of 86.2 Hz. If the experiment is repeated with the lower end of the tube open, resonance first occurs for a frequency of 171.0 Hz. Calculate the speed of sound in the column of air and the end correction for the tube. [L]

E78 (a) A note of frequency 600 Hz is sounded continuously over the open upper end of a vertical tube filled with water. As the water is slowly run out of the bottom the air in the tube resonates, first when the water level is 130 mm below the top of the tube and next when the level is 413 mm below the top of the tube. Calculate:
(i) the speed of sound in the air in the tube,
(ii) the position of the water level when the third resonance occurs.
(b) Describe the motion of the air particles at various points along the axis of the tube when the air first resonates. [J]

E79 (a) (i) Explain what is meant by *displacement, amplitude* and *wavelength* for a longitudinal progressive wave in air.
(ii) In terms of the motion of the particles give *one* similarity and *two* differences between a progressive wave and a stationary wave for sound waves in air.
(b) The air in a uniform pipe, closed at one end, is vibrating in its fundamental mode. Describe, with the aid of diagrams, the variation along the length of the tube of **(i)** the displacement amplitude of the air molecules, **(ii)** the pressure amplitude.
(c) A student is performing an experiment to measure the speed of sound in air using a uniform tube 800 mm long which stands vertically and is initially full of water. The level of water in the tube can be slowly lowered. A small loudspeaker connected to a signal generator is held over the open end of the tube.
With the frequency emitted by the loudspeaker set at 600 Hz the student lowers the water level and finds resonance for the first time when the length of the air column above the water is 130 mm. He misses the second resonance and finds the third resonance when the air column is 698 mm long.
(i) Show in a sketch the displacement nodes and antinodes for the third resonance.
(ii) Calculate the speed of sound in air and the end correction for the tube.
(iii) What would be the fundamental frequency for this tube if it were open at both ends? [J, '91]

E80 Explain qualitatively the difference between *progressive waves* and *stationary waves.*

Explain, in some detail, each of the following phenomena:

(a) A tuning fork is struck and then held vertically near the ear; as the fork is slowly rotated about a vertical axis maxima and minima of sound are heard.

(b) When two notes of almost equal frequency are sounded together periodic variations in intensity are heard.

(c) A tuning fork is sounded over the open end of a vertical tube the other end of which is closed by a movable piston. As the piston is moved up the tube it is found that, on passing through certain positions, the sound intensity rises to a maximum and then dies away. [L]

E81 **(a)** Explain what is meant by **(i)** the *pitch* and **(ii)** the *quality* of a musical note.

(b) Explain why the quality of the note from a 'closed' pipe differs from that given by an 'open' pipe. [J]

E82 **(a)** Describe a laboratory experiment to determine the speed of sound in air. Show how you would calculate the speed from the readings which you would take.

(b) State the equation relating the speed of sound, v, to the pressure, p, and density, ρ, of the gas through which it is passing. Use this equation to show how the speed of sound depends upon:

(i) the pressure of the air at a given temperature,

(ii) the presence of water vapour in the air.

(c) If the speed of sound in air is $336\ \text{m s}^{-1}$, what would be the length of an open organ pipe giving a fundamental frequency of 96 Hz? If these pipes were sounded together with another open pipe of length 2.10 m, what would be the beat frequency? (Ignore end corrections.) [AEB, '79]

E83 Definc *frequency* and explain the term *harmonics*. How do harmonics determine the *quality* of a musical note?

It is much easier to hear the sound of a vibrating tuning fork if it is **(a)** placed in contact with a bench, or **(b)** held over a *certain* length of air in a tube.Explain why this is so in both these cases and give two further examples of the phenomenon occurring in (b).

Describe how you would measure the wavelength in the air in a tube of the note emitted by the fork. How would the value obtained be affected by changes in **(i)** the temperature of the air, and **(ii)** the pressure of the air? [L]

E84 **(a)** Describe in detail a laboratory method of determining the speed of sound in *free air.* Explain what effect you would expect a change in atmospheric pressure to have on the result.

(b) The wire of a sonometer, of mass per unit length $1.0 \times 10^{-3}\ \text{kg m}^{-1}$, is stretched over the two bridges by a load of 40 N. When the wire is struck at its centre point so that it executes its fundamental vibration and at the same time a tuning fork of frequency 264 Hz is sounded, beats are heard and found to have a frequency of 3 Hz. If the load is slightly increased the beat frequency is lowered. Calculate the separation of the bridges.

By what amount must the load be increased to produce a beat frequency of 10 Hz if the same tuning fork is used? [L]

E85 A source of sound of frequency 2500 Hz is placed in front of a flat wall. If the microphone is moved from the source directly towards the wall a series of minimum values in its output is observed at equally spaced points.

Why does this effect occur? Calculate the separation of these points if the speed of sound in air at this temperature is $330\ \text{m s}^{-1}$. [L]

E86 L is a loudspeaker emitting sound waves of a single frequency and P is a metal plate. A small microphone, M, is positioned between L and P and connected to the Y-input of a CRO with the time base turned off. As M is moved

towards L, the trace height passes through a series of minimum and maximum readings. The distance moved by M between adjacent minima is 0.100 m.

(a) (i) Explain the variation in trace height and calculate the frequency of the sound waves.

(ii) The frequency of the signal is doubled and the experiment repeated. What will be the distance between adjacent minima? Explain your answer.
(Speed of sound waves in air = 340 m s^{-1}.)

(b) P is replaced by a different material which reflects a smaller proportion of the incident waves. Describe how the variation in the trace would be different from that described in (a) (i) when M is moved towards L. [J, '92]

E87 A source of sound frequency 550 Hz emits waves of wavelength 600 mm in air at 20 °C. What is the velocity of sound in air at this temperature? What would be the wavelength of the sound from this source in air at 0 °C? [L]

E88 The speed of sound in air is 341 m s^{-1} at a temperature of 290 K and a pressure of 1.00×10^5 Pa. If the temperature falls to 285 K, and the pressure rises to 1.02×10^5 Pa, does the speed of sound increase or decrease? Justify your statement.

What would be the effect of such a change of atmospheric conditions on the frequency of sound emitted by an organ pipe? [S]

E89 Suggest why the speed of sound in a gas is of the same order of magnitude as the average speed of the molecules of the gas. How does the speed of sound in a gas depend on pressure? [C(O)]

E90 Explain what is meant by the *displacement* of a wave, and distinguish between transverse and longitudinal waves. What is the nature of the disturbance produced by waves **(i)** in air and **(ii)** in a sonometer wire? Describe, with the aid of sketch graphs, the main features of a transverse progressive sinusoidal wave motion. Use these graphs to define *amplitude*, *period* and *wavelength*.

Sketch a curve to show the variation of particle displacement at a point with time for a sound wave in air. Explain why pressure variations develop, and mark on the curve a time at which the instantaneous pressure at the point: **(a)** is equal to the static air pressure and **(b)** reaches its maximum value.

Write down an expression, in terms of density and pressure, for the speed of sound in a gas. Define any other quantity appearing in this expression and explain its significance. Consider how the speed depends on the relative molecular mass of the gas.

The speed of sound in dry air at 20 °C is 356 m s^{-1}. Determine the speed of sound when the air is saturated with water vapour. Assume that the only factor affecting the speed is the relative molecular mass.

The effective relative molecular masses of dry air and saturated air may be taken as 29 and 28 respectively. [W]

E91 **(a)** describe in detail how to determine, by experiment, the speed of sound in air.

(b) The speed of sound in a gas is given by

$$c = \sqrt{\frac{\gamma p}{\rho}}$$

(i) Define each term in the equation.

(ii) Use the equation to deduce how the speed of sound in a gas depends on (1) the temperature, (2) the pressure, (3) the humidity.

(iii) Show that the speed of sound in a gas is of the order of the average molecular speed in the gas. [W]

E92 **(a)** Distinguish between transverse and longitudinal waves. To which of these categories do **(i)** sound waves in air, **(ii)** light waves belong?

Describe an experiment you could carry out in a school laboratory to justify your classification of light waves.

(b) (i) The speed, c, of a transverse progressive wave travelling along a stretched string of mass per unit length μ is given by

$$c = \sqrt{\frac{T}{\mu}}$$

where T is the tension in the string. Show that this equation is dimensionally correct.

(ii) An undamped progressive transverse wave travels along a uniform stretched string. Describe the motion of different points on the path of the wave with reference to frequency, amplitude and phase.

(iii) A uniform heavy rope hangs vertically from a fixed point. If a transverse pulse is generated at the bottom of the rope, discuss qualitatively how you would expect the speed of the pulse to change as it travels up the rope. [J]

DOPPLER EFFECT (Chapter 35)

E93 A train, sounding its whistle, is travelling along a long straight section of track and passes under a low bridge. Explain how the frequency of the note heard by an observer standing on the bridge differs from the frequency emitted. [L]

E94 Deduce expressions for the frequency heard by an observer **(a)** when he is stationary and a source of sound is moving towards him and **(b)** when he is moving towards a stationary source of sound. Explain your reasoning carefully in each case.

Give an example of change of frequency due to motion of source or observer from some other branch of physics, explaining either a use which is made of it, or a deduction from it.

A car travelling at $10\,\mathrm{m\,s^{-1}}$ sounds its horn, which has a frequency of 500 Hz, and this is heard in another car which is travelling behind the first car, in the same direction, with a velocity of $20\,\mathrm{m\,s^{-1}}$. The sound can also be heard in the second car by reflection from a bridge ahead. What frequencies will the driver of the second car hear?

(Speed of sound in air = $340\,\mathrm{m\,s^{-1}}$.) [L]

E95 **(a)** Describe one method by which the speed of light has been measured, explaining how the result is calculated from the observations. Describe one practical situation in which a knowledge of the speed of light, or other electromagnetic radiation, is needed.

(b) An observer, travelling with a constant velocity of $20\,\mathrm{m\,s^{-1}}$, passes close to a stationary source of sound and notices that there is a change of frequency of 50 Hz as he passes the source. What is the frequency of the source?

(Speed of sound in air = $340\,\mathrm{m\,s^{-1}}$.) [L]

E96 A source emitting a note of a certain frequency f approaches a stationary observer at a constant speed of one-tenth the speed of sound in air. The source is then maintained stationary and the observer moves towards it at the same constant speed. Determine from first principles the frequency of the note heard by the observer in each case. [J]

E97 **(a)** Explain what is meant by the *Doppler shift* in the case of light waves and give an expression for its magnitude.

(b) The Sun rotates with a period of 24.7 days and has a radius of 7.00×10^{8} m. For a terrestrial observer, calculate the resultant Doppler shift of light of wavelength 500 nm which is emitted from the solar equator at:

(i) each side of the disc,

(ii) the centre of the solar disc.

Explain briefly, *one* other possible source of Doppler shift to be expected in observations of the solar spectrum.

The speed of light, $c = 3.00 \times 10^{8}\,\mathrm{m\,s^{-1}}$. [J]

E98 What is the *Doppler effect?*

A police radar set emits a parallel beam of electromagnetic radiation at wavelength λ_0 and velocity c, which falls on a motor-car moving directly towards the set with a velocity u.

(a) Derive an expression for the time it takes a wavefront of the radiation initially a distance λ_0 from the car, to reach the car.

(b) Derive an expression for the wavelength λ of the radiation reflected from the car.

(c) if $\lambda_0 = 0.10$ m, $c = 3.0 \times 10^{8}\,\mathrm{m\,s^{-1}}$, and $u = 33\,\mathrm{m\,s^{-1}}$, calculate the change in wavelength of the radiation received at the set after reflection from the car. [J]

E99 The mean wavelength of the sodium D-lines as measured using a laboratory source is 5.893×10^{-7} m. The same line in the light from a particular star has a mean wavelength of 5.895×10^{-7} m when observed from the

Earth. Say whether the Earth and the star are approaching each other or receding from each other. Deriving from first principles any expressions that you use, calculate the velocity of the star with respect to the Earth. You may assume that the velocity of the star is not so great that a relativistic treatment is necessary.
(The speed of light $= 3.0 \times 10^8\,\mathrm{m\,s^{-1}}$.)

E100 **(a)** A source which is emitting sound waves of frequency f_0 is travelling at a speed u towards an observer who is travelling with a speed v in the same direction. Derive an expression for the frequency f heard by the observer.

(b) An engine travelling at constant speed towards a tunnel emits a short burst of sound of frequency 400 Hz which is reflected from the tunnel entrance. The engine driver hears an echo of frequency 500 Hz two seconds after the sound is emitted. Assuming the speed of sound is $340\,\mathrm{m\,s^{-1}}$ calculate the speed of the engine, and its distance from the tunnel when the driver hears the echo.

(c) Explain why the formula derived in (a) should not be applied to electromagnetic waves, and give an expression for the observed frequency in the case of electromagnetic waves. [J]

E101 **(a)** Describe a simple laboratory demonstration of the Doppler effect for sound waves.

(b) Many bats use the Doppler effect for detecting obstacles and prey. One species sends out high frequency sound waves and locates the objects in front of it from an analysis of the reflected waves. If the bat flies at a steady speed of $4\,\mathrm{m\,s^{-1}}$ and emits waves of frequency 90.0 kHz, what is the frequency of the wave detected by the bat after reflection from a stationary obstacle directly ahead of the bat? Derive any equation for the Doppler effect used in the calculation.

(c) State what could be deduced about the obstacle if a bat detected a reflected wave of frequency less than that emitted.

(d) Give an equation for the frequency change observed by an observer moving with a source of radio waves towards a stationary reflector.

Speed of sound $= 340\,\mathrm{m\,s^{-1}}$. [J]

E102 **(a)** Examination of the spectra of light emitted from many galaxies shows spectral lines characteristic of the elements known on Earth but with *wavelengths* longer than those produced on Earth.

(i) How does the Doppler effect explain these observations?

(ii) Write down an equation for the approximate wavelength shift, $\Delta\lambda$, of one particular spectral line, of wavelength λ.
(Define any other symbols in the formula you quote.)

(iii) What conditions are necessary for the formula quoted to be valid?

(iv) A line of wavelength 4.3×10^{-7} m in a spectrum produced on Earth appears at a wavelength 4.5×10^{-7} m in a corresponding spectrum from a particular galaxy. Make quantitative deductions from this information.

(b) When the spectrum of an incandescent gas is examined by a high resolution spectrometer each line of the spectrum is observed to consist of a narrow continuous range of wavelengths. This range increases as the temperature of the gas increases. Explain this effect.
(Speed of light in vacuo $= 3.0 \times 10^8\,\mathrm{m\,s^{-1}}$.) [J]

E103 **(a)** **(i)** Show from first principles that the frequency f_o of sound in still air, heard by a stationary observer as a source of sound of frequency f_s approaches the observer with a velocity v_s is given by

$$f_o = f_s\left(\frac{1}{1-\dfrac{v_s}{c}}\right)$$

where c is the velocity of sound in still air.

(ii) When $f_s = 1.0 \times 10^3$ Hz and $c = 300\,\mathrm{m\,s^{-1}}$, what is the percentage change in the frequency heard by the stationary observer when the source velocity changes from $30\,\text{m}\,\text{s}^{-1}$ to $35\,\text{m}\,\text{s}^{-1}$?

(iii) A source of sound of frequency f_s moves with simple harmonic motion along the x-axis of a co-ordinate system, the motion being symmetrical about the point $x = 0$. Sketch the variation with time of the frequency heard by an observer at the point $x = 0$. Label the points on your sketch corresponding to $x = 0$ and $x = \pm a$, where a is the amplitude of the oscillation. *Do not attempt a mathematical solution.*

(b) The Doppler broadening of a line in the spectrum of light emitted by a gaseous source is due to the motion of the atoms emitting the light.

(i) State *two* factors on which the speed of the atoms in the source depends.

(ii) Determine which gaseous source would have less Doppler broadening, a mercury lamp at 200 °C or a krypton lamp at 0 °C.
(Relative atomic mass of mercury = 200, relative atomic mass of krypton = 84.) [J, '90]

E104 **(a)** **(i)** Describe a simple laboratory demonstration of the Doppler effect for sound waves.

(ii) A source of sound of frequency f_s moves with a speed v_s along a straight line between an observer and a source. The frequency f_o of sound in still air heard by the observer, moving with a speed v_o is given by

$$f_o = \frac{(c \pm v_o)}{(c \pm v_s)} f_s$$

where c is the speed of sound in still air.
State the significance of the ± signs.

(b) A motor-cyclist travels at a constant speed of $72\,\text{km}\,\text{h}^{-1}$ along a straight, level road towards an observer standing at the centre of a bridge over the road. When the motor-cyclist is distant, the observer hears a sound of frequency 65.0 Hz from the engine of the motor-cycle.
What would be the frequency of the sound heard by the observer after the motor-cycle has passed under the bridge and is along way from the observer? You may assume that the air is still, and that the speed of sound, c is $330\,\text{m}\,\text{s}^{-1}$.

(c) A second motor-cyclist now rides alongside the one mentioned in part (b), at the same constant speed of $72\,\text{km}\,\text{h}^{-1}$. The frequency of the sound of the second motor cycle is lower than that of the first motor cycle. When the two machines are travelling towards the observer, from the same direction, the intensity of the sound received varies with a frequency of 3.0 Hz.
Name the effect causing the variation of sound intensity and explain how it arises in this case. What would be the frequency of the variation of sound intensity after the motor-cycles have passed under the bridge and are moving away from the observer?

(d) A *red shift* is observed in the light received on Earth from some galaxies.

(i) Explain what a red shift is and how it occurs.

(ii) What useful information can be obtained from the red shift of a particular galaxy? [J, '92]

E105 Two stars of equal mass m move in a circular orbit of radius a about their common centroid C, as shown below. Observations in the plane of the orbit show that the wavelength of a spectral line from one of the stars varies between 599.9 nm and 600.1 nm in the course of one revolution.

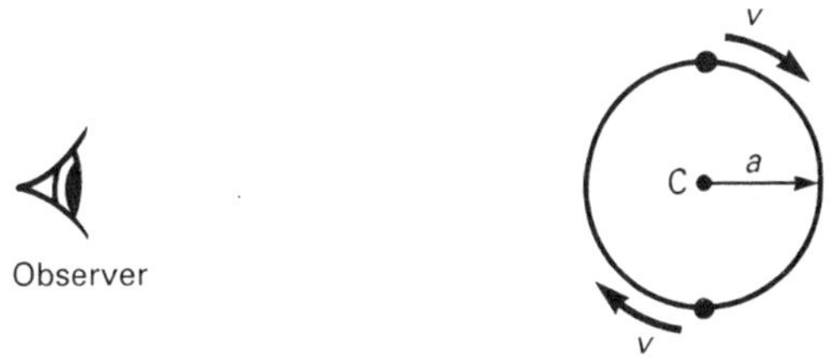

(a) Explain why the observed wavelength λ varies.

(b) Calculate the speed v of the star in its orbit. You may use the relation for the change in wavelength $\Delta\lambda$

$$\frac{\Delta\lambda}{\lambda} = \frac{v}{c}$$

where v is the speed of the star in the line of sight.
(Take the speed of light in free space c to be $3.0 \times 10^8\ \mathrm{m\,s^{-1}}$.)

(c) If the orbital period T of the stars is 3.5×10^6 s, calculate the orbital radius a.

(d) Write down, in terms of m, a and the gravitational constant G, an expression for the gravitational force between the stars.

(e) By equating the centripetal and gravitational forces, derive an expression for the mass m of one of the stars in terms of a, v, and G. [O, '92]

SECTION F

ELECTRICITY AND MAGNETISM

36

CHARGE, CURRENT, POTENTIAL DIFFERENCE AND POWER

36.1 BASIC CONCEPTS

Charge (Q)

In order to explain what are commonly called electrical effects, it is necessary to ascribe to certain 'particles' the property of charge. There are two types of charge – positive and negative. In any process in which charge is transferred from one body to another, the total charge is constant throughout, i.e. **charge is conserved**. The unit of charge is the coulomb (C).

> **The coulomb** is defined as the quantity of charge which passes any section of a conductor in one second when a current of one ampere is flowing, i.e. $1\,\text{C} = 1\,\text{A}\,\text{s}$.

Current (I)

An electric current consists of a flow of charged particles. In metals the charge is carried mainly by electrons. Ions (see Chapter 46) and holes (see Chapter 55) act as charge carriers in other types of conductor. The unit of current is the **ampere** (A).

The ampere is defined in section 41.11 and is the fundamental electrical unit on which the others are based.

The magnitude of the current in a circuit is equal to the rate of flow of charge through the circuit. Thus, for a steady current, I

$$I = \frac{Q}{t}$$

where

I = the (steady) current (in amperes)

Q = the charge (in coulombs) passing some point in time t (in seconds)

When the current is not steady, the instantaneous current, I, is given by

$$I = \frac{\mathrm{d}Q}{\mathrm{d}t}$$

Potential Difference (V)

Whenever a current flows from one point to another, it does so because the electrical potentials at the two points are different. **If two points are at the same potential, no current can flow between them.** The unit of potential difference is the **volt** (V); it is defined in section 39.6.

The action of the chemicals within a cell (or battery) causes each of its two terminals to have a different electrical potential. When a conductor is connected between the terminals of a cell, each end of the conductor acquires the potential of the terminal to which it is connected. Thus, a cell is capable of providing the potential difference (PD) needed to drive current through a conductor. The direction of current flow is conventionally taken to be from points at higher potential to points at lower (i.e. less positive) potential. This is always so, though when the current is being carried by electrons, the electrons themselves actually flow in the opposite direction. Positively charged current carriers, on the other hand, flow in the same direction as conventional current.

Resistance (R)

The electrical resistance of a conductor is defined by

$$R = \frac{V}{I} \qquad [36.1]$$

where I is the current flowing through the conductor when the PD across it is V.

The unit of electrical resistance is the ohm (Ω).

The ohm is defined as being the resistance of a conductor through which a current of one ampere is flowing when the PD across it is one volt, i.e. $1\,\Omega = 1\,\mathrm{V\,A^{-1}}$.

Some conductors have resistances which depend on the currents flowing through them. But the resistances of many conductors (notably metals), depend only on their physical circumstances (e.g. temperature or mechanical strain). This was discovered by Ohm, and such conductors are known as **ohmic conductors** and are said to obey **Ohm's law**.

It follows from equation [36.1] that when R is constant

$$\frac{V}{I} = \text{a constant}$$

i.e. $\quad I \propto V$

Therefore, Ohm's law may be stated as:

The current through an ohmic conductor is directly proportional to the potential difference across it, provided there is no change in the physical conditions (e.g. temperature) of the conductor.

The current–voltage relationships of various non-ohmic conductors are shown in Fig. 36.1 together with that of an ohmic conductor. (Note, in particular, that for the ohmic conductor the graph is a straight line through the origin, i.e. $I \propto V$.)

Fig. 36.1
Current–voltage relationships for various devices

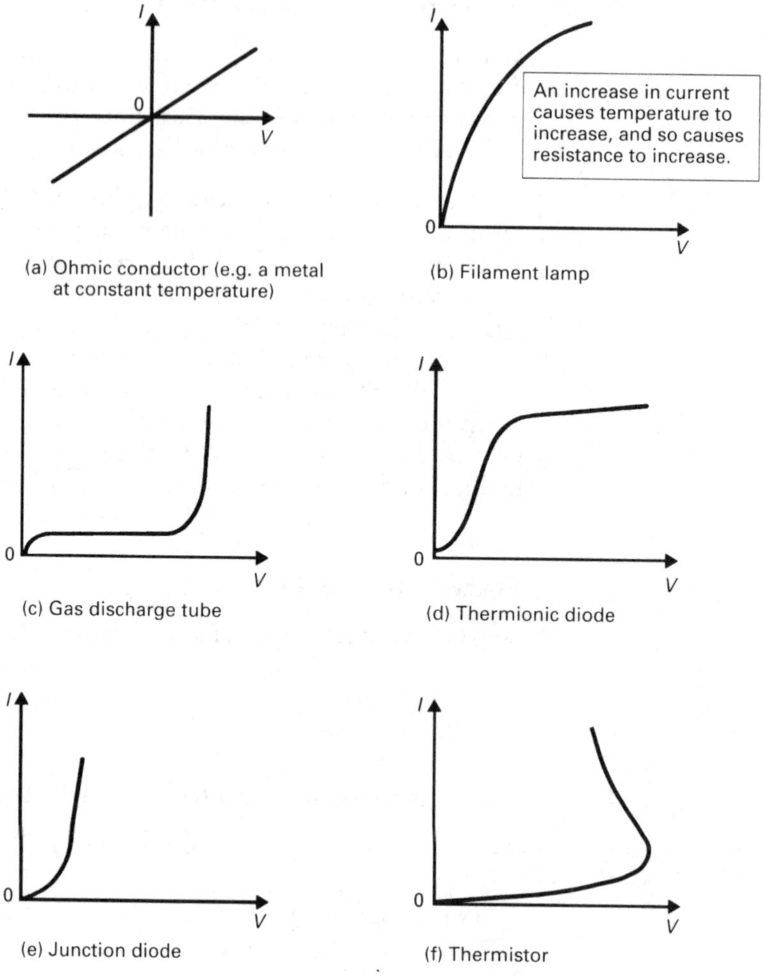

Resistivity (ρ)

The electrical resistivity of a material is defined by

$$R = \frac{\rho L}{A} \qquad [36.2]$$

where

R = resistance of some conductor (Ω)

L = length of the conductor (m)

A = area of cross-section of the conductor (m^2)

ρ = the resistivity of the material of which the conductor is made (Ω m).

The resistivities of various materials are given in Table 36.1. Note the wide range of values.

The experimental determination of the resistivity of a material involves measuring the resistance of a specimen of the material. The specimen must be regularly shaped in order that its dimensions (L and A) can be measured and used in equation [36.2]. If the specimen is in the form of a wire, its diameter should be measured at about six different points.

Table 36.1
Resistivities of various materials at room temperature

Material	*Resistivity*/Ω m
Silver	1.6×10^{-8}
Copper	1.7×10^{-8}
Aluminium	2.8×10^{-8}
Iron	10×10^{-8}
Constantan	49×10^{-8}
Mercury	96×10^{-8}
Germanium	~ 0.5
Alumina	10^{9}–10^{12}
Pyrex	10^{12}
Fused quartz	$>10^{16}$

Conductance (G)

The electrical conductance of a conductor is the reciprocal of its resistance, i.e.

$$G = \frac{1}{R}$$

where

G = electrical conductance (siemens (S) = Ω^{-1})

R = electrical resistance (Ω).

Conductivity (σ)

The electrical conductivity of a material is the reciprocal of its resistivity, i.e.

$$\sigma = \frac{1}{\rho}$$

where

σ = electrical conductivity (S m^{-1} = Ω^{-1} m^{-1})

ρ = electrical resistivity (Ω m).

Current Density (J)

The current density at a point in a conductor is defined as the current per unit cross-sectional area at that point, i.e.

$$J = \frac{I}{A}$$

where

J = current density (A m^{-2})

I = current (A)

A = cross-sectional area (m^2).

QUESTIONS 36A

1. Find the (steady) current in a circuit when a charge of 40 C passes in 5.0 s.

2. Find the (steady) current in a circuit when a charge of 20 μC passes in 5.0 ms.

3. What is the resistance of a copper cylinder of length 12 cm and cross-sectional area 0.40 cm^2? (Resistivity of copper $= 1.7 \times 10^{-8}\,\Omega\,m$.)

4. Find: **(a)** the current, **(b)** the current density when a potential difference of 24 V is connected between the ends of the cylinder described in question 3.

36.2 THE MECHANISM OF CONDUCTION IN METALS

When atoms combine, their outer electrons are used in chemical bonding. The structures of metals are such that each atom (on average) has one outer electron which is not required for bonding and which need not remain localized on its atom. When no current is flowing these 'free' electrons move randomly throughout the structure (typically at $10^6\,m\,s^{-1}$). When a potential difference is put across the metal it produces an electric field in the region occupied by the metal. The 'free' electrons are affected by this field and are urged to move in the opposite direction to it, i.e. to the point of higher potential. (Positively charged 'particles' would move in the same direction as the field, i.e. to the point of lower potential.) Thus, the application of a potential difference superimposes a **small drift velocity** (typically $\sim 10^{-3}\,m\,s^{-1}$) on the random motions of the free electrons. There is now a net flow of charge, i.e. an electric current.

As the electrons move through the metal, they collide with the positive ions of the lattice. (Typically, the average distance travelled between collisions is a hundred atomic diameters.) On collision, the kinetic energy which an electron has gained as a result of being accelerated by the field is transferred to the ion with which it has collided. This increases the vibrational energy of the lattice, and so increases the temperature of the metal. This, then, is the origin of the heating effect of an electric current. If there were no collisions, the electrons would never lose their drift velocity once they had acquired it. In such circumstances, a potential difference would be required only to establish a current in the first place. Once the current had been established, the potential difference could be removed and the current would remain.

The electrical resistance of a conductor is simply a measure of the potential difference required to maintain a current (1 ohm $=$ 1 volt per ampere). Thus, both the heating effect of a current and the electrical resistance are results of electron collisions with the lattice. It is not too surprising that the two phenomena have the same basic cause, for there can be no heating effect if there is no resistance, and when electrical energy is dissipated in this way it is known as **ohmic heating**.

If the temperature of a metal is increased, the amplitudes of vibration of the lattice ions increase and therefore the ions present a larger collision cross-section to the electrons. As a result, the electrons collide more frequently, and this accounts for the observation that the resistivity of a metal increases with temperature.

36.3 DERIVATION OF $I = nAve$ AND $J = nve$

Consider a section of a metallic conductor in which a current is flowing (Fig. 36.2). Let:

I = current flowing in the conductor (A)

L = length of the section considered (m)

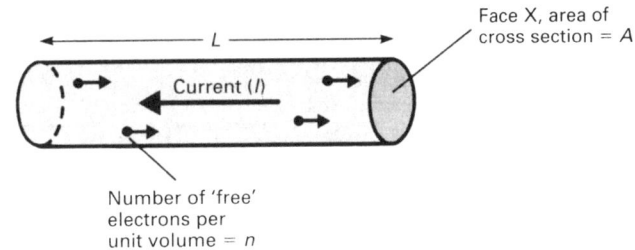

A = cross-sectional area (m^2)

n = number of 'free' electrons per unit volume (m^{-3})

e = charge on each electron (C)

v = average (drift) velocity of the electrons ($m\,s^{-1}$).

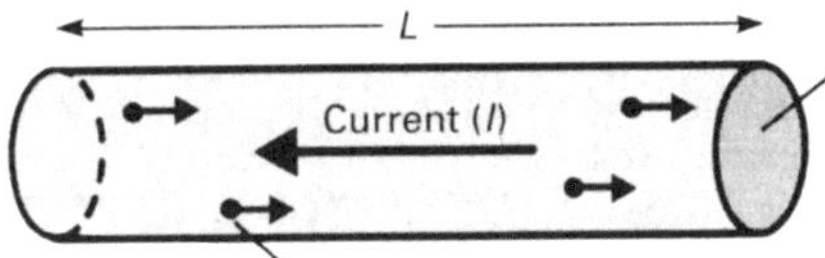

Fig. 36.2 Electron and current motion in a metallic conductor

It follows that

	Volume of section	$= LA$
∴	Number of free electrons in the section	$= nLA$
∴	Total quantity of charge which is free to move	$= nLAe$

The time required for all these electrons to emerge from face X is L/v, and therefore

$$\text{Rate of flow of charge past X} = \frac{nLAe}{L/v}$$

Thus, the current, I, is given by

$$I = nAve$$

Also, since the current density J is given by $J = I/A$ (section 36.1),

$$J = nve$$

36.4 TEMPERATURE COEFFICIENT OF RESISTANCE (α)

The temperature coefficient of resistance, α, is defined by

$$\alpha = \frac{\text{Increase in resistance per °C}}{\text{Resistance at 0 °C}} \qquad [36.3]$$

For any given material the value of α depends on temperature, but the variation is slight and it is meaningful to think in terms of the average value of α between two temperatures. Thus, from equation [36.3], the average value of α between two temperatures θ_1 and θ_2 is given by

$$\alpha = \frac{R_{\theta_2} - R_{\theta_1}}{R_0(\theta_2 - \theta_1)} \qquad [36.4]$$

where

R_{θ_2} = resistance of specimen at θ_2

R_{θ_1} = resistance of specimen at θ_1

R_0 = resistance of specimen at 0 °C.

Note Since it is the difference of θ_2 and θ_1 that is involved in equation [36.4], θ_2 and θ_1 may be measured in either kelvins or degrees Celsius, and values of α are numerically the same whether they are expressed in K^{-1} or $°C^{-1}$. Thus, for copper, $\alpha = 0.0043\,K^{-1} = 0.0043\,°C^{-1}$.

36.5 RESISTORS IN SERIES

Refer to Fig. 36.3.

Fig. 36.3
Resistors in series

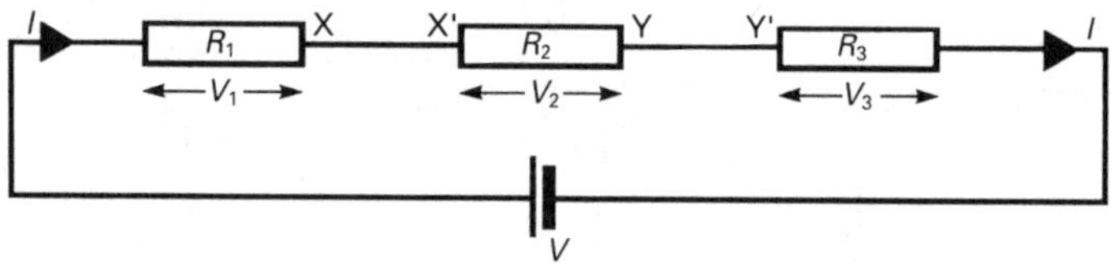

(i) **The same current, *I*, flows through each resistor.**

(ii) Assuming that the connecting wires have no resistance, the potential at X = the potential at X′, and the potential at Y = the potential at Y′, and therefore the total potential difference, *V*, is given by

$$V = V_1 + V_2 + V_3 \quad [36.5]$$

(iii) From the definition of resistance (equation [36.1]), $V_1 = IR_1$, $V_2 = IR_2$, $V_3 = IR_3$ and $V = IR$, where R is the equivalent resistance of the network (i.e. the single resistance which has the same effect as R_1, R_2 and R_3). Substituting for V, V_1, V_2 and V_3 in equation [36.5] gives

$$IR = IR_1 + IR_2 + IR_3$$

i.e.

$$R = R_1 + R_2 + R_3 \quad [36.6]$$

36.6 RESISTORS IN PARALLEL

Refer to Fig. 36.4.

Fig. 36.4
Resistors in parallel

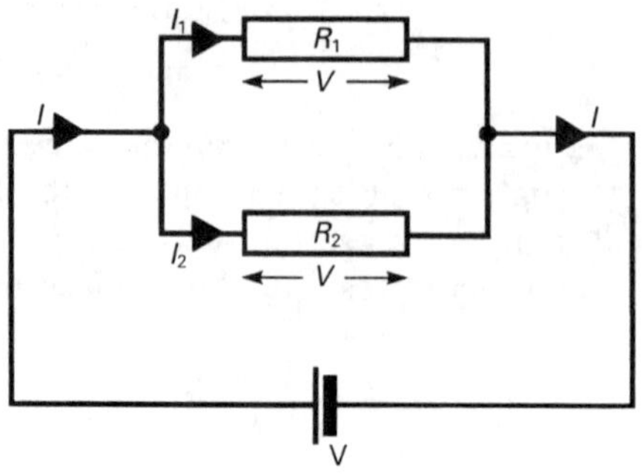

(i) **There is the same potential difference, V, across each resistor.**

(ii) Charge is conserved, and therefore

$$I = I_1 + I_2 \qquad [36.7]$$

(iii) From the definition of resistance (equation [36.1]), $I_1 = V/R_1$, $I_2 = V/R_2$ and $I = V/R$, where R is the equivalent resistance of the network. Substituting in equation [36.7] gives

$$\frac{V}{R} = \frac{V}{R_1} + \frac{V}{R_2}$$

i.e.

$$\frac{1}{R} = \frac{1}{R_1} + \frac{1}{R_2} \qquad [36.8]$$

Note For three resistors $\frac{1}{R} = \frac{1}{R_1} + \frac{1}{R_2} + \frac{1}{R_3}$.

QUESTIONS 36B

In order to answer the questions that follow, make use of equations [36.5] to [36.8] and be prepared to apply $V = IR$ to any section of a circuit, or all of it, as necessary.

1. For the circuit shown below, calculate: **(a)** the total resistance of the circuit, **(b)** I, **(c)** V_1, **(d)** V_2.

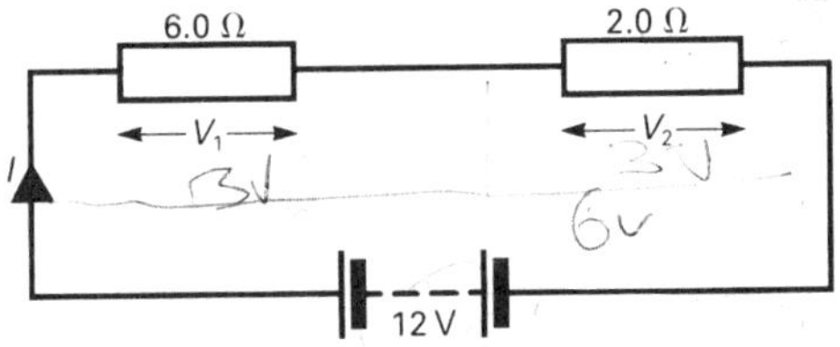

2. For the circuit shown below calculate in the order asked for: **(a)** the total resistance of the circuit, **(b)** I.

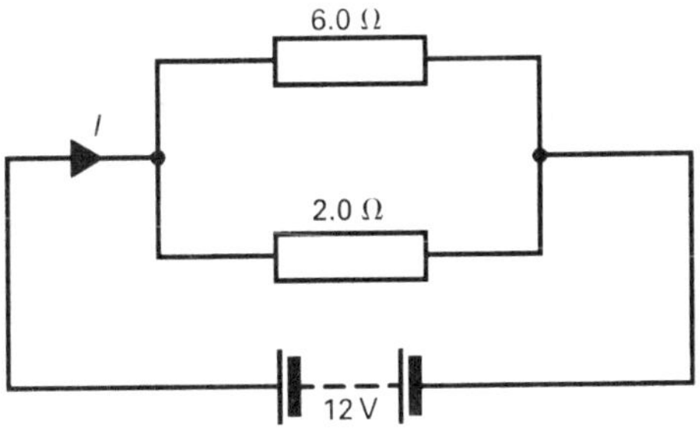

3. For the circuit shown below, calculate, without finding the total resistance of the circuit: **(a)** I_1, **(b)** I_2, **(c)** I.

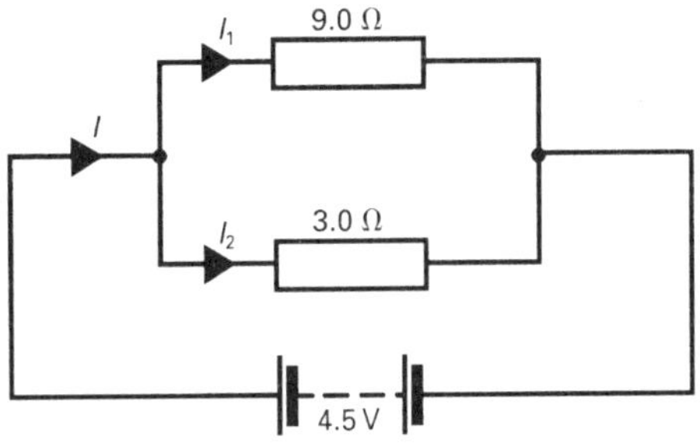

4. For the circuit shown below, calculate: **(a)** the combined resistance of the 12 Ω and 4.0 Ω resistors, **(b)** the total resistance of the circuit, **(c)** I, **(d)** V_1, **(e)** V_2, **(f)** I_1, **(g)** I_2.

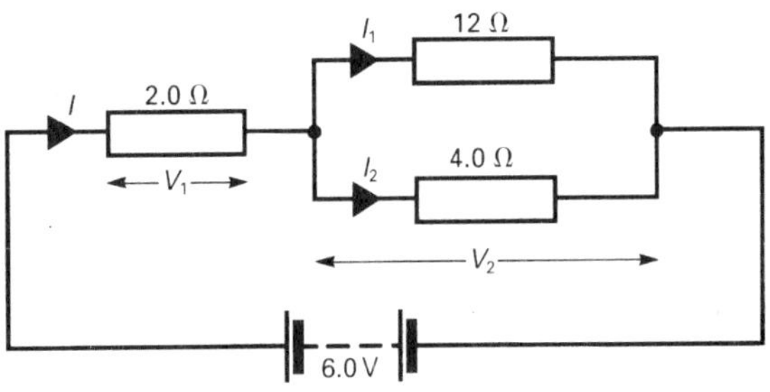

5. Find the value of I in each of circuits **(a)** and **(b)**

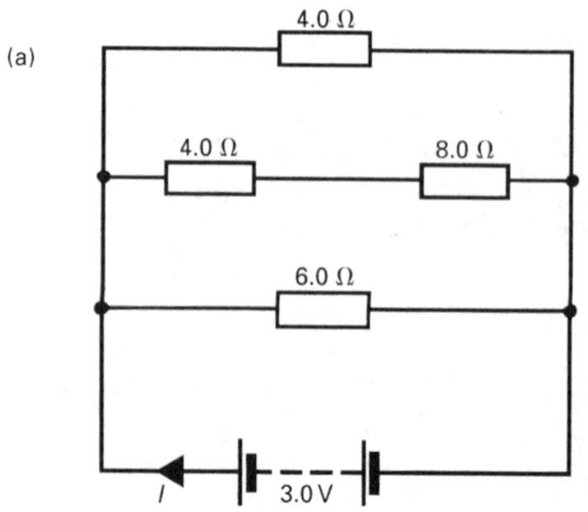

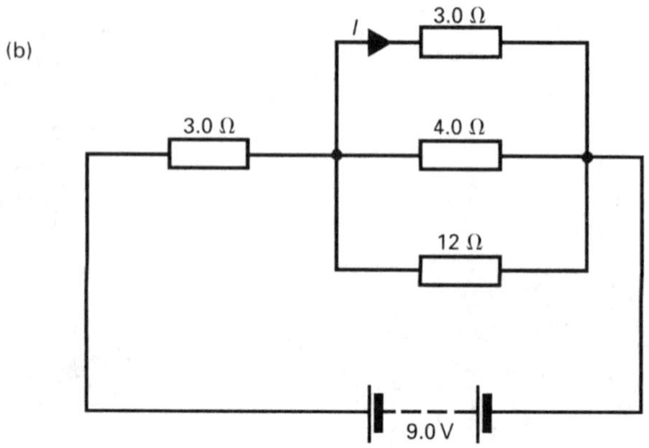

EXAMPLE 36.1

Calculate: (a) the total resistance, (b) I, for the circuit shown in Fig. 36.5.

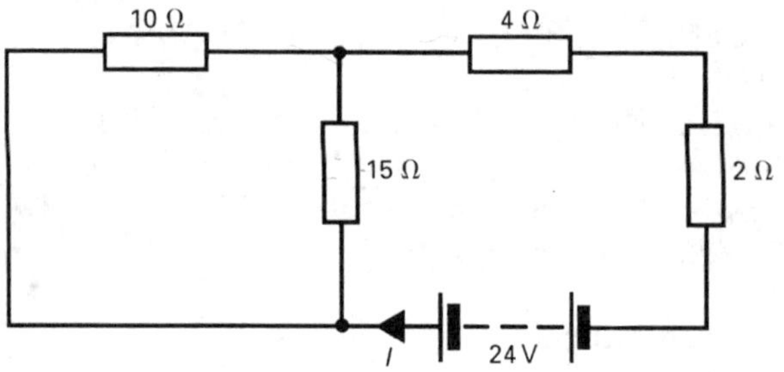

Fig. 36.5 Circuit diagram for Example 36.1

Solution

The problem is made easier by first redrawing the circuit in a more familiar form. To do this we note that the 4 Ω and 2 Ω resistors are in series, and that the 10 Ω and 15 Ω resistors are in parallel; the circuit becomes as shown in Fig. 36.6.

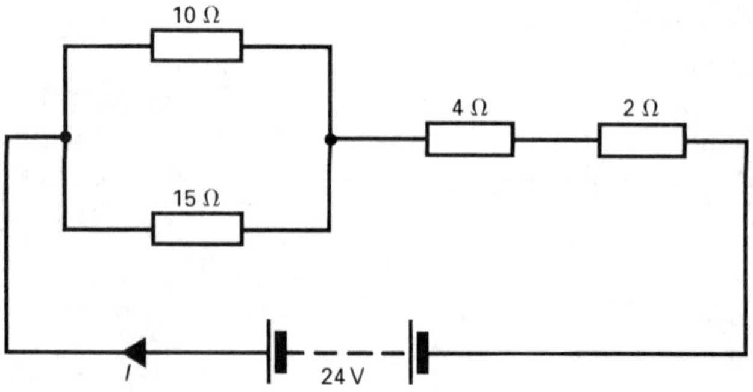

Fig. 36.6 Circuit diagram for Example 36.1 redrawn

(a) The resistance of the parallel combination of 10 Ω and 15 Ω is 6 Ω (from equation [36.8]), and therefore the total resistance of the circuit is $6 + 4 + 2 = 12\,\Omega$.

(b) It follows from $I = V/R$ that $I = 24/12 = 2\,\text{A}$✓

QUESTIONS 36C

1. Find the value of the current, I, in each of the circuits **(a)** to **(e)**.

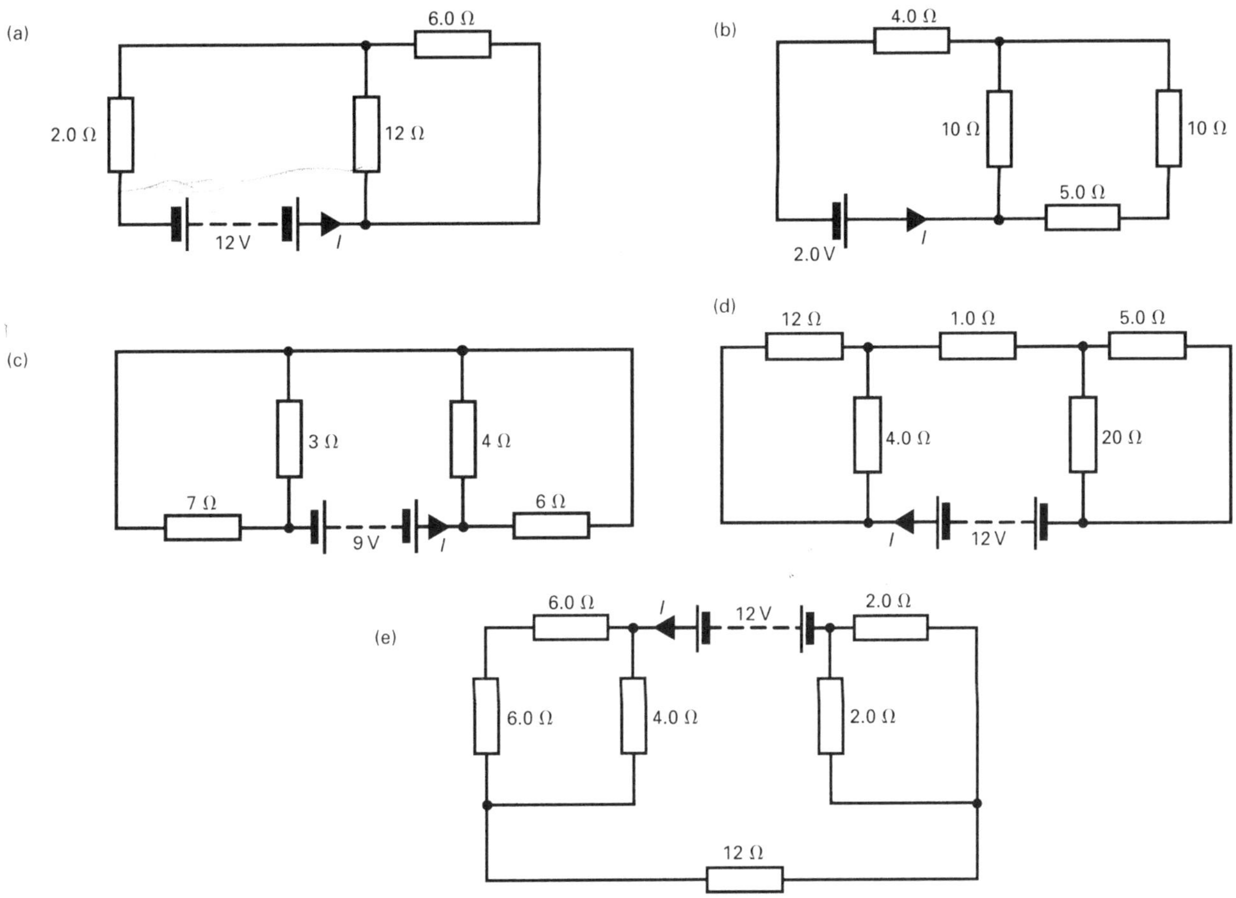

36.7 USE OF A MILLIAMMETER AS AN AMMETER

A moving coil milliammeter (or galvanometer) can be used as an ammeter, provided the extra current is caused to bypass the meter by placing a suitable resistance in parallel with it. Such a resistance is known as a **shunt**.

Consider a milliammeter which has a full-scale deflection of 5 mA (i.e. 0.005 A) and a coil resistance of 10 Ω (Fig. 36.7). Suppose that the meter is to be used as an ammeter with a full-scale deflection of 2 A. Since the maximum current that can be allowed to pass through the coil of the meter is 0.005 A, the rest of the current, 1.995 A, has to go through the shunt. Suppose that the resistance of the shunt is r.

The coil and the shunt are in parallel and therefore there is the same potential difference across each, in which case

$$1.995r = 0.005 \times 10$$

i.e. $\quad r = 0.0251\,\Omega$

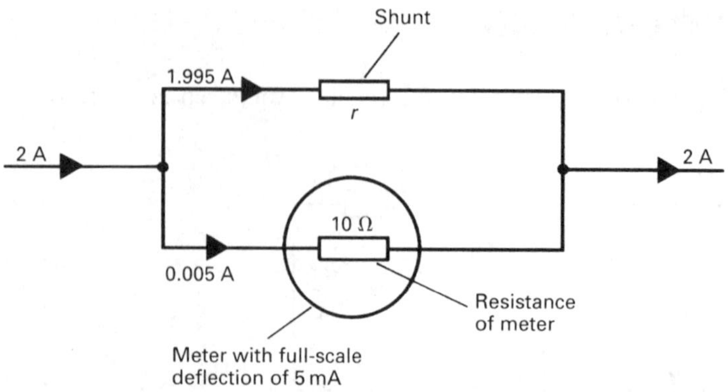

Fig. 36.7
Milliammeter with a shunt in parallel, used as an ammeter

36.8 USE OF A MILLIAMMETER AS A VOLTMETER

Consider, again, a milliammeter with a full-scale deflection of 0.005 A and a coil whose resistance is 10 Ω. When the meter is fully deflected, the potential difference across it is, by $V = IR$, $0.005 \times 10 = 0.05$ V. Suppose that the meter is to be used as a voltmeter with a full-scale deflection of 3 V. Since the maximum potential difference there can be across the coil of the meter is 0.05 V, the rest of the voltage ($3.00 - 0.05 = 2.95$ V) has to be across a resistance which is in series with the meter. Such a resistance is called a **multiplier** or **bobbin**. Suppose that the resistance of the multiplier is R (Fig. 36.8). Since the multiplier is in series with the meter, the current passing through it (when the meter is fully deflected) is 0.005 A. Therefore, applying $V = IR$ to the multiplier gives

$$2.95 = 0.005R$$

i.e. $$R = 590\,\Omega$$

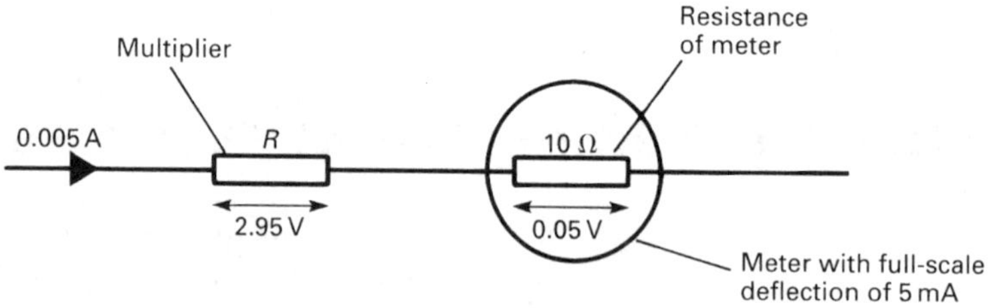

Fig. 36.8
Milliammeter with a multiplier in series, used as a voltmeter

QUESTIONS 36D

1. A milliammeter with a full-scale deflection of 5.0 mA and a coil resistance of 50 Ω is to be used as a voltmeter with a full-scale deflection of 2.0 V. What size resistance needs to be placed in series with the meter?

2. A milliammeter with a full-scale deflection of 20 mA and a coil resistance of 40 Ω is to be used as an ammeter with a full-scale deflection of 500 mA. What size shunt is required?

36.9 MEASUREMENT OF RESISTANCE USING AN AMMETER AND A VOLTMETER

In principle, the resistance, R, of a resistor can be found by measuring the current, I, which flows through it as a result of there being a potential difference, V, across it, and substituting the measured values of I and V into equation [36.1] (i.e. $R = V/I$). If a moving coil ammeter and a moving coil voltmeter are used to measure I and V, the result cannot be very accurate, even if the meters themselves are accurately calibrated. This is because one or other of the meters is bound to register a value which is higher than that which is associated with the resistor.

Suppose that the circuit shown in Fig. 36.9(a) is used. The voltmeter, as required, registers the potential difference across the resistor, but the ammeter records the current through the resistor plus that drawn by the voltmeter. The significance of the error is reduced if the resistance of the voltmeter (typically 50 000 Ω) is much greater than R, because the current drawn by the voltmeter is then much less than that through the resistor.

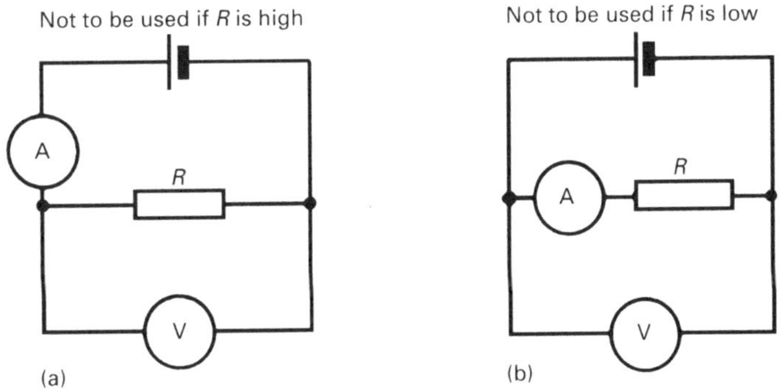

Fig. 36.9 Measurement of: (a) a low resistance, (b) a high resistance

If R is high, the circuit shown in Fig. 36.9(b) should be used, but even so, there is an error. In this case the ammeter, as required, registers the current through the resistor, but the voltmeter records the potential difference across the resistor plus that across the ammeter. The error in the voltmeter reading is small if the resistance of the ammeter (typically less than 1 Ω)* is much less than R, because the potential difference across the ammeter is much less than that across the resistor.

It might be thought that the problem can be overcome by using the meters separately – it cannot, for the removal of one of the meters from the circuit and the subsequent inclusion of the other, alters the potential difference across the resistor and the current through it.

It should be clear to the reader that if the resistance of the ammeter is zero and the resistance of the voltmeter is infinite, each circuit gives an accurate result for all values of R. There are, in fact, no circumstances where it is not desirable that ammeter resistances are zero and that voltmeter resistances are infinite. Whenever an ammeter is put into an existing circuit, it is put in series with the circuit components, and therefore it must have zero resistance if it is not to affect the current in the circuit. A voltmeter, on the other hand, is always placed in parallel with a circuit component, and therefore must have infinite resistance if it is not to draw current through itself and alter the current in the circuit. Thus: **an ideal ammeter has zero resistance; an ideal voltmeter has infinite resistance**.

*The resistance of a milliammeter is higher – typically 50 Ω.

36.10 ELECTROMOTIVE FORCE AND INTERNAL RESISTANCE

The potential difference across the terminals of a cell (the **terminal potential difference**) depends on the size of the current being drawn from the cell. If no current is being drawn (i.e. if the cell is on open circuit), the terminal potential difference has its maximum value – known as the **electromotive force** (EMF) of the cell.

The chemicals within a cell offer a resistance to current flow, known as the **internal resistance** of the cell. When a cell is connected across an external circuit (a load) some of the EMF is used to drive current through the load and the rest of the EMF drives the same current through the internal resistance.

The internal resistance behaves as if it is in series with the cell and the external circuit. Fig. 36.10 represents a cell of EMF, E, and internal resistance, r, connected across a load of resistance, R.

The EMF of a cell (and that of any other source of electrical energy) can be defined as the energy converted into electrical energy from other forms (e.g. chemical, mechanical) when unit charge passes through it.

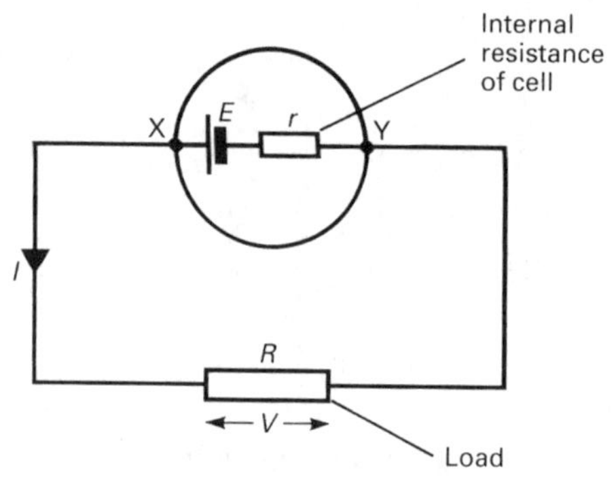

Fig. 36.10
Effect of a cell's internal resistance

It can be shown that, for the circuit of Fig. 36.10, this definition of EMF leads to

$$E = I(r + R) \qquad [36.9]$$

(I: Current delivered by cell; $(r + R)$: Total resistance of circuit)

Equation [36.9] can be rewritten as

$$E = Ir + IR$$

(Ir: PD across internal resistance; IR: PD across external resistance)

i.e. $E = Ir + V$ [36.10]

where V is the potential difference across the external resistance. From equation [36.10], when $\boldsymbol{I = 0}$, $V = E$. V is equal to the potential difference between X and Y – the terminal potential difference, and therefore **when $\boldsymbol{I = 0}$ the terminal potential difference is equal to the EMF**, as previously stated.

Notes (i) The internal resistance of an accumulator is low (typically 0.01 Ω). The internal resistance of a Leclanché dry cell is high (typically 1 Ω), and therefore the maximum current which this type of cell can deliver is low.

(ii) Any source of electrical energy has an internal resistance. For example, the coils in a generator have a resistance, as do the wires which constitute a thermocouple.

Measurement of EMF and Internal Resistance of a Cell

Connect the cell whose EMF (E) and internal resistance (r) are to be measured in the circuit shown in Fig. 36.11.

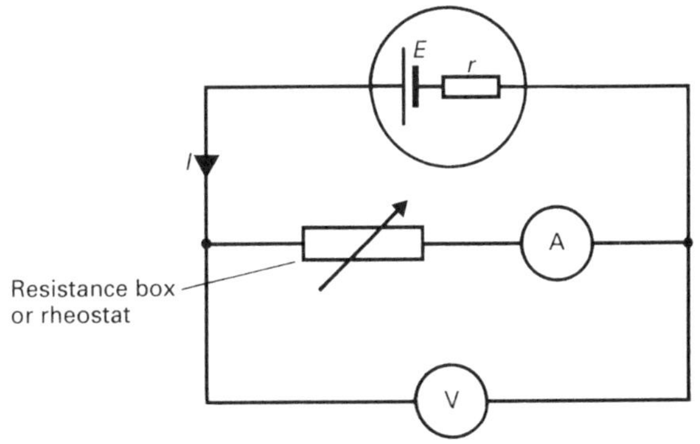

Fig. 36.11 Measurement of EMF and internal resistance of a cell

Record the readings on the voltmeter and ammeter. The voltmeter reading is the terminal potential difference (V) of the cell. Provided the resistance of the voltmeter is much greater than the combined resistance of the ammeter and the variable resistor, the ammeter reading can be taken to be the current (I) delivered by the cell. Obtain about six pairs of values of V and I by making suitable adjustments to the variable resistor.

By analogy with equation [36.10]

$$E = Ir + V$$

$$\therefore \quad V = -rI + E$$

It follows that a graph of V against I (Fig. 36.12) has a gradient of $-r$ and a y-intercept of E; hence r and E.

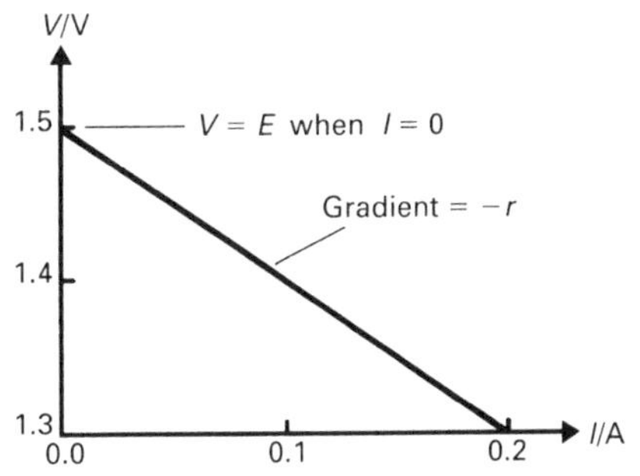

Fig. 36.12 Plot to obtain ***E*** and ***r***

Notes (i) The chemical reactions that occur in a cell when a current is drawn from it can cause the EMF to drop. This is due to an effect known as **polarization**. If it is significant, it leads to misleading results. It can be reduced by breaking the circuit between readings and limiting the current to a maximum of 0.2 A, say. The internal resistance of a typical 1.5 V dry cell is about 1 Ω. If the current is limited to 0.2 A, the terminal PD(V) will range from 1.5 V (when $I = 0$) to 1.3 V (when $I = 0.2$ A). Thus V changes by only 0.2 V and is best measured with a digital voltmeter.

(ii) Alternative methods of measuring EMF and internal resistance are described in section 38.3 and section 38.7 respectively.

EXAMPLE 36.2

When a 10 Ω resistor is connected across the terminals of a cell of EMF E and internal resistance r a current of 0.10 A flows through the resistor. If the 10 Ω resistor is replaced with a 3.0 Ω resistor the current increases to 0.24 A. Find E and r.

Solution

The circuit of Fig. 36.10 applies, initially with $R = 10\,\Omega$ and then with $R = 3.0\,\Omega$. It follows that

$$E = 0.10\,(r + 10) \quad \text{and} \quad E = 0.24\,(r + 3.0)$$

Solving these equations simultaneously gives

$$0.10r + 1.0 = 0.24r + 0.72$$

$$\therefore \quad 0.28 = 0.14r \quad \text{i.e.} \quad r = 2.0\,\Omega$$

Substituting for r in either of the original equations gives

$$E = 1.2\,\text{V}$$

QUESTIONS 36E

1. A battery of EMF 9.0 V and internal resistance 6.0 Ω is connected across a 30 Ω resistor. Find: **(a)** the current through the resistor, **(b)** the terminal PD. **(c)** What would the terminal PD be if the cell were on open circuit?

2. When a 12 V battery (i.e. a battery of EMF 12 V) is connected across a lamp with a resistance of 6.8 Ω the PD across the lamp is 10.2 V. Find: **(a)** the current through the lamp, **(b)** the internal resistance of the battery.

3. A voltmeter with a resistance of 10 kΩ is connected across a power supply which is known to have an EMF of 3.6 kV. The reading on the voltmeter is only 3.0 kV. Find the internal resistance of the power supply.

4. A battery of EMF E and internal resistance r is connected across a variable resistor. When the resistor is set at 21 Ω the current through it is 0.48 A; when it is set at 36 Ω the current is 0.30 A. Find E and r.

5. A voltmeter with a resistance of 20 kΩ connected across a power supply gives a reading of 44 V. A voltmeter with a resistance of 50 kΩ connected across the same power supply gives a reading of 50 V. Find the EMF of the power supply.

36.11 COMBINATIONS OF CELLS

Cells in Series

Refer to Fig. 36.13. The total EMF, E, and the total internal resistance, r, are given by

$$E = E_1 + E_2$$
$$r = r_1 + r_2$$

Fig. 36.13
Total EMF and internal resistance of two cells in series

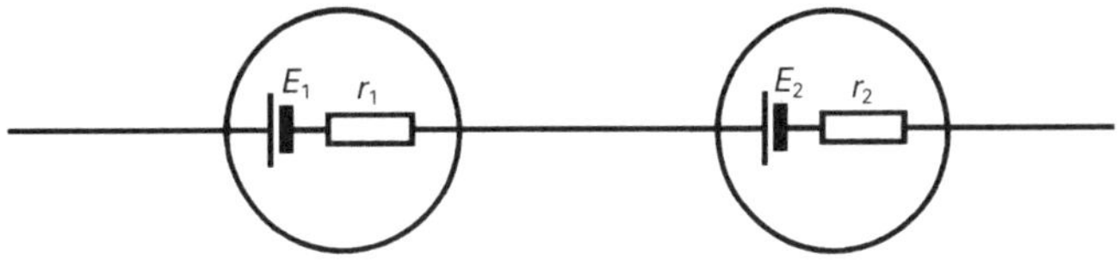

Cells in Parallel

For like cells, refer to Fig. 36.14. The total EMF, E, and the total internal resistance, r, are given by

$$E = E_1$$
$$\frac{1}{r} = \frac{1}{r_1} + \frac{1}{r_1}$$

When unlike cells are connected in parallel, there are no simple relationships and Kirchhoff's rules have to be used.

Fig. 36.14
Total EMF and internal resistance of two cells in parallel

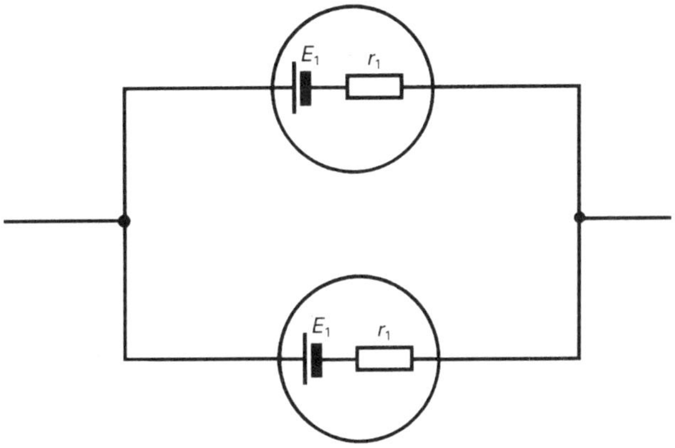

36.12 KIRCHHOFF'S RULES

The rules are useful in solving circuit problems.

Rule 1

The algebraic sum of the currents flowing into a junction is zero, i.e.

$$\Sigma I = 0 \qquad [36.11]$$

For an example of the use of this rule consider Fig. 36.15. If the current flowing into the junction is regarded as being positive, then the current flowing out must be taken to be negative. Thus, from equation [36.11]

$$I_1 + (-I_2) + (-I_3) = 0$$

i.e. $$I_1 = I_2 + I_3$$

Fig. 36.15
Kirchoff's first rule

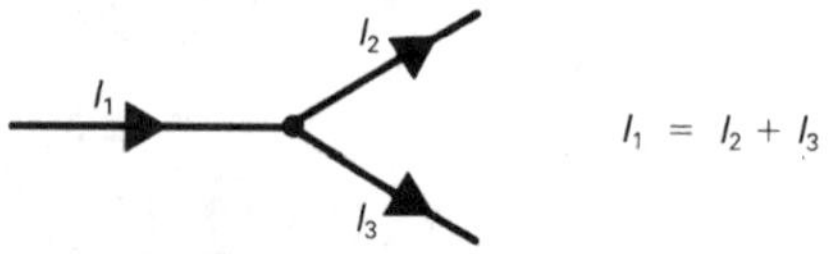

Note The same result could have been derived on the basis of the conservation of charge.

Rule 2

In any closed loop, the algebraic sum of the EMFs is equal to the algebraic sum of the products of current and resistance, i.e.

$$\Sigma E = \Sigma IR \qquad [36.12]$$

When using this rule, **each resistor within a particular loop must be traversed in the same sense** (either clockwise or anticlockwise). Any circuit component which has a current flowing through it in the opposite direction to that in which the loop is being traversed must be regarded as having a negative IR. Example 36.3 illustrates the use of Kirchhoff's rules.

EXAMPLE 36.3

Calculate the values of I_1, I_2 and I_3 in Fig. 36.16(a).

Fig. 36.16
Diagram for Example 36.3: (a) question, (b) redrawn for answer, using Kirchhoff's first rule

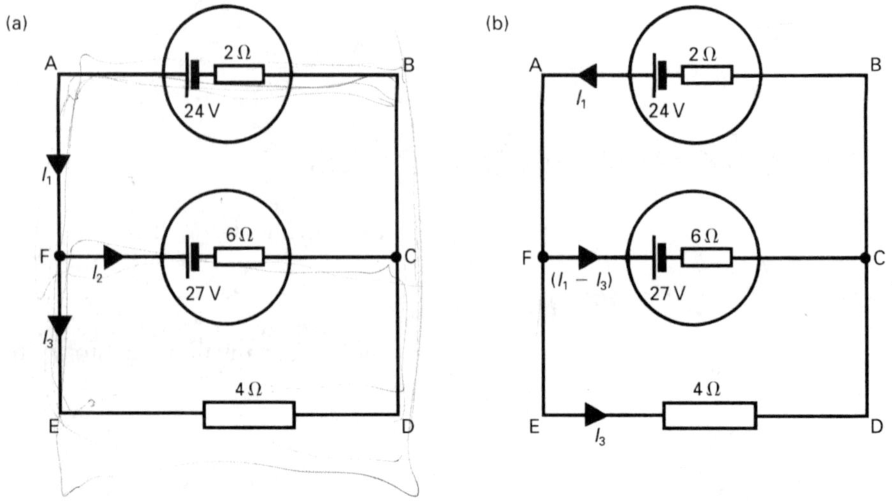

Solution

We can eliminate one of the unknown currents by making use of Kirchhoff's first rule, in which case the circuit may be redrawn as shown in Fig. 36.16(b). Traversing AEDBA in an anticlockwise sense we have, by Kirchhoff's second rule

$$24 = I_3 \times 4 + I_1 \times 2$$

i.e. $$24 = 4I_3 + 2I_1 \qquad [36.13]$$

Traversing FEDCF in an anticlockwise sense, we have

$$27 = I_3 \times 4 - (I_1 - I_3) \times 6$$

Because $(I_1 - I_3)$ is flowing from F to C and we are traversing from C to F

i.e. $$27 = 10I_3 - 6I_1 \qquad [36.14]$$

Solving equations [36.13] and [36.14] simultaneously we have

$$I_1 = 3\,\text{A} \quad \text{and} \quad I_3 = 4.5\,\text{A}$$

Also, $I_2 = I_1 - I_3$, therefore

$$I_2 = -1.5\,\text{A}$$

The presence of the minus sign indicates that the current of 1.5 A flows from C to F, and not from F to C as shown.

Note We could have used AFCBA in place of FEDCF (say) in which case equation [36.14] would have been replaced by

$$24 - 27 = (I_1 - I_3) \times 6 + I_1 \times 2$$

i.e. $$3 = 6I_3 - 8I_1$$

EXAMPLE 36.4

Find the PD between A and C in Fig. 36.17.

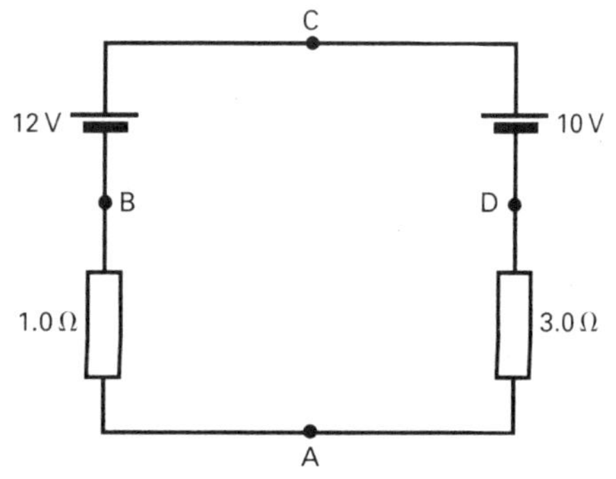

Fig. 36.17 Diagram for Example 36.4

Solution

Traversing ABCDA in a clockwise sense we have, by Kirchhoff's second rule,

$$12 - 10 = I \times 1.0 + I \times 3.0$$

i.e. $$I = 0.5\,\text{A}$$

$\therefore$ PD between A and B $= 0.5 \times 1.0 = 0.5\,\text{V}$

and PD between A and D $= 0.5 \times 3.0 = 1.5\,\text{V}$

The potential decreases by 0.5 V from A to B (because current is flowing from A to B, and current flows from high potential to low potential) and then increases by 12 V from B to C. The PD between A and C is therefore $12 - 0.5 = 11.5$ V.

Alternatively, the potential increases by 1.5 V from A to D and then increases by 10 V from D to C, giving a total increase from A to C of $10 + 1.5 = 11.5$ V.

QUESTIONS 36F

1. Find the current through the 10 Ω resistor in the circuit opposite.

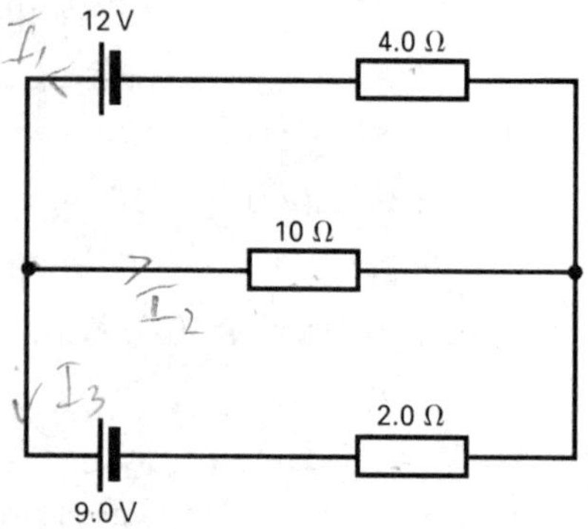

36.13 ENERGY AND POWER IN DC CIRCUITS

Consider a simple circuit in which a steady current, I, is flowing through a load. The load may be an electric motor for example, or an accumulator on charge, or simply a resistor. As the current flows through the load it dissipates energy in it. The energy dissipated is equal to the potential energy lost by the charge as it moves through the potential difference that exists between the input and output terminals of the device.

It follows from the definition of potential difference (section 39.6) that the energy dissipated is given by

$$W = QV$$

where

W = the energy dissipated in some time t

Q = the charge which flows in time t

V = the potential difference across the load.

The current is steady, and therefore

$$Q = It$$

Therefore

$$W = VIt \qquad [36.15]$$

The rate of dissipation of energy, $\mathrm{d}W/\mathrm{d}t$, is known as the **power**, P, and therefore since I and V are constant

$$P = VI \qquad [36.16]$$

The unit of power is the **watt** (W). $1\,\text{W} = 1\,\text{J s}^{-1}$. Note that in equations [36.15] and [36.16], W and P are in joules and watts respectively when V, I and t are expressed in volts, amperes and seconds respectively.

Note Energy values are sometimes expressed in kilowatt hours. One **kilowatt hour** is the energy consumed by a rate of working of one kilowatt for one hour, i.e.

$$1\,\text{kW h} = 1\,\text{kW} \times 1\,\text{h}$$
$$= 1000\,\text{W} \times 3600\,\text{s}$$

i.e. $$\mathbf{1\,kW\,h = 3.6 \times 10^6\,J}$$

When the load is a resistor the energy is dissipated as heat in the resistor. If its resistance is R, then from the definition of resistance, $R = V/I$, equations [36.15] and [36.16] may be rewritten as

$$W = I^2Rt = \frac{V^2}{R}t \qquad [36.17]$$

and

$$P = I^2R = \frac{V^2}{R} \qquad [36.18]$$

Note Equations [36.15] and [36.16] apply to any load. **Equations [36.17] and [36.18] apply only to resistors** (and to devices such as lamps where all the energy being consumed is being dissipated as heat). An example should illustrate this. Suppose that a current of 2 A flows through an electric motor when the potential difference across it is 30 V, and suppose that the resistance of the motor windings is 4 Ω. Notice that $R \neq V/I$ ($4 \neq 30/2$). The rate of production of heat in the windings is given by $P = I^2R$ as $2^2 \times 4 = 16$ W. The total power is given by $P = VI$ as $30 \times 2 = 60$ W. Thus $60 - 16 = 44$ W is the rate of production of mechanical energy.

QUESTIONS 36G

1. A battery of EMF 6.0 V and internal resistance 2.0 Ω is connected across a torch bulb with a resistance of 10 Ω. Calculate: **(a)** the current supplied by the battery, **(b)** the power consumed by the bulb, **(c)** the power consumed in the internal resistance of the battery. Verify that the total power supplied by the battery is equal to the product of its EMF and the current it supplies.

36.14 THE MAXIMUM POWER THEOREM

Consider the circuit shown in Fig. 36.18, in which a cell of EMF, E, and internal resistance, r, is driving a current, I, through a load of resistance, R.

The power, P, delivered to the load is given by

$$P = I^2R$$

But $$I = \frac{E}{R+r}$$

Therefore

$$P = \frac{E^2R}{(R+r)^2}$$

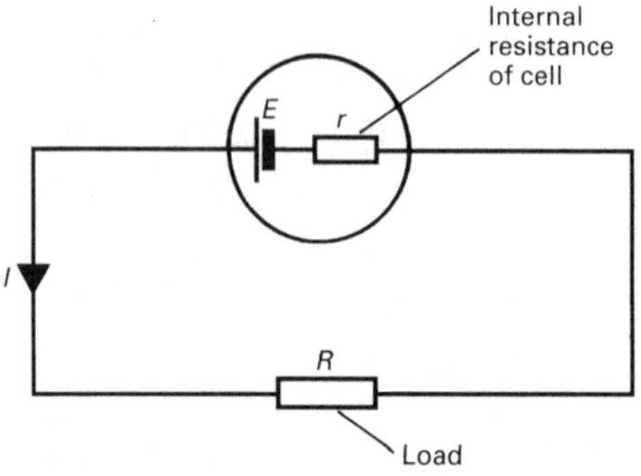

Fig. 36.18 Circuit to illustrate maximum power theorem

To discover the value of R which makes P a maximum we need to differentiate with respect to R and equate to zero. Thus

$$\frac{\mathrm{d}P}{\mathrm{d}R} = \frac{(R+r)^2E^2 - E^2R\cdot 2(R+r)}{(R+r)^4}$$

i.e.

$$\frac{\mathrm{d}P}{\mathrm{d}R} = \frac{E^2(r-R)}{(R+r)^3}$$

When $\mathrm{d}P/\mathrm{d}R = 0$, $R = r$ or $R = \infty$. When $R = \infty$ no current flows, and therefore no power is transferred to the load. This leaves $R = r$ as the condition for maximum power transfer. (If the reader is not convinced that P is a maximum when $R = r$, he should verify it by the usual calculus method or by sketching a simple graph.)

The result applies to any source of EMF and can be stated as:

> A given source of EMF delivers the maximum amount of power to a load when the resistance of the load is equal to the internal resistance of the source.

Note $R = r$ is the condition that the power delivered by a given source is a maximum. If a source with the same EMF but a lower internal resistance is used, it transfers a greater amount of power to the load even though it is not the maximum power it is capable of providing. (A 6 V accumulator can provide more power than a 6 V dry battery.)

36.15 CURRENT–VOLTAGE CHARACTERISTIC OF A THERMISTOR

The resistance of a thermistor decreases with increasing temperature (i.e. it has a negative temperature coefficient of resistance)*. Passing a current through a thermistor generates heat in it and therefore increases its temperature. Thus, **the greater the current, the greater the temperature and therefore the lower the resistance**.

The current–voltage characteristic of a commonly used thermistor (type TH-3) is shown in Fig. 36.19. It can be obtained using the circuit of Fig. 36.20.

The current through the thermistor is increased (in steps) by decreasing the resistance of the resistance box (from 10 000 Ω down to about 20 Ω). For the

*Thermistors with positive temperature coefficients of resistance are also available but are less common.

Fig. 36.19 Current–voltage characteristic of a thermistor

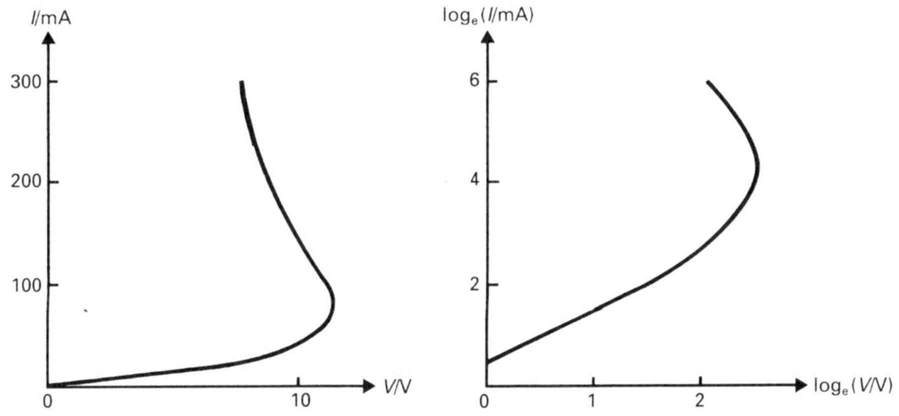

Fig. 36.20 Circuit to obtain current–voltage characteristic of a thermistor

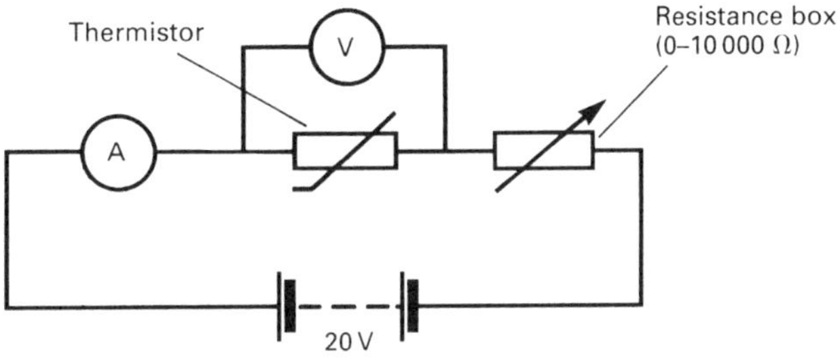

higher values of current it is necessary to wait a few minutes each time the resistance box is adjusted in order to allow the thermistor to reach thermal equilibrium, and therefore for the current and voltage readings to become steady.

Notes

(i) The resistance box makes it possible to alter the current through the thermistor whilst at the same time limiting the current. It is important that the current is limited in some way. If it is not, the situation is unstable because an increase in current causes the resistance of the thermistor to drop, which causes the current to increase even more, and so on – a case of positive feedback. The behaviour of a thermistor can be contrasted with that of a filament lamp. The lamp is inherently stable because any increase in current is accompanied by an increase in resistance which prevents any further current increase – a case of negative feedback.

(ii) The wide range of values of current makes a digital ammeter particularly useful.

CONSOLIDATION

Current flows from high potential to low potential.

$V = IR$ applies to all conductors, but $V = IR$ where R is a constant applies only to ohmic conductors, i.e. to conductors which obey Ohm's law.

Ohm's law. The current through an ohmic conductor is directly proportional to the PD across it, provided there is no change in the physical conditions (e.g. temperature) of the conductor.

An ideal ammeter has zero resistance.

An ideal voltmeter has infinite resistance.

The terminal PD of a cell is the PD across the terminals. If no current is being drawn from the cell, the terminal PD has its maximum value – the EMF of the cell.

37

THE WHEATSTONE BRIDGE

37.1 DERIVATION OF BALANCE CONDITION

The Wheatstone bridge circuit (Fig. 37.1) provides an accurate means of comparing an unknown resistance (X) with known resistances (P, Q and R).

Fig. 37.1
Wheatstone bridge circuit

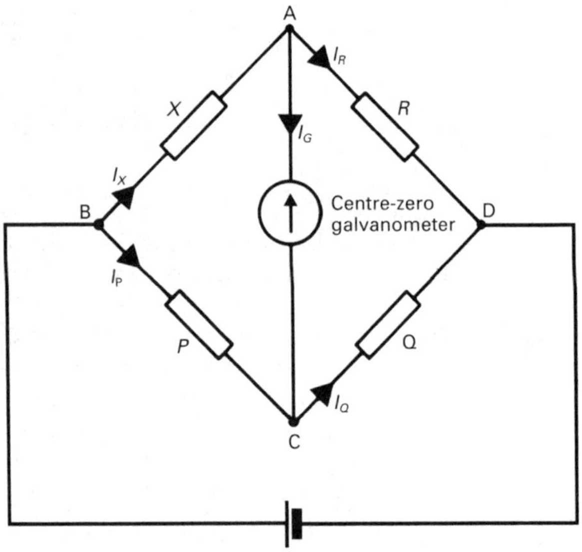

If the resistances X, P, Q and R are such that no current flows through the galvanometer, the bridge is said to be balanced.

At balance, $I_G = 0$, and therefore

$$\text{Potential at A} = \text{Potential at C}$$

$$\therefore \quad V_{BA} = V_{BC} \quad \text{and} \quad V_{DA} = V_{DC}$$

$$\therefore \quad I_X X = I_P P \quad \text{and} \quad I_R R = I_Q Q$$

Dividing gives

$$\frac{I_X X}{I_R R} = \frac{I_P P}{I_Q Q}$$

But, because $I_G = 0$,

$$I_X = I_R \quad \text{and} \quad I_P = I_Q$$

$$\therefore \quad \frac{X}{R} = \frac{P}{Q} \qquad [37.1]$$

37.2 THE METRE BRIDGE

It is clear from equation [37.1] that if R is known, it is sufficient to know the ratio of P and Q in order to determine X. The metre bridge is a practical form of the Wheatstone bridge circuit which makes use of this.

In Fig. 37.2, AB is a resistance wire (e.g. constantan) of uniform diameter, and therefore its resistance per unit length is the same at all points along its length. Its resistance is typically 10 Ω, and it is usually 1 m long – hence the name.

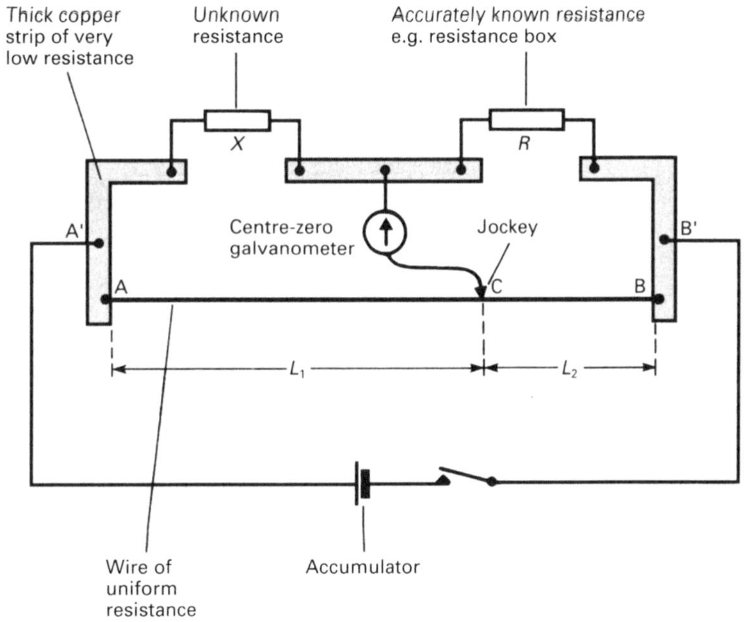

Fig. 37.2 Metre bridge

Let

$$R_{AC} = \text{resistance of AC}$$

$$R_{CB} = \text{resistance of CB}$$

Therefore, from the theory of the circuit (equation [37.1]), at balance

$$\frac{X}{R} = \frac{R_{AC}}{R_{CB}}$$

Therefore, since the wire is uniform,

$$\frac{X}{R} = \frac{L_1}{L_2} \quad [37.2]$$

where L_1 and L_2 are the lengths of AC and CB respectively.

Procedure

The switch is closed and the jockey (a knife-edged movable contact) is placed first at A and then at B. If the resulting galvanometer deflections are in opposite senses, it can reasonably be assumed that the circuit is properly connected. With the switch still closed, the balance point is found by trial and error, by placing the jockey at different points along AB. (The jockey should not be scraped along AB, because doing so could spoil the uniformity of the wire.) Once the balance point has been located, the values of L_1 and L_2 are noted and used in equation [37.2] together with the known value, R, in order to determine X.

The greatest accuracy is achieved when the balance point is close to the middle of AB. There are two distinct reasons for this.

(i) When the balance point is near the middle, both L_1 and L_2 may be measured to a reasonable degree of accuracy. The error in the ratio, L_1/L_2, as a result of errors in measuring the lengths L_1 and L_2, is then (close to) a minimum.

(ii) If (say) L_1 were small, the error introduced by any contact resistance (for example, as a result of poor soldering) at A would then be very significant. Such errors are known as **end-errors**; corrections applied to take account of them are **end-corrections**.

Notes (i) If R is variable, it may be adjusted so that the balance point is near the middle of AB.

(ii) If there is an end-error, repeating the determination with X and R interchanged yields a different value of X from that obtained originally. If the difference is slight, the average may be taken, but if the difference is large, the measurements should be repeated using different apparatus.

Further Points

(i) It is a **null-deflection** method (i.e. the operator looks for zero deflection – not for a reading of zero) and therefore the balance condition can be found with a high degree of sensitivity.

(ii) At balance, the galvanometer takes no current and therefore the result is not affected by faulty calibration of the galvanometer.

(iii) The error inherent in the voltmeter–ammeter method of measuring resistance (section 36.9) is not present.

(iv) Variation in the EMF of the supply does not affect the balance point.

(v) It cannot be used to measure very low resistances ($<1\ \Omega$), because the unknown resistances of the wires used to connect X and R into the bridge are then significant.

(vi) If a very high resistance ($>1\ \text{M}\Omega$) is being measured, a highly sensitive galvanometer is required. This is because the presence of high resistances in the circuit causes the current through the galvanometer to be very low even when far from balance. The inclusion of a suitable galvanometer protection circuit is then necessary (see section 38.3).

37.3 MEASUREMENT OF TEMPERATURE COEFFICIENT OF RESISTANCE

The temperature coefficient of resistance of (say) copper can be measured by using the circuit shown in Fig. 37.2. The unknown resistance, X, is a length of insulated copper wire, wrapped around the lower end of a thermometer and contained in a beaker of water (Fig. 37.3) so that its temperature may be varied. The wire should be both long and thin, for it needs to have a resistance of about $5\ \Omega$ so that the resistances of the connecting wires are not significant. The resistance is measured at a number of different temperatures and a graph of resistance against temperature is plotted (Fig. 37.4).

Fig. 37.3
Environment for copper wire to measure its temperature coefficient of resistance

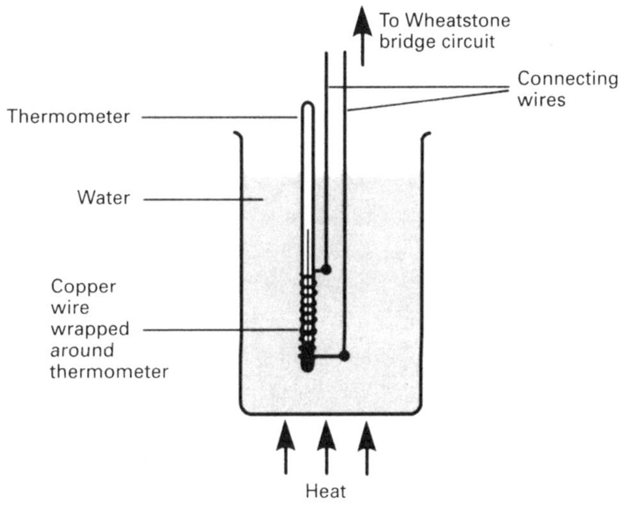

Fig. 37.4
Plot to determine temperature coefficient of resistance

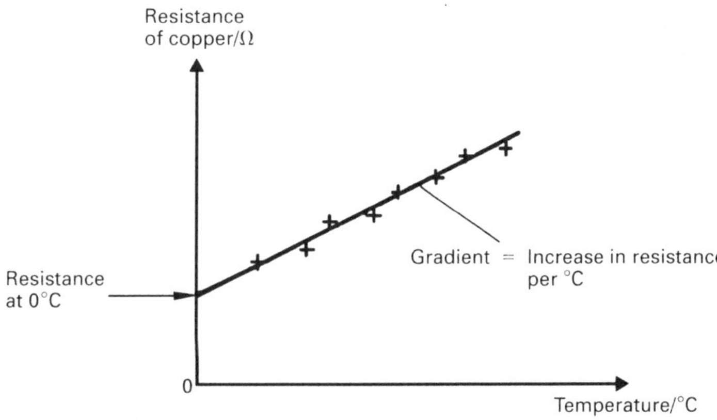

The temperature coefficient of resistance, α, is given by equation [36.3] as

$$\alpha = \frac{\text{Increase in resistance per } ^\circ\text{C}}{\text{Resistance at } 0^\circ\text{C}}$$

i.e. $$\alpha = \frac{\text{Gradient of graph}}{\text{Intercept on } y\text{-axis}}$$

37.4 THE STRAIN GAUGE

One common form of strain gauge consists of a small sheet of polyester to which a thin metal foil is firmly bonded in a zig-zag fashion as shown in Fig. 37.5. The foil is a copper–nickel alloy and typically has a resistance of about 100 Ω. Two low-resistance flying leads allow for electrical connection to a resistance measuring device such as a Wheatstone bridge.

Fig. 37.5
Strain gauge

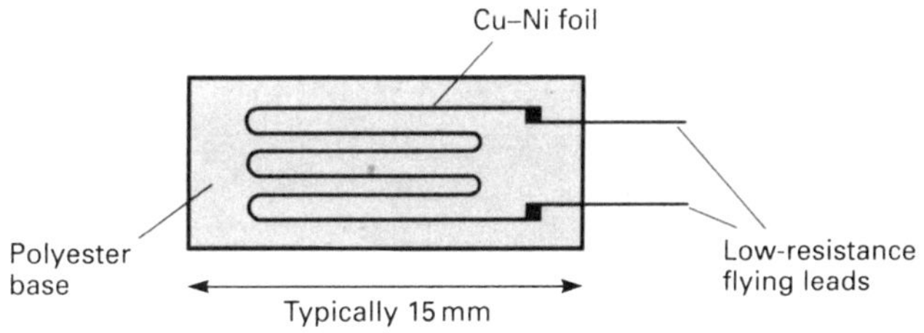

The polyester base is attached to the component under test with a very strong adhesive. Any subsequent deformation of the test component, such as a change in length, for example, alters the length and cross-sectional area of the foil, and so changes its resistance. Measuring this change in resistance gives the strain in the sample provided the gauge has been previously calibrated.

38

THE POTENTIOMETER

38.1 THE POTENTIAL DIVIDER CIRCUIT

In Fig. 38.1, on the assumption that the cells have negligible internal resistances,

$$V = I(R_1 + R_2) \qquad \text{and} \qquad V_1 = IR_1$$

Therefore

$$\frac{V_1}{V} = \frac{R_1}{(R_1 + R_2)}$$

i.e.

$$V_1 = \left(\frac{R_1}{R_1 + R_2}\right)V$$

Fig. 38.1
Potential divider circuit

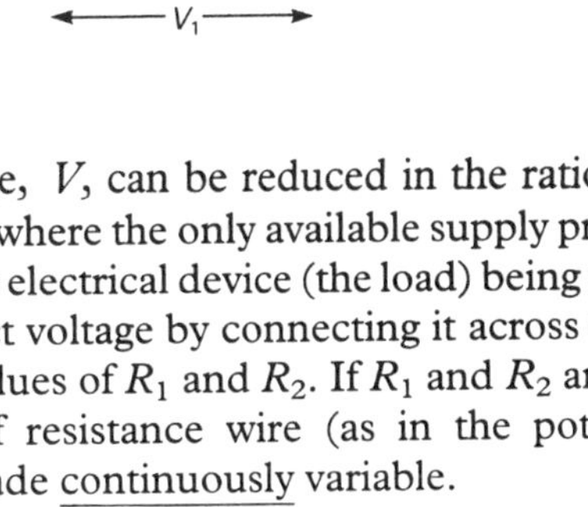

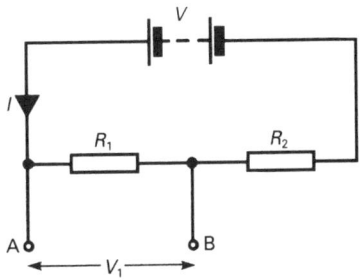

Thus, the supply voltage, V, can be reduced in the ratio $R_1/(R_1 + R_2)$. This is useful in circumstances where the only available supply provides a greater voltage than that required by the electrical device (the load) being used. The device can be supplied with the correct voltage by connecting it across terminals A and B, and suitably choosing the values of R_1 and R_2. If R_1 and R_2 are replaced by a (single) rheostat or a length of resistance wire (as in the potentiometer), the ratio, $R_1/(R_1 + R_2)$ can be made continuously variable.

Note If the resistance of the load is low compared with that of R_1, the load draws an appreciable current. In such circumstances the simple theory breaks down, for the current through R_1 is then not equal to that through R_2. The problem can be overcome by reducing the value of R_1 (and, therefore, that of R_2).

38.2 THE PRINCIPLE OF THE POTENTIOMETER

The potentiometer primarily measures potential difference, but on this basis it can be used to:

(i) compare EMFs (since it does not draw any current from the PD it is being used to measure),

(ii) compare resistances,

(iii) measure currents,

(iv) measure PDs in general.

The potentiometer circuit is shown in Fig. 38.2. Terminals X and Y are connected across the potential difference being measured in the same way as those of a moving coil voltmeter would be. If the positive terminal of the driver cell is connected to X (as shown here), then X must be connected to the positive side of the potential difference being measured. The potentiometer is said to be balanced when the jockey (sliding contact) is at such a position on AB that there is no current through the galvanometer.

Fig. 38.2 Potentiometer circuit

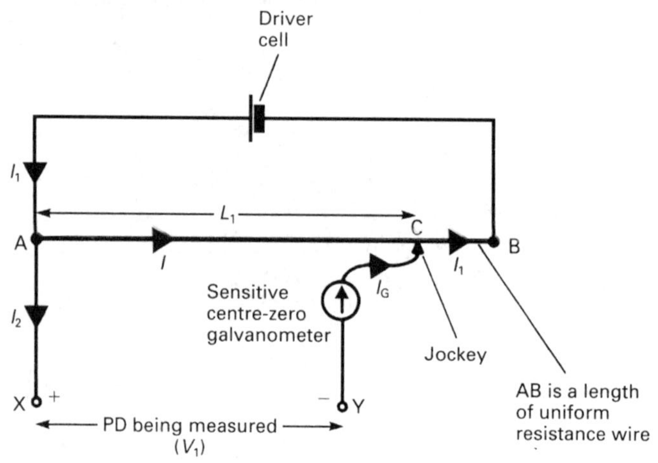

At balance, $I_G = 0$, and therefore

$$I = I_1 \quad \text{(by applying Kirchhoff's 1st rule at C)}$$

and so

$$I_2 = 0 \quad \text{(by applying Kirchhoff's 1st rule at A)}$$

Therefore

$$\text{Potential at X} = \text{Potential at A} \qquad (\text{since } I_2 = 0)$$

and

$$\text{Potential at Y} = \text{Potential at C} \qquad (\text{since } I_G = 0)$$

Therefore

$$\text{PD between X and Y} = \text{PD between A and C}$$

i.e.

$$V_1 = IrL_1 \tag{38.1}$$

where

r = resistance per unit length of AC.

If V_1 is replaced by another PD, V_2, and a new balance length, L_2, is found, then

$$V_2 = IrL_2 \qquad [38.2]$$

(**Note**. Replacing V_1 by V_2 does not affect the current, I, in AC because, at balance, no current flows in either AX or CY, i.e. the current through AC is provided solely by the driver cell.)

Dividing equation [38.1] by equation [38.2] gives

$$\frac{V_1}{V_2} = \frac{L_1}{L_2} \qquad [38.3]$$

Thus, by measuring the respective balance lengths, two potential differences can be compared. At balance, $I_2 = I_G = 0$, and therefore **no current is drawn from the potential differences being compared.**

38.3 TO COMPARE THE EMFs OF TWO CELLS

The circuit is shown in Fig. 38.3. With the cell of EMF E_1 in circuit, a balance point is found such that AC = L_1. With the cell of EMF E_2 in place of the first cell, a new balance point is found such that AC = L_2. No current is drawn from either cell, at balance, and therefore the potential differences across their terminals (and therefore across the potentiometer) are the EMFs of the cells. From the theory of the potentiometer (equation [38.3])

$$\frac{E_1}{E_2} = \frac{L_1}{L_2} \qquad [38.4]$$

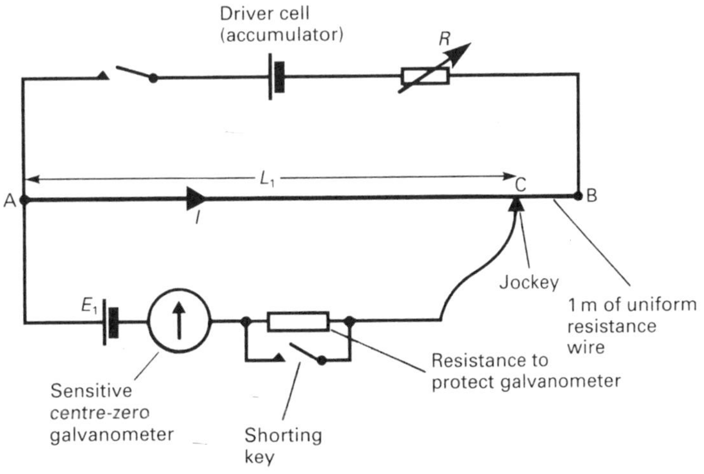

Fig. 38.3 Potentiometer used to compare the EMFs of two cells (E_1 and L_1 replaced by E_2 and L_2 in second measurement)

If the EMF of one of the cells is to be measured, the other cell should be one which has an accurately known EMF. Such a cell is known as a **standard cell** (see section 38.8). The standard Weston cell, which has an EMF of 1.0183 V, is often used.

Practical Details

(i) The accuracies of the measurements of length are greatest when L_1 and L_2 are large. Also, any end-error (see section 37.2) which is present is less significant when the balance lengths are large. Therefore, **it is necessary that both L_1 and L_2 are large if the result is to be accurate.** A preliminary measurement should be made to determine which of L_1 and L_2 is the larger. The value of R should then be adjusted so that the larger of the two balance lengths is close to B. (Compare this with the Wheatstone bridge, where the optimum position is close to the middle of the wire.) Then, provided that L_1 and L_2 are not very different, both measurements of length will be accurate. The value of R must be the same for the measurement of L_1 as it is for L_2, otherwise the current through AB (I) would be different in each case and the theory would not apply.

(ii) The EMF of the driver cell should be constant throughout. If it is not, the currents through AB are not the same for the two determinations of balance length, and the theory does not apply. A lead-acid accumulator, which is neither freshly charge nor run down, is suitable. As a precaution, the measurement which has been made first, say L_1, should be repeated after the other length has been measured. If the two values of L_1 are not very different, their average can be taken to be the value that would have been obtained for L_1 had it been measured at the same time as L_2.

(iii) A very sensitive galvanometer is needed if precise locations of the balance points are to be made. Such a galvanometer should be protected, by a large (typically 1 MΩ) series resistance (see Fig. 38.3), from the relatively high currents that would otherwise flow through it when in off-balance situations. The shorting key is left open until an approximate balance point has been found. The key is then closed, to short out the protective resistance and allow the full current to flow through the galvanometer, so that an accurate balance point can be found.

38.4 TO MEASURE A SMALL EMF (<10 mV)

It is not possible to measure a small EMF, such as that of a thermocouple, by comparing it directly with that of a standard Weston cell using the circuit of Fig. 38.3. An example will illustrate this.

Suppose that the balance length is L_1 when a thermocouple EMF of 5 mV is in circuit, and is L_2 when a standard cell whose EMF is 1.0183 V is in circuit. From equation [38.4]

$$\frac{5 \times 10^{-3}}{1.0183} = \frac{L_1}{L_2}$$

i.e. $L_1 = 0.0049L_2$

Thus, even if L_2 were 100 cm, L_1 would be a mere 0.49 cm. The determination of E_1 would be extremely inaccurate, because:

(i) end-error would be significant, and

(ii) the actual measurement of L_1 would be inaccurate.

The circuit shown in Fig. 38.4 should be used for small EMFs. With K connected to M, so that the small EMF, E_1, is in circuit, a balance length, L_1, is found. Next, by connecting K to N, a balance length, L_2 is found when the standard cell EMF, E_2, is in circuit.

At balance, no current is drawn from either the standard cell or the source of the small EMF. Therefore, no matter which of the two EMFs is in circuit, the current through R_1 is the same as that through AB. Let this current be I, and let the resistance per unit length of AB be r.

Fig. 38.4
Potentiometer used to measure a small EMF

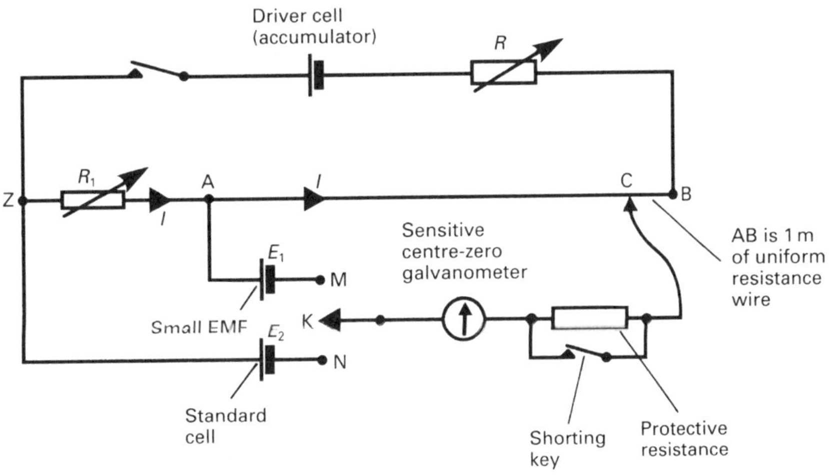

At balance with E_1 in circuit

$$E_1 = IrL_1 \qquad \text{(by using Kirchhoff's 2nd rule in CMAC)}$$

At balance with E_2 in circuit

$$E_2 = I(R_1 + rL_2) \qquad \text{(by using Kirchhoff's 2nd rule in CNZAC)}$$

Dividing gives

$$\frac{E_1}{E_2} = \frac{rL_1}{(R_1 + rL_2)} \qquad [38.5]$$

Equation [38.5] can be used to calculate E_1 if r and R_1 are known. A resistance box can be used to provide R_1, and r can be found by carrying out a preliminary Wheatstone bridge experiment.

Practical Details

(i) Details (i), (ii) and (iii) of section 38.3 apply.

(ii) Initially, one of R and R_1 should be put equal to zero. The other should be adjusted to give a suitably large value for L_1 (say 90 cm) when E_1 is in circuit. Next, with E_2 in circuit, both R and R_1 are adjusted, in such a way that their sum stays the same (and therefore preserves the value of L_1), until L_2 has a suitably large value. Once these preliminary adjustments of R and R_1 have been made, accurate measurements of L_1 and L_2 should be carried out.

Note The inclusion of R_1 in series with AB is equivalent to increasing the length of AB by many metres.

38.5 TO COMPARE TWO RESISTANCES

Refer to Fig. 38.5. The resistances being compared are R_X and R_Y. The accumulator, E', drives a steady current, I', through both R_X and R_Y. The size of this current may be adjusted by means of R'.

The potentiometer is connected first across R_X (as shown here) and then across R_Y (dotted connection). The potentiometer draws no current from the potential difference it is being used to measure, whether it be that across R_X, or that across

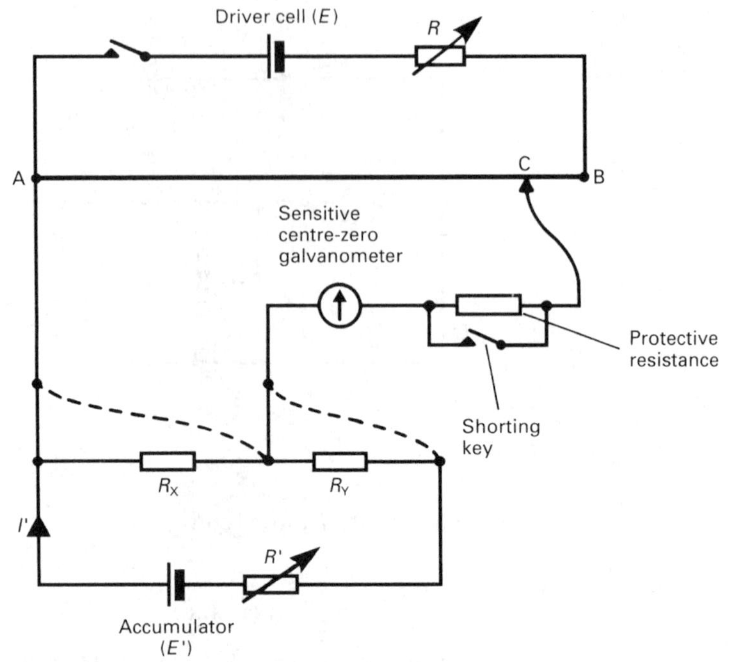

Fig. 38.5 Potentiometer used to compare two resistances

R_Y. Therefore, the replacement of R_X by R_Y in the potentiometer circuit does not affect the current through R_X and R_Y. Let

L_X = the balance length (AC) when the PD ($I'R_X$) across R_X is being measured

and

L_Y = the balance length (AC) when the PD ($I'R_Y$) across R_Y is being measured.

Therefore, from the theory of the potentiometer (equation [38.3]),

$$\frac{I'R_X}{I'R_Y} = \frac{L_X}{L_Y}$$

i.e.

$$\frac{R_X}{R_Y} = \frac{L_X}{L_Y}$$

Thus, the ratio of the resistances is the ratio of the respective balance lengths, and if the value of one of the resistances is known (e.g. if it has been provided by a resistance box), that of the other may be found. Unlike the Wheatstone bridge method, this method can be used to measure small resistances. This is because, at balance, the wires connecting the resistances into the potentiometer carry no current. There is, therefore, no potential difference across the connecting wires, and consequently their resistance does not affect the result.

Practical Details

(i) A preliminary measurement should be carried out in order that the values of R and R' may be adjusted to give suitably large balance lengths (see section 38.3(i)).

(ii) One of the balance lengths should be redetermined after the other has been measured, as a check on any variations in the values of E and E' (see section 38.3(ii)).

(iii) The protective resistance is made use of in the usual way (see section 38.3 (iii)).

38.6 TO MEASURE CURRENT

If the resistance, R_1, of a circuit component (Fig. 38.6) is known, the current flowing through it, I_1, can be measured with a potentiometer. The procedure is to obtain a balance length, L_1, when the component potential difference, I_1R_1, is across the potentiometer, and a second balance length, L_2, when the EMF, E_2, of a standard cell is across the potentiometer.

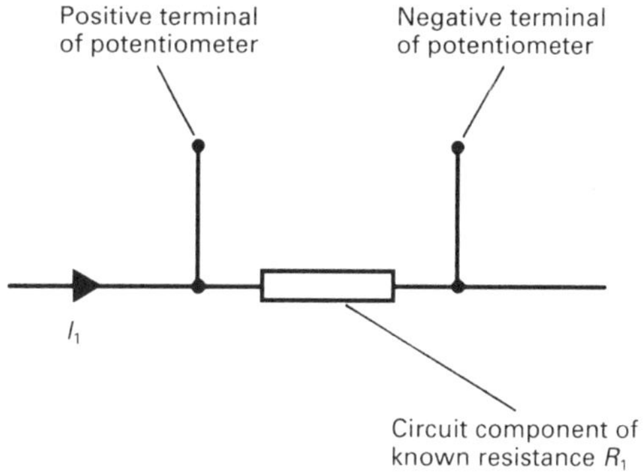

Fig. 38.6 Section of a potentiometer circuit to measure current

From equation [38.3],

$$\frac{I_1R_1}{E_2} = \frac{L_1}{L_2}$$

Hence we obtain I_1, because R_1 and E_2 are known, and L_1 and L_2 are measured.

A potentiometer draws no current from the potential difference it is used to measure, and therefore, when it is used to measure current in this way, it does not affect the size of the current. If the same technique were to be employed using a moving coil voltmeter, the measurement would be inaccurate as a result of the current drawn by the meter. (There would also, of course, be errors resulting from inaccuracies in the calibration of the voltmeter.)

Practical Details

Details (i), (ii) and (iii) of section 38.3 apply

38.7 TO MEASURE THE INTERNAL RESISTANCE OF A CELL

The internal resistance, r, of a cell can be found by using a potentiometer to compare its terminal potential difference on open circuit with that when it is driving a current, I_1, through a known resistance, R_1.

The circuit is shown in Fig. 38.7. When K is open, the cell whose internal resistance is required is on open circuit and its EMF, E, is across the potentiometer. When K is closed, the terminal potential difference, I_1R_1, of the cell is across the potentiometer. If the balance lengths are L_1 and L_2 when K is open and closed respectively, then from the theory of the potentiometer (equation [38.3])

$$\frac{E}{I_1R_1} = \frac{L_1}{L_2}$$

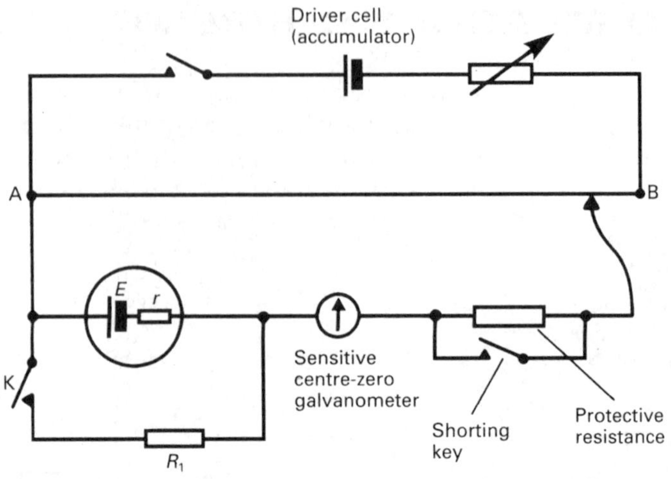

Fig. 38.7 Potentiometer used to measure the internal resistance of a cell

But

$$E = I_1(R_1 + r)$$

Therefore

$$\frac{I_1(R_1 + r)}{I_1 R_1} = \frac{L_1}{L_2}$$

i.e.

$$r = \left(\frac{L_1}{L_2} - 1\right)R_1$$

Practical Details

Details (i), (ii) and (iii) of section 38.3 apply.

38.8 USE OF STANDARD CELL

A standard cell is one which provides a constant and accurately known EMF. (The EMF of the standard Weston cell is 1.0183 V at 20 °C.) If a current of more than about 10 μA is drawn from such a cell, at any stage of its life, it is unlikely that the cell will continue to furnish the standard EMF. The resistance which protects the galvanometer from overload (see, for example, Fig. 38.3) also serves to limit the current drawn from the standard cell in the off-balance position. (A resistance of 1 MΩ limits the current to 1 μA.)

38.9 MEASUREMENT OF LARGE POTENTIAL DIFFERENCES

If a potentiometer is to be used to measure a potential difference which is larger than that of the driver cell of the potentiometer, the potential difference must first be reduced in a known ratio by using a potential divider circuit (section 38.1).

38.10 ADVANTAGES AND DISADVANTAGES OF POTENTIOMETER

By now the reader should be acquainted with the various features of the potentiometer, but it is useful to list a number of the more important ones.

Advantages

(i) It draws no current from the potential difference it is used to measure and therefore can be used to measure EMFs.

(ii) Results are dependent only on measurements of length and the values of standard resistances and standard EMFs.

(iii) It is a **null-deflection** method (see section 37.2) and therefore balance points can be found with high sensitivity and are not dependent on the calibration of the galvanometer.

Disadvantages

(i) It is slow in operation. In particular, it cannot be used to monitor rapidly changing voltages.

(ii) The resistance wire must be uniform.

(iii) There may be end-errors. (It is possible to eliminate the effects of these – if the reader wishes to know how, he should consult a more advanced text.)

39

ELECTROSTATICS

39.1 INTRODUCTION

If a polythene rod is rubbed with a woollen cloth, it becomes capable of repelling another polythene rod which is lightly suspended and has also been rubbed. The rods are said to have acquired an electric charge. A rubbed perspex rod is attracted by a rubbed polythene rod. It can be concluded that there are two* kinds of charge, and that:

(i) **like charges repel** (since there is no reason to suppose that the charge on the first polythene rod is of a different type from that on the second), and

(ii) **unlike charges attract.**

Early experiments of this nature made use of rods of ebonite and glass. Arbitrarily, the charge acquired by ebonite (rubbed with fur) was said to be negative, and that acquired by glass (rubbed with silk) was said to be positive. This convention makes rubbed polythene and rubbed Perspex negative and positive respectively, and is the reason that electrons and protons are, respectively, said to be negative and positive.

It is important to realize that the charges are not created by the rubbing action. When polythene is rubbed with a woollen cloth, some of the electrons of the surface atoms of the cloth become transferred to the rod. Thus, although the polythene becomes negatively charged, the cloth acquires an equal amount of positive charge, i.e. charge is conserved.

When a charge is placed on an insulator (say by rubbing) it stays in the region in which it has been placed. A charge placed on a metal, on the other hand, is quickly redistributed over the surface of the metal. The difference in behaviour results from metals having electrons which can move freely about the structure, whereas insulators have no charge carriers.

A hand-held metal rod cannot be charged by rubbing. Any charge acquired by the rod flows along it and through the body to become redistributed over the surface of the Earth. (The human body and the Earth are relatively good conductors of electricity.)

*Further investigation shows that there are only two types of charge.

39.2 COULOMB'S LAW

The law is based on experiments which Coulomb* concluded in 1785.

It states:

The magnitude of the force (F) between two electrically charged bodies, which are small compared with their separation (r), is inversely proportional to r^2, and proportional to the product of their charges (Q_1 and Q_2).

Thus

$$F \propto \frac{Q_1 Q_2}{r^2}$$

or

$$F = k\frac{Q_1 Q_2}{r^2} \qquad [39.1]$$

The value of the constant of proportionality, k, depends on the medium in which the charges are situated. The particular property of the medium which determines the value of k is known as the **permittivity** (ε) of the medium. It is convenient, in respect of units, to put $k = 1/4\pi\varepsilon$, in which case equation [39.1] becomes

$$F = \frac{1}{4\pi\varepsilon}\frac{Q_1 Q_2}{r^2} \qquad [39.2]$$

where

F = force (N)

Q_1 and Q_2 = charge (C)

r = separation (m)

ε = permittivity in farads per metre ($\mathrm{F\,m^{-1}}$). (The farad is the name given to the coulomb per volt – see section 40.2.)

Notes (i) The permittivity of free space (vacuum) is denoted by ε_0 and its value is determined by experiment. A method which can be used in a school laboratory is outlined in section 40.12; the most accurate method is based on a result of Maxwell's electromagnetic theory. The theory predicts that $c = 1/\sqrt{\mu_0\varepsilon_0}$, where c is the velocity of light in vacuum and μ_0 is a constant called the permeability of free space. The value of c is found by experiment to be $2.998 \times 10^8\,\mathrm{m\,s^{-1}}$ and μ_0 has a value of $4\pi \times 10^{-7}\,\mathrm{H\,m^{-1}}$ exactly (see section 41.4); it follows that

$$\varepsilon_0 = 8.854 \times 10^{-12}\,\mathrm{F\,m^{-1}}$$

and

$$\frac{1}{4\pi\varepsilon_0} = 8.988 \times 10^9\,\mathrm{m\,F^{-1}}$$

(ii) The permittivity of air at STP is $1.0005\varepsilon_0 \approx \varepsilon_0$.

(iii) $1\,\mathrm{F\,m^{-1}} = 1\,\mathrm{C^2\,N^{-1}\,m^{-2}}$.

(iv) The experimental investigation of Coulomb's law is dealt with in section 39.16.

*Electrostatic forces are often referred to as Coulomb forces.

39.3 ELECTRIC FIELDS

An electric field exists in a region if electrical forces are exerted on charged bodies in that region. **The direction of an electric field** at a point is the direction in which a small positive charge would move (under the influence of the field) if placed at that point.

The electric field intensity or electric field strength (E) at a point is defined as the force exerted by the field on a unit charge placed at that point. (Unit $= \text{N C}^{-1} = \text{V m}^{-1}$.)

It follows from the definition of electric field intensity that the force, F, exerted on a charge, Q, at a point where the field intensity is E is given by

$$F = EQ$$

Field Intensity Due to a Point Charge

In Fig. 39.1, the force, F, on the test charge, Q', due to the point charge, Q, in a medium of permittivity, ε, is given by equation [39.2] as

$$F = \frac{1}{4\pi\varepsilon}\frac{QQ'}{r^2} \qquad [39.3]$$

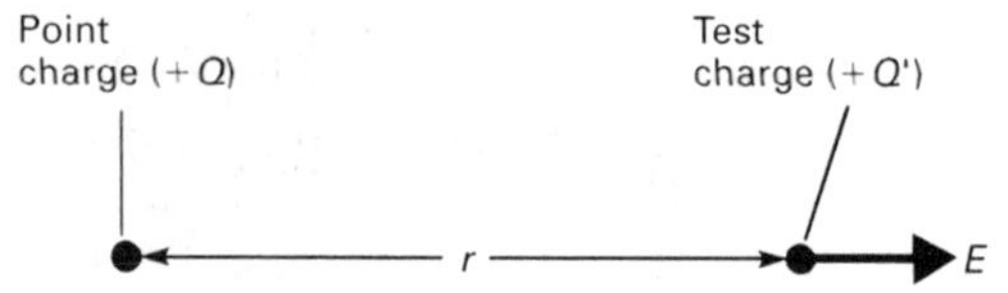

Fig. 39.1
To calculate the field intensity due to a point charge

By definition, electric field intensity is force per unit charge and therefore the field intensity, E, at the site of Q' is given by

$$E = \frac{F}{Q'}$$

Therefore, from equation [39.3]

$$E = \frac{1}{4\pi\varepsilon}\frac{Q}{r^2} \qquad [39.4]$$

39.4 ELECTRIC POTENTIAL

In order to move a charge from one point in an electric field to another, work may well have to be done on the charge. The two points must, therefore, have some property associated with them which is different at the two points. This property is called **electric potential**. The concept is analogous to that of gravitational potential (section 8.11). If two points have different electric potentials, then the potential energy of a charge changes as a result of moving from one point to the other. **The potential is a property of the field, the potential energy depends on both the field and the size of the charge.**

The potential at a point in an electric field is defined as being numerically equal to the work done in bringing a unit positive charge from infinity to the point. The unit of potential is the volt (V).

Notes (i) **It follows that the potential of a point at infinity is zero.** For practical purposes, though, the zero of potential is taken to be the potential of the Earth. This is possible because the potential of the Earth is constant and because we are usually concerned with differences of potential.

(ii) It also follows that the **potential energy** of a charge Q at a point where the potential is V is given by

$$\text{PE} = QV$$

39.5 THE POTENTIAL DUE TO A POINT CHARGE

Refer to Fig. 39.2. The point charge at A exerts a repulsive force F on the charge at B where (from equation [39.2])

$$F = \frac{QQ'}{4\pi\varepsilon x^2}$$

and ε is the permittivity of the medium in which the charges are situated. If the charge at B is moved by some external agent a small distance δx towards A where δx is so small that F can be considered constant, the work done δW by the external agent is given by

$$\delta W = -F\delta x$$

(The minus sign is necessary because the motion is in the opposite direction to F.) Substituting for F gives

$$\delta W = -\frac{QQ'\delta x}{4\pi\varepsilon x^2}$$

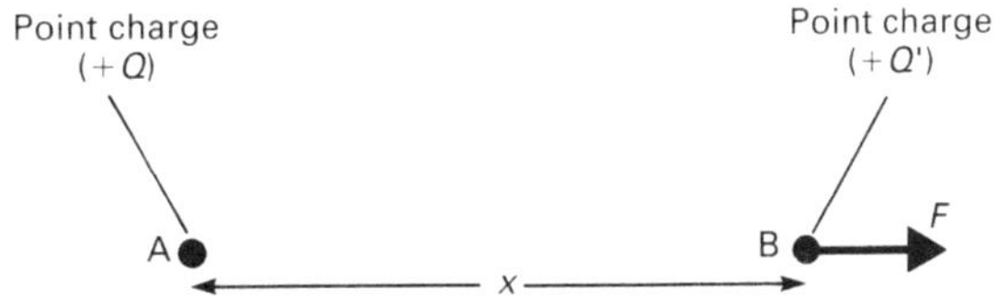

Fig. 39.2
To calculate the potential due to a point charge

The total work done W in bringing the charge at B from infinity (where $x = \infty$) to some point a distance r from A (where $x = r$) is therefore given by

$$W = -\frac{QQ'}{4\pi\varepsilon}\int_{\infty}^{r}\frac{1}{x^2}\,\mathrm{d}x = -\frac{QQ'}{4\pi\varepsilon}\left[\frac{-1}{x}\right]_{\infty}^{r} = \frac{QQ'}{4\pi\varepsilon r}$$

If Q' is a unit charge, the work done is $Q/4\pi\varepsilon r$, and therefore by the definition of potential (section 39.4) the potential V at a distance r from a point charge Q in a medium of permittivity ε is given by

$$V = \frac{1}{4\pi\varepsilon}\frac{Q}{r} \qquad [39.5]$$

Notes (i) Potential is a scalar quantity and therefore the potential at a point due to a number of point charges is the algebraic sum of the (separate) potentials due to each charge.

(ii) It should be clear from equation [39.5] that the potential due to a positive charge is positive; that due to a negative charge is negative.

(iii) It is clear from equation [39.5] that all points which are equidistant from a point charge are at the same potential. (A surface over which potential is constant is called an **equipotential surface**. **If a point lies on an equipotential surface, the electric field at that point is perpendicular to the surface.**)

(iv) It follows from note (ii) of section 39.4 that the **potential energy** of a charge Q' at a distance r from a point charge Q is given by

$$\text{PE} = \frac{1}{4\pi\varepsilon}\frac{QQ'}{r}$$

EXAMPLE 39.1

Find expressions for (a) the potential, (b) the electric field intensity at a point C on a line ACB in a medium of permittivity ε where AC = CB = r given that there is a positive point charge Q_1 at A and a negative point charge $-Q_2$ at B.

Solution

(a) From equation [39.5]

$$\text{Potential at C due to } Q_1 = \frac{1}{4\pi\varepsilon}\frac{Q_1}{r}$$

$$\text{Potential at C due to } -Q_2 = \frac{1}{4\pi\varepsilon}\left(\frac{-Q_2}{r}\right)$$

Potential is a scalar quantity and therefore the total potential is simply the algebraic sum of the individual potentials.

$$\therefore \quad \text{Potential at C} = \frac{1}{4\pi\varepsilon}\left(\frac{Q_1 - Q_2}{r}\right)$$

(b) The positive charge at A would cause a small positive charge at C to move in the direction A to B. The negative charge at B would also cause a small positive charge at C to move in the direction A to B. Thus both contributions to the field at C are in the direction A to B. From equation [39.4]

$$\text{Field at C due to } Q_1 = \frac{1}{4\pi\varepsilon}\frac{Q_1}{r^2} \qquad \text{(from A to B)}$$

$$\text{Field at C due to } -Q_2 = \frac{1}{4\pi\varepsilon}\frac{Q_2}{r^2} \qquad \text{(from A to B)}$$

Since the two fields are in the same direction

$$\text{Field at C} = \frac{1}{4\pi\varepsilon}\left(\frac{Q_1 + Q_2}{r^2}\right) \qquad \text{(from A to B)}$$

QUESTIONS 39A

1. Find: **(a)** the potential, **(b)** the electric field intensity at a point C on a line ABC in a vacuum where AB = BC = 5.0 cm given that there are point charges of 6.0 μC and -4.0 μC at A and B respectively. State the direction of the field.
(Assume $1/4\pi\varepsilon_0 = 9.0 \times 10^9 \text{ m F}^{-1}$.)

2. ABC is an equilateral triangle of side 4.0 cm in a vacuum. There are point charges of 8.0 μC at A and B. Find **(a)** the potential, **(b)** the electric field intensity at C. State the direction of the field.
(Assume $1/4\pi\varepsilon_0 = 9.0 \times 10^9 \text{ m F}^{-1}$.)

39.6 ELECTRICAL POTENTIAL DIFFERENCE

The potential difference between two points in an electric field is numerically equal to the work done in moving a unit positive charge from the point at the lower potential to that at the higher potential. (This follows from the definition of potential.)

The unit of potential difference is the **volt** (V). It is defined as follows:

If the work done in causing one coulomb of electric charge to flow between two points is one joule, then the PD between the points is one volt, i.e. $1 \text{ V} = 1 \text{ J C}^{-1}$.

It follows that the work done in moving a charge through a potential difference is given by

$$W = QV \qquad [39.6]$$

where

W = the work done (J)

Q = the charge (C)

V = the potential difference (V).

Note **The electric potential energy** of a positively charged particle increases when it moves to a point of higher potential.

The electric potential energy of a negatively charged particle increases when it moves to a point of lower potential.

The increase in potential energy results from:

(i) work being done by some external agent, in which case the increase in potential energy is equal to the work done, or

(ii) a decrease in the kinetic energy of the particle, in which case the increase in potential energy is equal to the decrease in kinetic energy.

The situation is analogous to that of a mass in a gravitational field, (i) is equivalent to the mass being lifted, (ii) is equivalent to the mass being thrown upwards.

When any given charge moves from one point to another, the work done depends only on the difference in the potentials of the starting and finishing points, i.e. **the work done is independent of the path taken.** It follows that the work done in taking a charged particle around a closed path in an electric field is zero (see Fig. 39.3).

Fig. 39.3
To calculate the work done in taking a point charge around a closed path

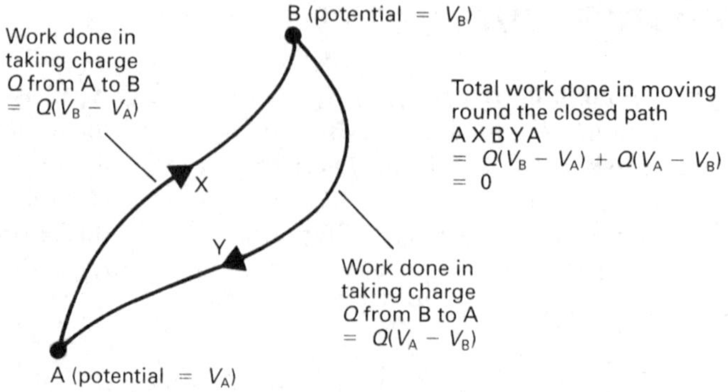

39.7 RELATION BETWEEN *E* AND *V*

Suppose that a charge, $+Q$, (Fig. 39.4) is moved by a force, F, from some point, A, in an electric field of intensity, E, to some other point, B. Suppose also, that the distance, δx, from A to B is so small that the force can be considered to be constant between A and B. The work done, δW, by the force is given (by equation [5.1]) as

$$\delta W = F\delta x \quad [39.7]$$

It follows from the definition of electric field intensity that

$$F = -EQ \quad [39.8]$$

Fig. 39.4
To establish the relationship between ***E*** and ***V***

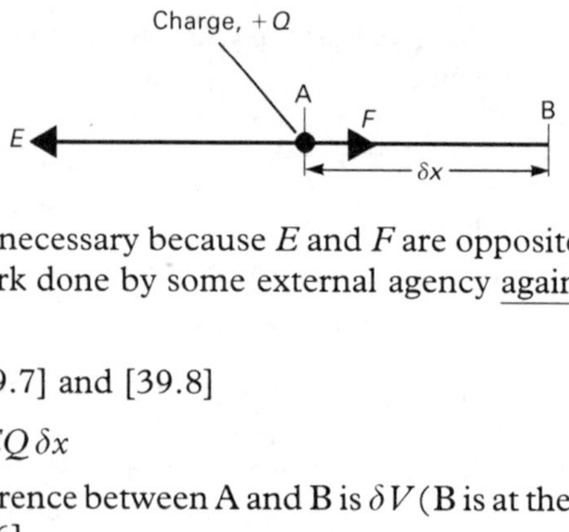

(The minus sign is necessary because E and F are oppositely directed, i.e. we are considering the work done by some external agency against the influence of the field.)

From equations [39.7] and [39.8]

$$\delta W = -EQ\,\delta x$$

If the potential difference between A and B is δV (B is at the higher potential), then from equation [39.6]

$$\delta W = Q\delta V$$

$$\therefore \quad -EQ\,\delta x = Q\delta V$$

and therefore in the limit

$$E = -\frac{dV}{dx} \quad [39.9]$$

Notes (i) dV/dx is the rate of change of potential with distance at a point in an electric field, and is called the **potential gradient** at that point. The minus sign expresses the fact that the field is in the negative direction of x if the potential increases in the positive direction of x. (V and x increase in going from A to B, E is directed from B to A.)

(ii) It should now be obvious why we were able to give $V\,m^{-1}$ as an alternative to $N\,C^{-1}$ as a unit of field strength in section 39.3.

Uniform Field

Consider two parallel conducting plates, A and B, a distance d apart and between which there is a potential difference V (Fig. 39.5).

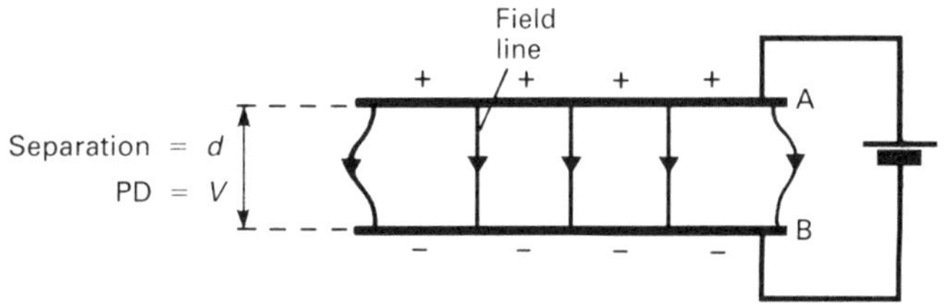

Fig. 39.5
Electric field between parallel plates

Except at the edges, the field between the plates is uniform (field lines parallel to each other). It follows from equation [39.9], therefore, that the potential gradients at all points between A and B (away from the edges) are the same. The potential gradient at any one of these points is therefore equal to the average potential gradient, the magnitude of which is V/d. Therefore, the magnitude of the electric field intensity, E, is given by equation [39.9] as

$$E = \frac{V}{d} \quad \text{(uniform field)} \qquad [39.10]$$

QUESTIONS 39B

1. The diagram below shows the electric field lines between two parallel metal plates, A and B.

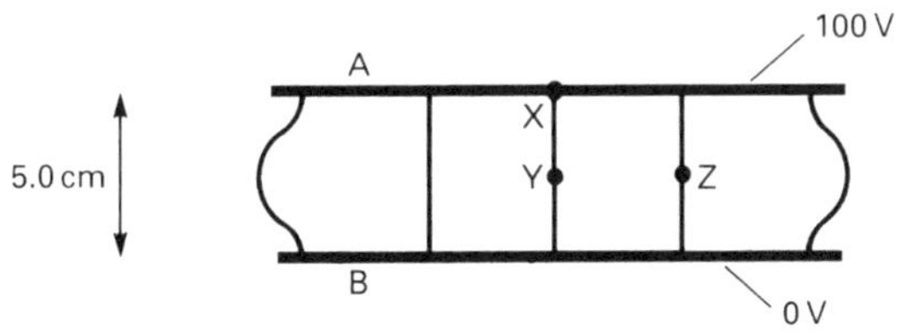

What is:

(a) the direction of the field (A to B or B to A),

(b) the field strength at X,

(c) the field strength at Y,

(d) the work done in moving an electron from X to Y (XY = 2.5 cm),

(e) the work done in moving an electron from Y to Z (YZ = 3.0 cm)?

(f) If a charge of 2.0 μC were to be placed at Y, what force would be exerted on it?
($e = 1.6 \times 10^{-19}$ C.)

39.8 CHARGED CONDUCTORS

In any conductor there are charge carriers which are free to move, and if two regions of a conductor are at different electric potentials, charge moves from one region to the other until the two potentials are the same. The act of charging a conductor alters its potential in the region where it is charged. It follows, therefore, that whenever a conductor becomes charged, the charge redistributes itself, and that, at equilibrium, **every part of the material of a conductor is at the same potential**.

Since the potential is the same at all points, $dV/dx = 0$, and therefore, since $E = -dV/dx$, **there can be no electric field within the material of a conductor.** It can be shown* that this requires that:

(i) **there is no charge within the material of a conductor,** and

(ii) **the charge resides entirely on the surface of a solid conductor, and entirely on the outer surface of an empty hollow conductor.** (Even when there is a charged body on an insulating support inside a hollow conductor, there is no net charge inside the conductor – see section 39.9.)

The field at the surface of a conductor is perpendicular to the surface. If this were not so, there would be a component of the field in the surface of the conductor. There can be no such component because, as stated above, the field is zero within the material of the conductor.

The amount of charge on unit area of surface is known as the **charge density** (σ). It can be shown (by using Gauss's theorem) that **the field strength at a point on the surface of a conductor is proportional to the charge density there.** It can also be shown (see, for example, section 39.13) that **the charge density (and therefore the field strength) is greatest at highly curved convex regions of the surface of a charged conductor, and is least at concave regions.** (This is not necessarily true if there are other charged objects nearby.) Figs. 39.6 and 39.7 illustrate many of these results.

Fig. 39.6
A positively charged solid conductor

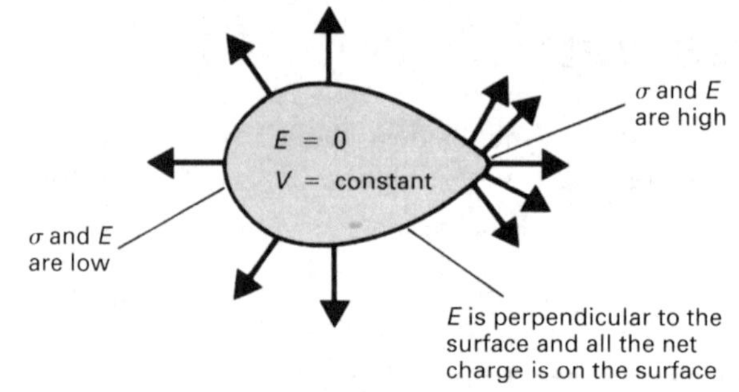

Fig. 39.7
A positively charged hollow conductor

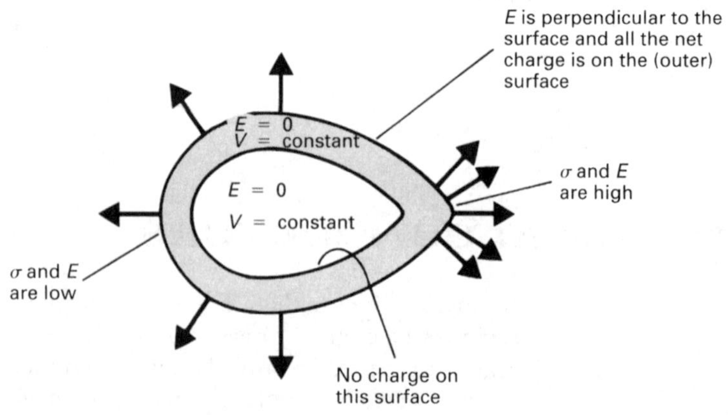

*The proof involves Gauss's theorem and is beyond the scope of this book.

The particular situation of a charged sphere is shown in Fig. 39.8. Note that outside the sphere, the field and the potential are as if all the charge is concentrated at the centre.

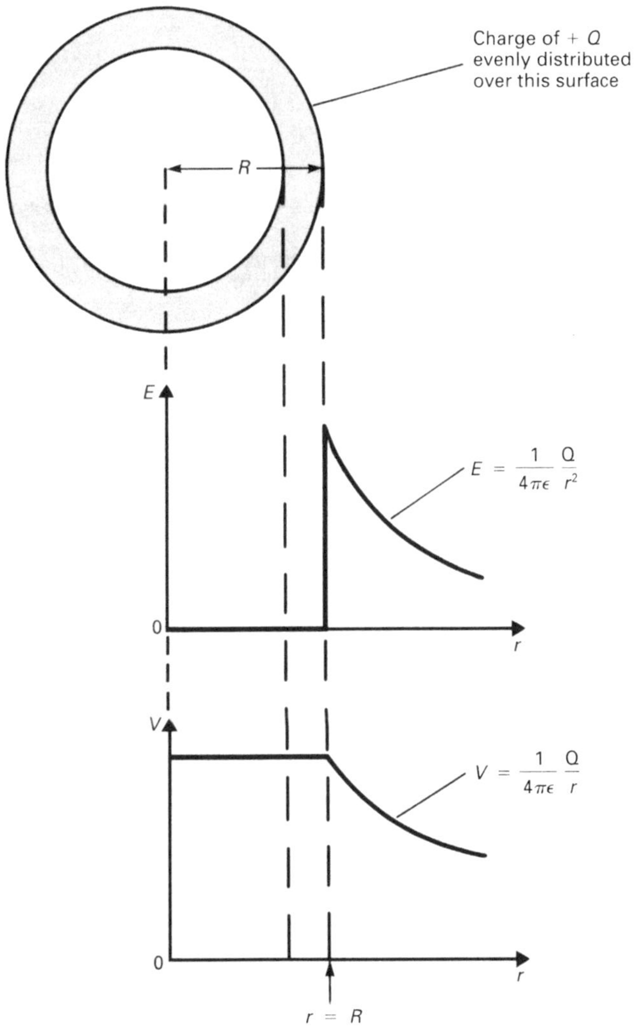

Fig. 39.8
A positively charged hollow sphere

39.9 HOLLOW CONDUCTOR CONTAINING A CHARGE

Fig. 39.9 illustrates the result of placing two uncharged hemispherical shells around a positively charged body on an insulating support.

Note:

(i) The net charge within the sphere is zero.

(ii) The total charge is the same as it was before the body was enclosed – an example of the conservation of charge.

(iii) Outside the sphere, the field and the potential are as if the total charge ($+Q$) were concentrated at the centre of the sphere.

(iv) The position of the enclosed charge does not affect (i), (ii) and (iii).

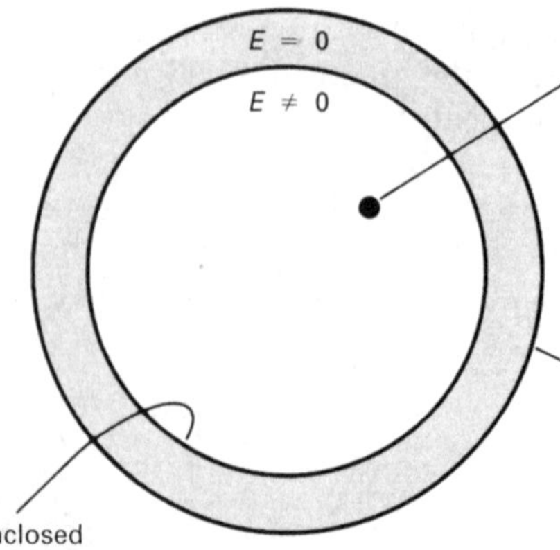

Fig. 39.9
A hollow (spherical) conductor enclosing a charge

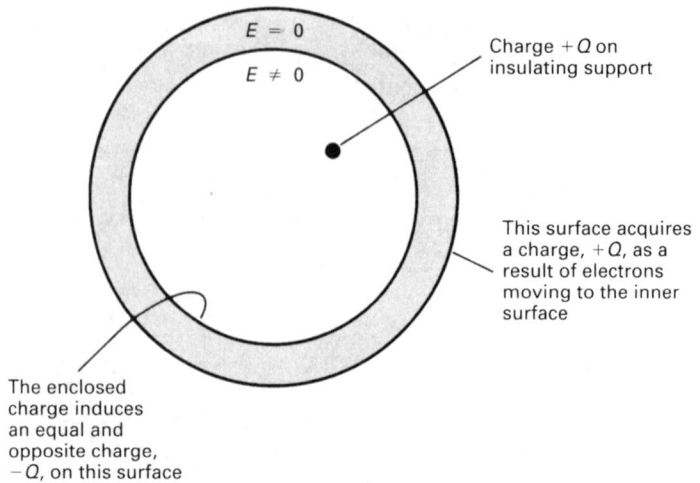

QUESTIONS 39C

1. A conducting sphere of radius 5.0 cm has a charge of 4.0 μC. Find the potential: **(a)** 6.0 cm, **(b)** 5.0 cm, **(c)** 4.0 cm from the centre of the sphere
(Assume $1/4\pi\varepsilon_0 = 9.0 \times 10^9 \text{ m F}^{-1}$.)

39.10 THE GOLD-LEAF ELECTROSCOPE

A gold-leaf electroscope is shown in Fig. 39.10. It is a device which measures **electric potential,** but it can also be used to detect the presence of charge and determine whether the charge is positive or negative.

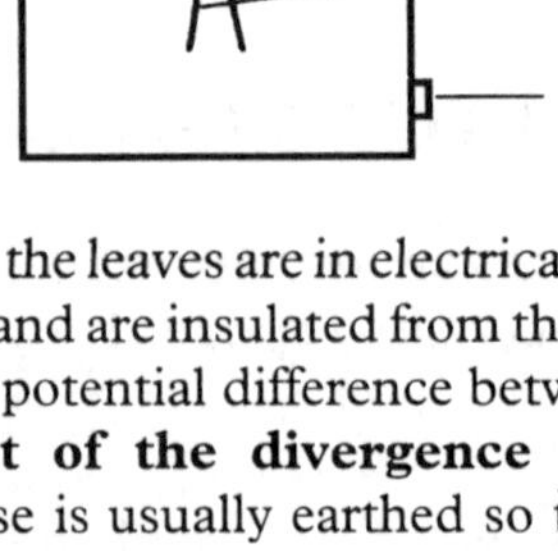

Fig. 39.10
A gold-leaf electroscope

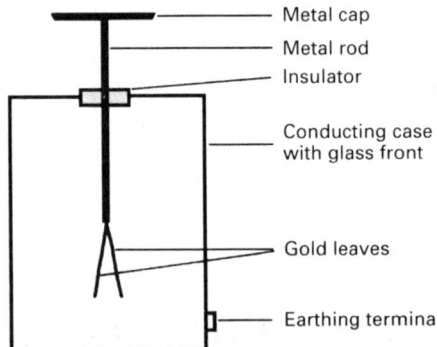

The cap, the rod and the leaves are in electrical contact (and therefore are always at the same potential), and are insulated from the conducting case of the instrument. Whenever there is a potential difference between the cap and the case the leaves diverge, **the extent of the divergence is a measure of the potential difference.** The case is usually earthed so that it is at zero potential, in which case the divergence of the leaves is a measure of the potential of the cap.

Suppose that a positively charged body is brought near to (but not in contact with) the cap of a gold-leaf electroscope (Fig. 39.11(a)). The presence of the positive charge draws electrons to the cap from the leaves. Thus, a negative charge is **induced** on the cap and the leaves acquire a positive charge. The positive charge on the leaves induces a negative charge on the case by causing electrons to flow to it from earth. The leaves are charged and are in an electric field (due to both themselves and the case) and therefore experience a force. It is this force which causes the leaves to diverge. Alternatively, the divergence can be thought of as being the result of mutual repulsion between the 'like' charges on the leaves and attraction by the charges of opposite sign on the case. The leaves diverge to such an extent that the restoring torques due to their weights are equal to those due to the electrostatic forces.

If the cap is earthed at this stage, the leaves fall (Fig. 39.11(b)). This happens because the case and the cap are now at the same potential (earth potential). Electrons have flowed from earth onto the leaves and have cancelled the positive charge that was previously there. Note that there is still a negative charge on the cap, owing to the presence of the positively charged body. This result highlights the fact that the divergence of the leaves is a measure of the potential difference between the cap and the case, and is not necessarily a measure of the charge on the cap.

If the earth lead to the cap is removed, the cap stays at zero potential and the leaves remain undeflected (Fig. 39.11(c)).

At this stage, removal of the positively charged body creates a negative potential at the cap (Fig. 39.11(d)). There is once again a potential difference between the cap and the case, and the leaves diverge. Removing the charged body has allowed electrons to flow from the cap to the leaves, under the influence of the repulsive forces between the electrons. Note that if the stages shown in Fig. 39.11(b) and (c) are omitted, removal of the charged body causes the leaves to fall – because there has been no opportunity for electrons to flow onto the cap from earth.

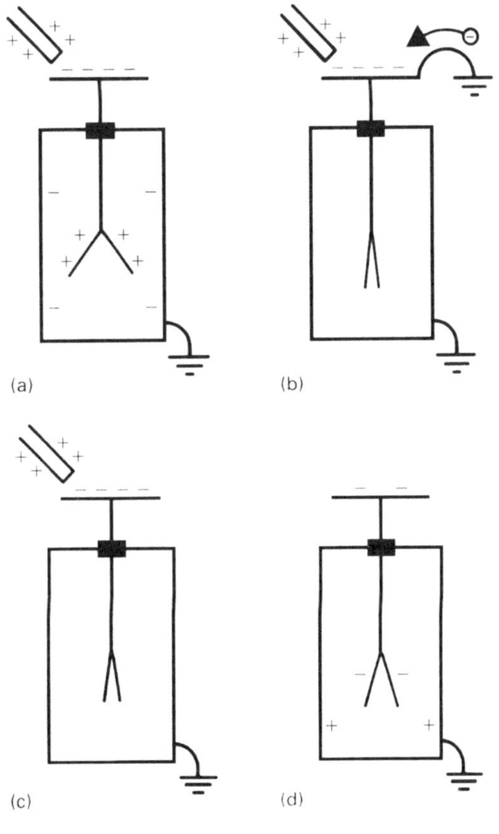

Fig. 39.11
Charging an electroscope by induction

In Fig. 39.12(a) a positively charged body has been brought into contact with the cap of an electroscope. Electrons flow onto the charged body from the cap and the leaves, and they become positively charged. If the charged body is now removed, positive charge remains on the leaves and they stay deflected.

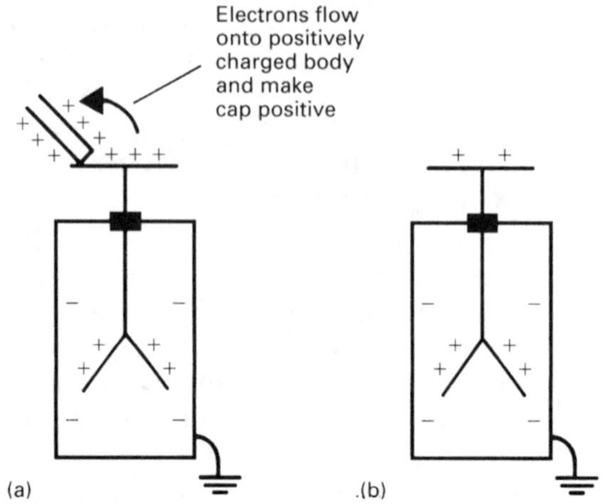

Fig. 39.12 Charging an electroscope by contact

Note Compare the negatively charged leaves of Fig. 39.11(d) (where the positively charged body never came into contact with the cap) with the positively charged leaves on Fig. 39.12(b) (where the charged body did come into contact with the cap).

Determining the Sign of a Charge

Suppose that a gold leaf electroscope has been charged so that the leaves are deflected and are known to have a positive charge. (It is preferable that the electroscope is charged by induction, as charging by contact often gives misleading results.) Table 39.1 lists the effects of bringing a positively charged body, a negatively charged body and an uncharged body near to (but not in contact with) the cap of the electroscope.

Table 39.1 The effects of bringing variously charged bodies near to a positively charged electroscope

Charge on body brought near	*Effect on divergence of leaves*
Positive	Increases
Negative	Decreases
Uncharged	Decreases

Thus, if a body of unknown charge is brought near to the cap and the divergence increases, the body must be positively charged. Note that there are two possibilities if the divergence decreases. In order to test for negative charge it is necessary to have a negatively charged electroscope.

39.11 FARADAY'S ICE-PAIL EXPERIMENT

The experiment derives its name from the ice-pail used in Faraday's original experiment in 1843 – a tall metal can does just as well.

The can is stood on the cap of a gold-leaf electroscope in order to detect any charge that appears on the outside of the can. A small charged sphere on an insulating thread is lowered into the can without touching it (Fig. 39.13(a)). (Fig. 39.13 has been drawn on the assumption that the sphere is positively charged.) The leaves of the electroscope diverge, showing that a charge has been induced on the outside of the can. The size of the deflection does not depend on the position of the sphere, as long as it is well inside the can. It follows, therefore, that **the potential of the can is independent of the position of the sphere.**

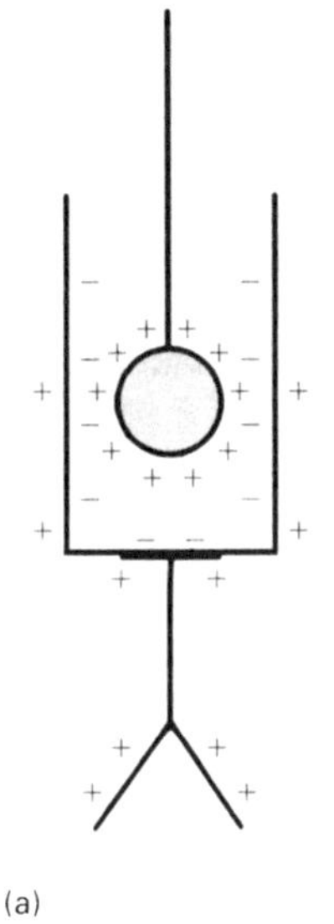

Fig. 39.13 Faraday's ice-pail experiment

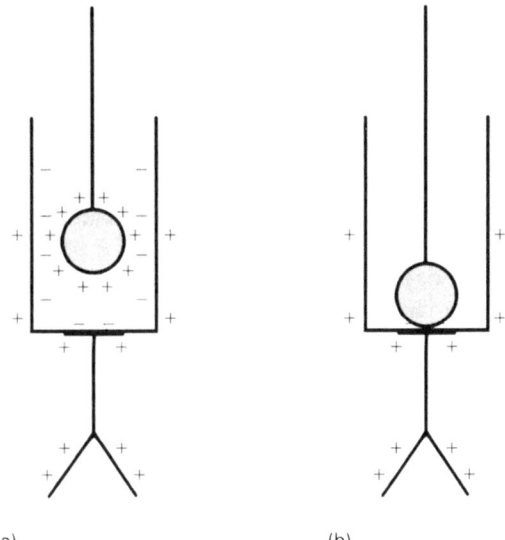

If the sphere is allowed to touch the can (Fig. 39.13(b)), there is no change in the deflection of the electroscope. If the sphere is now removed from the can, it can be shown that there is no charge on either the sphere or the inside surface of the can. These three observations allow it to be concluded that:

> **When a charged body is enclosed by a hollow conductor, a charge is induced on the outside of the conductor which is equal to that on the charged body.**

An extension to the experiment shows that **the charge induced on the outside of the can is equal and opposite to that on the inside.**

39.12 THE ELECTROPHORUS

The electrophorus is a device which can supply large quantities of charge. It consists of an insulating disc (of polythene say) and a metal disc – the latter being mounted on an insulating handle. The successive stages in the charging process are shown in Fig. 39.14.

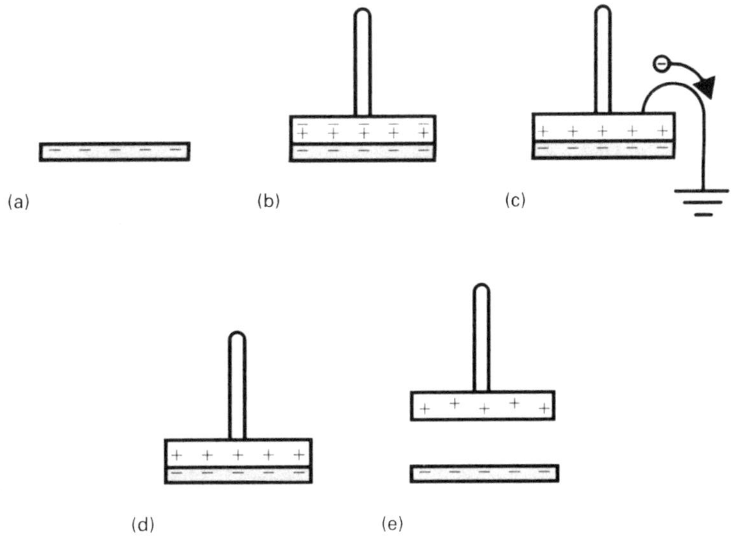

Fig. 39.14 Electrophorus

First, the polythene disc is rubbed vigorously with a woollen cloth so that its upper surface becomes negatively charged (Fig. 39.14(a)). Next, the metal disc is placed on top of the polythene (Fig. 39.14(b)). Although the discs appear to have flat surfaces, there are in fact very few points at which the two discs make contact. The result is that a positive charge is induced on the lower surface of the metal disc, and a negative charge is induced on its upper surface. The metal disc is momentarily earthed at this stage (say by touching it with a finger), and the electrons on its upper surface flow to earth (Fig. 39.14(c)). Once the earth connection has been removed (Fig. 39.14(d)), the disc can be lifted off the polythene (Fig. 39.14(e)), and can be used to provide a positive charge. The charge is almost equal in magnitude to the original charge on the polythene.

The charge remaining on the polythene is (apart from that transferred to the metal at the small number of contact points) equal to that produced originally by rubbing. Once the metal disc has been discharged, it can be charged again by returning it to the polythene. The process can be repeated many times before the charge on the polythene leaks away and has to be replenished by rubbing.

Notes

(i) Each time the metal disc is lifted off the polythene, work has to be done to overcome electrostatic forces of attraction.

(ii) Compare Fig. 39.14 with Fig. 39.11, each depicts a process of charging by induction. The charged body, the cap of the electroscope and the leaves in Fig. 39.11 are respectively equivalent to the polythene disc, the lower surface of the metal disc and the upper surface of the metal disc in Fig. 39.14.

39.13 EXPERIMENTAL INVESTIGATION OF CHARGE DISTRIBUTION OVER THE SURFACE OF A CONDUCTOR

Suppose it is required to investigate the charge distribution over the surface of a pear-shaped conductor. (The conductor may be charged by touching it with an electrophorus.)

A proof plane (a small metal disc on an insulating handle)is placed in contact with a section of the surface of the conductor (Fig. 39.15). The proof plane acquires a charge which is proportional to the charge density on that section of the conductor. The proof plane is now removed and placed in contact with the inner surface of a metal can standing on the cap of a gold-leaf electroscope. The leaves diverge by an amount which is proportional to the charge on the proof plane (and therefore to the charge density on the pear-shaped conductor).

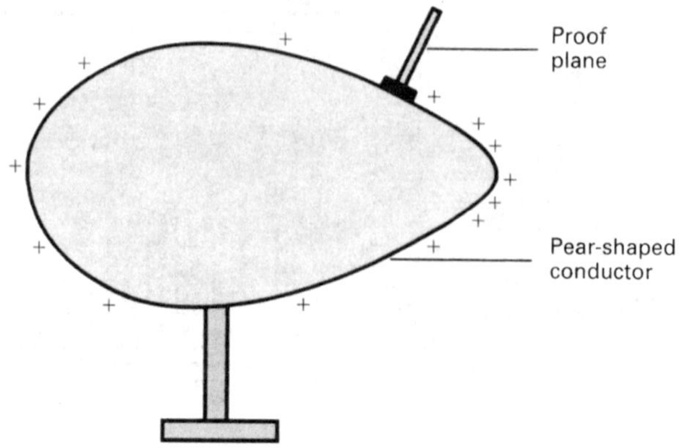

Fig. 39.15 Investigation of surface charge distribution on a conductor

Since the proof plane has touched the inside of the can, it loses all of its charge (see section 39.11). It is therefore ready to be used to test some other region of the conductor.

The experiment shows that the charge density is greatest where the surface is highly curved. Detailed experiments show that, for a convex surface, the charge density is inversely proportional to the radius of curvature.

39.14 FIELDS AROUND POINTS

The electric field strength at the surface of a charged conductor is greatest at highly curved (convex) regions of the surface, corresponding to the high charge density there (see also sections 39.8 and 39.13). It follows that if there are any sharp points on a highly charged conductor, there are extremely intense electric fields around them. A charged particle in the vicinity of such a point is accelerated by the field and ionizes the air around the point by making energetic collisions with the air molecules. Thus, the insulation of the air breaks down around sharp points on charged conductors, and the charge on the conductor is likely to leak away at such points. It follows that the presence of sharp points on the surface of a conductor limits the amount of charge that can be put on it. The ideal shape, as far as storing large amounts of charge is concerned, is a sphere of large radius. (A large flat sheet is no use because its edges act as points.)

39.15 THE VAN DE GRAAFF GENERATOR

Van de Graaff generators are electrostatic devices which are capable of producing very high potential differences. Large versions provide PDs in excess of 10^7 V, and are used to accelerate charged particles in the study of nuclear processes. The reader should be familiar with the ideas of the last section before reading on.

The essential parts of a Van de Graaff generator are shown in Fig. 39.16. The belt is driven at high speed past a series of metal points at X. The points are at a high positive potential with respect to earth, and the insulation of the air around them

Fig. 39.16
Van de Graaff generator

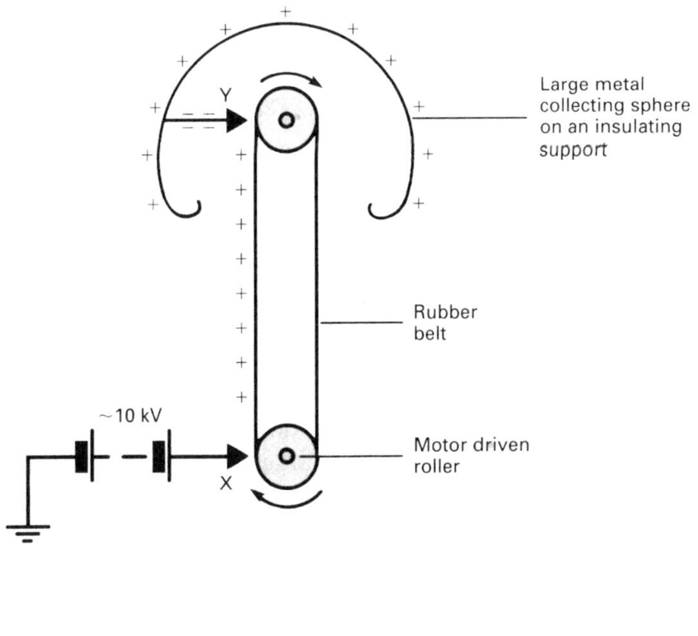

breaks down. Positively charged ions are 'sprayed' on to the belt and these are carried up into the collecting sphere. This induces a negative charge on the inner surface of the sphere and a corresponding positive charge on its outer surface. The sphere is in electrical contact with a second set of metal points, and the induced negative charge is sprayed from these on to the belt at Y. This neutralizes the positive charge on the belt so that the whole process can be repeated. In this way a considerable amount of positive charge builds up on the outer surface of the collecting sphere.

The size of the potential difference attainable is determined by the insulating properties of the air around the machine.

If a large Van de Graaff generator is discharged steadily (say by providing the potential difference necessary to operate a high-voltage X-ray tube), the current which flows is of the order of 100 μA. This corresponds to a power output of 100 W if the PD is 10^6 V. The origin of this power is the work done by the motor in driving the positive charge on the belt upwards against the repulsive force of the positive charge on the sphere.

The current available from a school version of the Van de Graaff generator can be measured by connecting a suitable microammeter in series with the collecting sphere and the (earthed) base of the device. This provides a useful demonstration of the equivalence of current and static electricity.

39.16 EXPERIMENTAL INVESTIGATION OF COULOMB'S LAW

Refer to Fig. 39.17(a). A and B are small expanded polystyrene balls wrapped in aluminium foil or coated with Aquadag (a graphite-based paint) to make them conducting. The balls are charged by being touched with a rubbed polythene rod or electrophorus. Initially A hangs vertically, and the position of the centre of its shadow on the graph paper screen is noted. Bringing B close to A deflects it and the values of a and r (Fig. 39.17(b)) are found. The measurements are then repeated for different values of r.

Refer to Fig. 39.17(b). Ball A is in equilibrium under the action of its weight, W, the tension, T, in the support thread, and the Coulomb repulsion, F. Resolving horizontally and vertically gives

$$F = T\sin\theta \quad \text{and} \quad W = T\cos\theta$$

Eliminating T between these equations gives

$$F = W\tan\theta$$

If the length, L, of the thread is large compared with a, $\tan\theta \approx \sin\theta = a/L$ and therefore to a good approximation

$$F = \frac{Wa}{L}$$

Thus F is proportional to a. From Coulomb's law, $F \propto 1/r^2$ and therefore a graph of a against $1/r^2$ should be a straight line through the origin.

It follows from Coulomb's law that for any given value of r, F is proportional to the product of the charges on A and B. In order to test this, B is touched against an identical but uncharged metallized polystyrene ball, C, on an insulating support. Since B and C are identical, this halves the charge on B. According to Coulomb's

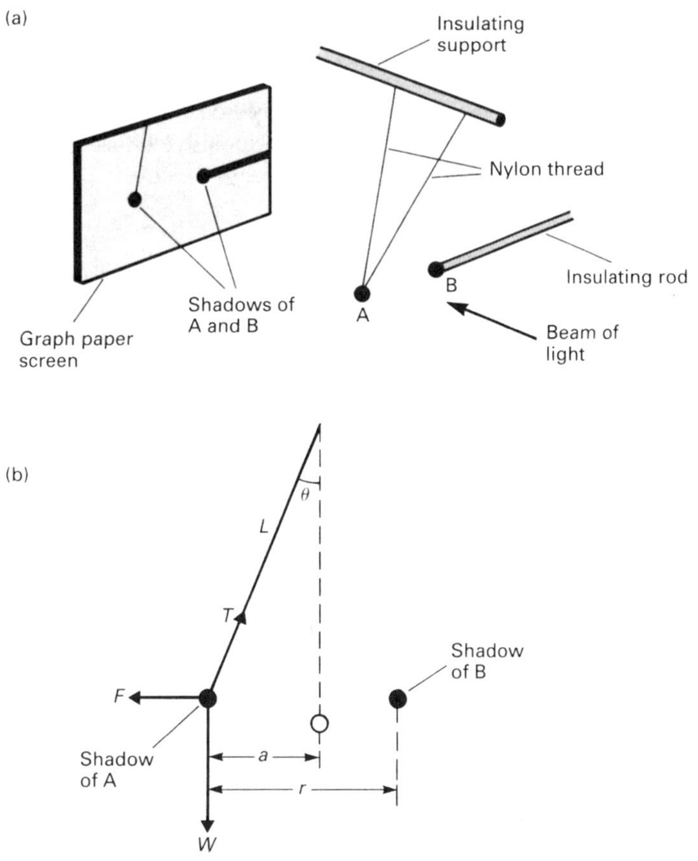

Fig. 39.17 Investigation of Coulomb's law: (a) experimental arrangement, (b) theory

law the force between A and B should now be one half of what it was previously for any particular value of r. This is easily tested by selecting some suitable value of r and measuring the new value of a. If C is discharged and touched against either A or B, the force (and therefore a) should halve again, and so on.

The method just described gives only a crude test of the validity of Coulomb's law. It is best tested by indirect methods involving the experimental verification of results deduced from it.

CONSOLIDATION

Coulomb's law The force between two charged bodies, which are small compared to their separation (r), is inversely proportional to r^2, and proportional to the product of their charges (Q_1 and Q_2).

$$F = \frac{1}{4\pi\varepsilon}\frac{Q_1 Q_2}{r^2}$$

Electric field intensity (or **field strength**) at a point is defined as the force exerted by the field on a unit charge placed at that point. (Unit = $\mathrm{N\,C^{-1}}$ or $\mathrm{V\,m^{-1}}$.)

Thus

$$F = EQ$$

The electric potential at a point is defined as being numerically equal to the work done in bringing a unit positive charge from infinity to the point. (Unit = V.)

The potential difference between two points is numerically equal to the work done in moving a unit positive charge from the point at the lower potential to the point at the higher potential. (Unit = V.) Thus

$$W = QV$$

Electric field intensity is a vector quantity. The direction of the field is the direction in which a positive charge would move under the influence of the field

Electric potential is a scalar quantity. The potential at infinity is zero

At a distance r from a point charge Q

$$E = \frac{1}{4\pi\varepsilon}\frac{Q}{r^2} \qquad V = \frac{1}{4\pi\varepsilon}\frac{Q}{r}$$

These expressions also apply at a point a distance r from the centre of a sphere with charge Q providing the point is outside the sphere.

$$E = -\frac{dV}{dx} \qquad \text{(In general)}$$

It follows that $E = 0$ whenever $V =$ constant. In particular, $E = 0$ and $V =$ constant throughout the material of a conductor.

$$E = \frac{V}{d} \qquad \text{(For a uniform field)}$$

An equipotential surface is a surface over which the potential is constant.

Electric field lines are perpendicular to equipotential surfaces.

The surface of a conductor is an equipotential surface and therefore the field lines at the surface of a conductor are perpendicular to the surface.

For a charge Q at a point where the potential is V

$$\text{PE} = QV$$

In particular, the potential energy of a charge Q' at a distance r from a point charge Q is given by

$$\text{PE} = \frac{1}{4\pi\varepsilon}\frac{QQ'}{r}$$

40

CAPACITORS

40.1 THE ACTION OF A CAPACITOR

A capacitor (the old name is **condenser**) is a device which is used to store electric charge.* Effectively, all capacitors consist of a pair of conducting plates separated by an insulator. The insulator is called a **dielectric** and is often air, oil or paper.

Fig. 40.1 shows a (parallel-plate) capacitor connected to a battery. When the battery is first connected there is a momentary flow of current. This is the result of electrons being drawn from plate A by the positive terminal of the battery, whilst at the same time, electrons are being deposited on B by the action of the negative terminal. After a very short time the potentials of A and B become equal to those of the positive and negative terminals of the battery, respectively. The potential difference across the capacitor is now equal to that across the battery and oppositely directed to it, so that there can be no further current flow. The capacitor is said to be (fully) charged. The charging process leaves A and B with equal amounts of charge of opposite sign.

Fig. 40.1
Charging a capacitor

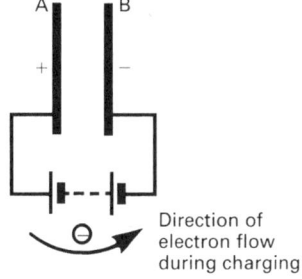

40.2 CAPACITANCE

The measure of the extent to which a capacitor can store charge is called its **capacitance**. It is defined by

$$C = \frac{Q}{V} \qquad [40.1]$$

*The gold-leaf electroscope is essentially a capacitor, though its prime purpose, of course, is not to store charge.

where

C = the capacitance (Unit = farad, symbol F.)

Q = the magnitude of the charge on either plate (C)

V = the PD between the plates (V).

Notes (i) The farad is a very large unit, capacitances are often measured in microfarads (μF).

(ii) 1 farad = 1 coulomb volt^{-1}, i.e. $1\,\text{F} = 1\,\text{C}\,\text{V}^{-1}$.

The capacitance of any particular capacitor depends on the size and separation of its plates, and on the permittivity (see section 39.2) of the material (the dielectric) between the plates. It is possible to derive an expression which relates the capacitance of a capacitor to its dimensions if its geometry is simple. Some of the more important expressions of this type are given below.

Parallel-plate Capacitor

Refer to Fig. 40.2

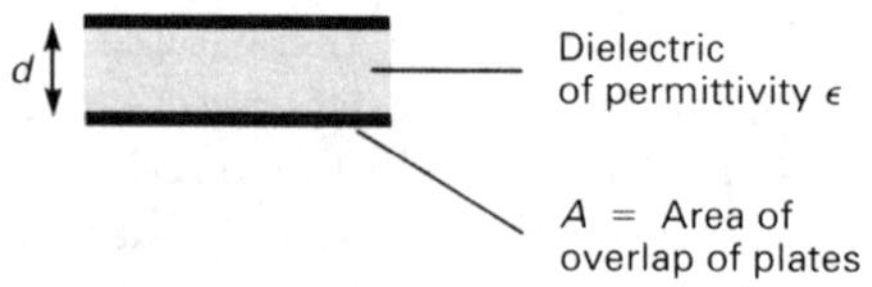

Fig. 40.2 Components of a parallel-plate capacitor

$$C = \frac{\varepsilon A}{d} \qquad [40.2]$$

Proof: To show that $C = \varepsilon A/d$

Consider a parallel-plate capacitor (such as that in Fig. 40.2) where the charge on either plate is Q and the PD between them is V. Let the area of each plate be A, and let their separation be d.

It follows from Gauss's theorem* that the field strength E between the plates of a parallel-plate capacitor is given by

$$E = \sigma/\varepsilon$$

where σ is the charge density (i.e. the charge per unit area) on either plate.

From the definition of capacitance (equation [40.1]) the capacitance C of the capacitor is given by

$$C = Q/V$$

i.e. $$C = \sigma A/V$$

*Gauss's theorem is beyond the scope of this book.

Substituting for σ gives

$$C = \varepsilon EA/V$$

But $E = V/d$ (equation [39.10]). Therefore

$$C = \varepsilon A/d$$

Note Strictly, $E = \sigma/\varepsilon$ for an infinite conducting plate, but it is a good approximation at points close to a finite conducting plate provided they are away from the edges. Strictly, then, $C = \varepsilon A/d$ is an approximate relationship.

Coaxial Cylindrical Capacitor

Refer to Fig. 40.3

Fig. 40.3
Components of a coaxial cylindrical capacitor

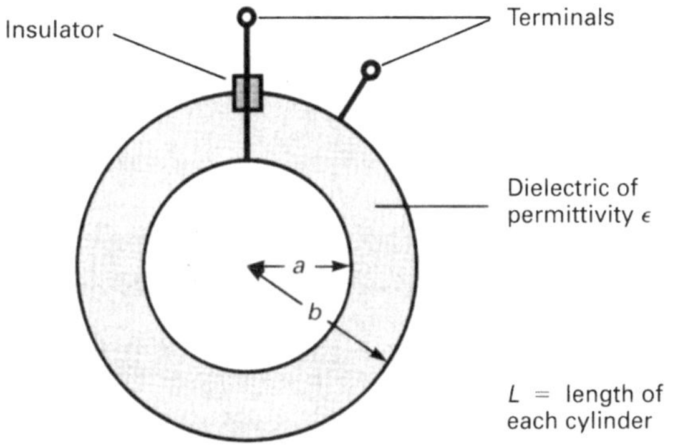

$$C = \frac{2\pi\varepsilon L}{\log_e(b/a)}$$

Concentric Spheres

The capacitance of two concentric metal spheres whose radii are a and b and which are separated by a material of permittivity, ε, is given by

$$C = \frac{4\pi\varepsilon ab}{(b-a)}$$

Isolated Sphere

An isolated metal sphere acts as a capacitor. The sphere itself is one plate, the Earth is the other. For a sphere of radius, r

$$C = 4\pi\varepsilon r \qquad [40.3]$$

where ε is the permittivity of the dielectric between the sphere and the Earth – usually air. The collecting sphere of a Van de Graaff generator is this type of capacitor.

Note In each case $C \propto \varepsilon$, showing that the capacitance of a capacitor can be increased by replacing its dielectric with one of greater permittivity.

40.3 RELATIVE PERMITTIVITY

The relative permittivity (or **dielectric constant**), ε_r, of a dielectric material is defined by

$$\varepsilon_r = \frac{C}{C_0} \qquad [40.4]$$

where

C = the capacitance of some capacitor when the dielectric is between its plates, and

C_0 = the capacitance of the same capacitor when there is free space (vacuum) between its plates.

The capacitance of any capacitor is proportional to the permittivity of the material between its plates (see section 40.2), and therefore

$$\frac{C}{C_0} = \frac{\varepsilon}{\varepsilon_0}$$

where

ε = the permittivity of the dielectric

ε_0 = the permittivity of free space.

Therefore, from equation [40.4]

$$\varepsilon_r = \frac{\varepsilon}{\varepsilon_0}$$

i.e.

$$\varepsilon = \varepsilon_0 \varepsilon_r \qquad [40.5]$$

Notes
(i) Relative permittivity is dimensionless and has no units.
(ii) The relative permittivity of free space is equal to unity, by definition.
(iii) Equation [40.5] <u>defines</u> the (absolute) permittivity ε of a material.

40.4 EXPLANATION OF THE EFFECT OF A DIELECTRIC IN A CAPACITOR

If a dielectric is placed between the plates of a charged capacitor, the field between the plates distorts the molecules of the dielectric. The (positive) nuclei are shifted slightly in the direction of the field (i.e. away from the positive plate), and the (negative) electrons are shifted in the opposite direction. The molecules are said to be **polarized**, there being an excess positive charge at one end of each molecule and an excess negative charge at the other end. As a result of this polarization, that surface of the dielectric which is adjacent to the positive plate of the capacitor acquires a negative charge, the opposite surface acquires a positive charge. The situation is shown in Fig. 40.4.*

*Polarization accounts for the ability of a charged comb to attract small pieces of paper. The side of the paper which is close to the comb acquires an induced charge of opposite sign to that on the comb and so is attracted to it.

Fig. 40.4
(a) Effect of a dielectric on an isolated capacitor.
(b) Effect of dielectric with battery connected

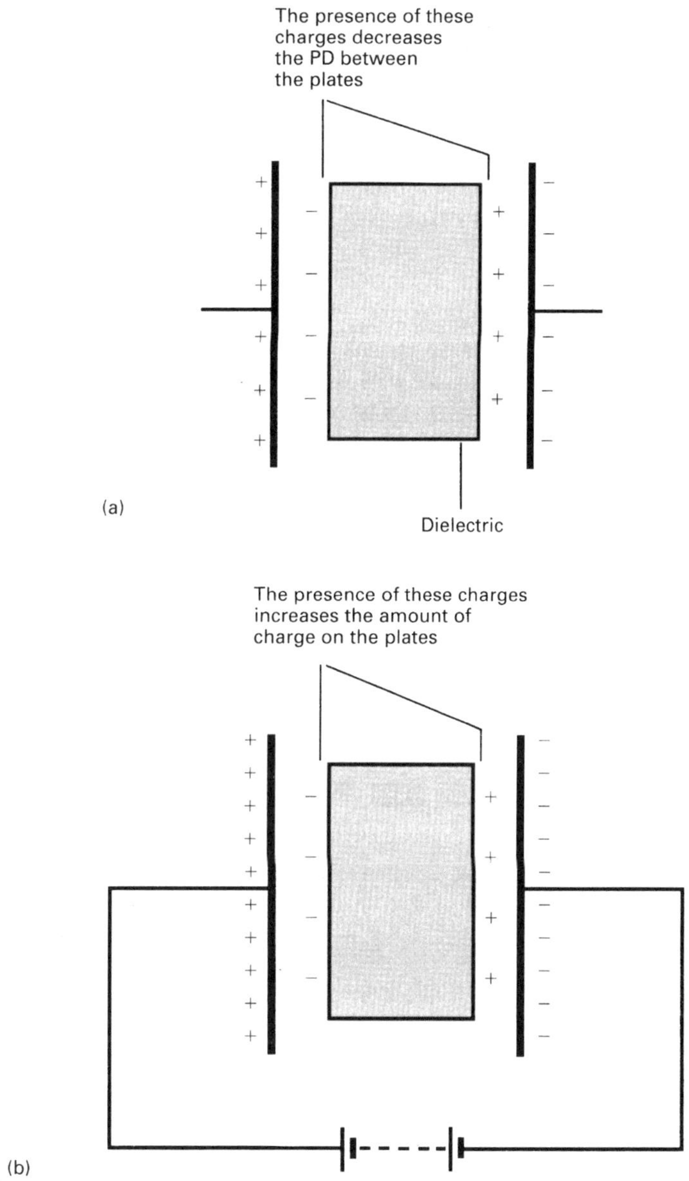

The presence of these charges decreases the PD between the plates
Dielectric
(a)
The presence of these charges increases the amount of charge on the plates
(b)

If the capacitor is isolated (Fig. 40.4(a)), the potential of the positive plate is decreased by the negative charge on that surface of the dielectric which is near to it; the potential of the negative plate is increased (i.e. made less negative). Thus, the presence of the dielectric decreases the potential difference between the plates. Since the capacitor is isolated, there can be no change in the amount of charge on either plate. For a charged capacitor, $Q = CV$ (equation [40.1]), and therefore since Q is unchanged and V decreases, C must have increased.

If the capacitor is connected to a battery (Fig. 40.4(b)), the potential difference between the plates cannot change when the dielectric is introduced. The battery maintains the potential of each plate by drawing electrons off the positive plate and depositing electrons on the negative plate. In this case then, V is unchanged and Q is increased. It follows from $Q = CV$ that C must have increased.

Thus, in each case, the introduction of the dielectric increases the capacitance of the capacitor.

Some molecules (known as **polar molecules**) are polarized even in the absence of an electric field. (The water molecule is an example.) If a dielectric which contains polar molecules is placed between the plates of a charged capacitor, the molecules become partially aligned in the direction of the field. (Random thermal motion prevents there being complete alignment.) As might be expected, dielectrics of this type have a much greater effect on capacitance than non-polar dielectrics – see Table 40.1.

Table 40.1 Relative permittivity

Substance	*Relative Permittivity* ($\varepsilon_r = \varepsilon/\varepsilon_0$)
Vacuum	1 (by definition)
Air (STP)	1.000 576
Polythene	2.3
Ebonite	3
Oiled paper	~4
Glass	~5
Mica	7
Water*	80

*Polar dielectric.

40.5 CAPACITORS IN PARALLEL

Fig. 40.5
Capacitors in parallel

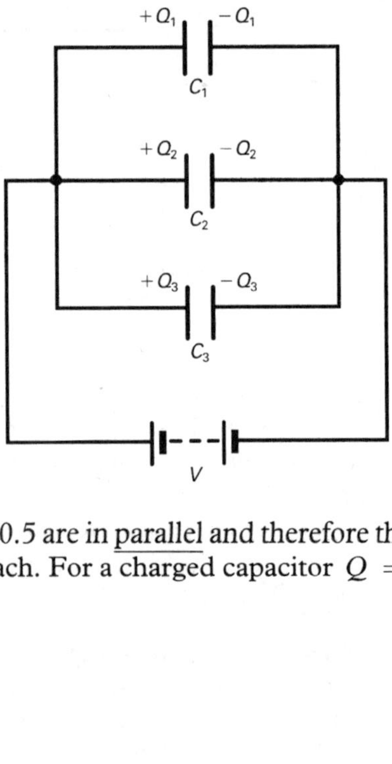

The capacitors in Fig. 40.5 are in parallel and therefore there is the same potential difference (V) across each. For a charged capacitor $Q = CV$ (equation [40.1]) and therefore

$$Q_1 = C_1 V$$

$$Q_2 = C_2 V$$

$$Q_3 = C_3 V$$

Adding gives

$$Q_1 + Q_2 + Q_3 = (C_1 + C_2 + C_3)V$$

If the total charge $(Q_1 + Q_2 + Q_3)$ is written as Q, this becomes

$$Q = (C_1 + C_2 + C_3)V \qquad [40.6]$$

A single capacitor which has the same effect as these three must store charge, Q, when the potential difference across its plates is V, and therefore has a capacitance, C, given by

$$Q = CV \qquad [40.7]$$

It follows from equations [40.6] and [40.7] that three capacitors, whose capacitances are C_1, C_2 and C_3 and which are in parallel, have a total (effective) capacitance, C, given by

$$C = C_1 + C_2 + C_3 \qquad [40.8]$$

40.6 CAPACITORS IN SERIES

The capacitors in Fig. 40.6 are in series. On connection, the battery draws electrons from plate A and leaves it with a positive charge, $+Q$. This induces a charge, $-Q$, on B. B and M together with their connecting wire constitute an isolated conductor, and the charge which appears on B results from electrons moving from M to B. M is left with a charge, $+Q$, and this induces a charge, $-Q$, on N and so on, so that the situation is as shown.

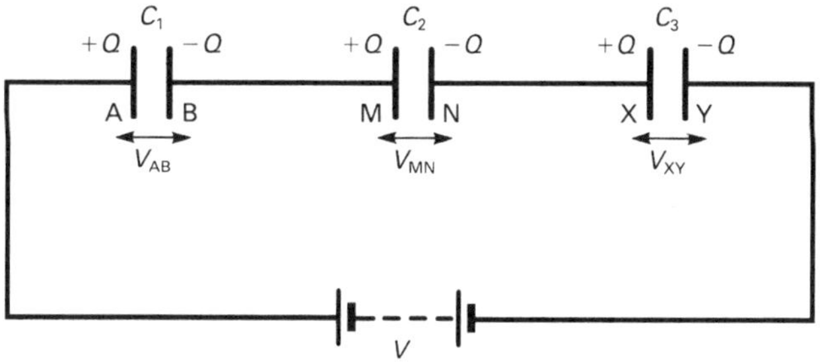

Fig. 40.6
Capacitors in series

From equation [40.1]

$$V_{AB} = \frac{Q}{C_1}$$

$$V_{MN} = \frac{Q}{C_2}$$

$$V_{XY} = \frac{Q}{C_3}$$

Adding gives

$$V_{AB} + V_{MN} + V_{XY} = Q\left(\frac{1}{C_1} + \frac{1}{C_2} + \frac{1}{C_3}\right)$$

i.e. $$V = Q\left(\frac{1}{C_1} + \frac{1}{C_2} + \frac{1}{C_3}\right) \qquad [40.9]$$

A single capacitor which has the same effect as these three would have a capacitance, C, given by

$$V = \frac{Q}{C} \qquad [40.10]$$

It follows from equations [40.9] and [40.10] that three capacitors, whose capacitances are C_1, C_2 and C_3 and which are connected in series, have a total (effective) capacitance, C, given by

$$\frac{1}{C} = \frac{1}{C_1} + \frac{1}{C_2} + \frac{1}{C_3} \qquad [40.11]$$

40.7 COMPARISON OF CAPACITOR AND RESISTOR NETWORKS

See Table 40.2. Note, in particular, that the capacitance of a series combination of capacitors is less than the smallest individual capacitance, and the resistance of a parallel combination of resistors is less than the smallest individual resistance.

Table 40.2 Comparison of capacitor and resistor networks

	Capacitor network	*Resistor network*
Series connection	Same charge $\frac{1}{C} = \frac{1}{C_1} + \frac{1}{C_2} + \frac{1}{C_3}$ $Q = VC$	Same current $R = R_1 + R_2 + R_3$ $V = IR$
Parallel connection	Same PD $C = C_1 + C_2 + C_3$ $Q = VC$	Same PD $\frac{1}{R} = \frac{1}{R_1} + \frac{1}{R_2} + \frac{1}{R_3}$ $V = IR$

EXAMPLE 40.1

Calculate the charges on the capacitors shown in Fig. 40.7, and the potential differences across each.

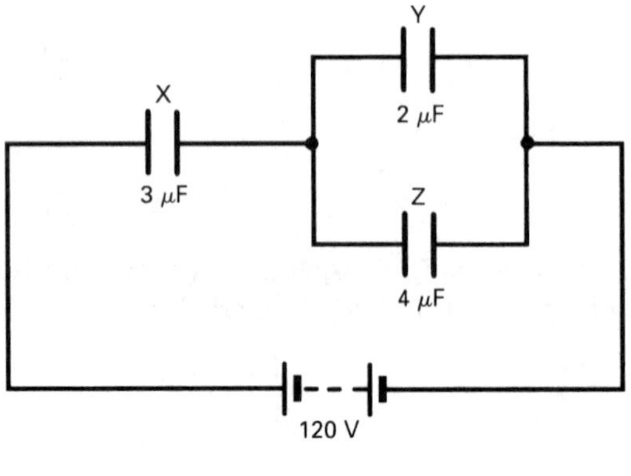

Fig. 40. 7 Diagram for Example 40.1

Solution

The capacitance C_{ZY} of the parallel combination of Z and Y is given (by equation [40.8]) as

$$C_{ZY} = 2\,\mu F + 4\,\mu F = 6\,\mu F$$

The circuit is therefore equivalent to a $3\,\mu$F capacitor in series with a $6\,\mu$F capacitor, and the total capacitance, C, is given by equation [40.11], as

$$\frac{1}{C} = \frac{1}{3 \times 10^{-6}} + \frac{1}{6 \times 10^{-6}} = \frac{3}{6 \times 10^{-6}}$$

i.e. $C = 2 \times 10^{-6}\,F$

For the circuit as a whole, the total charge, Q, is given, by equation [40.1], as

$$Q = VC$$

i.e. $Q = (120) \times (2 \times 10^{-6})$

i.e. $Q = 240 \times 10^{-6}$ coulombs

Therefore

Charge on X $= 240 \times 10^{-6}$ coulombs

and

Charge on Z + Y $= 240 \times 10^{-6}$ coulombs

For X:

If the PD across X is V_X, then by $Q = VC$ (equation [40.1]) we have

$$240 \times 10^{-6} = (V_X) \times (3 \times 10^{-6})$$

i.e. $V_X = 80\,V$

It follows that the PD across each of Y and Z is $120 - 80 = 40\,V$.

For Y:

If the charge on Y is Q_Y, then by $Q = VC$, we have

$$Q_Y = (40) \times (2 \times 10^{-6})$$

i.e. $Q_Y = 80 \times 10^{-6}$ coulombs

For Z:

If the charge on Z is Q_Z, then by $Q = VC$, we have

$$Q_Z = (40) \times (4 \times 10^{-6})$$

i.e. $Q_Z = 160 \times 10^{-6}$ coulombs

Alternatively,

$$Q_Z = Q - Q_Y$$

i.e. $Q_Z = 240 \times 10^{-6} - 80 \times 10^{-6}$

i.e. $Q_Z = 160 \times 10^{-6}$ coulombs

QUESTIONS 40A

1. A 3.0 μF capacitor and a 5.0 μF capacitor are each in parallel with a 12 V supply. Calculate the charge on each capacitor.

2. A 4.0 μF capacitor and a 6.0 μF capacitor are connected in series with a 20 V supply. Calculate: **(a)** the charge on each capacitor, **(b)** the PD across each capacitor.

3. Calculate: **(a)** the charge on each capacitor, **(b)** the PD across each capacitor in the circuit shown below.

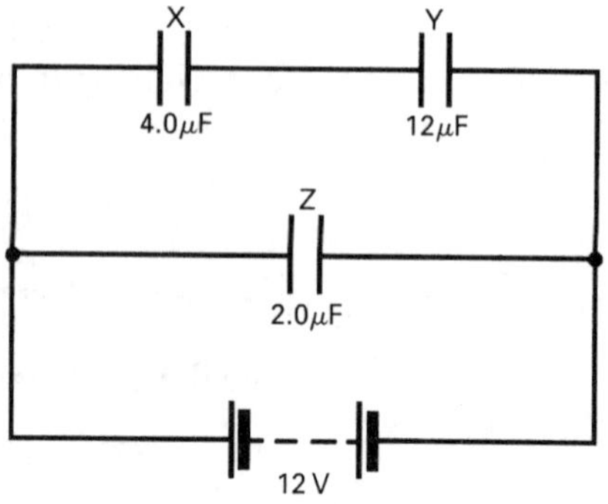

40.8 ENERGY STORED IN A CHARGED CAPACITOR

Once the charging of a capacitor has begun, the addition of electrons to the negative plate involves doing work against the repulsive forces of the electrons which are already there. Equally, the removal of electrons from the positive plate requires that work is done against the attractive forces of the positive charges on that plate. The work which is done is stored in the form of electrical potential energy.

Consider a capacitor whose capacitance is C. Suppose that it is partially charged, so that the charge on its plates is Q and the potential difference between them is V. Suppose now, that the charge increases to $(Q+\delta Q)$. This involves moving a charge δQ from one plate to the other. If δQ is small, V can be considered to be unchanged by this process, in which case the work done, δW, is given by equation [39.6] as

$$\delta W = V\delta Q$$

Therefore by equation [40.1]

$$\delta W = \frac{Q}{C}\delta Q$$

The total work done, W, in increasing the charge from 0 to Q_0 is therefore given by

$$W = \int_0^{Q_0} \frac{Q}{C}\,\mathrm{d}Q$$

i.e.

$$W = \frac{Q_0^{\,2}}{2C}$$

Writing Q for Q_0 and making use of $Q = VC$ leads to

$$W = \frac{Q^2}{2C} = \tfrac{1}{2}CV^2 = \tfrac{1}{2}QV \qquad [40.12]$$

where

W = the energy stored by a charged capacitor (J)

Q = the charge on the plates (C)

V = the PD across the plates (V)

C = the capacitance (F).

40.9 JOINING TWO CHARGED CAPACITORS

Fig. 40.8 shows two charged capacitors whose capacitances are C_1 and C_2. On closing S, charge flows until the potential difference across each capacitor is the same. The final charge, potential difference and energy of each capacitor can be calculated by making use of:

(i) there is no change in the total amount of charge,

(ii) the two capacitors acquire the same potential difference,

(iii) the capacitors are in parallel and therefore the capacitance, C, of the combination is given by $C = C_1 + C_2$.

Fig. 40.8
Joining two charged capacitors

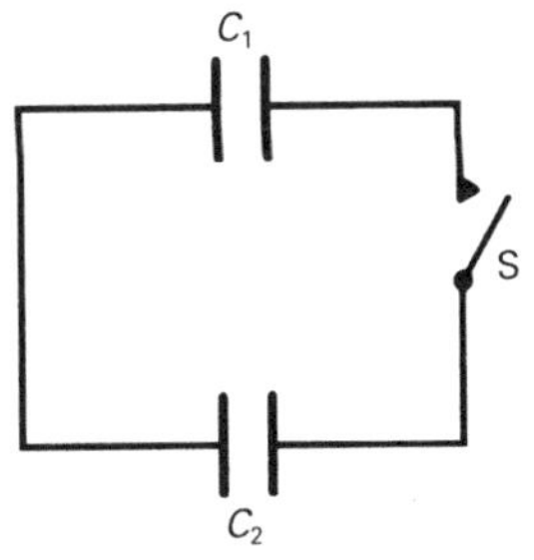

It turns out that unless the initial potential differences of the capacitors are the same, the total energy stored by the capacitors decreases when they are joined together. Energy is dissipated as heat in the connecting wires when charge flows from one capacitor to the other, and this accounts for the decrease in stored energy. (If two capacitors are placed in contact without using connecting wires, the 'lost' potential energy goes into producing a spark as the terminals of the capacitors approach each other.)

EXAMPLE 40.2

A 5 μF capacitor (X) is charged by a 40 V supply, and is then connected across an uncharged 20 μF capacitor (Y). Calculate:

(a) the final PD across each,

(b) the final charge on each,

(c) the initial and final energies stored by the capacitors.

Solution

If the initial charge on X is Q_0, then by $Q = VC$ (equation [40.1]),

$$Q_0 = (40) \times (5 \times 10^{-6})$$

i.e. $Q_0 = 200 \times 10^{-6}$ coulombs

The situation after the capacitors have been connected is shown in Fig. 40.9, where Q_X and Q_Y are the final charges on X and Y respectively, and V is the final PD.

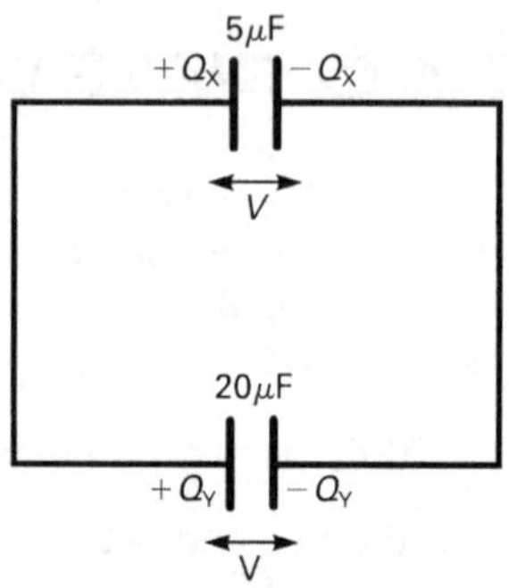

Fig. 40.9 Diagram for Example 40.2

The capacitors are in parallel, and therefore the total capacitance is given (by equation [40.8]) as

$$\text{Total capacitance} = 5 + 20 = 25\mu\text{F}$$

The total charge is unchanged, and therefore

$$\text{Total charge} = 200 \times 10^{-6} \text{ coulombs}$$

Applying $Q = VC$ to the combination gives

$$200 \times 10^{-6} = (V) \times (25 \times 10^{-6})$$

i.e. $V = 8\,\text{V}$

For X:

By $Q = VC$ we have

$$Q_X = (8) \times (5 \times 10^{-6})$$

i.e. $Q_X = 40 \times 10^{-6}$ coulombs

For Y:

Since $Q = VC$ we have

$$Q_Y = (8) \times (20 \times 10^{-6})$$

i.e. $Q_Y = 160 \times 10^{-6}$ coulombs

The energy of a charged capacitor is given by $\frac{1}{2}CV^2$ (equation [40.12]), and therefore

$$\text{Initial energy} = \tfrac{1}{2}(5 \times 10^{-6}) \times (40)^2 = 4 \times 10^{-3}\,\text{J}$$

$$\text{Final energy} = \tfrac{1}{2}(5 \times 10^{-6}) \times (8)^2 + \tfrac{1}{2}(20 \times 10^{-6}) \times (8)^2 = 0.8 \times 10^{-3}\,\text{J}$$

EXAMPLE 40.3

Calculate the effect of doubling the separation of the plates of a parallel-plate capacitor on the energy stored by the capacitor:

(i) when the capacitor is isolated,

(ii) when the capacitor is connected to a battery.

In each case account for the energy changes which occur.

Solution

The capacitance of a parallel-plate capacitor is given by $C = \varepsilon A/d$ (equation [40.2]), and therefore doubling the separation of the plates halves the capacitance.

Isolated capacitor When the capacitor is isolated the charge on the plates is constant. Since $Q = CV$, halving C causes V to double. The energy stored by a charged capacitor is given by $W = Q^2/2C$ (equation [40.12]) and therefore, since C is halved and Q is constant, the energy stored by the capacitor doubles.

Battery connected When the capacitor is connected to a battery the potential difference across the capacitor is constant. In this case it is useful to make use of $W = \frac{1}{2}CV^2$ (equation [40.12]). Thus, since C is halved and V is constant, the energy stored by the capacitor halves.

When the capacitor is isolated work has to be done in order to pull the positive charge on one plate away from the negative charge on the other. The work which is done is equal to the increase in the electrical energy stored by the capacitor.

When the capacitor is connected to a battery the decrease in capacitance results in a decrease in the amount of charge stored by the capacitor ($Q = VC$ and V is constant). This charge is returned to the battery. The decrease in energy is the result of the capacitor discharging.

40.10 DISCHARGING A CAPACITOR THROUGH A RESISTOR

Refer to Fig. 40.10. Suppose that initially the charge on the capacitor is Q_0 and the potential difference across it is V_0. On closing S the capacitor starts to discharge. The resistance, R, limits the current flow, and therefore the discharge is not instantaneous.

Consider the situation some time t after the discharge has commenced. Let

V = the PD across the capacitor at time t

V_R = the PD across the resistor at time t.

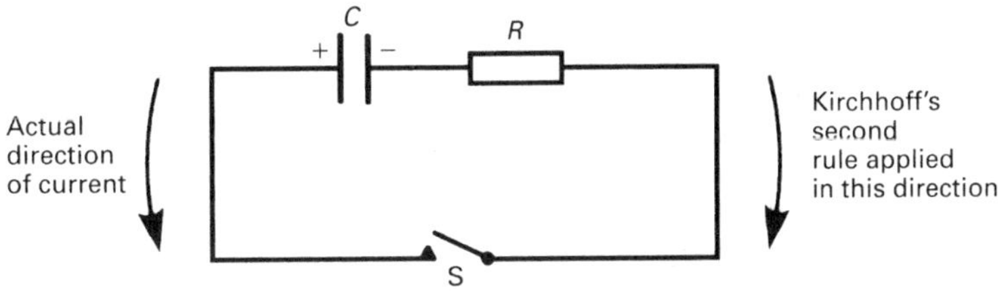

Fig. 40.10
Discharge of a capacitor through a resistor

There is no EMF in the circuit and therefore by Kirchhoff's second rule (equation [36.12])

$$0 = V + V_R \quad [40.13]$$

i.e. $$0 = \frac{Q}{C} + IR$$

where

I = current flowing at time t

Q = charge remaining on the capacitor at time t.

But $I = dQ/dt$, and therefore

$$0 = \frac{Q}{C} + \frac{dQ}{dt}R$$

i.e. $$\frac{dQ}{Q} = -\frac{dt}{CR}$$

Since $Q = Q_0$ when $t = 0$ and $Q = Q$ when $t = t$, integrating this expression gives

$$\int_{Q_0}^{Q} \frac{dQ}{Q} = -\frac{1}{CR}\int_0^t dt$$

i.e. $$\left[\log_e Q\right]_{Q_0}^{Q} = -\frac{1}{CR}\left[t\right]_0^t$$

i.e. $$\log_e Q - \log_e Q_0 = -\frac{t}{CR}$$

i.e. $$\log_e\left(\frac{Q}{Q_0}\right) = -\frac{t}{CR}$$

In exponential form

$$\frac{Q}{Q_0} = e^{-t/CR}$$

i.e. $$Q = Q_0 e^{-t/CR} \quad [40.14]$$

Since $Q = VC$ and $Q_0 = V_0C$, it follows from equation [40.14] that

$$V = V_0 e^{-t/CR} \quad [40.15]$$

Equations [40.14] and [40.15] show that when a capacitor is discharged through a resistor, both the charge on the capacitor and the potential difference across it decrease exponentially. It follows from equation [40.14] that when $t = CR$

$$Q = Q_0 e^{-1}$$

i.e. $$Q = \frac{Q_0}{e} = 0.368Q_0$$

CR is known as the **time constant** of the circuit and is the time taken for the charge on the capacitor to fall to $1/e$ (i.e. 36.8%) of its initial value (see Fig. 40.11).

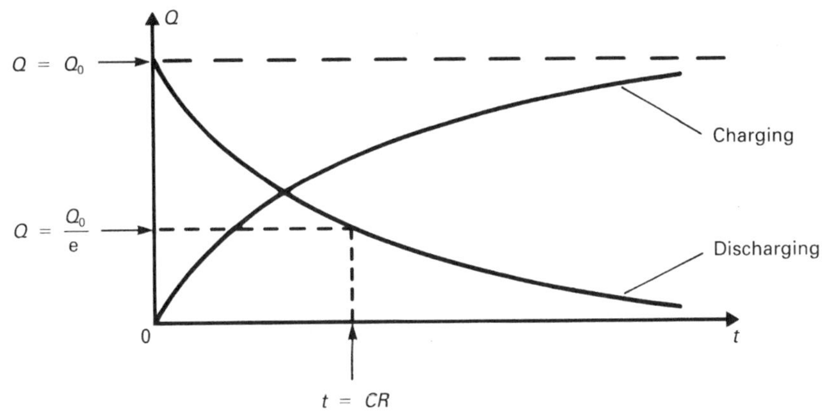

Fig. 40.11
Charge against time for a capacitor charging and discharging

Notes

(i) The time constant of a circuit containing a given capacitor is proportional to R, and therefore the discharge takes place slowly when there is a large series resistance.

(ii) From equation [40.13], $V = -V_R$, i.e. the PD across the capacitor is equal and opposite to that across the resistor.

(iii) It can be shown that when a capacitor is charged whilst in series with a resistance R, $Q = Q_0(1 - e^{-t/CR})$ where Q_0 is the final charge, i.e. when $t = \infty$ (see Fig. 40.11).

(iv) From equations [40.13] and [40.15]

$$V_R = -V_0 e^{-t/CR}$$

i.e. $$I = -(V_0/R)e^{-t/CR}$$

i.e. the current during discharge also decreases exponentially. (The minus sign arises because in applying Kirchhoff's second rule we traversed the circuit in the opposite direction to that in which the current is actually flowing. Note that, since $I = dQ/dt$, this makes dQ/dt negative and is therefore consistent with Q decreasing as t increases.) Note in particular that the size of the current is greatest when $t = 0$. This is also true in the case of the charging process. The current is normally taken to be positive in problems requiring numerical answers.

EXAMPLE 40.4

A 5.0 μF capacitor is charged by a 12 V supply and is then discharged through a 2.0 MΩ resistor.

(a) What is the charge on the capacitor at the start of the discharge?

(b) What is: (i) the charge on the capacitor, (ii) the PD across the capacitor, (iii) the current in the circuit 5.0 s after the discharge starts?

Solution

(a) The PD across the capacitor at the start of the discharge is 12 V and therefore the initial charge Q_0 is given by $Q = VC$ as

$$Q_0 = 12 \times 5.0 \times 10^{-6} = 60 \times 10^{-6}\ \text{C}$$

i.e. Initial charge $= 60\ \mu$C

(b) (i) By $Q = Q_0 e^{-t/CR}$ with $Q_0 = 60 \times 10^{-6}$ C, $t = 5.0$ s and $CR = 5.0 \times 10^{-6} \times 2.0 \times 10^6 = 10$ s

$$Q = 60 \times 10^{-6} e^{-5.0/10.0} = 3.64 \times 10^{-5} \text{ C}$$

i.e. Charge after 5.0 s = 36 μC

(ii) By $V = Q/C$

$$V = \frac{3.64 \times 10^{-5}}{5.0 \times 10^{-6}} = 7.28 \text{ V}$$

i.e. PD after 5.0 s = 7.3 V

(iii) The PD across the resistor is equal to the PD across the capacitor*, and therefore the current, I, through the resistor (i.e. the current in the circuit) is given by $I = V/R$ as

$$I = \frac{7.28}{2.0 \times 10^6} = 3.64 \times 10^{-6} \text{ A}$$

i.e. Current after 5.0 s = 3.6 μA

QUESTIONS 40B

1. A 500 μF capacitor with a charge of 3000 μC is discharged through a 200 kΩ resistor. What is **(a)** the initial discharge current, **(b)** the current after 20 s?

40.11 PRACTICAL CAPACITORS

For a parallel-plate capacitor $C = \varepsilon A/d$, and therefore large capacitances are achieved by having: (i) plates of large area, (ii) plates which are very close together, i.e. a thin layer of dielectric, (iii) a dielectric which has a high permittivity. Note, however, that there is no point in choosing to use a particular dielectric simply because its permittivity is high if it cannot be produced in the form of thin sheets.

Paper Capacitors

A paper capacitor is shown in Fig. 40.12. It consists of two long strips of metal foil (the electrodes) between which are two thin strips of paper (the dielectric). The

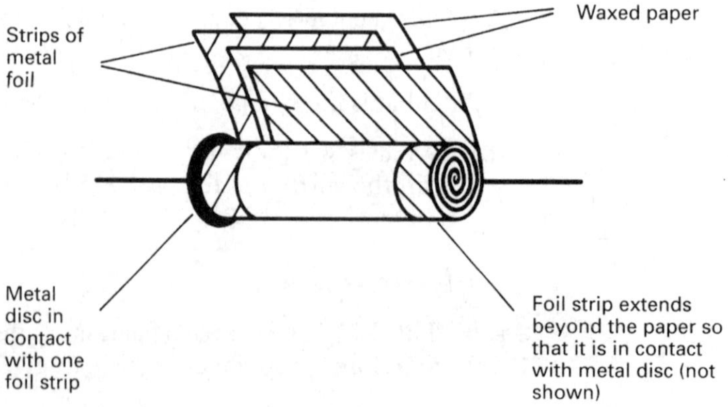

Waxed paper
Strips of metal foil
Metal disc in contact with one foil strip
Foil strip extends beyond the paper so that it is in contact with metal disc (not shown)

Fig. 40.12
Paper capacitor

*Strictly, they are equal and opposite (see note (ii) on p. 603) but there would be no point including a minus sign here.

'sandwich' is tightly rolled to form a small cylinder, so that the arrangement is essentially a parallel-plate capacitor of large surface area, occupying only a small volume.

There is a maximum electric field strength at which any given insulator can operate without breaking down (i.e. without ceasing to act as an insulator). There is therefore a maximum potential difference which can be applied across a given thickness of paper, and the paper in any particular capacitor must be thick enough to withstand the voltage at which the device is intended to operate. The insulating properties can be improved by impregnating the paper with paraffin wax.

Paper capacitors have capacitances ranging from about $10^{-3}\,\mu F$ to $10\,\mu F$. Their stability is poor but they are cheap, and they are extensively used. Paper capacitors are suitable for use in circuits where the frequency is between 100 Hz and 1 MHz. Capacitors of this type which make use of thin sheets of polystyrene rather than paper have greater stability and can be used at higher frequencies.

Electrolytic Capacitors

An electrolytic capacitor is shown in Fig. 40.13 together with its circuit symbol. Electrolytic capacitors are made by passing a current through the aluminium borate soaked paper using the two strips of aluminium foil as electrodes. This produces a thin film of aluminium oxide on the electrode which is acting as the anode, and it is this oxide film which is the dielectric of the capacitor. The capacitor plates are the aluminium anode and the electrolyte (i.e. the aluminium borate). The connection to the electrolyte is made via the cathode.

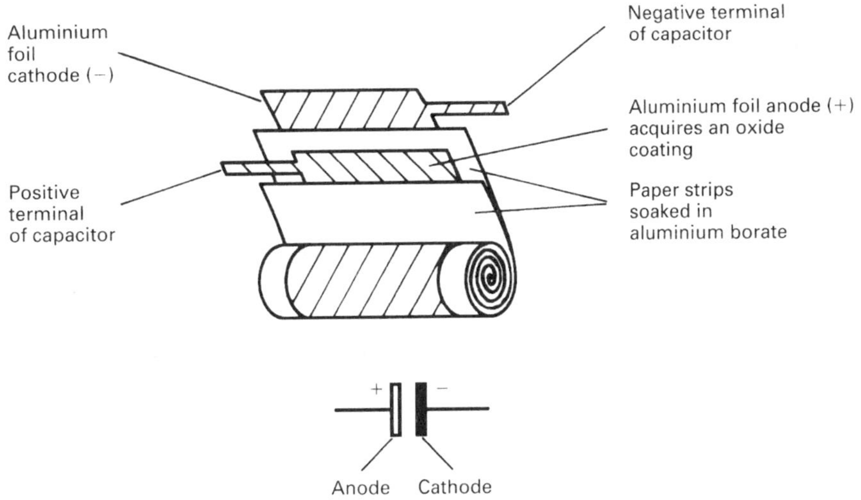

Fig. 40.13 Electrolytic capacitor and its circuit symbol

The oxide layer dielectric can withstand very high electric fields and therefore can be extremely thin (as little as 10^{-7} m for a capacitor operating at 100 V). As a result, electrolytic capacitors with capacitances of as much as $10^5\,\mu F$ are still physically small.

In order to maintain the oxide layer, a small leakage current of about 1 mA has to flow through the capacitor whilst it is being used. It is important that this current always flows in a particular direction. Consequently the anode terminal is usually marked with a plus sign or is coloured red to indicate that it is to be connected to the positive side of the circuit in which the capacitor is being used.

The stability of electrolytic capacitors is very poor and they cannot be used at frequencies above about 10 kHz.

Variable Air Capacitor

This type of capacitor (Fig. 40.14) is much used in the tuning circuits of radio receivers (see section 43.13). The capacitor consists of two interleaved sets of metal plates which are parallel to each other. One set of plates is attached to a spindle and can be moved in or out of the other set. Since the capacitance of a parallel-plate capacitor is proportional to the area of overlap of its plates, this alters the capacitance.

Fig. 40.14
Variable air capacitor and circuit symbol

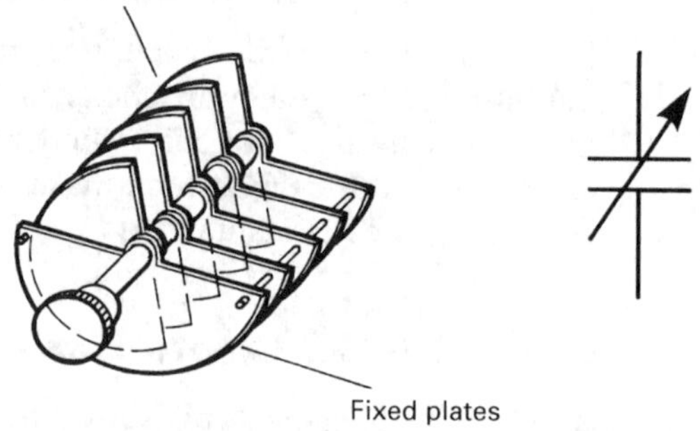

40.12 EXPERIMENTAL DETERMINATION OF CAPACITANCE AND RELATED EXPERIMENTS

Measurement of Capacitance

The circuit is shown in Fig. 40.15(a) and involves a **vibrating reed switch** (Fig. 40.15(b)). (Alternative methods are given in sections 40.13 and 42.21.) The switch contains two contacts, X and Y, one of which, X, becomes magnetized

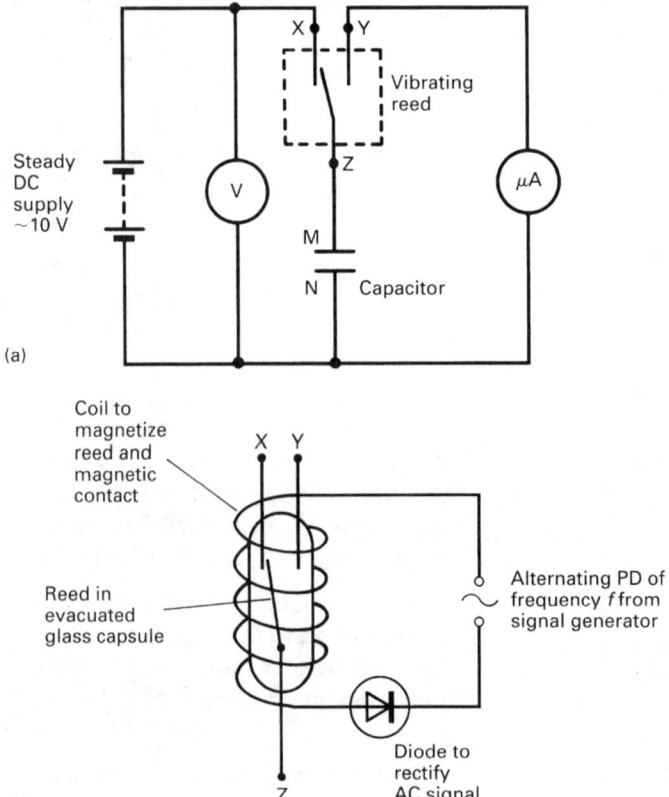

Fig. 40.15
Measurement of capacitance using a vibrating reed

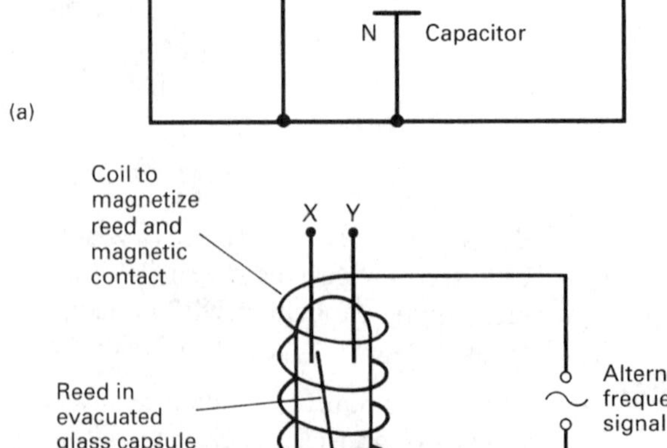

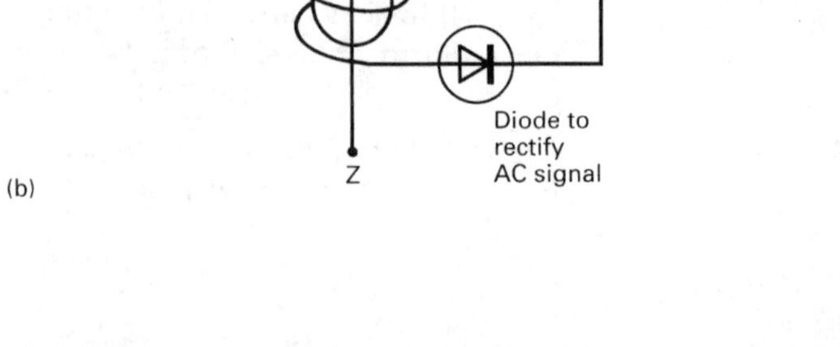

when it is in a magnetic field. The reed itself is also magnetized when in a magnetic field. The switch is surrounded by a coil connected to an alternating supply via a rectifier. On that half of the cycle that current is flowing through the coil, the reed and X become oppositely magnetized and the reed moves across to X providing a current path from X to Z. On the non-conducting half of the cycle the reed springs back to Y so that the current path is now from Y to Z. The reed therefore oscillates between X and Y at the frequency of the alternating supply.

As the reed vibrates, the capacitor is repeatedly charged by the DC supply and then discharged through the microammeter. If the charge on the capacitor is Q when it is fully charged, then Q is the quantity of charge which flows through the microammeter each time the capacitor discharges. If the frequency of the alternating supply is f, the charge which flows through the meter in one second is Qf, i.e. the current, I, through the microammeter is given by

$$I = Qf$$

If the capacitance of the capacitor is C, then

$$I = VCf$$

where V is the PD across the capacitor when it is fully charged, i.e. the voltage registered by the voltmeter. Hence C.

Notes

(i) It is sometimes necessary to protect the microammeter from current surges by putting a suitably large resistance in series with it. (This resistance should not be so large that the capacitor does not have time to discharge completely.)

(ii) It is desirable that both V and f are large so that I is not so small that its measurement presents difficulties. There is, however, an upper limit to f which is determined by the maximum speed of response of the reed; this is typically 1000 Hz.

To Verify that $Q \propto V$

This involves measuring I as a function of V whilst the frequency is kept constant. When the frequency is constant I is proportional to Q. It can be verified that Q is proportional to V, therefore, by plotting a graph of I against V and obtaining a straight line which passes through the origin.

Measurement of ε (and ε_0)

For a parallel-plate capacitor $C = \varepsilon A/d$. It follows that the permittivity, ε, of the material between the plates of such a capacitor can be determined by measuring C together with A and d.

With air between the plates the method gives a value for ε_0, the permittivity of free space. (There is no point in taking the trouble to evacuate the region between the plates because at the level of accuracy obtainable ε_0 is indistinguishable from the permittivity of air.)

To Verify that $C \propto A/d$ for a Parallel-plate Capacitor

For a parallel-plate capacitor $C = \varepsilon A/d$. By altering the area of overlap and the separation of two large metal plates it is possible to measure C as a function of A and d. A graph of C against A/d is a straight line through the origin and verifies that $C \propto A/d$. (The gradient of the graph is ε.)

Measurement of the Capacitance of an Isolated Sphere

When the capacitance being measured is that of an isolated sphere M and N in Fig. 40.15(a) are respectively the sphere and an earth.

40.13 THE DC AMPLIFIER

A DC amplifier (Fig. 40.16) is an electronic device which acts as a voltmeter with a very high input resistance (typically $10^{13}\,\Omega$). The instrument is designed to produce an output current which is proportional to the PD across its input terminals. The PD to be measured is applied to the input terminals and produces a reading on a microammeter connected across the output terminals. A typical instrument gives a full scale deflection on a 0–100 μA meter when the PD across the input is 1 V. Before the DC amplifier is used, it must be calibrated by applying a known PD of 1 V to the input and adjusting a sensitivity control (not shown) to give a reading of 100 μA on the microammeter.

Fig. 40.16
DC amplifier

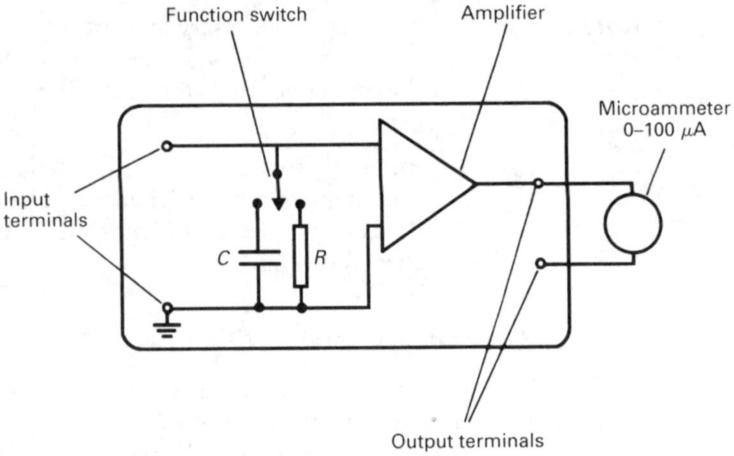

Although it is basically a voltmeter, a DC amplifier can be adapted to measure very small charges ($<10^{-9}$ C) and very small currents ($<10^{-11}$ A).

Measurement of Charge and Capacitance

By use of the function switch a capacitor, C, whose capacitance is known, is connected (internally) across the input. Capacitances of 10^{-7} F, 10^{-8} F and 10^{-9} F are normally available. Suppose that the 10^{-8} F capacitor is selected and that the charge to be measured, Q, is on the plates of a capacitor, C'. (It might equally well be on a proof plane or the dome of a Van de Graaff generator, etc.) In order that it can be measured, Q has to be transferred to the plates of C (the 10^{-8} F capacitor). Accordingly, C' is connected across the input terminals of the DC amplifier so that C and C' are in parallel, in which case, provided the capacitance of C' is much less than that of C, practically the whole of the charge on C' is transferred to C. Suppose that the charge transfer produces a (steady) reading of 20 μA on the microammeter. Since the instrument is calibrated to register 100 μA for an input PD of 1 V, it follows that the PD across C is 0.2 V and that (by $Q = CV$) the charge on C is $10^{-8} \times 0.2 = 2 \times 10^{-9}$ C. As long as the capacitance of C' is much less than that of C, 2×10^{-9} C can be taken to be the charge that was originally on C'. Further, if the PD across C' before it was connected to C was 20 V, say, then the capacitance of C' is $2 \times 10^{-9}/20 = 10^{-10}$ F.

Notes (i) The capacitance of C' is only 1% of that of C, in which case 99% of the charge on C' will have been transferred to C.

(ii) C discharges through the DC amplifier, but because its resistance is very high ($\sim 10^{13}\,\Omega$) the discharge takes place very slowly and the microammeter reading is to all intents and purposes steady; there is, therefore, ample time to take the reading.

Measurement of Current

By use of the function switch a resistor, R, whose resistance is known, is connected (internally) across the input of the DC amplifier. Resistances of $10^8\,\Omega$, $10^9\,\Omega$, $10^{10}\,\Omega$ and $10^{11}\,\Omega$ are normally available. Suppose that the $10^{10}\,\Omega$ resistor is selected. The current to be measured, I, is caused to flow through this resistor by connecting the source of the current (e.g. an ionization chamber) across the input terminals. Suppose that this produces a reading of 40 μA on the microammeter. Since the DC amplifier is calibrated to register 100 μA for an input PD of 1 V, it follows that the PD across the $10^{-10}\,\Omega$ resistor is 0.4 V and that (since $I = V/R$) $I = 0.4/10^{10} = 4 \times 10^{-11}$ A.

Note The input resistance of the DC amplifier ($\sim 10^{13}\,\Omega$) is much higher than that of R ($10^{10}\Omega$ in this example) and therefore, although the two resistances are in parallel, only an insignificant fraction (0.1%) of I flows through the $10^{13}\,\Omega$ resistance.

40.14 ANALOGY BETWEEN A CHARGED CAPACITOR AND A STRETCHED SPRING

It is possible to draw an analogy between a <u>charged</u> capacitor and a <u>stretched</u> spring which obeys Hooke's law.

(i) They both store energy.

(ii) The more a spring is stretched, the harder it becomes to stretch it any further because work has to be done against the ever increasing tension in the spring. Similarly, the more a capacitor is charged, the harder it becomes to charge it further because work has to be done against the ever increasing potential difference across the plates of the capacitor.

(iii) The tension, F, in a spring whose extension is x and which obeys Hooke's law is given by $F = kx$, where k is the spring constant (sections 7.5 and 7.7). It can be shown* that F, k and x are respectively analogous to V, $1/C$ and Q and therefore that $F = kx$ is analogous to $V = (1/C)Q$. The three expressions for the energy of a charged capacitor (section 40.8) are analogous to those derived in section 7.7 for a stretched spring – see Table 40.3.

*We stated in (ii) that in stretching a spring work has to be done against tension, and that in charging a capacitor work is done against potential difference. This gives us a clue that F and V are analogous. It is no more than a clue, however, for we could just as easily have said that in charging a capacitor work is done against the ever-increasing <u>charge</u> on the plates, and this would make F analogous to Q, in which case k would be analogous to $\overline{C}$ rather than to $1/C$. We show in (iv) that it is k and $1/C$ that are analogous to each other.

Table 40.3 Comparison of Springs and Capacitors

Stretched spring which obeys Hooke's law	*Analogous expressions for a charged capacitor*
$F = kx$	$V = \frac{1}{C}Q$
$W = \frac{1}{2}Fx$	$W = \frac{1}{2}VQ$
$W = \frac{1}{2}kx^2$	$W = \frac{Q^2}{2C}$
$W = \frac{F^2}{2k}$	$W = \frac{1}{2}CV^2$

(iv) If a charged capacitor is connected across an inductor (see section 42.14), the capacitor repeatedly discharges and charges up again. The current in the circuit oscillates sinusoidally. This may be compared with a mass on an oscillating spring. The respective periods of oscillation are given by

$$\text{Period} = 2\pi\sqrt{LC} \quad \text{and} \quad \text{Period} = 2\pi\sqrt{m/k}$$

This confirms our previous assertion that k is analogous to $1/C$, and also shows that mass (m) is analogous to inductance (L). There is bound to be some resistance in the circuit (mainly that of the inductor). This causes the oscillations to die away and corresponds to the way in which the motion of an oscillating spring is damped by any viscous forces that are present. Thus electrical resistance is analogous to viscosity. The reader may find it useful to compare Fig. 29.4 and Fig. 43.10.

CONSOLIDATION

$$C = \frac{Q}{V}$$ This equation defines capacitance.

$$\textbf{Energy stored} = \frac{Q^2}{2C} = \tfrac{1}{2}CV^2 = \tfrac{1}{2}QV$$

Parallel Plate Capacitor

$$C = \frac{\varepsilon A}{d} \quad \text{where} \quad \varepsilon = \varepsilon_0\varepsilon_r$$

Capacitors in Parallel

Same PD across each.

$$C = C_1 + C_2 + C_3 \ldots$$

Capacitors in Series

Same charge on each.

$$\frac{1}{C} = \frac{1}{C_1} + \frac{1}{C_2} + \frac{1}{C_3} \ldots$$

Discharging Through a Resistor

$$Q = Q_0\,\mathrm{e}^{-t/CR} \quad \text{and} \quad V = V_0\,\mathrm{e}^{-t/CR}$$

$$\textbf{Time constant} = CR = \text{Time to fall to } \frac{Q_0}{\mathrm{e}} = \frac{V_0}{\mathrm{e}}$$

$$I = \frac{V_R}{R} = -\frac{V_0}{R}\,\mathrm{e}^{-t/CR}$$

(**Note.** Current is normally taken to be positive in problems requiring numerical answers.)

41

MAGNETIC EFFECTS OF ELECTRIC CURRENTS

41.1 MAGNETIC FIELDS

The region around a magnet where magnetic effects can be experienced is called the magnetic field of the magnet. The direction of a field, at a point, is taken to be the direction in which a north magnetic pole would move under the influence of the field if it were placed at that point. The path which such a pole would follow is called a **magnetic field line** (or **line of force**). Field lines are directed away from the north poles of magnets because 'like' poles repel each other.

41.2 MAGNETIC FLUX DENSITY (*B*)

The magnitude and direction of a magnetic field can be represented by its **magnetic flux density** (B). (This is sometimes called **magnetic induction.**) Flux density is proportional to **magnetic field strength** (H) (see section 41.4). The unit of flux density is the **tesla** (T). (The tesla is defined in section 41.7.)

The direction of the flux density at a point is that of the tangent to the field line at the point. The magnitude of the flux density is high where the number of field lines per unit area is high (see Fig. 41.1).

Fig. 41.1
Field lines and flux density

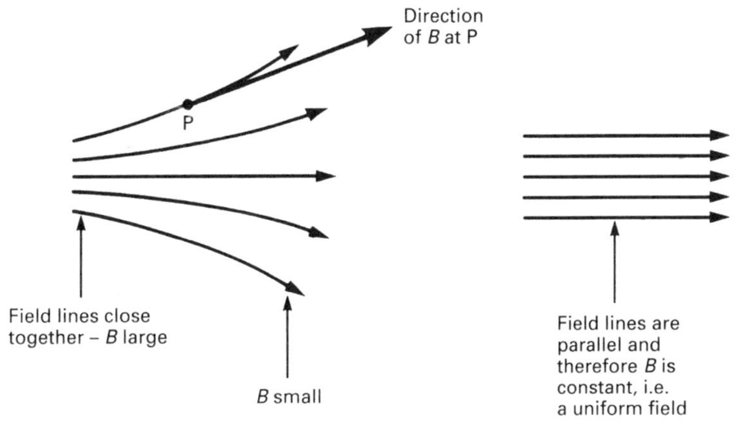

41.3 MAGNETIC FLUX (Φ)

The magnetic flux (Φ) through a region is a measure of the number of magnetic field lines passing through the region. The flux through an area A (Fig. 41.2), the normal to which lies at angle θ to a field of flux density B, is given by

$$\Phi = AB\cos\theta \qquad [41.1]$$

Fig. 41.2
Magnetic flux

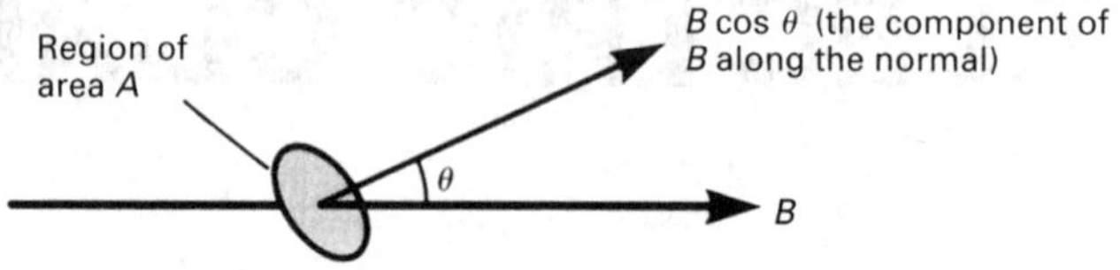

The unit of magnetic flux is the **weber** (Wb).

Note One tesla = 1 weber per square metre, i.e. $1\,\text{T} = 1\,\text{Wb}\,\text{m}^{-2}$

41.4 MAGNETIC FIELDS OF CURRENT-CARRYING CONDUCTORS

In 1820 Oersted discovered that a wire carrying an electric current has an associated magnetic field. For a straight wire the field lines are a series of concentric circles centred on the wire (Fig. 41.3). The direction of the field can be found by using the **right-hand grip rule**:

> Grip the wire using the right hand with the thumb pointing in the direction of the current – the fingers then point in the direction of the field.

Fig. 41.3
Magnetic field due to a wire carrying a current

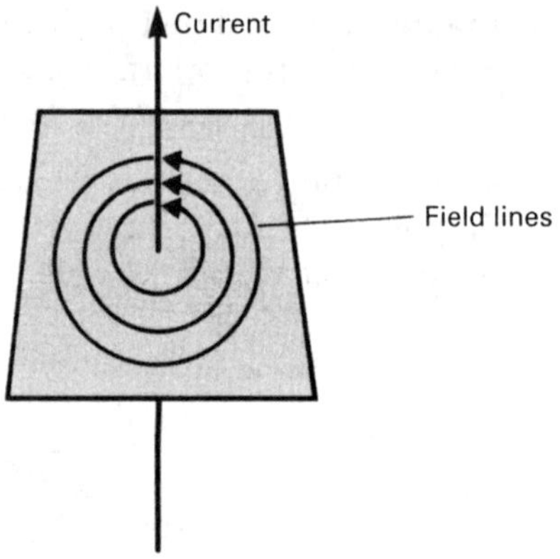

The field of a wire can be intensified by coiling it to form a solenoid. The field of a solenoid is shown in Fig. 41.4 together with those of a plane circular coil and a bar magnet. The directions of the fields associated with both the solenoid and the coil can be found by applying the right-hand grip rule to small sections of them, or from Ⓢ and Ⓝ (Fig. 41.4(a)).

Expressions for the magnetic flux density in four important cases are given below. Three of the expressions are derived in section 41.5.

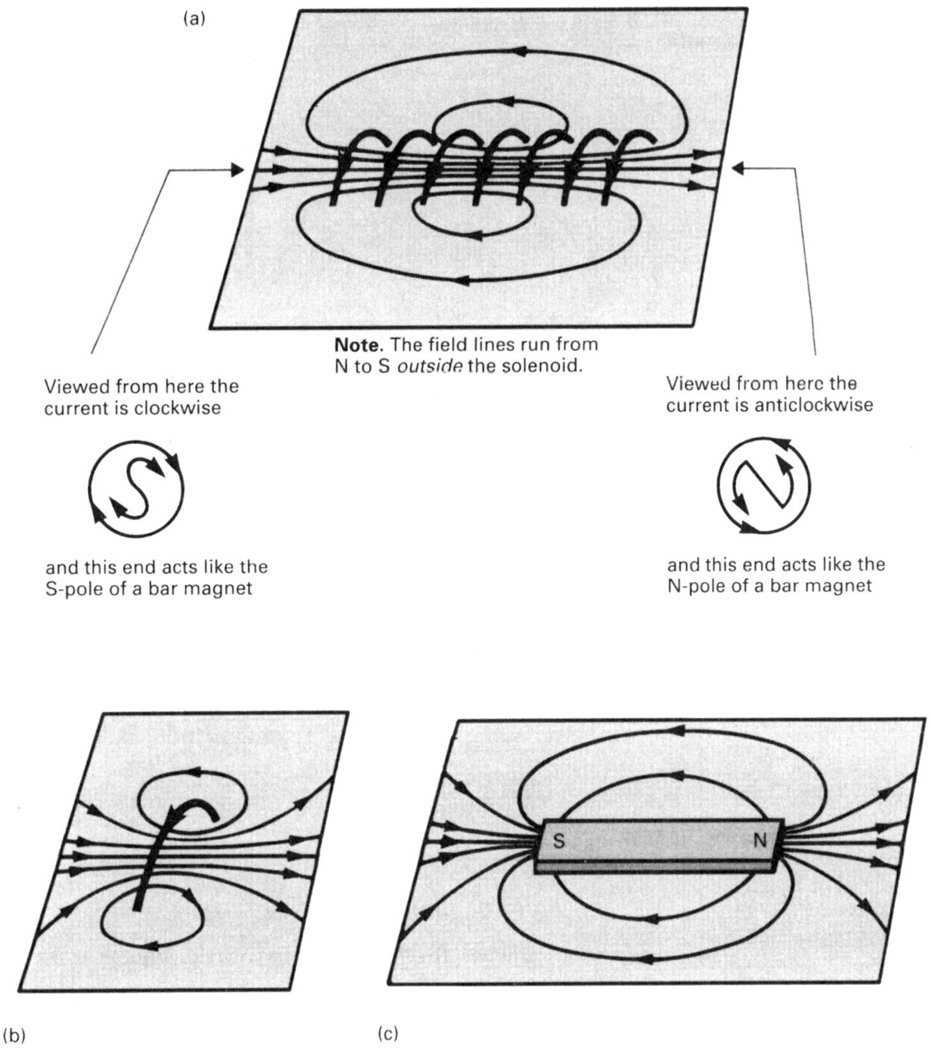
(a)
Note. The field lines run from N to S *outside* the solenoid.
Viewed from here the current is clockwise
and this end acts like the S-pole of a bar magnet
Viewed from here the current is anticlockwise
and this end acts like the N-pole of a bar magnet
S
N
(b)
(c)

Fig. 41.4
Magnetic field due to: (a) a solenoid, (b) a plane circular coil, (c) a bar magnet

Flux Density Due to an Infinitely Long Straight Wire

The magnetic flux density at P (Fig. 41.5) is directed into the paper (by the right-hand grip rule) and its magnitude, B, is given by

$$B = \frac{\mu_0 I}{2\pi a} \qquad [41.2]$$

where

B = the flux density (T)

I = the current in the wire (A)

a = the perpendicular distance of P from the wire (m)

μ_0 = a constant of proportionality known as the **permeability** of free space (vacuum). It is assigned the value of $4\pi \times 10^{-7}$ henrys per metre (i.e. $4\pi \times 10^{-7}\,\mathrm{H\,m^{-1}}$).

Fig. 41.5
To calculate the magnetic flux density near a current-carrying wire

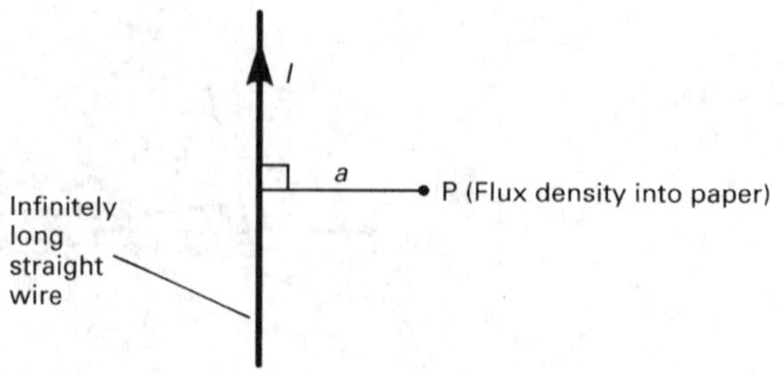

Notes (i) The permeability of air is very nearly the same as that of vacuum and can be taken to be equal to μ_0. Equation [41.2] and the expressions that follow therefore apply to conductors in both vacuum and air.

(ii) Real conductors are not infinitely long; however equation [41.2] is usually a good enough approximation if a is less than about one-twentieth of the length of the conductor and B is being calculated at the middle.

Flux Density on the Axis of an Infinitely Long Solenoid

The magnetic flux density on the axis of an infinitely long solenoid (Fig. 41.6) is directed along the axis and its magnitude, B, is given by

$$B = \mu_0 nI \qquad [41.3]$$

where

B = flux density (T)

μ_0 = the permeability of free space ($= 4\pi \times 10^{-7}\,\mathrm{H\,m^{-1}}$)

n = the number of turns per unit length ($\mathrm{m^{-1}}$)

I = the current through the solenoid (A).

Fig. 41.6
To calculate the magnetic flux density in an infinitely long solenoid

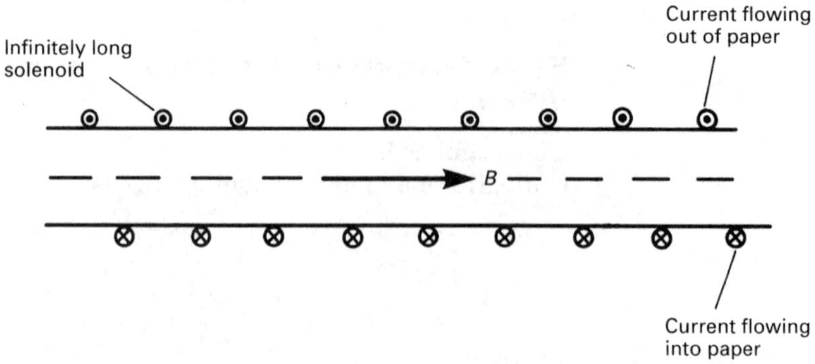

Notes (i) nI is called the **ampere–turns per metre**. It is equal to the magnetic field strength, H, and therefore by equation [41.3] $B = \mu_0 H$. **The unit of magnetic field strength is the ampere per metre** ($\mathrm{A\,m^{-1}}$).

(ii) Real solenoids are not infinitely long; however equation [41.3] is usually a good enough approximation if the length of the solenoid is at least ten times its diameter and B is being calculated at the middle.

(iii) At either end of a long thin solenoid the flux density along the axis is $\mu_0 nI/2$.

Flux Density at the Centre of a Plane Circular Coil

The flux density at P, the centre of the plane (flat) circular coil shown in Fig. 41.7 is directed into the paper (by the right-hand grip rule) and its magnitude is given by

$$B = \frac{\mu_0 NI}{2r} \qquad [41.4]$$

where

B = the flux density (T)

μ_0 = the permeability of free space ($= 4\pi \times 10^{-7}\,\mathrm{H\,m^{-1}}$)

N = the number of turns on the coil (pure number)

I = the current in the coil (A)

r = the radius of the coil (m).

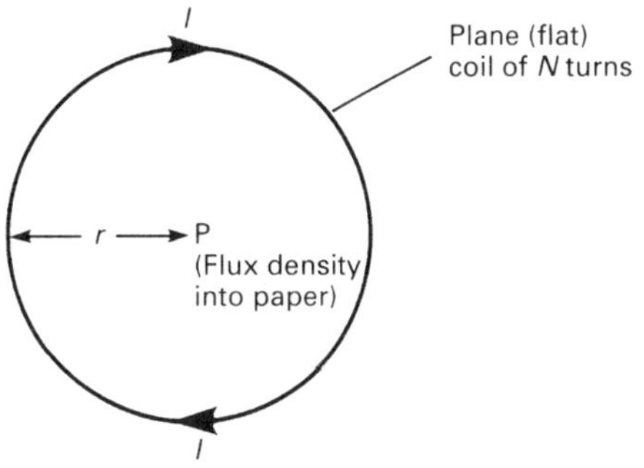

Fig. 41.7
To calculate the magnetic flux density in a plane circular coil

Flux Density of Helmholtz Coils

Two identical coils connected in series so that they each carry the same current produce a uniform field over some distance close to their common axis if the separation of the coils is equal to their radius (Fig. 41.8(a)). The field pattern is shown in Fig. 41.8(b). The magnitude of the flux density in the region of uniform field is given by

$$B \approx 0.72 \frac{\mu_0 NI}{r}$$

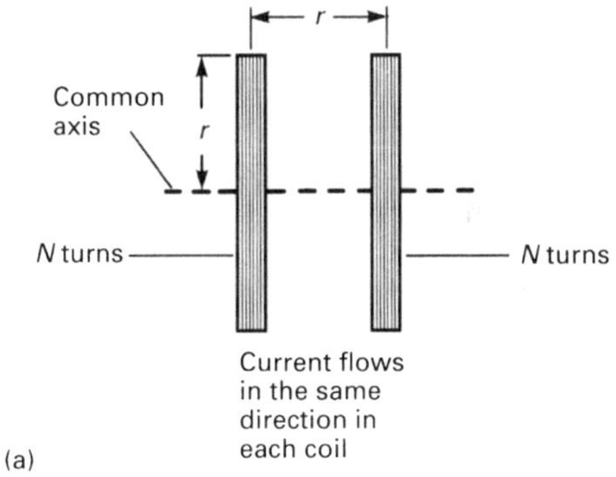

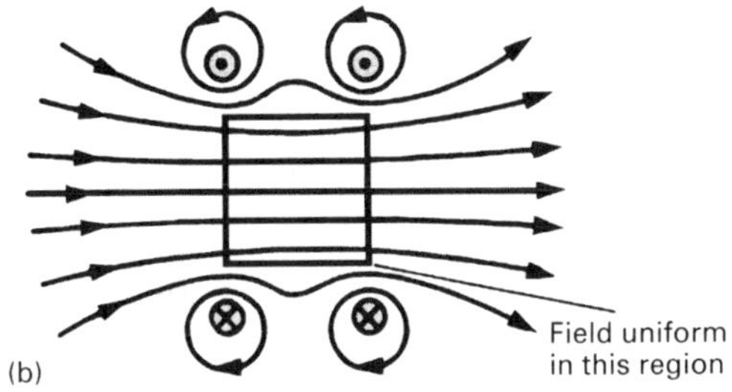

Fig. 41.8
(a) Helmholtz coils.
(b) Field pattern due to Helmholtz coils

where

B = flux density (T)

μ_0 = the permeability of free space ($= 4\pi \times 10^{-7}\,\mathrm{H\,m^{-1}}$)

N = the number of turns on the coil (pure number)

I = the current in each coil (A)

r = the radius of each coil (m).

QUESTIONS 41A

1. A 2000 turn solenoid of length 40 cm and resistance 16 Ω is connected to a 20 V supply. What is the flux density at the mid point of the axis of the solenoid? ($\mu_0 = 4\pi \times 10^{-7}\,\mathrm{H\,m^{-1}}$.)

2. A long wire (X) carrying a current of 30 A is placed parallel to, and 3.0 cm away from, a similar wire (Y) carrying a current of 6.0 A. What is the flux density midway between the wires: **(a)** when the currents are in the same direction, **(b)** when they are in opposite directions? **(c)** When the currents are in the same direction there is a point somewhere between X and Y at which the flux density is zero. How far from X is this point? ($\mu_0 = 4\pi \times 10^{-7}\,\mathrm{H\,m^{-1}}$.)

41.5 THE BIOT–SAVART LAW

The Biot–Savart law can be stated as:

The magnetic flux density δB at a point P which is a distance r from a very short length δl of a conductor carrying a current I is given by

$$\delta B \propto \frac{I\,\delta l \sin\theta}{r^2} \qquad [41.5]$$

where θ is the angle between the short length and the line joining it to P (Fig. 41.9).

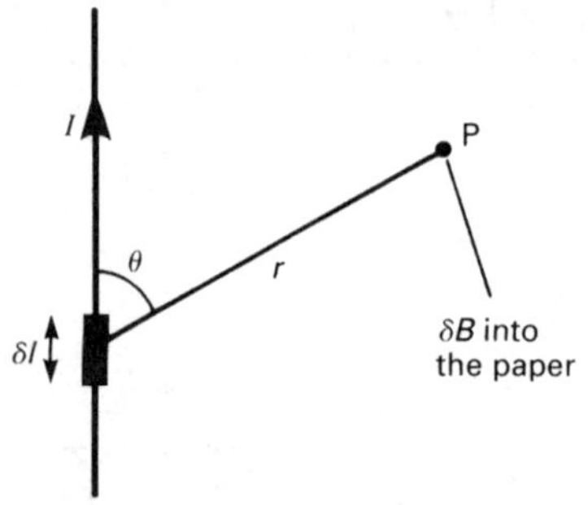

Fig. 41.9 Parameters in the Biot–Savart law

The direction of δB is given by the right-hand grip rule. $I\,\delta l$ is called a **current element**. The constant of proportionality depends on the medium around the conductor. In vacuum (or air) the constant is written as $\mu_0/4\pi$ and equation [41.5] becomes

$$\delta B = \frac{\mu_0}{4\pi}\frac{I\,\delta l \sin\theta}{r^2} \qquad [41.6]$$

The law cannot be tested directly, because it is not possible to have a current-carrying conductor of length δl. However, it can be used to derive expressions for the flux densities of real conductors and these give values which are in agreement with those determined by experiment.

To Show that $B = \mu_0 NI/2r$ at the Centre of Plane Circular Coil

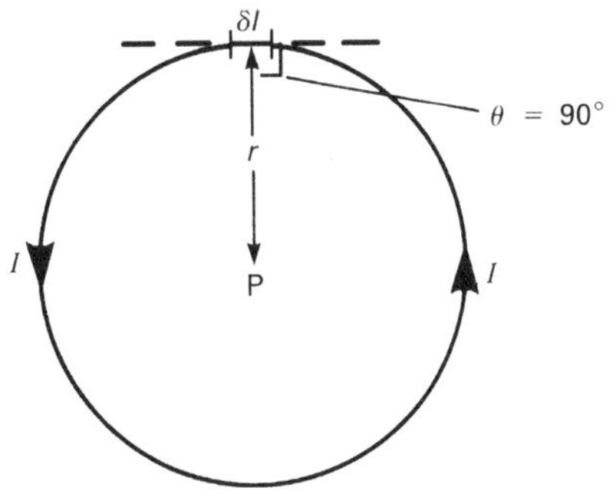

Fig. 41.10
To calculate ***B*** at the centre of a plane circular coil

The flux density δB at P (Fig. 41.10) due to the short length δl is given by equation [41.6] as

$$\delta B = \frac{\mu_0}{4\pi}\frac{I\,\delta l \sin\theta}{r^2}$$

The total flux density, B, at P is the sum of the flux densities of all the short lengths, i.e.

$$B = \sum \frac{\mu_0}{4\pi}\frac{I\,\delta l \sin\theta}{r^2}$$

Every section of the coil is at a distance r from P and makes an angle of 90° with the line joining it to P, and therefore

$$B = \frac{\mu_0 I \sin 90^\circ}{4\pi r^2}\sum \delta l$$

i.e. $$B = \frac{\mu_0 I}{4\pi r^2}\sum \delta l$$

Since $\sum \delta l$ is the total length of the coil, i.e. its circumference, $2\pi r$, this becomes

$$B = \frac{\mu_0 I}{4\pi r^2} 2\pi r$$

i.e. $$B = \frac{\mu_0 I}{2r}$$

If the coil has N turns each carrying current in the same sense, the contribution of each turn adds to that of every other and therefore

$$B = \frac{\mu_0 NI}{2r}$$

To Show that $B = \mu_0 I/2\pi a$ for an Infinitely Long Conductor

Fig. 41.11
To calculate ***B*** due to an infinitely long conductor

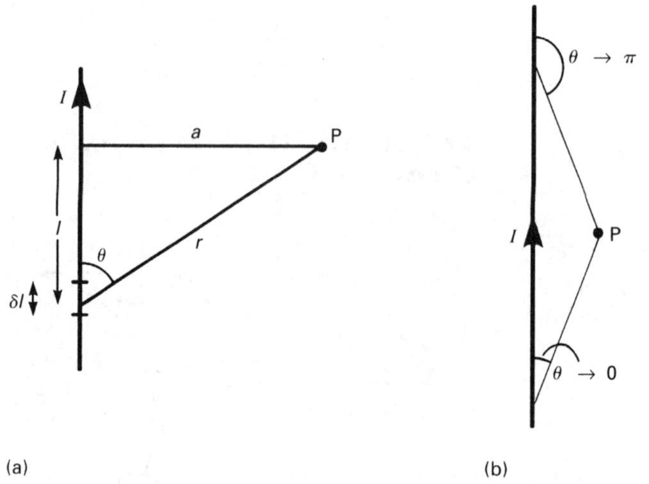

The flux density δB at P (Fig. 41.11(a)) due to the short length δl given by equation [41.6] as

$$\delta B = \frac{\mu_0}{4\pi}\frac{I\,\delta l \sin\theta}{r^2}$$

From Fig. 41.11(a)

$$r = a/\sin\theta$$

and

$$l = a\cot\theta \qquad \therefore \quad \delta l = -a\,\mathrm{cosec}^2\theta\,\delta\theta$$

Substituting for r and δl gives

$$\delta B = \frac{\mu_0}{4\pi}\frac{I(-a\,\mathrm{cosec}^2\theta\,\delta\theta)\sin\theta}{(a/\sin\theta)^2}$$

i.e. $$\delta B = \frac{-\mu_0 I}{4\pi a}\sin\theta\,\delta\theta$$

The total flux density, B, at P is the sum of the flux densities of all the short lengths and can be found by letting $\delta\theta \to 0$ and integrating over the whole length of the conductor. Thus

$$B = \frac{-\mu_0 I}{4\pi a}\int_{\pi}^{0}\sin\theta\,\mathrm{d}\theta$$

(The limits of the integration are π and 0 because these are the values of θ at the ends of the conductor – see Fig. 41.11(b).) Therefore,

$$B = \frac{-\mu_0 I}{4\pi a}\Big[-\cos\theta\Big]_{\pi}^{0}$$

i.e. $$B = \frac{\mu_0 I}{2\pi a}$$

Flux Density at Any Point on the Axis of a Plane Circular Coil

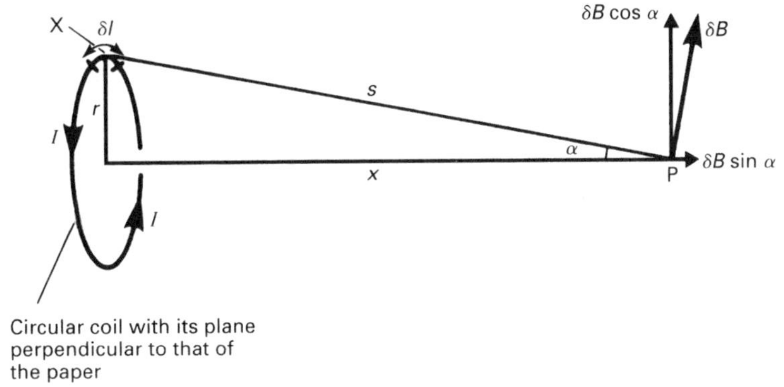

Fig. 41.12 To calculate ***B*** on the axis of a plane circular coil

The flux density δB at P (Fig. 41.12) due to the short length δl of the coil at X, where X is in the plane of the paper, is given by equation [41.6] as

$$\delta B = \frac{\mu_0}{4\pi}\frac{I\,\delta l \sin\theta}{s^2}$$

and is in the plane of the paper and in the direction shown.

By symmetry, when all the short lengths δl are taken into account, the components of magnitude $\delta B \cos\alpha$ sum to zero. Each short length produces a component of magnitude $\delta B \sin\alpha$ parallel to the axis, and all these components are in the direction shown. The total flux density is therefore in the direction of $\delta B \sin\alpha$, and its magnitude B is given by

$$B = \sum \delta B \sin\alpha$$

i.e. $$B = \sum \frac{\mu_0}{4\pi}\frac{I\,\delta l \sin\theta}{s^2}\sin\alpha$$

The radius vector XP of each small length is perpendicular to it, so that $\theta = 90^\circ$ and therefore $\sin\theta = 1$ in every case. Also, the length of each radius vector has the constant value s, and $\sin\alpha$ is the same for each small length. Therefore

$$B = \frac{\mu_0 I \sin\alpha}{4\pi s^2}\sum \delta l$$

Since $\sum \delta l = 2\pi r$ (the circumference of the coil), this becomes

$$B = \frac{\mu_0 I r \sin\alpha}{2s^2}$$

But $\sin\alpha = r/s$, and therefore

$$B = \frac{\mu_0 I r^2}{2s^3}$$

and for a coil of N turns

$$B = \frac{\mu_0 N I r^2}{2s^3} \qquad [41.7]$$

Note that when $s = r$, P is at the centre of the coil and equation [41.7] reduces to equation [41.4]. It is clear from Fig. 41.12 that $s^2 = (r^2 + x^2)$, and therefore equation [41.7] may be written in the alternative form

$$B = \frac{\mu_0 N I r^2}{2(r^2 + x^2)^{3/2}}$$

To show that $B = \mu_0 nI$ on the Axis of an Infinitely Long Solenoid

Consider the magnetic flux density δB at P (Fig. 41.13) due to a section of the solenoid of length δx. Since δx is small, the section can be treated as a plane circular coil of N turns in which case δB is given by equation [41.7] as

$$\delta B = \frac{\mu_0 N I r^2}{2s^3}$$

and is directed along the axis of the solenoid.

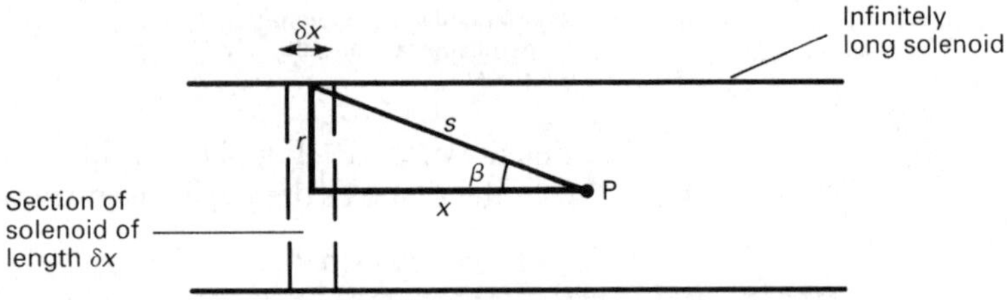

Fig. 41.13 To calculate ***B*** on the axis of an infinitely long solenoid

If the solenoid has n turns per unit length, then $N = n\,\delta x$, and therefore

$$\delta B = \frac{\mu_0 n I r^2 \delta x}{2s^3}$$

From Fig. 41.13

$$\sin\beta = r/s \qquad \therefore \quad s^3 = r^3 \operatorname{cosec}^3\beta$$

and

$$\tan\beta = r/x \qquad \therefore \quad x = r\cot\beta \qquad \therefore \quad \delta x = -r \operatorname{cosec}^2\beta\,\delta\beta$$

Substituting for s^3 and δx gives

$$\delta B = \frac{\mu_0 n I r^2(-r \operatorname{cosec}^2\beta\,\delta\beta)}{2r^3 \operatorname{cosec}^3\beta}$$

i.e. $$\delta B = \frac{-\mu_0 n I}{2}\sin\beta\,\delta\beta$$

The flux densities at P due to every section of the solenoid are all in the same direction and therefore the total flux density B can be found by letting $\delta\beta \to 0$ and integrating over the whole length of the solenoid. Thus,

$$B = \frac{-\mu_0 n I}{2}\int_{\pi}^{0} \sin\beta\,\mathrm{d}\beta$$

(The limits of integration are π and 0 because these are the values of β at the ends of the solenoid.) Therefore

$$B = \frac{-\mu_0 n I}{2}\Big[-\cos\beta\Big]_{\pi}^{0}$$

i.e. $$B = \mu_0 n I$$

41.6 AMPÈRE'S LAW

Ampère's law is an alternative to the Biot–Savart law but it can be used only in certain situations. For conductors in a vacuum (or air) the law may be stated as

$$\oint B\cos\theta\,\mathrm{d}l = \mu_0 I \qquad [41.8]$$

$\oint B\cos\theta\,\mathrm{d}l$ is called the **line integral of the flux density around a closed path.** It is the sum of the terms $B\cos\theta\,\delta l$ for every very short length δl of the closed path.

B is the magnitude of the flux density.

θ is the angle between a section of the path and the direction of the field at that section (see Fig. 41.14).

I is the current enclosed by the path.

Fig. 41.14
Parameters in Ampère's law

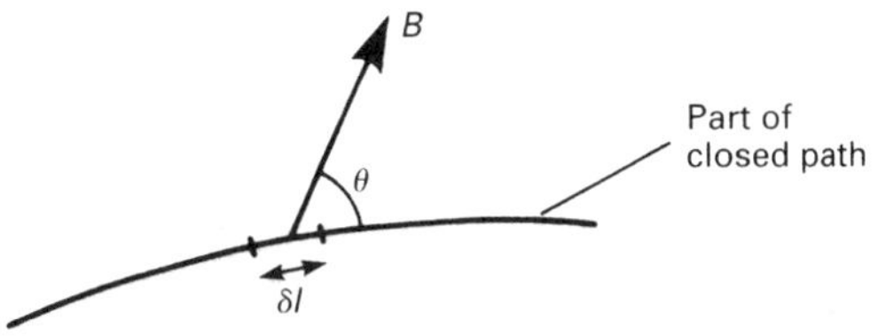

In order to use the law it is necessary to choose a path for which it is possible to determine the value of the line integral. It is because there are many situations where there is no such path that the law is of limited use. In particular, it can be used to determine flux densities at points close to long straight conductors and along the axes of long solenoids; it cannot be used in the case of a plane circular coil.

Ampère's Law Applied to a Long Straight Conductor

We choose as the closed path along which to carry out the integration a circle of radius a centred on the conductor (Fig. 41.15). The conductor is long and therefore by symmetry the flux density has a constant magnitude, B, at all points on the path, and is everywhere parallel to the path and therefore $\cos\theta = 1$ at every point. It follows that

$$\text{Line integral} = \oint B\cos\theta\,\mathrm{d}l = B\oint \mathrm{d}l$$

Fig. 41.15
To calculate ***B*** close to a conductor using Ampère's law

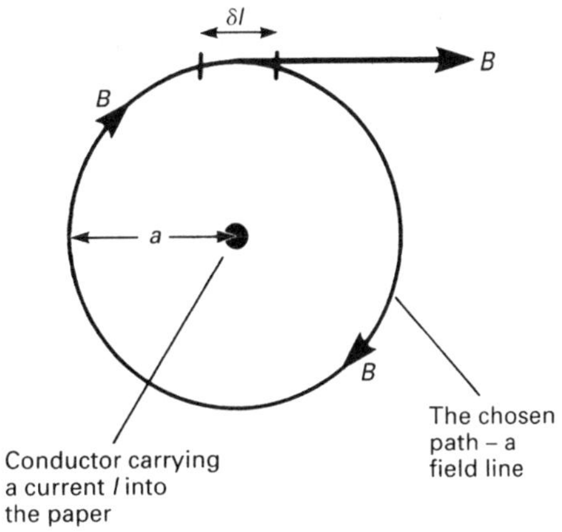

But

$$\oint \mathrm{d}l = 2\pi a$$

(Since $2\pi a$ is the path length, i.e. the sum of the lengths δl for the whole path.) Thus

$$\text{Line integral} = B2\pi a$$

Also

$$\text{Current enclosed} = I$$

Therefore by Ampère's law (equation [41.8])

$$B2\pi a = \mu_0 I$$

i.e. $$B = \frac{\mu_0 I}{2\pi a}$$

Ampère's Law Applied to a Long Thin Solenoid

In Fig. 41.16, WXYZW is the closed path around which the integration is to take place. The flux density outside a solenoid is very much less than that inside. Therefore, in comparison with its value along WX, the flux density can be taken to be zero along YZ. Also, if the solenoid is long and thin, WX can be chosen to be very much greater than XY and ZW, in which case the contributions which XY and ZW make to the line integral can be taken to be zero. This leaves only WX to be considered, and the flux density is parallel to every section δl of WX. Therefore

$$\text{Line integral} = \oint B\cos\theta\,\mathrm{d}l = B\,\mathrm{WX}$$

where B is the magnitude of the flux density along WX, i.e. along the axis of the solenoid.

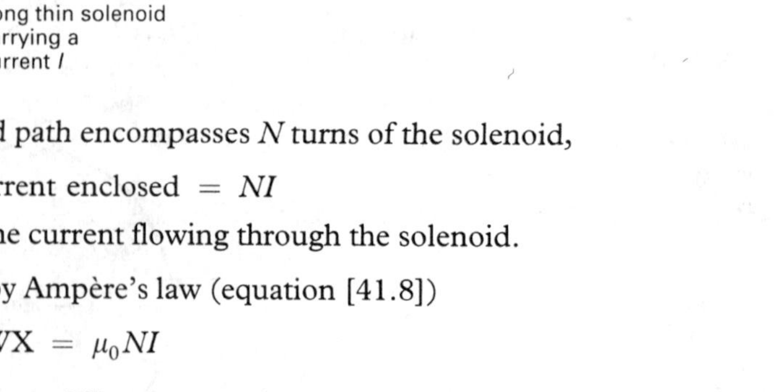

Fig. 41.16 To calculate ***B*** in a solenoid using Ampère's law

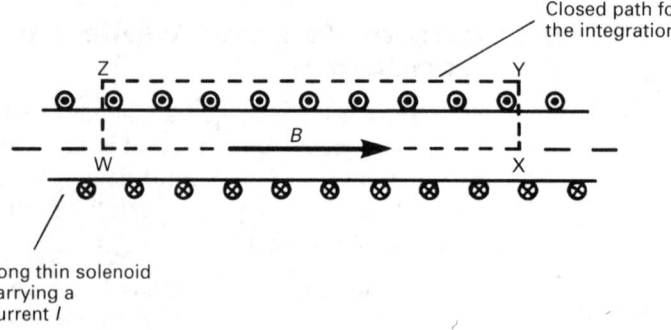

If the closed path encompasses N turns of the solenoid,

$$\text{Current enclosed} = NI$$

where I is the current flowing through the solenoid.

Therefore by Ampère's law (equation [41.8])

$$B\,\mathrm{WX} = \mu_0 NI$$

i.e. $$B = \frac{\mu_0 NI}{\mathrm{WX}}$$

But N/WX is the number of turns per unit length, n, of the solenoid and therefore

$$B = \mu_0 nI$$

41.7 THE FORCE ON A CONDUCTOR IN A MAGNETIC FIELD

A current-carrying conductor in a magnetic field experiences a force. One can think of the force as being the result of the magnetic field of the conductor interacting with that in which it is situated.

The Magnitude of the Force

In Fig. 41.17

$$F = BIL \sin\theta \qquad [41.9]$$

where

$F =$ the force on the conductor (N)

$B =$ the magnitude of the magnetic flux density of the field (T)

$I =$ the current in the conductor (A)

$L =$ the length of the conductor (m).

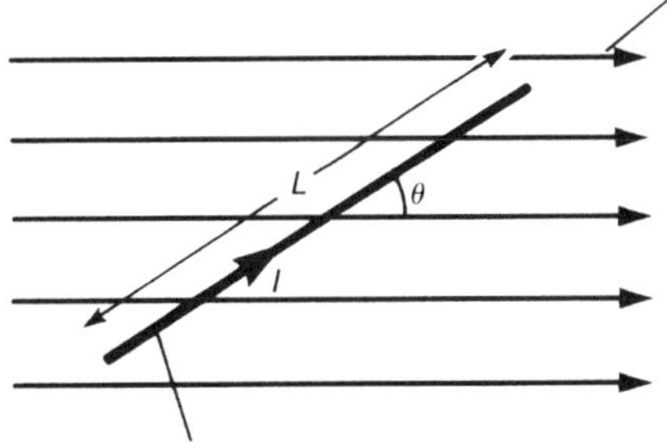

Fig. 41.17 Parameters in equation for the force on a conductor in a magnetic field

Notes

(i) It is clear from equation [41.9] that the force on the conductor has its maximum value when the conductor (and therefore the current) and the external field are at right angles to each other ($\theta = \pi/2$), and is zero when the conductor is parallel to the field ($\theta = 0$).

(ii) $F \propto IL\sin\theta$ is an experimental result. $F \propto BIL\sin\theta$ follows from the definition of magnetic flux density. $F = BIL\sin\theta$ follows from the definition of the unit of magnetic flux density – the tesla (T). Thus,

> **One tesla** is the magnetic flux density of a field in which a force of 1 newton acts on a 1 metre length of a conductor which is carrying a current of 1 ampere and is perpendicular to the field.

The Direction of the Force

Experiment shows that the force is always perpendicular to the plane which contains both the current and the external field at the site of the conductor. The direction of the force can be found by using **Fleming's left-hand (motor) rule:**

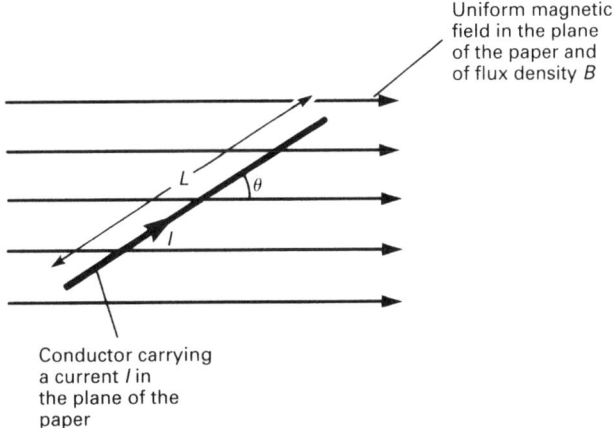

If the first and second fingers and the thumb of the left hand are placed comfortably at right angles to each other, with the **F**irst finger pointing in the direction of the **F**ield and the se**C**ond finger pointing in the direction of the **C**urrent, then the thu**M**b points in the direction of the force, i.e. in the direction in which **M**otion takes place if the conductor is free to move (Fig. 41.18).

Fig. 41.18
Fleming's left-hand rule

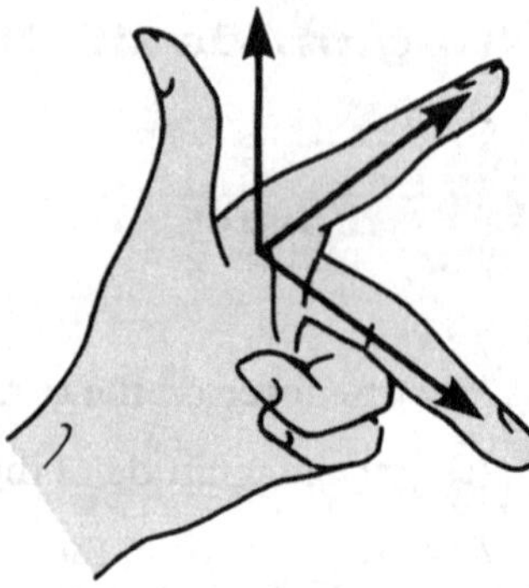

Applying the rule to the situation in Fig. 41.19 reveals that the forces act in the directions shown. The force experienced by the conductor shown in Fig. 41.17 is perpendicular to the paper and is directed into the paper.

Fig. 41.19
Application of Fleming's left-hand rule to a current carrying conductor between the poles of a magnet

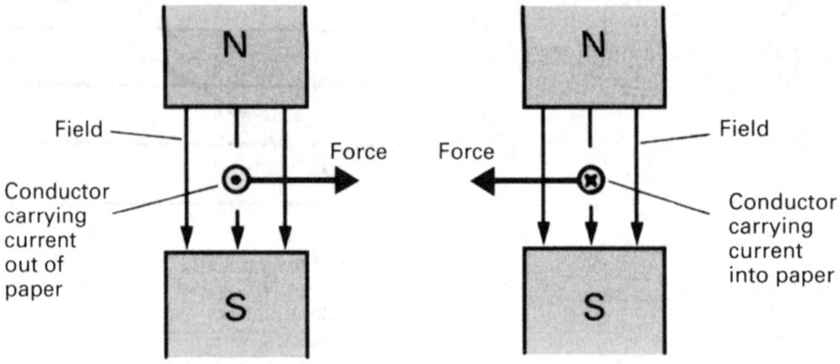

41.8 THE FORCE ON A CHARGED PARTICLE IN A MAGNETIC FIELD

The Magnitude of the Force

In Fig. 41.20

$$F = BQv \sin\theta \qquad [41.10]$$

where

F = the force of the particle (N)

B = the magnitude of the magnetic flux density of the field (T)

Q = the charge on the particle (C)

v = the magnitude of the velocity of the particle ($\mathrm{m\,s^{-1}}$).

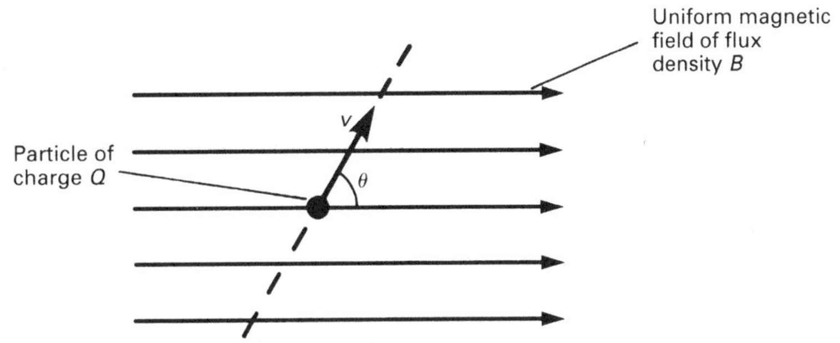

Fig. 41.20 Parameters in the equation for the force on a charged particle in a magnetic field

Note It is clear from equation [41.10] that **a magnetic field cannot exert a force on a stationary charged particle.**

The Direction of the Force

The force on a positively charged particle is in the same direction as that on a conductor which is carrying a current in the same direction as that in which the particle is moving. It follows that the direction of the force can be found by using Fleming's left-hand rule. Thus, if the particle shown in Fig. 41.20 is positively charged, the force acting on it is directed perpendicularly into the paper. A negatively charged particle feels a force in the opposite direction.

QUESTIONS 41B

1. Diagrams **(a)** to **(c)** show a magnetic field of flux density 0.20 T directed perpendicularly into the paper. In each of **(a)** and **(b)** a conductor of length 0.30 m is entirely within the field and is carrying a current of 4.0 A in the plane of the paper. In **(c)** an electron is moving in the plane of the paper at $2.0 \times 10^6\ \text{m s}^{-1}$. Copy the diagrams and show the direction of the force in each case. Also find the magnitude of the forces.
(Charge on the electron $= 1.6 \times 10^{-19}$ C.)

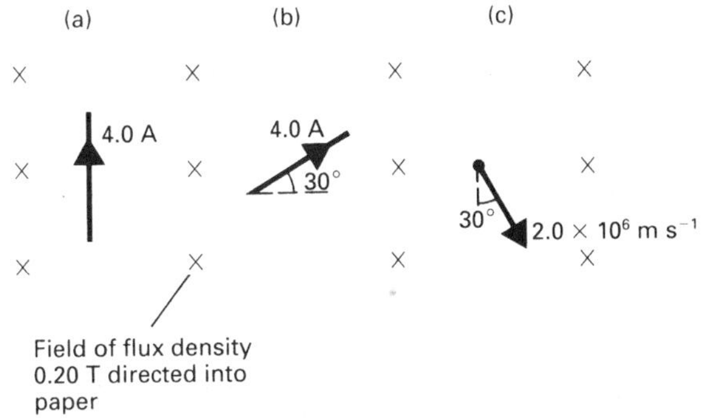

41.9 RELATIONSHIP BETWEEN THE FORCE ON A CONDUCTOR AND THE FORCE ON A CHARGED PARTICLE

An electric current is a flow of charge. It is therefore reasonable to suppose that the force experienced by a current-carrying conductor in a magnetic field is simply the resultant of the forces felt by the moving charges which constitute the current. If this supposition is correct, then it should be possible to derive the expression for the force acting on a current-carrying conductor (equation [41.9]) from the expression for the force acting on a charged particle (equation [41.10]). This is done below for the particular case of an electron current.

Suppose that in a length L of a conductor there are N electrons moving with an average (drift) velocity v (Fig. 41.21). The time required for all these electrons to emerge from face X is L/v, and therefore the rate of flow of charge past X is

$$\frac{Ne}{L/v}$$

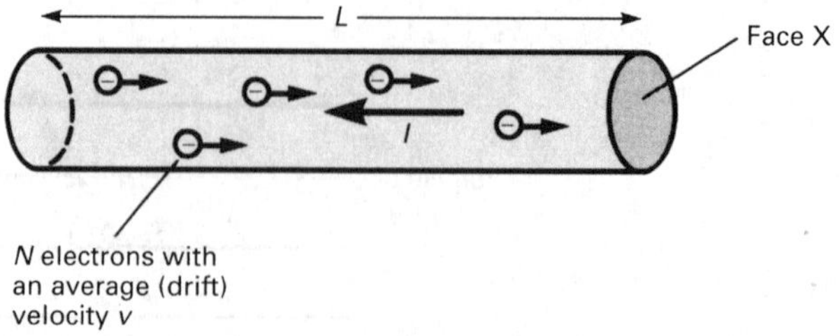

Fig. 41.21
A metallic conductor carrying a current

where e is the charge on each electron. Thus, the current, I, in the conductor is given by

$$I = \frac{Ne}{L/v} = \frac{Nev}{L} \quad [41.11]$$

If the conductor is at an angle θ to a magnetic field of flux density B, then it follows from equation [41.10] that the total force, F, felt by (all) the electrons is given by

$$F = NBev \sin\theta$$

Substituting for Nev from equation [41.11] gives

$$F = BIL \sin\theta$$

This is equation [41.9] and therefore it can be concluded that equation [41.9] and equation [41.10] are equivalent and that the original supposition is correct.

We have seen that the force experienced by a current-carrying conductor is fundamentally a force on the moving charges within the conductor. It is also reasonable to suppose that the magnetic field set up by the current-carrying conductor is itself due to the moving charges which constitute the current. Thus, a moving charge sets up a magnetic field and this exerts a force on any other charge which is moving in the field, i.e. a moving charge exerts a magnetic force on another moving charge. This force is quite distinct from the electrostatic (Coulomb) force which exists between charges whether they are moving or not.

41.10 THE FORCES BETWEEN TWO CURRENT-CARRYING CONDUCTORS

If two current-carrying conductors are close together, then each is in the magnetic field of the other and therefore each experiences a force.

In Fig. 41.22, X and Y are two infinitely long parallel conductors carrying currents I_1 and I_2 respectively. The conductors are in vacuum and their separation is a. From equation [41.2], the magnitude of the magnetic flux density, B, at any point, P, on Y due to the current in X is given by

$$B = \frac{\mu_0 I_1}{2\pi a} \quad [41.12]$$

and, by the right-hand grip rule, the field is directed into the paper.

From equation [41.9], a length L of Y experiences a force, F, which is given by

$$F = BI_2L \sin\theta$$

where θ is the angle between the direction of I_2 and the field along Y, and is therefore 90°, in which case

$$F = BI_2L$$

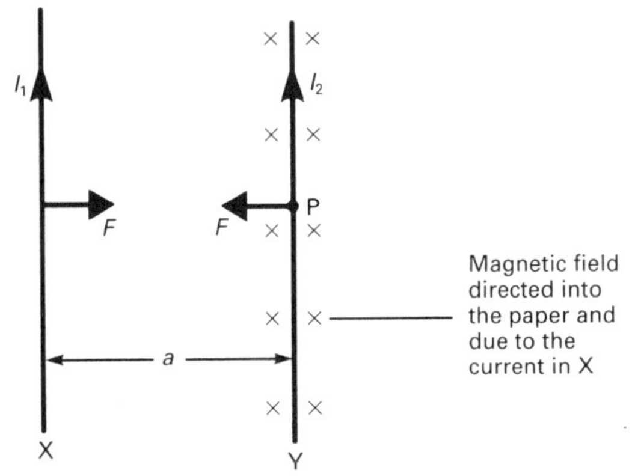

Fig. 41.22 Forces between two conductors

Substituting for B from equation [41.12] gives

$$F = \frac{\mu_0 I_1 I_2 L}{2\pi a} \qquad [41.13]$$

By Fleming's left-hand rule, the force is directed towards X as shown.

Similar reasoning reveals that X is subject to a force of the same magnitude. (Alternatively, this result can be deduced by noting that equation [41.13] is symmetrical in I_1 and I_2.) The force on X is in the opposite direction to that on Y because the field along X is directed out of the paper.

Thus, the forces are such that the wires attract each other. It can be shown that when the currents are in opposite directions the wires repel each other. Thus: **like currents attract; unlike currents repel.**

41.11 DEFINITION OF THE AMPERE AND THE VALUE OF μ_0

Equation [41.13] defines the ampere. Thus:

The ampere is that steady current which, when it is flowing in each of two infinitely long, straight, parallel conductors which have negligible areas of cross-section and are 1 metre apart in a vacuum, causes each conductor to exert a force of 2×10^{-7} N on each metre of the other.

Notes

(i) The choice of a force of 2×10^{-7} N makes the value of the ampere defined in this way very close to the value it had according to a previous definition based on the chemical effect of a current.

(ii) From the definition of the ampere, in equation [41.13] $I_1 = I_2 = 1$ A when $F = 2 \times 10^{-7}$ N, $a = 1$ m and $L = 1$ m, and therefore

$$2 \times 10^{-7} = \frac{\mu_0 \times 1 \times 1 \times 1}{2\pi \times 1}$$

i.e. $\mu_0 = 4\pi \times 10^{-7}\ \mathrm{H\,m^{-1}}$

Thus, the definition of the ampere assigns μ_0 a value of $4\pi \times 10^{-7}\ \mathrm{H\,m^{-1}}$.

41.12 THE TORQUE ON A COIL IN A MAGNETIC FIELD

In Fig 41.23 a rectangular coil is suspended in such a way that it can turn about a vertical axis PQ. The coil is in a magnetic field of flux density B and the plane of the coil is parallel to the field. A current, I, is flowing round the coil in the direction shown.

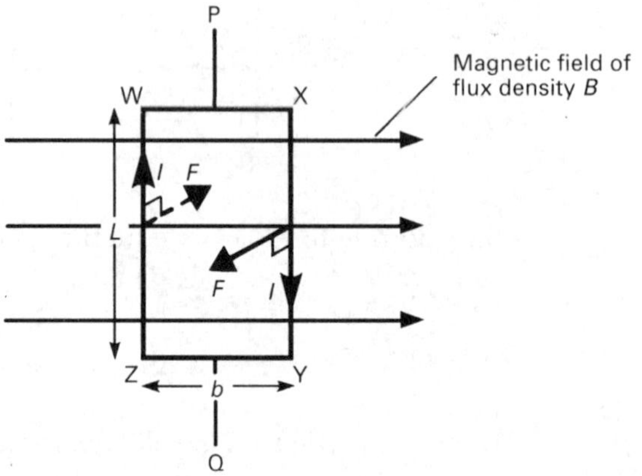

Fig. 41.23
Torque on a coil

By Fleming's left-hand rule, the vertical side WZ of the coil experiences a force F which is directed perpendicularly into the paper. There is an equal and opposite force on XY. (Since the plane of the coil is parallel to the field there is no force on either WX or ZY.) The forces on WZ and XY constitute a couple whose torque T is given by

$$T = Fb \qquad [41.14]$$

where b is the width of the coil. The directions of the currents in WZ and XY are each at 90° to the magnetic field, and therefore from equation [41.9]

$$F = BIL \sin 90^\circ$$

i.e.

$$F = BIL \qquad [41.15]$$

where L is the length of each vertical side of the coil. Substituting for F in equation [41.14] gives

$$T = BILb$$

i.e.

$$T = BIA$$

where A = the area of the coil.

For a coil of N turns

$$T = BIAN \qquad [41.16]$$

In a **radial** field (see section 41.14) the plane of the coil is always parallel to the field, in which case equation [41.16] applies for all orientations of the coil. On the other hand, if the coil is in a **uniform** field, then as soon as it turns under the influence of the torque it ceases to be parallel to the field. Fig. 41.24 shows the situation of the coil from above when it is at some angle ϕ to the field. Note that even though the coil has turned, its vertical sides WZ and XY are still perpendicular to the field. The forces acting on WZ and XY therefore have the same magnitude and the same directions as they had before the coil turned. However, the separation of the forces alters so that the torque, T, has a reduced value given by

$$T = Fb \cos \phi$$

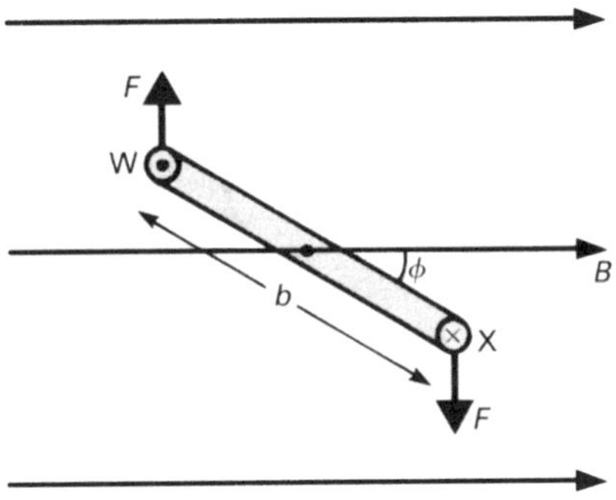

Fig. 41.24
Torque on a coil at an angle to the magnetic field

Therefore from equation [41.15], in general

$$T = BIAN\cos\phi \qquad [41.17]$$

Notes

(i) For values of ϕ other than zero the field exerts forces on WX and ZY, but these forces are always parallel to the axis about which the coil is turning and therefore make no contribution to the torque. For $0 < \phi < \pi$ the forces on WX and ZY tend to increase the height of the coil, for $\pi < \phi < 2\pi$ they tend to decrease its height.

(ii) Equation [41.17] can be shown to hold for a coil of any shape provided that its area is A.

41.13 ELECTROMAGNETIC MOMENT

It has been found convenient to define a quantity known as the **electromagnetic moment** m (sometimes called **magnetic moment**) of a current-carrying coil. It can be thought of as that property which determines:

(i) the magnitude of the torque that acts on the coil when it is at a given angle to a given magnetic field, and

(ii) the angle at which the coil ultimately comes to rest in the field.

Electromagnetic moment is a vector quantity. Its magnitude m can be defined as being numerically equal to the torque acting on the coil when it is parallel to a uniform field whose flux density is one tesla.

Thus, in equation [41.17], $T = m$ when $B = 1\,\text{T}$ and $\phi = 0$, and therefore (by equation [41.17])

$$m = IAN \qquad [41.18]$$

The electromagnetic moment is along the normal to the coil and is in the direction in which a 'right-handed' corkscrew would advance if it were to be rotated in the same sense as that of the current flowing in the coil (see Fig. 41.25).

Thus, the magnitude of m is equal to IAN and its direction depends on the direction of the current in the coil, and so m is a property of the coil itself and the current it is carrying.

Fig. 41.25
Electromagnetic moment of a coil

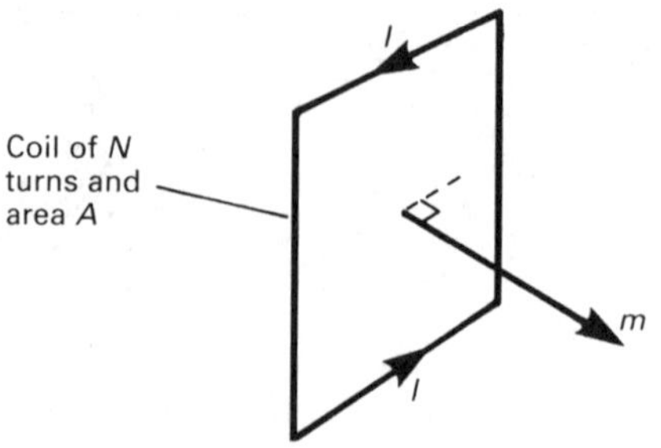

Notes (i) It follows from equation [41.18] that the unit of m is $\mathrm{A\,m^2}$.

(ii) From equations [41.17] and [41.18] the magnitude T of the torque acting on the coil in Fig. 41.24 can be written as

$$T = Bm\cos\phi$$

(iii) The torque acts so as to align the electromagnetic moment with the field direction. This is illustrated in Fig. 41.26 – when m is parallel to B the torque is zero, i.e. the coil is in its equilibrium position.

Fig. 41.26
Torque on a coil aligning magnetic moment with magnetic field

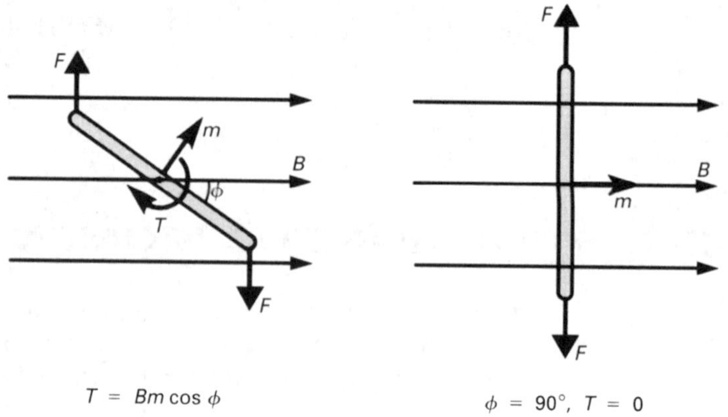

(iv) Area (surprisingly) is a vector quantity and equation [41.18] can be written in terms of vectors as $\boldsymbol{m} = IAN$. The direction associated with $\boldsymbol{A}$ is the direction of $\boldsymbol{m}$.

A bar magnet also has an electromagnetic moment $\boldsymbol{m}$. Its magnitude is numerically equal to the torque acting on the magnet when it is perpendicular to a uniform field whose flux density is 1 T. The direction of $\boldsymbol{m}$ is along the axis of the magnet from the S pole to the N pole (Fig. 41.27).

Note that for both a coil and a bar magnet the magnitude of $\boldsymbol{m}$ is numerically equal to the maximum torque that acts in a field of one tesla.

Fig. 41.27
Electromagnetic moment of a bar magnet

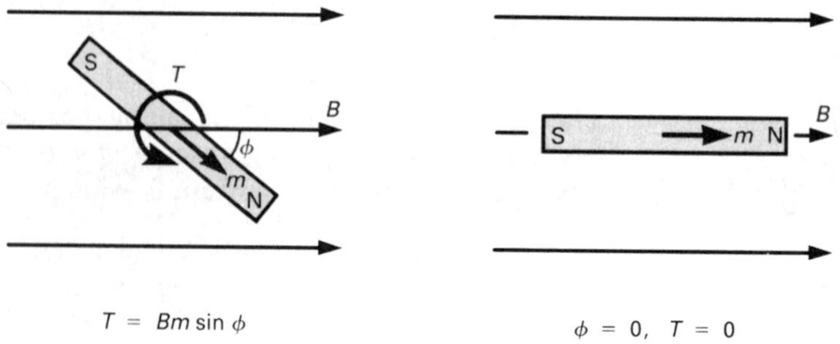

41.14 THE MOVING-COIL GALVANOMETER

The moving coil galvanometer makes use of the torque exerted on a current-carrying coil in a magnetic field. The coil is rectangular and consists of many turns of fine wire. In the most sensitive forms of the instrument the coil is suspended by a fine wire (Fig. 41.28(a)). A small mirror is attached to the coil and this reflects a beam of light on to a scale, the beam of light acting as a weightless pointer. Current is fed into and out of the coil by way of the support wires.

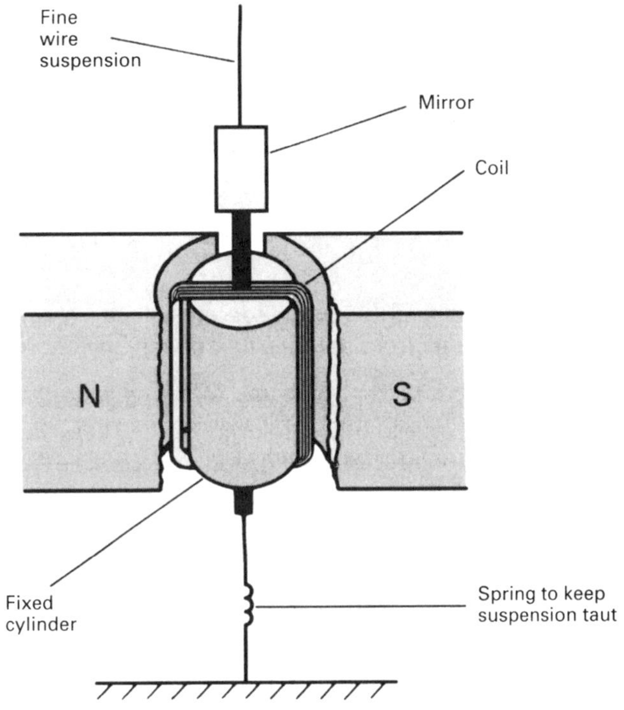

(a)

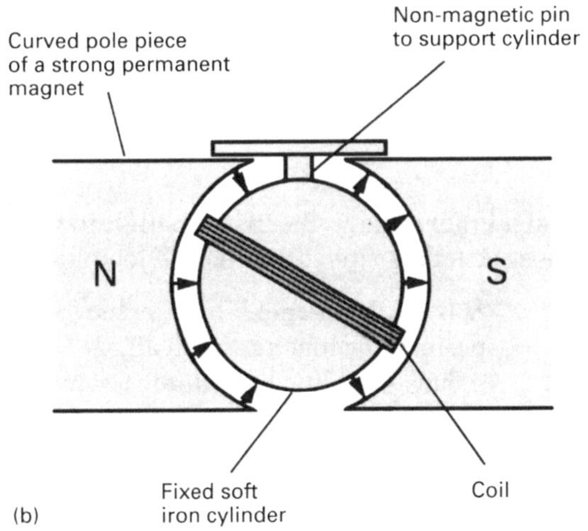

(b)

Fig. 41.28
Principle of the moving-coil galvanometer

The combination of the curved pole pieces and the soft iron cylinder (Fig. 41.28(b)) produces a radial magnetic field in the air gap. Therefore, no matter what its orientation, the coil is always parallel to the field and experiences a constant torque whose magnitude T is given by equation [41.16] as

$$T = BIAN$$

where B is the flux density of the field in the air gap, I is the current in the coil, A is the area of the coil and N is the number of turns on it.

The coil turns under the influence of this torque and, as it does, it twists the suspension. This causes the suspension to exert an oppositely directed torque of magnitude T', say. T' is proportional to the angle θ through which the suspension has been twisted, and therefore we can write

$$T' = c\theta$$

where c is a constant of proportionality known as the **torsion constant** of the suspension.

The coil turns until the restoring torque is of the same magnitude as that due to the magnetic field, and therefore at equilibrium

$$BIAN = c\theta$$

i.e. $$\theta = \left(\frac{BAN}{c}\right)I \qquad [41.19]$$

Thus, the deflection of the coil, θ, is proportional to the current through it and therefore the instrument can be calibrated with a linear scale.

More robust forms of the instrument have jewelled bearings and hair-springs instead of the fine wire suspension, and use a pointer in place of the mirror. These pointer-type instruments are not sufficiently sensitive to detect currents which are less than about 1 μA. (The corresponding full-scale deflection would be much greater than this.)

Current Sensitivity S_I

The current sensitivity S_I is defined by

$$S_I = \frac{\theta}{I}$$

Therefore from equation [41.19]

$$S_I = \frac{BAN}{c}$$

and it follows that if the instrument is to have high current sensitivity, B, A and N must be large, c must be small. Each of these requirements is dealt with below.

(i) ***B* must be large**. This is achieved by using a narrow air gap and a strong permanent magnet (typically $B = 0.4\,$T). In addition the high value of B means that the instrument is almost totally uninfluenced by external magnetic fields (cf. the Earth's field where $B \approx 4 \times 10^{-5}\,$T).

(ii) ***N* must be large**. There is, however, an upper limit to N because the sides of the coil have to be narrow enough to fit in the air gap.

(iii) ***A* must be large**. There is an upper limit to A because it must not be so large that the instrument is unwieldy. Also, a coil of large area swings about its equilibrium position for a long time.

(iv) **c must be small**. Thus a weak suspension is required. It must not be too weak, for this would cause the coil to swing about its equilibrium position for a long time.

Voltage Sensitivity S_V

The voltage sensitivity S_V is defined by

$$S_V = \frac{\theta}{V}$$

where V is the PD across the galvanometer which produces a deflection of θ. For a coil of resistance, R, $V = IR$, where I is the current through the instrument. Therefore

$$S_V = \frac{\theta}{IR}$$

in which case, by equation [41.19]

$$S_V = \frac{BAN}{cR}$$

It follows that high voltage sensitivity requires the same features as high current sensitivity, together with low coil resistance.

In deciding which galvanometer to use in a particular circuit it is necessary to take account of the resistance of the circuit as a whole. It can be shown that it is desirable to transfer the maximum amount of power to the meter. This occurs when the resistance of the meter is equal to that of the rest of the circuit. It follows that in circuits of low resistance (Wheatstone bridge and potentiometer circuits often have low resistances) a galvanometer with a high voltage sensitivity is required.

Notes

(i) Moving-coil galvanometers can be adapted for use as ammeters and voltmeters by the addition of shunts and multipliers, respectively (see sections 36.7 and 36.8).

(ii) Moving-coil galvanometers can be used to measure alternating currents and voltages if a rectifier is placed in series with the coil (see section 43.15).

(iii) The ballistic use of the galvanometer is dealt with in section 42.20.

(iv) Galvanometer damping is discussed in section 42.12.

41.15 MEASUREMENT OF MAGNETIC FLUX DENSITY BY USING A CURRENT BALANCE

A simple form of current balance is shown in Fig. 41.29. It consists of a wire frame (PQNM) pivoted about a horizontal axis (XY). The pivots are such that current can be fed into the frame at one of them and out at the other.

Suppose that it is desired to measure the magnitude, B, of the flux density inside a solenoid. The solenoid is arranged so that its field is perpendicular to PQ, as shown. With no current flowing, the frame is made horizontal (pointer indicating zero) by adding small riders to MN or PQ as necessary. A current I is then passed through the frame in the direction shown. This produces a downward directed force on PQ. The direction of the field is perpendicular to PQ, and therefore from equation [41.9]

$$\text{Force due to field} = BIL$$

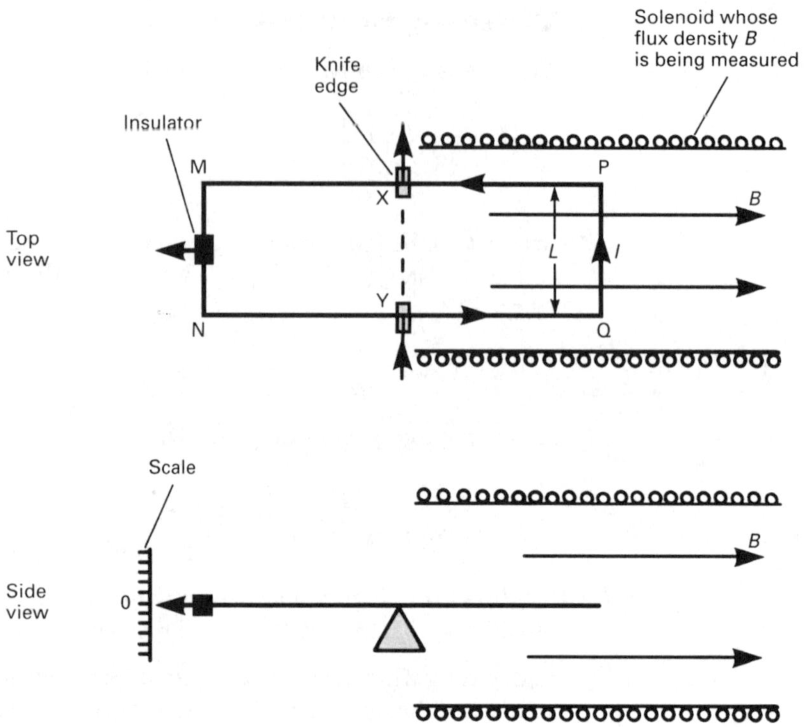

Fig. 41.29
Current balance

where L is the length of PQ. The frame is restored to the horizontal by adding more riders, of mass m and weight mg, to MN. If MN and PQ are equidistant from the axis XY, then

Force due to riders = Force due to field

$$\therefore \quad mg = BIL$$

i.e. $$B = \frac{mg}{IL} \qquad [41.20]$$

41.16 ABSOLUTE MEASUREMENT OF CURRENT BY USING A CURRENT BALANCE

The arrangement shown in Fig. 41.29 can be used to make an absolute determination of the current in PQ if the frame is connected in series with the solenoid. Such a connection ensures that the same current (I) flows through the solenoid as through PQ. The magnitude, B, of the flux density inside the solenoid is therefore given by equation [41.3] as

$$B = \mu_0 nI^\star$$

where n is the number of turns per unit length of the solenoid and μ_0 is the permeability of vacuum (air).

*Strictly, this expression applies only if the solenoid is infinitely long. It is, however, accurate enough for most purposes if the length of the solenoid is at least ten times its diameter and PQ is at the middle.

Substituting for B in equation [41.20] gives

$$\mu_0 n I = \frac{mg}{IL}$$

i.e. $$I = \sqrt{\frac{mg}{\mu_0 n L}} \qquad [41.21]$$

Equation [41.21] allows I to be calculated from the values of the mechanical quantities m, g, n and L, and μ_0 whose value is defined as $4\pi \times 10^{-7}\,\mathrm{H\,m^{-1}}$.

An even simpler, but less accurate, arrangement is shown in Fig. 41.30. P′Q′ is a long, fixed straight wire which is parallel to PQ and vertically below it. The two wires are in series so that each carries the same current, I. If the currents in PQ and P′Q′ are in opposite directions, PQ feels an upward directed force whose magnitude, F, is given by equation [41.13] as

$$F = \frac{\mu_0 I^2 L}{2\pi a}$$

where a is the separation of PQ and P′Q′ and L is the length of PQ. If m is the mass of the rider which has to be placed on PQ to keep the frame horizontal, then

$$\frac{\mu_0 I^2 L}{2\pi a} = mg$$

i.e. $$I = \sqrt{\frac{2\pi a m g}{\mu_0 L}}$$

Fig. 41.30
Simplified current balance

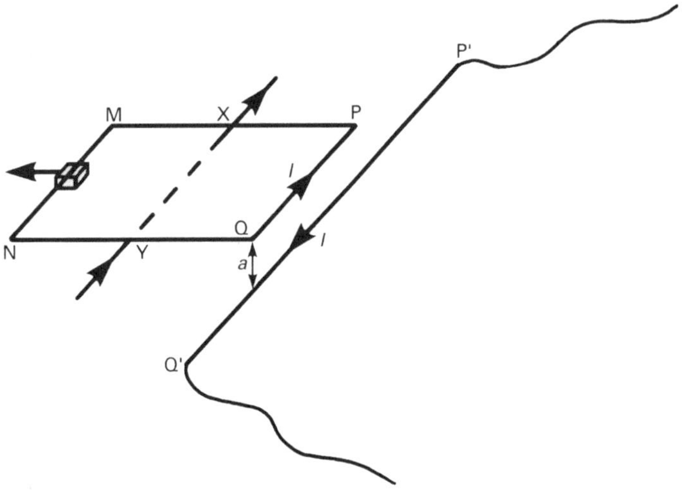

This arrangement is normally less accurate than that involving a solenoid because a solenoid is likely to exert a much greater force than a straight wire such as P′Q′. (The force per unit length on PQ when P′Q′ is 5 mm below it and is carrying a current of 1 A is only $4 \times 10^{-5}\,\mathrm{N\,m^{-1}}$. The force per unit length due to a solenoid of 2000 turns per metre and carrying 1 A is approximately $2.5 \times 10^{-3}\,\mathrm{N\,m^{-1}}$.)

The current balance is not normally used to measure currents directly, but is used to calibrate more convenient instruments such as the moving-coil ammeter. Elaborate versions of the current balance have accuracies of about 1 part in 10^6.

41.17 THE HALL EFFECT

Consider a piece of conducting material in a magnetic field of flux density B (Fig. 41.31). Suppose that the field is directed (perpendicularly) into the paper, and that there is a current flowing from right to left. If the material is a metal, the current is carried by electrons moving from left to right.

Fig. 41.31
Hall effect in a conductor

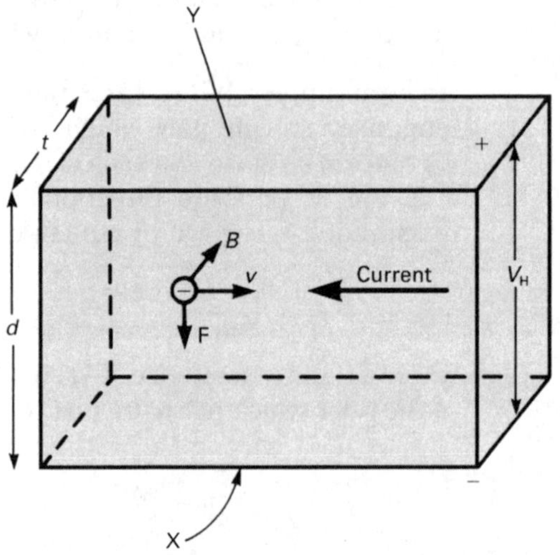

Consider the situation of one of these electrons and suppose that it has a velocity v. The electron feels a force, F, which, by Fleming's left-hand rule, is directed downwards as shown. Thus, in addition to the electron flow from left to right, electrons are urged away from face Y and towards face X. A negative charge builds up on X, leaving a positive charge on Y, so that a potential difference is established between X and Y. The build-up of charge continues until the potential difference becomes so large that it prevents any further increase. This, maximum, potential difference is called the **Hall voltage**.

Let F_V be the force exerted on an electron by the PD between X and Y. Therefore, when the build-up of charge on X and Y has ceased

$$F = F_V \quad [41.22]$$

By equation [41.10], for an electron whose component of velocity at right angles to the field is v

$$F = Bev$$

where e is the charge on the electron. Also by equations [39.8] and [39.10]

$$F_V = eE = \frac{eV_H}{d}$$

where

E = the strength of the uniform electric field between X and Y due to the Hall voltage

V_H = the Hall voltage

d = the separation of X and Y.

Therefore, by equation [41.22]

$$Bev = \frac{eV_H}{d}$$

i.e. $$V_H = Bvd \qquad [41.23]$$

It has been shown (section 36.3) that the current, I, in a material is given by

$$I = nAve$$

where

n = the number of charge carriers (electrons in metals) per unit volume of the material

A = the cross-sectional area of the material

v = the drift velocity of the charge carriers*

e = the charge on each charge carrier.

Substituting for v in equation [41.23] gives

$$V_H = \frac{BdI}{nAe}$$

In Fig. 41.31, $A = dt$ and therefore

$$V_H = \frac{BI}{net} \qquad [41.24]$$

It is clear from equation [41.24] that the Hall voltage is inversely proportional to the number of charge carriers per unit volume. There are typically 10^4 times as many charge carriers in metals as there are in extrinsic semiconductors (see Chapter 55). Therefore, whereas a typical Hall voltage for a metal might be as little as 1 μV, for a semiconductor under similar circumstances it would be about 10 mV.

The Hall effect can be used to determine whether the charge carriers in a material are negative or positive (for example, whether a semiconductor is n-type or p-type). If the direction of the conventional current is to the left, as it is in Fig. 41.31, it must be due to either negative charge moving to the right or positive charge moving to the left. In each case the magnetic field exerts a downward directed force. This results in X becoming negative if the charge carriers are negative, and positive if the charge carriers are positive. Thus, by determining the polarity of the Hall voltage, it is possible to determine the nature of the charge carriers.

41.18 MEASUREMENT OF MAGNETIC FLUX DENSITY BY MAKING USE OF THE HALL EFFECT

The flux density, B, of a magnetic field can be measured by making use of a **Hall probe**. One version of the device has a small wafer of germanium (a semiconductor) mounted on the end of a long, narrow handle, so that it can conveniently be used to probe the magnetic field being examined.

*A little thought should convince the reader that when large numbers of electrons are involved the v of equation [41.23] should be interpreted as drift velocity.

A current, I, is passed between two opposite faces of the germanium wafer. When the wafer is suitably orientated in a magnetic field, a Hall voltage, V_H, is established across two other faces. From the theory of the Hall effect (equation [41.24]),

$$V_H = \frac{BI}{net}$$

i.e. $$B = \frac{netV_H}{I} \qquad [41.25]$$

The value of *net* in equation [41.25] is normally supplied by the manufacturer of the probe, in which case B can be found by measuring I and V_H. A high-impedance voltmeter is used to measure V_H. An ammeter with a full-scale deflection of 1 A is suitable for measuring I.

41.19 THE DC MOTOR

A simple DC motor is shown in Fig. 41.32. It follows from Fleming's left-hand rule that at the instant shown side X is experiencing a downward directed force and side Y is experiencing an upward directed force. The coil therefore turns in an anticlockwise sense. When the plane of the coil is vertical the gaps in the commutator are facing the brushes and, momentarily, there is no current in the coil. However, its inertia carries it beyond this position so that side Y comes into contact with brush M and side X comes into contact with brush N. It follows that the current in the coil always flows around the coil in the same direction (clockwise as seen from above), and therefore the coil rotates in an anticlockwise sense no matter what its orientation.

Fig. 41.32
Principle of a DC motor

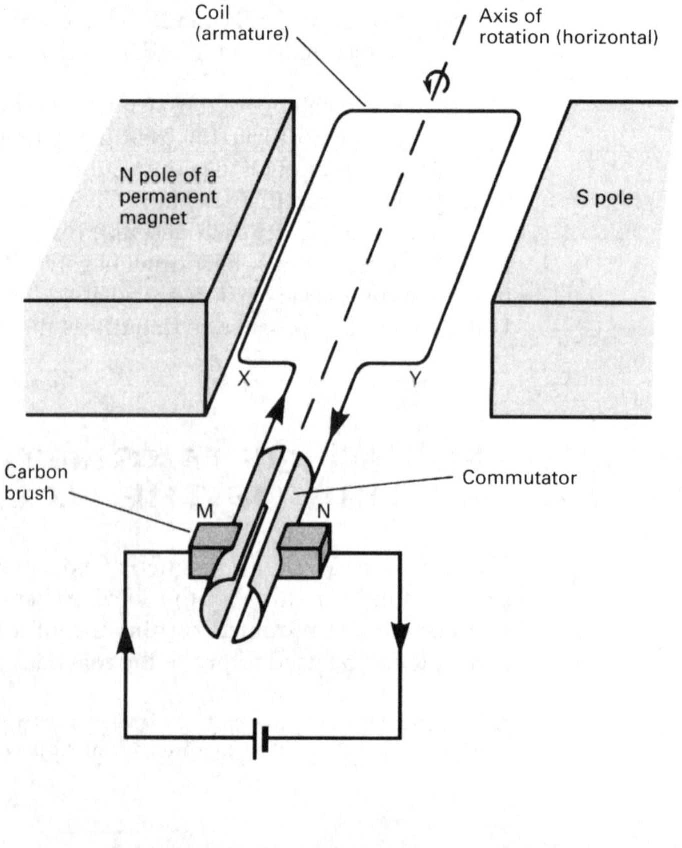

The rotating part of an electric motor is called the **armature** (or **rotor**) of the motor. In practice the armature consists of several equally spaced coils wound on a soft iron core and connected to a commutator which has a corresponding number of sections. The main advantage of using several coils, rather than just one, is that the motor provides an almost constant torque. The use of curved pole pieces together with the soft iron core produces a radial field and this also serves to make the torque constant.

If the core were a solid piece of soft iron, power would be wasted as a result of eddy current heating (see section 42.11). To avoid this the core is constructed from thin sheets of iron whose planes are perpendicular to the axis of rotation and which are insulated from each other by varnish, i.e. the core is **laminated**.

As the armature rotates in the magnetic field an EMF is induced in its windings. This EMF opposes the applied PD and so is known as a back EMF (see section 42.13). If E and V represent the magnitudes of the back EMF and the applied PD respectively and I is the current in the coils and R is their total resistance, then

$$V - E = IR \qquad [41.26]$$

The value of E is proportional to the speed at which the armature is rotating. When the motor is first switched on E is zero, but as the motor speeds up E increases and therefore the current, I, falls. It follows that the torque, which is proportional to I (see section 41.12), also falls. In the absence of a load a frictionless motor would carry on accelerating until the torque had become zero. This would occur when E had become equal to V. The motor would now be taking no current and consuming no energy and, of course, would be doing no work either. When a motor is loaded its speed of rotation falls to some new steady value. This is because a loaded motor has to exert a torque in order to perform mechanical work and this requires there to be a current flowing, and therefore the speed must not be so great that E is equal to V.

Multiplying equation [41.26] by I and rearranging gives

$$VI = I^2R + EI$$

VI is the power supplied to the motor and I^2R is the rate at which energy is dissipated as heat in the coils. The difference, EI, is the rate at which the motor is performing mechanical work. The efficiency of the motor is defined by

$$\text{Efficiency} = \frac{\text{Mechanical power obtained}}{\text{Power supplied}} \times 100$$

$$= \frac{EI}{VI} \times 100$$

$$= \frac{E}{V} \times 100$$

The efficiency is high when the coil resistance is small. As a consequence electric motors usually have coil resistances of less than 1 Ω.

When a motor is first switched on $E = 0$, and were it not for the inclusion of a so-called **starting resistance** the full supply voltage, V, would be across the coils and would probably cause them to burn out. The starting resistance is a rheostat and is in series with the coils. As the motor gathers speed, the resistance of the rheostat is gradually reduced. By the time the motor is running at its full operating speed the rheostat resistance is zero and the current is being limited solely by the back EMF, which by now will be only slightly less than the supply PD. The rheostat is operated automatically by an electromagnet energized by the same current as that which passes through the motor. Small motors tend to have higher resistances than large ones and do not normally need a starting resistance.

In large motors the magnetic field is usually provided by electromagnets. In the so-called **series-wound motor** the field coils are in series with the armature. In a **shunt-wound motor** the field coils and the armature are in parallel. Series-wound motors produce large torques at low speeds and are used, for example, as the power units of electric locomotives. The speed of a shunt-wound motor is relatively insensitive to variations in load. Motors of this type are used in record players for example.

41.20 EXPERIMENTAL INVESTIGATION OF THE EFFECTS OF VARYING THE TORQUE ON A DC MOTOR

The investigation about to be described is concerned with a shunt-wound motor, i.e. one in which the field coils are in parallel with the armature (rotor).

Method

Refer to Fig. 41.33.

Fig. 41.33
(a) Experimental arrangement, (b) circuit diagram

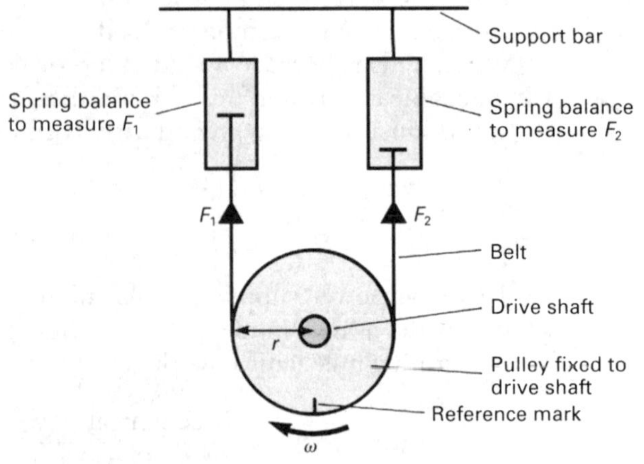

(a)

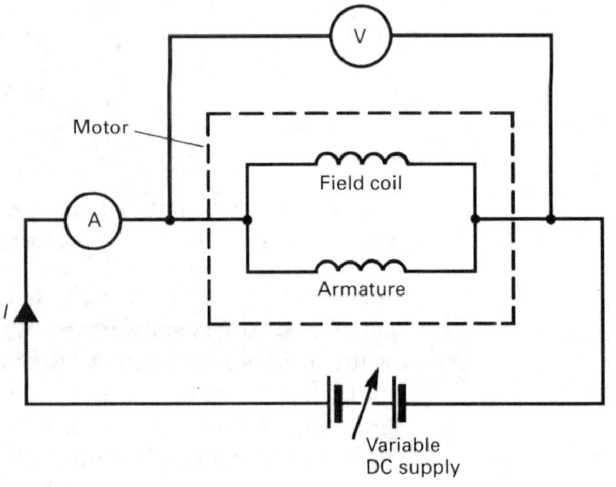

(b)

(a) Adjust the DC supply to some suitable value (e.g. the normal operating PD of the motor).

(b) Illuminate the pulley with a stroboscope set to a high frequency and then decrease the frequency until the reference mark on the pulley appears to be stationary. The stroboscope frequency (f) is now equal to the frequency of rotation of the motor.

(c) Record the values of F_1, F_2 and f together with the readings of the voltmeter and ammeter. The voltmeter reading is the PD (V) across the motor; the ammeter reading is (to a good approximation) the total current (I) taken by the motor.

(d) Raise the support bar. This increases the torque on the pulley by increasing the value of $(F_2 - F_1)$. If necessary, reset V to its previous value.

(e) Adjust the stroboscope frequency until the reference mark again appears to be stationary.

(f) Record the new values of I, F_1, F_2 and f.

(g) Repeat (d), (e) and (f) for increased values of $(F_2 - F_1)$.

Applied torque, $T = (F_2 - F_1)\,r$

Angular velocity of motor, $\omega = 2\pi f$

Power input to motor, $P = VI$

Efficiency of motor, $\eta = T\omega/P$

Plot ω against T, P against T and η against T to obtain the characteristics shown in Fig. 41.34.

Fig. 41.34
Motor characteristics

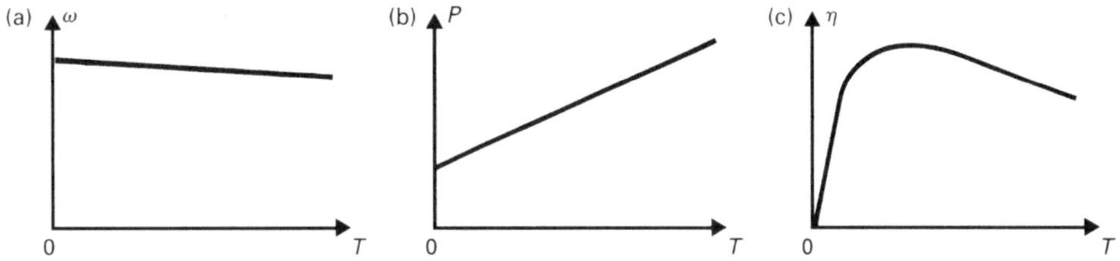

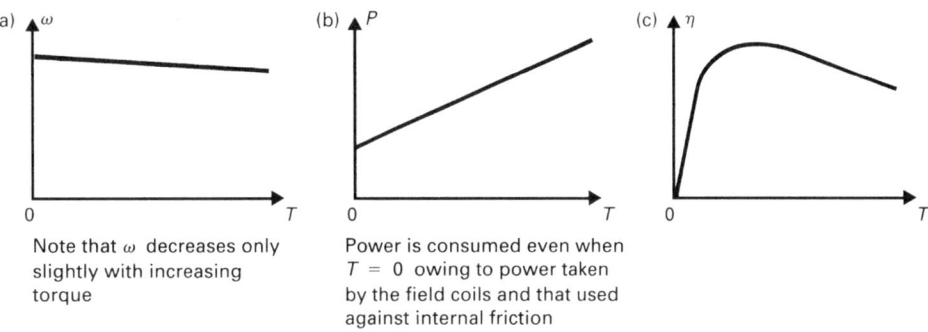

Note that ω decreases only slightly with increasing torque

Power is consumed even when $T = 0$ owing to power taken by the field coils and that used against internal friction

Theory

Refer to Fig. 41.35. If the back EMF in the armature is E, then

$$V - E = I_A R_A \qquad [41.27]$$

The back EMF is proportional to the rate of cutting of flux, and therefore to ω, in which case we may put

$$E = k\omega \qquad [41.28]$$

where k is a constant of proportionality. The rate at which the motor works in order to overcome the back EMF is EI_A and is equal to the rate at which the motor is performing mechanical work. This is given by $(T + T_0)\,\omega$ where T_0 is the constant torque used to overcome friction within the motor.

Thus

$$EI_A = (T + T_0)\,\omega \qquad [41.29]$$

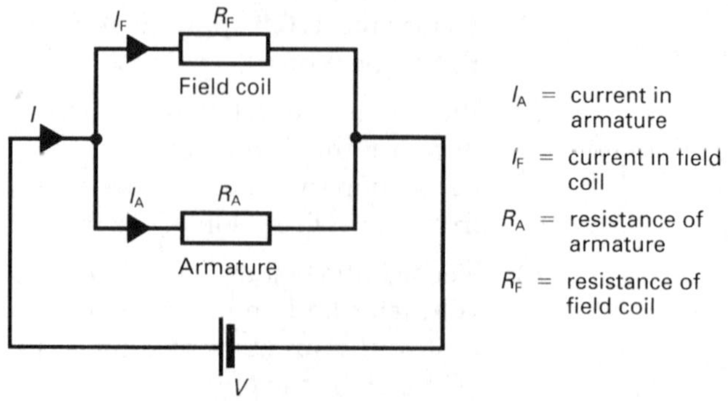

Fig. 41.35
Equivalent circuit

Eliminating E and I_A from equations [41.27], [41.28] and [41.29] gives

$$V - k\omega = \left(\frac{T + T_0}{k}\right)R_A$$

i.e.
$$\omega = -\frac{R_A}{k^2}T + \left(\frac{V}{k} - \frac{T_0 R_A}{k^2}\right)$$

hence Fig. 41.34(a), since R_A, k, V and T_0 are constant.

The power, P, consumed by the motor is given by

$$P = V(I_A + I_F) \qquad [41.30]$$

Eliminating E and I_A from equations [41.28], [41.29] and [41.30] gives

$$P = V\left(\frac{T + T_0}{k} + I_F\right)$$

i.e.
$$P = \frac{V}{k}T + \left(\frac{VT_0}{k} + VI_F\right)$$

hence Fig. 41.34(b), since k, V, T_0, and I_F are constant.

41.21 THE ELECTROMAGNETIC RELAY

This device is an electromagnetic switch (Fig. 41.36). When a current passes through the coil the soft-iron core becomes magnetized, attracting the soft-iron armature and closing the contacts together so that a current may flow in the circuit being switched. Although this current might be very large, the current through the coil may be only a few milliamps. In this way the small current from a

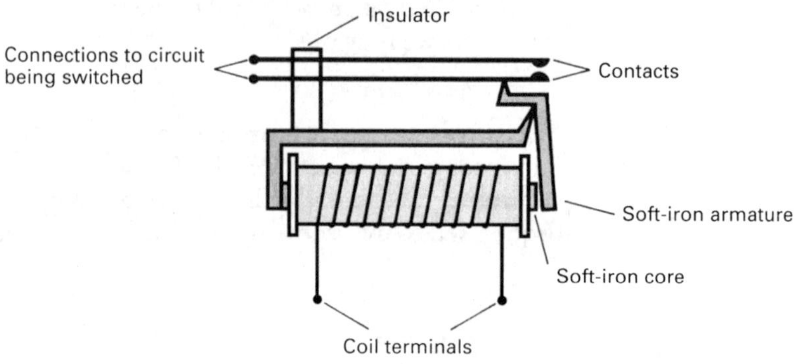

Fig. 41.36
The electromagnetic relay

photoemissive cell, for example, can be caused to control the very much larger current necessary to open or close a garage door, say. When a passenger in a lift pushes the operating button a small current flows through the coil of a relay. The relay switches on the large current required to operate the lift motor. The system has the advantage that the passenger is electrically isolated from the potentially dangerous high-current circuit.

A minor modification to the circuit shown in Fig. 41.36 would allow the current in the main circuit to be switched off, rather than on, when current flows through the coil.

41.22 THE MOVING-COIL LOUDSPEAKER

A moving-coil loudspeaker is shown in Fig. 41.37. The turns on the coil are at right angles to the radial field between the pole-pieces of the permanent magnet. It follows from Fleming's left-hand rule that if a current flows through the coil, the coil will move in or out of the magnet according to which way the current is flowing. If the current is an alternating current (from a signal generator or the amplifier of a record-player, for example) the coil and the stiff paper cone attached to it oscillate at the frequency of the current. The motion of the cone disturbs the air around it and generates a sound wave whose frequency is the same as that of the alternating current. Increasing the amplitude of the current increases the loudness of the sound produced. The cone must be rigid – if it were to flex as it moved to and fro, it would distort the output.

Fig. 41.37 The moving-coil loudspeaker

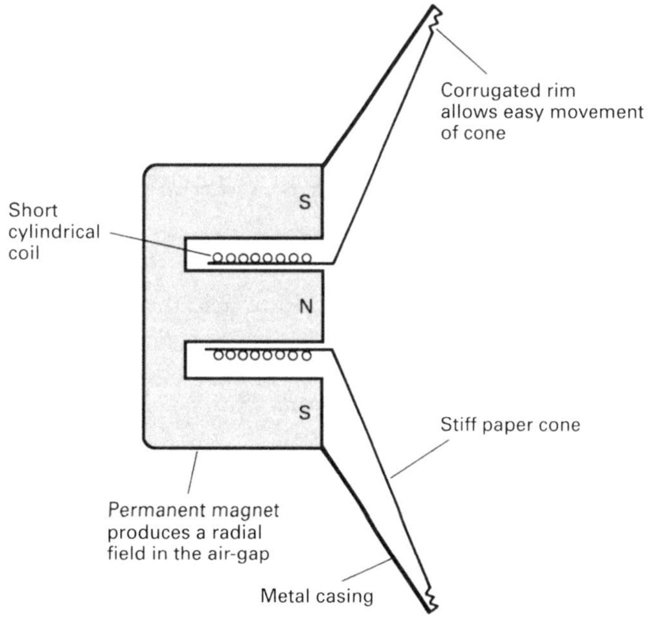

It is desirable that the maximum amount of available power is transferred to a loudspeaker. This requires that the impedance (section 43.10) of the speaker is equal to the output impedance of the amplifier or signal generator with which it is being used. This is known as **impedance matching**.

The sound waves radiating from a loudspeaker cone are diffracted as if they have passed through a circular hole with the same diameter as the cone. The sound therefore spreads sideways to some extent – the lower the frequency and the smaller the cone, the greater the spreading. A highly directional beam is normally undesirable (the position of the listener would be too critical) and therefore a small cone is required for high frequencies. In any case, a speaker designed to reproduce high (treble) frequencies (a 'tweeter') has to change direction very rapidly, and a large cone would have too much inertia for this to be possible. On the other hand, if a cone is small compared with the wavelength it is emitting, most of the energy simply pulses to and fro in front of the speaker rather than being radiated away from it*. It follows that a speaker designed to handle low (base) frequencies (a 'woofer') has a large cone or, if space will not allow, a small cone vibrating with large amplitude.

CONSOLIDATION

The direction of a magnetic field is the direction in which a north pole would move under the influence of the field.

The right-hand grip rule gives the direction of the field of a current-carrying conductor.

Fleming's left-hand (motor) rule gives the direction of the force due to a magnetic field acting on a current-carrying conductor or on a moving charged particle.

Like currents attract; unlike currents repel.

At the mid point of the axis of a long solenoid.

$$B = \mu_0 n I \qquad (n = \text{number of turns per unit length})$$

At the centre of a plane coil

$$B = \frac{\mu_0 N I}{2r} \qquad (N = \text{number of turns})$$

At a distance a from a long straight wire

$$B = \frac{\mu_0 I}{2\pi a}$$

For a coil in a uniform field the torque T is given by

$$T = BIAN\cos\phi \qquad (\phi = \text{angle between field and plane of coil})$$

For a coil in a radial field

$$T = BIAN$$

*The reason for this is beyond the scope of this book.

42

ELECTROMAGNETIC INDUCTION

42.1 THE PHENOMENON OF ELECTROMAGNETIC INDUCTION

An EMF is induced in a coil in a magnetic field whenever the flux (Φ) through the coil changes. Fig. 42.1 illustrates some ways of achieving this. (**Note**. The EMF ceases to exist once the change has taken place.) The effect is called electromagnetic induction and if the coil forms part of a closed circuit, the induced EMF causes a current to flow in the circuit. The effect was discovered by Faraday, and independently at about the same time, by Henry, in 1831 – eleven years after Oersted's discovery that a current-carrying conductor has an associated magnetic field.

Fig. 42.1
Three demonstrations of electromagnetic induction

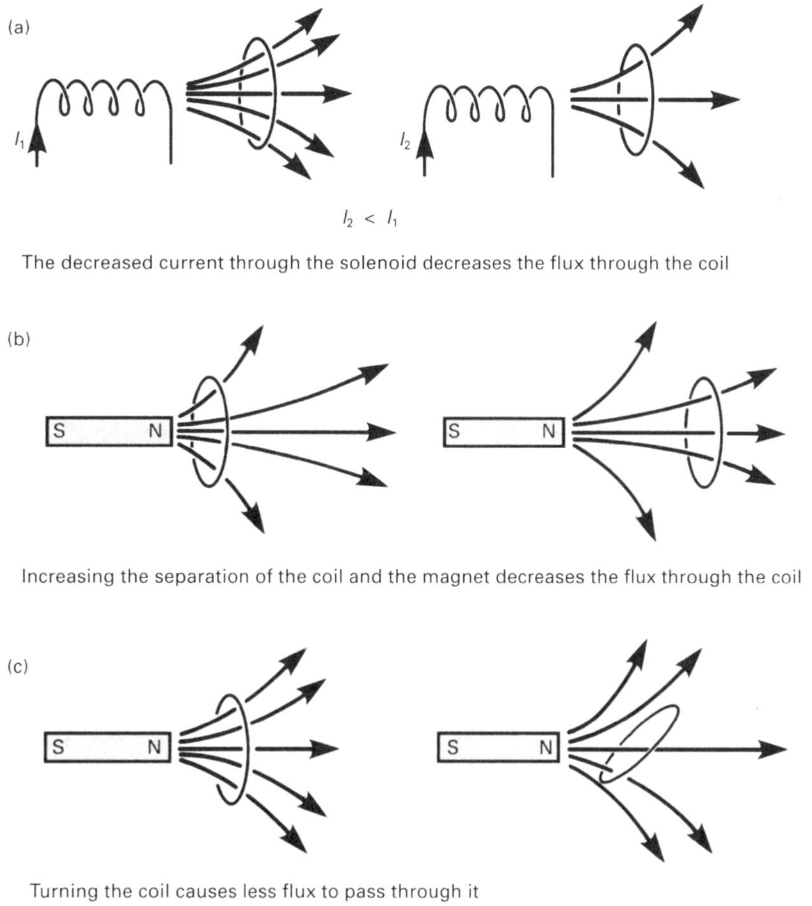

(a) The decreased current through the solenoid decreases the flux through the coil

(b) Increasing the separation of the coil and the magnet decreases the flux through the coil

(c) Turning the coil causes less flux to pass through it

Experiments show that the magnitude of the EMF depends on the rate at which the flux through the coil changes. It also depends on the number of turns, N, on the coil, and it is useful to define a quantity called **flux-linkage** as being the product of the number of turns and the flux through the coil, i.e.

$$\text{Flux-linkage} = N\Phi$$

It is not necessary that a conductor is in the form of a coil in order for it to be able to acquire an induced EMF. Experiment shows that **an EMF can be induced in a straight conductor whenever it is caused to cut across magnetic field lines**. The magnitude of the EMF is proportional to the rate of cutting.

We have seen that EMFs are induced when the flux through a coil changes, and when a conductor cuts magnetic flux. At first sight these appear to be two distinct ways of inducing EMFs – they are, in fact, equivalent. For example in Fig. 42.1(b) and (c) the motion of the coil causes its perimeter to cut across magnetic flux lines (see also section 42.7).

42.2 THE LAWS OF ELECTROMAGNETIC INDUCTION

A detailed investigation of electromagnetic induction leads to two laws.

(i) The magnitude of the induced EMF in a circuit is directly proportional to the rate of change of flux-linkage or to the rate of cutting of magnetic flux (**Faraday's law**).

(ii) The direction of the induced EMF is such that the current which it causes to flow (or would flow in a closed circuit) opposes the change which is producing it (**Lenz's law**).

The two laws can be expressed as **Neumann's equation**:

$$E = -\frac{\mathrm{d}}{\mathrm{d}t}(N\Phi) \quad [42.1]$$

where

E = the induced EMF in volts

$\frac{\mathrm{d}}{\mathrm{d}t}(N\Phi)$ = the rate of change of flux-linkage in webers per second.

Notes (i) If a non-consistent set of units were being used, a constant of proportionality with a value other than unity would be included in equation [42.1].

(ii) The minus sign in equation [42.1] takes account of Lenz's law. According to Lenz's law, the induced current flows in such a sense as to create a flux in the opposite direction to that in which the external flux has increased, i.e. the current flows in such a direction as to oppose the change which has taken place. The relative directions of current flow and increased external flux are shown in Fig. 42.2. (The reader should confirm, by using the right-hand grip rule, that current flowing in the direction shown gives rise to a magnetic field which is in the opposite direction to that in which the external flux has increased.) It can be seen that the induced current flows in

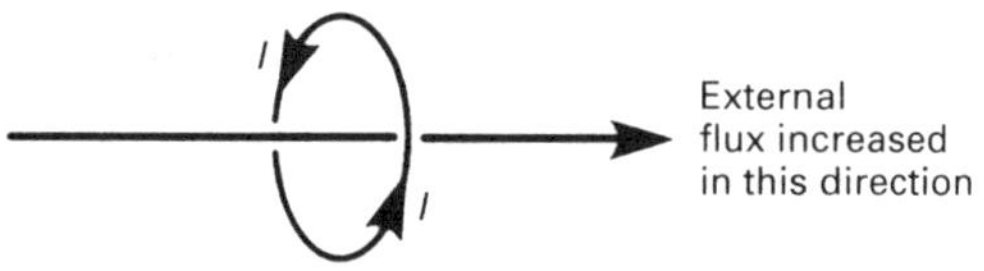

Fig. 42.2
Principle of Lenz's law

the same sense as a left-handed corkscrew would have to turn in order to advance in the direction of the increased external flux. It follows that the EMF which has produced the current also has a left-handed relationship with the external flux. There is a convention running through SI which demands that a minus sign be included to take account of the left-handedness.

(iii) When working in terms of rate of cutting of flux rather than rate of change of flux-linkage equation [42.1] becomes

$$E = -N\frac{d\Phi}{dt}$$

where

$\frac{d\Phi}{dt}$ = the rate of cutting of flux

N = the number of conductors cutting the flux.

42.3 AN ILLUSTRATION OF LENZ'S LAW

In Fig. 42.3 the strength of the magnetic field at the solenoid increases as the magnet is moved towards it. An EMF is induced in the solenoid and the galvanometer indicates that a current is flowing. If a preliminary experiment has been performed to determine the direction of the current through the galvanometer which corresponds to a deflection in a particular sense, then it is seen that the current through the solenoid is in the direction that makes end A a north pole. This opposes the motion of the magnet (like poles repel), i.e. the direction of the current is such as to oppose the change which has induced it – Lenz's law.

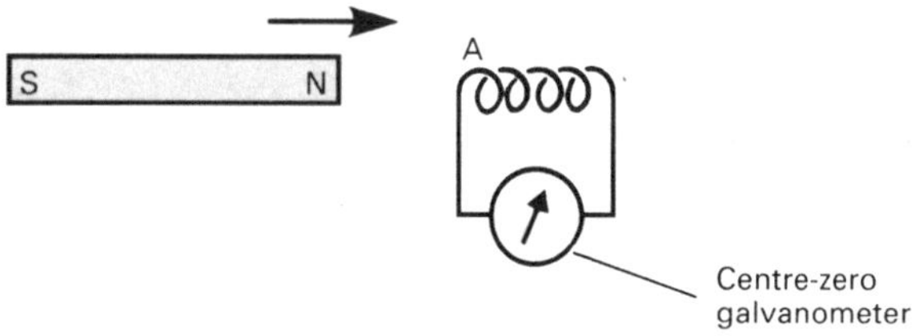

Fig. 42.3
Demonstration of Lenz's law

The presence of the north pole at A means that work has to be done in order to move the magnet and cause the current to flow. The work done is converted to electrical energy, some of which is dissipated as heat in the circuit and some of which provides the mechanical energy to deflect the galvanometer.

If the magnet is moved away from the solenoid, current flows in the opposite direction and there is a south pole at A which opposes the motion of the magnet.

Lenz's law is the law of conservation of energy expressed in such a way as to apply specifically to electromagnetic induction.

42.4 FLEMING'S RIGHT-HAND (DYNAMO) RULE

The direction of an induced current can always be found by using Lenz's law, but if the current is being induced by the motion of a straight conductor, it is more convenient to use **Fleming's right-hand (dynamo) rule:**

If the first and second fingers and the thumb of the right hand are placed comfortably at right angles to each other, with the **F**irst finger pointing in the direction of the **F**ield and the thu**M**b pointing in the direction of the **M**otion, then the se**C**ond finger points in the direction of the induced **C**urrent (Fig. 42.4).

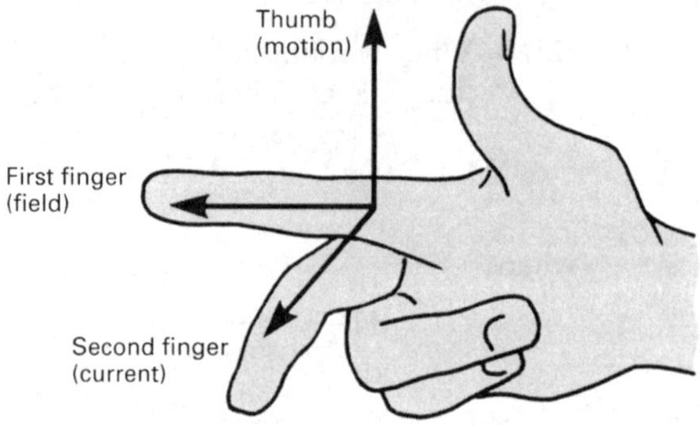

Fig. 42.4
Fleming's right-hand rule

Applying the rule to the conductor shown in Fig. 42.5 reveals that the induced current is directed into the paper.

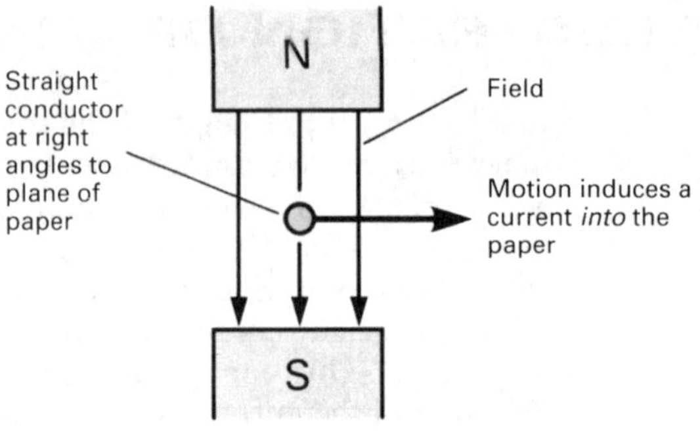

Fig. 42.5
Application of Fleming's right-hand rule

42.5 THE EQUIVALENCE OF LENZ'S LAW AND FLEMING'S RIGHT-HAND RULE

In Fig. 42.6 the electrons within the conductor are being moved to the right, and so constitute a current to the left. By Fleming's left-hand rule, therefore, the electrons are subject to a downward directed force (F) and so move towards P, i.e. the motion induces a current which flows towards Q. This current gives rise to a force which, by the left-hand rule, is directed to the left. This force opposes the motion of the conductor and so is consistent with Lenz's law.

Fleming's right-hand rule also predicts that the induced current is directed towards Q.

Note that the motion of the conductor exerts a force on the positive ions within it, but since they are not free to move they cannot give rise to an induced current.

Fig. 42.6
To illustrate the equivalence of Lenz's law and Fleming's right-hand rule

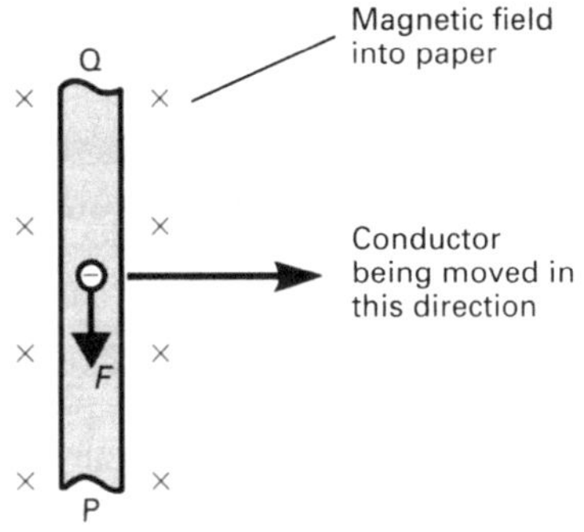

42.6 EMF INDUCED IN A STRAIGHT CONDUCTOR

Fig. 42.7 illustrates five distinct ways in which a straight conductor of length L can be moved with velocity v through a magnetic field of flux density B. The only motion which induces an EMF between the ends of the conductor is that shown in (a), where the conductor cuts across the field lines. The EMF, E, is given by

$$E = BLv \qquad [42.2]$$

(We are concerned only with EMFs between the ends of the conductor, but it should not be overlooked that real conductors are of finite width in which case the motions shown in (b) and (c) induce EMFs across the width of the conductor.)

Fig. 42.7
Conductor moving in a magnetic field

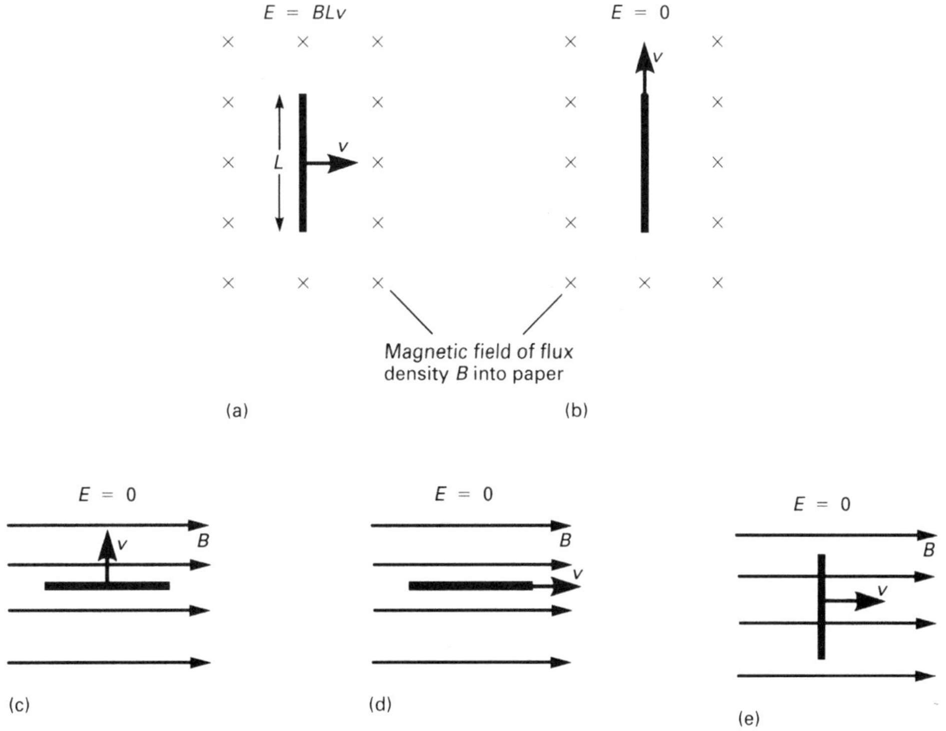

Derivation of $E = BLv$ from Faraday's Law

In Fig. 42.7(a)

The area swept out by the conductor in 1 s $= Lv$

$\therefore$ The flux cut by the conductor in 1 s $= BLv$

i.e. The rate of cutting of flux $= BLv$

In consistent units the relevant version of Faraday's law can be written

EMF = Rate of cutting of flux

i.e. $E = BLv$

Derivation of $E = BLv$ from Conservation of Energy

Suppose that the conductor in Fig. 42.8 forms part of a closed circuit and that the induced EMF causes a current, I, to flow through it. By Fleming's right-hand rule the current is in the direction shown. Since the conductor is carrying a current and is in a magnetic field, there will be a force acting on it, the magnitude of which is given by equation [41.9] as BIL. By Fleming's left-hand rule the force is directed towards the left, i.e. it opposes the applied force.

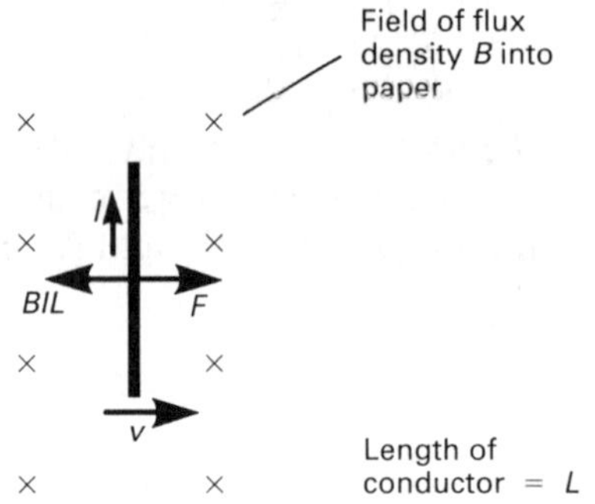

Fig. 42.8
Forces on a conductor moving with velocity ***v*** and carrying a current ***I***

The conductor moves at <u>constant</u> velocity and therefore the net force on it is zero, i.e.

$$F = BIL$$

where F is the magnitude of the applied force.

The applied force is moving its point of application and therefore is doing work (against BIL). The rate at which this work is being done is given by equation [5.7] as

Rate of working $= Fv$

i.e. Rate of working $= BILv$

This work is being converted to electrical energy at a rate EI (the electrical power). By the principle of conservation of energy the rate of working is equal to the rate of production of electrical energy, i.e.

$$EI = BILv$$

i.e. $E = BLv$

QUESTIONS 42A

1. Diagrams **(a)** to **(c)** below show a metal rod, AB, of length 8.0 cm being moved in the plane of the paper at $3.0\,\mathrm{m\,s^{-1}}$ through a magnetic field of flux density $4.0 \times 10^{-2}\,\mathrm{T}$ which is directed into the paper. Find the magnitude of the induced EMF in each case.

2. Two small bar magnets, X and Y, are released from rest at the same height above the ground. X falls directly to the ground, but Y passes through a metal ring which is fixed with its plane horizontal. Does X reach the ground: (A) at the same time as Y, (B) before Y, (C) after Y? Explain your answer.

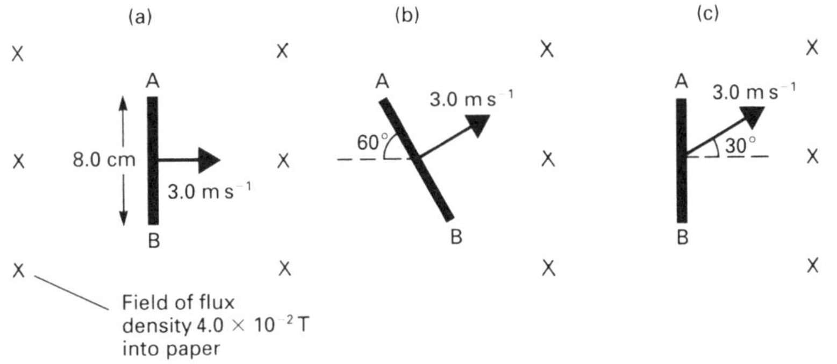

42.7 EMF INDUCED IN A RECTANGULAR COIL

Consider a rectangular coil PQRS of one turn situated in a magnetic field of flux density B (Fig. 42.9). The plane of the coil is perpendicular to the field and its dimensions are as shown.

Fig. 42.9
Rectangular coil moving in a magnetic field

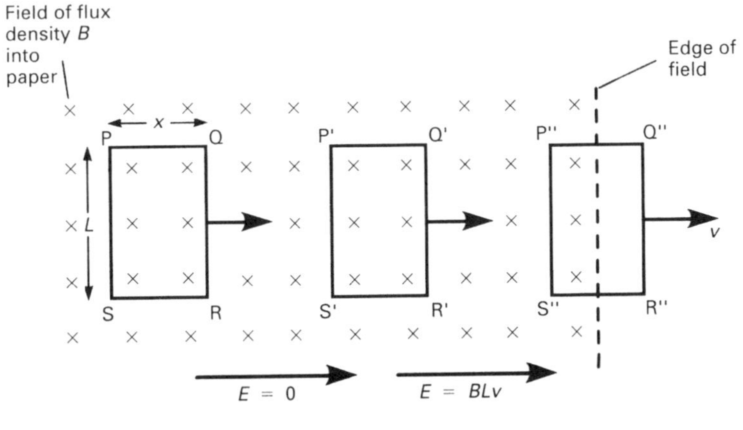

Suppose that the coil is moved sideways with velocity v. As the coil moves to P′Q′R′S′, both PS and QR cut across the field lines and equal and opposite EMFs are induced in them. The net EMF in the coil is therefore zero – there has been no change in the flux through the coil. (There is no induced EMF in either PQ or SR because each of these sections is moving parallel to its length.)

As the coil moves to P″Q″R″S″, QR leaves the field before PS. Once QR has left the field there is no EMF across it to oppose that across PS, and the coil as a whole has an induced EMF, E, where

$$E = BLv$$

Derivation of $E = BLv$ on the Basis of Changing Flux-linkage

When all of the coil is in the field the flux-linkage is BLx; when all of it is out of the field the flux-linkage is zero. Therefore,

$$\text{Change in flux-linkage} = BLx$$

The flux-linkage starts to change when QR leaves the field, and stops changing when PS leaves the field. The time taken for this is the time taken for the coil to travel a distance equal to its width, x. Therefore,

$$\text{Time for flux-linkage to change} = x/v$$

Therefore,

$$\text{Rate of change of flux-linkage} = \frac{BLx}{x/v} = BLv$$

In consistent units the relevant version of Faraday's law can be written

$$\text{EMF} = \text{Rate of change of flux-linkage}$$

i.e.

$$E = BLv$$

Derivation of $E = BLv$ on the Basis of Flux-cutting

The induced EMF results from the motion of PS once QR has left the field. PS is a straight conductor of length L moving with velocity v at right angles to a field of flux density B. It has been shown in section 42.6 that, on the basis of cutting of flux, the EMF, E, in such a situation is given by $E = BLv$.

Thus, the treatment based on flux-cutting gives the same result as that based on changing flux-linkage, i.e. the two treatments are equivalent.

42.8 EMF INDUCED IN A ROTATING COIL

Consider a rectangular coil of N turns, each of area A, being rotated with constant angular velocity ω in a uniform magnetic field of flux density B about an axis which is perpendicular to the paper (Fig. 42.10).

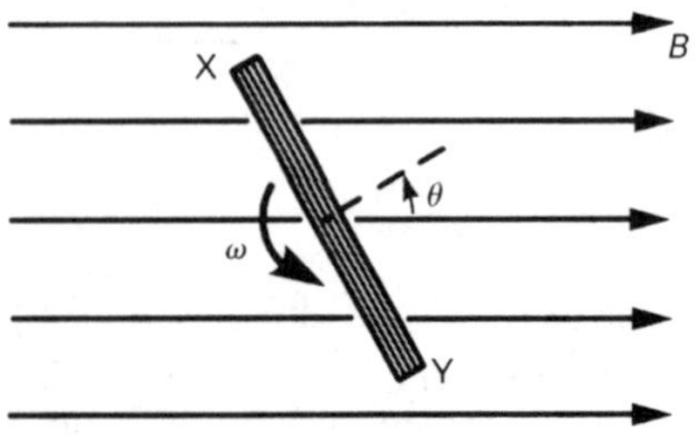

Fig. 42.10
Rectangular coil rotating in a magnetic field

When the normal to the coil is at an angle θ to the field the flux Φ through each turn of the coil is given by

$$\Phi = AB\cos\theta$$

But

$$\theta = \omega t$$

where

$$t = \text{the time that has elapsed since } \theta = 0.$$

Therefore

$$\Phi = AB \cos \omega t$$

The coil has N turns and therefore the flux-linkage $N\Phi$ is given by

$$N\Phi = NAB \cos \omega t$$

By Neumann's equation, the induced EMF, E, is given by

$$E = -\frac{d}{dt}(N\Phi)$$

i.e. $$E = -\frac{d}{dt}(NAB \cos \omega t)$$

$$= -NAB\frac{d}{dt}\cos \omega t$$

i.e. $$E = NAB\omega \sin \omega t \qquad [42.3]$$

Thus, a coil rotating with constant angular velocity in a uniform magnetic field produces a sinusoidally alternating EMF (see Fig. 42.11).

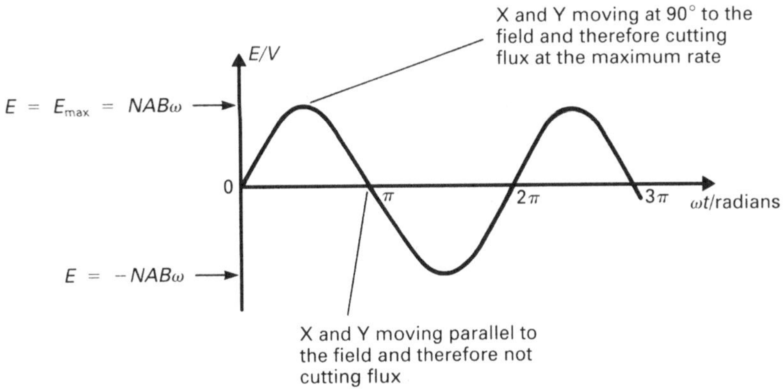

Fig. 42.11 EMF induced in a coil rotating with constant angular velocity

42.9 THE AC GENERATOR (ALTERNATOR)

In its simplest form an AC generator consists of a rectangular coil rotating in a uniform magnetic field (Fig. 42.12).

The generated current is fed from the coil in such a way that each side of the coil is always connected to the same output terminal, no matter what the orientation of the coil. This is achieved by the use of **slip rings**. Side W of the coil is attached to slip ring X. This rubs against brush Y which is connected to output terminal Z. Thus, W is always in electrical contact with Z. Similarly, side M of the coil is always in electrical contact with output terminal N, and therefore the alternating EMF of the rotating coil appears at the output terminals.

In practice the coil is wound on a soft iron core. The coil and its core are collectively called the **armature** of the generator. The core is laminated to reduce power losses due to eddy current heating.

Fig. 42.12
Principle of an AC generator

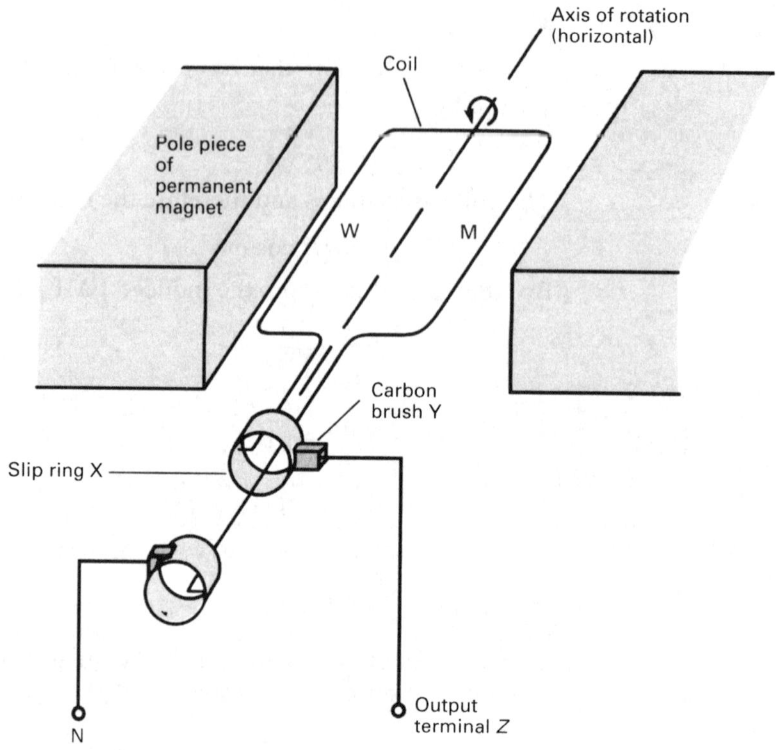

In a large generator the magnetic field is provided by an electromagnet. The coils of the electromagnet are called **field coils** and are energized by the current from a separate DC generator. The current in the field coils is much less than that in the armature coils. In view of this it is normal practice to keep the armature fixed and to rotate the electromagnet so that the rotating slip rings need only be capable of handling the smaller of the two currents. Whatever the arrangement, the stationary component is called the **stator** and the rotating component is called the **rotor**.

42.10 DC GENERATORS

Rotating Coil

A simple DC generator is shown in Fig. 42.13. At the instant shown, side X of the coil is connected by way of the **commutator** to output terminal Z. When the coil is turned through 90° from this position, the gaps between the two sections of the commutator come to face the brushes, and therefore, momentarily, there is no electrical contact between the coil and the output terminals. As the coil is moved beyond this position side Y comes into contact with Z. Terminal Z, then, is always connected to the side of the coil which is moving down. The polarity of the EMF at the output terminals never changes therefore, and takes the form shown in Fig. 42.14, and gives rise to varying, yet unidirectional, current.

In practice several equally spaced coils are wound on a soft iron core, and are connected to a multisectioned commutator in such a way that the EMFs of the individual coils add to each other. The whole assembly is called the **armature** of the generator. This arrangement gives a larger and steadier EMF than that produced by a single coil. As in the case of the electric motor (section 41.19) the core of the armature is laminated to cut down power losses due to eddy current heating.

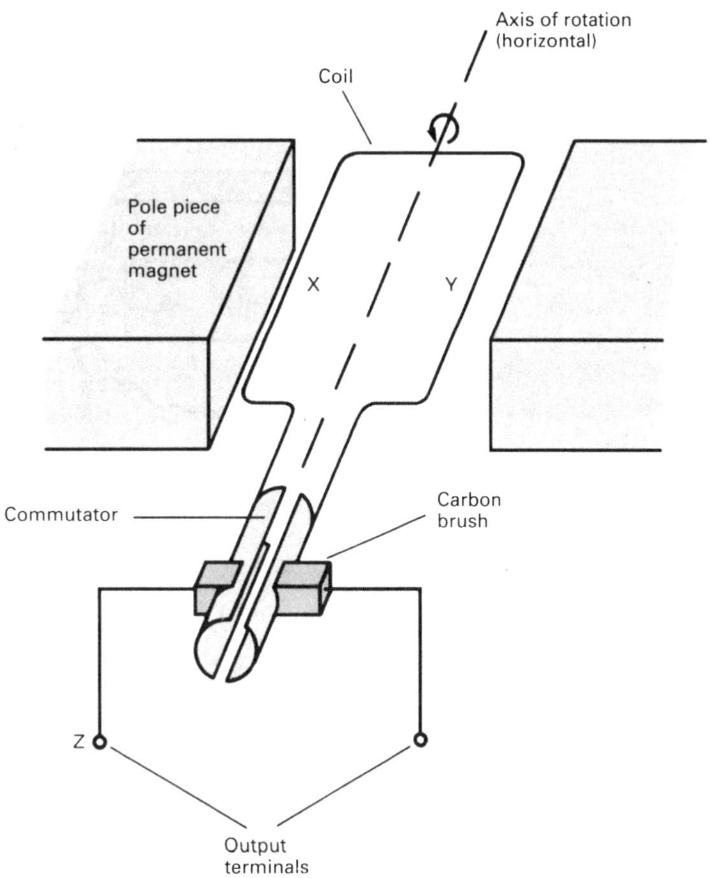

Fig. 42.13
Principle of a DC generator

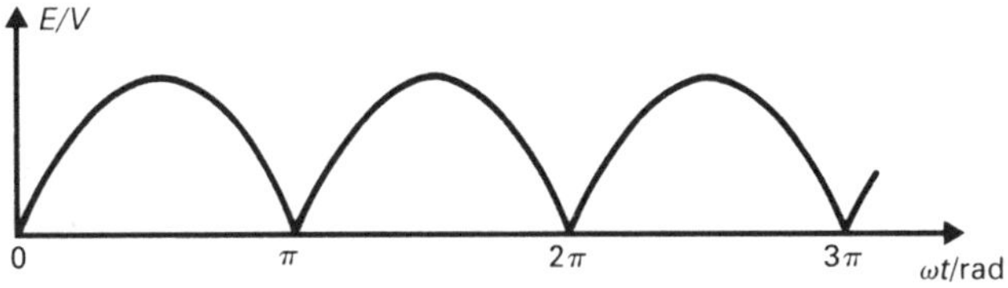

Fig. 42.14
EMF generated by a DC generator

In large generators the magnetic field is provided by electromagnets. The current which energizes the electromagnets is provided by the generator itself. This would not be possible if it were not for the fact that the cores of the electromagnets always retain a little magnetism, for otherwise there would be no possibility of generating a current in the first place.

The Homopolar Generator (Faraday's Disc)

This consists of a metal (usually copper) disc which is rotated between the poles of a magnet. The magnetic field is perpendicular to the plane of the disc and parallel to its axis of rotation (Fig. 42.15). Brushes form sliding contacts with the axle and the rim at X and Y respectively.

In one revolution all the radii of the disc cut through the flux between X and Y. It is as if a single radius has cut through a magnetic field which passes through the whole of the area of the disc, and therefore the flux cut in one revolution $= B\pi\,(r_2{}^2 - r_1{}^2)$, where B is the magnetic flux density between X

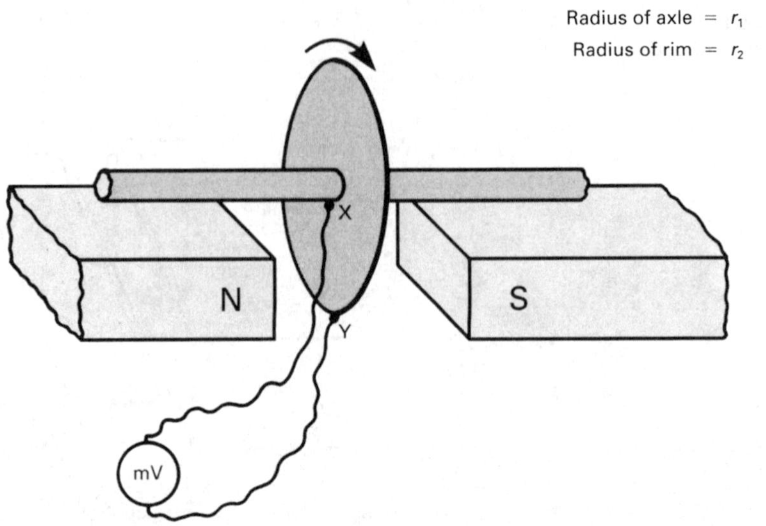

Fig. 42.15
Homopolar generator (Faraday's disc)

and Y. If the disc makes f revolutions in one second, then the rate of cutting of flux $= fB\pi(r_2^2 - r_1^2)$. Therefore by Faraday's law the induced EMF, is given by

$$E = fB\pi(r_2^2 - r_1^2)$$

If the angular velocity of the disc is constant, the induced EMF is also constant. The EMF is small (of the order of millivolts).

42.11 EDDY CURRENTS

Whenever a block of metal moves in a non-uniform magnetic field, or is sited in a changing magnetic field, EMFs are induced in it. These give rise to induced currents, **eddy currents**, which circulate within the body of the metal. The eddy currents follow low-resistance paths and therefore may be large, even though the induced EMFs are often small. As a result, eddy currents can produce quite considerable heating and magnetic effects.

The Heating Effect

Eddy current heating is made use of in the **induction furnace**. A water-cooled coil carrying high-frequency alternating current surrounds the sample to be heated. The rapidly changing magnetic field of the coil induces large eddy currents in those parts of the sample which are conducting – any insulators which are present are unaffected. The technique is particularly useful when a metal object is to be heated in close proximity to an insulator which cannot be allowed to reach such a high temperature as the metal.

The Magnetic Effect

In accordance with Lenz's law, eddy currents always flow in such a direction as to oppose the motion which has produced them. They therefore act as a brake on any solid metal object which is moving through a non-uniform magnetic field. The effect can be demonstrated by comparing the motions of two specially designed pendulums swinging between the poles of a large magnet (Fig. 42.16). Large eddy

Fig. 42.16
Demonstration of the effect of eddy currents

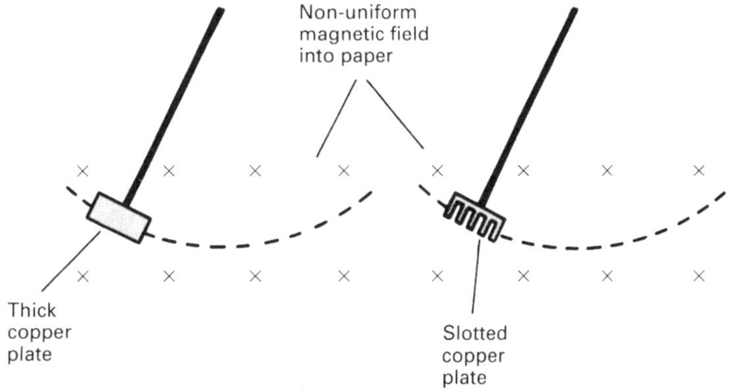

currents circulate in the solid plate and the motion very quickly ceases. Eddy currents induced in the slotted pendulum are free to circulate only in the narrow regions between the slits, so the motion is largely undamped and continues for some time.

42.12 GALVANOMETER DAMPING

The coil of a moving-coil meter comes to rest only after it has used up its energy in overcoming the damping forces which are present. If the coil is lightly damped, it overshoots its equilibrium position and then oscillates about it for a considerable time. If the damping is large, the coil creeps slowly to its equilibrium position. Ideally, a meter which is used to measure current or voltage (rather than charge) should be **critically damped.**

A meter is said to be critically damped when its coil reaches its equilibrium position as quickly as possible **without overshooting** (see Fig. 42.17).

Fig. 42.17
Types of damping in a galvanometer

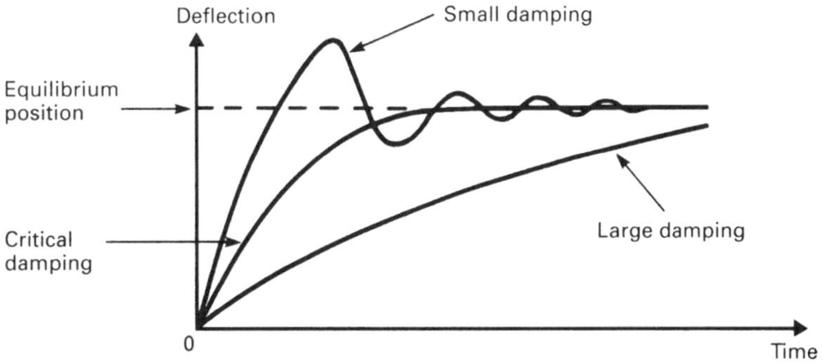

The coil of a pointer-type meter is usually wound on a metal frame. As the coil swings in the instrument's magnetic field, eddy currents circulate in the low-resistance circuit provided by the frame. These currents oppose the motion of the coil – a feature which is known as **electromagnetic damping** (or **eddy current damping**). The ideal design of frame is that which produces critical damping, in which case the movement of the instrument is said to be **dead-beat**.

The coil of a mirror-type galvanometer is usually wound on a non-conducting frame or is simply coated with glue to give mechanical strength. As a consequence, this type of instrument can be used ballistically (see section 42.20). When one of these instruments is used to measure current or voltage a suitably sized shunt is connected in parallel with the coil, and eddy currents flowing around the coil and shunt provide the necessary electromagnetic damping. In the Scalamp-type mirror galvanometer, commonly used in schools, these shunts are connected internally on all ranges except 'direct'. On the direct setting the damping, apart from that due to air resistance and internal friction in the suspension, is provided by the induced current which circulates through the coil via the external circuit. If the resistance of the external circuit is low, the damping may be excessive, in which case a large resistance should be connected in series with the instrument.

42.13 SELF-INDUCTION

A coil through which a current is flowing has an associated magnetic field. If, for any reason, the current changes, then so too does the magnetic flux and an EMF is induced in the coil. Since this EMF has been induced in the coil by a change in the current through the same coil, the process is known as **self-induction**. In accordance with Lenz's law, the EMF opposes the change that has induced it and it is therefore known as a back EMF. If the current is increasing, the back EMF opposes the increase; if the current is decreasing, it opposes the decrease.

The measure of the ability of a coil to give rise to a back EMF is known as the **self-inductance** of the coil. It is defined by

$$E = -L\frac{dI}{dt} \qquad [42.4]$$

where

E = the back EMF induced in the coil (V)

L = the self-inductance of the coil. The unit of self-inductance is the **henry** (H)

$\frac{dI}{dt}$ = the rate of change of current in the coil ($A\,s^{-1}$).

Notes

(i) The back EMF opposes the current change, and therefore the inclusion of the minus sign in equation [42.4] makes L a positive constant.

(ii) The value of L depends on the dimensions of the coil, the number of turns and the permeability of the core material.

(iii) If Φ is the flux through a coil of N turns and self-inductance L when it is carrying a current I, then from equations [42.1] and [42.4]

$$-\frac{d}{dt}(N\Phi) = -L\frac{dI}{dt}$$

$$\therefore \quad N\int_{\Phi=0}^{\Phi=\Phi}\frac{d\Phi}{dt}\,dt = L\int_{I=0}^{I=I}\frac{dI}{dt}\,dt$$

i.e.

$$N\Phi = LI \qquad [42.5]$$

Equation [42.5] provides an alternative definition of L.

(iv) Equation [42.4] is used to define **the henry**. Thus:

A coil has a self-inductance of one henry (H) if the back EMF in it is one volt when the current through it is changing at one ampere per second.

42.14 THE *L–R* DC CIRCUIT

Consider a circuit containing an iron-cored coil, a resistor of resistance R and a cell of EMF E (Fig. 42.18). The iron core causes the coil to have a high associated magnetic flux when it is carrying a current, and therefore a high self-inductance, L. A coil such as this is called an **inductor**. The resistance of the inductor, and the inductance of the resistor are assumed to be zero.

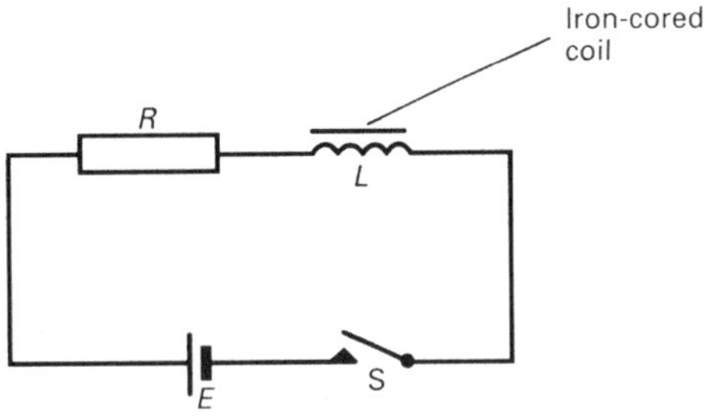

Fig. 42.18 DC circuit with an inductor and a resistor in series

When S is closed, the current build-up is opposed by the back EMF induced in the coil, and the current takes a considerable time to reach its equilibrium value (Fig. 42.19). The current build-up can be conveniently displayed by connecting an oscilloscope across the resistor.

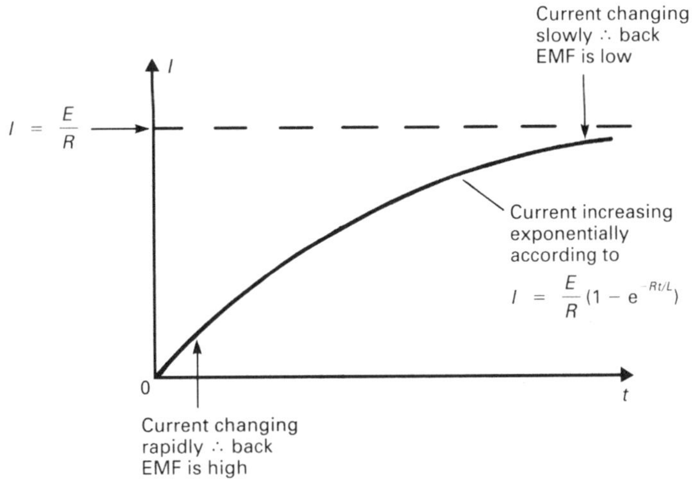

Fig. 42.19 Current build-up in the circuit of Fig. 42.18 after the switch is closed

If at time t the current in the circuit is I, the back EMF is given by equation [42.4] as $-L\,\mathrm{d}I/\mathrm{d}t$, and the resultant EMF in the circuit is $E - L\,\mathrm{d}I/\mathrm{d}t$. Therefore, by Kirchhoff's second rule

$$E - L\frac{\mathrm{d}I}{\mathrm{d}t} = IR \qquad [42.6]$$

The solution to equation [42.6] involves time and therefore depends on what is taken to be the starting point. Two situations are of particular interest.

(i) **Make**. In this case $I = 0$ when $t = 0$ and the solution to equation [42.6] can be shown to be

$$I = \frac{E}{R}(1 - \mathrm{e}^{-Rt/L}) \qquad [42.7]$$

(ii) **Break.** In this case, if we assume the current has already reached its equilibrium value, $I = E/R$ when $t = 0$, then the solution to equation [42.6] can be shown to be

$$I = \frac{E}{R} e^{-Rt/L} \qquad [42.8]$$

Equations [42.7] and [42.8] show that both the current build-up and the current decay are exponential functions of time. It follows from equation [42.8] that when $t = L/R$,

$$I = \frac{E}{R} e^{-1}$$

i.e. $$I = 0.368 \frac{E}{R}$$

L/R is known as the **time-constant** of the circuit, and is the time taken for the current to fall to $1/e$ (i.e. 36.8%) of its initial value (see Fig. 42.20). From equation [42.7], when $t = L/R$ the current has built up to 63.2% of its maximum value.

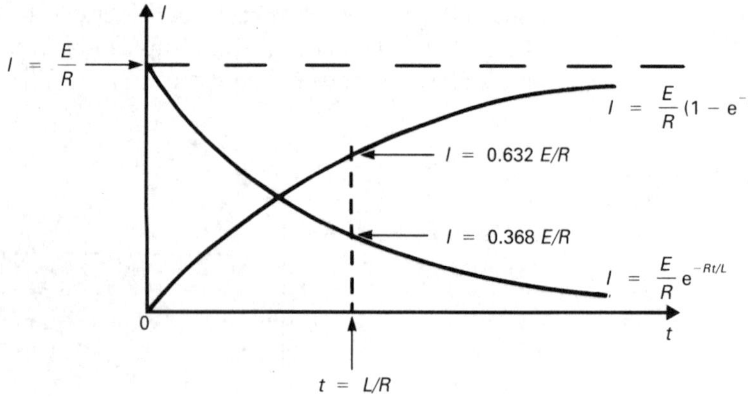

Fig. 42.20 Illustration of time-constant

Notes

(i) In a real circuit the inductor has some resistance and the resistor may well have some inductance, in which case L and R represent the total inductance and the total resistance respectively.

(ii) Equation [42.8] applies to a circuit in which the resistance is R. It does not apply, therefore, if the circuit of Fig. 42.18 is broken by opening S, because this makes the circuit resistance infinite.

The relevant condition can be achieved by first shorting out the cell and then removing it.

High Back EMF at Break

In Fig. 42.18, opening S causes the current to fall rapidly to zero and, in accordance with equation [42.4], a very high back EMF is induced. This EMF may be much higher than that of the supply and may cause a spark to jump between the switch contacts.

When a circuit containing a large inductor is made, the current has to do work against the back EMF. The work done in this way is stored in the form of the magnetic field of the coil. (The energy stored can be shown to be $LI^2/2$ – see section 42.15.) It is this energy which produces the spark when the current falls to zero and the magnetic field ceases to exist.

The circuit shown in Fig. 42.21 provides a convincing demonstration. The minimum PD required to ionize the neon and cause the lamp to strike is 60 V. When the switch is opened, the lamp flashes, showing that the back EMF is at least 60 V – very much greater than the supply PD of 2 V. Putting a suitable resistor in parallel with the coil before opening S prevents the lamp striking, because the current decay, and therefore the magnitude of the induced EMF, is limited by the presence of the resistor.

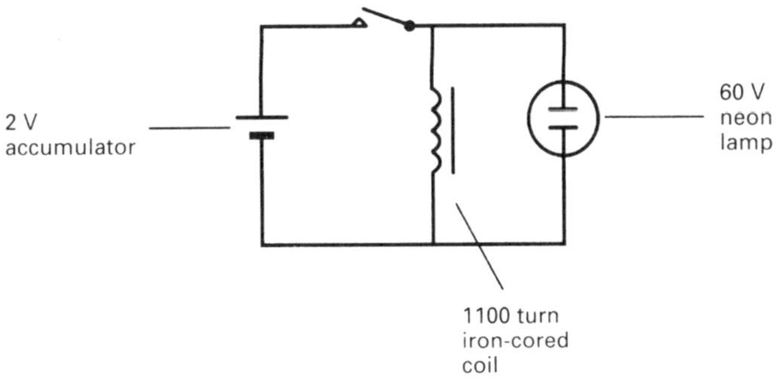

Fig. 42.21
Circuit to demonstrate the high back-EMF on breaking an inductive circuit

EXAMPLE 42.1

A coil with an inductance of 20 H and a resistance of 10 Ω is connected in series with a battery of EMF 12 V and a switch. What is: (a) the rate of change of current immediately after the switch is closed, (b) the final current, (c) the current after 3.0 s? (d) How long after the switch is closed will the current be 0.40 A?

Solution

(a) $$E - L\frac{dI}{dt} = IR \qquad \text{(equation [42.6])}$$

where $E = 12\,\text{V}$, $L = 20\,\text{H}$, $R = 10\,\Omega$ and, since the switch has just been closed, $I = 0$.

$$\therefore \quad 12 - 20\frac{dI}{dt} = 0$$

$$\therefore \quad \frac{dI}{dt} = \frac{12}{20} = 0.60$$

i.e. Initial rate of change of current $= 0.60\,\text{A s}^{-1}$

(b) When the current has reached its final value the inductance plays no part and $I = E/R$

$$\therefore \quad I = \frac{12}{10} = 1.2$$

i.e. Final current $= 1.2\,\text{A}$

(c) By equation [42.7]

$$I = \frac{E}{R}(1 - e^{-Rt/L})$$

$$\therefore \quad I = \frac{12}{10}(1 - e^{-10\times 3.0/20})$$

$$= 1.2(1 - e^{-1.5})$$

$$= 1.2(1 - 0.223) = 0.932$$

∴ Current after 3.0 s = 0.93 A

(d)

$$I = \frac{E}{R}(1 - e^{-Rt/L})$$

$$\therefore \quad 0.40 = \frac{12}{10}(1 - e^{-10t/20})$$

$$\therefore \quad 0.40 = 1.2 - 1.2e^{-0.5t}$$

$$\therefore \quad 1.2e^{-0.5t} = 0.8$$

$$\therefore \quad e^{-0.5t} = 0.667$$

$$\therefore \quad \ln(e^{-0.5t}) = \ln(0.667)$$

$$\therefore \quad -0.5t = -0.405 \qquad \therefore \quad t = 0.810$$

i.e. Current is 0.40 A after 0.81 s

QUESTIONS 42B

1. A coil with an inductance of 40 H and a resistance of 10 Ω is connected in series with a battery of EMF 9.0 V and a switch. What is: **(a)** the current, **(b)** the rate of change of current 2.0 s after closing the switch? **(c)** How long after closing the switch will the current reach 0.60 A?

42.15 ENERGY STORED IN AN INDUCTOR

Consider a coil of self-inductance L. Suppose that at time t the current in the coil is in the process of building up to its equilibrium value I_0 at a rate $\mathrm{d}I/\mathrm{d}t$. The magnitude, E, of the back EMF is given by equation [42.4] as

$$E = L\frac{\mathrm{d}I}{\mathrm{d}t}$$

The current works at a rate P in overcoming this back EMF, where

$$P = IE$$

and therefore

$$P = LI\frac{\mathrm{d}I}{\mathrm{d}t}$$

The work done δW in a small time interval δt is therefore given by

$$\delta W = LI\left(\frac{\mathrm{d}I}{\mathrm{d}t}\right)\delta t$$

The total work done, W, as the current increases from 0 to I_0 can be found by letting $\delta t \to 0$ and integrating. Thus,

$$W = \int_{I=0}^{I=I_0} LI \left(\frac{dI}{dt}\right) dt$$

i.e.

$$W = \int_0^{I_0} LI \, dI$$

i.e.

$$W = \tfrac{1}{2} LI_0^2 \qquad [42.9]$$

QUESTIONS 42C

1. A coil with a resistance of 6.0 Ω and an inductance of 30 mH is connected to a 12 V supply. What is the energy stored in the coil when the current has reached its equilibrium value?

42.16 NON-INDUCTIVE COILS

A coil which has been wound with wire which has been doubled back on itself, or which has been wound first one way and then the other, is non-inductive. There are equal and opposite currents in each section of the coil and therefore the coil has no magnetic field and no induced EMF.

The coils used in resistance boxes are non-inductively wound. This has two advantages:

(i) there is no delay in reaching equilibrium when, for example, balancing a Wheatstone bridge;

(ii) for any given alternating PD, the current through such a coil is not frequency-dependent (see section 43.4).

42.17 MUTUAL INDUCTION

If two coils are close together, then a changing current in one coil (the **primary**) sets up a changing magnetic field at the site of the other (the **secondary**) and so induces an EMF in it. The effect is known as mutual induction and the so-called **mutual inductance,** M, of the pair of coils is defined by

$$E_s = -M\frac{dI_p}{dt} \qquad [42.10]$$

where

E_s = the back EMF induced in the secondary coil (V)

$\frac{dI_p}{dt}$ = the rate of change of current in the primary ($A\,s^{-1}$)

M = the mutual inductance of the pair of coils. The unit of mutual inductance is the **henry** (H).

Notes (i) It can be shown (both theoretically and experimentally) that equation [42.10] gives the same value for M no matter which of a given pair of coils is taken to be the primary.

(ii) It can be shown that if all the flux produced by the primary passes through the secondary, then

$$M = \sqrt{L_p L_s}$$

where L_p and L_s are the respective self-inductances of the primary and the secondary.

(iii) If the rate of change of flux-linkage in the secondary is $d(N_s\Phi_s)/dt$ as a result of the current in the primary changing at a rate dI_p/dt, then by equations [42.10] and [42.1] the EMF E_s in the secondary is given by

$$E_s = -M\frac{dI_p}{dt}$$

and

$$E_s = -\frac{d}{dt}(N_s\Phi_s)$$

It follows that

$$N_s\Phi_s = MI_p$$

This equation provides an alternative definition of M and in some circumstances is useful in calculating M (see Example 42.2).

QUESTIONS 42D

1. The mutual inductance of a pair of coils is 0.36 H. Find the EMF induced across the secondary when the current in the primary is reduced from 3.5 A to zero at a steady rate over a period of 3.0 s.

EXAMPLE 42.2

A short coil of 20 turns is wound on a long air-cored solenoid which has 1000 turns per metre and a cross-sectional area of $1.2 \times 10^{-3}\,m^2$ (Fig. 42.22). Calculate the mutual inductance of the arrangement. ($\mu_0 = 4\pi \times 10^{-7}\,H\,m^{-1}$.)

Fig. 42.22 Diagram for Example 42.2

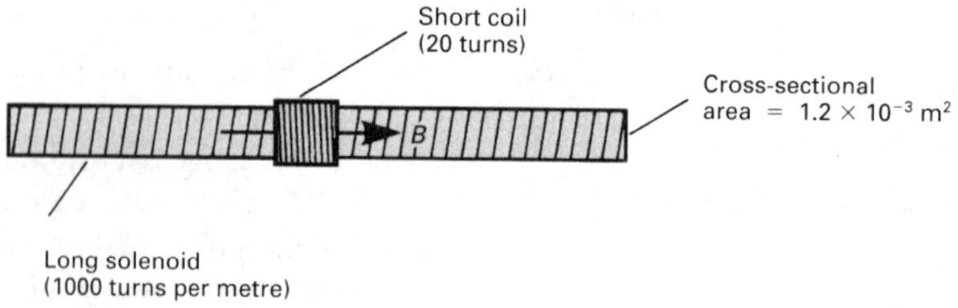

Solution

In order to calculate the mutual inductance we shall assume that the solenoid is carrying a current I_p, in which case the flux density B along the axis of the solenoid is given by equation [41.3] as

$$B = \mu_0 n I_p$$

where n is the number of turns per unit length of the solenoid. Thus

$$B = 4\pi \times 10^{-7} \times 1000 \times I_p$$

i.e. $$B = 4\pi \times 10^{-4} I_p$$

If we assume that the flux density is the same across the entire cross-section, then the flux Φ_s through the coil is given by

$$\Phi_s = 4\pi \times 10^{-4} I_p \times 1.2 \times 10^{-3}$$

i.e. $$\Phi_s = 4.8\pi \times 10^{-7} I_p$$

The coil has 20 turns and therefore the flux-linkage $N_s\Phi_s$ is given by

$$N_s\Phi_s = 20 \times 4.8\pi \times 10^{-7} I_p$$

i.e. $$N_s\Phi_s = 96\pi \times 10^{-7} I_p$$

From section 42.17

$$N_s\Phi_s = MI_p$$

where M is the mutual inductance. It follows that

$$M = 96\pi \times 10^{-7}$$

i.e. $$M = 3.0 \times 10^{-5}\,\text{H}$$

42.18 THE TRANSFORMER

The transformer is a device which makes use of mutual induction to produce a large alternating EMF from a small one, or a small alternating EMF from a large one. Fig. 42.23 shows a common design. Two coils, the **primary** and **secondary**, are wound, one on top of the other, on a soft iron core. There is no electrical connection between the coils but they are linked magnetically – the presence of the soft iron ensures that all the flux associated with one coil also passes through the other.

Fig. 42.23
Transformer

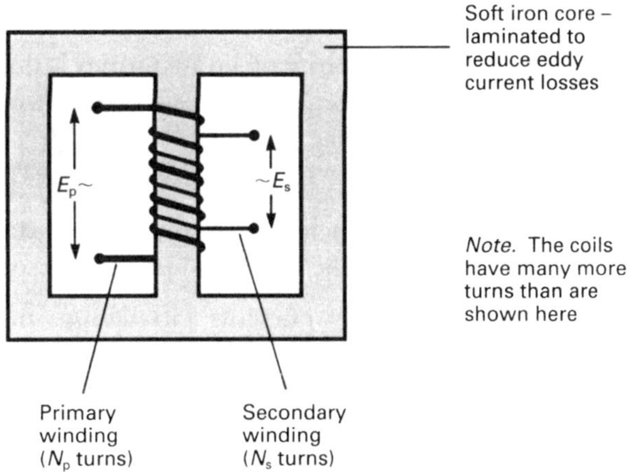

Suppose that an EMF, E_p, is applied to the primary coil. If at some instant the flux in the primary is Φ, then there will be a back EMF in it given by Neumann's equation (equation [42.1]) as $-\,\mathrm{d}(N_p\Phi)/\mathrm{d}t$, where N_p is the number of turns on the primary. Applying Kirchhoff's second rule to the primary circuit gives

$$E_p - \frac{\mathrm{d}}{\mathrm{d}t}(N_p\Phi) = IR$$

where I is the current in the primary and R is the resistance. In order to simplify the theory we shall assume that $R = 0$, in which case

$$E_p = N_p \frac{\mathrm{d}\Phi}{\mathrm{d}t} \qquad [42.11]$$

The flux through the primary also passes through the secondary and therefore the rate of change of flux in the secondary is also $\mathrm{d}\Phi/\mathrm{d}t$. It follows that there will be an EMF induced in the secondary. Its magnitude, E_s, is given by

$$E_s = N_s \frac{\mathrm{d}\Phi}{\mathrm{d}t} \qquad [42.12]$$

where N_s is the number of turns on the secondary. By equations [42.11] and [42.12]

$$\frac{E_s}{E_p} = \frac{N_s}{N_p} \qquad [42.13]$$

Notes (i) E_p must be an alternating EMF. If it were not, the flux in the primary would not change (except for a short time immediately after the EMF is first applied) and there would be no induced EMF in the secondary. The induced EMF (E_s) has the same frequency as the applied EMF (E_p).

(ii) If $N_s > N_p$, the transformer is called a **step-up transformer** because then $E_s > E_p$. A **step-down transformer** has $N_s < N_p$.

When a load (a resistance) is connected across the secondary, a current, I_s say, flows in the secondary. Suppose that the current in the primary is I_p. If the transformer is 100% efficient,

$$\text{Power output} = \text{Power input}$$

i.e. $$I_sE_s = I_pE_p$$

in which case, by equation [42.13]

$$\frac{E_s}{E_p} = \frac{I_p}{I_s} = \frac{N_s}{N_p} \qquad [42.14]$$

The efficiency of a transformer is defined by

$$\text{Efficiency} = \frac{\text{Power output}}{\text{Power input}}$$

The efficiencies of commercial transformers are very high – typically in the range 95–99%. The four most important sources of power loss are listed below.

(i) Eddy currents circulating in the soft iron core produce heating and therefore reduce the amount of power that can be transferred to the secondary. The core is **laminated,** i.e. made up of thin sheets of soft iron each separated from the next by a layer of insulating varnish. This very nearly eliminates eddy current heating.

(ii) Each time the direction of magnetization of the core is reversed, some energy is wasted in overcoming internal friction. This is known as **hysteresis loss** (see section 45.4), and it produces heating in the core. It is minimized by using special alloys (e.g. Permalloy) for the core material.

(iii) Some energy is dissipated as heat in the coils (I^2R). This is reduced to an acceptable level by using suitably thick wire. (The coil which has the smaller number of turns carries the larger current (equation [42.14]) and therefore is wound from thicker wire than the other.)

(iv) Some loss of energy occurs because a small amount of the flux associated with the primary fails to pass through the secondary.

42.19 TRANSMISSION OF ELECTRICAL ENERGY. THE NATIONAL GRID

Suppose that power is to be transmitted from a power station to a home or a factory. Since power is the product of current and voltage, a given amount of power can be transmitted either at high voltage and low current or at low voltage and high current. The cables that transmit the power have resistance, and therefore some of the power is bound to be wasted by producing heat in the cables as the current flows through them. If the resistance of the cables is R, the heat energy produced in time t when a current I is flowing through them is I^2Rt. Thus the amount of energy that is wasted is proportional to the square of the current in the cables. **The most efficient way to transmit power is therefore at high voltage and low current.** This is known as **high tension** transmission.

A second advantage of high voltage/low current transmission of electrical power is that low currents require thinner and therefore cheaper cables. A disadvantage is the high cost of the substantial insulation needed when employing high voltages.

The National Grid (Fig. 42.24) is the network of cables (transmission lines) which connects Britain's power stations to their consumers. Modern power stations generate alternating current at a precise frequency of 50 Hz and about 25 kV. This is stepped up, using transformers, to 275 kV or 400 kV for efficient transmission over long distances. It is subsequently stepped down by other transformers to 33 kV, 11 kV or 240 V before being supplied to the various consumers.

The National Grid uses AC rather than DC because the most efficient way to convert high voltages to low voltages and vice versa is by using transformers, and these cannot operate on DC.

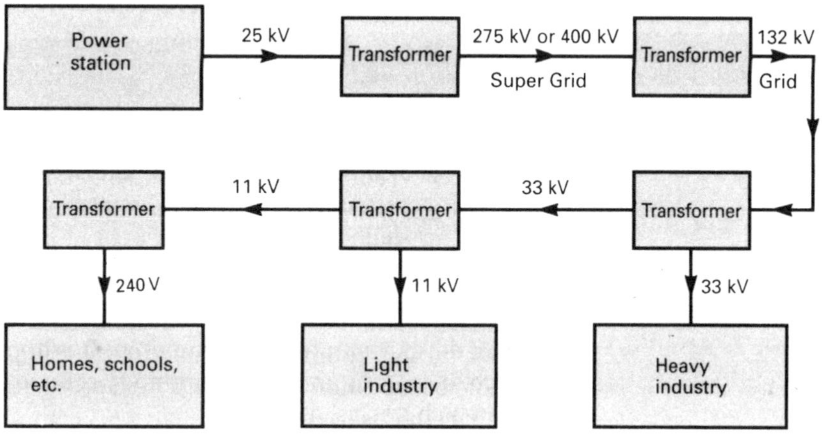

Fig. 42.24
The National Grid

Notes (i) The Heysham power station supplies 1320 MW of electricity to the National Grid at 400 kV – this corresponds to a current of 3300 A.

(ii) The CEGB uses multistrand aluminium cables reinforced with steel and with a resistance of 0.067 Ω/km.

QUESTIONS 42E

1. The total resistance of the cables connecting a power station to a small factory is 0.40 Ω. The power station output PD is 11 000 V and the current supplied to the factory is 100 A. Calculate: **(a)** the PD across the cables, **(b)** the PD at the factory, **(c)** the power wasted in heating the cables, **(d)** the power consumed by the factory, **(e)** the total power supplied by the power station, **(f)** the percentage of the power which is usefully consumed.

2. Suppose that the power station in question 1 were to output the same total amount of power (i.e. 1100 kW) but at 1100 V and 1000 A. What percentage of the power output would now be usefully consumed now?

42.20 THE BALLISTIC GALVANOMETER

A moving coil galvanometer which is being used to measure charge is called a **ballistic galvanometer**. If a galvanometer is to be used ballistically, two conditions need to be fulfilled.

(i) All of the charge being measured should be delivered to the instrument before its coil has moved appreciably. (The coil moves as a result of the momentary current that flows.)

(ii) The damping should be as small as possible. This is achieved by winding the coil on a non-conducting frame so that the only electromagnetic damping which is present (see section 42.12) is that due to the induced current which flows through both the coil and the external circuit. It may be necessary to include a large resistance in the circuit to limit this current, and so reduce the associated damping.

When the charge passes through the instrument its coil is deflected and then oscillates, with decreasing amplitude, about its initial (zero) position (see Fig. 42.25). It can be shown that the maximum angular deflection, θ_m, is proportional to the charge delivered, i.e.

$$\theta_m = aQ \qquad [42.15]$$

where

a = a constant of proportionality known as the **charge sensitivity** of the instrument. It may be expressed in rad C^{-1}, or, in terms of the scale deflection, as mm C^{-1}.

The value of a can be found by carrying out a subsidiary calibration experiment. Its value depends on the amount of damping present, and this depends on the circuit in which the galvanometer is being used. It is important, therefore, that the circuit in which the galvanometer is calibrated produces the same amount of damping as that in which it is used.

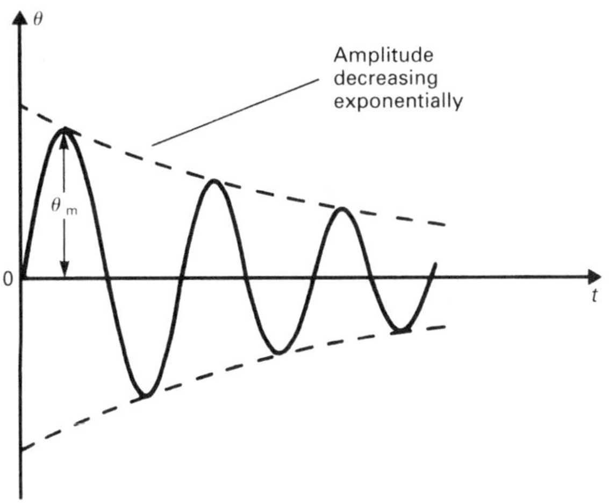

Fig. 42.25
Damping in a ballistic galvanometer

42.21 COMPARISON OF CAPACITANCES BY USING A BALLISTIC GALVANOMETER

Suppose that it is required to compare the capacitances, C_1 and C_2, of two capacitors. With X connected to Y (Fig. 42.26), the capacitor whose capacitance is

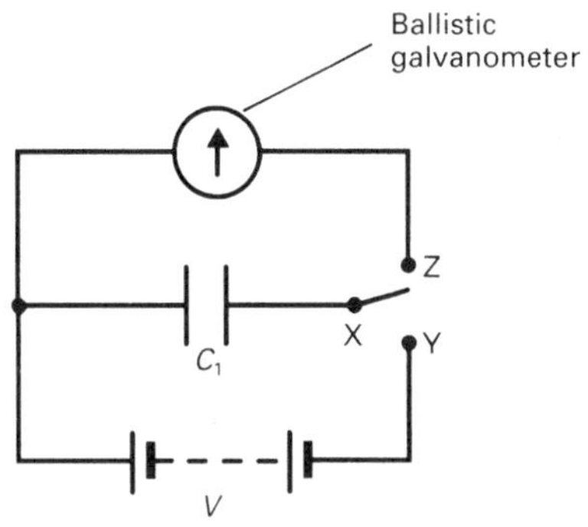

Fig. 42.26
Ballistic galvanometer used to compare capacitances (***C*₁** subsequently replaced by ***C*₂**)

C_1 is charged to a PD V, so that the charge on the capacitor is (by equation [40.1]) VC_1. Then with X connected to Z, the capacitor is discharged through the ballistic galvanometer. If this produces a maximum deflection θ_1, then by equation [42.15]

$$\theta_1 = aVC_1 \qquad [42.16]$$

If repeating this procedure with the second capacitor produces a maximum deflection θ_2, then

$$\theta_2 = aVC_2$$

Dividing gives

$$\frac{\theta_1}{\theta_2} = \frac{C_1}{C_2}$$

42.22 MEASUREMENT OF MAGNETIC FLUX DENSITY BY USING A BALLISTIC GALVANOMETER AND A SEARCH COIL

Consider a small flat coil which has N turns of area A and whose plane is perpendicular to a magnetic field of flux density B. The flux-linkage is therefore NAB. Suppose now that the coil is removed from the field to a point where the field is zero. The flux-linkage changes from NAB to zero, and therefore an EMF is induced in the coil, the magnitude of which, E, is given by Neumann's equation (equation [42.1]) as

$$E = -\frac{d}{dt}(N\Phi)$$

where $N\Phi$ is the flux-linkage at some time, t. It should be noted that $d(N\Phi)/dt$ is not necessarily constant, and that E is the instantaneous EMF at time t. If the search coil is connected to a ballistic galvanometer and the total resistance of the circuit is R, the instantaneous current, I, through the ballistic galvanometer is given by

$$I = \frac{E}{R}$$

i.e. $$I = -\frac{1}{R}\frac{d}{dt}(N\Phi)$$

The current is equal to the rate of flow of charge dQ/dt through the galvanometer, and therefore

$$\frac{dQ}{dt} = -\frac{1}{R}\frac{d}{dt}(N\Phi)$$

Integrating with respect to time gives

$$\int_0^t \frac{dQ}{dt}\,dt = -\frac{1}{R}\int_0^t \frac{d(N\Phi)}{dt}\,dt$$

When $t = 0$, $Q = 0$ and $N\Phi = NAB$. When $t = t$ the search coil is completely out of the field and therefore all of the charge has passed through the galvanometer, i.e. $Q = Q$ and $N\Phi = 0$. Therefore

$$\int_0^Q dQ = -\frac{1}{R}\int_{NAB}^0 d(N\Phi)$$

i.e. $$\Big[Q\Big]_0^Q = -\frac{1}{R}\Big[N\Phi\Big]_{NAB}^0$$

i.e. $$Q = \frac{NAB}{R} \qquad [42.17]$$

Note that the charge delivered to the galvanometer does not depend on how long it takes to remove the search coil from the field. However, the proper use of a ballistic galvanometer requires that all the charge passes through it before its coil moves appreciably (see section 42.20), and therefore the search coil has to be pulled out of the field as quickly as possible. If the subsequent maximum deflection is θ_m, then from equations [42.15] and [42.17]

$$\theta_m = a\frac{NAB}{R}$$

where a is the charge sensitivity of the galvanometer. Rearranging this equation gives

$$B = \frac{\theta_m R}{aNA} \qquad [42.18]$$

If N and A are known, B can be calculated once θ_m has been measured and the charge sensitivity, a, has been determined. (If a is determined by the method described below, it is not necessary to know the value of R.)

Determination of Charge Sensitivity

The ballistic galvanometer is calibrated by using a standard mutual inductance in the circuit of Fig. 42.27. (It is this same circuit, with S open, that is used to determine θ_m.)

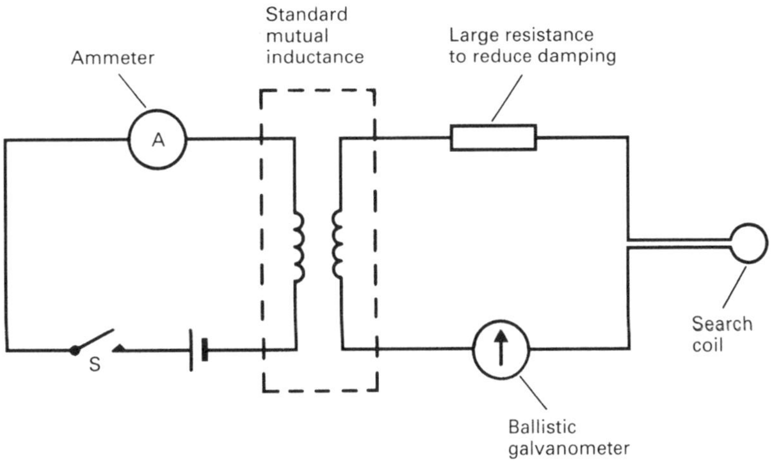

Fig. 42.27 Circuit to calibrate a ballistic galvanometer

S is closed so that after a short time a known, steady current, I_p, passes through the primary of the mutual inductance. Suppose that this produces a flux-linkage of $N_s\Phi'$ with the secondary. On opening S, the current falls to zero and the flux-linkage with the secondary also falls to zero. If the resulting charge through the galvanometer is Q', then by comparison with equation [42.17]

$$Q' = \frac{N_s\Phi'}{R}$$

where R is the total resistance of the galvanometer circuit, and has the same value as it has in equation [42.18].

It can be shown (see section 42.17) that

$$N_s\Phi' = MI_p$$

and therefore

$$Q' = \frac{MI_p}{R} \qquad [42.19]$$

If the maximum deflection of the ballistic galvanometer as a result of opening S is θ'_m, then from equation [42.15]

$$\theta'_m = aQ'$$

Therefore, from equation [42.19]

$$\theta'_{\mathrm{m}} = \frac{aMI_{\mathrm{p}}}{R}$$

Substituting for a in equation [42.18] gives

$$B = \frac{MI_{\mathrm{p}}}{NA}\frac{\theta_{\mathrm{m}}}{\theta'_{\mathrm{m}}}$$

A standard mutual inductance is used and therefore M is known, I_{p} is measured by using the ammeter in the primary circuit.

Note If the ballistic galvanometer is being used on open circuit, equation [42.16] can be used to determine the charge sensitivity, provided that C_1 is known. This method of calibration should not be employed when the galvanometer is being used with a search coil, because such a circuit does not have infinite resistance.

CONSOLIDATION

There is an induced EMF whilst the flux through a coil is changing or whilst a straight conductor is cutting across field lines.

The induced EMF is directly proportional to the rate of change of flux-linkage or to the rate of cutting of flux. (**Faraday's law**)

The direction of the induced EMF is such that the current which it causes to flow (or would flow in a closed circuit) opposes the change which is producing it. (**Lenz's law**)

$$E = -\frac{\mathrm{d}}{\mathrm{d}t}(N\Phi) \qquad \text{(for a coil)}$$

$$E = -N\frac{\mathrm{d}\Phi}{\mathrm{d}t} \qquad \text{(for a straight conductor)}$$

The direction of the induced current is found by using Lenz's law in the case of a coil and by using Fleming's right-hand (dynamo) rule in the case of a straight conductor.

The coil of a meter which is **critically damped** reaches its equilibrium position **as quickly as possible without overshooting**.

A galvanometer used to measure current or voltage should be critically damped; one used to measure charge (i.e. a ballistic galvanometer) should have as little damping as possible.

Self Inductance

$$E = -L\frac{\mathrm{d}I}{\mathrm{d}t} \qquad \text{(This equation defines self-inductance and the henry)}$$

The *L–R* DC Circuit

$$E - L\frac{\mathrm{d}I}{\mathrm{d}t} = IR \qquad \begin{pmatrix}\text{Make: } I = 0 \text{ when } t = 0 \\ \text{Break: } I = E/R \text{ when } t = 0\end{pmatrix}$$

$$I = \frac{E}{R}(1 - \mathrm{e}^{-Rt/L}) \qquad \text{(Make)}$$

$$I = \frac{E}{R}\,\mathrm{e}^{-Rt/L} \qquad \text{(Break)}$$

$$\text{Time constant} = \frac{L}{R}$$

Mutual Inductance

$$E_\mathrm{s} = -M\frac{\mathrm{d}I_\mathrm{p}}{\mathrm{d}t}$$

43

ALTERNATING CURRENTS

43.1 INTRODUCTION

If the polarity of an EMF changes with time, it is known as an alternating EMF. The current that such an EMF causes to flow repeatedly changes its direction and is known as **alternating current** (AC).

The most commonly encountered type of alternating EMF varies sinusoidally with time, like that generated by the mains, and can be represented by

$$E = E_0 \sin \omega t \qquad [43.1]$$

where

E = the value of the EMF at time t (V)

E_0 = the **peak value** (i.e. the maximum value) of E (V)

ω = the **angular frequency** of the supply (rad s^{-1}).
(**Note**. $\omega = 2\pi f$, where f is the frequency of the supply in hertz (Hz).)

The corresponding alternating current, I, is given in terms of its maximum value, I_0, as

$$I = I_0 \sin \omega t \qquad [43.2]$$

Note that the current has the same frequency as the EMF that produces it.

43.2 ROOT MEAN SQUARE VALUES

If an alternating current flowing in a circuit produces the same heating effect as a direct current of (say) 3 A flowing in the same circuit, then the **effective value** of the alternating current is also 3 A.

In general, the **effective value** of an alternating current is equal to that direct current which results in the same expenditure of energy under the same conditions.

The energy, W, supplied in time, t, by an alternating current, I, flowing through a resistance, R, is (by equation [36.17]) equal to the product of t and the average value of I^2R, i.e.

$$W = (I^2R)_{\text{avge}}\, t$$

Since R is constant, this becomes

$$W = (I^2)_{\text{avge}} Rt \qquad [43.3]$$

If (anticipating the result) the effective value of I is denoted by I_{RMS}, then W is equal to the energy supplied by a steady current of magnitude I_{RMS} flowing through a resistance R for time t, i.e.

$$W = (I_{\text{RMS}})^2 Rt \qquad [43.4]$$

Therefore, by equations [43.3] and [43.4]

$$(I_{\text{RMS}})^2 Rt = (I^2)_{\text{avge}} Rt$$

i.e. $$I_{\text{RMS}} = \sqrt{(I^2)_{\text{avge}}}$$

i.e. $$I_{\text{RMS}} = \sqrt{\text{average value of } I^2} \qquad [43.5]$$

Thus, **the effective value (denoted by I_{RMS}) is the root mean square (RMS) value** and the reason for the notation is now clear. Equation [43.5] holds for any alternating current but the relationship between I_{RMS} and the peak value, I_0, depends on the nature of the AC. Two cases are of particular interest.

Sinusoidal AC

Refer to Fig. 43.1. It can be shown that

$$I_{\text{RMS}} = I_0/\sqrt{2} = 0.707\,I_0$$

and

$$E_{\text{RMS}} = E_0/\sqrt{2} = 0.707\,E_0$$

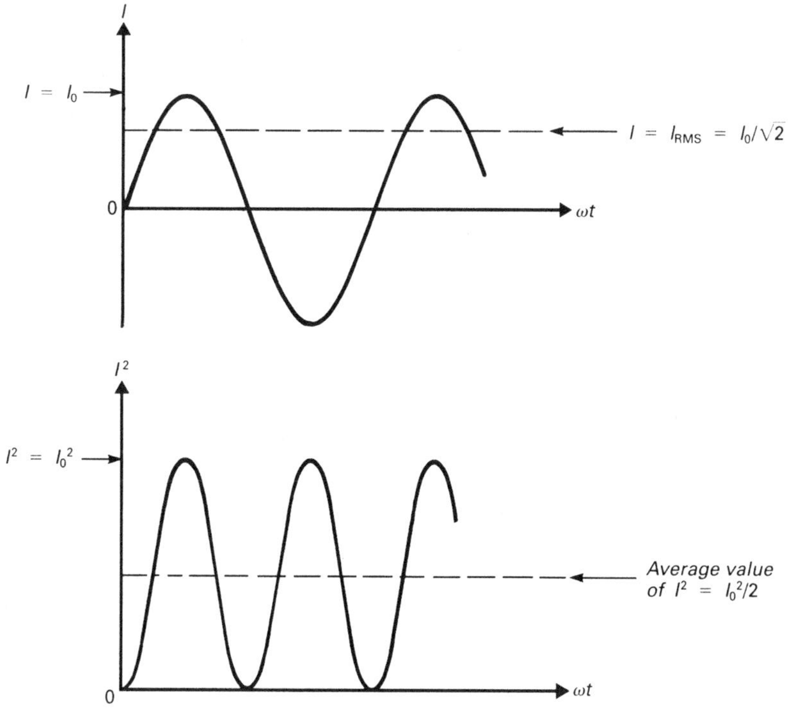

Fig. 43.1
To illustrate the relationship between sinusoidal alternating current and its RMS value

Square-wave AC

Refer to Fig. 43.2. In this case I^2 has the constant value of $I_0{}^2$, and therefore

$$\text{Average value of } I^2 = I_0{}^2$$

Fig. 43.2
Square wave alternating current

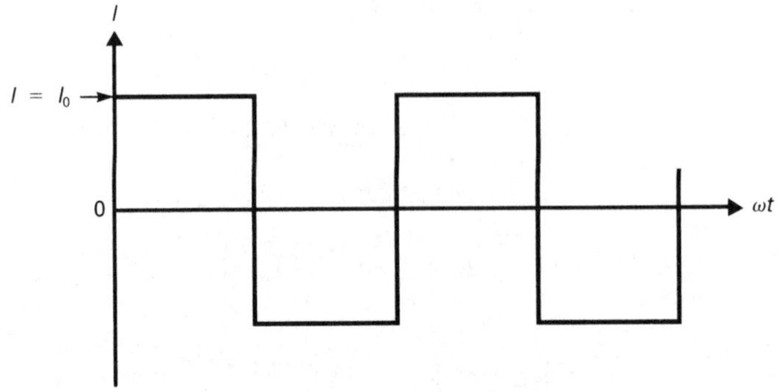

Therefore, from equation [43.5]

$$I_{RMS} = I_0$$

Similarly

$$E_{RMS} = E_0$$

EXAMPLE 43.1

Find the RMS value of the alternating current, *I*, shown in Fig. 43.3.

Fig. 43.3
Diagram for Example 43.1

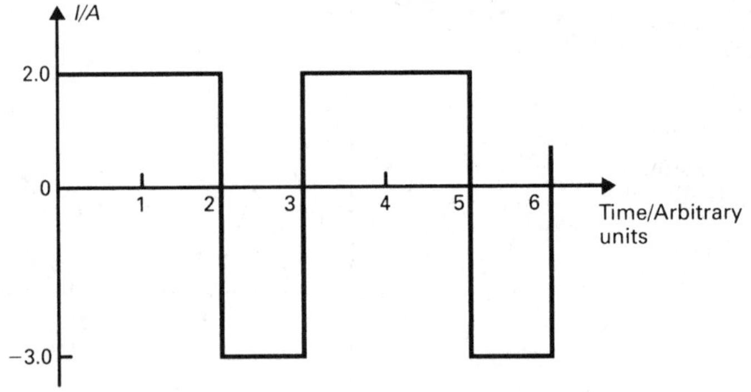

Solution

Consider the current variation over one complete cycle.

Interval	I/A	I^2/A^2
0–1	2.0	4.0
1–2	2.0	4.0
2–3	−3.0	9.0

$$\text{Average value of } I^2 = \frac{4.0 + 4.0 + 9.0}{3} = 5.67$$

$$\therefore \quad \text{RMS value of current} = \sqrt{5.67} = 2.4\,\text{A}$$

QUESTIONS 43A

1. Find the value of the RMS current in the following cases:
   (a) a sinusoidally varying current with a peak value of 4.0 A,
   (b) a square wave current which has a constant value of 4.0 A for the first 3 ms and −2.4 A for the next 2 ms of each 5 ms cycle,
   (c) an alternating current which has the same effect as a steady DC current of 2.4 A,
   (d) a 240 V RMS supply driving current through a 16 Ω resistor.

43.3 AC THROUGH A PURE INDUCTANCE

Consider an inductor of inductance L and zero resistance (i.e. $R = 0$) connected across an alternating supply. (An inductor which has zero resistance is called a pure inductance.) An alternating current ($I = I_0 \sin \omega t$) flows through the inductor and sets up a changing magnetic flux. This induces a back EMF given by equation [42.4] as $-L\,\mathrm{d}I/\mathrm{d}t$ at some time t. Suppose that the value of the applied PD at time t is V.

By Kirchhoff's second rule $\Sigma E = \Sigma IR$ and therefore, since $R = 0$,

$$V - L\frac{\mathrm{d}I}{\mathrm{d}t} = 0$$

i.e.
$$V - L\frac{\mathrm{d}}{\mathrm{d}t}(I_0 \sin \omega t) = 0$$

i.e.
$$V - \omega L I_0 \cos \omega t = 0$$

Writing

$$V_0 = \omega L I_0 \qquad [43.6]$$

gives

$$V = V_0 \cos \omega t$$

Thus, V is a cosine function whereas I is a sine function, and therefore there is a phase difference of $\pi/2$ radians between the current and the applied PD (see Fig. 43.4). The voltage reaches its maximum value before the current, i.e. the voltage leads the current. Thus:

In a purely inductive circuit the applied PD leads the current by $\pi/2$ radians.

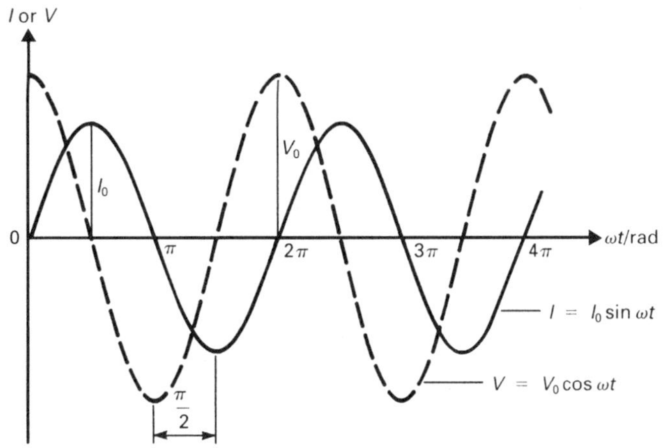

Fig. 43.4 Phase difference between PD and current for a pure inductance

43.4 INDUCTIVE REACTANCE X_L

The opposition (note: we do not use the word 'resistance') which an inductor offers to current flow is called its **reactance**, X_L, and is defined by

$$X_L = \frac{V_0}{I_0} \qquad [43.7]$$

where V_0 is the maximum PD across the inductor and I_0 is the maximum current through it.

Therefore, from equation [43.6]

$$X_L = \frac{\omega L I_0}{I_0}$$

i.e. $$X_L = \omega L \qquad [43.8]$$

Notes (i) If ω is in radians per second and L is in henrys, X_L is in ohms.

(ii) $V_{RMS} = V_0/\sqrt{2}$ and $I_{RMS} = I_0/\sqrt{2}$, and therefore

$$\frac{V_{RMS}}{I_{RMS}} = \frac{V_0}{I_0}$$

This provides an alternative definition of X_L, namely

$$X_L = \frac{V_{RMS}}{I_{RMS}} \qquad [43.9]$$

(iii) X_L does not equal V/I. V and I are the instantaneous values of the applied PD and the current and their ratio ranges from 0 to ∞.

QUESTIONS 43B

1. A 240 V RMS supply with a frequency of 50 Hz causes an RMS current of 3.0 A to flow through an inductor which can be taken to have zero resistance. Calculate: **(a)** the reactance of the inductor, **(b)** the inductance of the inductor.

43.5 EXPLANATION OF THE BEHAVIOUR OF AN INDUCTOR IN AN AC CIRCUIT

We need to explain:

(i) why there is a phase difference between the applied PD and the current through the inductor;

(ii) why the reactance, X_L, is proportional to ωL. (The fact that in equation [43.8] X_L is equal to ωL is due to the choice of units.)

The back EMF induced in an inductor is greatest when the current is varying at its maximum rate, and this occurs when the current is zero. Since the magnitude of the back EMF is equal to that of the applied PD, it follows that the magnitude of the applied PD is a maximum when the current is zero. When the current has reached its maximum value its rate of change is momentarily zero. As a consequence, the back EMF and the applied PD are zero at such times. Thus, when $I = 0$, $V = V_0$, and when $I = I_0$, $V = 0$, i.e. current and voltage are out of phase by $\pi/2$ radians.

Suppose that the RMS value of the applied PD is kept constant but that its frequency is increased. Since the RMS PD is constant, so too is the average back EMF, and since $E \propto \mathrm{d}I/\mathrm{d}t$, the average rate of change of current must also be constant. The increase in frequency, though, causes the current to change direction more frequently, and the only way this can happen, if the average value of $\mathrm{d}I/\mathrm{d}t$ is to be constant, is for the average value of the current to decrease. (A current of 10 A falling to zero in 2 s is changing at the same rate as a current of 5 A falling to zero in 1 s.) Thus, when V_{RMS} (and L) is constant, $I_{\mathrm{RMS}} \propto 1/\omega$.

Suppose now that the inductance, L, is increased whilst, as before, the average value of the applied PD is kept constant. The increase in L tends to increase the average back EMF, and therefore the average value of $\mathrm{d}I/\mathrm{d}t$ must decrease in proportion. The only way this can happen at constant frequency is if the average current decreases. Thus, when V_{RMS} and ω are constant, $I_{\mathrm{RMS}} \propto 1/L$.

Combining these results gives

$$I_{\mathrm{RMS}} \propto \frac{1}{\omega L}$$

when V_{RMS} is constant. Therefore

$$X_L = \frac{V_{\mathrm{RMS}}}{I_{\mathrm{RMS}}} \propto \omega L$$

43.6 AC THROUGH A PURE CAPACITANCE

Consider a capacitor of capacitance C connected across an alternating supply, so that the PD V across the capacitor at time t is given by

$$V = V_0 \sin \omega t$$

The charge Q on the capacitor at time t is given by equation [40.1] as

$$Q = CV$$

i.e. $$Q = CV_0 \sin \omega t$$

Differentiating gives

$$\frac{\mathrm{d}Q}{\mathrm{d}t} = \omega C V_0 \cos \omega t$$

The current, I, in the circuit is equal to the rate of flow of charge, and therefore

$$I = \omega C V_0 \cos \omega t$$

Writing

$$I_0 = \omega C V_0 \qquad [43.10]$$

gives

$$I = I_0 \cos \omega t$$

Thus:

In a purely capacitative circuit the applied PD lags the current by $\pi/2$ radians.

43.7 CAPACITATIVE REACTANCE X_C

The opposition which a capacitor offers to current flow is called its **reactance,** X_C, and is defined by

$$X_C = \frac{V_0}{I_0} = \frac{V_{RMS}}{I_{RMS}} \qquad [43.11]$$

Substituting for I_0 from equation [43.10] gives

$$X_C = \frac{V_0}{\omega C V_0}$$

i.e.

$$X_C = \frac{1}{\omega C} \qquad [43.12]$$

Note If ω is in radians per second and C is in farads, X_C is in ohms.

QUESTIONS 43C

1. A 240 V RMS supply with a frequency of 50 Hz is connected across a 4700 pF capacitor. Find: **(a)** the reactance of the capacitor, **(b)** the RMS current in the circuit.
2. When a capacitor is connected across a 200 Hz alternating supply with a peak value of 100 V the maximum amount of charge on the capacitor in each cycle is 500 μC. Find: **(a)** the reactance of the capacitor, **(b)** the peak value of the current in the circuit.

43.8 EXPLANATION OF THE BEHAVIOUR OF A CAPACITOR IN AN AC CIRCUIT

Current cannot, of course, actually flow through a capacitor; the current flows only in the circuit on either side of it. If a capacitor is connected across a steady DC supply, current flows only at make or break, i.e. for the short periods of time during which the charge on the capacitor is building up to its maximum value or decaying to zero. A capacitor in an AC circuit is continually being charged and discharged and therefore there is always a current in such a circuit.

It follows that if a capacitor is connected across a supply which has a steady DC component and an AC component, **the capacitor blocks the DC component but passes the AC**.

The phase difference between the applied PD and the current in an AC circuit is accounted for as follows. When the PD is at its maximum value the capacitor is fully charged and therefore the rate of flow of charge, i.e. the current, is zero. When the PD of the supply starts to fall, charge has to flow off the capacitor in order that the PD across it can stay the same as that of the supply. The more rapidly the PD falls, the more rapidly the charge flows. Since the rate at which a sinusoidally varying PD falls is greatest as it reaches zero, the current has its maximum value when the PD is zero. Thus, when $V = V_0$, $I = 0$, and when $V = 0$, $I = I_0$, i.e. current and voltage are out of phase by $\pi/2$ radians.

For a capacitor, $Q = CV$. In an AC circuit, therefore, for any given value of V_{RMS}, the charge that has to flow on to the plates of the capacitor or off them is proportional to C. It follows, therefore, that when V_{RMS} is constant, the root mean square current, I_{RMS}, is proportional to C.

Increasing the frequency of the supply, whilst keeping V_{RMS} and C constant, increases the rate at which the capacitor is charged and discharged and therefore increases the current, i.e. I_{RMS} is proportional to ω.

Combining these results gives

$$I_{RMS} \propto \omega C$$

when V_{RMS} is constant. Therefore

$$X_C = \frac{V_{RMS}}{I_{RMS}} \propto \frac{1}{\omega C}$$

43.9 ROTATING VECTOR REPRESENTATION

In Fig. 43.5 ON is the projection of the line OP on Oy. If OP rotates with a constant angular velocity ω, the length of ON varies sinusoidally according to ON = OP $\sin \omega t$. If OP is regarded as a vector of magnitude y_0, then the projection of OP on Oy represents the instantaneous value y (say) of a sinusoidally alternating quantity whose peak value is y_0. This is known as the **rotating vector** (or **phasor**) representation. The representation is particularly useful for dealing with two or more sinusoidally varying quantities which have the same frequency but different phases, and we make use of it in section 43.10.

Fig. 43.5
Rotating vector representation of AC

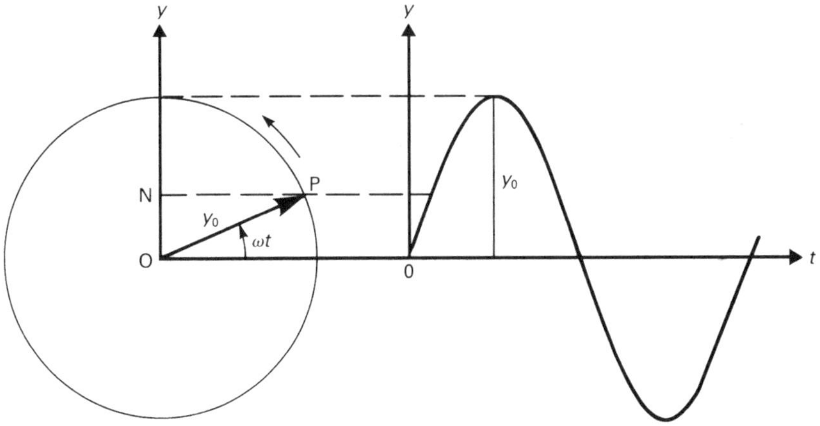

43.10 *R*, *L* AND *C* IN SERIES

Consider a circuit in which a resistance R, an inductance L and a capacitance C are in series, and across which there is an alternating PD V (Fig. 43.6). If at some time t the PDs across R, L and C are V_R, V_L and V_C respectively, then

$$V = V_R + V_L + V_C \tag{43.13}$$

The circuit components are in series, and therefore at any instant there is the same current, I, flowing through each of R, L and C, and this is equal to the current being supplied by the source at the same time. V_R is in phase with I, V_L leads I by 90° and V_C lags I by 90°, and therefore V_L leads V_R by 90° and V_C lags V_R by 90°

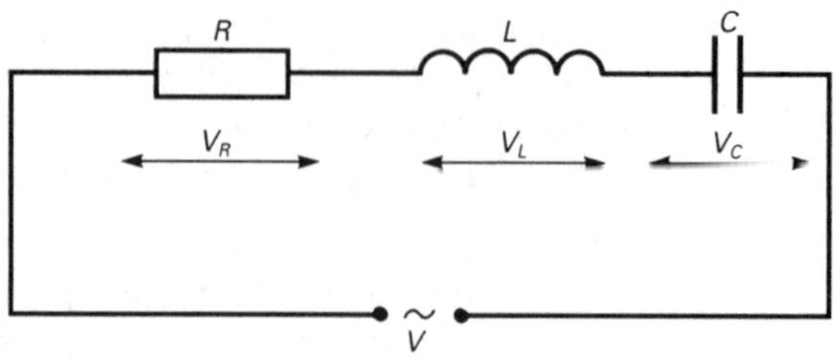

Fig. 43.6
AC circuit with a resistor, an inductor and a capacitor in series

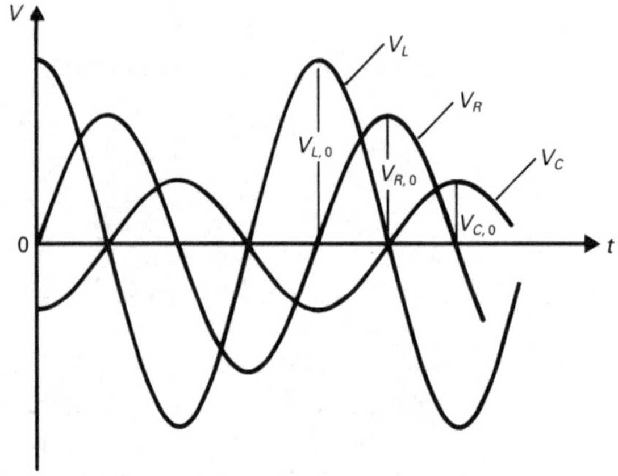

Fig. 43.7
To illustrate the phase relationships of the PDs across the components in the circuit of Fig. 43.6

(see Fig. 43.7). In particular, when V_R has its maximum value $V_{R,0}$, $V_L \neq V_{L,0}$ and $V_C \neq V_{C,0}$. When working in terms of peak values, therefore, it is necessary to take account of the phase differences. Thus, although the instantaneous value of the PD is equal to the algebraic sum of V_R, V_L and V_C (see equation [43.13]), the peak value of the applied PD is equal to the vector sum of the peak values $V_{R,0}$, $V_{L,0}$ and $V_{C,0}$. It is useful to represent the relative magnitudes and phases of the PDs by a vector diagram (the phasor representation).

If we let $V_R = 0$ when $t = 0$, then the relevant diagram for $t = 0$ is as shown in Fig. 43.8(a). After some time t, each vector has turned through an angle ωt and the situation is shown in Fig. 43.8(b). The instantaneous PDs V_R, V_L and V_C are the projections of their respective peak values $V_{R,0}$, $V_{L,0}$ and $V_{C,0}$ on Oy. As the vectors rotate, so does their resultant, V_0, the peak value of the applied PD. The projection of V_0 on Oy is the instantaneous value V of the total (i.e. the supply) potential difference. V_0 is the vector sum of $V_{R,0}$, $V_{L,0}$ and $V_{C,0}$, and therefore

$$V_0{}^2 = V_{R,0}{}^2 + (V_{L,0} - V_{C,0})^2$$

At time t the projection of V_0 on Oy is $V_0 \sin(\omega t + \phi)$, i.e.

$$V = V_0 \sin(\omega t + \phi)$$

where

$$V_0 = \sqrt{V_{R,0}{}^2 + (V_{L,0} - V_{C,0})^2} \qquad [43.14]$$

The opposition which the circuit as a whole offers to current flow is called the **impedance**, Z, of the circuit. It is defined by

$$Z = \frac{V_0}{I_0} \qquad [43.15]$$

The unit of impedance is the ohm.

Fig. 43.8
Vector representation of the PDs in the circuit of Fig. 43.6

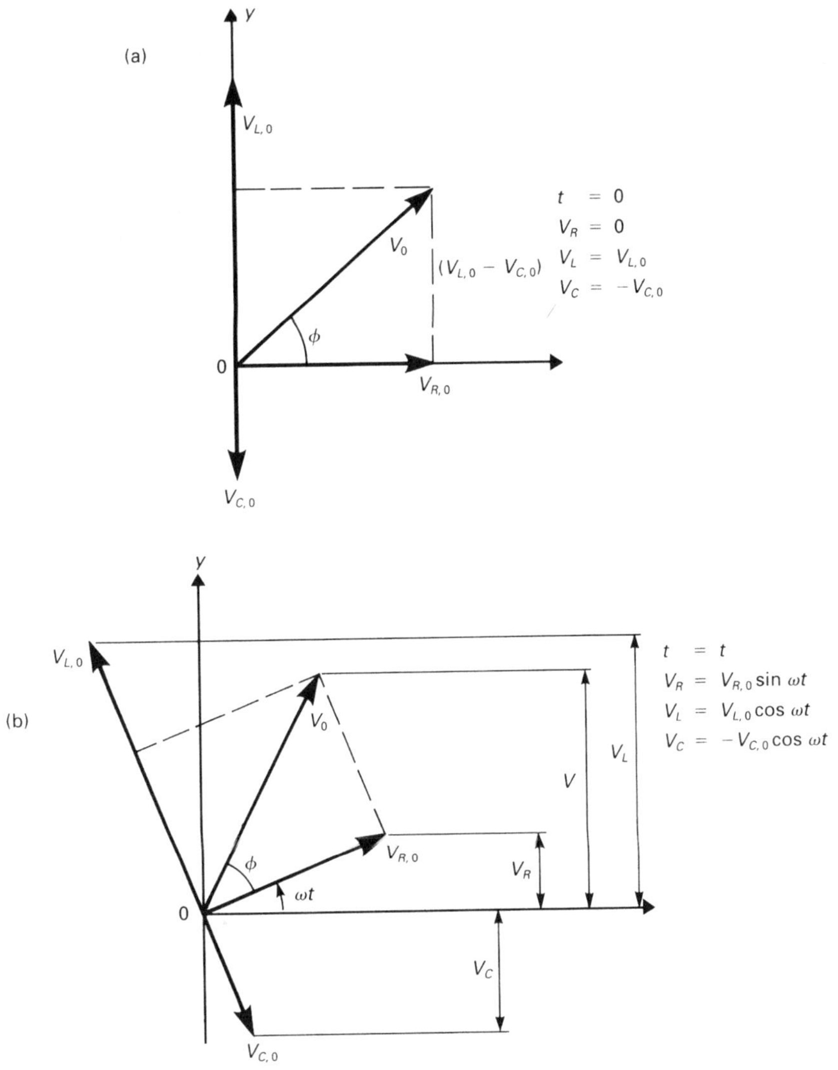

Since

$$V_{R,0} = I_0 R \qquad V_{L,0} = I_0 X_L \qquad V_{C,0} = I_0 X_C$$

equation [43.14] can be written as

$$V_0 = I_0 \sqrt{R^2 + (X_L - X_C)^2}$$

Therefore, by equation [43.15]

$$Z = \sqrt{R^2 + (X_L - X_C)^2} \qquad [43.16]$$

From Fig. 43.8

$$\tan \phi = \frac{V_{L,0} - V_{C,0}}{V_{R,0}}$$

i.e. $$\tan \phi = \frac{I_0 X_L - I_0 X_C}{I_0 R}$$

i.e. $$\tan \phi = \frac{X_L - X_C}{R} \qquad [43.17]$$

where ϕ is called the **phase angle**, and is the angle by which the applied PD, V, leads V_R, i.e. the angle by which V leads I, and, in particular, the angle by which V_0 leads I_0.

Note Equation [43.14] can be restated in terms of the corresponding RMS voltages. Thus,

$$V_{RMS} = \sqrt{(V_{R,RMS})^2 + (V_{L,RMS} - V_{C,RMS})^2}$$

EXAMPLE 43.2

A coil which has an inductance of 0.4 H and a resistance of 5 Ω is in series with a resistor whose resistance is 25 Ω. The pair are connected across a 200 V RMS supply alternating at $50/\pi$ Hz. Calculate:

(a) the RMS current,

(b) the RMS voltage across the resistor,

(c) the RMS voltage across the coil.

Solution

$$\begin{aligned}\text{Inductive reactance } X_L &= \omega L \\ &= 2\pi f L \\ &= 2\pi \times (50/\pi) \times 0.4 \\ &= 40\,\Omega\end{aligned}$$

$$\begin{aligned}\text{Total resistance } R &= 5 + 25 \\ &= 30\,\Omega\end{aligned}$$

In general (equation [43.16])

$$Z = \sqrt{R^2 + (X_L - X_C)^2}$$

Here, $R = 30\,\Omega$, $X_L = 40\,\Omega$, $X_C = 0$, and therefore

$$Z = \sqrt{30^2 + 40^2}$$

i.e. $$Z = 50\,\Omega$$

Since $Z = V_{RMS}/I_{RMS}$ (this follows from equation [43.15]),

$$I_{RMS} = 200/50$$

i.e. $$I_{RMS} = 4\text{ A}$$

$$\begin{aligned}\text{RMS PD across resistor} &= I_{RMS} \times \text{resistance} \\ &= 4 \times 25 \\ &= 100\text{ V}\end{aligned}$$

$$\begin{aligned}\text{RMS PD across coil} &= I_{RMS} \times \text{impedance of coil} \\ &= 4 \times \sqrt{5^2 + 40^2} \\ &= 4 \times \sqrt{1625} \\ &= 161.2\text{ V}\end{aligned}$$

Alternatively

PD across resistive component of coil $= 4 \times 5 = 20\,\text{V}$

PD across reactive component of coil $= 4 \times 40 = 160\,\text{V}$

$$\text{Total PD across coil} = \text{Vector sum} = \sqrt{20^2 + 160^2} = 161.2\,\text{V}$$

QUESTIONS 43D

1. A 47.0 μF capacitor and a resistor of resistance 30.0 Ω are connected in series. The pair are connected across a 50 Hz supply with a peak value of 100 V. Find: **(a)** the reactance of the capacitor, **(b)** the impedance of the circuit, **(c)** the peak value of the current in the circuit, **(d)** the peak value of the PD across the capacitor, **(e)** the peak value of the PD across the resistor.

2. A resistor of resistance 120 Ω, a capacitor of capacitance 22.0 μF and a coil with a resistance of 10 Ω and an inductance of 300 mH are connected in series with a 100 Hz alternating supply. Find: **(a)** the impedance of the circuit, **(b)** the angle by which the applied PD leads the current.

43.11 POWER IN AC CIRCUITS

No power is absorbed by either inductors or capacitors over a complete cycle,* and therefore **the average power absorbed by both inductors and capacitors is zero.**

A capacitor absorbs energy during that part of the cycle that the charge on its plates is increasing, but returns the energy to the source in the subsequent section of the cycle as the charge falls to zero. Similarly, an inductor absorbs energy as its magnetic field increases but returns the energy to the supply when the field collapses.

It follows that in a circuit in which there is resistance, inductance and capacitance, the total average power consumed, $\overline{P}$, is equal to that dissipated in the resistance, i.e.

$$\overline{P} = I_{\text{RMS}}{}^2 R \qquad [43.18]$$

The RMS voltage $V_{R,\text{RMS}}$ across the resistance is given by

$$V_{R,\text{RMS}} = I_{\text{RMS}} R$$

and therefore equation [43.18] can be rewritten as

$$\overline{P} = I_{\text{RMS}} V_{R,\text{RMS}} \qquad [43.19]$$

But

$$\frac{V_{R,\text{RMS}}}{V_{\text{RMS}}} = \frac{V_{R,0}}{V_0}$$

*A mathematical justification of this statement is given in section 43.12.

and from Fig. 43.8

$$\frac{V_{R,0}}{V_0} = \cos\phi$$

$$\therefore \quad \frac{V_{R,\text{RMS}}}{V_{\text{RMS}}} = \cos\phi$$

i.e. $V_{R,\text{RMS}} = V_{\text{RMS}} \cos\phi$

Substituting for $V_{R,\text{RMS}}$ in equation [43.19] gives

$$\overline{P} = I_{\text{RMS}} V_{\text{RMS}} \cos\phi \qquad [43.20]$$

Eliminating I_{RMS} between equations [43.18] and [43.20] gives

$$\overline{P} = \frac{{V_{\text{RMS}}}^2}{R} \cos^2\phi \qquad [43.21]$$

Notes

(i) $\cos\phi$ is called the **power factor** of the circuit. From Fig. 43.8 $\cos\phi = V_{R,0}/V_0 = (I_0 R/I_0 Z)$

i.e. $\cos\phi = R/Z$ [43.22]

(ii) Equations [43.18], [43.20] and [43.21] are analogous to the DC expressions $P = I^2R$, $P = IV$ and $P = V^2/R$ in which, in particular, V has been replaced by $V_{\text{RMS}} \cos\phi$. This arises because the applied PD leads the current by ϕ and $V_{\text{RMS}} \cos\phi$ can be thought of as the component of V_{RMS} which is in phase with I_{RMS}.

EXAMPLE 43.3

A coil which has an inductance of 0.4 H and negligible resistance is in series with a resistor whose resistance is 120 Ω. The pair are connected across a 100 V RMS supply alternating at $200/\pi$ Hz. Calculate:

(a) the total impedance of the circuit,

(b) the power factor,

(c) the phase angle,

(d) the average power.

Solution

$$\begin{aligned}\text{Inductive reactance } X_L &= \omega L \\ &= 2\pi f L \\ &= 2\pi \times (200/\pi) \times 0.4 \\ &= 160\,\Omega\end{aligned}$$

$$\text{Resistance } R = 120\,\Omega$$

In general, the impedance Z is given (by equation [43.16]) as

$$Z = \sqrt{R^2 + (X_L - X_C)^2}$$

Here, $R = 120\,\Omega$, $X_L = 160\,\Omega$, $X_C = 0$, and therefore

$$Z = \sqrt{120^2 + 160^2}$$

i.e. $$Z = 200\,\Omega$$

The power factor, $\cos\phi$, is given by $\cos\phi = R/Z$, and therefore

$$\cos\phi = \frac{120}{200} = 0.6$$

i.e. Power factor $= 0.6$

The phase angle ϕ is therefore $\cos^{-1} 0.6 = 53^\circ$ (approx.).

The average power consumed $\overline{P}$ is given by

$$\overline{P} = \frac{{V_{\text{RMS}}}^2}{R}\cos^2\phi$$

$$\therefore \quad \overline{P} = \frac{(100)^2(0.6)^2}{120}$$

i.e. $$\overline{P} = 30\,\text{W}$$

Note The phase angle can also be calculated from $\tan\phi = X_L/R$ (equation [43.17] with $X_C = 0$), and $\overline{P}$ can be calculated from $\overline{P} = {I_{\text{RMS}}}^2 R$ (equation [43.18]) or from $\overline{P} = I_{\text{RMS}} V_{\text{RMS}} \cos\phi$ (equation [43.20]). The reader is advised to confirm this by calculation.

QUESTIONS 43E

1. A $20.0\,\mu$F capacitor and a $100\,\Omega$ resistor are connected in series with a 240 V RMS 50 Hz supply. Calculate: **(a)** the reactance of the capacitor, **(b)** the impedance of the circuit, **(c)** the RMS current, **(d)** the average power.

43.12 MATHEMATICAL TREATMENT OF THE POWER ABSORBED BY RESISTANCES, INDUCTANCES AND CAPACITANCES

Resistance

If $I = I_0 \sin\omega t$, then $V = V_0 \sin\omega t$ and therefore the instantaneous power P being absorbed is given by

$$P = (I_0 \sin\omega t)(V_0 \sin\omega t)$$

i.e. $$P = I_0 V_0 \sin^2\omega t$$

Thus, P varies as shown in Fig. 43.9(a) and is non-zero over a complete cycle.

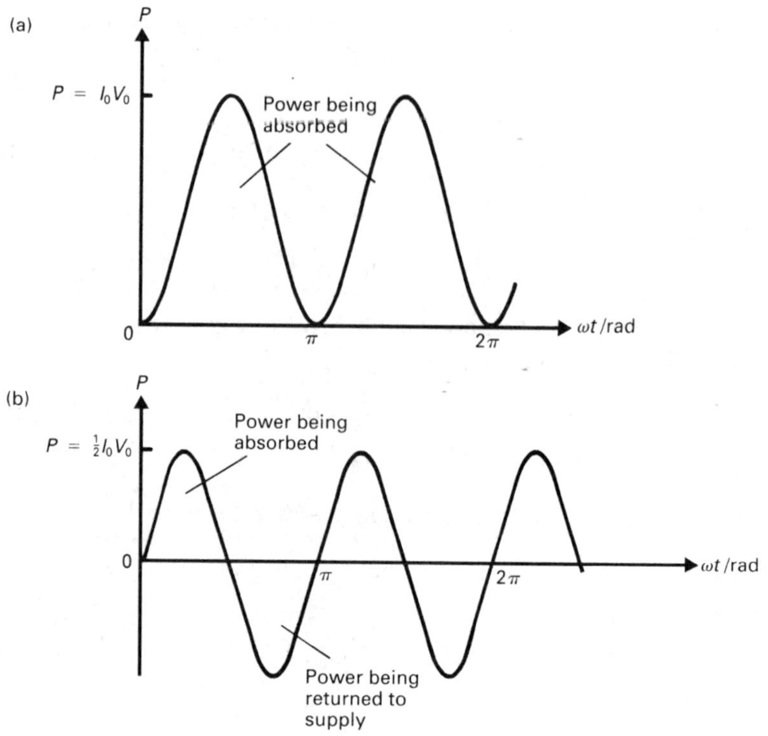

Fig. 43.9
Power as a function of time in the circuit of Fig. 43.6: (a) for the resistor, (b) for the inductor

Inductance

If $I = I_0 \sin \omega t$, then $V = V_0 \cos \omega t$ (see section 43.3) and therefore the instantaneous power P being absorbed is given by

$$P = (I_0 \sin \omega t)(V_0 \cos \omega t)$$

i.e. $$P = \tfrac{1}{2} I_0 V_0 \sin 2\omega t$$

Thus, P varies as shown in Fig. 43.9(b) and is zero over a complete cycle. Note that the angular frequency of the power variation is 2ω – twice that of the applied PD.

Capacitance

As with an inductance, the current and voltage are out of phase by 90° and the power variation is proportional to $\sin 2\omega t$, i.e. P varies as shown in Fig. 43.9(b) and is zero over a complete cycle.

43.13 RESONANCE IN *R–L–C* SERIES CIRCUITS

Consider the circuit of Fig. 43.6. The RMS current I_{RMS} is given by

$$I_{RMS} = V_{RMS}/Z$$

where V_{RMS} is the RMS applied PD and Z is the impedance of the circuit. Therefore, by equation [43.16]

$$I_{RMS} = \frac{V_{RMS}}{\sqrt{R^2 + (X_L - X_C)^2}} \qquad [43.23]$$

Both X_L and X_C depend on frequency and therefore the RMS current depends on the frequency f of the supply (Fig. 43.10). The frequency, f_0, which causes I_{RMS} to be a maximum (for a given value of V_{RMS}, R, L and C) is called the **resonant frequency** of the circuit. It is clear from equation [43.23] that I_{RMS} is a maximum when

$$X_L = X_C$$

i.e. $$2\pi f_0 L = \frac{1}{2\pi f_0 C}$$

i.e. $$f_0 = \frac{1}{2\pi\sqrt{LC}} \qquad [43.24]$$

At resonance

$$X_L = X_C$$

$\therefore$ $$I_0 X_L = I_0 X_C$$

i.e. $$V_{L,0} = V_{C,0}$$

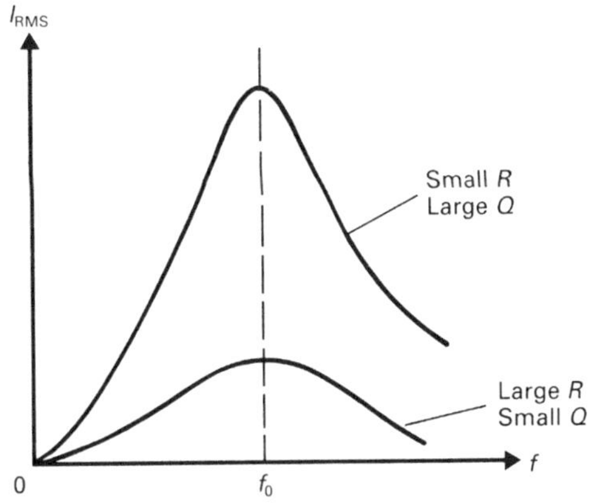

Fig. 43.10
Variation of I_{RMS} with frequency in the circuit of Fig. 43.6

Thus, at resonance, the peak values of the PDs across L and C are equal, and therefore, since V_L and V_C are always out of phase by 180°, at every instant the total PD across L and C is zero. It follows, therefore, that at resonance the applied PD is equal to the PD across R (see Fig. 43.11), and the applied PD and the current are in phase.

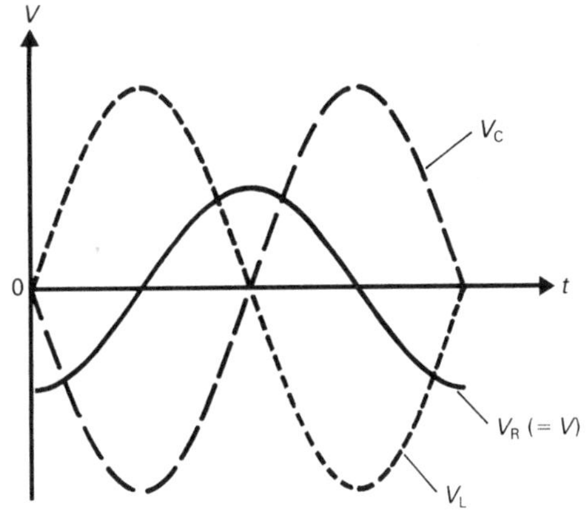

Fig. 43.11
Conditions at resonance in the circuit of Fig. 43.6

Note that $V_{R,0} = I_0 R$, $V_{L,0} = I_0 X_L$ and $V_{C,0} = I_0 X_C$, and therefore if X_L and X_C are much larger than R, $V_{L,0}$ and $V_{C,0}$ are much larger than the peak value of the applied PD ($V_{R,0}$). The **Q-factor** (or **quality factor**) of the circuit is defined by

$$Q = \frac{V_{L,0}}{V_0} = \frac{X_L}{R}$$

It can be shown that a circuit which has a high Q-factor gives rise to a sharp resonance.

Series resonance circuits in which C is variable are used to tune radio receivers (see Fig. 43.12). Radio waves induce currents of many different frequencies in the aerial coil, and these induce currents of the same frequencies in L. By altering C (a variable air capacitor) one can 'tune' the circuit to resonate at the frequency of the desired signal. The currents due to the unwanted signals are negligibly small in comparison.

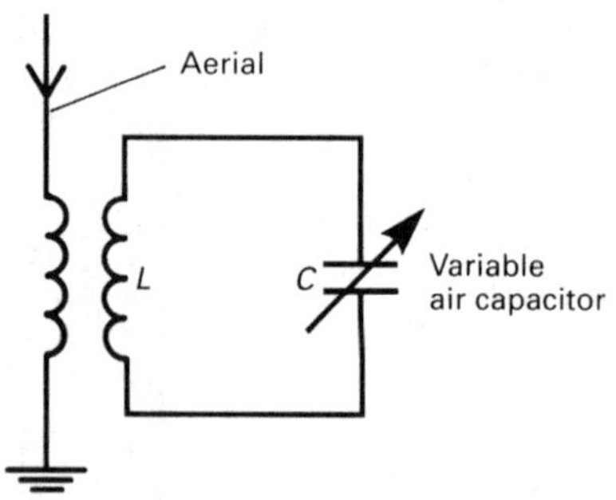

Fig. 43.12
Resonance circuit for tuning a radio receiver

In a circuit of this type, R is due only to the resistance of the wire forming the coil and that of the connecting wires. It is likely, therefore, to be very much less than X_L, in which case the Q-factor of the circuit is high and the circuit is highly selective.

The energy stored by the capacitor is associated with the charge on its plates and therefore with the PD across it. The energy stored by the inductor is associated with its magnetic field and therefore with the current flowing through it. When the PD across the capacitor is a maximum the current is zero, and therefore at such times the energy of the system is stored as the electric field of the capacitor, none is stored in the inductor. A quarter of a cycle later, the current is a maximum and the PD across the capacitor is zero so that the energy is now stored entirely as the magnetic field of the inductor. Thus, the energy continually passes back and forth between the electric field and the magnetic field. This may be compared with the continual interchange of kinetic energy and potential energy of an oscillating pendulum.

In order to maintain the oscillation of a pendulum, energy has to be supplied to offset that lost through friction; in the R–L–C series circuit energy has to be supplied to offset that dissipated as heat in the resistive component of the circuit.

QUESTIONS 43F

1. A 47 μF capacitor and a 2.0 mH inductor with a resistance of 100 Ω are connected in series with a 50 V RMS supply. What is: **(a)** the resonant frequency of the circuit, **(b)** the average power consumed at this frequency? (Note, you should be able to do part **(b)** in your head!)

2. Classify the following statements concerning an R–L–C series circuit as **A**, **B** or **C** where:

   A = true only at resonance,
   B = always true,
   C = never true.

   (a) V_R in phase with I,
   (b) V in phase with I,
   (c) $Z \geqslant R$,
   (d) $Z = R$,
   (e) $Z < R$,
   (f) magnitude of V_L = magnitude of V_C for the whole of every cycle,
   (g) $V_{L.0} = V_{C.0}$,
   (h) $V_R = V$,
   (i) V_L and V_C out of phase by π radians,
   (j) $\overline{P} = I_{RMS}\ V_{R.RMS}$,
   (k) $\overline{P} = I_{RMS}\ V_{RMS}$,
   (l) power factor = 1,
   (m) V_L in phase with I.

43.14 THE THERMOCOUPLE METER

A thermocouple meter is shown in Fig. 43.13. The alternating current to be measured passes through the fine wire XY. The 'hot' junction of a thermocouple is attached to XY at Z. The current heats XY and a thermoelectric EMF is generated at Z. This produces a direct current which is measured by the moving-coil milliammeter. The milliammeter reading depends on the temperature of the thermocouple junction and therefore on the RMS value of the alternating current. An evacuated bulb surrounds Z to shield it from draughts.

Fig. 43.13
Thermocouple meter

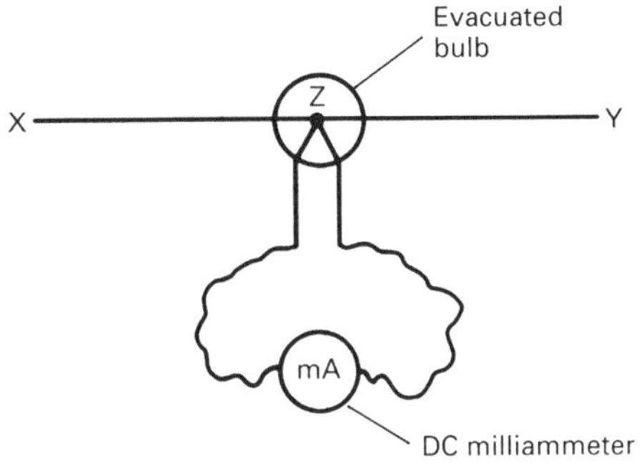

43.15 THE RECTIFIER METER

A moving-coil ammeter in series with a rectifier (see section 44.1) can be used to measure alternating currents (Fig. 43.14(a)). The current passing through the meter is varying but unidirectional (Fig. 43.14(b)). The meter cannot (except for very low frequencies) respond quickly enough to follow the current variation and it registers the mean value of the current. The calibration is such that the value of the RMS current is displayed.

Fig. 43.14
Rectifier meter

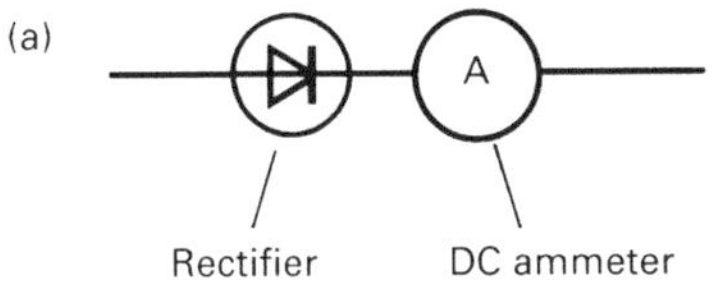

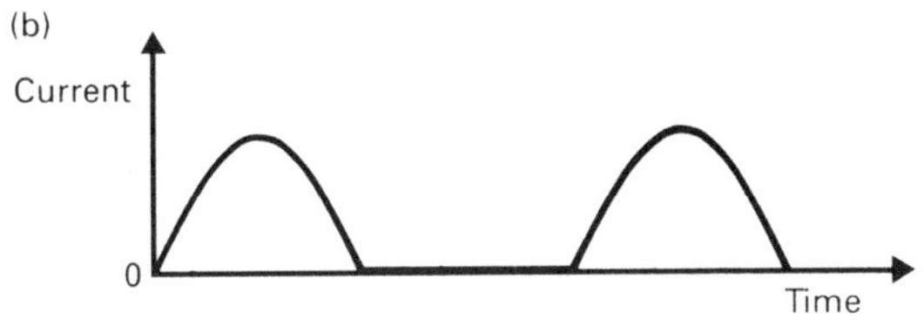

CONSOLIDATION

The effective value of an alternating current is the steady DC current which produces the same effect. It is always equal to the square root of the average value of the square of the current, i.e. it is the root mean square current, I_{RMS}.

Sinusoidal AC

$$I_{RMS} = \frac{I_0}{\sqrt{2}} \qquad E_{RMS} = \frac{E_0}{\sqrt{2}}$$

Purely Inductive Circuit

V leads I by $\pi/2$ radians

$$X_L = \frac{V_0}{I_0} = \frac{V_{RMS}}{I_{RMS}} \qquad \text{(Definitions of } X_L\text{)}$$

where V_0 and V_{RMS} are respectively the maximum PD and the RMS PD across the inductor. When there are circuit components other than an inductor present the symbols $V_{L,0}$ and $V_{L,RMS}$ have been used.

$$X_L = \omega L = 2\pi f L$$

Purely Capacitative Circuit

I leads V by $\pi/2$ radians

$$X_C = \frac{V_0}{I_0} = \frac{V_{RMS}}{I_{RMS}} \qquad \text{(Definitions of } X_C\text{)}$$

where V_0 and V_{RMS} are respectively the maximum PD and the RMS PD across the capacitor.

$$X_C = \frac{1}{\omega C} = \frac{1}{2\pi f C}$$

R–L–C Series Circuit

V_R is always in phase with I

$$V_0 = \sqrt{V_{R,0}^{\,2} + (V_{L,0} - V_{C,0})^2} = \sqrt{V_{R,RMS}^{\,2} + (V_{L,RMS} - V_{C,RMS})^2}$$

$$Z = \frac{V_0}{I_0} \qquad \text{(Definition of } Z\text{)}$$

$$Z = \sqrt{R^2 + (X_L - X_C)^2}$$

$$\tan\phi = \frac{X_L - X_C}{R} \qquad \cos\phi = \frac{R}{Z}$$

Power is consumed only by the resistive component of the circuit.

$$\overline{P} = I_{\mathrm{RMS}}{}^2 R = I_{\mathrm{RMS}} V_{\mathrm{RMS}} \cos\phi = \frac{V_{\mathrm{RMS}}{}^2}{R} \cos^2\phi$$

$$V_{\mathrm{RMS}} \cos\phi = V_{R,\mathrm{RMS}}$$

Power factor $= \cos\phi$

Phase angle $= \phi$ (the angle by which the applied PD leads the current)

R–L–C Series Circuit at Resonance

(i) $X_L = X_C$

(ii) $f_0 = \dfrac{1}{2\pi\sqrt{LC}}$

(iii) Magnitude of V_L = magnitude of V_C (at every instant). In particular, $V_{L,0} = V_{C,0}$ and $V_{L,\mathrm{RMS}} = V_{C,\mathrm{RMS}}$

44

RECTIFICATION

44.1 INTRODUCTION

Alternating current can be converted to direct current (i.e. **rectified**) by making use of devices which conduct appreciable amounts of current in one direction only. Such devices are called **rectifiers** and include:

(i) thermionic diodes,

(ii) metal rectifiers,

(iii) semiconductor diodes.

A rectifier is said to be **forward-biased** when it is connected to a supply in such a way that it conducts. If connected the other way round, the rectifier is **reverse-biased** (see Fig. 44.1). The current–voltage curve of a typical rectifier is shown in Fig. 44.2.

Fig. 44.1
Rectifier, allowing appreciable current in one direction only

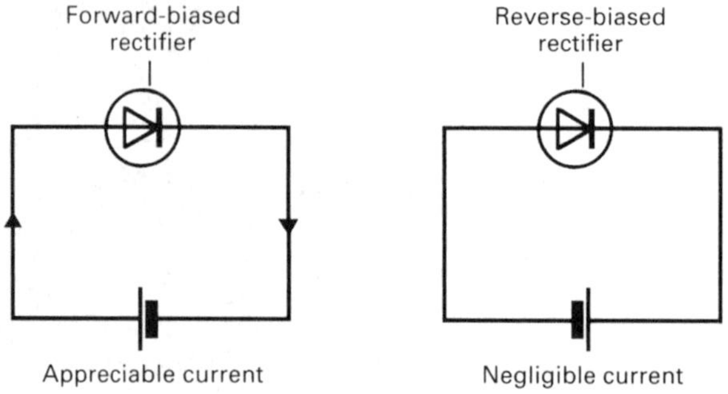

Note. The arrowhead in the rectifier symbol indicates the direction in which current can flow through the rectifier

Fig. 44.2
Current–voltage curve of a rectifier

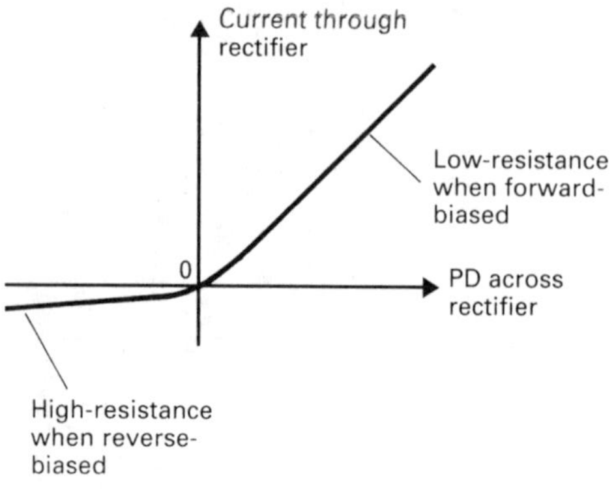

44.2 HALF-WAVE RECTIFICATION

Refer to Fig. 44.3. The rectifier conducts only during the half of the cycle which makes X positive (see Fig. 44.4). Although the output is pulsating, it is unidirectional, i.e. direct current. (**Note**. The output is exactly sinusoidal, only if the positive section of the *I*–*V* curve of the rectifier is linear.)

Fig. 44.3
Rectifier in an AC circuit

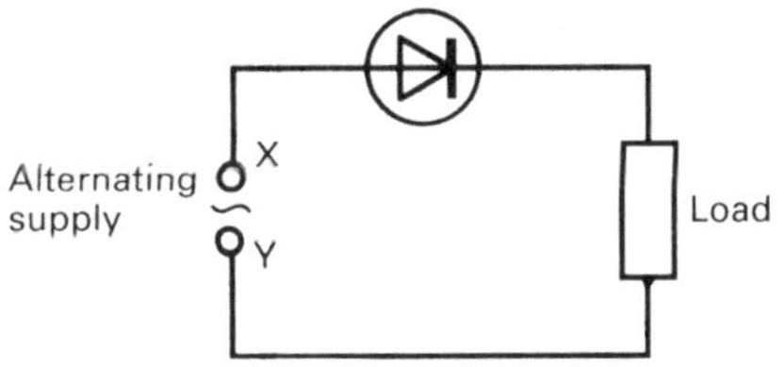

Fig. 44.4
PD variations in the circuit of Fig. 44.3

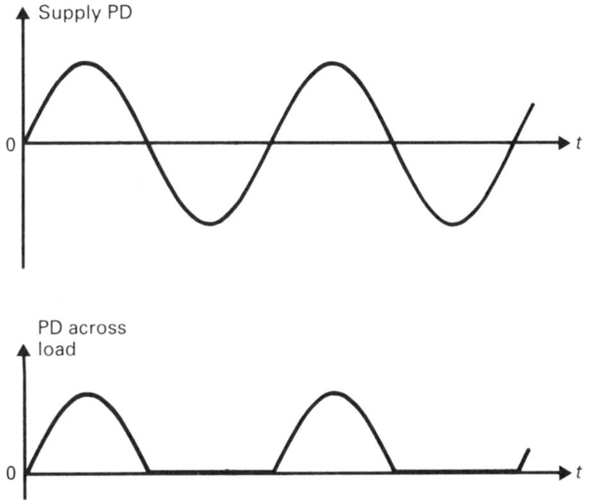

44.3 FULL-WAVE RECTIFICATION

This can be achieved by using an arrangement of four rectifiers known as a **bridge rectifier** (Fig. 44.5). When X is positive, B and D conduct; when Y is positive, A and C conduct. In each case the current through the load is in the same direction – from N to M. The PD across the load has the form shown in Fig. 44.6.

Fig. 44.5
Circuit for full-wave rectification

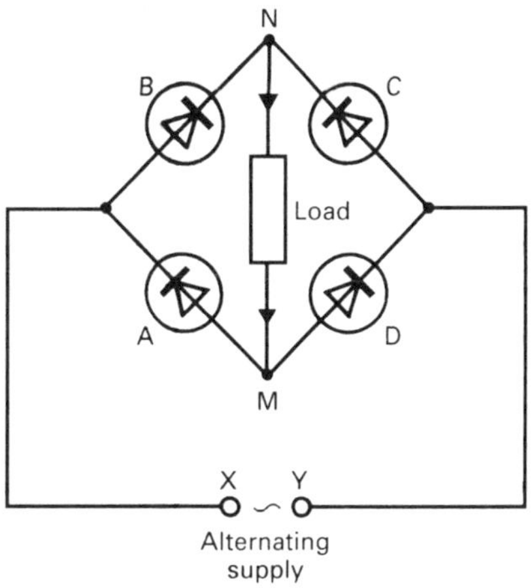

Fig. 44.6
Variation of PD across load in Fig. 44.5

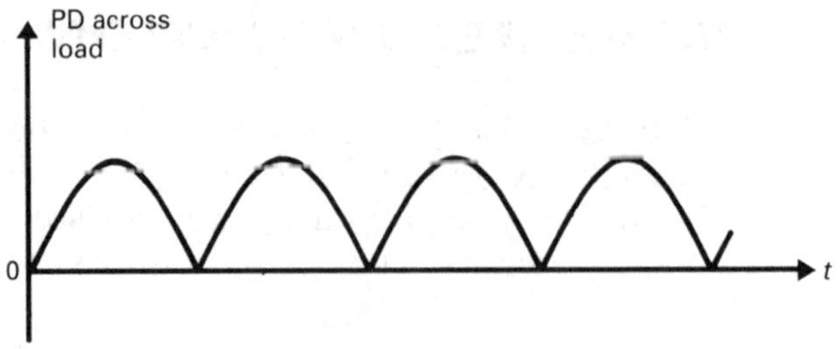

Thus, full-wave rectification allows the load to draw current from the supply on each half of each cycle and therefore the power that can be utilized is double that achieved with half-wave rectification.

44.4 SMOOTHING

The pulsating output produced by both half-wave and full-wave rectifiers can be made more steady (**smoothed**) by putting a suitable capacitor in parallel with the load (see Fig. 44.7). The 'smoothed' outputs for both the half-wave and the full-wave situations are shown in Fig. 44.8.

Fig. 44.7
Use of a capacitor to smooth PD across load

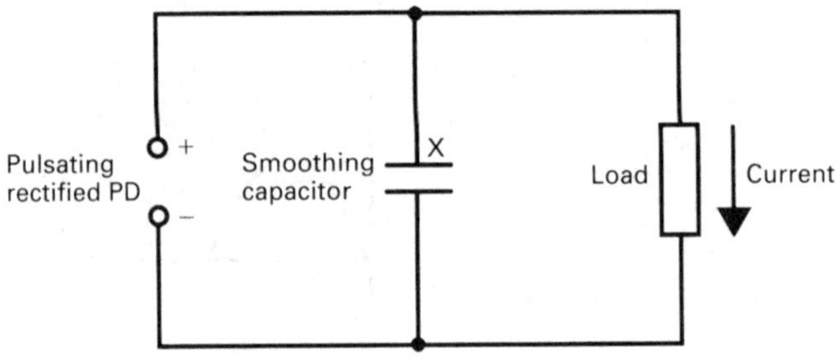

Fig. 44.8
Variation of PD across load in Fig. 44.7 for:
(a) half-wave rectification, (b) full-wave rectification

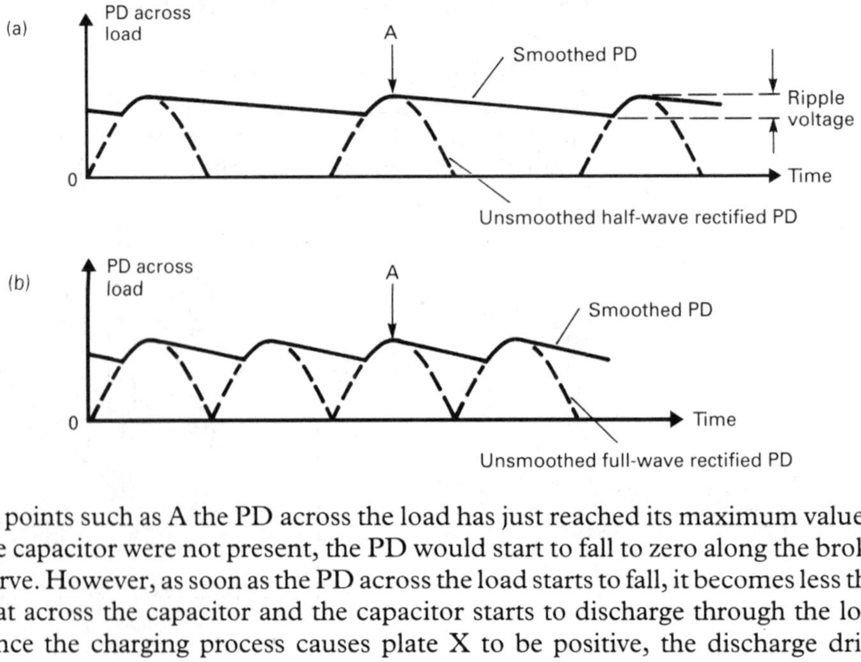

At points such as A the PD across the load has just reached its maximum value. If the capacitor were not present, the PD would start to fall to zero along the broken curve. However, as soon as the PD across the load starts to fall, it becomes less than that across the capacitor and the capacitor starts to discharge through the load. Since the charging process causes plate X to be positive, the discharge drives current through the load in the same direction as it flowed during charging. If the time constant (see section 40.10) of the capacitor-load combination is suitably large, the PD across the load falls by only a small amount before it starts to rise again.

If further smoothing is required, an inductor and a second capacitor are used (see Fig. 44.9). **The inductor (*L*) has a low DC resistance but high AC impedance. The capacitor (*C*), on the other hand, has a low AC impedance but a high DC resistance.** Therefore, the bulk of the DC component of the output from the smoothing capacitor appears across *C*. The bulk of the ripple voltage is across *L*. Thus, when a load is connected across *C* (as shown) it is in parallel with a very steady PD as required.

Fig. 44.9
Additional components for further smoothing

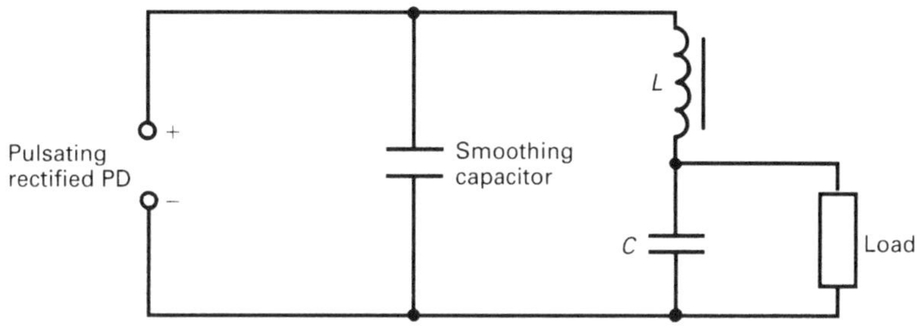

45

MAGNETIC MATERIALS

45.1 RELATIVE PERMEABILITY (μ_r)

Suppose that a substance is introduced into a vacuum in a region where the magnetic flux density is B_0. The flux density takes on a new value, B, and it is useful to define a quantity called the **relative permeability** μ_r of the substance by

$$\mu_r = \frac{B}{B_0}$$

It follows that μ_r is a pure number, and that the relative permeability of vacuum is unity.

45.2 CLASSIFICATION OF MAGNETIC MATERIALS

All substances are affected by magnetic fields and are classified as being diamagnetic, paramagnetic or ferromagnetic according to how they are affected.

For **diamagnetic** materials μ_r is very slightly less than one (typically 0.99999) and for **paramagnetic** materials it is slightly greater than one (typically 1.001). Thus, diamagnetism and paramagnetism are very weak forms of magnetism.

Those materials which exhibit very strong magnetic effects are said to be **ferromagnetic** and have very large values of μ_r (typically 10^4). The only elements which are ferromagnetic are iron, nickel, cobalt, gadolinium and dysprosium. A number of oxides and alloys are also ferromagnetic. For ferromagnetic materials μ_r depends on:

(i) the (magnetic) history of the sample and the strength of the magnetizing field, i.e. on B_0 (see section 45.3), and

(ii) the temperature of the material. The value of μ_r decreases with increasing temperature, and at a critical temperature which is known as the **Curie temperature** a ferromagnetic material becomes paramagnetic. The Curie temperature of iron is 1043 K.

45.3 THE MAGNETIZATION CURVE

Fig. 45.1 illustrates how B (the flux density at the site of a sample of initially unmagnetized ferromagnetic material) varies with B_0 (the flux density of the field in which the sample is situated). (Some of the practical considerations involved in obtaining such a curve are outlined in section 45.6.)

Fig. 45.1
A typical hysteresis curve

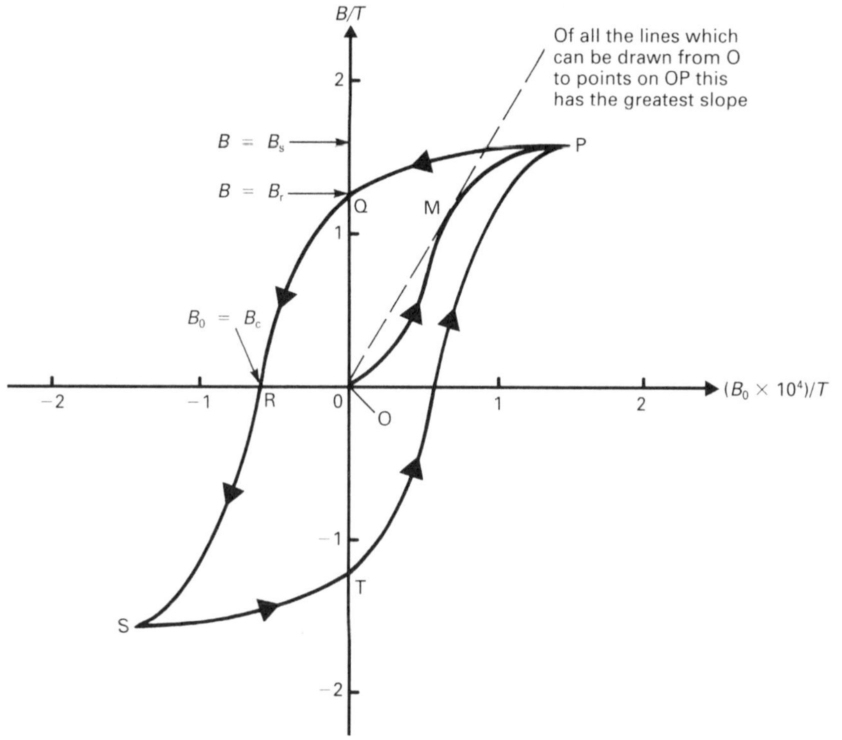

The sample is initially unmagnetized and its situation is represented by O. As B_0 is increased B increases along OP and reaches its **saturation value** B_s at P. If B_0 is now reduced to zero, B falls only slightly (to B_r), and the situation of the sample is represented by Q. The sample has retained some of its magnetism; the flux density which remains in the sample is called its **remanence** or **retentivity**, B_r. In order to reduce B to zero it is necessary to apply a reverse field of flux density B_c (point R). B_c/μ_0 is called the **coercivity** of the sample. (μ_0 is the permeability of vacuum, $4\pi \times 10^{-7}\,\mathrm{H\,m^{-1}}$.) If the reverse field is increased, the sample saturates at S. When the reverse field is reduced to zero the sample again retains some of its magnetism (point T). If B_0 is increased to its original maximum (forward) value, B increases to B_s along TP. When the magnitude of B_0 is being reduced (i.e. from P to Q and from S to T) the magnitude of B is greater than it was for the same value of B_0 whilst B_0 was being increased. Thus the sample shows a reluctance to being demagnetized. The phenomenon is called hysteresis – a term which is derived from a Greek word meaning 'lagging behind'. The sample can be taken around PQRSTP indefinitely; the curve is called a **hysteresis loop**.

Notes (i) $\mu_r = B/B_0$, and therefore since the plot of B against B_0 is not a straight line through the origin, the curve shows that μ_r is not constant for a ferromagnetic material. Bearing in mind that $\mu_r = B/B_0$, we see that during the initial magnetization the sample has its maximum permeability at M.

(ii) It can be shown that the area of the hysteresis loop is proportional to the energy used in taking unit volume of the material once round the loop. The energy is dissipated as heat in the material.

(iii) Ferromagnetic materials exhibit hysteresis even when they are not taken to saturation.

(iv) For values of B_0 which are greater than that required to produce saturation the B–B_0 curve continues to rise with a gradient of 1*. This is because $B = B_M + B_0$, where B_M is the flux density of the material being investigated and although B_M has stopped increasing, B continues to increase because of the increasing value of B_0.

45.4 PROPERTIES AND USES OF FERROMAGNETIC MATERIALS

Ferromagnetic materials are classified as being either **soft** or **hard**. The names arise because the properties associated with each type are respectively typified by the behaviour of soft iron and hard steel. Fig. 45.2 compares the hysteresis loop of a magnetically soft material with that of a magnetically hard material. Note that the soft magnetic material has: (i) a narrow hysteresis loop, (ii) low coercivity, (iii) high saturation flux density, (iv) high remanence. (High remanence is not a characteristic of all magnetically soft materials.)

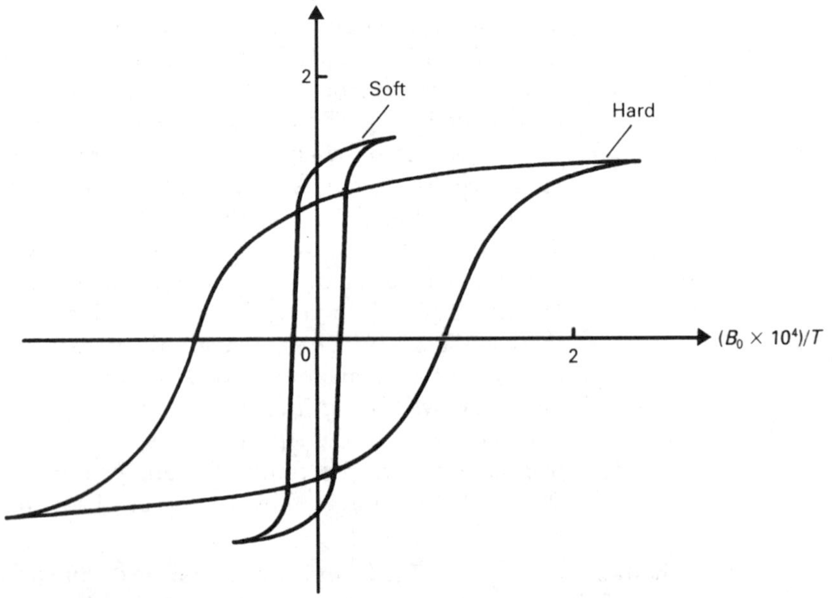

Fig. 45.2
Hysteresis loops for 'hard' and 'soft' ferromagnetic materials

*The gradient of the curve at P appears to be very much less than 1, this is because the scale on one axis is 10^4 times bigger than that on the other.

Permanent Magnets

These are made from hard magnetic materials. It is necessary that the material has:

(i) high remanence (so that the magnet is 'strong'), and

(ii) high coercivity (so that it is unlikely to be demagnetized by stray magnetic fields).

Table 45.1 lists the remanence and coercivity values of some of the materials which are used to make permanent magnets.

Table 45.1 Remanence and coercivity of some hard magnetic materials

Material	*Remanence*/T	*Coercivity*/kA m^{-1}	*Comment*
Cobalt steel	1.0	6.0	
Alnico	0.65	43	Alloy of Fe, Ni, Al, Co
Ticonal GX	1.35	58	Alloy of Fe, Co, Ni, Al, Cu
Magnadur	0.36	110	Ceramic. Oxides of Ba and Fe

Electromagnets

The most important requirement is that the material has a high magnetic flux density at saturation (so that the magnet will be strong). For reasons which are explained in section 45.6, it is necessary that a magnetically soft material is used.

Transformer Cores

Soft magnetic materials are used. When a transformer is in use its core is taken through many complete cycles of magnetization. Energy is dissipated in the core in the form of heat during each cycle. The energy dissipated is known as **hysteresis loss** and is proportional to the area of the hysteresis loop. It is necessary therefore that the hysteresis loop of the core material has a small area. Silicon iron is used at low frequencies. At high frequencies eddy current losses (see section 42.11) tend to be a serious problem and high resistivity ferrites are used.

45.5 DEMAGNETIZATION

Refer to Fig. 45.1. Applying a reverse flux density of B_c to a sample of ferromagnetic material reduces its flux density to zero. This does not mean that the sample has been permanently demagnetized, for as soon as the reverse field is removed B becomes positive. In order to produce true demagnetization it is necessary to take the sample around a series of hysteresis loops of gradually decreasing amplitude (see Fig. 45.3). This can be achieved, for example, by placing the sample inside a solenoid through which alternating current is flowing, and then reducing the current to zero.

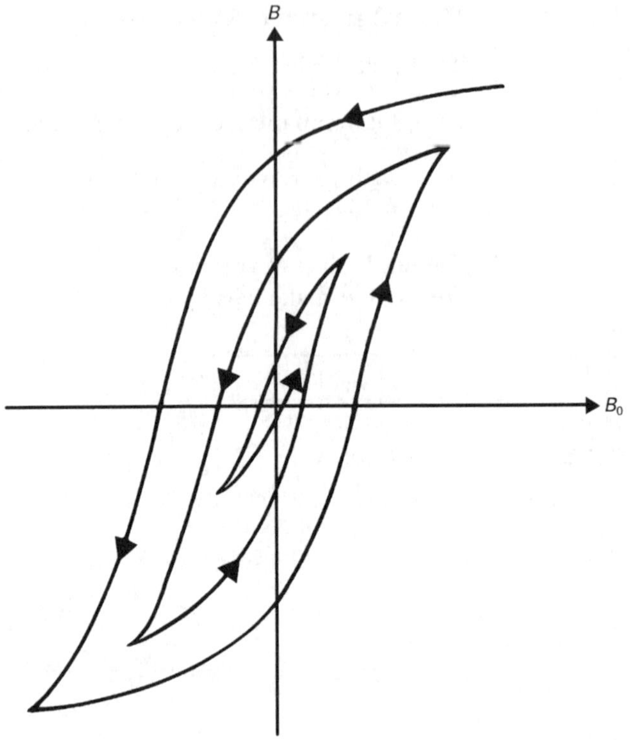

Fig. 45.3 Demagnetization curve for a ferromagnetic material

45.6 SELF-DEMAGNETIZATION

The hysteresis curves of Figs. 45.1 and 45.2 are obtained with closed loops of magnetic material. Any attempt to produce such curves with (say) an iron bar inside a solenoid would be unsuccessful. This is because forces exist within a piece of magnetized material which tend to demagnetize it. The effects of these forces are particularly marked near the ends of a rod-shaped specimen, and the magnetization at the ends is appreciably less than it is elsewhere. This non-uniformity of the magnetization gives a hysteresis curve of a different shape from that which is obtained with a uniformly magnetized sample. In the case of a magnetically soft material this self-magnetization is capable of almost completely destroying the flux density within the material as soon as the magnetizing field is removed. With hard magnetic materials the effect takes place much more slowly. Nevertheless, it is necessary to store permanent magnets with soft iron 'keepers' in order to form a magnetically closed loop and reduce the self-demagnetization.

The core of an electromagnet is rarely in the form of a magnetically closed loop, and therefore the self-demagnetizing effect reduces the flux density very nearly to zero as soon as the magnetizing current is switched off. The fact that many soft magnetic materials have high remanence is therefore of no consequence.

45.7 THE DOMAIN THEORY OF FERROMAGNETISM

Paramagnetic and ferromagnetic materials contain atoms which have magnetic fields resulting from the motions of the electrons within the atoms. The atoms of diamagnetic materials, on the other hand, have no such fields because their electron configurations are such that the fields of the individual electrons cancel each other. In paramagnetic materials the fields of the individual atoms are oriented randomly, owing to thermal agitation, and there is no overall magnetization.

In ferromagnetic materials there are very strong interactions between neighbouring atoms which cause the magnetic fields of groups of them to line up in the same direction. These groups are called **domains**, and have volumes of 10^{-3} mm^3 or more. The magnetic field within a domain is perhaps a hundred times stronger than the strongest fields that can be produced in laboratories.

The direction of magnetization varies from one domain to another, and in an unmagnetized sample of ferromagnetic material the fields of the various domains cancel (Fig. 45.4(a)). When an external field is applied to the sample those domains whose directions of magnetization are in (or close to) the direction of the field grow at the expense of others and there is a general movement of the domain walls (Fig. 45.4(b)). The sample now has a resultant magnetization in the direction of the external field. If the strength of the field is increased, the extent to which the walls move increases. In addition, there comes a point when the magnetic axes of some of the domains suddenly rotate, and line up with the external field (Fig. 45.4(c)). Further increases in field strength cause more and more domains to rotate, and eventually all the domain axes are in line with the applied field. The sample is now saturated.

Fig. 45.4
The effect of an external field on the domain structure of a ferromagnetic material

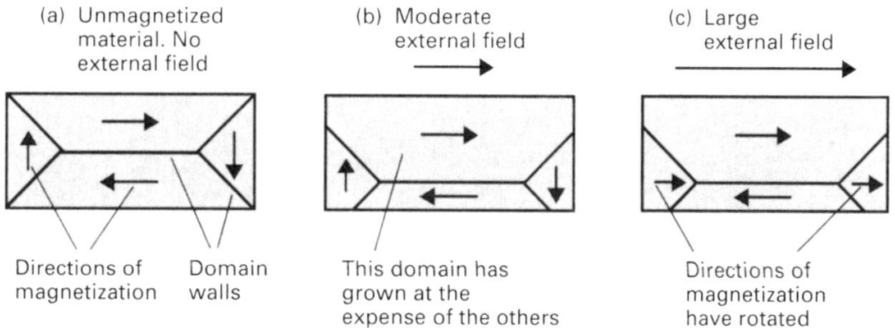

The wall movements which occur in weak fields are small and (almost completely) reversible – the sample loses its magnetism when the external field is removed. The larger wall movements and rotations produced by stronger fields are mainly irreversible (Fig. 45.5).

Fig. 45.5
Magnetization curve showing the dominant processes in the different regions of the curve

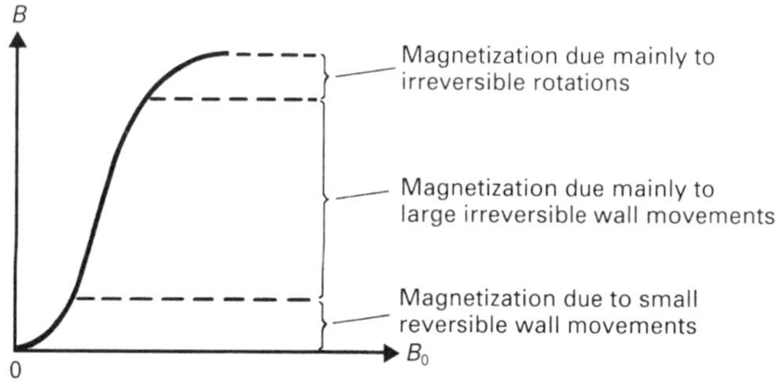

Once a particular arrangement of domains has been established by a strong external field, simply removing the field has little effect. It is necessary to apply a large reverse field in order to rearrange the domains and destroy the magnetism. This explains why ferromagnetic materials exhibit hysteresis. Magnetically hard materials are more difficult to magnetize and demagnetize than magnetically soft materials.

This is because the hard materials have more crystal imperfections and grain boundaries (see section 9.10) which hamper the movements of the domain walls.

Perhaps the most convincing evidence for the existence of magnetic domains is provided by what are known as **Bitter patterns**. These can be produced by placing a colloidal suspension of magnetite on the surface of a single crystal of ferromagnetic material which has been polished electrolytically to remove surface irregularities. The magnetite tends to settle along the domain walls creating a pattern which can be observed with a microscope to reveal the domain structure.

46

ELECTROLYSIS

46.1 THE PHENOMENON OF ELECTROLYSIS

Electrolysis is the name given to the process in which chemical changes are caused to occur by passing an electric current through a liquid. The liquid is called an **electrolyte**.

Liquid metals are not electrolytes since they pass current without there being any associated chemical change. Molten salts, and solutions of most inorganic compounds in water are electrolytes. Some solutions do not conduct electricity and are said to be **non-electrolytes** – sugar solution is an example.

Fig. 46.1 illustrates a simple arrangement for producing electrolysis. The plates by which the current enters and leaves are called **electrodes**. That through which the (conventional) current enters, i.e. that which is connected to the positive terminal of the battery, is called the **anode**; the other electrode is called the **cathode**. The apparatus as a whole is called a **voltameter**.

Fig. 46.1
The principle of electrolysis

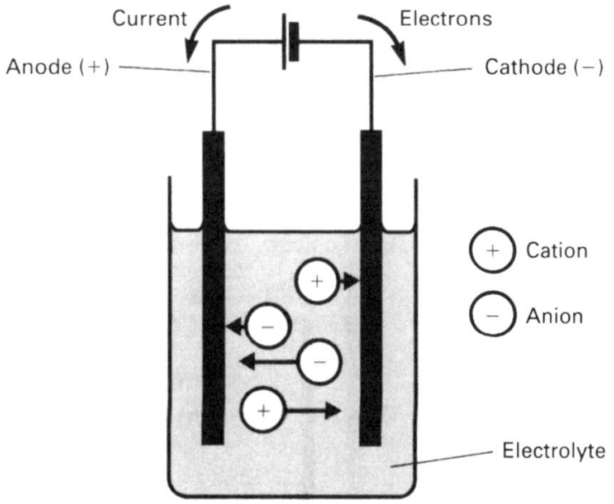

The current flow in an electrolyte is due, not to electrons, but to positive and negative ions. The positive ions move towards the cathode and are called **cations**; the negative ions move towards the anode and are known as **anions**. (The current in the connecting wires and the electrodes is carried by electrons; it follows that the electrodes must be metals or graphite.)

As an example of an electrolyte we shall consider sodium chloride (NaCl). This contains Na^+ ions and Cl^- ions. If the sodium chloride is molten or is in aqueous solution, these ions are free to move and it can conduct an electric current. Solid sodium chloride cannot conduct a current because, although it too contains Na^+ and Cl^- ions, they are not free to move.

We stated at the beginning of this section that electrolysis results in chemical change. The chemical reactions occur only at the electrodes. When an anion reaches the anode it gives up its excess electron(s) to the anode and becomes a neutral atom.* The electrons received by the anode in this way (effectively) move through the external circuit to the cathode. When cations arrive at the cathode they combine with these electrons and, like the anions, become neutral atoms.

When there is more than one type of cation or anion present only one type of each is neutralized. Just which cation and which anion are neutralized is determined by the relative positions of the ions in the so-called electrochemical series, the concentration of the solution, and the nature of the electrodes. In some cases the anode dissolves in the electrolyte as a result of the electrolysis. For example, in electrolysis of copper sulphate solution using copper electrodes, copper atoms are deposited on the cathode and an equal number go into solution at the anode. If platinum electrodes are used, though copper is still deposited at the cathode, oxygen is evolved at the anode and the anode remains intact.

46.2 FARADAY'S LAWS OF ELECTROLYSIS

Faraday summarized the results of many detailed experiments on electrolysis in two laws.

The First Law

The mass of any given substance which is liberated or dissolved as a result of electrolysis is proportional to the quantity of electric charge that has flowed.

The first law can be expressed as

$$m = zQ \qquad [46.1]$$

where

m = the mass of a substance which has been liberated or dissolved by a charge Q, and

z = a constant for any given substance. It is called the **electrochemical equivalent** (ECE) of the substance. (Unit = $kg\,C^{-1}$.)

Notes (i) Equation [46.1] defines ECE. Thus, **the electrochemical equivalent of a substance is the mass of it which is liberated or dissolved in electrolysis by the passage of one unit of electric charge**.

(ii) If the charge which passes is due to a steady current I flowing for time t, equation [46.1] can be written as

$$m = zIt$$

*Often a radical e.g. OH, rather than an atom.

The Second Law

The ratio of the numbers of moles of any two different substances liberated or dissolved in electrolysis as a result of equal quantities of charge flowing is equal to the reciprocal of the ratio of their respective valencies.

For example, the valency of oxygen is twice that of hydrogen and therefore the number of moles of hydrogen liberated by any given quantity of charge is twice the number of moles of oxygen liberated by the same charge.

46.3 THE FARADAY CONSTANT (*F*)

The Faraday constant (F) is numerically equal to the quantity of charge which liberates one mole of monovalent (i.e. singly charged) ions in electrolysis. It is found by experiment that

$$F = 9.649 \times 10^4 \,\mathrm{C\,mol^{-1}}$$

The magnitude of the charge on a monovalent ion is equal to the electronic charge e. The total charge on a mole of monovalent ions is therefore $N_A e$, where N_A is the Avogadro constant. It follows that

$$F = N_A e$$

This equation can be used to obtain a value for e which is much more accurate than that from a 'Millikan-type' experiment (section 50.4). (N_A is obtained from X-ray diffraction measurements; F is found from electrolysis experiments.)

46.4 TWO USEFUL RELATIONSHIPS

$$\text{Total charge on one mole of ions of valency } v = N_A v e = Fv$$

$$\text{Mass in kg of one mole of ions (of any valency)} = A_r \times 10^{-3}$$

where A_r is the relative atomic mass.

It follows that the specific charge (charge-to-mass ratio) in $\mathrm{C\,kg^{-1}}$ of an ion of valency v is given by

$$\text{Specific charge} = \frac{Fv}{A_r \times 10^{-3}}$$

The electrochemical equivalent z (in $\mathrm{kg\,C^{-1}}$) is the number of kilograms of a substance liberated by one coulomb of electric charge, and as such, it is the mass of that quantity of ions which have a total charge of one coulomb. It follows that the reciprocal ($1/z$) of the electrochemical equivalent is the specific charge of the ions, i.e.

$$\text{Specific charge} = \frac{1}{z}$$

(This, of course, assumes that each ion of the substance has the same mass and carries the same charge as every other.)

Combining these two results, we have

$$\underset{(\text{in C kg}^{-1})}{\text{Specific charge}} = \frac{1}{z} = \frac{Fv}{A_r \times 10^{-3}}$$

QUESTIONS ON SECTION F

Assume $\varepsilon_0 = 8.85 \times 10^{-12}\,\mathrm{F\,m^{-1}}$ and $\mu_0 = 4\pi \times 10^{-7}\,\mathrm{H\,m^{-1}}$ unless otherwise stated.

DC ELECTRICITY (Chapter 36)

F1 For the circuit shown below, calculate **(a)** I, **(b)** V_1, **(c)** V_2, **(d)** V_3.

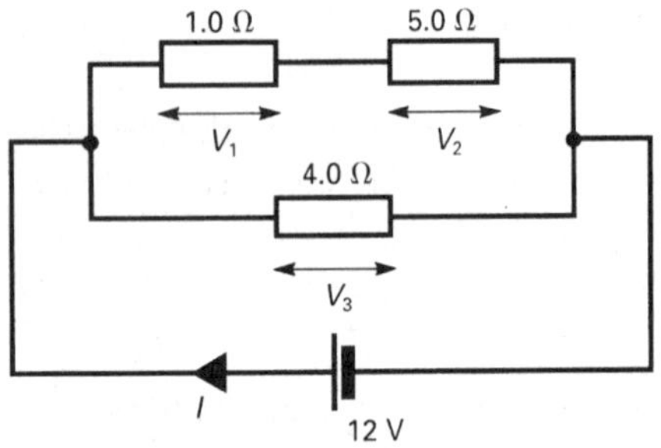

F2 When 81 cm of a wire with a cross-sectional area of $3.1 \times 10^{-7}\,\mathrm{cm^2}$ is connected across a 2.0 V cell a current of 1.6 A flows in the wire. Find the resistivity of the material of the wire.

F3 In the circuit below a voltmeter with a resistance of 80 kΩ is connected across a 10 kΩ resistor. Calculate the voltmeter reading.

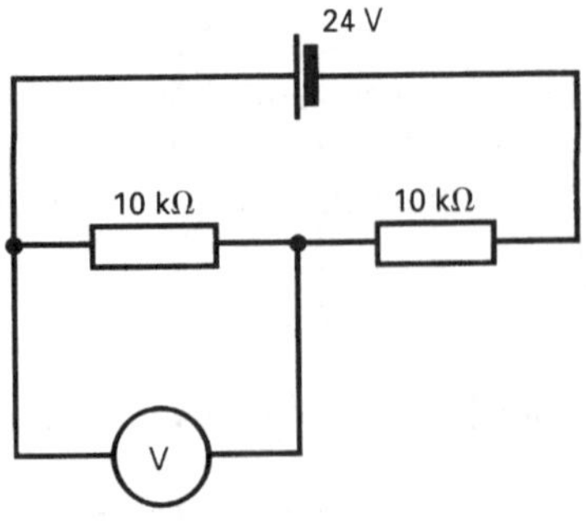

F4 Find the current drawn from the cell in the following circuit.

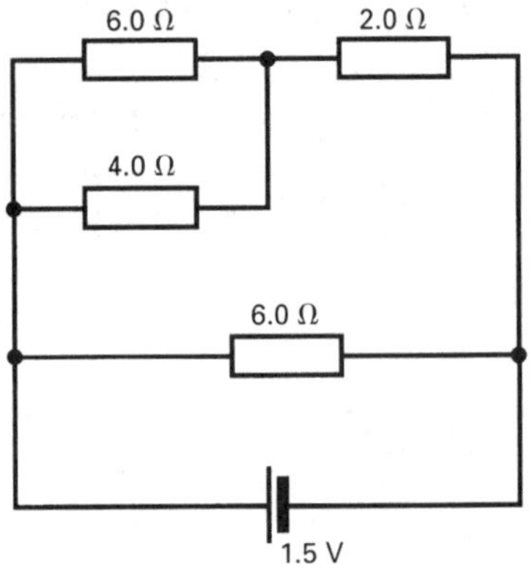

F5 **(a)** State Ohm's law.

(b) The following are four electrical components:

A a component which obeys Ohm's law

B another component which obeys Ohm's law but which has higher resistance than A

C a filament lamp

D a component, other than a filament lamp, which does not obey Ohm's law.

(i) For each of these components, sketch current–voltage characteristics, plotting current on the vertical axis, and showing both positive and negative values. Use one set of axes for A and B, and separate sets of axes for C and for D. Label your graphs clearly.

(ii) Explain the shape of the characteristic for C.

(iii) Name the component you have chosen for D. [J, '92]

F6 There are n free electrons in unit volume of a wire and each electron carries a charge e.

(a) Show that when a current flows, the current density J is given by

$$J = nev$$

where v is the drift velocity of the electrons.

(b) A PD of 4.5 V is applied to the ends of a 0.69 m length of manganin wire of cross-sectional area $6.6 \times 10^{-7}\,\text{m}^2$. Calculate the drift velocity of the electrons along the manganin wire.

The resistivity of manganin is $4.3 \times 10^{-7}\,\Omega\,\text{m}$ and n for manganin is $10^{28}\,\text{m}^{-3}$. ($e = 1.6 \times 10^{-19}\,\text{C}$.) [W, '92]

F7 Calculate the resistance of the network shown below:
(a) between A and B,
(b) between A and C.

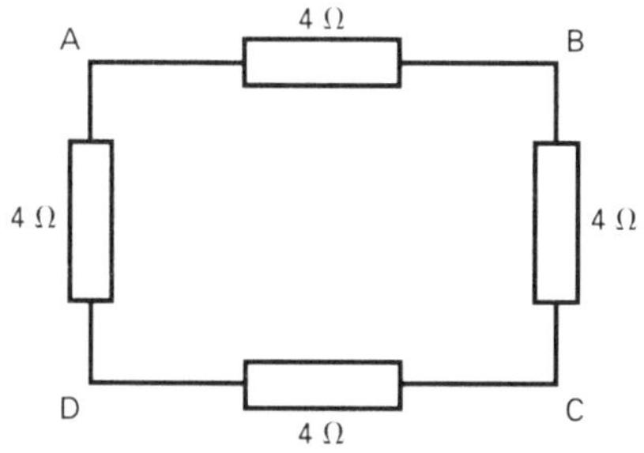

F8 Deduce the value of the current I as shown in the circuit below. [C]

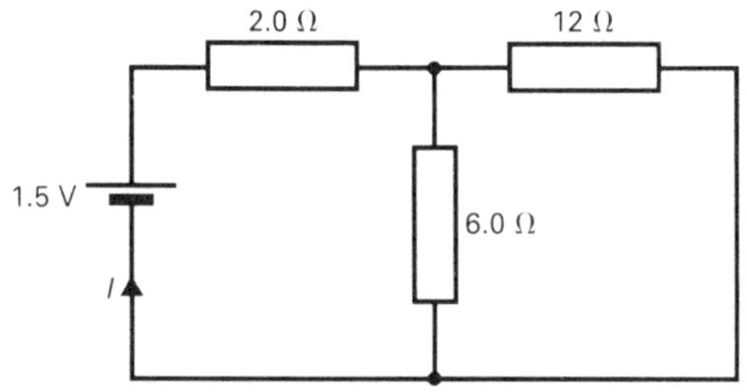

F9

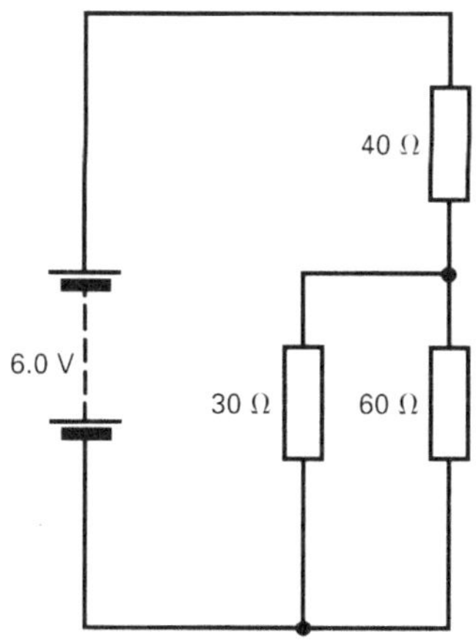

Fig. 1

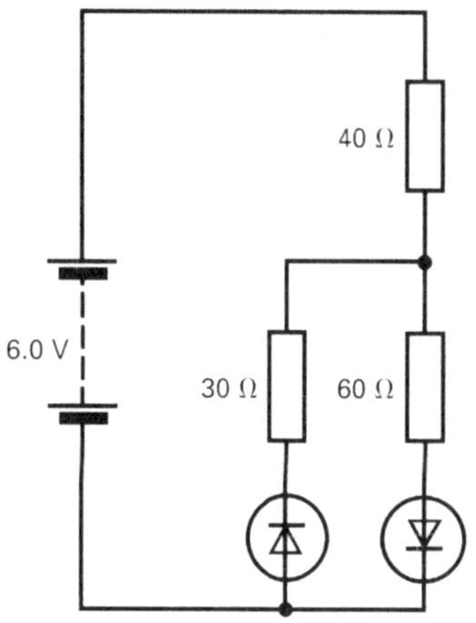

Fig. 2

When answering the questions below, you may assume that the source resistance is negligible and that each diode has a potential difference of 0.6 V across it when conducting.

(a) For the circuit shown in Fig. 1, calculate
 (i) the current in the 30 Ω resistor,
 (ii) the current in the 60 Ω resistor,
 (iii) the PD across the 40 Ω resistor,

(b) For the circuit shown in Fig. 2, calculate
 (i) the current in the 30 Ω resistor,
 (ii) the current in the 60 Ω resistor,
 (iii) the PD across the 40 Ω resistor.

[J, '91]

F10 The light-dependent resistor (LDR) in the circuit below is found to have resistance 800 Ω in moonlight and resistance 160 Ω in daylight. Calculate the voltmeter reading, V_m, in moonlight with the switch S open.

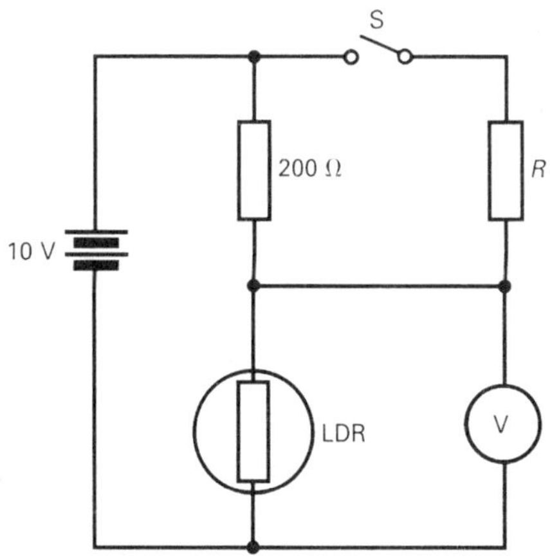

If the reading of the voltmeter in daylight with the switch S closed is also equal to V_m what is the value of the resistance R? [L, '91]

F11 Find the value of the current, I, in the circuit shown below.

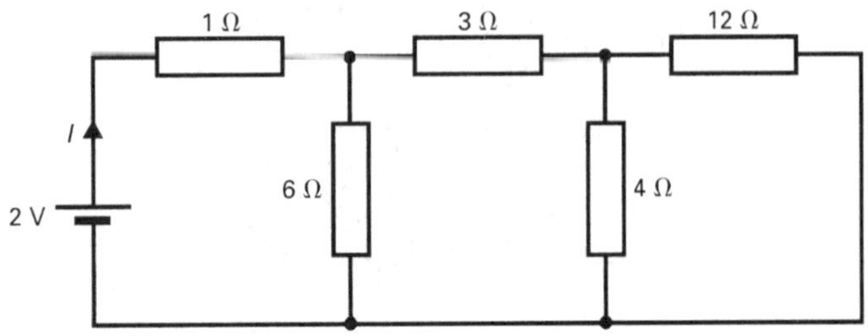

F12

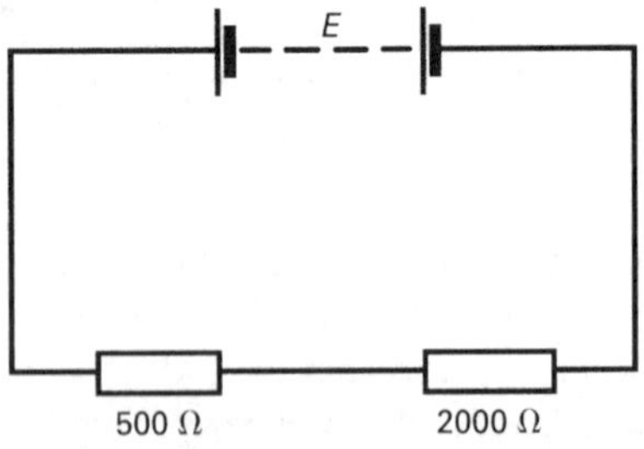

In the above circuit E is a DC source of negligible internal resistance driving a current through two resistors of resistance 500 Ω and 2000 Ω respectively arranged in series. The PD across the 500 Ω resistor is measured using a voltmeter of resistance 2000 Ω and is found to be 10 V. What is the EMF of the source and what is the PD across the 2000 Ω resistor when the voltmeter is not being used? [AEB, '79]

F13 A semiconductor diode and a resistor of constant resistance are connected in some way inside a box having two external terminals, as shown in the diagram below. When a potential difference of 1.0 V is applied across the terminals the ammeter reads 25 mA. If the same potential difference is applied in the reverse direction the ammeter reads 50 mA.

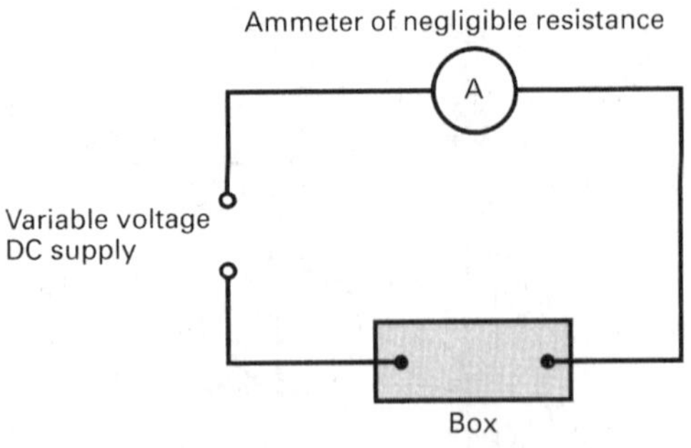

What is the most likely arrangement of the diode and the resistor? Explain your deduction. Calculate the resistance of the resistor and the forward resistance of the diode. [L]

F14 Two resistors, each of resistance 6 kΩ, are connected in series across a 12 V battery which has negligible internal resistance. What potential difference is indicated by a moving coil meter with resistance 15 kΩ connected in parallel with one of the 6 kΩ resistors? [S]

F15 A moving-coil meter has a resistance of 5.0 Ω and full-scale deflection is produced by a current of 1.0 mA. How can this meter be adapted for use as:

(a) a voltmeter reading up to 10 V,
(b) an ammeter reading up to 2 A? [S]

F16 For the values shown in the figure below, calculate

(a) the current through ammeter A_1,
(b) the current through ammeter A_2,
(c) the PD across R_2.

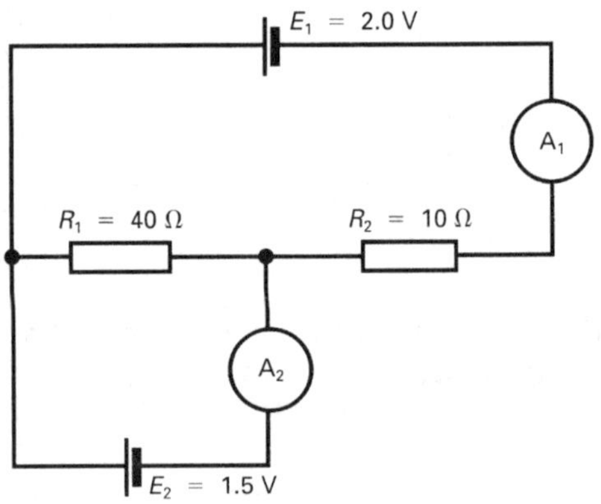

Assume that the internal resistance of each cell and the resistance of each ammeter is negligible. [AEB, '79]

F17 A DC milliammeter has a full-scale deflection of 10 mA and a resistance of 50 Ω. How would you adapt this to serve as a voltmeter with a full-scale deflection of 150 V? Comment on whether this voltmeter would be suitable for accurately measuring the potential difference across a resistor of about 100 kΩ carrying a current of about 1 mA. [J]

F18 A galvanometer of resistance 40 Ω requires a current of 10 mA to give a full-scale deflection. A shunt is put in position to convert it to a meter reading up to 1.0 A full-scale deflection. A resistance bobbin, intended to convert the galvanometer to one reading up to 1.0 V full-scale deflection, is now attached to the instrument but the shunt is inadvertently left in position. By considering the potential difference across the bobbin and the meter, calculate the voltage which would produce full-scale deflection of the meter. [L]

F19

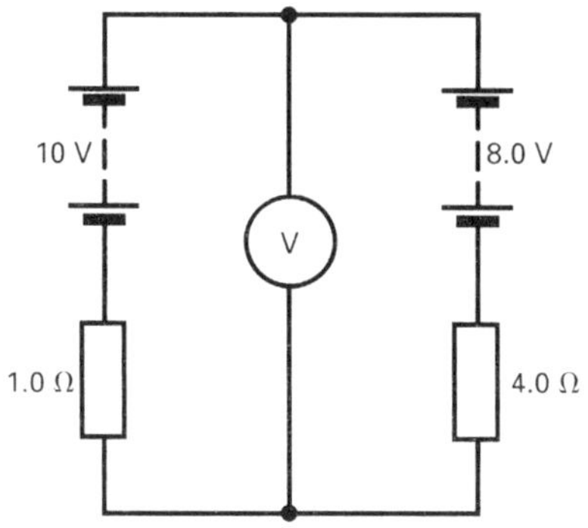

In the above circuit the batteries have negligible internal resistance and the voltmeter V has a very high resistance. What would be the reading of the voltmeter? [L]

F20 The 4.00 V cell in the circuits shown below has zero internal resistance.
An accurately calibrated voltmeter connected across YZ records 1.50 V. Calculate
(a) the resistance of the voltmeter,
(b) the voltmeter reading when it is connected across Y′Z′.

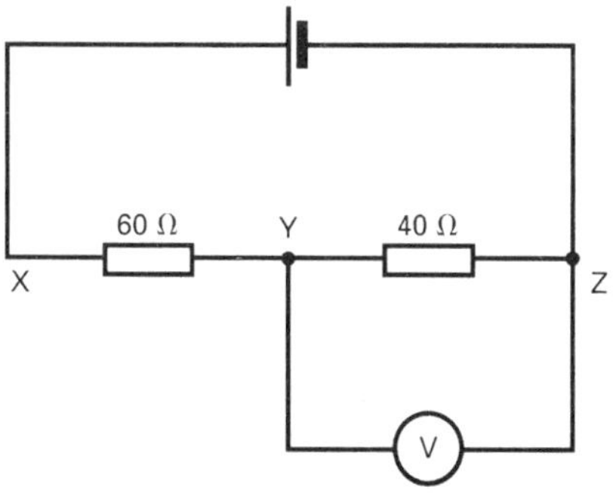

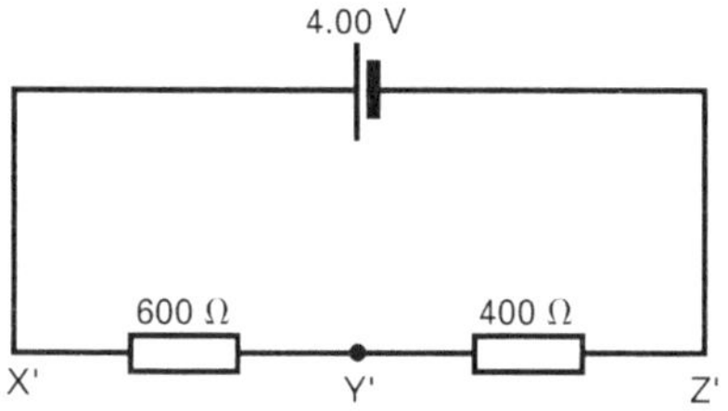

What do your results suggest concerning the use of voltmeters? [L]

F21 **(a)** Define
(i) *potential difference* and the *volt*;
(ii) *resistance* and the *ohm*.
(b) Two resistors having resistances of 1.8 kΩ and 4.7 kΩ are connected in series with a battery of EMF 12 V and negligible internal resistance as shown below:

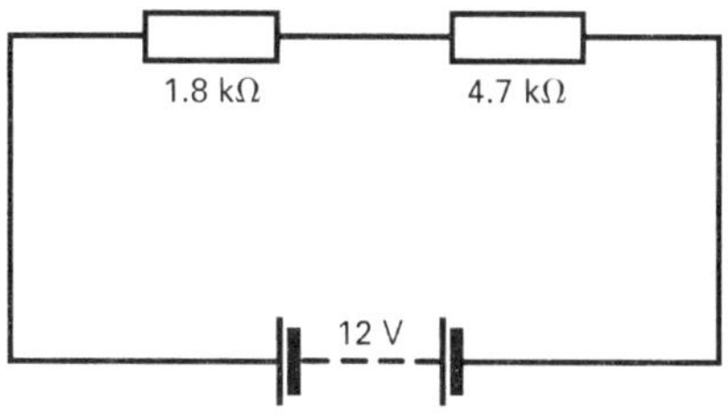

(i) What is meant by the expression an EMF of 12 V?
(ii) What is the potential difference across each of the resistors?
(c) When a particular voltmeter of fixed resistance R, which is known to be accurately calibrated, is placed across the 1.8 kΩ resistor in the diagram above it reads 2.95 V. When placed across the 4.7 kΩ resistor it reads 7.70 V.
(i) Why do these two readings not add up to 12 V?
(ii) Calculate the resistance R of the voltmeter.
(d) A second, identical, voltmeter is used so that a voltmeter is placed across each resistor. What will each voltmeter read? [C, '91]

F22 Explain why the PD between the terminals of a cell is not always the same as its EMF.

A cell, a resistor and an ammeter of negligible resistance are connected in series and a current of 0.80 A is observed to flow when the resistor has a value of 2.00 Ω. When a resistor of 5.00 Ω is connected in parallel with the 2.00 Ω resistor, the ammeter reading is 1.00 A. Calculate the EMF of the cell. [J]

F23 A power supply used in a laboratory has an EMF of 5000 V. When, however, a voltmeter of resistance 20 kΩ is connected to the terminals of the power supply a reading of only 40 V is obtained.
(a) Explain this observation.
(b) Calculate the current flowing in the meter and the internal resistance of the power supply. [S]

F24 A battery of internal resistance 0.50 Ω is connected (as shown below) through a switch S to a resistor X, which is initially at 0 °C. When S is closed, the voltmeter reading falls immediately from 12.0 V to 10.0 V. The reading then rises gradually to a steady value of 10.5 V.

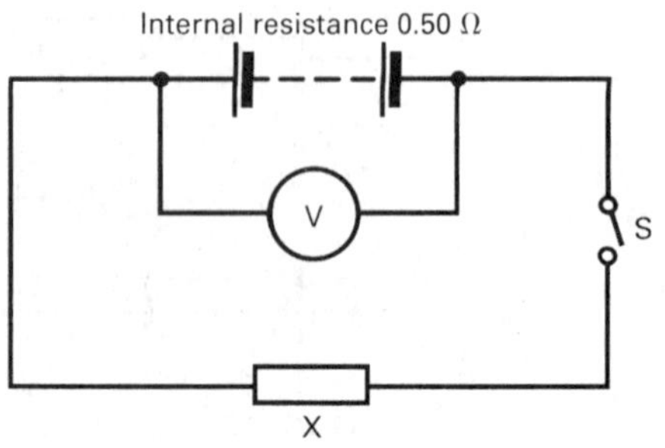

(a) Explain these observations.
(b) Calculate:
 (i) the initial current when S is closed;
 (ii) the initial resistance of X;
 (iii) the resistance of X when the steady state has been reached. [O, '92]

F25 A battery of EMF 12.6 V and internal resistance 0.1 Ω is being charged from a DC source of EMF 24.0 V and internal resistance 1.0 Ω using the circuit shown in the figure below. V_1 and V_2 are high resistance voltmeters and R is a fixed resistor.

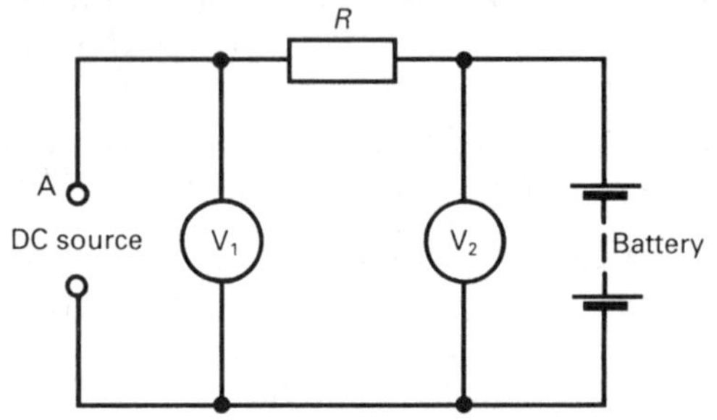

(a) What is the polarity of terminal A of the source?
(b) If the charging current is 5.0 A, determine the resistance of the resistor R.
(c) If the resistance of R were changed to 0.9 Ω what would be the reading on each voltmeter? [AEB, '79]

F26 Explain why the *potential difference* between the terminals of a battery is not always the same as its *EMF*.

A battery is known to have an EMF of 5.0 V but when a certain voltmeter is connected to it the reading is 4.9 V. The battery can deliver a current of 0.40 A when connected to a resistance of 12 Ω. What is the resistance of the voltmeter? [L]

F27 (a) State Kirchhoff's second law.
(b)

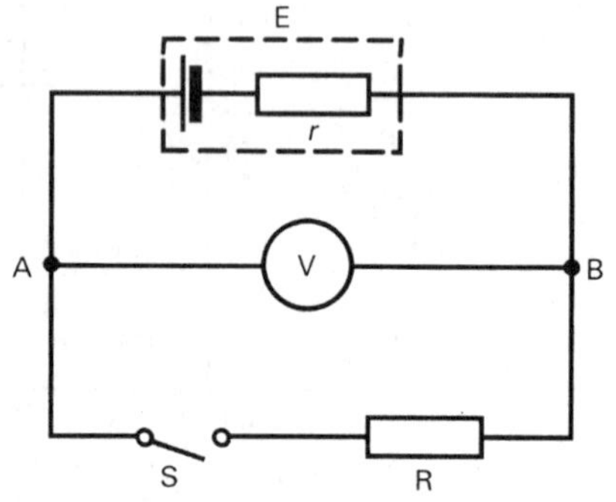

 (i) In the circuit shown E is a cell of source (internal) resistance r and the resistance of R is 4.0 Ω. With the switch S open, the high resistance voltmeter reads 10.0 V and with S closed the voltmeter reads 8.0 V. Show that $r = 1.0\,\Omega$.
 (ii) If R were replaced by a cell of e.m.f. 4.0 V and source resistance 1.0 Ω with its negative terminal connected to B, what would be the reading of the voltmeter with S closed? [J, '90]

F28 A source of EMF E and internal resistance r is connected to a load of resistance R.
(a) What current flows in the load?
(b) Find the power P dissipated in the load.
(c) For what value of R is P a maximum? [C]

F29 Find the current I in the circuit shown.

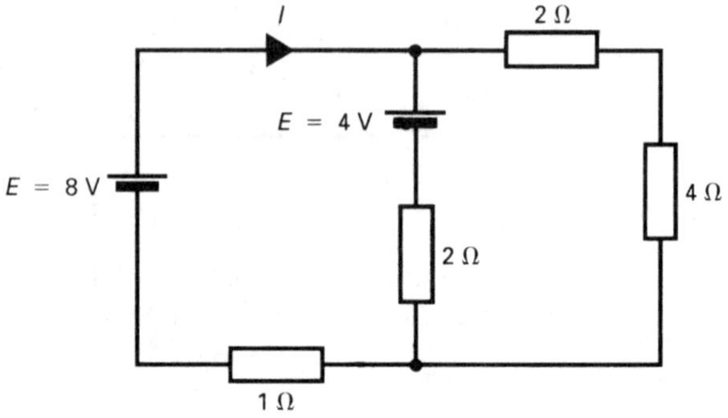

[W, '91]

F30

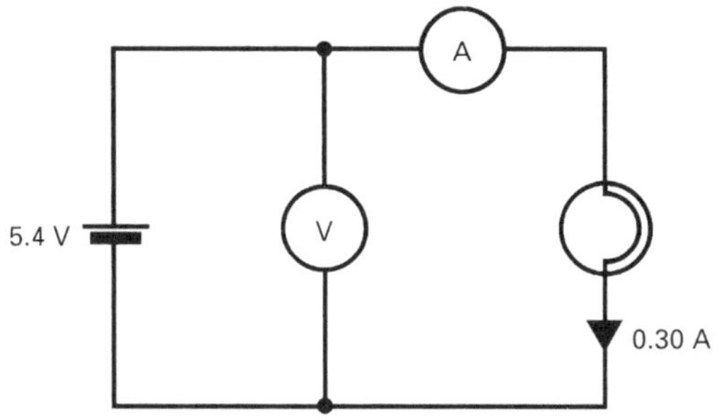

The battery in the circuit above has EMF 5.4 V and drives a current of 0.30 A through a lamp. The voltmeter reading is 4.8 V.

Explain why the voltmeter reading is less than the EMF of the cell.

Calculate values for

(a) the internal resistance of the battery, and

(b) the energy transformed per second in the lamp.

State *two* assumptions you made in order to complete these calculations. [L, '92]

F31 A string of electric lamps for decorating a Christmas tree consists of 20 12 V lamps connected in series across the 240 V mains. The power consumption of the whole string is 24 W.

(a) What is the resistance of each lamp?

(b) If one lamp becomes short-circuited, what is approximately the new power consumption of the string?

(c) Explain why, when one of these lamps is tested by applying a potential difference of 0.1 V, it passes a current of 10 mA. [S]

F32 **(a)** Explain what is meant by the *electromotive force* and the *terminal potential difference* of a battery.

(b) A bulb is used in a torch which is powered by two identical cells in series each of EMF 1.5 V. The bulb then dissipates power at the rate of 625 mW and the PD across the bulb is 2.5 V. Calculate **(i)** the internal resistance of each cell and **(ii)** the energy dissipated in each cell in one minute. [J]

F33 A steady uniform current of 5 mA flows axially along a metal cylinder of cross-sectional area 0.2 mm^2, length 5 m and resistivity $3 \times 10^{-5}\ \Omega$ m. Find:

(a) the potential difference between the ends of the cylinder,

(b) the rate of production of heat. [W]

F34 A heating coil is to be made, from nichrome wire, which will operate on a 12 V supply and will have a power of 36 W when immersed in water at 373 K. The wire available has an area of cross-section of 0.10 mm^2. What length of wire will be required?

(Resistivity of nichrome at 273 K $= 1.08 \times 10^{-6}\ \Omega$ m. Temperature coefficient of resistivity of nichrome $= 8.0 \times 10^{-5}\ \mathrm{K}^{-1}$.) [L]

F35

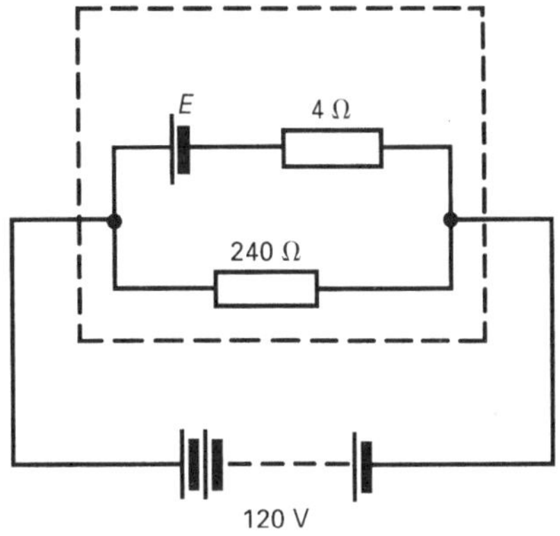

Fig. 1

In this question you are asked to analyse the simple circuit shown in Fig. 1. The region within the box represents the circuit of a DC motor running at a steady speed. The motor behaves as a source of EMF E (opposing that of the 120 V battery), and an internal resistance 4 Ω. These are in parallel with a fixed 240 Ω resistor. A current of 5.5 A is drawn from the 120 V battery (which has negligible internal resistance).

(a) Calculate

(i) the current through the 4 Ω internal resistance,

(ii) the EMF E.

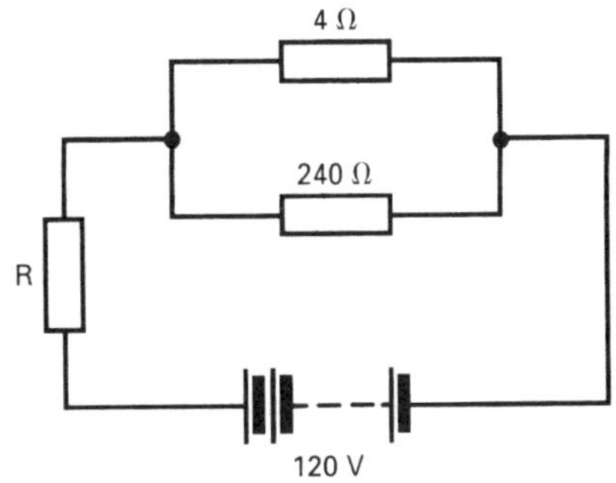

Fig. 2

(b) When the motor is not running, the EMF E is zero. To limit the current on starting, an extra series resistor R is included as shown in Fig. 2. Calculate a value for its resistance which will limit the starting current drawn from the battery to 20 A. [O & C, '90]

F36 An electrical heating element is to be designed so that the power dissipated will be 750 W when connected to the 240 V mains supply.

(a) Calculate the resistance of the wire needed.

(b) The element is to be made from nichrome ribbon 1.0 mm wide and 0.050 mm thick. The resistivity of nichrome = $1.1 \times 10^{-6}\,\Omega\,\text{m}$. Calculate the length of ribbon required.

(c) Draw a circuit diagram to show how a second heating element would be connected to increase the power dissipated to 1.5 kW.

(d) State one important property of a conductor used to make heating elements. [AEB, '87]

F37 A 12 V, 24 W lamp and a resistor of fixed value are connected in some way inside a box with two external terminals. In order to discover the circuit arrangement inside the box a student connects a variable DC power source and an ammeter in series with the box and obtains the following results:

Applied potential difference	1.0 V	12 V
Current	1.17 A	4.00 A

(a) Draw a circuit diagram of the most likely arrangement inside the box, giving your reason.

(b) Use the 12 V, 4.00 A reading to deduce the value of the fixed resistor.

(c) What is the percentage increase in the resistance of the lamp as the applied potential difference changes from 1.0 V to 12 V?

(d) When the applied potential difference is increased to 24 V the current is again found to be 4.00 A. Explain this observation.

(You may neglect the internal resistance of the power supply and the resistance of the ammeter in all your calculations.) [L]

F38 A student is provided with a 2 V cell, a lamp, a switch and a thermistor with a negative temperature coefficient of resistance. The lamp, which is in series with the cell as in Figure 1, lights immediately the switch is turned on.

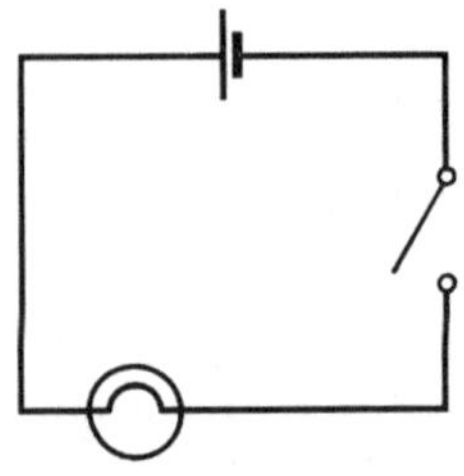

Fig. 1

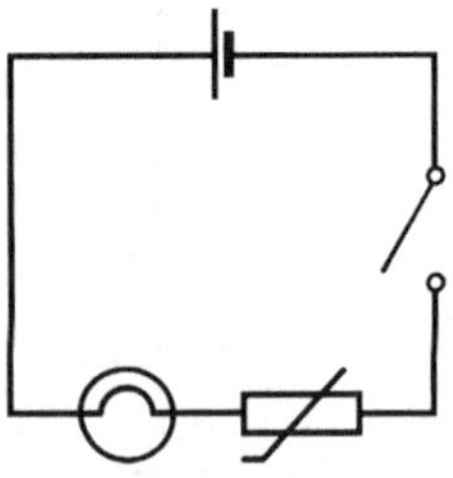

Fig. 2

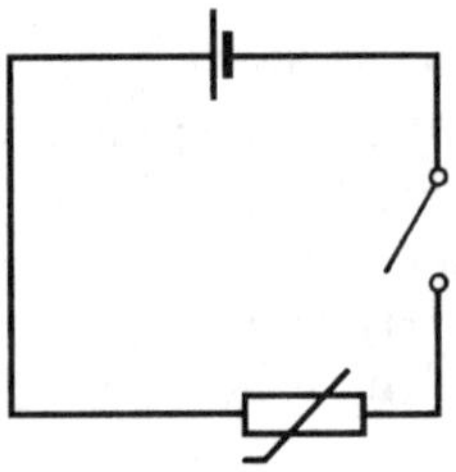

Fig. 3

Explain why

(a) when the thermistor is connected in series with the lamp, as in Figure 2, and the switch is turned on, the lamp lights up slowly, and

(b) if the lamp is omitted, as in Figure 3, and the switch is turned on, the cell is soon destroyed by overheating. [L]

F39 State *Kirchhoff's laws* for circuit networks.

In the following circuit, cell A has an EMF of 10 V and an internal resistance of 2 Ω; cell B has an EMF of 3 V and an internal resistance of 3 Ω.

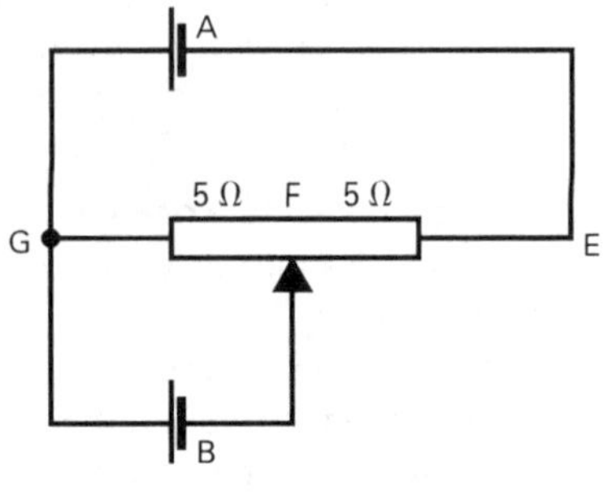

(a) Show that the currents through A and B are $\frac{65}{71}$ amps and $\frac{14}{71}$ amps respectively. What is the magnitude of the current through GF?

(b) Determine the power dissipated as heat in the resistor FE. If the circuit is switched on for 30 minutes, calculate the energy dissipated in FE in kilowatt-hours.

(c) What is the potential difference **(i)** across the terminals of cell A, **(ii)** across the terminals of cell B?

(d) Calculate the rate at which energy is being supplied (or absorbed) by cells A and B.

(e) If the contact F can be moved along the resistor GE, find the value of the resistance GF when no current is flowing through cell B.

(f) At what setting of F would cell B **(i)** be discharging at the maximum possible rate and **(ii)** be charging at the maximum possible rate? [W]

F40 **(a)** Give an expression for P, the power dissipated in a resistor of resistance R, in terms of V, the potential difference across the resistor, and I, the current through the resistor. Hence show that P is given by the expression

$$P = \frac{V^2}{R}$$

A certain electric hotplate, designed to operate on a 250 V supply, has two coils of nichrome wire of resistivity 9.8×10^{-7} Ω m. Each coil consists of 16 m of wire of cross-sectional area 0.20 mm^2.

(b) For one of the coils calculate

(i) its resistance,

(ii) the power dissipation when a 250 V supply is connected across the coil, assuming its resistance does not change with temperature.

(c) Show, by means of diagrams, how these coils may be arranged so that the hotplate may be made to operate at three different powers. In each case, calculate the power rating.

(d) The hotplate is connected to the 250 V supply by means of cable of total resistance 3.0 Ω.

(i) Calculate the power loss in the connecting cable when the hotplate is being used on its middle power rating.

(ii) Comment qualitatively on any change in power loss in the cable when the hotplate is operating at each of its other power ratings.

(e) Different connecting cables are available for use with the hotplate. The maximum safe current which can be used in any one of the cables is 1 A or 3 A or 6 A or 12 A. State which is the most appropriate cable to use and briefly explain one possible danger of using cable with a lower maximum safe current. [C, '92]

THE WHEATSTONE BRIDGE (Chapter 37)

F41 Draw a circuit diagram of a metre bridge circuit being used to compare two resistances. Why is the method only suitable for two resistances of the same order of magnitude? [L]

F42 The slide-wire of a Wheatstone bridge is 1 m long, and has a resistance of 2 Ω. The bridge has a 2 V power supply with negligible internal resistance, and is used with a galvanometer of resistance 1000 Ω. The bridge is used to compare a 1 Ω with a 3 Ω resistor.

(a) Where is the balance point on the bridge wire?

(b) If the balance point can be determined to within ±1 mm, what is the smallest current which the galvanometer can detect? [S]

F43 In an experiment to investigate the variation of resistance with temperature, a nickel wire and a 10 Ω standard resistor were connected in the gaps of a metre bridge. When the nickel wire was at 0 °C a balance point was found 40 cm from the end of the bridge wire adjacent to the nickel wire. When it was at 100 °C the balance point occurred at 50 cm. Calculate:

(a) the temperature of the nickel wire (on its resistance scale) when the balance point was at 42 cm,

(b) the resistivity of nickel at this temperature if the wire was then 150 cm long and of cross-sectional area 2.5×10^{-4} cm^2.

Explain the advantage of using a 10 Ω standard resistor in preference to a 100 Ω standard in this experiment. [J]

F44 **(a)** Explain what is meant by *electrical resistivity*, and show that its unit is Ω m.

(b) 3.00 m of iron wire of uniform diameter 0.80 mm has a potential difference of 1.50 V across its ends.

(i) Calculate the current in the wire.

(ii) If E is the uniform electric field strength along the length of the wire, and J is the uniform current per unit cross-sectional area of the wire, calculate the magnitude of E and J. Calculate the ratio E/J and comment on the result.

Resistivity of iron $= 10.2 \times 10^{-8}\,\Omega\,\text{m}$.

(c) In the circuit shown below, each resistor has a resistance of $10.0\,\Omega$; the battery has an emf of $12.0\,\text{V}$ and is of negligible internal resistance.

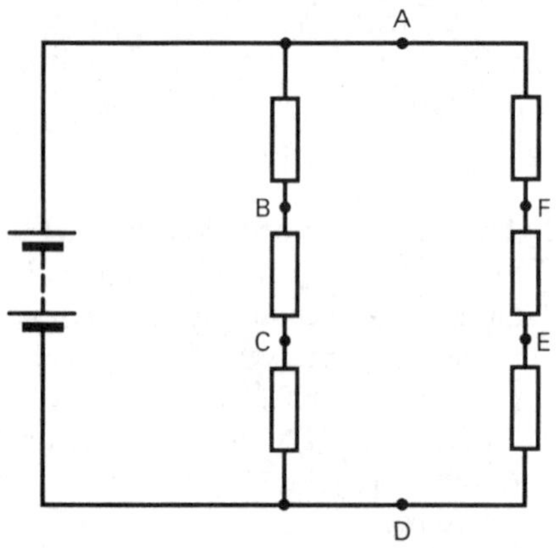

(i) Calculate the potential difference between C and F.

(ii) When a certain resistor is connected between D and F no current flows in a galvanometer connected between C and F. Calculate the resistance of this resistor and the total power delivered by the battery in this case.

[J]

F45 For the Wheatstone bridge shown derive a relation between the resistances P, Q, R and X when the bridge is balanced.

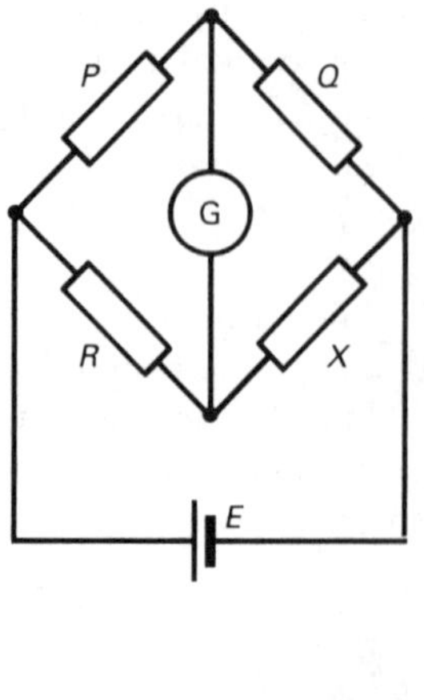

In an experiment in which X was a torch bulb the following results were obtained for the balanced bridge:

	P	Q	R
(a)	$1000\,\Omega$	$0.5\,\Omega$	$1000\,\Omega$
(b)	$1000\,\Omega$	$1000\,\Omega$	$5\,\Omega$

What conclusions can you draw about the resistance of X and how are they explained?

[J]

F46 **(a)** Define *resistivity* and *temperature coefficient of resistance*.

(b) Outline briefly how you would measure the temperature coefficient of resistance of copper in the laboratory.

(c) The temperature at which the tungsten filament of a 12 V, 36 W lamp operates is $1750\,°\text{C}$. Taking the temperature coefficient of resistance of tungsten to be $6 \times 10^{-3}\,\text{K}^{-1}$, find the resistance of the filament at room temperature, $20\,°\text{C}$.

(d) The table gives readings for two filament lamps A and B of different ratings.

Current, I in A	0	0.05	0.10	0.15	0.20
PD across lamp A, V_A *in* V	0	0.40	1.1	2.8	6.5
PD across lamp B, V_B *in* V	0	1.25	2.6	5.0	9.1

(i) On the same graph, with I as y-axis, draw the graph of I against V for each lamp.

(ii) The two lamps A and B are connected in parallel. Find and tabulate the corresponding values of the current I and the PD V across the lamps up to 6 V, and draw the I–V graph (on the same graph as (i)). [O]

F47 Explain what is meant by a Wheatstone Bridge network, and derive a relation between the resistances when the bridge is balanced.

The potential difference V across a filament lamp is related to the current I by $V = 2I + 8I^2$. The lamp is connected in one arm of a Wheatstone bridge and the other arms are each of constant resistance $4\,\Omega$. Determine the potential difference which must be applied to the bridge so that it is balanced. [W*]

THE POTENTIOMETER (Chapter 38)

F48

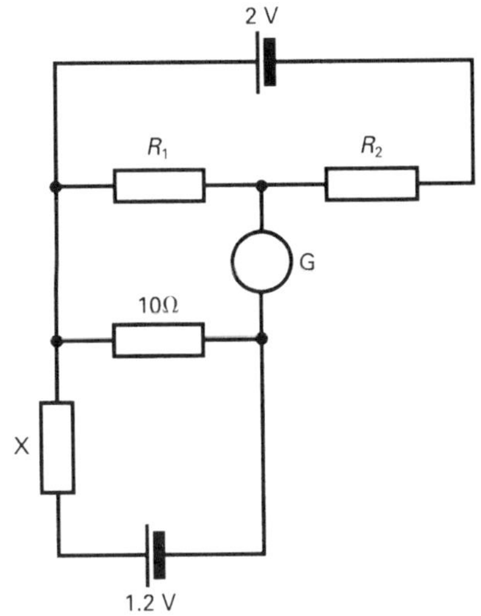

A circuit is set up as shown above. When $R_1 = 10$ ohms and $R_2 = 90$ ohms there is no current through the galvanometer G. What is the resistance of X? (The resistances of the cells are negligible.) [W, '90]

F49

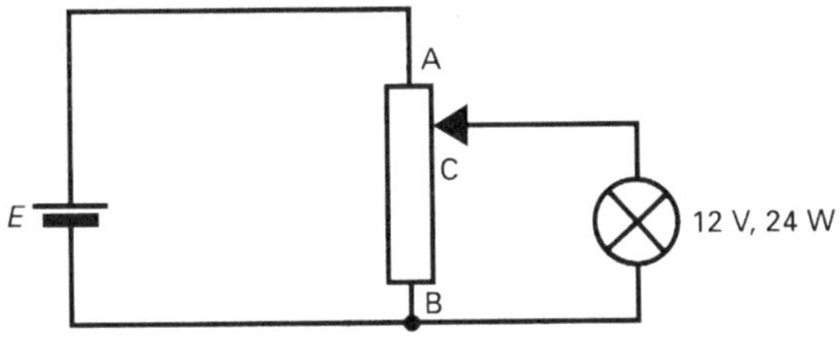

In the diagram above AB is a variable resistor with moveable contact C. The total resistance of AB is 16 Ω. The resistor is used as a potential divider which, with a cell of EMF E and no internal resistance provides a PD to a bulb rated 12 V, 24 W. When the resistance of AC is 4 Ω (and of CB is 12 Ω) the bulb operates at 24 W.

Calculate the value of E. [W, '92]

F50 A simple potentiometer circuit is set up as shown, using a uniform wire AB, 1.0 m long, which has a resistance of 2.0 Ω. The resistance of the 4 V battery is negligible. If the variable resistor R were given a value of 2.4 Ω, what would be the length AC for zero galvanometer deflection?

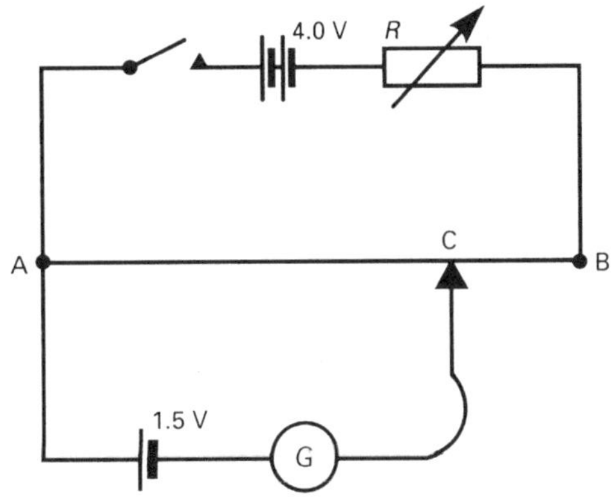

If R were made 1.0 Ω and the 1.5 V cell and galvanometer were replaced by a voltmeter of resistance 20 Ω, what would be the reading of the voltmeter if the contact C were placed at the mid-point of AB? [L]

F51 Draw a labelled circuit diagram of a potentiometer arranged to compare potential differences. Explain why no current flows through the galvanometer when the potentiometer is balanced. Why is a standard cell needed in order to measure the EMF of another cell using a potentiometer? Why is this measurement unaffected by the internal resistance of the cell?

Describe an experiment, using a potentiometer, to determine the internal resistance of a cell. Give the theory of the method and show how the observations may be displayed in the form of a straight line graph from which the internal resistance may be deduced. [L]

F52 What is the advantage of the slide-wire potentiometer over the moving coil voltmeter when used to compare two EMFs?

A battery of EMF E volt and internal resistance r ohm is joined in series with two resistances X ohm and Y ohm in a closed circuit. A standard cell of EMF 1.06 V and a galvanometer are joined in series and the combination is connected across X. The galvanometer shows no deflection when $X = 40.0\,\Omega$ and $Y = 149\,\Omega$, nor when $X = 60.0\,\Omega$ and $Y = 224\,\Omega$. Calculate the values of E and r. [J]

F53 **(a)** Explain what is meant by **(i)** the electromotive force, **(ii)** the internal resistance, of a cell.

(b) Describe how you would use a potentiometer to determine the internal resistance of a cell. Give full experimental details of your procedure and a labelled circuit diagram. Explain how you would use your results to plot a linear graph and determine the internal resistance from it.

(c) A series circuit is formed consisting of a battery with considerable internal resistance and two resistors of resistance 10 Ω and 990 Ω respectively. The potential difference across the 10 Ω resistor is balanced against the potential drop across 715 mm of uniform potentiometer wire carrying a steady current. When the resistors in the circuit are replaced by two others of resistance 1 Ω and 99 Ω respectively, the length of the same potentiometer wire carrying the same current required to balance the PD across the 1 Ω resistor is 500 mm. Calculate the internal resistance of the battery. [J]

F54 **(a)** A cell has an EMF of 1.5 V and an internal resistance of 10 Ω.

(i) Explain what is meant by an *EMF* of 1.5 V.

(ii) Explain why it is necessary, when measuring the EMF of the cell, that it should not be supplying current to a load.

(iii) What will be the voltage recorded by a voltmeter of resistance 100 Ω when it is used in an attempt to measure the EMF of the cell by connecting it between the terminals of the cell?

(iv) What power would be dissipated in the voltmeter?

(v) Explain why the power delivered by a cell is small for both very low resistance loads and for very high resistance loads.

(b) A slide-wire potentiometer can be used to give an accurate value of the EMF of the cell in (a) provided that a suitable driver cell and standard cell are available.

(i) Draw a diagram of the circuit you would use. Give a suitable value for the EMF of the driver cell and state a suitable range for any meter you would include.

(ii) Explain how you would proceed to determine the EMF of the unknown cell showing clearly how the value should be determined from your measurements.

(iii) State two advantages of using a potentiometer compared with that of using a moving coil voltmeter such as that in (a) (iii).

(c) Assuming that the potentiometer wire in the circuit you have described in (b) has a resistance of 20 Ω, explain how you would modify the circuit to enable measurement of the EMF of a thermocouple known to give an EMF of about 10 mV. [AEB, '92]

F55 **(a)** Describe how you would use a potentiometer incorporating a wire of resistance about 2 Ω and other necessary apparatus to calibrate an ammeter in the range 0.2 to 1.0 A. Give a labelled circuit diagram; specify, with the reason, the magnitude of any resistance components used; and show how the results are calculated from the observations.

(b) If the current in the potentiometer wire slowly decreased during the above experiment, how would you detect this change and what effect would the change have on your results?

(c) A standard cell has an EMF of 1.0186 V. When a 0.5 MΩ resistor is connected across it the potential difference between the terminals is 1.0180 V. Calculate a value for the internal resistance of the cell. If a standard cell with this internal resistance were used in your experiment would it cause an error in the results? Give a reason for your answer. [J]

F56 **(a)** Explain, with the aid of circuit diagrams, how a potentiometer can be used to measure:

(i) a current known to be of the order of 10 A,

(ii) an EMF known to be of the order of 5 mV.

You may assume that any apparatus you require is available.

(b) In an experiment to measure the EMF of a thermocouple, a potentiometer is used in which the slide wire is 2.000 m long and the resistance of the wire is 6.000 Ω. The current through the wire is 2.000 mA and the balance point is 1.055 m from one end. Calculate the EMF of the thermocouple. [AEB, '79]

F57 **(a)** Explain the principle of the slide-wire potentiometer.

(b) Discuss the relative merits of using **(i)** a potentiometer, **(ii)** a moving-coil voltmeter for determining the EMF of a cell of EMF approximately 1.5 V.

(c) A potentiometer wire is 1.00 m long and has a resistance of about 4 Ω. Explain how you could obtain a potential difference of about 4 mV across the wire using a 2 V accumulator of negligible internal resistance and any other essential apparatus. Support your explanation with a suitable calculation.

(d) A thermocouple generates an EMF which is proportional to the temperature difference between its junctions and, for a temperature difference of 100 K, the EMF is approximately 4 mV. Draw a labelled circuit diagram showing how you could use the potentiometer adapted in (c) to obtain temperature readings in the range 0 °C to 100 °C so that each centimetre of wire corresponds to 1 K and describe how you would use the apparatus to determine room temperature. You may assume that apparatus is available for maintaining fixed temperatures of 0 °C and 100 °C.

(e) Give *two* reasons (other than faulty apparatus and poor electrical contact) why it may not be possible to obtain a balance point in a potentiometer experiment and indicate in each case how the fault can be remedied. [J]

ELECTRIC FIELDS AND CAPACITORS (Chapters 39 and 40)

F58 **(a)** For each of the following, state whether it is a scalar or a vector and give an appropriate unit:

(i) electric potential,

(ii) electric field strength.

(b) Points A and B are 0.10 m apart. A point charge of $+3.0 \times 10^{-9}$ C is placed at A and a point charge of -1.0×10^{-9} C is placed at B.

(i) X is the point on the straight line through A and B, between A and B, where the electric potential is zero. Calculate the distance AX.

(ii) Show on a diagram the approximate position of a point, Y, on the straight line through A and B where the electric field strength is zero. Explain your reasoning, but no calculation is expected. [J, '92]

F59

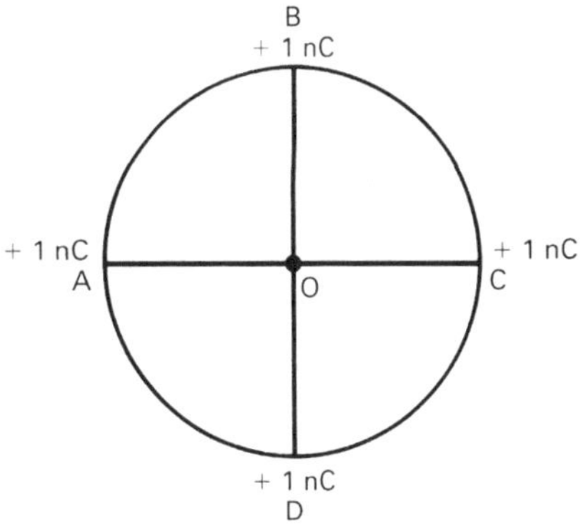

Charges of +1 nC are situated at each of A, B, C and D as shown. Given that AOC = 6 m, calculate the work done in bringing a charge of +5 nC from a distant point to the centre O.

(Assume that $1/4\pi\varepsilon_0 = 9 \times 10^9 \text{ m F}^{-1}$.) [W, '90]

F60 Two point charges of +5 μC and −3 μC are placed at the points A and B as shown in the diagram below.

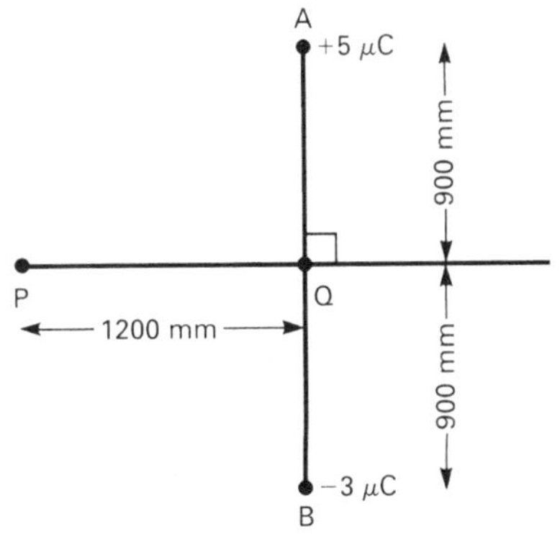

Calculate the work done in moving a charge of −3 μC from P to Q.

(Take $1/4\pi\varepsilon_0 = 9 \times 10^9 \text{ m F}^{-1}$.) [W, '91]

F61 **(a)** State the relationship between electric field intensity and potential gradient.

(b) The diagram below shows a scale drawing of the equipotential lines in the vicinity of a collection of charges.

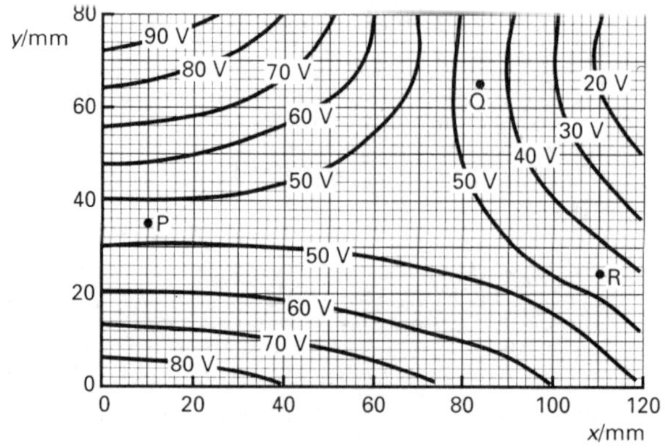

Use your expression from part (a) to *estimate* the magnitudes of the electric field intensities at points P, Q and R on the scale drawing. Indicate, where appropriate, the approximate field directions by arrows on [a simplified copy of] the diagram. [W, '91]

F62

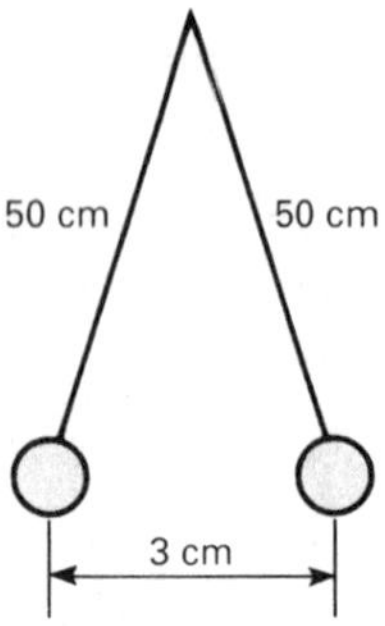

Two light, conducting spheres, each 6 mm diameter and having a mass of 10 mg, are suspended from the same point by fine insulating fibres 50 cm long. Due to electrostatic repulsion, the spheres are in equilibrium when 3 cm apart. What is:

(a) the force of repulsion between the spheres,
(b) the charge on each sphere,
(c) the potential of each sphere?

(Acceleration due to gravity $= 10\,\text{m}\,\text{s}^{-2}$.) [S]

F63 Two large horizontal metal plates are arranged one above the other a short distance apart in a vacuum. A small, negatively charged sphere introduced between them may be held stationary if an appropriate potential difference is applied to the plates. Explain how (if at all) the equilibrium of the sphere might be affected if the plates were slowly pulled apart while remaining connected to the source of EMF. [C]

F64 (a) Define the terms **(i)** electric field strength, **(ii)** electric potential, both at a point in an electric field.

(b) An electric field is established between two parallel plates as shown below.

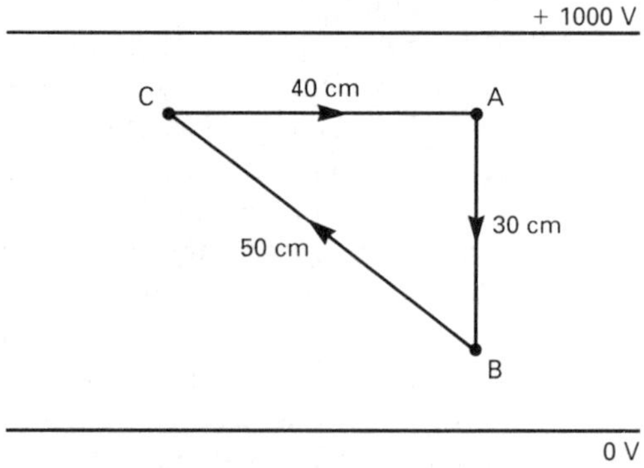

The plates are 50 cm apart and a PD of 1000 V is applied between them.

A point charge of value $+1.0\,\mu\text{C}$ is held at point A. It is moved first to B then to C and finally back to A. The distances are shown in the diagram. Calculate:

(i) the force experienced by the charge at A,
(ii) the force experienced by the charge at B,
(iii) the energy involved in moving the charge from A to B,
(iv) the energy required to move the charge from C to A,
(v) the net energy needed to move the charge along the route ABCA. [S*]

F65 (a) Define the terms *potential* and *field strength* at a point in an electric field. The diagram shows two horizontal parallel conducting plates in a vacuum.

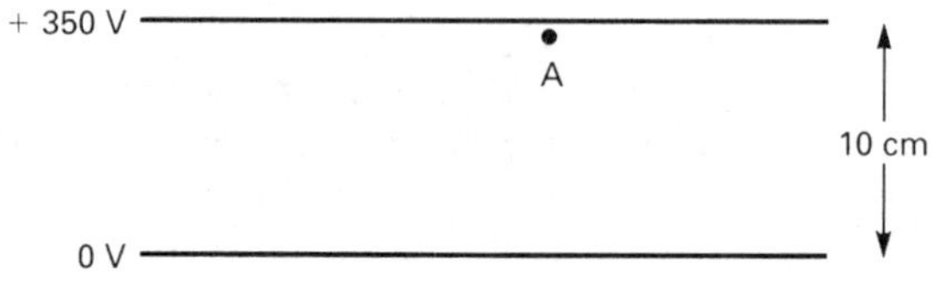

A small particle of mass 4×10^{-12} kg, carrying a positive charge of 3.0×10^{-14} C is released at A close to the upper plate. What *total* force acts on this particle?

Calculate the kinetic energy of the particle when it reaches the lower plate.

(b) The diagram shows a positively charged metal sphere and a nearby uncharged metal rod.

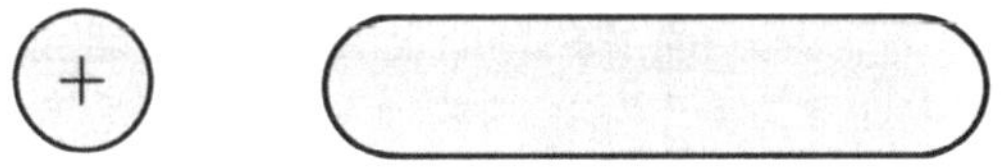

Explain why a redistribution of charge occurs on the rod when the charged metal sphere is brought close to the rod.

Copy this diagram and show on it the charge distribution on the rod. Sketch a few electric field lines in the region between the sphere and the rod.

Sketch graphs which show how **(i)** the potential relative to earth, and **(ii)** the field strength vary along the axis of the rod from the centre of the charged sphere to a point beyond the end of the rod furthest from the sphere. How is graph (i) related to graph (ii)?

How will the potential distribution along this axis be changed if the rod is now earthed? [L]

F66 **(a)** The diagrams show a hollow metal sphere supported on an insulating stand. In (i) a large positive charge is near to the sphere; in (ii) the sphere is earthed; in (iii) the earth connection has been removed and finally in (iv) the positive charge has been removed.

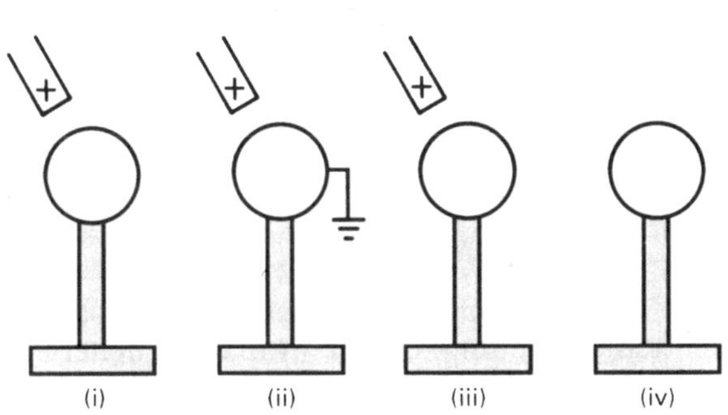

Sketch the distribution of charge on the sphere which you would expect at each of the four stages.

(b) A large, hollow, metal sphere is charged positively and insulated from its surroundings. Sketch graphs of (i) the electric field strength, and (ii) the electric potential, from the centre of the sphere to a distance of several diameters. [AEB, '85]

F67 In a demonstration of Faraday's ice-pail experiment a conducting hollow container C is placed on the cap of an uncharged gold-leaf electroscope. A conducting sphere A, on an insulated handle, is charged positively.

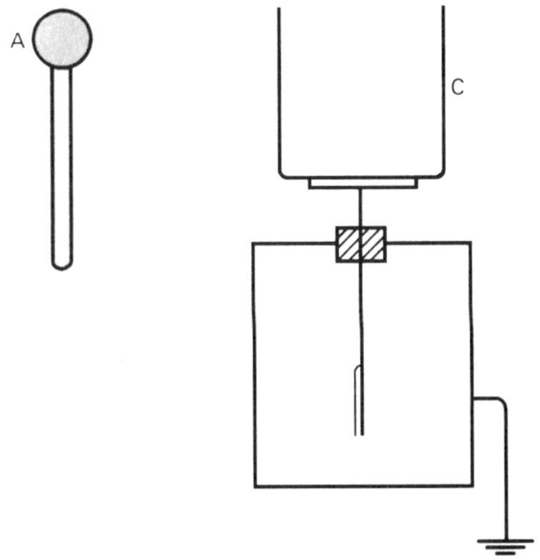

The following sequence of operations is then performed.

(i) The sphere A is placed well inside the container C and moved around without touching the inside of the container.
(ii) The sphere is touched against the inside of the container.
(iii) The sphere is removed from the container.
(iv) The container C is touched with a finger.
(v) The sphere is again placed inside the container and touched against the inside.

(a) State what happens to the leaves of the electroscope in each of the above operations.
(b) Draw a diagram showing the distribution of the charge during operation (i).
(c) State a conclusion which may be drawn from the experiment. [AEB, '79]

F68 **(a)** Explain what is meant by electrostatic induction. In Faraday's ice-pail experiment, a charged conducting sphere is (i) lowered into a deep, insulated metal can, (ii) lowered further so that it touches the can, and (iii) removed from the can. Describe the distribution and magnitude of the charge induced on the can at each stage.

(b) Describe the construction and explain the mode of action of a Van de Graaff electrostatic generator. Discuss briefly the factors which limit the maximum voltage attainable.

The belt of a Van de Graaff generator is of width 10^{-1} m and travels at a speed of $12\,\mathrm{m\,s^{-1}}$. The charge density on the belt is $3 \times 10^{-5}\ \mathrm{C\,m^{-2}}$. The generator is connected to a resistor of $4 \times 10^{10}\,\Omega$. Determine:

(i) the maximum steady current that can be drawn from the generator,

(ii) the maximum potential difference across the resistor, and

(iii) the minimum possible power output of the motor which drives the belt. [W]

F69 Describe the Van de Graaff electrostatic generator, and explain how it operates. Why is the 'live' terminal of the generator made in the form of a large sphere?

A large Van de Graaff generator has a top terminal in the form of a sphere of diameter 4 m. When the terminal is at the operating potential of 5×10^6 V, what is **(a)** the stored charge, **(b)** the stored energy, **(c)** the electric field (potential gradient) at the surface of the sphere? [S]

F70

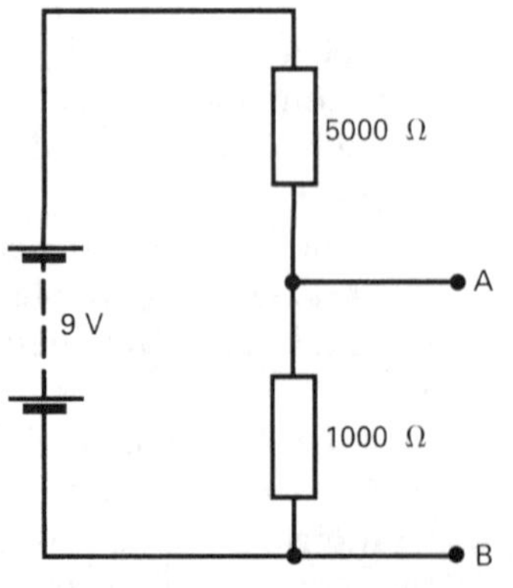

What is the final potential difference between A and B in the above circuit:

(a) in the circuit as shown,

(b) if an additional 500 Ω resistor were connected from A to B,

(c) If the 500 Ω resistor were replaced by a 2 μF capacitor?

For what purposes would the circuit in (a) be useful? [L]

F71 A 12 V DC source is connected in a series circuit containing two capacitors (A and B) and a switch, as shown below.

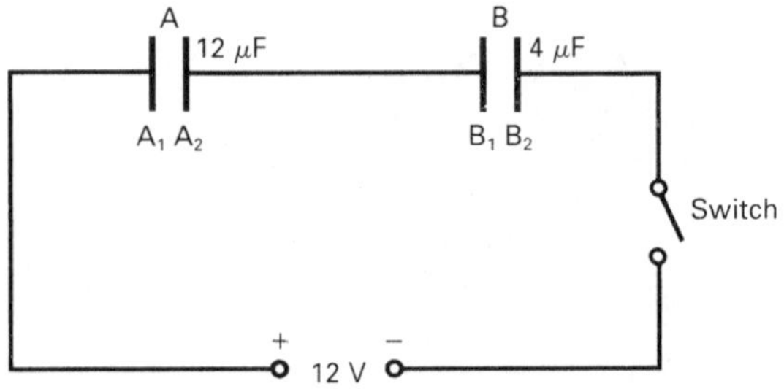

(a) Describe the charge movements (in terms of electrons) that occur throughout the circuit when the switch is first closed.

(b) Calculate the combined capacitance of the two capacitors.

(c) Calculate the charge on each of the four plates (A_1, A_2, B_1, B_2) after the switch has been closed for some time.

(d) How would your answers (i) in (a), (ii) in (c), be affected, if at all, if a 1 megohm resistor were connected in series between A_2 and B_1? [O, '91]

F72

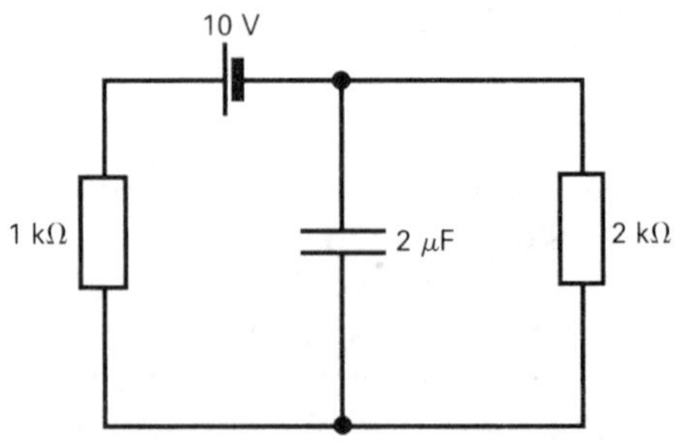

Calculate the charge on the capacitor in the above circuit. [L]

F73 Explain what is meant by **(a)** the *potential*, and **(b)** the *electric field strength*, at a point in an electric field. State and explain how these quantities are related.

A hollow metal container with a small opening, standing on an insulating material, is given an electric charge. Explain how, using such other apparatus as you may require, you would show that there is no charge on the inside of the container.

A parallel plate capacitor consists of two plates of area $1.00 \times 10^{-2}\,m^2$ placed a distance $2.00 \times 10^{-2}\,m$ apart in air. The capacitor is charged so that the potential difference between the plates is 1000 V.

Calculate:

(i) the electric field between the plates (neglect edge effects),

(ii) the capacitance of the capacitor,

(iii) the energy stored in the capacitor.

What limits the potential difference which may exist between the plates?

(Permittivity of free space, $\varepsilon_0 = 8.84 \times 10^{-12}\ F\,m^{-1}$; relative permittivity of air = 1.00.) [L]

F74

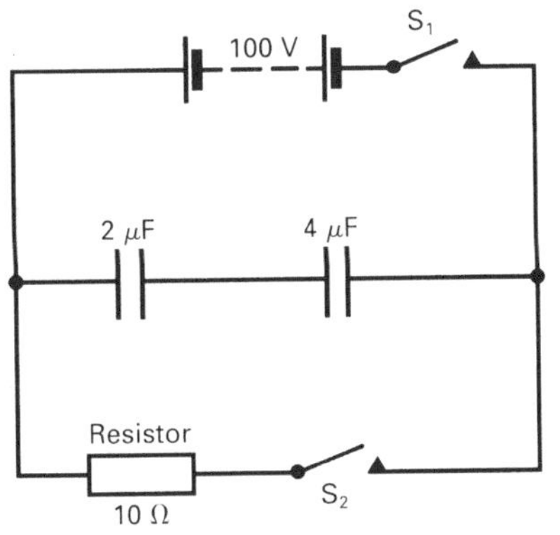

If S_2 is left open and S_1 is closed, calculate the quantity of charge on each capacitor.

If S_1 is now open and S_2 is closed, how much charge will flow through the 10 Ω resistor?

If the entire process were repeated with the 10 Ω resistor replaced by one of much larger resistance what effect would this have on the flow of charge? [L]

F75 A thundercloud has a horizontal lower surface (area 25 km^2) which is 750 m above the earth. Treating the arrangement as a capacitor, calculate the electrical energy stored when its potential is 10^5 V above earth. If the cloud rises to 1250 m what change in electrical energy occurs? [J]

F76 Two horizontal parallel plates, each of area 500 cm^2, are mounted 2 mm apart in a vacuum. The lower plate is earthed and the upper one is given a positive charge of 0.05 μC. Neglecting edge effects, find the electric field strength between the plates and state in what direction the field acts.

Deduce values for **(a)** the potential of the upper plate, **(b)** the capacitance between the two plates and **(c)** the electrical energy stored in the system.

If the separation of the plates is doubled, keeping the lower plate earthed and the charge on the upper plate fixed, what is the effect on the field between the plates, the potential of the upper plate, the capacitance and the electrical stored energy?

Discuss how the change in energy can be accounted for. [O & C]

F77

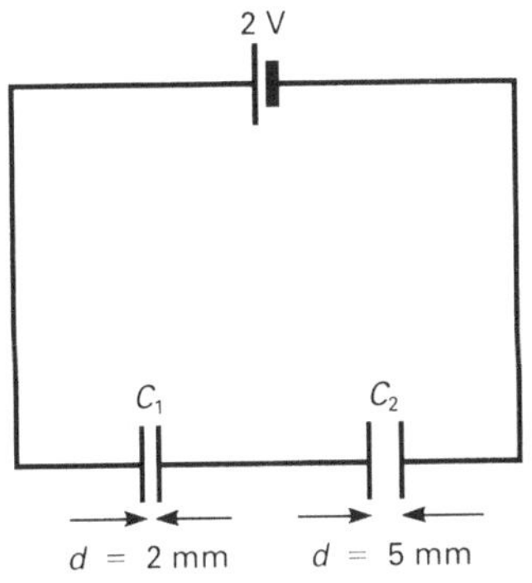

In the above circuit, the parallel-plate capacitors are identical except that the distances apart of the plates, d, are as shown. Find the potential difference across each capacitor and the electric field intensity between each pair of plates. [W]

F78 (a) Two capacitors of capacitance C_1 and C_2 respectively are connected in series. Derive an expression for the capacitance of a single equivalent capacitor.

(b)

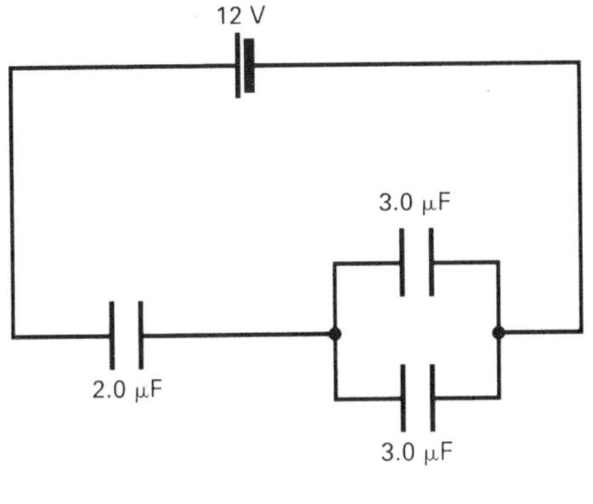

For the circuit shown, calculate

(i) the capacitance of this combination of capacitors,

(ii) the total energy stored in the capacitors,

(iii) the PD across the 2.0 μF capacitor. [J, '92]

F79 **(a)** Define *capacitance.*

Describe an experiment to show that the capacitance of an air-spaced parallel-plate capacitor depends on the separation of the plates.

State three desirable properties of a dielectric material for use in a high-quality capacitor.

(b) Two plane parallel-plate capacitors, one with air as dielectric and the other with mica as dielectric, are identical in all other respects. The air capacitor is charged from a 400 V DC supply, isolated, and then connected across the mica capacitor which is initially uncharged. The potential difference across this parallel combination becomes 50 V. Assuming the relative permittivity of air to be 1.00, calculate the relative permittivity of mica.

(c) Compare the energy stored by the single charged capacitor in part (b) with the energy stored in the parallel combination.

Comment on these energy values. [AEB, '79]

F80 A capacitor charged from a 50 V DC supply is discharged across a charge-measuring instrument and found to have carried a charge of 10 μC. What is the capacitance of the capacitor and how much energy was stored in it? [L]

F81 What is the electrical energy stored in an isolated capacitor A of capacitance 10 μF when the potential difference across it is 300 V?

A second, initially uncharged capacitor B of capacitance 20 μF is now connected across A. What is the energy stored in A and in B?

How do you reconcile your result with the principle of conservation of energy? [S]

F82 A capacitor consists of two parallel metal plates in air. The distance between the plates is 5.00 mm and the capacitance is 72 pF. The potential difference between the plates is raised to 12.0 V with a battery.

(a) Calculate the energy stored in the capacitor.

(b) The battery is then disconnected from the capacitor; the capacitor retains its charge. Calculate the energy stored in the capacitor if the distance between the plates is now increased to 10.00 mm.

The answers to (a) and (b) are different. Why is this so?

Use your answers to estimate the average force of attraction between the plates while the separation is being increased. [L, '91]

F83 Explain the meaning of the terms *capacitor* and *dielectric*. Give a brief qualitative explanation, based on the behaviour of molecules, for the change in capacitance of a capacitor when the space between its plates is filled with a dielectric.

Give *one* practical example of the use of a capacitor, explaining how the properties of the capacitor are utilized in the example you choose.

Two capacitors, of capacitance 2 μF and 3 μF, are connected as shown to batteries A and B which have EMF 6 V and 10 V respectively.

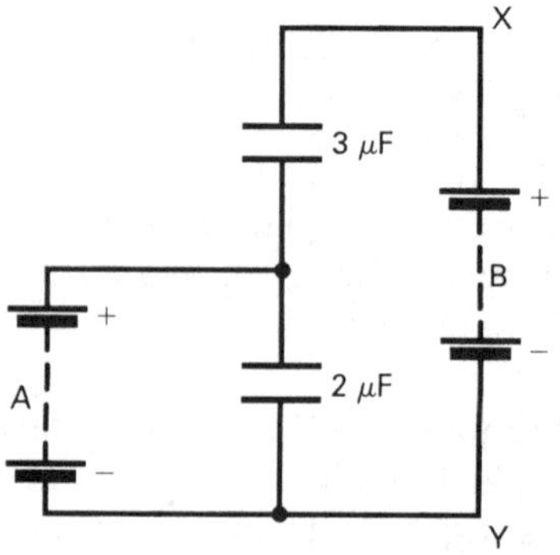

What is the energy stored in each of the capacitors? Calculate also the stored energy in each capacitor **(a)** when the terminals of battery A are reversed, and **(b)** when the battery B is disconnected and the points X and Y are connected together. [O & C]

F84 Two capacitors, of capacitance 2.0 μF and 4.0 μF respectively, are each given a charge of 120 μC. The positive plates are now connected together, as are the negative plates. Calculate the new potential difference between the plates of the capacitors.

Calculation shows that the total energy stored by the capacitors has decreased. Explain this change in energy. [AEB, '79]

F85 A 100 μF capacitor is charged from a supply of 1000 V, disconnected from the supply and then connected across an uncharged 50 μF capacitor. Calculate the energy stored initially and finally in the two capacitors. What conclusion do you draw from a comparison of these results? [J]

F86 The capacitance of a certain variable capacitor may be varied between limits of 1×10^{-10} F and 5×10^{-10} F by turning a knob attached to the movable plates. The capacitor is set to 5×10^{-10} F, and is charged by connecting it to a battery of EMF 200 V.

(a) What is the charge on the plates?

The battery is then disconnected and the capacitance changed to 1×10^{-10} F.

(b) Assuming that no charge is lost from the plates, what is now the potential difference between them?

(c) How much mechanical work is done against electrical forces in changing the capacitance? [C]

F87 The following circuit is set up with both switches open.

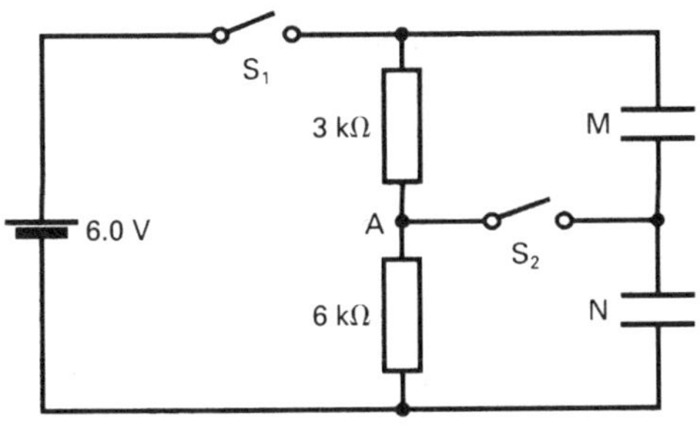

Assume that the internal resistance of the 6.0 V cell is negligible. Calculate the charges stored on the capacitors M and N:

(a) with S_1 closed but S_2 left open,

(b) with both switches closed.

In which direction is the initial flow of current in AB when S_2 is closed a few seconds later than S_1? [L]

F88 **(a)** A *capacitor* is charged through a *resistor* using a battery of constant *EMF.*

(i) Explain the meaning of the terms in italics.

(ii) Draw sketch graphs on the same time axis showing how the charge on the capacitor and how the current through the circuit vary with time. Qualitatively explain their shapes.

(b) **(i)** Derive an expression for the electrical energy stored in a capacitor of capacitance C when charged to a potential difference V.

(ii) If $C = 2\,\mu$F and $V = 4$ V, calculate

(I) the final energy stored in the capacitor,

(II) the work done by the battery in the charging process

(c) Account for any difference between your answers in parts (I) and (II) above. [W, '90]

F89 **(a)** Define *capacitance.*

(b) A capacitor of capacitance 10 μF is fully charged from a 20 V DC supply.

(i) Calculate the charge stored by the capacitor.

(ii) Calculate the energy delivered by the 20 V supply.

(iii) Calculate the energy stored by the capacitor.

(iv) Account for the difference between the answers for (ii) and (iii).

(c) The 10 μF capacitor in part (b) was charged from the supply through a resistor of resistance 2.0 kΩ.

(i) Calculate the time constant for this circuit.

(ii) When the capacitor was charged from zero charge, how long did it take for V, the potential difference across the capacitor, to reach 99% of its final value?

(You may use the equation $V = V_0\,(1 - \exp(-t/CR))$ if you wish.) [C, '92]

F90 **(a)** **(i)** Define capacitance.
(ii) Write down an expression for the capacitance of a parallel plate capacitor. Identify *each* term in your expression.

(b) Derive expressions for the effective capacitance of two capacitors C_1 and C_2
(i) in series,
(ii) in parallel.

(c) If a potential difference V is applied across the two capacitors C_1 and C_2 in series as shown below, show that V_1 is given by the expression

$$\frac{1}{V_1} = \left(\frac{1}{C_1} + \frac{1}{C_2}\right)\frac{C_1}{V}.$$

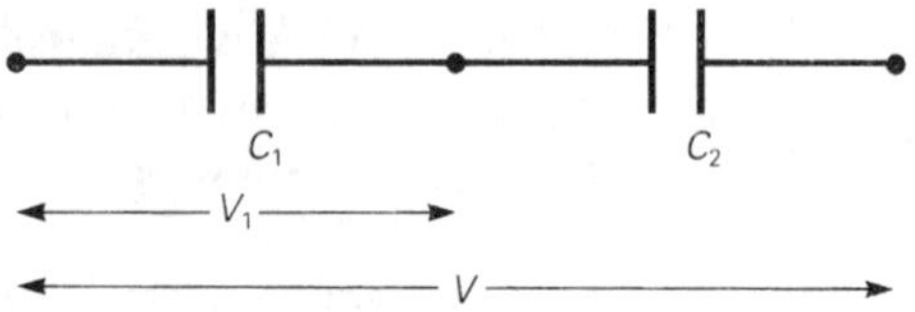

(d) The movement of a thin metal diaphragm in a pressure gauge is determined by the arrangement shown where the two electrodes E_1, E_2 and the diaphragm D form two capacitors C_1 and C_2 in series.

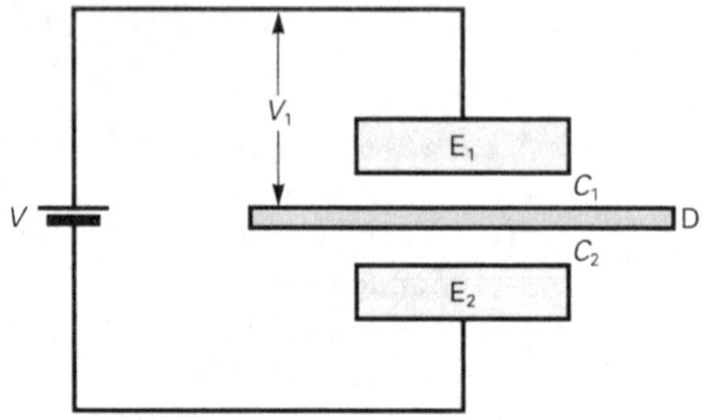

(Assume equilibrium distances E_1 to D and E_2 to D to be both equal to d, and the effective area of C_1 and C_2 to be A.)

Use the expression in part (c) to show that when the diaphragm has been moved down a distance x from its central equilibrium position

$$V_1 = \frac{V(d+x)}{2d}$$ [W,' 92]

F91 **(a)** List the factors which determine the capacitance of a parallel plate capacitor.
(b) Explain what is meant by the time constant of the RC circuit in Fig. 1.

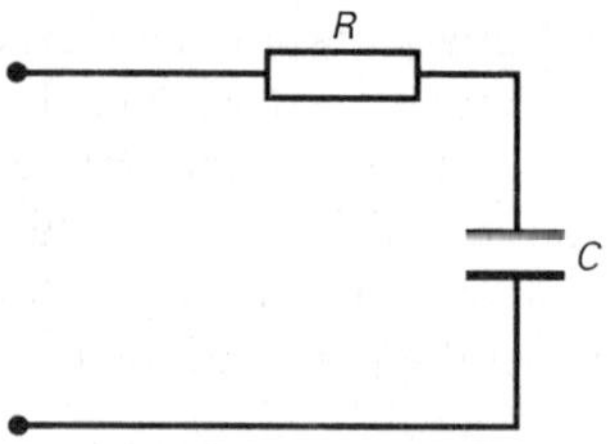

Fig. 1

(c) The switch S of the circuit shown in Fig. 2 is moved according to the timing diagram shown in Fig. 3. Sketch the variation in the voltage V_0 with time for
(i) $R_1 = 0$ and $R_2 = 10\,\text{k}\Omega$
(ii) $R_1 = 10\,\text{k}\Omega$ and $R_2 = 10\,\text{k}\Omega$

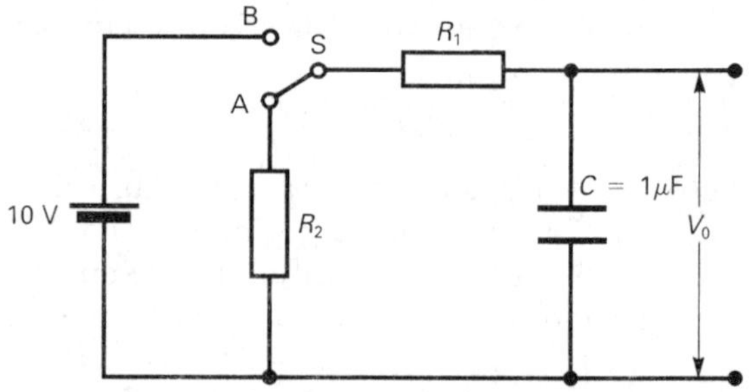

Fig. 2

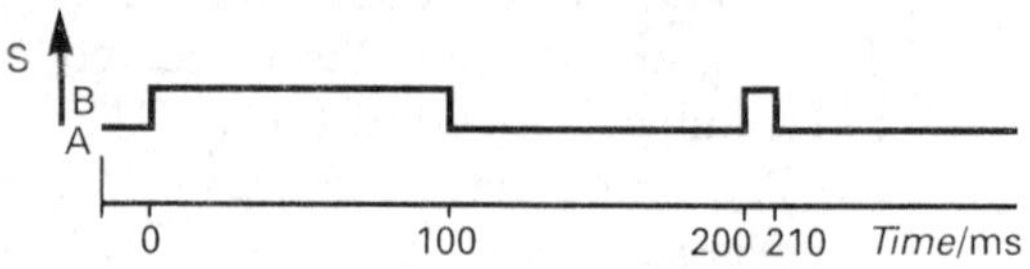

Fig. 3

Assume that the switch has been at position A for a long time before the movement starts. Explain how you arrived at the answer. [W, '91]

F92 **(a)** In the circuit shown in Fig. 1, both switches are open and both capacitors discharged. The capacitor C_1 of value 2 μF, is charged by closing switch S_1.

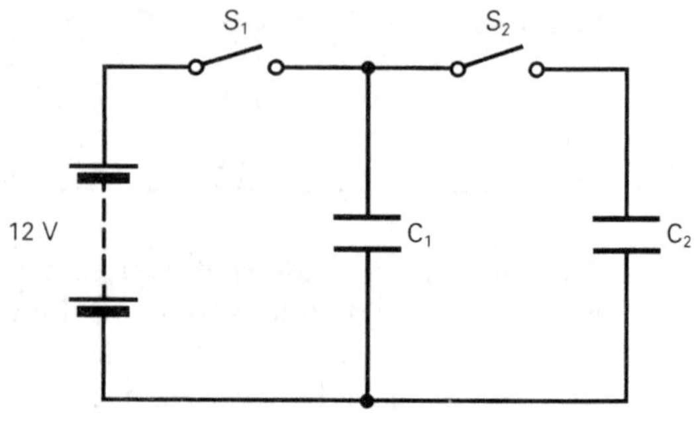

Fig. 1

(i) Calculate the charge on C_1 and the energy stored on it.

S_1 is opened after which S_2 is closed so connecting capacitor C_2, of value 3 μF, across C_1.

(ii) Calculate the final voltage across C_2 and the total energy stored on the two capacitors. Account for the difference between this energy and the energy calculated in (i).

(b) In another experiment using capacitor C_1, a resistor R is placed in series, Fig. 2.

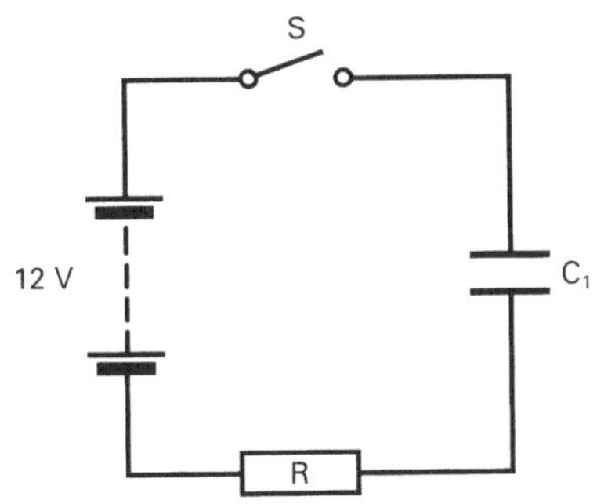

Fig. 2

The voltage, V_C, across C_1 is plotted against time from the closing of the switch, Fig. 3.

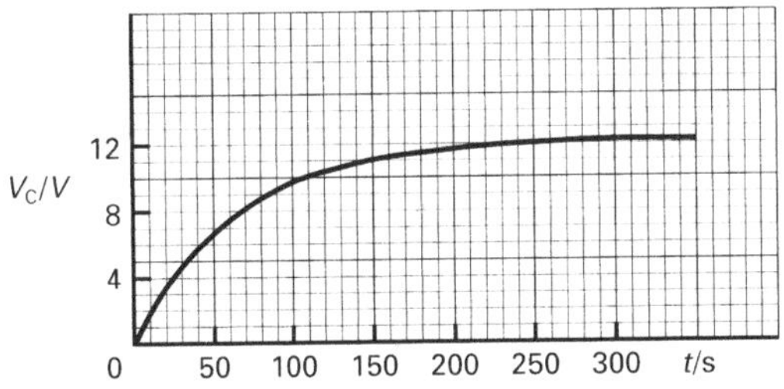

Fig. 3

(i) Sketch a graph of the voltage, V_R, across R, over the same time interval. Estimate from the graph, Fig. 3, the time at which $V_C = V_R$.

(ii) What is the significance of the quantity RC in this circuit, Fig. 2? Explain the shape of the V_C against time graph, shown in Fig. 3, using the quantity RC in your explanation.

(iii) Find the value of R. [O & C, '92]

F93 **(a)** **(i)** Define *capacitance*.

(ii) Describe, with the aid of a circuit diagram, an experiment to determine the ratio of two capacitances each roughly equal to 0.1 μF.

(b) A parallel-plate capacitor is charged by connecting it to the terminals of a battery. Explain how the energy stored in the capacitor is changed when the distance between the plates is increased if **(i)** the battery is left connected, **(ii)** the battery is first disconnected.

(c) A capacitor and a resistor are joined in series and connected across the terminals of a battery of negligible internal resistance. Show that, when the capacitor is fully charged, the energy stored in the capacitor is equal to the energy dissipated in the resistor during the charging process. [J]

F94 Three identical capacitors are discharged and then connected in series across a potential difference V. The *total* energy stored by them is E_s and the charge *on each* is Q_s. They are then discharged and again connected across the same potential difference, but this time they are in parallel. The *total* energy stored this time is E_p and the charge *on each* is Q_p. Find **(a)** the ratio E_p/E_s, and **(b)** the ratio Q_p/Q_s. [S]

F95 Define electric field strength and potential at a point in an electric field.

Explain what is meant by the *relative permittivity* of a material. How may its value be determined experimentally?

A capacitor of capacitance 9.0 μF is charged from a source of EMF 200 V. The capacitor is now disconnected from the source and connected in parallel with a second capacitor of capacitance 3.0 μF. The second capacitor is now removed and discharged. What charge remains on the 9.0 μF capacitor? How many times would the process have to be performed in order to reduce the charge on the 9.0 μF capacitor to below 50% of its initial value? What would the PD between the plates of the capacitor now be? [L]

F96 Define *capacitance* of a capacitor. Describe experiments you would perform to demonstrate qualitatively how the capacitance of a parallel-plate capacitor is affected by the area and separation of the plates and the dielectric between them.

A capacitor is charged to a voltage V and discharged through a resistor.

(a) Sketch the graph of voltage against time during the discharge.

(b) On the same axes sketch the graph obtained if the operation were repeated with a resistor having a much lower resistance than in (a). Label the graphs (a) and (b) respectively.

Two capacitors, each of capacitance 500 pF (1 pF $= 10^{-12}$ F) and with air as dielectric, are connected in parallel and charged by a 6 V battery. The battery is removed, the capacitors remaining connected to each other, and an insulating liquid is poured into one of the capacitors increasing its capacitance to 2500 pF. Calculate the charge on each capacitor before and after the liquid is poured in and the change in total energy of the capacitors which results. [L]

F97 A capacitor of capacitance C is fully charged by a 200 V battery. It is then discharged through a small coil of resistance wire embedded in a thermally insulated block of specific heat capacity 2.5×10^2 J kg^{-1} K^{-1} and of mass 0.1 kg. If the temperature of the block rises by 0.4 K, what is the value of C? [L]

F98

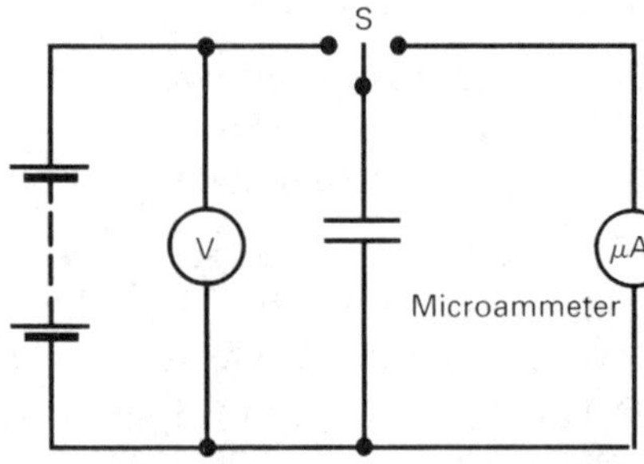

In the circuit shown above, S is a vibrating reed switch and the capacitor consists of two flat metal plates parallel to each other and separated by a small air-gap. When the number of vibrations per second of S is n and the potential difference between the battery terminals is V, a steady current I is registered on the microammeter.

(a) Explain this and show that $I = nCV$, where C is the capacitance of the parallel plate arrangement.

(b) Describe how you would use the apparatus to determine how the capacitance C depends upon **(i)** the area of overlap of the plates, **(ii)** their separation, and show how you would use your results to demonstrate the relationships graphically.

(c) Explain how you could use the measurements made in (b) to obtain a value for the permittivity of air.

(d) In the above arrangement, the microammeter records a current I when S is vibrating. A slab of dielectric having the same thickness as the air-gap is slid between the plates so that one-third of the volume is filled with dielectric. The current is now observed to be $2I$. Ignoring edge effects, calculate the relative permittivity of the dielectric. [J]

MAGNETIC EFFECTS OF CURRENTS (Chapter 41)

F99 A long air-cored solenoid has two windings wound on top of each other. Each has N turns per metre and resistance R. Deduce expressions for the flux density at the centre of the solenoid when the windings are connected **(a)** in series, and **(b)** in parallel, to a battery of EMF E and of negligible internal resistance. (In each case the magnetic fields produced by the currents in the two windings reinforce.) [L]

F100 A current of 3 A flows down each of two long, vertical wires, which are mounted side by side 5 cm apart. Show on a diagram the magnetic field pattern in a horizontal plane, indicating clearly the direction of the magnetic field at any point. What is the magnitude and direction of the force on a 25 cm length of a wire? [S]

F101 The diagram represents a cylindrical aluminium bar AB resting on two horizontal aluminium rails which can be connected to a battery to drive a current through AB. A

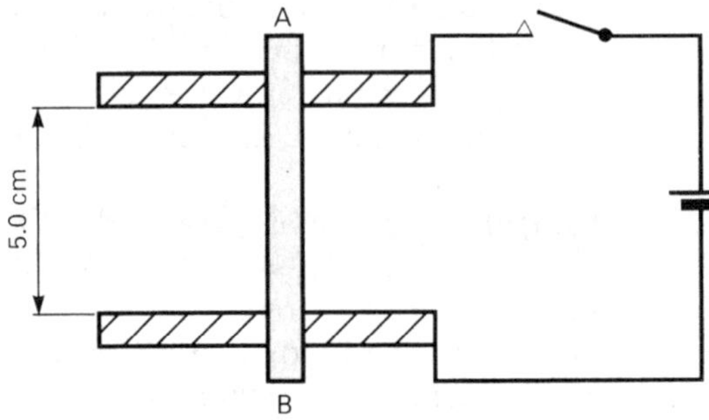

magnetic field, of flux density 0.10 T, acts perpendicularly to the paper and into it. In which direction will AB move if the current flows from A to B?

Calculate the angle to the horizontal to which the rails must be tilted to keep AB stationary if its mass is 5.0 g, the current in it is 4.0 A and the direction of the field remains unchanged.

(Acceleration of free fall, $g = 10\,\text{m}\,\text{s}^{-2}$.)[L]

F102

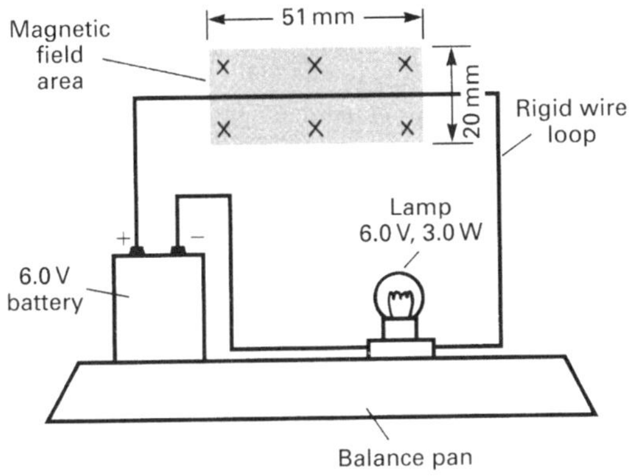

The diagram shows a rigid conducting wire loop connected to a 6.0 V battery through a 6.0 V, 3.0 W lamp. The circuit is standing on a top-pan balance. A uniform horizontal magnetic field of strength 50 mT acts at right angles to the straight top part of the conducting wire in the direction indicated in the diagram, i.e. into the paper. This magnetic field extends over the shaded area in the diagram. The balance reads 153.860 g. Calculate

(a) the force exerted on the conducting wire by the magnetic field, and
(b) the new balance reading if the direction of the magnetic field is reversed.

[L, '91]

F103 Two parallel wires have currents passed through them which are in the same direction. Draw a diagram showing the directions of the currents and of the forces on the wires.

Alternating currents are now passed through the wires. Explain what forces would act if the currents were:

(a) in phase, and
(b) out of phase, by π rad. [L]

F104 **(a)** A long straight wire of radius a carries a steady current. Sketch a diagram showing the lines of magnetic flux density (B) near the wire and the relative directions of the current and B. Describe, with the aid of a sketch graph, how B varies along a line from the surface of the wire at right-angles to the wire.

(b) Two such identical wires R and S lie parallel in a horizontal plane, their axes being 0.10 m apart. A current of 10 A flows in R in the opposite direction to a current of 30 A in S. Neglecting the effect of the Earth's magnetic flux density calculate the magnitude and state the direction of the magnetic flux density at a point P in the plane of the wires if P is **(i)** midway between R and S, **(ii)** 0.05 m from R and 0.15 m from S.

[J]

F105 Write down expressions for:

(a) the magnetic field B at a distance r from a long straight wire carrying a current I,
(b) the force on a straight wire of length l placed perpendicular to a magnetic field B and carrying a current I.

Define the *ampere*, and show that the expressions you have given above are consistent with this definition. [S]

F106 The diagram shows a uniform magnetic flux density B in the plane of the paper. Q and R mark the points where two long, straight and parallel wires carry the *same* current, I, in the same direction and perpendicular to the paper. The line through QR is at right angles to the direction of B.

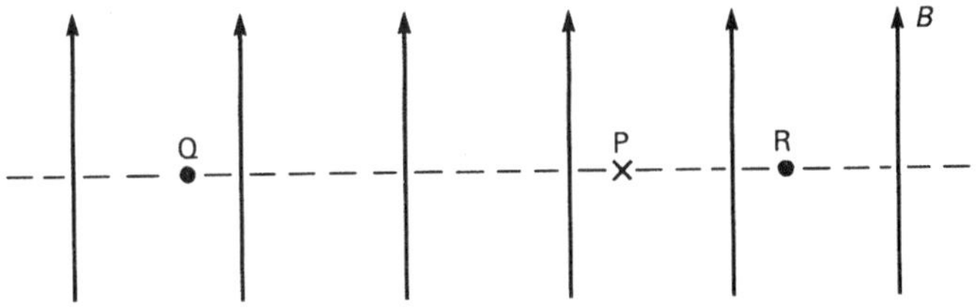

P is a point where the resultant magnetic flux density is zero, i.e. it is a neutral point. P is closer to R than to Q.

(a) Explain whether the direction of the current I is into or out of the paper and sketch a diagram which shows the directions of the different magnetic flux densities present at P.

(b) If I is increased slightly, will the neutral point at P move towards Q or towards R?

(c) There is a second neutral point on the line through QR. State whether it is to the left of Q, between Q and R or to the right of R. [L]

F107 A square coil of side a and consisting of N turns is free to rotate about a vertical axis through the mid-points of two opposite sides. It is situated in a uniform horizontal magnetic field of flux density B so that the plane of the coil makes an angle θ with the field. Draw a diagram of this arrangement as seen from above and show the couple acting on the coil when a current I flows through it. Write down an expression for the magnitude of this couple.

Explain how and why this simple arrangement is modified in most moving-coil galvanometers. [J]

F108 A moving-coil galvanometer consists of a rectangular coil of N turns each of area A suspended in a radial magnetic field of flux density B. Derive an expression for the torque on the coil when a current I passes through it. (You may assume the expression for the force on a current-carrying conductor in a magnetic field.)

If the coil is suspended by a torsion wire for which the couple per unit twist is C, show that the instrument will have a linear scale.

How may the current sensitivity of the instrument be made as large as possible? What practical considerations limit the current sensitivity?

Two galvanometers, which are otherwise identical, are fitted with different coils. One has a coil of 50 turns and resistance 10 ohms while the other has 500 turns and a resistance of 600 ohms. What is the ratio of the deflections when each is connected in turn to a cell of EMF 2.5 V and internal resistance 50 ohms? [L]

F109 Describe with the aid of diagrams the structure and mode of action of a moving-coil galvanometer having a linear scale and suitable for measuring small currents. If the coil is rectangular, derive an expression for the deflecting couple acting upon it when a current flows in it, and hence obtain an expression for the current sensitivity (defined as the deflection per unit current).

If the coil of a moving-coil galvanometer having 10 turns and of resistance 4 Ω is removed and replaced by a second coil having 100 turns and of resistance 160 Ω calculate:

(a) the factor by which the current sensitivity changes and

(b) the factor by which the voltage sensitivity changes.

Assume that all other features remain unaltered. [J]

F110 Explain what is meant by the Hall effect. Show how the sign of the Hall voltage depends on the sign of the charge of the current carriers. [AEB, '79]

F111 You are given a small slice of doped semiconductor material but you do not know whether it is n-type or p-type. Explain with the aid of a diagram the physical principles of an investigation which would enable you to identify the type. [S]

F112 This question is about investigating the variation in magnetic field along the axis of a plane circular coil.

(a) (i) Copy the representation of the coil shown in Fig. 1. The plane of the coil is at right angles to the page. On your diagram sketch the field associated with the coil alone when it is carrying a current in the direction shown.

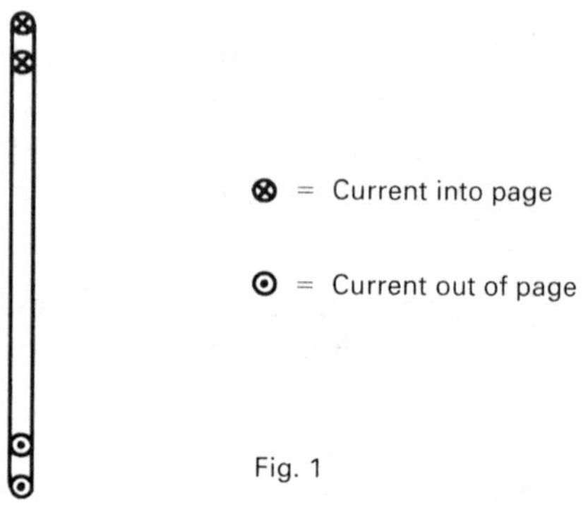

Fig. 1

(ii) The coil has 600 turns and a diameter of 10.0 cm. Show that the magnetic flux density at its centre is 9.0 mT when it is carrying a direct current of 1.2 A.
(Permeability of free space $\mu_0 = 4\pi \times 10^{-7}\,\text{H m}^{-1}$.)

(b) Magnetic flux densities can be investigated using a current balance or a Hall probe.

When there is no magnetic field the current balance shown in Fig. 2 is balanced horizontally with the rider in the position shown.

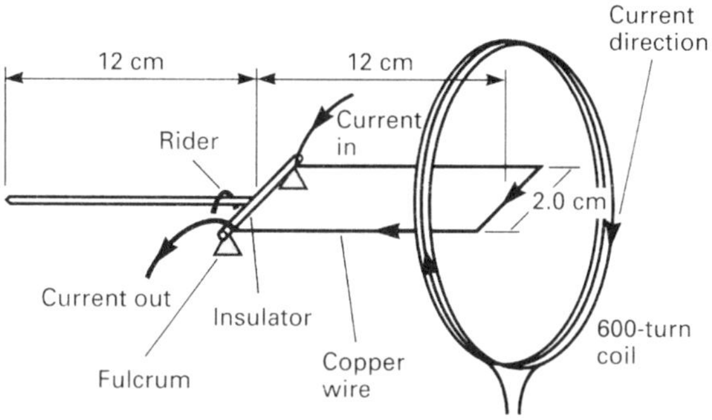

Fig. 2

(i) State the measurements you would make to confirm that the flux density at the centre of the coil is 9.0 mT.

(ii) Show clearly how you would use your measurements and suggest suitable values for the mass of the rider and the current in the current balance.

(c) A Hall probe is to be used to investigate how the field varies with the distance, a, between the centre of the coil and a point along the axis of the coil.

In a calibration experiment the Hall probe was found to give a Hall voltage of 5 mV when it was in a magnetic field of flux density 0.10 T and when it was carrying a current of 50 mA. The maximum permissible current through the probe is 100 mA.

(i) Assuming that the maximum permissible current through the Hall probe is used, state the range of meter you would select to measure *the Hall voltage*, giving your reasoning.

(ii) Describe how you would carry out the investigation. Your account should include
- a circuit diagram for the Hall probe
- a clear indication of how you would position the Hall probe in the field and its orientation relative to the field
- the measurements made and the apparatus used. [AEB, '90]

F113

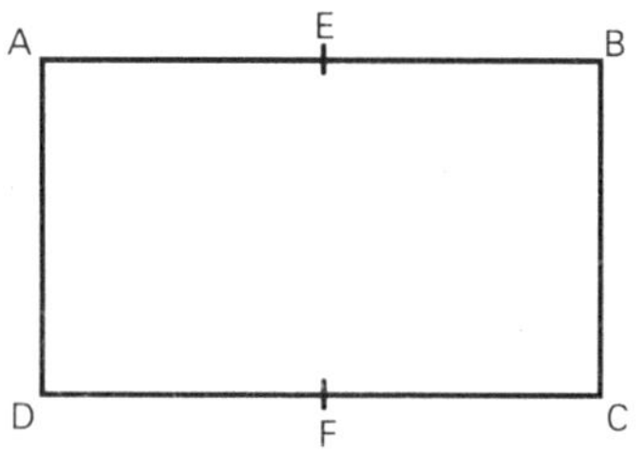

ABCD is a plane rectangular strip of conducting material of uniform thickness, with a steady current flowing uniformly from AD to BC. The potential difference between E and F, the mid-points respectively of AB and CD, is zero, but when a magnetic field is set up at right angles to ABCD, into the plane of the diagram, a small, steady potential difference appears between E and F. Explain these observations and briefly describe a practical application of the phenomenon.

When the rectangle is made of copper, of density $9 \times 10^3\,\text{kg m}^{-3}$, relative atomic mass 63, with thickness 1 mm, breadth 20 mm, carrying a current of 10 A and with an applied field of flux density 1.67 T, the potential difference between E and F is found to be $1\,\mu\text{V}$, F being positive with respect to E. Taking the electronic charge to be 1.6×10^{-19} C and the Avogadro constant to be $6 \times 10^{23}\,\text{mol}^{-1}$, find:

(a) the drift velocity of the current carriers in the copper;

(b) the number of charge carriers per unit volume in the copper and the sign of their charge;

(c) the mobility (drift velocity per unit electric field) of these carriers, given the resistivity of copper to be $1.7 \times 10^{-8}\,\Omega\,\text{m}$;

(d) the ratio of the number of atoms per unit volume to the number of charge carriers per unit volume in the copper. [O & C]

F114 **(a)** The diagram shows a rectangular piece of semiconductor material with leads attached to metal end faces.

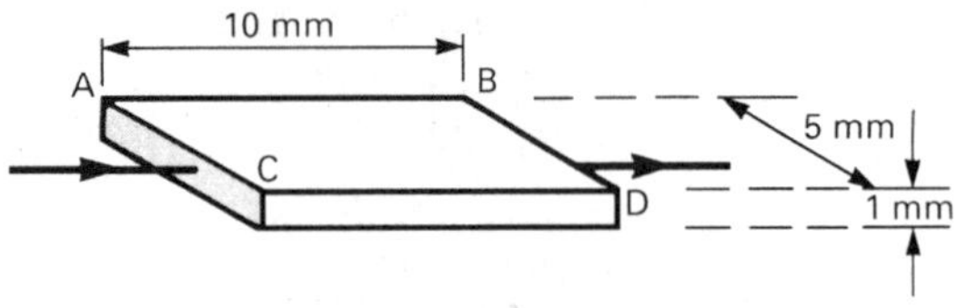

(i) If the resistance of the specimen is approximately 100 Ω, show that the resistivity of the material, is about 0.050 Ω m.

(ii) Describe how you would determine experimentally the resistivity of the material of a similar specimen using apparatus normally available in a school or college laboratory. You can assume that the specimen is mounted in such a way that its dimensions can be measured.

(b) For a specimen similar to that shown in the diagram, a magnetic field is applied perpendicular to the face ABCD. When a current flows, a potential difference, called the Hall PD, develops across the specimen and is perpendicular to both field and current.

(i) Explain how this effect occurs.

(ii) Explain why the Hall PD is much larger in a semiconductor than in a metal specimen of the same dimensions under the same conditions. You may assume that the current, I, in a specimen of cross-sectional area, A, is given by

$$I = nAvq$$

where n is the number of charge carriers per unit volume, q is the charge of each carrier and v is the drift velocity of the charge carriers.

(c) You are given two pieces of semiconductor material which are identical in appearance. One is p-type material and the other is n-type. Outline an experiment using the Hall effect to distinguish between them, explaining carefully how you would use the observations to identify the n-type material. [J, '89]

F115 A rectangular coil of n turns and area A, carrying a current I, is free to rotate about an axis in its plane in a uniform magnetic field of flux density B perpendicular to the axis of rotation. Draw a diagram showing the position of the coil with respect to the field when the torque acting upon it is **(a)** zero, and **(b)** a maximum. Derive an expression for the maximum value of the torque.

Explain, with the aid of a diagram, how the coil can be made to rotate continuously when connected to a direct current supply. Describe briefly how the torque may be made almost constant in value as the coil rotates.

Explain what is meant by back EMF in a motor and indicate its magnitude relative to the supply voltage when the armature is **(i)** at rest, and **(ii)** rotating at maximum speed. [L]

F116 Explain why, when a DC motor is switched on with no load, it accelerates until a certain speed is reached and then continues to run at that speed.

Explain why, if it is now loaded, the speed falls to a lower steady value. (The motor uses a permanent magnet to produce the field.) [L]

ELECTROMAGNETIC INDUCTION (Chapter 42)

F117 **(a)** A wire of length l is horizontal and oriented North–South. It moves East with velocity v through the Earth's magnetic field which has a downward vertical component of flux density B. Write down an expression for the potential difference between the two ends of the wire. Which end of the wire is at the more positive potential?

(b) A horizontal square wire frame ABCD, of side d, moves with velocity v parallel to sides AB, DC from a field-free region into a region of uniform magnetic field of flux density B. The boundaries of the field are parallel to the sides BC, AD of the frame and the field is directed vertically downward. Write down expressions for the electromotive force induced in the frame:

(i) when side BC has entered the field but side AD has not,
(ii) when the frame is entirely within the field region,
(iii) when side BC has left the field but side AD has not.

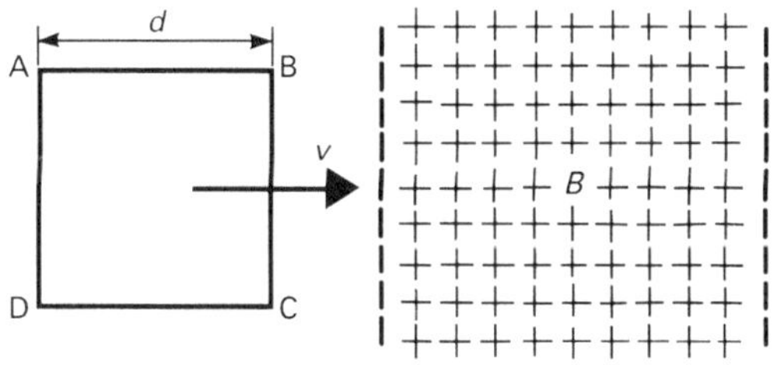

For each position derive an expression for the magnitude and direction of the current in the frame and the resultant force acting on the frame due to the current. The total resistance of the wire frame is R, and its self inductance may be neglected. [O & C]

F118 A closed wire loop in the form of a square of side 4 cm is placed with its plane perpendicular to a uniform magnetic field, which is increasing at the rate of $0.3\,\mathrm{T\,s^{-1}}$. The loop has negligible inductance, and a resistance of $2 \times 10^{-3}\,\Omega$. Calculate the current induced in the loop, and explain with the aid of a clear diagram the relation between the direction of the induced current and the direction of the magnetic field. [S]

F119

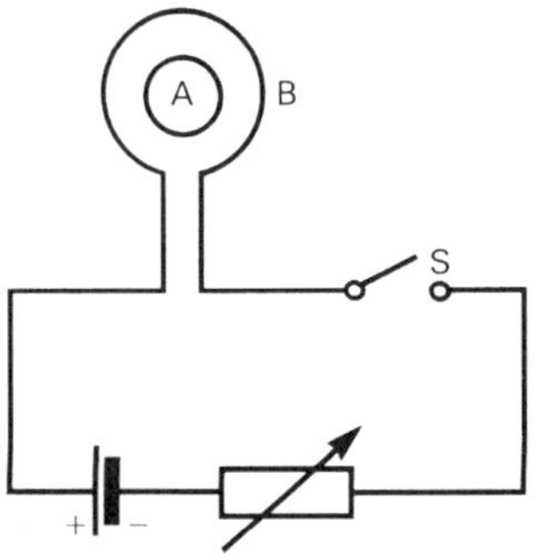

A closed circular wire loop A lies in the plane of a larger loop B, which is connected as shown to a battery. Show on diagrams the direction of the current induced in A when:
(a) the switch S is closed,
(b) when S is opened,
(c) with S closed, the loop A is raised, parallel to itself, out of the plane of the figure. Justify your conclusions. [S]

F120 A coil of 100 turns each of cross-sectional area $3.0 \times 10^{-4}\,\mathrm{m^2}$ is placed with its plane perpendicular to a uniform magnetic field of flux density B. The terminals of the coil are connected together.

B increases steadily with time from zero to 0.2 T in 2.0 ms. It then remains constant for 1.0 ms and decreases uniformly to zero in 1.0 ms.
(a) **(i)** Calculate the maximum flux through one turn of the coil.
(ii) Calculate the EMF induced in the coil during the first 2.0 ms.
(iii) Sketch a graph to show how the EMF induced in the coil varies with time during the 4.0 ms. Give numerical values on both axes.
(b) In a sketch show two turns of the coil, the magnetic field B and the direction of the induced current when B is decreasing. [J, '90]

F121 A closed wire loop in the form of a square of side 4.0 cm is mounted with its plane horizontal. The loop has a resistance of $2.0 \times 10^{-3}\,\Omega$, and negligible self inductance. The loop is situated in a magnetic field of strength 0.70 T directed vertically downwards. When the field is switched off, it decreases to zero at a uniform rate in 0.80 s. What is:
(a) the current induced in the loop,
(b) the energy dissipated in the loop during the change in the magnetic field?
Show on a diagram, justifying your statement, the direction of the induced current. [S]

F122 State the laws of electromagnetic induction.

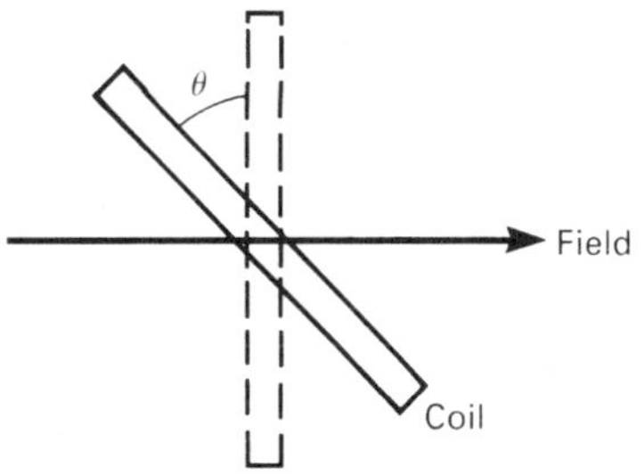

A plane coil of wire of 20 turns and area $0.040\,\mathrm{m^2}$ is placed with its plane at right-angles to a uniform magnetic field of flux-density 0.30 T.

(a) Write down an expression for the total flux [flux-linkage] through the coil when it is at an angle θ from its original position (see diagram).

(b) Hence deduce the maximum EMF induced in the coil if it is rotated steadily at 10 revolutions every second.

The axis of rotation of the coil is parallel to the plane of the coil and perpendicular to the field. [W, '92]

F123 **(a)** Explain how the laws of electromagnetic induction predict the magnitude and direction of induced EMFs. Describe two simple experiments which you could carry out in the laboratory to illustrate the laws.

(b) A powerful electromagnet produces a uniform field in the gap between its poles, each of which measures 0.10 m × 0.08 m. There is no field outside the gap. A circular coil of 80 turns and radius 0.09 m is placed so that it encloses all the flux of the magnetic field (Fig. 1).

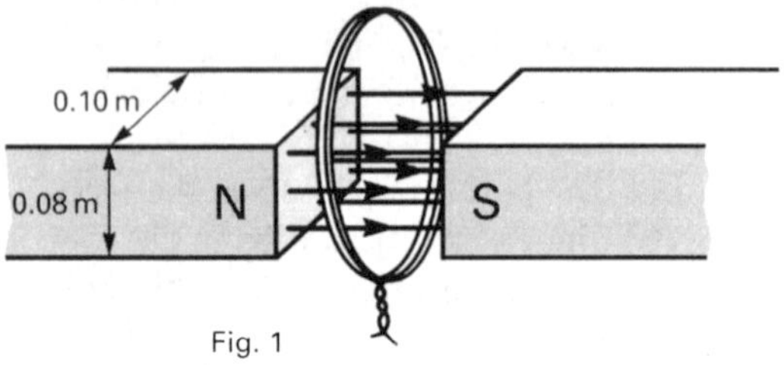

Fig. 1

(i) The current in the electromagnet is reduced so that the field falls linearly from 0.20 T to zero in 5.0 s. Calculate the initial total flux [flux-linkage] in the gap and hence the EMF generated in the coil during this time.

(ii) The coil is part of a circuit in which the total resistance is 24 Ω. Calculate the current in the circuit while the field is collapsing and the magnetic field which this current produces at the centre of the coil.

(iii) Fig. 2 shows the poles and coil. Copy this diagram and show on your copy the direction of the induced current in the coil and the direction of the magnetic field which this current produces. Explain how your application of the laws led you to these deductions about the directions of field and current. [O & C, '92]

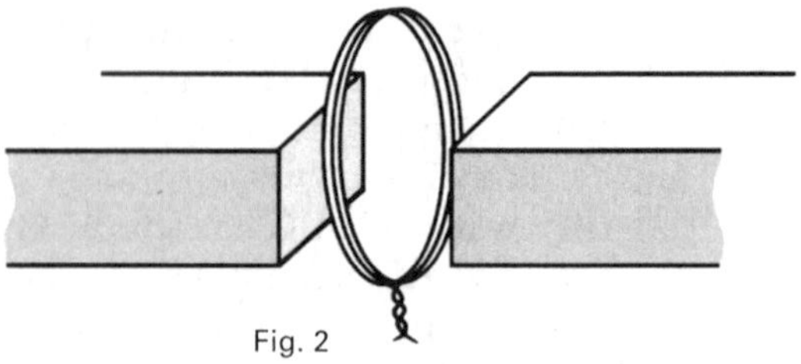

Fig. 2

F124 **(a)** P, Q and R in the diagram are circular coils each with the same number of turns placed in a magnetic field of flux density B. The area of each coil is as shown.

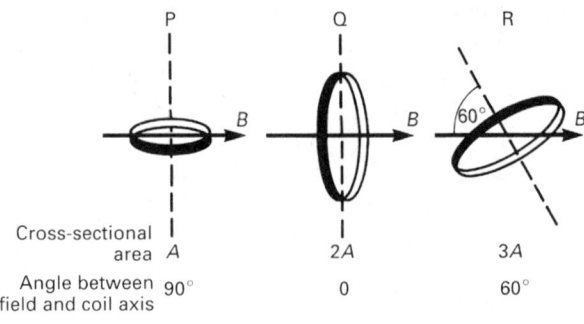

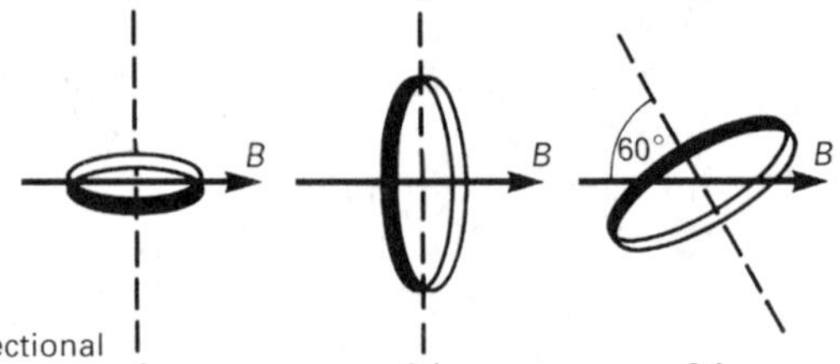

In which case is the flux passing through the coil **(i)** least, **(ii)** greatest? Explain your answers.

(b) A coil is connected to a datalogger, which is a device which records voltages at regular time intervals. The coil is positioned with its axis vertical and a bar magnet is held above the coil with its axis lined up with that of the coil. It is dropped so that it falls through the coil and the datalogger, triggered by the approaching magnet, records the induced EMF. The results are shown in the graph opposite.

(i) Identify the time at which the rate of change of flux through the coil has its greatest magnitude. Give the magnitude of this quantity and its unit.

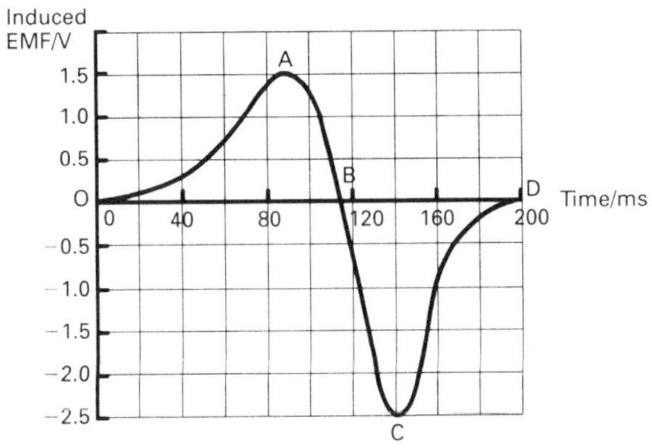

(ii) Explain how you would expect the graph to be different if the magnet had been dropped from a greater height.

(iii) Explain the significance of the area under the curve. Discuss whether or not you would expect the positive and negative areas to be equal.

(c) The coil in (b) has a resistance of 20 Ω. If it had been short-circuited when the magnet was dropped instead of being connected to the datalogger, calculate

(i) the current which would have flowed corresponding to point A on the graph.

(ii) the power being dissipated at A due to this current and hence make an estimate of the energy dissipated in the solenoid [coil] in the time represented by OB. [J, '92]

F125 A rectangular coil of N turns, each of dimensions a metres by b metres, has its ends short-circuited and is rotated at constant angular speed ω rad s^{-1} in a uniform magnetic flux density of B tesla (Wb m^{-2}). The axis of rotation passes through the midpoints of the sides, of length a, of the coil and is at right-angles to the direction of the magnetic field.

(a) Explain why there is an EMF in the coil and derive an expression which shows how its magnitude varies with time.

(b) What is the frequency of the EMF in Hz?

(c) Derive an expression for the maximum value of the EMF and state, with the reason, the position of the coil relative to the field when this occurs.

(d) Apart from any mechanical resistance to the motion, why does the coil slow down when it is disconnected from the device which drives it?

(e) With the aid of a diagram, show how the arrangement could be modified to act as a generator which causes an alternating current to flow in an external load. [J]

F126 A metal disc rotates anticlockwise at an angular velocity ω in a uniform magnetic field which is directed into the paper in the diagram, and covers the whole of the disc. A and B are brushes.

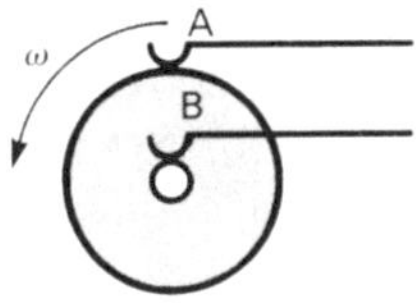

By considering an electron between the brushes show that a potential difference exists between A and B.

Explain what changes, if any, will be needed in the torque required to keep the disc rotating at the same angular velocity ω if A and B are connected by a resistor. [L]

F127 A copper disc of radius 10 cm is situated in a uniform field of magnetic flux density 1.0×10^{-2} T with its plane perpendicular to the field.

The disc is rotated about an axis through its centre parallel to the field at 3.0×10^3 rev min^{-1}. Calculate the EMF between the rim and centre of the disc.

Draw a circle to illustrate the disc. Show the direction of rotation as clockwise and consider the field directed into the plane of the diagram.

Explaining how you obtain your result, state the direction of the current flowing in a stationary wire whose ends touch the rim and centre of the disc. [J]

F128 **(a)** What is meant by *critical damping* of an oscillatory system?

(b) The diagram shows the arrangement by which a modern laboratory balance is critically damped. The aluminium beam supporting the balance pan moves in the field of powerful magnets.

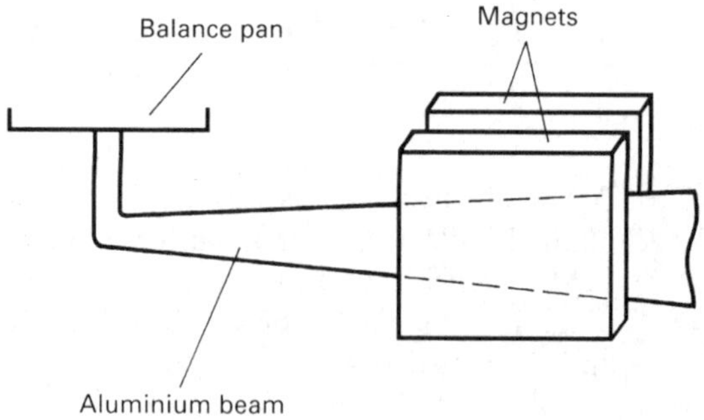

(i) Explain, in terms of electromagnetic principles, how the damping works.

(ii) After long use the magnets will weaken. What change would occur in the performance of the balance? [O, '91]

F129 A choke of large self-inductance and small resistance, a battery and a switch are connected in series. Sketch and explain a graph illustrating how the current varies with time after the switch is closed. If the self-inductance and resistance of the coil are 10 henries and 5 ohms respectively and the battery has an EMF of 20 volts and negligible resistance, what are the greatest values after the switch is closed of **(a)** the current, **(b)** the rate of change of current? [J]

F130 In the circuit shown, the voltmeter has a high resistance and both the ammeter and cell have negligible resistance.

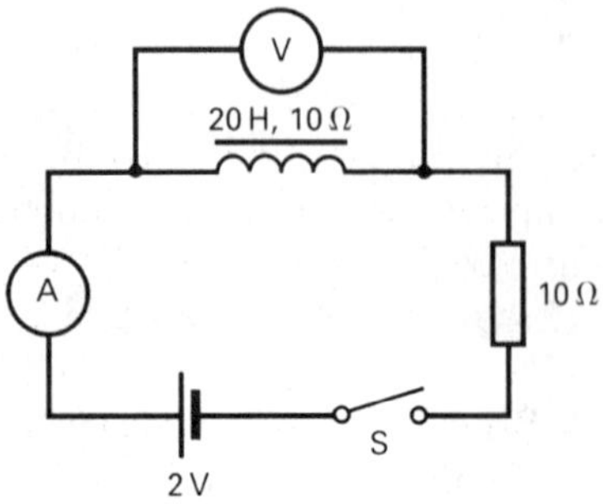

(a) Calculate the following:

(i) the reading on the ammeter immediately after switch S is closed,

(ii) the reading on the voltmeter immediately after switch S is closed,

(iii) the steady reading on the ammeter,

(iv) the steady reading on the voltmeter,

(v) the time constant of the circuit.

(b) Sketch a graph of the voltmeter reading against time for the first 10 seconds after the switch is closed, indicating the scales on the axes. [AEB, '85]

F131 Write down an equation relating E, the EMF induced in an inductor, to dI/dt, the rate of change of current through the inductor, and L, its self-inductance. Hence express the henry, the unit of self-inductance, in terms of the joule and the ampere. [C]

F132 It can be shown that the energy stored in an inductor of inductance L carrying a current I is $\frac{1}{2}LI^2$. In what form is this energy stored. How may it be converted into some other form of energy? [C]

F133 Explain in terms of the laws of electromagnetic induction why a 'back EMF' is developed in a coil when an alternating potential difference is applied across it. [L]

F134 Fig. 1 shows a circuit consisting of a 40 H inductor connected in series with a 50 Ω resistor and a 12 V supply. The supply and inductor have negligible resistance. Two wires are connected to the inductor with a small gap between them, as shown in Fig. 1.

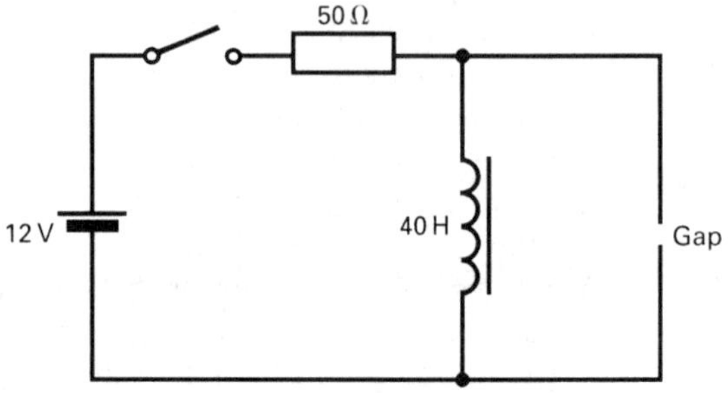

Fig. 1

(a) When the switch is closed the current does not rise instantly to the final value owing to the presence of the inductor. The variation of current with time is shown in Fig. 2.

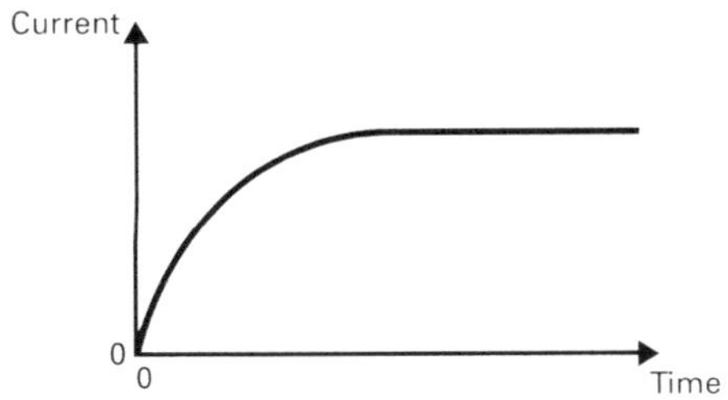

Fig. 2

(i) What is the initial rate of change of current?
(ii) What is the final steady value of the current?
(iii) Explain qualitatively why the current does not rise instantly to its final steady value.

(b) Outline the energy changes which take place as the current rises.

(c) When the switch is opened a spark jumps the gap. What is the maximum energy which could be dissipated in the spark? [AEB, '91]

F135 (a) What is meant by the statement that a solenoid has an inductance of 2.0 H?

A 2.0 H solenoid is connected in series with a resistor, so that the total resistance is 0.50 Ω, to a 2.0 V DC supply. Sketch the graph of current against time when the current is switched on.

What is:
(i) the final current,
(ii) the initial rate of change of current with time, and
(iii) the rate of change of current with time when the current is 2.0 A?

Explain why an EMF greatly in excess of 2.0 V will be produced when the current is switched off.

(b) A long air-cored solenoid has 1000 turns of wire per metre and a cross-sectional area of 8.0 cm^2. A secondary coil, of 2000 turns, is wound around its centre and connected to a ballistic galvanometer, the total resistance of coil and galvanometer being 60 Ω. The sensitivity of the galvanometer is 2.0 divisions per microcoulomb. If a current of 4.0 A in the primary solenoid were switched off, what would be the deflection of the galvanometer? [L]

F136 Two coils, X and Y, are fixed near to each other, and when the current in X changes at a steady rate of 5 A s^{-1} an EMF of 20 mV is induced in Y. What is the mutual inductance between X and Y?

What rate of change of current in coil Y would be needed to induce an EMF of 5 mV in X?

How might it be possible, without altering the windings of either of the coils, **(a)** to reduce the mutual inductance between X and Y, and **(b)** to increase it considerably? [S]

F137 A farmer installs a private hydroelectric generator to provide power for equipment rated at 120 kW 240 V AC. The generator is connected to the equipment by two conductors which have a total resistance of 0.20 Ω. The system is shown schematically in Fig. 1.

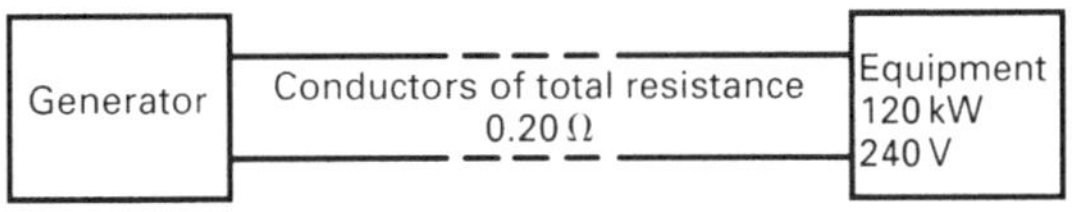

Fig. 1

(a) The equipment is operating at its rated power. Calculate:
(i) the power loss in the cables,
(ii) the voltage which must be developed by the generator,
(iii) the efficiency of the transmission system.

(b) An engineer suggests that the farmer uses a transformer to convert the generator output to give a PD of 2400 V at the end of the transmission line, as shown in Fig. 2. A second transformer is to be used to step down this PD to 240 V.

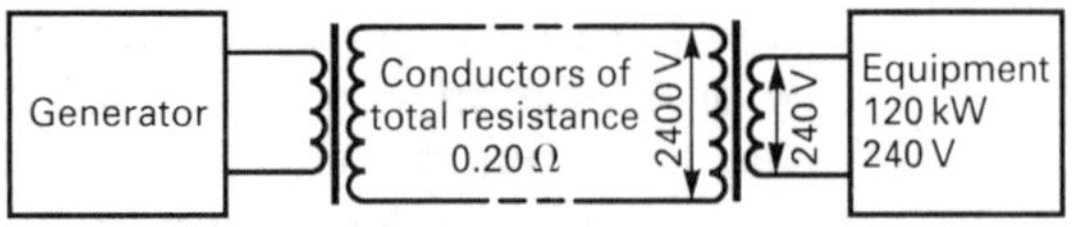

Fig. 2

(i) Explain briefly how a transformer makes use of electromagnetic induction to produce an output voltage several times bigger than the input voltage.

(ii) The transformers are 100% efficient. Calculate the power loss in the new transmission system.

[AEB, '92]

F138 Describe and account for *two* constructional differences between a moving-coil galvanometer used to measure current and the ballistic form of the instrument.

An electromagnet has plane-parallel pole faces. Give details of an experiment, using a search coil and ballistic galvanometer of known sensitivity, to determine the variation in the magnitude of the magnetic flux density (magnetic induction) along a line parallel to the pole faces and mid-way between them. Indicate in qualitative terms the variation you would expect to get.

A coil of 100 turns each of area $2.0 \times 10^{-3}\,\text{m}^2$ has a resistance of 12 Ω. It lies in a horizontal plane in a vertical magnetic flux density of $3.0 \times 10^{-3}\,\text{Wb m}^{-2}$. What charge circulates through the coil if its ends are short-circuited and the coil is rotated through 180° about a diametral axis? [J]

F139 A flat search coil containing 50 turns each of area $2.0 \times 10^{-4}\,\text{m}^2$ is connected to a galvanometer; the total resistance of the circuit is 100 Ω. The coil is placed so that its plane is normal to a magnetic field of flux density 0.25 T.

(a) What is the change in magnetic flux linking the circuit when the coil is moved to a region of negligible magnetic field?

(b) What charge passes through the galvanometer? [C]

F140 (a) What quantity is measured by a ballistic galvanometer?

(b) State *two* essential requirements of a galvanometer intended for ballistic use, giving the reason for each.

(c) When using a ballistic galvanometer with a search coil, why is a series resistor often included in the circuit?

[AEB, '79]

F141 Describe experiments (one in each case) involving the use of a moving coil ballistic galvanometer to **(a)** compare two capacitances of approximately the same magnitude, **(b)** compare the magnetic induction (flux density) between the poles of one electromagnet with that between the poles of another. In each case justify the method used to calculate the result.

Explain *two* special features of a galvanometer suitable for use in these experiments.

[J]

F142 Write down an expression for the magnetic field B at a distance r in vacuo from a long, straight wire carrying a current I. State the units in which B is measured.

A *small* air-cored coil of several hundred turns of wire is connected to a ballistic galvanometer and placed at a distance r from such a wire. When the current in the wire is switched off, the galvanometer gives a kick θ. In what direction should the axis of the coil lie to get the largest possible kick θ_{max} for a given value of r? How does θ_{max} vary with r?

The Earth's magnetic field is comparable in magnitude with some parts of the field due to the wire, but it does not affect the magnitude of the kick θ. Why? [S]

F143 (a) (i) State the laws of electromagnetic induction.

(ii) Use the laws to show that a charge

$$\frac{n}{R}(\phi_1 - \phi_2)$$

flows through a coil of n turns contained in a closed circuit of total resistance R, if the magnetic flux being investigated changes from ϕ_1 to ϕ_2.

(b) The magnetic flux density in the vicinity of a large air-cored electromagnet is determined by measuring the induced flow of charge in a small coil as the current in the electromagnet is switched on – see the diagram below.

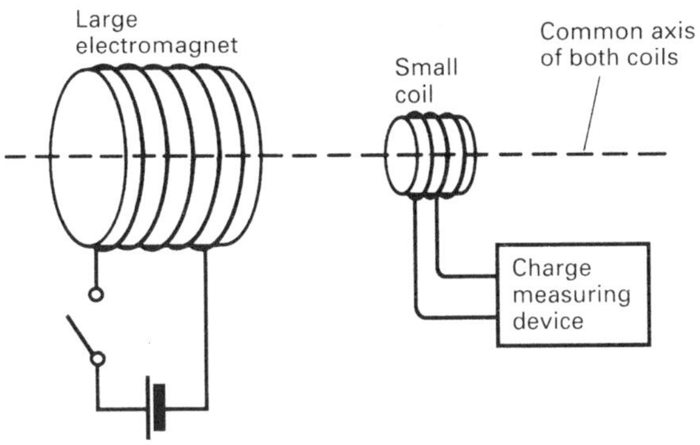

(i) Calculate the magnetic flux density at the position of the small coil, due to the electromagnet from the following data:

Radius of small coil = 10 mm.
Resistance per unit length of wire of small coil = $0.10\,\Omega\,\mathrm{m}^{-1}$.
Total flow of charge through the small coil due to switching on the electromagnet = 0.01 C.

(ii) Comment on the expected total flow of charge when the electromagnet has a ferrous core instead of an air core. [W, '91]

AC THEORY AND RECTIFICATION (Chapters 43 and 44)

F144 If a sinusoidal current, of peak value 5 A, is passed through an AC ammeter the reading will be $5/\sqrt{2}$ A. Explain this.

What reading would you expect if a square-wave current, switching rapidly between +0.5 and −0.5 A were passed through the instrument? [L]

F145 The figure shows (in part) the variation with time of a periodic current.

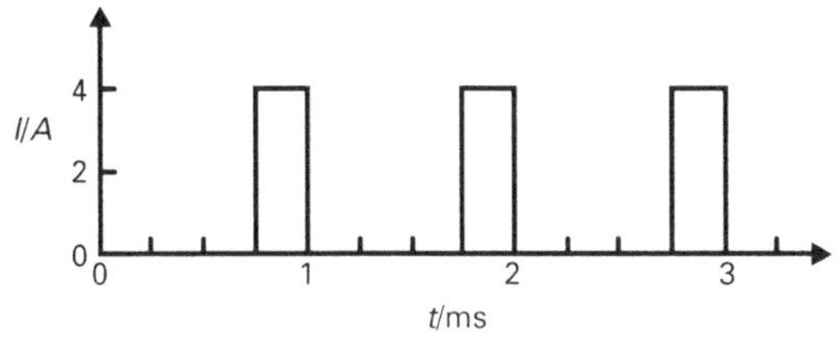

(a) What is the average value of the current?
(b) Find the root-mean-square current.

The periodic current passes through a resistor, producing heat at a certain rate.

(c) What steady current, passing through the same resistor, would have an identical heating effect? [C]

F146 **(a)** **(i)** Explain what is meant by root mean square voltage.
(ii) Write down an expression relating RMS and peak voltage for a sinusoidal electrical supply.

(b) A voltage given by

$$V = 339.4 \sin(100\,\pi t)$$

is applied in turn across the terminals of the devices shown in diagrams A and B below.

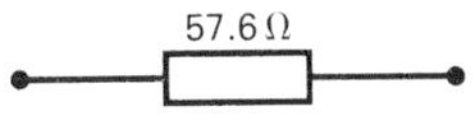

Diagram A

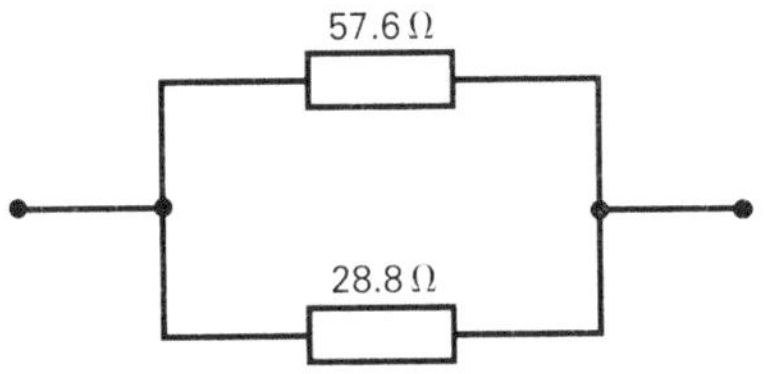

Diagram B

(i) In *each* case, what is the total energy dissipated by the device over a period of 100 s?
(ii) Suggest a practical device whose operation may be described by the application of the above voltage to A or B. [W, '91]

F147 A coil is rotated at a constant rate in a uniform magnetic field. The peak value of the EMF induced in the coil is 10 V. Find:
(a) the RMS value of the EMF,
(b) the instantaneous value of the EMF one-quarter of a period after the EMF is a maximum. [C(O)]

F148 A 1.2 μF capacitor is connected across a variable frequency 20 V RMS AC supply. Calculate the RMS value of the current in the circuit if the supply frequency is 50 Hz. (Neglect the internal impedance of the supply.)

What will the current be if the frequency of the supply is raised to 5 kHz? [L]

F149 **(a)** In an alternating current demonstration (see Figure 1), a capacitor of large capacitance C is connected to an AC signal generator G producing very low frequency signals. The potential difference V across the generator and the current I in the circuit are recorded and their simultaneous values plotted (Figure 2).

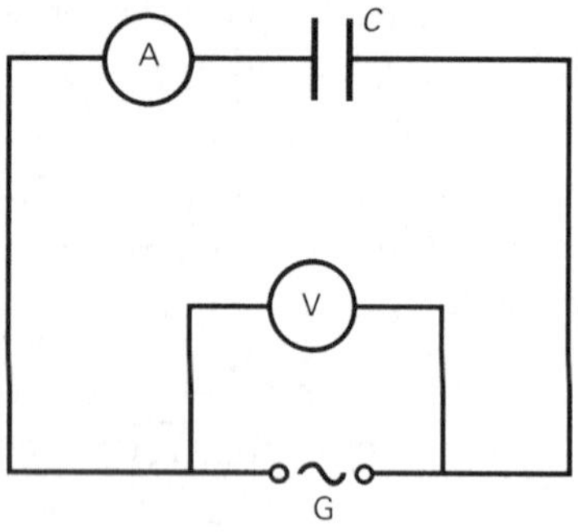

Fig. 1

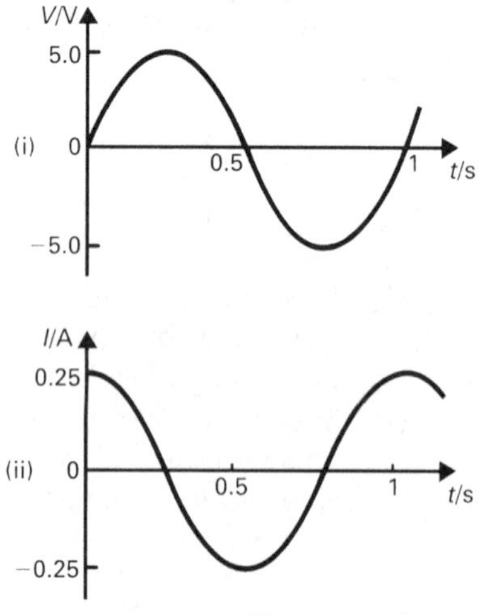

Fig. 2

Using the data from the graphs, calculate:

(i) the RMS voltage produced by the signal generator,

(ii) C.

(b) **(i)** What electrical term, applicable to the capacitor, is given by the quotient

$$\frac{\text{peak value of } V}{\text{peak value of } I}?$$

(ii) State the phase relation of V and I. [O, '91]

F150 The graphs show how the voltage across a coil varied with the current flowing through it. One graph was obtained using DC and the other using AC of frequency 200 Hz.

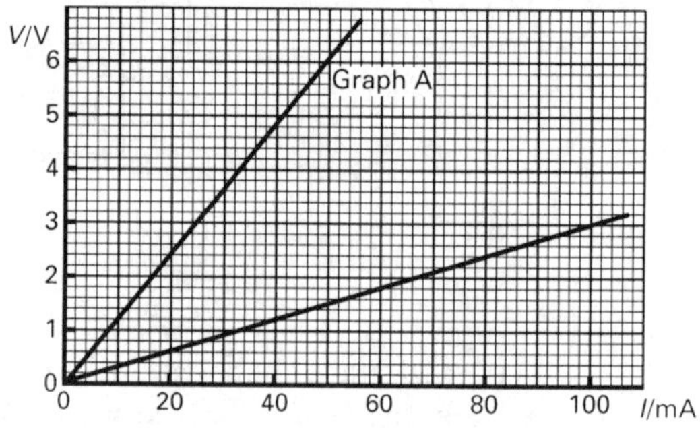

(a) Explain why graph A must be the one obtained when AC was used.

(b) Determine the resistance of the coil.

(c) Determine the inductance of the coil. [AEB, '90]

F151 A generator which produces a potential difference $V = V_0 \cos \omega t$ where $V_0 = 10$ V and $\omega = 1000 \text{ rad s}^{-1}$ is connected in series with a capacitance of 1 μF. There is no resistance in the circuit.

Draw *five* sketch graphs with quantitative values on both axes, of the variation with time of:

(a) the voltage across the capacitor,

(b) the charge on *one* plate of the capacitor,

(c) the current in the circuit,

(d) the energy stored by the capacitor,

(e) the power provided by the generator.

Use the same time scales for each graph starting from $t = 0$, ($V = V_0$) and extend the time over two cycles. Carefully label + and − directions on the various scales.

(f) Describe the relationship between graphs (d) and (e). [W]

F152 Explain the terms *reactance* and *impedance* in relation to AC circuits. Distinguish between mean and root mean square values of potential difference in such circuits.

A small non-inductively wound immersion heater rated at 20 W, 10 V RMS is connected in series with an inductance of 10 mH. A 10 V RMS supply of variable frequency is connected across the two. At what frequency will the power dissipated in the heater be one half of its value when the supply is 10 V DC? (The resistance of the heater may be assumed constant.)

What capacitance needs to be added in series at this frequency to restore the power dissipated by the heater to its DC value? Illustrate this result by means of a suitable vector diagram. [L]

F153 Define the *impedance* of an AC circuit.

A 2.5 μF capacitor is connected in series with a non-inductive resistor of 300 ohms across a source of PD of RMS value 50 volts alternating at $\frac{1000}{2\pi}$ Hz. Calculate:

(a) the RMS value of the current in the circuit and the PD across the capacitor,
(b) the mean rate at which energy is supplied by the source. [J]

F154 A 60 W light bulb, designed for use with a 120 V supply, may be operated at the correct rating from a 240 V, 50 Hz supply by connecting it in series with a resistor, as shown in Fig. 1.

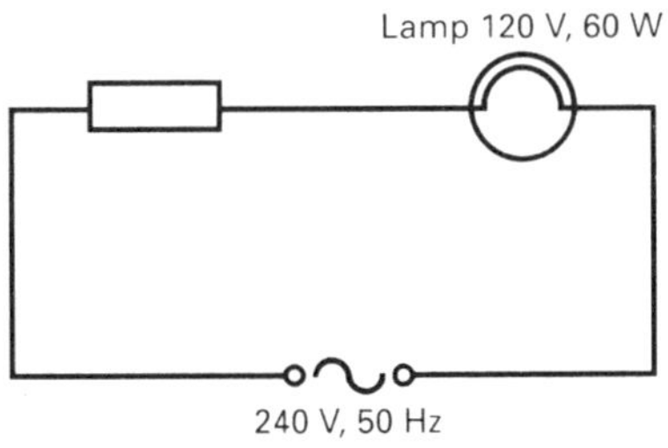

Fig. 1

(a) Calculate, for normal working conditions,
  (i) the current flowing in the lamp
  (ii) the resistance R of the lamp
  (iii) a value for a suitable series resistor.
(b) The same lamp may also be operated at the correct rating from the 240 V, 50 Hz supply by connecting it in series with a capacitor as shown in Fig. 2.

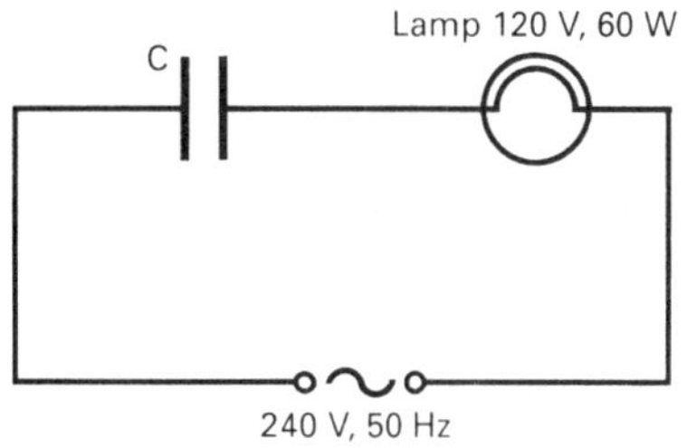

Fig. 2

  (i) Write down an expression for the impedance of the circuit containing the lamp and the capacitor of reactance X_C.
  (ii) Calculate a value for X_C.
  (iii) Calculate a value for the capacitance of the capacitor.
(c) By considering the power dissipated in each circuit explain which circuit provides the better solution. [AEB, '89]

F155 Define the *self-inductance* of a coil.

A coil carries a steady current of 50 mA when the steady PD across its ends is 2.0 V. When the steady PD is replaced by a PD alternating at 50 Hz and of magnitude 4.1 V RMS, the current flowing through the coil is 100 mA RMS.

Calculate:

(a) the resistance of the coil,
(b) the self-inductance of the coil,
(c) the phase difference between the alternating current and the alternating PD. [J]

F156

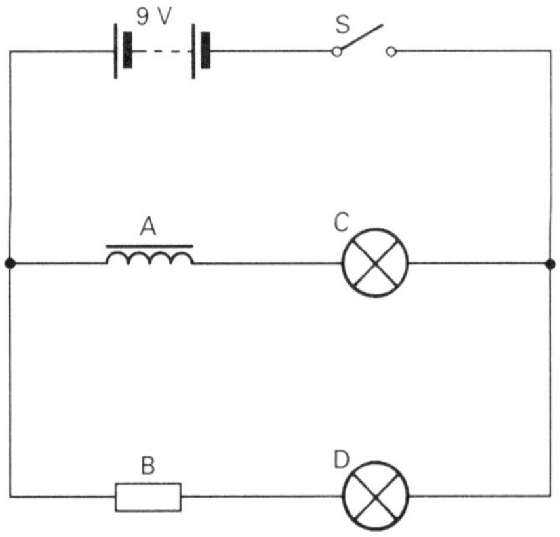

A is a coil having a low resistance and a high inductance. B is a resistor having the same resistance as A, but negligible inductance. C and D are identical filament lamps and the battery can be assumed to have negligible resistance.

(a) Describe and explain how the appearance of each lamp changes in the period after the switch S is closed.

(b) The DC battery is replaced by a sinusoidal alternating source of RMS potential difference 9 V and negligible impedance. Describe and explain the appearance of each lamp after S is closed. [J]

F157 What is meant by the root mean square value of a sinusoidal alternating current? Why is this a useful measure of alternating current?

A 3.0 Ω resistor is joined in series with a 10 mH inductor of negligible resistance, and a potential difference V (= 5.0 volts RMS) alternating at $\frac{200}{\pi}$ Hz is applied across the combination.

(a) Calculate the PD V_R across the resistor and that of V_L across the inductor.

(b) Showing clearly your procedure draw a vector diagram representing the relation between V, V_R and V_L.

(c) Determine the phase difference between V and V_L.

(d) How would you use a cathode ray oscilloscope to show that there is a phase difference between V and V_L? [J]

F158

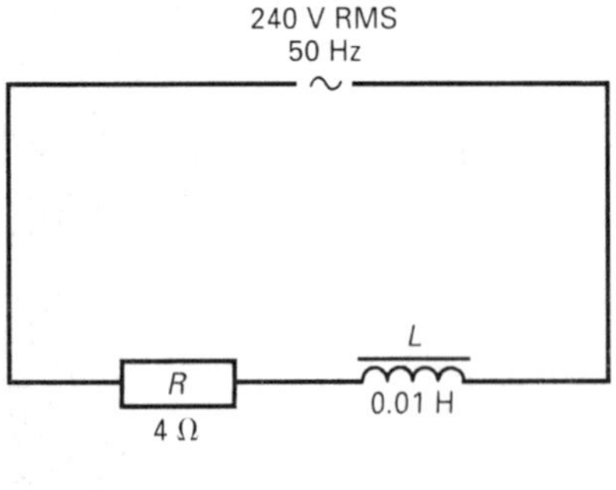

In the above circuit, the source has negligible internal impedance. Find:

(a) the RMS current in the circuit,

(b) the mean rate of production of heat. [W]

F159 **(a)** The operations of resistors and inductors are described by the following equations relating voltage drop to current:

$$V_R = IR \qquad \text{(resistor)}$$

$$V_L = -L\frac{\mathrm{d}I}{\mathrm{d}t} \qquad \text{(inductor)}$$

Use these equations to determine relationships between peak voltage and peak current for both devices, when the current through each device is given by $I = I_0 \sin \omega t$. By using the current I_0 as a reference vector, draw (and explain) vectors representing peak voltage V_0 for both devices.

(b) **(i)** By drawing a suitable vector diagram derive a relationship between peak supply voltage V_S and peak output voltage V_R of the following circuit:

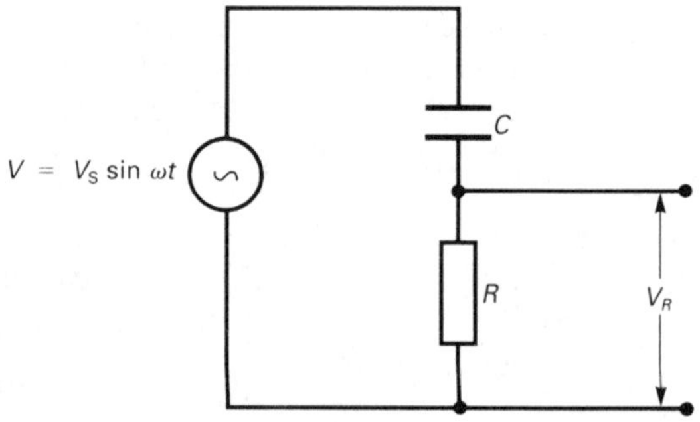

(ii) Calculate the frequency at which $\frac{V_S}{V_R} = \sqrt{2}$ when $C = 10$ nF and $R = 1$ kΩ. [W, '91]

F160 A sealed box with two external terminals is known to contain a resistor and a capacitor connected in some way.

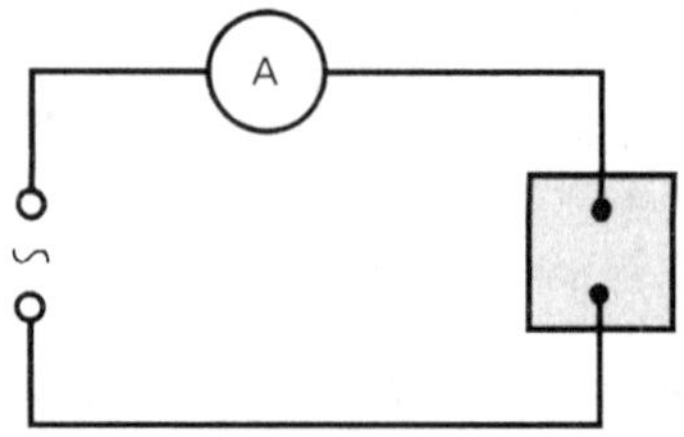

When a potential difference of 32 V RMS at a frequency of 100 Hz is applied across the terminals of the box as shown above a current of 1.0 mA RMS flows into it. If the frequency of the applied potential difference is gradually increased, the current at first rises and then reaches a steady value of 2.0 mA RMS, no matter how high the frequency. There is no current when a DC power supply is used instead.

What is the most likely arrangement of the two components inside the box? Explain.

Calculate the resistance of the resistor. [L]

F161 When an inductance L, a capacitance C and a resistance R are connected in series with a source providing an alternating potential difference V, a current I flows in the circuit.

Explaining your procedure, sketch a vector (phasor) diagram to represent the magnitudes and phase relationships of the PDs across L, C and R.

Explain with the help of your diagram:
(a) how the magnitude of V is related to the PDs across L, C and R,
(b) how the phase of V is related to that of I.
[J]

F162 Define the terms *impedance*, *reactance*, *frequency* and *phase angle* as used in AC theory.

Derive an expression for the resulting potential difference across the component when a sinusoidal current flows through **(i)** a resistor, **(ii)** a capacitor, and **(iii)** an inductor, and give the phase angle between the potential difference and the current in *each* case. Write down expressions for the reactance of a capacitor and an inductor.

A 600 ohm resistor, a 5 μF capacitor and a 0.8 H inductor are connected in series with the 240 V, 50 Hz mains supply. Deriving any expressions used, determine the current and the potential difference across each component. Explain why the sum of the potential differences is not 240 V. [W]

F163 What do you understand by the impedance of an electrical circuit?

A series circuit consists of a resistor R and a pure capacitor C fed by an alternating EMF of frequency f. Show, by means of a vector diagram or otherwise, but with clear explanation, how you could find the impedance of this circuit.

If the applied EMF has peak value 100 V, $f = 50$ Hz, $C = 5\,\mu$F and $R = 500\,\Omega$ find:
(a) the peak current,
(b) the peak PD across C.
The PDs across C and R are simultaneously displayed on a double-beam oscilloscope. Sketch and explain what you would expect to find.

Show that no energy is dissipated in C. [W]

F164 A heater is rated 1.00 kW 200 V. It is required to use it on a 250 V 50 Hz supply. This may be done by placing either **(a)** a resistor or **(b)** an inductor in series with the heater. Calculate:
(i) the resistance of the resistor,
(ii) the inductance of the inductor if its resistance is 5.0 Ω.
State an advantage of using an inductor rather than a resistor. [AEB, '82]

F165 **(a)** Define *resistance*, *reactance* and *impedance*. Give an expression for the reactance of **(i)** a capacitor and **(ii)** an inductor.

Write an account of the vector method of finding the relationship between current and voltage in AC circuits.

(b) A lamp is rated at 120 V, 2 A. Calculate, deriving any expressions used, the value of **(i)** a resistor, **(ii)** an inductor, and **(iii)** a capacitor, which, when individually connected in series with the lamp, will enable the lamp to be run at full rating from the 240 V, 50 Hz mains. Determine the power drawn from the mains in each case. State, with reasons, which of the three items (resistor, inductor, capacitor) you would choose for running the lamp. [W]

F166 Explain the meaning of the term *reactance*, and derive an expression for the reactance of an inductor L when alternating current of frequency f is flowing through it.

Show that no power is dissipated in an inductor when a current passes through it.

A 2 H pure inductor is connected in series with a 500 Ω resistor across AC mains which have a frequency of 50 Hz. A high resistance voltmeter connected across the mains reads 240 V. What does it read when connected across **(a)** the resistor, **(b)** the inductor? What is the time interval between the instants at which the potential differences across the generator and the resistor reach their respective maxima? [S]

F167 A 500 Ω resistor and a capacitor C are connected in series across the 50 Hz AC supply mains. The RMS potential differences recorded on high impedance voltmeters V_1 and V_2 are: $V_1 = 120$ V; $V_2 = 160$ V.

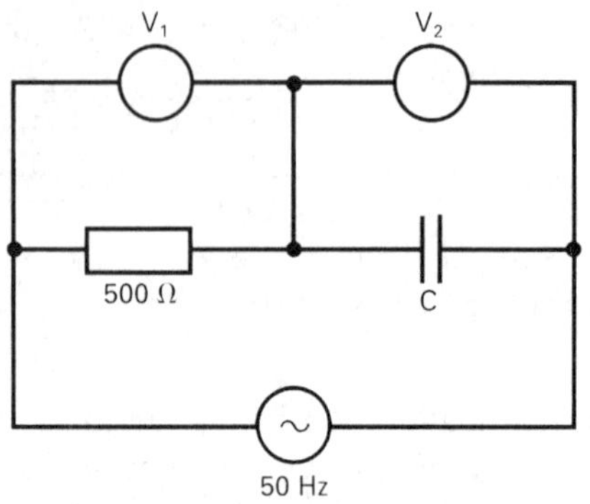

(a) What is the current flowing in the circuit?
(b) What is the power taken from the supply mains?
(c) What is the capacitance of C? [S]

F168 Explain what is meant by the *peak value* and the *root mean square value* of an alternating current.

When a sine wave alternating current is passed through a 100 Ω resistor whose ends are connected to the Y-plates of a cathode ray tube the trace on the screen is a vertical line 30 mm high. A steady potential difference applied between these plates gives a displacement of 0.5 mm per volt. Calculate the root mean square value of the alternating current.

A long solenoid connected to a source of 12 V DC passes a steady current of 2 A. When the solenoid is connected to a supply of AC at 12 V RMS and 50 Hz the current flowing is 1 A RMS. Calculate the inductance of the solenoid.

State and explain how you would expect the values of current for both the AC and DC cases to differ from the above:

(a) when the solenoid is replaced by one of the same length and diameter wound with more turns of the same gauge of wire,
(b) as a bar of soft iron is introduced slowly into the original solenoid while the solenoid is connected to the supply. [O & C]

F169 A coil having inductance and resistance is connected to an oscillator giving a fixed sinusoidal output voltage of 5.00 V RMS. With the oscillator set at a frequency of 50 Hz, the RMS current in the coil is 1.00 A and at a frequency of 100 Hz, the RMS current is 0.625 A.

(a) Explain why the current through the coil changes when the frequency of the supply is changed.
(b) Determine the inductance of the coil.
(c) Calculate the ratio of the powers dissipated in the coil in the two cases. [J]

F170 A constant voltage AC generator, of 20 V RMS and variable frequency, is connected in series with a resistor of resistance 2.0 Ω, a coil of inductance 5.0 H and a capacitor of capacitance 2.0 μF. The frequency is adjusted until the current in the circuit has a maximum value of 2.0 A RMS. Calculate the resistance of the wire of the inductor and the value of this frequency.

(Assume $\pi^2 = 10$.) [L]

F171 Sketch a curve to show how the current in an *LCR* series circuit varies with the frequency of the sinusoidal supply. Show the shape of the curve at very low and very high frequencies. Also illustrate the effect of increasing the resistance in the circuit.

If $L = 100$ mH, $C = 10.0\,\mu$F, $R = 50\,\Omega$ and the RMS voltage of the supply is 1.50 V, calculate values for:

(a) the resonant frequency,
(b) the maximum RMS current in the circuit. [AEB, '79]

F172 A tuning capacitor is to be used with a 10 mH inductor in an *LC* circuit of a radio to provide tuning of all stations broadcasting in the band from 500 kHz to 1.50 MHz.

(a) The largest value of capacitance required is 10.1 pF. What is the smallest value used?

(b) The capacitor is to be made as shown in the diagram using two semi-circular plates which are separated by an air gap of 1.00 mm.

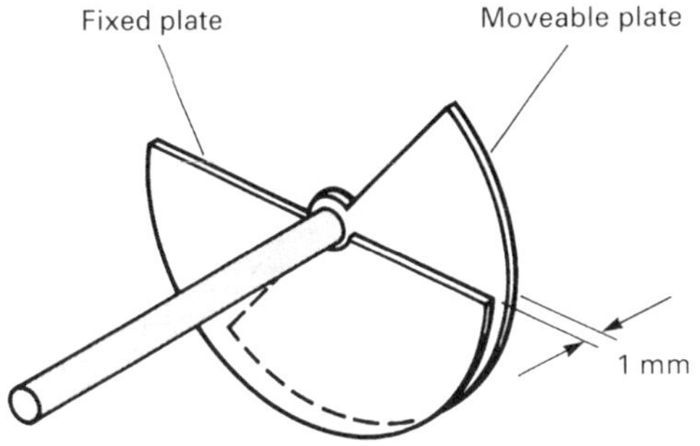

Calculate the diameter of the plates required.

Permittivity of air $= 8.9 \times 10^{-12}\,\mathrm{F\,m^{-1}}$

(c) State two ways of designing the capacitor so that the diameter can be reduced whilst giving the same maximum capacitance.

(d) To change the range of tuning capacitance a 22 pF capacitor is connected in series with the variable capacitor. What is the new maximum capacitance?

[AEB, '92]

F173 **(a)** A cathode ray oscilloscope is set up with equal X and Y voltage sensitivities and the time base is switched off. In *each* of the following cases, sketch the trace you would expect to see on the screen. Explain why the traces have the shapes you have drawn.

(i) A sinusoidal voltage of frequency 150 Hz is connected across the X plates and a sinusoidal voltage of the same amplitude but of frequency 50 Hz is connected across the Y plates.

(ii) A resistor, R, and a capacitor, C, are connected in series with a 50 Hz sinusoidal supply, which is not earthed. The values of the resistance and the capacitance are such that the voltages across R and C are of the same amplitude. The CRO is connected as shown in the diagram.

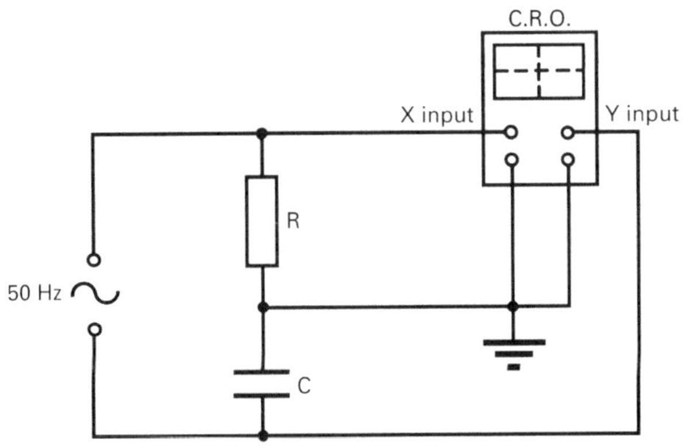

(b) A microphone is connected through an amplifier to the Y plates of a CRO whose time base is on and set to $2\,\mathrm{ms\,cm^{-1}}$. The length of the trace is 10 cm. Two sources of sound, of frequencies 500 Hz and 600 Hz, which have the same loudness are placed near the microphone.

(i) Calculate the beat frequency.

(ii) Draw the trace you would expect to see on the screen. Show the corresponding time scale on the horizontal axis. [J, '89]

F174 A resistance of 80 Ω, a capacitance of 25 μF and an inductance of 0.10 H are connected in series to a 250 V (RMS) supply of pulsatance (ω) of 300 radians $\mathrm{s^{-1}}$.

(a) By drawing a vector diagram for an R, L, C circuit, derive the equations for determining the impedance of the circuit.

(b) Find

(i) the values of the capacitative reactance (X_C) and the inductive reactance (X_L),

(ii) the circuit impedance,

(iii) the RMS current,

(iv) the phase shift between the peak current and the peak potential difference,

(v) the power factor.

(c) At what frequency would the circuit resonate?

(d) The PDs across L and C are displayed on a double beam oscilloscope. Describe how these would appear when **(i)** $\omega \ll \omega_r$, **(ii)** $\omega = \omega_r$, **(iii)** $\omega \gg \omega_r$, where ω_r is the resonant frequency.

[W]

F175 The diagram shows a resistor of resistance R in series with an inductor of inductance L, a capacitor of capacitance C and an alternating supply of RMS potential difference 5.0 V and of frequency 1.00 kHz. The RMS potential differences across R, L and C are each 5.0 V and the RMS current is 100 mA.

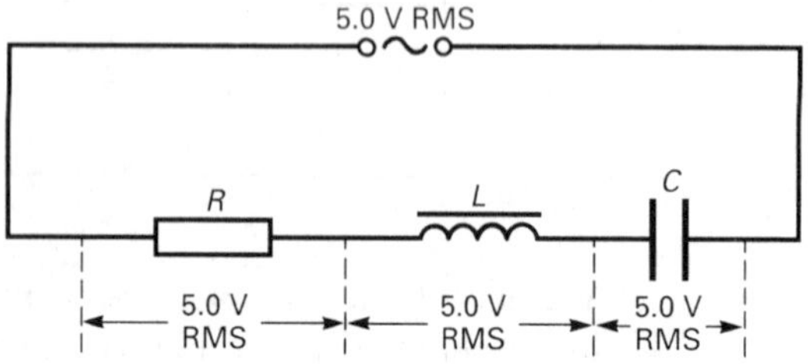

(a) Calculate R, L and C.
(b) Calculate *peak* values for:
  (i) the potential difference across C,
  (ii) the charge on C,
  (iii) the potential difference across L,
  (iv) the rate of change of current in L.
(c) Explain why the rate of change of current in the inductor is a maximum when the charge on the capacitor has its maximum value.
(d) Any change in the value of either L alone or C alone is accompanied by a drop in the peak value of the current. Explain this. [L]

F176

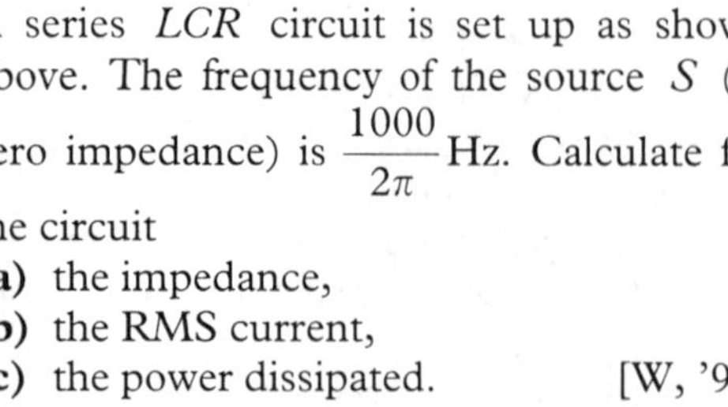

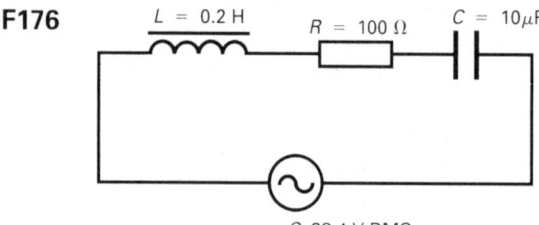

A series LCR circuit is set up as shown above. The frequency of the source S (of zero impedance) is $\frac{1000}{2\pi}$ Hz. Calculate for the circuit
(a) the impedance,
(b) the RMS current,
(c) the power dissipated. [W, '92]

F177 Explain the term *phase difference* as applied to circuits carrying sinusoidally alternating currents.

In a series circuit containing an inductance L, a capacitance C and a resistance R, what are the phase relationships between the voltage and the current for each of the components? Using an appropriate vector diagram obtain an expression for the impedance Z of such a circuit when the frequency of the applied EMF is f. What is the phase relationship between the current and the voltage across the whole circuit?

A circuit consists of an inductor of inductance 1.0 H, a capacitor of capacitance 15 μF and a resistor of resistance 50 Ω in series. Calculate the voltage necessary to pass an RMS current of 2.0 A through the circuit if the frequency of this voltage is 50 Hz. If the voltage is kept constant but the frequency varied, at what frequency will the current have its maximum value? [L]

F178 **(a)** The *impedance* of a circuit containing a capacitor C and a resistor R connected to an alternating voltage supply is given by $\sqrt{R^2 + X^2}$, where X is the *reactance* of the capacitor. Define the two terms in italics.

The current in the circuit leads the voltage by a phase angle ϕ, where $\tan\phi = X/R$. Explain, using a vector diagram, why this is so.

An inductor is put in series with the capacitor and resistor and a source of alternating voltage of constant value but variable frequency. Sketch a graph to show how the current will vary as the frequency changes from zero to a high value.

(b) An alternating voltage of 10 V RMS and 5.0 kHz is applied to a resistor, of resistance 4.0 Ω, in series with a capacitor of capacitance 10 μF. Calculate the RMS potential differences across the resistor and the capacitor. Explain why the sum of these potential differences is not equal to 10 V.

(Assume $\pi^2 = 10$.) [L]

F179 **(a)** **(i)** Explain why power is transmitted by the National Grid at high voltages and why alternating currents are used.

(ii) The power output of a power station is 500 MW at an RMS PD of 132 kV. If the total resistance of the power line is 2.0 Ω, calculate the RMS current in the line and the percentage power loss in transmission due to heat produced in the power line.

(b) In the circuit shown in the diagram resistor R has resistance 100 Ω, capacitor C has capacitance 20 μF and the pure inductor L has inductance 0.20 H.

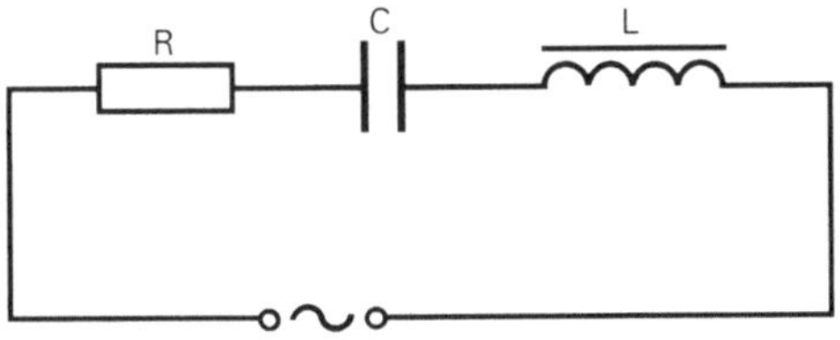

The supply PD is sinusoidal. RMS potential differences and currents are referred to throughout.

When the current in the circuit is 0.50 A, V_L, the PD measured across the inductor, is 25 V.

(i) Calculate the reactance of L and the frequency of the supply.
(ii) Calculate V_R, the PD across R, and V_C, the PD across C.
(iii) Draw a phasor diagram showing V_R, V_L, V_C and V_S, the PD of the supply.
(iv) Calculate V_S and the phase difference between the current and the supply voltage. [J, '92]

F180 In the circuit shown below, the output of the generator is a 200 Hz, 2.5 V RMS sinusoidal signal. The diode may be assumed to be ideal. The Y-plate sensitivity and time-base of the cathode-ray oscilloscope (CRO) are set at 2.0 V cm^{-1} and 1.0 ms cm^{-1} respectively.

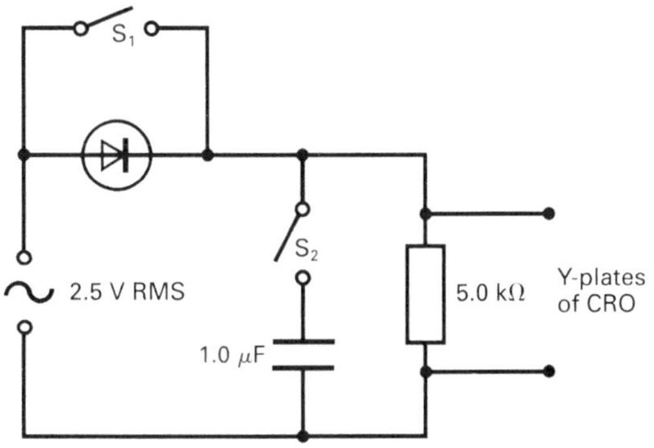

Sketch full-scale diagrams to show the waveform observed when
(a) S_1 is closed and S_2 is open,
(b) S_1 and S_2 are both open,
(c) S_1 is open and S_2 is closed. [C, '91]

F181 A certain type of ammeter uses the heating effect of the current to produce the deflection of the pointer. Discuss whether the same calibration can be used for measuring direct currents and alternating currents. [C]

F182 **(a)** A sinusoidal alternating potential difference of which the peak value is 20 V is connected across a resistor of resistance 10 Ω. What is the mean power dissipated in the resistor?

(b) A sinusoidal alternating potential difference is to be rectified using the circuit shown, which consists of a diode D, a capacitor C and a resistor R. Sketch the variation with time of the potential difference between A and B which you would expect and explain why it has that form.

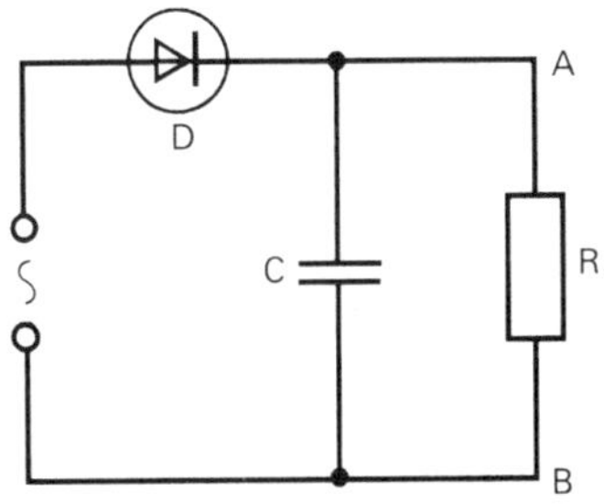

(c) A sinusoidal alternating potential difference of constant amplitude is applied across the resistor R and an inductor L as shown below.

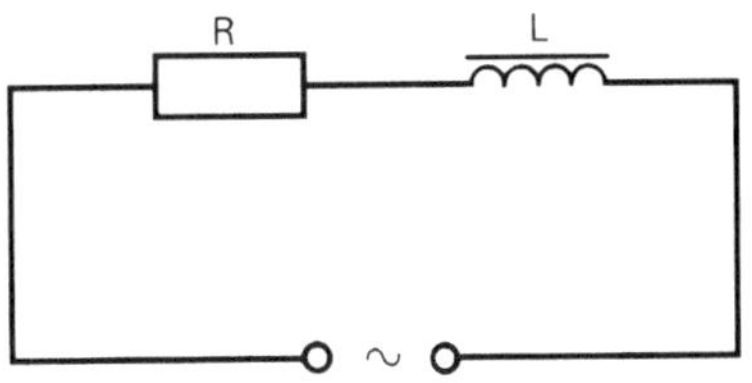

Explain why the amplitude of the current through the circuit decreases as the frequency of the alternating potential difference is increased.

(d)

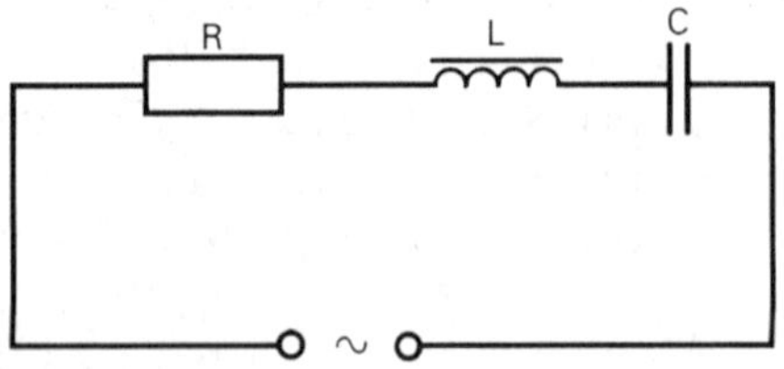

A sinusoidal alternating potential difference of constant frequency and amplitude is applied to the above circuit. Describe and explain how the amplitude of the current through the circuit changes as the capacitance of C is increased slowly from a very small value to a very large value. [O & C]

F183 Draw a circuit of a full-wave rectifier being used to drive a direct current through a resistor and explain its mode of action. (The mode of action of the individual diodes is *not* required.) [AEB, '79]

F184

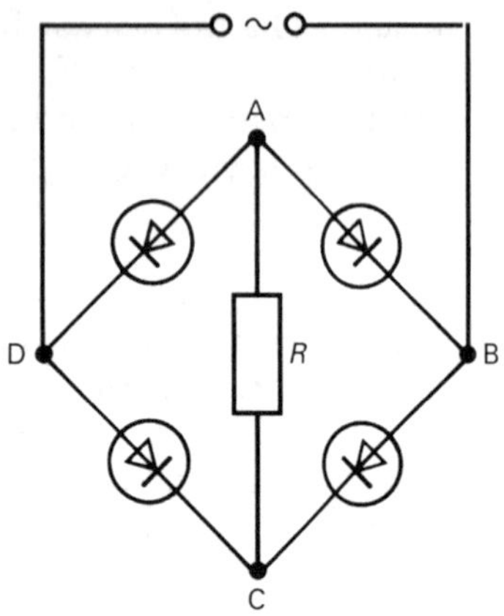

The circuit shows four junction diodes and a resistor, R, connected to a sinusoidally alternating supply. Sketch graphs showing the variation with time over two cycles of the supply of:

(a) the potential of C with respect to A,
(b) the potential of B with respect to A, and
(c) the potential of B with respect to D. [L]

F185 The diagram shows three stages A, B and C required to produce a 5 V DC power supply from a 240 V, 50 Hz alternating mains supply. This question is about different aspects of its design.

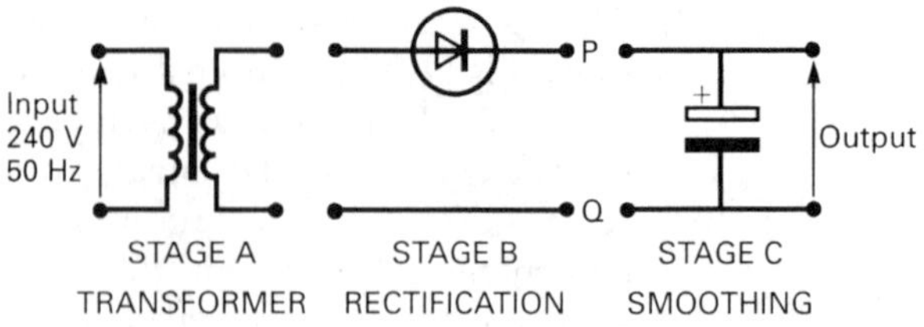

(a) This section concerns stage A on its own.
(i) Describe briefly the operation of the transformer.
(ii) The transformer used is ideal and has 2700 turns on the primary coil and 40 turns on the secondary. Show that the peak voltage across the secondary is about 5.0 V.
(iii) Given that the maximum power output permissible without damaging the transformer is 10 W, determine the maximum current drawn from the mains. How would you protect the transformer against overloading?

(b) You are now to consider what happens when stage B is added to stage A, and a load resistor is connected between P and Q.

Assuming that the diode is ideal, sketch graphs to show the corresponding variations of voltage with time between the output of stage A and that of stage B indicating the time scale and the peak voltages in each case.

(c) In this section you are to consider the operation when stage C, the smoothing capacitor, is connected between P and Q firstly with no load connected and then together with a load.
(i) The capacitor used is marked 500 μF, 10 V. Explain the significance of the 10 V rating.
(ii) Calculate the energy stored by the capacitor when the output voltage is 5 V and no load is connected.
(iii) A load resistor of 100 Ω is now connected across the output terminals. Assuming that the capacitor charges to the peak voltage during each cycle and that the charging time is negligible so that the capacitor discharges for a time equal to the period of the AC, calculate the minimum voltage across the capacitor.

(iv) Estimate the average power delivered to the load by the circuit. [AEB, '90]

MAGNETIC MATERIALS (Chapter 45)

F186 Explain what is meant by *relative permeability*.

How do the relative permeabilities of diamagnetic, paramagnetic and ferromagnetic materials differ from each other?

State two ways in which the relative permeability of a ferromagnetic material can be caused to change.

F187 The data in the table below refer to two ferromagnetic materials:

Material	A	B
Average value of relative permeability	8000	500
Saturation flux density/T	2.0	1.7
Coercive force/A m^{-1}	80	1200
Remanence/T	1.4	1.2

(a) With the aid of a suitable sketch graph, explain what is meant by **(i)** a hysteresis loop and **(ii)** the four quantities (remanence, etc.) in the table.

(b) (i) From the data given in the table, sketch on a common graph the likely shapes of the hysteresis loops for the two materials.

(ii) The two materials are soft iron and a steel; which is which?

(iii) State, with reasons, which of the materials A and B you would use to make a permanent magnet. [W]

F188 Name materials, one in each case, which are suitable for **(a)** a permanent magnet, **(b)** the core of an electromagnet, **(c)** the core of an electric transformer. Explain, in terms of their magnetic properties, why these materials are used. [J]

ELECTROLYSIS (Chapter 46)

F189 A current of 1.5 A is passed through copper sulphate solution, of resistivity 0.4 Ω m, between two plane parallel copper electrodes, each of area 10^{-3} m^2, 4 mm apart.

(a) What is the potential difference across the electrolytic cell?

(b) What power is dissipated in the cell?

(c) How long is required for 2 g of copper (electro-chemical equivalent 3.3×10^{-7} kg C^{-1}) to be deposited on the cathode?

[S]

F190 Calculate the volume at STP of the oxygen liberated in a water voltammeter by a current of 3 A flowing for 20 minutes.

(The ECE of oxygen is 8.3×10^{-8} kg C^{-1} and the density of oxygen at STP is 1.4 kg m^{-3}.)

F191 Calculate the specific charge of a Cu^{++} ion.

(Faraday constant = 9.649×10^4 C mol^{-1}, relative atomic mass of copper = 63.54.)

F192 Calculate the steady current which must flow through an electroplating tank in order that a layer of silver with a uniform thickness of 20 μm may be deposited in 30 minutes on an article which has a surface area of 40 cm^2.

(Density of silver = 1.05×10^4 kg m^{-3}. Silver is monovalent and its relative atomic mass is 108. The Faraday constant = 9.649×10^4 C mol^{-1}.)

SECTION G

MODERN PHYSICS

47

THE PHOTOELECTRIC EFFECT, WAVE-PARTICLE DUALITY

47.1 THE ELECTRONVOLT

The electronvolt (eV) is a unit of energy. It is equal to the kinetic energy gained by an electron in being accelerated by a potential difference of one volt.

The work done when a particle of charge Q moves through a PD V is given by equation [39.6] as QV. The charge on the electron is 1.6×10^{-19} C, and therefore when an electron is accelerated through a PD of 1 V the work done is

$$(1.6 \times 10^{-19}) \times (1)$$

i.e. 1.6×10^{-19} J

The work done is equal to the kinetic energy gained by the electron, and therefore the kinetic energy gained by an electron in being accelerated through one volt is 1.6×10^{-19} J,

i.e.

$$1 \text{ eV} = 1.6 \times 10^{-19} \text{ J}$$

EXAMPLE 47.1

Write down the kinetic energy, in eV, of (a) an electron accelerated from rest through a PD of 10 V, (b) a proton accelerated from rest through a PD of 20 V, (c) a doubly charged calcium ion accelerated from rest through a PD of 30 V.

Solution

(a) 10 eV (Since an electron accelerated through 1 V would gain 1 eV, it follows from $W = QV$ that an electron accelerated through 10 V gains 10 eV.)

(b) 20 eV (Since the charge on the proton has the same magnitude as the charge on the electron, and an electron accelerated through 20 V would gain 20 eV.)

(c) 60 eV (Since the charge on the calcium ion is twice that on an electron, and an electron accelerated through 30 V would gain 30 eV.)

QUESTIONS 47A

1. An electron is accelerated from rest through a PD of 1000 V. What is **(a)** its kinetic energy in eV, **(b)** its kinetic energy in joules, **(c)** its speed?
($1 \text{eV} = 1.602 \times 10^{-19}$ J, mass of electron $= 9.110 \times 10^{-31}$ kg.)

2. An electron accelerated from rest through a PD of 50 V acquires a speed of $4.2 \times 10^6 \text{ m s}^{-1}$. Without performing a detailed calculation, write down the speed that a PD of 200 V would produce.

3. What is the kinetic energy, in eV, of a triply charged ion of iron (Fe^{3+}) which has been accelerated from rest through a PD of 100 V?

4. An electron moves between a pair of electrodes in a vacuum tube. The first electrode is at a potential of 50 V, and the electron has kinetic energy of 20 eV as it leaves it. What is the potential of the second electrode if the electron just reaches it?

5. The kinetic energy of an α-particle from a radioactive source is 4.0 MeV. What is its speed?
(Charge on electron $= 1.6 \times 10^{-19}$ C, mass of α-particle $= 6.4 \times 10^{-27}$ kg.)

47.2 THE PHENOMENON OF PHOTOELECTRIC EMISSION

Electromagnetic radiation (usually visible light or ultraviolet) incident on a metal surface can cause electrons to be emitted from the surface. The phenomenon is called the **photoelectric effect**. (A broad outline of the experiments which established the nature of the effect is given in section 47.5.)

The conclusions which can be drawn from detailed investigations of the effect are summarized below.

(i) Emission occurs only if the frequency of the incident radiation is above some minimum value called the **threshold frequency,** and this depends on the particular metal being irradiated. (For example, the threshold frequency of sodium is in the yellow region of the visible spectrum, that of zinc is in the ultraviolet.)

(ii) Emission commences at the instant the surface starts to be irradiated.

(iii) If the incident radiation is of a single frequency (above the threshold frequency), the number of electrons emitted per second is proportional to the intensity of the radiation.

(iv) The emitted electrons have various kinetic energies, ranging from zero up to some maximum value. Increasing the frequency of the incident radiation increases the energies of the emitted electrons and, in particular, increases the maximum kinetic energy.

(v) The intensity of the radiation has no effect on the kinetic energies of the emitted electrons.

47.3 THE INABILITY OF THE WAVE THEORY TO ACCOUNT FOR PHOTOELECTRIC EMISSION

The photoelectric effect is due to electrons absorbing energy from the incident radiation and so becoming able to overcome the attractive forces of the nuclei. According to the wave theory of light the energy of the incident radiation is distributed uniformly over the wavefront. On the basis of the wave theory, therefore, each electron in the surface of an irradiated metal would absorb an equal share of the radiant energy. It is to be expected, therefore, that if the intensity of the radiation were very low, no single electron would gain sufficient energy to escape, or at least, that a considerable time would elapse before any electron did escape. Neither of these predictions is consistent with observation. Furthermore, an increase in intensity increases the energy falling on the surface and would be expected to increase the energies of the emitted electrons. This also is inconsistent with observation. The wave theory can offer no explanation of the frequency dependence of the kinetic energies of the emitted electrons, nor why there should be a minimum frequency at which emission occurs.

47.4 EINSTEIN'S THEORY OF THE PHOTOELECTRIC EFFECT

In 1901 Max Planck had shown that the energy distribution in the black-body spectrum (section 17.10) could be accounted for by assuming that the radiation was emitted as discrete (separate) packets of energy known as **quanta**, the energy, E, of a **quantum** being given by

$$E = hf \qquad [47.1]$$

where

h = a constant, now called **Planck's constant**, equal to 6.626×10^{-34} J s

f = the frequency of the radiation.

In 1905 Einstein extended this idea by suggesting that the quantum of energy emitted by an atom continues to exist as a concentrated packet of energy. A beam of light of frequency f can therefore be considered to be a stream of particles (called **photons**), each of energy hf. At large distances from a point source of light the intensity is low because the photons are spread over a large area, but there is no diminution in the energy associated with each photon. (**Note**. The intensity of a beam of light is proportional to the number of photons per unit cross-section of the beam per unit time.)

Einstein proposed that when a photon collides with an electron, it must either be reflected with no reduction in energy, or it must give up all its energy to the electron. The energy of a single photon cannot be shared amongst the electrons – no more than one electron can absorb the energy of one photon. It follows that (in a given time) **the number of electrons emitted by a surface is proportional to the number of incident photons, i.e. to the intensity of the radiation.** Furthermore, an electron can be emitted as soon as a photon reaches the surface, explaining why photoemission begins instantaneously.

Einstein reasoned that some of the energy imparted by a photon is actually used to release an electron from the surface (i.e. to overcome the binding forces) and that the rest appears as the kinetic energy of the emitted electron. This is summed up by **Einstein's photoelectric equation**

$$hf = W + \frac{1}{2}mv^2 \qquad [47.2]$$

where

hf = the energy of each incident photon of frequency f,

W = the **work function** of the surface, i.e. the minimum amount of energy that has to be given to an electron to release it from the surface,

$\frac{1}{2}mv^2$ = the **maximum** kinetic energy of the emitted electrons. (Many of the emitted electrons are involved in collisions on their way out of the surface and therefore emerge with energy which is less than the maximum.)

That there should be a minimum frequency which causes emission follows immediately. If $hf < W$, there is not sufficient energy to release an electron, i.e. the threshold frequency, f_0, is given by

$$hf_0 = W$$

The corresponding maximum wavelength, λ_0, is given by

$$h\frac{c}{\lambda_0} = W$$

47.5 MILLIKAN'S VERIFICATION OF EINSTEIN'S PHOTOELECTRIC EQUATION AND MEASUREMENT OF *h*

The photoelectric effect was discovered by Hertz in 1887. He had observed that a spark passes more easily between two electrodes if they are illuminated by ultraviolet radiation. Within a year of Hertz announcing his discovery, Hallwachs found that a negatively charged zinc plate loses its charge when it is illuminated by ultraviolet radiation, but that there is no effect with a positively charged plate. He concluded that the illumination causes the negatively charged plate to emit negatively charged particles. In 1899 Lenard established that these particles were electrons.* By 1902 he had shown that although the maximum kinetic energy of the emitted electrons was independent of the intensity of the incident radiation, it was affected by the frequency being used and by the nature of the illuminated surface. He also showed that the number of electrons emitted was proportional to the intensity of the radiation. It was these experimental results that Einstein explained in 1905.

Lenard's results were not sufficiently accurate to be considered a thorough test of Einstein's theory. Since the theory had such far-reaching consequences, it was of vital importance that it be tested rigorously, and therefore in 1916 Millikan carried out a series of refined experiments – these completely verified Einstein's theory.

*The electron had been discovered two years earlier by J. J. Thomson.

A simplified form of the apparatus used by Millikan is shown in Fig. 47.1. He chose to test Einstein's equation by irradiating sodium, potassium and lithium, each of which has a loosely bound electron and therefore exhibits the photoelectric effect over a large range of visible frequencies. Photoelectric emission is a surface phenomenon, and therefore meaningful results can be obtained only if the emitting surfaces are chemically clean. The metals which Millikan used are very reactive and oxidize rapidly in the presence of air, and therefore the apparatus was evacuated. Any oxide that formed on the samples could be removed with the knife (B). The position of the knife was adjusted immediately before a sample was illuminated, so that by rotating the table (T) the surface of the metal could be pared off by moving it against the knife.

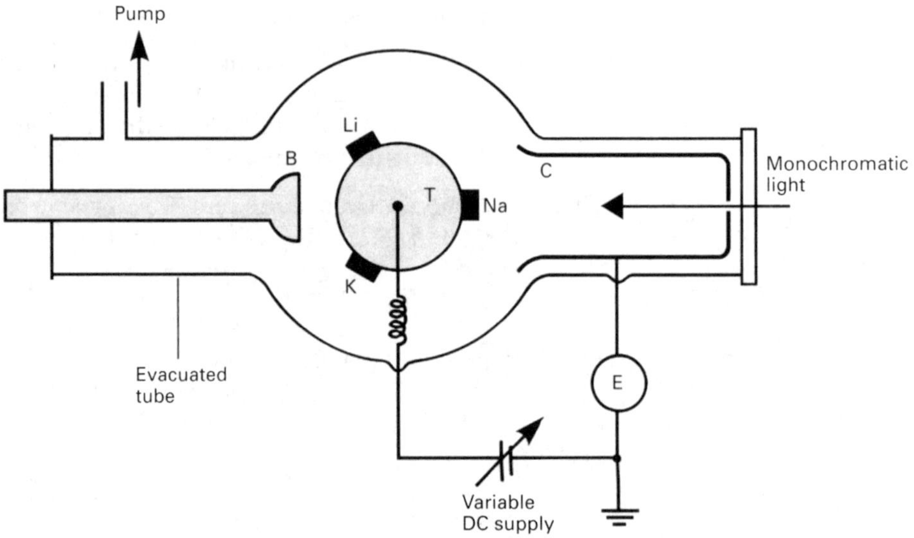

Fig. 47.1
Millikan's apparatus to verify Einstein's photoelectric equation

The target metal was irradiated with monochromatic light. The emitted electrons were collected by the electrode (C) and flowed to earth through the electrometer (E) – the electrometer therefore registered a current. (The electrode was coated with copper oxide, a material which shows no photoelectric emission with visible light.)

By applying a positive potential to the target metal, Millikan was able to slow down the electrons. This prevented those that had been emitted with low values of kinetic energy from reaching the electrode, and the current through the electrometer decreased. He increased the potential until even those electrons which had been emitted with the maximum kinetic energy were unable to reach the electrode, whereupon the current fell to zero. The minimum potential which reduces the current to zero is called the **stopping potential** V. The work done by an electron in moving against the stopping potential is given by equation [39.6] as eV, where e is the charge on the electron. Since this work is done at the expense of the kinetic energy of the electron, it follows that

$$eV = \frac{1}{2}mv^2$$

where $\frac{1}{2}mv^2$ = the maximum KE of an emitted electron.

By Einstein's photoelectric equation

$$hf = W + \frac{1}{2}mv^2$$

$$\therefore \quad hf = W + eV$$

$$\text{i.e.} \quad V = \frac{h}{e}f - \frac{W}{e}$$

Thus, for any given target material, a plot of V against f is linear if Einstein's photoelectric equation is correct. Millikan irradiated each sample with beams of different frequencies and measured the corresponding stopping potentials. For each metal he obtained a graph of the form shown in Fig. 47.2, thus verifying Einstein's equation. The results also provided an accurate value of h.

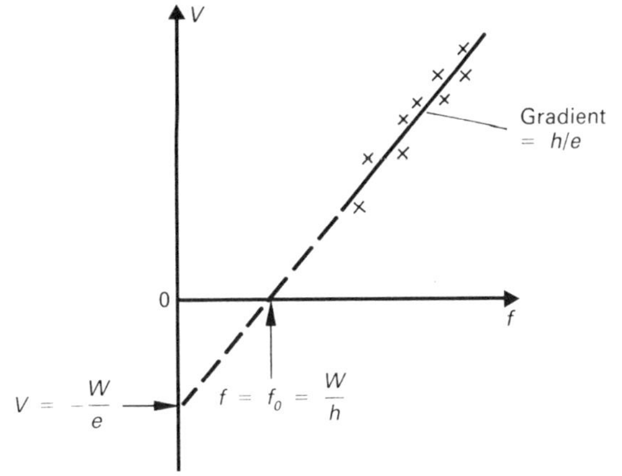

Fig. 47.2 Graph to verify Einstein's photoelectric equation

Notes (i) Changing the target metal gives a graph with the same gradient (h/e) but different intercepts.

(ii) The measurements also provide values of W. These are in good agreement with those found from thermionic emission (see Table 50.1).

EXAMPLE 47.2

Sodium has a work function of 2.3 eV. Calculate: (a) its threshold frequency, (b) the maximum velocity of the photoelectrons produced when the sodium is illuminated by light of wavelength 5×10^{-7} m, (c) the stopping potential with light of this wavelength. ($h = 6.6 \times 10^{-34}$ J s, $c = 3.0 \times 10^{8}$ m s^{-1}, 1eV $= 1.6 \times 10^{-19}$ J, mass of electron $m = 9.1 \times 10^{-31}$ kg.)

Solution

(a) The threshold frequency f_0 is given by

$$hf_0 = W$$

Therefore, since $2.3\,\text{eV} = 2.3 \times 1.6 \times 10^{-19}\,\text{J}$

$$6.6 \times 10^{-34} f_0 = 2.3 \times 1.6 \times 10^{-19}$$

i.e. $f_0 = 5.6 \times 10^{14}$ Hz

(b) By Einstein's photoelectric equation

$$hf = W + \tfrac{1}{2}mv^2$$

$$\therefore \quad h\frac{c}{\lambda} = W + \tfrac{1}{2}mv^2$$

$$\therefore \quad \frac{6.6 \times 10^{-34} \times 3 \times 10^8}{5 \times 10^{-7}} = 2.3 \times 1.6 \times 10^{-19} + \tfrac{1}{2}mv^2$$

$$\therefore \quad 3.96 \times 10^{-19} = 3.68 \times 10^{-19} + \tfrac{1}{2}mv^2$$

i.e. $\tfrac{1}{2}mv^2 = 0.28 \times 10^{-19}$

Therefore the maximum velocity v is given by

$$v = \sqrt{\frac{2 \times 0.28 \times 10^{-19}}{m}}$$

i.e. $v = \sqrt{\dfrac{0.56 \times 10^{-19}}{9.1 \times 10^{-31}}}$

i.e. $v = 2.5 \times 10^5\ \text{m s}^{-1}$

(c) The stopping potential V is given by

$$eV = \tfrac{1}{2}mv^2$$

i.e. $1.6 \times 10^{-19}\ V = 0.28 \times 10^{-19}$

i.e. $V = 0.18\ \text{V}$

QUESTIONS 47B

1. Calculate the energy of **(a)** a photon of frequency 7.0×10^{14} Hz, **(b)** a photon of wavelength 3.0×10^{-7} m.
($h = 6.6 \times 10^{-34}$ J s, $c = 3.0 \times 10^8$ m s^{-1}.)

2. Calcium has a work function of 2.7 eV. **(a)** What is the work function of calcium expressed in joules? **(b)** What is the threshold frequency for calcium? **(c)** What is the maximum wavelength that will cause emission from calcium?
($e = 1.6 \times 10^{-19}$ C, $h = 6.6 \times 10^{-34}$ J s, $c = 3.0 \times 10^8$ m s^{-1}.)

3. Gold has a work function of 4.9 eV. **(a)** Calculate the maximum kinetic energy, in joules, of the electrons emitted when gold is illuminated with ultraviolet radiation of frequency 1.7×10^{15} Hz. **(b)** What is this energy expressed in eV? **(c)** What is the stopping potential for these electrons?
($e = 1.6 \times 10^{-19}$ C, $h = 6.6 \times 10^{-34}$ J s.)

4. Calculate the stopping potential for a platinum surface irradiated with ultraviolet light of wavelength 1.2×10^{-7} m. The work function of platinum is 6.3 eV.
($h = 6.6 \times 10^{-34}$ J s, $c = 3.0 \times 10^8$ m s^{-1}, $e = 1.6 \times 10^{-19}$ C.)

47.6 THE PHOTOEMISSIVE CELL

A photoemissive cell is shown in Fig. 47.3. If light of a frequency which is greater than the threshold frequency of the cathode-coating is incident on the cathode, electrons are emitted and move to the anode. A current ($\sim \mu$A) flows in the external circuit. The size of this current increases with the intensity of the light. If the light beam is interrupted, the current flow ceases, and when the device is used with a suitable relay circuit it can be used to open doors, act as a burglar alarm, etc.

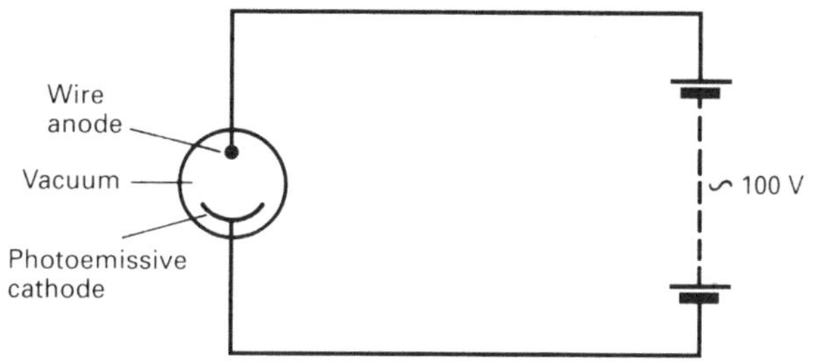

Fig. 47.3
Photoemissive cell and circuit

47.7 WAVE–PARTICLE DUALITY

The reflection and refraction of light are satisfactorily explained on the basis of light being a wave motion. Furthermore, light can be diffracted and can produce interference effects – convincing evidence that it behaves as a wave motion. Light can be polarized; it is therefore a transverse wave motion. Towards the end of the nineteenth century Maxwell showed, on entirely theoretical grounds, that electromagnetic waves could propagate through space; the velocity of these waves is exactly the same as that of light.

It seems paradoxical therefore that it is necessary to attribute particle properties to light in order to explain the photoelectric effect. What the reader must accept, though, is that the idea of light being a wave motion and the idea of it being a particle motion are merely two different models which help us explain the behaviour of light; neither is necessarily a literal description of what light is.

In a thesis presented in 1924 Louis de Broglie, having reflected on the wave–particle duality of light, suggested that matter might also have a dual nature. He proposed that any particle of momentum p has an associated wavelength λ (now called the **de Broglie wavelength**) given by

$$\lambda = \frac{h}{p} = \frac{h}{mv}$$

where m is the relativistic mass* of the particle and v is its velocity. This relationship was confirmed in 1927 when Davisson and Germer succeeded in diffracting electrons. They used a single crystal of nickel as a diffraction grating, the regularly spaced atoms within the crystal acting as diffraction centres in much the same way as they do for X-rays. (The de Broglie wavelengths of electrons which have been accelerated by potentials in the range 1.5 V to 15 000 V are in the range 10^{-9} m to 10^{-11} m, similar to the wavelengths of X-rays.)

Neutrons, protons, hydrogen atoms and helium atoms have also been diffracted. Neutron diffraction is used to study crystal structures. The wavelengths associated with macroscopic bodies are very much less than the width of any aperture through which such a body might pass, and any diffraction which occurs is too small to be observable. (The de Broglie wavelength of a billiard ball moving at 1 m s^{-1} is of the order of 10^{-33} m.)

The de Broglie waves, though often referred to as **matter waves**, are not composed of matter. The intensity of the wave at a point represents the probability of the associated particle being there.

*The special theory of relativity distinguishes between the mass m_0 of a stationary particle (the rest mass) and the mass m of a particle moving with velocity v. It can be shown that $m = m_0(1 - v^2/c^2)^{1/2}$, where c is the velocity of light. The distinction is unimportant if $v \ll c$.

QUESTIONS 47C

1. Calculate the de Broglie wavelength of an electron moving at $3.0 \times 10^6\,\mathrm{m\,s^{-1}}$.
($h = 6.6 \times 10^{-34}\,\mathrm{J\,s}$, mass of electron $= 9.1 \times 10^{-31}$ kg.)

2. Calculate **(a)** the speed, **(b)** the de Broglie wavelength of an electron which has been accelerated from rest through a PD of 250 V.
($e = 1.6 \times 10^{-19}$ C, $h = 6.6 \times 10^{-34}\,\mathrm{J\,s}$, mass of electron $= 9.1 \times 10^{-31}$ kg.)

3. Calculate the de Broglie wavelength of an α-particle of energy 4.0 MeV.
($e = 1.6 \times 10^{-19}$ C, $h = 6.6 \times 10^{-34}\,\mathrm{J\,s}$, mass of α-particle $= 6.4 \times 10^{-27}$ kg.)

CONSOLIDATION

The electronvolt is a unit of energy equal to the kinetic energy gained by an electron in being accelerated through a PD of 1 volt.

In order to explain the photoelectric effect a beam of light (of frequency f and wavelength λ) is regarded as a stream of **photons**, each of energy E where

$$E = hf \qquad \text{or} \qquad E = h\frac{c}{\lambda}$$

The Photoelectric Effect

Electromagnetic radiation incident on a metal surface causes electrons to be emitted from the surface.

The maximum KE of the emitted electrons increases when the **frequency** of the radiation increases.

The number of emitted electrons is proportional to the **intensity** of the radiation.

The work function of a surface is the minimum amount of energy that has to be given to an electron to release it from the surface.

The threshold frequency is the minimum frequency that will cause emission.

Einstein's Photoelectric Equation

$$hf = W + \tfrac{1}{2}mv^2$$

Since hf can be replaced by hc/λ, W by hf_0, and $\frac{1}{2}mv^2$ by eV, where V = stopping potential and f_0 = threshold frequency

$$\begin{pmatrix} hf \\ \text{or} \\ h\dfrac{c}{\lambda} \end{pmatrix} = \begin{pmatrix} W \\ \text{or} \\ hf_0 \end{pmatrix} + \begin{pmatrix} \frac{1}{2}mv^2 \\ \text{or} \\ eV \end{pmatrix}$$

Millikan's Experiment

(i) Verified Einstein's photoelectric equation.

(ii) Provided an accurate value of h.

(iii) Provided values of the work functions of the metals used.

$$hf = W + \tfrac{1}{2}mv^2$$

$$\therefore \quad hf = W + eV \qquad \text{where } V = \text{ stopping potential}$$

$$\therefore \quad V = \frac{h}{e}f - \frac{W}{e}$$

The graph of V against f is a straight line (hence (i)) of gradient h/e (hence (ii)) and y-intercept $-W/e$ (hence (iii)).

Changing the target material changes the intercept, but not the gradient.

Wave-Particle Duality

A particle of mass m and velocity v has an associated wavelength λ where

$$\lambda = \frac{h}{mv}$$

48

THE STRUCTURE OF THE ATOM, ENERGY LEVELS

48.1 INTRODUCTION

By 1900, the idea that matter was composed of atoms was well established. It was known that atoms contained negatively charged particles (**electrons**). It was also known that atoms as a whole are electrically neutral, and therefore they must have a positively charged component. No positively charged equivalent of an electron was known (positive ions are much more massive), and this led J. J. Thomson to propose the so called **'plum-pudding' model** of the atom. In this model, the positive charge was supposed to be continuously distributed throughout a sphere in which the electrons were embedded – rather like plums in a pudding. The model had very limited success and, in particular, was totally incapable of accounting for the results of Rutherford's α-particle scattering experiments.

48.2 THE SCATTERING OF α-PARTICLES

In 1909 Rutherford investigated the scattering of α-particles (positively charged particles resulting from radioactive decay) by thin films of heavy metals, notably gold. (The experiments were actually carried out by Geiger and Marsden under Rutherford's direction.) The experimental arrangement is shown in Fig. 48.1.

Fig. 48.1
Rutherford's apparatus to investigate scattering of α-particles

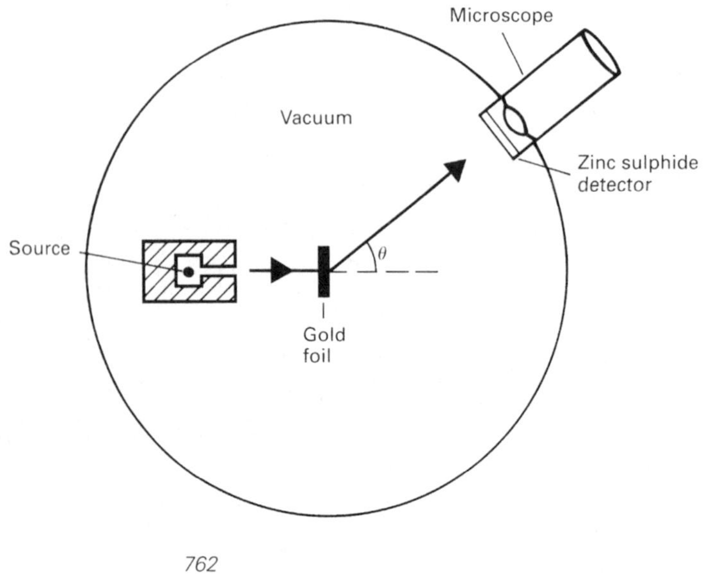

A narrow pencil of α-particles from a radon source inside a metal block was incident on a thin metal foil. A glass screen coated with zinc sulphide was used to detect the scattered α-particles. Whenever a particle hit the screen it produced a faint flash of light (a **scintillation**). The experiment was carried out in a darkened room and the scintillations were observed through a microscope. The screen could be rotated about the metal foil, and by counting the number of scintillations produced in various positions in equal intervals of time, the angular dependence of the scattering was determined. The range of α-particles in air is limited to about 5 cm, and therefore the apparatus was evacuated so that the particles would not be prevented from reaching the screen.

The majority of the α-particles were scattered through small angles, but a few (about 1 in 8000) were deviated by more than 90°.

The most popular model of the atom at the time was the plum-pudding model (section 48.1). The net charge at any point in such an atom is small, and therefore it would not be able to scatter an α-particle through a large angle. The large deflections which were observed might be thought to be due to successive deflections through small angles. However, Rutherford was able to show that the number of large deflections, though small, was far too high to be accounted for in this way, and he suggested that the large-angle scattering was due to a single encounter between an α-particle and an intense positive electric charge. In view of this, Rutherford proposed in 1911 that **an atom has a positively charged core (now called the nucleus) which contains most of the mass of the atom and which is surrounded by orbiting electrons**. On the basis of this model, Rutherford calculated that the number of particles scattered through an angle, θ, should be proportional to $\operatorname{cosec}^4(\theta/2)$. Geiger and Marsden performed a second series of experiments and verified this prediction in 1913.

Large-angle scattering occurred whenever an α-particle was incident almost head-on to a nucleus (see Fig. 48.2). Since very few of the particles were scattered through large angles, it follows that the probability of a head-on approach is small and indicated that **the nucleus occupies only a small proportion of the available space**. (The nuclear radius is of the order of 10^{-15} m; that of an atom as a whole is about 10^{-10} m.)

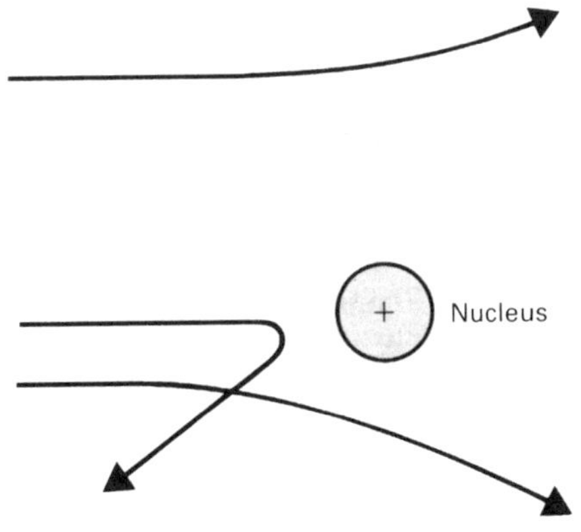

Fig. 48.2
Scattering of α-particles by a nucleus

An atom is electrically neutral and therefore the positive charge on the nucleus must be equal to Ze, where e is the charge on the electron and Z is an integer equal to the number of electrons in the atom. According to Rutherford, the number of α-particles scattered in any given direction should be proportional to Z^2. Geiger and Marsden found that the scattering was proportional to the square of the atomic weight (now called relative atomic mass) of the scatterer. It had already been shown by Barkla (1911) that the atomic weight of an element was, to a first approximation, proportional to the number of electrons in each of its atoms, i.e. to Z. Therefore, Geiger's and Marsden's observation that the scattering was proportional to the square of the atomic weight was consistent with Rutherford's prediction that it should be proportional to Z^2 – further verification of the Rutherford model.

48.3 THE BOHR MODEL OF THE ATOM

Although Rutherford's model of the atom was completely consistent with the results of the α-particle scattering experiments, there was considerable opposition to it on theoretical grounds. An orbiting electron is constantly changing its direction and therefore is accelerating. It was well established (classical electromagnetic theory) that when a charged particle is accelerated it emits electromagnetic radiation. The orbiting electrons, therefore, could be expected to emit radiation continuously. This is not possible, for if an electron were to emit radiation, it would have to do so at the expense of its own energy and as a consequence it would slow down and spiral into the nucleus, in which case the atom would cease to exist.

The problem was resolved by Neils Bohr in 1913. Bohr assumed that each electron moves in a circular orbit which is centred on the nucleus, the necessary centripetal force being provided by the electrostatic force of attraction between the positively charged nucleus and the negativey charged electron. On this basis he was able to show that the energy of an orbiting electron depends on (among other factors) the radius of its orbit. This much was obvious, but Bohr made two revolutionary proposals.

(i) The angular momenta of the electrons are whole number multiples of $h/2\pi$, where h is a constant known as **Planck's constant**. (Thus, the angular momentum does not have a continuous range of values, i.e. it is **quantized**.) This means that the electrons can have only certain orbital radii, which in turn means that **the electrons are allowed to have only certain values of energy (called energy levels)**. This nullifies the idea that the electrons should continuously emit radiation, for if they were to do so, they would lose energy continuously and would need to have a continuous range of energies available to them. The allowed energy levels are often referred to as **stationary states**, since an electron can remain in an energy level indefinitely without radiating any energy.

(ii) An electron can jump from an orbit in which its energy is E_2, say, to one which is closer to the nucleus and of lower energy, E_1 say. In doing so, the electron gives up the energy difference of the two levels by emitting an electromagnetic wave whose frequency, f, is given by

$$E_2 - E_1 = hf \qquad [48.1]$$

where h = Planck's constant = 6.626×10^{-34} J s.

The principal justification for the Bohr model is that it predicts, to a high degree of accuracy, the wavelengths emitted by atomic hydrogen. It fails, however, when it is applied to more complex atoms. A second objection is that the model involves the arbitrary assumption that the allowed values of angular momentum are integral multiples of $h/2\pi$. (Over ten years later it was shown that electrons can be regarded as waves, and that the allowed values of angular momentum were consistent with the allowed orbits being exactly the right size to accommodate a stationary (standing) electron wave.)

The current model of the atom is based on **wave mechanics**. The electrons are no longer considered to move in definite orbits, but exist as an 'electron cloud' throughout the volume of the atom – the Bohr radii are the most probable positions of the electrons. The concepts of discrete energy levels and transitions between them giving rise to the emission or absorption of radiation are retained. The existence of discrete energy levels (the energies of which are the same as those given by Bohr) is a natural consequence of the theory and there is no need to make any arbitrary assumptions.

48.4 MATHEMATICAL TREATMENT OF THE HYDROGEN ATOM ACCORDING TO THE BOHR MODEL

Consider an electron of mass m and charge e moving with velocity v in a circular orbit of radius r about a hydrogen nucleus.* The charge on the nucleus is also e.

There is an inward directed Coulomb force acting on the electron, the magnitude of which, F, is given by equation [39.2] as

$$F = \frac{1}{4\pi\varepsilon_0}\frac{e^2}{r^2}$$

The centripetal acceleration of the electron is v^2/r (see section 6.3). Therefore, by Newton's second law

$$F = m\frac{v^2}{r}$$

Combining these equations gives

$$\frac{e^2}{4\pi\varepsilon_0 r^2} = \frac{mv^2}{r} \qquad [48.2]$$

Multiplying each side of equation [48.2] by mr^3 gives

$$\frac{me^2 r}{4\pi\varepsilon_0} = (mvr)^2 \qquad [48.3]$$

According to Bohr's first proposal, the angular momentum mvr is given by

$$mvr = \frac{nh}{2\pi} \qquad (n = 1, 2, 3, \ldots)$$

Therefore, from equation [48.3]

$$\frac{me^2 r}{4\pi\varepsilon_0} = \left(\frac{nh}{2\pi}\right)^2$$

i.e.
$$r = \frac{n^2 h^2 \varepsilon_0}{\pi m e^2} \qquad [48.4]$$

The total energy, E, of the system is given by

$$E = E_k + E_p \qquad [48.5]$$

*A hydrogen nucleus is much more massive than an electron and therefore it is a sufficiently good approximation to assume that the electron rotates about the nucleus, rather than about the centre of mass of the system.

where

$$E_k = \text{the kinetic energy of the electron}$$

$$= \tfrac{1}{2}mv^2 = \frac{1}{2}\frac{e^2}{4\pi\varepsilon_0 r} \qquad \text{(from equation [48.2])}$$

and

$$E_p = \text{the potential energy of the electron}$$

If the nucleus is considered to be a point charge, the electric potential at a distance r is given by equation [39.5] as

$$\frac{e}{4\pi\varepsilon_0 r}$$

Therefore, the work done in bringing an electron from infinity to a point a distance r from the nucleus is given by equation [39.6] as

$$-\frac{e^2}{4\pi\varepsilon_0 r}$$

(The minus sign arises because the nucleus attracts the electron.)

If the potential energy of the electron is taken to be zero when it is at infinity, then

$$E_p = \frac{-e^2}{4\pi\varepsilon_0 r}$$

Substituting for E_k and E_p in equation [48.5] gives

$$E = \frac{1}{2}\frac{e^2}{4\pi\varepsilon_0 r} - \frac{e^2}{4\pi\varepsilon_0 r}$$

i.e. $$E = -\frac{e^2}{8\pi\varepsilon_0 r}$$

Therefore, from equation [48.4]

$$E = -\frac{me^4}{8\varepsilon_0{}^2 n^2 h^2} \qquad (n = 1, 2, 3, \ldots) \qquad [48.6]$$

Notes (i) The energy is always negative (equation [48.6]). Work has to be done to remove the electron to infinity, where it is considered to have zero energy, i.e. the electron is 'bound' to the atom.

(ii) Increasing values of r are associated with increasing values of n and therefore with increasing (i.e. less negative) values of E.

48.5 ENERGY LEVELS

The energies of the electrons in an atom can have only certain values. These values are called the energy levels of the atom. All atoms of a given element have the same set of energy levels and these are characteristic of the element, i.e. they are different from those of every other element. The energies of the various levels can be calculated by using wave mechanics (and, in the specific case of the hydrogen atom, by using the Bohr model). It is convenient to express energy level values in **electronvolts** (see section 47.1), where one electronvolt (i.e. 1 eV) is equal to 1.6×10^{-19} joules.

The energy levels of an atom are usually represented as a series of horizontal lines. Fig. 48.3 shows the energy level diagram of the hydrogen atom. Hydrogen has only one electron and this normally occupies the lowest level and has an energy of −13.6 eV. When the electron is in this level the atom is said to be in its **ground state**. If the atom absorbs energy in some way (for example by being involved in a collision or by absorbing electromagnetic radiation), the electron may be promoted into one of the higher energy levels. (On the Bohr model the electron has moved into an orbit which is farther away from the nucleus.) The atom is now unstable – it is said to be in an **excited state** – and after a short, but random, interval the electron 'falls' back into the lowest level so that the atom returns to its ground state. The energy that was originally absorbed is emitted as an electromagnetic wave.

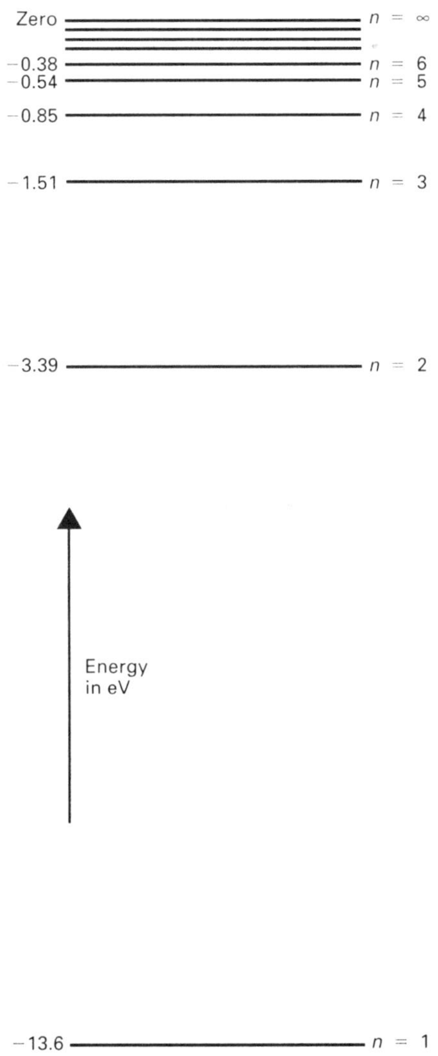

Fig. 48.3
Energy levels of the hydrogen atom

Each energy level is characterized by what is called a **quantum number**, n (see Fig. 48.3). The lowest level has $n = 1$, the next has $n = 2$, etc. The energy of the level which has $n = \infty$ is zero. If the electron is raised to this level, it becomes free of the atom. An atom which has lost an electron is said to be **ionized**, and therefore the energy required to ionize a hydrogen atom which is in its ground state is 13.6 eV.

48.6 THE OPTICAL LINE SPECTRUM OF ATOMIC HYDROGEN

Suppose that a hydrogen atom has acquired energy in some way, and that as a result its electron is in the energy level characterized by $n = 4$. After a short time the electron will return to the level with $n = 1$. There are four possible 'routes'.

(a) $n = 4 \longrightarrow n = 3 \longrightarrow n = 2 \longrightarrow n = 1$

(b) $n = 4 \longrightarrow n = 3 \longrightarrow n = 1$

(c) $n = 4 \longrightarrow n = 2 \longrightarrow n = 1$

(d) $n = 4 \longrightarrow n = 1$

This involves six different transitions, namely:

$$n = 4 \longrightarrow n = 3 \qquad n = 4 \longrightarrow n = 2$$

$$n = 4 \longrightarrow n = 1 \qquad n = 3 \longrightarrow n = 2$$

$$n = 3 \longrightarrow n = 1 \qquad n = 2 \longrightarrow n = 1$$

($n = 4 \rightarrow n = 3$ appears in both (a) and (b), $n = 2 \rightarrow n = 1$ appears in both (a) and (c)). Each transition involves the emission of an electromagnetic wave whose frequency depends on the difference in energy of the two levels involved. When there are large numbers of atoms the different transitions take place simultaneously and radiation of many different frequencies is emitted. The line spectrum (see section 28.3) of hydrogen is composed of light of these frequencies.

When an electron moves from a level with energy E_2 to one of lower energy, E_1, the frequency, f, of the emitted radiation is given by

$$E_2 - E_1 = hf \qquad [48.7]$$

where h = Planck's constant. When using this equation it is necessary to bear in mind that h and f are normally expressed in the relevant SI units (J s and Hz respectively), in which case E_1 and E_2 must be expressed in joules.

Suppose it is required to calculate the frequency, f, and the wavelength, λ, of the radiation emitted as a result of an electron transition from $n = 4$ to $n = 3$. From Fig. 48.3 the energies involved are -0.85 eV and -1.51 eV, i.e.

$$E_2 = -0.85\,\text{eV} = -0.85 \times 1.6 \times 10^{-19} = -1.36 \times 10^{-19}\,\text{J}$$

and

$$E_1 = -1.51\,\text{eV} = -1.51 \times 1.6 \times 10^{-19} = -2.42 \times 10^{-19}\,\text{J}$$

If the value of h is taken to be 6.6×10^{-34} J s, then by equation [48.7]

$$(-1.36 \times 10^{-19}) - (-2.42 \times 10^{-19}) = 6.6 \times 10^{-34} \times f$$

i.e. $$1.06 \times 10^{-19} = 6.6 \times 10^{-34} \times f$$

i.e. $$f = \frac{1.06 \times 10^{-19}}{6.6 \times 10^{-34}}$$

i.e. $$f = 1.6 \times 10^{14}\,\text{Hz}$$

Also, $\lambda = c/f$, where c = the velocity of light = $3 \times 10^8\ \mathrm{m\,s^{-1}}$. Therefore

$$\lambda = \frac{3 \times 10^8}{1.6 \times 10^{14}}$$

i.e. $$\lambda = 1.9 \times 10^{-6}\ \mathrm{m}$$

Wavelengths calculated in this way are in excellent agreement with those observed in the line spectrum of atomic hydrogen – providing convincing evidence of the existence of energy levels.

Fig. 48.4
The main spectral transitions of atomic hydrogen

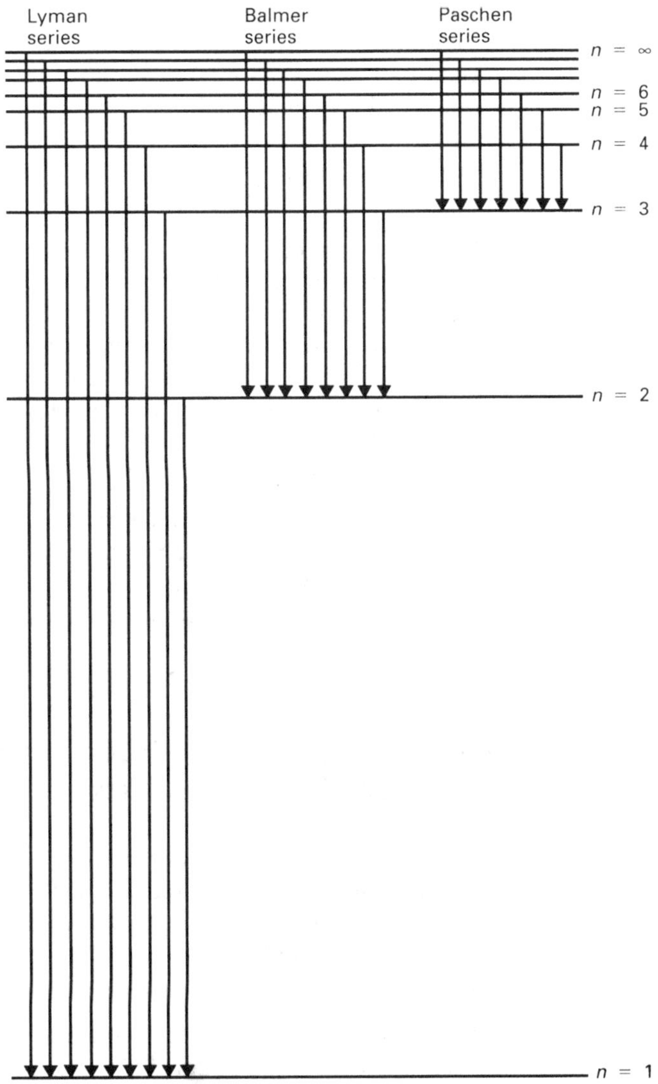

The spectrum of atomic hydrogen contains distinct groups of lines. The three most obvious groups are the Lyman series, the Balmer series and the Paschen series. The wavelengths of the lines in the Lyman series are in the ultraviolet and each is associated with a transition involving the level with $n = 1$ (see Fig. 48.4). The Balmer series involves transitions to the level with $n = 2$, and as a consequence smaller energy differences are involved and the wavelengths are in the visible. The lines of the Paschen series are in the infrared.

QUESTIONS 48A

1. Refer to Fig. 48.3. Calculate the frequency and wavelength of the radiation resulting from the following transitions: **(a)** $n = 4$ to $n = 2$, **(b)** $n = 2$ to $n = 1$. ($h = 6.6 \times 10^{-34}$ J s, $c = 3.0 \times 10^{8}$ m s^{-1}.)

48.7 IONIZATION AND EXCITATION POTENTIALS

The minimum amount of energy required to ionize an atom which is in its ground state, i.e. to remove its most loosely bound electron, is called the **first (or principal) ionization energy** of the atom. For example, the first ionization energy of hydrogen is 13.6 eV, i.e. 2.18×10^{-18} J (see Fig. 48.3). The energy to remove the next most loosely bound electron is called the second ionization energy, etc. The first ionization energy is often referred to simply as **the ionization energy**. This is particularly so in the case of hydrogen which has only one electron and therefore no higher ionization energies than the first.

It follows from the definition of the electronvolt that 13.6 eV is the kinetic energy gained by an electron in being accelerated through a PD of 13.6 V. Therefore, if an electron which has been accelerated from rest by a PD of 13.6 V collides with a hydrogen atom, it has exactly the right amount of energy to produce ionization. This is a common method of producing ionization, and therefore the term ionization potential is often used. **Ionization potential** is expressed in volts and is numerically equal to the ionization energy. Thus, the ionization potential of hydrogen is 13.6 V.

The energy required to excite an atom which is in its ground state is called an **excitation energy** of the atom. For example, the first and second excitation energies of hydrogen are (approximately) 10.2 eV and 12.1 eV respectively (see Fig. 48.3). The corresponding **excitation potentials** are 10.2 V and 12.1 V.

48.8 MEASUREMENT OF EXCITATION POTENTIALS BY ELECTRON COLLISION

The spectroscopic evidence for the existence of discrete energy levels (see section 48.6) is supported by the results of experiments in which electrons are caused to collide with gas atoms. The first successful experiments of this type were carried out by Franck and Hertz in 1914. A schematic form of the apparatus used in a Franck–Hertz type experiments is shown in Fig. 48.5.

Electrons which have been emitted thermionically by the filament (F) are accelerated towards the grid (G) and pass through it. The anode (A) is slightly negative with respect to G and therefore the electrons are retarded as they move from G to A. If V_1 is slightly larger than V_2, the electrons have sufficient energy to reach A and a current flows through the galvanometer. If V_1 is increased, the current increases at first (see Fig. 48.6). As the electrons move across the tube they collide with gas atoms, and at this stage the collisions are elastic, the electrons bounce off the atoms without losing any energy. As V_1 is increased further, eventually a point is reached when the electrons have exactly the right amount of energy to promote the atomic electrons to higher energy levels and inelastic collisions result. All the energy of a bombarding electron is given up to the atom

Fig. 48.5
Apparatus for a Franck–Hertz type experiment

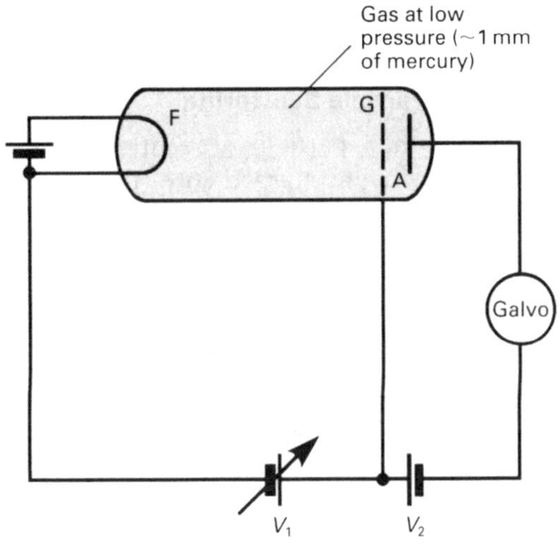

Fig. 48.6
Typical results of a Franck–Hertz type experiment

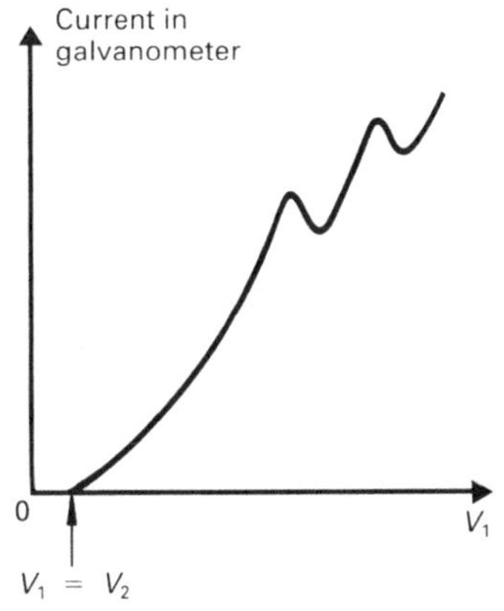

with which it has collided. The electrons no longer have sufficient energy to reach A and the galvanometer current falls. By determining the value of V_1 at which this happens it is possible to calculate the excitation potential of the transition which has taken place. Increasing V_1 beyond this value causes the current to rise until V_1 is again such that the bombarding electrons can produce excitation, whereupon the current falls.

48.9 ENERGY LEVELS IN ATOMS WITH MORE THAN ONE ELECTRON

The energy levels of many electron atoms are arranged in groups, known as shells, the levels within any one group having similar energies. There is a limit to how many electrons can occupy a shell. That known as the K shell can contain no more than two electrons, the L shell has a maximum of eight, the M shell has a maximum of eighteen, etc. (These numbers are given by $2n^2$, where $n = 1, 2, 3, \ldots$.) When an atom is in its ground state the electrons occupy the lowest available energy levels. Thus sodium, which has eleven electrons, has two in the K shell, eight in the L shell and one in the M shell. (This simple situation breaks down for atoms with more than eighteen electrons because one of the energy levels in the N shell has less energy than one of those in the M shell.)

CONSOLIDATION

α-Particle Scattering

Some α-particles are scattered through large angles – suggesting that an atom has a positively charged core (the nucleus).

Only a few α-particles are scattered through large angles – suggesting that the nucleus is very much smaller than the atom as a whole.

Bohr Model of Atom

Electrons are allowed to have only certain orbital radii and therefore only certain values of energy (called energy levels).

An electron may 'jump' from a level with energy E_2 to a lower level of energy E_1 causing the emission of electromagnetic radiation of frequency f where

$$E_2 - E_1 = hf$$

The principal justification for the Bohr model is that it predicts, to a high degree of accuracy, the wavelengths emitted by atomic hydrogen.

An atom is in its **ground state** when all the electrons are in the lowest available energy levels.

The **ionization energy** of an atom is the minimum amount of energy required to remove its most loosely bound electron when the atom is in its ground state.

49

X-RAYS

49.1 INTRODUCTION

X-rays are short (typically $\sim 10^{-10}$ m) wavelength electromagnetic radiation. They were discovered by Röntgen in 1895.

49.2 PRODUCTION OF X-RAYS

A modern form of X-ray tube is shown in Fig. 49.1. A focused beam of electrons is accelerated towards the target during the half-cycles that it is positive with respect to the filament. On collision the electrons decelerate rapidly and X-rays are produced. Over 99% of the kinetic energy of the electrons goes into producing heat. The target is a high-melting point metal such as tungsten or molybdenum embedded in a copper rod, the purpose of which is to conduct heat away from the target. The rod is cooled by circulating oil through it or by the use of cooling fins.

The intensity of an X-ray beam increases with the number of electrons hitting the target and therefore with filament current. It is also increased by increasing the PD across the tube because this increases the energy with which the electrons hit the target and so makes more energy available for X-ray production.

The penetrating power (or quality) of an X-ray beam increases with the PD across the tube. X-rays with low penetrating power are called **soft X-rays**, those with high penetrating power are called **hard X-rays**.

Fig. 49.1
A Coolidge type X-ray tube

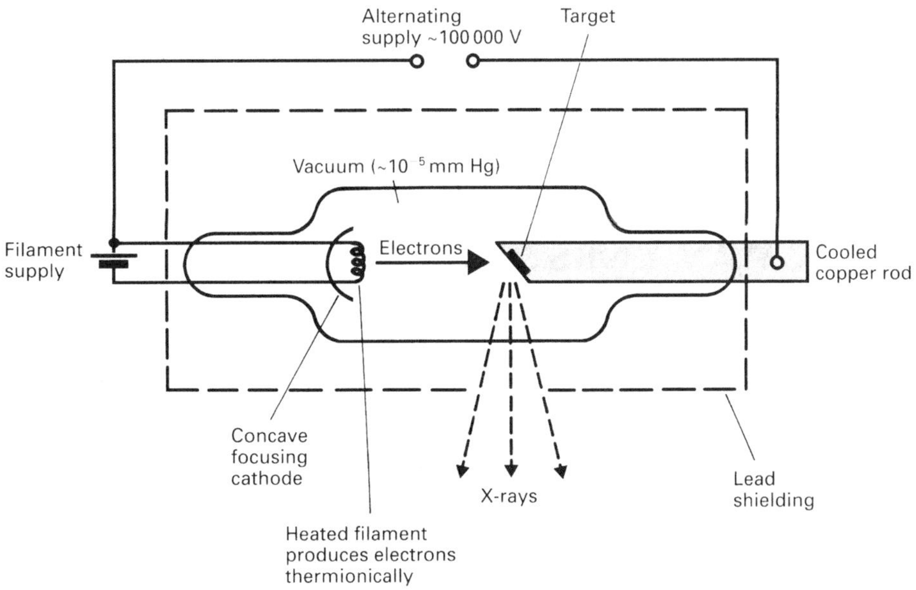

49.3 PROPERTIES OF X-RAYS

(i) They travel in straight lines at the velocity of light.

(ii) They cannot be deflected by electric or magnetic fields. (This is convincing evidence that they are not charged particles.)

(iii) They penetrate matter. Penetration is least with materials of high density.

(iv) They can be reflected, but only at very large angles of incidence.

(v) Refractive indices of all materials are very close to unity for X-rays so that very little bending occurs when they pass from one material to another. They cannot be focused by lenses.

(vi) They can be diffracted.

The following properties, (vii) to (x), are used to detect X-rays.

(vii) They ionize gases through which they pass.

(viii) They affect photographic film.

(ix) They can produce fluorescence.

(x) They can produce photoelectric emission.

49.4 USES OF X-RAYS

(i) They are used in medicine to:

(a) locate bone fractures, etc.
(b) destroy cancer cells.

(ii) They are used to locate internal imperfections in welded joints and castings.

(iii) The spacings of the regularly arrayed atoms within crystals are such that crystals act as diffraction gratings for X-rays. Analysis of the diffraction pattern provides detailed information about the crystal structure. X-ray diffraction has been used to determine the structure of complex organic molecules.

49.5 X-RAY EMISSION SPECTRA

A typical X-ray spectrum is shown in Fig. 49.2. X-ray spectra have two distinct components.

(i) A background of continuous radiation, the minimum wavelength of which depends on the operating voltage of the tube, i.e. on the energy of the bombarding electrons.

(ii) Very intense emission at a few discrete wavelengths (an X-ray line spectrum). These wavelengths are characteristic of the target material and are independent of the operating voltage.

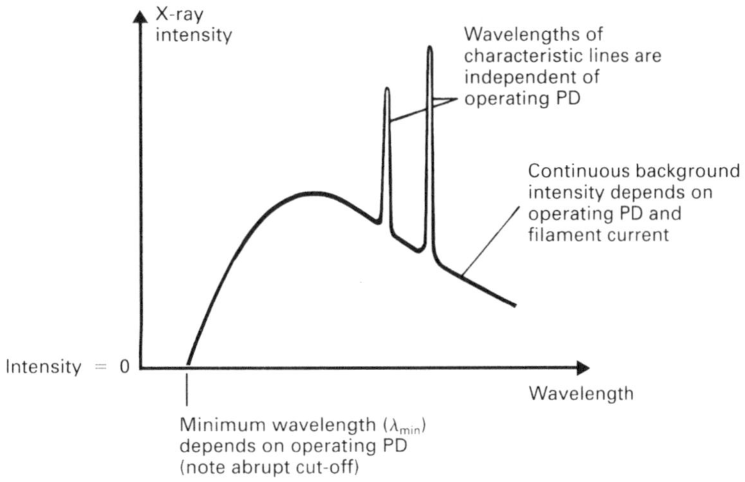

Fig. 49.2
X-ray spectrum

The continuous background is produced by electrons colliding with the target and being decelerated. The energy of the emitted X-ray quantum is equal to the energy lost in the deceleration. An electron may lose any fraction of its energy in this process. The most energetic X-rays (i.e. those whose wavelength is λ_{min}) are the result of bombarding electrons losing all their energy at once. Since the energy of the electrons depends on the operating voltage, so also does λ_{min} (see Example 49.1). X-rays with longer wavelengths are the result of electrons losing less than their total energy.

The line spectrum is the result of electron transitions within the atoms of the target material. The electrons which bombard the target are very energetic ($\sim$ 100 000 eV) and are capable of knocking electrons out of deep-lying energy levels of the target atoms. (This corresponds to removing an electron from an inner orbit on the Bohr model.) An outer electron may 'fall' into the vacancy created in its atom, and in doing so causes a high-energy quantum of electromagnetic radiation, i.e. an X-ray, to be emitted. The frequency of the X-ray is given by $E = hf$, where E is the difference in energy of the levels involved and h is Planck's constant. Since the energy levels are characteristic of the target atoms, so too are the X-rays produced in this way.

The essential difference between the transitions which give rise to X-rays and those which give rise to visible light is that the X-ray transitions involve deep-lying energy levels, the optical transitions do not.

EXAMPLE 49.1

Calculate the wavelength of the most energetic X-rays produced by a tube operating at 1.0×10^5 V. ($h = 6.6 \times 10^{-34}$ J s, $e = 1.6 \times 10^{-19}$ C, $c = 3.0 \times 10^8$ m s^{-1}.)

Solution

The most energetic X-rays are those produced by electrons which lose all their kinetic energy on impact.

$$\text{KE on impact} = \text{work done by accelerating PD}$$
$$= 1.6 \times 10^{-19} \times 10^5 \qquad \text{(from equation [39.6])}$$
$$= 1.6 \times 10^{-14} \text{ joules}$$

$$\therefore \quad \text{Maximum KE lost} = 1.6 \times 10^{-14} \text{ joules}$$

The energy of the corresponding X-ray quantum is hc/λ_{min} and therefore

$$\frac{hc}{\lambda_{\text{min}}} = 1.6 \times 10^{-14}$$

i.e. $$\lambda_{\text{min}} = \frac{6.6 \times 10^{-34} \times 3.0 \times 10^{8}}{1.6 \times 10^{-14}}$$

i.e. $$\lambda_{\text{min}} = 1.24 \times 10^{-11}\ \text{m}$$

EXAMPLE 49.2

The current in a water-cooled X-ray tube operating at 60 kV is 30 mA. 99% of the energy supplied to the tube is converted into heat at the target and is removed by water flowing at a rate of 0.060 kg s^{-1}. Calculate: (a) the rate at which energy is being supplied to the tube, (b) the increase in temperature of the cooling water. (Specific heat capacity of water $= 4.2 \times 10^3$ J kg^{-1} °C^{-1}.)

Solution

(a) Rate of supply of electrical energy $= 60 \times 10^3 \times 30 \times 10^{-3}$

$= 1.8 \times 10^3\ \text{J s}^{-1}$

(b) Rate of production of heat $= 0.99 \times 1.8 \times 10^3$

$= 1782\ \text{J s}^{-1}$

Let $\Delta\theta$ = increase in temperature of water

Heat gained by water in 1 s $= 0.060 \times 4.2 \times 10^3\ \Delta\theta$

$= 252\ \Delta\theta$

Since all the heat produced is removed by the water,

$$252\,\Delta\theta = 1782$$

$$\therefore \quad \Delta\theta = \frac{1782}{252} = 7.1\text{°C}$$

QUESTIONS 49A

1. The most energetic X-rays produced by a particular X-ray tube have a wavelength of 2.1×10^{-11} m. What is the operating PD of the tube?
($e = 1.6 \times 10^{-19}$ C, $h = 6.6 \times 10^{-34}$ J s, $c = 3.0 \times 10^{8}$ m s^{-1}.)

2. An X-ray tube which is 1% efficient produces X-ray energy at a rate of 20 J s^{-1}. Calculate the current in the tube if the operating PD is 50 kV.

3.
$n = 3$ ———— -11×10^3 eV
$n = 2$ ———— -26×10^3 eV
$n = 1$ ———— -98×10^3 eV

The diagram shows the three lowest energy levels of an atom of the target material of an X-ray tube. What is the minimum PD at which the tube can operate: **(a)** if the transition $n = 3$ to $n = 1$ is to be possible, **(b)** if $n = 2$ to $n = 1$ is to be possible? **(c)** What is the wavelength corresponding to the transition $n = 3$ to $n = 1$?
($e = 1.6 \times 10^{-19}$ C, $h = 6.6 \times 10^{-34}$ J s, $c = 3.0 \times 10^{8}$ m s^{-1}.)

49.6 X-RAY ABSORPTION SPECTRA

In Fig. 49.3 a monochromatic beam of X-rays of intensity I_0 is incident on a thickness x of some absorber. The intensity I of the emergent beam is given by

$$I = I_0 e^{-\mu x}$$

where

μ = a constant for a given material and a given wavelength. It is called the **linear absorption coefficient** of the material at the wavelength concerned (m^{-1}).

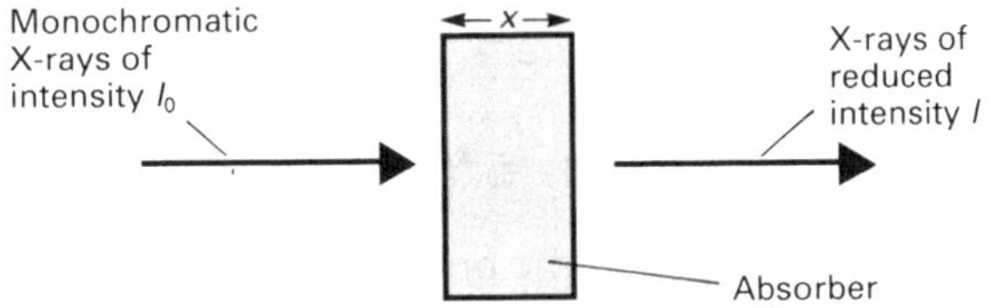

Fig. 49.3
X-rays passing through an absorber

The way in which μ varies with wavelength (λ) for a given absorber is illustrated in Fig. 49.4. As λ is increased μ (and therefore the absorption) rises rapidly until $\lambda = \lambda_K$. Up to this point (which is called the **K absorption edge**) absorption has mainly been due to the ejection of electrons from the K shell of the absorber (see section 48.9). The sudden drop in μ at $\lambda = \lambda_K$ occurs because X-rays of longer wavelength are not sufficiently energetic to eject K shell electrons. As λ is increased beyond λ_K the absorption increases again; the absorption now being due mainly to the ejection of electrons from the higher lying L shell.

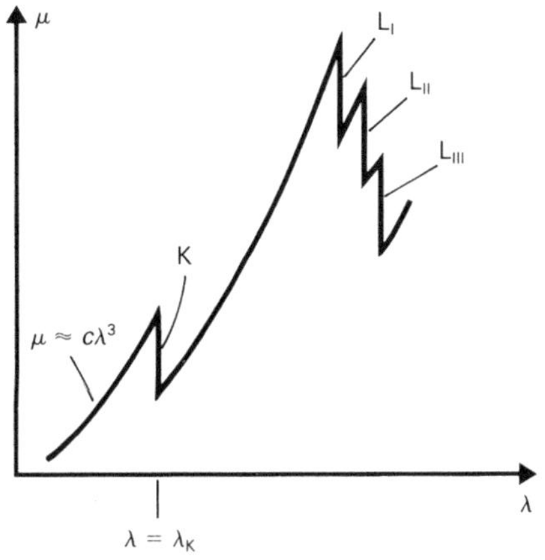

Fig. 49.4
Variation of linear absorption coefficient of X-rays with wavelength

The three absorption edges designated as L_I, L_{II} and L_{III} lie very close together and are taken as evidence that the L shell consists of three energy levels. Indeed, it is the fact that the energy levels are grouped in this way that leads us to think in terms of a shell structure. A group of five absorption edges (not shown) at longer wavelengths indicates that the M shell contains five levels, etc.

CONSOLIDATION

X-rays are short wavelength electromagnetic radiation ($\lambda \sim 10^{-10}$ m).

The X-ray Tube

The intensity of the X-ray beam can be increased by increasing the filament current and/or by increasing the PD across the tube.

The penetrating power (or **quality**) of the X-rays increases with the PD across the tube.

The X-ray Emission Spectrum

The energy of an X-ray quantum in the **continuous background** is equal to the energy lost by an electron as a result of being slowed down by hitting the target. The wavelength λ_{min} of the most energetic X-rays is given by

$$\frac{hc}{\lambda_{min}} = eV$$

where V is the operating PD.

The intense peaks are the result of electron transitions involving **deep-lying energy levels** of the target atoms. Their wavelengths are characteristic of the target material.

50

THE ELECTRON

50.1 DEFLECTION OF ELECTRONS

In an Electric Field

It follows from the definition of electric field intensity (section 39.3) that an electron in a field of intensity E is subject to a force F, given by

$$F = eE$$

where

e = the charge on the electron ($= 1.6 \times 10^{-19}$ C).

Since electrons are negatively charged, the force is in the opposite direction to the field.

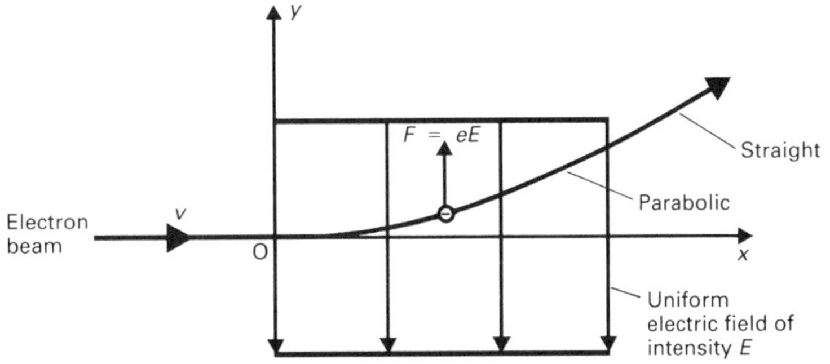

Fig. 50.1
Deflection of an electron beam in an electric field

Consider a beam of electrons, each moving with velocity v, entering a uniform electric field of intensity E which is perpendicular to their direction of motion (Fig. 50.1). Once in the field each electron is subject to a force eE in the positive direction of the y-axis, and therefore by Newton's second law acquires an acceleration eE/m in this direction, where m is the mass of the electron. At the instant an electron enters the field its y-component of velocity is zero, and therefore after it has spent a time t in the field it will have undergone a vertical displacement, y, given by equation [2.8] as

$$y = 0 + \frac{1}{2}\left(\frac{eE}{m}\right)t^2$$

i.e. $$y = \frac{1}{2}\left(\frac{eE}{m}\right)t^2 \qquad [50.1]$$

The electron's x-component of velocity is unaffected by the field and therefore x, its horizontal displacement from O, is given by

$$x = vt \qquad [50.2]$$

Eliminating t between equations [50.1] and [50.2] gives

$$y = \left(\frac{eE}{2mv^2}\right)x^2$$

This is the equation of a parabola, i.e. **the path of the electron whilst in the field is parabolic**. Once the electron has left the field its path is linear. Since the electron has gained a y-component of velocity whilst in the field and there has been no change in its x-component, its kinetic energy has increased.

The situation is analogous to that of a particle in a uniform gravitational field. In each case the magnitude and direction of the force are constant.

In a Magnetic Field

An electron moving with velocity v at right angles to a magnetic field of flux density B experiences a force F, which is given by equation [41.10] as

$$F = Bev$$

The force is perpendicular to both the field direction and the velocity and its direction is given by Fleming's left-hand rule.

Consider an electron moving with velocity v into a uniform magnetic field of flux density B which is at right angles to its direction of motion (Fig. 50.2).

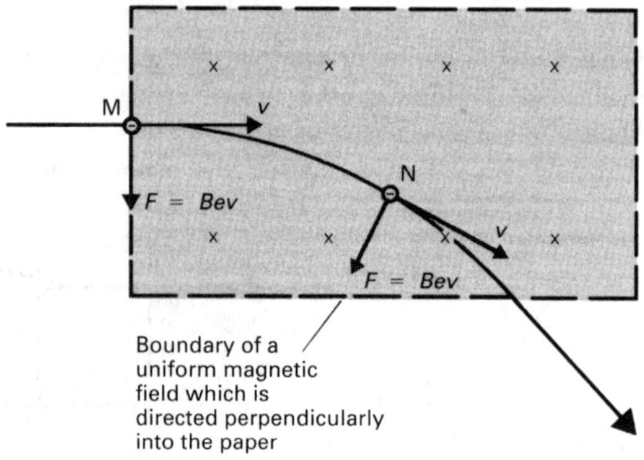

Fig. 50.2
Deflection of an electron beam in a magnetic field

On entering the field at M the electron feels a force F as shown and is deflected. The force is at right angles to the direction of motion of the electron and can neither speed it up nor slow it down. When the electron reaches some other point N, the magnitude of the force acting on it is the same as it was at M (since none of B, e and v has changed) but the direction of the force is different. Thus, the force is perpendicular to the direction of motion at all times and has a constant magnitude, and therefore the electron travels with constant speed along a circular arc. **The magnetic field does not change the kinetic energy of the electron.**

If the electron moves along an arc of radius r, then its centripetal acceleration v^2/r is given by Newton's second law as

$$Bev = \frac{mv^2}{r}$$

Crossed Fields

If a uniform electric field and a uniform magnetic field are perpendicular to each other in such a way that they produce deflections in opposite senses, they are known as **crossed fields**. If the forces exerted by each field are of the same size, then

$$Bev = eE$$

i.e. $$v = E/B \qquad [50.3]$$

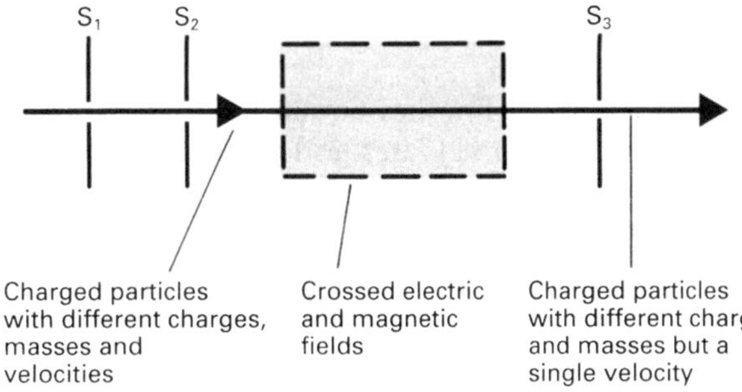

Fig. 50.3 Principle of a velocity selector

Fig. 50.3 shows how crossed fields can be used as a **velocity selector**, i.e. to select charged particles of a single velocity from a beam containing particles with a range of different velocities. The particles need not be electrons, and may have a range of charge-to-mass ratios. Slits S_1 and S_2 confine the particles to a narrow beam. The only particles which are undeflected, and therefore which emerge from slit S_3, are those whose velocity v is given by $v = E/B$.

The Bainbridge mass spectrograph (section 51.4) uses a velocity selector of this type.

The determination of the charge-to-mass ratio (specific charge) of the electron by the method described in section 50.2 also makes use of crossed fields.

EXAMPLE 50.1

Refer to Fig. 50.4. A beam of electrons is accelerated through a PD of 500 V and then enters a uniform electric field of strength $3.00 \times 10^3\ \text{V m}^{-1}$ created by two parallel plates each of length 2.00×10^{-2} m. Calculate: (a) the speed, v, of the electrons as they enter the field, (b) the time, t, that each electron spends in the field, (c) the angle, θ, through which the electrons have been deflected by the time they emerge from the field. (Specific charge (e/m) for electron $= 1.76 \times 10^{11}\ \text{C kg}^{-1}$.)

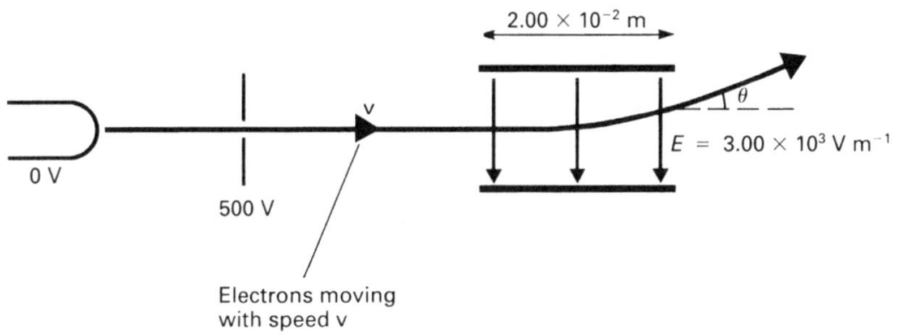

Fig. 50.4 Diagram for Example 50.1

Solution

(a) The kinetic energy gained by an electron is equal to the work done by the PD. Therefore

$$\tfrac{1}{2}mv^2 = eV$$

i.e. $$v = \sqrt{\frac{2eV}{m}}$$

$$= \sqrt{2 \times 1.76 \times 10^{11} \times 500} = 1.327 \times 10^7$$

i.e. Speed on entering field $= 1.33 \times 10^7\,\text{m s}^{-1}$

(b) The horizontal component of velocity is unaffected by the field between the plates and is therefore constant. It follows that t is given by

$$t = \frac{2.00 \times 10^{-2}}{1.327 \times 10^7} = 1.507 \times 10^{-9}$$

i.e. Time between plates $= 1.51 \times 10^{-9}\,\text{s}$

(c) The electric field, E, exerts a force, F, on each electron where

$$F = eE$$

By Newton's second law this gives each electron an acceleration, a, towards the top of the page, where

$$a = \frac{F}{m} = \frac{eE}{m}$$

On emerging from the field the electron will have gained a vertical component of velocity, v_y, given by $v = u + at$ (equation [2.6]) as

$$v_y = 0 + \left(\frac{eE}{m}\right)t$$

i.e. $$v_y = \left(\frac{e}{m}\right) \times E \times t$$

$$= 1.76 \times 10^{11} \times 3.00 \times 10^3 \times 1.507 \times 10^{-9} = 7.957 \times 10^5$$

Since $\tan\theta = v_y/v$

$$\tan\theta = \frac{7.957 \times 10^5}{1.327 \times 10^7} = 5.996 \times 10^{-2}$$

$$\therefore \quad \theta = 3.4^\circ$$

EXAMPLE 50.2

Suppose that in an arrangement of the type described in Example 50.1, particles of charge, Q, and mass, M, are accelerated by a PD, V, and then enter a field of strength, E, between plates of length, d. Obtain an expression for the angle, θ, through which the particles will have been deflected by the time they leave the plates.

Solution

The speed, v, on entering the field between the plates is given by

$$\tfrac{1}{2}Mv^2 = QV \qquad \therefore \qquad v = \sqrt{\frac{2QV}{M}}$$

The time, t, between the plates is given by

$$t = \frac{d}{v}$$

The vertical acceleration, a, is given by

$$a = \frac{QE}{M}$$

The vertical component of velocity, v_y, is given by $v = u + at$ as

$$v_y = \left(\frac{QE}{M}\right)\frac{d}{v}$$

$$\therefore \qquad \tan\theta = \frac{v_y}{v} = \frac{QEd}{Mv^2}$$

Substituting for v gives

$$\tan\theta = \frac{QEd}{M}\frac{M}{2QV} = \frac{Ed}{2V}$$

$$\text{i.e.} \qquad \theta = \tan^{-1}\left(\frac{Ed}{2V}\right)$$

Note that θ depends only on E, d and V – it does not depend on either Q or M. This result is interesting in itself, but it also serves to illustrate that we could have obtained the answer to Example 50.1 even if we had not been given the value of the specific charge of the particles involved. It further illustrates the effort that can be saved by not putting in numerical values until it is absolutely necessary!

QUESTIONS 50A

1. Calculate the speed of a proton which has been accelerated through a PD of 400 V. (Mass of proton $= 1.67 \times 10^{-27}$ kg, charge on proton $= 1.60 \times 10^{-19}$ C.)

2. An electron is moving in a circular path at $3.0 \times 10^6\ \text{m s}^{-1}$ in a uniform magnetic field of flux density 2.0×10^{-4} T. Find the radius of the path. (Mass of electron $= 9.1 \times 10^{-31}$ kg, charge on electron $= 1.6 \times 10^{-19}$ C.)

50.2 DETERMINATION OF THE SPECIFIC CHARGE (*e/m*) OF THE ELECTRON AND ITS DISCOVERY

One form of apparatus used to determine e/m is shown in Fig. 50.5. Electrons which have been emitted thermionically by the filament are accelerated towards the cylindrical anode and pass through it. Two small holes on the axis of the anode confine the electrons to a narrow beam. When both fields (E and B) are zero the electrons reach the screen at X and produce fluorescence there.

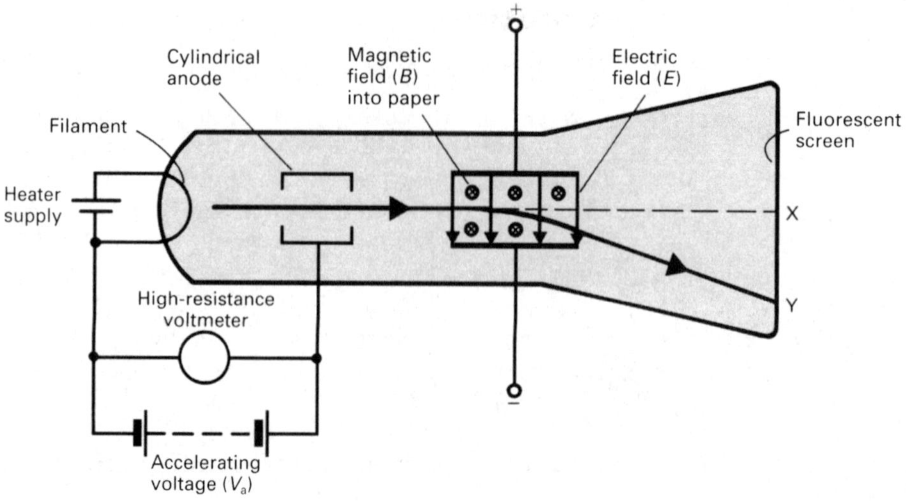

Fig. 50.5 Apparatus to determine ***e/m*** for the electron

If the velocity of the electrons on emerging from the anode is v, then

$$eV_a = \tfrac{1}{2}mv^2$$

i.e. $$\frac{e}{m} = \frac{v^2}{2V_a} \qquad [50.4]$$

where V_a = the accelerating voltage.

The position of X is noted and the magnetic field is switched on, deflecting the beam to Y. The electrons can be brought back to X by using the electric field, because it is arranged so that it exerts a force in the same region as the magnetic field but in the opposite direction. The electric field is switched on and is adjusted until the beam is again at X. The forces being exerted by each field must now be of equal size, and therefore

$$Bev = eE$$

i.e. $$v = E/B$$

Substituting for v in equation [50.4] gives

$$\frac{e}{m} = \frac{E^2}{2V_a B^2}$$

The value of E is found from $E = V/d$,

where V = the PD between the deflecting plates

d = their (known) separation.

The magnetic field is normally produced by a pair of Helmholtz coils (see section 41.4), in which case the value of B can be found from the current through them, their radius and the number of turns. The value of V_a is read off directly from the voltmeter.

Discovery of the Electron

In 1897 J. J. Thomson used this method (but with a cold cathode rather than a heated filament) to determine the specific charge of **cathode rays** – the 'rays' which move from cathode to anode when an electric discharge is passed through a gas at low pressure. (The 'rays' were already known to be particles and there was evidence to suggest that in any particular discharge they all had the same specific charge, but there was no knowledge of the size of the charge, or of whether it was different under different conditions.) Thomson found that the value of the specific charge did not depend on the nature of the gas nor on the electrode material, suggesting that the 'rays' were composed of previously unknown particles which are a basic constituent of all matter. The idea that the particles were fundamentally different from any particle known at the time was strongly supported by Thomson's value for their specific charge – of the order of a thousand times that of the hydrogen ion as found from experiments in electrolysis, and this was the highest of any known particle. The relatively high value of the specific charge of the cathode ray particles could be due to them having a large charge, or a small mass, or both. Thomson assumed that they had the same charge as a monovalent ion, in which case they must be particles of very small mass. There is now no doubt that Thomson's assumption was correct. The particles whose specific charge he had measured are what are now called electrons, and in view of this, Thomson is normally credited with having discovered the electron.

50.3 DETERMINATION OF THE SPECIFIC CHARGE (e/m) OF THE ELECTRON BY USING A FINE-BEAM TUBE

The fine-beam tube (Fig. 50.6) is a glass bulb containing hydrogen at low pressure. Electrons are produced by an electron-gun arrangement (see section 50.7) at one side of the tube. The electrons collide with hydrogen atoms causing the atoms to emit light and so reveal the path of the electrons. A pair of Helmholtz coils (see section 41.4) provides a uniform magnetic field (directed perpendicularly out of the paper) which, provided it is sufficiently strong, deflects the electrons so that they travel in a complete circle. This circular path shows up as a luminous ring.

Fig. 50.6
The fine-beam tube

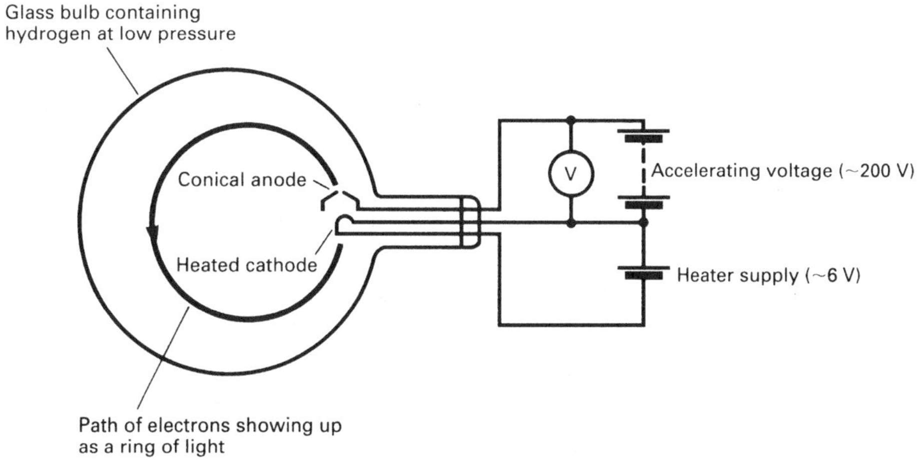

If r = the radius of the electron path,

v = the velocity of the electrons on leaving the electron gun, and

B = the magnetic flux density,

then the force on the electron is Bev, the centripetal acceleration (see section 6.3) is v^2/r, and therefore by Newton's second law

$$Bev = mv^2/r \quad [50.5]$$

Also, if the accelerating voltage (i.e. the PD between the anode and cathode of the electron gun) is V, then

$$\frac{1}{2}mv^2 = eV \qquad [50.6]$$

because (by equation [39.6]) eV is the work done by the accelerating PD, and $\frac{1}{2}mv^2$ is the kinetic energy gained by the electrons as a result. By equation [50.5]

$$v = Br(e/m) \qquad [50.7]$$

By equation [50.6]

$$v^2 = 2V(e/m) \qquad [50.8]$$

By equations [50.7] and [50.8]

$$(Br)^2(e/m)^2 = 2V(e/m)$$

$$\therefore \quad \frac{e}{m} = \frac{2V}{B^2r^2}$$

Hence e/m may be determined.

A high-resistance voltmeter is used to measure V, and B can be found from the current through the Helmholtz coils, their radius and number of turns. The diameter, and therefore the radius, of the electron path can be measured by placing a mirror with a scale on it behind the tube and lining up the luminous ring with its image.

50.4 MILLIKAN'S DETERMINATION OF *e* (1909)

A Note on Terminal Velocity

An object falling through air experiences a viscous drag (see section 12.9). Initially the downward force due to gravity is greater than the drag force and the object accelerates. Drag forces increase with velocity, and therefore as the object accelerates, the upward directed force due to the drag increases and eventually becomes equal to the gravitational force. Once this happens there is no further acceleration and the object is said to have reached its **terminal velocity**.

Theory of Millikan's Experiment

The principle of Millikan's experiment is to measure the terminal velocity of a small, charged oil drop falling under gravity, and then to oppose its motion with an electric field in such a way that it remains stationary.

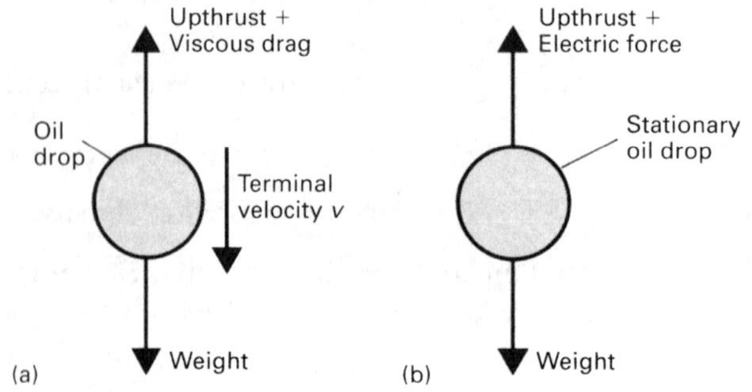

Fig. 50.7
Forces acting on an oil drop in Millikan's oil-drop experiment: (a) without electric field, (b) with electric field

When there is no electric field the forces acting on the drop are as shown in Fig. 50.7(a). Once the drop has reached its terminal velocity it has no acceleration, and therefore

$$\text{Weight} = \text{Upthrust due to air} + \text{Viscous drag} \quad [50.9]$$

But

$$\text{Weight} = \text{Volume of drop} \times \text{Density of oil} \times g$$

Therefore for a spherical drop

$$\text{Weight} = \tfrac{4}{3}\pi r^3 \rho_o g$$

where

r = radius of drop

ρ_o = density of oil.

Also

$$\text{Upthrust} = \text{Weight of air displaced by drop}$$
$$= \text{Volume of drop} \times \text{Density of air} \times g$$

i.e. $$\text{Upthrust} = \tfrac{4}{3}\pi r^3 \rho_a g$$

where

ρ_a = density of air.

By Stokes' law (section 12.9)

$$\text{Viscous drag} = 6\pi r \eta v$$

where

η = coefficient of viscosity of air.

Substituting in equation [50.9] gives

$$\tfrac{4}{3}\pi r^3 \rho_o g = \tfrac{4}{3}\pi r^3 \rho_a g + 6\pi r \eta v \quad [50.10]$$

When an electric field has been applied such that the drop is stationary, the forces acting on the drop are as shown in Fig. 50.7(b). The drop has no velocity and no acceleration and therefore

$$\text{Weight} = \text{Upthrust} + \text{Electric force}$$

i.e. $$\tfrac{4}{3}\pi r^3 \rho_o g = \tfrac{4}{3}\pi r^3 \rho_a g + QE \quad [50.11]$$

where

Q = the charge on the drop

E = the electric field strength.

Subtracting equation [50.10] from equation [50.11] gives

$$0 = QE - 6\pi r \eta v$$

i.e. $$Q = \frac{6\pi r \eta v}{E} \quad [50.12]$$

Millikan measured E and v and did a separate experiment to find η. He was not able to measure r directly, but by equation [50.10]

$$\tfrac{4}{3}\pi r^3 (\rho_o - \rho_a) g = 6\pi r \eta v$$

i.e. $$r = \left(\frac{9\eta v}{2(\rho_o - \rho_a)g}\right)^{1/2}$$

Substituting for r in equation [50.12] gives

$$Q = \frac{6\pi\eta}{E}\left(\frac{9\eta v}{2(\rho_o - \rho_a)g}\right)^{1/2} v \qquad [50.13]$$

Note The density of air at room temperature and pressure is less than one thousandth of that of oil, and except for very accurate calculations equation [50.13] can be replaced by

$$Q = \frac{6\pi\eta}{E}\left(\frac{9\eta v}{2\rho_o g}\right)^{1/2} v$$

This of course is the result that would have been obtained had the upthrust due to the displaced air been ignored in the first place.

Experimental Procedure

The apparatus is shown in schematic form in Fig. 50.8. A and B are two metal plates which are accurately parallel to each other. (In Millikan's apparatus the plates had a diameter of 20 cm and a separation of 1.5 cm.) An atomizer is used to create a fine mist of oil drops in the region of the small hole in the upper plate. The drops are charged, either positively or negatively, as a result of losing or gaining electrons through frictional effects on emerging from the atomizer. Some of the drops fall through the holes and are observed by reflected light through a low-power microscope. The eyepiece of the microscope incorporates a calibrated graticule so that the terminal velocity v of any particular oil drop can be determined by timing its fall through a known distance.

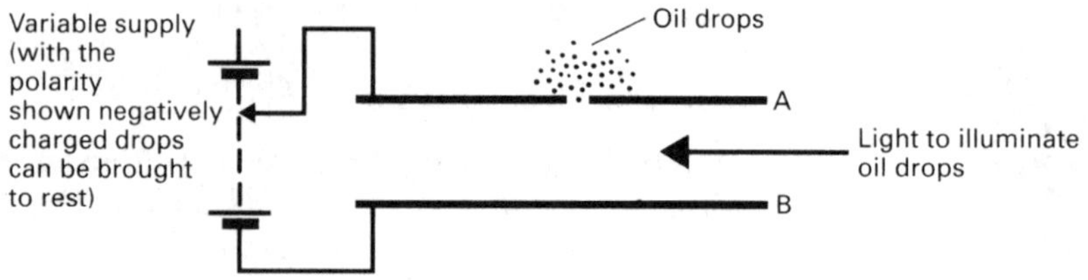

Fig. 50.8 Millikan's apparatus to determine the charge on the electron

An electric field ($\sim 10^5\,\text{V m}^{-1}$) is applied at this stage and is adjusted so that the drop whose velocity has just been determined is held stationary. The strength E of this field is given by $E = V/d$, where V is the PD between the plates and d is their separation.

Millikan measured the charges on hundreds of drops. **The charges were always integral multiples of 1.6×10^{-19} C, and he concluded that electric charge can never exist in fractions of this amount and that the magnitude of the electronic charge e is 1.6×10^{-19} C,** i.e. Millikan established the **quantization of electric charge**.

Notes (i) For some measurements Millikan used X-rays to ionize the air through which the drops fell. When this was done the speeds of some of the drops changed suddenly (though not all at the same time). Millikan interpreted this as being due to the drops colliding with ions and acquiring positive or negative charges according to whether they had collided with a positive ion or a negative ion. He found that the change in charge was always plus or minus a small whole-number multiple of *e*.

(ii) A constant-temperature enclosure surrounded Millikan's apparatus in order to eliminate convection currents. It also served to shield the apparatus from draughts.

(iii) Millikan used a low-vapour-pressure oil to reduce problems due to evaporation.

(iv) Millikan's actual technique was to cause the drops to move upwards under the influence of the electric field – we have considered the drops to be at rest in order to simplify the theory.

(v) Millikan's early results showed that it is necessary to use a modified form of Stokes' law to account for very small drops.

50.5 THERMIONIC EMISSION

All metals contain some electrons which are free to move about within the lattice. Though the attractive forces exerted on these electrons by the atomic nuclei are not strong enough to bind them to particular atoms, they do prevent them from leaving the surface. When a metal is heated the energies of its electrons increase and some of them acquire sufficient energy to escape from the surface. The process is called **thermionic emission**. The rate at which electrons are emitted increases rapidly with temperature. It can be shown on theoretical grounds that if the electrons are drawn towards a positively charged electrode so that they constitute a current I, then

$$I = AT^2 e^{-W/kT} \qquad [50.14]$$

This is known as **Richardson's equation**; A and W are constants which are characteristic of the emitter, k is Boltzmann's constant and T is the temperature of the emitter in kelvins.

In equation [50.14] W is the minimum amount of energy which has to be supplied to the metal to remove an electron from the surface of the metal; it is called the **work function** of the metal. There is good agreement between work function values estimated on the basis of the thermionic effect (i.e. from equation [50.14]) and those estimated on the basis of the photoelectric effect (see Table 50.1).

Table 50.1 Thermionic and photoelectric work functions compared

Metal	W/eV *Thermionic*	W/eV *Photoelectric*
Caesium	1.81	1.9
Tungsten	4.52	4.49

50.6 THE THERMIONIC DIODE

Action and Construction

The thermionic diode is used to rectify (Chapter 44) alternating currents. It has two electrodes – the anode and the cathode. The cathode is heated and emits electrons by the process of thermionic emission. When the anode is positive with respect to the cathode, the electrons emitted by the cathode are drawn to the anode and a current flows. When the anode is negative with respect to the cathode, the electrons are unable to reach the anode and there is no current flow. The device is often referred to as a diode valve because it allows current to pass in one direction only.

The anode is usually in the form of a nickel cylinder which surrounds the cathode, and the two electrodes are sealed in a highly evacuated glass bulb (Fig. 50.9(a)).

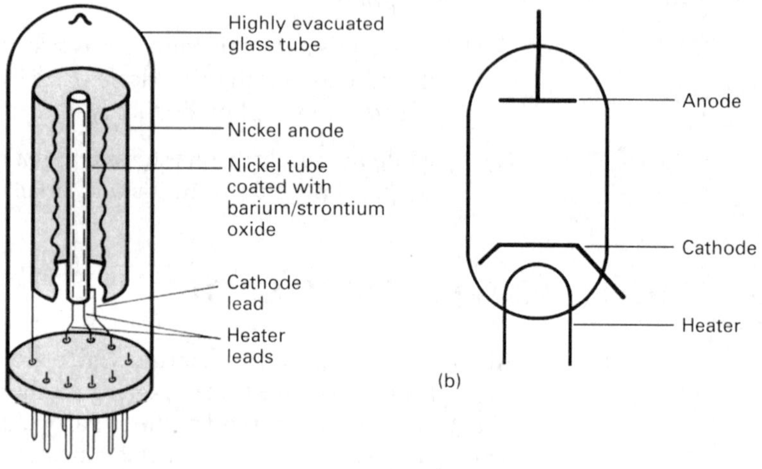

Fig. 50.9
(a) Thermionic diode, and (b) its circuit symbol

Two types of cathode are used. Most diodes are of the type illustrated and have an **indirectly heated cathode**. This is a nickel tube with a tungsten filament, called the **heater**, inside it. The tube is packed with a refractory powder such as alumina, the purpose of which is to insulate it electrically from the filament. The outer surface of the tube is coated with a mixture of barium oxide and strontium oxide. This mixture has a low work function (1.8 eV) and gives considerable thermionic emission at the relatively low temperature of 1100 K. The circuit symbol is shown in Fig. 50.9(b). **A directly heated cathode** is a simple filament and is usually made of thoriated tungsten (i.e. a tungsten wire with a monatomic layer of thorium on its surface). Thoriated tungsten has a work function of 2.6 eV (that of tungsten is 4.5 eV) and gives good emission at 1900 K.

Both types of cathode are heated electrically. It follows that the potential at one end of a directly heated cathode is higher than at the other, and therefore the PD between the anode and the cathode varies from one part of the cathode to another. This is undesirable because it means that there can be no sharp cut-off in the electron flow to the anode. Another disadvantage of the directly heated cathode is that if AC is used to heat it, there is a similar alternating component superimposed on the current flowing to the anode. The indirectly heated cathode has neither of these disadvantages but it requires a slightly longer warm-up time.

Characteristics

The current which flows through a diode is called the **anode current** (I) and depends on the PD between the anode and cathode – the **anode voltage** (V). A plot of I against V for any particular cathode temperature (i.e. for any particular filament current, I_f) is called the **characteristic** of the valve at that temperature. Typical characteristics are shown in Fig. 50.10. The curves can be obtained using the circuit of Fig. 50.11.

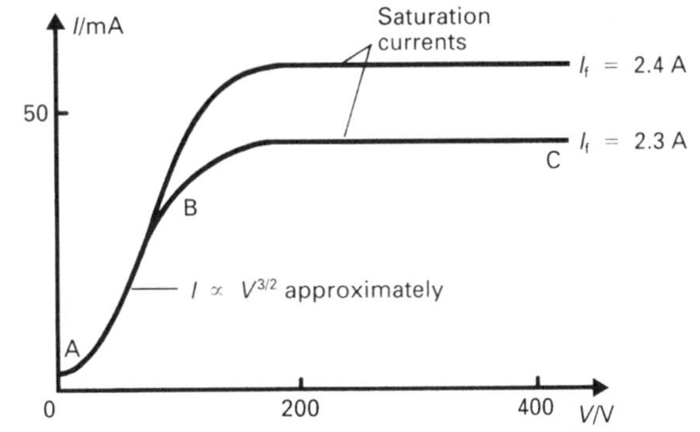

Fig. 50.10
Current–voltage characteristics of a thermionic diode

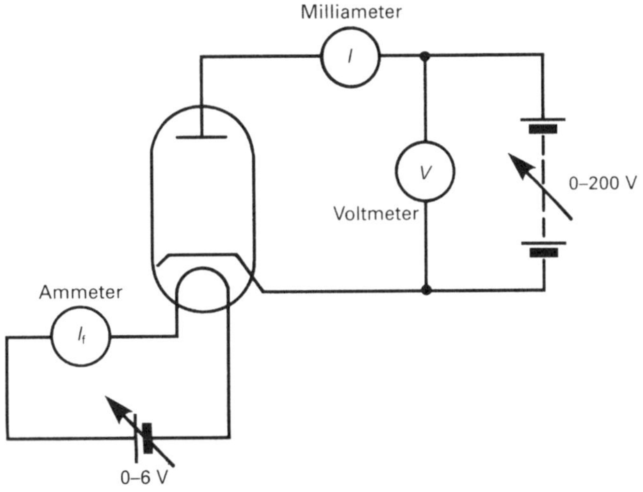

Fig. 50.11
Circuit for investigating diode characteristics

Consider the situation when $V = 0$. Electrons are emitted by the cathode with a range of velocities, and some of the electrons have sufficient energy to reach the anode. Those emitted with low kinetic energies are prevented from reaching the anode because the electrons repel each other in the region between the anode and cathode, and as a result some are actually turned back towards the cathode. Large numbers of electrons gather together close to the cathode and exist there as an almost stationary cloud of negative charge called the **space charge**. A dynamic equilibrium exists in which the rate at which electrons enter the space charge is equal to the rate at which electrons leave it by being returned to the cathode and by moving to the anode.

As V is increased the electron density of the space charge decreases so that its effect becomes less marked and an increasing number of electrons reach the anode. When V reaches a sufficiently large value the space charge ceases to exist and all the electrons reach the anode. Any further increase in V has very little effect on I and the current is said to have reached its **saturation value**. The current is said to be **space-charge-limited** along AB in Fig. 50.10; it is **temperature-limited** along BC. At higher filament currents the cathode temperature is greater and more electrons are emitted, resulting in an increased saturation current.

Theory shows that under space-charge-limited conditions

$$I = kV^{3/2}$$

where k is a constant which depends on the temperature of the cathode and the nature of its surface. The relationship is known as the **three-halves power law** or the **Langmuir–Child's law**. In practice the law is obeyed only approximately.

Notes (i) Modern valves emit so many electrons that saturation cannot be reached without causing damage to the valve.

(ii) The saturation current is not quite constant but increases slightly with increasing anode voltage. This is because the increased electric field at the cathode decreases the work function slightly.

(iii) Because the electrons have finite velocities when they are emitted, some can reach the anode even when it is slightly negative with respect to the cathode. This reverse current exists for negative anode potentials of up to about 0.5 V.

50.7 THE CATHODE RAY OSCILLOSCOPE (CRO)

The principal component of a cathode ray oscilloscope is a cathode ray tube (Fig. 50.12). **The electron gun** (i.e. the indirectly heated cathode, the grid and the anodes A_1 and A_2) provides a beam of electrons which converge to a point on the screen. Typical potentials are:

Cathode	−1000 V
Grid	−1000 V to − 1050 V
A_1 (the focusing anode)	−800 V to − 700 V
A_2 (the accelerating anode)	0 V (i.e. earthed)

The screen is coated with a phosphor such as zinc sulphide and fluoresces under the impact of the electrons. The grid is always negative with respect to the cathode but by making it less so, by use of the **brightness control**, the rate at which electrons pass through it can be increased in order to increase the brightness of the display. The anodes (A_1 and A_2) are at different potentials and the non-uniform field between them focuses the electron beam. The focusing can be adjusted by use of the **focus control** which alters the potential of A_1. The beam is accelerated primarily by A_2.

Fig. 50.12
Cathode ray oscilloscope

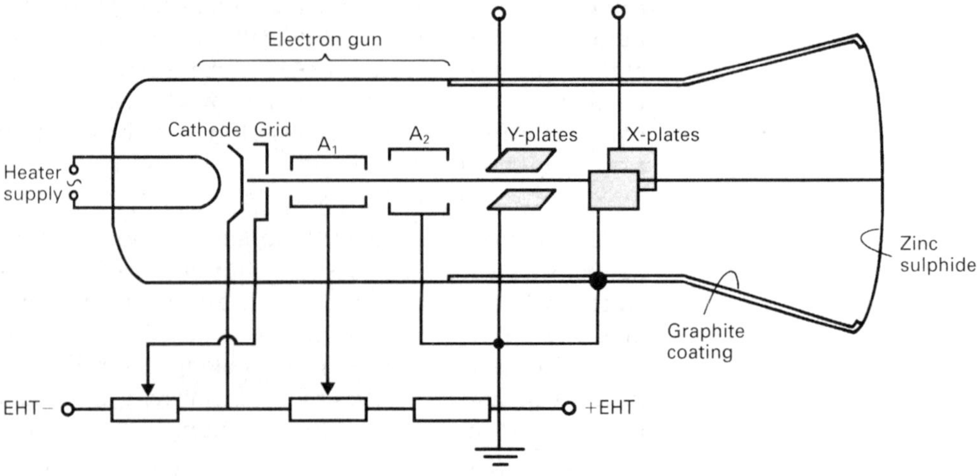

There must be a return path for the electrons that hit the screen. This is provided by coating the inner walls of the tube with graphite which is electrically connected to A_2 and therefore is at earth potential. When the electrons hit the zinc sulphide they knock electrons from it. These **secondary electrons** are collected by the graphite coating and flow to earth.

Apart from the intentionally applied electric fields between the X and Y deflecting plates, the region between A_2 and the screen must be field-free to avoid spurious deflections of the beam. The graphite coating provides an equipotential surface and excludes external electric fields. A mumetal screen around the outside of the tube shields it from stray magnetic fields. One of each pair of deflecting plates is connected to A_2 (i.e. earthed) to reduce the effect of the field between it and the plates.

When the **time base** is on, an internally generated saw-tooth voltage (Fig. 50.13) is automatically applied to the X-plates. This sweeps the electron beam from left to right at a constant rate. The particular sweep-rate is determined by the setting of the **time base control** (e.g. 100 ms, 10 ms, . . . ,1 μs for 1 cm of horizontal travel). The saw-tooth then returns the beam to the initial position at the extreme left of the screen almost instantaneously. The time taken for this right to left sweep is called the **fly-back time**. The brightness is automatically reduced to zero at the start of the fly-back.

Fig. 50.13
Time base voltage (saw-tooth)

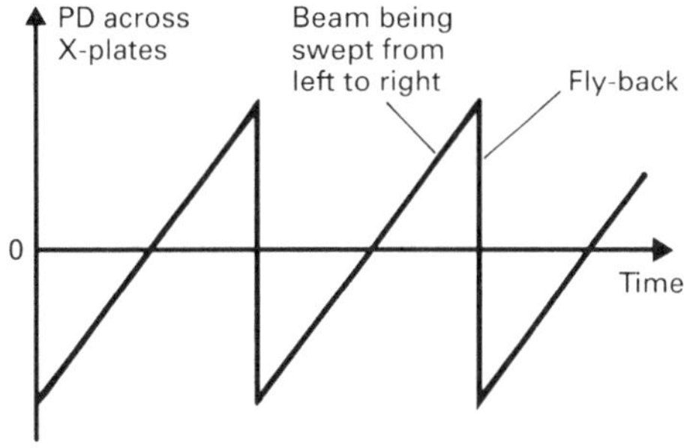

If a time-varying voltage V is applied to the Y-plates whilst the time base is on, the variation of V as a function of time can be investigated. The sweep time has to be synchronized with the frequency of V. If it is not, the waveform appears in a different position each time a sweep is made and the display drifts. To achieve synchronization a small fraction of V is automatically fed to a trigger circuit which produces a voltage pulse to initiate the fly-back at the same point on V each time.

50.8 USES OF THE CRO

In addition to displaying waveforms, the CRO can be used to:

(i) measure voltages (AC or DC),

(ii) measure frequencies,

(iii) measure phase differences,

(iv) measure small time intervals.

Measurement of Voltage

The voltage to be measured is applied to the Y-plates. If the time base is on, a sinusoidally alternating voltage produces a display of the form shown in Fig. 50.14(a); a steady DC voltage appears as shown in Fig. 50.14(b). Voltage measurements can be made with the time base off, in which case it is normal practice to centralize the beam by using the **X-shift control**. When this is done

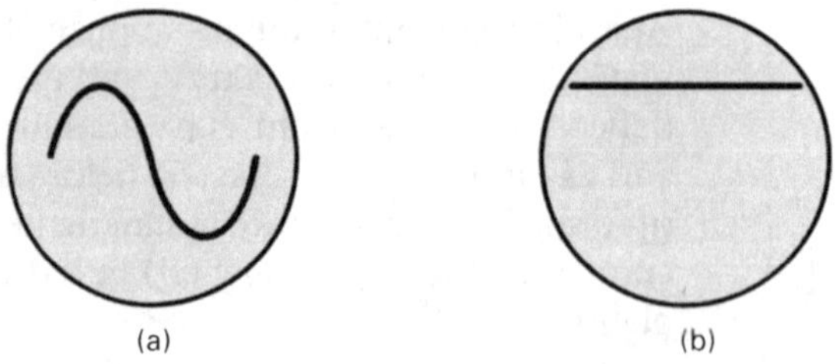

Fig. 50.14
Display of voltages on a CRO with time base on: (a) AC, (b) DC

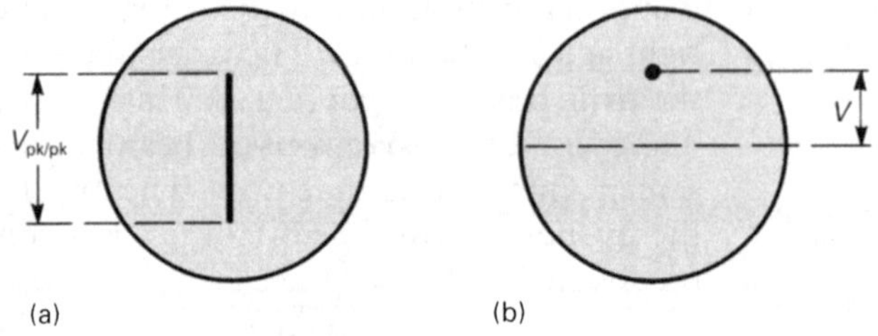

Fig. 50.15
Display of voltages on a CRO with time base off: (a) AC, (b) DC

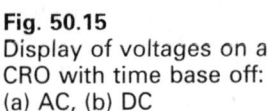

AC and DC signals appear as a vertical line and as a deflected spot respectively (Fig. 50.15). In the AC case the vertical height of the line represents the peak-to-peak value of the voltage ($= 2\sqrt{2}V_{RMS}$). In the DC case the magnitude of the voltage is represented by the vertical displacement of the spot from the zero position. In each case it is necessary to adjust the **Y-sensitivity** control (calibrated in $V\,cm^{-1}$) to give a display of a suitable size. For accurate work the calibration should be checked by applying a signal of known voltage.

The CRO is a particularly useful instrument for measuring voltage because:

(i) it has nearly infinite resistance to DC and very high (AC) impedance, and therefore draws very little current,

(ii) it can be used for both AC and DC,

(iii) it has no coil to burn out,

(iv) it has instantaneous response.

Measurement of Frequency

This may be achieved by making use of the calibrated time base. (It is normally necessary to set the fine time base control to the CAL position and for the **X-gain control** to be set to minimum gain for the calibration to hold.) The signal whose frequency is to be measured is applied to the Y-plates. Suppose that a trace of the form shown in Fig. 50.16 is obtained when the time base setting is $10\,ms\,cm^{-1}$, and that three complete cycles occupy 8.7 cm. It follows that one cycle corresponds to $8.7/3 = 2.9\,cm$. The period T of the signal is therefore $2.9 \times 10 = 29\,ms = 29 \times 10^{-3}\,s$. Therefore

$$\text{Frequency} = \frac{1}{T} = \frac{1}{29 \times 10^{-3}} \approx 34.5\,Hz$$

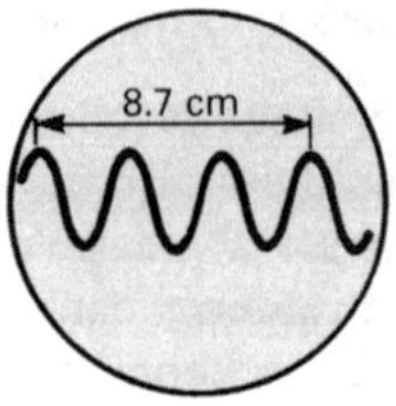

Fig. 50.16
Measurement of AC frequency on a CRO

If the CRO does not have a calibrated time base, an unknown frequency can be determined by comparing it with a known one. If one cycle of the signal of unknown frequency f_1 occupies a horizontal distance d_1 and, with the same time base setting, one cycle of the signal of known frequency f_2 occupies a distance d_2, then

$$\frac{f_1}{f_2} = \frac{d_2}{d_1}$$

Frequencies may also be compared by making use of Lissajous' figures as described in Measurement of Phase Difference.

Measurement of Phase Difference

Suppose the phase difference of two sinusoidal signals is to be measured. With the time base off and the beam centralized, one signal is applied to the Y-plates in the usual way and the other is applied to the X-plates. (This provides a sinusoidally varying time base rather than the usual linear time base.) A series of patterns which are known as **Lissajous' figures** results. Their shapes are determined by the frequency ratio and the phase difference of the two signals. Some examples are shown in Fig. 50.17. The frequency ratio corresponding to any particular figure can be determined by counting the number of times the figure crosses imaginary horizontal and vertical lines. For example, in Fig. 50.17 the Lissajous' figure with $f_X/f_Y = 2/1$ and a phase difference of $\pi/4$ rad crosses the vertical 4 times and the horizontal twice, corresponding to $f_X/f_Y = 4/2 = 2/1$.

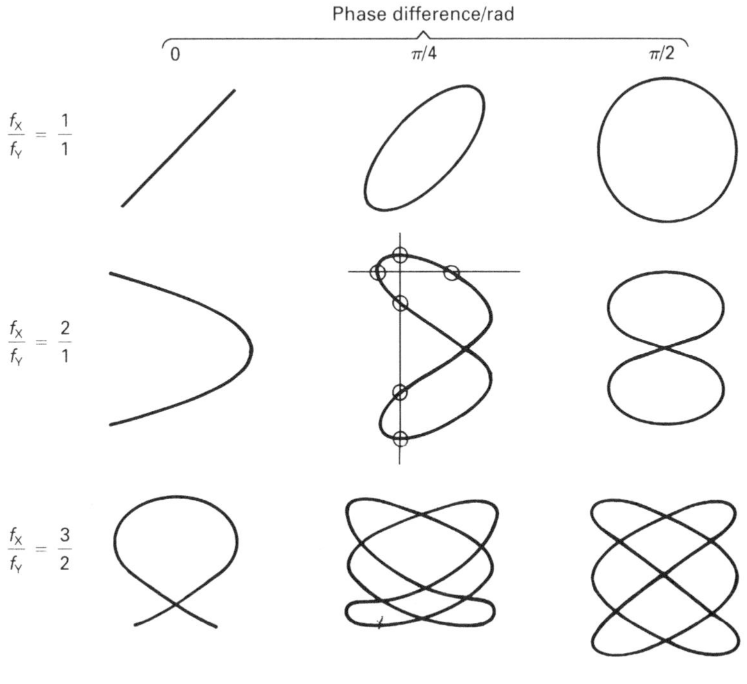

Fig. 50.17
Lissajous' figures on a CRO produced by signals of equal strength

For each frequency ratio the phase difference of the signals affects the shape of the figure. Thus, two signals which have the same frequency and are in phase produce a line at 45° to the horizontal; a phase difference of $\pi/2$ rad gives rise to a circle.

Measurement of Small Time Intervals

If two events which are separated by a small time interval are each caused to produce a voltage pulse on a CRO, the horizontal separation of the pulses can be used together with the value of the time base setting to estimate the time interval between the events.

CONSOLIDATION

A charged particle follows a parabolic path in a uniform electric field (except when it moves parallel to the field lines).

A charged particle moving in a plane which is at right angles to a uniform magnetic field follows a circular path.

An electric field changes the kinetic energy of a charged particle.

A magnetic field cannot change the kinetic energy of a charged particle (because the force it exerts is always at right angles to the motion).

For an electron in an **electric field**

$$F = eE \qquad eV = \tfrac{1}{2}mv^2$$

For an electron in a **magnetic field**

$$F = Bev \qquad Bev = \frac{mv^2}{r}$$

51

THE NUCLEUS

51.1 STRUCTURE OF THE NUCLEUS, THE NUCLEONS

Every atom has a central, positively charged nucleus. Nuclear diameters are $\sim 10^{-15}$ m, atomic diameters are $\sim 10^{-10}$ m. Over 99.9% of the mass of an atom is in its nucleus. Atomic nuclei are unaffected by chemical reactions.

Nuclei contain **protons** and **neutrons*** which, because they are the constituents of nuclei, are collectively referred to as **nucleons**. Their properties are compared with those of the electron in Table 51.1. The charge on the proton is equal and opposite to that on the electron, and it follows that (neutral) atoms contain equal numbers of protons and electrons.

Table 51.1
The nucleons compared with the electron

	Electron	*Proton*	*Neutron*
Mass	$m_e = 9.110 \times 10^{-31}$ kg	$m_p = 1836\, m_e$	$m_n = 1839\, m_e$
Charge	-1.602×10^{-19} C	$+1.602 \times 10^{-19}$ C	Zero

The nucleons are held together by one of the four fundamental forces – the **strong interaction**. (The other three are weaker than this, and in descending order of effectiveness are the electromagnetic force, the weak interaction and the gravitational force.) The strong interaction (sometimes called the **nuclear force**) is a very strong short-range force, and it more than offsets the considerable electrostatic repulsion of the positively charged protons.

51.2 ISOTOPES: DEFINITIONS

Two atoms which have the same number of protons but different numbers of neutrons are said to be **isotopes** of each other. It follows that each atom contains the same number of electrons as the other and, therefore, that their chemical properties are identical. Isotopes cannot be separated by chemical methods. Some elements have only one naturally occurring isotope (gold and cobalt are examples); tin has the largest number – ten.

The atomic number (or **proton number**) ***Z*** of an element is the number of protons in the nucleus of an atom of the element.

The atomic number of an element was originally used to represent its position in the periodic table; it still does, but is now more meaningfully defined as above.

*The nucleus of the common isotope of hydrogen is an exception: it has a single proton and no neutron.

The mass number (or **nucleon number**) A of an atom is the number of nucleons (i.e. protons + neutrons) in its nucleus.

The various isotopes of an element whose chemical symbol is represented by X are distinguished by using a symbol of the form

$$^{A}_{Z}\text{X}$$

where A and Z are respectively the mass number and atomic number of the isotope. The most abundant isotope of lithium (lithium 7) has 3 protons and 4 neutrons, i.e. $Z = 3$ and $A = 3+4 = 7$, and therefore it is represented by $^{7}_{3}\text{Li}$. Lithium 6 has only 3 neutrons and is written as $^{6}_{3}\text{Li}$. The Z-value is sometimes omitted because it gives the same information as the chemical symbol (for example, all lithium atoms have 3 protons).

Hydrogen is exceptional in that its three isotopes are given different names. The most abundant isotope has one proton and no neutron and is actually called hydrogen ($^{1}_{1}\text{H}$). The other isotopes are **deuterium** ($^{2}_{1}\text{D}$ or $^{2}_{1}\text{H}$) and **tritium** ($^{3}_{1}\text{T}$ or $^{3}_{1}\text{H}$). A deuterium nucleus is called a **deuteron**.

The relative atomic mass A_r of an atom is defined by

$$A_r = \frac{\text{Mass of atom}}{\text{One twelfth the mass of a } ^{12}_{6}\text{C atom}} \qquad [51.1]$$

It follows that the relative atomic mass of $^{12}_{6}\text{C}$ is exactly twelve, that of $^{1}_{1}\text{H}$ is 1.008, that of $^{16}_{8}\text{O}$ is 15.995, etc. Note that **the relative atomic mass of an atom is approximately equal to the mass number of the atom.**

The unified atomic mass unit (u) is defined such that the mass of $^{12}_{6}\text{C}$ is 12 u (exactly). It follows that the mass of an atom expressed in unified atomic mass units is numerically equal to its relative atomic mass. The mass of a $^{12}_{6}\text{C}$ atom is found by experiment to be 1.993×10^{-26} kg, and therefore

$$1\,\text{u} = 1.661 \times 10^{-27}\,\text{kg}$$

Note The definition of relative atomic mass given above (equation [51.1]) applies to a single atom or single isotope. An alternative definition, which takes account of the natural isotopic composition of the element concerned, is

$$A_r = \frac{\text{Average mass of an atom of the element}}{\text{One twelfth the mass of a } ^{12}_{6}\text{C atom}} \qquad [51.2]$$

When dealing with a single isotope (which is usually the case in nuclear physics) equation [51.1] should be used. Equation [51.2] is relevant in situations such as chemical reactions, where an element (normally) has its natural isotopic composition. Consider, for example, the case of uranium. On the basis of equation [51.1] the relative atomic mass of $^{235}_{92}\text{U} = 235.04$ and that of $^{238}_{92}\text{U} = 238.05$. The relative atomic mass of uranium, on the other hand, is given by equation [51.2] as 238.03. This is very close to the value for $^{238}_{92}\text{U}$, because over 99% of naturally occurring uranium is in the form of $^{238}_{92}\text{U}$.

The term **nuclide** is used to specify an atom with a particular number of protons and a particular number of neutrons. Thus $^6_3$Li, $^7_3$Li, $^{16}_8$O and $^{18}_8$O are four different nuclides.

Isotopes are nuclides with the same number of protons.

Isotones are nuclides with the same number of neutrons.

Isobars are nuclides with the same number of nucleons.

QUESTIONS 51A

1. How many of each of the following particles are there in a single atom of iron 56 ($^{56}_{26}$Fe)?
   (a) electrons,
   (b) protons,
   (c) neutrons,
   (d) nucleons,
   (e) negatively charged particles,
   (f) positively charged particles,
   (g) neutral particles.

2. The two naturally occurring isotopes of copper have relative atomic masses of 62.9296 and 64.9278. **(a)** What are the mass numbers of these isotopes? **(b)** What are their masses in unified atomic mass units? **(c)** Calculate the relative atomic mass of naturally occurring copper, given that the isotope with a relative atomic mass of 62.9296 is 69.0% abundant.

51.3 THE MASS SPECTROMETER

Mass spectrometers can be used to determine the amounts of different gases present in a vacuum chamber. (The gases are often introduced purposely for analysis.)

Suppose that one of the gases being analysed is neon. A beam of electrons (Fig. 51.1) is caused to collide with the neon atoms and dislodges electrons from them producing positively charged neon ions (Ne^+ and Ne^{2+}). The positive ions are accelerated towards a slit in a negatively charged plate (X). The ions pass through the slit and are further accelerated by an even more negatively charged plate (Y). The velocities of the ions on reaching Y depend on their charge-to-mass ratios and on the accelerating voltage. As the ions pass through the slit in Y they enter a uniform magnetic field which is directed into the paper. Under the influence of the field the ions move along semi-circular paths.

Fig. 51.1
Mass spectrometer

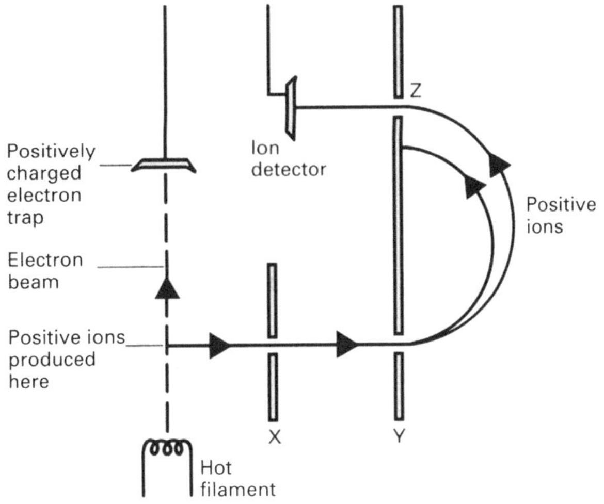

If the accelerating voltage is V, then an ion of mass M and charge Q reaches Y with a velocity v given by

$$\tfrac{1}{2}Mv^2 = QV$$

If the magnetic flux density beyond Y is B, such an ion moves along a semicircular path of radius r, where

$$BQv = \frac{Mv^2}{r}$$

Eliminating v between these equations gives

$$\frac{Q}{M} = \frac{2V}{B^2r^2}$$

It follows that for any particular value of V and any given value of B only those ions which have a particular charge-to-mass ratio (Q/M) will travel along a path whose radius is such that they can pass through the narrow slit at Z and reach the detector. Ions of a different charge-to-mass ratio can be caused to reach the detector by altering V. (The magnetic field is constant.) The current produced by the ion detector is proportional to the number of ions present which have the relevant charge-to-mass ratio. By measuring the current and the accelerating voltage used to produce it, it is possible to determine the number of ions that are present with any particular charge-to-mass ratio. Since isotopes are characterized by their charge-to-mass ratio, the quantities of the various isotopes in the chamber can be found.

Mass spectrometers were originally used to determine the proportions of the various isotopes of the elements. Nowadays they are used principally for chemical analysis and for leak detection. They are capable of detecting the presence of a gas which is exerting a pressure of as little as 10^{-12} mm Hg.

51.4 THE BAINBRIDGE MASS SPECTROGRAPH

The mass spectrograph, like the mass spectrometer, is used to separate ions with different charge-to-mass ratios. Unlike the mass spectrometer, though, it uses photographic detection. There are various types; the essential features of that due to Bainbridge are shown in Fig. 51.2. Positive ions, produced by electron bombardment of the gas under investigation, emerge from the velocity selector (see section 50.1) with a range of different charge-to-mass ratios but the same velocity.

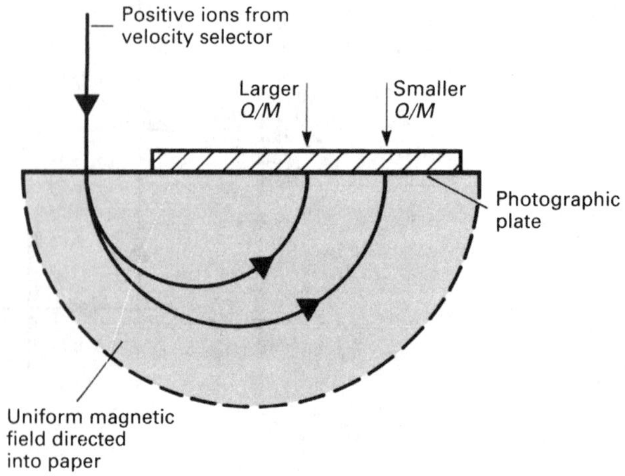

Fig. 51.2
Mass spectrograph

On entering the semicircular region the ions come under the influence of a uniform magnetic field which is directed into the paper and is of flux density B. The ions move along semicircular paths and strike the photographic plate. For an ion of charge Q and mass M moving on a path of radius r

$$BQv = \frac{Mv^2}{r}$$

i.e.

$$r = \frac{M}{Q}\frac{v}{B}$$

Since v and B are constant, r depends only on the charge-to-mass ratio. It follows that the position at which an ion strikes the photographic plate depends only on its charge-to-mass ratio. The instrument is easily capable of distinguishing between different isotopes of a single element. The relative abundances of the various isotopes can be determined from the extents to which they have exposed the photographic plate.

CONSOLIDATION

Nuclei contain protons and neutrons.

A nucleon is a proton or a neutron.

The strong interaction (or **nuclear force**) is the force that holds nucleons together inside the nucleus.

Atoms which have the same number of protons are atoms of the same **element**.

Two atoms which have the same number of protons but different numbers of neutrons are **isotopes** of each other.

Atomic number (or **proton number**) Z of an element is the number of protons in the nucleus of an atom of the element.

Mass number (or **nucleon number**) A of an atom is the number of nucleons (i.e. protons + neutrons) in its nucleus.

Relative atomic mass (A_r)

$$A_r = \frac{\text{Mass of atom}}{\text{One twelfth the mass of a } {}^{12}_{6}\text{C atom}} \qquad [51.1]$$

or

$$A_r = \frac{\text{Average mass of an atom of the element}}{\text{One twelfth the mass of a } {}^{12}_{6}\text{C atom}} \qquad [51.2]$$

Equation [51.1] applies to a single atom or single isotope. Equation [51.2] applies when different isotopes of the element are present.

The relative atomic mass of an atom is approximately equal to the mass number of the atom.

The unified atomic mass unit (u) is one twelfth the mass of a ${}^{12}_{6}$C atom.

The mass of an atom in unified atomic mass units is numerically equal to its relative atomic mass.

52

RADIOACTIVITY

52.1 RADIOACTIVE DECAY

In 1896 Becquerel noticed that some photographic plates which had been stored close to a uranium compound had become fogged. He showed that the fogging was due to 'radiations'* emitted by the uranium. The phenomenon is called **radioactivity**, or **radioactive decay**; and the 'radiations' are emitted when an unstable nucleus disintegrates to acquire a more stable state. **The disintegration is spontaneous** and most commonly involves the emission of an **α-particle** or a **β-particle**. In both α-emission and β-emission the **parent nucleus** (i.e. the emitting nucleus) undergoes a change of atomic number and therefore becomes the nucleus of a different element. This new nucleus is called the **daughter nucleus** or the **decay product**. It often happens that the daughter nucleus is in an excited state when it is formed, in which case it reaches its ground state by emitting a third type of radiation called a **γ-ray**. The emission of a γ-ray simply carries away the energy released when the daughter nucleus undergoes its transition to the ground state. Some radioactive substances decay in such a way that the daughter nuclei are produced in their ground states and therefore do not give any γ-emission. Though most nuclides emit either α-particles or β-particles, some emit both. For example 64% of $^{212}_{83}Bi$ nuclei emit β-particles and 36% emit α-particles.

52.2 α-PARTICLES (SYMBOL $^4_2\alpha$ OR ^4_2He)

An α-particle consists of two protons and two neutrons, i.e. it is identical to a helium nucleus. The velocity with which an α-particle is emitted depends on the species of nucleus which has produced it and is typically 6% of the velocity of light. This corresponds to a kinetic energy of 6 MeV, and the α-particles are the most energetic form of 'radiation' produced by radioactive decay. Many α-emitters produce α-particles of one energy only. Others emit α-particles with a small number of nearly equal, discrete values of energy. For example, $^{212}_{83}Bi$ emits α-particles with energies of 6.086, 6.047, 5.765, 5.622, 5.603, and 5.481 MeV; those with 6.086 MeV and 6.047 MeV account for over 97% of the emission. α-particles are emitted by heavy nuclei.

*Many of the 'radiations' are in fact particles; the term was applied before this was realized and is still in use.

Since α-particles are charged and move relatively slowly, they produce considerable ionization ($\sim 10^5$ ion-pairs* per cm in air at atmospheric pressure). As a consequence they lose their energy over a short distance and, for example, are capable of penetrating only a single piece of paper or about 5 cm of air.

When a nucleus undergoes α-decay it loses four nucleons, two of which are protons. Therefore

(i) its mass number (A) decreases by 4, and

(ii) its atomic number (Z) decreases by 2.

Thus, if a nucleus X becomes a nucleus Y as a result of α-decay, then

$$\underset{\text{(Parent)}}{{}^{A}_{Z}\mathrm{X}} \longrightarrow \underset{\text{(Daughter)}}{{}^{A-4}_{Z-2}\mathrm{Y}} + \underset{(\alpha\text{-particle})}{{}^{4}_{2}\alpha}$$

For example, uranium 238 decays by α-emission to thorium 234 according to

$${}^{238}_{92}\mathrm{U} \longrightarrow {}^{234}_{90}\mathrm{Th} + {}^{4}_{2}\alpha$$

Note that the number of nucleons is conserved ($238 \rightarrow 234 + 4$) and the charge is conserved ($92 \rightarrow 90 + 2$).

The specific charge (i.e. the charge-to-mass ratio) of the α-particles was measured, soon after their discovery, by deflecting them in electric and magnetic fields. This showed that the particles were positively charged, and that their specific charge was the same as that of a doubly ionized helium atom (i.e. a helium nucleus). Confirmation that α-particles are helium nuclei was provided by Rutherford and Royds in 1909. Their apparatus is shown schematically in Fig. 52.1. Radon, an α-emitting gas, was contained in a thin-walled glass tube A. The walls of this tube were less than 0.01 mm in thickness and could be penetrated by the α-particles produced by the radon. The particles were incapable of passing through the thick outer wall B and so were trapped in C. Although C was evacuated, traces of air

Fig. 52.1
Rutherford and Royd's apparatus to confirm α-particles as helium nuclei

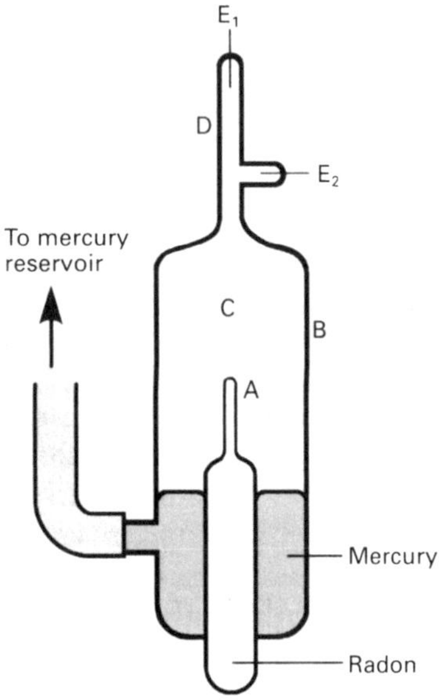

*Ionization results in the release of two charged particles – an electron and a positive ion, collectively known as an **ion-pair**.

remained, and the α-particles picked up electrons as a result of ionizing air molecules and by colliding with the walls. After about a week the gas which had accumulated in C was compressed and forced into the narrow tube D by raising the level of the mercury. An electrical discharge passed between the electrodes E_1 and E_2 caused the gas to produce its emission spectrum. When this was examined it was found to be the spectrum of helium – final proof that α-particles are helium nuclei.

52.3 β-PARTICLES (SYMBOL ${}^{0}_{-1}\beta$ OR ${}^{0}_{-1}e$)

These are very fast electrons (up to 98% of the velocity of light). In spite of their great velocities they have less energy than α-particles on account of their much smaller mass. **Any given species of nucleus emits β-particles with a continuous range of energies.** The maximum energies of the β-particles emitted by the naturally occurring nuclides vary from 0.05 MeV in the case of ${}^{228}_{88}Ra$ to 3.26 MeV in the case of ${}^{214}_{83}Bi$.

β-particles are emitted by nuclei which have too many neutrons to be stable. Such a nucleus attains a more stable state (i.e. a lower energy state) when one of its neutrons changes into a proton and an electron. When this happens the electron is immediately emitted as a β-particle. The proton remains in the nucleus so that the nucleus has effectively lost a neutron and gained a proton, i.e. it has become a different element. There is no change in the total number of nucleons. Thus, when a nucleus undergoes β-decay:

(i) its mass number (A) does not change, and

(ii) its atomic number (Z) increases by 1, i.e.

$$\underset{\text{(Parent)}}{{}^{A}_{Z}X} \longrightarrow \underset{\text{(Daughter)}}{{}^{A}_{Z+1}Y} + \underset{(\beta\text{-particle})}{{}^{0}_{-1}e}$$

For example, carbon 14 decays by β-emission to nitrogen 14 according to

$${}^{14}_{6}C \longrightarrow {}^{14}_{7}N + {}^{0}_{-1}e$$

Note that, as in α-decay, the number of nucleons is conserved ($14 \rightarrow 14 + 0$) and the charge is conserved ($6 \rightarrow 7 - 1$).

β-particles are much less massive than α-particles and are much more easily deflected. The path of a β-particle through matter is therefore tortuous. Because they move quickly and are easily deflected, β-particles spend very little time in the vicinity of a single atom and therefore produce much less ionization than α-particles ($\sim 10^3$ ion-pairs per cm in air at atmospheric pressure). The most energetic have ranges which are about 100 times those of α-particles.

By deflecting β-particles in electric and magnetic fields Kaufman (1902) was able to establish that the particles were negatively charged and that their charge-to-mass ratio decreased with increasing velocity. A later series of measurements by Bucherer (1909) showed that the variation was in excellent agreement with the supposition that the particles were electrons whose mass was varying with velocity in the manner predicted by the Special Theory of Relativity.

52.4 γ-RAYS (SYMBOL ${}^0_0\gamma$ OR γ)

γ-rays are electromagnetic radiation of very short wavelength. The wavelength of the radiation is characteristic of the nuclide which produces it. Many nuclides produce γ-rays of more than one wavelength; these wavelengths do not form a continuous spectrum, but are limited to a few discrete values. The wavelengths of the γ-rays produced by naturally occurring radioactive nuclides are typically in the range 10^{-10} m to 10^{-12} m, corresponding to energies of about 0.01 MeV to about 1 MeV. (The γ-rays produced by cosmic rays may have a wavelength of less than 10^{-15} m, i.e. an energy of more than about 10^3 MeV.) It is not uncommon for X-rays to have a wavelength of 10^{-11} m. These differ from γ-rays of the same wavelength only in the manner in which they are produced; **γ-rays are a result of nuclear processes, whereas X-rays originate outside the nucleus**.

In comparison with α-particles and β-particles γ-rays produce very little ionization and are very penetrating. γ-rays cannot be deflected by electric and magnetic fields.

The properties of the 'radiations' are summarized in Table 52.1.

Table 52.1 Summary of the properties of the 'radiations'

Property	*α-particle*	*β-particle*	*γ-ray*
Nature	Helium nucleus	Fast electron	Electromagnetic radiation
Charge	$+3.2 \times 10^{-19}$ C	-1.6×10^{-19} C	0
Rest mass	6.4×10^{-27} kg $= 4.0015$ u	9.1×10^{-31} kg $= 0.000\,55$ u	0
Velocity	$\sim 0.06\,c$	Up to $0.98\,c$	c
Energy	~ 6 MeV	~ 1 MeV	$hf \sim 0.1$ MeV
Number of ion-pairs per cm of air	$\sim 10^5$	$\sim 10^3$	~ 10
Penetration	~ 5 cm of air	~ 500 cm of air ~ 0.1 cm of aluminium	~ 4 cm of lead reduces intensity to 10%
Path through matter	Straight	Tortuous	Straight
Ability to produce fluorescence	Yes (strong)	Yes	Yes (weak)
Ability to affect a photographic plate	Yes	Yes	Yes

QUESTIONS 52A

1. The following questions refer to α-particles, β-particles and γ-rays. Which: **(a)** are emitted by parent nuclei, **(b)** are emitted by daughter nuclei, **(c)** cause the nuclei emitting them to change **(i)** mass number, **(ii)** atomic number, **(d)** are electrons, **(e)** are helium nuclei, **(f)** typically have the most energy, **(g)** typically have the least energy, **(h)** have a continuous range of energies, **(i)** are most easily absorbed, **(j)** produce the least ionization, **(k)** have the greatest specific charge (i.e. the greatest charge-to-mass ratio), **(l)** have tortuous paths through matter, **(m)** can be deflected by both electric and magnetic fields, **(n)** have a range of about 500 cm in air?

2. Write down the numerical values of the letters a to i in the following equations.

$$^{224}_{88}\text{Ra} \longrightarrow {}^{a}_{b}\alpha + {}^{c}_{d}\text{Rn}$$

$$^{e}_{83}\text{Bi} \longrightarrow {}^{f}_{-1}\beta + {}^{210}_{g}\text{Po} + {}^{h}_{i}\gamma$$

52.5 THE ABSORPTION OF α, β AND γ BY MATTER

Absorption of α-particles

Many α-emitters produce α-particles of one energy only. The α-particles are said to be **monoenergetic**, and they have very nearly equal ranges in any particular absorber (Fig. 52.2).

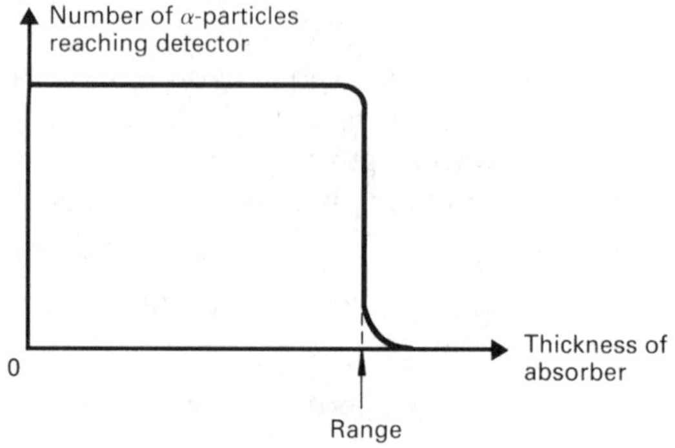

Fig. 52.2
Range of α-particles in an absorber

The range in air of the particles coming from a source can be found by determining the number* of α-particles that reach a detector when it is placed at different distances from the source. As the detector is moved away from the source the number of α-particles reaching it stays constant at first and then falls rapidly. An α-particle loses some of its energy each time it ionizes an air molecule. Since all the α-particles have the same energy initially, they each produce the same amount of ionization, but because of the randomness associated with the chance of an α-particle encountering an air molecule some travel a little further than others before giving up all their energy. This accounts for the tail on the curve and is known as **straggling**. The range is taken to be the maximum thickness that the majority of the α-particles can penetrate.

The range in aluminium (say) is found by placing successively thicker sheets of very thin aluminium foil between the source and the detector. (The source and detector are as close as possible so that there is no absorption due to air.) This also gives a curve of the type shown in Fig. 52.2. The range is reduced in aluminium, by approximately the ratio of the density of aluminium to that of air.

Absorption of β-particles

The absorption of β-particles is complicated by the fact that the particles emitted by any nuclide have a continuous range of energies. Even if monoenergetic particles are selected by some means, it is still difficult to assess their range. This is because:

(i) when β-particles are absorbed they sometimes eject high-speed electrons from the atoms of the absorber and these are easily confused with genuine β-particles, and

(ii) β-particles are easily deflected and therefore many are scattered away from the detector and therefore are not counted even though they have not been absorbed.

Over a limited thickness of absorber, however, the absorption of β-particles is approximately exponential.

*In practice it is usually some quantity which is proportional to the number of particles that is measured.

Absorption of γ-rays

Many nuclides emit γ-rays of more than one wavelength. If γ-rays of a single wavelength are selected, their absorption is an exponential function of absorber thickness, i.e.

$$I = I_0 \, e^{-\mu d} \qquad [52.1]$$

where

I = the intensity transmitted by a thickness d of absorber (Fig. 52.3)

I_0 = the intensity of the γ-rays incident on the absorber

μ = the **linear absorption coefficient** (or **attenuation coefficient**) of the absorber. The value of μ depends on the nature of the absorber and the wavelengths of the γ-rays. (Unit = m^{-1}.)

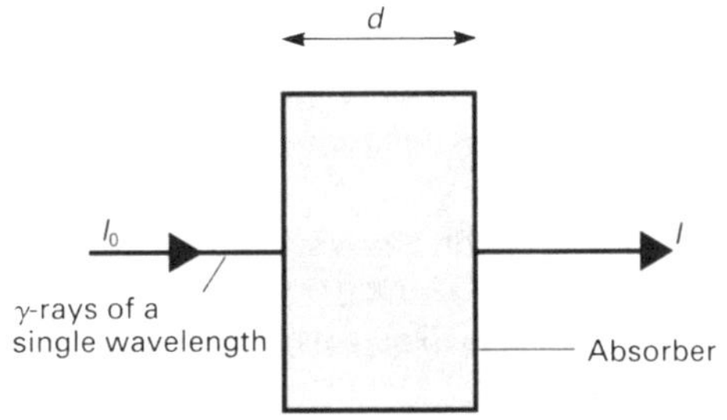

Fig. 52.3
γ-rays passing through an absorber

It follows from equation [52.1] that

$$\log_e I = \log_e (I_0 \, e^{-\mu d})$$

$$= \log_e I_0 + \log_e (e^{-\mu d})$$

i.e. $$\log_e I = -\mu d + \log_e I_0$$

Plots of I against d, and $\log_e I$ against d are shown in Fig. 52.4. The gradient of the latter gives μ.

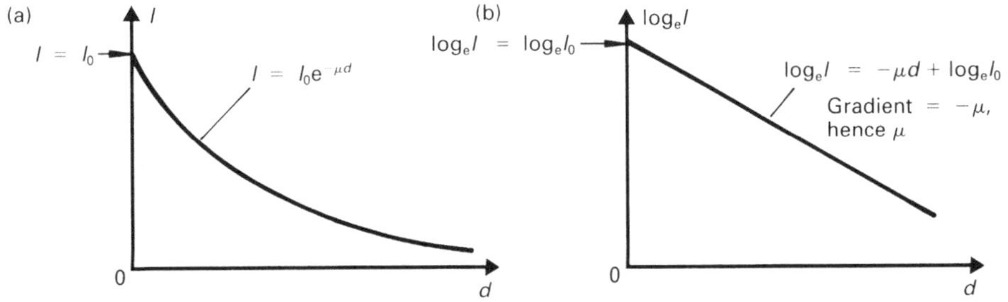

Fig. 52.4
(a) The effect of absorber thickness on the intensity of γ-rays transmitted.
(b) Plot to determine the linear absorption coefficient

Notes

(i) The absorption of γ-rays increases with the atomic number of the material of the absorber. Atomic number has little effect on the absorption of α-particles and β-particles.

(ii) The exponential nature of γ-ray absorption arises because, in most cases, a γ-ray quantum loses all its energy in a single event, and therefore the fractional intensity of the beam falls by a fixed amount each time it traverses any given small thickness of absorber.

(iii) Equation [52.1] also holds for X-rays (see section 49.6).

52.6 THE INVERSE SQUARE LAW FOR γ-RAYS

A point source of γ-rays emits in all directions about the source. It follows that the intensity of the γ-radiation decreases with distance from the source because the rays are spread over greater areas as the distance increases. This decrease in intensity is distinct from that produced by absorption.

Consider a point source of γ-rays, situated in a vacuum so that there is no absorption. The radiation spreads in all directions about the source, and therefore when it is a distance d from the source it is spread over the surface of a sphere of radius d and area $4\pi d^2$. If E is the energy radiated per unit time by the source, then the intensity of the radiation (= energy per unit time per unit area) is given by I, where

$$I = E/(4\pi d^2)$$

i.e. $$I \propto 1/d^2$$

Thus the intensity varies as the inverse square of the distance from the source. The law is entirely true in vacuum. The absorption of γ-rays by air at atmospheric pressure is very slight, and therefore the inverse square law can be taken to hold over large distances in air.

Note The inverse square law holds for α-particles and β-particles in vacuum providing they are coming from a point source. The law does not apply to these particles in air because air absorbs them.

52.7 EXPERIMENTAL VERIFICATION OF THE INVERSE SQUARE LAW FOR γ-RAYS

The experimental arrangement is shown in Fig. 52.5. γ-rays may be absorbed at any point in the Geiger–Müller tube. Nevertheless, it is as if they are all absorbed at a single point (B). Neither the location of this point nor that of the source (A) are known and this makes it impossible to measure d directly. There is no real difficulty though, since, as can be seen from Fig. 52.5, $d = x + c$ and x is measurable and c, though unknown, is constant.

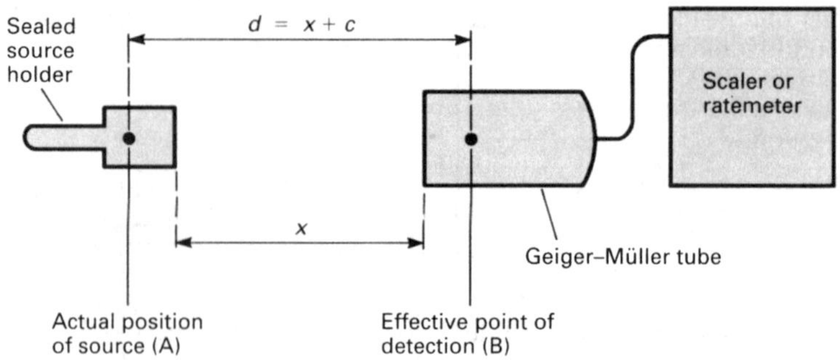

Fig. 52.5
Apparatus to verify the inverse square law for γ-rays

The aim is to verify that

$$I \propto \frac{1}{d^2}$$

i.e. $$I \propto \frac{1}{(x+c)^2} \qquad [52.2]$$

Since I is proportional to the **corrected count rate** R (i.e. the actual count rate minus the background count rate), equation [52.2] can be rewritten as

$$R \propto \frac{1}{(x+c)^2}$$

i.e. $$x + c = \frac{k}{R^{1/2}}$$

where k is a constant of proportionality

i.e. $$x = kR^{-1/2} - c$$

If a plot of x against $R^{-1/2}$ turns out to be linear, the inverse square law has been verified. The intercept when $R^{-1/2}$ is zero gives c – see Fig. 52.6.

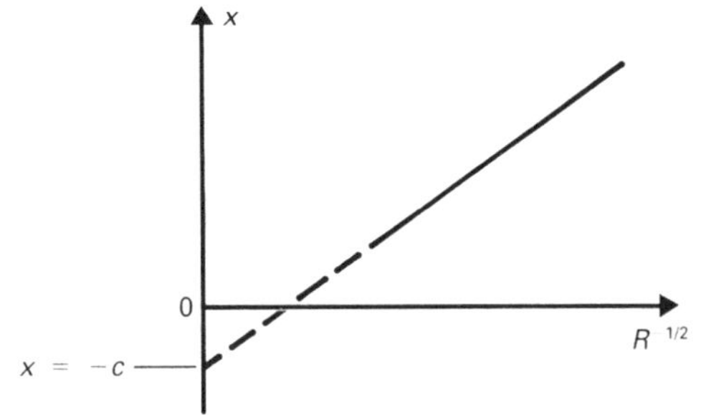

Fig. 52.6 Graph to verify the inverse square law for γ-rays

52.8 THE EXPONENTIAL LAW OF RADIOACTIVE DECAY

Radioactive nuclei disintegrate spontaneously; the process cannot be speeded up or slowed down. It follows that **for large numbers of any particular species of nuclei the rate of decay is proportional to the number of parent nuclei present**. If there are N parent nuclei present at time t, the rate of increase of N is $\mathrm{d}N/\mathrm{d}t$ and therefore the rate of decrease, i.e. the rate of decay, is $-\mathrm{d}N/\mathrm{d}t$. It follows that

$$-\frac{\mathrm{d}N}{\mathrm{d}t} = \lambda N$$

or $$\frac{\mathrm{d}N}{\mathrm{d}t} = -\lambda N \qquad [52.3]$$

where

λ = a (positive) constant of proportionality called the **decay constant** (unit = s^{-1})

$-\mathrm{d}N/\mathrm{d}t$ = the rate of decay and is called the **activity** of the source. When used in equation [52.3] the activity must be expressed in the relevant SI unit – the becquerel. One **becquerel** (Bq) is equal to an activity of one disintegration per second. Until recently activity has been expressed in curies. One **curie** (Ci) is defined as (exactly) 3.7×10^{10} disintegrations per second, i.e.

$$1\ \mathrm{Ci} = 3.7 \times 10^{10}\ \mathrm{s}^{-1}$$

Equation [52.3] can be rearranged as

(**Note.** The variables, N and t, are now on opposite sides of the equation.)

$$\frac{dN}{N} = -\lambda\, dt$$

Hence

$$\int \frac{dN}{N} = -\lambda \int dt$$

i.e. $$\log_e N = -\lambda t + c \qquad [52.4]$$

If the initial number of nuclei is N_0, i.e. if $N = N_0$ when $t = 0$, then by equation [52.4]

$$\log_e N_0 = c$$

Substituting for c in equation [52.4] gives

$$\log_e N = -\lambda t + \log_e N_0$$

i.e. $$\log_e N - \log_e N_0 = -\lambda t$$

i.e. $$\log_e \left(\frac{N}{N_0}\right) = -\lambda t$$

i.e. $$\frac{N}{N_0} = e^{-\lambda t}$$

i.e. $$N = N_0\, e^{-\lambda t} \qquad [52.5]$$

Equation [52.5] expresses the exponential nature of radioactive decay, i.e. that the number of nuclei remaining after time t (i.e. the number of parent nuclei) decreases exponentially with time (see Fig. 52.7). It is known as the exponential law of radioactive decay.

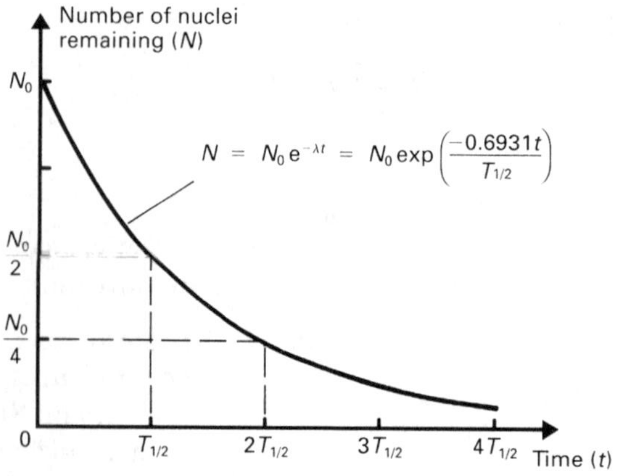

Fig. 52.7
Graph to illustrate the exponential nature of radioactive decay

Thus, although it is impossible to predict when any particular nucleus will disintegrate, it is possible to say what proportion of a large number of nuclei will disintegrate in any given time.

52.9 HALF-LIFE ($T_{1/2}$)

If the life of a radioactive nuclide is taken to mean the time that elapses before all the nuclei present disintegrate, then it is clear from equation [52.5] (or from Fig. 52.7) that the life of any radioactive nuclide is infinite, i.e. $N = 0$ when $t = \infty$. It is not very useful, therefore, to talk about the life of a radioactive nuclide, and instead we refer to its half-life. **The half-life of a radioactive nuclide is the time taken for half the nuclei present to disintegrate.** If the half-life is represented by $T_{1/2}$, then when $t = T_{1/2}$, $N = N_0/2$, and therefore by equation [52.5]

$$\frac{N_0}{2} = N_0\,e^{-\lambda T_{1/2}}$$

i.e. $$\tfrac{1}{2} = e^{-\lambda T_{1/2}}$$

i.e. $$\log_e\left(\tfrac{1}{2}\right) = -\lambda T_{1/2}$$

i.e. $$-0.6931 = -\lambda T_{1/2}$$

i.e. $$\boldsymbol{T_{1/2} = \frac{0.6931}{\lambda}} \qquad [52.6]$$

The concept of half-life is illustrated in Fig. 52.7. The reader should verify, by selecting any point on the curve, that the number of nuclei halves whenever t increases by $T_{1/2}$. The half-life of any given nuclide is constant, in particular, it does not depend on the number of nuclei present. Half-lives have a very wide range of values; this can be confirmed by inspection of Table 52.2.

The experimental determination of half-lives is discussed in sections 52.17 and 52.18.

EXAMPLE 52.1

A sample of a radioactive material contains 10^{18} atoms. The half-life of the material is 2.000 days. Calculate:

(a) the fraction remaining after 5.000 days,

(b) the activity of the sample after 5.000 days.

Solution

(a) Since $N = N_0\,e^{-\lambda t}$, the fraction, N/N_0, remaining after time t is given by

$$\frac{N}{N_0} = e^{-\lambda t}$$

Here $t = 5.000$ days and $\lambda = 0.6931/2.000$ day^{-1}.

$$\therefore \quad \lambda t = \frac{0.6931}{2.000} \times 5.000 = 1.7328$$

$$\therefore \quad \frac{N}{N_0} = e^{-1.7328} = 0.1768$$

i.e. Fraction remaining after 5.000 days $= 0.1768$

(There has been no need to express t in s, nor to express λ in s^{-1}. We are concerned with λt, which is a pure number and therefore any unit of time can be used for t as long as the reciprocal of the same unit is used for λ.)

(b) $$\frac{dN}{dt} = -\lambda N$$

Here

$$N = 0.1768 \times 10^{18} \quad \text{and} \quad \lambda = \frac{0.6931}{2.000 \times 24 \times 3600}\,\text{s}^{-1}$$

$$\therefore \quad \frac{dN}{dt} = -\frac{0.6931 \times 0.1768 \times 10^{18}}{2.000 \times 24 \times 3600}$$

$$= -7.092 \times 10^{11}\,\text{s}^{-1}$$

i.e. Activity after 5.000 days $= 7.092 \times 10^{11}$ Bq

EXAMPLE 52.2

An isotope of krypton $({}^{87}_{36}\text{Kr})$ has a half-life of 78 minutes. Calculate the activity of 10 μg of ${}^{87}_{36}$Kr. (The Avogadro constant, $N_A = 6.0 \times 10^{23}\,\text{mol}^{-1}$.)

Solution

To a good approximation, the relative atomic mass of an atom is equal to the mass number of the atom. It therefore follows (see section 14.3) that

87 g of ${}^{87}_{36}$Kr contain 6.0×10^{23} atoms

$$\therefore \quad 10\,\mu\text{g of } {}^{87}_{36}\text{Kr contain } \frac{10 \times 10^{-6} \times 6.0 \times 10^{23}}{87} \text{ atoms}$$

$$= 6.90 \times 10^{16} \text{ atoms}$$

$$\frac{dN}{dt} = -\lambda N$$

Here $N = 6.90 \times 10^{16}$ and $\lambda = \dfrac{0.6931}{78 \times 60}\,\text{s}^{-1}$

$$\therefore \quad \frac{dN}{dt} = -\frac{0.6931}{78 \times 60} \times 6.90 \times 10^{16}$$

$$= -1.02 \times 10^{13}\,\text{s}^{-1}$$

i.e. Activity $= 1.02 \times 10^{13}$ Bq

EXAMPLE 52.3

A sample of radioactive material has an activity of 9.00×10^{12} Bq. The material has a half-life of 80.0 s. How long will it take for the activity to fall to 2.00×10^{12} Bq?

Solution

Since activity is proportional to the number of parent nuclei present (see section 52.8), it follows from

$$N = N_0\,e^{-\lambda t}$$

that

$$A = A_0\,e^{-\lambda t}$$

where A = activity at time t, and A_0 = activity at $t = 0$. Rearranging gives

$$\frac{A}{A_0} = e^{-\lambda t}$$

$$\therefore \quad \log_e\left(\frac{A}{A_0}\right) = -\lambda t$$

$$\therefore \quad \log_e\left(\frac{2.00 \times 10^{12}}{9.00 \times 10^{12}}\right) = -\frac{0.6931}{80.0} \times t$$

$$\therefore \quad -1.504 = -8.664 \times 10^{-3}\,t$$

$$\therefore \quad t = 174\,\text{s}$$

i.e. Time for activity to fall to 2.00×10^{12} Bq is 174 s.

QUESTIONS 52B

1. The half-life of a particular radioactive material is 10 minutes. Without using a calculator, determine what fraction of a sample of the material will decay in 30 minutes.

2. A Geiger counter placed 20 cm away from a point source of γ-radiation registers a count rate of 6000 s^{-1}. What would the count rate be 1.0 m from the source?

3. A radioactive source has a half-life of 20 s, and an initial activity of 7.0×10^{12} Bq. Calculate its activity after 50 s have elapsed.

4. A sample of radioactive waste has a half-life of 80 years. How long will it take for its activity to fall to 20% of its current value?

5. Potassium 44 $\left({}^{44}_{19}\text{K}\right)$ has a half-life of 20 minutes and decays to form ${}^{14}_{20}\text{Ca}$, a stable isotope of calcium.
   (a) How many atoms would there be in a 10 mg sample of potassium 44?
   (b) What would be the activity of the sample?
   (c) What would the activity be after one hour?
   (d) What would the ratio of potassium atoms to calcium atoms be after one hour?
   ($N_A = 6.0 \times 10^{23}\ \text{mol}^{-1}$.)

6. What mass of radium 227 would have an activity of 1.0×10^6 Bq. The half-life of radium 227 is 41 minutes. ($N_A = 6.0 \times 10^{23}\ \text{mol}^{-1}$.)

52.10 THE RADIOACTIVE SERIES

Most of the radioactive nuclides which occur naturally have atomic numbers which are greater than that of lead. Each of these nuclides can be arranged in one or other of three series – the radioactive series. The series are known as the uranium series, the thorium series, and the actinium series. The first eight members of the uranium series are listed in Table 52.2. Thus ${}^{238}_{92}\text{U}$ decays by α-emission to produce ${}^{234}_{90}\text{Th}$ which is itself unstable and undergoes β-decay to give ${}^{234}_{91}\text{Pa}$, etc. The final member of each series is a stable isotope of lead – a different isotope in each case.

Table 52.2
The first eight members of the uranium series

Element	*Nuclide*	*Half-life*	*Radiation*	*Energy of α or β in* MeV
Uranium	$^{238}_{92}U$	4.51×10^9 years	α γ	4.2
Thorium	$^{234}_{90}Th$	24.1 days	β γ	0.19
Protactinium	$^{234}_{91}Pa$	6.75 hours	β γ	2.3
Uranium	$^{234}_{92}U$	2.47×10^5 years	α γ	4.77
Thorium	$^{230}_{90}Th$	8.0×10^4 years	α γ	4.68
Radium	$^{226}_{88}Ra$	1620 years	α γ	4.78
Radon	$^{222}_{86}Rn$	3.82 days	α	5.49
Polonium	$^{218}_{84}Po$	3.05 minutes	α	6.0

Of the elements whose atomic numbers are less than that of lead, indium and rhenium are the only ones whose most abundant isotopes are radioactive. The half-life is very long in each case – 6×10^{14} years for $^{145}_{49}In$ and $>10^{10}$ years for $^{187}_{75}Re$.

52.11 RADIOACTIVE EQUILIBRIUM

Consider a freshly produced sample of uranium 238 (the parent of the uranium series). Uranium 238 decays to produce thorium 234. Initially the rate of production of thorium will exceed the rate at which it is decaying, and the thorium content of the sample will increase. As the amount of thorium increases, its activity increases and eventually a situation is reached in which the rate of production of thorium is equal to its rate of decay. The half-life of uranium 238 is very much greater than the half-lives of its decay products, and therefore to a good approximation the rate of production of thorium is constant. It follows, therefore, that once the rate of decay of the thorium has become equal to its rate of production, the quantity of thorium in the sample will remain constant. Thorium decays to produce protactinium 234, and some time after the thorium content has become constant, the protactinium content will also stabilize. Eventually the amounts of each of the decay products in the series will be constant. The situation is known as radioactive equilibrium. At equilibrium the rate of decay of each nuclide is equal to its rate of production and it follows that all the rates of decay are equal. From equation [52.3] therefore

$$\lambda_1 N_1 = \lambda_2 N_2 = \dots$$

where the subscripts $_1$, $_2$, ... relate to the first, second, ... members of the series. Since $\lambda_1 = 0.6931/T_1$ and $\lambda_2 = 0.6931/T_2$, where T_1 and T_2 are the half-lives of the first and second members of the series, then

$$\frac{N_1}{T_1} = \frac{N_2}{T_2} = \dots \qquad [52.7]$$

52.12 ARTIFICIAL RADIOACTIVITY. POSITRON DECAY

Radioactive nuclides which do not occur in nature can be produced by bombarding naturally occurring nuclides with atomic particles – notably with neutrons inside a nuclear reactor. For example, neutron bombardment of the common isotope of beryllium (^9_4Be) results in the creation of ^9_3Li – a radioactive isotope of lithium which does not occur in nature. The reaction is

$$^9_4Be + ^1_0n \longrightarrow ^9_3Li + ^1_1p$$

Other particles used in this way include protons, deuterons and α-particles.

A number of artificially produced radioactive nuclides decay by emitting a **positron**. A positron is a positively charged particle which has the same mass as the electron and whose charge is numerically equal to that of the electron. For example, α-particle bombardment of ${}^{19}_{9}F$ produces ${}^{22}_{11}Na$ – an artificial isotope of sodium. This decays by positron emission (sometimes called positive β-decay) to produce ${}^{22}_{10}Ne$ – a stable isotope of neon. The reactions are:

$${}^{19}_{9}F + {}^{4}_{2}\alpha \longrightarrow {}^{22}_{11}Na + {}^{1}_{0}n$$

and

$${}^{22}_{11}Na \longrightarrow {}^{22}_{10}Ne + \underbrace{{}^{0}_{1}e}_{\text{(Positron)}}$$

There are no naturally occurring positron emitters.

52.13 SCHOOL LABORATORY SOURCES

The radioactive sources which are commonly used in schools are listed in Table 52.3. Activities of 5 μCi are typical.

Table 52.3
Sources used in schools

Element	*Nuclide*	*Radiation*	*Comment*
Americium	${}^{241}_{95}Am$	α	Also emits γ-rays. These are of low energy and of no importance.
Plutonium	${}^{239}_{94}Pu$	α	Also emits γ-rays. These are of low energy and of no importance.
Strontium	${}^{90}_{38}Sr$	β	The β-particles actually come from yttrium – a decay product of strontium.
Cobalt	${}^{60}_{27}Co$	γ	Also emits low energy β-particles. These are absorbed by the foil surrounding the source.
Radium	${}^{226}_{88}Ra$	α β γ	The β-particles are actually produced by some of the decay products of the radium.

52.14 RADIOACTIVE DATING

Uranium Dating

The presence of ${}^{238}_{92}U$ in some rocks allows estimates of their ages to be made.

From the ratio of ${}^{206}_{82}Pb$ to ${}^{238}_{92}U$

${}^{206}_{82}Pb$ is the stable end product of the uranium series (section 52.10). ${}^{238}_{92}U$ is the parent of the series and therefore for every uranium atom that has decayed since the rock containing the uranium was formed, one atom of ${}^{206}_{82}Pb$ will have been produced.* The ratio of ${}^{206}_{82}Pb$ to ${}^{238}_{92}U$ can be used to determine the age of the rock.

*This ignores the small number of uranium atoms which have disintegrated but which are still in the process of becoming lead.

As an example of the method suppose that in a particular sample of rock the ratio of $^{206}_{82}\mathrm{Pb}$ to $^{238}_{92}\mathrm{U}$ is 0.6. If

$$N_{\mathrm{U}} = \text{number of uranium atoms present and}$$

$$N_{\mathrm{Pb}} = \text{number of lead atoms present, then}$$

$$N_{\mathrm{U}} + N_{\mathrm{Pb}} = \text{number of uranium atoms present initially.}$$

Therefore, from equation [52.5]

$$N_{\mathrm{U}} = (N_{\mathrm{U}} + N_{\mathrm{Pb}})\,\mathrm{e}^{-\lambda t} \qquad [52.8]$$

where

λ = the decay constant of $^{238}_{92}\mathrm{U}$

t = the time for which the uranium has been decaying.

Rearranging equation [52.8] gives

$$\frac{N_{\mathrm{U}} + N_{\mathrm{Pb}}}{N_{\mathrm{U}}} = \mathrm{e}^{\lambda t}$$

i.e.

$$1 + \frac{N_{\mathrm{Pb}}}{N_{\mathrm{U}}} = \mathrm{e}^{\lambda t}$$

But $N_{\mathrm{Pb}}/N_{\mathrm{U}} = 0.6$, and therefore

$$1.6 = \mathrm{e}^{\lambda t}$$

in which case, from calculator

$$\lambda t = 0.4700$$

The half-life of $^{238}_{92}\mathrm{U}$ is 4.5×10^9 years, and therefore

$$\lambda = \frac{0.6931}{4.5 \times 10^9}$$

i.e.

$$t = \frac{0.4700 \times 4.5 \times 10^9}{0.6931}$$

i.e.

$$t = 3.1 \times 10^9 \text{ years}$$

If it is assumed that there was no lead present when the rock was formed, then the age of the rock is 3.1×10^9 years.

From the ratio of helium to $^{238}_{92}\mathrm{U}$

The decay process which eventually converts $^{238}_{92}\mathrm{U}$ to $^{206}_{82}\mathrm{Pb}$ involves the emission of eight α-particles and six β-particles. If it is assumed that the α-particles remain in the uranium-bearing rock as helium atoms, then the ratio of helium to uranium can be used to determine the age of the rock.

Carbon 14 Dating

The common isotope of carbon is the stable isotope $^{12}_{6}\mathrm{C}$. A radioactive form of carbon, $^{14}_{6}\mathrm{C}$, is formed in the upper atmosphere by neutron bombardment of $^{14}_{7}\mathrm{N}$, the common isotope of nitrogen. The reaction is

$$^{14}_{7}\mathrm{N} + {}^{1}_{0}\mathrm{n} \rightarrow {}^{14}_{6}\mathrm{C} + {}^{1}_{1}\mathrm{p}$$

The neutrons are produced by the interaction of cosmic rays with atmospheric nuclei. If it is assumed that cosmic ray activity has been constant for a period which is long compared with the half-life of $^{14}_{6}C$ (5730 years), then for some considerable time the rate of decay of $^{14}_{6}C$ must have been equal to its rate of production. This means that the ratio of $^{14}_{6}C$ to $^{12}_{6}C$ in the atmosphere will also have been constant for a considerable time.

Living matter takes in carbon in the form of carbon dioxide from the atmosphere. When an organism dies it ceases to take in atmospheric carbon and its $^{14}_{6}C$ content starts to decrease as a result of radioactive decay. The $^{12}_{6}C$ content, on the other hand, stays constant and therefore from the moment of death the ratio of $^{14}_{6}C$ to $^{12}_{6}C$ decreases.

The radioactivity of carbon can be used to date archaeological samples. For example, the activity of a given mass of carbon taken from an ancient piece of wood can be compared with that of an equal mass of carbon from (say) a living plant. From this comparison it is possible to estimate the time that has passed since the wood was part of a living tree.

As an example of the method suppose that an archaeological sample has an activity of 7.5 disintegrations per minute, and that an equal mass of carbon from a living plant has an activity of 15 disintegrations per minute. The activity of the sample is one half that of the present-day level and therefore its age is equal to the half-life of $^{14}_{6}C$, i.e. the sample is 5730 years old.

Matter which has been dead for more than about 12 000 years cannot be dated in this way since its $^{14}_{6}C$ content is too low to allow an accurate analysis.

52.15 USES OF RADIOACTIVITY

(i) Cancer cells can be destroyed by γ-radiation from a high-activity source of cobalt 60. Deep-lying tumours can be treated by planting radium 226 or caesium 137 inside the body close to the tumour.

(ii) The thickness of metal sheet can be monitored during manufacture by passing it between a γ-ray source and a suitable detector. The thicker the sheet the greater the absorption of γ-rays.

(iii) The exact position of an underground pipe can be located if a small quantity of radioactive liquid is added to the liquid being carried by the pipe. This also allows leaks to be detected; the soil close to the leak becomes radioactive. Provided a short-lived radioisotope is used there is no permanent contamination of the soil.

(iv) A radioisotope is chemically identical to a non-active isotope of the same element, and therefore takes part in the same chemical reactions. Thus the rate at which iodine passes through the thyroid can be determined by feeding radioactive iodine to a patient and externally monitoring the subsequent radioactivity of the thyroid. Radioactive phosphorus is used to assess the different abilities of plants to take up phosphorus from different types of phosphate fertilizer.

(v) Radioactive dating: see section 52.14.

52.16 RADIATION HAZARDS

The cells of the body may undergo dangerous physical and chemical changes as a result of exposure to radiation. The extent of the damage depends on:

(i) the nature of radiation,

(ii) the part of the body exposed to the radiation,

(iii) the dose received.

α-particles are absorbed in the dead surface layers of the skin and therefore do not constitute a serious hazard unless their source is taken into the body. γ-rays can penetrate deeply into the body and are a serious hazard.

The energy absorbed by unit mass of irradiated material is called the **absorbed dose**. The unit is the **gray (Gy)** and 1 Gy $=$ 1 J kg^{-1}. Until recently the unit was the **rad**. (1 Gy $=$ 100 rad.)

In order to take account of the different biological effects of the different radiations it is useful to define the **dose equivalent** as

$$\text{Dose equivalent} = Q \times \text{Absorbed dose}$$

where Q is the **quality factor** of the radiation concerned (see Table 52.4). The unit of dose equivalent is the **sievert (Sv)**. (1 Sv $=$ 1 J kg^{-1}.) Until recently the unit was the **rem**. (1 Sv $=$ 100 rem.).

Table 52.4
Quality factors of some radiations

Radiation	*Quality factor (Q)*
β, γ, X	1
n	5 to 20
α	20

52.17 EXPERIMENTAL DETERMINATION OF THE HALF-LIFE OF RADON 220*

If the half-life of a substance is short enough for its activity to decrease by a measurable amount in a reasonable time, its half-life may be determined directly by monitoring its activity as a function of time. It must be borne in mind, however, that the product of a radioactive decay is often itself radioactive, and care must be taken to avoid confusing the decay of such a product with that of the substance being investigated.

The half-life of radon 220 ($^{220}_{86}$Rn) is very different from the half-lives of its decay products, and therefore the observed decay rate is a very good approximation to that of the radon itself. Its half-life is also suitably short (55 s), and since it is a gas, it is easily isolated from its parent.

$^{220}_{86}$Rn is an α-emitter and therefore any instrument which is capable of detecting α-particles can be used to monitor the activity. In the method outlined here, a Geiger–Müller tube (with a suitably thin window) is used (see Fig. 52.8). An alternative method of detection is described in section 54.2. $^{220}_{86}$Rn is one of the

*The original name for the element radon was **emanation**; the term is now obsolete. The particular isotope radon 220 is often called **thoron**.

Fig. 52.8
Apparatus to determine the half-life of radon

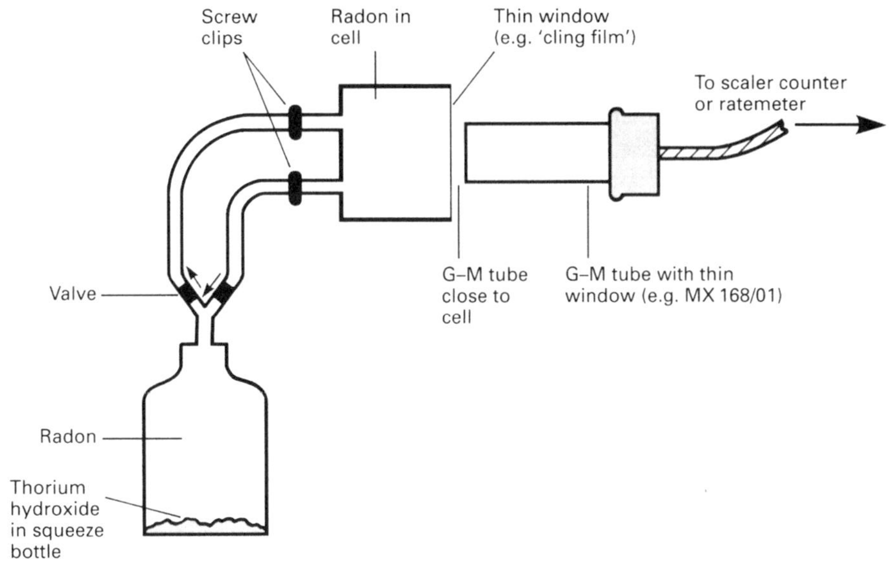

decay products of $^{232}_{90}Th$, the common isotope of thorium. The $^{220}_{86}Rn$ is conveniently obtained by squeezing a plastic bottle containing thorium hydroxide powder. The bottle is fitted with two valved tubes so that the radon can be fed into the cell without any escaping into the atmosphere. (It is very important that none of the radon and none of the powder is inhaled by the experimenter.)* The cell has a thin window through which the α-particles given off by the radon can pass.

Method

(i) Determine the background count-rate.

(ii) Open the clips and squeeze the bottle. This forces radon into the cell.

(iii) Close the clips. This isolates the radon in the cell from that being freshly produced by the thorium hydroxide.

(iv) Immediately after closing the clips, start the scaler and a stopwatch.

(v) Count for 10 s. During the next 10 s interval reset the counter and record the previous count.

(vi) Repeat (v) so that counts are being made over 10 s intervals every other 10 s.

(vii) Continue for about 5 minutes.

(viii) The count-rates obtained in this way can be taken to correspond to 5 s, 25 s, 45 s, etc.

(ix) Subtract the background count-rate to obtain the true count-rates (R).

At any time t the count-rate (R) is proportional to the activity ($-dN/dt$) which, from equation [52.3], is proportional to the number of undisintegrated nuclei (N), and therefore by equation [52.5]

$$R = R_0 e^{-\lambda t}$$

where

R = count-rate after a time t has elapsed

R_0 = initial count-rate (i.e. when $t = 0$)

λ = the decay constant of $^{220}_{86}Rn$

*The experiment should be carried out in a fume-cupboard.

Taking logs to base e gives

$$\log_e R = \log_e R_0 - \lambda t$$

Therefore, from equation [52.6]

$$\log_e R = \frac{-0.6931}{T_{1/2}} t + \log_e R_0$$

where $T_{1/2}$ is the half-life of $^{220}_{86}\text{Rn}$. Thus the gradient of a plot of $\log_e R$ against t is $-0.6931/T_{1/2}$, hence $T_{1/2}$ (see Fig. 52.9).

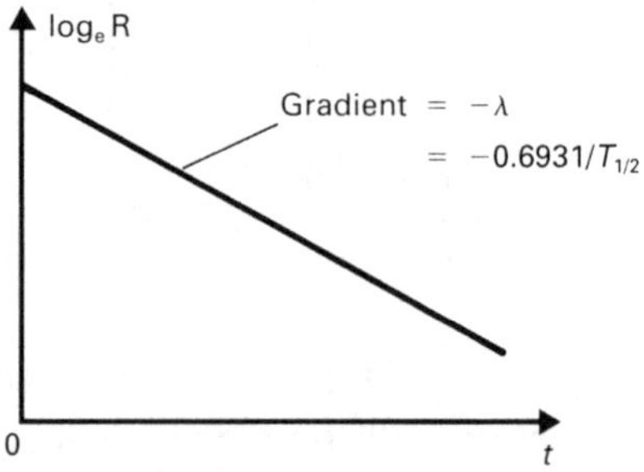

Fig. 52.9
Plot to determine half-life

52.18 DETERMINATION OF LONG HALF-LIVES

The half-life of a long-lived nuclide cannot be determined by monitoring its decrease in activity as a function of time because the decrease is slight and cannot be measured with sufficient accuracy. Instead, long half-lives can be determined by measuring the activity of a known mass of the nuclide concerned.

If m is the mass of the nuclide in grams, then the number of nuclei present, N, is given by

$$N = \frac{mN_A}{A_r}$$

where N_A is the Avogadro constant ($= 6.022 \times 10^{23}\,\text{mol}^{-1}$) and A_r is the relative atomic mass of the nuclide. Substituting for N in equation [52.3] gives

$$\frac{dN}{dt} = -\lambda \frac{mN_A}{A_r}$$

i.e. $$\frac{dN}{dt} = \frac{-0.6931}{T_{1/2}} \frac{mN_A}{A_r}$$

Hence the half-life, $T_{1/2}$, by measuring m and the activity $(-dN/dt)$ in disintegrations per second.

If a long-lived nuclide is in radioactive equilibrium (see section 52.11), its half-life can be determined from equation [52.7] provided that the half-life and relative amount of another member of the series are also known.

CONSOLIDATION

α-particles ($^4_2\alpha$ or $^4_2$He) are identical to helium nuclei.

β-particles ($_{-1}^{\ 0}\beta$ or $_{-1}^{\ 0}$e) are (fast) electrons.

γ-rays ($^0_0\gamma$ or γ) are very short wavelength electromagnetic radiation.

Properties of α, β and γ are summarized in Table 52.1.

α-emission A new element is formed,

$$\underset{\text{(Parent)}}{^A_Z\text{X}} \longrightarrow \underset{\text{(Daughter)}}{^{A-4}_{Z-2}\text{Y}} + \underset{(\alpha\text{-particle})}{^4_2\text{He}}$$

β-emission A new element is formed,

$$\underset{\text{(Parent)}}{^A_Z\text{X}} \longrightarrow \underset{\text{(Daughter)}}{^{A}_{Z+1}\text{Y}} + \underset{(\beta\text{-particle})}{^{\ 0}_{-1}\text{e}}$$

γ-emission γ-rays are emitted by the daughter nucleus – no new element is formed.

α-particles and γ-rays are emitted with a single energy, or with a small number of discrete values of energy.

β-particles are emitted with a continuous range of energies.

Radioactive decay is spontaneous and therefore the rate of decay is proportional to the number of parent nuclei present. This leads to

$$\frac{dN}{dt} = -\lambda N$$

where λ is a constant of proportionality called the **decay constant**.

Solving this equation gives **the exponential law of radioactive decay**

$$N = N_0 e^{-\lambda t}$$

Because activity, A, and count-rate, R, are proportional to N, it follows that

$$A = A_0 e^{-\lambda t} \qquad \text{and} \qquad R = R_0 e^{-\lambda t}$$

Activity is the rate of decay, i.e. the number of disintegrations per second. The unit of activity is the becquerel.

One becquerel (Bq) = 1 disintegration per second.

$$\text{Activity} = -\frac{dN}{dt} = \lambda N$$

Half-life ($T_{1/2}$) is the time taken for half the nuclei present to disintegrate.

$$T_{1/2} = \frac{\log_e 2}{\lambda} = \frac{0.6931}{\lambda}$$

Inverse square law for γ-rays

The intensity, I, at a distance, d, from a point source of γ-rays is given by

$$I \propto \frac{1}{d^2} \qquad \text{i.e.} \qquad I = \frac{\text{constant}}{d^2}$$

The law is entirely true in vacuum, and is a good approximation in air.

53

NUCLEAR STABILITY. FISSION AND FUSION

53.1 EINSTEIN'S MASS–ENERGY RELATION

According to the Special Theory of Relativity a mass m is equivalent to an amount of energy E, where

$$E = mc^2 \tag{53.1}$$

c being the speed of light ($\approx 3 \times 10^8\ \text{m s}^{-1}$).

It follows that whenever a reaction results in a release of energy there is an associated decrease in mass. For example, when 1 kg of $^{235}_{92}\text{U}$ undergoes fission (see section 53.7) the energy released is approximately 8×10^{13} J, and therefore according to equation [53.1] there is a decrease in mass of $8 \times 10^{13}/(3 \times 10^8)^2 \approx 9 \times 10^{-4}$ kg. This is a significant fraction of the initial mass of $^{235}_{92}\text{U}$ and can be measured. Chemical reactions, on the other hand, release relatively small amounts of energy and the associated decrease in mass is too small to be measured. For example, when 1 kg of petrol is burned the energy released is only 5×10^7 J and, by equation [53.1], this corresponds to a decrease in mass of a mere 5.5×10^{-10} kg.

The reader should be left in no doubt that no matter how a change in energy arises there is a change in mass. For example, an increase in temperature is accompanied by an increase in mass, as is an increase in velocity.

The unified atomic mass unit (u) is defined in section 51.2, and

$$1\ \text{u} = 1.661 \times 10^{-27}\ \text{kg}$$

Therefore, from equation [53.1]

$$1\ \text{u} = 1.661 \times 10^{-27} \times (2.998 \times 10^8)^2\ \text{J}$$

$$= \frac{1.661 \times 10^{-27} \times (2.998 \times 10^8)^2}{1.602 \times 10^{-19}}\ \text{eV}$$

i.e.

$$1\ \text{u} = 932\ \text{MeV} \tag{53.2}$$

53.2 BINDING ENERGY

The mass of a nucleus is always less than the total mass of its constituent nucleons. The difference in mass is called the **mass defect** of the nucleus, i.e.

$$\text{Mass defect} = \text{Mass of nucleons} - \text{Mass of nucleus}$$

The reduction in mass arises because the act of combining the nucleons to form the nucleus causes some of their mass to be released as energy (in the form of γ-rays). Any attempt to separate the nucleons would involve them being given this same amount of energy – it is therefore called the **binding energy** of the nucleus. It follows from equation [53.1] that

$$\underset{(\text{J})}{\text{Binding energy}} = \underset{(\text{kg})}{\text{Mass defect}} \times \underset{(\text{m s}^{-1})^2}{c^2}$$

It follows from equation [53.2] that

$$\underset{(\text{MeV})}{\text{Binding energy}} = 932 \times \underset{(\text{u})}{\text{Mass defect}}$$

Tables normally give atomic masses rather than nuclear masses and it is useful to redefine the mass defect as

$$\begin{matrix}\text{Mass}\\ \text{defect}\end{matrix} = \begin{pmatrix}\text{Mass of nucleons}\\ \text{and electrons}\end{pmatrix} - \begin{pmatrix}\text{Mass of}\\ \text{atom}\end{pmatrix}$$

Consider, as an example of the calculation of binding energies, the case of the helium atom. It consists of two protons (each of mass 1.007 28 u), two neutrons (each of mass 1.008 67 u) and two electrons (each of mass 0.000 55 u). The total mass of the particles is

$$2 \times 1.007\,28 + 2 \times 1.008\,67 + 2 \times 0.000\,55 = 4.033\,00\,\text{u}$$

The mass of a helium atom is 4.002 60 u, and therefore the mass defect is

$$4.033\,00 - 4.002\,60 = 0.030\,4\,\text{u}$$

From equation [53.2], therefore, the binding energy of a helium atom is

$$0.0304 \times 932 = 28.3\,\text{MeV}$$

Note The binding energy of a nucleus is the energy required to break it up into its component neutrons and protons. The binding energy of an atom, on the other hand, is the energy required to break it up into its component neutrons, protons and electrons. The difference between the two is negligible, because the energy required to remove the electrons is very much less than that required to remove the neutrons and protons. For example, the binding energy of a helium nucleus is also 28.3 MeV.

A useful measure of the stability of a nucleus is its **binding energy per nucleon** (i.e. binding energy divided by mass number), since this represents the (average) energy which needs to be supplied to remove a nucleon. Fig. 53.1 shows the way this quantity varies with mass number for the naturally occurring nuclides with mass numbers in the range 2–238. It can be seen that the nuclides of intermediate mass numbers have the largest values of binding energy per nucleon. $^{56}_{26}\text{Fe}$ has a value of 8.8 MeV and is one of the most stable nuclides. Three nuclides, $^{4}_{2}\text{He}$, $^{12}_{6}\text{C}$

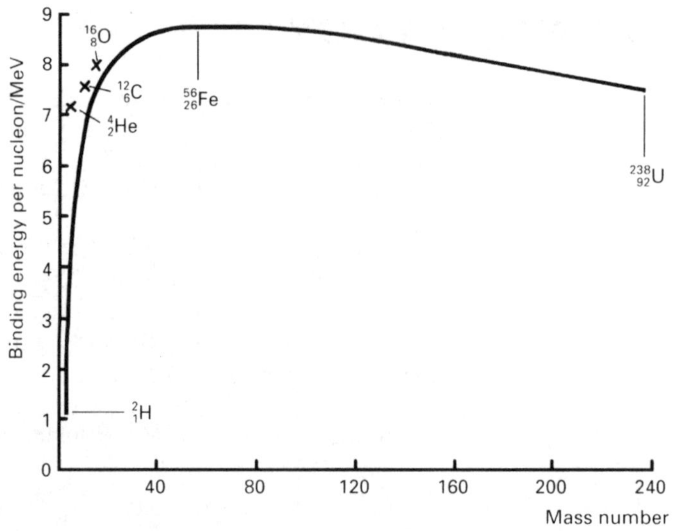

Fig. 53.1
Variation of binding energy per nucleon with mass number

and $^{16}_{8}O$, lie significantly above the main curve. Though these are not the most stable of all nuclides, they are considerably more stable than those of adjacent mass. Note that $^{12}_{6}C$ and $^{16}_{8}O$ are respectively combinations of three and four α-particles. A combination of two α-particles would be $^{8}_{4}Be$. The binding energy per nucleon of $^{8}_{4}Be$ is slightly less than that of an α-particle, and therefore it is unstable and disintegrates to form two α-particles.

Nuclides of intermediate mass number have the greatest binding energy per nucleon and therefore energy is released when two light nuclides are fused to produce a heavier one (**fusion**) and when a heavy nucleus splits into two lighter ones (**fission**).

QUESTIONS 53A

1. Calculate:
(a) the mass defect,
(b) the binding energy per nucleon for $^{238}_{92}U$.

(Atomic mass of $^{238}_{92}U$ = 238.050 76 u,
mass of neutron = 1.008 67 u,
mass of proton = 1.007 28 u,
mass of electron = 0.000 55 u,
1 u = 932 MeV.)

53.3 STABILITY AGAINST α-PARTICLE EMISSION

A nucleus which undergoes radioactive decay by emitting an α-particle is able to do so because its mass is greater than the sum of the masses of the daughter nucleus and the emitted α-particle.*

In order to illustrate this we shall consider $^{210}_{84}Po$, which decays according to

$$^{210}_{84}\mathrm{Po} \longrightarrow \, ^{206}_{82}\mathrm{Pb} + \, ^{4}_{2}\mathrm{He}$$

The various terms in this equation can be taken to represent atoms rather than nuclei because each side of the equation involves the same number of electrons. From tables

*Fulfilment of this condition does not guarantee that an α-particle is emitted.

Atomic mass of $^{210}_{84}Po = 209.983\,u$

Atomic mass of $^{206}_{82}Pb = 205.974\,u$

Atomic mass of $^{4}_{2}He = 4.003\,u$

The mass of the parent (209.983 u) is greater than the total mass of the products ($205.974 + 4.003 = 209.977\,u$) – as required.

It can be shown that about 98% of the energy provided by the decrease in mass is carried away as the kinetic energy of the α-particle; the remaining 2% is the recoil energy of the nucleus. (The nucleus recoils in order that momentum is conserved.)

53.4 STABILITY AGAINST (NEGATIVE) β-PARTICLE EMISSION

If a nucleus is to decay by emitting a β-particle, the mass of the nucleus must be greater than the total mass of the decay products. Consider the possibility of $^{14}_{6}C$ decaying by (negative) β-emission. If it does, the relevant nuclear equation is

$$^{14}_{6}C \longrightarrow {}^{14}_{7}N + {}^{0}_{-1}e$$

Adding six electrons to each side of the equation gives

$$^{14}_{6}C + 6\,{}^{0}_{-1}e \longrightarrow {}^{14}_{7}N + 7\,{}^{0}_{-1}e$$

Bearing in mind that a carbon atom has six electrons and that a nitrogen atom has seven electrons, we can rewrite this equation as

$$^{14}_{6}C \longrightarrow {}^{14}_{7}N$$

where the terms now represent atoms rather than nuclei. From tables:

Atomic mass of $^{14}_{6}C = 14.0032\,u$

Atomic mass of $^{14}_{7}N = 14.0031\,u$

Thus the atomic mass of $^{14}_{7}N$ is less than that of $^{14}_{6}C$ and the decay is possible, and in fact does occur.

EXAMPLE 53.1

Calculate the energy released (i.e. the **Q-value**) when gallium 70 ($^{70}_{31}Ga$) undergoes (negative) β-decay to produce germanium 70 ($^{70}_{32}Ge$). (Atomic mass of $^{70}_{31}Ga = 69.92605\,u$, of $^{70}_{32}Ge = 69.92425\,u$. $1\,u = 932\,MeV$.)

Solution

The nuclear equation is

$$^{70}_{31}Ga \longrightarrow {}^{70}_{32}Ge + {}^{0}_{-1}e$$

Adding 31 electrons to each side of the equation gives

$$^{70}_{31}Ga \longrightarrow {}^{70}_{32}Ge$$

where the terms now represent atoms.

Decrease in mass $= 69.92605 - 69.92425 = 0.00180\,u$

$\therefore$ Energy released $= 0.00180 \times 932 = 1.68\,MeV$

QUESTIONS 53B

1. Radium 224 decays by α-emission to produce radon 220 according to

$$^{224}_{88}\text{Ra} \longrightarrow {}^{220}_{86}\text{Rn} + {}^{4}_{2}\text{He}$$

Calculate: **(a)** the decrease in mass, **(b)** the energy released (the **Q-value**).
(Atomic mass of $^{224}_{88}$Ra = 224.020 22 u, of $^{220}_{86}$Rn = 220.011 40 u, of $^{4}_{2}$He = 4.002 60 u. 1 u = 932 MeV.)

2. Nitrogen 13 decays by positron emission (see section 52.12) to produce carbon 13. The relevant nuclear equation is

$$^{13}_{7}\text{N} \longrightarrow {}^{13}_{6}\text{C} + {}^{0}_{1}\text{e}$$

Calculate: **(a)** the decrease in mass, **(b)** the energy released (the **Q-value**).
(Atomic mass of $^{13}_{7}$N = 13.005 74 u, of $^{13}_{6}$C = 13.003 35 u. Mass of $^{0}_{1}$e and of $^{0}_{-1}$e = 0.000 55 u. 1 u = 932 MeV.) Hint – add 7 electrons to each side of the nuclear equation.

3. A nucleus decays to produce an α-particle of mass 4.00 u and a daughter of mass 204 u, releasing 5.21 MeV in the process. **(a)** Bearing in mind that the daughter nucleus recoils in order that momentum is conserved, calculate the value of the ratio: kinetic energy of α-particle/kinetic energy of daughter. **(b)** Hence find the kinetic energy of the α-particle.

53.5 STABILITY AGAINST FISSION

The fission process is discussed in section 53.7. The ideas of the last two sections apply, i.e. the total mass of the fission products is less than that of the nucleus which has undergone fission.

53.6 STABILITY AND NEUTRON–PROTON RATIO

Fig. 53.2 shows a plot of neutron number against proton number for all the known stable nuclides. It can be seen that among the light nuclei the tendency is for there

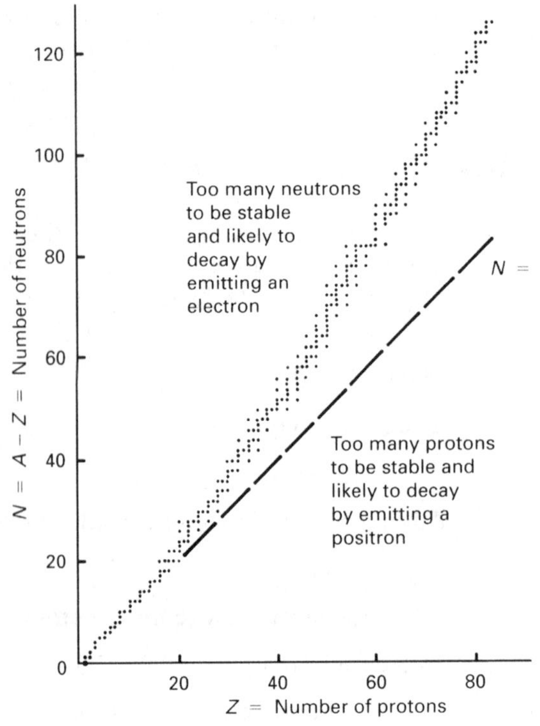

Fig. 53.2
Variation of neutron number with proton number for stable nuclides

to be equal numbers of neutrons and protons. Heavy nuclei, on the other hand, have more neutrons than protons, the neutron–proton ratio reaching about 1.6 with $^{208}_{82}Pb$. A nuclide which would be represented by a point above the stability belt has too many neutrons to be stable and can acquire a more stable state by converting one of its neutrons into a proton, i.e. it emits a (negative) β-particle. A nuclide which lies below the stable region has too many protons and can correct this by converting one of its protons to a neutron, i.e. it emits a positron.

53.7 NUCLEAR FISSION

Nuclear fission is the disintegration of a heavy nucleus into two lighter nuclei. Energy is released by the process because the average binding energy per nucleon of the fission products is greater than that of the parent.

As an example of a fission reaction, consider the bombardment of $^{235}_{92}U$ by slow neutrons. This can result in the capture of a neutron and the formation of $^{236}_{92}U$ which is unstable and undergoes fission. Many different pairs of nuclei can be produced by the fission of $^{236}_{92}U$, but in over 95% of the cases the mass number of the heavier fragment is between 130 and 149. One possible reaction is

$$^{235}_{92}U + ^{1}_{0}n \longrightarrow ^{236}_{92}U \longrightarrow ^{141}_{56}Ba + ^{92}_{36}Kr + 3\,^{1}_{0}n + \text{Energy}$$

The energy released by the fission of a single uranium atom is about 200 MeV, and about 80% of this goes into providing the kinetic energy of the two fission fragments. These are often radioactive, and their subsequent decay accounts for a further 10% of the energy released. The remaining 10% appears as the kinetic energy of the neutrons which are ejected, and in the form of those γ-rays which are a result of the fission process itself.

Nuclear reactors (section 53.9) make use of controlled fission reactions to provide energy. The atom bomb makes use of an uncontrolled fission reaction.

EXAMPLE 53.2

Calculate the energy released when 10 kg of $^{235}_{92}U$ undergoes fission according to

$$^{235}_{92}U + ^{1}_{0}n \longrightarrow ^{141}_{56}Ba + ^{92}_{36}Kr + 3\,^{1}_{0}n$$

(Mass of $^{235}_{92}U$ = 235.04 u, of $^{141}_{56}Ba$ = 140.91 u, of $^{92}_{36}Kr$ = 91.91 u, of $^{1}_{0}n$ = 1.01 u, 1 u = 932 MeV, $N_A = 6.02 \times 10^{23}\,\text{mol}^{-1}$.)

Solution

$$\text{Mass difference} = (235.04 + 1.01) - (140.91 + 91.91 + 3 \times 1.01)$$
$$= 0.20\,\text{u}$$

$\therefore$ Energy released $= 0.20 \times 932 = 186.4\,\text{MeV}$

235.04×10^{-3} kg of $^{235}_{92}U$ contains 6.02×10^{23} atoms

$\therefore$ 10 kg of $^{235}_{92}U$ contains $\dfrac{10 \times 6.02 \times 10^{23}}{235.04 \times 10^{-3}}$ atoms

$= 2.56 \times 10^{25}$ atoms

$\therefore$ Energy released by 10 kg of $^{235}_{92}U$ $= 2.56 \times 10^{25} \times 186.4$

$= 4.77 \times 10^{27}$ MeV

EXAMPLE 53.3

Calculate the energy released (i.e. the **Q-value**) when a uranium 236 nucleus undergoes fission according to

$$^{236}_{92}U \longrightarrow ^{146}_{57}La + ^{87}_{35}Br + 3^{1}_{0}n$$

(Binding energy per nucleon of $^{236}_{92}U$ = 7.59 MeV, of $^{146}_{57}La$ = 8.41 MeV, of $^{87}_{35}Br$ = 8.59 MeV.)

Solution

Binding energy of $^{146}_{57}La = 146 \times 8.41 = 1227.86$ MeV

Binding energy of $^{87}_{35}Br = 87 \times 8.59 = 747.33$ MeV

$\therefore$ Total binding energy after fission $= 1227.86 + 747.33 = 1975.19$ MeV

Binding energy of $^{236}_{92}U = 236 \times 7.59 = 1791.24$ MeV

$\therefore$ Increase in binding energy $= 1975.19 - 1791.24 = 183.95$ MeV

$\therefore$ Energy released $= 184$ MeV

53.8 NUCLEAR FUSION

Nuclear fusion is the combining of two light nuclei to produce a heavier nucleus. Energy is released by the process. An example is the fusion of two deuterium nuclei to produce helium 3:

$$^{2}_{1}H + ^{2}_{1}H \longrightarrow ^{3}_{2}He + ^{1}_{0}n + 3.27\ \text{MeV}$$

Reactions of this type (the conversion of hydrogen to helium) are the source of the Sun's energy. Temperatures in excess of 10^8 K are required to provide the nuclei which are to fuse with the energy needed to overcome their mutual electrostatic repulsion. To date, this has been achieved only in an uncontrolled way, in the hydrogen bomb. The high temperature required for the fusion reaction is provided by the explosion of an atom bomb.

The energy released by the fusion of two nuclei is very much less than that which results from the fission of, say, a uranium nucleus. However, it should be borne in mind that fusion involves very much less massive nuclei and, in fact, the energies provided <u>per unit mass of reactants</u> by the two processes are much the same.

53.9 THE THERMAL REACTOR

When a $^{235}_{92}U$ nucleus captures a neutron and undergoes fission (section 53.7) an average of about 2.5 neutrons is released. (The actual number depends on just which pair of fission products is formed.) The principle of the thermal reactor is to cause these neutrons to produce more fission by being captured by other $^{235}_{92}U$ nuclei so that a chain reaction occurs.

In natural uranium only about 1 atom in 140 is a $^{235}_{92}U$ atom – the rest are $^{238}_{92}U$. $^{238}_{92}U$ can be fissioned, but only by being bombarded with <u>very fast neutrons</u>. On the other hand, slow neutrons are required to produce fission in $^{235}_{92}U$. The neutrons released by the fission of $^{235}_{92}U$ are not fast enough to produce fission in $^{238}_{92}U$, but need to be slowed down before they can cause fission with $^{235}_{92}U$.

The neutrons are slowed down by the use of a material called a **moderator** – commonly graphite, water or heavy water (D_2O). In a graphite-moderated reactor the uranium fuel is in sealed tubes which are arranged inside a block of graphite (Fig. 53.3). The neutrons released by the fission of $^{235}_{92}U$ collide with the atoms of the moderator and are slowed to such an extent that they are far more likely to cause fission of $^{235}_{92}U$ than to be unproductively captured by $^{238}_{92}U$.

Fig. 53.3
The advanced gas-cooled reactor (AGR) – an example of a graphite-moderated thermal reactor

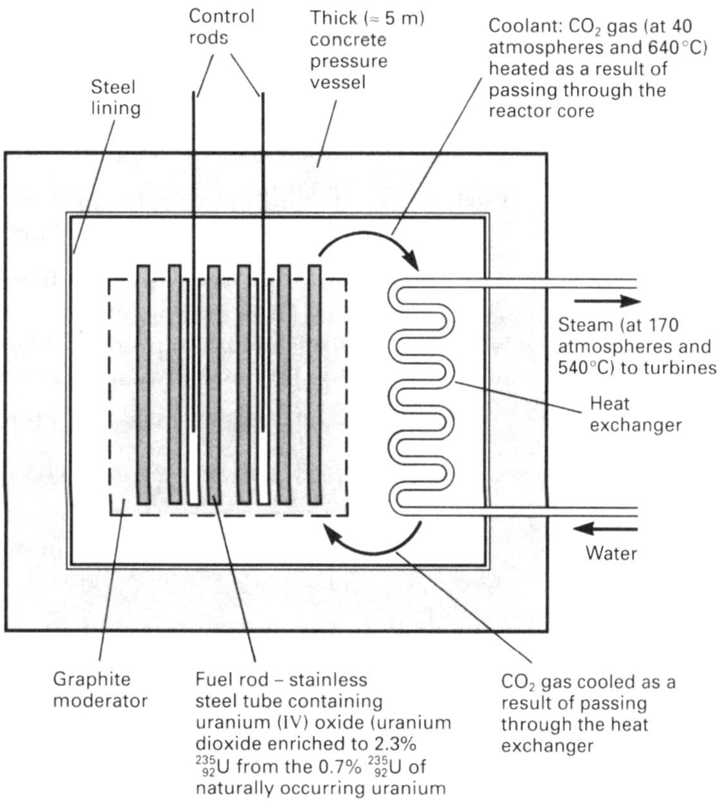

A typical reactor of this type generates 660 MW of electrical power, and has 324 fuel rods and 81 control rods

Ideally, just over one neutron per fission is required to sustain the reaction; much higher rates than this would release energy too quickly and the reaction would go out of control. **Control rods** of boron-coated steel are used to keep the net rate of production of neutrons to the required level by capturing the necessary proportion before they can initiate fission. When the control rods are moved upwards out of the heart of the reactor, the number of neutrons left to produce fission is increased; when the rods are lowered, the number of neutrons is decreased.

The heat energy produced by the fission reaction is removed by passing a coolant such as carbon dioxide or water through the reactor. The coolant then passes through some form of heat exchanger, producing steam to drive turbines which in turn generate electricity. A thick concrete shield prevents potentially harmful radiation from reaching the operators.

CONSOLIDATION

$$E = mc^2 \qquad 1\,\text{u} = 932\,\text{MeV}$$

$$\begin{matrix}\text{Mass defect}\\ \text{of an atom}\end{matrix} = \begin{pmatrix}\text{Mass of neutrons}\\ \text{protons and electrons}\end{pmatrix} - \begin{pmatrix}\text{Mass of}\\ \text{atom}\end{pmatrix}$$

The binding energy of an atom is the energy required to split it up into its component neutrons, protons and electrons.

$$\underset{(\text{MeV})}{\text{Binding energy}} = 932 \times \underset{(\text{u})}{\text{Mass defect}}$$

$$\underset{(\text{J})}{\text{Binding energy}} = \underset{(\text{kg})}{\text{Mass defect}} \times \underset{(\text{m s}^{-1})^2}{c^2}$$

Fission The disintegration of a heavy nucleus into two lighter ones, accompanied by a release of energy and an associated decrease in mass.

Fusion The joining of two light nuclei to produce a heavier one, accompanied by a release of energy and an associated decrease in mass.

The plot of binding energy per nucleon against mass number (Fig. 53.1) shows that there is an increase in binding energy, and therefore a release of energy, when two light nuclei fuse to produce a heavier one, and when a heavy nucleus undergoes fission producing two lighter ones.

$$\text{Energy released} = \text{Increase in binding energy}$$

$$\underset{(\text{MeV})}{\text{Energy released}} = 932 \times \underset{(\text{u})}{\text{Decrease in mass}}$$

$$\underset{(\text{J})}{\text{Energy released}} = \underset{(\text{kg})}{\text{Decrease in mass}} \times \underset{(\text{m s}^{-1})^2}{c^2}$$

To calculate the energy released in α-decay When working with atomic masses the terms in the nuclear equation can be taken to represent atoms.

To calculate the energy released in β-decay When working with atomic masses remember to add a suitable number of electrons to each side of the nuclear equation.

The Thermal Reactor

Moderator – slows the neutrons so that they are more likely to cause fission.

Control rods – absorb neutrons so that the reaction does not go out of control.

Coolant – carries away the heat generated in the reactor core.

54

DETECTORS OF RADIATION

54.1 THE GEIGER–MÜLLER TUBE (G–M TUBE)

A Geiger–Müller tube (Fig. 54.1) can be used to detect the presence of X-rays, γ-rays and β-particles. Tubes with very thin mica windows can also detect α-particles.

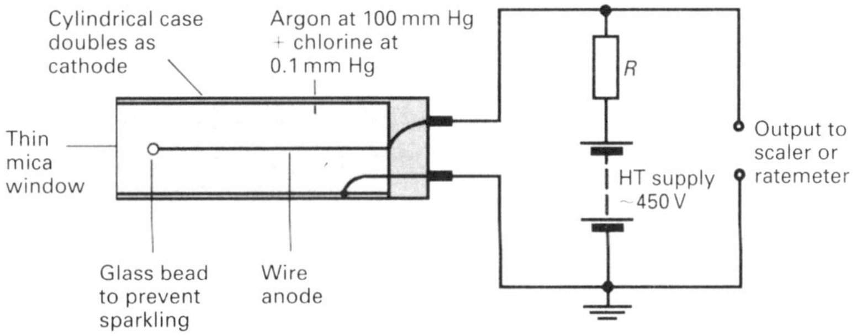

Fig. 54.1 A Geiger–Müller tube and circuit

When one of these ionizing 'particles' enters the tube, either through the window or the wall, some of the argon atoms become ionized. The free electrons and positive ions which result are accelerated towards the anode and cathode respectively by the PD across the tube. The geometry of the tube is such that the electric field near the anode is very intense, and as electrons approach the anode they gain sufficient kinetic energy to produce further ionization. The electrons released by this 'secondary' ionization produce even more ionization so that there is soon a large number of electrons moving towards the anode – the resistance of the gas is said to have broken down. The positive ions are much more massive, and move much more slowly, than the electrons, and after about 10^{-6} s there are so many positive ions near the anode that the electric field around it is cancelled out. This prevents further ionization, and the **electron avalanche** and the associated anode current, cease to exist. Thus, the effect of a single ionizing 'particle' entering the tube is to produce a relatively large current pulse. The process is called **gas amplification** and as many as 10^8 electrons can be released as a result of a single ionizing event.

The positive ions move slowly towards the cathode. Some of the ions would release electrons from the cathode surface if they were allowed to collide with it. These electrons would initiate a second avalanche, and this would give rise to a third, and so on, so that a whole series of current pulses would be produced. This would make it impossible to know whether a second ionizing 'particle' had entered the tube. In order to ensure that only one pulse is produced by each 'particle' that enters it the tube contains a **quenching agent** – chlorine. (Bromine is used in some cases.)

The argon ions are neutralized as a result of collisions with chlorine molecules before they reach the cathode, and, in effect, their energy is used to dissociate the chlorine molecules rather than to release electrons from the cathode.

A resistor, R, of about 1 MΩ, is connected in series with the tube and the HT supply. The current pulse from the tube creates a voltage pulse of about 1 V across R, and this can be amplified and fed to a **scaler counter** or a **ratemeter**. A scaler registers the number of pulses it receives whilst it is switched on; a ratemeter indicates the rate at which it receives pulses and registers it in counts per second.

Immediately after a pulse has been registered there is a period of about 300 μs during which the tube is insensitive to the arrival of further ionizing 'particles'. This can be divided into two parts – the **dead time** and the **recovery time**. During the dead time the tube does not respond at all to the arrival of an ionizing 'particle'. The recovery time is the second stage of the period of insensitivity, and during this time pulses are produced but they are not large enough to be detected. The dead time is the time taken by the positive ions to move far enough away from the anode for the electric field there to return to a level which is large enough for an avalanche to start. The recovery time is the time which elapses while the argon ions are being neutralized by the quenching gas. The period of insensitivity limits the count rate to a maximum of about 1000 counts per second.

Apart from those which arrive during the period of insensitivity, almost every α-particle and β-particle that enters a Geiger–Müller tube is counted. γ-rays and X-rays are more likely to be detected indirectly, as a result of being absorbed by the walls of the tube and releasing electrons in the process, than by direct ionization of argon atoms. They are only weakly absorbed and only about 1% are detected.

Fig. 54.2 shows the way in which the count rate varies with the PD applied to a typical Geiger–Müller tube. When the applied PD is less than the **threshold voltage** there is not sufficient gas amplification to produce pulses which are large

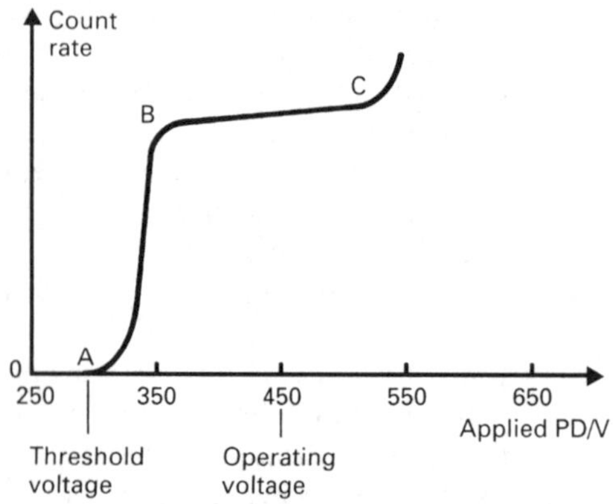

Fig. 54.2
Variation of count rate with applied PD in a G–M tube

enough to be detected. Between A and B, the **proportional region**, the size of any particular pulse depends on the 'strength' of the initial ionization; some of the 'particles' which enter the tube produce less ionization than others and go undetected. In the **plateau region** (B to C) all the pulses have the same amplitude, irrespective of the 'strength' of the initial ionization. Every particle which produces any ionization at all is detected. This is the region in which the tube should be operated. If the voltage is increased beyond C, the quenching process becomes less and less effective and eventually a continuous discharge occurs.

54.2 THE IONIZATION CHAMBER

A common form of ionization chamber is shown in Fig. 54.3. It is essentially a metal can containing a small brass platform mounted on the upper end of a metal rod. The can forms one electrode of the device; the metal rod is the other. The simplest types contain air at atmospheric pressure and are capable of detecting the ionization produced by α-particles and intense sources of β-particles. The source of the radiation may be either outside the chamber, or inside it, in which case the platform provides a convenient means of supporting the source. When the source is outside the chamber a gauze 'lid' is used so that the particles can enter without appreciable absorption.

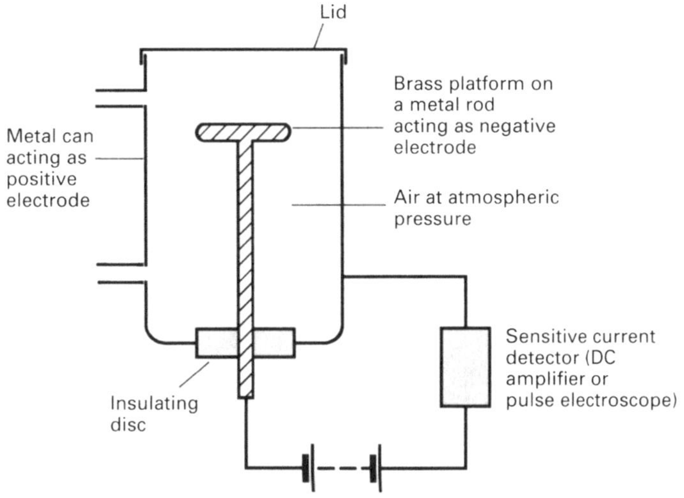

Fig. 54.3 Ionization chamber

A typical α-particle produces 10^5–10^6 ion-pairs as it passes through the air in the chamber. The electrons are attracted to the can and the positive ions move to the central rod causing a current to flow in the external circuit. Fig. 54.4 illustrates the way in which this **ionization current** depends on the PD across the chamber. Between O and A the PD is not large enough to draw all the electrons and positive ions to their respective electrodes before some recombination has occurred. Between A and B the PD is large enough to prevent recombination but is not so high that it produces secondary ionization. The ionization current is said to have reached its **saturation value** (I_s). Beyond B the PD is large enough to cause secondary ionization. The PD at which an ionization chamber is operated should be such that the ionization current has its saturation value. Under such conditions:

(i) the ionization current is independent of fluctuations in supply voltage, and

(ii) the ionization current is proportional to the rate at which ionization is being produced in the chamber. (The reader should contrast this with the case of the Geiger–Müller tube, where the output is proportional to the number of ionizing particles.)

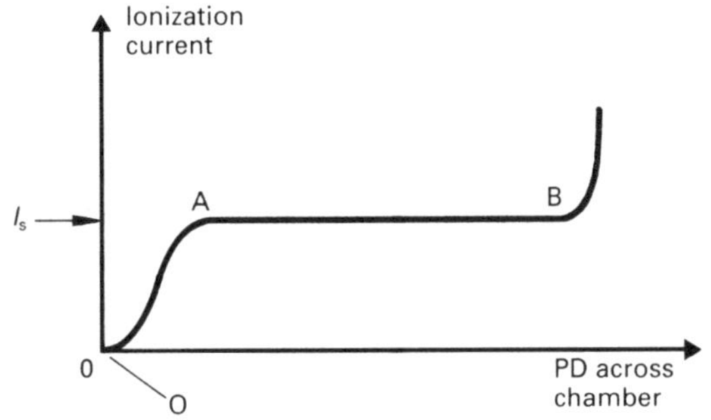

Fig. 54.4 Variation of ionization current with PD across an ionization chamber

The ionization currents produced by the sources commonly used in schools are small – 10^{-9} A or less. This means that very sensitive current detectors are required; DC amplifiers and pulse electroscopes are suitable. The use of a DC amplifier is described in section 40.13.

An ionization chamber and DC amplifier is a convenient alternative to a Geiger–Müller tube and scaler counter in the determination of the half-life of radon 220 (section 52.17). The ionization chamber (with suitable inlet and outlet tubes) is substituted for the cell (Fig. 52.8). The correct operating voltage for the chamber being used can be found by carrying out a preliminary experiment in which the PD across the chamber is increased gradually until the ionization current saturates.

54.3 THE DIFFUSION CLOUD CHAMBER

The cloud chamber (Fig. 54.5) displays the tracks of any ionizing agents which pass through it. It is superior to Wilson's earlier expansion cloud chamber (section 54.4) in that it does not have to be re-set before it can display a second track.

Fig. 54.5
Diffusion cloud chamber

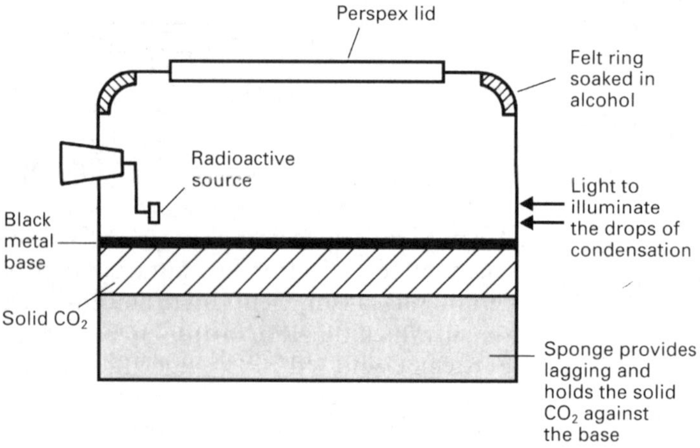

The base of the chamber is maintained at about −80°C by the solid carbon dioxide there. The top of the chamber is at room temperature and so there is a temperature gradient between top and bottom. The air at the top of the chamber is saturated with alcohol vapour from the felt ring. The vapour continually diffuses downwards into the cooler regions so that the air there becomes supersaturated with alcohol vapour. The excess vapour in the supersaturated regions can condense only if there are nucleating sites present. Condensation occurs on ionized atoms in preference to neutral atoms, and so if an ionizing agent passes through the supersaturated air, the ions produced along its path act as nucleating sites. The path therefore shows up as a series of small drops of condensation.

α-particles leave dense, straight tracks; β-particles produce less ionization and give thinner tracks. The tracks of fast β-particles are straight; those travelling more slowly are easily deflected and leave tortuous tracks. γ-rays are uncharged and therefore must actually collide with an atom in order to ionize it. Since such collisions are rare, there is very little ionization along the path of the rays. However, when a collision does occur, the electron which is ejected has sufficient energy to ionize atoms along its path. All such paths originate on the path of the γ-rays (Fig. 54.6). X-rays produce a similar effect.

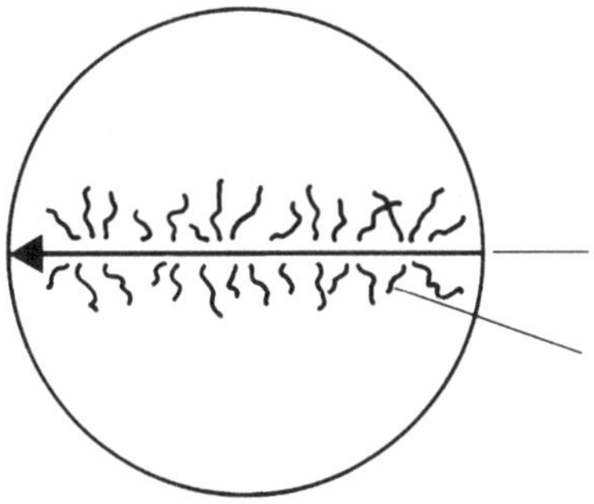
γ-rays
Path of an ejected electron shows up as droplets of condensed vapour

Fig. 54.6
The effect of γ-rays passing through a cloud chamber

54.4 THE WILSON CLOUD CHAMBER

The Wilson cloud chamber, like the diffusion cloud chamber, shows up the paths of ionizing 'particles' which pass through it. Both types of chamber make use of the fact that a supersaturated vapour condenses more readily on ions than on neutral atoms. In the Wilson cloud chamber, a gas saturated with vapour is caused to undergo an adiabatic expansion; the gas cools and so becomes supersaturated with vapour. Various combinations of gas and vapour are used, e.g. air–water, air–alcohol, argon–water.

54.5 THE BUBBLE CHAMBER

The bubble chamber was invented by Glaser in 1951 and makes use of a superheated liquid. (A superheated liquid is one which is at a higher temperature than that at which it normally exists as a liquid under the prevailing pressure, and as such is unstable.) The ions produced in a superheated liquid by the passage of a charged particle act as nucleating sites on which bubbles form. The path of such a particle (or ionizing radiation) therefore shows up as a train of bubbles.

Bubble chambers are superior to cloud chambers in that the detecting medium is a liquid which, being more dense than the air (and other gases) used in cloud chambers, has a greater stopping power. This is particularly useful when particles with very high energies are being studied – such particles are likely to pass through a cloud chamber without producing any ionization at all.

One of the most commonly used liquids is hydrogen. This provides an opportunity to bombard protons (hydrogen nuclei) with high-energy particles and to study the particles which result from the collisions. By using magnetic fields to deflect the particles, and by making use of the laws of conservation of energy and momentum, the charges, masses and speeds of the particles can be deduced.

54.6 PHOTOGRAPHIC EMULSIONS

Ionizing radiation passing through a photographic emulsion (silver halide grains in gelatin) has the same effect on the emulsion as light does. Plates which are intended for the detection of ionizing radiations are now produced commercially and have special emulsions with a high density of silver halide grains. The path of an ionizing particle which has passed through such a plate shows up as a well-defined track when the plate is developed. The tracks are short (typically 1 mm for a β-particle) and microscopes have to be used to study them.

The method automatically provides a permanent record of the 'particle' tracks. The plates are easily portable making them convenient for the study of cosmic ray activity as balloons can carry them to considerable heights where cosmic ray activity is much greater than it is on the ground. A disadvantage is that the path lengths are too short to allow the measurement of any curvature due to a magnetic field.

55

SEMICONDUCTORS AND ELECTRONICS

55.1 INTRODUCTION

Semiconductors are materials whose electrical conductivities are higher than those of insulators but less than those of conductors. Though we normally think of semiconductors as being solids, some liquids are semiconducting. Commonly used semiconducting materials include silicon, germanium, gallium arsenide, indium antimonide and cadmium sulphide, all of which are solids.

Semiconductors have negative temperature coefficients of resistance, i.e. their electrical resistivities decrease with increasing temperature. Insulators also have negative temperature coefficients of resistance, and the distinction between semiconductors and insulators is only one of degree – all insulators are semiconductors at high temperatures and all semiconductors are insulators at low temperatures. It is on the basis of their room-temperature resistivities that they are categorized.

The extent to which a semiconductor conducts electricity is considerably affected by the presence of impurities. Very pure semiconductors are called **intrinsic** semiconductors; those to which impurities have been added are called **extrinsic** semiconductors.

55.2 INTRINSIC SEMICONDUCTORS

Both silicon and germanium are tetravalent, i.e. each has four electrons (valence electrons) in its outermost shell. Both elements crystallize with a diamond-like structure, i.e. in such a way that each atom in the crystal is inside a tetrahedron formed by the four atoms which are closest to it. Fig. 55.1 shows one of these tetrahedral units. Each atom shares its four valence electrons with its four immediate neighbours on a one to one basis, so that each atom is involved in four covalent bonds. A two-dimensional representation of the crystal structure is shown in Fig. 55.2, and can be taken to represent either silicon or germanium.

Fig. 55.1
The basic tetrahedral unit of silicon and germanium

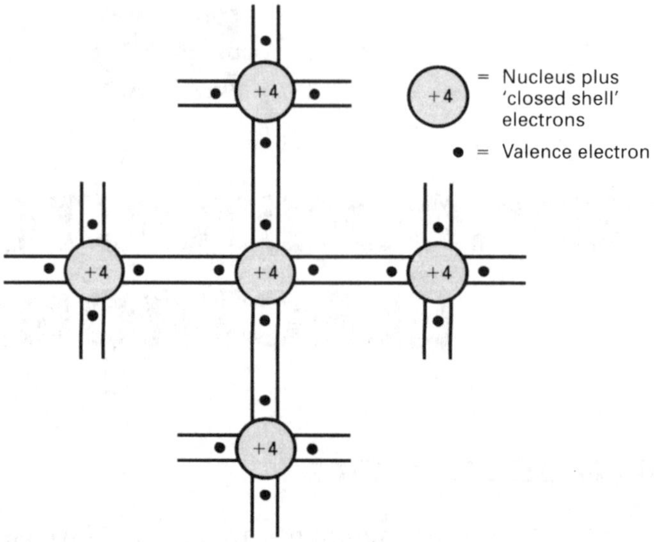

Fig. 55.2 Two-dimensional representation of the crystal structure of germanium or silicon at 0 K

At 0 K all the valence electrons are involved in bonding, and so the crystal is a perfect insulator because there are no electrons available for conduction. At higher temperatures, however, some of the valence electrons have sufficient energy to break away from the bonds and move about the structure. The higher the temperature, the greater the number of 'free' electrons, accounting for the observation that the electrical resistivity decreases with increasing temperature.

When an electron escapes from a bond it leaves behind a vacancy in the lattice. The vacancy is called a **hole**. Clearly, a hole is a region in which there is an excess of positive charge. If, in the course of its random motion through the solid, an electron comes near a hole, it is likely to be captured, in which case the hole ceases to exist. Since holes are continually being filled in this way and, at the same time, are continually being created as more electrons are freed from their bonds, it is as if the holes are moving randomly through the structure. Thus, in semiconductors there are two types of charge carrier – the electrons (negative) and the holes (positive). When a battery is connected the electrons drift towards the positive plate; the holes drift in the opposite direction. (The reader should not confuse positive holes with positive ions; the concept of electrical conduction by positive holes is merely a convenience.)

Each hole in an intrinsic semiconductor is produced by the thermal excitation of a bound electron and therefore there is an equal number of free electrons and holes. However, the **mobility** of an electron, i.e. its average drift velocity per unit electric field intensity, is usually greater than that of a hole, and because of this the contribution of the electrons to the total current is greater than that of the holes.

55.3 EXTRINSIC SEMICONDUCTORS

Small amounts ($\sim$1 part in 10^6) of certain elements can be added to germanium (and silicon) without producing any distortion of the basic crystal structure. The process is known as **doping** and produces a considerable increase in conductivity. The atoms which are added are often referred to as **impurities**, but this is not to be taken to imply that there is anything haphazard or accidental about their inclusion, the process is very carefully controlled. The atoms which are added are either pentavalent (e.g. antimony, arsenic and phosphorus) or trivalent (e.g. indium and gallium).

We shall illustrate the effect of adding a pentavalent impurity by considering the particular case of antimony incorporated into a crystal of germanium. The antimony atoms occupy sites at which there would otherwise be germanium atoms (Fig. 55.3). Antimony has five valence electrons, only four of which are required for bonding. The fifth electron is very loosely bound, and needs only about 0.01 eV to become free. This is less than the average thermal energy of the lattice ions and therefore almost all the surplus electrons are liberated from their parent nuclei. The energy required to release an electron from a germanium–germanium bond, on the other hand, is about 0.75 eV and therefore at ordinary temperatures only about 1 in 10^{10} of these electrons is free. If germanium is doped to the extent of one part per million with antimony, then since each antimony atom contributes an electron which is available for conduction, the conductivity is increased by a factor of 10^4. Since antimony and the other pentavalent elements which are used in doping donate conduction electrons to the structure, they are known as **donor impurities**. Germanium which has been doped in this way is called an **n-type** semiconductor because it is primarily negative charge carriers (electrons) that are involved when it conducts electricity. Since in an n-type semiconductor there are many more free electrons than holes, the electrons are called **majority carriers** and the holes are called **minority carriers**. It is worth stressing that the addition of a donor impurity leaves the crystal electrically neutral; the excess negative charge associated with the charge carriers is balanced by the excess positive charge at the sites of the antimony nuclei.

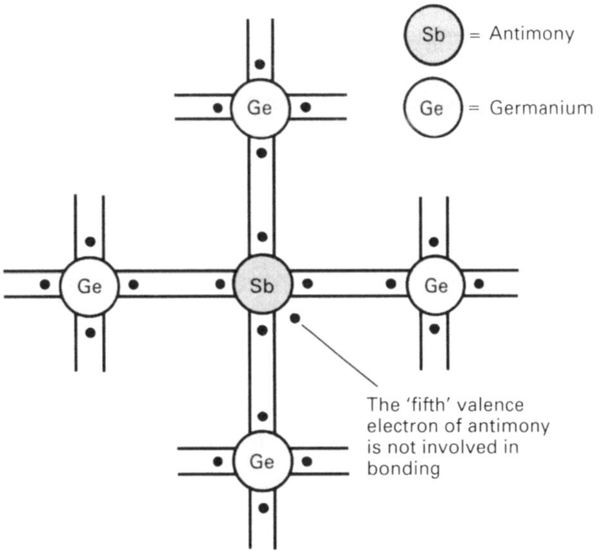

Fig. 55.3
Effect of adding a pentavalent impurity into a germanium crystal

The addition of a trivalent impurity (e.g. indium or gallium) to a crystal of germanium or silicon also produces a marked increase in conductivity. In this case the majority charge carriers are holes, and semiconductors of this type are called **p-type** semiconductors. Consider the effect of adding indium to a crystal of germanium. The indium atoms substitute for germanium atoms (Fig. 55.4). Each indium atom has only three valence electrons and therefore can form covalent bonds with only three of its four germanium neighbours. It requires very little energy for an electron in a nearby germanium–germanium bond to move across and fill the vacancy on the indium. At room temperature, lattice vibrations readily provide this energy, thus creating a hole on one of the germanium atoms. Because trivalent impurities accept electrons from the structure in this way they are called **acceptors**. The majority carriers in a p-type material are holes. Like n-type semiconductors the crystal as a whole is electrically neutral.

Fig. 55.4
Effect of adding a trivalent impurity into a germanium crystal

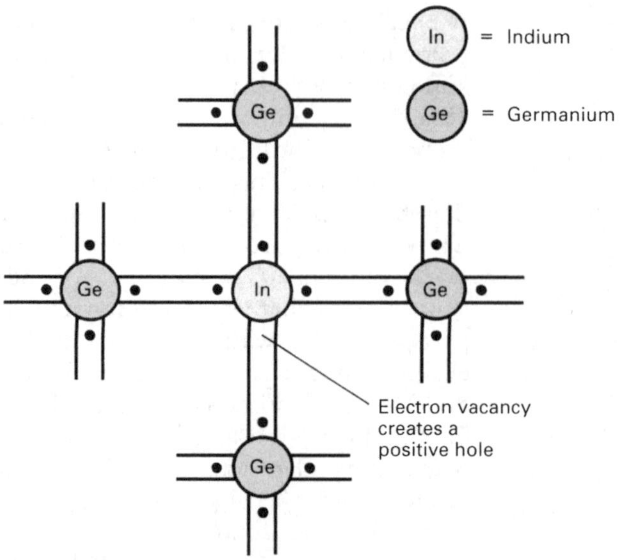

55.4 BAND THEORY TREATMENT OF CONDUCTION

The electrons in an isolated atom have a well-defined set of energy levels (section 48.5). When two identical atoms are close together their electrons move under the influence of the combined electric fields of the two atoms and each previously single energy level splits into two levels, one higher and one lower than the corresponding level of the isolated atoms (Fig. 55.5). When large numbers of atoms are together, as they are in a crystal, the energy levels spread into bands (Fig. 55.6). Each band contains a large number of levels, and these are so close together that, in effect, there is a continuous range of energies available to the electrons. The energy bands are separated by gaps in which there are no available energy levels.

Fig. 55.5
Energy levels of a pair of atoms

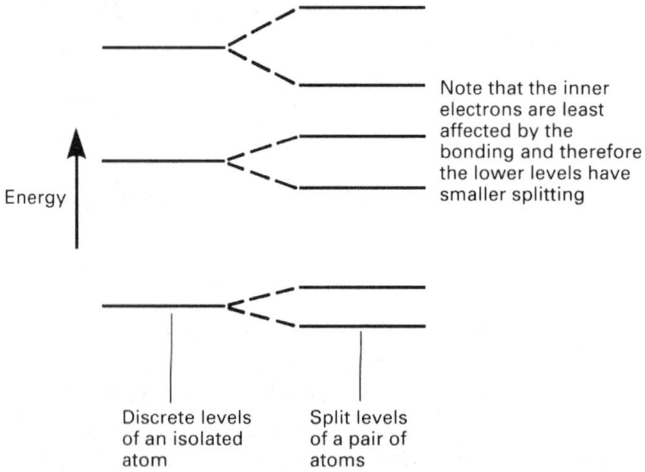

If an electron is to take part in the conduction of an electric current, it must be capable of being accelerated by an applied PD and so must be capable of being raised to a slightly higher energy level. It follows that a material can conduct electricity only if some of its electrons are in a band which does not contain its full quota of electrons, for otherwise the only energy levels which are available to the electron are in higher bands, in which case the energy differences involved are prohibitive.

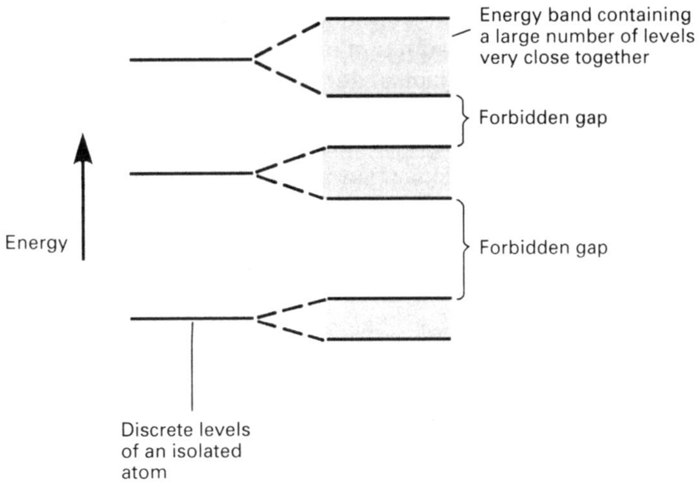

Fig. 55.6
Energy bands of a large number of atoms

The energy band diagrams of an intrinsic semiconductor and an insulator are compared in Fig. 55.7; in each case only the two highest bands are shown. (Bands which are lower than these are full and are of no interest.)

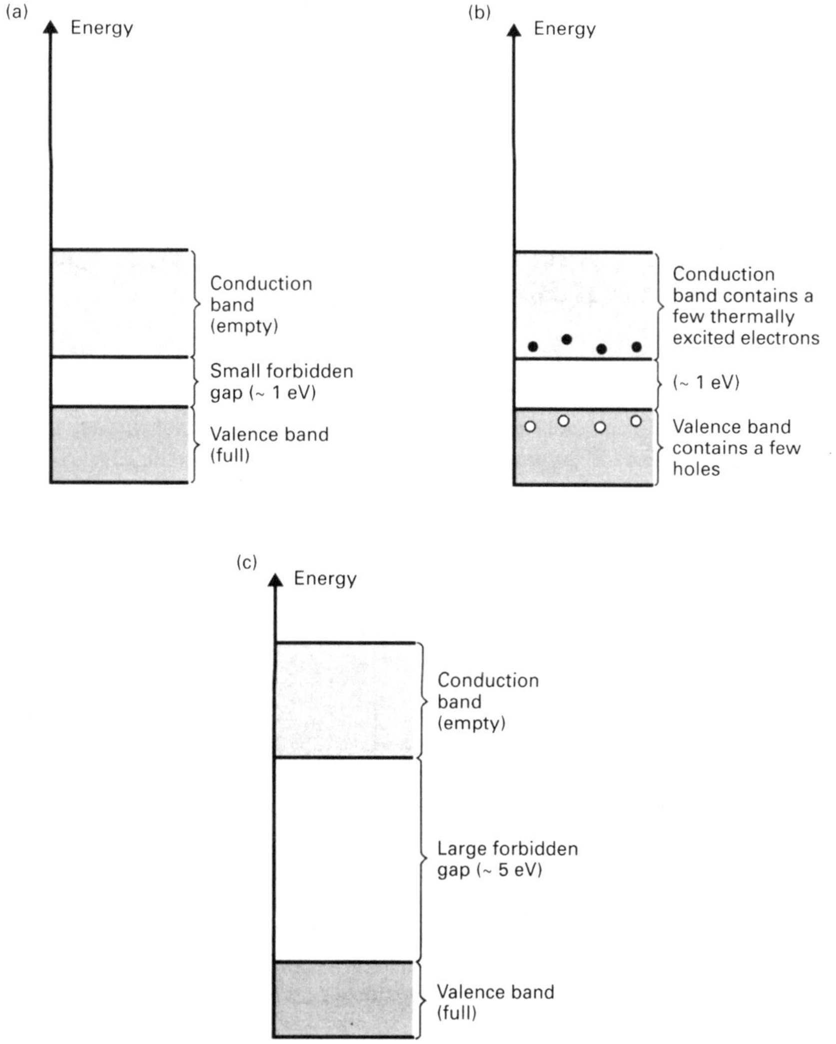

Fig. 55.7
Energy bands of (a) an intrinsic semiconductor at 0 K, (b) an intrinsic semiconductor at room temperature, (c) an insulator at room temperature

In a semiconductor at 0 K (Fig. 55.7(a)) the valence band is completely full and the conduction band is totally empty. The semiconductor therefore behaves like a perfect insulator because it has no electrons which can conduct a current. However, the forbidden gap is narrow (0.75 eV in the case of germanium, 1.1 eV in the case of silicon) and at room temperature (Fig. 55.7(b)) some of the electrons in the valence band gain enough energy from lattice vibrations to move up into the conduction band. This electron promotion creates an equal number of holes in the valence band. If the temperature is increased, there is even more thermal excitation of electrons into the conduction band and the conductivity increases.

Insulators, like semiconductors at 0 K, have completely full valence bands and totally empty conduction bands (Fig. 55.7(c)). However, in the case of an insulator the forbidden gap is large and, at room temperature, the number of electrons that can acquire sufficient energy to cross it is so small as to be insignificant. At very high temperatures, or in very strong electric fields, a significant number of electrons may gain enough energy to be raised to the conduction band, in which case the insulator is said to have broken down. At even higher temperatures there are increased amounts of electron promotion and therefore, like semiconductors, insulators have negative temperature coefficients of resistance.

Monovalent metals have valence bands in which only half the available energy levels are occupied. The valence bands of divalent metals are full, but are overlapped by their conduction bands. In each case, therefore, higher energy levels are available to the electrons – hence metals are good conductors.

55.5 BAND THEORY TREATMENT OF EXTRINSIC SEMICONDUCTORS

The addition of a donor impurity to an intrinsic semiconductor creates extra energy levels just below the bottom of the conduction band (Fig. 55.8). At room temperature, lattice vibrations are easily capable of providing the small amount of energy (~0.1 eV) needed to raise the electrons in these levels to the conduction band. Once in the conduction band the electrons can conduct an electric current.

Fig. 55.8
Energy bands of an n-type semiconductor

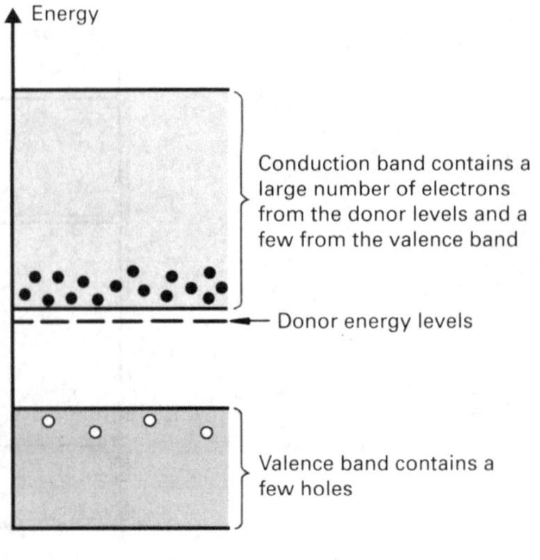
Energy
Conduction band contains a large number of electrons from the donor levels and a few from the valence band
Donor energy levels
Valence band contains a few holes

The addition of an acceptor impurity to an intrinsic semiconductor creates extra energy levels just above the top of the valence band (Fig. 55.9). At room temperature these levels are occupied by electrons which have been thermally excited from the valence band. This leaves a large number of holes in the valence band and so increases the conductivity.

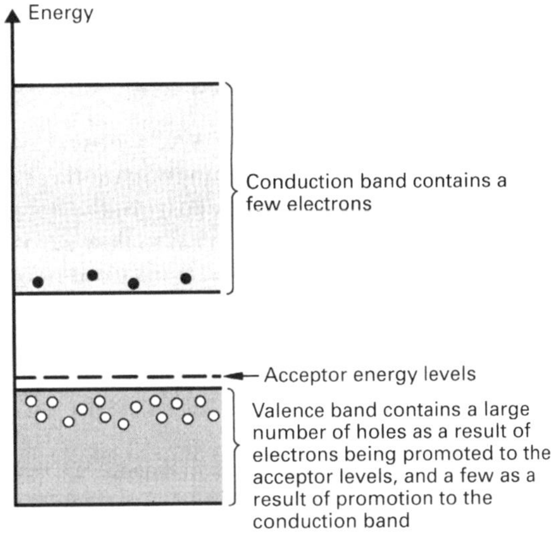
Energy
Conduction band contains a few electrons
Acceptor energy levels
Valence band contains a large number of holes as a result of electrons being promoted to the acceptor levels, and a few as a result of promotion to the conduction band

Fig. 55.9
Energy bands of a p-type semiconductor

55.6 THE p–n JUNCTION DIODE

A single crystal of silicon or germanium which has been doped in such a way that one half of it is p-type and the other is n-type can be used as a rectifier. It is the existence of the junction between the two types of semiconducting material which gives the device its ability to rectify; it is therefore called a p–n junction diode. (**Note**. A junction formed simply by putting a p-type crystal in contact with an n-type would have too many imperfections to behave in a reliable fashion.)

As soon as such a junction is formed, electrons from the electron-rich n-type material on one side of the junction diffuse into the p-type side and fill some of the holes there. At the same time, holes from the p-type side diffuse into the n-type material and are filled by electrons. This exchange takes place in a narrow ($\sim 1\ \mu m$) region known as the **depletion layer** (Fig. 55.10). That part of the depletion layer which is in the n-type material has lost electrons and gained holes and is therefore left with a positive charge; the p-type side of the depletion layer becomes negative.

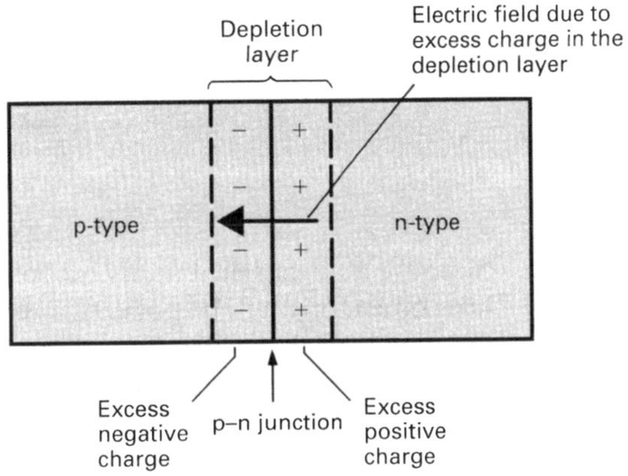
Depletion layer
Electric field due to excess charge in the depletion layer
p-type
n-type
Excess negative charge
p–n junction
Excess positive charge

Fig. 55.10
Depletion layer in a p–n junction

Note that although the p-type material now has a net negative charge, it still contains many more holes than free electrons; the excess charge is due to the presence of negatively charged ions, and as such is immobile. Positive ions are responsible for the excess charge in the n-type material. Thus the diffusion establishes a potential difference across the junction (the so-called **contact potential**), and within a very short time of the junction being formed this becomes large enough to prevent any further diffusion. The size of the contact potential depends on the nature of the crystal, its temperature and the amount of doping, but it is typically a few tenths of a volt.

Suppose that a battery, whose EMF is bigger than the contact potential, is connected across a junction diode as in Fig. 55.11(a) – the so-called **forward bias connection**. The polarity of the battery is such as to urge the majority carriers on each side of the junction to flow across it and so constitute a current. Since it is the majority carriers (electrons from n to p and holes from p to n) that are carrying the current, the current is appreciable (~mA). (The current in the connecting wires is of course carried by electrons only.) A typical forward bias characteristic (current against applied PD) is shown in Fig. 55.11(b).

Fig. 55.11
(a) A junction diode with forward bias, and (b) its current–voltage characteristic

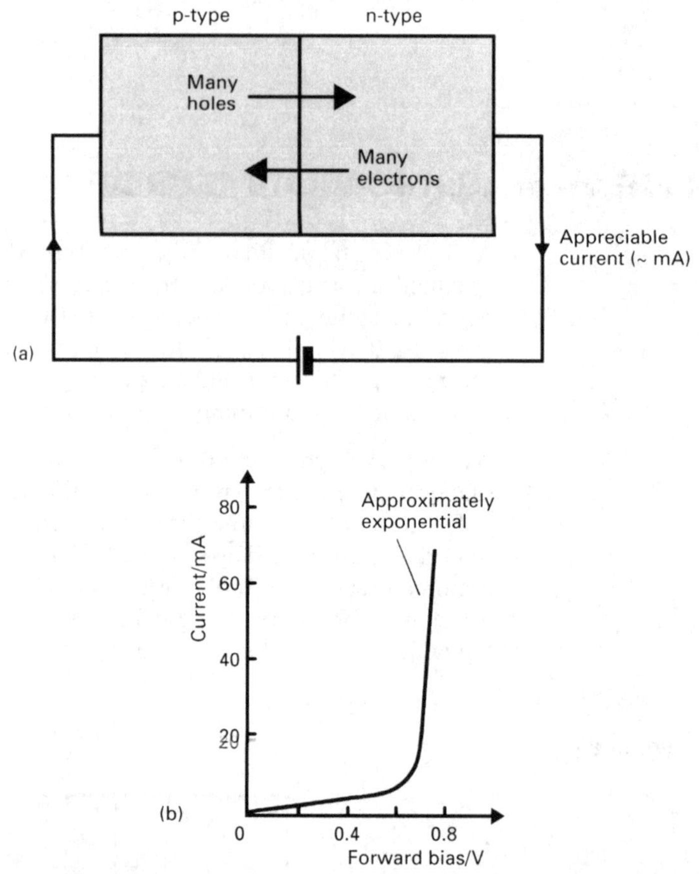

Suppose now that the battery is connected as shown in Fig. 55.12(a) – the **reverse bias connection**. The polarity is such that only the minority carriers can cross the junction and therefore the current which flows is small. This reverse current is often referred to as the **leakage current**. A typical reverse bias characteristic is shown in Fig. 55.12(b); note that the scales are different from those in Fig. 55.11(b).

Fig. 55.12
(a) A junction diode with reverse bias, and (b) its current–voltage characteristic

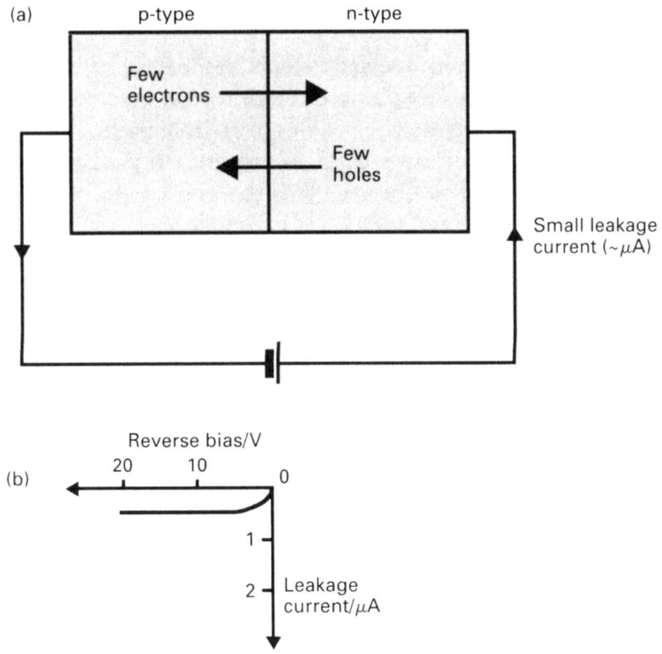

Thus a junction diode has low resistance to current flow when it is forward-biased, but has high resistance to current flow in the opposite direction and therefore can be used as a rectifier. The circuit symbol for a junction diode is shown in Fig. 55.13.

Fig. 55.13
Circuit symbol for a junction diode

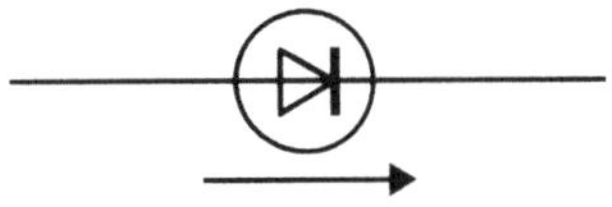

55.7 THE ZENER DIODE

Fig. 55.14 shows the effect of applying a large reverse PD to a junction diode. There is a marked increase in current as the PD is increased beyond a critical **breakdown voltage**, V_Z (sometimes called the **Zener voltage**). There are two distinct processes by which breakdown may occur – **Zener breakdown** and **avalanche breakdown**.

Fig. 55.14
Current–voltage characteristic and circuit symbol of a Zener diode

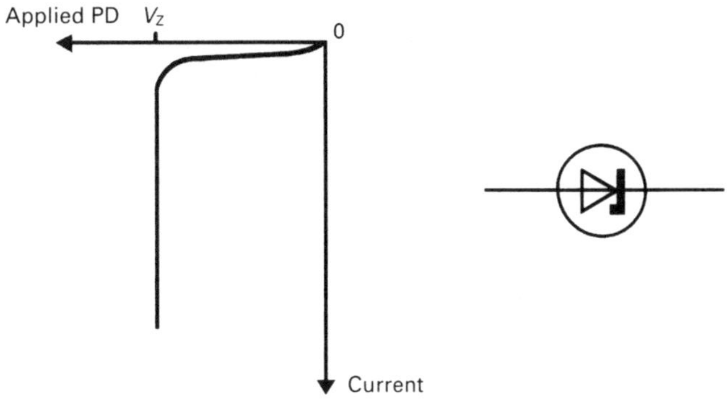

Zener Breakdown

Zener breakdown is the result of the electric field across the depletion layer becoming high enough to tear electrons away from their nuclei and so produce a large number of electron–hole pairs. This has a very marked effect on the number of minority carriers present (in particular) and leads to a massive increase in the reverse current. It is the predominant effect in diodes which have a high level of doping, and V_Z is typically less than 5 V. The process is reversible, the current returns to its pre-breakdown level when the applied PD is reduced below V_Z.

Avalanche Breakdown

Avalanche breakdown occurs in diodes which have a low level of doping. These have wide depletion layers, and reverse bias PDs of as much as 200 V can be applied without the electric field across the depletion layer becoming large enough to produce ionization. However, with high PDs the minority carriers can gain sufficient energy to produce ionization by collision. The electrons which are ejected can themselves produce ionization and a current avalanche occurs.

Zener Diodes

Zener diodes are p–n junction diodes which are specifically intended to make use of one or other of these breakdown effects. When a Zener diode is used it is operated at a reverse bias voltage which is slightly higher than its breakdown voltage. Under these conditions the voltage across the diode is very nearly independent of the current through it, i.e. it is a **voltage regulator**.

55.8 THE JUNCTION TRANSISTOR (BIPOLAR TRANSISTOR)

A junction transistor is a single crystal of semiconducting material doped in such a way that a piece of p-type material is sandwiched between two pieces of n-type material, or such that a piece of n-type is between two pieces of p-type. The two types are respectively called n–p–n transistors and p–n–p transistors, and are illustrated schematically in Fig. 55.15 together with their circuit symbols. The three regions of a junction transistor are called the **emitter**, the **base** and the **collector**. The current in an n–p–n transistor is due mainly to electrons flowing from the emitter to the collector; in a p–n–p type it is due mainly to the movement of holes, also flowing from the emitter to the collector. The arrowheads on the circuit symbols point in the direction of conventional current flow. In both types the base is very much thinner than the emitter and the collector, and is much more lightly doped.

A transistor is normally sealed inside a light-proof case through which protrude three metal leads so that connections can be made to the emitter, the base and the collector. Nowadays the most widely used transistors are silicon n–p–n types.* These can be used at higher frequencies than the p–n–p types. This is because the main charge carriers are electrons and have greater mobilities than holes. The forbidden gap in silicon is larger than that in germanium and because of this silicon transistors are less influenced by temperature variations.

*From now on we shall limit our discussion to the behaviour of n–p–n types. The reader should be able to apply this to p–n–p types by bearing in mind that (i) the polarities of the biasing voltages are the reverse of those used with n–p–n transistors, and (ii) the main charge carriers are holes rather than electrons.

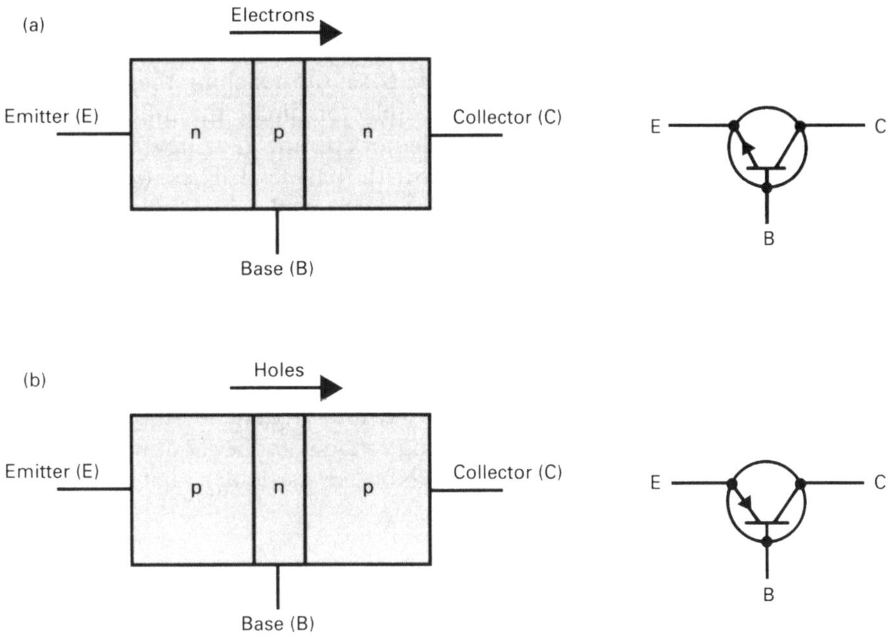

Fig. 55.15
Junction transistors: (a) n–p–n, (b) p–n–p

When a transistor is in use the base–emitter junction is normally forward-biased and the base–collector junction is reverse-biased. In the case of an n–p–n transistor this means that the base must be positive with respect to the emitter, and the collector must be positive with respect to the base.

Transistors can be connected into circuits in three different ways: the **common-emitter connection**, the **common-base connection** and the **common-collector connection**. The common-emitter connection is the most widely used and is the only one that we shall consider.

Fig. 55.16 illustrates the way in which an n–p–n transistor is biased when it is operating in the common-emitter mode. Note that the emitter is common to the base circuit and the collector circuit. Provided the PD between the emitter and the base is large enough to overcome the contact potential (section 55.6) at the base–emitter junction, electrons (the majority carriers in the emitter) cross into the base

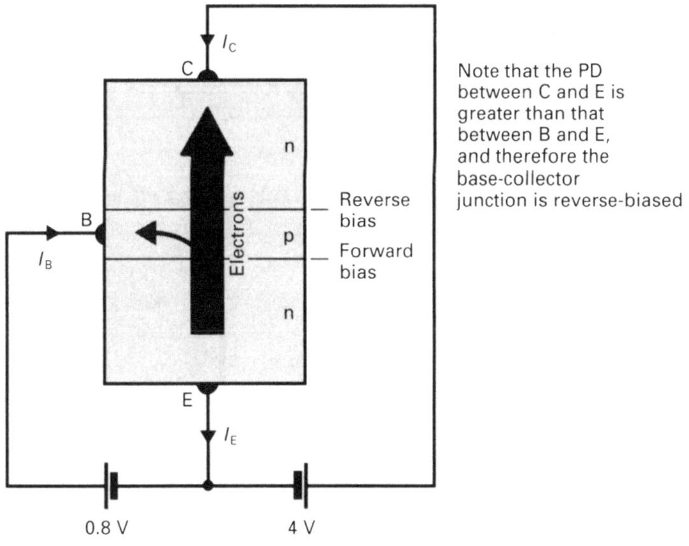

Fig. 55.16
Biasing of an n–p–n transistor in the common-emitter mode

region. Because the base is very thin and is only lightly doped most of the electrons that enter the base diffuse to the base–collector junction before they can combine with holes in the base. On reaching the base–collector junction they are pulled across into the collector under the influence of the base–collector PD and so constitute a collector current, I_C. A few electrons combine with holes in the base and a few diffuse to the base lead. These two effects give rise to a small base current, I_B, in the form of electrons flowing out of the base into the external circuit.

If the emitter current is represented by I_E, then by Kirchhoff's first rule

$$I_E = I_B + I_C$$

The values of I_E, I_B and I_C depend on the particular transistor being considered and on the biasing voltages applied, but typically $I_E = 2\,\text{mA}$, $I_C = 1.98\,\text{mA}$ and $I_B = 0.02\,\text{mA}$. Note, in particular, that I_B is much less than I_C.

55.9 TRANSISTOR CHARACTERISTICS IN COMMON-EMITTER CONNECTION

In order to be in a position to determine how any particular transistor will behave in a circuit, it is necessary to have knowledge of the relationships which exist among the currents in the three sections of the transistor and the biasing voltages. These relationships are called the **characteristics** of the transistor and the information is usually presented as a set of curves. The three most important relationships for a transistor which is operating in the common-emitter mode are those shown in Fig. 55.17.

Fig. 55.17 Transistor characteristics in the common-emitter mode

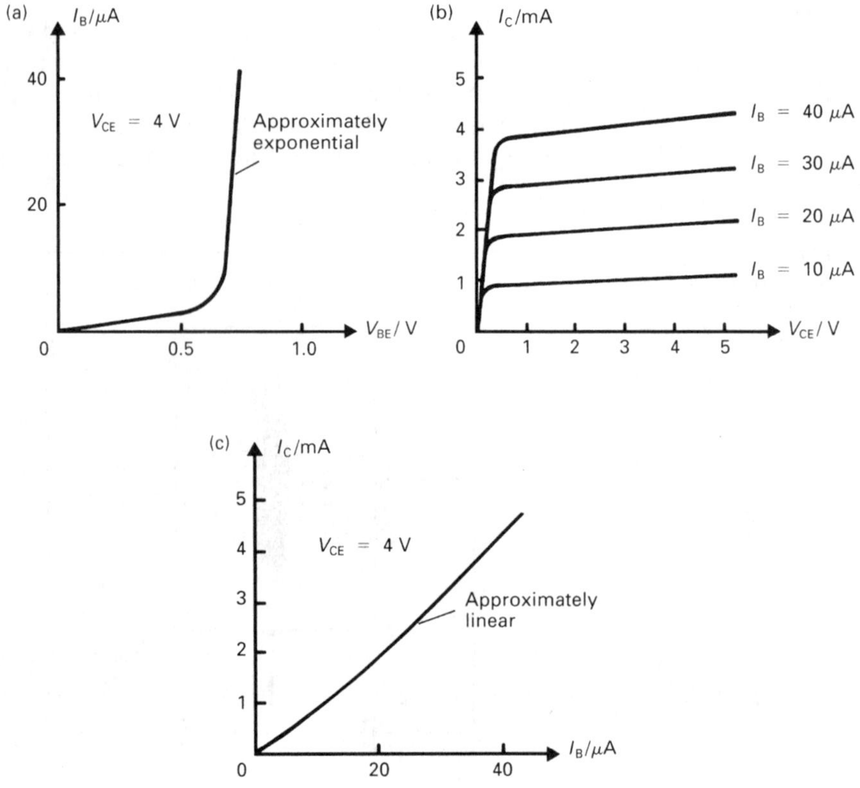

The Input (Base) Characteristics

These are plots of base current (I_B) against base–emitter voltage (V_{BE}) at fixed values of collector–emitter voltage (V_{CE}) – one such curve is shown in Fig. 55.17(a). The shape of the curve is very like that of a forward-biased junction diode (Fig. 55.11(b)). The similarity arises because I_B, though much smaller than the emitter current, I_E, is an almost constant fraction of it, and I_E is the current which, under the influence of V_{BE}, is flowing across the base–emitter junction, and this is essentially a forward-biased junction diode. Note that I_B is very small for values of V_{BE} which are less than the contact potential (~0.5 V for silicon).

The input (base) resistance is defined by

$$\text{Input resistance} = \frac{\Delta V_{BE}}{\Delta I_B} \qquad \text{(typically } 10^3\,\Omega\text{)}$$

where ΔI_B is the small increase in I_B which corresponds to a small increase, ΔV_{BE}, in V_{BE}. The curve is not linear, and therefore the value of the input resistance depends on the value of V_{BE} at which it is measured. It also has a slight dependence on V_{CE}. Under normal operating conditions, i.e. $V_{BE} > 0.5$ V, the input resistance is typically $10^3\,\Omega$.

The Output (Collector) Characteristics

These are shown in Fig. 55.17(b) and are plots of collector current (I_C) against collector–emitter voltage (V_{CE}) at fixed values of base current (I_B). For values of V_{CE} which are greater than about 1 V, I_C increases only slightly with V_{CE} but is strongly dependent on I_B. Transistors which are being used as amplifiers are operated at values of V_{CE} which are to the right of the knee. The curves are linear in this region and the amplifier produces an undistorted output.

The output resistance is defined by

$$\text{Output resistance} = \frac{\Delta V_{CE}}{\Delta I_C} \qquad \text{(often} > 10^5\,\Omega\text{)}$$

where ΔI_C is the increase in I_C which corresponds to an increase ΔV_{CE} in V_{CE} in the region to the right of the knee. The value of the output resistance depends on the particular value of I_B but, because I_C varies linearly with V_{CE} in this region, it is independent of V_{CE}. The output resistance is high (often $> 10^5\,\Omega$),* a feature which is brought about by the reverse bias across the collector–base junction.

The Transfer Characteristics

These are plots of collector current (I_C) against base current (I_B) at fixed values of collector–emitter voltage (V_{CE}). One such curve is shown in Fig. 55.17(c). The graph can be obtained from Fig. 55.17(b) and is approximately linear, reflecting the fact that I_B is an almost constant fraction of I_C (and I_E).

For a fixed value of V_{CE}, the **static value of the forward current transfer ratio,** h_{FE} (formerly called the **DC current gain**, β) is defined by

$$h_{FE}(=\beta) = \frac{I_C}{I_B}$$

Its value is typically in the range 20–200, but it can be much higher.

*The curves shown in Fig. 55.17(b) correspond to an output resistance of only about $10^4\,\Omega$ at $I_B = 40\,\mu$A.

The small signal forward current transfer ratio, h_{fe}, is defined by

$$h_{fe} = \frac{\Delta I_C}{\Delta I_B}$$

where ΔI_C is the change in collector current brought about by a change ΔI_B in base current at a fixed value of collector–emitter voltage.

Since the transfer characteristic approximates to a straight line through the origin

$$\frac{\Delta I_C}{\Delta I_B} \approx \frac{I_C}{I_B} = h_{FE}$$

55.10 EXPERIMENTAL DETERMINATION OF COMMON-EMITTER CHARACTERISTICS

The common-emitter characteristics of an n–p–n transistor can be obtained by using the circuit shown in Fig. 55.18. R_1 and R_3 may be high-resistance rheostats or wire-wound potentiometers. R_2 is a safety feature; its value should be such that even when R_1 is providing the maximum PD the base current does not exceed the maximum allowable for the particular transistor being studied. The voltmeters should have very high resistances – DC valve voltmeters are ideal. Alternatively the voltmeters may be connected between X and X′, and between Y and Y′ and corrections made for the PDs across the ammeters. (For example if the microammeter has a resistance of 1000 Ω and is registering a current of 20 μA, the PD across it is $20 \times 10^{-6} \times 10^3 = 0.02$ V, and it is necessary to subtract this from the voltmeter reading to obtain the true value of V_{BE}.)

Fig. 55.18
Circuit to investigate n–p–n transistor characteristics in the common-emitter mode

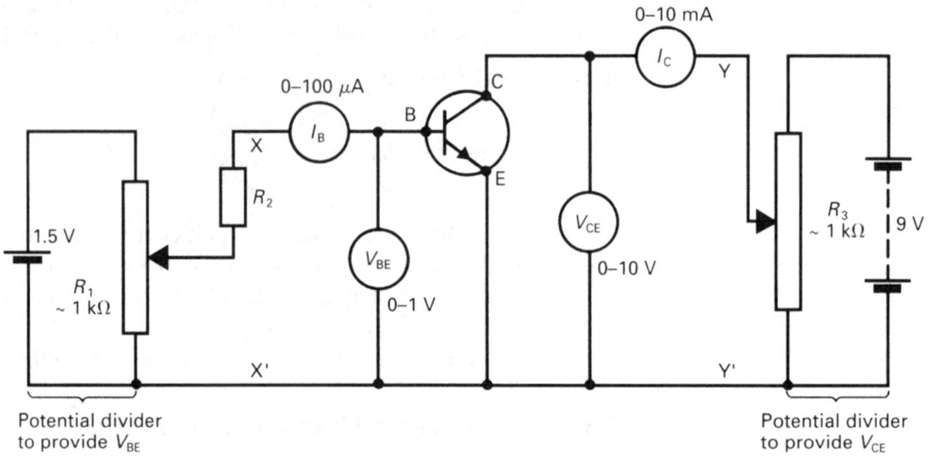

To Obtain the Input Characteristics (I_B–V_{BE})

Select a suitable value of V_{CE} by adjusting R_3. With V_{CE} fixed at this value vary R_1 to obtain a series of values of I_B at different values of V_{BE}.

To Obtain the Output Characteristics (I_C–V_{CE})

Select a suitable value of I_B by adjusting R_1. With I_B fixed at this value vary R_3 to obtain a series of values of I_C at different values of V_{CE}. Repeat for different values of I_B to give a set of curves.

To Obtain the Transfer Characteristics (I_C–I_B)

Once the output characteristics have been plotted, corresponding pairs of values of I_C and I_B can be read off for any particular value of V_{CE}.

55.11 THE TRANSISTOR AS AN AMPLIFIER IN THE COMMON-EMITTER CONNECTION

Examination of the transfer characteristics of a transistor in common-emitter connection (Fig. 55.17(c)) reveals that a small change in base current produces a large change in collector current. For example, in the particular case shown, a change of 1 mA in collector current is brought about by a change of less than 10 μA in base current. It follows that if a small alternating current is superimposed on the steady base current, it will cause a much larger alternating current to be superimposed on the steady collector current, i.e. the transistor can amplify a current change. Furthermore, since the transfer characteristic is approximately linear, the amplification produces very little distortion, i.e. the amplified current has nearly the same waveform as the original current.

The common-emitter circuit shown in Fig. 55.19 can be used to amplify small voltage changes. A single battery provides both the base–emitter bias and the collector–emitter bias – a feature which is made possible by the presence of the base bias resistor. The alternating PD which is to be amplified is applied across the input terminals. This causes a small alternating current to be superimposed on the steady current I_B flowing into the base from the bias resistor. This results in a large alternating current being superimposed on the collector current, I_C, and therefore a large alternating PD is superimposed on the steady PD across the load resistor. Since the PD between X and Y is constant ($= V_S$), there is an equal and opposite alternating PD superimposed on the steady PD between emitter and collector. C_2 acts as a blocking capacitor and ensures that only the alternating component appears at the output terminals. C_1 allows the alternating input current to pass to the base and, at the same time, ensures that the steady biasing current from the bias resistor flows into the transistor and not into the AC input.

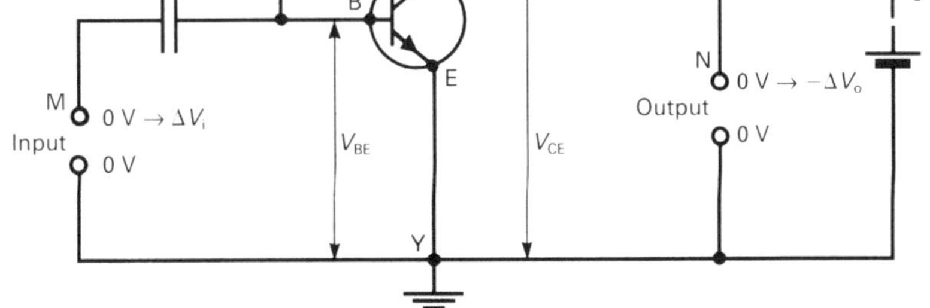

Fig. 55.19
Transistor used as a voltage amplifier

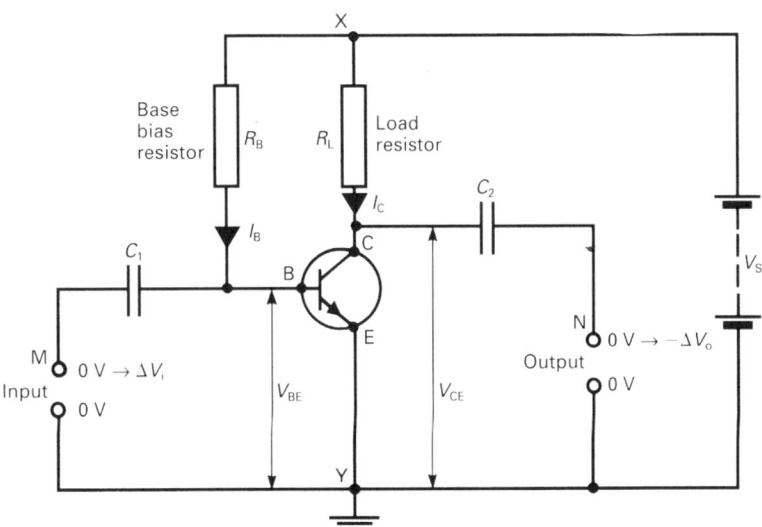

We shall now examine the circuit in more detail by considering a numerical example. In order to do so we shall assume that the characteristics of the transistor are those shown in Fig. 55.17 and that it is required to operate the amplifier such that in the absence of an input signal the steady collector current $I_C = 2$ mA, the collector–emitter voltage $V_{CE} = 4$ V and the supply PD $V_S = 9$ V. The resistance of the base bias resistor is R_B, that of the load resistor is R_L.

To Calculate R_L

From Fig. 55.19

$$V_S = I_C R_L + V_{CE}$$

$$\therefore \quad 9 = (2 \times 10^{-3}) R_L + 4$$

$$\text{i.e.} \quad R_L = 2.5 \times 10^3 \, \Omega$$

To Calculate R_B

From Fig. 55.17(b), when $I_C = 2$ mA and $V_{CE} = 4$ V, $I_B = 20\,\mu$A. From Fig. 55.17(a), when $I_B = 20\,\mu$A, $V_{BE} = 0.8$ V. From Fig. 55.19

$$V_S = I_B R_B + V_{BE}$$

$$\therefore \quad 9 = (20 \times 10^{-6}) R_B + 0.8$$

$$\text{i.e.} \quad R_B = 4.1 \times 10^5 \, \Omega$$

To Calculate the Voltage Amplification

Suppose that the PD across the input terminals increases from zero to ΔV_i and that its polarity is such that the upper terminal becomes positive. There will be a corresponding increase ΔI_B in the base current given by

$$\Delta I_B = \frac{\Delta V_i}{R_i + R_S}$$

where R_i is the input resistance of the base–emitter junction and R_S is the internal output resistance of the alternating supply. If the transfer characteristic is taken to be a straight line through the origin, then from section 55.9(c) the corresponding increase, ΔI_C, in the collector current is proportional to ΔI_B and is given by

$$\Delta I_C = h_{FE} \Delta I_B$$

$$\text{i.e.} \quad \Delta I_C = h_{FE} \frac{\Delta V_i}{R_i + R_S}$$

The increased collector current causes the PD across the load resistor to increase by $\Delta I_C R_L$, and therefore the voltage between the collector and the emitter decreases by ΔV_o, where

$$\Delta V_o = \Delta I_C R_L$$

$$\text{i.e.} \quad \Delta V_o = h_{FE} \frac{\Delta V_i}{R_i + R_S} R_L$$

i.e. the voltage amplification $\Delta V_o / \Delta V_i$ is given by

$$\frac{\Delta V_o}{\Delta V_i} = h_{FE} \frac{R_L}{R_i + R_S}$$

For this particular example $R_L = 2500\,\Omega$, $h_{FE} \approx 100$ (from Fig. 55.17(c)), $R_i \approx 1000\,\Omega$ (from Fig. 55.17(a) at $I_B = 20\,\mu$A), and therefore, if $R_S = 0$

$$\frac{\Delta V_o}{\Delta V_i} \approx 250$$

The current amplification $\Delta I_C / \Delta I_B$ is given by

$$\frac{\Delta I_C}{\Delta I_B} \approx h_{FE} \approx 100$$

It follows that the power amplification is approximately $100 \times 250 = 25\,000$.

Further Points

(i) The PD between collector and emitter has decreased and therefore the potential at the upper output terminal has changed from 0 V to $-\Delta V_o$. Thus, an increased potential at M has produced a decreased potential at N, i.e. there is a phase difference of π radians between the input and output signals.

(ii) R_i is the slope of the input characteristic (Fig. 55.17(a)) and depends on the PD between emitter and base. It follows that unless R_S is both constant and is much greater than R_i, the change in base current will not be even approximately proportional to the change in input voltage, in which case the waveform of the output voltage will not be a faithful reproduction of the input waveform. (In obtaining a numerical value for the voltage amplification we took R_S to be zero, and therefore the figure arrived at (250) could be achieved only at the expense of considerable distortion.)

(iii) The circuit shown in Fig. 55.19 is less complex than those actually used. In particular, there is no provision to prevent unwanted feedback of the amplified signal to the base emitter circuit, and there is nothing to prevent temperature changes producing undesirable effects through causing changes in base current.

55.12 THE TRANSISTOR AS A SWITCH

The common-emitter circuit shown in Fig. 55.20(a) acts as a switching circuit. The output voltage, V_o, is equal to V_{CE} (the PD between collector and emitter) and the way in which it depends on the input voltage, V_i, is shown in Fig. 55.20(b). When $V_i = 0$ there is no base–emitter bias (i.e. $V_{BE} = 0$) and therefore the base

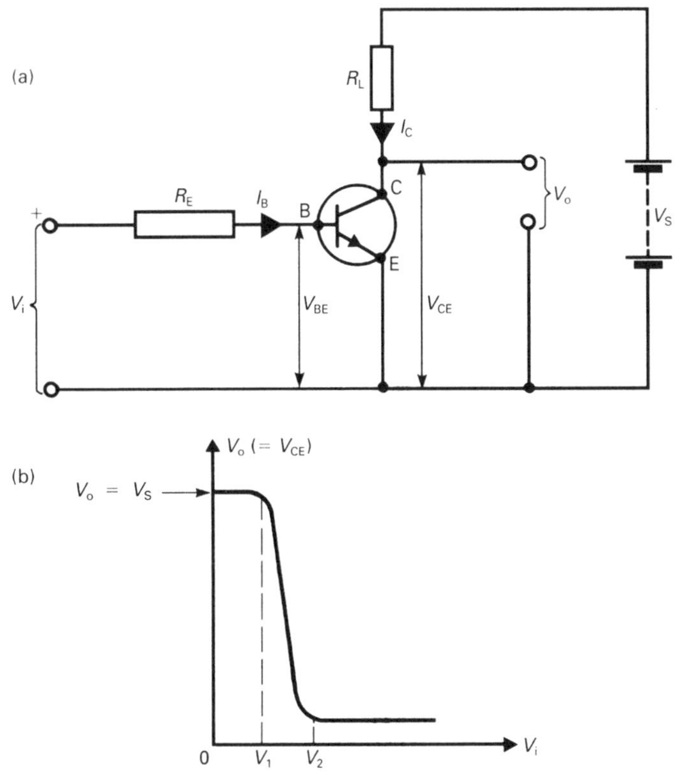

Fig. 55.20
(a) Transistor used as a switch, (b) variation of V_o with V_i

current, I_B, is zero. Since $I_B = 0$, the collector current, I_C, is also zero. From Fig. 55.20(a), under all circumstances,

$$V_o = V_S - I_C R_L \qquad [55.1]$$

and therefore since $I_C = 0$ when $V_i = 0$, it follows that when $V_i = 0$, $V_o = V_S$. If V_i is increased, then providing it does not exceed V_1 (Fig. 55.20(b)), there is very little increase in I_B (and therefore in I_C) and so V_o decreases only slightly. When it is in this state (i.e. $V_i \leq V_1$) the transistor is said to be **cut-off** and $I_C \approx 0$. If V_i is increased to V_2, I_C is appreciable and the PD across R_L is almost as large as the supply voltage, V_S. It follows from equation [55.1] that V_o is approximately zero when $V_i = V_2$. Increasing V_i beyond V_2 has very little effect on I_C, and therefore has very little effect on V_o. In this state (i.e. $V_i \geq V_2$) the transistor is said to be **saturated**.

Thus, by varying the input voltage, the transistor can be made to switch between two states – cut-off and saturation. The sharpness of the switching depends on the value of the resistance, R_E, in series with the base. Increasing R_E makes V_{BE} less sensitive to changes in V_i, and so increases the value of $V_2 - V_1$, i.e. it decreases the sharpness of the switching.

The ability of transistors to switch between two distinct states is made use of in digital computers. A single transistor can be switched many millions of times in one second. The output from one switch can be used as the input of a second switch, and interlinking a large number of these switches in this way makes it possible to carry out complex arithmetical calculations at high speed. The switching circuits are called logic gates, and these are discussed briefly in section 55.13.

55.13 LOGIC GATES

NOT Gate (Inverter)

The circuit shown in Fig. 55.20(a) acts as a NOT gate because, as explained in section 55.12, its output (V_o) is high only when its input (V_i) is not high. Thus, if an output (or input) voltage V_S is taken to represent the binary digit '1' and an output (or input) voltage of (approximately) zero is taken to represent the binary digit '0', then when the input = 1 the output = 0 and when the input = 0 the output = 1. (This system, in which the higher of the two voltages represents 1 and the lower voltage represents 0, is known as **positive logic**.) The circuit symbol for a NOT gate is shown in Fig. 55.21(a)* and the function of the gate is summarized in the table in Fig. 55.21(b), which is known as a **truth table**.

Fig. 55.21
(a) NOT gate, (b) its truth table

(a)

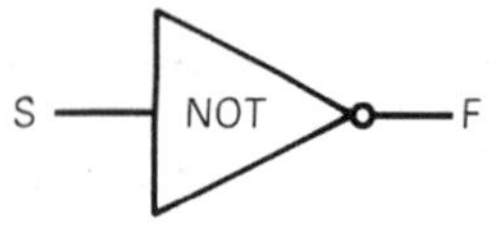

(b)

Input S	Output F
0	1
1	0

NOR Gate

The circuit of Fig. 55.22(a) acts as a NOR gate. It has two inputs (A and B) and the output is high only when both inputs are not high, i.e. when neither A nor B is high. The circuit symbol and truth table are shown in Fig. 55.22(b) and (c).

*The symbols for the various logic gates in Figs. 55.21 to 55.25 show identifying labels. This is not strictly necessary but has been done to assist the reader.

Fig. 55.22
(a) Circuit for NOR gate, (b) circuit symbol for NOR gate, (c) its truth table

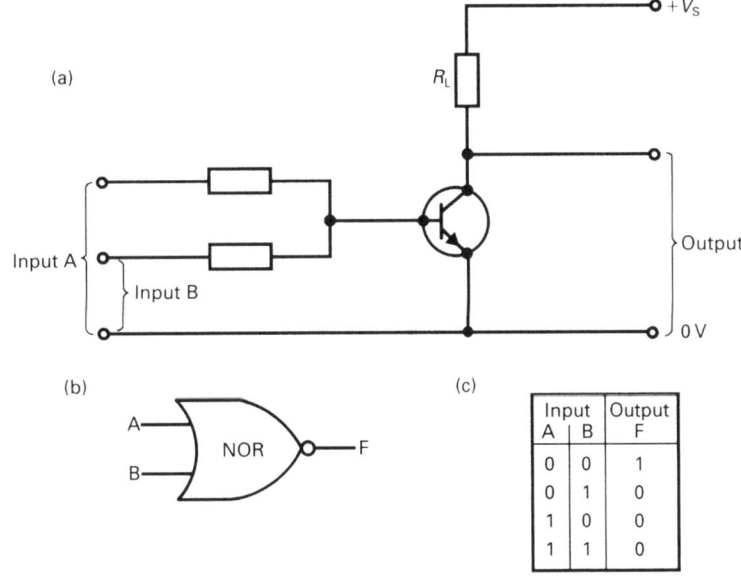

(a)

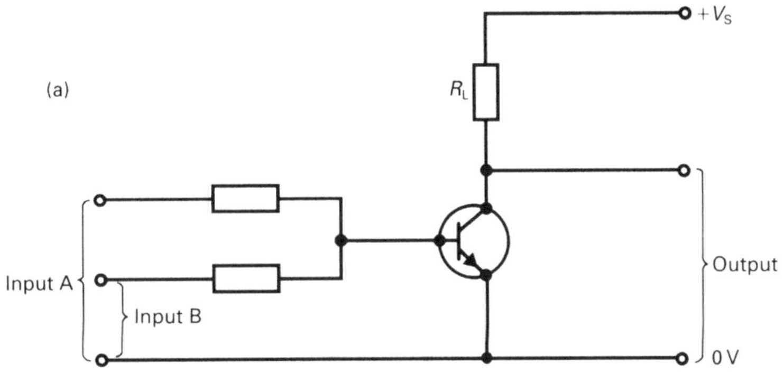

(b)

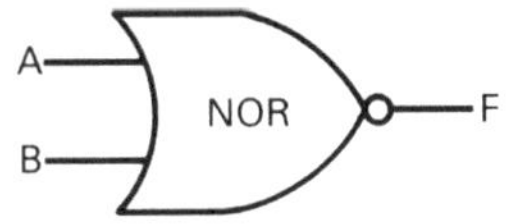

(c)

Input A	Input B	Output F
0	0	1
0	1	0
1	0	0
1	1	0

OR Gate

This has two inputs and operates in such a way that its output is high if one input or the other (or both) is high. It is actually a NOR gate followed by a NOT gate (Fig. 55.23(a)) and as such is the opposite of a NOR gate. The circuit symbol and truth table are shown in Fig. 55.23(b) and (c).

Fig. 55.23
(a) OR gate, (b) its circuit symbol, (c) its truth table

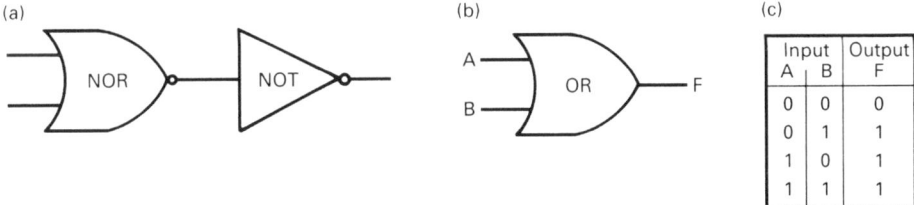

(a)

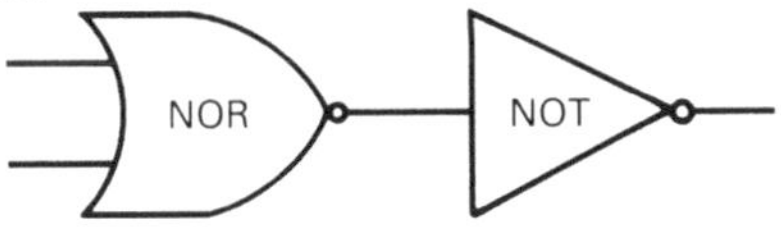

(b)

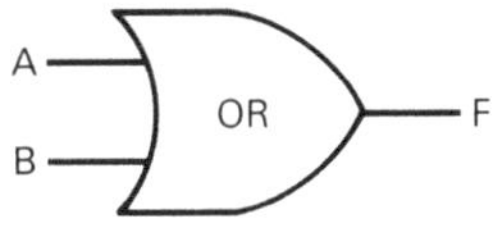

(c)

Input A	Input B	Output F
0	0	0
0	1	1
1	0	1
1	1	1

AND Gate

This has two inputs and operates in such a way that the output is high only if one input is high and the other is also high. It consists of two NOT gates followed by a NOR gate (Fig. 55.24(a)). The circuit symbol and truth table are shown in Fig. 55.24(b) and (c).

Fig. 55.24
(a) AND gate, (b) its circuit symbol, (c) its truth table

(a)

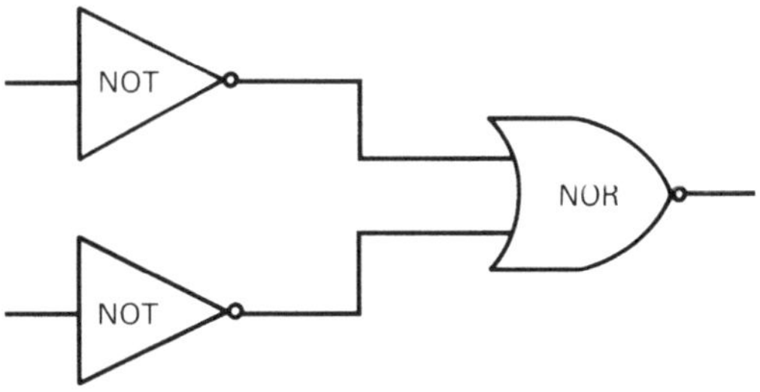

(b)

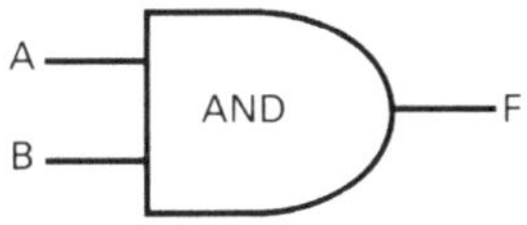

(c)

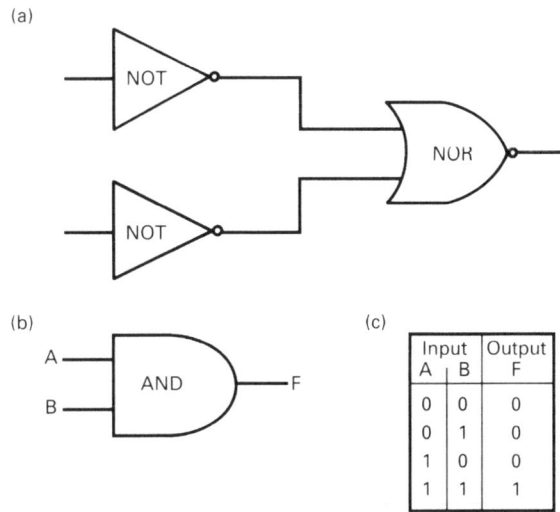

Input A	Input B	Output F
0	0	0
0	1	0
1	0	0
1	1	1

NAND Gate

This has two inputs and operates in such a way that the output is high if either input or both inputs are low. It is an AND gate followed by a NOT gate (Fig. 55.25(a)) and as such is the opposite of an AND gate. The circuit symbol and truth table are shown in Fig. 55.25(b) and (c).

Fig. 55.25
(a) NAND gate, (b) its circuit symbol, (c) its truth table

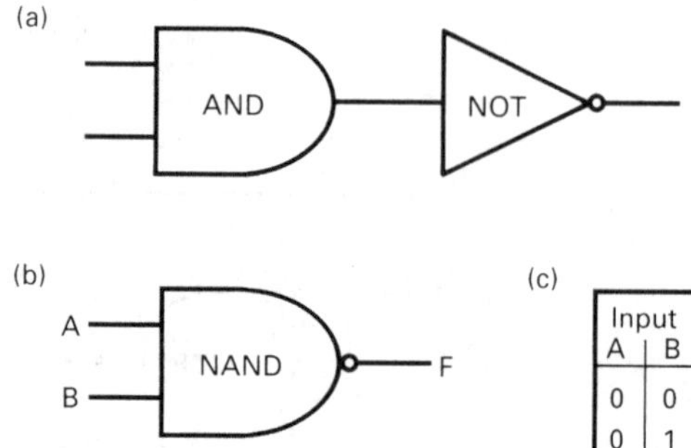

(c)

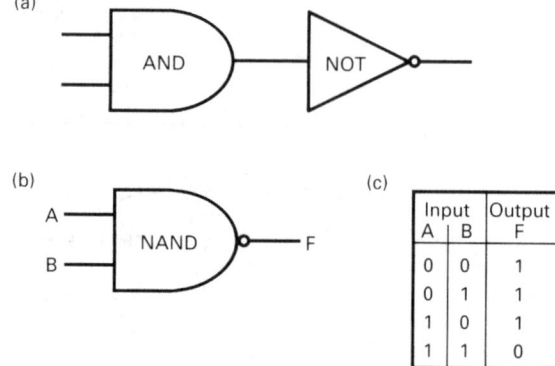

Input A	Input B	Output F
0	0	1
0	1	1
1	0	1
1	1	0

55.14 THE BISTABLE MULTIVIBRATOR (FLIP-FLOP)

A bistable multivibrator circuit is shown in Fig. 55.26. It consists of two transistors X and Y, each of which is connected in the common-emitter mode and which are arranged such that the collector of X is coupled to the base of Y via a resistance R_X, and such that the collector of Y is coupled to the base of X via a resistance R_Y. Thus the output from X is linked to the input of Y and the output from Y is linked to the input of X. R_{LX} and R_{LY} are the load resistances of X and Y respectively.

Fig. 55.26
Bistable multivibrator (or flip-flop) circuit

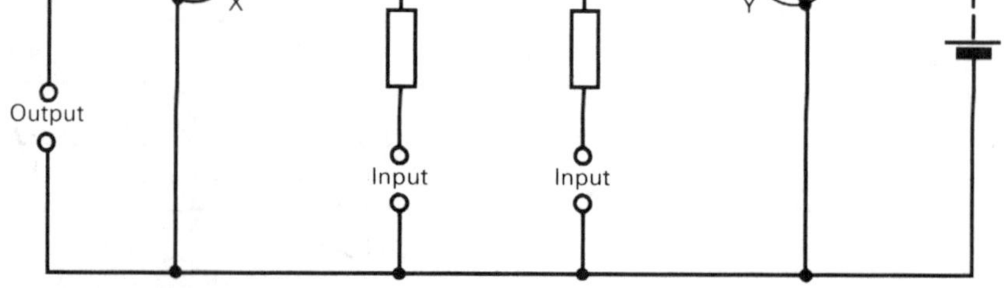

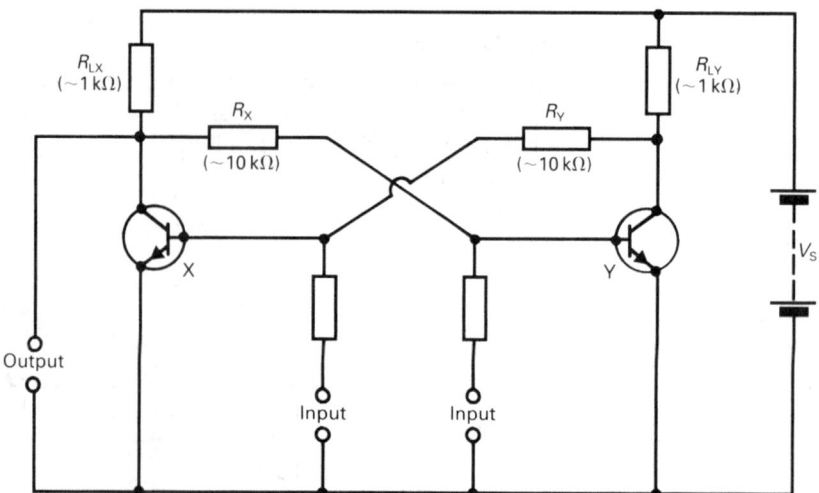

Suppose that initially the circuit conditions are such that there is no current flowing into (or out of) the base of X. Recalling what was said in section 55.12 we note that X is cut-off and that its output voltage is V_S. Providing R_X is not too high, the fraction of V_S which is across the input of Y will be sufficient to saturate it. Thus the output from Y is (approximately) zero, and so although this output is linked to the input of X, it is not sufficient to prevent X remaining cut-off.

Suppose now that an external signal is applied to the input on the left so that a current pulse flows into the base of X. The pulse is amplified by X and fed to Y where it is amplified even more. Since the output of Y is connected to the input of X, the amplified pulse is returned to X. Each transistor reverses the phase of the pulse and therefore the returned pulse has the same phase as the original – an effect which is known as **positive feedback**. The process continues with ever larger pulses repeatedly being passed from X to Y and then from Y to X until X saturates. When this happens the output from X is zero, and therefore Y is cut-off. Thus the circuit has been caused to 'flip' from its original state in which X was cut-off and Y was saturated to the opposite state – Y cut-off and X saturated. The transition takes place very rapidly. Both states are stable, and the circuit remains in whichever state it happens to be until it is triggered into changing by an external signal applied to the base of the transistor which is cut-off. The output, which can be taken from the collector of either transistor, is high when the circuit is in one state and is low when it is in the other state.

Bistable multivibrators are used as binary counters and as frequency dividers.

55.15 THE ASTABLE MULTIVIBRATOR

An astable multivibrator circuit is shown in Fig. 55.27. It consists of two transistors, X and Y, each of which is connected in the common-emitter mode. They are arranged such that the output from X is coupled to the input of Y via a capacitance C_X, and such that the output from Y is coupled to the input of X via a capacitance C_Y. R_{LX} and R_{LY} are the load resistances of X and Y respectively.

Fig. 55.27 Astable multivibrator circuit

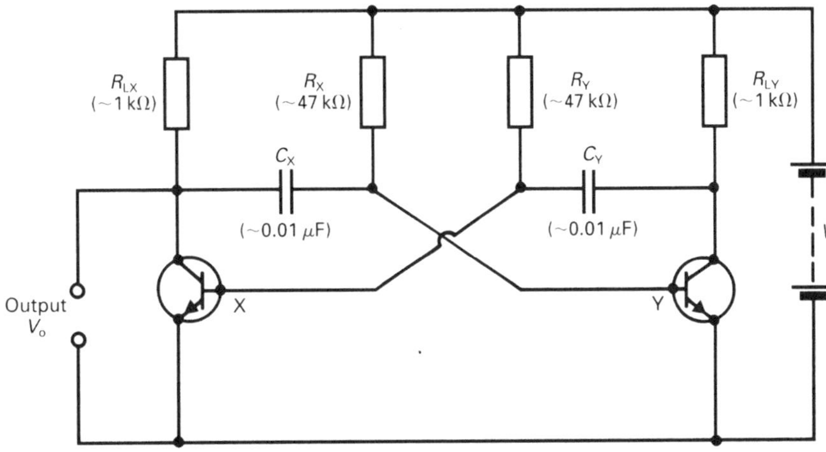

Whenever the circuit is on, one of the transistors is conducting (i.e. saturated) and the other is cut-off. When X is conducting C_X discharges through R_X and turns Y on, whereupon C_Y starts to discharge through R_Y and eventually turns X on again. This sequence repeats indefinitely with a frequency which is determined by the time constants $C_X R_X$ and $C_Y R_Y$. Thus the circuit has two quasi-stable states and switches periodically from one to the other. If the time constants are equal, the output voltage, V_o, has the form shown in Fig. 55.28. The output can be taken from the collector of either transistor; the output from X differs in phase with that from Y by π radians. Because the waveform is approximately rectangular the circuit is often referred to as a **square-wave oscillator**.

Astable multivibrators are used in the timing circuits of computers.

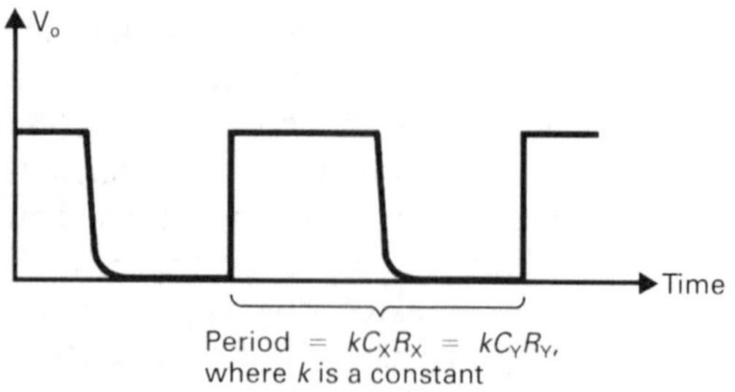

Fig. 55.28
Voltage output (approximately square-wave) from an astable multivibrator

55.16 THE MONOSTABLE MULTIVIBRATOR

A monostable multivibrator circuit is shown in Fig. 55.29. It consists of two transistors, X and Y, each of which is connected in the common-emitter mode. The output of X is connected to the input of Y via a resistance R_X, and the output of Y is connected to the input of X via a capacitance C_Y. The circuit has one stable state only, and this is such that X is conducting and Y is cut-off. The circuit can be switched into its unstable state (Y conducting, X cut-off) by applying a positive voltage pulse to the base of Y. Once Y is conducting, C_Y starts to discharge through R_Y, and eventually turns X on again, i.e. the circuit reverts to its stable state. The time that the circuit spends in the unstable state is proportional to C_YR_Y and, in particular, is independent of the size and shape of the triggering pulse. Fig. 55.30 shows how the output varies with time as a result of a trigger pulse being applied at the input. Since the shape of the output pulse is independent of the shape of the trigger pulse, the circuit can be used as a pulse-shaping circuit.

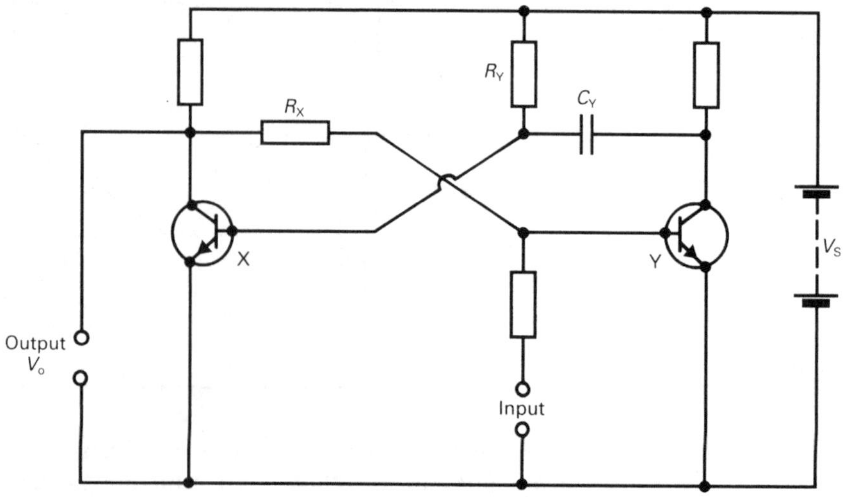

Fig. 55.29
Monostable multivibrator circuit

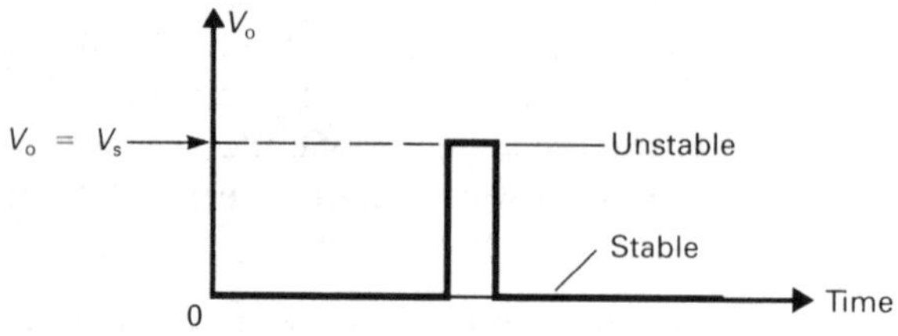

Fig. 55.30
The effect of a trigger pulse on the output voltage from a monostable multivibrator

55.17 THE OPERATIONAL AMPLIFIER

The term 'operational amplifier' was originally used in the field of analogue computing* to describe an amplifier circuit whose characteristics were such that it could be used to carry out mathematical operations such as addition, subtraction, differentiation and integration. Nowadays these amplifiers are available at modest prices (~50p) in integrated circuit form and because of this it is becoming increasingly popular to use them as general purpose amplifiers.

Operational amplifiers can amplify both DC and AC voltages and have high input resistances (typically 2 MΩ) and low output resistances (typically 200 Ω). They have very large voltage gains ($\sim 10^5$) at low frequencies.

The type 741 operational amplifier is commonly available and is manufactured as an 'eight-pin dual-in-line package' (see Fig. 55.31).

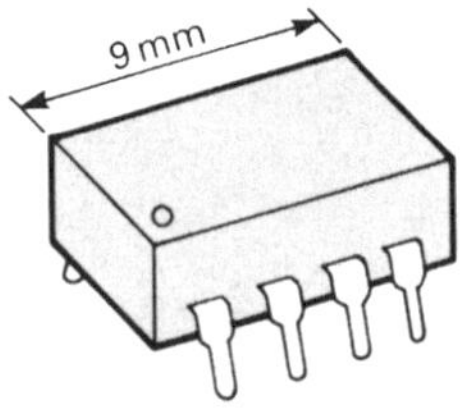

Fig. 55.31
A typical operational amplifier

One of the eight terminals is not used and two will not concern us; the other five are shown on the circuit symbol in Fig. 55.32 and are the inverting input (−), the non-inverting input (+), the output, and the positive and negative voltage supplies ($+V_S$ and $-V_S$ respectively).

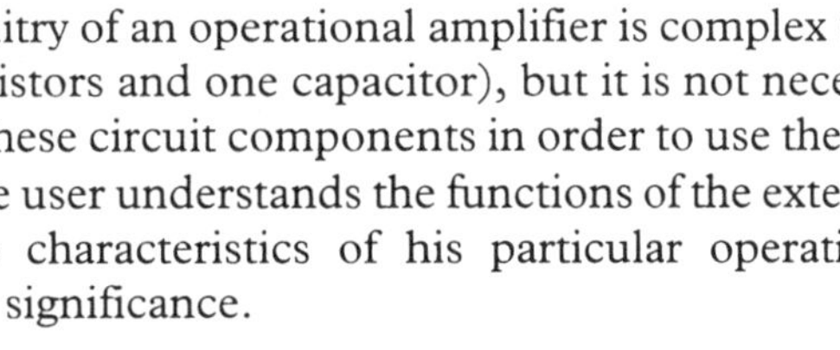

Fig. 55.32
Circuit symbol of an operational amplifier

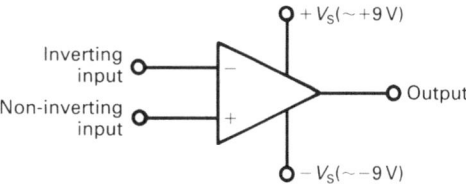

The internal circuitry of an operational amplifier is complex (the 741 contains 20 transistors, 11 resistors and one capacitor), but it is not necessary to understand the behaviour of these circuit components in order to use the amplifier. All that is required is that the user understands the functions of the external terminals and is familiar with the characteristics of his particular operational amplifier and understands their significance.

The complex circuitry of an operational amplifier can be represented by what is called its **equivalent circuit**. (An equivalent circuit is a theoretical circuit which is much simpler than the circuit it represents but which has the same input and

*__An analogue computer__ is distinct from a **digital computer** in that the data which it processes is input as a continuously varying quantity (usually a continuously varying voltage) rather than as a set of binary digits. For example, when an analogue computer is used to solve a differential equation the magnitudes of the variables in the equation are input as voltages and the output voltage is proportional to the solution of the equation.

output characteristics. The reader will already be familiar with the idea, though perhaps is unaware of it, in which case reference to Fig. 55.33 should convince.) The equivalent circuit of an operational amplifier is shown in Fig. 55.34. Thus an operational amplifier can be regarded as being a device which generates a voltage E given by

$$E = A(V_2 - V_1) \tag{55.2}$$

where A is known as the **open-loop gain** of the amplifier, and V_1 and V_2 are the voltages applied to the inverting and non-inverting inputs respectively. The value of A depends on the frequency of the input voltage(s) and is very high at low frequencies. (The DC open-loop gain of a 741 is 10^5.) Note that V_o, the PD appearing at the output, will be less than E because of the non-zero output resistance, R_o, of the amplifier.

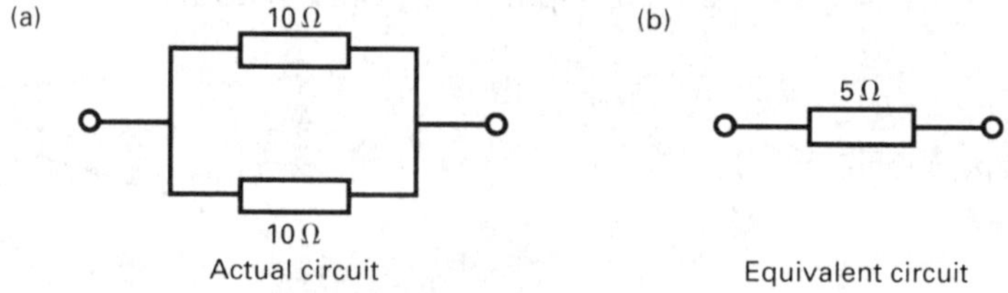

Fig. 55.33
(a) An actual circuit,
(b) its equivalent circuit

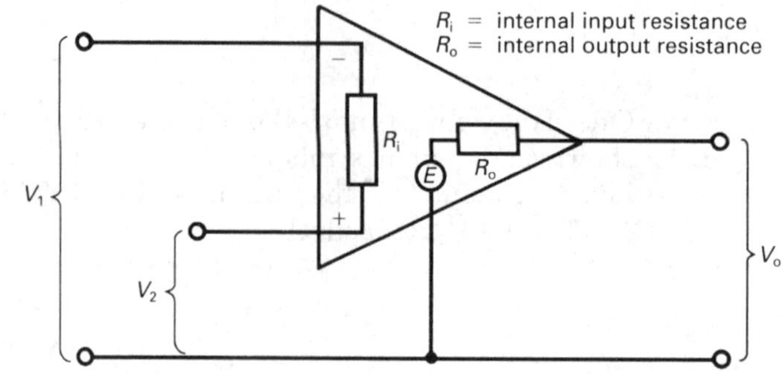

Fig. 55.34
Equivalent circuit of an operational amplifier

It follows from equation [55.2] that when $V_2 = 0$, $E = -AV_1$, i.e. **the output voltage is out of phase with the voltage applied to the inverting input** (hence the name). When $V_1 = 0$, $E = AV_2$, i.e. the output is in phase with the voltage applied to the non-inverting input.

The maximum value of the output cannot exceed about 80% of either the positive or the negative supply voltages. It follows that with supply voltages of +9 V and −9 V and a gain of 10^5 the output saturates when the PD between the input terminals is greater than about 70 μV.

55.18 THE OPERATIONAL AMPLIFIER AS A VOLTAGE AMPLIFIER WITH NEGATIVE FEEDBACK

Inverting Amplifier

In the circuit shown in Fig. 55.35 a fraction of the output signal of the operational amplifier is fed back to the inverting input by way of the feedback resistor. Because it is the inverting input which is being used the feedback is out of phase with the input signal and so the process is known as **negative feedback** because it results in

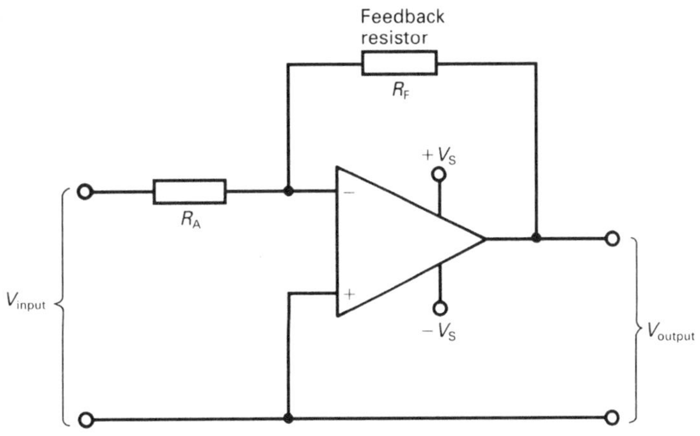

Fig. 55.35 Operational amplifier used with negative feedback as an inverting amplifier

a reduction in the overall voltage gain. Although the gain is reduced, the system has the advantage of having a gain which is independent of the characteristics of the amplifier itself. As such the gain is highly stable and is constant over a large range of input voltages and frequencies. It can be shown that providing both R_F and R_A are small in comparison with the input resistance of the operational amplifier (2 MΩ in the case of a 741) and that the open-loop gain is very large

$$\frac{V_{output}}{V_{input}} = -\frac{R_F}{R_A} \qquad [55.3]$$

The minus sign takes account of the phase inversion, i.e. V_{output} is negative when V_{input} is positive and vice versa. Equation [55.3] holds for positive and negative inputs which are not so large that the amplifier saturates. If $V_S = 9\,V$, the maximum output voltage will be plus or minus about 7 V, in which case if the overall gain is −10 (e.g. $R_F = 100\,k\Omega$, $R_A = 10\,k\Omega$), saturation occurs for input voltages of plus or minus 0.7 V. A typical output/input characteristic is shown in Fig. 55.36. Note that the gain is exactly linear between the two saturated states. The characteristic can be obtained by using the circuit of Fig. 55.37; this is basically the same as that of Fig. 55.35 but shows the necessary power supplies, etc. In practice the positive and negative saturation voltages may not be equal. The asymmetry is associated with the fact that the meter used to measure V_{output} acts as a highly resistive load.

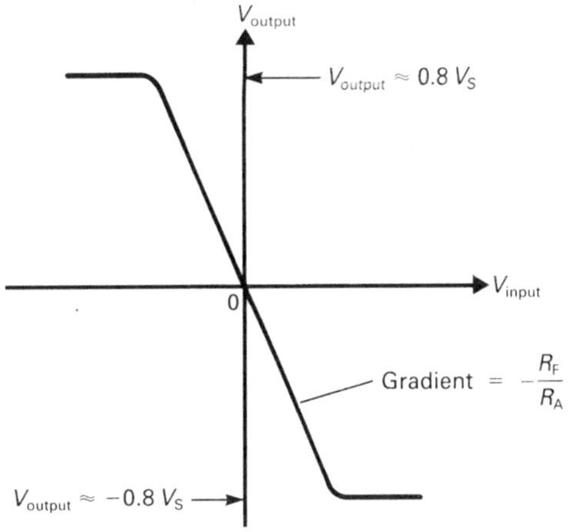

Fig. 55.36 Output–input characteristics for the circuit of Fig. 55.35

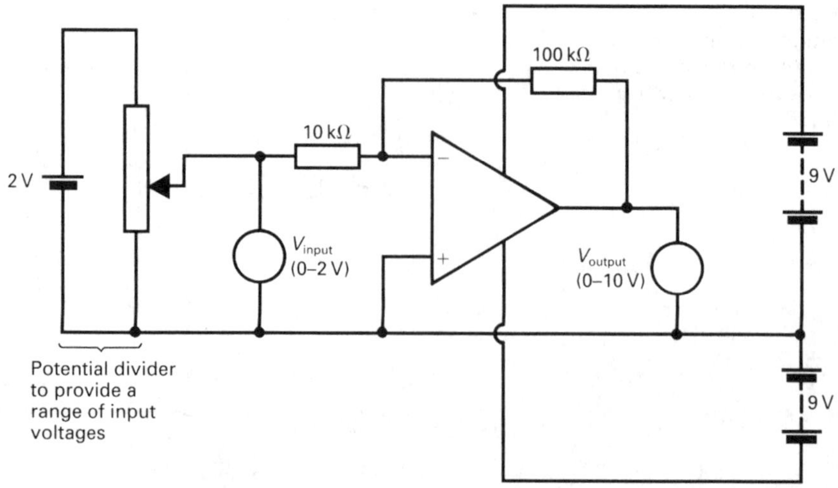

Fig. 55.37
Circuit to obtain characteristics of Fig. 55.36

For a 741 with a DC input, equation [55.3] is valid for values of R_F/R_A which are less than about 10^3. The equation is valid for AC inputs with frequencies of less than about 10 kHz provided R_F/R_A is limited to about 10.

Non-inverting Amplifier

In the circuit shown in Fig. 55.38 the two resistors form a potential divider so that a fraction $R_A/(R_A + R_F)$ of the output voltage is fed back to the inverting input by way of R_F. As with the circuit shown in Fig. 55.35, the signal is being fed back to the inverting input and therefore the process is one of negative feedback. In this case, though, the input signal is applied to the non-inverting input and this gives an output voltage which is in phase with the input voltage. It can be shown that provided both R_F and R_A are small in comparison with the input resistance of the operational amplifier (2 MΩ in the case of a 741) and that the open-loop gain is very large:

$$\frac{V_{output}}{V_{input}} = \frac{R_A + R_F}{R_A} = \frac{1}{\beta}$$

where β is the fraction of the output that is fed back to the input. Thus the gain depends only on the resistances R_F and R_A. It is therefore independent of the characteristics of the amplifier itself, and so the amplifier is highly stable and the gain is constant over a large range of voltages and frequencies. A typical output–input characteristic is shown in Fig. 55.39 and may be obtained by using a suitably modified form of Fig. 55.37.

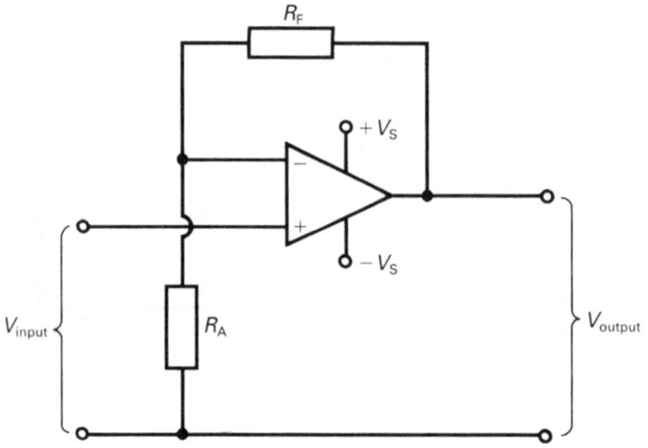

Fig. 55.38
Operational amplifier used with negative feedback as a non-inverting amplifier

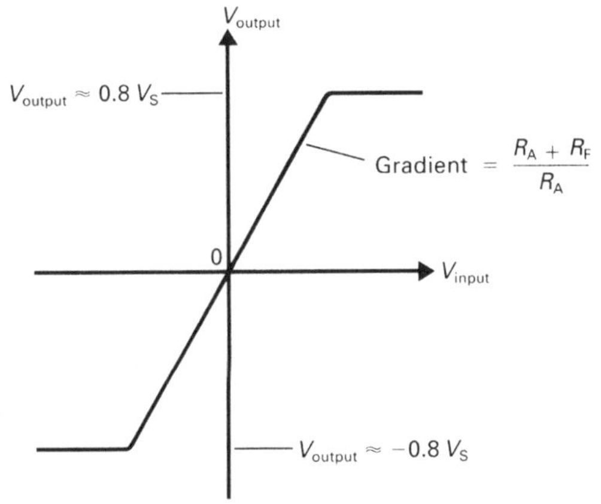

Fig. 55.39 Output–input characteristic for the circuit in Fig. 55.38

55.19 THE OPERATIONAL AMPLIFIER AS A VOLTAGE COMPARATOR

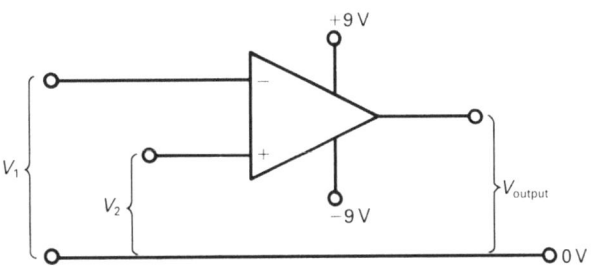

A voltage comparator circuit is used to compare the size of one voltage with that of another. A typical application is to sense when a varying input signal has reached the level of a steady reference voltage. Operational amplifiers can be used in such circuits, in which case the voltages being compared (V_1 and V_2) are connected to the two inputs of the amplifier (see Fig. 55.40). The operational amplifier is being used in the open-loop configuration, and therefore a difference of about 70 μV between V_1 and V_2 is sufficient to saturate it if a 9 V supply is being used (see section 55.17). Suppose that $V_1 = 1$ V and that V_2 is a sinusoidally varying voltage whose maximum value is 2 V. When V_2 is more than about 70 μV greater than V_1 (i.e. > 1.000 07 V) the amplifier saturates and the output voltage has a constant value of about 7 V. When V_2 is more than about 70 μV less than V_1 (i.e. < 0.999 93 V) the amplifier is again saturated and the output has a constant value of about −7 V. Thus the circuit switches from one saturated state to the other as the input swings above and below the reference level. The situation is illustrated in Fig. 55.41 and can be investigated by using the circuit of Fig. 55.42. The alternating signal and the output are viewed simultaneously on a double-beam CRO. The 10 kΩ potential divider enables V_1 to be set to the desired value. The 2.7 kΩ resistor is necessary to provide a DC path to ground if the signal generator is AC-coupled.

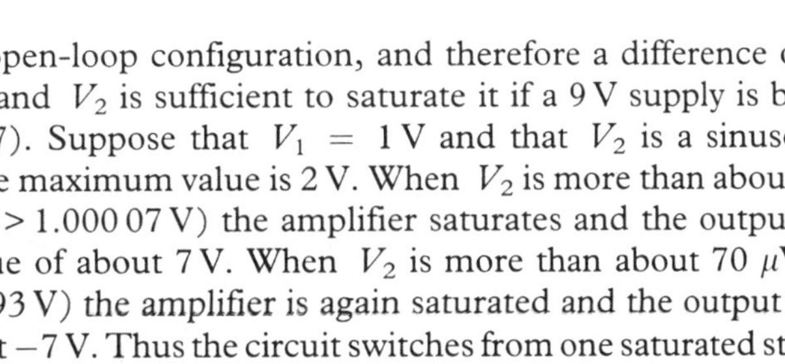

Fig. 55.40 Operational amplifier as a voltage comparator

The circuit shown in Fig. 55.43 is a voltage comparator circuit. The PD across the series combination $R_1 + R_2$ is the same as that across $R_3 + R_4$, and it therefore follows that if $R_1/R_2 > R_3/R_4$, $V_1 > V_2$, in which case the output voltage V_o is negative. If $R_1/R_2 < R_3/R_4$, $V_1 < V_2$, and V_o is positive.

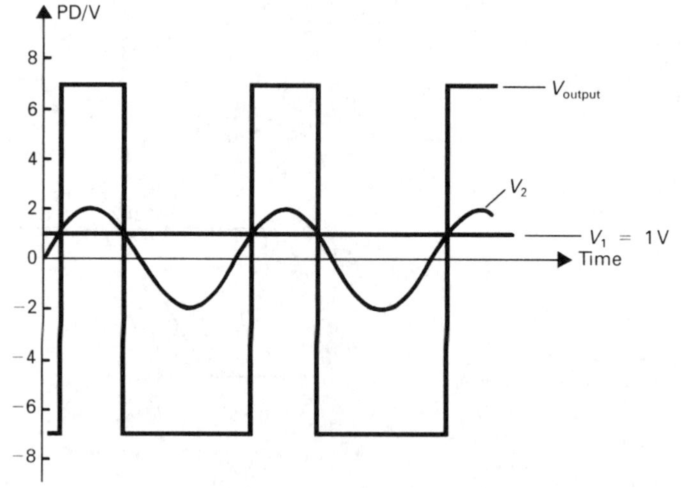

Fig. 55.41
To illustrate the action of the circuit in Fig. 55.40

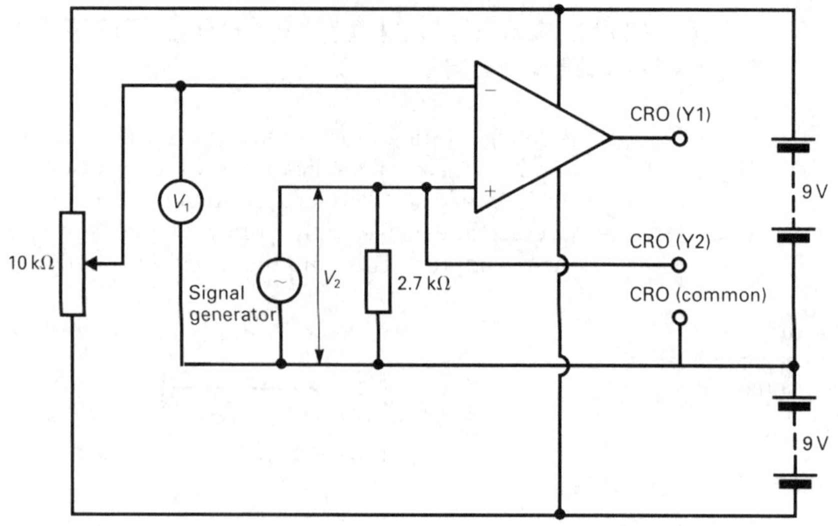

Fig. 55.42
Circuit to obtain Fig. 55.41

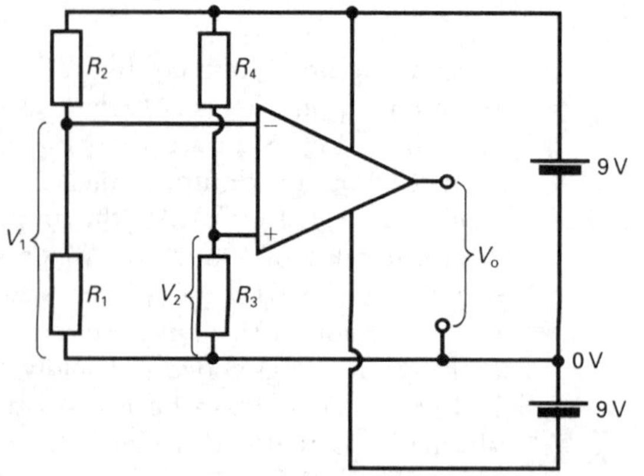

Fig. 55.43
A voltage comparator circuit

If R_1 (say) is a light-dependent resistor (LDR) or a thermistor, the output voltage will flip from one saturated state to the other as the light level or temperature rises or falls above or below some critical value determined by the values of R_2, R_3, R_4 and the characteristics of the particular LDR or thermistor. Used in conjunction with a relay the circuit can be caused to switch on or off some auxiliary lighting or heating system.

The circuit also provides a very sensitive means of measuring the value of an unknown resistance. If R_1, R_2, R_3 and R_4 are the four resistance arms of a Wheatstone bridge circuit, $V_o = 0$ only when $R_1/R_2 = R_3/R_4$ exactly. Since V_1 and V_2 need differ by only about 70 μV to saturate the amplifier, any slight deviation from this balance condition will cause the output to change markedly. Thus the operational amplifier and the meter used to measure V_o are taking the place of the galvanometer in a conventional Wheatstone bridge circuit.

55.20 THE OPERATIONAL AMPLIFIER AS AN ASTABLE MULTIVIBRATOR

The circuit is shown in Fig. 55.44. It produces a square-wave output of period T, where

$$T = 2CR\log_e\left(1 + \frac{2R_2}{R_1}\right)$$

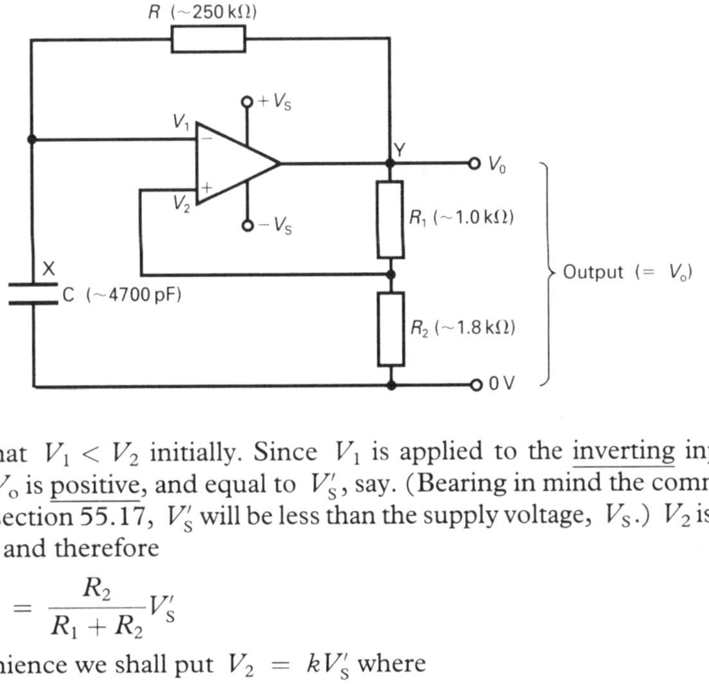

Fig. 55.44
Operational amplifier as an astable multivibrator

Suppose that $V_1 < V_2$ initially. Since V_1 is applied to the inverting input and $V_1 < V_2$, V_o is positive, and equal to V'_S, say. (Bearing in mind the comments at the end of section 55.17, V'_S will be less than the supply voltage, V_S.) V_2 is the PD across R_2, and therefore

$$V_2 = \frac{R_2}{R_1 + R_2} V'_S$$

For convenience we shall put $V_2 = kV'_S$ where

$$k = \frac{R_2}{R_1 + R_2}$$

At any stage, the potential at X (the upper plate of the capacitor) is V_1. The potential at Y is V'_S, and therefore the potential at X is less than that at Y.* The capacitor will therefore be being charged through R, and but for the action of the amplifier it would continue to be charged until the potential at X became equal to that at Y. However, as soon as the potential at X exceeds kV'_S, V_1 is greater than V_2 and the output voltage V_o becomes negative and equal to $-V'_S$, whereupon V_2 becomes $-kV'_S$. The potential at X is now greater than that at Y. The capacitor starts to discharge through R and the potential at X falls. As soon as it reaches $-kV'_S$, V_1 is again less than V_2, and therefore the output flips back to its original value of $+V'_S$ and the cycle starts again. The situation is shown graphically in Fig. 55.45.

*$V_1 < V_2 = kV'_S$, where $k < 1$; $\therefore$ $V_1 < V'_S$, i.e. the potential at X is less than the potential at Y.

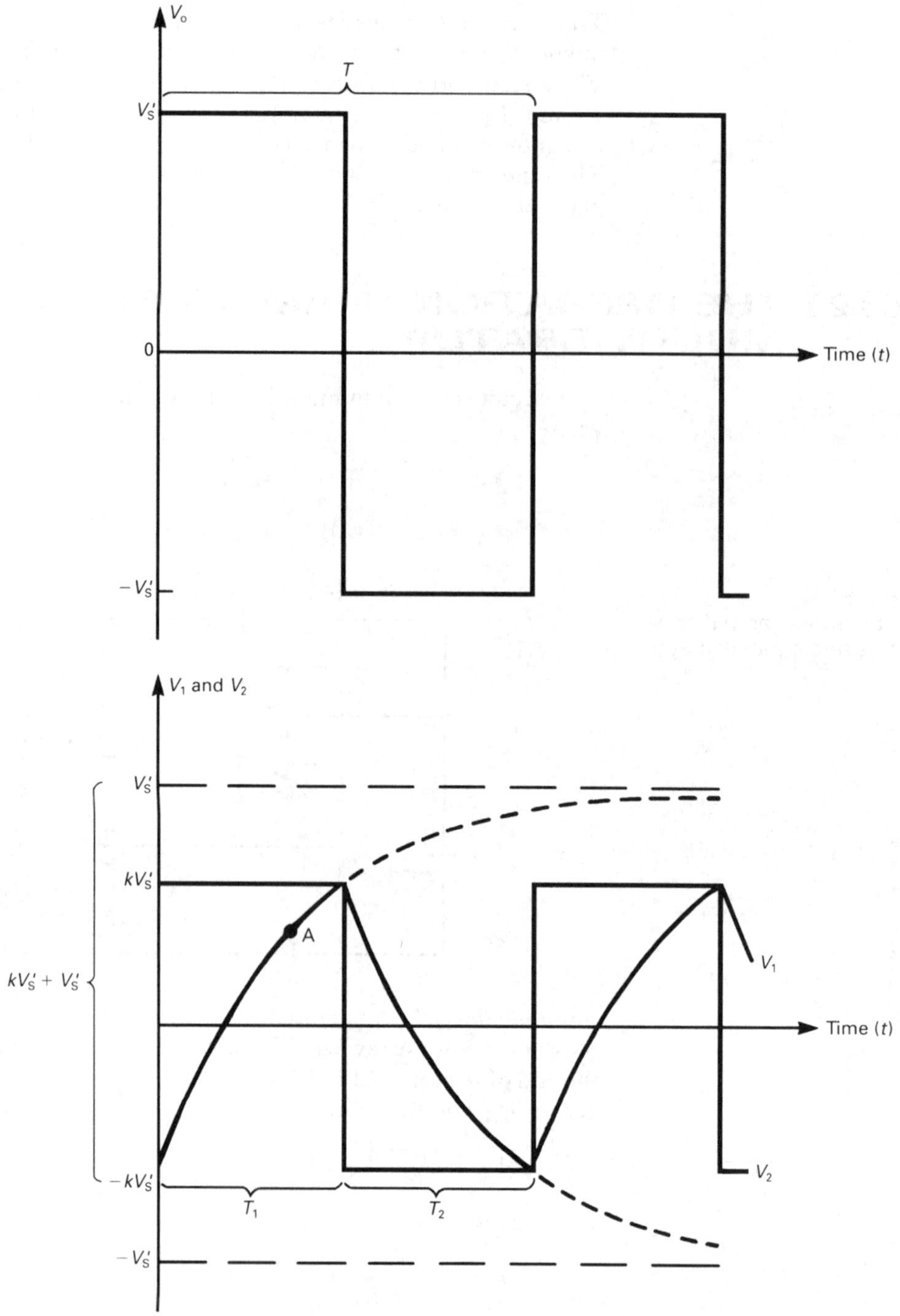

Fig. 55.45 Waveforms for circuit of Fig. 55.43

To show that $T = 2CR\log_e(1 + 2R_2/R_1)$

Consider the situation at some time t when the capacitor is charging, e.g. a point such as A (Fig. 55.45). The potential at the upper plate of the capacitor is in the process of changing from $-kV_S'$ to V_S', a change of $(1 + k)V_S'$, and therefore, since at $t = 0$, $V_1 = -kV_S'$,

$$V_1 = -kV_S' + (1 + k)V_S'(1 - e^{-t/CR})$$

Switching occurs when $V_1 = kV'_S$ and $t = T_1$, and therefore

$$kV'_S = -kV'_S + (1+k)V'_S(1 - e^{-T_1/CR})$$

$$\therefore \quad \frac{2k}{1+k} = 1 - e^{-T_1/CR}$$

$$\therefore \quad e^{-T_1/CR} = 1 - 2k/(1+k)$$

$$= (1-k)/(1+k)$$

$$\therefore \quad e^{T_1/CR} = (1+k)/(1-k)$$

$$\therefore \quad T_1 = CR\log_e\left(\frac{1+k}{1-k}\right)$$

The discharge time T_2 is equal to T_1 because the charging and discharging processes take place under equivalent conditions. The total period T is therefore equal to $2T_1$, i.e.

$$T = 2CR\log_e\left(\frac{1+k}{1-k}\right)$$

Since $k = R_2/(R_1+R_2)$, $(1+k) = (R_1+2R_2)/(R_1+R_2)$ and $(1-k) = R_1/(R_1+R_2)$; therefore

$$T = 2CR\log_e\left(\frac{R_1+2R_2}{R_1}\right)$$

i.e. $$T = 2CR\log_e(1 + 2R_2/R_1)$$

55.21 THE WIEN BRIDGE OSCILLATOR

The frequency of oscillation of the multivibrator circuit of section 55.20 is determined by the time-constant of the R–C combination. In the oscillator which is about to be described the frequency of oscillation is determined by the resonant frequency of the feedback circuit.

Fig. 55.46(a) shows an operational amplifier version of a Wien bridge oscillator. The circuit produces an output voltage V_o which varies sinusoidally with frequency f where

$$f = \frac{1}{2\pi RC}$$

The oscillator makes use of positive feedback, i.e. a fraction of the output from the operational amplifier is fed back to its non-inverting input in phase with the voltage already at the input. This is then amplified to offset the attenuation introduced by the feedback loop and reappears at the output.

An integral part of the circuit is the combination of resistors and capacitors shown in Fig. 55.46(b) and known as a **Wien network**. It will be useful to discuss the properties of this network before explaining the operation of the oscillator circuit itself.

The Wien network is an AC potential divider. In Fig. 55.46(b) an alternating voltage V_o is applied across the whole network and an alternating voltage V_f appears across the parallel R–C combination. In general there will be a phase difference between V_f and V_o (see Chapter 43) but it can be shown that at one frequency only, namely $1/(2\pi RC)$, V_f and V_o are in phase with each other. It can also be shown that at this frequency V_f has its maximum value of $V_o/3$.

Fig. 55.46
(a) Operational amplifier version of a Wien bridge oscillator. (b) The Wien network

(a)

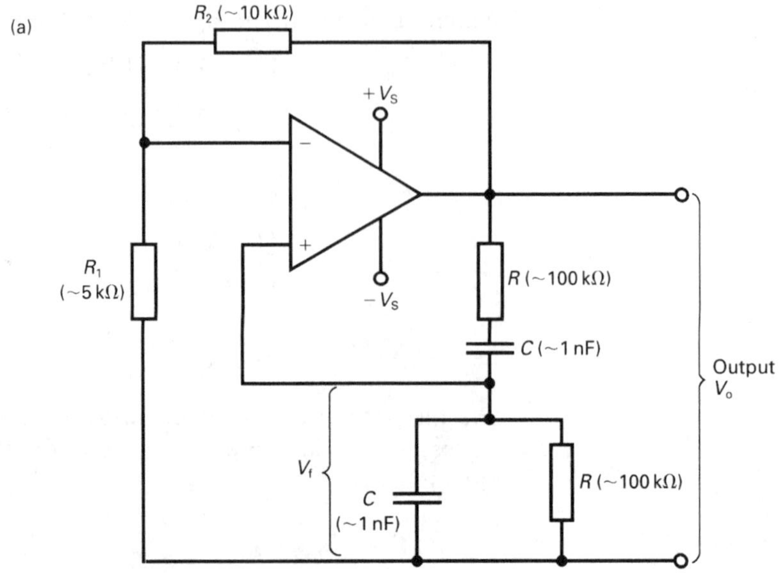

Note. The ratio R_2/R_1 is critical and in practice is unlikely to be exactly 2 since the values of R and C in the Wien network are unlikely to be exactly as stated by the manufacturer. At least one of R_1 and R_2 should therefore be provided by a resistance box so that the required ratio (close to 2) can be found by trial and error.

(b)

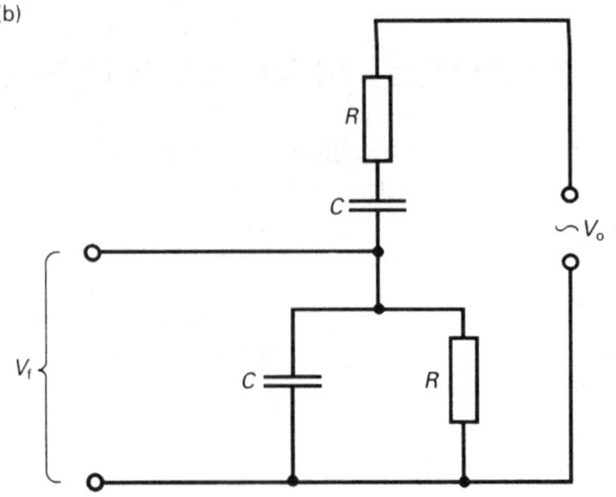

In Fig. 55.46(a) the input to the Wien network is the output from the operational amplifier. When the oscillator circuit is first switched on small stray currents (noise) of many different frequencies start to flow and produce an output voltage V_o which is itself comprised of voltages of a whole range of frequencies. A fraction V_f/V_o of this output is fed back to the non-inverting input of the operational amplifier. The feedback is in phase with the signal at the input (i.e. positive) only for that component of the output voltage which is at a frequency of $1/(2\pi RC)$. It follows that providing the amplifier is producing sufficient gain, oscillations of this frequency increase in amplitude; those of other frequencies do not.

At a frequency of $1/(2\pi RC)$, $V_f = \frac{1}{3}V_o$, i.e. only one-third of the output is fed back to the non-inverting input. In order to offset this attenuation the operational amplifier needs to produce a voltage gain of 3. The gain is provided by way of the negative feedback applied to the inverting input. Bearing in mind that the signal being amplified is applied to the non-inverting input, we see that the operational amplifier is being used as a non-inverting amplifier with negative feedback (see section 55.18). The gain is therefore $(R_1 + R_2)/R_1$ and so we require

$$\frac{R_1 + R_2}{R_1} = 3$$

i.e. $R_2 = 2R_1$

55.22 THE LIGHT-DEPENDENT RESISTOR

The light-dependent resistor (or LDR) is a semiconductor device whose resistance depends on the amount of light falling on it – the greater the intensity of the light (visible, ultraviolet or infrared), the lower the resistance. The circuit symbol is shown in Fig. 55.47.

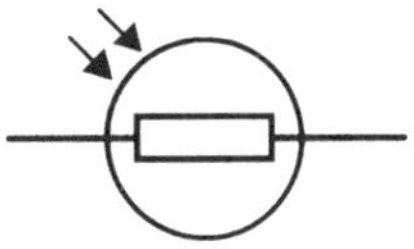

Fig. 55.47
Circuit symbol of LDR

A typical LDR consists of a pair of grid-like electrodes evaporated on to the surface of a semiconductor such as cadmium sulphide. The resistance of the cadmium sulphide decreases when light falls on it because the light provides energy to release bound electrons from their atoms, and so increases the number of electrons and holes available for conduction.

A commonly used LDR (the ORP12) has a diameter of 14 mm and a resistance which ranges from about 100 Ω in bright light to 10 MΩ in the dark. The response time for a change of this magnitude is of the order of 100 ms.

Light-dependent resistors can be used as photographic exposure meters or, in conjunction with relays, as light-activated switches.

55.23 THE LIGHT-EMITTING DIODE

The light-emitting diode (or LED) is a p–n junction diode which emits light when a current (typically 20 mA) flows through it in the forward direction. The circuit symbol is shown in Fig. 55.48(a). The current provides a continuous supply of

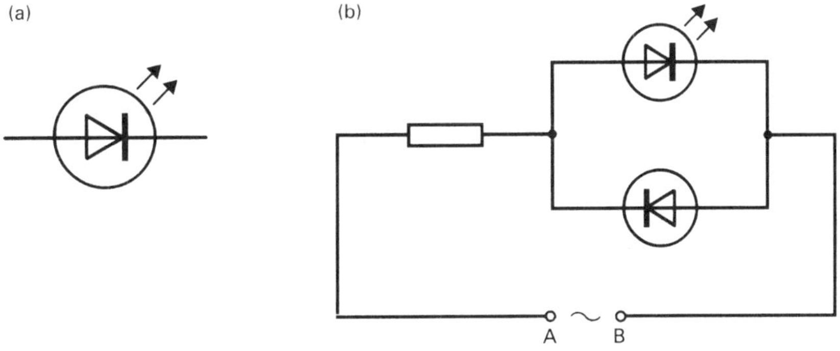

Fig. 55.48
The light-emitting diode: (a) circuit symbol, (b) AC operation

electrons and holes moving across the junction in opposite directions. Some electron–hole recombination occurs in the region around the junction, and whenever it does a quantum of energy is released. (In the band theory model this corresponds to an electron 'falling' from the conduction band or one of the donor levels just beneath it in the n-type region to the valence band or one of the acceptor levels in the p-type region.) In a normal junction diode (one made from silicon or germanium, for example) this energy is all dissipated as heat, but light-emitting diodes are made from semiconducting materials which are such that some of the energy is emitted as light. The colour of the light depends on the composition of the semiconductor – gallium phosphide gives green light and gallium arsenide phosphide gives red or yellow light according to the relative proportions of gallium, arsenic and phosphorus. In each case the light is emitted as a narrow band of wavelengths. Light-emitting diodes made from gallium arsenide emit infrared radiation.

These materials do not transmit light readily. In order that the light can escape, either the n-type or the p-type region has to be extremely thin ($\sim 1\,\mu$m) so that the junction (where the light is produced) is very close to the surface.

Light-emitting diodes are used as indicator lamps, and in the seven-segment displays commonly used on radio alarm clocks and early pocket calculators. They are small, cheap, reliable, mechanically robust, consume very little power and can be switched on and off at high speed. As with an ordinary diode, the current through an LED increases rapidly with voltage when it is forward-biased. Unless the current is being provided by some form of constant current source, the LED should be connected in series with a resistor to limit the current and prevent damage to the junction. They should also be protected from reverse bias – a reverse bias of as little as 5 V can destroy them. Fig. 55.48(b) shows an LED being operated from an AC supply. The resistor limits the current through the LED when it is forward-biased, i.e. when A is positive. When A is negative the (ordinary) diode is forward-biased and its resistance is then very much lower than that of the resistor. As a consequence all but a small fraction of the supply PD is across the resistor and there is no danger of damaging the LED.

55.24 THE PHOTODIODE

This is a p–n junction diode in a case which has a transparent region or 'window' so that the junction can be exposed to light (visible or infrared). It is operated under reverse bias conditions, and in the dark the current is the normal leakage (minority carrier) current which might be as little as 1 nA. When light falls on the junction extra electron–hole pairs are created and the current increases, very nearly in direct proportion to the intensity of the light.

A photodiode has very much faster response ($\sim 0.1\,\mu$s) than an LDR. The circuit symbol is shown in Fig. 55.49.

Fig. 55.49
Circuit symbol of photodiode

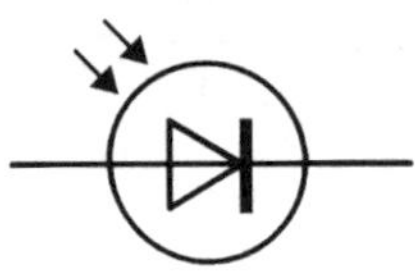

55.25 ELECTRICAL TRANSDUCERS

These are devices which convert electrical energy into non-electrical energy or vice versa.

An input transducer (or sensor) converts a non-electrical signal (such as sound or light) into an electrical signal that can be input into an electronic system such as an amplifier. An output transducer converts the electrical signal which is the output of the electronic system into a non-electrical signal. (See Fig. 55.50.) Some examples of transducers are given in Table 55.1.

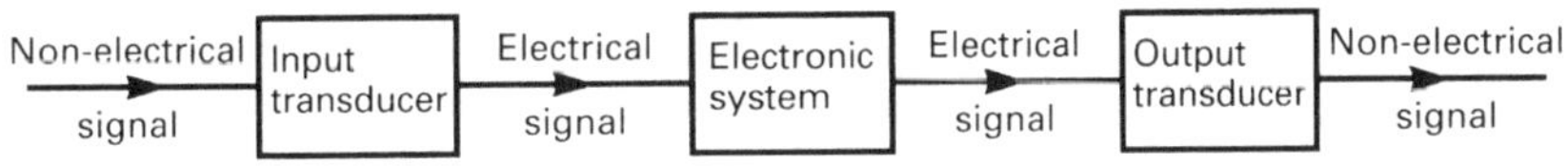

Fig. 55.50 To illustrate the function of transducers

Table 55.1 Examples of electrical transducers

Input transducer	*Output transducer*
Microphone	Loudspeaker
LDR	Lamp
Thermistor	LED
Strain gauge	Motor
Record-player pick-up	Relay

55.26 MODULATION

If a radio wave is to carry useful information, such as speech or music, it needs to be modified in some way. The process is known as modulation and is commonly amplitude modulation or frequency modulation.

Amplitude Modulation (AM)

In amplitude modulation (Fig. 55.51) the amplitude of the high frequency radio signal (the **carrier wave**) is made to vary in sympathy with the amplitude and frequency of the much lower frequency information signal (e.g. the audio frequency signal from a microphone). The information signal is known as the **modulating signal**.

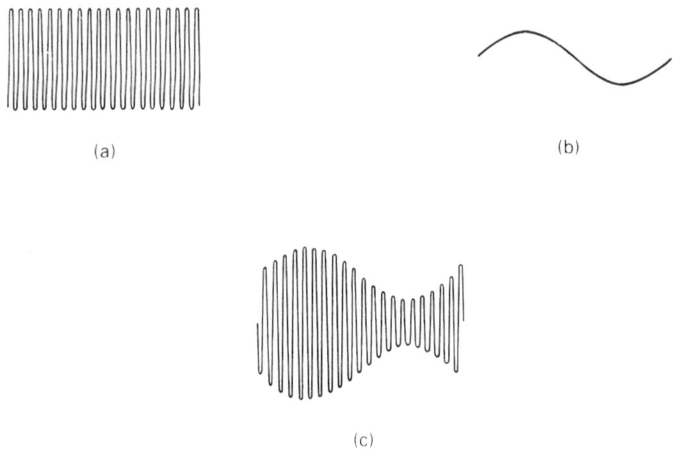

Fig. 55.51 Amplitude modulation (a) unmodulated (RF) carrier wave, (b) audio frequency modulating signal, (c) amplitude modulated signal

Frequency Modulation (FM)

In frequency modulation the amplitude of the carrier wave is kept constant, but its frequency is increased or decreased by an amount proportional to the amplitude of the modulating signal and at a rate which is proportional to the frequency of the signal – see Fig. 55.52. FM is a more complicated process than AM but it has the advantage of not being susceptible to changes in amplitude produced by extraneous signals (noise) and so gives better reception.

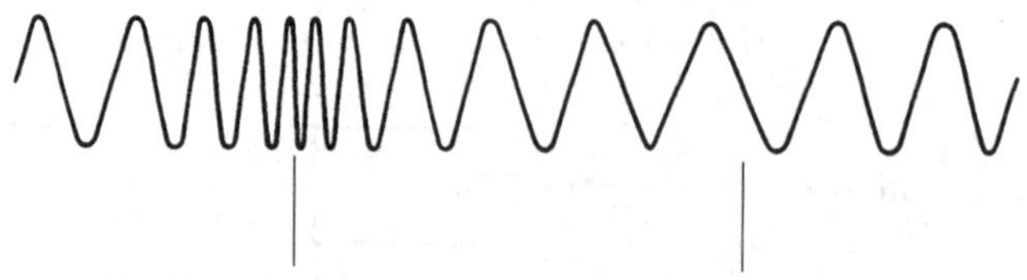
Maximum frequency corresponds to maximum positive value of modulating signal
Minimum frequency corresponds to maximum negative value of modulating signal

Fig. 55.52 Frequency modulation

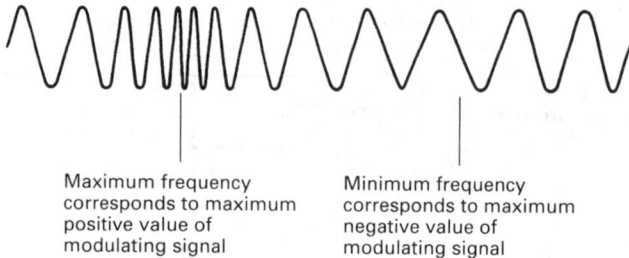

Bandwidth

It can be shown that the amplitude modulated signal of Fig. 55.51(c) may be regarded as being made up of the carrier wave (of frequency f_c) together with a signal of lower frequency, $f_c - f_m$, and one of higher frequency, $f_c + f_m$ (where f_m is the frequency of the modulating signal). The signals of frequency $f_c - f_m$ and $f_c + f_m$ are known as **side frequencies**. In practice, of course, the modulating signal has a whole range of frequencies, each of which produces its own pair of side frequencies. There are therefore two bands of frequencies, known as **sidebands**, lying above and below the carrier frequency. If the highest frequency in the modulating signal is 4 kHz, for example, then the total spread in frequency is 8 kHz. This is called the **bandwidth** of the signal. In general, **the bandwidth of an AM signal is twice the highest frequency contained in the (audio frequency) modulating signal**. The bandwidth of a frequency modulated signal is somewhat greater than that of an amplitude modulated signal carrying the same range of audio frequencies.

The BBC uses a bandwidth of 9 kHz for its amplitude modulated medium waveband, the frequencies of which lie between 540 kHz and 1600 kHz. This is a range of 1060 kHz and the band can therefore accommodate $1060/9 \approx 118$ different signals without any overlap. The frequency modulated, VHF band (from 88 MHz to 108 MHz) uses a bandwidth of 200 kHz (= 0.2 MHz) and so can carry $20/0.2 = 100$ simultaneous transmissions.

55.27 PULSE CODE MODULATION (PCM)

This is a form of **analogue to digital (A/D) conversion** in which the analogue signal (an audio frequency sine wave for example) is 'sampled' at regular intervals and represented as a binary number which can be transmitted as a series of voltage pulses.

As an example of the process, consider the analogue signal shown in Fig. 55.53(a). The signal shown is sampled at times $t_1, t_2, t_3, \ldots$, and its amplitudes at these times are then allocated to one or other of eight voltage levels designated as 0, 1, 2, ..., 7 (Fig. 55.53(b)). The corresponding binary numbers are then transmitted

Fig. 55.53
Illustration of pulse code modulation (a) analogue signal, (b) digital coding, (c) reconstructed analogue signal

(a)

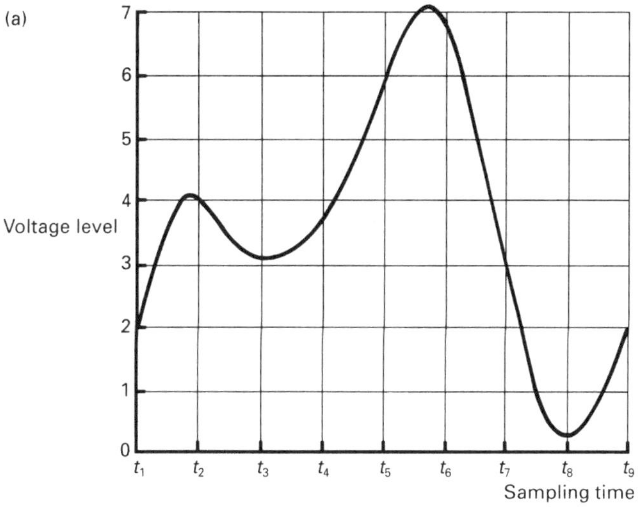

Sample time	Voltage level	Binary code	Voltage pulse
t_1	2	0 1 0	
t_2	4	1 0 0	
t_3	3	0 1 1	
t_4	4	1 0 0	
t_5	6	1 1 0	
t_6	7	1 1 1	
t_7	3	0 1 1	
t_8	0	0 0 0	
t_9	2	0 1 0	

(b)

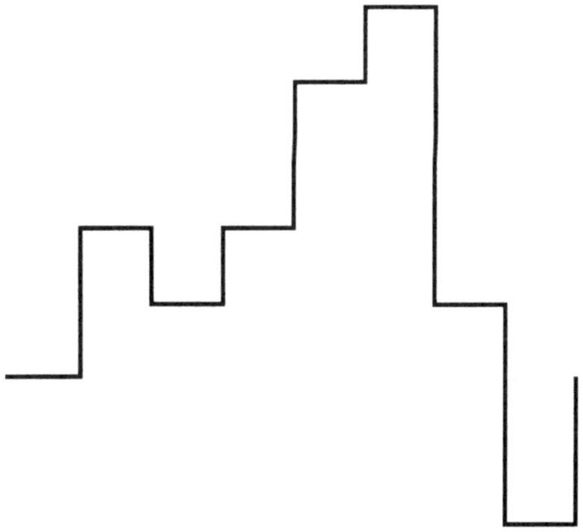

(c)

as a train of voltage pulses using a 3-bit code. When the signal is received it is converted back to analogue form using a **digital to analogue (D/A) converter**. The reconstituted form of the signal of Fig. 55.53(a) is shown in Fig. 55.53(c). This 'stepped' waveform is only a poor copy of the original signal. A much more faithful reproduction would have been obtained had the signal been sampled more frequently and had a greater number of voltage levels been used.

British Telecom's System X uses PCM and samples the signal at 125 μs intervals (i.e. 8000 times per second) and represents it as an 8-bit binary code using 256 different voltage levels.

A major advantage of PCM is that the signal is protected from the effects of noise and from any distortions introduced by the various components of the transmission system. However distorted the pulses might become, as long as their presence can be detected, the code can still be recognized. A second advantage of the system is that a number of different signals can be interleaved. Each 8-bit number requires only about 4 μs of sampling time, and therefore since each signal is sampled only once every 125 μs, about 30 different signals can be interleaved and sent down a single transmission line – a system known as **Multiplexing**.

55.28 OPTICAL COMMUNICATION SYSTEMS

Communication systems in which signals are transmitted along optical fibres as a series of pulses of light are now widespread.

An optical fibre consists of a glass core ($\sim$ 0.1 mm in diameter) surrounded by a glass cladding of lower refractive index, and which is protected by a plastic coating. Most of the light that enters the fibre hits the core/cladding boundary at such an angle that it undergoes total internal reflection and so is trapped in the core (Fig. 55.54).

Fig. 55.54
Light propagation in an optical fibre.

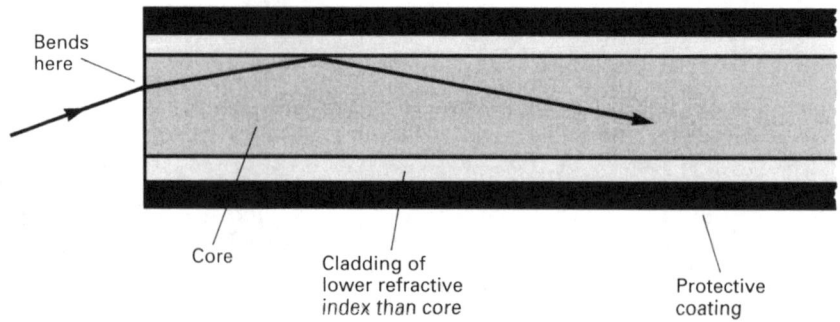

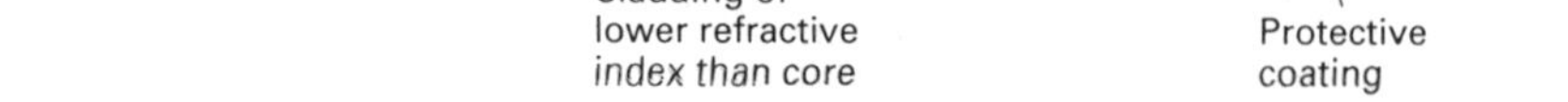

Note. There is only a small difference in refractive index (typically 0.07) between the core and the cladding. The critical angle for the core/cladding boundary is therefore large (typically 72°) in which case all the light that undergoes total internal reflection and travels along the fibre is approximately parallel to the fibre axis. All the possible light paths are therefore of approximately the same length and so produce very little increase in the duration of each pulse.

The light source must be capable of being switched on and off very rapidly so that high data transmission rates can be achieved. (Systems currently in use carry 140 Mbits s^{-1}, the equivalent of nearly 2000 simultaneous telephone channels, on a single fibre.) Furthermore, the light should have only a narrow spread of wavelengths so that problems due to dispersion (i.e. due to the different wavelengths travelling at different speeds) are avoided. Low power lasers and LEDs have been specially developed for the purpose. The 'light' is actually infrared, at a wavelength where the core glass produces very little absorption (0.84 μm, 1.3 μm or 1.55 μm). A photodiode is used to detect the signal.

An optical fibre has a number of advantages over the traditional copper cable.

(i) It is cheaper, lighter and easier to handle than a copper cable with the same signal carrying capacity.

(ii) There is much less attentuation of the signal and therefore fewer repeater stations are required.

(iii) It is free from electrical interference.

(iv) An alternating current in a copper wire can induce a similar current in an adjacent wire so that the two signals become superimposed – this is known as 'crosstalk' and is not a problem with optical systems.

(v) An electrical signal radiates electromagnetic waves which can be detected along the entire length of the cable. This cannot happen with an optical signal and therefore total security of transmission is much more likely.

QUESTIONS ON SECTION G

PHOTOELECTRIC EFFECT AND WAVE–PARTICLE DUALITY (Chapter 47)

G1 An α-particle accelerated between a pair of parallel plates in a vacuum tube acquires a kinetic energy of 10^3 eV. What is the potential difference between the plates? [W]

G2 Describe the principal experimental facts concerning the photoelectric effect, and show how they are explained by the quantum theory.

Describe how Planck's constant h can be determined by experiments on the photoelectric effect. [S]

G3 Light of frequency 6.0×10^{14} Hz incident on a metal surface ejects photoelectrons having a kinetic energy 2.0×10^{-19} J.

Calculate the energy needed to remove an electron from the metal (work function).

Very briefly indicate how you would determine experimentally the kinetic energy of the photoelectrons.
(The Planck constant $= 6.6 \times 10^{-34}$ J s.) [S]

G4 Light of wavelength $0.50\,\mu$m incident on a metal surface ejects electrons with kinetic energies up to a maximum value of 2.0×10^{-19} J. What is the energy required to remove an electron from the metal? If a beam of light causes no electrons to be emitted, however great its intensity, what condition must be satisfied by its wavelength?

(The Planck constant $= 6.6 \times 10^{-34}$ J s, the speed of light $= 3.0 \times 10^8$ m s^{-1}.) [S]

G5 The maximum kinetic energy of photoelectrons ejected from a tungsten surface by monochromatic light of wavelength 248 nm was found to be 8.6×10^{-20} J. Find the work function of tungsten.

(The Planck constant, $h = 6.6 \times 10^{-34}$ J s; speed of light, $c = 3.0 \times 10^8$ m s^{-1}; electronic charge, $e = -1.6 \times 10^{-19}$ C.) [C(O)]

G6 When a metallic surface is exposed to monochromatic electromagnetic radiation electrons may be emitted. Apparatus is arranged so that **(a)** the intensity (energy per unit time per unit area) and **(b)** the frequency of the radiation may be varied. If each of these is varied in turn whilst the other is kept constant what is the effect on **(i)** the number of electrons emitted per second, and **(ii)** their maximum speed? Explain how these results give support to the quantum theory of electromagnetic radiation.

The photoelectric work function of potassium is 2.0 eV. What potential difference would have to be applied between a potassium surface and the collecting electrode in order just to prevent the collection of electrons when the surface is illuminated with radiation of wavelength 350 nm? What would be **(iii)** the kinetic energy, and **(iv)** the speed, of the most energetic electrons emitted in this case?

(Speed of electromagnetic radiation *in vacuo* $= 3.0 \times 10^8$ m s^{-1}. The electronic charge $= -1.6 \times 10^{-19}$ C. Mass of an electron $= 9.1 \times 10^{-31}$ kg. The Planck constant $= 6.6 \times 10^{-34}$ J s.) [L]

G7 List the important experimental facts relating to the photoelectric effect, and explain how Einstein's equation accounts for them.

A clean surface of potassium in a vacuum is irradiated with light of wavelength 5.5×10^{-7} m and electrons are found just to emerge, but when light of wavelength 5×10^{-7} m is incident, electrons emerge each with energy 3.62×10^{-20} J. Estimate the value for Planck's constant h.

Deduce the effect of irradiating in vacuum **(a)** a copper surface, and **(b)** a caesium surface, with light of wavelength 5×10^{-7} m, given that the work functions of copper and caesium are, respectively, 6.4×10^{-19} J and 3.2×10^{-19} J.

(Velocity of light $= 3 \times 10^8\,\mathrm{m\,s^{-1}}$.) [W]

G8 Einstein's equation for the photoelectric emission of electrons from a metal surface under radiation of frequency v can be written as

$$hv = \tfrac{1}{2}m_e v^2 + \phi,$$

where m_e is the mass of an electron, v the greatest speed with which an electron can emerge and ϕ is a quantity called the work function of the metal.

(a) Explain briefly the physical process with which this equation is concerned.

(b) Describe briefly an experiment by which you could determine the values h/e and of ϕ.

(c) For sodium the value of ϕ is 3.12×10^{-19} J, and the wavelength of sodium yellow light is 590 nm.

- **(i)** Explain why electrons are emitted when a sodium surface is irradiated with sodium yellow light, and calculate the greatest speed of the emitted electrons.
- **(ii)** Estimate the 'stopping potential' for these electrons, assuming that no contact potential differences are involved.

(Take the value of the speed of light in vacuum, c, to be $3.00 \times 10^8\,\mathrm{m\,s^{-1}}$, the Planck constant, h, to be 6.63×10^{-34} J s, the electronic charge, e, to be -1.60×10^{-19} C, and m_e to be 9.11×10^{-31} kg.) [O]

G9 **(a)** When electromagnetic radiation falls on a metal surface, electrons may be emitted. This is the photoelectric effect.

- **(i)** State Einstein's photoelectric equation, explaining the meaning of each term.
- **(ii)** Explain why, for a particular metal, electrons are emitted only when the frequency of the incident radiation is greater than a certain value.
- **(iii)** Explain why the maximum speed of the emitted electrons is independent of the intensity of the incident radiation.

(b) A source emits monochromatic light of frequency 5.5×10^{14} Hz at a rate of 0.10 W. Of the photons given out, 0.15% fall on the cathode of a photocell which gives a current of 6.0 μA in an external circuit. You may assume this current consists of all the photoelectrons emitted. Calculate:

- **(i)** the energy of a photon,
- **(ii)** the number of photons leaving the source per second,
- **(iii)** the percentage of the photons falling on the cathode which produce photoelectrons.

(c) **(i)** Calculate the wavelength associated with electrons which have been accelerated from rest through 3000 V.

- **(ii)** Indicate *one* situation where you would expect electrons of about this energy to behave as waves. Give a reason for your answer.

(The Planck constant $= 6.6 \times 10^{-34}$ J s, the electron charge $= 1.6 \times 10^{-19}$ C, the electron mass $= 9.1 \times 10^{-31}$ kg.) [J, '91]

G10 **(a)** Write down an expression for the energy of a photon, explaining the meanings of the symbols used in your expression and giving the units of the physical quantities involved.

(b) In an experiment with a vacuum photocell the maximum kinetic energy of the electrons emitted was measured for different wavelengths of the illuminating radiation. The following results were obtained.

Maximum kinetic energy/10^{-19} J	*Wavelength*/ 10^{-7} m
3.26	3.00
2.56	3.33
1.92	3.75
1.25	4.29
0.58	5.00

Use these results to plot a linear graph and derive a value for Planck's constant.

(c) If the experiment were repeated with radiation of wavelength **(i)** 7.5×10^{-7} m, **(ii)** 2.8×10^{-7} m, would photoelectrons be emitted and, if so, what would be their maximum kinetic energy?

(d) Describe and explain how the graph might change if a different metal were used for the surface of the photo-cathode.

Speed of light, $c = 3.00 \times 10^8\,\mathrm{m\,s^{-1}}$. [J]

G11 Light of photon energy 3.5 eV is incident on a plane photocathode of work function 2.5 V. Parallel and close to the cathode is a plane collecting electrode. The cathode and collector are mounted in an evacuated tube.

(a) Find the maximum kinetic energy E_{max} of photoelectrons emitted from the cathode. (Express your answer in eV.)

(b) Find the minimum value of the potential difference which should be applied between collector and cathode in order to prevent electrons of energy E_{max} from reaching the collector for electrons emitted **(i)** normal to the cathode, **(ii)** at an angle of 60° to the cathode. [C]

G12 (a) (i) Explain what is meant by *photoelectric emission*.

(ii) Briefly describe a simple experiment to demonstrate this effect qualitatively.

(b) Write down Einstein's photoelectric equation and explain the meaning of each term in it.

(c) In an experiment in photoelectricity, the maximum kinetic energy of the photoelectrons was determined for different wavelengths of the incident radiation. The following results were obtained:

Wavelength/nm	300	375	500
Maximum kinetic energy/eV	2.03	1.20	0.36

Use the results to determine

(i) the work function for the metal,

(ii) a value for Planck's constant.

(d) (i) Describe how a photocell functions.

(ii) Describe *one* practical application of a photocell.

($c = 3 \times 10^8\,\text{m s}^{-1}$, $e = 1.6 \times 10^{-19}$ C.)

[W, '90]

G13 (a) (i) Electrons may be emitted from a surface by *thermionic emission* or by *photoelectric emission*. Distinguish between the two.

(ii) Describe the construction of a simple photocell.

(b) (i) Write down Einstein's photoelectric equation relating the maximum kinetic energy E_{max} of the photoelectrons with the frequency f of the incident radiation and with the work function ϕ for the emitting surface.

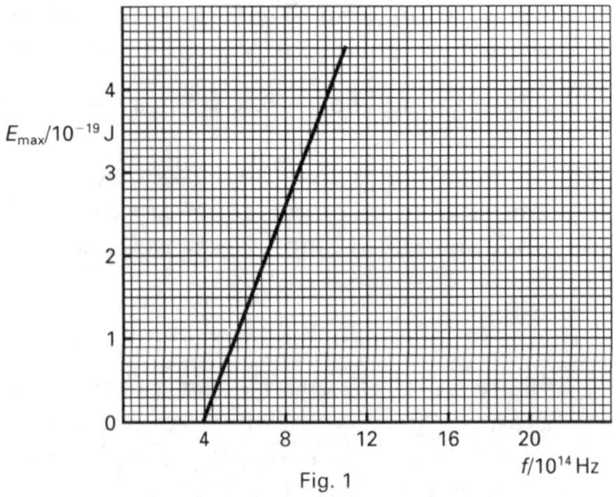

Fig. 1

(ii) The graph shows how E_{max} varies with f for a particular surface. Use the graph to find the value of

(I) ϕ,

(II) the threshold wavelength of the radiation.

(iii) Describe and explain the effect of increasing the intensity of the incident radiation.

(c) A photocell of work function $\phi = 1.0$ eV is connected to the circuit shown in Fig. 2. The wavelength of radiation incident on the cell is gradually decreased and the output of the amplifier continuously monitored. Fig. 3 gives the variation of wavelength and amplifier output with time. Using Figures 2 and 3 calculate the wavelength at time T.

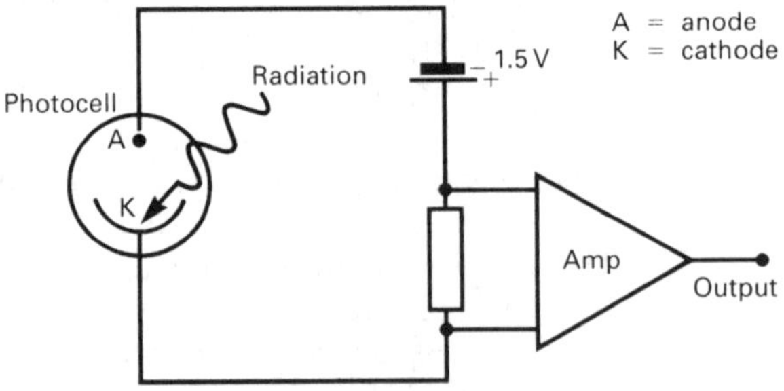

Fig. 2

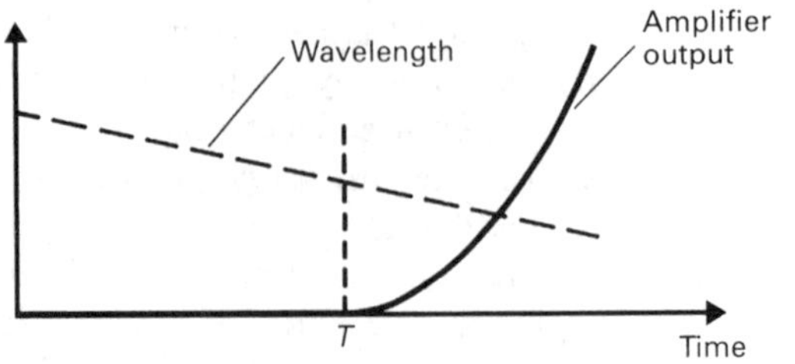

Fig. 3

(Charge on an electron $e = -1.60 \times 10^{-19}$ C, Speed of light $c = 3.00 \times 10^{8}\,\mathrm{m\,s^{-1}}$, the Planck constant $h = 6.63 \times 10^{-34}$ J s.) [W, '92]

G14 Explain the physical processes described by the Einstein photoelectric equation $h\nu = \frac{1}{2}mv^2 + h\nu_0$, and state the significance of each term.

Describe briefly how the values of h and ν_0 can be determined.

An argon laser emits a beam of light of wavelength 4.88×10^{-7} m, the power in the beam being 100 mW. How many photons per second are emitted by the laser? If the beam falls on the caesium cathode of a photocell, what photoelectric current would be observed, assuming 10% of the photons are able to eject an electron? Given that the limiting frequency ν_0 of caesium is 5.2×10^{14} Hz, what reverse potential difference between the cell electrodes is needed to suppress the photocell current?

(The Planck constant $= 6.6 \times 10^{-34}$ J s, the speed of light $= 3.0 \times 10^{8}\,\mathrm{m\,s^{-1}}$, the electronic charge $= 1.6 \times 10^{-19}$ C.) [S]

G15 What is meant by *photoelectricity*? Describe the main features of photoelectric emission. Give an expression for the kinetic energy of the photoelectrons emitted from a surface, explaining what is meant by *work function*, the *threshold frequency* (or *cut-off frequency*) and *Planck's constant*.

Describe an experiment to verify the equation for the kinetic energy of the photoelectrons, and show how the work function of the surface and Planck's constant can be obtained.

A monochromatic light source provides a 5 W beam of radiation of wavelength 4.5×10^{-7} m and this beam liberates 10^{10} photoelectrons per second from the surface of a sodium block. The threshold wavelength of sodium is 5.5×10^{-7} m.

(a) Calculate the magnitude of the photoelectric emission current (in amperes) given by **(i)** this arrangement, **(ii)** an otherwise identical one with a 10 W beam, and **(iii)** a 5 W beam of wavelength 6×10^{-7} m.

(b) If, in the original arrangement, the sodium block were electrically isolated, the emission of photoelectrons would cause it to acquire a positive potential. Find the steady value which it would eventually reach.

(Electronic charge $= 1.6 \times 10^{-19}$ C; the Planck constant $= 6.6 \times 10^{-34}$ J s; speed of light $= 3 \times 10^{8}\,\mathrm{m\,s^{-1}}$.) [W]

G16 (a) (i) What is a *photon*?

(ii) Show that E, the energy of a photon, is related to λ, its wavelength, by

$$E\lambda = 1.99 \times 10^{-16}$$

where E is measured in J and λ is measured in nm.

(b) Two metal electrodes A and B are sealed into an evacuated glass envelope and a potential difference V, measured using the voltmeter, is applied between them as shown in Fig. 1.

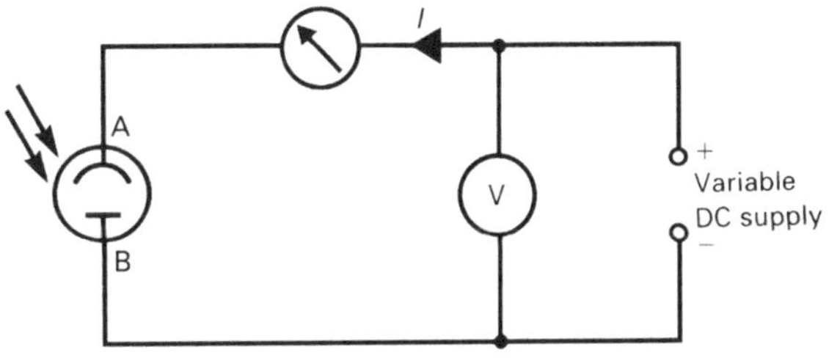

Fig. 1

B is then illuminated with monochromatic light of wavelength 365 nm and I, the current in the circuit, is measured for various values of V. The results are shown in Fig. 2.

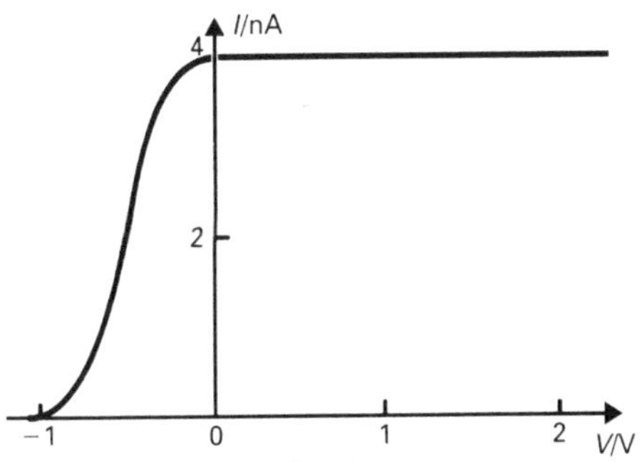

Fig. 2

(i) From this graph, deduce the PD required to stop photoelectric emission from B.

(ii) Calculate the maximum kinetic energy of the photoelectrons.

(iii) Deduce the work function energy of B.

(The Planck constant $= 6.6 \times 10^{-34}$ J s, the electron charge $= 1.6 \times 10^{-19}$ C, velocity of light $= 3.0 \times 10^{8}$ m s^{-1}.)

[C, '91]

G17 (a) When electromagnetic radiation falls on a metal surface, electrons may be emitted. Use the quantum theory to explain the following experimental observations.

(i) For a particular metal, electrons are emitted only when the wavelength of the radiation is less than a certain value.

(ii) The number of electrons emitted per second increases as the intensity of the radiation increases.

(b) Indicate *one* piece of experimental evidence showing that particles have wave properties. [J, '89]

G18 (a) What are the dimensions of **(i)** energy, **(ii)** momentum, **(iii)** the Planck constant, h?

It can be shown that the total energy E of a particle is related to its momentum p and its mass m_0 when at rest, by the equation

$$E^2 = p^2c^2 + m_0^2c^4$$

where c is the speed of electromagnetic radiation in vacuo. Show this equation to be dimensionally correct.

(b) Given that m_0 is zero for a photon, use the above equation to derive an expression for the momentum of a photon in terms of its wavelength.

(c) Estimate the potential difference across which an electron should be accelerated if it is to have the same energy as a photon of ultraviolet radiation. Support your estimate with a calculation, by choosing a suitable value for the wavelength of the electromagnetic radiation. You may assume that the value of the product hc is 2.0×10^{-25} J m.

(Charge on electron $= -1.6 \times 10^{-19}$ C.)

[J]

G19 Find an expression for the de Broglie wavelength of an electron in terms of its kinetic energy E, the electron mass m_e, and the Planck constant h. [C]

G20 Calculate the de Broglie wavelength of an electron which has been accelerated from rest through a PD of 10^4 V.

(Electron charge $= 1.6 \times 10^{-19}$ C, electron mass $= 9.1 \times 10^{-31}$ kg, Planck's constant $= 6.6 \times 10^{-34}$ J s.)

ATOMIC STRUCTURE AND ENERGY LEVELS (Chapter 48)

G21 The atomic nucleus may be considered to be a sphere of positive charge with a diameter very much less than that of the atom. Discuss the experimental evidence which supports this view. Describe briefly how the experimental evidence was obtained. [J]

G22 What do you understand by the term *nuclear atom*?

Alpha particles in a narrow parallel beam are scattered by a thin metallic foil. State the results of such an experiment and explain why they are evidence for the nuclear atom.

Explain why:

(a) such an experiment is carried out *in vacuo*,

(b) the incident alpha particles are confined in a narrow parallel beam,

(c) the foil is thin. [J]

G23 Experiment shows that if a fine beam of alpha particles is directed normally on to a thin gold film

(a) most of the alpha particles go straight through the film with undiminished speed, and

(b) a small proportion of the alpha particles is strongly deflected.

How do you explain these two observations?

Write down an approximate value for the ratio of the diameters of a gold atom and its nucleus.

Use this ratio to calculate a value for the density of the nucleus of a gold atom given that the density of gold is 19.3×10^3 kg m^{-3}.

[L, '92]

G24 Describe *the principle* of the experiment which established the nuclear model of the atom, explaining how the deduction is made from the observations.

The emission spectrum of the hydrogen atom consists of a series of lines. Explain why this suggests the existence of definite energy levels for the electron in the atom.

By considering the intervals between the energy levels explain the spacing of the lines in the visible hydrogen spectrum.

The ionisation potential of the hydrogen atom is 13.6 V. Use the data below to calculate:

(a) the speed of an electron which could just ionize the hydrogen atom,

(b) the minimum wavelength which the hydrogen atom can emit.

(Charge on an electron $= -1.60 \times 10^{-19}$ C. Mass of an electron $= 9.11 \times 10^{-31}$ kg. The Planck constant $= 6.63 \times 10^{-34}$ J s. Speed of light $= 3.00 \times 10^{8}\,\mathrm{m\,s^{-1}}$.) [L]

G25 **(a)** Sketch, on the same diagram, the paths of three alpha particles of the same energy which are directed towards a nucleus so that they are deflected through **(i)** about 10°, **(ii)** 90°, **(iii)** 180° respectively.

(b) For the deflection of 180°, describe in qualitative terms how **(i)** the kinetic energy, **(ii)** the potential energy of the alpha particle varies during its path, assuming the nucleus remains stationary.

(c) If, in (b) above, the alpha particle has an initial kinetic energy of 1.60×10^{-13} J and the nucleus has a charge of $+50e$, calculate the nearest distance of approach of the alpha particle to the nucleus.

(Magnitude of the electronic charge: $e = 1.60 \times 10^{-19}$ C.

Permittivity of free space: $\varepsilon_0 = 10^{-9}/36\pi$ or $8.85 \times 10^{-12}\,\mathrm{F\,m^{-1}}$.) [J]

G26 **(a)** What are the chief characteristics of a line spectrum?

Explain briefly how line spectra are used:

(i) in analysis for the identification of elements present;

(ii) in astronomy for estimating the component in the line of sight of the velocity of a star relative to the Earth.

(b) The figure below representing the lowest energy levels of the electron in the hydrogen atom, gives the principal quantum number n associated with each, and the corresponding value of the energy, measured in joules.

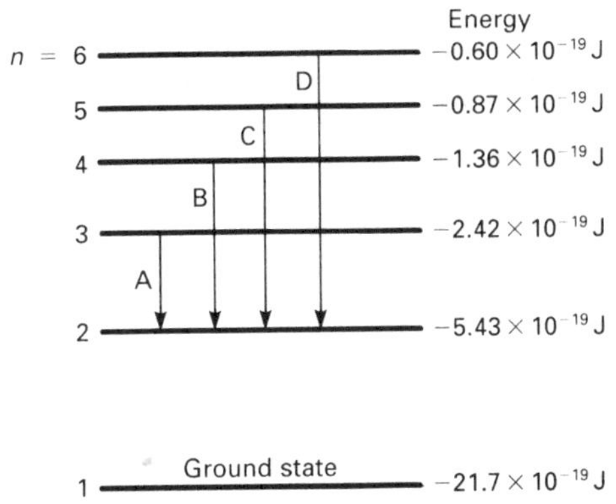

(i) Calculate the wavelengths of the lines arising from the transitions marked A, B, C, D on the figure.

(ii) Show that the other transitions that can occur give rise to lines which are in either the ultraviolet or the infrared regions of the spectrum.

(iii) The level $n = 1$ is the 'ground state' of the unexcited hydrogen atom. Explain why hydrogen in its ground state is quite transparent to light emitted by the transitions A, B, C, D, and also what happens when 21.7×10^{-19} J of energy is supplied to a hydrogen atom in its ground state.

(Take the value of the speed of light in vacuum, c, to be $3.00 \times 10^{8}\,\mathrm{m\,s^{-1}}$, and that of the Planck constant, h, to be 6.63×10^{-34} J s.) [O]

G27 **(a)** Calculate the energy of one photon of light emitted within the *D*-lines of a sodium lamp if the wavelength of the *D*-lines is 589 nm.

(b) In a 200 W sodium street lamp, 30% of input electrical energy is emitted within the *D*-lines. How many photons of light are emitted within the *D*-lines per second?

($h = 6.6 \times 10^{-34}$ J s, $c = 3.0 \times 10^{8}\,\mathrm{m\,s^{-1}}$.) [W, '91]

G28 The ionisation energy for a hydrogen atom is 13.6 eV if the atom is in its ground state. It is 3.4 eV if the atom is in the first excited state.

Explain the terms *ionisation energy* and *excited state.*

Calculate the wavelength of the photon emitted when a hydrogen atom returns to the ground state from the first excited state. Name the part of the electromagnetic spectrum to which this wavelength belongs.

(Electronic charge, $e = -1.60 \times 10^{-19}$ C, the Planck constant, $h = 6.63 \times 10^{-34}$ J s, speed of light, $c = 3.00 \times 10^{8}$ m s^{-1}.)

[L, '92]

G29 The energy levels of the hydrogen atom are given by the expression

$$E_n = -2.16 \times 10^{-18}/n^2 \text{ J}$$

where n is an integer.

(a) What is the ionization energy of the atom?
(b) What is the wavelength of the H_α line, which arises from transitions between $n = 3$ and $n = 2$ levels?

(The Planck constant $= 6.6 \times 10^{-34}$ J s, the speed of electromagnetic radiation $= 3.0 \times 10^{8}$ m s^{-1}.)

[S]

G30 The lowest energy level in a helium atom (the ground state) is -24.6 eV. There are a number of other energy levels, one of which is at -21.4 eV.

(a) Define an eV.
(b) **(i)** Explain the significance of the negative signs in the values quoted.
(ii) What is the energy, in J, of a photon emitted when an electron returns to the ground state from the energy level at -21.4 eV?
(iii) Calculate the wavelength of the radiation emitted in this transition. The electronic charge, $e = 1.6 \times 10^{-19}$ C. The speed of electromagnetic radiation, $c = 3.0 \times 10^{8}$ m s^{-1}. The Planck constant, $h = 6.6 \times 10^{-34}$ J s.
(c) Helium was first discovered from observations of the absorption spectrum produced by helium in the Sun's atmosphere.

What is an *absorption spectrum*?

[AEB, '89]

G31 The frequencies f of the spectral lines emitted by atomic hydrogen can be represented by the expression

$$f = a\left(\frac{1}{n_1{}^2} - \frac{1}{n_2{}^2}\right)$$

where a is a constant and n_1 and n_2 are integers with $n_2 > n_1$.

Given that the ionisation potential for atomic hydrogen is 13.6 V:

(a) Calculate the energies (in eV) of the quanta emitted by electronic transitions between the levels given by
(i) $n_1 = 2, n_2 = 3$
(ii) $n_1 = 1, n_2 = 2$.
(b) State in which parts of the spectrum these quanta are found.

[W, '92]

G32 Hydrogen atoms in a discharge tube emit spectral lines whose frequencies f are given by

$$f = cR_H\left(\frac{1}{n_1{}^2} - \frac{1}{n_2{}^2}\right)$$

where $c = 3 \times 10^{8}$ m s^{-1}, $R_H = 1.10 \times 10^{7}$ m^{-1} and n_1 and n_2 are any positive whole numbers.

(a) Calculate **(i)** the highest and **(ii)** the lowest frequencies in the Lyman series of spectral lines.
($n_1 = 1$ in the Lyman series.)
(b) Multiply each answer in (a) by h (Planck's constant $h = 6.6 \times 10^{-34}$ J s) and state what each answer represents.

[W, '90]

G33 Some of the energy levels of the hydrogen atom are shown (not to scale) in the diagram.

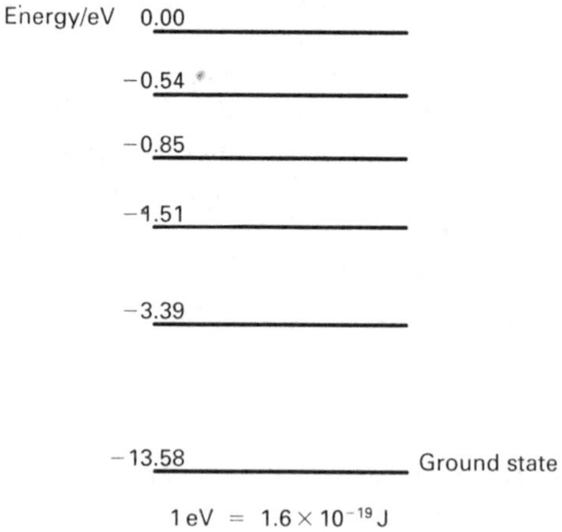

(a) Why are the energy levels labelled with negative energies?
(b) State which transition will result in the emission of radiation of wavelength 487 nm. Justify your answer by suitable calculation.
(c) What is likely to happen to a beam of photons of energy **(i)** 12.07 eV, **(ii)** 5.25 eV, when passed through a vapour of atomic hydrogen?

(1 eV $= 1.6 \times 10^{-19}$ J.)

[O & C, '92]

G34 **(a)** **(i)** Describe briefly the Bohr model for the hydrogen atom, and

(ii) state the important assumptions that Bohr made.

(b) The table shows some of the energy levels for the hydrogen atom.

	Energy/eV
a	0
b	−0.54
c	−0.85
d	−1.51
e	−3.39
f	−13.6

(i) Define the electron volt.

(ii) Explain why the energy levels are given negative values.

(iii) How might the atom be changed from state e to state c?

(iv) State which level corresponds to the ground state.

(v) Calculate the ionization energy of atomic hydrogen in joules.

(vi) The result of the Bohr theory for the hydrogen atom can be expressed by

$$\frac{1}{\lambda} = R_H\left[\frac{1}{n_1^2} - \frac{1}{n_2^2}\right]$$

where n_1 and n_2 are whole numbers: for the ground state $n_1 = 1$.

Calculate the value of R_H.

($h = 6.6 \times 10^{-34}$ J s; $e = -1.6 \times 10^{-19}$ C; $c = 3 \times 10^8$ m s^{-1}.) [W]

G35 **(a)** **(i)** Explain what is meant by 'electron energy levels' in an atom.

(ii) How does this concept account for the characteristic emission line spectrum of an element?

(iii) Describe a simple laboratory demonstration of line spectrum emission.

(b) The diagram below represents the lowest energy levels of the electron in the hydrogen atom, giving the principal quantum number n associated with each level and the corresponding values of the energy.

(i) Why are the energies quoted with negative values?

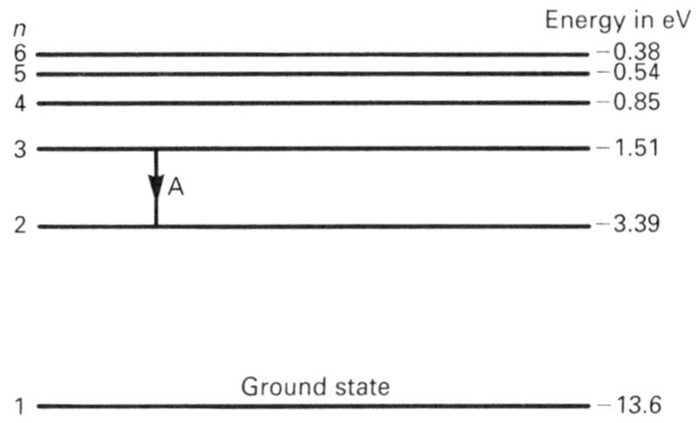

(ii) Calculate the wavelength of the line arising from the transition A, indicating in which region of the electromagnetic spectrum this occurs.

(iii) What happens when 13.6 eV of energy is absorbed by a hydrogen atom in its ground state?

(c) **(i)** Explain, in terms of electron energy levels, how X-ray line spectra are emitted, pointing out differences between production of X-ray line spectra and optical line spectra.

(ii) Electrons are accelerated from rest through a potential difference of 10 kV in an X-ray tube. Calculate the minimum wavelength of the X-rays emitted. Why do the remainder of the X-rays emitted have longer wavelengths than this?

($e = 1.6 \times 10^{-19}$ C, $h = 6.6 \times 10^{-34}$ J s, $c = 3.0 \times 10^8$ m s^{-1}.) [W, '91]

X-RAYS (Chapter 49)

G36 What are X-rays and how are they produced? Do you think it would be possible, in principle, to obtain an X-ray line spectrum from hydrogen? Give reasons in support of your answer. [W]

G37 Explain how the radiation from an evacuated X-ray tube is affected by changing **(a)** the filament current, **(b)** the filament-target potential difference, **(c)** the target material. [J]

G38 Give a labelled diagram showing the essential components of a modern X-ray tube.

State briefly how you would control electrically **(a)** the intensity, **(b)** the penetrating power of the emitted X-rays. [S]

G39 **(a)** When atoms absorb energy colliding with moving electrons, light or X-radiation may subsequently be emitted. For each type of radiation, state typical values of the energy per atom which must be absorbed and explain in atomic terms how each type of radiation is emitted.

(b) State *one* similarity and *two* differences between optical atomic emission spectra and X-ray emission spectra produced in this way.

(c) Electrons are accelerated from rest through a potential difference of 10 000 V in an X-ray tube. Calculate:

(i) the resultant energy of the electrons in eV;

(ii) the wavelength of the associated electron waves;

(iii) the maximum energy and the minimum wavelength of the X-radiation generated.

(Charge of electron = 1.60×10^{-19} C. Mass of electron = 9.11×10^{-31} kg. Planck's constant = 6.62×10^{-34} J s. Speed of electromagnetic radiation in vacuo = 3.00×10^{8} m s^{-1}.) [J]

G40 The potential difference between the target and cathode of an X-ray tube is 50 kV and the current in the tube is 20 mA. Only 1% of the total energy supplied is emitted as X-radiation.

(a) What is the maximum frequency of the emitted radiation?

(b) At what rate must heat be removed from the target in order to keep it at a steady temperature

(The Planck constant, $h = 6.6 \times 10^{-34}$ J s; electron charge, $e = -1.6 \times 10^{-19}$ C.) [C]

G41 **(a)** Explain briefly why a modern X-ray tube can be operated directly from the output of a step-up transformer.

(b) An X-ray tube works at a DC potential difference of 50 kV. Only 0.4% of the energy of the cathode rays is converted into X-radiation and heat is generated in the target at a rate of 600 W. Estimate **(i)** the current passed through the tube, **(ii)** the velocity of the electrons striking the target.

(Electron mass = 9.00×10^{-31} kg. Electron charge = -1.60×10^{-19} C.) [J]

G42 **(a)** A 900 W X-ray tube operates at a DC potential difference of 30 kV. Calculate the minimum wavelength of the X-rays produced.

(b) Calculate the current through the tube.

(c) If 99% of the power is dissipated as heat, estimate the number of X-ray photons produced per second.

($h = 6.6 \times 10^{-34}$ J s, $c = 3.0 \times 10^{8}$ m s^{-1}, $e = 1.6 \times 10^{-19}$ C.) [W, '92]

G43 **(a)** Place the following radiations in ascending order of photon energy, giving approximate values to these energies: X-rays, visible, infrared, ultraviolet.

(b) Describe *briefly* the atomic or molecular processes involved in the production of X-rays and infrared radiation. [W, '91]

THE ELECTRON (Chapter 50)

G44 A beam of electrons travelling with speed 1.2×10^{7} m s^{-1} in an evacuated tube is made to move in a circular path of radius 0.048 m by a uniform magnetic field of flux density $B = 1.4$ mT.

(a) Calculate, in electronvolts, the kinetic energy of an electron in the beam. (The charge on an electron = 1.6×10^{-19} C and 1 eV = 1.6×10^{-19} J.)

(b) A similar technique is used to accelerate protons to very high speeds. Protons with energies of 500 GeV can be held by magnetic fields in circular orbits of radius 2 km.

Suggest why such a large radius orbit is necessary for high energy protons. [L]

G45 **(a)** Give the equation for the force on a particle carrying a charge e:

(i) in an electric field of intensity E,

(ii) while moving with velocity v at right angles to a magnetic field of flux density B.

(iii) Draw diagrams to indicate the directions of the forces. (Make sure that you show the direction of the fields in relation to + and −, and N and S.)

(b) A beam of electrons is directed into a region of uniform magnetic field of flux density B along a path at right angles to the direction of B.

(i) What is the path followed by the electrons when they enter the field?
(ii) Explain why this path is followed.

(c) Hydrogen ions moving at various speeds are directed at a region of combined electric and magnetic fields as shown in the diagram below. The electric field is between two parallel plates 10 mm apart with a potential difference V across them, while the magnetic field of flux density 0.1 T is at right angles to the electric field.

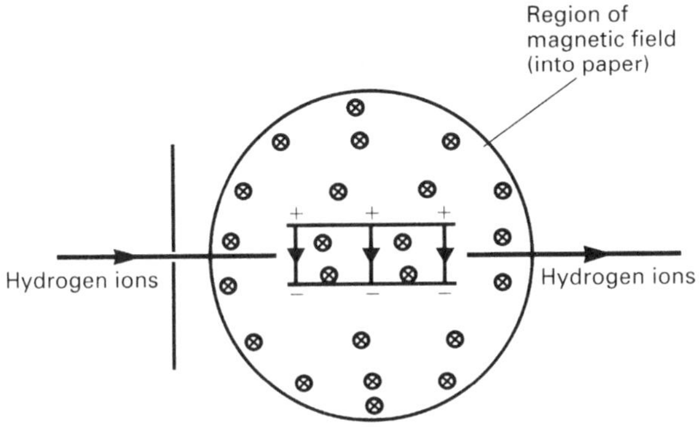

(i) Calculate the value of V required so that ions of speed $100\,\text{m}\,\text{s}^{-1}$ pass through the region of the two fields without being deviated.
(ii) What is the kinetic energy per ion as they leave the combined fields?

($1\,\text{u} = 1.67 \times 10^{-27}\,\text{kg}$.) [W, '90]

G46 A beam of protons is accelerated from rest through a potential difference of 2000 V and then enters a uniform magnetic field which is perpendicular to the direction of the proton beam.

If the flux density is 0.2 T calculate the radius of the path which the beam describes.

(Proton mass $= 1.7 \times 10^{-27}\,\text{kg}$, electronic charge $= -1.6 \times 10^{-19}\,\text{C}$.) [L]

G47 Electrons, accelerated from rest through a potential difference of 3000 V, enter a region of uniform magnetic field, the direction of the field being at right angles to the motion of the electrons. If the flux density is 0.010 T, calculate the radius of the electron orbit.

(Assume that the specific charge, e/m for electrons $= 1.8 \times 10^{11}\,\text{C}\,\text{kg}^{-1}$.) [AEB, '79*]

G48 (a) An electron (mass m, charge e) travels with speed v in a circle of radius r in a plane perpendicular to a uniform magnetic field of flux density B.
(i) Write down an algebraic equation relating the centripetal and electromagnetic forces acting on the electron.
(ii) Hence show that the time for one orbit of the electron is given by the expression $T = 2\pi m/Be$.

(b) If the speed of the electron changed to $2v$, what effect, if any, would this change have on:
(i) the orbital radius r,
(ii) the orbital period T?

(c) Radio waves from outer space are used to obtain information about interstellar magnetic fields. These waves are produced by electrons moving in circular orbits. The radio wave frequency is the same as the electron orbital frequency.

(The mass of an electron is $9.1 \times 10^{-31}\,\text{kg}$, and its charge is $-1.6 \times 10^{-19}\,\text{C}$.)

If waves of frequency 1.2 MHz are observed, calculate:
(i) the orbital period of the electrons;
(ii) the flux density of the magnetic field.

[O, '92]

G49 (a) A beam of singly ionized carbon atoms is directed into a region where a magnetic and an electric field are acting perpendicularly both to each other and to the beam. The fields have intensities 0.10 T and $1.0 \times 10^4\,\text{N}\,\text{C}^{-1}$ respectively. If the beam is able to pass undeviated through this region, what is the velocity of the ions?

(b) The beam then enters a region where a magnetic field alone is acting. As a result the beam describes an arc of radius 0.75 m. Calculate the flux density of this magnetic field.

(Mass of carbon atom $= 2.0 \times 10^{-26}\,\text{kg}$; $e = 1.6 \times 10^{-19}\,\text{C}$.) [S]

G50 Describe, giving the theory and a labelled diagram of the apparatus, a method of determining e/m for the electron.

In an evacuated tube, electrons are accelerated through a potential difference of 500 V. Calculate their final speed, and consider

whether this depends on the accelerating field being uniform. After this acceleration, the electrons pass through a uniform electric field which is perpendicular to the direction of travel of the electrons as they enter the field. This electric field is produced by applying a potential difference of 10 V to two parallel plates which are 0.06 m long and 0.02 m apart. Assume that the electric field is uniform and confined to the space between the two plates. Determine the angular deflection of the electron beam produced by the field.
(e/m for the electron $= 1.76 \times 10^{11}\,\mathrm{C\,kg^{-1}}$.)
[W]

G51 Two parallel metal sheets of length 10 cm are separated by 20 mm in a vacuum. A narrow beam of electrons enters symmetrically between them as shown.

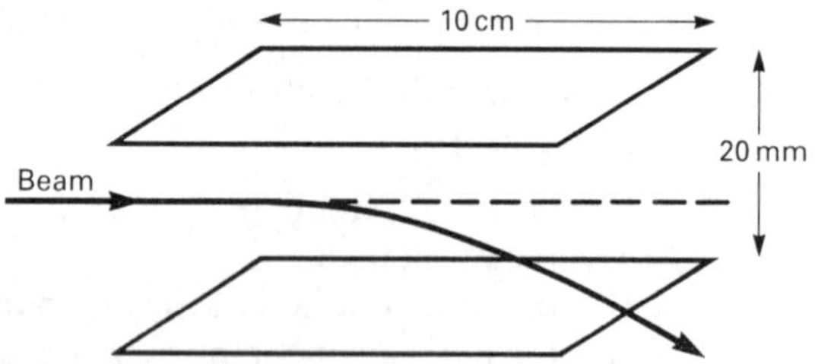

When a PD of 1000 V is applied between the plates the electron beam just misses one of the plates as it emerges.

Calculate the speed of the electrons as they enter the gap. (Take the field between the plates to be uniform.)

($e/m = 1.8 \times 10^{11}\,\mathrm{C\,kg^{-1}}$.) [W, '92]

G52 An ion source supplies ions of different velocities to an arrangement of three apertures X, Y and Z, which are arranged on the same straight line as shown in the figure below. The region between apertures Y and Z contains a uniform magnetic field of flux density B and a uniform electric field of intensity E. Both fields are at right angles to each other and also at right angles to the line joining the apertures. Derive the condition for ions to pass through all three apertures.
[W, '91]

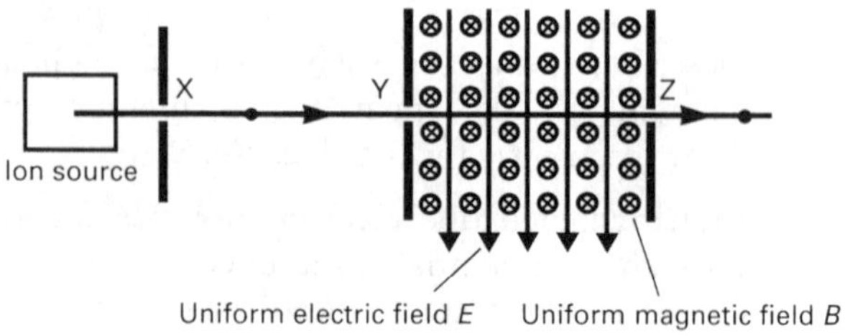

G53 **(a)** Describe by means of a diagram and a simple equation the force due to **(i)** an electric field, **(ii)** a magnetic field, acting on an electron moving at right angles to each field.

Hence explain how an electric field and a magnetic field may be used in the selection of the velocity of negatively charged particles.

(b) Ions having charge $+Q$ and mass M are accelerated from rest through a potential difference V. They then move into a region of space where there is a uniform magnetic field of flux density B, acting at right angles to the direction of travel of the ions, as shown below. [C, '92]

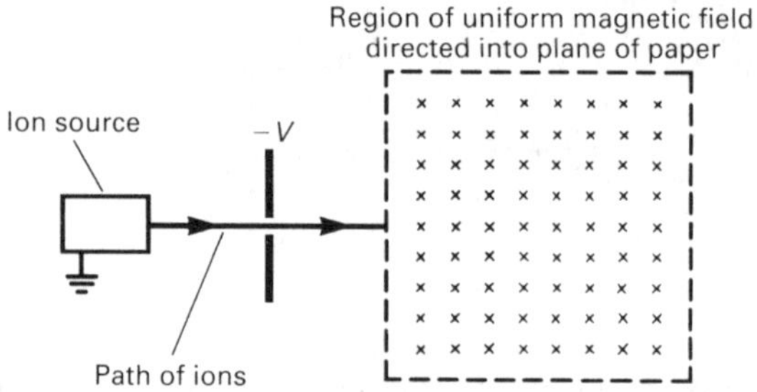

(i) Show that v, the speed with which the ions enter the magnetic field, is given by

$$v = \sqrt{\frac{2QV}{M}}$$

(ii) Hence derive an expression, in terms of M, Q, B and V, for the radius of the path of the ion in the magnetic field.

(iii) Briefly describe and explain any change in the path in the magnetic field of an ion of twice the specific charge (i.e. for which the ratio Q/M is doubled). [C, '92]

G54 Describe a method by which the charge/mass ratio e/m of the electron has been determined.

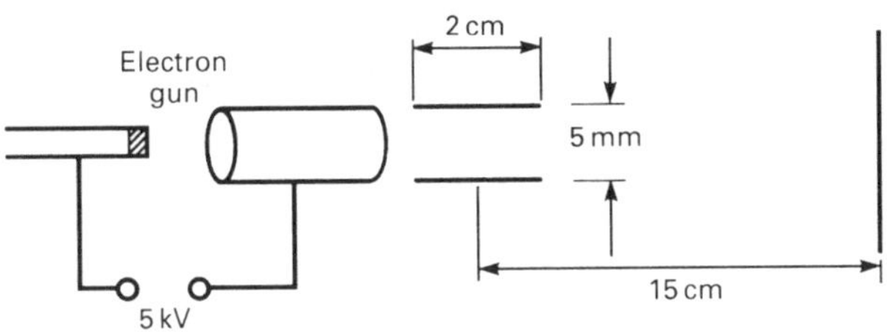

Calculate the deflection sensitivity (deflection of spot in mm per volt potential difference) of a cathode ray tube from the following data:

Electrons are accelerated by a potential difference of 5.0 kV between cathode and anode;

(Length of deflector plates = 2.0 cm; Separation of deflector plates = 5.0 mm; Distance of mid point of deflector plates from screen = 15.0 cm.) [S]

G55 A heated filament and an anode with a small hole in it are mounted in an evacuated glass tube so that a narrow beam of electrons emerges vertically upwards from the hole in the anode. A uniform magnetic field is applied so that the electrons describe a circular path in a vertical plane.

(a) Draw a diagram showing the path of the electrons and indicate the direction of the magnetic field which will cause the beam to curve in the direction you have shown. Explain why the path is circular.

(b) Derive an expression for the specific electronic charge (e/m) of the electrons in terms of the PD between the anode and filament, V, the radius of the circular path, r, and the magnetic flux density, B.

(c) What value of B would be required to give a radius of the electron path of $2r$, assuming that V remains constant? If B is now held constant at its new value, what value of V will restore the beam to its former radius?

(d) Describe and account for the changes in **(i)** kinetic energy, **(ii)** momentum which an electron undergoes from the instant it leaves the heated filament with negligible velocity until it has completed a full circle in the magnetic field. [J]

G56 The diagram shows a type of cathode ray tube containing a small quantity of gas. Electrons from a hot cathode emerge from a small hole in a conical shaped anode, and the path subsequently followed is made visible by the gas in the tube.

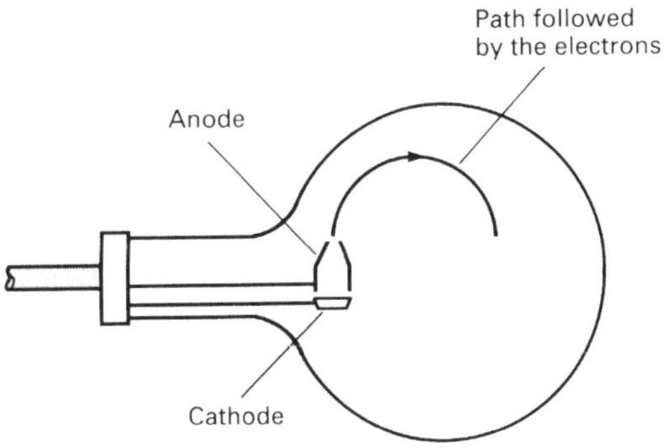

(a) The accelerating voltage is 5.0 kV. Calculate the speed of the electrons as they emerge from the anode.

(b) The apparatus is situated in a uniform magnetic field acting into the plane of the diagram. Explain why the path followed by the beam is circular. Calculate the radius of the circular path for a flux density of 2.0×10^{-3} T.

(c) Suggest a possible process by which the gas in the tube might make the path of the beam visible.

(Specific charge of an electron = 1.8×10^{11} C kg^{-1}.) [AEB, '87]

G57 **(a)** Explain what is meant by quantization of charge.

(b) A cloud of oil droplets is formed between two horizontal parallel metal plates. Explain the following observations:

(i) In the absence of an electric field between the plates, all the oil droplets fall slowly at uniform speeds.

(ii) On applying a vertical electric field, some droplet speeds are unaltered, some are increased downwards, whereas some droplets move upwards. [W, '90]

G58 **(a)** A charged oil drop falls at constant speed in the Millikan oil drop experiment when there is no PD between the plates. Explain this.

(b) Such an oil drop, of mass 4.0×10^{-15} kg, is held stationary when an electric field is applied between the two horizontal plates.

If the drop carries 6 electric charges each of value 1.6×10^{-19} C, calculate the value of the electric field strength.

(Assume $g = 9.8\,\text{m}\,\text{s}^{-2}$.) [L]

G59 **(a)** A charged drop of oil is held stationary in the space between two metal plates across which an electric field is applied. Explain with the aid of a diagram how the forces acting enable the drop to remain stationary.

(b) In an investigation using the above arrangement the voltage across the plates is recorded. The drop is then given a different charge and the voltage adjusted until the drop is stationary again. This procedure is repeated several times and the voltage recorded on each occasion.

The following voltage values were obtained:

142, 425, 569, 709, 999.

Discuss whether these results support the notion that the charge on the drop is a multiple of some fundamental value. [S]

G60 In a measurement of the electron charge by Millikan's method, a potential difference of 1.5 kV can be applied between horizontal parallel metal plates 12 mm apart. With the field switched off, a drop of oil of mass 10^{-14} kg is observed to fall with constant velocity $400\,\mu\text{m}\,\text{s}^{-1}$. When the field is switched on, the drop rises with constant velocity $80\,\mu\text{m}\,\text{s}^{-1}$. How many electron charges are there on the drop? (You may assume that the air resistance is proportional to the velocity of the drop, and that air buoyancy may be neglected.)

(The electronic charge $= 1.6 \times 10^{-19}$ C, the acceleration due to gravity $= 10\,\text{m}\,\text{s}^{-2}$.) [S]

G61 **(a)** In an experiment to attempt to confirm Millikan's conclusion that the electron carries a discrete charge, a charged oil droplet of known mass was held stationary between a pair of parallel horizontal plates by an electric field. The PD producing the field was measured. The charge on the drop was changed and the new voltage to maintain equilibrium was measured. The experiment was repeated and the following results were obtained.

Measurement	*PD required/V*
1	225
2	110
3	150
4	50

(i) Draw a diagram which shows the forces acting on the oil drop when it is held stationary in the field. State the origins of the forces.

(ii) Suggest a means by which the charge on the oil drop could be changed.

(iii) Explain clearly how these measurements suggest that Millikan's conclusion was correct.

(iv) The separation of the parallel plates was 0.020 m. Making the assumption that the oil drop in the first set of observations carried an excess of two electrons, calculate the force acting on the oil drop when the PD was switched off.

The charge on an electron, $e = -1.6 \times 10^{-19}$ C.

(b) In order to determine the mass of an electron the specific charge of an electron has first to be found.

(i) State what is meant by the term *specific charge* and state the unit in which it is measured.

(ii) Describe an experiment to determine a value for the specific charge. Your account should include

—a diagram of the apparatus showing essential circuitry

—the experimental procedure stating clearly the measurements you would make

—an explanation of the theory of the method. [AEB, '90]

G62 In Millikan's experiment an oil drop of mass 1.92×10^{-14} kg is stationary in the space between the two horizontal plates which are 2.00×10^{-2} m apart, the upper plate being earthed and the lower one at a potential of -6000 V. State, with the reason, the sign of the electric charge on the drop. Neglecting the buoyancy of the air, calculate the magnitude of the charge.

With no change in the potentials of the plates, the drop suddenly moves upwards and attains a uniform velocity. Explain why **(a)** the drop moves, **(b)** the velocity becomes uniform.

(The acceleration due to gravity $= 10\,\text{m s}^{-2}$.) [J]

G63 A small oil drop, carrying a negative electric charge, is falling in air with a uniform speed of $8.00 \times 10^{-5}\,\text{m s}^{-1}$ between two horizontal parallel plates. The upper plate is maintained at a positive potential relative to the lower one. Draw a diagram showing all the forces acting on the drop, stating the cause of each force.

Use the following data to determine the charge on the oil drop.

(Radius of drop	$= 1.60 \times 10^{-6}\,\text{m}$.
Density of oil	$= 800\,\text{kg m}^{-3}$.
Density of air	$= 1.30\,\text{kg m}^{-3}$.
Viscosity of air	$= 1.80 \times 10^{-5}\,\text{N s m}^{-2}$.
Distance between plates	$= 1.00 \times 10^{-2}\,\text{m}$.
PD between plates	$= 2.00 \times 10^{3}\,\text{V}$.
Acceleration of free fall, g	$= 10\,\text{m s}^{-2}$.)

[L*]

G64 Describe, with the necessary theory, some experiment by which the charge on the electron has been determined.

An oil drop, of mass $3.2 \times 10^{-15}\,\text{kg}$, falls vertically with uniform velocity through the air between vertical parallel plates 3 cm apart. When a potential difference of 2000 V is applied between the plates, the drop moves with uniform velocity at an angle of 45° to the vertical. Calculate the charge on the drop.

The path of the drop suddenly changes, becoming inclined at 18°26′ to the vertical; later, the path changes again and becomes inclined at 33°42′ to the vertical. Estimate from these data the elementary unit of charge (electron charge). [S]

G65 **(a)** Describe, giving a labelled diagram, the essential constructional features of the cathode ray tube and outline how the electron beam is focused, varied in intensity, and deflected.

(b) In a cathode ray tube the electrons are accelerated through a potential difference of 500 V and then pass between deflecting plates which are 0.05 m long.

(i) Calculate the time it takes an electron to pass between the plates.

(ii) If the PD across the plates is 10 V DC and the plates are 1 cm apart, calculate the angle through which the electrons are deflected.

(iii) Explain why the deflection produced by an alternating voltage, applied to the deflecting plates, can be zero at some high frequency.

($e/m_e = 1.76 \times 10^{11}\,\text{C kg}^{-1}$.) [W]

G66 Two signals of identical voltage, amplitude and frequency, but different phase, are fed to a cathode ray oscilloscope, one to the X plates and one to the Y plates.

Sketch and briefly explain what form of trace you would get if the phase difference was **(a)** zero, **(b)** 90°, and **(c)** 180°. [W]

G67 **(a)** Draw a labelled diagram of a cathode ray tube showing to which of the electrodes the following controls would be connected: brilliance, focus, X and Y shifts, time base. (Details of the circuitry are not required.)

(b) Calculate the accelerating potential which would be required to give the electrons a velocity of $2.0 \times 10^{7}\,\text{m s}^{-1}$. Indicate on your diagram between which electrodes this potential would be applied and show the polarity.

(c) Explain how you would use an oscilloscope to measure an alternating potential difference of the order of 10 V peak-to-peak. (You may assume that an alternating PD of 5 V peak-to-peak is available.) What is the RMS value of a peak-to-peak voltage of 10.0 V?

(d) Sketch the waveform which you would expect to see if the Y plates of an oscilloscope were connected to a sinusoidally varying PD of frequency 100 Hz and the X plates were connected:

(i) to a PD with a saw-tooth waveform of frequency 50 Hz,

(ii) to a PD with a sinusoidal waveform of frequency 50 Hz,

(iii) together and earthed.

(Specific electronic charge $e/m = 1.76 \times 10^{11}\,\text{C kg}^{-1}$.) [AEB, '79]

G68 **(a)** With the aid of sketch graphs, one in each instance, describe how the displacement of the spot from the centre of the screen of a cathode-ray oscilloscope varies with time when the time-base is **(i)** linear, **(ii)** sinusoidal, there being no potential difference between the Y plates.

(b) Describe how you would calibrate a sinusoidal signal generator at frequencies of 100, 200 and 300 Hz, using a CRO and a sinusoidal signal generator providing a standard frequency of 100 Hz.

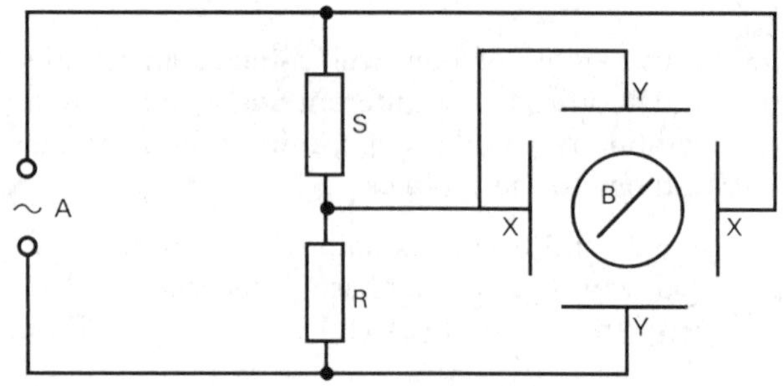

(c) In the diagram A is an alternating voltage supply, S is a standard resistor of 10 000 Ω resistance, R is a resistor of comparable resistance and B is a CRO. With the internal time-base switched off, the trace is as shown. Account for the form of the trace and explain how the resistance of R could be determined from measurements on this trace if the X and Y deflection sensitivities, in V mm^{-1}, are known.

(d) State and explain one possible advantage of this method of measuring the resistance of R. [J]

NUCLEAR PHYSICS (Chapters 51–54)

G69 Chlorine has a *proton number* (*atomic number*) 17 and, as occurring in nature, a *relative atomic mass* 35.5. There are two naturally occurring *isotopes* of chlorine, with *nucleon numbers* (*mass numbers*) 35 and 37. Explain the meaning of the terms in italics. What further information can be obtained from the data given? Calculate the relative abundance of the two isotopes in naturally occurring chlorine.

Explain why it is impossible to separate the different isotopes of an element by ordinary chemical means.

Describe *in outline* one method by which the existence of different isotopes of an element can be demonstrated. [L]

G70 Describe, and give the theory of, a mass spectrometer for measuring the charge/mass ratio e/M of positive ions.

Mention some of the uses of the mass spectrometer.

An examination of copper (atomic number 29) in the mass spectrometer shows two lines corresponding to relative atomic masses 62.93 and 64.93 respectively. The relative atomic mass of copper is found by chemical methods to be 63.55. What is the proportion of each isotope of copper? What elementary particles, and how many of each, comprise the atoms of the two copper isotopes? [S]

G71 Describe and give the theory of, a mass spectrometer for measuring the charge/mass ratio e/M of positive ions.

The mass of the singly charged neon isotope $^{20}_{10}Ne^{+}$ is 3.3×10^{-26} kg. A beam of these ions enters a uniform transverse magnetic field of 0.3 T, and describes a circular orbit of radius 0.22 m. What is:

(a) the velocity of the ions,

(b) the potential difference which has been used to accelerate them to this velocity?

(The electronic charge $= 1.6 \times 10^{-19}$ C.)[S]

G72 The diagram shows a mass spectrometer used for measuring the masses of isotopes. It consists of an ion generator and accelerator, a velocity selector and an ion separator, all in a vacuum.

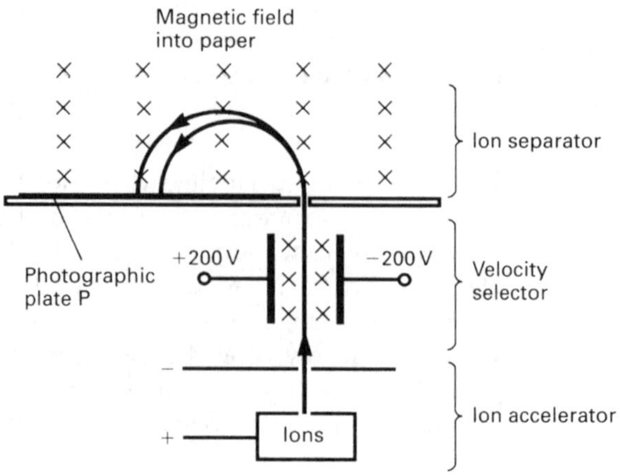

In one experiment, tin ions, each of which carries a charge $+1.6 \times 10^{-19}$ C, are produced in the ion generator and are then accelerated by a p.d. of 20 000 V. Tin has a number of isotopes, two of which are tin–118 (^{118}Sn) and tin–120 (^{120}Sn).

(a) **(i)** State one similarity and one difference between the isotopes of tin.

(ii) Assuming that an ion of tin–120 is at rest before being accelerated, show that the final speed after acceleration is 177 km s^{-1}.

Mass of a nucleon $= 1.7 \times 10^{-27}$ kg

(iii) What will be the final speed of an ion of tin–118?

(b) In practice all ions produced by the ion generator have a range of speeds. A velocity selector is used to isolate ions with a single speed. In the velocity selector the force produced by the electric field is balanced by that due to the magnetic field which is perpendicular to the plane of the paper.

(i) The plates producing the electric field have a separation of 2.0 cm. The potentials of the plates are marked on the diagram. What is the magnitude of the force on an ion due to this electric field in the velocity selector?

(ii) Write down the equation which must be satisfied if the ions are to emerge from the exit hole of the velocity selector. Define the terms in the equation.

(iii) What magnetic flux density is required if ions travelling with a speed of 177 km s^{-1} are to be selected?

(c) After selection the ions are separated using a magnetic field on its own, as shown in the diagram.

(i) Explain why the ions move in circular paths in this region.

(ii) Show that the radius of the path is directly proportional to the mass of the ion.

(iii) The ions are detected using the photographic plate P. Determine the distance between the points of impact on the photographic plate of the two isotopes of tin when a magnetic flux density of 0.75 T is used in the ion separator.

(d) Explain whether the distance between the points of impact of the ions would be the same, greater or smaller for two isotopes of uranium, one with a nucleon number of 236 and the other 238, assuming that they have the same velocity as the tin ions.

[AEB, '92]

G73 Define *nucleon number* (*mass number*) and *proton number* (*atomic number*) and explain the term *isotope*. Describe a simple form of mass spectrometer and indicate how it could be used to distinguish between isotopes.

In the naturally occurring radioactive decay series there are several examples in which a nucleus emits an α particle followed by two β particles. Show that the final nucleus is an isotope of the original one. What is the change in mass number between the original and final nuclei? [L]

G74 Part of the actinium radioactive series can be represented as follows:

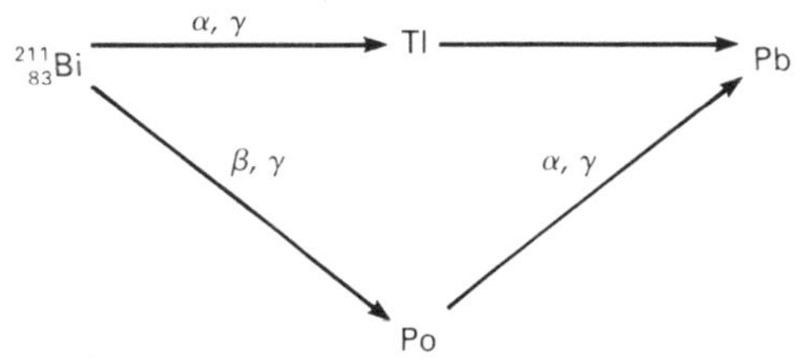

The symbols above the arrows indicate the modes of decay.

(a) Write down the atomic numbers and mass numbers of Tl, Po and Pb in this series.

(b) What is a possible mode of decay for the stage Tl to Pb? [C(O)]

G75 Explain the following observations:

(a) A radioactive source is placed in front of a detector which can detect all forms of radioactive emission. It is found that the activity registered is noticeably reduced when a thin sheet of paper is placed between the source and the detector.

(b) A brass plate with a narrow vertical slit is now placed in front of the radioactive source and a horizontal magnetic field, normal to the line joining source and detector, is applied. It is found that the activity recorded is further reduced.

(c) The magnetic field in (b) is removed and a sheet of aluminium is placed in front of the source. The activity recorded is similarly reduced.

(d) The aluminium sheet in (c) is replaced by a sheet of lead and the detector records much less activity. This activity is not affected by the reintroduction of the magnetic field. [AEB, '83]

G76 A certain α-particle track in a cloud chamber has a length of 37 mm. Given that the average energy required to produce an ion pair in air is 5.2×10^{-18} J and that α-particles in air produce on average 5.0×10^3 such pairs per mm of track, find the initial energy of the α-particle. Express your answer in MeV.

(Electron charge $= 1.6 \times 10^{-19}$ C.) [C]

G77 **(a)** You are required to carry out an experiment to determine the count rate measured when different thicknesses of aluminium sheet are placed in turn between a radioactive source and a G–M tube. How could you obtain a reliable set of results so that you could plot a graph of the measured count rate *due to the source* against the thickness of the aluminium sheet? You may assume that a set of absorbers is available as well as a ratemeter or scaler and the usual apparatus of a physics laboratory.

(b) The table below refers to an experiment similar to that described in (a). Mean count rates due to the source are shown with the thicknesses of different aluminium absorbers.

Absorber thickness/ mm	*Mean count rate/* count min^{-1}
0.11	988
0.26	786
0.37	679
0.50	542
0.73	417
0.90	323
1.23	202
1.57	105

(i) Plot a graph of mean count rate against absorber thickness.

(ii) When a sheet of aluminium foil is folded into 16 thicknesses and positioned between the source and detector, the mean count rate is 815 count min^{-1}. Use the graph to determine the thickness of a single sheet of the foil.

(iii) The uncertainty in the count rate due to the randomness of radioactive emission is ± 30 when the count rate is 815 count min^{-1}. Use this information and the graph to find the uncertainty in the thickness of the aluminium you have calculated in (ii).

(c) The following results were also obtained with the same experimental arrangement as that described in (b):

aluminium absorber, 3.50 mm thick, count rate due to source = 12 count min^{-1}.
lead absorber, 3.50 mm thick, count rate due to source = 10 count min^{-1}.

Explain the significance of these results. [J, '92]

G78 What is gamma-radiation? Explain *one* way in which it originates.

An experiment was conducted to investigate the absorption by aluminium of the radiation from a radioactive source by inserting aluminium plates of different thicknesses between the source and a Geiger tube connected to a ratemeter (or scaler). The observations are summarised in the following table:

Thickness of aluminium /cm	*Corrected mean count rate* /min^{-1}
2.3	1326
6.9	802
11.4	496
16.0	300

Use these data to plot a graph and hence determine for this radiation in aluminium the *linear absorption coefficient*, μ (defined by $\mu = -\frac{dI}{I}\frac{1}{dx}$ where I is the intensity of the incident radiation and dI is the part of the incident radiation absorbed in thickness dx).

Draw a diagram to illustrate the arrangement of the apparatus used in the experiment and describe its preliminary adjustment.

What significance do you attach to the words 'corrected' and 'mean' in the table? [J]

G79 **(a)** What are gamma-rays? Mention one way in which they originate.

(b) Describe, with the aid of a labelled diagram, how you would verify experimentally the inverse square law for gamma-rays using a Geiger–Müller tube. Discuss the precautions you would take and explain how you would present your results graphically.

(c) The window of a gamma-ray detector has an area of $4.0 \times 10^{-4}\,m^2$ and is placed horizontally so that it lies 2.0 m vertically above and on the axis of an effective point source of gamma-rays. When a sheet of gamma-ray absorber is introduced between source and detector, the initial rate of arrival of gamma-rays at the widow, 60 photon min^{-1}, can be maintained only by moving the gamma-ray detector vertically down through 0.20 m. Estimate **(i)** the rate of emission of gamma-rays from the source and **(ii)** the percentage of gamma-rays effectively absorbed by the sheet. [J]

G80 A sample of iodine contains 1 atom of the radioactive isotope iodine 131 (^{131}I) for every 5×10^7 atoms of the stable isotope iodine-127. Iodine has a proton number of 52 and the radioactive isotope decays into xenon 131 (^{131}Xe) with the emission of a single negatively charged particle.

(a) State the similarities and differences in composition of the nuclei of the two isotopes of iodine.

(b) What particle is emitted when iodine 131 decays? Write the nuclear equation which represents the decay.

(c) The diagram shows how the activity of a freshly prepared sample of the iodine varies with time. Use the graph to determine the decay constant of iodine 131. Give your answer in s^{-1}.

(d) Determine the number of iodine atoms in the original sample. [AEB, '91]

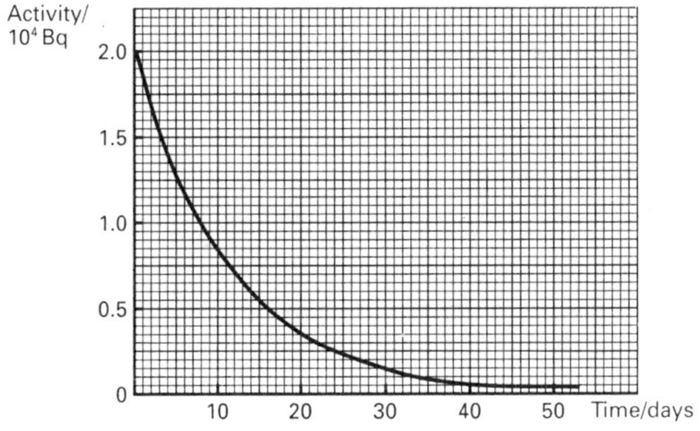

G81 A radioactive source has a half-life of 20 days. Calculate the activity of the source after 70 days have elapsed if its initial activity is 10^{10} Bq.

G82 The radioactive isotope $^{218}_{84}Po$ has a half-life of 3 min, emitting α-particles according to the equation

$$^{218}_{84}Po \rightarrow \alpha + ^{x}_{y}Pb.$$

What are the values of x and y?

If N atoms of $^{218}_{84}Po$ emit α-particles at the rate of $5.12 \times 10^4\,s^{-1}$, what will be the rate of emission after $\frac{1}{2}$ hour? [S]

G83 **(a)** In the uranium decay series $^{238}_{92}U$ decays by stages to $^{234}_{92}U$.

(i) Why is the new nucleus given the symbol U?

(ii) Identify the total number and types of particles that have been emitted in this transformation.

(b) An isotope of the element radon has a half-life of 4 days. A sample of radon originally contains 10^{10} atoms.

(Take 1 day to be 86×10^3 s.)

Calculate:

(i) the number of radon atoms remaining after 16 days;

(ii) the radioactive decay constant for radon;

(iii) the rate of decay of the radon sample after 16 days. [O, '92]

G84 The half-life of $^{30}_{15}P$ is 2.5 minutes. Calculate the mass of $^{30}_{15}P$ which has an activity of 10^{15} Bq.

(The Avogadro constant $= 6.0 \times 10^{23}\,mol^{-1}$.)

G85 The activity of a particular radioactive nuclide falls from 1.0×10^{11} Bq to 2.0×10^{10} Bq in 10 hours. Calculate the half-life of the nuclide.

G86 Calculate the activity of 2.0 μg of $^{64}_{29}$Cu.

(The half-life of $^{64}_{29}$Cu = 13 hours.
The Avogadro constant = 6.0×10^{23} mol^{-1}.)

G87 The radioactive isotope of iodine ^{131}I has a half-life of 8.0 days and is used as a tracer in medicine. Calculate:
(a) the number of atoms of ^{131}I which must be present in the patient when she is tested to give a disintegration rate of 6.0×10^5 s^{-1},
(b) the number of atoms of ^{131}I which must have been present in a dose prepared 24 hours before. [J, '91]

G88 The activity of a mass of $^{14}_{6}$C is 5×10^8 Bq and the half-life is 5570 years. Estimate the number of $^{14}_{6}$C nuclei present.

($\ln 2 \approx 0.69$.) [J]

G89 **(a)** A sample initially consists of N_0 radioactive atoms of a single isotope. After a time t the number N of radioactive atoms of the isotope is given by:

$$N = N_0 e^{-\lambda t}$$

(i) Sketch a graph of this equation and show on the graph the time equal to the half-life of the sample, $T_{1/2}$.
(ii) Explain what is meant by the disintegration rate of the sample and represent this on the graph at zero time and at time $T_{1/2}$. State the ratio of these two disintegration rates.
(iii) Explain the physical significance of the constant λ in the equation above.
(b) If you are provided with a small gamma ray source of very long half-life, describe the arrangement you would use, and the measurements you would make to investigate the inverse-square law for gamma rays. Show how you would use your measurements to verify the law. [J]

G90 **(a)** Discuss the assumption on which the law of radioactive decay is based. What is meant by the *decay constant* λ and the *half-life* $T_{1/2}$ of a radioactive isotope? Show, *from first principles*, that

$$\lambda T_{1/2} = 0.69$$

(b) Outline an experiment to verify the decay law if a fairly weak radioactive source with $T_{1/2}$ about an hour is available. Make clear the readings you would take and the way in which you would use them.
(c) At a certain time, two radioactive sources R and S contain the same number of radioactive nuclei. The half-life is 2 hours for R and 1 hour for S. Calculate:
(i) the ratio of the rate of decay of R to that of S at this time.
(ii) the ratio of the rate of decay of R to that of S after 2 hours.
(iii) the proportion of the radioactive nuclei in S which have decayed in 2 hours. [O & C, '92]

G91 **(a)** The various *isotopes* of an element X are distinguished by using the notation $^{A}_{Z}$X. Explain the meaning of A, Z and of the term *isotope*.
(b) Radioactive sources which might be used in schools are ^{226}Ra which emits α-, β- and γ-rays, and ^{90}Sr which emits β-rays only.
(i) List *three* safety precautions which need to be taken into account when using such sources.
(ii) The half-life of ^{90}Sr is 28 years. When its activity falls to 25% of its original value it should be replaced. After how many years should it be replaced?
(c) **(i)** Explain briefly how β-particles are emitted by the nucleus.
(ii) When $^{226}_{88}$Ra emits an α-particle it decays to radon (Rn). Write down a balanced equation for this change.
(d) Radioactive isotopes have many applications merely by virtue of being isotopes. Describe and explain *one* such application.
(e) **(i)** Describe briefly Rutherford's α-particle scattering experiment and summarise the evidence that it provided for the nuclear model of the atom.
(ii) How would the results be different if aluminium foil were used instead of the gold foil in such an experiment? [W, '90]

G92 In 420 days, the activity of a sample of polonium, Po, fell to one-eighth of its initial value. Calculate the half-life of polonium. Give the numerical values of a, b, c, d, e, f in the nuclear equation

$$^{a}_{b}\text{Po} \rightarrow {}^{c}_{d}\alpha + {}^{206}_{82}\text{Pb} + {}^{e}_{f}\gamma$$

G93 A point source of γ-radiation has a half-life of 30 minutes. The initial count rate, recorded by a Geiger counter placed 2.0 m from the source, is $360\,s^{-1}$. The distance between the counter and the source is altered. After 1.5 hour the count rate recorded is $5\,s^{-1}$. What is the new distance between the counter and the source? [L]

G94 **(a)** **(i)** What are alpha, beta and gamma rays?

(ii) Describe briefly one method whereby they may be distinguished from one another experimentally.

(b) Explain what is meant by:

(i) radioactive decay,

(ii) radioactive decay constant,

(iii) half-life,

(iv) the becquerel.

(c) **(i)** A newspaper article stated that the NASA Galileo space probe to Jupiter 'contained 49 lb of plutonium to provide 285 watts of electricity through its radioactive thermonuclear generator (RTG)'.

(Note: An RTG is a device for converting thermal energy produced by fission into electrical energy.)

Assuming that the plutonium is ^{239}Pu, which is built into a small nuclear reactor and that the efficiency of the RTG is 10%, what is the maximum time for which the RTG will supply the required energy output?

(Take the energy emitted for each nuclear disintegration of the ^{239}Pu to be 32 pJ, $N_A = 6.0 \times 10^{23}\,mol^{-1}$, 1 lb = 0.45 kg.)

(ii) What factors will tend to: (I) increase, (II) decrease your estimate of the time? [W, '91]

G95 When iron is irradiated with *neutrons* an *isotope* of iron is formed. This isotope is *radioactive* with a *half-value period* (*half-life*) of 45 days. Give the meanings of the terms printed in italics.

A steel piston ring of mass 16 g was irradiated with neutrons until its activity due to the formation of this isotope was 10 microcurie. Ten days after the irradiation the ring was installed in an engine and after 80 days continuous use the crankcase oil was found to have a total activity of 1.85×10^3 disintegrations per second. Determine the average mass of iron worn off the ring per day assuming that all the metal removed from the ring accumulated in the oil and that one curie is equivalent to 3.7×10^{10} disintegrations per second. [J]

G96 Discuss the assumptions on which the law of radioactive decay is based.

What is meant by the *half-life* of a radioactive substance?

A small volume of a solution which contained a radioactive isotope of sodium had an activity of 12 000 disintegrations per minute when it was injected into the bloodstream of a patient. After 30 hours the activity of $1.0\,cm^3$ of the blood was found to be 0.50 disintegrations per minute. If the half-life of the sodium isotope is taken as 15 hours, estimate the volume of blood in the patient. [J]

G97 A tube containing an isotope of radon, $^{222}_{86}Rn$, is to be implanted in a patient. The radon has an initial activity of 1.6×10^4 Bq, a half-life of 4 days and it decays by alpha emission. To provide the correct dose, the tube, containing a freshly prepared sample of the isotope, is to be implanted for 8 days.

(a) **(i)** What are the proton (atomic) number and the nucleon (mass) number of the daughter nucleus produced by the decay of the radon?

(ii) State one reason why an alpha emitter is preferred to a beta or gamma emitter for such purposes.

(b) Determine:

(i) the decay constant for radon in s^{-1}

(ii) the initial number of radioactive radon atoms in the tube.

(c) The operation to implant the tube has to be delayed.

Ignoring the effects of any daughter products of the decay, determine the maximum delay possible if the patient is to receive the prescribed dose using the source. [AEB, '92]

G98 $^{210}_{83}Bi$ is a *radioactive isotope* of bismuth with a *half-life* period of 5.00 days, which emits *negative beta particles*. Explain the italicized terms in the above statement and the significance of the numbers 210 and 83.

Describe experiments you would do to verify that alpha and gamma radiations from $^{210}_{83}$Bi are negligible. (You are not required to explain the principles underlying the action of any detector you may use.)

This isotope appears to be an ideal source of beta particles for an experiment you wish to perform. Assuming that for your experiments, which run continuously for 300 hours, the strength of the source must not fall below 10 μCi, what strength of source is required at the start of an experiment? [O & C]

G99 A decay sequence for a radioactive atom of radon 219 to a stable lead 207 atom is as shown below.

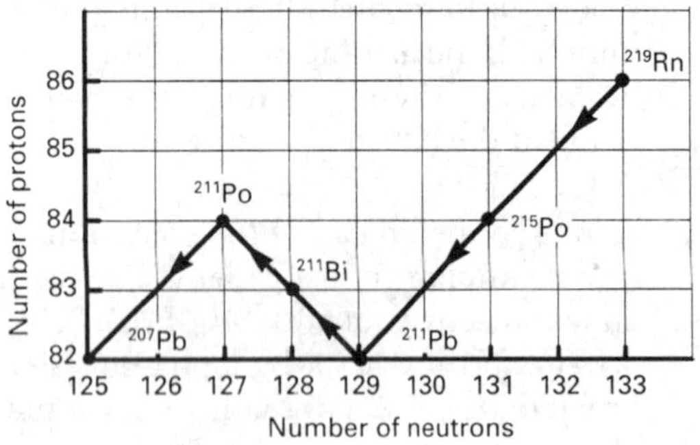

(a) What do the numbers on the symbol $^{207}_{82}$Pb represent?

(b) (i) Write down a nuclear equation representing the decay of $^{219}_{86}$Rn to $^{215}_{84}$Po.

(ii) Write down the name of the particle which is emitted in this decay.

(c) (i) What particle is emitted when $^{211}_{83}$Bi decays?

(ii) What happens within the nucleus to cause this decay?

(d) The half-life of $^{219}_{86}$Rn is 4.0 s. At time $t = 20$ s, what fraction of the radon atoms present at time $t = 0$, will be undecayed? [C, '92]

G100 The potassium isotope $^{42}_{19}$K has a half-life of 12 hr, and disintegrates with the emission of a γ-ray to form the calcium isotope $^{42}_{20}$Ca. What other radiation besides γ-rays must be emitted? How many electrons, protons, and neutrons are there in an atom of the calcium isotope?

The amount of radiation received in unit time by a person working near a radioactive source, commonly called the dose rate, is measured in rem hr^{-1}. The safety regulations forbid dose rates in excess of 7.5×10^{-4} rem hr^{-1}. The γ-ray dose rate from the $^{42}_{19}$K source is found to be 3×10^{-3} rem hr^{-1} at a distance of 1 m. What is the minimum distance from this source at which it is safe to work?

After how long will it be safe to work at a distance of 1 m from this source? [S]

G101 (a) Why is radioactive decay described as a 'random process'?

(b) The equation for the rate of decay of a radioactive nuclide is

$$-\frac{dN}{dt} = \lambda N$$

where N is the number of atoms surviving at time t, and λ is the constant for the particular decay.

(i) Obtain the expression for the value of N at time t, given that $N = N_0$ when $t = 0$.

(ii) Define the half-life, $T_{1/2}$, for the decay, and express it in terms of λ.

(iii) Obtain the expression for the number of nuclei that have disintegrated during the period up to time t.

(c) The hydrogen isotope $^{3}_{1}$H (tritium) is a β-emitter of half-life 12.3 years.

(i) Outline briefly how you would identify the emission as β-particles, and how you would measure the half-life of the decay.

(ii) What is the difference in structure between a tritium atom and an ordinary hydrogen atom, and what is the product of the tritium decay? [O]

G102 Explain what is meant by **(a)** *radioactivity*, **(b)** *half-life*.

Why is it necessary to specify the half-life of a radioactive substance rather than the full life?

Describe how you would determine experimentally the half-life of a radioactive substance of comparatively short half-life explaining how you would calculate the result from your observations.

Describe an experiment to demonstrate the sign of the charge carried by one form of radiation from radioactive substances. [L]

G103 How are **(a)** the atomic number, **(b)** the atomic mass number of a radioactive nucleus changed by the emission of **(i)** an α-particle, **(ii)** a β-particle, **(iii)** a γ-ray?

Very briefly explain how the nature of α-particles has been established.

How may the half-life of a radioactive material, of which the half-life is known to be about 10 minutes, be determined? [S]

G104 **(a)** Explain what is meant by *radioactive decay*, and state the nuclear changes that accompany the emission of an α-particle, a β-particle and a γ-ray photon.

(b) What do you understand by the *half-life* of a particular decay process? Given that the activity of a source undergoing a single type of decay is I_0 at time $t = 0$, obtain an expression in terms of the half-life $T_{1/2}$ for the activity I at any subsequent time t.

(c) A demonstrator has a thorium hydroxide preparation which produces thoron gas with a half-life of 56 seconds. Describe how he would demonstrate:

(i) that radioactivity involves decay, with a half-life that can be measured;

(ii) that radioactive decay is a random process. [O]

G105 **(a)** Explain what are meant by the *half-life* and the *decay constant* of a radioactive isotope.

(b) At the start of an experiment a mixture of radioactive materials contains 20.0 μg of a radioisotope A, which has a half-life of 70 s, and 40.0 μg of radioisotope B, which has a half-life of 35 s.

(i) After what period of time will the mixture contain equal masses of each isotope? What is the mass of each isotope at this time?

(ii) Calculate the rate at which the atoms of isotope A are decaying when the masses are the same.

(Molar mass of isotope A = 234 g, the Avogadro constant = 6.0×10^{23} mol^{-1}.) [AEB, '87]

G106 **(a)** Radioactive decay is a random process in which the time of decay of an individual nucleus cannot be predicted. The decay of a radioactive substance follows an exponential law with a well-defined decay constant. Explain how these two statements are consistent.

(b) You are given a small radioactive source believed to emit two different types of radiation. Describe tests you might carry out to identify the radiations present.

(c) The isotope of bismuth of mass number 200 has a half-life of 5.4×10^3 s. It emits alpha particles with an energy of 8.2×10^{-13} J.

(i) State the meaning of the term *half-life*.

Calculate for this isotope:

(ii) the decay constant,

(iii) the initial activity of 1.0×10^{-6} mole of the isotope,

(iv) the initial power output of this quantity of the isotope.

($N_A = 6.0 \times 10^{23}$ mol^{-1}.) [S]

G107 A Geiger–Müller tube was fixed with its axis horizontal at a place on a bench well removed from all known radioactive sources. Background count rate measurements were made by recording counts over several ten-minute periods. The following counts were obtained for five such periods: 290, 277, 273, 263 and 247. Find the mean background count rate in counts per minute.

A weak radioactive source was then mounted on the axis of the tube with its protective grille facing the end window of the tube, so that it could be moved along the axis to give various distances s between the grille and the end-window, as in the Figure below.

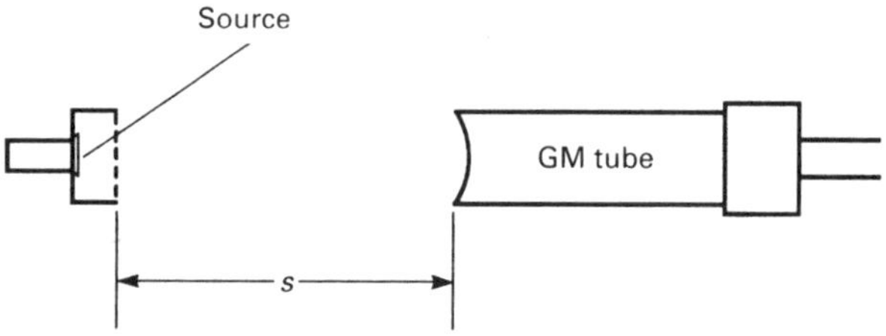

Counts, N, were taken over four-minute periods for the various values of s, with results as follows:

s/mm	N
10	*
15	17 820
30	7980
45	4536
60	2942
75	2076
90	1554
105	1215

*Too rapid for the counter.

Copy this table adding further columns for values of n and $1/\sqrt{n}$, where n is the corrected count over four minutes, i.e. counts recorded minus background counts.

It is thought that the relationship between n and s is likely to be of the form $1/\sqrt{n} = k(s+x)$, where k and x are constants. Plot a graph of $1/\sqrt{n}$ against s/mm, and use it to obtain values for k and x.

What practical significance can be attached to x?

Plot a further graph of $\lg n$ against $\lg((s+x)/\text{mm})$ and find its gradient. [S]

G108 Uranium ores contain one radium 226 atom for every 2.8×10^6 uranium 238 atoms. Calculate the half-life of ${}^{238}_{92}U$ given that the half-life of ${}^{226}_{88}Ra$ is 1600 years. (${}^{226}_{88}Ra$ is a decay product of ${}^{238}_{92}U$.)

G109 A uranium-bearing rock is found to contain 9 uranium 238 atoms for every 8 helium atoms present in the rock. Assuming that the decay process which eventually converts a uranium atom to lead involves the emission of 8 α-particles, calculate the age of the rock.

(Half-life of ${}^{238}_{92}U = 4.5 \times 10^9$ years.)

G110 Wood from a buried ship has a specific activity of 1.2×10^2 Bq kg^{-1} due to ${}^{14}C$, whereas comparable living wood has an activity of 2.0×10^2 Bq kg^{-1}. What is the age of the ship?

(Half life of ${}^{14}C = 5.7 \times 10^3$ years.)

[W, '90]

G111 Living wood takes in radioactive carbon 14 from the atmosphere during the process of photosynthesis, the proportion of carbon 14 to carbon 12 atoms being 1.25 to 10^{12}. When the wood dies the carbon 14 decays, its half-life being 5600 years. 4 g of carbon from a piece of dead wood gave a total count rate of 20.0 disintegrations per minute. Estimate the age of the piece of wood.

(Avogadro constant $= 6.0 \times 10^{23}$ mol^{-1}.
One year $= 3.16 \times 10^7$ seconds.
$\ln 2 = \log_e 2 = 0.693$.) [L*]

G112 **(a)** The radioactive isotope ${}^{228}_{90}Th$ emits an alpha particle to become the isotope ${}^{a}_{b}Ra$. Explain the meanings of the superscript and subscript numbers 228 and 90 and state the values of a and b.

(b) The graph shows how the activity in counts per minute (c.p.m) of 1.00 g of processed carbon decreases as it ages.

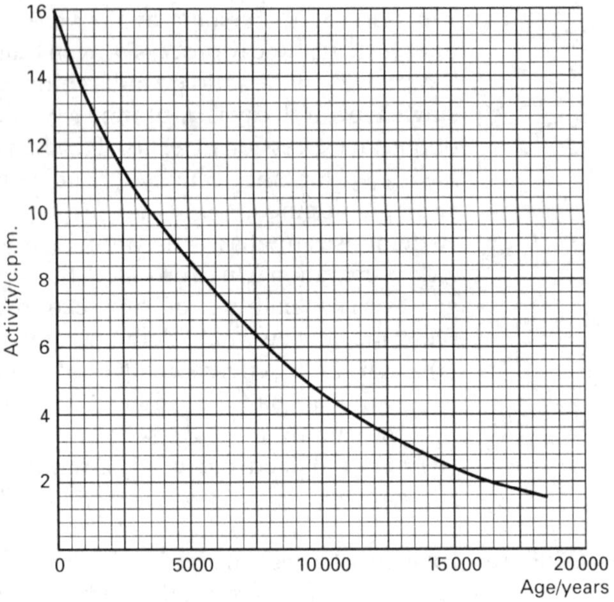

(i) State the half-life of the radioactive isotope present in the carbon.

(ii) 0.50 g of similarly processed carbon taken from some ancient charcoal gave a measured activity of 5.0 c.p.m. Estimate the age of the charcoal. [S]

G113 **(a)** Explain the significance of the characters H, 1 and 2 in the expression

$${}^{2}_{1}H$$

and hence state what is meant by an isotope.

(b) Copy and complete the following equations:
(i) $^{238}_{92}U \rightarrow ^{234}_{90}Th + \ldots$.
(ii) $^{234}_{90}Th \rightarrow ^{234}_{91}Pa + \ldots$.
(c) Sketch a graph of the variation of binding energy per nucleon with atomic mass number. Use your sketch to explain why **(i)** fusion and **(ii)** fission, occur. [W, '91]

G114 Explain what is meant by the binding energy of a nucleus. Using the following data calculate the binding energy per nucleon, in MeV, of an alpha particle.

(Mass of proton = 1.0080 u. Mass of neutron = 1.0087 u. Mass of alpha particle = 4.0026 u. 1 u is equivalent to 930 MeV.) [J]

G115 In the fusion reaction $^2_1H + ^3_1H = ^4_2He + ^1_0n$, how much energy, in joules, is released? (Mass of $^2_1H = 3.345 \times 10^{-27}$ kg; $^3_1H = 5.008 \times 10^{-27}$ kg; $^4_2He = 6.647 \times 10^{-27}$ kg; $^1_0n = 1.675 \times 10^{-27}$ kg.
Speed of light $= 3.0 \times 10^8$ m s^{-1}.) [L]

G116 $^{234}_{92}U$ decays by spontaneous emission to $^{234}_{93}Np$.
(a) What decay process has occurred?
(b) How does the nucleus of U-234 differ from the nucleus of Np-234?
(c) The particle emitted in the decay process has kinetic energy. What is the source of this energy? [L]

G117 **(a)** **(i)** Explain what is meant by the *mass defect* of an atomic nucleus.
(ii) Describe how the binding energy of an atomic nucleus can be found with the aid of this quantity.
(b) **(i)** Sketch a graph to show how the binding energies per nucleon of atomic nuclei vary with nuclear mass.
(ii) Explain how details of your graph can be used to identify an isotope that can undergo nuclear fission.
(c) **(i)** Two isotopes of lead, $^{208}_{82}Pb$ and $^{210}_{82}Pb$, have atomic masses 207.977 u and 209.984 u respectively. Calculate the mass defect (in u) of an atom of each of these isotopes.
(The mass of a neutron is 1.009 u, and that of a hydrogen atom is 1.008 u.)
(ii) One (only) of these isotopes is radioactive. What further calculation would identify the unstable isotope? Outline the steps in this calculation and give your conclusion.
(d) A radioactive gas G (contained in a thick glass flask) emits alpha radiation accompanied by gamma radiation. G has a half-life of a few days and decays to a substance S that emits alpha radiation only. S decays to products with negligible activity.

Describe an experimental procedure to establish an accurate value for the half-life of G. [O, '92]

G118 Explain the term *nuclear binding energy*. Sketch a graph showing the variation of binding energy per nucleon number (mass number) and show how both nuclear fission and nuclear fusion can be explained from the shape of this curve.

Calculate in MeV the energy liberated when a helium nucleus (^4_2He) is produced **(a)** by fusing two neutrons and two protons, and **(b)** by fusing two deuterium nuclei (^2_1H). Why is the quantity of energy different in the two cases?

(The neutron mass is 1.008 98 u, the proton mass is 1.007 59 u, the nuclear masses of deuterium and helium are 2.014 19 u and 4.002 77 u respectively. 1 u is equivalent to 931 MeV.) [L]

G119 **(a)** Explain the meaning of the symbol $^{238}_{92}U$. What is meant by an isotope?
(b) Certain types of nucleus may spontaneously lose a small amount of mass by the process known as radioactivity.
(i) Describe the nature of the radiations which may be emitted during this process.
(ii) Define *radioactive decay* and *briefly* explain how this leads to the equation $\frac{dN}{dt} = -\lambda N$.
Why is there a negative sign?

(iii) Sketch the variation of N with time and explain what is meant by half-life.

(c) (i) Draw a sketch of the variation of binding energy per nucleon with mass number. Use the sketch to explain why nuclear fusion occurs in some circumstances and nuclear fission in others.

(ii) Explain why very high temperatures are required for nuclear fusion.

(iii) A future fusion reactor might use the reaction

$$^2_1\text{H} + ^2_1\text{H} \rightarrow ^4_2\text{He} + \text{energy}$$

to produce useful energy. From the following data calculate the number of reactions required to produce 1 J of energy.

(iv) Calculate the mass of ^2_1H required to provide 1 J of energy.

(Mass of ^2_1H = 2.0136 amu, Mass of ^4_2He = 4.0015 amu, 1 amu = 1.661×10^{-27} kg, Velocity of light $c = 3.00 \times 10^8\,\text{m s}^{-1}$) [W, '92]

G120 (a) Explain the meaning of the term *mass difference* and state the relationship between the mass difference and the *binding energy* of a nucleus.

(b) Sketch a graph of nuclear binding energy per nucleon versus mass number for the naturally occurring isotopes and show how it may be used to account for the possibility of energy release by nuclear fission and by nuclear fusion.

(c) The Sun obtains its radiant energy from a thermonuclear fusion process. The mass of the Sun is 2×10^{30} kg and it radiates 4×10^{23} kW at a constant rate. Estimate the life time of the Sun, in years, if 0.7% of its mass is converted into radiation during the fusion process and it loses energy only by radiation. (1 year may be taken as 3×10^7 s.)

(The speed of light, $c = 3 \times 10^8\,\text{m s}^{-1}$.) [J]

G121 An electron and a positron (a particle of equal mass to an electron but with positive charge) may annihilate one another, producing two γ-ray photons of equal energy. What is the minimum energy of each of these photons?

(Mass of electron, $m_e = 9.1 \times 10^{-31}$ kg; speed of light, $c = 3.0 \times 10^8\,\text{m s}^{-1}$.) [C]

G122 Radioactive decay occurs by either α or β emission. Write down the general equation for the decay of an isotope X with nucleon number A and proton number Z to an isotope Y by **(a)** α-emission, **(b)** β-emission.

Part of a radioactive series consists of the following sequence of emissions: α, α, β, α, β, α. Draw this part of the series on a N against Z graph. (N is the number of neutrons in a nucleus.) What are the total changes in N and Z during this part of the series?

How does the ratio N:Z vary for stable nuclei throughout the Periodic Table? Discuss the relevance of this variation to **(i)** the fact that in a radioactive decay series the α-emissions are interspersed with β-emissions, and **(ii)** the likely radioactivity of the fission fragments in a nuclear fission.

The isotopes with the longest half-lives occurring in the four natural radioactive decay series have half-lives of 1.4×10^{10} years, 4.5×10^9 years, 7.1×10^8 years and 2.2×10^4 years respectively. What is the significance of the fact that only those series containing the three longest half-lives occur naturally in the earth? [L]

G123 (a) (i) Explain what is meant by the *nucleon number* and the *nuclear binding energy* of a nucleus.

(ii) Sketch a fully labelled graph to show the variation with nucleon number of the binding energy per nucleon. Hence explain why fusion of nuclei having high nucleon numbers is not associated with a release of energy.

(b) A rolling mill produces sheet aluminium, the thickness of which must be kept within certain limits. In order to achieve this, the thickness of the sheet is monitored as it leaves the final set of rollers. The monitor consists of a β-particle source and a detector placed as shown below.

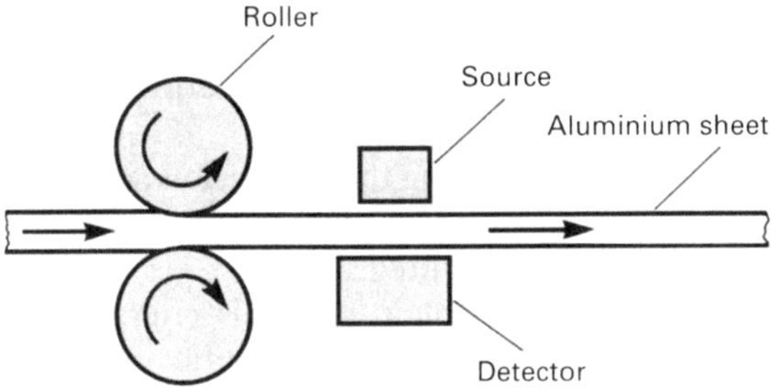

(i) Discuss the advisability, or otherwise, of using a β-emitting source which has
(1) a high activity, giving a count rate many times that of background,
(2) a short half-life.

(ii) For sheets of aluminium about 1 mm thick, the β-particle count rate C/s^{-1} is known to vary with thickness x/mm according to the expression

$$C = C_0 e^{-0.62x}$$

where C_0/s^{-1} is the count rate when $x = 0$.

Calculate the ratio

$$\frac{\text{maximum count rate}}{\text{minimum count rate}}$$

when $x = 1.00$ mm and x varies by $\pm 5.0\%$. [C, '92]

G124 Two reactions which occur in the upper atmosphere are:

$$^{14}_{7}\text{N} + ^{1}_{0}\text{n} \rightarrow ^{14}_{6}\text{C} + ^{1}_{1}\text{H}$$
$$^{14}_{7}\text{N} + ^{1}_{0}\text{n} \rightarrow ^{12}_{6}\text{C} + ^{3}_{1}\text{H}$$

Explain the meaning of the subscript and superscript numbers. Calculate in MeV the energy involved in each reaction. Explain the significance of these quantities. The data given below is for *atomic* masses although the reactions are *nuclear*. Why is this acceptable?

The $^{14}_{6}\text{C}$ produced in the atmosphere in the first reaction above is radioactive. Carbon from carbon dioxide in the atmosphere is absorbed by living material as long as it is alive. The half-life of $^{14}_{6}\text{C}$ is 5600 years.

Explain the meaning of the last statement. How can the above information be used to establish the age of a piece of ancient wood?

(Mass of neutron = 1.008 67 u.
Mass of $^{1}_{1}\text{H}$ = 1.007 83 u.
Mass of $^{3}_{1}\text{H}$ = 3.016 05 u.
Mass of $^{12}_{6}\text{C}$ = 12.000 00 u.
Mass of $^{14}_{6}\text{C}$ = 14.003 24 u.
Mass of $^{14}_{7}\text{N}$ = 14.003 07 u.
1 u ≡ 931 MeV.) [L]

G125 Explain the meanings of the superscript 238 and the subscript 92 in $^{238}_{92}\text{U}$.

How, if at all, would these numbers become altered by **(a)** the emission of one α-particle, **(b)** the emission of one β-particle, and **(c)** the absorption of one neutron?

Give an example of a fusion reaction and explain where reactions of this kind occur in nature.

The mass of the isotope $^{7}_{3}\text{Li}$ is 7.018 u. Find its binding energy given that the mass of $^{1}_{1}\text{H}$ is 1.008 u, the mass of the neutron = 1.009 u and 1 u = 931 MeV. [W]

G126 Explain why the mass of a nucleus is always less than the combined masses of its constituent particles.

Naturally occurring chlorine is a mixture of two isotopes, $^{35}_{17}\text{Cl}$ with a relative abundance of 75% and $^{37}_{17}\text{Cl}$ with a relative abundance of 25%. Explain fully what is meant by this statement, and describe briefly how the relative abundance of the two isotopes could be verified experimentally.

When natural chlorine is irradiated with slow neutrons from a reactor, another isotope of chlorine of mass number 38 is produced which decays by β-emission. What nuclear reaction would you expect to be responsible for producing this isotope, and what are the mass number and atomic number of the nucleus remaining after the β-decay? [O & C]

G127 In a controlled thermal fission reactor what is the function of **(a)** the moderator, **(b)** the control rods, and **(c)** the coolant?

A typical fission reaction is

$$^{235}_{92}\text{U} + ^{1}_{0}\text{n} \rightarrow ^{95}_{42}\text{Mo} + ^{139}_{57}\text{La} + 2^{1}_{0}\text{n} + 7\,^{0}_{-1}\text{e}$$

Calculate the total energy released by 1 g of $^{235}_{92}U$ undergoing fission by this reaction, neglecting the masses of the electrons.

(Mass of neutron = 1.009 u.
Mass of $^{95}_{42}Mo$ = 94.906 u.
Mass of $^{139}_{57}La$ = 138.906 u.
Mass of $^{235}_{92}U$ = 235.044 u.
1 u ≡ 1.66×10^{-27} kg.
Number of atoms in one mole of atoms = 6.02×10^{23}.
Speed of light = 3.00×10^{8} m s^{-1}.) [L*]

G128 A control rod to limit the rate at which a nuclear reactor is working is made from boron which is sealed in a casing. A boron atom ($^{10}_{5}B$) is able to capture a neutron; an atom of lithium (Li) and an alpha particle being produced in the process. As a result helium gas is produced which occupies the spaces between the atoms in the rods which may be assumed to have a crystalline structure.

Each cubic metre of the control rod can absorb 1.5×10^{27} neutrons before it must be replaced.

(a) State the difference between a *crystalline structure* and an *amorphous structure*.
(b) Write down an equation for the nuclear reaction which takes place in the control rods.
(c) How many moles of helium are liberated in each cubic metre of control rod?
(d) The boron atoms themselves occupy 75% of the total volume occupied by the rods. Calculate the pressure inside the casing, at a temperature of 300 K, just before the rod is replaced.

(Molar gas constant $R = 8.3$ J mol^{-1} K^{-1}, Avogadro constant $N_A = 6.0 \times 10^{23}$ mol^{-1}.) [AEB, '90]

G129 **(a)** In terms of the constituents of atomic nuclei, explain the meaning of **(i)** atomic number, **(ii)** mass number, **(iii)** isotope.
(b) Account for the fact that, although nuclei do not contain electrons, some radioactive nuclei emit beta particles.
(c) Cobalt has only one stable isotope, ^{59}Co. What form of radioactive decay would you expect the isotope ^{60}Co to undergo? Give a reason for your answer.
(d) The radioactive nuclei $^{210}_{84}Po$ emit alpha particles of a single energy, the product nuclei being $^{206}_{82}Pb$.
- **(i)** Using the data below, calculate the energy, in MeV, released in each disintegration.
- **(ii)** Explain why this energy does not all appear as kinetic energy, E_α, of the alpha particle.
- **(iii)** Calculate E_α, taking integer values of the nuclear masses.

Nucleus	Mass (u)
$^{210}_{84}Po$	209.936 730
$^{206}_{82}Pb$	205.929 421
α-particle	4.001 504

(1 atomic mass unit, u = 931 MeV.) [J]

G130 Describe the structure of a Geiger–Müller tube. Why are some tubes fitted with thin end windows? Why does the anode of a Geiger–Müller tube have to be made of a *thin* wire?

Explain the principle of operation of a cloud chamber. Describe and explain the differences between the tracks formed in such a chamber by alpha and beta particles.

A radioactive source has decayed to 1/128th of its initial activity after 50 days. What is its half-life? [L]

G131 Draw a labelled diagram of an ionization chamber. How is the ionization current measured? [L]

G132 **(a)** The path of a charged particle through a cloud chamber appears as a thin white line. What does this line consist of and how is it formed?

How may the tracks of beta particles be distinguished from those of alpha particles in the absence of any deflecting electric or magnetic fields?

(b) The diagram shows an exaggerated view of the track of a beta particle as it passes through a thin sheet of lead in a cloud chamber. A uniform magnetic field acts *into* the paper, through the whole region shown in the diagram, and the beta particle moves in the plane of the paper.

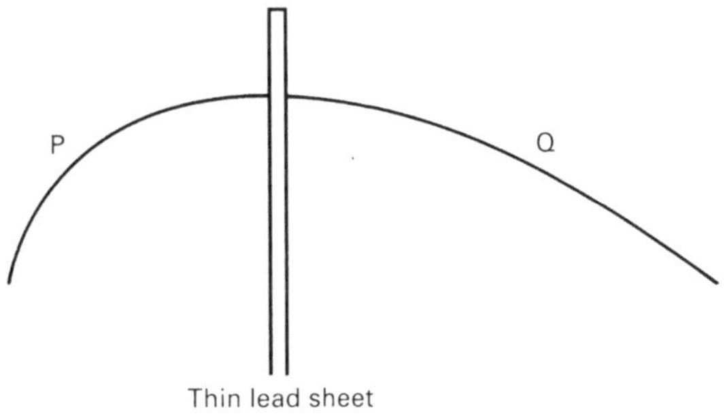

(i) Is the beta particle travelling in the direction P to Q or Q to P? Explain your reasoning.
(ii) Does this beta particle carry a positive or negative charge? Explain your reasoning.

(c) A $^{40}_{19}$K atom has mass 39.964 001 u and a $^{40}_{20}$Ca atom has mass 39.962 582 u where 1 u equals 1.6604×10^{-27} kg. The speed of light in vacuum is 3.00×10^{8} m s^{-1}.
(i) Write down an equation which describes the decay of a $^{40}_{19}$K atom into a $^{40}_{20}$Ca atom and identify the particle emitted by the nucleus.
(ii) Why, in the decay process, does the number of extranuclear electrons increase by one?
(iii) Calculate the energy released from the $^{40}_{19}$K nucleus during the decay.
(iv) Describe one possible mode of decay of a $^{40}_{19}$K nucleus into a $^{40}_{18}$Ar nucleus. [L]

G133 (a) A G–M tube is exposed to a constant flux of alpha particles. The graph below shows how the recorded count rate depends on the potential difference across the tube.

Draw and label a diagram of a G–M tube. Outline its working principle with reference to what happens when an alpha particle enters the tube. Explain why there is an upper limit to the rate at which a G–M tube can detect α-particles.

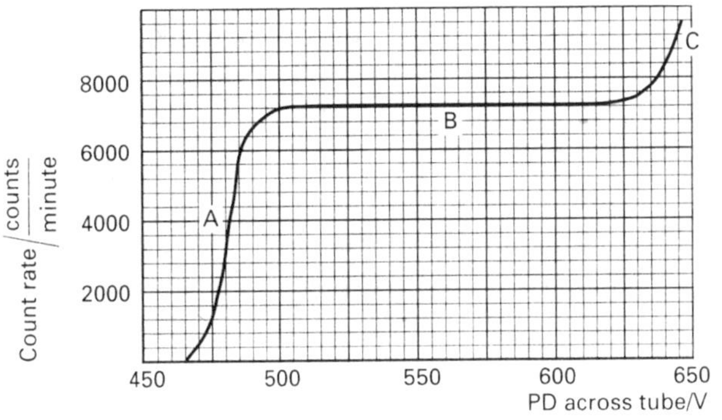

How do you account for:
(i) the sharp rise in the recorded count rate at A,
(ii) the 'plateau' at B, and
(iii) the uncontrolled rise in the recorded count rate at C?

State what potential difference you would choose for the Geiger counter whose response is shown in the graph. Give one good reason for your choice.

(b) A small amount of ^{24}Na is smeared on to a card and its activity falls by 87.5% in 45 h. What is the half-life of ^{24}Na? Describe how you would use a G–M tube in conjunction with a suitable counter to measure the half-life of ^{24}Na. Explain carefully how the result is found from the measurements.

$$\left(\text{Decay constant} = \frac{0.693}{\text{half-life}}.\right)$$ [L]

SEMICONDUCTORS AND ELECTRONICS (Chapter 55)

G134 Distinguish between *intrinsic* and *extrinsic* conduction in semiconductors. Explain the terms *donor* impurity and *acceptor* impurity. [AEB, '79]

G135 What is the difference between *intrinsic* and *extrinsic* conduction in semiconductors? Explain the effect of an increase of temperature on intrinsic conduction. How does this differ from the effect of an increase in temperature on a metallic conductor? [AEB, '79]

G136 (a) Give a typical sketch of the energy bands or levels at which electrons may reside in a pure semiconductor, such as silicon.
(b) Use your sketch to explain briefly:
(i) electric conduction in pure silicon due to holes and electrons,
(ii) the variation of semiconductor resistivity with temperature. [W, '91]

G137 Give an account of the mechanism of the flow of electric current in n-type and p-type semiconductors.

Explain what is meant by the *Hall effect*, and show how it can be used to distinguish between n-type and p-type semiconductors.

What are the important differences between **(a)** the resistivities of metals and semiconductors, **(b)** the variation of these resistivities with temperature? [S]

G138 Describe how the energy levels of electrons in solids differ from those in free atoms. Distinguish, in terms of the filling of such energy levels, between metals, insulators and intrinsic semiconductors.

Explain why the addition of small quantities of suitable impurities to an intrinsic semiconductor may result in a considerable decrease in its resistivity.

The mobility μ of charge carriers in a conductor is defined by the equation $v = \mu E$, where v is the drift velocity produced by an electric field E. A rod of p-type germanium of length 10 mm and cross-section area 1 mm^2 contains 3×10^{21} holes per m^3, the electron density being negligible. Given that the mobility of the holes is $0.35\, m^2\, V^{-1}\, s^{-1}$, what is the resistance between the ends of the rod?

(Electronic charge $e = 1.6 \times 10^{-19}$ C.)
[O & C]

G139 Give a brief account of electrical conduction in solids, pointing out the basic differences between insulators, conductors and semiconductors.

Explain what is meant by n-type and p-type semiconducting material. How are these used in a semiconductor diode? Discuss and explain the action of this diode as a rectifier. [L]

G140 Silicon has a valency of four (i.e. its electronic structure is 2:8:4). Explain the effect of doping it with an element of valency three (i.e. of electronic structure 2:8:3).

Explain the process by which a current is carried by the doped material.

Describe the structure of a solid-state diode.

Draw a circuit diagram showing a reverse-biased diode and explain why very little current will flow.

Suggest why a suitable reverse-biased diode could be used to detect alpha-particles. [L]

G141 **(a)** **(i)** What are meant by electron energy bands in a solid?
(ii) Draw diagrams to show how these bands are arranged in conductors and in semiconductors.
(iii) Explain how the conduction process in an intrinsic semiconductor gives rise to an increase in conductivity with temperature.
(b) Draw a diagram of a circuit used to investigate the characteristics of a junction diode and draw the graph which would be obtained.
(c) Describe, with reference to the motion of electrons and holes, how a potential barrier is set-up at a p–n junction. How is a p–n junction able to rectify an alternating current? What effect does the presence of minority carriers have on this process? [AEB, '79]

G142 **(a)** Explain the difference between a p-type and an n-type semiconductor. If you were given samples of each type of material, what experiments could you perform in order to identify them?
(b) Explain how a p–n junction acts as a rectifier, and describe how it is used in a practical rectifying device.
(c) Draw a diagram of a circuit in which an n–p–n transistor acts as a common emitter voltage amplifier. Explain how the circuit responds to a small input voltage change, and show how the output voltage can be calculated. [O]

G143 What is intrinsic semiconduction? Discuss in terms of band theory the effect of temperature upon the conductivity of intrinsic semiconductors.

What are meant by 'n-type' and 'p-type' semiconducting materials? Discuss these materials in terms of valence electrons. Explain the operation of a rectifying device made from n-type and p-type semiconducting materials.

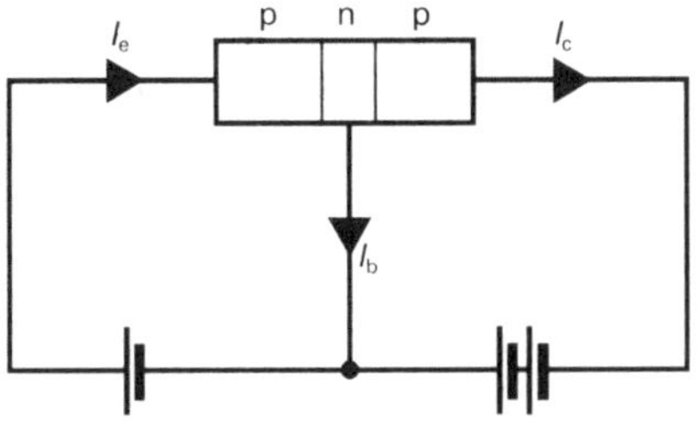

The diagram shows a p–n–p transistor and its associated power supplies. Explain why the current I_c is considerably greater than the current I_b. [L]

G144 What is meant by a *semiconductor*? Explain how the conductivity of such a material changes with **(a)** temperature, and **(b)** the presence of impurities.

Describe the structure of a solid state diode, explaining the nature of the semiconducting materials from which it is made. Explain the action of the diode in rectifying an alternating current.

(i)

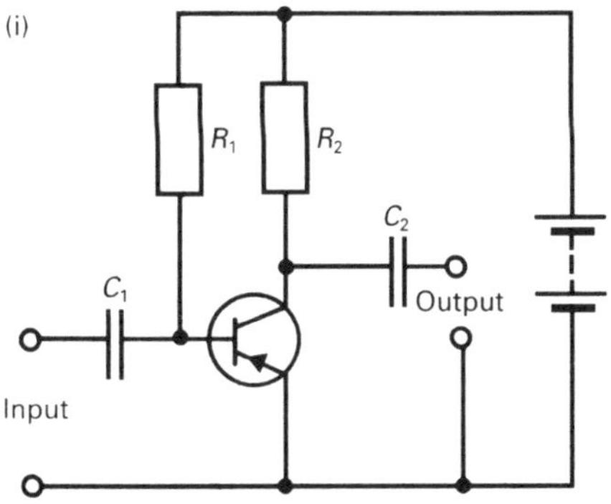

(ii)

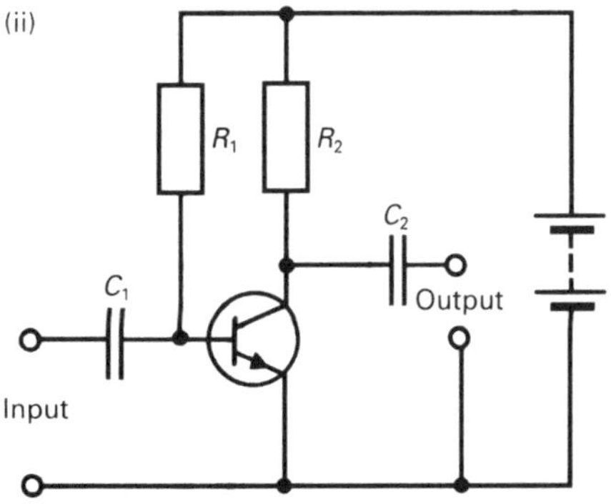

The diagrams above show simple forms of transistor voltage amplifiers using **(i)** a p–n–p transistor, and **(ii)** an n–p–n transistor. Choose *one* of these circuits and explain the functions of the components R_1, R_2, C_1 and C_2. (State which circuit you are considering.) [L]

G145 In the transistor circuit shown in (a), suitable potential differences are applied to the terminals AB and CD respectively.

(a)

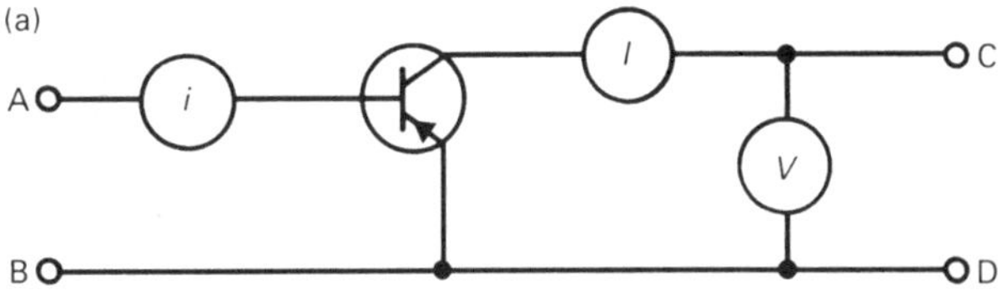

(b)

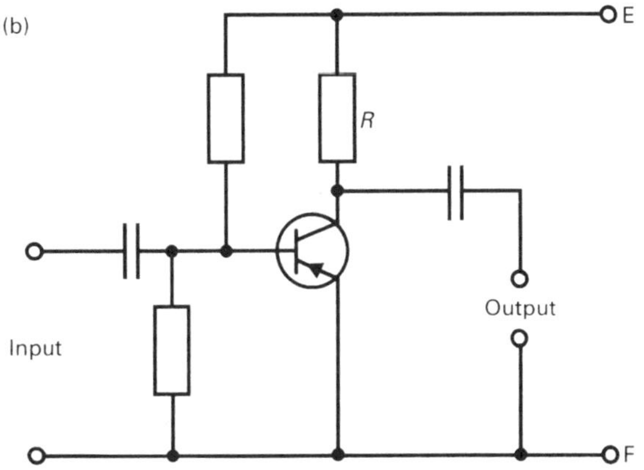

(a) Is the transistor n–p–n or p–n–p?

(b) Sketch graphs to show how the current I in the collector circuit varies with the potential difference V for different values of the current i in the base circuit of the top figure above.

If the transistor is used in the simple amplifier circuit shown in Figure (b) above.

(c) to which of the terminals, E, F is the positive terminal of the power supply connected?

(d) What would be the effect on the gain of the amplifier if the resistor R were increased in value by about 10%? [S]

G146 **(a)** What is meant by **(i)** an n-type semiconductor, **(ii)** a p-type semiconductor? Give one example of each, and explain how its characteristic property arises.

(b) The diagram represents a junction transistor (either p–n–p or n–p–n) in which a thin region of one material (shaded) separates two regions of the other. Electrodes E (emitter), B (base) and C (collector) are attached as shown.

(i) Draw a diagram showing suitable biasing of the electrodes for one of these types of transistor, and

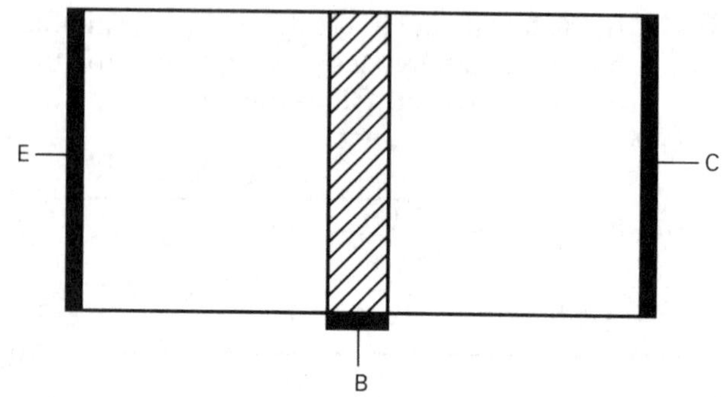

explain why the collector current I_c is much greater than the base current I_b.

(ii) Sketch typical transfer characteristics showing the collector current I_c against the collector-emitter voltage V for different base currents. Show how the current amplification factor β can be derived from these.

(c) Draw a diagram of a simple common-emitter amplifier circuit for small alternating voltages, and state suitable values for the components, given that β is approximately 50. [O]

G147 **(a)** For an *npn* transistor in the common-emitter configuration draw a circuit diagram of an arrangement for measuring the collector current as a function of the collector-emitter voltage for several values of the base current.

By reference to your diagram show how you would vary the base and collector currents and indicate the purpose and range of any meters in the circuit.

(b) State the observations you would make and draw a family of curves representing the results you would expect to obtain.

(c) The diagram shows an *npn* transistor circuit in which the base current, I_B, is 0.20 mA; the collector current, I_C, is

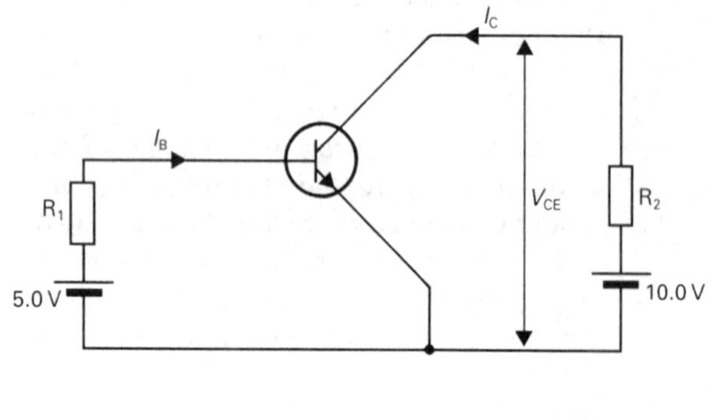

10.0 mA; and the collector-emitter voltage, V_{CE}, is 5.0 V. Neglecting the base–emitter voltage, calculate the resistance of each of the resistors R_1 and R_2.

If by subsequently altering R_1 only, V_{CE} were adjusted to 2.5 V what would then be the value of I_C? [J]

G148 Fig. 1 shows an arrangement for investigating the characteristics of a transistor circuit. The input voltage V_i is varied using the potentiometer, P. The corresponding output voltage V_o is shown graphically in Fig. 2.

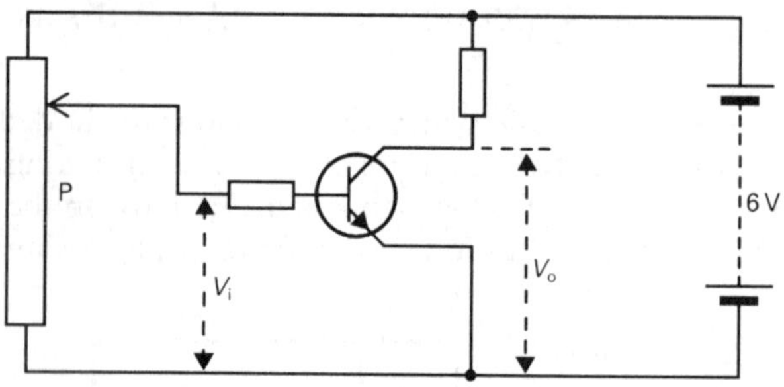

Fig. 1

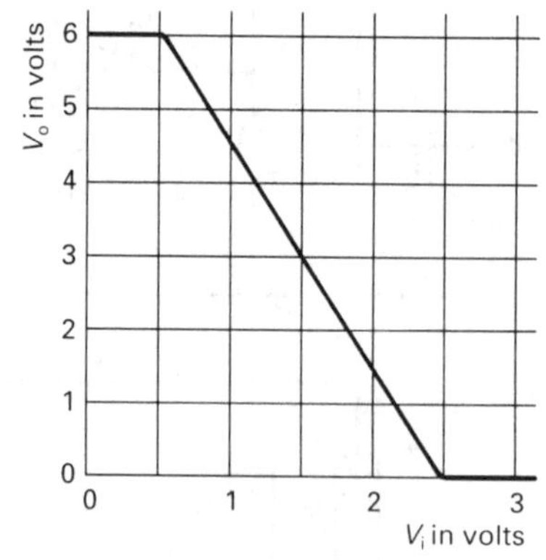

Fig. 2

The circuit is to be used as an alternating voltage amplifier. The input voltage must first be fixed at a suitable value by adjusting P.

(a) Suggest the most suitable value for this fixed input voltage, explaining your answer.

(b) A sinusoidally alternating voltage of amplitude 0.5 V is superimposed on this fixed voltage. What will be the amplitude of the output voltage variations? Will the output variations be sinusoidal? Justify your answers.

(c) Sketch one complete cycle of the output voltage which would be obtained if the amplitude of the superimposed sinusoidal voltage were increased to 1.5 V. [AEB, '83]

G149 (a) (i) In the context of semiconductors, what is meant by p-type and n-type materials? What conduction processes occur in such materials when a potential difference is applied across them?

(ii) Explain why a junction between a p-type and n-type material can act as a rectifier.

(b) The figure shows a circuit incorporating an npn transistor whose current amplification factor h_{fe} (β) is 50.

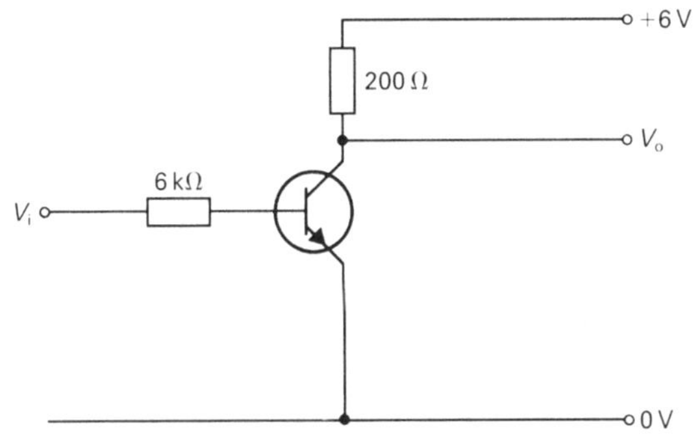

(i) Ignoring the base-emitter voltage, calculate the base current and emitter current when the input voltage V_i is 1.5 V. What is the output voltage V_o?

(ii) Draw a sketch-graph to show how the output voltage V_o changes as the input voltage V_i is increased from 0 V up to 6 V. Explain briefly.

(c) The resistance of a cadmium sulphide light-dependent resistor (LDR) decreases when it is illuminated. Design transistor circuits using an LDR:

(i) to switch on a 6 V filament lamp when darkness falls;

(ii) to switch off a 6 V filament lamp when darkness falls. [O]

G150 (a) Explain why, when a p–n junction diode is connected in a circuit and is reverse biased, there is a very small leakage current across the junction. How will the size of this current depend on the temperature of the diode?

(b) Sketch the output characteristic (I_C against V_{CE}) for a typical n–p–n transistor in *common-emitter* mode. Suggest typical values for I_C and V_{CE} on the axes and indicate how the base current varies over the family of curves.

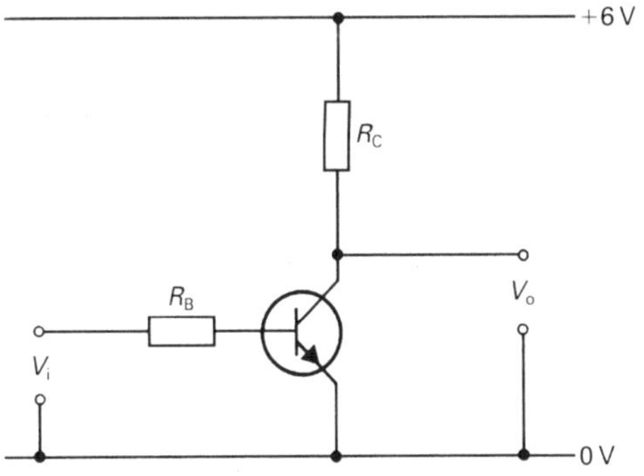

Explain with the aid of the characteristic the meaning of the terms **(i)** saturation, and **(ii)** cut-off, applied to the silicon transistor in the circuit shown. What bias conditions exist at the transistor junctions when it is in each of these states?

(c) The input voltage to the circuit in part (b) is varied in the way shown on the graph below. Copy this graph and on a sketch graph drawn to the same scale show how the output voltage, V_o, varies with time.

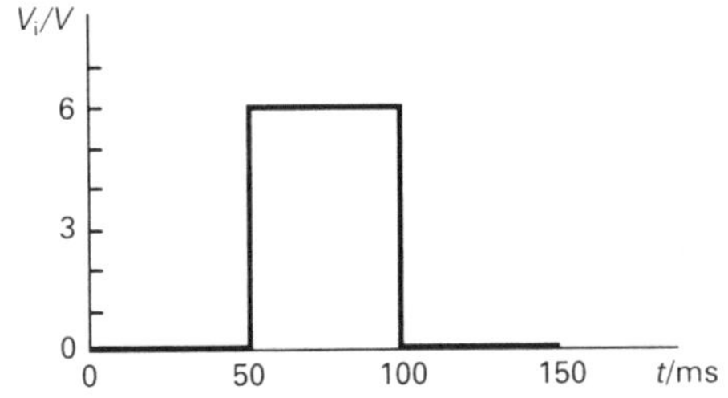

Given that the DC current gain for this silicon transistor, h_{fe}, is 80 and that $R_C = 1.5\,k\Omega$, determine a value for R_B if the transistor is to saturate for a minimum input V_i equal to 3.0 V. [L]

G151 (a) A semiconductor material may have an extrinsic conductivity due to the presence of *donor impurities* or *acceptor impurities*. Explain the meanings of each of the terms in italics.

(b) The circuit shown provides a visual warning when the temperature rises

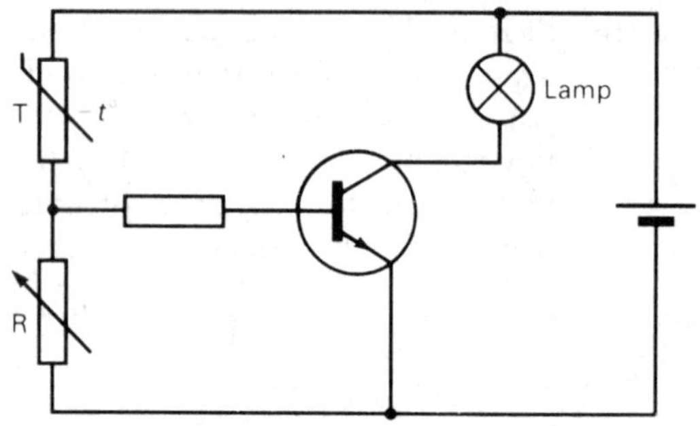

above a predetermined level. It incorporates a thermistor T, the resistance of which decreases as temperature rises. When the temperature is low the lamp is off. Explain the action of the circuit as the temperature rises from a low value.
[AEB, '86]

G152

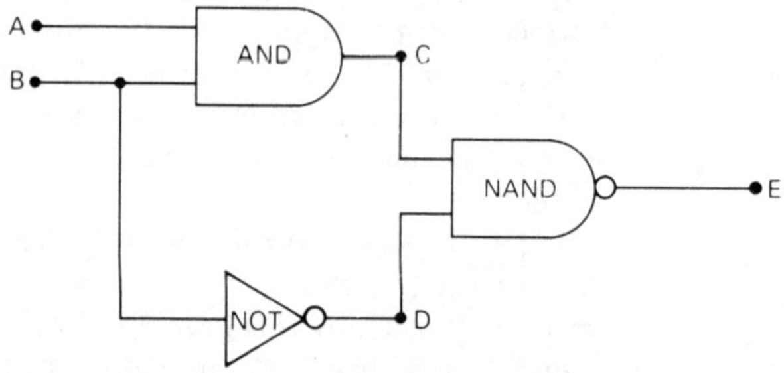

Complete the truth table for the simple combination of logic gates shown above.

A	B	C	D	E
0	0			
0	1			
1	0			
1	1			

G153

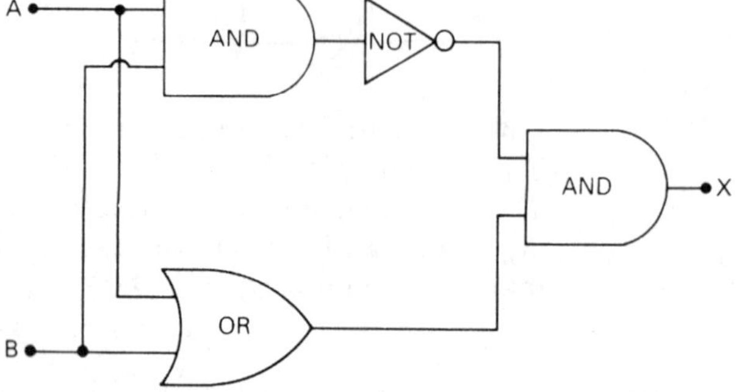

Construct a truth table for the combination of logic gates shown above.

G154 (a)

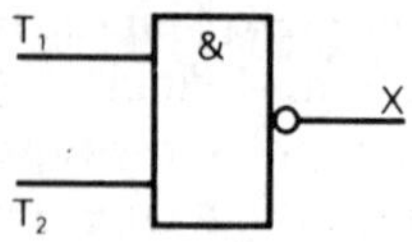

T_1	T_2	X
0	0	1
0	1	
1	0	
1	1	

The diagram represents a NAND gate with two inputs, T_1 and T_2, and an output X. Copy the truth table above and complete it.

Show how a NAND gate can become a NOT gate.

(b)

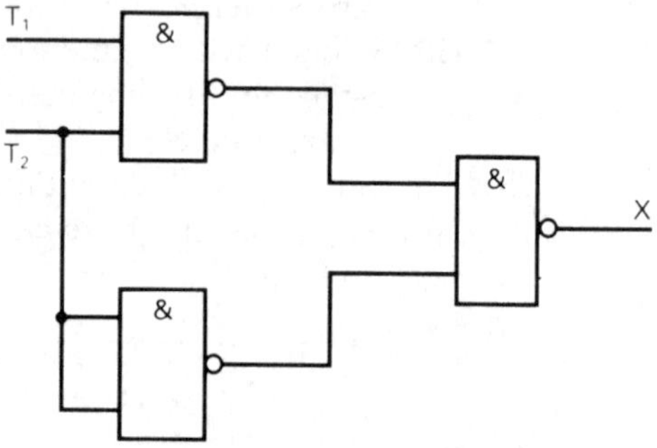

Draw up a truth table for the combination of NAND gates shown above. [L]

G155

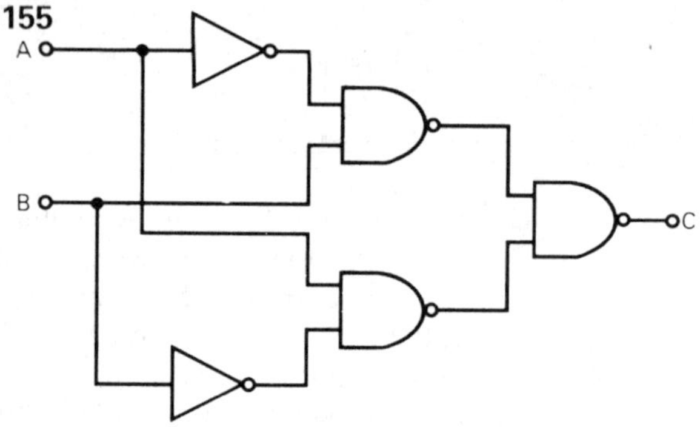

(a) Determine the truth table for the above circuit.

(b) Which single gate would provide exactly the same function? [W, '91]

G156 (a)

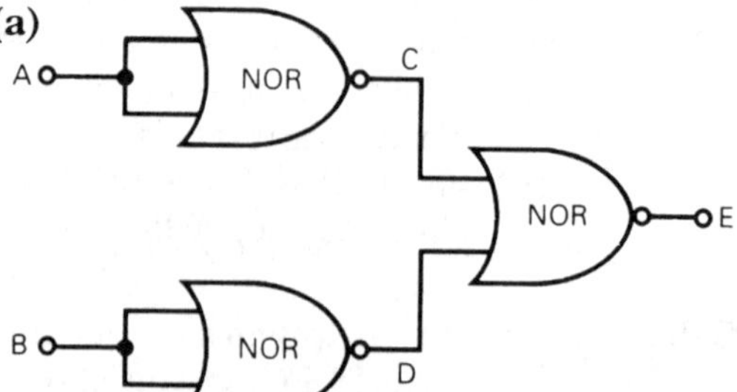

The NOR gate combination in Figure 1 acts as an AND gate. Copy and complete the truth table.

A	B	C	D	E
0	0			
0	1			
1	0			
1	1			

(b) An aircraft door is locked by two bolts A and B each of which operates a sensor giving a logic 1 when the bolt is fully inserted. A further sensor C gives a logic 0 output when the door is shut.

A circuit using logic gates is required to give logic output 1 only when both bolts are inserted and the door is shut.

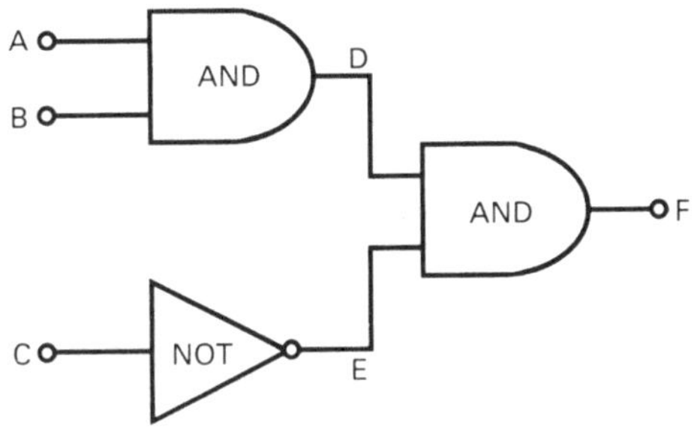

By completing a copy of its truth table, show that the circuit of Figure 2 meets this requirement.

A	B	C	D	E	F
0	0	0			
0	0	1			
0	1	0			
0	1	1			
1	0	0			
1	0	1			
1	1	0			
1	1	1			

(c) Draw a circuit using two-input NOR gates *only* that performs the same logic function. [O, '92]

G157 **(a)** Write down the complete truth table for:
(i) an AND gate,
(ii) an OR gate.
(b) Three square-wave generators are connected to a system of gates, as shown in Fig. 1. The output of square-wave generator S_2 is shown in Fig. 2. The outputs of the other two generators have

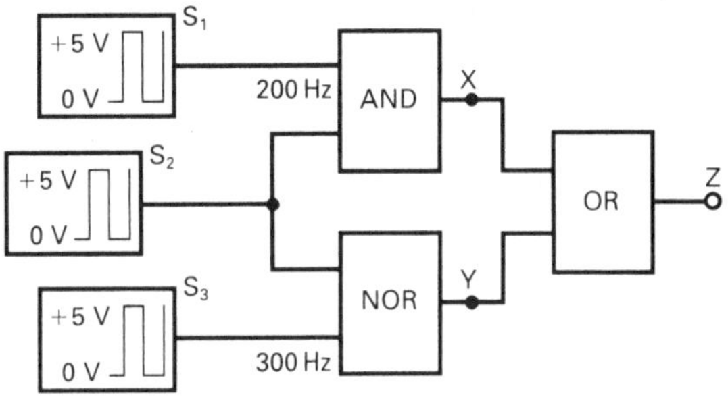

Fig. 1

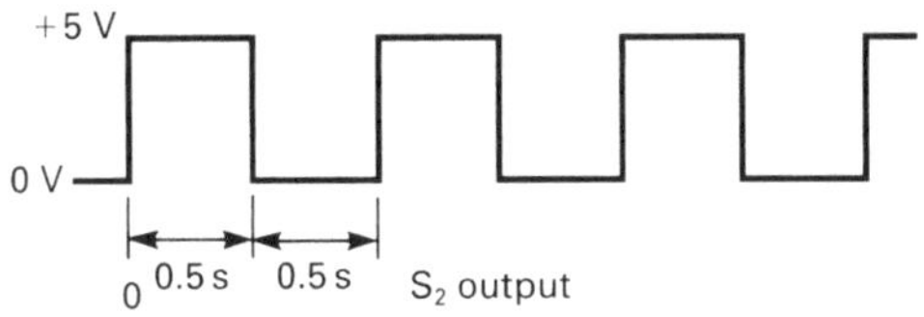

Fig. 2

a similar shape and they oscillate between the same voltage levels.
(i) In the time interval from $t = 0$ to $t = 0.5$ s, how often does the output go high at X? State your reasoning.
(ii) In the same time interval how often does the output go high at Y?
(iii) What happens at the output Z over a time interval of 2 s? Justify your answer. [AEB, '92]

G158

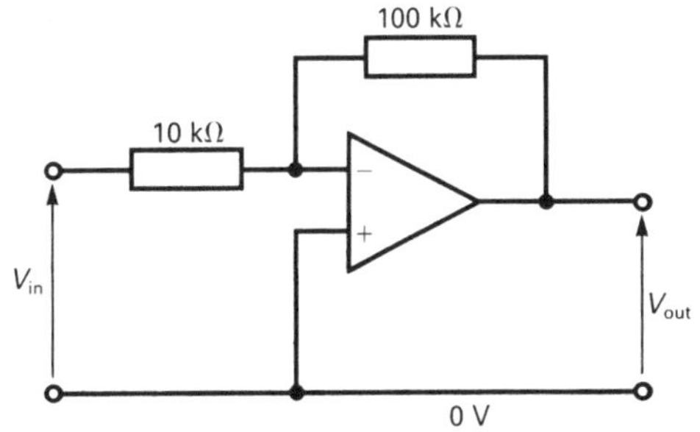

The diagram above shows an inverting amplifier operated from a ± 15 V supply.
(a) Calculate the voltage gain of the amplifier.
(b) A sinusoidal PD of 0.50 V RMS and frequency 1.0 kHz is applied to the input.
(i) Using the same set of axes showing suitable voltage and time scales, sketch the variation with time of the input PD and of the output PD

(ii) On a separate set of labelled axes, sketch the variation with time of the output PD if the input PD were 2.0 V RMS and the frequency remained unchanged. [J, '90]

G159 **(a)** The operational amplifier shown in the diagram can be used as a voltage amplifier. Describe how you would

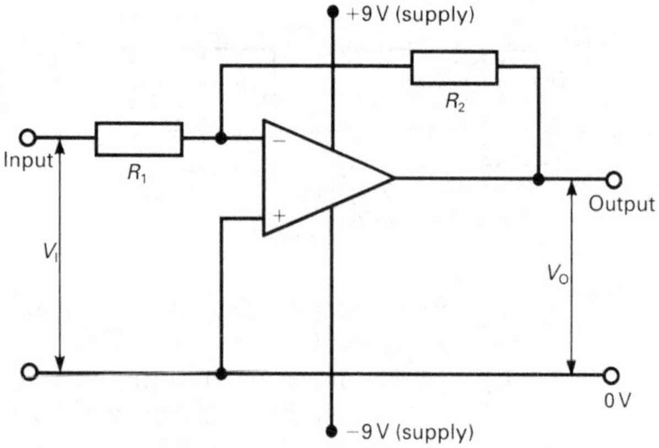

determine experimentally its DC input/output characteristic for positive and negative input voltages. Include a labelled circuit diagram showing how the amplifier is connected to a suitable power supply and explain how you obtain different input voltages using a potential divider. Suggest suitable ranges of voltmeters you would use to measure V_I and V_O given that the resistance of $R_1 = 10\,k\Omega$ and the resistance of $R_2 = 100\,k\Omega$.

(b)

Input voltage V_1/V	*Output voltage* V_O/V
+2.0	−8.0
+1.5	−8.0
+1.0	−8.0
+0.5	−4.0
0	0
−0.5	+4.0
−1.0	+8.0
−1.5	+8.0
−2.0	+8.0

The table above shows typical results for a voltage amplifier similar to that shown above. Draw the input/output characteristic and, by reference to it, explain what is meant by **(i)** *voltage gain*, **(ii)** *saturation* and **(iii)** *inversion*. State the range of input voltages for which the amplifier has a linear response and calculate the voltage gain within this range.

(c) A sinusoidal voltage of frequency 50 Hz is applied to the input terminals of the amplifier described in (b). Sketch graphs on one set of axes showing how the output voltage varies with time when the input voltage is **(i)** 0.5 V RMS, **(ii)** 1.0 V RMS. In each case indicate the peak value of the output voltage and comment on the waveform. [J]

G160 In the operational amplifier circuit of Fig. 1, $V_i = +2.0\,V$.

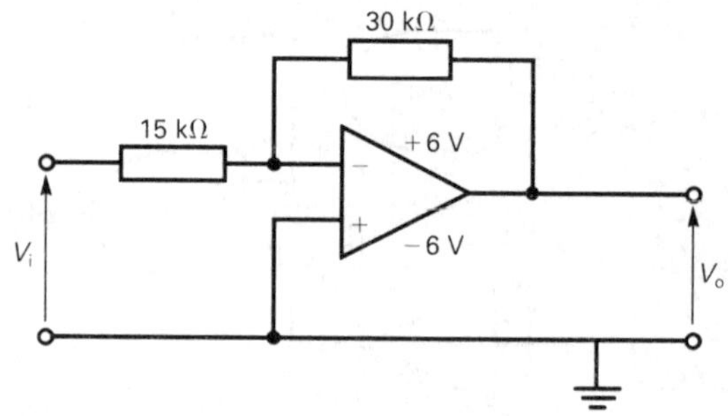

Fig. 1

(a) Calculate the output potential V_o.
(b) The input signal V_i is then replaced by the signal shown in Fig. 2. Copy this diagram and add to it the corresponding output potential V_o.

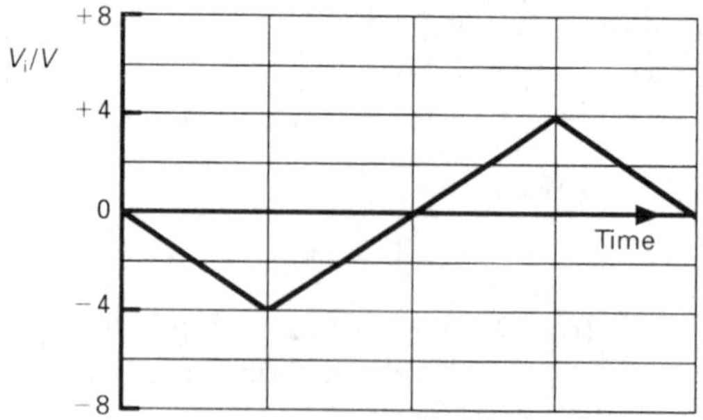

Fig. 2

[C, 91]

G161 The diagram on the next page shows an operational amplifier used as an astable multivibrator.

(a) Assuming that initially the capacitor is uncharged and that V_{out} is positive, explain how a square wave output is produced.

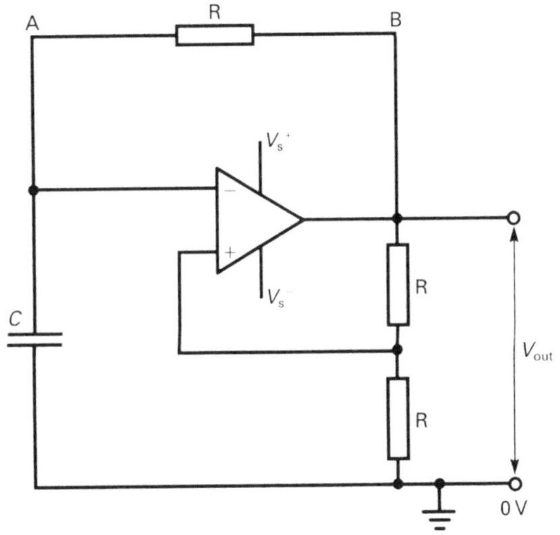

(b) Explain how the form of the output would change if the resistance of the resistor between the points A and B were increased. [J]

G162 This question is about the design of an experiment to investigate the frequency response of an amplifier. In the experiment the gain of an amplifier is to be investigated over a range of frequencies.

The diagram shows the circuit of the amplifier to be investigated. The operational amplifier may be assumed to be ideal. It is to be operated using a −15 V–0–+15 V supply which is not shown.

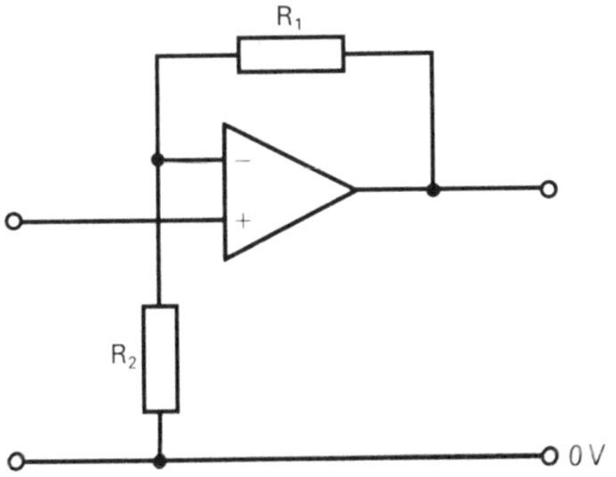

(a) **(i)** Select from the following list the two resistors you would use for R_1 and R_2 so that the gain of the amplifier at low frequencies would be as near to 30 as is possible.
1 kΩ, 2.7 kΩ, 4.7 kΩ, 10 kΩ, 39 kΩ, 150 kΩ

(ii) Calculate the expected low frequency gain of the amplifier using the resistors you have chosen.

(b) A sinusoidal input signal is to be provided by an uncalibrated oscillator which has a variable frequency with a range 1 Hz to 1 MHz. The output has a peak value which is constant at 2 V for all frequencies.

(i) Explain why this voltage is too large for investigating the frequency response of the amplifier.

(ii) Draw a diagram to show how you would reduce this voltage to a suitable magnitude using components from the list in (a)(i) and calculate the new peak output voltage.

(iii) Explain how you would proceed to measure the DC gain of the amplifier given that a 2 V DC supply is available. Indicate clearly the instrument(s) you would use, where the instruments would be connected and the measurements you would make.

(iv) Describe how you would use an oscilloscope to calibrate the oscillator and use the calibrated oscillator to investigate the frequency response of the amplifier.

(v) Draw a graph indicating the shape of the frequency response graph you would expect. [AEB, '89]

G163

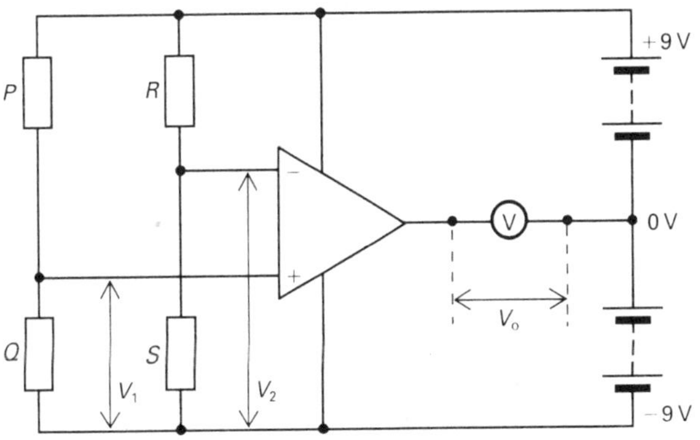

In the above circuit P, Q, R and S form part of a Wheatstone bridge network in which an operational amplifier circuit is used to detect the difference between the voltages V_1 and V_2.

(a) The output voltage in this circuit, V_o, is given by $V_o = A(V_1 - V_2)$ where A is the open-loop gain of the amplifier and V_o must lie between 9 V and −9 V. Show that, if $A = 90\,000$, then V_o should be either 9 V or −9 V if the difference between V_1 and V_2 exceeds 100 μV.

(b) If the resistances P and Q are each 10 kΩ and R is the resistance of a variable standard resistor, outline how you would use the arrangement to determine the resistance S, which is of the order of a few kΩ.

(c) A typical centre zero galvanometer has a resistance of 40 Ω and is graduated in divisions of 0.1 mA. If you were to choose between such a meter and the above operational amplifier as a null detector, which would you choose, and why? [J]

G164 (a) For the circuit shown in Fig. 1, calculate **(i)** the PD across Q, **(ii)** the PD across Y.

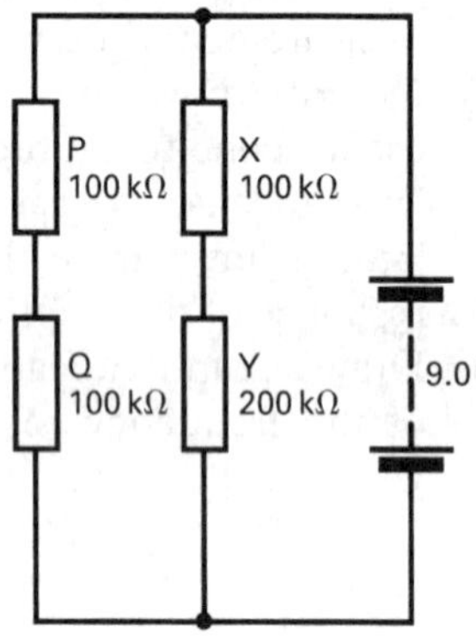

Fig. 1

(b) In Fig. 2, capacitor C is initially uncharged and switch S is closed.

(i) Sketch a graph of the variation of the PD across C with time from the instant when the switch is opened.

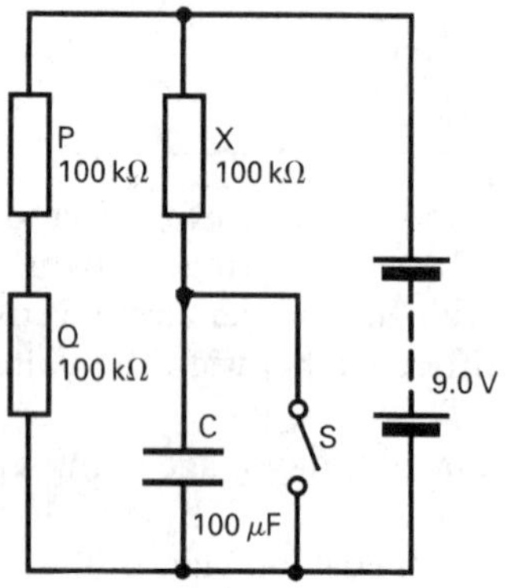

Fig. 2

Give a qualitative explanation for the shape of the curve.

(ii) Calculate the PD across C 10 s after the switch is opened.

(c) The components P, Q, X and C shown in Fig. 2 are connected in a circuit with an operational amplifier as shown in Fig. 3. The capacitor is uncharged, switch S is closed and the light emitting diode (LED) is off (unlit).

(i) Explain why the LED comes on 6.9 s after the switch is opened and then remains lit.

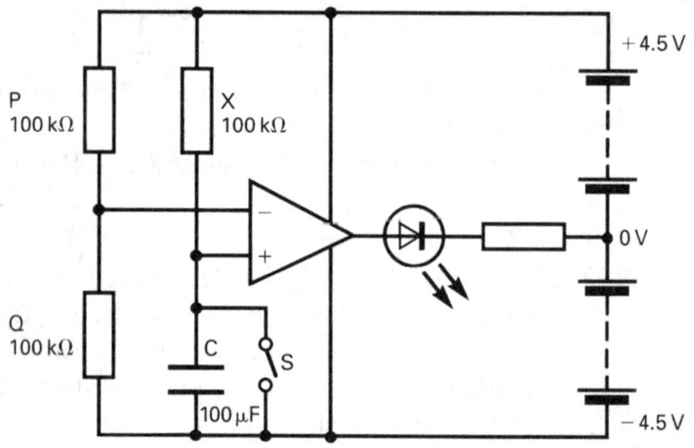

Fig. 3

(ii) Why is a resistor connected in series with the LED? What change, if any, would be observed if the resistor were replaced with one with a higher resistance?

(iii) State *two* ways of increasing the time delay between closing the switch and the LED lighting. Explain how *one* of these ways produces the desired effect.

[J, '91]

G165 In the circuit shown the light emitting diode (LED) is turned ON when the moisture content in the soil, in which the probes are placed, falls below a certain value. The moisture content at which the circuit switches is determined by the position of the wiper W of the potentiometer PQ.

The moisture affects the resistance between the probes.

(a) Explain why the conductivity of the soil increases when it is wet.

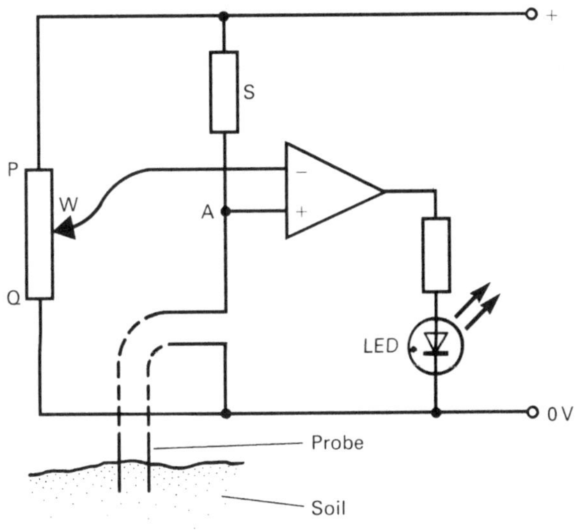

(b) Explain how the circuit operates so that the LED is ON when the moisture content of the soil is below a certain value but OFF when the moisture content is above that value.

(c) State two ways in which the circuit may be changed so that the circuit switches at a lower moisture content. [AEB, '90]

A1

SI UNITS

A1.1 BASIC UNITS

The international system of units known as SI units (*Système International d'Unités*) is based on the seven units listed in Table A1.1. These are called **basic** units, and the particular seven used in the system are chosen for convenience – not out of necessity. Three of the basic units are defined below; some of the others are defined at relevant places in the text.

Table A1.1
The basic units

Basic quantity	*Unit*	
	Name	*Symbol*
Mass	kilogram	kg
Length	metre	m
Time	second	s
Electric current	ampere	A
Temperature	kelvin	K
Amount of substance	mole	mol
Luminous intensity	candela	cd

The metre (m) is the unit of length and is equal to 1/299 792 458 of the distance travelled by light in vacuum in one second.

The kilogram (kg) is the unit of mass and is equal to the mass of the **International Prototype kilogram** (a platinum–iridium cylinder) kept at Sèvres, Paris.

The second (s) is the unit of time and is the duration of exactly 9 192 631 770 periods of the radiation corresponding to the transition between the two hyperfine levels of the ground state of a caesium 133 atom.

A1.2 DERIVED UNITS

Apart from the radian and the steradian,* all the other units used in the system are called **derived** units. Derived units are formed by multiplication and/or division of one or more basic units without the inclusion of any numerical factors (e.g., one coulomb = one ampere × one second). Some derived units are relatively complex when expressed in terms of the basic units, and, for convenience, are given special names (e.g. the kg m^2 s^{-3} A^{-2} is called the ohm, Ω). Much used units also have special names (e.g. the A s is called the coulomb, C). Those derived units which have special names and are used in this book are listed in Table A1.2. **The symbol for a unit which is named after a person has a capital letter.**

*The unit of angle (the radian) and the unit of solid angle (the steradian) are officially designated as **subsidiary** units and can be treated as being either basic or derived as convenience dictates.

Table A1.2
Derived units with special names

Derived quantity	*Unit*	
	Name	*Symbol*
Force	newton	N $= \mathrm{kg\,m\,s^{-2}}$
Pressure	pascal	Pa $= \mathrm{kg\,m^{-1}\,s^{-2}}$
Energy, work	joule	J $= \mathrm{kg\,m^2\,s^{-2}}$
Power	watt	W $= \mathrm{kg\,m^2\,s^{-3}}$
Frequency	hertz	Hz $= \mathrm{s^{-1}}$
Charge	coulomb	C $= \mathrm{A\,s}$
Electromotive force	volt	V $= \mathrm{kg\,m^2\,s^{-3}\,A^{-1}}$
Resistance	ohm	Ω $= \mathrm{kg\,m^2\,s^{-3}\,A^{-2}}$
Conductance	siemens	S $= \mathrm{kg^{-1}\,m^{-2}\,s^3\,A^2}$
Inductance	henry	H $= \mathrm{kg\,m^2\,s^{-2}\,A^{-2}}$
Capacitance	farad	F $= \mathrm{kg^{-1}\,m^{-2}\,s^4\,A^2}$
Magnetic flux	weber	Wb $= \mathrm{kg\,m^2\,s^{-2}\,A^{-1}}$
Magnetic flux density	tesla	T $= \mathrm{kg\,s^{-2}\,A^{-1}}$

A1.3 PREFIXES

Prefixes are used with the unit symbols to indicate decimal multiples or submultiples. Most of the standard prefixes are listed in Table A1.3.

Table A1.3
Standard prefixes

Submultiple	*Prefix*	*Symbol*	*Multiple*	*Prefix*	*Symbol*
10^{-2}	centi	c	10^3	kilo	k
10^{-3}	milli	m	10^6	mega	M
10^{-6}	micro	μ	10^9	giga	G
10^{-9}	nano	n	10^{12}	tera	T
10^{-12}	pico	p			
10^{-15}	femto	f			
10^{-18}	atto	a			

Note

$1\,\mathrm{cm} = 1 \times 10^{-2}\,\mathrm{m}$ $\quad 1\,\mathrm{mm} = 1 \times 10^{-3}\,\mathrm{m}$

$1\,\mathrm{cm^2} = 1 \times 10^{-4}\,\mathrm{m^2}$ $\quad 1\,\mathrm{mm^2} = 1 \times 10^{-6}\,\mathrm{m^2}$

$1\,\mathrm{cm^3} = 1 \times 10^{-6}\,\mathrm{m^3}$ $\quad 1\,\mathrm{mm^3} = 1 \times 10^{-9}\,\mathrm{m^3}$

A2

DIMENSIONS AND DIMENSIONAL METHODS

A2.1 DIMENSIONS

The dimensions of a physical quantity indicate how it is related to the basic quantities listed in Table A1.1 of Appendix 1. For example, area is obtained by multiplying one length by another (terms such as breadth, width, distance, radius are merely convenient ways of saying length) and therefore the dimensions of area are those of length squared. We represent the underlined statement by using the notation

$$[\text{Area}] = \text{L}^2$$

Volume is the product of three lengths and therefore

$$[\text{Volume}] = \text{L}^3$$

Density is mass divided by volume and therefore

$$[\text{Density}] = \text{M}\text{L}^{-3}$$

where M denotes 'dimensions of mass'. Speed is distance divided by time and therefore

$$[\text{Speed}] = \text{L}\text{T}^{-1}$$

where T denotes 'dimensions of time'. Force is mass multiplied by acceleration and therefore

$$[\text{Force}] = \text{M}\text{L}\text{T}^{-2}$$

Taking the dimensions of current, temperature and amount of substance to be represented by I, Θ and N respectively, the reader should convince himself of the following:*

$$[\text{Specific heat capacity}] = \text{L}^2\text{T}^{-2}\Theta^{-1}$$

$$[\text{Molar heat capacity}] = \text{M}\text{L}^2\text{T}^{-2}\Theta^{-1}\text{N}^{-1}$$

$$[\text{Thermal conductivity}] = \text{M}\text{L}\text{T}^{-3}\Theta^{-1}$$

$$[\text{Electrical potential}] = \text{M}\text{L}^2\text{T}^{-3}\text{I}^{-1}$$

$$[\text{Electrical resistance}] = \text{M}\text{L}^2\text{T}^{-3}\text{I}^{-2}$$

*Assuming, of course, that he is already familiar with the way in which the quantities in square brackets are defined.

Quantities which have no units associated with them are dimensionless (e.g. refractive index). Some quantities (e.g. angle) are dimensionless even though they have an associated unit. Note also:

$$[\text{Frequency}] = \mathrm{T}^{-1}$$

$$[\text{Angular frequency}] = \mathrm{T}^{-1}$$

A2.2 DIMENSIONAL HOMOGENEITY

If an equation is correct, each term in it must have the same dimensions as every other. Suppose we are told that a force F moves a body of mass m a distance s in time t, where $t = 2ms/F$. The dimensions of the left-hand side of the equation are simply T. Bearing in mind that 2 is a number and therefore is dimensionless and that $[m] = \mathrm{M}$, $[s] = \mathrm{L}$ and $[F] = \mathrm{MLT}^{-2}$, we see that the dimensions of the right-hand side are $\mathrm{ML}/(\mathrm{MLT}^{-2}) = \mathrm{T}^2$. Thus, the dimensions of the right-hand side are not the same as those of the left-hand side and therefore the equation cannot be correct.

Notes (i) The method does not tell us where the equation is wrong.

(ii) If the dimensions had been the same on each side of the equation, we would know only that it might be correct, for the method does not provide a check on any numerical factors.

A2.3 DIMENSIONAL ANALYSIS

The fact that an equation must be dimensionally homogeneous enables predictions to be made about the way in which physical quantities are related to each other. An example of the method is given below.

Period of a Simple Pendulum

Experiment shows that the period of a simple pendulum is independent of the amplitude of oscillation providing it is small. Therefore it may reasonably be supposed that for small oscillations the period depends only on the mass m of the bob, the length l of the string and the acceleration due to gravity g. If we express the relationship as

$$\text{Period} = km^x l^y g^z$$

where k is a dimensionless constant and x, y, and z are unknown indices, then since each side of the equation must have the same dimensions

$$[\text{Period}] = [m^x][l^y][g^z]$$

$$\therefore \quad \mathrm{T} = \mathrm{M}^x\mathrm{L}^y(\mathrm{LT}^{-2})^z$$

i.e. $$\mathrm{T} = \mathrm{M}^x\mathrm{L}^{y+z}\mathrm{T}^{-2z}$$

Equating the indices of M, L and T on each side of the equation gives

(for M) $0 = x$

(for L) $0 = y + z$

(for T) $1 = -2z$

Solving gives $x = 0$, $z = -\frac{1}{2}$, $y = \frac{1}{2}$. The relationship is therefore

$$\text{Period} = km^0 l^{1/2} g^{-1/2}$$

i.e.

$$\text{Period} = k\sqrt{\frac{l}{g}}$$

Dimensional analysis does not provide the value of the dimensionless constant k; a mathematical analysis of the type carried out in section 7.3 does. However, there are situations where mathematical analysis is too difficult; dimensional analysis is particularly useful in such cases and we make use of it in sections 12.7 and 12.9.

A3

RELEVANT MATHEMATICS

A3.1 SYMBOLS

$=$ equal to

$\neq$ not equal to

$\approx$ approximately equal to

$\sim$ of the order of, i.e. within a factor of ten of. (**Note**: $99.7 \approx 100$, $170 \sim 100$)

$>$ greater than

$\gg$ much greater than

$\geq$ greater than or equal to

$\not>$ not greater than

$<$ less than

$\ll$ much less than

$\leq$ less than or equal to

$\not<$ not less than

$\propto$ proportional to

$\bar{x}$ means the average value of x

$|x|$ means the modulus of x, i.e. the value of x without regard to whether it is negative. For example, $|-3| = |3| = 3$.

A3.2 INDICES

$$x^{1/a} = \sqrt[a]{x}$$

$$x^{-a} = 1/x^a$$

$$x^a \times x^b = x^{(a+b)}$$

$$x^a \div x^b = x^{(a-b)}$$

$$(x^a)^b = x^{ab}$$

A3.3 THE BINOMIAL EXPANSION

If $-1 < x < 1$ (i.e. if $|x| < 1$), then for all values of n

$$(1+x)^n = 1 + nx + \frac{n(n-1)}{1 \times 2}x^2 + \frac{n(n-1)(n-2)}{1 \times 2 \times 3}x^3 + \ldots$$

If $|x| \ll 1$, then terms in $x^2, x^3, \ldots$ may be ignored, and therefore

$$(1+x)^n \approx 1 + nx$$

and

$$(1-x)^n \approx 1 - nx$$

For example, if $|x| \ll 1$

$$(1+x)^{1/2} \approx 1 + \tfrac{1}{2}x \qquad (1+x)^{-1/2} \approx 1 - \tfrac{1}{2}x$$

$$(1-x)^{1/2} \approx 1 - \tfrac{1}{2}x \qquad (1-x)^{-1/2} \approx 1 + \tfrac{1}{2}x$$

Also

$$\frac{(1+\alpha)}{(1+\beta)} = (1+\alpha)(1+\beta)^{-1}$$

$$\approx (1+\alpha)(1-\beta) \quad \text{if } |\beta| \ll 1$$

If, in addition, $|\alpha| \ll 1$, then the term in $\alpha\beta$ may be ignored, and therefore

$$\frac{(1+\alpha)}{(1+\beta)} \approx 1 + \alpha - \beta$$

A3.4 QUADRATIC EQUATIONS

If

$$ax^2 + bx + c = 0$$

then

$$x = \frac{-b \pm \sqrt{b^2 - 4ac}}{2a}$$

A3.5 DEGREES AND RADIANS

Refer to Fig. A3.1. The value of the angle θ expressed in radians is given by

$$\theta = s/r$$

Fig. A3.1

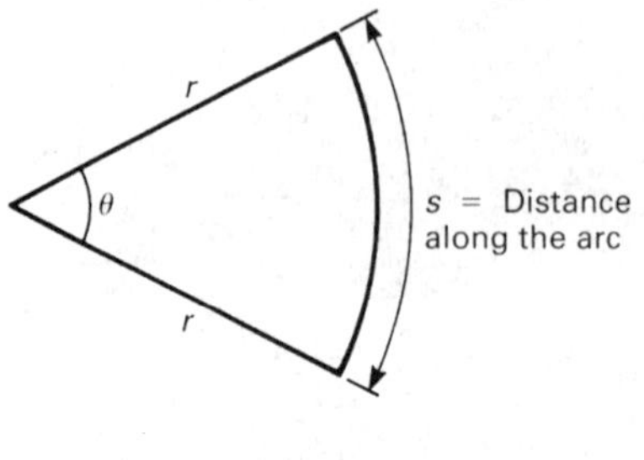

For a complete circle $s = 2\pi r$, and therefore the angle θ at the centre (i.e. the angle for a complete revolution) is given by

$$\theta = 2\pi r / r = 2\pi \text{ radians}$$

The angle at the centre of a circle is 360°, and therefore

$$2\pi \text{ rad} = 360^\circ$$

i.e. $\pi \text{ rad} = 180^\circ$

i.e. $1 \text{ rad} = 57.3^\circ$ (approx.)

A3.6 TRIGONOMETRIC RELATIONS

Refer to Fig. A3.2.

$$\sin\theta = a/c \qquad \operatorname{cosec}\theta = 1/\sin\theta$$

$$\cos\theta = b/c \qquad \sec\theta = 1/\cos\theta$$

$$\tan\theta = a/b \qquad \cot\theta = 1/\tan\theta$$

$$\tan\theta = \frac{\sin\theta}{\cos\theta} \qquad \sin(90^\circ - \theta) = \cos\theta$$

$$\cos(90^\circ - \theta) = \sin\theta$$

$$\sin^2\theta + \cos^2\theta = 1 \qquad \tan(90^\circ - \theta) = \cos\theta$$

Fig. A3.2

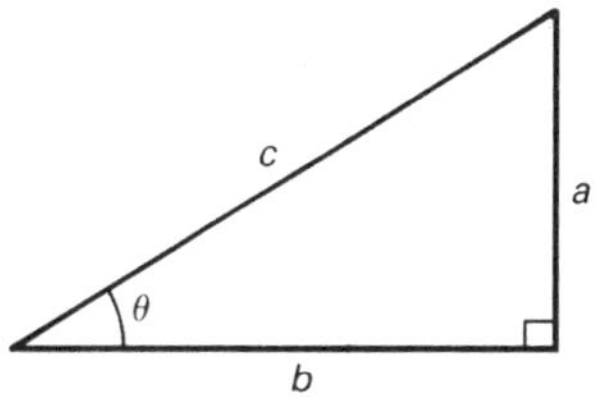

The values of the sines and cosines listed below are worth remembering:

θ/deg	θ/rad	$\sin\theta$	$\cos\theta$
0	0	0	1
30	$\pi/6$	$\frac{1}{2}$	$\frac{1}{2}\sqrt{3}$
45	$\pi/4$	$1/\sqrt{2}$	$1/\sqrt{2}$
60	$\pi/3$	$\frac{1}{2}\sqrt{3}$	$\frac{1}{2}$
90	$\pi/2$	1	0
180	π	0	−1

$$\sin(\alpha + \beta) = \sin\alpha\cos\beta + \sin\beta\cos\alpha$$

$$\sin(\alpha - \beta) = \sin\alpha\cos\beta - \sin\beta\cos\alpha$$

$$\cos(\alpha + \beta) = \cos\alpha\cos\beta - \sin\alpha\sin\beta$$

$$\cos(\alpha - \beta) = \cos\alpha\cos\beta + \sin\alpha\sin\beta$$

$$\sin\alpha + \sin\beta = 2\sin\left(\frac{a+\beta}{2}\right)\cos\left(\frac{\alpha-\beta}{2}\right)$$

$$\sin\alpha - \sin\beta = 2\sin\left(\frac{a-\beta}{2}\right)\cos\left(\frac{\alpha+\beta}{2}\right)$$

$$\cos\alpha + \cos\beta = 2\cos\left(\frac{a+\beta}{2}\right)\cos\left(\frac{\alpha-\beta}{2}\right)$$

$$\cos\alpha - \cos\beta = -2\sin\left(\frac{a+\beta}{2}\right)\sin\left(\frac{\alpha-\beta}{2}\right)$$

A3.7 THE SINE AND COSINE RULES

For any triangle, such as that in Fig. A3.3

$$\frac{a}{\sin A} = \frac{b}{\sin B} = \frac{c}{\sin C} \qquad \text{(sine rule)}$$

$$a^2 = b^2 + c^2 - 2bc\cos A \qquad \text{(cosine rule)}$$

Fig. A3.3

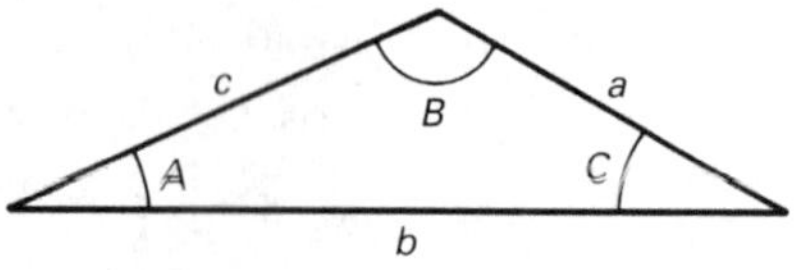

Note that when $A = 90^\circ$ this reduces to

$$a^2 = b^2 + c^2 \qquad \text{(Pythagoras' theorem)}$$

A3.8 SMALL ANGLES

If θ is small

$\sin\theta \approx \theta$ (expressed in radians)

$\tan\theta \approx \theta$ (expressed in radians)

$\cos\theta \approx 1$

These relationships are examined in the table below.

θ/deg	θ/rad	$\sin\theta$	$\tan\theta$	$\cos\theta$
0	0.0000	0.0000	0.0000	1.0000
1	0.0175	0.0175	0.0175	0.9998
2	0.0349	0.0349	0.0349	0.9994
5	0.0873	0.0872	0.0875	0.9962
10	0.1745	0.1736	0.1763	0.9848
15	0.2618	0.2588	0.2679	0.9659
20	0.3491	0.3420	0.3640	0.9397

Note how the relationships become less accurate as θ increases. Note also the exact relationships

$$\lim_{\theta \to 0} \frac{\sin\theta}{\theta} = 1$$

and

$$\lim_{\theta \to 0} \frac{\tan\theta}{\theta} = 1$$

A3.9 PERIMETERS, AREAS AND VOLUMES

Area of a triangle	$= \frac{1}{2}ab\sin C$	(Fig. A3.4)
Circumference of a circle	$= 2\pi r$	
Area of a circle	$= \pi r^2$	
Surface area of a sphere	$= 4\pi r^2$	
Volume of a sphere	$= \frac{4}{3}\pi r^3$	
Area of curved surface of a cylinder	$= 2\pi rh$	
Volume of a cylinder	$= \pi r^2 h$	
Area of curved surface of a cone	$= \pi rl$	(l = slant height)
Volume of a cone	$= \frac{1}{3}\pi r^2 h$	(h = height)

Fig. A3.4

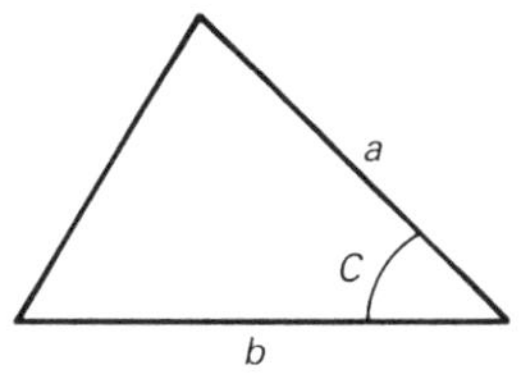

A3.10 THEOREM OF INTERSECTING CHORDS

Refer to Fig. A3.5. For any two chords

$$AX.XB = CX.XD$$

Fig. A3.5

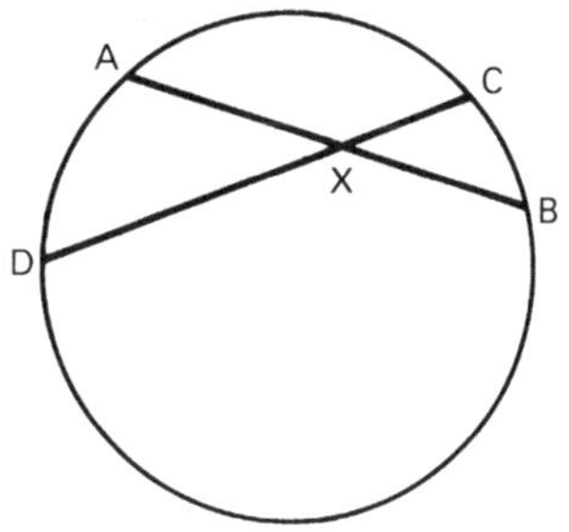

A3.11 LOGARITHMS

Logarithms to base 10 are called **common logarithms** and are represented by $\log_{10}$ or lg. If

$$\log_{10}x = y \qquad \text{then} \qquad x = 10^y$$

Logarithms to base e are called **natural (or hyperbolic or Napierian) logarithms** and are represented by $\log_e$ or ln. If

$$\log_e x = y \qquad \text{then} \qquad x = e^y$$

For logarithms to any base

$$\log(A \times B) = \log A + \log B$$

$$\log(A \div B) = \log A - \log B$$

$$\log(A^p) = p \log A$$

Two useful relations are

$$\log_{10}10^x = x$$

and

$$\log_e e^x = x$$

A logarithm to some base a can be expressed in terms of a logarithm to some other base b by

$$\log_a x = \log_a b \times \log_b x$$

In particular

$$\log_e x = \log_e 10 \times \log_{10} x$$

i.e.

$$\log_e x = 2.3026 \log_{10} x$$

For $-1 < x < 1$ (i.e. $|x| < 1$)

$$\log_e(1+x) = x - \frac{x^2}{2} + \frac{x^3}{3} - \dots$$

Therefore if $|x| \ll 1$

$$\log_e(1+x) \approx x - x^2/2$$

A3.12 DIFFERENTIATION

y	$\dfrac{dy}{dx}$
x^n	$nx^{(n-1)}$
$\log_e x$	$1/x$
$\sin x$	$\cos x$
$\cos x$	$-\sin x$
e^{ax}	$a\,e^{ax}$

A3.13 INTEGRATION

y	$\int y.dx$
x^n $(n \neq -1)$	$\frac{x^{(n+1)}}{n+1} + C^\star$
$1/x$	$\log_e x + C$
$\sin x$	$-\cos x + C$
$\cos x$	$\sin x + C$
e^{ax}	$\frac{e^{ax}}{a} + C$

*C is the constant of integration.

A3.14 LINEAR GRAPHS

If there is a linear relationship between two **variables** x and y, then they are related by

$$y = mx + c$$

where m and c are **constants**, i.e. $y = mx + c$ is the equation of a straight line. The gradient of the line is m; the intercept on the y-axis is c (see Fig. A3.6). Note that although the graph is linear, it does not pass through the origin ($x = 0$, $y = 0$), and therefore y is not proportional to x. The equation can be rearranged to give $(y - c) = mx$, i.e. $(y - c)$ is proportional to x.

Fig. A3.6

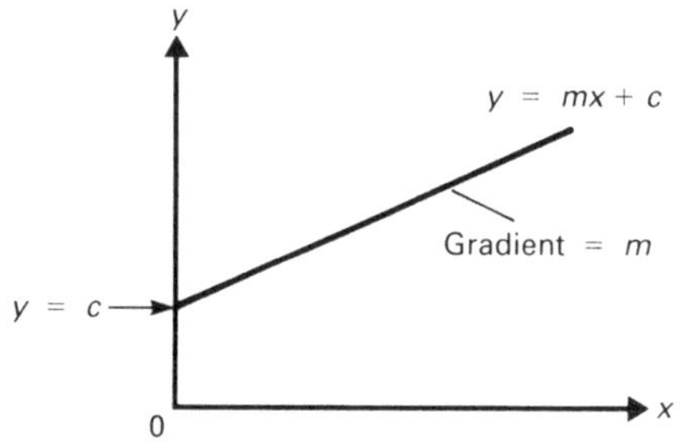

Suppose that in some experiment a series of pairs of values of two measured quantities x and y have been obtained. Suppose also that it is suspected that x and y are related by

$$y^2 = \frac{a}{x} + b$$

where a and b are constants. The relationship between x and y is not of the form $y = mx + c$ and therefore a graph of y against x would not be linear. A graph of y^2 against $1/x$, however, would be linear and would have a gradient of a and an intercept on the y-axis of b. Thus if y^2 is plotted against $1/x$ and a straight line results, the original suspicion (i.e. that $y^2 = a/x + b$) will have been verified. The advantage of a straight line graph is that it is easy to see whether the graph is linear or not; it is not easy to see whether a curved graph curves in the required manner.

Refer to the table below. The reader is advised to confirm (by rearranging the equations in the first column so that they are in the form $y = mx + c$) that the plots listed in the fourth column do indeed produce linear graphs and that the gradients of these graphs are as listed in the fifth column.

Relationship	*Variables*	*Unknown constants*	*Plot*	*Gradient*
$T = k\sqrt{m + m_0}$	T, m	k, m_0	T^2 vs m	k^2
$I = aV^{3/2}$	I, V	a	I vs $V^{3/2}$	a
$I = aV^b$	I, V	a, b	$\log I$ vs $\log V$	b
$ax + y = bx^3$	x, y	a, b	y/x vs x^2	b

A3.15 USEFUL MATHEMATICAL CONSTANTS

π	$= 3.141\,59$	$\sqrt{2}$	$= 1.414\,21$
π^2	$= 9.869\,60 \approx 10$	$\sqrt{3}$	$= 1.732\,05$
e	$= 2.718\,28$	$1/\sqrt{2}$	$= 0.707\,11$

A4

VALUES OF SELECTED PHYSICAL CONSTANTS

Quantity	*Symbol*	*Value*
Speed of light in vacuum	c	$2.998 \times 10^{8}\,\mathrm{m\,s^{-1}}$
Planck's constant	h	$6.626 \times 10^{-34}\,\mathrm{J\,s}$
Electronic charge	e	$1.602 \times 10^{-19}\,\mathrm{C}$
Mass of electron	m_e	$9.110 \times 10^{-31}\,\mathrm{kg}$
Charge-to-mass ratio of electron	e/m_e	$1.759 \times 10^{11}\,\mathrm{C\,kg^{-1}}$
Mass of proton	m_p	$1.673 \times 10^{-27}\,\mathrm{kg}$
Mass of neutron	m_n	$1.675 \times 10^{-27}\,\mathrm{kg}$
Unified atomic mass unit	u	$1.661 \times 10^{-27}\,\mathrm{kg}$
Permittivity of vacuum	ε_0	$8.854 \times 10^{-12}\,\mathrm{F\,m^{-1}}$
Permeability of vacuum	μ_0	$4\pi \times 10^{-7}\,\mathrm{H\,m^{-1}}$
Faraday constant	F	$9.649 \times 10^{4}\,\mathrm{C\,mol^{-1}}$
Avogadro constant	N_A	$6.022 \times 10^{23}\,\mathrm{mol^{-1}}$
Molar gas constant	R	$8.314\,\mathrm{J\,K^{-1}\,mol^{-1}}$
Boltzmann's constant	k	$1.381 \times 10^{-23}\,\mathrm{J\,K^{-1}}$
Stefan's constant	σ	$5.670 \times 10^{-8}\,\mathrm{W\,m^{-2}\,K^{-4}}$
Gravitational constant	G	$6.672 \times 10^{-11}\,\mathrm{N\,m^{-2}\,kg^{-2}}$
Density of water (0 °C)		$9.998 \times 10^{2}\,\mathrm{kg\,m^{-3}}$
Density of water (20 °C)		$9.982 \times 10^{2}\,\mathrm{kg\,m^{-3}}$
Density of mercury (0 °C)		$1.360 \times 10^{4}\,\mathrm{kg\,m^{-3}}$
Density of mercury (20 °C)		$1.355 \times 10^{4}\,\mathrm{kg\,m^{-3}}$
Density of air (0 °C)*		$1.293\,\mathrm{kg\,m^{-3}}$
Density of air (20 °C)*		$1.204\,\mathrm{kg\,m^{-3}}$
Speed of sound in air (0 °C)†		$3.315 \times 10^{2}\,\mathrm{m\,s^{-1}}$
Speed of sound in air (20 °C)†		$3.436 \times 10^{2}\,\mathrm{m\,s^{-1}}$
Standard atmospheric pressure		$7.600 \times 10^{2}\,\mathrm{mmHg}$
Standard atmospheric pressure		$1.013 \times 10^{5}\,\mathrm{Pa}$

* = dry and at a pressure of 760.0 mmHg.
† = dry + 0.03% CO_2 by volume.

ANSWERS TO QUESTIONS AND INDEX

ANSWERS TO QUESTIONS 1A–53B

QUESTIONS 1A

1. **(a)** 25 N at 16.3° to 24 N and 73.7° to 7 N
   (b) 43.6 N at 23.4° to 30 N and 36.6° to 20 N
   (c) 52.3 N at 46.0° to 50 N and 64.0° to 40 N
   (d) 43.8 N at 136.7° to 20 N and 13.3° to 60 N
2. 76.2 m at E 66.8° S
3. **(a)** 10 m s^{-1} **(b)** 50 m s^{-1} due N
4. 13.0 m s^{-1} at N 25.9° W

QUESTIONS 1B

1. **(a)** 26.0 N, 15.0 N **(b)** 25.0 m s^{-1}, 43.3 m s^{-1}
2. **(a)** 153 N **(b)** 129 N
3. 48.8 N at 44.2° to 20.0 N and 45.8° to 60.0 N

QUESTIONS 2A

1. 5.0 m s^{-2}
2. 7.0 m s^{-2} in direction of 60 N force
3. 2.0 m s^{-2} at 37° to 40 N and 53° to 30 N
4. Ç
5. **(a)** 5.0 m s^{-2} **(b)** 2.0 m s^{-2}
6. 2.0×10^2 N
7. 5.0 kg
8. 4.5×10^3 N
9. **(a)** 4.0 m s^{-2} **(b)** 15.4 N
10. **(a)** 1.6 kN **(b)** zero
11. **(a)** 3.02 kN **(b)** 1.12 kN

QUESTIONS 2B

1. 12 N
2. 20 N
3. 3.0×10^2 N
4. 9.4×10^2 kg
5. 5.2×10^2 N
6. 36 N
7. 1.7 m s^{-2}

QUESTIONS 2C

1. 80 m s^{-1}
2. 90 m
3. 12 m s^{-1}
4. **(a)** 4.0 m s^{-2} **(b)** 20 m s^{-1}
5. 125 m
6. 4.0 s
7. **(a)** 3 s **(b)** 2 s, 4 s **(c)** 1.5 s, 4.5 s
8. 25 m s^{-1}, 3.0 s
9. 3.0 s
10. **(a)** 20 m s^{-1} **(b)** 20 m s^{-1}
11. 25.6 kN

QUESTIONS 2D

1. **(a)** 1.25 s **(b)** 22.9 m s^{-1} at 19.1° below horizontal
2. **(a)** 3.9 s **(b)** 89 m
3. **(a)** 150 m **(b)** 52 m
4. 0.90 m s^{-1}
5. **(a)** 5.0 m **(b)** 40 m **(c)** 22.4 m s^{-1} at 26.6° below horizontal
6. 1500 m

QUESTIONS 2E

1. 3 m s^{-1}
2. mv/M
3. 80 kg
4. 9.2 m s^{-1}
5. $4v$

QUESTIONS 2F

1. A: 1.75 m s^{-1} in original direction of B,
   B: 1.85 m s^{-1} in original direction of A
2. $-5u/3$ (mass m), $u/3$ (mass $2m$)
3. **(a)** 4.41 m **(b)** 2.17 m

QUESTIONS 2G

1. **(a)** 13 m s^{-1} due N **(b)** 3 m s^{-1} due N
2. **(a)** 1.5 N s **(b)** 50 N
3. **(a)** 160 N s **(b)** 80 m s^{-1}
4. 80 m s^{-1}

QUESTIONS 3A

1. **(a)** zero **(b)** 120 N m anti-clockwise **(c)** 120 N m anti-clockwise **(d)** zero **(e)** 60 N m anti-clockwise
2. **(a)** 69 N m clockwise **(b)** 51 N m anti-clockwise **(c)** 120 N m anti-clockwise

QUESTIONS 4A

2. $P = 49.5$ N, $Q = 21.2$ N
3. **(a)** 61.0 N **(b)** 35.0 N
4. **(a)** 77.8 N **(b)** 59.6 N
5. 66.4°, 91.7 N
6. 5.5 W (C), 8.5 W (D)
7. **(a)** 4 N **(b)** 28 N
8. **(a)** 0.58 W **(b)** 0.5 W upwards **(c)** 0.29 W away from B
9. **(a)** 50 N **(b)** 30 N at 90° to wall
   Hint – make use of the fact that there are three forces acting on the sphere.

QUESTIONS 4C

1. **(a)** $7a$ **(b)** $5a$
2. 9.5 cm
3. **(a)** 1.6 cm **(b)** 5.9 cm
4. 0.25 cm

QUESTIONS 5A

1. 2.4×10^5 J
2. $\sqrt{v^2 - 2gh}$

3. 1.6×10^5 J, 1.6×10^3 N
4. **(a)** 2.9×10^3 J **(b)** 32 N
5. 2.0 m s^{-1}
6. 2.5×10^3 N
7. **(a)** 70 J **(b)** 14 N
8. 4.5 m s^{-1}
9. 0.4 J

QUESTIONS 5B

1. 125 W
2. 7.5×10^2 kg
3. 1.0×10^2 W
4. 30 m s^{-1}
5. 8.0 kW
6. 2.5×10^2 W

QUESTIONS 6A

1. **(a)** 60 m s^{-1} **(b)** 3.2 rev s^{-1} **(c)** 0.31 s **(d)** 1.2×10^3 m s^{-2}
2. 0.8 m
3. 3.8 N
4. 4 rad s^{-1}
5. 7.5 rad s^{-1}
6. 24.1 hours
7. **(a)** $\sqrt{5gr}$ **(b)** $\sqrt{4gr}$ **(c)** $\sqrt{2gr}$ **(d)** $\sqrt{5gr}$
8. 160 N
9. $\sqrt{gr}$

QUESTIONS 6B

1. 37°
2. $m(g - 0.4\omega^2)$, $\sqrt{5g/2}$
3. 4 rad s^{-1}
4. 6.7 N, 22 cm

QUESTIONS 6C

1. **(a)** 5.0 rad s^{-2} **(b)** 1.5 N m **(c)** 57
2. **(a)** 0.60 rad s^{-2} **(b)** 8.7 rad s^{-1}

QUESTIONS 6D

1. **(a)** 4.5 N m **(b)** 283 J **(c)** 53 rad s^{-1}
2. **(a)** Hoop **(b)** No **(c)** Disc

QUESTIONS 7A

1. 3.1 m s^{-1}
2. 5.6 m, 3.3 m s^{-1}
3. 64 cm s^{-1}
4. **(a)** 0.46 s **(b)** 1.54 s **(c)** 0.54 s **(d)** 0.38 s
5. **(a)** 5.0 m **(b)** 10 m s^{-1} **(c)** 20 m s^{-2}

QUESTIONS 8A

1. 2.7×10^{-6} N
2. 1.9 years
3. 6.0×10^{24} kg

QUESTIONS 10A

1. **(a)** 2.20×10^3 Pa **(b)** 1.02×10^5 Pa
2. 1.03×10^5 Pa
3. 860 mmHg
4. 19.6 m
5. **(a)** 30 mmHg **(b)** 800 mmHg **(c)** 1.96×10^3 Pa

QUESTIONS 11A

1. **(a)** 4.0×10^{-3} **(b)** 2.8×10^8 Pa **(c)** 1.3×10^{-7} m^2 **(d)** 35 N

QUESTIONS 13A

1. 87 °C
2. 369.2 K
3. **(a)** 369.5 K **(b)** 369.2 K (because it is measured at a lower pressure) **(c)** 368.9 K

QUESTIONS 13B

1. 693 J
2. 40.1 °C
3. 9.00×10^3 J
4. **(a)** **(i)** $(70 - \theta)$°C **(ii)** $(\theta - 18)$°C **(b)** **(i)** $36(70 - \theta)$ J **(ii)** $840(\theta - 18)$ J **(c)** 20.1 °C

QUESTIONS 13C

1. 6.8×10^4 J
2. 3.8×10^5 J
3. 1.56×10^6 J

QUESTIONS 14A

1. 102 °C
2. 66.2 cm^3
3. **(a)** 2.31 **(b)** 1.39×10^{24} **(c)** 4.62 g **(d)** 1.93 kg m^{-3}

QUESTIONS 14B

1. **(a)** 2 **(b)** $\sqrt{2}$
2. 707 m s^{-1}
3. **(a)** 4.0 **(b)** 3.6

QUESTIONS 18A

1. 35°
2. **(a)** 70° **(b)** 26°
3. 24.4°

QUESTIONS 19A

1. **(a)** 120 cm, virtual **(b)** 13.3 cm, virtual **(c)** 8.6 cm, virtual **(d)** 100 cm, real **(e)** 33.3 cm, real

QUESTIONS 19B

1. **(a)** 2× **(b)** 0.6×
2. **(a)** 60 cm **(b)** 15 cm, 30 cm
3. **(a)** 60 cm **(b)** 30 cm

QUESTIONS 20A

1. **(a)** 20 cm **(b)** 100 cm, real
2. **(a)** 16 cm **(b)** concave

QUESTIONS 21A

1. 72 cm
2. **(a)** 85.0 cm **(b)** 16 **(c)** 0.40 cm
3. **(a)** 60.0 mm **(b)** **(i)** 75.0 mm **(ii)** 15.0 mm **(c)** 1.2 mm **(d)** 33.3

QUESTIONS 21B

1. 0.88 cm
2. **(a)** 5.0 cm **(b)** 0.7 cm

QUESTIONS 21C

1. **(a)** long sight **(b)** converging **(c)** 27.8 cm
   (d) 27.8 cm

QUESTIONS 23A

1. **(a)** π **(b)** $\frac{\pi}{2}$ **(c)** $\frac{\pi}{2}\left(\text{or } \frac{3\pi}{2}\right)$ **(d)** 0 **(e)** $\frac{\pi}{3}$ **(f)** π
   (g) 0 **(h)** $\frac{\pi}{2}$

QUESTIONS 25A

1. 5.4×10^{-7} m
2. 0.96 mm (**Not** 1.0 mm!)
3. **(a)** A,B **(b)** A,E **(c)** D,G **(d)** A,Ç **(e)** A,D

QUESTIONS 26A

1. **(a)** 16.6° **(b)** 34.8°
2. 600 mm^{-1}

QUESTIONS 30A

1. 515 Hz
2. 34 Hz

QUESTIONS 32A

1. **(a)** 250 Hz **(b)** 750 Hz
2. $4f$

QUESTIONS 33A

1. **(a)** 1200 Hz **(b)** 800 Hz
2. **(a)** 425 Hz **(b)** 850 Hz

QUESTIONS 35A

1. **(a)** 529.4 Hz **(b)** 470.6 Hz **(c)** 531.3 Hz
   (d) 472.7 Hz **(e)** 629.0 Hz **(f)** 467.7 Hz
   (g) 527.0 Hz **(h)** 391.9 Hz
2. **(a)** 0.680 m **(b)** 0.680 m **(c)** 0.640 m **(d)** 0.720 m
   (e) 0.620 m **(f)** 0.620 m **(g)** 0.740 m **(h)** 0.740 m

QUESTIONS 35B

1. 7.1×10^{-10} m

QUESTIONS 36A

1. 8.0 A
2. 4.0 mA
3. $5.1 \times 10^{-5}\,\Omega$
4. **(a)** 4.7×10^{5} A **(b)** 1.2×10^{10} A m^{-2}

QUESTIONS 36B

1. **(a)** 8.0 Ω **(b)** 1.5 A **(c)** 9.0 V **(d)** 3.0 V
2. **(a)** 1.5 Ω **(b)** 8.0 A
3. **(a)** 0.5 A **(b)** 1.5 A **(c)** 2.0 A
4. **(a)** 3.0 Ω **(b)** 5.0 Ω **(c)** 1.2 A **(d)** 2.4 V **(e)** 3.6 V
   (f) 0.3 A **(g)** 0.9 A
5. **(a)** 1.5 A **(b)** 1.0 A

QUESTIONS 36C

1. **(a)** 2.0 A (b) 0.20 A **(c)** 2 A **(d)** 1.5 A **(e)** 0.75 A

QUESTIONS 36D

1. $3.5 \times 10^{2}\,\Omega$
2. 1.7 Ω

QUESTIONS 36E

1. **(a)** 0.25 A **(b)** 7.5 V **(c)** 9.0 V
2. **(a)** 1.5 A **(b)** 1.2 Ω
3. 2.0 kΩ
4. 12 V, 4.0 Ω
5. 55 V

QUESTIONS 36F

1. 0.88 A

QUESTIONS 36G

1. **(a)** 0.50 A **(b)** 2.5 W **(c)** 0.50 W

QUESTIONS 39A

1. **(a)** -1.8×10^{5} V **(b)** 9.0×10^{6} N Ç$^{-1}$ from C to A
2. **(a)** 3.6×10^{6} V **(b)** 7.8×10^{7} N Ç$^{-1}$ at 90° to AB directed away from AB

QUESTIONS 39B

1. **(a)** A to B **(b)** 2.0×10^{3} V m^{-1} **(c)** 2.0×10^{3} V m^{-1}
   (d) 8.0×10^{-18} J **(e)** 0 **(f)** 4.0×10^{-3} N

QUESTIONS 39C

1. **(a)** 6.0×10^{5} V **(b)** 7.2×10^{5} V **(c)** 7.2×10^{5} V

QUESTIONS 40A

1. 36 μC (3.0 μF), 60 μC (5.0 μF)
2. **(a)** 48 μC on each **(b)** 12 V (4.0 μF), 8.0 V (6.0 μF)
3. **(a)** 36 μC (X), 36 μC (Y), 24 μC (Z)
   (b) 9 V (X), 3 V (Y), 12 V (Z)

QUESTIONS 40B

1. **(a)** 30.0 μA **(b)** 24.6 μA

QUESTIONS 41A

1. 7.9×10^{-3} T
2. **(a)** 3.2×10^{-4} T **(b)** 4.8×10^{-4} T **(c)** 2.5 cm

QUESTIONS 41B

1. **(a)** 0.24 N **(b)** 0.24 N **(c)** 6.4×10^{-14} N
   Note. The mention of 30° is a red herring in (b) and (c). In (b) the field lines which pass through the conductor are at 90° to it. In (c) the field lines which pass through the path of the electron are at 90° to the path.

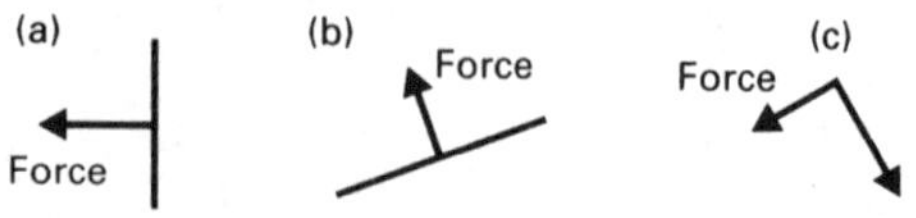

QUESTIONS 42A

1. **(a)** 9.6 mV **(b)** 9.6 mV **(c)** 8.3 mV

QUESTIONS 42B

1. **(a)** 0.35 A **(b)** 0.14 A s^{-1} **(c)** 4.4 s

QUESTIONS 42C

1. 60 mJ

QUESTIONS 42D

1. 0.42 V

QUESTIONS 42E

1. **(a)** 40 V **(b)** 10 960 V **(c)** 4000 W **(d)** 1096 kW **(e)** 1100 kW **(f)** 99.6%
2. 63.6%

QUESTIONS 43A

1. **(a)** 2.8 A **(b)** 3.3 A **(c)** 2.4 A **(d)** 15 A

QUESTIONS 43B

1. **(a)** 80 Ω **(b)** 0.25 H

QUESTIONS 43C

1. **(a)** 6.8×10^5 Ω **(b)** 35 mA
2. **(a)** 159 Ω **(b)** 0.628 A

QUESTIONS 43D

1. **(a)** 67.7 Ω **(b)** 74.1 Ω **(c)** 1.35 A **(d)** 91.4 V **(e)** 40.5 V
2. **(a)** 174 Ω **(b)** 41.8°

QUESTIONS 43E

1. **(a)** 159 Ω **(b)** 188 Ω **(c)** 1.28 A **(d)** 163 W

QUESTIONS 43F

1. **(a)** 519 Hz **(b)** 25 W
2. **(a)** B **(b)** A **(c)** B **(d)** A **(e)** C **(f)** A **(g)** A **(h)** A **(i)** B **(j)** B **(k)** A **(l)** A **(m)** C

QUESTIONS 47A

1. **(a)** 1000 eV **(b)** 1.602×10^{-16} J **(c)** 1.875×10^7 m s^{-1}
2. 8.4×10^6 m s^{-1}
3. 300 eV
4. 30 V
5. 1.4×10^7 m s^{-1}

QUESTIONS 47B

1. **(a)** 4.6×10^{-19} J **(b)** 6.6×10^{-19} J
2. **(a)** 4.3×10^{-19} J **(b)** 6.5×10^{14} Hz **(c)** 4.6×10^{-7} m
3. **(a)** 3.4×10^{-19} J **(b)** 2.1 eV **(c)** 2.1 V
4. 4.0 V

QUESTIONS 47C

1. 2.4×10^{-10} m
2. **(a)** 9.4×10^6 m s^{-1} **(b)** 7.7×10^{-11} m
3. 7.3×10^{-15} m

QUESTIONS 48A

1. **(a)** 6.2×10^{14} Hz, 4.9×10^{-7} m **(b)** 2.5×10^{15} Hz, 1.2×10^{-7} m

QUESTIONS 49A

1. 59 kV
2. 40 mA
3. **(a)** 87 kV **(b)** 72 kV **(c)** 1.4×10^{-11} m

QUESTIONS 50A

1. 2.77×10^5 m s^{-1}
2. 8.5 cm

QUESTIONS 51A

1. **(a)** 26 **(b)** 26 **(c)** 30 **(d)** 56 **(e)** 26 **(f)** 26 **(g)** 30
2. **(a)** 63, 65 **(b)** 62.9296 u, 64.9278 u **(c)** 63.5490

QUESTIONS 52A

1. **(a)** α, β **(b)** γ **(c)** **(i)** α **(ii)** α, β **(d)** β **(e)** α **(f)** α **(g)** γ **(h)** β **(i)** α **(j)** γ **(k)** β **(l)** β **(m)** α, β **(n)** β
2. $a = 4$, $b = 2$, $c = 220$, $d = 86$, $e = 210$, $f = 0$, $g = 84$, $h = 0$, $i = 0$

QUESTIONS 52B

1. $\frac{7}{8}$
2. 240 s^{-1}
3. 1.2×10^{12} Bq
4. 186 years
5. **(a)** 1.4×10^{20} **(b)** 7.9×10^{16} Bq **(c)** 9.8×10^{15} Bq **(d)** 1:7
6. 1.3×10^{-12} g

QUESTIONS 53A

1. **(a)** 1.935 42 u **(b)** 7.58 MeV

QUESTIONS 53B

1. **(a)** 0.006 22 u **(b)** 5.80 MeV
2. **(a)** 0.001 29 u **(b)** 1.20 MeV
3. **(a)** 51 **(b)** 5.11 MeV

ANSWERS TO SECTION QUESTIONS

The Examination Boards accept no responsibility whatsoever for the accuracy or method of working in the answers given. These are the sole responsibility of the author.

SECTION A

A1 38 N horizontally, 32 N vertically
A2 36 N at 56° to the horizontal
A3 54.1 N to the right at 20° below the horizontal
A5 33.7°, 3.25×10^{-6} J
A7 **(b)** **(ii)** 11.3° **(c)** 17.0°
A8 8.75 m s^{-1}, **(a)** 385 N **(b)** 770 N
A9 17.2 m s^{-1}
A10 **(a)** 1.92×10^{-2} kg s^{-1} **(b)** 2.22×10^{-2} m s^{-2}
A11 0.051 kg
A12 5 s, 50 m s^{-1}, 11.25 m s^{-1}
A13 **(a)** 45 m **(b)** 63.4°
A15 **(a)** 15 s **(b)** 1181 m
A17 80.0 m, 2.90×10^3 J
A24 1 m s^{-1}, 2.25 J
A25 0.087 m s^{-1}
A26 2/3
A27 **(b)** **(i)** 0.60 m s^{-2} **(ii)** 7508 m **(iii)** 1.2×10^4 N **(iv)** 50 kg
A28 **(a)** 0.60 m s^{-2} **(b)** 225 J (before), 81.4 J (after)
A30 0.75
A31 2.57×10^6 m s^{-1} at 43.2° to the original direction of the proton
A33 **(a)** 1.6×10^2 m s^{-1} **(c)** 0.027°
A34 **(b)** **(i)** 57 N
A36 $\sqrt{2}$ kg (A), 1 kg (C)
A37 23.1 N (CD), 11.6 N (AB)
A38 520 N (X), 380 N (Y)
A39 16 cm
A40 2.0 cm
A41 **(a)** 506 N at 90° to the wall **(b)** 1121 N at 63° to the ground
A43 **(b)** **(i)** 410 N
A44 **(d)** 500 N **(e)** 15 kW
A45 **(b)** **(ii)** 550 N **(iii)** 2.75×10^6 Pa
A46 9×10^{-15} J
A47 **(a)** 10 m s^{-1} **(b)** 24 J, 12.65 m s^{-1}
A48 **(a)** 300 m s^{-1} **(b)** 89.64 J
A49 319.4 m s^{-1}
A50 **(a)** 0.15 m s^{-1} (before), 0.060 m s^{-1} (after) **(c)** 4.6×10^{-3} J (before), 1.8×10^{-4} J (after)
A51 2.5×10^5 N, 1.54×10^5 N
A54 **(a)** 4.3×10^{-2} J **(b)** 0.66 m s^{-1} **(c)** 0.44 m s^{-1}
A55 5.0 m s^{-1}
A56 2 m s^{-1} (A), 4 m s^{-1} (B), 0.3 m from A
A57 2 m s^{-1} (P), 6 m s^{-1} (Q), 15 m s^{-1}, 60 750 W
A58 **(a)** 125 m **(b)** $6\frac{2}{3}$ N
A59 **(a)** 45 m **(b)** 400 N
A60 **(a)** **(i)** 250 m **(ii)** 4.0×10^3 N **(iii)** 2.0×10^4 kg m s^{-1}, 2.0×10^3 N **(b)** 8.0 m s^{-1}, 2.0×10^5 J, 8.0×10^4 J
A61 **(a)** 4.0 m s^{-2} **(b)** 7.1 m s^{-1}
A63 **(c)** **(i)** 5.0 N **(ii)** 5.0 N **(iii)** 3.6 m s^{-1}
A64 **(a)** 3.0 m s^{-2} **(b)** 45 kJ **(c)** 30 kW
A65 **(a)** 1.1×10^2 m s^{-1} **(b)** 0.16 m
A67 1.85×10^3 N, 3.3 m s^{-2}
A68 **(a)** 2.0 m s^{-2} **(b)** 1.2×10^4 kg m s^{-1} **(c)** 2.7×10^5 J
A69 **(a)** 1.4×10^4 W **(b)** 7.0×10^3 W, 3.0×10^4 W
A70 **(a)** **(i)** 100 N **(ii)** 10 J **(b)** 57.7 N **(c)** 57.7 N
A71 **(a)** 2.5×10^4 kg s^{-1} **(b)** 7.1×10^5 J s^{-1} **(c)** 2025
A73 90 N
A74 **(a)** 0.13 rad s^{-1} **(b)** 2.7×10^5 N towards centre of circle
A75 0.68 rev s^{-1}
A77 31 rev min^{-1}
A79 **(a)** **(i)** 3.2 m s^{-1} **(ii)** 2.2 m s^{-1} **(b)** 0.20 N
A80 **(a)** 9.0×10^{-2} J **(b)** **(i)** 2.7 m s^{-1} **(ii)** 1.8 m s^{-1} **(iii)** 0.55 N
A81 **(a)** 8 N **(b)** 8 N
A82 **(a)** 0.01 J **(b)** 2 m s^{-1} **(c)** 20 m s^{-2} **(d)** 0.15 N
A83 **(b)** 18.5 N **(c)** 12.25 m
A85 **(b)** 157 J
A86 8.9 rev s^{-1}
A88 **(b)** **(i)** 2.0×10^7 J **(ii)** 10 km
A89 58.9 rev, 235.6 s
A90 **(b)** **(i)** 2.9×10^7 J, 29 km
A91 **(c)** 1.07 rad s^{-1}, 0.161 m s^{-1}
A92 **(a)** 49 rad s^{-2} **(b)** 5.4×10^{-2} N m
A93 **(c)** **(i)** 2.0×10^2 rad s^{-1} **(iii)** 80 N m
A95 **(a)** 0.08 m s^{-1} **(b)** $\pi/2$ s
A96 **(b)** **(i)** 0.10 m s^{-1} **(ii)** 0.25 m s^{-2} **(iii)** 1.5×10^{-3} J
A97 **(a)** 0.026 m s^{-1} **(b)** 0.014 m s^{-2}; P = 0.05, Q = $\pi/6$
A98 168 m s^{-1}
A99 0.42 m s^{-1}
A101 0.2 J
A102 **(d)** 9.63×10^{11} Hz
A103 **(a)** 0.25 m s^{-1}, 0.016 J
A105 **(b)** 0.12 J
A106 **(a)** **(i)** 0.05 m **(ii)** 1.0 s, 1.8 m s^{-2}
A107 $a = 15$ mm, $\omega = 2\pi/3$ s^{-1}, $\varepsilon = \pi/2$ rad
A108 **(c)** **(i)** 3.38×10^{-3} J **(ii)** 3.45 N, 2.55 N
A109 **(d)** 139 N
A112 **(a)** 0.2π s **(b)** 1.25×10^{-4} J
A113 **(a)** 1.2 N **(b)** **(i)** 2.9 Hz **(ii)** 3.3 m s^{-2}
A115 16 Hz
A116 6.3 cm
A117 **(c)** **(i)** 100 kg **(ii)** 5.03 Hz
A118 **(a)** **(i)** 3.1 rad s^{-1} **(ii)** 0.30 m s^{-2}
A119 **(b)** **(i)** 1.0 N **(ii)** 0.79 s, 0.04 J
A120 889 N
A121 14.4 N, 24.5 h
A122 **(d)** 3.91×10^{-4} N **(e)** 2.65×10^{-4} N **(f)** 0.001%
A125 6.0×10^{24} kg
A127 9.0 m
A128 5×10^{24} kg
A130 **(a)** **(ii)** -5.36×10^9 J kg^{-1}

(b) **(i)** 2.68×10^{28} N **(ii)** 2.59×10^{4} m s^{-1}
(iii) 2.43×10^{7} s
A131 2×10^{30} kg
A133 3.37×10^{5}
A134 1.1×10^{3} kg m^{-3}
A136 0.35%, 27 days
A138 5.3×10^{6} J; 3.9×10^{8} m, 2.7×10^{-3} m s^{-2}
A141 **(b)** 6.1 km
A143 **(a)** -6.3×10^{7} J kg^{-1} **(b)** -8.9×10^{8} J kg^{-1}
A144 **(d)** 1.07×10^{11} J
A145 **(c)** **(i)** 1.26×10^{7} J kg^{-1} **(ii)** 1.26×10^{14} J
A146 **(b)** 5.4×10^{26} kg **(c)** 6.9×10^{-11} N m^{2} kg^{-2}
A147 **(a)** **(i)** -40.0 MJ kg^{-1}, -26.7 MJ kg^{-1}
(ii) 5.33×10^{11} J, 1.07×10^{5} N **(iii)** 11 km s^{-1}
A148 **(b)** **(i)** 4.4×10^{8} J, 2.1 km s^{-1}
A149 **(a)** **(iv)** 7.9×10^{3} m s^{-1}, 5.1×10^{3} s
(b) **(i)** 4.2×10^{7} m **(ii)** 0.28 s **(iii)** 81.3°

SECTION B

B2 **(c)** 500 N m^{-1} **(d)** 1.22×10^{14} Hz
B4 2×10^{-21} J, 4×10^{4} J kg^{-1}
B5 **(a)** 4.5×10^{-2} eV
B6 1.3×10^{-20} J
B7 1.25×10^{-28} m^{3}, 5×10^{-10} m
B8 **(a)** 3.1×10^{-10} m **(b)** 2.8×10^{-21} J
B9 **(a)** 3.7×10^{3} kg m^{-3} **(b)** 3.5×10^{3} kg m^{-3}
B10 1.6 kg (X), 1.2 kg (Y)
B11 1.81×10^{5} Pa
B12 36.8 cm
B13 3.2
B14 **(a)** 4.7×10^{3} kg m^{-3}
B15 15 N
B16 16 g
B17 9.0×10^{2} kg m^{-3}
B18 **(a)** 600 kg **(b)** 0.91 m s^{-2}
B19 **(a)** 4.0 mm
B22 1.6×10^{15}
B24 $\alpha = -\frac{1}{2}, \beta = \frac{1}{2}, \gamma = -\frac{1}{2}$
B25 2.74×10^{-5} J, 0.35 m s^{-1}
B26 5 cm
B27 761.9 mmHg
B28 30 mm
B29 7.5×10^{-2} N m^{-1}. The water would rise to only 4.3 cm
B30 **(b)** 7 mm
B32 **(c)** 4.5×10^{-4} J, $10\sqrt{3}$ m s^{-1}
B33 **(b)** **(i)** 7.3×10^{-2} N m^{-1}
B34 **(a)** 1 mm **(b)** 5.5×10^{-2} J
B35 2.6×10^{-4} m^{2}
B36 7×10^{6} N m^{-2}, 0.14 J
B37 **(c)** 0.45 m^{2}
B38 1.5×10^{8} Pa
B39 **(a)** 8:1 **(b)** $D/9$
B40 **(c)** **(ii)** 1.2×10^{-2} J
B41 6.25×10^{11} N m^{-2}
B42 **(b)** **(i)** 50 N **(ii)** 1.8×10^{-3} m, 4.4×10^{-2} J
(iii) 0.85 mm **(iv)** 0.084 m from B
B43 **(b)** **(i)** 2.0×10^{11} Pa **(ii)** 9.6×10^{2} N
(iii) 3.6×10^{6} J m^{-3}
B44 5.1×10^{6} N m^{-2}, 0.63 s
B45 **(a)** 50 N **(b)** 0.5 J
B47 5 mm
B48 **(a)** 3:2 (Cu:Fe) **(b)** 6.0 mm, 4.0 mm **(c)** 780 N
B49 **(a)** 589 N, 589 N **(b)** 10.046 m **(c)** 25.2 J
B51 **(a)** 3.0×10^{-4} **(b)** 6.0×10^{7} Pa **(c)** 1.44×10^{3} J, 1.56×10^{7} J
B52 **(a)** 1020 °C **(b)** 1.8×10^{8} Pa
B53 **(c)** **(i)** 2.5×10^{8} Pa **(ii)** 0.10 J **(iii)** 0.20 J
(v) 1.9×10^{11} Pa
B55 **(b)** **(i)** 1.1 mm **(ii)** 1.7×10^{-2} J
B56 2.8×10^{-10} m **(a)** 2.0×10^{8} Pa **(b)** 3.0 J
(c) 1.4×10^{10} Pa, 3.0 mm
B59 **(b)** 1.9×10^{11} Pa
B61 **(a)** 0.1 m (A), 0.2 m (B) **(b)** 2 N m^{-1}
B62 **(c)** **(ii)** 5.0 mm
B63 **(c)** **(i)** 2.0 m s^{-1} **(ii)** 3.9×10^{-2} kg s^{-1}
B64 **(c)** **(i)** 2.8 m s^{-1} **(ii)** 5.7×10^{-3} m^{3} s^{-1}
B65 2.97×10^{4} N
B66 **(a)** 2.0 m s^{-1} **(b)** 0.8 m. The second hole must be 20 cm above the base of the tank
B67 **(b)** **(ii)** 1.1×10^{3} Pa
B68 **(a)** 133 m s^{-1} **(c)** 120 m s^{-1}, 1.82×10^{5} Pa
B71 **(c)** 15.8 m s^{-1}
B74 7.6 cm
B75 6000 N m^{-2}
B76 **(b)** **(iii)** 386 mm
B80 **(a)** **(ii)** 5.0×10^{-7} m **(b)** **(iii)** 169 m s^{-1}
B82 2, 2.19
B83 **(b)** 24 m s^{-1}
B84 $A = 15$ cm^{2} g^{-1} s^{-1}, $B = 15$ cm^{-1} s^{-1}, radius = 0.2 cm
B85 **(b)** **(ii)** 0.45 N s m^{-2}
B86 **(b)** **(i)** 5.0×10^{-4} m^{3}

SECTION C

C4 290 K
C6 **(a)** 37 °C **(b)** 50 °C
C7 16.6 °C, 17.0 °C
C8 49.04 °C
C9 **(b)** **(ii)** 23 °C (approx) **(iii)** 0 °C, 100 °C
C10 **(f)** 364.21 K; 367.86 K
C11 0.45 W
C12 279 J kg^{-1} K^{-1}
C13 5.7 V
C14 6000 J kg^{-1} K^{-1}, 10%
C15 1700 J kg^{-1} K^{-1}
C16 **(d)** 2.91×10^{3} J kg^{-1} K^{-1}, 12.3 W
C17 4
C18 **(b)** **(iii)** 57.6 °C
C19 0.67 kg
C20 **(a)** 250 W **(b)** 2.50×10^{4} J kg^{-1} **(c)** 167 J kg^{-1} K^{-1}
C21 0.1225 kg
C22 1.67×10^{3} J kg^{-1} K^{-1}
C23 88%
C24 0.0396 kg
C25 7.4%
C26 **(c)** 1.6×10^{2} J, 7.0%
C27 2.4×10^{4} J kg^{-1}, 2.0×10^{3} J kg^{-1} K^{-1}
C28 **(a)** **(i)** 3.6×10^{4} J **(ii)** 15.8 g
C29 1.1×10^{5} Pa
C30 30 m
C31 745 mm
C32 **(b)** 4.6 mol **(c)** 517 m s^{-1}
C33 **(a)** 1.33×10^{5} Pa **(b)** 2 g in A, 1 g in B
C34 **(a)** 2.1×10^{3} mol **(b)** 60 kg
C35 40
C36 **(b)** **(i)** 2.3 kg **(ii)** 1.4 kg **(iii)** 3.1×10^{17} s^{-1}
C37 0.33 kg
C38 **(a)** 8.3 J K^{-1} mol^{-1} **(b)** 1.4×10^{-23} J K^{-1}
(c) 6.3×10^{-21} J
C39 **(a)** 0.32 g **(b)** 4.8×10^{22} **(c)** 1.4×10^{3} m s^{-1}
C40 $1.81(2) \times 10^{5}$ J, $1.85(5) \times 10^{5}$ J, 6.1×10^{2} J kg^{-1} K^{-1}

C42 **(d)** $552\,m\,s^{-1}$
C43 **(b)** **(i)** 2.57×10^{-4} kg **(ii)** 240 J
C44 **(a)** **(i)** 40.2 **(ii)** 14.1 kg **(iii)** $141\,kg\,m^{-3}$ **(iv)** $146\,m\,s^{-1}$
C46 **(b)** **(i)** 6.0×10^{-3} **(ii)** 3.6×10^{21} **(iii)** $483\,m\,s^{-1}$
C47 **(a)** **(i)** 2.4×10^{22} **(ii)** 150 J
C48 **(a)** 1.07:1 **(b)** 4:1 **(c)** 1.15:1
C51 **(b)** **(i)** $1.2 \times 10^{-3}\,m^3$ **(ii)** 1.9×10^2 J **(c)** **(i)** 31 J **(ii)** 21 J **(iii)** 52 J
C52 **(a)** 2.26×10^6 J **(b)** 1.69×10^5 J **(c)** 2.09×10^6 J
C54 **(a)** 1.2×10^{-2} **(b)** 2.6×10^6 Pa **(c)** 91 J
C55 **(b)** **(ii)** 75.0 J **(d)** **(i)** 50.0 J **(ii)** 83.3 J
C56 **(b)** 7.2×10^2 J **(c)** **(i)** 8.0 kN **(ii)** 1.6 MPa
C59 **(d)** **(i)** $4.75 \times 10^{-3}\,m^3$ **(ii)** 984 K **(iii)** **(1)** $29.1\,J\,K^{-1}\,mol^{-1}$ **(2)** $4.92 \times 10^{-4}\,m^3$ **(3)** 1.76×10^3 J **(4)** 4.39×10^3 J
C60 $1.44 \times 10^4\,J\,kg^{-1}\,K^{-1}$, $1.03 \times 10^4\,J\,kg^{-1}\,K^{-1}$
C61 **(a)** 3000 K **(b)** $600\,J\,kg^{-1}\,K^{-1}$, $783\,J\,kg^{-1}$; $268\,m\,s^{-1}$
C63 $1.07 \times 10^3\,J\,kg^{-1}\,K^{-1}$
C67 **(a)** 2.4×10^3 J **(b)** **(i)** 1.1×10^5 Pa **(ii)** 8.8×10^4 Pa
C68 **(c)** **(i)** 4.0×10^2 N **(ii)** 5.7×10^2 N **(iii)** 6.6×10^2 N
C69 **(b)** **(iii)** 150 K
C70 1.7×10^{-2} kg, 60 K, $7.3 \times 10^2\,J\,kg^{-1}\,K^{-1}$
C71 **(b)** $3.76 \times 10^3\,m\,s^{-1}$
C72 **(a)** $1250\,cm^3$ **(b)** 1.38×10^5 Pa
C73 **(b)** $7.44 \times 10^{-3}\,m^3$ **(d)** 1.66
C74 **(b)** **(i)** 301.2 K at A and C, 150.6 K at B **(ii)** 4375 J **(iii)** 3125 J **(iv)** 625 J **(v)** 625 J
C75 **(c)** $2.4(5) \times 10^5$ Pa **(d)** 2.5×10^5 Pa, 2.1×10^5 Pa, 31
C76 **(a)** 1.47×10^5 J **(b)** 1.69×10^6 J
C77 **(c)** **(i)** 2.2×10^4 J **(ii)** 3.1×10^5 J
C82 **(c)** $2.3 \times 10^{-3}\,m^3$
C84 **(b)** **(ii)** $a = 0.362\,N\,m^4\,mol^{-2}$, $b = 4.27 \times 10^{-5}\,m^3$
C85 **(c)** **(i)** 127 K, 36.9 K **(d)** 1.05×10^3 J, 3.06×10^2 J
C86 **(c)** **(i)** 9.9×10^{-2} kg **(ii)** 1.9×10^{24} **(iii)** $4.7 \times 10^2\,m\,s^{-1}$
C87 **(c)** **(i)** 1024 kPa **(ii)** 533 K
C91 **(c)** **(i)** 540 J **(ii)** 60 J
C94 1.01×10^5 Pa
C96 9×10^2 Pa
C97 $1.7 \times 10^4\,N\,m^{-2}$, $3.0 \times 10^3\,N\,m^{-2}$, 82%
C99 **(c)** **(ii)** 2.0 kPa **(iii)** 1.5 kPa
C100 **(d)** **(i)** 80 kPa (nitrogen), 20 kPa (oxygen) **(ii)** $0.27\,kg\,m^{-3}$ **(iii)** $1.2\,kg\,m^{-3}$
C101 $3.9 \times 10^5\,J\,kg^{-1}$
C103 **(a)** 1.60×10^6 J **(b)** 1.33×10^6 J
C104 5.03%
C105 **(a)** 90 J **(b)** **(i)** 70 J **(ii)** 20 J
C106 $1.22 \times 10^3\,J\,K^{-1}$
C107 **(a)** 0 **(b)** $40\,J\,K^{-1}$
C108 **(c)** 19.6%
C110 **(a)** 0 **(b)** 4.2×10^3 J **(c)** -3.0×10^3 J
C111 $3.0 \times 10^2\,°C\,m^{-1}$ (copper), $5.5 \times 10^2\,°C\,m^{-1}$ (aluminium)
C112 0.2 °C, $2.4 \times 10^6\,J\,kg^{-1}$
C113 1.9×10^3 W
C114 240 W, 232 min
C115 48.1 W, 2.16 g
C116 **(d)** 2.6 °C
C117 6400 W, 66:1
C120 102.5 °C
C121 **(a)** 4.0 kW
C122 **(a)** 0.02 **(b)** 0.008
C123 94%
C128 **(a)** **(ii)** $1.5 \times 10^2\,°C\,m^{-1}$, 29 W
C129 **(b)** 10 °C **(c)** **(ii)** $3.0 \times 10^{-3}\,K\,W^{-1}$, 1.0×10^3 W, 9.8 W
C130 $6.5 \times 10^2\,J\,s^{-1}\,m^{-2}$, $2.0 \times 10^{-3}\,mm\,s^{-1}$
C132 **(e)** 9.0 kW **(f)** 88 W
C133 **(b)** **(ii)** 507 °C
C134 $6.78 \times 10^{-2}\,K\,s^{-1}$
C137 1105 K
C138 1073 K
C139 2021 K
C141 $19\,W\,m^{-1}$
C143 **(a)** 0.19 nm **(b)** 1.07 mm
C144 378 K

SECTION D

D1 **(a)** 19.5° **(b)** 10.5°
D2 18.6°
D3 38.5°
D4 26.4°
D5 1.6
D6 40.2°
D8 27.9°
D9 **(b)** **(i)** 60°4′ **(ii)** 48°27′
D10 39.6°
D11 37.2°, 48.6°
D12 32.77°, 41.05°
D13 **(b)** 1.50
D14 3.17°
D15 **(a)** 1.5 cm high, 30 cm from lens on opposite side from object **(b)** 1.0 cm high, 20 cm from lens on same side as object
D16 20 cm; 40 cm from lens
D18 100 cm
D19 30 cm above lens, diameter = 6 mm
D20 5 mm, 5 mm, 25.0 mm
D21 100 cm from diverging lens on same side as object
D23 2.6 mm
D24 100 mm, 95 mm
D25 0.2 m (diverging)
D27 **(a)** 70 cm **(b)** 135 cm
D30 **(b)** 100 cm
D31 **(a)** 6 cm behind the mirror, 1.5× **(b)** 3.3 cm behind the mirror, 0.83×
D32 40 cm
D33 **(a)** $41\frac{2}{3}$ mm to 50 mm **(b)** 6 to 5
D36 16.0 cm from the second lens and between the lenses; 27.8 cm away on the side remote from the object
D37 **(a)** 41.7 mm **(b)** 90 **(c)** 10.7 mm
D38 **(b)** 110 mm
D40 **(c)** 22.7 mm, 37.5
D42 **(a)** 550 mm **(d)** 10 **(e)** 55 mm from eyepiece lens
D44 400 mm (objective), 50 mm (eyepiece)
D45 **(b)** 60 mm
D46 4.0 cm
D47 112.4 cm, 15.6
D48 **(a)** 160 **(b)** 25.2 mm **(c)** 0.938 mm
D49 **(c)** **(i)** 1.84 cm
D50 0.18(2) cm; 20 cm from second lens on the same side as the first lens; 0.91 cm
D51 9.1×10^{-3} rad
D52 **(a)** 2.1 cm **(b)** 112.5 cm
D53 100

D54 **(b)** **(ii)** $104\frac{1}{6}$ cm, 24 **(iii)** 5.3 cm behind eyepiece **(c)** **(i)** $106\frac{2}{3}$ cm
D55 **(a)** 0.029 cm **(b)** 0.55 cm **(c)** 0.11 rad
D58 **(a)** **(i)** 50 mm **(b)** 2.6 mm away from the film
D59 0.065 s
D60 **(c)** 1.75 cm; 11, 32 ms
D61 $4\frac{16}{21}$ cm, $\frac{5}{21}$ cm, 100 cm (diverging)
D62 $f = 200$ cm diverging
D63 $f = 30$ cm converging
D64 $66\frac{2}{3}$ cm
D65 **(a)** $f = 400$ cm diverging **(b)** $44\frac{4}{9}$ cm to infinity
D66 **(a)** $42\frac{6}{7}$ cm converging **(b)** 40 cm
D67 **(b)** **(i)** $-2.0\,m^{-1}$, 0.28 m
D68 200 cm (diverging), 50 cm (converging)
D69 **(b)** 3.1×10^4 m, 300 rev s^{-1}
D71 $3.0(2) \times 10^8\,m\,s^{-1}$

SECTION E

E1 6.0×10^4 m
E4 **(a)** $3.3 \times 10^2\,m\,s^{-1}$ **(b)** $6.6 \times 10^{-4}\,m\,s^{-1}$
E5 **(a)** **(i)** 7.5° (light), 45.6° (sound)
E12 $342\,m\,s^{-1}$
E13 **(a)** 1.5 m
E14 5.0×10^{-7} m
E15 **(b)** 1.79×10^{-3} rad, 0.064 mm
E17 1.5×10^{-4} rad
E18 **(b)** 0.28 mm **(c)** **(i)** 0.23 mm
E19 6.43×10^{-7} m
E20 **(b)** 711 nm, 427 nm
E21 **(b)** 0.021°
E23 1.5×10^{-5} m
E25 **(b)** 9.82×10^{-8} m
E26 2.7 mm
E28 **(b)** **(iii)** 1.26×10^{-5} m
E29 **(a)** 25 cm from card **(b)** 20 cm
E31 2.4×10^{-6} m
E32 4.60×10^{-7} m (violet), 6.90×10^{-7} m (red); 66.9°
E33 640 nm, 480 nm; 28.7°
E34 **(a)** 0.96° **(b)** 0.34 cm
E36 600 nm, 2.85×10^5, 43.2°
E37 **(c)** 1.04 cm
E38 4.34×10^{-7} m
E39 **(a)** $5.5 \times 10^5\,m^{-1}$ **(b)** **(i)** 6.2×10^{-7} m **(ii)** 4
E40 **(a)** 13.6° **(b)** 70.5°
E44 53.1°
E45 **(a)** 1.6 **(b)** 32°
E55 **(a)** 1.83 s **(b)** 19%
E56 **(a)** $3.2 \times 10^2\,m\,s^{-1}$ **(b)** $1.6 \times 10^{-3}\,m\,s^{-1}$, 0.13 m
E57 $3\lambda/8$, 300 Hz
E58 $1.4l$
E59 **(b)** **(i)** 91.3 Hz **(ii)** 87.8 Hz
E60 **(a)** $1.5 \times 10^2\,m\,s^{-1}$ **(b)** $1.8 \times 10^8\,N\,m^{-2}$ **(c)** $1.8 \times 10^{11}\,N\,m^{-2}$
E61 **(a)** $f - 2.0$ **(b)** $f - 0.7$; 0.76%
E63 **(d)** 71.5 cm
E64 **(a)** $300\,m\,s^{-1}$ **(b)** 9.0 kg
E65 **(a)** 250 Hz **(b)** 1.4×10^2 N
E66 1.7×10^2 Hz
E67 **(b)** **(i)** 514 Hz **(ii)** 512 mm
E68 1.0×10^2 Hz
E69 $336\,m\,s^{-1}$
E70 **(a)** $320\,m\,s^{-1}$ **(b)** 10 cm **(c)** 40 cm **(d)** 1200 Hz
E71 5.1 Hz
E73 0.83 m
E74 **(b)** **(ii)** $340\,m\,s^{-1}$ **(iii)** 34.6 cm
E76 $336\,m\,s^{-1}$
E77 $348\,m\,s^{-1}$, 8.26 mm
E78 **(a)** **(i)** $340\,m\,s^{-1}$ **(ii)** 696 mm below top of tube
E79 **(c)** **(ii)** $341\,m\,s^{-1}$, 12 mm **(iii)** 207 Hz
E82 **(a)** 1.75 m, 16 Hz
E84 **(b)** 0.38 m, 4.1 N
E85 0.066 mm
E86 **(a)** **(i)** 1700 Hz **(ii)** 0.050 m
E87 $330\,m\,s^{-1}$, 579 mm
E90 $362\,m\,s^{-1}$
E94 514 Hz, 545 Hz
E95 425 Hz
E96 $1.11f$, $1.10f$
E97 **(b)** **(i)** ±0.003 43 nm **(ii)** zero
E98 **(c)** 2.2×10^{-8} m
E99 $1.02 \times 10^5\,m\,s^{-1}$
E100 **(b)** $37.8\,m\,s^{-1}$, 302 m
E101 **(b)** 92.1 kHz
E103 **(a)** **(ii)** 1.9%
E104 **(b)** 57.6 Hz **(c)** 2.7 Hz
E105 **(b)** $5.0 \times 10^4\,m\,s^{-1}$ **(c)** 2.8×10^{10} m

SECTION F

F1 **(a)** 5.0 A **(b)** 2.0 V **(c)** 10 V **(d)** 12 V
F2 $4.8 \times 10^{-11}\,\Omega m$
F3 11.3 V
F4 0.59 A
F6 **(b)** $9.5 \times 10^{-3}\,m\,s^{-1}$
F7 **(a)** 3 Ω **(b)** 4 Ω
F8 0.25 A
F9 **(a)** **(i)** 0.067 A **(ii)** 0.033 A **(iii)** 4.0 V **(b)** **(i)** zero **(ii)** 0.054 A **(iii)** 2.16 V
F10 8.0 V, 50 Ω
F11 0.5 A
F12 60 V, 48 V
F13 40 Ω, 40 Ω
F14 5 V
F15 **(a)** 9995 Ω in series **(b)** $2.5 \times 10^{-3}\,\Omega$ in parallel
F16 **(a)** 0.05 A **(b)** 0.0125 A **(c)** 0.5 V
F17 14 950 Ω in series
F18 60.4 V
F19 9.6 V
F20 **(a)** 360 Ω **(b)** 0.96 V
F21 **(b)** **(ii)** 3.3 V (1.8 kΩ), 8.7 V (4.7 kΩ) **(c)** **(ii)** 10.3 kΩ **(d)** 3.9 V, 8.1 V
F22 2.29 V
F23 **(b)** 2.0 mA, 2.48 MΩ
F24 **(b)** **(i)** 4.0 A **(ii)** 2.5 Ω **(iii)** 3.5 Ω
F25 **(b)** 1.18 Ω **(c)** 18.3 V (V_1), 13.2 V (V_2)
F26 24.5 Ω
F27 **(b)** **(ii)** 7.0 V
F29 2.0 A
F30 **(a)** 2.0 Ω **(b)** 1.44 W
F31 **(a)** 120 Ω **(b)** 25.3 W
F32 **(b)** **(i)** 1 Ω **(ii)** 3.75 J
F33 **(a)** 3.75 V **(b)** 19 mW
F34 0.367 m
F35 **(a)** **(i)** 5.0 A **(ii)** 100 V **(b)** 2.1 Ω
F36 **(a)** 76.8 Ω **(b)** 3.49 m
F37 **(b)** 6.0 Ω **(c)** 500%
F39 **(a)** 0.718 A **(b)** 4.19 W, 2.095×10^{-3} kW h **(c)** 8.17 V, 3.59 V **(d)** 9.15 W, 0.592 W **(e)** 3.60 Ω
F40 **(b)** **(i)** 78 Ω **(ii)** 8.0×10^2 W **(d)** **(i)** 28 W
F42 **(a)** 0.25 m from one end **(b)** 2 μA

F43 **(a)** 17.2 °C **(b)** $1.21 \times 10^{-7}\,\Omega\,m$
F44 **(b)** **(i)** 2.46 A **(ii)** $0.500\,V\,m^{-1}$, $4.90 \times 10^6\,A\,m^{-2}$, $10.2 \times 10^{-8}\,\Omega\,m$
(c) **(i)** 4.0 V **(iii)** 6.7 Ω, 14.4 W
F46 **(c)** 0.390 Ω
F47 2 V
F48 50 Ω
F49 24 V
F50 82.5 cm, 1.29 V
F52 1 Ω, 5.04 V
F53 **(c)** 50.2 Ω
F54 **(a)** **(iii)** 1.36 V **(iv)** 18.6 mW
F55 **(c)** 294.7 Ω
F56 **(b)** 6.330 mV
F58 **(b)** **(i)** 0.075 m
F59 $6.0 \times 10^{-8}\,J$
F60 −24 mJ
F62 **(a)** $3.0 \times 10^{-6}\,N$ **(b)** $5.5 \times 10^{-10}\,C$
(c) $1.8 \times 10^3\,V$
F64 **(b)** **(i)** $2.0 \times 10^{-3}\,N$ **(ii)** $2.0 \times 10^{-3}\,N$
(iii) $6.0 \times 10^{-4}\,J$ **(iv)** 0 **(v)** 0
F65 **(a)** $1.45 \times 10^{-10}\,N$, $1.45 \times 10^{-11}\,J$
F68 **(i)** 36 μA **(ii)** $1.44 \times 10^6\,V$ **(iii)** 51.8 W
F69 **(a)** $1.1 \times 10^{-3}\,C$ **(b)** $2.8 \times 10^3\,J$
(c) $2.5 \times 10^6\,V\,m^{-1}$
F70 **(a)** 1.5 V **(b)** 0.56 V **(c)** 1.5 V
F71 **(b)** 3 μF **(c)** 36 μC (A_1 and B_1), −36 μC (A_2 and B_2)
F72 13.3 μC
F73 **(i)** $5 \times 10^4\,V\,m^{-1}$ **(ii)** $4.4 \times 10^{-12}\,F$
(iii) $2.2 \times 10^{-6}\,J$
F74 133 μC on each; 133 μC
F75 1475 J; an increase of 983 J
F76 $1.13 \times 10^5\,V\,m^{-1}$; **(a)** 226 V **(b)** $2.2 \times 10^{-10}\,F$
(c) $5.65 \times 10^{-6}\,J$
F77 $\frac{4}{7}$ V; $1\frac{3}{7}$ V, $\frac{2}{7} \times 10^3\,V\,m^{-1}$
F78 **(b)** **(i)** 1.5 μF **(ii)** 108 μJ **(iii)** 9.0 V
F79 7
F80 0.2 μF; 0.25 mJ
F81 0.45 J; 0.15 J
F82 **(a)** $5.18 \times 10^{-9}\,J$ **(b)** $1.56 \times 10^{-8}\,J$, $1.04 \times 10^{-6}\,N$
F83 36 μJ, 24 μJ; **(a)** 36 μJ, 384 μJ **(b)** 36 μJ, 54 μJ
F84 40 V
F85 50 J, $33\frac{1}{3}$ J
F86 **(a)** $1 \times 10^{-7}\,C$ **(b)** 1000 V **(c)** $4 \times 10^{-5}\,J$
F87 **(a)** 4 μC on each **(b)** 2 μC (M), 8 μC (N)
F88 **(b)** **(ii)** **(I)** 16 μJ **(II)** 32 μJ
F89 **(b)** **(i)** $2.0 \times 10^{-4}\,C$ **(ii)** $4.0 \times 10^{-3}\,J$
(iii) $2.0 \times 10^{-3}\,J$
(c) **(i)** 20 ms **(ii)** 92 ms
F92 **(a)** **(i)** 24 μC, 144 μJ **(ii)** 4.8 V, 57.6 μJ
(b) **(iii)** 33 MΩ (approx.)
F95 $1.35 \times 10^{-3}\,C$; 3 times; 84.4 V
F96 Initially $3 \times 10^{-9}\,C$ on each; finally $1 \times 10^{-9}\,C$ and $5 \times 10^{-9}\,C$; $1.2 \times 10^{-8}\,J$
F97 500 μF
F98 4
F100 $9 \times 10^{-6}\,N$
F101 21.8°
F102 **(a)** $1.3 \times 10^{-3}\,N$ **(b)** 1.54; 115 g
F104 **(b)** **(i)** $1.6 \times 10^{-4}\,T$ **(ii)** zero
F108 1.08:1
F109 **(a)** 10 **(b)** 0.25
F113 **(a)** $3.0 \times 10^{-5}\,m\,s^{-1}$ **(b)** $1.0(4) \times 10^{29}\,m^{-3}$
(c) $3.5 \times 10^{-3}\,m^2\,s^{-1}\,V^{-1}$ **(d)** 0.82
F118 0.24 A

F120 **(a)** **(i)** $6.0 \times 10^{-5}\,Wb$ **(ii)** 3.0 V
F121 **(a)** 0.7 A **(b)** $7.8 \times 10^{-4}\,J$
F122 **(a)** 0.24 cos θ Wb **(b)** 15 V
F123 **(b)** **(i)** 0.128 Wb, $2.6 \times 10^{-2}\,V$ **(ii)** 1.1 mA, $6.0 \times 10^{-7}\,T$
F124 **(b)** **(i)** 140 ms, 2.5 V **(c)** **(i)** 75 mA **(ii)** 0.11 W
F127 16 mV
F129 **(a)** 4.0 A **(b)** $2.0\,A\,s^{-1}$
F130 **(a)** **(i)** 0 **(ii)** 2 V **(iii)** 0.1 A **(iv)** 1 V **(v)** 1 s
F134 **(a)** **(i)** $0.30\,A\,s^{-1}$ **(ii)** 0.24 A **(c)** 1.2 J
F135 **(a)** **(i)** 4.0 A **(ii)** $1.0\,A\,s^{-1}$ **(iii)** $0.5\,A\,s^{-1}$
(b) 268 divisions
F136 4 mH, $1.25\,A\,s^{-1}$
F137 **(a)** **(i)** 50 kW **(ii)** 340 V **(iii)** 70.6%
(b) **(ii)** 500 W
F138 100 μC
F139 **(a)** $2.5 \times 10^{-3}\,Wb$ **(b)** 25 μC
F143 **(b)** **(i)** 0.2 T
F144 0.5 A
F145 **(a)** 1 A **(b)** 2 A **(c)** 2 A
F146 **(b)** **(i)** 100 kJ, 300 kJ
F147 **(a)** 7.1 V **(b)** zero
F148 7.5 mA, 750 mA
F149 **(a)** **(i)** 3.5 V **(ii)** $8.0 \times 10^{-3}\,F$
F150 **(b)** 30 Ω **(c)** 92 mH
F152 80 Hz; 400 μF
F153 **(a)** 0.10 A, 40 V **(b)** 3.0 W
F154 **(a)** **(i)** 0.50 A **(ii)** 240 Ω **(iii)** 240 Ω
(b) **(ii)** 416 Ω **(iii)** 7.7 μF
F155 **(a)** 40 Ω **(b)** $2.9 \times 10^{-2}\,H$ **(c)** 0.22 rad
F157 **(a)** $V_R = 3.0$ V RMS, $V_L = 4.0$ V RMS **(c)** 0.64 rad
F158 **(a)** 47.2 A **(b)** $8.91 \times 10^3\,W$
F159 **(b)** **(ii)** $1.6 \times 10^4\,Hz$
F160 $1.6 \times 10^4\,\Omega$
F162 0.337 A; 202.2 V, 214.3 V, 84.6 V
F163 **(a)** 0.124 A **(b)** 78.6 V
F164 **(i)** 10 Ω **(ii)** 69 mH
F165 **(b)** **(i)** 60 Ω, 480 W **(ii)** 0.33 H, 240 W
(iii) 30.6 μF, 240 W
F166 **(a)** 149 V **(b)** 188 V, 2.9 ms
F167 **(a)** 0.240 A RMS **(b)** 28.8 W **(c)** 4.77 μF
F168 0.21 A, 33 mH
F169 **(b)** 11.5 mH **(c)** 2.56:1
F170 8 Ω, 50 Hz
F171 **(a)** 159 Hz **(b)** 0.03 A
F172 **(a)** 1.13 pF **(b)** 5.4 cm **(d)** 6.9 pF
F173 **(b)** **(i)** 100 Hz
F174 **(b)** **(i)** $X_C = 133\,\Omega$, $X_L = 30.0\,\Omega$ **(ii)** 131 Ω
(iii) 1.91 A **(iv)** 52° **(v)** 0.61 **(c)** 101 Hz
F175 **(a)** 50.0 Ω, 7.96 mH, 3.18 μF
(b) **(i)** 7.07 V **(ii)** 22.5 μC **(iii)** 7.07 V
(iv) $8.9 \times 10^2\,A\,s^{-1}$
F176 **(a)** 141 Ω **(b)** 0.20 A **(c)** 4.0 W
F177 227 V, 41.1 Hz
F178 7.8 V, 6.2 V
F179 **(a)** **(ii)** $3.79 \times 10^3\,A$, 5.7%
(b) **(i)** 50 Ω, 40 Hz **(ii)** $V_R = 50$ V, $V_C = 100$ V **(iv)** 90 V, 56°
F182 **(a)** 20 W
F185 **(a)** **(iii)** 0.042 A (RMS)
(c) **(ii)** $6.25 \times 10^{-3}\,J$ **(iii)** 3.4 V **(iv)** 0.17 W
F189 **(a)** 2.4 V **(b)** 3.6 W **(c)** $4.0 \times 10^3\,s$
F190 $2.1 \times 10^{-4}\,m^3$
F191 $3.037 \times 10^6\,C\,kg^{-1}$
F192 0.42 A

SECTION G

G1 500 V
G3 1.96×10^{-19} J
G4 1.96×10^{-19} J, $>1.01 \times 10^{-6}$ m
G5 4.45 eV
G6 1.5 V, 2.5×10^{-19} J, 7.3×10^{5} m s^{-1}
G7 6.64×10^{-34} J s
G8 **(c)** **(i)** 2.35×10^{5} m s^{-1} **(ii)** 0.157 V
G9 **(b)** **(i)** 3.6×10^{-19} J **(ii)** 2.8×10^{17} s^{-1} **(iii)** 9.1%
(c) **(i)** 2.2×10^{-11} m
G10 **(b)** 6.7×10^{-34} J s **(c)** **(i)** no emission
(ii) electrons emitted with a KE of 3.74×10^{-19} J
G11 **(a)** 1 eV **(b)** **(i)** 1 V **(ii)** 0.75 V
G12 **(c)** **(i)** 3.4×10^{-19} J **(ii)** 6.68×10^{-34} J s
G13 **(b)** **(ii)** **(I)** 2.6×10^{-19} J **(II)** 7.5×10^{-7} m
(c) 1.24×10^{-6} m
G14 2.5×10^{17} s^{-1}, 3.9 mA, 0.39 V
G15 **(a)** **(i)** 1.6×10^{-9} A **(ii)** 3.2×10^{-9} A **(iii)** zero
(b) 0.5 V
G16 **(b)** **(i)** 1.0 V **(ii)** 1.6×10^{-19} J **(iii)** 3.8×10^{-19} J
G20 1.2×10^{-11} m
G24 **(a)** 2.19×10^{6} m s^{-1} **(b)** 9.14×10^{-8} m
G25 **(c)** 1.44×10^{-13} m
G26 **(b)** **(i)** 661 nm, 489 nm, 436 nm, 412 nm
G27 **(a)** 3.4×10^{-19} J **(b)** 1.8×10^{20} s^{-1}
G28 1.22×10^{-7} m
G29 **(a)** 2.16×10^{-18} J **(b)** 6.6×10^{-7} m
G30 **(b)** **(ii)** 5.1×10^{-19} J **(iii)** 3.9×10^{-7} m
G31 **(a)** **(i)** 1.9 eV **(ii)** 10.2 eV
G32 **(a)** **(i)** 3.3×10^{15} Hz **(ii)** 2.5×10^{15} Hz
G34 **(b)** **(vi)** 1.1×10^{7} m^{-1}
G35 **(b)** **(ii)** 6.6×10^{-7} m **(c)** **(ii)** 1.2×10^{-10} m
G39 **(c)** **(i)** 10 000 eV **(ii)** 1.23×10^{-11} m
(iii) 1.60×10^{-15} J, 1.24×10^{-10} m
G40 **(a)** 1.2×10^{19} Hz **(b)** 9.9×10^{2} W
G41 **(b)** **(i)** 12 mA **(ii)** 1.33×10^{8} m s^{-1}
G42 **(a)** 4.1×10^{-11} m **(b)** 30 mA **(c)** 1.9×10^{15} s^{-1}
G44 **(a)** 4.0×10^{2} eV
G45 **(c)** **(i)** 0.1 V **(ii)** 8.4×10^{-24} J
G46 3.3 cm
G47 1.8 cm
G48 **(c)** **(i)** 8.3×10^{-7} s **(ii)** 4.3×10^{-5} T
G49 **(a)** 1.0×10^{5} m s^{-1} **(b)** 1.7×10^{-2} T
G50 1.33×10^{7} m s^{-1}, 1.71°
G51 6.7×10^{7} m s^{-1}
G54 6×10^{-2} mm V^{-1}
G56 **(a)** 4.2×10^{7} m s^{-1} **(b)** 0.12 m
G58 **(b)** 41 kV m^{-1}
G60 6
G61 **(a)** **(iv)** 3.6×10^{-15} N
G62 6.4×10^{-19} C
G63 4.68×10^{-19} C
G64 4.8×10^{-19} C; 1.6×10^{-19} C
G65 **(b)** **(i)** 3.8×10^{-9} s **(ii)** 2.9°
G67 **(b)** 1.14×10^{3} V **(c)** 3.53 V
G69 75% $^{35}_{17}Cl$, 25% $^{37}_{17}Cl$
G70 69% (62.93), 31% (64.93)
G71 **(a)** 3.2×10^{5} m s^{-1} **(b)** 10.6 kV
G72 **(a)** **(iii)** 179 m s^{-1} **(b)** **(i)** 3.2×10^{-15} N
(iii) 0.11 T **(c)** **(iii)** 0.50 cm
G76 6.0 MeV
G77 **(b)** **(ii)** 0.015 cm **(iii)** ±0.001 cm
G78 0.11 cm^{-1}
G79 **(c)** **(i)** 7.5×10^{6} min^{-1} **(ii)** 19%
G80 **(c)** 1.0×10^{-6} s^{-1} **(d)** 1.0×10^{18}
G81 8.8×10^{8} Bq
G82 50 s^{-1}
G83 **(b)** **(i)** 6.3×10^{8} **(ii)** 2.0×10^{-6} s^{-1}
(iii) 1.3×10^{3} Bq
G84 11 μg
G85 4.3 hours
G86 2.8×10^{11} Bq
G87 **(a)** 6.0×10^{11} **(b)** 6.5×10^{11}
G88 1.27×10^{20}
G90 **(c)** **(i)** 1:2 **(ii)** 1:1 **(iii)** 75%
G91 **(b)** **(ii)** 56 years
G92 140 days; $a = 210$, $b = 84$, $c = 4$, $d = 2$, $e = 0$, $f = 0$
G93 6.0 m
G94 **(c)** **(i)** 6.2×10^{11} s
G95 4.0 mg per day
G96 6.0×10^{3} cm^{3}
G97 **(b)** **(i)** 2.0×10^{-6} s^{-1} **(ii)** 8.0×10^{9} **(c)** 40 hours
G98 >57 μCi
G99 **(d)** $\frac{1}{32}$
G100 2.0 m; 24 hours
G105 **(b)** **(i)** 70 s, 10.0 μg **(ii)** 2.5×10^{14} s^{-1}
G106 **(c)** **(ii)** 1.3×10^{-4} s^{-1} **(iii)** 7.7×10^{13} s^{-1}
(iv) 63 W
G108 4.5×10^{9} years
G109 6.8×10^{8} years
G110 4.2×10^{3} years
G111 8700 years
G112 **(b)** **(ii)** 3.8×10^{3} years
G114 7.16 MeV
G115 2.79×10^{-12} J
G117 **(c)** **(i)** 1.813 u, 1.824 u
G118 **(a)** 28.3 MeV **(b)** 23.8 MeV
G119 **(c)** **(iii)** 2.6×10^{11} **(iv)** 1.7×10^{-15} kg
G120 **(c)** 1.1×10^{11} years
G121 8.2×10^{-14} J
G123 **(b)** **(iii)** 1.06
G124 +0.624 MeV, −4.01 MeV
G125 39.1 MeV
G127 8.53×10^{10} J
G128 **(c)** 2.5×10^{3} mol **(d)** 2.5×10^{7} Pa
G129 **(d)** **(i)** 5.40(4) MeV **(iii)** 5.30(2) MeV
G130 7.1 days
G132 **(c)** **(iii)** 2.1×10^{-13} J
G133 **(b)** 15 hours
G138 60 Ω
G147 2.5×10^{4} Ω (R_1), 5.0×10^{2} Ω (R_2), 15 mA
G148 **(a)** 1.5 V **(b)** 1.5 V
G149 **(b)** **(i)** 250 μA, 12.75 mA, 3.5 V
G150 **(c)** 60 kΩ
G157 **(b)** **(i)** 100 times **(ii)** never
G158 **(a)** 10
G159 **(b)** 8
G160 **(a)** −4.0 V
G162 **(a)** **(i)** 4.7 kΩ, 150 kΩ **(ii)** 33
G164 **(a)** **(i)** 4.5 V **(ii)** 6.0 V **(b)** **(ii)** 5.7 V

INDEX

Parentheses indicate a page where there is a minor reference.